I0064387

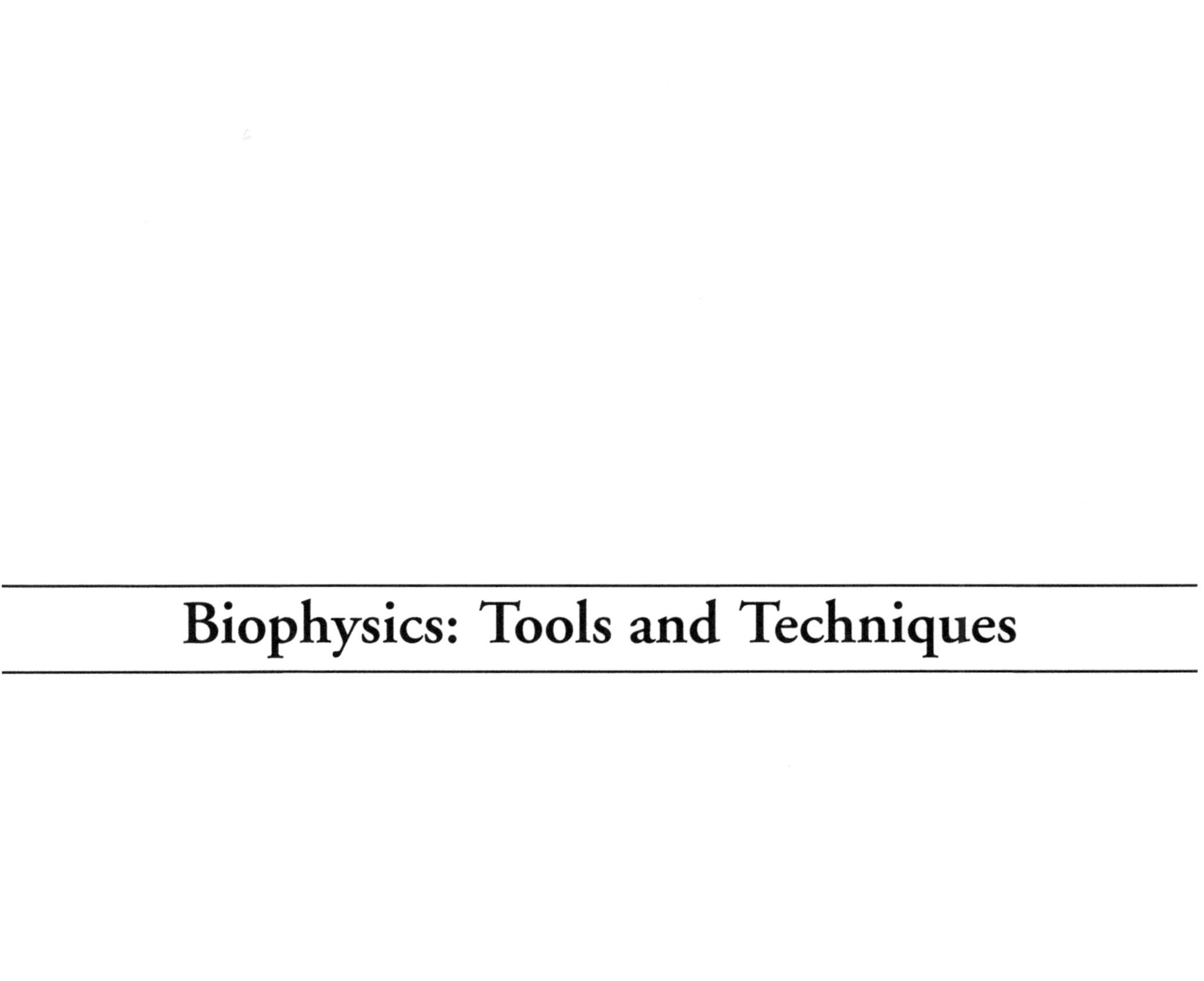

Biophysics: Tools and Techniques

Biophysics: Tools and Techniques

Editor: Betty Karasek

New York

Published by NY Research Press
118-35 Queens Blvd., Suite 400,
Forest Hills, NY 11375, USA
www.nyresearchpress.com

Biophysics: Tools and Techniques
Edited by Betty Karasek

International Standard Book Number: 978-1-63238-544-4 (Hardback)

Cataloging-in-Publication Data

Biophysics : tools and techniques / edited by Betty Karasek.
p. cm.
Includes bibliographical references and index.
ISBN 978-1-63238-544-4
1. Biophysics. 2. Biology. 3. Physics. I. Karasek, Betty.
QH505 .B56 2017
571.4--dc23

Printed in the United States of America.

Contents

Preface

Biophysics can be termed as the application of the principles and methods of physics to biological structures. This book on biophysics is interdisciplinary in its approach. This book presents the complex subject of biophysics in the most comprehensible and easy to understand language. Some of the diverse topics covered herein address the varied branches that fall under this category. The text presents the current practices that are being developed in this field. The various studies that are constantly contributing towards advancing technologies and evolution of this field are examined in detail. This book will be of great use to researchers in the fields of bioengineering, systems biology and computational biology.

This book is a comprehensive compilation of works of different researchers from varied parts of the world. It includes valuable experiences of the researchers with the sole objective of providing the readers (learners) with a proper knowledge of the concerned field. This book will be beneficial in evoking inspiration and enhancing the knowledge of the interested readers.

In the end, I would like to extend my heartiest thanks to the authors who worked with great determination on their chapters. I also appreciate the publisher's support in the course of the book. I would also like to deeply acknowledge my family who stood by me as a source of inspiration during the project.

Editor

1

The Shine-Dalgarno sequence of riboswitch-regulated single mRNAs shows ligand-dependent accessibility bursts

Arlie J. Rinaldi[1,*,†], Paul E. Lund[1,2,*], Mario R. Blanco[1,†] & Nils G. Walter[1]

In response to intracellular signals in Gram-negative bacteria, translational riboswitches—commonly embedded in messenger RNAs (mRNAs)—regulate gene expression through inhibition of translation initiation. It is generally thought that this regulation originates from occlusion of the Shine-Dalgarno (SD) sequence upon ligand binding; however, little direct evidence exists. Here we develop Single Molecule Kinetic Analysis of RNA Transient Structure (SiM-KARTS) to investigate the ligand-dependent accessibility of the SD sequence of an mRNA hosting the 7-aminomethyl-7-deazaguanine (preQ$_1$)-sensing riboswitch. Spike train analysis reveals that individual mRNA molecules alternate between two conformational states, distinguished by 'bursts' of probe binding associated with increased SD sequence accessibility. Addition of preQ$_1$ decreases the lifetime of the SD's high-accessibility (bursting) state and prolongs the time between bursts. In addition, ligand-jump experiments reveal imperfect riboswitching of single mRNA molecules. Such complex ligand sensing by individual mRNA molecules rationalizes the nuanced ligand response observed during bulk mRNA translation.

[1] Single Molecule Analysis Group, Department of Chemistry, University of Michigan, Ann Arbor, Michigan 48109, USA. [2] Program in Chemical Biology, University of Michigan, Ann Arbor, Michigan 48109, USA. * These authors contributed equally to this work. † Present addresses: W.M. Keck Science Center, The Claremont Colleges, Claremont, California 91711, USA (A.J.R.); Division of Biology and Biological Engineering, California Institute of Technology, Pasadena, California 91125, USA (M.R.B.). Correspondence and requests for materials should be addressed to N.G.W. (email: nwalter@umich.edu).

Riboswitches are noncoding structural elements most commonly embedded in the 5′ untranslated region (UTR) of bacterial messenger RNAs (mRNAs) that regulate the expression of a downstream gene through the binding of an intracellular signal[1–5]. These signals include nucleobases[6,7], amino acids[8,9], cofactors of metabolic enzymes[10,11] and metal ions[12–14], among others. Genetic regulation is achieved through a multitude of mechanisms; however, the two most common modes are transcriptional attenuation and inhibition of translation initiation. In translational riboswitches, ligand binding sequesters the Shine-Dalgarno (SD) sequence of the mRNA through alternative base pairing, resulting in occlusion of the ribosomal binding site needed for efficient initiation of translation. In recent years, a plethora of biophysical techniques has been employed to understand the genetic regulation mechanism utilized by translational riboswitches[15–21]; however, these techniques have rarely involved the entire mRNA and have largely failed to provide direct mechanistic insight at a molecular level into the coupling of ligand-induced conformational changes with downstream regulatory effects.

The preQ$_1$ riboswitch from *Thermoanaerobacter tengcongensis* (*Tte*) is a strikingly small translational riboswitch that responds to the presence of the modified nucleobase preQ$_1$ (ref. 22). Crystal structures[18], in combination with single-molecule fluorescence resonance energy transfer (smFRET), NMR and computational studies[19], have indicated that this riboswitch achieves genetic regulation by the formation of a pseudoknot, in which the P2 helix containing the first two nucleotides of the SD sequence is formed. In the absence of ligand, this helix is only transiently closed[19], presumably revealing the SD sequence in its entirety as part of the downstream expression platform to bind the anti-SD sequence of the 16S ribosomal RNA (rRNA, Fig. 1a,b). These and many similar studies[21,22] have provided valuable insight into the conformational sampling of the preQ$_1$ riboswitch as well as other translational riboswitches as a function of ligand concentration. They have left largely unresolved, however, the molecular mechanism by which sequestration of just a small fraction of the SD sequence brings about the proposed ON/OFF gene-regulatory control of an entire mRNA through coupling between the aptamer and expression platform.

We here develop a technique termed Single Molecule Kinetic Analysis of RNA Transient Structure (SiM-KARTS), wherein a short, fluorescently labelled nucleic acid probe, whose sequence is complementary to a particular region of interest, is used to probe changes in structure of a longer target RNA through repeated binding and dissociation events. In the current implementation, an RNA probe corresponding to the 3′ sequence of *T. tengcongensis* (*Tte*) 16S rRNA (that is, an anti-SD probe) binds to preQ$_1$ riboswitch-containing, surface-immobilized, single *Tte* mRNA molecules, and thus directly reports on the accessibility of the SD sequence. Our results reveal unexpected complexities of ligand-induced riboswitching during translation initiation and add a new dimension to an emerging model, wherein stochastic single-molecule events contribute to fine-tuned temporal gene expression fluctuations in bacteria. We anticipate that SiM-KARTS will find broad application in probing dynamic RNA structural elements at the single-molecule level.

Results

preQ$_1$ regulates translation of the *Tte* mRNA *in vitro*. The expression platform of translational riboswitches contains the SD sequence, a short (three to eight nucleotides, nt), purine-rich sequence located ~5–9 nt upstream of the start codon of bacterial mRNAs[23]. It hybridizes with the 3′ end of 16S rRNA (Fig. 1b), and this interaction is important for canonical initiation and proper start codon selection by the 30S ribosomal subunit[24]. To date, a majority of riboswitch studies have focused on the properties of only the isolated aptamer domain[21]. Studies incorporating the full riboswitch including the downstream

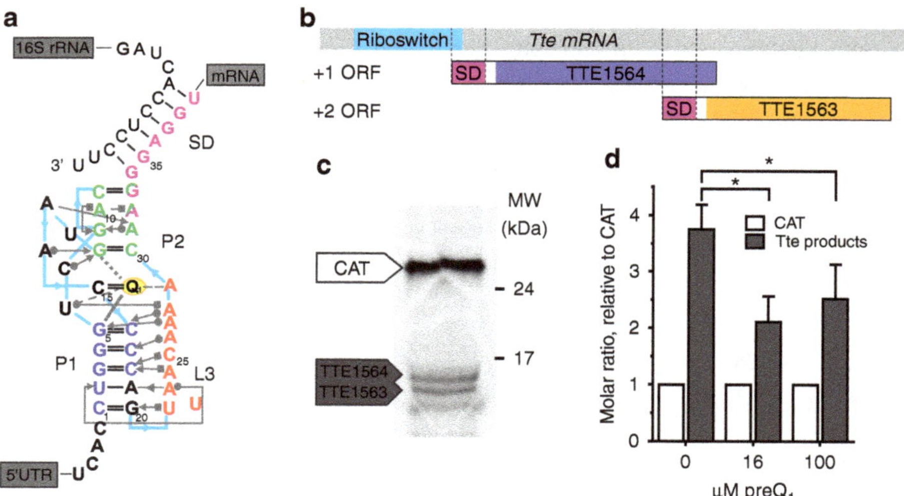

Figure 1 | *In vitro* translation of *Tte* mRNA. (**a**) Structural map of the *Tte* preQ$_1$ translational riboswitch displayed with Leontis-Westhof notations[57]. The SD sequence (purple) partially overlaps the P2 stem nucleotides (green). Formation of the P2 stem interferes with proper base pairing between the anti-SD sequence at the 3′ end of the 16S rRNA and the SD sequence of the *Tte* mRNA. (**b**) Schematic of *Tte* mRNA used for *in vitro* translation assays. The putative mRNA transcript is bicistronic, containing the overlapping reading frames for TTE1564 and TTE1563. The preQ$_1$ riboswitch aptamer (light blue) overlaps with a portion of the SD sequence (purple) of TTE1564. (**c**) Example autoradiograph of *in vitro* translation products. *Tte* mRNA was translated using L-[^{35}S]-Cys in the presence of the control mRNA encoding CAT at a 4:1 ratio of *Tte*:CAT mRNA. Molecular weight markers are indicated on the right (full-length gel is shown in Supplementary Fig. 11). (**d**) Quantification of *in vitro* translation products as a function of preQ$_1$ concentration. The total protein produced from the *Tte* mRNA (sum of TTE1564 and TTE1563 bands, grey bar) is reported relative to the intensity of the CAT product in the same lane, after normalizing for the cysteine content of each protein (5, 1 and 1 for CAT, TTE1564 and TTE1563, respectively). The results represent the mean ± s.d. of three replicates (*$P<0.05$).

expression platform have mostly involved reporter gene assays replacing the mRNA's native gene(s) with the exogenous reporter. Here we instead opted to test the functional activity of the $preQ_1$ riboswitch in the context of its native mRNA. In *T. tengcongensis*, the riboswitch is located in the 5′ UTR of a putative bicistronic operon as the *Tte* mRNA encodes two genes (Fig. 1b). *In vitro* translation using purified *Escherichia coli* translation factors and ribosomes, which share the anti-SD sequence of *T. tengcongensis* 16S rRNA with the exception of an additional 3′ single-nucleotide overhang (Supplementary Fig. 1a), produced the corresponding two proteins, TTE_RS07450 and TTE_RS07445 (subsequently referred to by their former locus tags TTE1564 and TTE1563, respectively), as expected (Fig. 1c and Supplementary Fig. 1b,c). We next performed competition experiments using a 4:1 molar ratio of *Tte* to chloramphenicol acetyltransferase (CAT) control mRNA, where the latter encodes CAT, does not contain the $preQ_1$ riboswitch and thus is not expected to be modulated in its translation by $preQ_1$ (Supplementary Fig. 2a). We observed an mRNA-specific, ~40% decrease in translation of the two *Tte* mRNA genes on addition of saturating concentrations (16 and 100 μM, see below) of $preQ_1$ (Fig. 1d; note that the quantification accounts for the higher number of labelled cysteines in CAT, see Methods). This result suggests that $preQ_1$ decreases the translational efficiency of *Tte* mRNA, and that the native mRNA is thus responsive to ligand-induced structural changes.

SiM-KARTS detects changes in the structure of single mRNAs. To observe changes in SD sequence accessibility as a function of ligand concentration, we developed SiM-KARTS (Fig. 2), utilizing a short, fluorescently (Cy5) labelled RNA anti-SD probe with the sequence of the 12 nt at the very 3′ end of *T. tengcongensis* 16S rRNA (Fig. 1a and Supplementary Note 1). Target mRNA molecules were hybridized with a high-melting-temperature TYE563-labelled locked nucleic acid (TYE563-LNA) for visualization, immobilized on a quartz slide at low density via a biotinylated capture strand and imaged with single-molecule sensitivity by total internal reflection fluorescence microscopy (TIRFM, Fig. 2a,b). To simplify our analysis in the context of the full-length mRNA, we chose the TYE563-LNA marker to also block the distinct SD sequence and start codon of the TTE1563 open reading frame (ORF), preventing the anti-SD probe from binding to the downstream TTE1563 SD (Fig. 2a and Supplementary Fig. 3). TYE563 fluorescence could only be observed once all three components (biotinylated capture strand, *Tte* mRNA and TYE563-LNA) were assembled on the surface (Fig. 2b and Supplementary Fig. 4), attesting to the high specificity of the experiment.

Because the interaction between the *Tte* mRNA and the anti-SD probe is limited to seven Watson–Crick base pairs and one wobble (Fig. 1a), binding of the probe to a single mRNA molecule under equilibrium conditions is reversible and transient (Fig. 2b). In addition, since the experiment is performed using TIRFM, only probe molecules transiently immobilized to the slide surface via the mRNA target will be observed within the evanescent field and co-localized with TYE563 in a diffraction-limited spot, whereas probes diffusing freely in solution will instead contribute to a modest background fluorescence. Repeated, transient

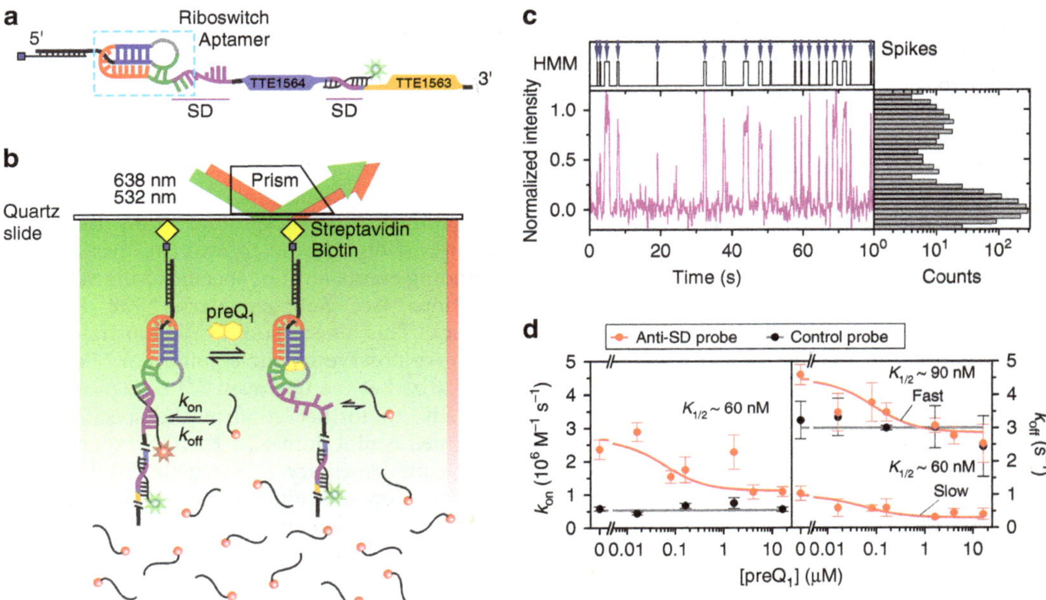

Figure 2 | SiM-KARTS measurements of $preQ_1$-dependent anti-SD-binding kinetics. (**a**) The *Tte* mRNA complex used in SiM-KARTS experiments. Full-length *Tte* mRNA molecules are immobilized to the slide surface via a biotinylated-capture strand that is hybridized to the 5′ end of the mRNA. Features of the riboswitch and associated reading frames are coloured as in Fig. 1a,b, respectively. A TYE563-LNA (with green star) is hybridized to the start of the downstream open reading frame to occlude this second SD sequence and to locate mRNAs on the slide surface. (**b**) Experimental prism-based TIRFM set-up. The *Tte* mRNA complexes shown in **a** are immobilized to a slide surface that has been passivated with biotinylated BSA (omitted for clarity). Repeated binding and dissociation of the anti-SD probe labelled with Cy5 (red sphere, red star) is monitored through co-localization of TYE563 and Cy5 fluorescence. (**c**) Representative anti-SD probe-binding fluorescence versus time trajectory and corresponding fluorescence intensity histogram for a single *Tte* mRNA molecule in the absence of $preQ_1$. Cy5 intensity from the anti-SD probe (magenta) and Hidden Markov idealization to a two-state model (HMM, grey) are plotted as a function of time. The TYE563 fluorescence trace used to identify and localize the *Tte* mRNA has been omitted for clarity. (**d**) Anti-SD (red) and control (black) probe binding and dissociation rate constants (k_{on}, left plot; k_{off}, right plot) were determined from exponential fits of dwell times in the unbound and bound states, respectively, as a function of $preQ_1$ concentration. Binding and dissociation rate constants for the control probe are unaffected by $preQ_1$ concentration. The corresponding $K_{1/2}$ value from the saturation curve fit of the anti-SD probe binding is indicated. The results represent the average ± s.e.m. of at least three independent experiments.

diffraction-limited co-localization of Cy5 and TYE563 fluorescence therefore unambiguously characterizes individual binding events of the anti-SD probe to a single target mRNA molecule (Fig. 2b). Since changes in the probe binding and dissociation time constants can be sensitively monitored over an arbitrarily long time window with high precision, these characteristic repeat signals are expected to quantitatively report on the accessibility of the SD sequence and thus secondary structure of individual mRNA molecules.

Previous SAXS[18], smFRET and MD simulation data[19] on the *Tte* $preQ_1$ riboswitch aptamer have found that helix P2 of the riboswitch pseudoknot is partially open in the absence of $preQ_1$, leaving the SD sequence more exposed than it is in the presence of ligand (Fig. 1a). To ask whether SiM-KARTS detects the expected difference in accessibility of the SD sequence under equilibrium conditions in the absence and presence of $preQ_1$, Cy5-labelled anti-SD probe was flowed on a slide with immobilized and TYE563-LNA-bound *Tte* mRNA first in the absence of ligand. Demonstrating the highly parallel nature of SiM-KARTS, thousands of transient binding events were observed in over 100 mRNA molecules per experiment. The resulting Cy5 emission trajectories were fit using a two-state Hidden Markov Model (HMM) to extract dwell times of the probe in the bound and unbound states, τ_{bound} and $\tau_{unbound}$, respectively (Fig. 2c). The use of HMMs allows us to objectively identify binding events in single-molecule fluorescence time trajectories that are inherently noisy because of the background of excess free Cy5-labelled probe in solution. HMMs filter out this noise and can detect binding events as short as a single camera integration time. HMMs also offer an advantage over simple signal thresholding or image spot finding algorithms because HMMs inherently consider the kinetics of binding events that are the key information sought from SiM-KARTS. Using HMM analysis, we found that a lower-than-average signal-to-noise ratio of some trajectories sometimes could lead to misidentification of the molecule's state (bound or unbound). To further improve the HMM analysis, refitting was performed on select trajectories where binding events were clearly missed (Methods). Although this additional refitting improved the fits of a small number of individual trajectories, the rate constants were only modestly affected (Supplementary Fig. 5). This suggests that the analysis of SiM-KARTS data is robust, provided sufficient numbers of molecules are sampled (Supplementary Note 2). The resulting cumulative τ_{bound} and $\tau_{unbound}$ dwell time distributions were fit with a single-exponential function to calculate the binding rate constant k_{on} and a double-exponential function to extract a fast and a slow dissociation rate constant k_{off}, respectively, based on an analysis of the residuals (Supplementary Fig. 6). In the absence of ligand, the anti-SD probe binds with a bimolecular rate constant k_{on} of $2.4 \pm 0.3 \times 10^6\,M^{-1}\,s^{-1}$ and dissociates with two unimolecular rate constants k_{off} of $4.6 \pm 0.2\,s^{-1}$ (relative amplitude = 76%) and $1.0 \pm 0.2\,s^{-1}$ (relative amplitude = 24%; Fig. 2d and Supplementary Table 1 and Supplementary Note 3). As our further analysis will demonstrate, the apparent biphasic nature of k_{off} is simply the result of the shortcomings of a phenomenological fit to a set of complex, kinetically broadly distributed molecular behaviours.

Next, mRNA molecules were folded in the presence of varying concentrations of $preQ_1$ and subjected to SiM-KARTS. The value of k_{on} of the anti-SD probe decreased as the concentration of $preQ_1$ increased, with a half-saturation point $K_{1/2}$ of $\sim$60 nM $preQ_1$ (Fig. 2d). Such a decrease in the binding rate indicates an occlusion of this target sequence in the presence of $preQ_1$, as expected. It is important to note that this $K_{1/2}$ is related to, but not a direct measure of, ligand affinity, which is known to be in the low nanomolar range[18]. Instead, it reflects $preQ_1$-induced structural changes in the expression platform, in contrast to previous studies performed on only the minimal aptamer[19]. Unexpectedly, increasing $preQ_1$ concentrations also resulted in a decrease in both the fast and slow k_{off} rate constants (Fig. 2d), indicating that high $preQ_1$ concentrations stabilize the SD:anti-SD interaction once formed. A plausible explanation for such an effect is the potential for $preQ_1$ to stabilize co-axial stacking of the anti-SD probe on a more fully formed P2 helix (Fig. 1a).

To test whether these changes are indeed due to conformational rearrangements near the SD sequence of the riboswitch, we performed equilibrium SiM-KARTS using a Cy3-labelled blocking strand that anneals to a 35-nt region, effectively sequestering the expression platform as well as the initial five nucleotides in the TTE1564 ORF and disrupting the riboswitch aptamer domain (Supplementary Fig. 7a and Supplementary Note 1). Performing equilibrium SiM-KARTS experiments in the presence of this blocking strand along with the TYE563-LNA strand results in a situation where both SD sequences are occluded from anti-SD probe binding. Accordingly, the probe dissociation constant, $K_{d,effective}$, for the mRNA at high ligand concentration is significantly lower than that for mRNA in the absence of ligand but annealed with the blocking strand (Supplementary Fig. 7e, red bar versus red striped bar). This indicates that the affinity of the anti-SD probe for the *Tte* mRNA is greatly reduced when the expression platform of the riboswitch is blocked. In addition, while the k_{on} of the anti-SD probe in the presence of the blocking strand ($0.82 \times 10^6\,M^{-1}\,s^{-1}$, Supplementary Table 2) is comparable to that of the probe in the presence of high ligand concentration ($1.1 \times 10^6\,M^{-1}\,s^{-1}$, Supplementary Table 1), the values of k_{off} differ substantially (8.3 versus $2.1\,s^{-1}$, respectively), indicating that any spurious, nonspecific events are characterized by significantly faster probe dissociation (Supplementary Fig. 8b). This suggests that, while both conditions (blocking strand or high ligand) induce a conformation in which the SD sequence is (partially) blocked, the ligand induces a distinct kinetic profile. Conservatively assuming that the entire bound dwell time distribution for *Tte* mRNA annealed with the blocking strand represents nonspecific probe-binding events, we can proportionally subtract this background from the distribution in the presence of high ligand to estimate that a majority (>60%) of all binding events must be specific to the SD sequence under these conditions (see Supplementary Note 1 and Supplementary Fig. 8c). This unequivocally demonstrates that most binding events we observe in our equilibrium SiM-KARTS experiments are localized at the upstream SD sequence.

Finally, to further confirm the specificity of our assay, we performed equilibrium SiM-KARTS experiments using a control probe complementary to a region distal to the riboswitch and within the open reading frame of the mRNA (Supplementary Fig. 3 and Supplementary Note 4), whose structure and, thus, accessibility to control probe binding should be unaffected by $preQ_1$. Indeed, the kinetics of the binding of this probe showed little change in response to $preQ_1$ concentration (Fig. 2d), indicating that the ligand-induced conformational changes are localized to the SD region of the riboswitch.

Periods of high SD sequence accessibility occur in bursts. Further inspection of individual probe-binding trajectories revealed that single molecules interconvert between periods of frequent probe-binding events and periods of more sporadic events, which can be interpreted as periods of high and low SD accessibility, respectively (Fig. 3a). Traditional analysis methods of single molecules in aggregate failed to detect these changes. For example, common scatter plots of the mean τ_{bound} and $\tau_{unbound}$ dwell times for individual molecules[25–30] in the presence of

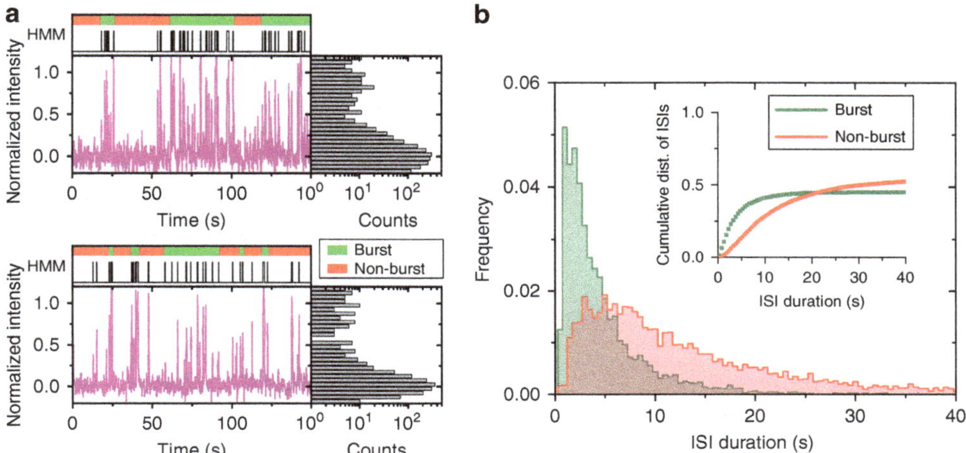

Figure 3 | Detection of burst behaviour through spike train analysis. (**a**) Representative trajectories as in Fig. 2b for two single *Tte* mRNA molecules in the absence of preQ$_1$, annotated with bursts (green bars) and non-burst periods (red bars) detected through spike train analysis. (**b**) Cumulative histogram displaying the distribution of ISIs during burst (green) and non-burst (red) periods in the absence of preQ$_1$.

saturating ligand concentration revealed a shift towards longer unbound times compared with the absence of ligand; however, all molecules generally fit within a single broad distribution (Supplementary Fig. 8a). This observation suggests that calculating average rate constants for each mRNA molecule, while revealing heterogeneity among a population of molecules[25–30], largely fails to detect time evolution in the equilibrium behaviour of a single molecule.

The probe-binding events detected via SiM-KARTS strongly resemble neuronal spike trains, where neuronal firing is monitored and detected as sharp, transient increases (or 'spikes') in electrical activity in response to external stimuli. A common feature of these spike trains is short intervals of high firing activity, or 'bursts', separated by periods of relative inactivity (non-bursts)[31]. This type of analysis previously has been used to describe transcription time series in *E. coli*[32,33], leading us to ask whether spike train analysis could detect and separate in an unbiased manner the periods of high and low frequency of probe-binding events observed within single molecules. To specifically justify its use in the context of SiM-KARTS, we first calculated the Fano factor[34] of the number of spikes within a certain time window (Supplementary Fig. 9). The Fano factor here is defined as the variance of the number of spikes within a certain time frame divided by the mean. For a random Poisson distribution, that is, one that is completely independent of the time window, the Fano factor is consistently equal to one (simulated data points in Supplementary Fig. 9). Our SiM-KARTS data clearly deviate from this expectation (Supplementary Fig. 9), rendering spike train analysis an appropriate tool. Next, exploiting its nonparametric approach we applied the Rank Surprise (RS) method of burst detection, which has been utilized to detect regions of high spike activity[31] that, in our case, represent periods of high SD accessibility. The RS method does not make any assumptions about the distribution of spikes and is based solely on the definition of bursts as representing many spikes in a comparably short amount of time[31]. Global burst analysis across the various concentrations of ligand tested found that individual molecules displayed detectable bursts of anti-SD probe binding, which were separated by non-bursting periods characterized by areas of low average binding activity (Fig. 3a). When the duration of interspike intervals (ISIs), equivalent to τ_{unbound} dwell times, in the bursting and non-bursting periods were plotted, two distinct, previously hidden intramolecular behaviours became evident (Fig. 3b). We found that single molecules typically interconvert between periods of bursting and non-bursting behaviour, rather than segregating into separate subpopulations of highly and poorly accessible molecules, with bursts of high SD accessibility identified even at saturating ligand concentrations (Fig. 4a). This finding suggests that *Tte* mRNA switches between (at least) two distinct conformational states: a bursting state with overall high SD accessibility and frequent binding events of the anti-SD probe (that is, shorter ISIs) and a non-bursting state characterized by low SD accessibility where the SD sequence is more sequestered away from the probe (longer ISIs). The latter non-bursting state is adopted even in the absence of ligand (Fig. 4a), in accord with previous studies indicating that the P2 helix, which partially sequesters the SD sequence, can form without ligand present[18,19].

preQ$_1$ decreases the number of bursts and the burst duration. Visual inspection of the trajectories at high ligand concentration suggested that the bursting state is shorter lived, and non-bursting periods longer lived, compared with low ligand concentration conditions (Fig. 4a). In global spike train analysis, this observation is reflected in an increasing bias towards non-burst-associated ISIs with increasing preQ$_1$ concentration (Fig. 4b and Supplementary Table 1). Furthermore, the cumulative burst duration distribution shifts towards shorter values (Fig. 4c), indicating that the conformation in which the SD sequence is more accessible becomes shorter-lived and destabilized by ligand. However, the duration of ISIs, and thus the binding rate constant k_{on}, within bursting states is largely unaffected by ligand, indicating that the bursting state conformation is similar in the absence and presence of ligand, just that its lifetime is shorter (Supplementary Fig. 10 and Supplementary Table 1). Overall, our results suggest that the *Tte* mRNA with embedded preQ$_1$ riboswitch transitions between two distinguishable equilibrium conformational states: a bursting state conformation with an exposed SD sequence that is available for frequent binding of the anti-SD sequence and a non-bursting conformation with a less accessible SD sequence. Both of these states coexist and interconvert in both the presence and absence of ligand. As more ligand is added, transitions to the bursting state become less frequent and shorter lived, yet remain a persistent feature of the mRNA-embedded riboswitch.

mRNA only partly adapts to changes in preQ$_1$ concentration. To assess the response of individual mRNAs and their SD

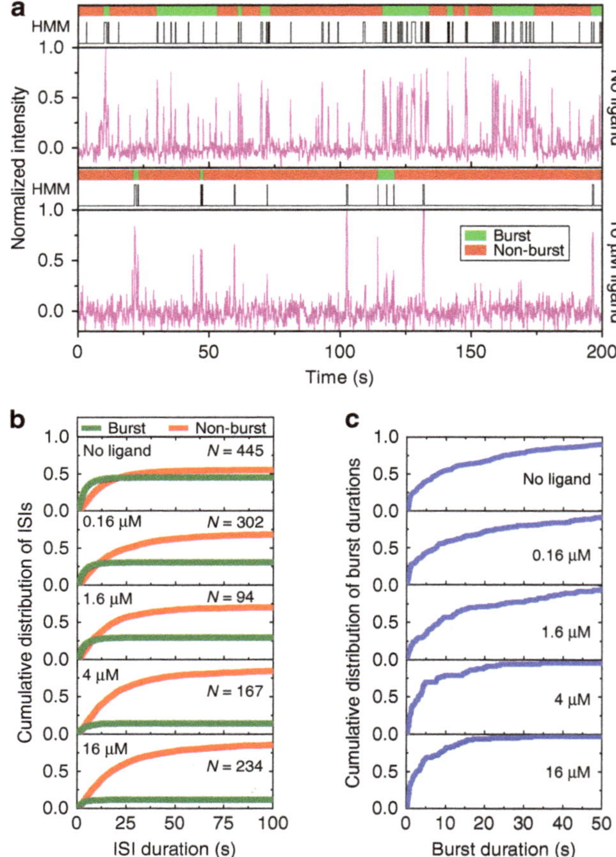

Figure 4 | Ligand-dependent changes in bursting behaviour of single riboswitches. (**a**) Single-molecule trajectories as in Fig. 3a but in the absence and presence of saturating $preQ_1$ (top and bottom, respectively). (**b**) Cumulative distribution plots indicating the distribution of ISIs during burst (green) and non-burst (red) periods at varying $preQ_1$ concentrations, where N is number of molecules included in the analysis. (**c**) Cumulative distribution plots of burst duration for the molecules in **b** as a function of $preQ_1$ concentration. As $preQ_1$ concentration increases, the average burst duration decreases.

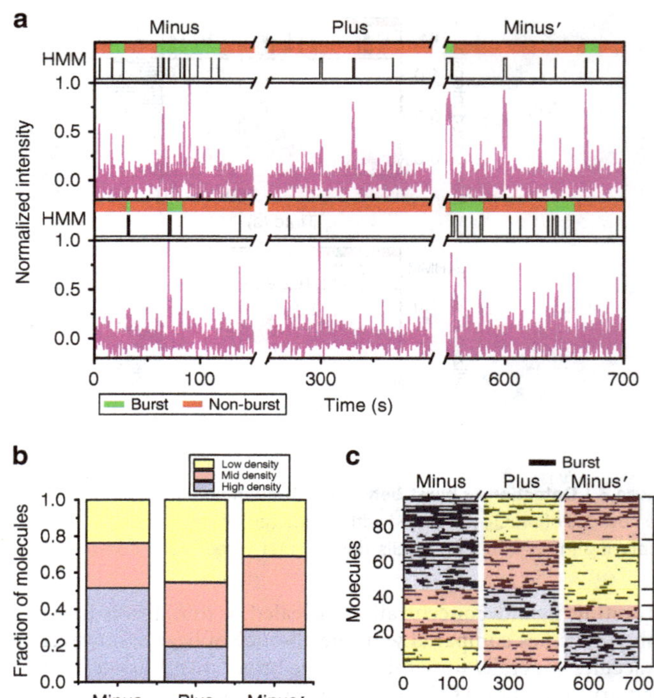

Figure 5 | Single mRNA molecules can undergo conformational switching depending on their environment. (**a**) Exemplary single-molecule trajectories from ligand-jump SiM-KARTS experiments composed of three time segments. Anti-SD probe binding to the same set of individual *Tte* mRNA molecules is monitored first in the absence of $preQ_1$ (Minus), then in the presence of 16 μM $preQ_1$ (Plus) and again in the absence of $preQ_1$ (Minus′). Each axis break represents a 50-s dark period between segments during which buffer was exchanged. (**b**) Distribution of burst density rankings for each segment of an individual molecule's trajectory, for all molecules in the ligand-jump SiM-KARTS experiment ($N = 97$). (**c**) Rastergram displaying the bursting behaviour of the *Tte* mRNA molecules in **b**. Bursts are displayed as black bars. Individual probe-binding events (spikes) are omitted for clarity.

accessibility to changing (nonequilibrium) ligand concentrations, as may occur in the bacterial cell, we devised a ligand-jump experiment that allowed us to apply SiM-KARTS to molecules tracked throughout a transition from no ligand, to saturating ligand and back to no ligand (Minus, Plus, Minus′ segments, respectively) in a set of contiguous fluorescence–time trajectories (Fig. 5a). We then applied global spike train analysis to all molecules we were able to track through this nonequilibrium ligand-jump experiment. To further characterize the evolution of the SD accessibility through the changes in ligand concentration, each of the three segments for a given molecule was ranked by time spent in the bursting state. An individual molecule's Minus, Plus or Minus′ segment with the highest density of bursts was ranked as High (H), the next highest as Mid (M) and the lowest as Low (L). We then quantified the overall distribution of burst ranks in the three different segments (Fig. 5b). For the majority of molecules, the highest burst density occurred in the first segment where ligand is absent (Minus), as expected. By contrast, the saturating ligand segment (Plus) exhibited mostly Low- and Mid-burst density rankings, again as expected. Finally, the Minus′ segment, where the ligand-containing buffer had been extensively washed out, exhibited a fairly equal distribution of ranks. These measures indicate that, as an ensemble, single mRNA molecules respond to ligand concentration with the expected modulation in SD accessibility. We further plotted the bursting behaviour of each of 97 molecules as a rastergram and organized them into six groups according to their per-segment burst density ranks (Fig. 5c), several of which were of particular interest. This comparison showed that ~24% of all molecules responded to the addition and removal of ligand with reduction and recovery of bursting behaviour (H–L–M in the Minus, Plus, Minus′ sequence; molecules 75–97 in Fig. 5c and upper trace Fig. 5a), respectively, as expected. However, ~30% of molecules seemed to not revert to greater burst density on the timescale of the experiment after the ligand was washed out (H–M–L, molecules 46–74 in Fig. 5c). This population of mRNAs reacts to initial ligand binding by SD sequence occlusion, but appears to remain in a ligand-bound conformation even after the $preQ_1$-containing buffer is removed, consistent with the known slow rate of $preQ_1$ dissociation[18]. Interestingly, we also observed that ~16% of molecules displayed the opposite behaviour, that is, displayed their highest bursting density in the Minus′ segment (L–M–H, molecules 1–16 in Fig. 5b, lower trace in Fig. 5a), after the introduction and removal of ligand. This behaviour suggests that in some cases the ligand may help promote refolding of an mRNA in which the SD sequence was occluded before the addition of ligand.

Discussion

It is generally thought that translational riboswitches achieve gene regulation through a ligand-mediated conformational change in the aptamer domain that is then transduced into the downstream expression platform to actuate an ON/OFF switch in gene expression[1–5]. However, the molecular underpinnings of this transduction, especially in the context of the native gene, are still poorly studied and understood. Here we present SiM-KARTS as a cost-effective and relatively non-invasive alternative technique to FRET that sensitively probes site-specific changes in secondary structure of arbitrarily large or complex single RNA molecules in real-time and without the need for covalently modifying the targeted RNA, which can be an inefficient and expensive undertaking. Utilizing an anti-SD probe mimicking the 3′ end of the corresponding 16S rRNA, we used SiM-KARTS to quantify the accessibility of the SD sequence in the expression platform of an mRNA hosting a small $preQ_1$ riboswitch over a significantly longer time window than a similar FRET experiment, since SiM-KARTS is inherently not limited by photobleaching. The resulting extended observation of single molecules is critical for long RNAs that demonstrate relatively slow changes in structure, such as the accessibility changes in the SD sequence of the mRNA studied here. As expected, we detected a decrease in SD accessibility on addition of $preQ_1$ as a marked decrease in the binding rate constant k_{on} of the probe. Less expectedly, however, probe-binding events to a single mRNA (spikes) typically occurred in bursts. Consequently, spike train analysis of these bursts provided evidence for two conformational states repeatedly interconverting within single molecules that are characterized by periods of high and low SD accessibility. Unexpectedly, these two states were observed not only in the absence of $preQ_1$, but also at saturating ligand concentration, indicating that the riboswitch continues to occasionally sample a conformation with high SD accessibility. Finally, nonequilibrium ligand-jump experiments indicated that single mRNA molecules only imperfectly switch between high and low SD accessibility. These findings rationalize the significant, but relatively modest impact of saturating $preQ_1$ concentrations on the *in vitro* translation output of the mRNA and lead us to propose the model for stochastic riboswitch-controlled gene expression depicted in Fig. 6.

Notably, previous smFRET studies of the isolated *Tte* riboswitch also found two conformations populated from zero to saturating ligand concentration, identified as pre-folded and folded, that both respond to (that is, 'sense') ligand[19]. It is tempting to speculate that these two local conformations of the aptamer give rise to the two mRNA conformations that differentially bind the anti-SD during SiM-KARTS; the pre-folded state appears to have an only partially formed P2 helix[19] that is expected to bind the probe more readily (resulting in a burst of spikes), whereas the folded state features a fully formed P2 (ref. 19) that disfavours probe binding and thus leads to only sparse spikes (non-bursts, Fig. 6). Consistent with the relatively modest binding free energy of $-11\,\mathrm{kcal\,mol^{-1}}$ available from this small-molecule ligand[19], the mode of action of $preQ_1$ then is to subtly remodel the aptamer and only modestly reduce the hosting mRNA's translation-initiation frequency, consistent with the observed reduction of *in vitro* translation product by ~40%. This seemingly moderate (approximately twofold) change *in vitro* that can be directly attributed to the ligand suggests that regulation by $preQ_1$ is more nuanced than previously appreciated and that it may be potentiated by other forces at work in the cell, including: co-transcriptional folding of the mRNA from 5′ to 3′ ends[35,36], competition between ribosome-catalysed translation and RNase-mediated mRNA decay[37,38] (leading to a shorter lifetime of a sparsely translated mRNA, Fig. 6) and repeated unfolding of the RNA resulting from close spacing between the transcription complex and the leading and sequentially loaded ribosomes[39].

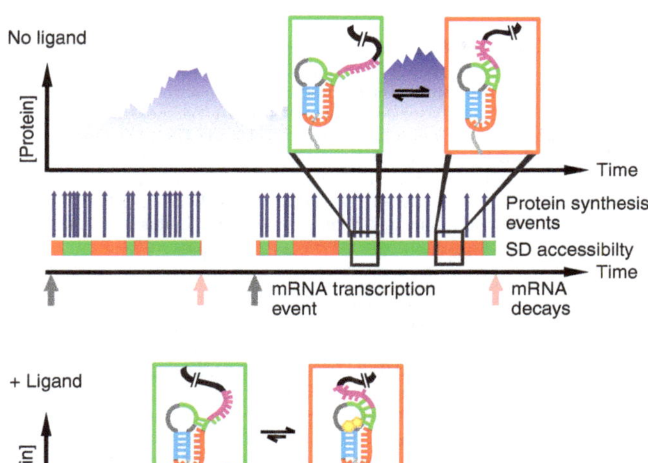

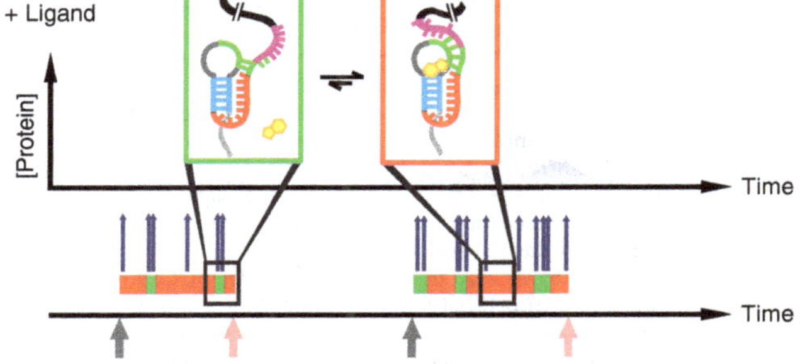

Figure 6 | A stochastic burst model for $preQ_1$-dependent expression of the *Tte* mRNA. In the absence of ligand (top), a transcription event (grey arrow) gives rise to an individual riboswitch-containing mRNA molecule that persists in the cell until it is degraded (light red arrow). During its lifetime, the mRNA transitions between burst (green bar segments) and non-burst (red bar segments) conformational states during which its SD sequence (purple nucleotides) is highly accessible and largely occluded, respectively. As reflected by increased anti-SD probe binding during SiM-KARTS, there are more potential opportunities for translation initiation events (blue arrows) in the burst state, leading to bursts of protein biosynthesis, than in the non-burst state. In the presence of the ligand (bottom), $preQ_1$ favours formation of the full P2 stem (green nucleotides) and occlusion of the SD sequence, resulting in shorter excursions to the burst state and, consequently, fewer opportunities for translation initiation. This may be accentuated by a decrease in the mRNA's lifetime (depicted as overall shorter burst/non-burst bars) resulting from its scarce occupancy with actively translating ribosomes, leading to significant downregulation of protein expression. Colouring of the riboswitch cartoon is as in Figs 1a and 2a.

The spike trains detected by SiM-KARTS resemble the transcriptional bursting that has been suggested as an underlying cause of genetic 'noise'[33,40,41]. This generally stochastic nature of biological systems results in cell-to-cell variability and has been shown to be beneficial to organisms, particularly during times of environmental stress[42–44]. Each time a single mRNA molecule is transcribed it gives rise to a few to tens of protein molecules[33]. We here show that translational bursting of an mRNA, which can be modulated by ligand binding to an embedded riboswitch, appears to add another layer of stochasticity to the gene expression cascade (Fig. 6). We anticipate that such translational bursting will turn out to be a widespread phenomenon among mRNAs that generally present structurally dynamic ribosome substrates whose SD region is known to significantly influence translation efficiency[45], and that our SiM-KARTS approach can detect changes in the secondary structure not just of single riboswitch-hosting mRNA molecules but of virtually any RNA under a wide range of conditions, poising it to find broad application.

Methods

Ligand and oligonucleotides. The $preQ_1$ ligand used in this study was generously provided by George Garcia; $preQ_1$ was synthesized following established methods[46] and additionally purified using reverse-phase HPLC as described[47]. DNA and LNA oligonucleotides were purchased from Integrated DNA Technologies Inc. (IDT) and Exiqon, respectively. Fluorophore-labelled DNA and LNA oligonucleotides were HPLC-purified by the respective manufacturer. The control and anti-SD probe RNAs were purchased from IDT with a 5′ aminohexyl-linker modification and labelled with Cy5-NHS ester (GE Healthcare) exactly as described[19]. The sequences of all oligonucleotides and cloning primers used in this study are listed in Supplementary Note 4.

RNA preparation. Reference genomic sequences for *T. tengcongensis* were downloaded from the National Center for Biotechnology Information (NC_003869.1, http://www.ncbi.nlm.nih.gov). The complete mRNA transcript, including the TTE_RS07450 and TTE_RS07445 (TTE1564 and TTE1563, respectively) ORFs, and its 3′ UTR as predicted from the FindTerm algorithm (SoftBerry), was amplified using PCR from *T. tengcongensis* genomic DNA, which was purchased from the NITE Biological Resource Center. The amplified region was cloned into the pUC19 plasmid between the BamHI and HindIII sites with an engineered upstream T7 promoter. For use as a control for *in vitro* translation assays, an FspI site was introduced through site-directed mutagenesis[48] into the control vector provided with the ActivePro In Vitro Translation Kit (Ambion) containing the CAT gene under the control of a T7 promoter (pAMB CAT). All of the plasmids used in this study are available through Addgene (www.addgene.org).

mRNA was produced by *in vitro* transcription. The *Tte* pUC19 plasmid was linearized with HindIII (AflII or XbaI for *in vitro* translation assays; New England Biolabs) for run-off transcription. Similarly, the pAMB CAT plasmid was linearized with FspI (New England Biolabs). Transcription reactions were performed in the presence of 120 mM HEPES-KOH (pH 7.5 at 25 °C), 25 mM $MgCl_2$, 2 mM spermidine, 40 mM dithiothreitol (DTT), 30 mM NTPs, 0.01% (w/v) Triton X-100, 200 nM linearized plasmid, 0.01 U μl^{-1} pyrophosphatase and 0.07 mg ml^{-1} T7 RNA polymerase in a total volume of 1 ml. Transcription reactions were incubated at 37 °C for 4 h. Enzyme was removed using phenol/chloroform extraction, and the resulting solution was spun in an Amicon 100 MWCO spin column to reduce the volume to ~100 µl. mRNA was purified by denaturing, 7 M urea, PAGE, detected using brief 254-nm ultraviolet radiation and gel-eluted overnight. mRNAs were ethanol-precipitated and resuspended in TE buffer at pH 7.0. The sequences of the *Tte* mRNAs used in this study are listed in Supplementary Note 4.

Equilibrium SiM-KARTS. An RNA mix containing 2 nM each of *Tte* mRNA, TYE563-LNA, biotin capture strand and Cy3-blocking strand (when present, see Supplementary Fig. 7 and Supplementary Note 1) were heat-annealed at 70 °C for 2 min in the presence of SiM-KARTS buffer containing 50 mM Tris-HCl (pH 7.5 at 25 °C), 0.6 M NaCl and 20 mM $MgCl_2$, and were allowed to cool to room temperature over 20 min in the presence or absence of $preQ_1$. Next, the RNA mix was diluted to 40 pM in the same buffer in the presence or absence of $preQ_1$, with an additional 12.5-fold excess of TYE563-LNA, biotin-capture strand and Cy3-blocking strand (when present) to ensure that the complex would stay intact during dilution. All sequences of mRNA, capture strand, TYE563-LNA, Cy3-blocking strand and Cy5 anti-SD probe can be found in Supplementary Note 4. The diluted complex was chilled on ice. The chilled solution was flowed over an assembled microfluidic channel on a quartz slide coated with biotinylated BSA and streptavidin, as previously described[49,50]. The chilled, 40 pM RNA complex solution (100 µL) was flowed over the slide and allowed to equilibrate for 5 min. Excess RNA was washed off the slide with SiM-KARTS buffer with or without $preQ_1$. An oxygen-scavenging system consisting of 5 mM protocatechuic acid and 50 nM protocatechuate-3,4-dioxygenase with or without $preQ_1$, to slow photobleaching, and 2 mM Trolox, to reduce photoblinking[51], as well as 50 nM Cy5 probe was flowed over the slide and allowed to equilibrate for 5 min. Both Cy5 and TYE563 dyes were directly and simultaneously excited using 638 nm red and 532 nm green diode lasers, respectively. Emission from both fluorophores was simultaneously recorded using an intensified charge-coupled device camera (I-Pentamax, Princeton Instruments) at 100-ms time resolution using the Micro-Manager software (https://www.micro-manager.org/). Fluorescence time traces were extracted from the raw movie files using IDL (Research Systems) and analysed using Matlab (The MathWorks, Inc.) scripts. Genuine traces exhibiting binding were manually selected using the following criteria: a single photobleaching step of the TYE563 signal to localize the mRNA molecule on the slide surface, TYE563 fluorescence intensity of >200 intensity units and at least two Cy5 co-localization signals per trajectory corresponding to anti-SD-binding events with a signal to noise ratio of at least 3:1. Suitable traces were compiled. Hidden Markov Modeling analysis was performed on the Cy5 intensity using the segmental k-means algorithm in the QuB software suite[52]. A two-state model was used with an unbound and bound state to idealize the data (for an additional discussion of the idealization procedure, see Supplementary Note 2). Transition density plots were constructed to extract the dwell times in the bound and unbound states, as described[53]. The normalized cumulative distributions of bound dwell times were fit with a double-exponential, and unbound dwell times were fit with a single-exponential association function (see text) in OriginLab 8.5 from which on- and off-rates were calculated. Rate constants for the anti-SD probe as a function of $preQ_1$ concentration were fit with a dose–response curve for an inhibitor with a standard Hill slope of -1. Linear regression of the data for the control probe in OriginLab showed the slope to not be significantly different from zero; these data were thus fit with regression lines of zero slope.

Ligand-jump SiM-KARTS. Fluorescently (TYE563) labelled $preQ_1$ riboswitches were immobilized as detailed above and first imaged in SiM-KARTS buffer without ligand. Initial co-localization of the TYE563 and Cy5 signals provided for unambiguous determination of the relative locations of single-mRNA molecules on the slide, even after TYE563 photobleaching. In some cases, the TYE563 signal persisted throughout both dark periods, further confirming that the same molecule was tracked throughout the duration of the experiment. Binding of anti-SD-labelled probes (Cy5) at these locations was continuously monitored to determine the accessibility of the SD sequence. SD accessibility was monitored for 150 s, and then a new solution of anti-SD probe in SiM-KARTS buffer was introduced in conjunction with $preQ_1$ at saturating concentration (16 µM). The process was repeated but with a final SiM-KARTS buffer solution without ligand. Because the fluorescence measured is of the anti-SD probe, which is in great excess and only excited briefly while near the surface because of the TIRFM illumination conditions, we can observe binding events to the same mRNA molecule throughout the change in ligand concentrations with limited risk of photobleaching the rapidly dissociating anti-SD probe molecules.

Ribosome preparation. Salt-washed ribosomes and separated ribosomal subunits were prepared using a previously described protocol with several modifications[54]. Briefly, *E. coli* MRE600 was grown in Luria-Bertani (LB) medium at 37 °C to an OD_{600} of 0.6–0.8, cooled at 4 °C for 45 min and then pelleted. All subsequent steps were performed on ice or at 4 °C. The cell pellet was resuspended in buffer A (20 mM Tris-HCl (pH 7.05 at 25 °C), 100 mM NH_4Cl, 10 mM $MgCl_2$, 0.5 mM EDTA and 6 mM β-mercaptoethanol), and the cells were lysed in a single pass using a M-110L Microfluidizer processor (Microfluidics). The lysate was cleared by centrifugation at 30,000*g*, and the clarified lysate was pelleted over a 35-ml sucrose cushion (1.1 M sucrose, 20 mM Tris-HCl (pH 7.05 at 25 °C), 500 mM NH_4Cl, 10 mM $MgCl_2$ and 0.5 mM EDTA) in a Beckman Ti-45 rotor for at least 16 h at 32,000 r.p.m. The pellet was washed with 1 ml of buffer B (20 mM Tris-HCl (pH 7.05 at 25 °C), 500 mM NH_4Cl, 10 mM $MgCl_2$ and 0.5 mM EDTA), resuspended in 10 ml of buffer B by gentle stirring and brought to a final volume of 35 ml in buffer B. This material was then pelleted again over a 35-ml sucrose cushion. The resulting pellet was washed with 1 ml of storage buffer (50 mM Tris-HCl (pH 7.5 at 25 °C), 70 mM NH_4Cl, 30 mM KCl, 7 mM $MgCl_2$ and 6 mM β-mercaptoethanol), resuspended in 2.5 ml storage buffer by gentle stirring and the dialysed against three changes of buffer C (50 mM Tris-HCl (pH 7.05 at 25 °C), 150 mM NH_4Cl, 1 mM $MgCl_2$ and 6 mM β-mercaptoethanol) to dissociate the subunits. A portion of the dialysed sample was adjusted to a final Mg^{2+} concentration of 7 mM, flash-frozen with liquid nitrogen in aliquots (salt-washed ribosomes) and stored at -80 °C. For the remaining dialysed sample, ~100 A_{260} units of material was loaded on each of six 10–40% sucrose gradients in buffer C and separated by zonal centrifugation in a Beckman SW-28 Ti rotor for 18 h at 20,000 r.p.m. Gradient fractions containing 30S or 50S ribosomal subunit peaks were pooled separately and pelleted in a Beckman Ti-70 rotor for 12 h at 61,500 r.p.m. Pelleted subunits were resuspended in storage buffer and flash-frozen with liquid nitrogen in aliquots (separated subunits) and stored at -80 °C.

***In vitro* translation assays.** Salt-washed ribosomes and separated subunits were found to perform similarly (Supplementary Fig. 12), and therefore salt-washed ribosomes were used for all *in vitro* translation assays unless otherwise noted. *In vitro* transcribed mRNAs were translated using the PURExpress Δ Ribosome Kit (New England Biolabs). For each reaction, 3 µl of a 4-µM mRNA solution (4 µM CAT, 4 µM *Tte* mRNA or a mixture of 0.8 µM CAT and 3.2 µM *Tte* mRNA) was re-folded in the presence of 0, 16 or 100 µM $preQ_1$ by heating to 70 °C for 2 min, followed by slow cooling to room temperature for 20 min, and then placed on ice. The remaining components required for translation were master-mixed and aliquoted to each reaction (1.5–9 µCi L-[^{35}S]-Cysteine, 4 µl PURExpress Solution A, 1.2 µl Factor Mix and 6 pmol salt-washed ribosomes or separated 30S and 50S subunits), along with additional $preQ_1$ required to maintain a final concentration of 0, 16 and 100 µM $preQ_1$ in the final reaction volume of 12 µl. Reactions were incubated at 37 °C for 2 h, frozen on dry ice and stored at -20 °C. The following day, reactions were thawed at room temperature and 2 µl of 1 M KOH was added to quench the reaction and cleave any remaining peptide from their tRNA. Protein products were precipitated by adding 5 volumes of cold acetone and pelleted by centrifugation at 14,000*g* for 10 min. Pellets were resuspended in 20 µl of 1 × loading buffer (45 mM Tris-HCl (pH 8.45 at 25 °C), 10% (v/v) glycerol, 50 mM DTT, 1% (w/v) SDS and 0.01% (w/v) bromophenol blue) and heated at 37 °C for 45 min. Protein products were resolved on 16% Tris-tricine SDS–PAGE gels[55] electrophoresed at 150 V for 2.5 h. Gels were then fixed for 45 min in 5% (v/v) glycerol, 40% (v/v) methanol and 10% (v/v) acetic acid and dried on a 3-mm Whatman paper using a Bio-Rad Model 583 Gel Dryer. Dried gels were imaged using a storage phosphorscreen and Typhoon 9410 Variable Mode Imager (GE Healthcare Life Sciences). Gel images were quantified using ImageQuant v5.2 (Molecular Dynamics). Unless otherwise noted, after background correction the intensities for CAT, TTE1564 and TTE1563 bands were divided by their respective number of cysteine residues (5, 1 and 1, respectively). For each lane, the intensities for TTE1564 and TTE1563 bands were summed and then divided by the intensity of the respective CAT band. Values at each $preQ_1$ concentration were graphed in Prism (GraphPad Software) and an unpaired, two-tailed t-test was used to assess statistical significance.

Burst analysis. Burst analysis was carried out using the RS method described by Gourévitch and co-workers[31]. We utilized a modified Matlab implementation of the RS method based on the Matlab script provided in the supplement to Gourévitch *et al.*[31] In the standard implementation, each molecule's ISIs are ranked independently. In our implementation, termed global burst analysis, we have extended this so that each ISI detected in all our experimental conditions is ranked simultaneously; therefore, a burst is defined as a global property of the molecules and is not biased by the number of binding events of an individual molecule. Briefly, ISIs were determined by calculating the time in between individual binding events for each molecule. ISIs for all molecules were collected and used as an input for global burst analysis, using the RS method to demarcate the start and end points of bursts, that is, the first and last binding events, respectively, in a sequential series of binding events occurring in quick succession ('burst'). Each start and end point was then reassigned to the corresponding molecule, preserving the single-molecule burst profile. Global burst analysis was carried out separately for equilibrium and ligand-jump data sets. The RS algorithm developed by Gourévitch *et al.*[31] requires two parameters, the maximal ISI between spikes to be considered part of a burst and a RS cutoff (alpha). These were set to 40 s and 3, respectively. Although the maximal ISI in a burst was set to 40 s, the distribution of ISIs we obtained after analysis is much lower than this, suggesting that we have provided enough flexibility in the algorithm to find the true distribution of ISIs in burst.

For the ligand-jump SiM-KARTS experiments, we further characterized the time evolution of SD accessibility throughout changes in ligand concentration by examining changes in the burst density for each molecule. We ranked the three segments (Minus, Plus, Minus′) for each molecule in the ligand-jump SiM-KARTS experiment by their burst density. For a given molecule, the segment with the highest density of bursts was ranked as High (H), the next highest as Mid (M) and lowest as Low (L). We then quantified the overall distribution of burst density rankings for the three segments. We plotted each individual molecule's burst behaviour as a rastergram and organized them such that their per-segment ranks were the same within a group. MATLAB scripts for global burst analysis are provided as part of the Supplementary Information.

Fano factor calculations. Matlab scripts were written to calculate and simulate the Fano factor from our experimental data and from a simulated Poisson distribution, respectively. The Fano factor is defined as the variance in spike counts divided by the mean spike count for a given time interval, T. For every molecule analysed in a particular condition, a time interval of length T was randomly selected from the molecule's fluorescence time trace. The Fano factor was calculated for T equal to 5, 10, 20 and 40 s. Each time window T was sampled 100 times with a different random seed for each molecule to generate an average Fano factor. The average Fano factor and the s.d. for each time window are presented in Supplementary Fig. 9. For the simulations, the Matlab Poisson random number generator was utilized to generate spike counts with an average firing rate equal to the average ISI in the burst from our experimental data and equal number of samplings. 95% Confidence intervals were calculated in Matlab utilizing the function:

bounds = gaminv([.025,.975],$(n-1)/2,2/(n-1)$), where n = sample size[34].

Measurement of the anti-SD probe diffusion coefficient. Fluorescence correlation spectroscopy was used to determine the diffusion coefficient of the Cy5 anti-SD probe. Glass coverslips were passivated before use as follows: a coverslip was pre-wet with 50 µl of T50 buffer (Tris-HCl (pH 8.0 at 22 °C), 50 mM NaCl), dried with a gentle stream of air and then incubated with $1.0\,\mathrm{mg\,ml^{-1}}$ of biotinylated BSA for 10 min. The biotinylated BSA solution was then removed, the coverslip was washed once with 50 µl T50 buffer, dried and then incubated with $0.2\,\mathrm{mg\,ml^{-1}}$ of streptavidin for 5 min. The streptavidin solution was then removed, and the coverslip was finally washed twice with SiM-KARTS buffer. The anti-SD probe was diluted to 2.5 nM in SiM-KARTS buffer, a small aliquot was placed on the washed and dried coverslip and measurements were performed at room temperature on an Olympus IX81 inverted microscope with an ISS ALBA 5 confocal system (Champaign, IL). Ten replicate traces of 30 s each were acquired and sampled at 50 kHz. The data from each replicate were averaged and fit in PyCorrFit v0.9.1 (http://pycorrfit.craban.de/) with a correlation function equation (1) for three-dimensional free diffusion with a Gaussian laser excitation profile (elliptical), including a triplet component, where n is the effective number of particles in the confocal volume, τ_{diff} is the characteristic residence time in the confocal volume, SP is the structural parameter, T is the fraction of particles in the non-fluorescent triple state and τ_{trip} is the characteristic residence time in the triplet state[56].

$$G(\tau) = \left(\frac{1}{n}\right)\left(\frac{1}{1+\frac{\tau}{\tau_{\text{diff}}}}\right)\left(\frac{1}{\sqrt{1+\frac{\tau}{SP^2\tau_{\text{diff}}}}}\right)\left(1+\frac{Te^{\left(-\frac{\tau}{\tau_{\text{trip}}}\right)}}{1-T}\right) \quad (1)$$

The average correlation curve is shown in Supplementary Fig. 14. A diffusion coefficient, D_{probe}, was calculated from the fit of the average according to the relationship:

$$\tau_{\text{diff}} = \frac{w_0^2}{4D_{\text{probe}}} \quad (2)$$

where w_0 is the dimension of the Gaussian detection volume transversal to the optical axis. The uncertainty for the diffusion coefficient was estimated by re-fitting 7 randomly selected subsets of the data, where each subset contained 4 of the 10 replicates, and taking the s.d.

References

1. Breaker, R. R. Prospects for riboswitch discovery and analysis. *Mol. Cell* **43,** 867–879 (2011).
2. Mandal, M. & Breaker, R. R. Gene regulation by riboswitches. *Nat. Rev. Mol. Cell Biol.* **5,** 451–463 (2004).
3. Nudler, E. & Mironov, A. S. The riboswitch control of bacterial metabolism. *Trends Biochem. Sci.* **29,** 11–17 (2004).
4. Winkler, W. C. & Breaker, R. R. Regulation of bacterial gene expression by riboswitches. *Annu. Rev. Microbiol.* **59,** 487–517 (2005).
5. Winkler, W. C., Nahvi, A., Roth, A., Collins, J. A. & Breaker, R. R. Control of gene expression by a natural metabolite-responsive ribozyme. *Nature* **428,** 281–286 (2004).
6. Batey, R. T., Gilbert, S. D. & Montange, R. K. Structure of a natural guanine-responsive riboswitch complexed with the metabolite hypoxanthine. *Nature* **432,** 411–415 (2004).
7. Mandal, M. & Breaker, R. R. Adenine riboswitches and gene activation by disruption of a transcription terminator. *Nat. Struct. Mol. Biol.* **11,** 29–35 (2004).
8. Blount, K. F., Wang, J. X., Lim, J., Sudarsan, N. & Breaker, R. R. Antibacterial lysine analogs that target lysine riboswitches. *Nat. Chem. Biol.* **3,** 44–49 (2007).
9. Mandal, M. *et al.* A glycine-dependent riboswitch that uses cooperative binding to control gene expression. *Science* **306,** 275–279 (2004).
10. Barrick, J. E. & Breaker, R. R. The distributions, mechanisms, and structures of metabolite-binding riboswitches. *Genome Biol.* **8,** R239 (2007).
11. Sudarsan, N. *et al.* Tandem riboswitch architectures exhibit complex gene control functions. *Science* **314,** 300–304 (2006).
12. Dann, 3rd C. E. *et al.* Structure and mechanism of a metal-sensing regulatory RNA. *Cell* **130,** 878–892 (2007).
13. Regulski, E. E. *et al.* A widespread riboswitch candidate that controls bacterial genes involved in molybdenum cofactor and tungsten cofactor metabolism. *Mol. Microbiol.* **68,** 918–932 (2008).
14. Shi, Y., Zhao, G. & Kong, W. Genetic analysis of riboswitch-mediated transcriptional regulation responding to Mn^{2+} in *Salmonella*. *J. Biol. Chem.* **289,** 11353–11366 (2014).
15. Anthony, P. C., Perez, C. F., Garcia-Garcia, C. & Block, S. M. Folding energy landscape of the thiamine pyrophosphate riboswitch aptamer. *Proc. Natl Acad. Sci. USA* **109,** 1485–1489 (2012).
16. Haller, A., Altman, R. B., Souliere, M. F., Blanchard, S. C. & Micura, R. Folding and ligand recognition of the TPP riboswitch aptamer at single-molecule resolution. *Proc. Natl Acad. Sci. USA* **110,** 4188–4193 (2013).
17. Haller, A., Rieder, U., Aigner, M., Blanchard, S. C. & Micura, R. Conformational capture of the SAM-II riboswitch. *Nat. Chem. Biol.* **7,** 393–400 (2011).
18. Jenkins, J. L., Krucinska, J., McCarty, R. M., Bandarian, V. & Wedekind, J. E. Comparison of a $preQ_1$ riboswitch aptamer in metabolite-bound and free states with implications for gene regulation. *J. Biol. Chem.* **286,** 24626–24637 (2011).
19. Suddala, K. C. *et al.* Single transcriptional and translational $preQ_1$ riboswitches adopt similar pre-folded ensembles that follow distinct folding pathways into the same ligand-bound structure. *Nucleic Acids Res.* **41,** 10462–10475 (2013).
20. Wilson, R. C. *et al.* Tuning riboswitch regulation through conformational selection. *J. Mol. Biol.* **405,** 926–938 (2011).
21. Henkin, T. M. Preface—riboswitches. *Biochim. Biophys. Acta* **1839,** 899 (2014).
22. Eichhorn, C. D., Kang, M. J. & Feigon, J. Structure and function of $preQ_1$ riboswitches. *Biochim. Biophys. Acta* **1839,** 939–950 (2014).
23. Shine, J. & Dalgarno, L. The 3′-terminal sequence of *Escherichia coli* 16S ribosomal RNA: complementarity to nonsense triplets and ribosome binding sites. *Proc. Natl Acad. Sci. USA* **71,** 1342–1346 (1974).
24. Laursen, B. S., Sørensen, H. P., Mortensen, K. K. & Sperling-Petersen, H. U. Initiation of protein synthesis in bacteria. *Microbiol. Mol. Biol. Rev.* **69,** 101–123 (2005).

25. Greenfeld, M., Solomatin, S. V. & Herschlag, D. Removal of covalent heterogeneity reveals simple folding behavior for P4-P6 RNA. *J. Biol. Chem.* **286,** 19872–19879 (2011).
26. Hyeon, C., Lee, J., Yoon, J., Hohng, S. & Thirumalai, D. Hidden complexity in the isomerization dynamics of Holliday junctions. *Nat. Chem. Biol.* **4,** 907–914 (2012).
27. Okumus, B., Wilson, T. J., Lilley, D. M. & Ha, T. Vesicle encapsulation studies reveal that single molecule ribozyme heterogeneities are intrinsic. *Biophys. J.* **87,** 2798–2806 (2004).
28. Tan, E. *et al.* A four-way junction accelerates hairpin ribozyme folding via a discrete intermediate. *Proc. Natl Acad. Sci. USA* **100,** 9308–9313 (2003).
29. deSilva, C. & Walter, N. G. Leakage and slow allostery limit performance of single drug-sensing aptazyme molecules based on the hammerhead ribozyme. *RNA* **15,** 76–84 (2009).
30. Johnson-Buck, A. *et al.* Kinetic fingerprinting to identify and count single nucleic acids. *Nat. Biotechnol.* **33,** 730–732 (2015).
31. Gourévitch, B. & Eggermont, J. J. A nonparametric approach for detection of bursts in spike trains. *J. Neurosci. Method.* **160,** 349–358 (2007).
32. Cai, L., Friedman, N. & Xie, X. S. Stochastic protein expression in individual cells at the single molecule level. *Nature* **440,** 358–362 (2006).
33. Yu, J., Xiao, J., Ren, X. J., Lao, K. Q. & Xie, X. S. Probing gene expression in live cells, one protein molecule at a time. *Science* **311,** 1600–1603 (2006).
34. Eden, U. T. & Kramer, M. A. Drawing inferences from Fano factor calculations. *J. Neurosci. Methods* **190,** 149–152 (2010).
35. Lai, D., Proctor, J. R. & Meyer, I. M. On the importance of cotranscriptional RNA structure formation. *RNA* **19,** 1461–1473 (2013).
36. Poot, R. A., Tsareva, N. V., Boni, I. V. & van Duin, J. RNA folding kinetics regulates translation of phage MS2 maturation gene. *Proc. Natl Acad. Sci. USA* **94,** 10110–10115 (1997).
37. Caron, M. P. *et al.* Dual-acting riboswitch control of translation initiation and mRNA decay. *Proc. Natl Acad. Sci. USA* **109,** E3444–E3453 (2012).
38. Laalami, S., Zig, L. & Putzer, H. Initiation of mRNA decay in bacteria. *Cell. Mol. Life Sci.* **71,** 1799–1828 (2014).
39. Picard, F. *et al.* Bacterial translational regulations: high diversity between all mRNAs and major role in gene expression. *BMC Genomics* **13,** 528 (2012).
40. Golding, I., Paulsson, J., Zawilski, S. M. & Cox, E. C. Real-time kinetics of gene activity in individual bacteria. *Cell* **123,** 1025–1036 (2005).
41. So, L. H. *et al.* General properties of transcriptional time series in *Escherichia coli*. *Nat. Genet.* **43,** 554–560 (2011).
42. Balazsi, G., van Oudenaarden, A. & Collins, J. J. Cellular decision making and biological noise: from microbes to mammals. *Cell* **144,** 910–925 (2011).
43. Beaumont, H. J., Gallie, J., Kost, C., Ferguson, G. C. & Rainey, P. B. Experimental evolution of bet hedging. *Nature* **462,** 90–93 (2009).
44. Hebenstreit, D., Deonarine, A., Babu, M. M. & Teichmann, S. A. Duel of the fates: the role of transcriptional circuits and noise in CD4+ cells. *Curr. Opin. Cell Biol.* **24,** 350–358 (2012).
45. Tuller, T. & Zur, H. Multiple roles of the coding sequence 5' end in gene expression regulation. *Nucleic Acids Res.* **43,** 13–28 (2015).
46. Akimoto, H., Imamiya, E., Hitaka, T., Nomura, H. & Nishimura, S. Synthesis of queuine, the base of naturally occurring hypermodified nucleoside (queuosine), and its analogues. *J. Chem. Soc.* **1,** 1637–1644 (1988).
47. Roth, A. *et al.* A riboswitch selective for the queuosine precursor $preQ_1$ contains an unusually small aptamer domain. *Nat. Struct. Mol. Biol.* **14,** 308–317 (2007).
48. Todd, G. C. & Walter, N. G. Secondary structure of bacteriophage T4 gene *60* mRNA: implications for translational bypassing. *RNA* **19,** 685–700 (2013).
49. Roy, R., Hohng, S. & Ha, T. A practical guide to single-molecule FRET. *Nat. Method* **5,** 507–516 (2008).
50. Rueda, D. *et al.* Single-molecule enzymology of RNA: essential functional groups impact catalysis from a distance. *Proc. Natl Acad. Sci. USA* **101,** 10066–10071 (2004).
51. Rasnik, I., McKinney, S. A. & Ha, T. Nonblinking and long-lasting single-molecule fluorescence imaging. *Nat. Method* **3,** 891–893 (2006).
52. Blanco, M. & Walter, N. G. Analysis of complex single-molecule FRET time trajectories. *Methods Enzymol.* **472,** 153–178 (2010).
53. Abelson, J. *et al.* Conformational dynamics of single pre-mRNA molecules during *in vitro* splicing. *Nat. Struct. Mol. Biol.* **17,** 504–512 (2010).
54. Shaw, J. J. & Green, R. Two distinct components of release factor function uncovered by nucleophile partitioning analysis. *Mol. Cell* **28,** 458–467 (2007).
55. Schagger, H. & Vonjagow, G. Tricine-sodium dodecyl sulfate-polyacrylamide gel electrophoresis for the separation of proteins in the range from 1 to 100-KDa. *Anal. Biochem.* **166,** 368–379 (1987).
56. Müller, P., Schwille, P. & Weidemann, T. PyCorrFit-generic data evaluation for fluorescence correlation spectroscopy. *Bioinformatics* **30,** 2532–2533 (2014).
57. Leontis, N. B. & Westhof, E. Geometric nomenclature and classification of RNA base pairs. *RNA* **7,** 499–512 (2001).

Acknowledgements

We thank J. Widom and K. Suddala for careful reading of the manuscript, and G. Garcia for the generous gift of $preQ_1$. We thank the Single Molecule Analysis in Real-Time (SMART) Center of the University of Michigan, seeded by NSF MRI-R2-ID award DBI-0959823 to N.G.W., as well as J.D. Hoff for performing the Fluorescence correlation spectroscopy measurements. This research was funded by NIH grants GM062357, GM098023 and a sub-contract on GM063162 (PI: J. E. Wedekind) to N.G.W.

Author contributions

A.J.R. and N.G.W. conceived of the project. A.J.R. designed, conducted and analysed the SiM-KARTS experiments. M.R.B. wrote and implemented the scripts for burst analysis. P.E.L. designed, conducted and analysed the *in vitro* translation experiments and SiM-KARTS experiments performed in the presence of blocking strand. M.R.B. and P.E.L. wrote and implemented scripts for burst analysis. All authors were involved in interpreting and discussing the results, and participated in writing the paper.

Additional information

Competing financial interests: The authors declare no competing financial interests.

2

Cyclophilin A stabilizes the HIV-1 capsid through a novel non-canonical binding site

Chuang Liu[1,2,*], Juan R. Perilla[3,*], Jiying Ning[1,2], Manman Lu[2,4], Guangjin Hou[2,4], Ruben Ramalho[5], Benjamin A. Himes[1], Gongpu Zhao[1,2], Gregory J. Bedwell[6], In-Ja Byeon[1,2], Jinwoo Ahn[1,2], Angela M. Gronenborn[1,2], Peter E. Prevelige[2,6], Itay Rousso[5], Christopher Aiken[2,7], Tatyana Polenova[2,4], Klaus Schulten[3] & Peijun Zhang[1,2]

The host cell factor cyclophilin A (CypA) interacts directly with the HIV-1 capsid and regulates viral infectivity. Although the crystal structure of CypA in complex with the N-terminal domain of the HIV-1 capsid protein (CA) has been known for nearly two decades, how CypA interacts with the viral capsid and modulates HIV-1 infectivity remains unclear. We determined the cryoEM structure of CypA in complex with the assembled HIV-1 capsid at 8-Å resolution. The structure exhibits a distinct CypA-binding pattern in which CypA selectively bridges the two CA hexamers along the direction of highest curvature. EM-guided all-atom molecular dynamics simulations and solid-state NMR further reveal that the CypA-binding pattern is achieved by single-CypA molecules simultaneously interacting with two CA subunits, in different hexamers, through a previously uncharacterized non-canonical interface. These results provide new insights into how CypA stabilizes the HIV-1 capsid and is recruited to facilitate HIV-1 infection.

[1] Department of Structural Biology, University of Pittsburgh School of Medicine, Pittsburgh, Pennsylvania 15260, USA. [2] Pittsburgh Center for HIV Protein Interactions, University of Pittsburgh School of Medicine, Pittsburgh, Pennsylvania 15260, USA. [3] Department of Physics and Beckman Institute, University of Illinois at Urbana-Champaign, Urbana, Illinois 61801, USA. [4] Department of Chemistry and Biochemistry, University of Delaware, Newark, Delaware 19716, USA. [5] Department of Physiology and Cell Biology, Ben-Gurion University of the Negev, Be'er-Sheva, 84105, Israel. [6] Department of Microbiology, University of Alabama at Birmingham, Birmingham Alabama 35294, USA. [7] Department of Pathology, Microbiology and Immunology, Vanderbilt University School of Medicine, Nashville, Tennessee 37232, USA. * These authors contributed equally to this work. Correspondence and requests for materials should be addressed to K.S. (email: kschulte@ks.uiuc.edu) or to P.Z. (email: pez7@pitt.edu).

Infection by retroviruses, including HIV-1, is critically dependent on the functions of the viral capsid[1–3], which plays multiple roles during replication, including the prevention of innate sensor triggering[4,5] and regulation of reverse transcription[1,6] and nuclear import[7–9]. Such functions are highly dependent on interactions between the viral capsid and cellular factors. The host cell protein cyclophilin A (CypA) binds directly to the HIV-1 capsid and modulates capsid uncoating and viral infectivity[10–12]. Interference with CypA-binding inhibits HIV-1 replication in cell culture[13–15]. In addition, host cell proteins containing the CypA domain, including Trim-Cyp and Nup358, interact directly with the viral capsid and control HIV-1 infection[8,16,17].

CypA-binding appears to stabilize or destabilize the HIV-1 capsid, depending on the cell type[6,18]. In addition, the interaction between CypA and HIV-1 CA promotes HIV-1 infection of human cells[19,20], yet, in non-human primates, the same interaction enhances the anti-HIV-1 restriction activity of Trim5α (refs 21–23). Furthermore, while CypA promotes HIV-1 reverse transcription in human cells, its cell-type specific effect is best correlated with nuclear entry, and probably involves an unknown CypA-dependent host restriction factor[24]. Owing to this complexity, understanding the role of CypA in HIV-1 infection has been difficult, especially given the limited structural information that is available.

The crystal structure of CypA in complex with the CA N-terminal domain (CA_{NTD}) shows that CA residues A88-G89-P90 and the configuration of the CypA-binding loop are important for the binding interaction[11]. Binding of CypA to the CypA-binding loop of the viral capsid is essential for viral infectivity[25,26], as substitutions in loop residues G89 or P90 in HIV-1 CA are deleterious to replication[14,21,27]. CypA binding to monomeric CA can be detected at high protein concentration[11], but binding is enhanced by CA multimerization[10], suggesting that CypA favours binding to an assembled capsid. Despite extensive studies on the interaction between CypA and HIV-1 CA, structural information on how CypA interacts with the assembled viral capsid, and an understanding of the consequences of the interaction for capsid function, are lacking.

To elucidate the molecular interactions between CypA and the HIV-1 capsid, we determined the structure of CypA in complex with an HIV-1 CA tubular assembly at 8-Å resolution by cryoEM. The density map, combined with structure-guided molecular dynamics (MD) simulations, unexpectedly revealed a novel, non-canonical, second capsid-binding site on CypA that is vital for stabilizing the viral capsid. The finding was confirmed experimentally by solid-state NMR.

Results

Binding of CypA to wild-type (wt) CA assemblies. We began our studies by characterizing the binding of CypA to HIV-1 CA tubular assemblies at various CypA:CA molar ratios. Binding was measured in a high-speed centrifugation assay, after incubation of CypA with a fixed concentration of pre-assembled wild-type (wt) CA tubes. Incubation of CypA with CA tubes, up to a 6:6 molar ratio, resulted in co-sedimentation of CypA and CA (Fig. 1a, top), with the CypA:CA binding ratio increasing as the CypA concentration increased, up to 40 μM, beyond which point, binding was saturated (Fig. 1b, solid line). Interestingly, while a relatively low amount of CypA (CypA:CA input ratios $\leq$ 2:6) slightly enhanced the amount of CA in the pelleted fraction (Fig. 1a top), CypA:CA ratios above 2:6 led to a reduction in the amount of pelleted CA (Fig. 1a, top). This finding is consistent with an earlier report that high molar ratios of CypA:CA alter CA assembly *in vitro*[28].

We also tested the effect of CypA on the process of CA assembly by including it in the CA assembly reaction under the high-salt condition (2.25 M NaCl) (Fig. 1a, bottom). In this case, CypA binding was consistently higher compared with when it was added to pre-assembled CA tubes (Fig. 1b, dotted lines). However, the presence of CypA during CA assembly induced marked morphological changes in the resulting CA assemblies (Fig. 1c): in the absence of CypA, long CA tubes were formed; at an intermediate level of CypA, short tubes and cones were generated; and, at high CypA concentrations (6:6 CypA:CA molar ratio input), very small cones and spheres were observed, indicating a significant increase in the curvature of the assemblies. These data suggest that CypA influences the curvature of CA assemblies possibly by preferentially binding to and stabilizing CA assembly conformations with a high degree of curvature, a distinctive feature that had not been previously appreciated but can be clearly explained by the structure presented here (further sections). This influence of CypA on the capsid curvature may shed new light on the maturation process, since CypA is packaged into viral particles and may facilitate the formation of cones by stablizing a high degree of curvature at the ends during maturation. Similar to the pre-assembled tubes, CA was reduced in the pelleted fraction with higher CypA:CA molar ratios in the co-assembly reaction (Fig. 1a).

CypA affects the stiffness of CA assemblies. To determine the effect of CypA on CA stability, we used atomic force microscopy (AFM) operated in the nano-indentation mode to measure the stiffness of wt CA assemblies as a function of CypA level. The point stiffness measured for CA without CypA was $0.043 \pm 0.006\,N\,m^{-1}$ ($n = 30$) (Fig. 1d). A gradual elevation in CypA level resulted in an increase in CA stiffness (1:6 CypA:CA ratio is $0.11 \pm 0.02\,N\,m^{-1}$ ($n = 31$) and 3:6 is $0.14 \pm 0.018\,N\,m^{-1}$ ($n = 24$)). The stiffness value represents the amount of force that is required to deform the CA structure or, alternatively, the ability of the structure to withstand an external force. In the case of a CA assembly, this ability is governed by the interaction strength between its building blocks, namely, CA hexamers. Thus, higher stiffness values represent stronger interactions between the building blocks of the CA lattice. Incorporation of CypA into the CA structure increased the effective interaction strength between the CA hexamers. This effect was maximal at a 3:6 CypA:CA ratio. Interestingly, further addition of CypA to a 6:6 ratio reduced the stiffness of CA by almost twofold ($0.088 \pm 0.011\,N\,m^{-1}$ ($n = 45$)), compared with the stiffness obtained at a 3:6 ratio. However, the CA stiffness at 6:6 remains significantly larger than CA lacking CypA. These results suggest that CypA incorporation in the capsid structure strengthens the stability of capsid in a stoichiometrically dependent manner, with the efficiency of this effect being notably decreased at high (6:6) CypA to CA molar ratios.

CryoEM structure of CypA-CA tubular complex. To determine the structure of the CypA-CA tubular assembly complex, we chose the CypA-CA ratio and assembly conditions that yielded homogeneous complexes suitable for cryoEM: pre-assembled CA tubes with CypA added at 2:6 CypA:CA molar ratio (Fig. 1a top). Addition of CypA resulted in well-separated, long helical tubes with a layer of density decorating the CA tube surface (Fig. 2a, inset, black arrows). Such tubes are well-ordered to at least 13.8-Å resolution, as indicated by the Fourier transform (Fig. 2b). We carried out three-dimensional (3D) reconstruction of the CypA-CA tubular complex using the iterative real-space helical reconstruction method[29]. Combining 19 high-resolution tubes belonging to a single helical symmetry of (− 8, 13) (see ref. 29 for

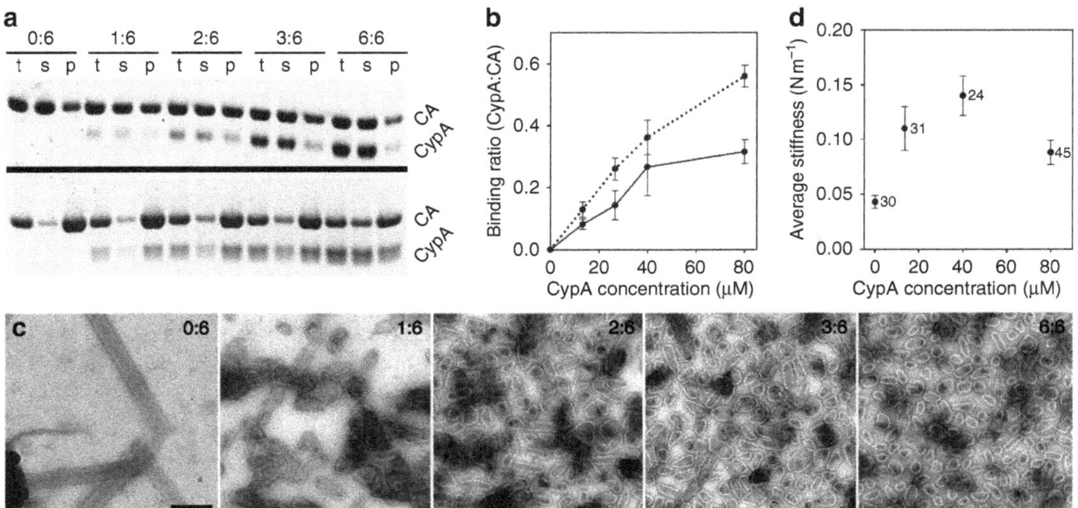

Figure 1 | Interaction of CypA with HIV-1 CA assemblies. (**a**) Binding of CypA to pre-assembled wt CA tubes in 1 M NaCl (top) and co-assembly of CypA and CA in 2.25 M NaCl (bottom); the molar ratio of CypA:CA in each condition is indicated. Reaction products before centrifugation (t) and after centrifugation, supernatant (s) and pellet (p), were analysed by 4–12% gradient gel. (**b**) The binding ratios of CypA:CA were quantified for the pre-assembled sample (solid line) and for the co-assembled sample (dotted line), calibrated using the input ratio as standard. (**c**) EM images of negative stained samples of CypA-CA complexes co-assembled in 2.25 M NaCl with the indicated CypA:CA molar ratios. Scale bar, 200 nm. (**d**) AFM measurements of the stiffness of CA assemblies in the presence of CypA at CypA:CA molar ratios of 0:6, 1:6, 3:6 and 6:6. At least 24 particles (numbers shown to the right of each data point), obtained from 2–3 individual assembly reactions, were measured for each stiffness value. Error bars represent the s.e.m.

information regarding helical symmetry), we determined the cryoEM structure of the CypA-CA tubular complex at 8-Å resolution, as indicated by the Fourier shell correlation of 0.5 (Supplementary Fig. 1).

The resulting density map displays a number of intriguing features. First, among the three CA helical arrays in the assembly (Fig. 2c; dashed black arrows denote the three directions along which helical tubular arrays are formed, each direction corresponding to a different curvature), CypA binds preferentially along the most highly curved array (Fig. 2c, Supplementary Fig. 2a). Very little CypA binds to the CA arrays that run in the other two helical directions. The same binding preference was observed in an independent reconstruction generated from tubes with a very different helical symmetry (− 12, 11) (Supplementary Fig. 2b). Such selective binding explains the sub-stoichiometric binding ratios observed in our CypA-binding assays (Fig. 1b). Second, while the CA region is highly ordered and all the α-helices are well resolved (Fig. 2d, Supplementary Fig. 1b,c), the resolution at the CypA region is poorer (Fig. 2e, Supplementary Fig. 1b,c). This could be because of the known flexibility of the CypA-binding loop[11,30], which would give rise to variable CypA orientations, and/or to partial occupancy by the CypA molecule. Indeed, when individual images were aligned based on the CA density (Supplementary Fig. 1a), the resolution for the CA region was improved, but the density corresponding to the CypA molecule became less well-defined (Supplementary Fig. 1b,c). Third, rigid body docking of the hexamer model (PDB: 3J4F) derived from our previous study[30] showed a near-perfect fit into the CA density (Fig. 2d), with a cross-correlation coefficient (CCC) of 0.84 between the atomic models of the hexamer and the density map, suggesting that CypA binding to a pre-assembled tube has little effect on the overall structure of the CA hexamer. Finally, aligning the CA_{NTD} from the crystal structure of the CypA-CA_{NTD} complex (PDB: 1AK4) (ref. 11) with the CA_{NTD} in our rigid body fitted model (3J4F) places the CypA model exactly into the CypA density region (Fig. 2e). Most interestingly, the CypA density appears to bridge two neighbouring CA hexamers and is located directly above the CA_{CTD} dimer interface (Fig. 2e) of two adjacent, non-CypA-binding CA molecules (Fig. 2c–e; Fig. 3e, red CA molecules).

An initial interpretation may be that the bridging density corresponds to a CypA dimer, with each CypA molecule binding to the CypA-binding loop from the adjacent CA hexamer. However, there is not enough space to accommodate two CypA molecules without extensive clashing. In addition, analytical centrifugation experiments suggest that CypA is primarily monomeric in our experimental conditions (Supplementary Fig. 3). Furthermore, focused classification using reference-based cross-correlation to separate CypA-binding modes into distinct classes, resulted in class averages that contain monomeric CypA bound to CA hexamers (Supplementary Fig. 2c). Thus, the bridging density likely corresponds to a single-CypA molecule, in which case, one may ask why CypA only occupies one particular helical array, rather than being equally distributed along all directions. One possibility is that a single CypA interacts with two CA molecules that are arranged in a specific geometry, with each CA molecule recognizing a different region of the enzyme. Supporting evidence for this idea comes from two sources: (1) a crystal structure of the CypA/CsA complex (PDB: 2RMA), in which the opposite faces of CypA each bind a molecule of CsA (ref. 31); and (2) an analysis of the inhibition of CypA-Gag interaction by CsA, showing a sigmoidal CsA inhibition curve, indicating synergy between two CsA in the inhibition[12]. For these reasons, we suggest that a single-CypA molecule spans two CA hexamers in the tubular complex (Fig. 2e).

Computational modelling of the CypA-CA lattice. For the binding of a single-CypA molecule to two adjacent CA hexamers in a helical array, two binding modes are possible: (1) CypA binds more strongly to the left hexamer, extending across to the right one; or (2) CypA binds the other way around. The density map is

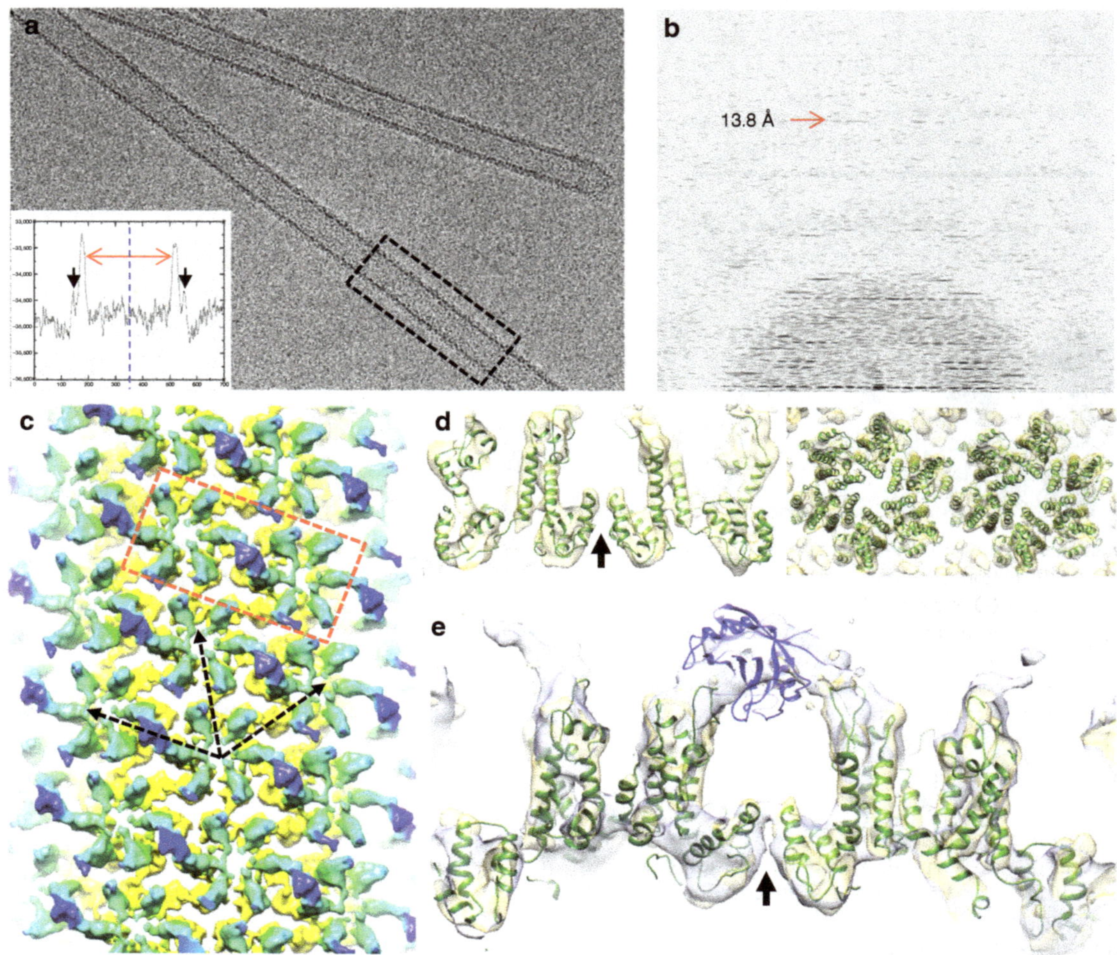

Figure 2 | CryoEM reconstruction of CypA /CA tubular assemblies. (**a**) CryoEM image of a CypA-CA tubular assembly at 2:6 CypA:CA input ratio. Inset, radial density profile from the boxed region showing the CypA density (black arrows) decorating the CA tubular assembly (red arrows). (**b**) Fourier transform of a CypA-CA tubular assembly. Arrow indicates a layer line at 13.8 Å resolution. (**c**) 3D density map of the CypA-CA assembly complex reconstructed at 8 Å resolution (see Supplementary Fig. 1). The density (contoured at 2σ) is coloured radially from yellow-green (CA) to blue (CypA). CypA selectively binds to the CA array that has the highest curvature. (**d**) Rigid body docking of CA hexamers (3J4F, green ribbon) into the CA region of the density map (yellow density), viewed from the side and top. The area shown is from the boxed region in C. (**e**) Aligning the CypA-CA complex (PDB: 1AK4) to the docked CA hexamer model, overlaid with the density maps contoured at 2σ (yellow) and at 1.5σ (blue). The CypA molecule (blue ribbon) is situated right at the bridging density, across two CA hexamers (green ribbons) and above the CA dimer interface (arrow).

an ensemble average of all possible binding modes from many tubes and, thus, cannot distinguish between the binding modes. Therefore, we carried out MD flexible fitting (MDFF) to model the interaction between CypA and the CA tubes (simulations 1and 2—Supplementary Table 1), monitoring, during the simulation, the cross-correlation between the model and the experimental density (Supplementary Fig. 4). An ensemble of 10 fully saturated CypA-CA tubular complex models (Fig. 3a) was constructed as described in Supplementary Information. Overlapping or clashing CypA molecules were removed with no bias from the saturated models, resulting in 10 tubular complex models with thousands of statistically independent arrangements of CypA over the CA lattice (Supplementary Fig. 5a). Interestingly, in all cases, clashes between CypA and adjacent CA molecules were only observed along the least curved helical direction of the tube (a typical HIV tube contains a CA lattice with three helical directions along which repeating arrays are found, each direction with different characteristic curvatures as illustrated in Fig. 2c and Supplementary Fig. 6), resulting in removal of CypA molecules along this direction (Supplementary Fig. 5a). Subsequently, the CypA-CA tubular complex models were equilibrated for 100 ns (Supplementary Fig. 7, simulation 2—Supplementary Table 1), and simulated electron density maps were generated for each model (Supplementary Fig. 5b). Mimicking the cryoEM real-space helical reconstruction process, a final, averaged density map with helical symmetry imposed was calculated from all these models.

The resulting average density map (Fig. 3b) matches the experimental map remarkably well, especially the selectivity of CypA for the most curved CA lattice array, with a cross-correlation value of 0.967 (Figs 2c and 3b). During the course of the MD simulations (simulation 2), the CA lattices remained rigid, as previously reported[30], whereas the CypA-binding loops and the CypA molecules exhibited a range of motions (as calculated from root-mean-square fluctuations of the atomic positions, mapped onto the structure of the tube in Supplementary Fig. 8), which can be considered a likely cause for the lower resolution of the cryoEM density map at the CypA region (Supplementary Fig. 1b,c). Indeed, principal component analysis of longer trajectories of CypA-CA subsystems extracted from the MDFF-derived

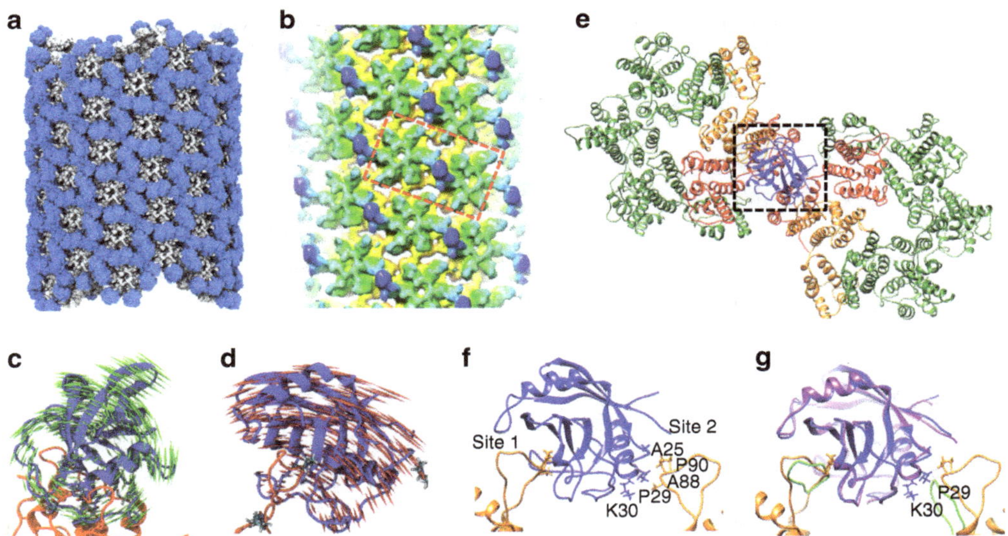

Figure 3 | All-atom MD simulations of the CypA-CA complex in a helical assembly. (**a**) A fully saturated CypA-CA tubular complex model with extensive clashes between CypA and CA and between adjacent CypA molecules. CypA is in blue and CA in grey. (**b**) A density map averaged from ten helically symmetrized MD models of CypA-CA tubes. Colour scheme is the same as that in Fig 2c. The density is contoured at 2σ. The red box encloses the two CA hexamers in complex with CypA that are shown in **e**. (**c**,**d**) Porcupine representation of the most dominant mode from principal component analysis performed on the trajectories of the bridging (**c**) and non-bridging (**d**) binding modes of CypA (blue). (**e**) Result from an all-atom MD simulation of the CypA-CA complex comprising two CA molecules (orange) and one CypA (blue) within a tubular configuration, shown with two CA hexamers (green). The CA molecules making the dimer interface are in red. (**f**) An enlarged view of the CypA-CA complex from the boxed region in E rotated 90°, illustrating the two CA binding sites on CypA: site 1 is the canonical binding site and site 2 is the non-canonical binding site (see text). Selective residues at the site 2 interface are labelled. (**g**) Overlay of the MD CypA-CA model with the crystal structure of the CypA-CsA complex (PDB: 2RMA, CypA in purple and CsA in green), illustrating that CypA interacts with a second CsA at site 2.

tube (simulations 4 and 6, Supplementary Fig. 9), revealed two distinct types of motion depending on whether CypA is bridging along the most curved CA array or not (Fig. 3c,d). Although the bridging CypA exhibits limited motion (1°) perpendicular to the array direction and appears locked onto the curved CA lattice (Fig. 3c), disjointed CypA is more mobile (20°) and interconverts between the two conformers observed in the crystal structure of the CypA-CA_{NTD} complex (Fig. 3d, Supplementary Movie 1)[11].

To investigate the detailed interactions between CypA and CA, a subsystem comprising two adjacent CA subunits with a corresponding bridging CypA (Fig. 3e, Supplementary Table 1 and Supplementary Movie 2). The MD simulation revealed, beside the first binding interface at the canonical CypA-binding site, a second binding interface between CypA and CA, comprising residues A88, G89 and P90 from CA, and A25, P29 and K30 from CypA (Fig. 3f, site 2). The interaction at the second, non-canonical CypA-binding site is weaker than at the canonical binding site, as shown from pair interaction energies -16.1 ± 7.3 and $-28.9 \pm 17.0\,\mathrm{kcal\,mol^{-1}}$, respectively, obtained from unrestrained CypA simulations (simulation 4—Supplementary Table 1). However, the second site, properly positioned, could yield a strong binding avidity once CypA binds to the first CA through the canonical site. Interestingly, a similar second binding site, involving P29 and K30 on CypA, was observed in an early crystal structure of the CypA/CsA complex (PDB: 2RMA), in which two CsA molecules bind CypA (Fig. 3g)[31]. We conclude, therefore, that CypA likely employs both binding sites to strengthen the CA dimer interface in the CA assembly, thereby stabilizing the viral capsid.

Solid-state NMR spectroscopy. To experimentally delineate the CA and CypA residues that form the intermolecular CypA-CA interface, we employed magic angle spinning (MAS) NMR spectroscopy. We previously showed that CA tubular assemblies yield unprecedented high-resolution MAS NMR spectra, permitting their in-depth characterization[32]. In complex with CypA, multiple chemical shift perturbations were detected in the two-dimensional (2D) COmbined R2-Symmetry Driven and NCACX spectra from isotopically labelled CA (U-^{13}C,^{15}N) tubular assemblies (Fig. 4a, Supplementary Fig. 10). The perturbed resonances belong mostly to residues in the CypA-binding loop (Fig. 4b), in agreement with the cryoEM results. Additional perturbations, observed for a few residues located in other CA regions (Fig. 4b), are likely due to allosteric effects. Our recent study revealed that binding of CypA to tubular assemblies of CA results in significantly attenuated mobility of CypA loop and other CA regions[33].

To infer the CypA residues that interact with the assembled CA, we prepared isotopically labelled CypA (U-^{13}C,^{15}N) with assembled CA tubes at three different CypA:CA input ratios of 1:4, 1:2 and 1:1, and recorded CORD and NCA spectra (Fig. 4c, Supplementary Fig. 10). Numerous chemical shift perturbations and/or intensity changes were observed as the CypA:CA ratio increased (Fig. 4c, Supplementary Table 2). Many of the spectral changes correspond to the CypA residues comprising the canonical binding site. In addition, we observed many residues, not associated with the canonical binding site, that experience spectral changes. Mapping these residues onto the CypA structure reveals that many are located at or in the vicinity of the non-canonical second binding site (Fig. 4d), consistent with the results of cryoEM and MD simulations. Interestingly, at a CypA:CA ratio of 6:6, additional spectral perturbations are found in residues distal to either the canonical or the non-canonical binding site (Fig. 4d). These perturbations are likely due to allosteric effects at high CypA concentrations rather than direct interactions with CA. We note that the NMR experiments were run for co-

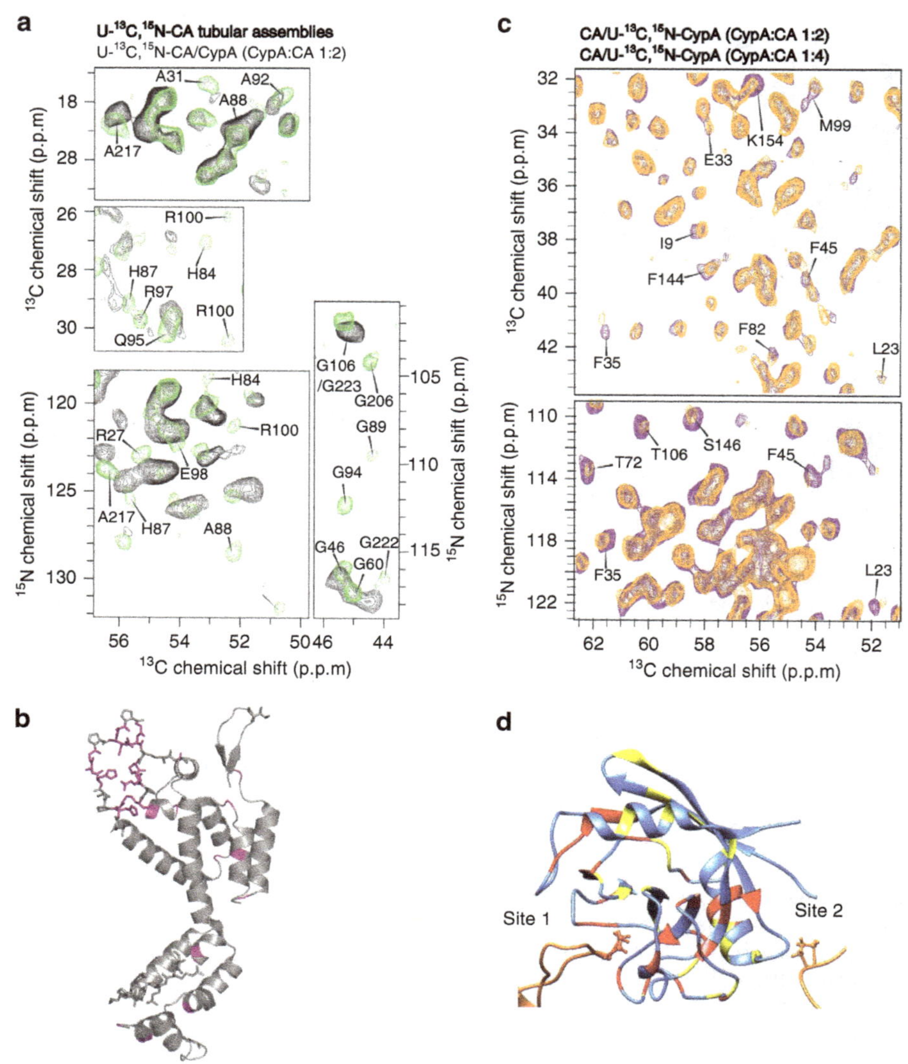

Figure 4 | Solid-state MAS NMR studies of the intermolecular interfaces formed between CypA and CA in CypA-CA complex assemblies. (**a**) Expansions of 2D MAS NMR spectra of U-^{13}C,^{15}N-CA tubular assemblies, free (green) and in complex with CypA (black), acquired at 19.9 T. Top two panels: ^{13}C-^{13}C CORD, bottom two panels: ^{15}N-^{13}C NCACX. The peaks whose chemical shifts or intensities change in the presence of CypA are labelled. (**b**) Mapping of chemical shift and/or intensity changes onto the 3D structure of CA (PDB: 3NTE). Purple indicates residues whose shifts or intensities are perturbed. (**c**) Expansions of 2D MAS NMR spectra of CA/U-^{13}C,^{15}N-CypA assemblies acquired at 19.9 T. Top: ^{13}C-^{13}C CORD, bottom: ^{15}N-^{13}C NCA. The spectra are shown for the varying CypA:CA ratios: 1:4 (purple), and 1:2 (orange). The peaks whose chemical shifts or intensities change as a function of the CypA:CA ratios are labelled. (**d**) Mapping of chemical shift and/or intensity changes on the 3D structure of CypA. Red and yellow are the changes when the CypA:CA ratio is shifted from 1:4 to 1:2 and 1:2 to 1:1, respectively.

assembled samples (Fig. 1a–c). The observed enhanced binding (not saturated at sub-stoichiometric conditions) in these highly curved assemblies (as opposed to the tubular assemblies) is consistent with our model that one CypA binds the CA hexamer along the curved direction of CA tubes. The MAS NMR results corroborate the cryoEM findings and computer simulations, revealing that CypA binds to assembled CA through canonical and non-canonical binding sites.

Discussion

A significant result from this study is the discovery of a non-canonical, second site of CypA for binding to the assembled CA lattice. Through this site, CypA bridges two CA molecules from adjacent hexamers, thus stabilizing the CA lattice (Fig. 1a). The additional interaction interface that is established by the new site further strengthens the binding of CypA to the CA lattice through avidity. CypA is known to have different effects on viral infectivity in different host cells; some are enhancive while others are restrictive[24]. Whether these effects correlate with CypA's cellular expression level or are associated with an unknown cell-specific cellular factor is still not clear. The fact that CypA, via its second binding site, binds selectively along the most curved CA array (Fig. 2c) and promotes curved assembly (Fig. 1c), further implies that CypA and proteins containing CypA domains are best suited to interact with a conical capsid exhibiting a continuous curvature.

We found that sub-stoichiometric amounts of CypA (up to 3:6 CypA:CA) stabilize the CA assembly (Fig. 1d). On the other hand, a high level of CypA reduces capsid stability (Fig. 1d) and disrupts the capsid assembly (Fig. 1a)[28]. These seemingly contradictory observations can be explained by steric hindrance

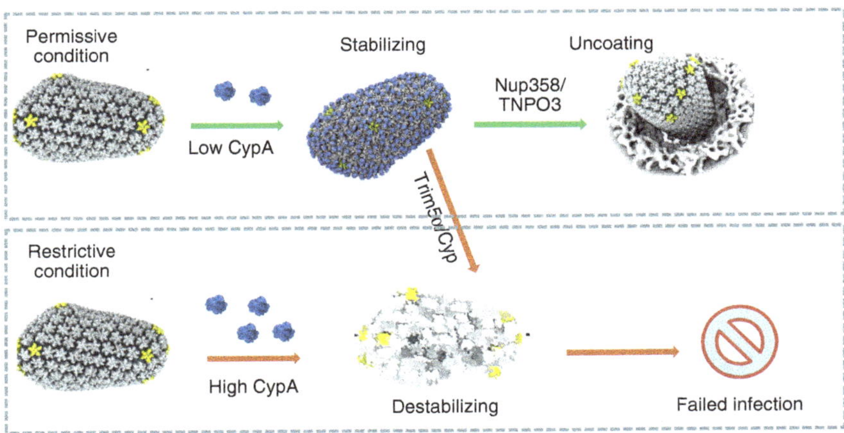

Figure 5 | Model for CypA's role in HIV-1 infection. In permissive cells, sub-stoichiometric levels of CypA stabilize and protect the viral capsid moving in the cytoplasm as it moves towards the nuclear pore, where Nup358 additionally binds to the capsid and promotes uncoating. In non-permissive cells, either high levels of CypA or CypA in combination with host restriction factors such as Trim5α and Trim-Cyp facilitate premature uncoating and result in viral restriction.

between two CypA molecules upon simultaneous binding to neighbouring CA hexamers, thus weakening, at high concentration, the dimer assembly interface. This effect is utilized by the host restriction factors Trim-Cyp and Trim5α to carry out their function[34–36]; both proteins contain stable dimeric capsid-binding motifs, CC-Cyp and CC-SPRY, in addition to the B-box domain for higher-order oligomerization, to enhance the affinity of interaction through avidity effects[37]. The CC-SPRY dimer, not the SPRY monomer, effectively binds to and breaks the capsid between CA hexamers[38,39]. Our data further suggest a plausible mechanism for Trim-mediated premature uncoating and viral restriction that involves direct binding of Trim-Cyp and Trim5α at the inter-hexamer interface and subsequent destabilization of the viral capsid[37,40,41]. Since CypA prefers curvature and stabilizes curved assemblies, the variable curvature in the mature core would help CypA binding. CypA binding may also have a greater effect on a continuously curved capsid. CypA binding to the HIV-1 capsid also enhances the antiviral activity of some small molecule inhibitors, such as PF74 (ref. 42). Based on our observation that CypA binding hardens the capsid lattice, in our AFM experiments, we suggest that PF74, which appears to act at the uncoating step of infection, acts more effectively on CypA-stabilized capsid.

What are the functional consequences of CypA binding to the HIV-1 core? Lentiviruses, such as HIV-1, are differentiated from other retroviruses by having evolved the ability to traverse the nuclear pore complex, allowing replication in non-dividing cells. HIV-1 has been suggested to use CypA and CPSF6 to cloak its genome during migration through the cytoplasm, allowing evasion of innate immune sensing in primary human macrophages and dendritic cells[4,5,43,44]. Our data support this notion and suggest that the second interaction interface is important for the protective role of CypA to the viral capsid. On the basis of previous studies by others[8] and our current results, we propose a simplified model for the function of CypA in HIV-1 infection (Fig. 5). In permissive cells, CypA affects HIV-1 transduction in two stages. The first stage is within the cytoplasm during reverse transcription. Since a single CypA bridges two CA molecules from adjacent hexamers, a sub-stoichiometric level of CypA stabilizes the capsid and protects the viral core from premature uncoating, thus promoting reverse transcription in all cell types[34]. At the second stage, once the HIV-1 core reaches the nuclear membrane, the CA subunits that are not bound by CypA could be available for binding to Nup358 (ref. 24), thus facilitating HIV-1 infection by coordinating proper uncoating of the core in target cells[6,45]. On the other hand, a high level of CypA binding or addition of Trim-Cyp/Trim5α could destabilize the capsid[18,46,47]; CypA enhances Trim-mediated restriction at a stage after reverse transcription in restrictive cells[48], and productive restriction can occur even when a fraction of CA in the capsid is available for Trim-Cyp binding[49]. Taken together, our results provide new insights into how CypA engages in modulating HIV-1 infectivity.

In summary, we used an integrative multi-disciplinary approach, combining cryoEM, computer modelling and MD simulations, solid-state NMR and biochemical analysis, to investigate the interactions between the host cell factor CypA and HIV-1 capsid assembly. We uncovered a novel interaction interface of CypA which may provide a new avenue for the development of therapeutic interventions that target CypA interactions with the HIV-1 capsid.

Methods

Protein expression and purification (CA-FL and CypA). Recombinant wt HIV-1 CA (HXB2) and CypA proteins were expressed in *Escherichia coli*, Rosetta 2 (DE3), cultured in Luria-Bertani media or modified minimal medium, and induced with 0.4 mM IPTG at 23 °C for 16 h (refs 50,51). Natural abundance and U-^{13}C, ^{15}N isotopically labelled CA proteins were purified over a 5 ml Hi-Trap QP column (GE Healthcare, Piscataway, NJ, USA) in 25 mM sodium phosphate (pH 7.0), 1 mM DTT and 0.02% sodium azide, followed by a 5 ml Hi-Trap SP columns (GE Healthcare) using a 0.1 M NaCl gradient in a 25 mM sodium phosphate (pH 5.8), 1 mM DTT and 0.02% sodium azide mixture. The final purification step comprised gel-filtration over Hi-Load Superdex 200 16/60 columns (GE Healthcare) in 25 mM sodium phosphate (pH 6.5), 100 mM NaCl, 1 mM DTT and 0.02% sodium azide. Proteins were concentrated to 10 mg ml^{-1} using Amicon concentrators (Millipore, Billerica, MA, USA), flash-frozen with liquid N2, and stored at − 80 °C. Cells expressing unlabelled and U-^{13}C, ^{15}N isotopically labelled CypA were lysed using sonication in 25 mM sodium phosphate buffer at pH 7.0, and centrifuged at 15,000 r.p.m. at 4 °C for 1 h. After adjusting the pH to 5.5 with acetic acid, the solution was centrifuged again at 15,000 r.p.m. at 4 °C for 1 h, and the supernatant was loaded onto a 5 ml HiTrap SP HP (GE Healthcare) column and eluted with a 0–1 M NaCl gradient. Fractions containing the target protein were combined and further purified with a Hi-Load Superdex 75 pg 26/600 (GE Healthcare) column in a buffer containing 25 mM sodium phosphate, 1 mM DTT and 0.02% NaN_3, pH 5.5.

CA assembly and CypA-binding assays. For CA and CypA co-assembly, 80 μM (2 mg ml^{-1}) wt CA was mixed with CypA at different molar ratios in a buffer containing 2.25 M NaCl and 50 mM Tris-Hcl (pH 8.0) and incubated at 37 °C for 1 h. For CypA binding to pre-assembled CA tubes, 10 mg ml^{-1} wt CA was first diluted to 2 mg ml^{-1} in a buffer containing 1 M NaCl, 50 mM Tris pH 8.0 and incubated at 37 °C for 1 h. The assembly mixture was then incubated with CypA at different molar ratios at room temperature for 2 h. At the end of incubation, 5 µl samples were immediately used for cryoEM analysis. An amount of 6 µl samples from the same reaction mixtures were mixed with 4 × LDS loading buffer (Invitrogen, NP-0007) supplemented with 10 mM DTT for SDS–polyacrylamide gel electrophoresis (SDS–PAGE) analysis (t). The remaining sample was pelleted at 20,000*g* with an Eppendorf centrifuge 5417R for 15 min, and supernatants (s) and resuspended pellets (p) were mixed with 4 × LDS loading buffer for gel analysis. Total, supernatant and pellet samples, without boiling, were fractionated on 10% SDS–PAGE and stained with Coomassie Blue. Each experiment was carried out at least three times.

To determine the binding ratio of CypA:CA, the SDS–PAGE gels were scanned using an Epson 4990 scanner. The integrated intensities of CA and CypA protein bands were measured using Image J 1.40 program (NIH). The molar ratios were calculated according to the formula (CypA intensity/CypA molecular weight)/(CA intensity/CA molecular weight) and calibrated using the input ratios as standards.

AFM indentation experiments. For AFM measurements, 160 µM wt CA was dialysed against a high-salt buffer (50 mM Tris-Hcl pH 8.0, 1 M NaCl) overnight at 4 °C. The assembled protein was diluted to 80 µM and incubated with varying concentrations of CypA (13, 40 and 80 µM) for 2 h at 4 °C. CA assemblies were attached to hexamethyldisilazane-coated microscope glass slides using previously described methods[52,53]. All measurements were conducted in a fluid environment (high-salt buffer). The experimental system consisted of a Nanowizard ULTRA Speed atomic force microscope (JPK, Berlin, Germany) mounted on an inverted optical microscope (AxioObserver, Carl Zeiss, Heidelberg, Germany). Images of capsid particles were acquired in AFM tapping mode. Silicon nitride probes (DNP, $k = 0.1 \pm 0.01\ \mathrm{N\,m^{-1}}$, Bruker) were used, their spring constants being determined experimentally by measuring the thermal fluctuations of the cantilevers (Hutter and Bechhoefer, 1993). To determine the stiffness of an individual capsid, an indentation experiment was performed. For each capsid, ~100 successive force-distance curves were collected at a scan rate of 20 Hz. To maintain a constant range of indentation depths, the maximal loading force was limited to a range of 0.2–1.5 nN, yielding maximal indentation depths of 4 nm. Capsid stiffness was calculated from the slope of the force-distance curve, according to Hooke's law, on the assumption that the capsid and the cantilever can be modelled as two springs arranged in series.

CryoEM specimen preparation and data collection. Full-length wt CA ($10\ \mathrm{mg\,ml^{-1}}$) was diluted to $2\ \mathrm{mg\,ml^{-1}}$ in high-salt buffer (1 M NaCl, 50 mM Tris pH 8.0) at 37 °C for 1 h for tubular assembly. The assembly mixture was then incubated with CypA (27 µM) at room temperature for 2 h. CryoEM grids were prepared according to the procedure described previously[30]. Low-dose ($\sim 15\ e^{-}\ \text{Å}^{-2}$) projection images were collected with a Tecnai TF20 microscope (FEI Corp., OR) operated at 200 kV, on Kodak SO163 films at a nominal magnification of 55,000, with under-focus values ranging from 1.0 to 3.5 µm. The best micrographs were selected and digitized using a Nikon super coolscan 9000 ED scanner (Nikon, Japan) at an optical resolution of 4,000 d.p.i.

3D EM reconstruction. More than 800 films of tube images possessing a wide range of diameters and helical symmetries were collected. Nineteen tubes with (−8, 13) symmetry were selected and boxed into segments of size 512 × 512 pixels with a shift of two subunits along the helical axis by the EMAN BOXER program (ref. 54). The defocus value of each micrograph was evaluated using the program CTFFIND3 (ref. 55). An initial density map was reconstructed from binned segments of 256 × 256 pixels using iterative helical real-space reconstruction[29]. The refined helical parameters corresponded to a rise of 7.393 Å and a rotation angle of 138.13°. The preliminary map was further refined using iterative helical real-space reconstruction without binning. During the refinement, helical symmetry and full-contrast transfer function (CTF) correction were applied. A total of 8,640 helical segments, containing 17,280 asymmetric units, were included in the final reconstruction. The Fourier shell correlation curve was calculated using gold standard procedure from two independent half data sets.

Focused classification. To accurately assess a given CypA-CA binding site for CypA occupancy, reference-based classification was used. Owing to the very large mass discrepancy between the rest of the CypA-CA tube and an individual CypA monomer or dimer, very accurate background subtraction was needed, essentially as described in ref. 56.

References were generated by simulating electron density maps for CypA from the CypA-$\mathrm{CA_{NTD}}$ crystal structure (PDB: 1AK4) filtered to 12 Å using Chimera[57]. These were then aligned to the final EM map in three binding modes: CypA monomer binding to the left hexamer; CypA monomer binding to the right hexamer; CypA composite dimer (referred to as CypA dimer hereon) binding to both, with steric clashes present. In addition, we included a negative control with a CypA dimer positioned above the centre of the CA hexamer in question; any significant density showing up in the results would indicate a bias selecting for noise. The cross-correlation between each of these four references was calculated for each particle, in replicates of three as described in the following.

To account for the large difference in signal-to-noise ratio between the re-projections from the final model and its individual data projections, we added noise and information from the particle's CTF for each comparison. Three hundred noise images were selected from different micrographs near the tube density to provide the most accurate combination of structural and detector related noise. Three noise images were selected at random for each comparison, convolved with the particles CTF, centred to a mean of zero, and normalized by the subsequent images intensity range. To each of these images, a re-projection of the final EM map at the given particles orientation was added (also convolved with the particles CTF), centred and normalized. These three replicates were individually subtracted from the data projection, everything being low-pass filtered to 12 Å. Finally, the three difference images were masked off using a binary mask derived from the projection of the corresponding CypA dimer reference (one for the dimer and both monomer-binding modes, and one for the negative control).

The average CCC from the three replicates was calculated for all four binding modes and used as the metric for classification. Those with a CCC > 0 with the dimer reference were initially separated from the full data set. This subset was then further split in two, based on the difference between the left ('up') and right ('down') binding modes, with a $\mathrm{CCC_{up}}$ - $\mathrm{CCC_{down}} > 0$ indicating an 'up' class. No particles were excluded from the analysis and no further alignment was introduced. The classified maps were directly derived from the pre-aligned tube segments that have been sorted into different classes.

Molecular dynamics flexible fitting. Ten tubular CypA-CA complexes were derived as described in Supporting Information, and further refined using MDFF, as follows. Ions near the surface of the protein were placed using cionize[58]. Bulk water and Na/Cl ions were then added using visual molecular dynamics (VMD)[58], setting the total concentration of NaCl to 1 M. Each of the resulting systems contained 1,900,000 protein atoms and ~25,600,000 total atoms, including solvent. The systems were equilibrated at 310 K and 1 atm for 5 ns while applying positional restraints on all heavy atoms of the protein. MDFF[59] was then applied for 10 ns with a grid scaling of $\varphi = 0.1$ coupled only to backbone heavy atoms. Domain restraints were applied to maintain the structural integrity of each individual CypA-$\mathrm{CA_{NTD}}$ complex. Additional restraints were applied in the form of extra bonds to maintain secondary structure integrity and to prevent transitions of *cis/trans* bonds and chirality errors[59]. MDFF was performed using NAMD 2.9 (ref. 60) with the CHARMM22-CMAP force field[61,62]. Simulated density maps needed for determining cross-correlation between model and EM data, were generated using VMD[58].

Molecular dynamics simulations of CypA-CA complexes. All-atom MD simulations of reduced systems for the two types of CypA-CA complexes, namely bridged and disjoined, were performed. For this purpose, ions in close proximity to the protein were placed using cionize. Bulk water and Na/Cl ions were then added using VMD, setting the total concentration of NaCl to 1 M. The systems were then equilibrated at 310 K and 1 atm for 10 ns using the NAMD 2.10 and CHARMM36FF (ref. 63) together with the TIP3P water model[64], corresponding to simulations 3 and 5 in Supplementary Table 1. Simulations of CypA-CA complexes were then performed, again employing NAMD 2.10, but with the polarizable CHARMM force field based on the classical Drude oscillator model[65,66] together with the SWM4-NDP water model[67] at 310 K and 1 atm (simulations 4 and 6 in Supplementary Table 1).

All simulations carried out in the present study used the r-RESPA integrator available in NAMD. Long-range electrostatic force calculations employed the PME method (particle-mesh Ewald), utilizing a grid spacing of 2.1 Å and eighth order interpolation, with a 1.2 nm cutoff. Simulations using the CHARMM additive force field, employed a time-step of 2 fs, with non-bonded interactions evaluated every 2 fs and electrostatics updated every 4 fs; all hydrogen bonds were constrained with the SHAKE algorithm. For the simulations employing the CHARMM Drude polarizable force field (simulations 4 and 6), the systems were first equilibrated using CHARMM36 for 10 ns at 310 K and 1 atm (simulations 3 and 5), then the Drude particles were added using 'drude prepper' as implemented in CHARMM-GUI[68]. All systems were minimized for 20 steps, and then equilibrated at 310 K and 1 atm for 1 ns while maintaining positional restraints for the proteins' heavy atoms with an integration time step of 0.5 fs. Screened Coulomb corrections of Thole were calculated within a cutoff of 5 Å. Restraints were applied to a Drude oscillator if its length exceeded 0.2 Å, assuming a force constant of $40000\ \mathrm{kcal}\ \text{Å}^{-2}$. Production runs were performed with a time step of 1 fs, with non-bonded interactions evaluated every 1 fs and electrostatics updated every 1 fs. Temperature of the Drude particles was maintained at 1 K.

Computational modelling of CypA-CA tubular complexes. A helical tube of CA was constructed by docking the MDFF-derived hexameric structure (PDB: 3J34), using Chimera, to each of the hexamers in the experimental tubular density with (−8, 13) helical symmetry. Reverse mutations, L6I, L83V, E92A and H120N, were applied to match the wt (HXB2) sequence. Protonation states of titratable groups were calculated using PDB2PQR (ref. 69). The tube was decorated with CypA employing VMD[70], as described below. Two binding modes of CypA to CA are observed in the CypA-CA_{NTD} crystal complex and, accordingly, in building the tube model for each of the CA monomers in the lattice, one of the two CypA conformers was selected, at random with equal probability. The location of the CypA molecule was derived by aligning the CA_{NTD} domain of each CA in the tubular lattice with the CA_{NTD} of the CypA-CA_{NTD} crystal structure (PDB: 1AK4) (ref. 11) using root-mean-square fitting, with excellent agreement (C_α-RMSD of 0.5 Å). Coordinates were taken from the selected conformer, including the coordinates of the CA loop binding to CypA. The procedure was repeated 10 times, to explore the combinatorial complexity of the binding of CypA to CA. The resulting 10 fully saturated CypA-CA tubes contain extensive clashes between neighbouring CypAs. To correct for steric clashes, CypAs were removed from each of the saturated models, following a random walk on a self-avoidance path over the lattice. At each step of the removal procedure, a CypA was chosen at random, and, if clashes between the selected CypA and adjacent CypAs were present, then the adjacent CypA molecule(s) was (were) removed from the model. The procedure was repeated over the set of remaining CypAs—those that had not been removed or selected on a previous step—yielding a model with no clashes.

Solid-state NMR specimen preparation. Natural abundance CA and U-^{13}C,^{15}N-labelled CypA proteins were concentrated to 30 mg ml^{-1} in 25 mM phosphate buffer (pH 5.5), and then mixed at varying CypA:CA ratios (1:4, 3:6 and 6:6). Assemblies of CypA-CA were prepared by adding NaCl to a final salt concentration of 2.4 M, followed by one-hour incubation at 37 °C. Assembled CypA-CA were pelleted by centrifugation at 50,000 r.p.m. at 4 °C and packed into three 3.2 mm thin-wall Bruker rotors. The amount of U-^{13}C,^{15}N-labelled CypA in each rotor was ~0.15 μmol.

MAS NMR spectroscopy. 2D MAS NMR spectra of CypA-CA assemblies were acquired on a Bruker 19.9 T spectrometer with a 3.2 mm HCN Efree probe. Larmor frequencies were 850.400 MHz for ^{1}H, 213.833 MHz for ^{13}C, and 86.170 MHz for ^{15}N. 2D CORD and NCA spectra were collected at the MAS frequency of 14.000 ± 0.001 kHz, regulated by a Bruker MAS controller, and the temperature was maintained at 4.0 ± 0.5 °C throughout the experiments by a Bruker temperature controller.

Typical pulse lengths were 2.6 μs (^{1}H), 3.4 μs (^{13}C) and 3.7 μs (^{15}N); ^{1}H-^{13}C cross-polarization was performed using a linear amplitude ramp, with the ^{1}H rf field of 88 kHz matched to the second spinning sideband. The ^{1}H-^{13}C and ^{1}H-^{15}N contact times were 0.6 and 1.1 ms, respectively. The ^{15}N-^{13}C DCP contact time was 6 ms. Small phase incremental alternation ^{1}H decoupling (95 kHz) was applied during the evolution and acquisition periods. The modulated ^{1}H field strengths during CORD were 14 and 7 kHz, and the CORD mixing time was 50 ms. Under these conditions predominantly one-bond correlations are observed in the spectra. Time-proportional phase incrementation was used for frequency discrimination in the indirect dimension.

Processing and analysis of solid-state NMR spectra. All spectra were processed with NMRpipe and analysed using Sparky (https://www.cgl.ucsf.edu/home/sparky/). 30°-shifted, for CORD spectra, or 45°-shifted, for NC spectra, sine bell apodization followed by a Lorentzian-to-Gaussian transformation was applied in both dimensions. Forward linear prediction to twice the number of the original data points was used in the indirect dimension, followed by zero filling to twice the total number of points.

Analytical centrifugation. Sedimentation velocity experiments were performed in an XL-A analytical ultracentrifuge (Beckman Coulter) in two-channel Epon centrepieces. The data were collected at 280 nm and a speed of 40,000 r.p.m. in an An60Ti four-hole rotor at 25 °C. The concentration of protein loaded was approximately 0.35 mg ml^{-1}. The data were analysed with the program SEDFIT using the c(s) and c(s,ff0) models (http://www.analyticalultracentrifugation.com/default.htm). The c(s) model was used for molecular weight estimation of the sedimenting species. The buffer density, buffer viscosity, and protein partial specific volume used in these calculations were estimated with the program SEDNTERP (http://bitcwiki.sr.unh.edu/index.php/Main_Page).

References

1. Forshey, B. M., von Schwedler, U., Sundquist, W. I. & Aiken, C. Formation of a human immunodeficiency virus type 1 core of optimal stability is crucial for viral replication. *J. Virol.* **76,** 5667–5677 (2002).
2. Rihn, S. J. *et al.* Extreme genetic fragility of the HIV-1 capsid. *PLoS Pathog.* **9,** e1003461 (2013).
3. von Schwedler, U. K., Stray, K. M., Garrus, J. E. & Sundquist, W. I. Functional surfaces of the human immunodeficiency virus type 1 capsid protein. *J. Virol.* **77,** 5439–5450 (2003).
4. Rasaiyaah, J. *et al.* HIV-1 evades innate immune recognition through specific cofactor recruitment. *Nature* **503,** 402–405 (2013).
5. Lahaye, X. *et al.* The capsids of HIV-1 and HIV-2 determine immune detection of the viral cDNA by the innate sensor cGAS in dendritic cells. *Immunity* **39,** 1132–1142 (2013).
6. Shah, V. B. *et al.* The host proteins transportin SR2/TNPO3 and cyclophilin A exert opposing effects on HIV-1 uncoating. *J. Virol.* **87,** 422–432 (2013).
7. Matreyek, K. A. & Engelman, A. The requirement for nucleoporin NUP153 during human immunodeficiency virus type 1 infection is determined by the viral capsid. *J. Virol.* **85,** 7818–7827 (2011).
8. Schaller, T. *et al.* HIV-1 capsid-cyclophilin interactions determine nuclear import pathway, integration targeting and replication efficiency. *PLoS Pathog.* **7,** e1002439 (2011).
9. Lee, K. *et al.* Flexible use of nuclear import pathways by HIV-1. *Cell Host Microbe* **7,** 221–233 (2010).
10. Colgan, J., Yuan, H. E., Franke, E. K. & Luban, J. Binding of the human immunodeficiency virus type 1 Gag polyprotein to cyclophilin A is mediated by the central region of capsid and requires Gag dimerization. *J. Virol.* **70,** 4299–4310 (1996).
11. Gamble, T. R. *et al.* Crystal structure of human cyclophilin A bound to the amino-terminal domain of HIV-1 capsid. *Cell* **87,** 1285–1294 (1996).
12. Luban, J., Bossolt, K. L., Franke, E. K., Kalpana, G. V. & Goff, S. P. Human immunodeficiency virus type 1 Gag protein binds to cyclophilins A and B. *Cell* **73,** 1067–1078 (1993).
13. Franke, E. K., Yuan, H. E. & Luban, J. Specific incorporation of cyclophilin A into HIV-1 virions. *Nature* **372,** 359–362 (1994).
14. Sokolskaja, E., Sayah, D. M. & Luban, J. Target cell cyclophilin A modulates human immunodeficiency virus type 1 infectivity. *J. Virol.* **78,** 12800–12808 (2004).
15. Thali, M. *et al.* Functional association of cyclophilin A with HIV-1 virions. *Nature* **372,** 363–365 (1994).
16. Nisole, S., Lynch, C., Stoye, J. P. & Yap, M. W. A Trim5-cyclophilin A fusion protein found in owl monkey kidney cells can restrict HIV-1. *Proc. Natl Acad. Sci. USA* **101,** 13324–13328 (2004).
17. Sayah, D. M., Sokolskaja, E., Berthoux, L. & Luban, J. Cyclophilin A retrotransposition into TRIM5 explains owl monkey resistance to HIV-1. *Nature* **430,** 569–573 (2004).
18. Li, Y., Kar, A. K. & Sodroski, J. Target cell type-dependent modulation of human immunodeficiency virus type 1 capsid disassembly by cyclophilin A. *J. Virol.* **83,** 10951–10962 (2009).
19. Ikeda, Y., Ylinen, L. M., Kahar-Bador, M. & Towers, G. J. Influence of gag on human immunodeficiency virus type 1 species-specific tropism. *J. Virol.* **78,** 11816–11822 (2004).
20. Song, C. & Aiken, C. Analysis of human cell heterokaryons demonstrates that target cell restriction of cyclosporine-resistant human immunodeficiency virus type 1 mutants is genetically dominant. *J. Virol.* **81,** 11946–11956 (2007).
21. Berthoux, L., Sebastian, S., Sokolskaja, E. & Luban, J. Cyclophilin A is required for TRIM5{alpha}-mediated resistance to HIV-1 in Old World monkey cells. *Proc. Natl Acad. Sci. USA* **102,** 14849–14853 (2005).
22. Towers, G. J. *et al.* Cyclophilin A modulates the sensitivity of HIV-1 to host restriction factors. *Nat. Med.* **9,** 1138–1143 (2003).
23. Stremlau, M., Song, B., Javanbakht, H., Perron, M. & Sodroski, J. Cyclophilin A: an auxiliary but not necessary cofactor for TRIM5alpha restriction of HIV-1. *Virology* **351,** 112–120 (2006).
24. De Iaco, A. & Luban, J. Cyclophilin A promotes HIV-1 reverse transcription but its effect on transduction correlates best with its effect on nuclear entry of viral cDNA. *Retrovirology* **11,** 11 (2014).
25. Yoo, S. *et al.* Molecular recognition in the HIV-1 capsid/cyclophilin A complex. *J. Mol. Biol.* **269,** 780–795 (1997).
26. Braaten, D. & Luban, J. Cyclophilin A regulates HIV-1 infectivity, as demonstrated by gene targeting in human T cells. *EMBO J.* **20,** 1300–1309 (2001).
27. Aiken, C. Mechanistic independence of Nef and cyclophilin A enhancement of human immunodeficiency virus type 1 infectivity. *Virology* **248,** 139–147 (1998).
28. Grattinger, M. *et al.* *In vitro* assembly properties of wild-type and cyclophilin-binding defective human immunodeficiency virus capsid proteins in the presence and absence of cyclophilin A. *Virology* **257,** 247–260 (1999).
29. Zhang, P., Meng, X. & Zhao, G. Tubular crystals and helical arrays: structural determination of HIV-1 capsid assemblies using iterative helical real-space reconstruction. *Methods Mol. Biol.* **955,** 381–399 (2013).
30. Zhao, G. *et al.* Mature HIV-1 capsid structure by cryo-electron microscopy and all-atom molecular dynamics. *Nature* **497,** 643–646 (2013).
31. Ke, H. *et al.* Crystal structures of cyclophilin A complexed with cyclosporin A and N-methyl-4-[(E)-2-butenyl]-4,4-dimethylthreonine cyclosporin A. *Structure* **2,** 33–44 (1994).
32. Han, Y. *et al.* Magic angle spinning NMR reveals sequence-dependent structural plasticity, dynamics, and the spacer peptide 1 conformation in HIV-1 capsid protein assemblies. *J. Am. Chem. Soc.* **135,** 17793–17803 (2013).
33. Lu, M. *et al.* Dynamic allostery governs cyclophilin A-HIV capsid interplay. *Proc. Natl Acad. Sci. USA* **112,** 14617–14622 (2015).

34. Sayah, D. M. & Luban, J. Selection for loss of Ref1 activity in human cells releases human immunodeficiency virus type 1 from cyclophilin A dependence during infection. *J. Virol.* **78**, 12066–12070 (2004).
35. Stremlau, M., Perron, M., Welikala, S. & Sodroski, J. Species-specific variation in the B30.2(SPRY) domain of TRIM5alpha determines the potency of human immunodeficiency virus restriction. *J. Virol.* **79**, 3139–3145 (2005).
36. Yap, M. W., Mortuza, G. B., Taylor, I. A. & Stoye, J. P. The design of artificial retroviral restriction factors. *Virology* **365**, 302–314 (2007).
37. Goldstone, D. C. *et al.* Structural studies of postentry restriction factors reveal antiparallel dimers that enable avid binding to the HIV-1 capsid lattice. *Proc. Natl Acad. Sci. USA* **111**, 9609–9614 (2014).
38. Zhao, G. *et al.* Rhesus TRIM5alpha disrupts the HIV-1 capsid at the inter-hexamer interfaces. *PLoS Pathog.* **7**, e1002009 (2011).
39. Yang, H. *et al.* Structural insight into HIV-1 capsid recognition by rhesus TRIM5alpha. *Proc. Natl Acad. Sci. USA* **109**, 18372–18377 (2012).
40. Sanchez, J. G. *et al.* The tripartite motif coiled-coil is an elongated antiparallel hairpin dimer. *Proc. Natl Acad. Sci. USA* **111**, 2494–2499 (2014).
41. Weinert, C., Morger, D., Djekic, A., Grutter, M. G. & Mittl, P. R. Crystal structure of TRIM20 C-terminal coiled-coil/B30.2 fragment: implications for the recognition of higher order oligomers. *Sci. Rep.* **5**, 10819 (2015).
42. Shi, J., Zhou, J., Shah, V. B., Aiken, C. & Whitby, K. Small-molecule inhibition of human immunodeficiency virus type 1 infection by virus capsid destabilization. *J. Virol.* **85**, 542–549 (2011).
43. Hilditch, L. & Towers, G. J. A model for cofactor use during HIV-1 reverse transcription and nuclear entry. *Curr. Opin. Virol.* **4**, 32–36 (2014).
44. Manel, N. *et al.* A cryptic sensor for HIV-1 activates antiviral innate immunity in dendritic cells. *Nature* **467**, 214–217 (2010).
45. Qi, M., Yang, R. & Aiken, C. Cyclophilin A-dependent restriction of human immunodeficiency virus type 1 capsid mutants for infection of nondividing cells. *J. Virol.* **82**, 12001–12008 (2008).
46. Ylinen, L. M. *et al.* Cyclophilin A levels dictate infection efficiency of human immunodeficiency virus type 1 capsid escape mutants A92E and G94D. *J. Virol.* **83**, 2044–2047 (2009).
47. Battivelli, E. *et al.* Strain-specific differences in the impact of human TRIM5alpha, different TRIM5alpha alleles, and the inhibition of capsid-cyclophilin A interactions on the infectivity of HIV-1. *J. Virol.* **84**, 11010–11019 (2010).
48. Lin, T. Y. & Emerman, M. Determinants of cyclophilin A-dependent TRIM5 alpha restriction against HIV-1. *Virology* **379**, 335–341 (2008).
49. Shi, J., Friedman, D. B. & Aiken, C. Retrovirus restriction by TRIM5 proteins requires recognition of only a small fraction of viral capsid subunits. *J. Virol.* **87**, 9271–9278 (2013).
50. Byeon, I. J. *et al.* Structural convergence between Cryo-EM and NMR reveals intersubunit interactions critical for HIV-1 capsid function. *Cell* **139**, 780–790 (2009).
51. Han, Y. *et al.* Solid-state NMR studies of HIV-1 capsid protein assemblies. *J. Am. Chem. Soc.* **132**, 1976–1987 (2010).
52. Kol, N. *et al.* A stiffness switch in human immunodeficiency virus. *Biophys. J.* **92**, 1777–1783 (2007).
53. Pang, H. B. *et al.* Virion stiffness regulates immature HIV-1 entry. *Retrovirology* **10**, 4 (2013).
54. Ludtke, S. J., Baldwin, P. R. & Chiu, W. EMAN: semiautomated software for high-resolution single-particle reconstructions. *J. Struct. Biol.* **128**, 82–97 (1999).
55. Mindell, J. A. & Grigorieff, N. Accurate determination of local defocus and specimen tilt in electron microscopy. *J. Struct. Biol.* **142**, 334–347 (2003).
56. Scheres, S. H. RELION: implementation of a Bayesian approach to cryo-EM structure determination. *J. Struct. Biol.* **180**, 519–530 (2012).
57. Pettersen, E. F. *et al.* UCSF Chimera--a visualization system for exploratory research and analysis. *J. Comput. Chem.* **25**, 1605–1612 (2004).
58. Stone, J. E., McGreevy, R., Isralewitz, B. & Schulten, K. GPU-accelerated analysis and visualization of large structures solved by molecular dynamics flexible fitting. *Faraday Discuss* **169**, 265–283 (2014).
59. Trabuco, L. G., Villa, E., Schreiner, E., Harrison, C. B. & Schulten, K. Molecular dynamics flexible fitting: a practical guide to combine cryo-electron microscopy and X-ray crystallography. *Methods* **49**, 174–180 (2009).
60. Phillips, J. C. *et al.* Scalable molecular dynamics with NAMD. *J. Comput. Chem.* **26**, 1781–1802 (2005).
61. Mackerell, Jr. A. D., Feig, M. & Brooks, 3rd C. L. Extending the treatment of backbone energetics in protein force fields: limitations of gas-phase quantum mechanics in reproducing protein conformational distributions in molecular dynamics simulations. *J. Comput. Chem.* **25**, 1400–1415 (2004).
62. MacKerell, A. D. *et al.* All-atom empirical potential for molecular modeling and dynamics studies of proteins. *J Phys. Chem.. B* **102**, 3586–3616 (1998).
63. Best, R. B. *et al.* Optimization of the additive CHARMM all-atom protein force field targeting improved sampling of the backbone phi, psi and side-chain chi(1) and chi(2) dihedral angles. *J. Chem. Theory Comput.* **8**, 3257–3273 (2012).
64. Jorgensen, W. L., Chandrasekhar, J., Madura, J. D., Impey, R. W. & Klein, M. L. Comparison of simple potential functions for simulating liquid water. *J. Chem. Phys.* **79**, 926–935 (1983).
65. Jiang, W. *et al.* High-performance scalable molecular dynamics simulations of a polarizable force field based on classical Drude oscillators in NAMD. *J. Phys. Chem. Lett.* **2**, 87–92 (2011).
66. Lopes, P. E. *et al.* Force field for peptides and proteins based on the classical Drude oscillator. *J. Chem. Theory Comput.* **9**, 5430–5449 (2013).
67. Lamoureux, G., Harder, E., Vorobyov, I. V., Roux, B. & MacKerell, A. D. A polarizable model of water for molecular dynamics simulations of biomolecules. *Chem. Phys. Lett.* **418**, 245–249 (2006).
68. Jo, S., Kim, T., Iyer, V. G. & Im, W. CHARMM-GUI: a web-based graphical user interface for CHARMM. *J. Comput. Chem.* **29**, 1859–1865 (2008).
69. Dolinsky, T. J. *et al.* PDB2PQR: expanding and upgrading automated preparation of biomolecular structures for molecular simulations. *Nucleic Acids Res.* **35**, W522–W525 (2007).
70. Humphrey, W., Dalke, A. & Schulten, K. VMD: visual molecular dynamics. *J. Mol. Graph.* **14**(): 27–38 (1996).

Acknowledgements

We thank Dr Teresa Brosenitsch for critical reading of the manuscript. This work was supported by the National Institutes of Health NIGMS Grant R01GM085043 (P.Z.), P50GM082251 (A.M.G.), P41GM104601 (K.S.), R01GM067887 (K.S.), the National Science Foundation (NSF) PHY-1430124 (K.S.) and the Israeli Science Foundation (grant 1115/13) (I.R.). Large-scale MD simulations were performed on the Blue Waters supercomputer, financed by the NSF (awards OCI-0725070 and ACI-1238993), and on Titan Supercomputers, supported by the Department of Energy (Contract DE-AC05-00OR22725). T.P. acknowledges the support of the NSF CHE0959496 grant, for acquisition of an 850 MHz NMR spectrometer, and the NIGMS P30GM110758 grant for support of core instrumentation infrastructure at the University of Delaware.

Author contributions

This study was conceived by P.Z. J.A. made CypA constructs and purified wt CypA. P.Z. designed *in vitro* binding and cryoEM experiments; C.L, J.N., B.A.H. and G.Z. performed experiments and data analysis; J.N. purified proteins and performed binding and screening; C.L. and G.Z. collected cryoEM data and performed 3D reconstruction. J.R.P. and K.S. developed computational methodology. Computer simulations were designed by J.R.P. and K.S.; J.R.P. constructed structural models and performed MD simulations. C.L., J.R.P., K.S. and P.Z. analysed the structural models. I.B and A.M.G. performed and analysed solution NMR experiments, M.L., G.H. and T.P. designed, performed and analysed SSNMR experiments. M.L. purified wt CypA for SSNMR studies. G.J.B. and P.E.P. designed and performed analytical centrifugation experiments. R.R. and I.R. designed and performed AFM experiments. C.L., J.R.P., T.P, C.A., P.E.P., I.R., K.S. and P.Z. wrote the paper with support from all the authors.

Additional information

Accession codes: CryoEM structural data have been deposited at the EMDB under accession code EMD-3075 and EMD-3076, and the MD atomic model of the CypA-CA complex has been deposited at the Protein Data Bank under accession code 5FJB.

Competing financial interests: The authors declare no competing financial interests.

3

Mechanism of FGF receptor dimerization and activation

Sarvenaz Sarabipour[1] & Kalina Hristova[1]

Fibroblast growth factors (fgfs) are widely believed to activate their receptors by mediating receptor dimerization. Here we show, however, that the FGF receptors form dimers in the absence of ligand, and that these unliganded dimers are phosphorylated. We further show that ligand binding triggers structural changes in the FGFR dimers, which increase FGFR phosphorylation. The observed effects due to the ligands fgf1 and fgf2 are very different. The fgf2-bound dimer structure ensures the smallest separation between the transmembrane (TM) domains and the highest possible phosphorylation, a conclusion that is supported by a strong correlation between TM helix separation in the dimer and kinase phosphorylation. The pathogenic A391E mutation in FGFR3 TM domain emulates the action of fgf2, trapping the FGFR3 dimer in its most active state. This study establishes the existence of multiple active ligand-bound states, and uncovers a novel molecular mechanism through which FGFR-linked pathologies can arise.

[1] Department of Materials Science and Engineering, Johns Hopkins University, Baltimore, Maryland 21218, USA. Correspondence and requests for materials should be addressed to K.H. (email: kh@jhu.edu).

The fibroblast growth factor receptor (FGFR) family includes four receptors that bind 18 ligands called fibroblast growth factors, using heparin as a co-factor[1–4]. These receptors play important roles in all cell types, but are best known for the critical role that they play in the development of the skeletal system[5]. Many pathogenic mutations of FGFR genes are linked to skeletal, cranial and other developmental abnormalities in humans[6,7]. Furthermore, FGFR overexpression and mutations have been reported in a variety of cancers[8–15].

FGF receptors are single-pass membrane proteins, with N-terminal extracellular (EC) domains consisting of three immunoglobulin-like subdomains (D1, D2 and D3), a transmembrane (TM) domain consisting of a single α-helix, and an intracellular (IC) region encompassing a tyrosine kinase domain[16–18]. FGFRs transduce biochemical signals via lateral dimerization in the plasma membrane. Receptor dimerization is necessary for activation, as it brings the two tyrosine kinase domains into close proximity, allowing them to cross-phosphorylate each other on tyrosines in their activation loops[2,19]. This activates the kinases, which then bind adaptor proteins and phosphorylate cytoplasmic substrates, triggering downstream signalling cascades that control cell growth and differentiation[20–22].

High-resolution crystal structures of isolated FGFR EC domains in the presence of different fgfs have provided detailed views of ligand–receptor and receptor–receptor interactions in the EC portion, as well as the role of the co-factor heparin[23–27]. However, there is little mechanistic understanding of how conformational changes are transmitted from the EC domains through the TM domains to the kinase domains, in response to ligand binding. Different fgf ligands can elicit distinctly different biological responses[28], but the mechanism behind the specificity is unknown. To gain insight into these issues, here we study the dimerization of FGFR1, FGFR2 and FGFR3, as well as the response of these receptors to the ligands fgf1 and fgf2. Our results show that ligand binding to unliganded FGFR dimers triggers a switch to ligand-specific configurations of the TM helices, which in turn increase receptor phosphorylation. We further show that a pathogenic FGFR mutant causes unregulated ligand-independent signalling by mimicking the most active ligand-bound configuration.

Results

Full-length FGF receptors dimerize in the absence of ligand. FGF receptors have been proposed to form dimers in the absence of ligand[29–32]. However, the dimerization propensities of unliganded FGF receptors have not been measured directly or quantified. We used a Förster resonance energy transfer (FRET)-based technique, 'Quantitative Imaging FRET' or QI-FRET[33], to quantify the dimerization of full length FGFR1, FGFR2 and FGFR3 in plasma membrane-derived vesicles obtained directly from mammalian cells. While they lack cell cytoplasm, the vesicles are composed of lipid and proteins that are native to the plasma membrane[34]. Two-dimensional receptor concentrations are readily measurable in these vesicles, allowing us to quantify the physical interactions between membrane receptors[34–36].

The monomeric fluorescent proteins YFP or mCherry are a FRET pair suitable for QI-FRET[37]. Single YFP or mCherry genes were fused to the C termini of the full-length FGFR1, FGFR2 and FGFR3 genes via a sequence encoding for a flexible GGS linker (Supplementary Fig. 1). For each of the FGF receptors, Chinese Hamster Ovary (CHO) cells were co-transfected with two plasmids, one encoding FGFR-YFP and one encoding FGFR-mCherry. After the receptors were expressed and trafficked to the plasma membrane, the cells were vesiculated using gentle osmotic stress method[38,39]. The vesicles were collected and imaged, in the absence of ligand, using laser scanning confocal microscopy[35,36,40] (Supplementary Fig. 3). For each vesicle, we used the images captured in the donor, acceptor and FRET channels to determine the donor concentration, the acceptor concentration, the total receptor concentration, the FRET efficiency and the receptor dimeric fraction. As shown in Fig. 1a,b, 800–1200 individual plasma membrane-derived vesicles were analysed in 3–5 independent experiments for each FGF receptor. The single-vesicle data were combined to yield dimerization curves for the receptors, as described in the Methods section under 'QI-FRET Data Analysis'.

The dimeric receptor fraction as a function of receptor concentration is shown in Fig. 1c. From this concentration dependence we obtained, by fitting, the two-dimensional dissociation constant K_{diss} and the structural parameter 'intrinsic FRET', $\tilde{E}$ (refs 33,41; Table 1). Intrinsic FRET does not depend on the dimerization propensities, and is directly related to the distance between the fluorescent proteins. As discussed below, measurements of intrinsic FRET allow us to capture structural

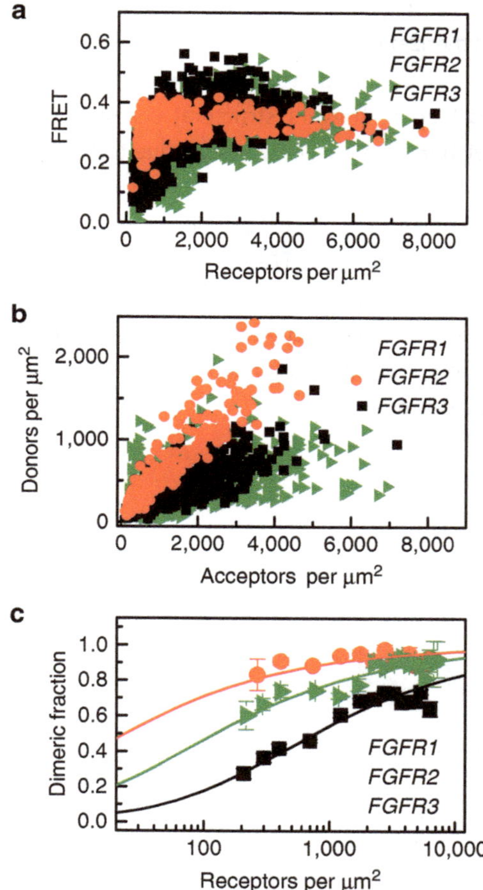

Figure 1 | FGF receptor dimerization in the absence of ligand. (**a**) Measured FRET in plasma membrane-derived vesicles, as a function of receptor concentration, for FGFR1 (black), FGFR2 (olive) and FGFR3 (red). Every data point represents a single vesicle. (**b**) The donor concentration is plotted as a function of the acceptor concentration, for each vesicle. (**c**) Dimeric fraction as a function of total receptor concentrations. The experimentally determined dimeric fractions are binned and are shown with the symbols, along with the standard errors. Each bin contains between 5 and 50 experimental points. The solid lines are the dimerization curves, plotted for the optimized dimerization parameters in Table 1.

Table 1 | Parameters describing FGFR unliganded dimerization.

	K_{diss} (rec μm^{-2})	ΔG (kcal mol^{-1})	Intrinsic FRET	d (Å)
Full FGFR1	710 (630 to 826)	−4.3 (−4.2 to −4.4)	0.66 (0.65 to 0.68)	47.6 (46.8 to 47.9)
Full FGFR2	111 (100 to 146)	−5.4 (−5.3 to −5.5)	0.43 (0.42 to 0.44)	55.7 (55.3 to 56.1)
Full FGFR3	24 (14 to 34)	−6.3 (−6.1 to −6.5)	0.55 (0.54 to 0.56)	51.4 (51.0 to 51.7)
ECTM FGFR1	428 (370 to 540)	−4.6 (−4.5 to −4.7)	0.5 (0.47 to 0.51)	53.1 (52.8 to 54.2)
ECTM FGFR2	3,235 (2,726 to 3,809)	−3.4 (−3.3 to −3.5)	0.57 (0.55 to 0.61)	50.7 (49.3 to 51.4)
ECTM FGFR3	3,235 (2,670 to 3,660)	−3.4 (−3.3 to −3.5)	0.52 (0.49 to 0.55)	52.4 (51.4 to 53.5)
TM FGFR1	40 (27 to 55)	−6.0 (−5.8 to −6.2)	0.5 (0.47 to 0.51)	53.1 (52.8 to 54.2)
TM FGFR2	67 (51 to 86)	−5.7 (−5.6 to −5.8)	0.52 (0.51 to 0.53)	52.4 (52.1 to 52.8)
TM FGFR3	156 (124 to 180)	−5.2 (−5.1 to −5.3)	0.65 (0.64 to 0.66)	47.9 (47.6 to 48.3)
ECTM FGFR3 A391E*	290 (251 to 360)	−4.8 (−4.7 to −4.9)	0.72 (0.7 to 0.73)	45.5 (45.0 to 46.1)

The parameters were obtained from least-square fits of a dimer model to the FRET data. K_{diss}: two-dimensional dissociation constants, ΔG: dimerization free energies (dimer stabilities) calculated using equation (10), d: effective distance between fluorescent proteins in a dimer, calculated from the intrinsic FRET efficiencies using equation (12). Best fits are shown, along with the 67% confidence intervals (standard errors). The values shown in parenthesis are the lower and upper bounds for the confidence intervals, determined from the fit.
*Data from ref. 36.

changes that occur on the cytoplasmic side of the receptor on ligand binding to the extracellular domains.

The values of the two-dimensional dissociation constants, K_{diss} are 710, 111 and 24 μm^{-2} for FGFR1, FGFR2 and FGFR3, respectively, corresponding to dimerization free energies of -4.3 ± 0.1, -5.4 ± 0.1 and -6.3 ± 0.1 kcal mol^{-1} (see equation (10) and Table 1; uncertainties are standard errors). The intrinsic FRET values for the unliganded FGFR1, FGFR2 and FGFR3 dimers are 0.66, 0.43 and 0.55, respectively (Table 1).

To evaluate the biological significance of the measured unliganded dimerization of the FGFRs, we note that physiological FGFR expression levels can be as high as ~80 000 receptors per cell, corresponding to ~80–100 receptors per μm^2 (ref. 42). The experimental dimerization curves that we measured for the three receptors, shown in Fig. 1c, predict substantial dimer populations, at least for FGFR2 (~20%) and FGFR3 (~50%), at receptor concentrations as low as 10 receptors per μm^2. Furthermore, we see a substantial increase in dimeric fraction with concentration, consistent with reports that FGFR overexpression is linked to cancer[10–15]. Thus, unliganded FGFR dimerization is important in physiological context.

Contributions of FGFR domains to unliganded dimerization. To determine the contribution of individual domains to the energetics of dimerization of the unliganded FGF receptors, we created two truncated versions of each receptor. In one truncated version (EC + TM), the intracellular domains were removed and the fluorescent proteins were attached to the cytoplasmic end of the TM domains via flexible, 15 residue $(GGS)_5$ linkers[43] (Supplementary Fig. 1). In the second truncated version (TM), both the IC and EC domains were removed, so that these constructs only contained the TM domains attached to the fluorescent proteins via the same flexible $(GGS)_5$ linkers (Supplementary Fig. 1). In earlier work, we showed that the attachment of the fluorescent protein to the TM domain via this linker does not have an effect on dimerization[36].

The dimerization of the truncated constructs in plasma membrane-derived vesicles was characterized by QI-FRET as above. The dimerization curves for the truncated receptors are shown in Fig. 2, along with the results for the full-length receptors for comparison. The dimerization constants, free energies of dimerization and intrinsic FRET values for all the variants are shown in Table 1. These results reveal several important aspects of FGFR unliganded dimerization. First, they show that the TM domains alone have a strong propensity for dimerization, with dimerization free energies between −5.2 and −6.0 kcal mol^{-1}. Second, they demonstrate that FGFRs that lack IC domains form dimers. It has been proposed previously that the IC domain is required for FGFR dimerization in the absence of ligand[44,45]. However, our results directly show that the IC domain is not necessary for FGFR dimerization. Third, the differences in stability of the two types of truncated receptors (EC + TM and TM) suggest that the contribution of the EC domains to unliganded FGFR dimerization is destabilizing for all three receptors, by 1.4–2.3 kcal mol^{-1}.

Last, by comparing the stabilities of the full-length receptors to the truncated receptors without the IC domains, we obtained directly, and for the first time, the thermodynamic contribution of the IC domain to FGFR unliganded dimerization. Surprisingly, there are large differences in the IC domain contribution among the three receptors. While the contributions of FGFR2 and FGFR3 IC domains are stabilizing by −2.0 and −2.9 kcal mol^{-1}, respectively, the contribution of FGFR1 IC domain is practically zero, suggesting that either FGFR1 IC domain does not engage in contacts that stabilize the full-length FGFR1 dimer, or that stabilizing contacts are balanced by repulsive ones.

Structural changes in FGFR dimers on fgf1 and fgf2 binding. In the experiments described above, we determined the intrinsic FRET for all the studied dimers in the absence of ligand (Table 1). The intrinsic FRET is a structural parameter that is directly related to the distance between the fluorescent proteins, and that does not depend on the dimerization propensities. By comparing the intrinsic FRET in the presence and absence of ligand, here we investigated structural changes that occur on the cytoplasmic side of the receptor on ligand binding to the extracellular domains.

The interaction interface of the TM domains has been proposed to play a role in FGFR activation[46,47], and we therefore looked at the effect of ligand on the truncated EC + TM FGFR constructs, in which the fluorescent proteins were attached to the TM domains via flexible, 15-residue $(GGS)_5$ linkers. The intrinsic FRET values for the unliganded EC + TM FGFR1, FGFR2 and FGFR3 dimers were 0.50, 0.57 and 0.52, respectively (Table 1). Assuming random orientation of the fluorophores (justified because they were attached via flexible 15 amino acid long linkers[43]), we calculated the effective distances between the fluorescent proteins in the unliganded EC + TM dimers (Table 1).

Since the fluorescent proteins were attached directly to the TM domains via the flexible linker, we could directly monitor changes in the structure of the TM domains in the receptor dimers in response to ligand binding. Experiments were performed at high, saturating ligand concentration (5 μg ml^{-1}) such that ligand concentration exceeded the ligand–receptor dissociation constants and the total FGFR concentration by at least two orders of

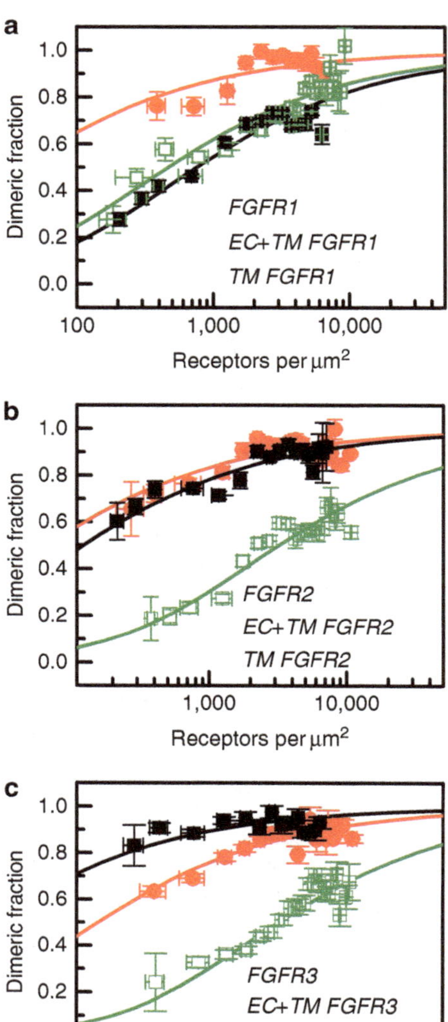

Figure 2 | FGFR domain contributions to unliganded dimerization. Dimerization curves are shown for the full-length receptors (black), for truncated receptors that lack the IC domain and thus contain only the EC and TM domains (olive), and for the TM domains only (red). (**a**) FGFR1. (**b**) FGFR2. (**c**) FGFR3. Data for EC + TM FGFR3 and TM FGFR3 are from ref. 36. The measured dimeric fractions are binned and are shown with the symbols, along with the standard errors. Each bin contains between 5 and 50 experimental points. The solid lines are the best fits of a monomer-dimer equilibrium model to the single-vesicle data. These data demonstrate that the TM domains have a strong propensity for dimerization. The EC domains, on the other hand, inhibit dimerization. The contribution of the IC domains is favourable, but it varies from zero to $-3\,\mathrm{kcal\,mol^{-1}}$ for the three receptors.

magnitude as shown previously[48-50]. This ensured that all receptors were in the ligand-bound dimeric state[48-50]. In the case of 100% dimeric receptors, the FRET signal does not depend on receptor concentration (Supplementary Fig. 5). Instead, it depends only on the intrinsic FRET and on the acceptor fraction, which is measured in each vesicle (equation (11)).

Histograms of intrinsic FRET values, measured in single vesicles for each receptor–ligand pair, are shown in Fig. 3a, and means and standard errors are given in Table 2. We see that intrinsic FRET depends only on the type of ligand, and not on the receptor identity. For all three receptors, FGFR1, FGFR2 and FGFR3, intrinsic FRET in the fgf1-bound dimer state was ~0.55, while the intrinsic FRET value in the fgf2-bound dimer state was ~0.73 (Table 2). These differences are statistically significant, $P<0.01$. The distances between the fluorescent proteins, calculated under the assumption of random fluorophore orientation using equation (12), are shown in Table 2. Note that these are the effective average distances as the linkers have been shown to behave as random coils and explore a variety of configurations[43]. In the presence of fgf1, the effective distance between the fluorescent proteins is 51.6 ± 0.4 Å, which is similar to the unliganded case ($P>0.1$, Table 1). In the presence of fgf2, the effective distance between the fluorescent proteins is 44.9 ± 0.6 Å, significantly smaller than in the fgf1-bound dimer ($P<0.01$). These measured differences in intrinsic FRET reflect differences in the separation of the C termini of the TM domains in the two ligand-bound states, diagrammed in Fig. 3d.

Changes in FGFR phosphorylation on fgf1 or fgf2 binding. To investigate the biological significance of the two different ligand-bound states that we observed in the QI-FRET experiments, we compared the phosphorylation of full-length FGFR1, FGFR2 and FGFR3 at saturating fgf1 or fgf2 concentrations ($5\,\mu\mathrm{g\,ml^{-1}}$) using western blotting. For detection of activated (phosphorylated) receptors we used anti-phospho-Tyr antibodies that are specific for phosphorylated tyrosines in the activation loop of the three receptors (anti-phospho-Y653/4) or other intracellular tyrosines. Typical western blot results are shown in Fig. 3b and Supplementary Fig. 6. Only the top bands, corresponding to the mature fully glycosylated receptors, were quantified. Comparing liganded FGFR1 and FGFR3, we consistently observe 20–40% higher phosphorylation in the presence of fgf2 than fgf1 in CHO cells. To test for statistical significance of this observation using a Student's *t*-test, we performed five independent FGFR3 experiments in two different cell lines, CHO and HEK 293T. In both cell lines, activation by saturating fgf2 is 40% higher than for fgf1 (Fig. 3c). The calculated P value is <0.01, showing that the difference is highly statistically significant. The western blot results therefore support the finding of two distinct ligand-bound active FGFR1 and FGFR3 states, and demonstrate that these different structural states correlate with biological activity.

We used the same method to measure phosphorylation in the absence of ligand. Supplementary Figure 6 shows typical western blot results in the absence of ligand, along with the fgf1 and fgf2 results. For FGFR1 and FGFR3, the phosphorylation in the absence of ligand was 35–65% of the phosphorylation in the presence of fgf1. Thus, the unliganded FGFR1 and FGFR3 dimers exhibit significant phosphorylation even in the absence of ligand, consistent with previous reports[48,49,51]. The phosphorylation increased by a factor of ~1.5 to ~3 on treatment with fgf1, and by a factor of ~2 to ~4 on treatment with fgf2.

Unlike the cases of FGFR1 and FGFR3, the phosphorylation of FGFR2 was the same whether the receptor was unliganded or was liganded by either fgf1 or fgf2 ($P>0.1$; Supplementary Fig. 6). This finding is consistent with the literature[45], and may be due to the fact that FGFR2 interacts with soluble adaptor proteins, such as Grb2, which can regulate its dimerization and activity[44,45]. These adapter proteins are present in the activation/western blot experiments that probe the overall biological response of the receptors to their ligands, but they are absent from the plasma membrane-derived vesicles, which do not contain cell cytoplasm[34].

Ligand binding triggers a switch in the FGFR3 TM dimer. The only FGFR TM domain high-resolution dimer structure reported thus far is the one for the TM domain of FGFR3 (ref. 46). In this

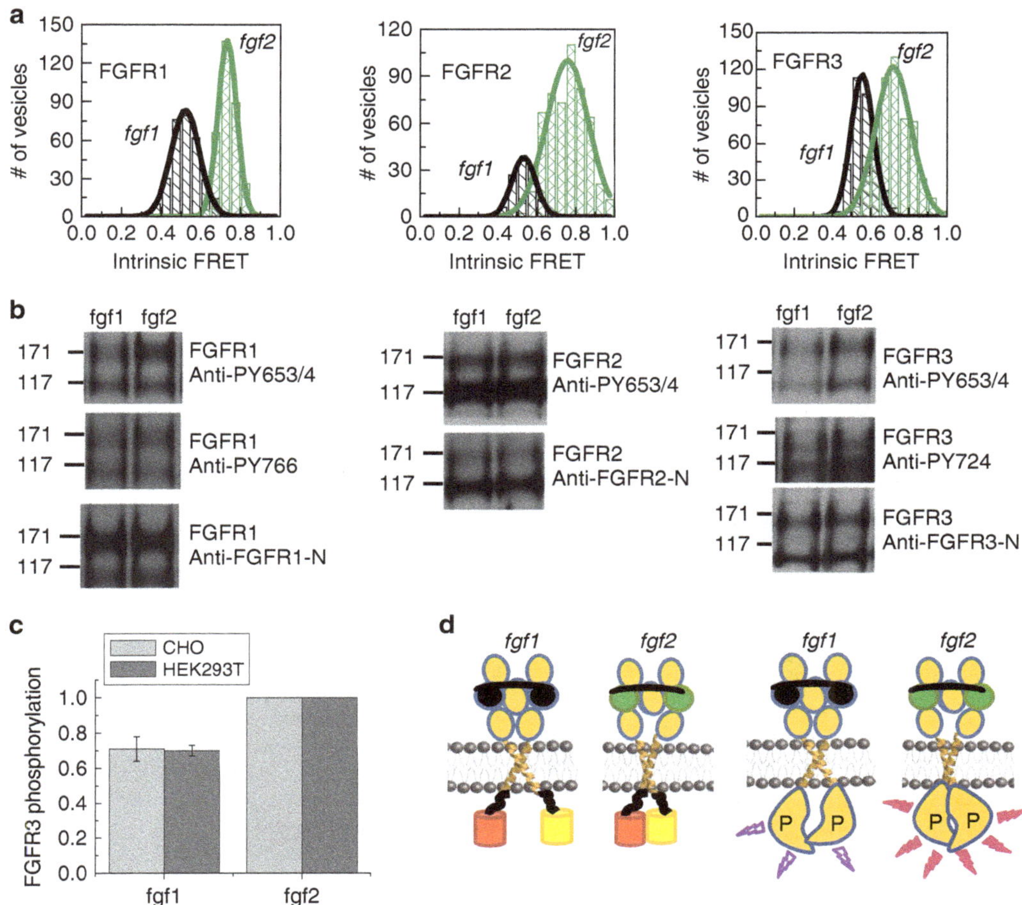

Figure 3 | Conformational changes and activation of the FGF receptors. (**a**) Histograms of single-vesicle intrinsic FRET values, measured for the three FGF EC + TM receptor constructs in the presence of saturating fgf1 (black) or fgf2 (olive) concentrations. Intrinsic FRET is a measure of the separation between the fluorescent proteins in the dimer. Two different intrinsic FRET values were measured for fgf1 and fgf2. Therefore, the binding of these two ligands to the extracellular domains leads to different separation of the fluorescent proteins on the cytoplasmic side of the membrane (Table 2). (**b**) Western blots, reporting on the phosphorylation of the full-length receptors in the presence of saturating concentrations of fgf1 and fgf2 (5 μg ml^{-1}). Expression of the receptors was probed with antibodies to the extracellular domains of the three receptors. Phosphorylation was assayed using antibodies against phosphorylated tyrosines in the activation loop of the three kinases (anti-phospho-Y653/4) or other phosphorylated tyrosine residues. Two bands are observed for all receptors. Only the top bands, corresponding to the fully glycosylated mature receptors that reside primarily in the plasma membrane, were considered in our analysis. There is a difference between the phosphorylation in response to fgf1 and fgf2 for FGFR1 and FGFR3, but not for FGFR2 (see text). (**c**) Relative FGFR3 phosphorylation in response to fgf1 and fgf2 is quantified and compared using a *t*-test. Five independent experiments were performed in two cell lines, CHO and HEK 293T. Phosphorylation was calculated by dividing the intensities of the anti-phospho-Y bands to the intensities of the anti-receptor bands, and scaled to the fgf2 case. The difference in FGFR3 phosphorylation in response to fgf1 and fgf2 is highly statistically significant ($P<0.01$). (**d**) Graphic representation of the findings that the fgf1- and fgf2-bound states are structurally and functionally distinct. Left: graphic (not to scale) representation of the finding that the average distance between the fluorescent proteins is larger when fgf1 is bound, as compared with the fgf2-bound case. Right: graphic representation of the finding that phosphorylation is higher when fgf2 is bound. The representation of the kinase domains is a cartoon, not based on structural data.

structure, solved by NMR in micelles, the FGFR3 TM helices form a left-handed dimer, with helix–helix interactions occurring along the entire TM domain. The TM helices are almost parallel, and wrap around each other in a tight, closed-packed configuration (Supplementary Fig. 7). Importantly, the TM domain of FGFR3 also contains several GxxxG-like motifs, sometimes called SMALLxxxSMALL motifs. They have small amino acids such as Gly, Ala, Thr and Ser in i, $i+4$ positions, and are capable of driving interactions between hydrophobic helices in membranes[52–56]. The FGFR3 TM domain has four such motifs in the N-terminal part of the TM helix (shown in colour in Supplementary Fig. 7C). While they do not participate in the NMR interface, they have been proposed to form an alternative dimer interface for FGFR3 (ref. 46).

To investigate if either the observed NMR interface or the putative GxxxG-like interfaces are related to the structures that we observed in our experiments, we created two sets of amino-acid mutations, each designed to destabilize one of the two TM dimer interfaces. First, we mutated residues L377, G380 and A391 to Ile. These residues mediate the interactions between the TM helices in the NMR structure[46]. Second, we mutated A374, G375 and S378 to Ile, to eliminate all GxxxG-like motifs in the N-terminal portion of the FGFR3 sequence (Supplementary Fig. 7). The comprehensive QI-FRET characterization of the dimerization of the EC + TM NMR interface L377I-G380I-A391I mutant and the GxxxG-like A374I-G375I-S378I mutant, in the absence and presence of fgf1 or fgf2, is shown in Supplementary Figs 8 and 9. Histograms of single-vesicle intrinsic FRET values

Table 2 | Intrinsic FRET efficiencies in the presence of saturating fgf1 or fgf2 concentrations.

	Intrinsic FRET	*d* (Å)
ECTM FGFR1 + fgf1	0.55 ± 0.01	51.4 ± 0.4
ECTM FGFR1 + fgf2	0.73 ± 0.01	45.0 ± 0.4
ECTM FGFR2 + fgf1	0.53 ± 0.01	52.0 ± 0.4
ECTM FGFR2 + fgf2	0.75 ± 0.01	44.2 ± 0.4
ECTM FGFR3 + fgf1	0.55 ± 0.01	51.4 ± 0.4
ECTM FGFR3 + fgf2	0.72 ± 0.01	45.4 ± 0.4
377I-380I-391I	0.39 ± 0.01	57.2 ± 0.4
377I-380I-391I + fgf1	0.52 ± 0.01	52.4 ± 0.4
377I-380I-391I + fgf2	0.53 ± 0.01	52.0 ± 0.4
374I-375I-378I	0.43 ± 0.01	55.7 ± 0.4
374I-375I-378I + fgf1	0.63 ± 0.01	48.6 ± 0.4
374I-375I-378I + fgf2	0.74 ± 0.01	44.6 ± 0.4
A391E + fgf1	0.75 ± 0.01	44.2 ± 0.4
A391E + fgf2	0.78 ± 0.01	43.0 ± 0.4

The intrinsic FRET efficiencies, $\tilde{E}$, were measured for the truncated EC + TM FGFR constructs studies here. In the fgf1-bound state, the measured distance between the fluorescent proteins, d, is 51–52 Å. In the fgf2-bound state, the measured distance between the fluorescent proteins, d, is 43–45 Å. Averages are shown together with standard errors.

a

L377I-G380I-A391I: DEAGSVYAGI**L**SY**G**VGFFLFILVV**A**AVTLCRLR
377 380 391

A374I-G375I-S378I: DEAGSVY**AG**IL**S**YGVGFFLFILVVAAVTLCRLR
374,5 378

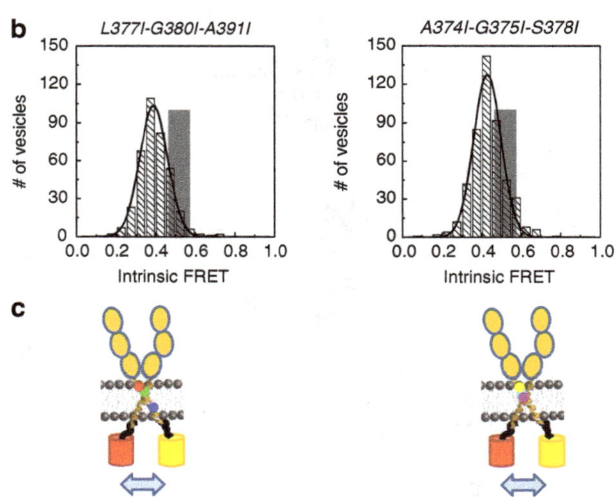

Figure 4 | The L377I-G380I-A391I and the A374I-G375I-S378I sets of mutations affect the unliganded dimer state. (**a**) Sequence of FGFR3 TM domain, with the mutations that were engineered in this study underlined. The L377I-G380I-A391I set of mutations (left) were engineered to destabilize the interface in the FGFR3 dimer structure, solved for the isolated TM domain in detergent micelles[46]. The A374I-G375I-S378I mutations (right) were engineered to destabilize a putative alternative dimer structure, mediated by GxxxG-like motifs[46]. (**b**) Intrinsic FRET values measured for the L377I-G380I-A391I (left) and A374I-G375I-S378I (right) mutants in the absence of ligand. Dark grey Gaussians: histograms of single-vesicle intrinsic FRET measured for the constitutively dimeric EC + TM L377I-G380I-A391I (left) and A374I-G375I-S378I (right) mutants in the absence of ligand (Supplementary Figs 8 and 9). Grey bars: intrinsic FRET for the wild-type EC + TM, obtained by fitting the FRET data to a dimerization model (Table 1). The width of the bar represents the standard error from the fit. The intrinsic FRET decreases due to both mutations, suggesting that the fluorescent proteins in the mutant dimers move away from each other due to the mutations. (**c**) Graphic representation of the effect of the mutations on structure, indicating an increase in separation between the fluorescent proteins. Cartoons are not drawn to scale.

for the two mutants are shown in Fig. 4 (unliganded) and Fig. 5 (liganded), and in Table 2.

In the absence of ligand, both sets of mutations significantly stabilized the EC + TM FGFR3 unliganded dimers, rendering them constitutively dimeric (Supplementary Figs 8 and 9). Both sets of mutations decreased the intrinsic FRET, indicating that the fluorescent proteins were further away from each other in the mutants, as compared with the wild type (Fig. 4). The fact that mutations in both the NMR interface, and the alternate GxxxG interface, had very significant effects on unliganded FGFR3 dimerization suggests that the dimerization interface in the unliganded EC + TM FGFR3 dimer in the plasma membrane does not map exclusively onto either one of these putative interfaces.

In the presence of saturating fgf1 concentrations, the NMR interface mutations did not affect the measured intrinsic FRET values (Fig. 5). The GxxxG-like interface mutations, however, had a significant effect on the intrinsic FRET. Thus, the TM dimer interface in the fgf1-bound dimer does not involve the amino acids in the NMR interface, but instead likely involves the alternative GxxxG-like motifs.

In the presence of saturating fgf2 concentrations, the NMR interface mutations lead to a decrease in intrinsic FRET. As a result, the intrinsic FRET for the NMR interface mutant is identical in the presence of fgf1 or fgf2, and is also identical to the intrinsic FRET value for the wild type in the presence of fgf1. On the other hand, mutating the GxxxG-like motifs had no effect on intrinsic FRET in the presence of fgf2. The mutagenesis results are consistent with the idea that the NMR interface is used by the FGFR3 dimer in the fgf2-bound state. This is supported by the highest intrinsic FRET observed in the fgf2-bound state, suggesting tight packing between the C termini of the TM helices, as in the NMR structure.

Effect of a pathogenic mutation. The pathogenic A391E mutation in the TM domain of FGFR3 has been linked to Crouzon syndrome with acanthosis nigricans, and to bladder cancer[8,57]. Previous QI-FRET studies have shown that this mutation stabilizes the unliganded EC + TM FGFR3 dimer by $-1.4\,\mathrm{kcal\,mol^{-1}}$ (ref. 36). The two-parameter fit of the QI-FRET data in the absence of ligand for this mutant yielded an intrinsic FRET value of 0.72 ± 0.02, the same as the value measured here in the fgf2-state for the wild type (0.72 ± 0.01). This fact prompted us to further investigate the behaviour of the mutant. Using QI-FRET, we characterized the intrinsic FRET of the A391E EC + TM FGFR3 construct in the presence of saturating fgf1 and fgf2 concentrations. We also characterized the phosphorylation of the full-length mutant receptor in the presence of saturating fgf1 and fgf2 concentrations.

The comprehensive characterization of the dimerization of the A391E pathogenic mutant, in the absence and presence of fgf1 and fgf2, is shown in Supplementary Fig. 10. The intrinsic FRET values as well as the western blots that report the phosphorylation of the mutants in the presence of fgf1 or fgf2 are shown in Fig. 6. The intrinsic FRET values in the fgf1 and fgf2 bound states were identical for this mutant, and were the same as the intrinsic FRET value for wild-type EC + TM in the presence of fgf2 (Fig. 6b). The phosphorylation of the full-length A391E mutant in the presence of saturating fgf1 or fgf2 concentrations was also the same, and was the same as in the wild-type fgf2-bound state (Fig. 6b). Specifically, the phosphorylation levels of fgf1- and fgf2-bound

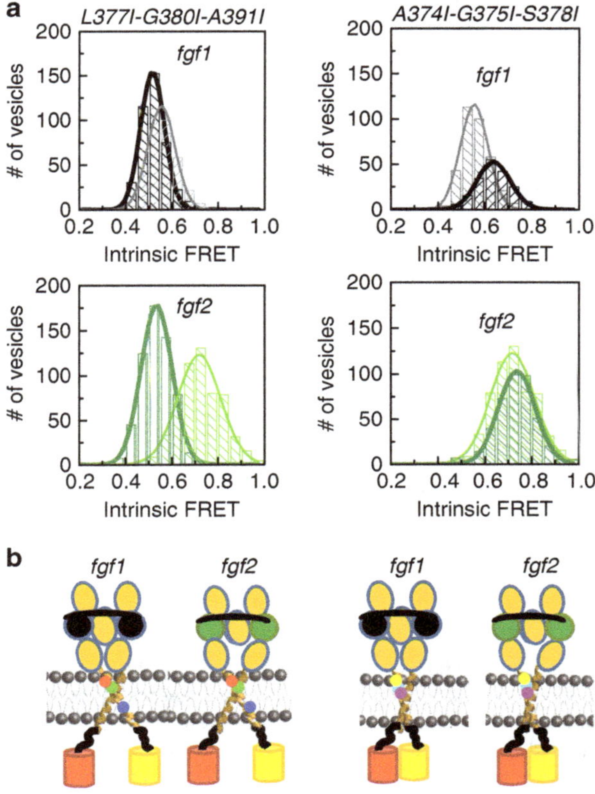

Figure 5 | Effect of the L377I-G380I-A391I and A374I-G375I-S378I mutations on the fgf1- and fgf2-bound FGFR3 dimer structures. Left: results for the L377I-G380I-A391I mutant. Right: results for the A374I-G375I-S378I mutant. (**a**) Intrinsic FRET values, measured for the truncated EC + TM FGFR3 mutants. The histograms of measured intrinsic FRET values in single vesicles for the wild type are shown in grey for the fgf1 case and in green for the fgf2 case. The histograms for the mutants are shown in black in the presence of fgf1 and in olive in the presence of fgf2. The L377I-G380I-A391I set of mutations (left) decreases the intrinsic FRET in the presence of fgf2, down to fgf1 wild-type levels. The A374I-G375I-S378I set of mutations (right) increases the intrinsic FRET in the presence of fgf1. (**b**) Graphic representation of the findings that the L377I-G380I-A391I mutations abolish the fgf2-bound state and induce a transition to the fgf1-bound state, while the A374I-G375I-S378I mutations abolish the fgf1 state. Cartoons are not drawn to scale.

A391E FGFR3 were 1.14 ± 0.07 and 1.04 ± 0.07, respectively, scaled to wild-type FGFR3 phosphorylation of 1.0. These differences were not statistically significant, $P = 0.38$. Thus, the A391E mutation increased FGFR3 phosphorylation in the presence of fgf1, up to fgf2 levels.

Taken together, published data[36] and the results reported here show that the A391E mutation abolishes the fgf1 state and traps the FGFR3 dimer in the fgf2 state even in the absence of ligand. This finding can be explained by the formation of a stabilizing hydrogen bond between the mutant Glu and the neighbouring helix, an idea that is supported by molecular modelling[58]. Once formed, this structure does not change significantly on binding fgf1 or fgf2. The A391E mutation therefore mimics the action of fgf2 in enforcing a close-packed TM dimer structure that leads to increased phosphorylation and thus disregulated signalling and disease.

Correlation between structure and phosphorylation. Finally, we sought to determine if a global correlation exists between the measured intrinsic FRET of the various EC + TM FGFR3 constructs and the phosphorylation of the equivalent full-length FGFR3. We therefore compared the intrinsic FRET and the phosphorylation of all FGFR3 variants that exhibited constitutive dimerization. These included all fgf1- and fgf2-bound FGFR3 variants (the wild type, the pathogenic A391E mutant, the L377I-G380I-A391I mutant and the A374I-G375I-S378I mutant), as well as the constitutive unliganded dimers of the L377I-G380I-A391I and the A374I-G375I-S378I mutants. Phosphorylation, measured using western blotting, was normalized to the phosphorylation of fgf2-bound wild type, which was assigned a value of 1. An example of a western blot, used to arrive at the relationship, is shown in Supplementary Fig. 11. The final results, shown in Fig. 7a, reveal a strong correlation between intrinsic FRET and phosphorylation ($P < 0.001$, $R^2 = 0.8$). In Fig. 7b, we show the same correlation, but we plot the calculated distance between fluorescent proteins in the EC + TM FGFR3 dimers, instead of their intrinsic FRET.

Discussion

Since the discovery of receptor tyrosine kinases (RTKs) in the 1970s, researchers have been searching for a model that captures the essence of RTK signal transduction across the plasma membrane. The most widely accepted model is the 'diffusion-based' or 'canonical' model of RTK activation[16]. It postulates that RTKs are monomers in the absence of ligand, but dimerize and cross-phosphorylate/activate each other on ligand binding. Here we demonstrate, however, that FGFR dimers exist in the absence of ligand at physiological concentrations (< 100 receptors per square micron, see Fig. 1). These unliganded dimers are stabilized through contacts between the TM domains and the IC domains (Fig. 2). We also show that the unliganded FGFR dimers are phosphorylated, providing an explanation of the fact that FGFR overexpression leads to cancer[10–15]. The dimers undergo structural changes in response to ligand binding, and these structural changes increase phosphorylation. These observations are consistent with the 'pre-formed dimer' model of RTK activation[44,59–61].

The dimerization propensities of FGFR1, FGFR2 and FGFR3 in the absence of ligand are very different. Indeed, FGFR1 is predominantly monomeric at low physiological expression levels (Fig. 1). This finding suggests that the difference between the 'canonical' and the 'pre-formed dimer' models of RTK activation may not be fundamental, but may simply lie in the magnitude of the dimerization constant for unliganded receptors. Thus, all RTKs may follow a universal model of activation, which includes unliganded dimers of various stabilities as intermediates. The hypothesis that such a universal model can describe the activation mechanism of the different RTKs is testable with the QI-FRET method used here.

We provide a direct experimental demonstration that ligand-induced structural changes occur in FGFR dimers within the plasma membrane. Ultimately, the ligand controls the structure of the TM domain by triggering a switch to a specific configuration, and the resulting structure of the TM dimer controls the activity of the receptor. The structural changes in response to fgf1 and fgf2 binding are very different, resulting in different distances between the intracellular domains, and different phosphorylation levels for the fgf1- and fgf2-bound dimers. Thus, there exist multiple active ligand-bound states for the FGF receptors.

The FRET experiments demonstrate that FGFR3 TM domains can form at least three different dimer structures, corresponding to the unliganded, fgf1- and fgf2-bound states. On fgf1 binding, FGFR3 TM domains change conformation and engage in interactions that likely involve small amino-acid residues

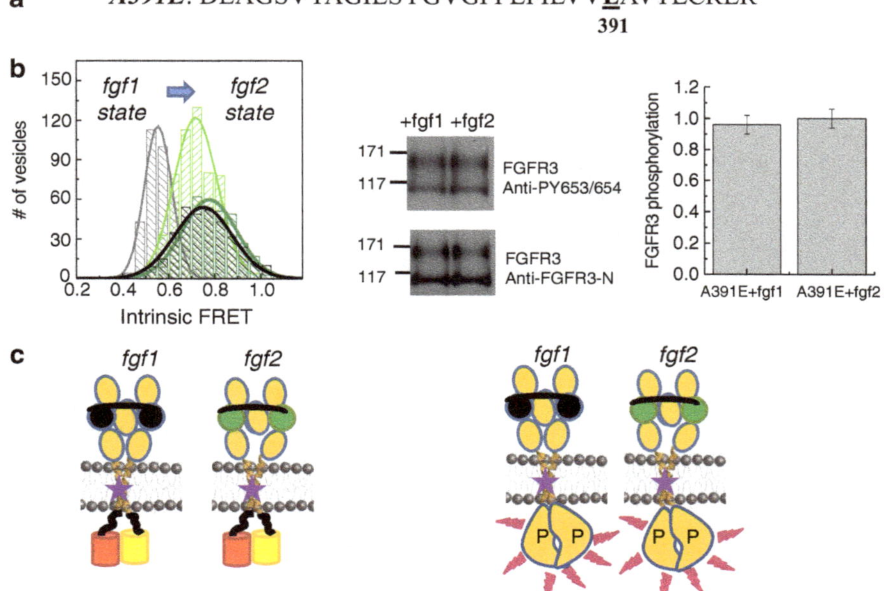

Figure 6 | Effect of the pathogenic A391E FGFR3 mutation on dimer structures in the fgf1- and fgf2-bound states. The A391E mutation is the genetic cause for Crouzon syndrome with acanthosis nigricans, a cranial abnormality[57], and has been linked to bladder cancer[8]. (**a**) Sequence of the A391E TM domain, with the mutation underlined. (**b**) Intrinsic FRET values, measured for the A391E EC + TM mutant in the presence of saturating concentrations of fgf1 or fgf2. The histograms for the wild type are shown in grey in the presence of fgf1 and in green in the presence of fgf2. The histograms for the A391E mutant are shown in black in the presence of fgf1 and in olive in the presence of fgf2. The intrinsic FRET values measured for the A391E mutant in the presence of fgf1 shift up, such that they overlap with the fgf2 wild-type values. Thus, the A391E mutation abolished the fgf1-bound state. (**b**) Western blots showing expression, as assayed by anti-FGFR3 antibodies, and phosphorylation of the tyrosines in the activation loop, as assayed by anti-phospho-Y653/4 antibodies. The phosphorylation of the mature fully glycosylated A391E mutant (top bands) is identical in the fgf1 and fgf2-bound states ($P > 0.01$), and is the same as the phosphorylation of the wild type in the fgf2-bound state. Data are from three independent experiments. Thus, the A391E mutation increases the phosphorylation in the fgf1 state to fgf2-state levels. (**c**) Graphic representation of the finding that the A391E mutation abolishes the fgf1-bound state and induces a transition to the fgf2 state. Left: graphic representation of the finding that the average distance between the fluorescent proteins is the same in the presence of both fgf1 and fgf2. Distances are not drawn to scale. Right: graphic representation of the finding that phosphorylation is also the same in the presence of fgf1 and fgf2. The representation of the kinase domains is not based on structural data.

in the N-terminal portion of the receptors (Fig. 8). FGFR3 phosphorylation increased by a factor of 1.5–3 on treatment with saturating fgf1 concentrations. We note that this cannot be explained by an increase in dimerization because FGFR3 dimerization is already very high under these conditions. Indeed, the dimeric fractions in QI-FRET experiments with full-length FGFR3 exceed 80% (Fig. 1), and phosphorylation experiments were done at similar expression levels (assessed by western blotting). We thus conclude that the structural change that occurs in the preformed FGFR3 dimer on fgf1 binding increases FGFR3 phosphorylation. Binding of fgf2 triggers a structural change towards a different TM dimer structure in which the interface likely involves contacts between L377, G380 and A391 (Fig. 8). The fgf2-bound FGFR3 dimer structure ensures the smallest separation between the TM domains and the highest FGFR3 phosphorylation, which increases by a factor of 2–4.

The difference between the fgf1- and fgf2-bound states observed here provides an explanation of the different biological roles of these two ligands. Hidai *et al.*[28] studied the effects of fgf1 and fgf2 on the differentiation of a multipotent embryonal carcinoma cell line, and found that fgf1 promotes differentiation into cardiac muscle cells, while fgf2 induces differentiation into skeletal muscle cells. Our study suggests that these profoundly different biological effects may originate in structural differences of the receptor–ligand complexes on the cell surface.

An important result of this study is the strong correlation between intrinsic FRET and kinase phosphorylation (Fig. 7). Since in the full-length receptors the kinase domains are attached to the TM domains via the juxtamembrane domains, these results suggest that a correlation exists between the separation of the kinase domains in the dimer and their phosphorylation: the smaller the distance between the kinases, the higher their phosphorylation. However, the distance between the TM domain C termini is not the only parameter that affects phosphorylation levels. For example, fgf1 binding to FGFR3 does not change the intrinsic FRET, but increases receptor phosphorylation. There appears to be a conformational change in the TM domain on fgf1 binding that likely affects TM helix rotation, not separation, and thus cannot be captured in the FRET experiments. These findings, and published work[62], suggest that the relative orientation of the kinases with respect to each other is another important parameter that determines phosphorylation efficiencies.

The activation mechanism discovered here sheds new light on the effect of pathogenic FGFR mutations on FGFR signalling[6,63–65]. We show that the A391E mutation in FGFR3, linked to Crouzon syndrome with acanthosis nigricans and to bladder cancer[8,57], mimics the structural and functional effects of fgf2 binding. In particular, the mutation prevents the FGFR3 dimer from exploring the unliganded and fgf1-bound conformations, and traps it in its most active state, the fgf2

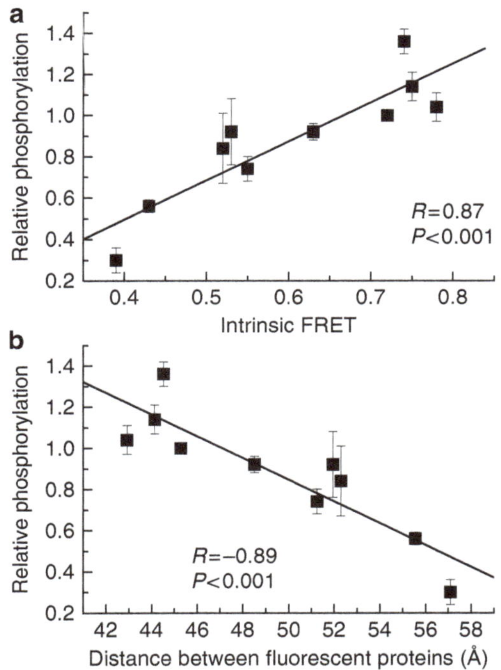

Figure 7 | Correlation between intrinsic FRET and phosphorylation. Results are shown for wild-type FGFR3 and the three studied FGFR3 mutants: the L377I-G380I-A391I mutant, the A374I-G375I-S378I mutant, and the A391E mutant, when 100% dimeric (see text). At least three independent experiments were performed for each mutant. The phosphorylation of the wild type in the fgf2-bound state is assigned a value of 1, and all other measured phosphorylation levels are scaled accordingly. (**a**) There is a strong correlation between the measured intrinsic FRET and phosphorylation ($P<0.001$). (**b**) Strong correlation between the distance between fluorescent proteins in the EC + TM FGFR3 constructs and full-length FGFR3 phosphorylation.

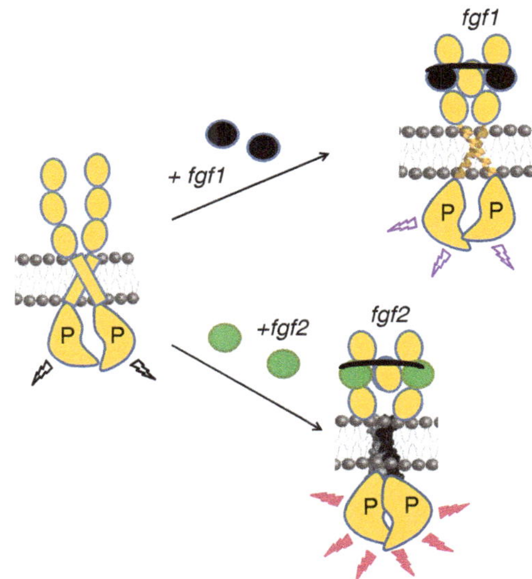

Figure 8 | The mechanism of FGFR3 activation. Top: on fgf1 binding to extracellular domains D2 and D3, and the D2-D3 linker[26,69], FGFR3 TM domains change their configuration and engage in interactions that involve small residues in the N-terminal portion of the TM domains. FGFR3 dimer phosphorylation increases by a factor of ~1.5 to ~3. Bottom: binding of fgf2 to D2, D3, and the linker, on the other hand, triggers a switch towards a closely packed TM dimer structure. Contacts between the TM helices are likely mediated by L377, G380 and/or A391, as in the case of a published NMR structure of the isolated TM domain[46]. FGFR3 dimer phosphorylation increases by a factor of ~2 to ~4.

state. This is a fundamentally novel mechanism through which FGFR-linked pathologies can arise.

Methods

Plasmids. The YFP plasmid was received from Dr M. Betenbaugh (Johns Hopkins University, Baltimore, MD) and the pRSET-mCherry plasmid was obtained from Dr R.Tsien (University of California, San Diego). The plasmids encoding human wild-type FGFR1 IIIc and FGFR2 IIIc in the pRK5 vector were received from Dr M. Mohammadi, NYU. The plasmid encoding human wild-type FGFR3 IIIc in the pcDNA3.1(+) vector was a gift from Dr D. J. Donoghue, UCSD. All primers were purchased from Invitrogen.

For this work, the full length *FGFR1* and *FGFR2* genes were cloned into the pcDNA3.1(+) vector. To accomplish this, the genes were first amplified using polymerase chain reaction (PCR) and then each gene was double digested using Hind III and XhoI restriction enzymes and ligated into the pcDNA3.1(+) vector. *YFP* or *mCherry* genes were subsequently fused to the C-terminal tail of each receptor via a 3-amino-acid (GGS) linker (Supplementary Fig. 1) between the XhoI and XbaI restriction sites in the vector. The A206K mutation was introduced in the *YFP* gene to render it monomeric.

Details on the cloning of *FGFR3 EC + TM-(GGS)$_5$-YFP* and *FGFR3 EC + TM-(GGS)$_5$-mCherry* into the pcDNA3.1(+) vector are given elsewhere[33]. For this work, the sequences encoding for the EC and TM domains of FGFR1 and FGFR2 were amplified by PCR, double digested using HindIII and EcoRV restriction enzymes and inserted in place of the *FGFR3 EC + TM* gene in the *FGFR3 EC + TM-(GGS)$_5$-YFP* and *FGFR3 EC + TM-(GGS)$_5$-mCherry* plasmid constructs, to produce the *FGFR1 EC + TM-(GGS)$_5$-YFP*, *FGFR1 EC + TM-(GGS)$_5$-mCherry*, *FGFR2 EC + TM-(GGS)$_5$-YFP* and *FGFR2 EC + TM-(GGS)$_5$-mCherry* plasmids (Supplementary Table 1).

All the mutant full-length *FGFR3* gene constructs were created from the wild type using QuickChange II XL Site-Directed Mutagenesis Kit (Stratagene, CA). These constructs are shown in Supplementary Fig. 2, and are used in the western blotting experiments.

The EC + TM (A374I-G375I-S378I) FGFR3 mutant and the EC + TM (L377I-G380I-A391I) FGFR3 mutant used in the FRET studies were generated from the full-length mutants. The complementary DNA (cDNA) encoding the EC and mutant TM domains was amplified using PCR and was double digested with HindIII and EcoRV. The *FGFR2 EC + TM-(GGS)$_5$-YFP* and *FGFR2 EC + TM-(GGS)$_5$-mCherry* plasmid constructs were also double digested with HindIII and EcoRV enzymes, and ligated with the PCR products.

Details on the cloning of *FGFR3 TM-(GGS)$_5$-YFP* and *FGFR3 TM-(GGS)$_5$-mCherry* into the pcDNA3.1(+) vector are given elsewhere[40]. For this work, the sequences encoding for the TM domains of FGFR1 and FGFR2 were amplified by PCR, double digested using KpnI and EcoRV restriction enzymes and ligated with *pcDNA-(GGS)$_5$-mCherry* and *pcDNA-(GGS)$_5$-YFP* vectors to yield the *FGFR1 TM-(GGS)$_5$-YFP*, *FGFR1 TM-(GGS)$_5$-mCherry*, *FGFR2 TM-(GGS)5-YFP* and *FGFR2 TM-(GGS)$_5$-mCherry* plasmids.

Cell culture and transfection for FRET experiments. CHO cells were cultures at 37 °C with 5% CO_2 for 24 h. Transfection was carried out using Fugene HD transfection reagent (Roche Applied Science), following the manufacturer's protocol. Cells were co-transfected with 3–7 µg of DNA encoding receptors tagged with either YFP or mCherry.

Production of mammalian plasma membrane vesicles. Vesiculation was performed using a chloride salt vesiculation buffer consisting of 200 mM NaCl, 5 mM KCl, 0.5 mM $MgSO_4$, 0.75 mM $CaCl_2$, 100 mM bicine and protease inhibitor cocktail (Complete mini EDTA-free tabs, Roche Applied Science) adjusted to pH of 8.5 (ref. 38). CHO cells were rinsed twice with 30% PBS (pH 7.4), and incubated with 1 ml of chloride salt vesiculation buffer overnight at 37 °C. A large number of vesicles were produced after 12 h, and the vesicles were transferred into four-well Nunc Lab-Tek II chambered coverslips for imaging. Images of vesicles with FGF receptors tagged with fluorescent proteins are shown in Supplementary Fig. 3. The cytoplasm in the vesicles is lost during vesicle production, as attested by the fact that soluble proteins of molecular weight up to 200 kDa are not retained inside the vesicles[34].

QI-FRET image acquisition. Vesicles were imaged using a Nikon Eclipse confocal laser scanning microscope using a × 60 water immersion objective. All the images were collected and stored at a 512 × 512 resolution. Three different scans were performed for each vesicle: (1) excitation at 488 nm, with a 500–530-nm emission filter (donor scan); (2) excitation at 488 nm, with a 565–615-nm emission filter

(FRET scan); and (3) excitation at 543 nm, with a 650-nm longpass filter (acceptor scan). Gains of 8.0 were used for all the three scans. The bleaching of the fluorescent proteins was minimized through the use of ND8 filters when exciting with the 488-nm laser, and low pixel dwell time (1.68 μs).

QI-FRET data analysis. The QI-FRET method has been described previously[33,41], but was applied here for the first time to full-length FGFRs. Purified solutions of YFP and mCherry[66]of known concentration were imaged in the donor, FRET and acceptor scans, to obtain the calibration constants for the donor and the acceptor, i_D and i_A, as well the bleed-through coefficients for the donor and the acceptor, β_D and β_A (ref. 37). A soluble linked YFP–mCherry protein was also imaged in the three channels to obtain the gauge factor G_F as described by Li *et al.*[37]

Each vesicle co-expressing FGFR–YFP and FGFR–mCherry was imaged in the donor, acceptor and FRET channels (Supplementary Fig. 3). The fluorescence intensities across the plasma membrane, I_D, I_{FRET} and I_A, in the three channels, were determined as described in detail elsewhere[33]. The acceptor concentration in each vesicle, C_A, was calculated according to ref. 33:

$$C_A = \frac{I_A}{i_A} \tag{1}$$

The sensitized emission of the acceptor in each vesicle was determined as[37]:

$$I_{SEN} = I_{FRET} - \beta_A I_A - \beta_D I_D \tag{2}$$

The donor intensity in the absence of the acceptor $I_{D,corr}$, and the donor concentrations (C_D) were calculated as:

$$I_{D,corr} = I_D + G_F I_{SEN} \tag{3}$$

$$C_D = \frac{I_{D,corr}}{i_D} \tag{4}$$

From equations (1) and (4), the total concentration, T, and the acceptor fraction, x_A, are calculated according to:

$$T = C_A + C_D \tag{5}$$

$$x_A = \frac{C_A}{C_A + C_D} \tag{6}$$

The FRET efficiency, E, was calculated using equation (7):

$$E = 1 - \frac{I_D}{I_{D,corr}} \tag{7}$$

The FRET efficiency was corrected for the so-called 'proximity FRET' contribution, which describes the close approach of donors and acceptors (within distances of 100 Å or so) in the membrane without specific interactions[67]. The dimeric fraction is determined from the corrected FRET efficiency according to:

$$f_D = \frac{E}{x_A \tilde{E}} \tag{8}$$

The constant $\tilde{E}$ in equation (8) is the 'intrinsic FRET', the FRET efficiency in a dimer containing a donor and an acceptor. This is a structural parameter, a constant for each receptor dimer, which depends only on the separation and the orientation of the two fluorescent proteins in the dimer, not on the dimerization propensity. x_A is the acceptor fraction.

Based on the law of mass action, the dimeric fraction can be written as a function of the total receptor concentration, T, and the dimerization constant K according to equation (9):

$$f_D = \frac{1}{T}\left(T - \frac{1}{4K}\left(\sqrt{1+8TK} - 1\right)\right) \tag{9}$$

Equations (8) and (9) are used to fit the dimerization model to the data while optimizing for two adjustable parameters: the dimerization constant K, and the intrinsic FRET, $\tilde{E}$ (Supplementary Fig. 4). The dissociation constant $K_{diss} = 1/K$ is reported in units of receptors per μm^2 in Table 1. The value of K_{diss} can be then directly compared with expressions levels in order to evaluate the biological significance of dimerization.

The free energy of dimerization (dimer stability) ΔG is calculated from the dimerization constant $K = 1/K_{diss}$. The standard state is defined as nm^2 per receptor[33], and therefore:

$$\Delta G = -RT\ln(\frac{10^6}{K_{diss}}) \tag{10}$$

In the case of 100% dimers ($f_D = 1$), equation (8) can be re-written as:

$$\tilde{E} = \frac{E}{x_A} \tag{11}$$

Thus, measurements of E and x_A for each vesicle in this case allows us to directly determine the value of the intrinsic FRET, $\tilde{E}$, in each vesicle. Histograms of the measured $\tilde{E}$ are shown throughout the manuscript, such as in Fig. 3, for example.

Finally, the dependence of the intrinsic FRET, $\tilde{E}$, on the distance between the fluorescent proteins in the dimer is given by equation (12).

$$\tilde{E} = \frac{1}{1 + (d/R_0)^6}. \tag{12}$$

Here d is the distance between the acceptor and the donor in the dimer, and R_o is the Förster radius of the FRET pair[33]. For eYFP and mCherry, R_o is 53.1 Å.

Western blots. CHO and HEK 293T cells were starved in serum-free medium for 24 h following transfection with 0.3 to 2.5 μg of DNA encoding the full-length receptors and then treated with lysis buffer (25 mM Tris-HCl, 0.5% Triton X-100, 20 mM NaCl, 2 mM EDTA, phosphatase inhibitor and protease inhibitor, Roche Applied Science). Lysates were collected following centrifugation at 15,000*g* for 15 min at 4 °C and loaded onto 3–8% NuPAGE Novex Tris-Acetatemini gels (Invitrogen, CA). The proteins were transferred onto a nitrocellulose membrane, and blocked using 5% milk in Tris-buffered saline. The phosphorylation of the tyrosines in the activation loop of the FGFR kinases was assessed first following protein transfer to the nitrocellulose membrane. Immunostaining was performed using specific anti-phospho-Tyr antibodies (Tyr653/654; #3471; Cell Signaling Technologies). These antibodies are raised against Tyr653/654 in the activation loop of FGFR1, but are reactive to all FGF receptors because of their identical activation loop sequences. Anti-phospho-Tyr766 (1E5; #2544; Cell Signaling Technology) was used to asses FGFR1 phosphorylation at Tyr766. Anti-phospho-Tyr724 (sc-33041; Santa Cruz Biotechnology) was used to detect the phosphorylation of Tyr724 in FGFR3. The membranes were then stripped and total FGFR protein levels were assessed using antibodies against FGFR3 (H-100; sc-9007), FGFR2 (H-80; sc-20735) and FGFR1 (H-76; sc-7945) from Santa Cruz Biotechnology. All the primary antibodies were followed by anti-rabbit HRP conjugated antibody (W4011, Promega). Dilutions were 1:1,000 for all primary antibodies and 1:2,500 for the secondary antibody. The proteins were detected using the Amersham ECL detection system (GE Healthcare). Uncropped versions of gels in Fig. 3b are shown in Supplementary Fig. 6.

Experimental protocols ensured that measurements were preformed in the 'linear regime', as described in detail in the Supplementary Data file in ref. 68.

Activation with fgf1 and fgf2. CHO and HEK 293T cells were cultured in normal medium for 24 h following transfection and then starved in serum-free medium for 24 h. Human fgf1 and human fgf2 were purchased from Cell Signaling Technologies (catalogue # 5234LC and # 8910LC, respectively). 5000 ng ml^{-1} of fgf1 or fgf2 were added to the serum-free medium. After incubating for 10 min at 37 °C with ligand, cells were placed on ice immediately and lysed as described above and analysed using western blotting.

References

1. Wilkie, A. O. M., Morriss-Kay, G. M., Jones, E. Y. & Heath, J. K. Functions of fibroblast growth factors and their receptors. *Curr. Biol.* **5,** 500–507 (1995).
2. Eswarakumar, V. P., Lax, I. & Schlessinger, J. Cellular signaling by fibroblast growth factor receptors. *Cytokine Growth Factor Rev.* **16,** 139–149 (2005).
3. Ornitz, D. M. FGFs, heparan sulfate and FGFRs: complex interactions essential for development. *BioEssays* **22,** 108–112 (2000).
4. Ornitz, D. M. & Itoh, N. Fibroblast growth factors. *Genome Biol.* **2** (2001).
5. Deng, C., Wynshaw-Boris, A., Zhou, F., Kuo, A. & Leder, P. Fibroblast growth factor receptor 3 is a negative regulator of bone growth. *Cell* **84,** 911–921 (1996).
6. Vajo, Z., Francomano, C. A. & Wilkin, D. J. The molecular and genetic basis of fibroblast growth factor receptor 3 disorders: the achondroplasia family of skeletal dysplasias, Muenke craniosynostosis, and Crouzon syndrome with acanthosis nigricans. *Endocr. Rev.* **21,** 23–39 (2000).
7. Wilkie, A. O. M. Bad bones, absent smell, selfish testes: The pleiotropic consequences of human FGF receptor mutations. *Cytokine Growth Factor Rev.* **16,** 187–203 (2005).
8. van Rhijn, B. W. G. *et al.* Novel fibroblast growth factor receptor 3 (FGFR3) mutations in bladder cancer previously identified in non-lethal skeletal disorders. *Eur. J. Hum. Genet.* **10,** 819–824 (2002).
9. van Rhijn, B. W. G., Montironi, R., Zwarthoff, E. C., Jobsis, A. C. & van der Kwast, T. H. Frequent FGFR3 mutations in urothelial papilloma. *J. Pathol.* **198,** 245–251 (2002).
10. Chesi, M. *et al.* The t(4;14) translocation in myeloma dysregulates both FGFR3 and a novel gene, MMSET, resulting in IgH/MMSET hybrid transcripts. *Blood* **92,** 3025–3034 (1998).
11. Chesi, M. *et al.* Frequent translocation t(4;14)(p16.3;q32.3) in multiple myeloma is associated with increased expression and activating mutations of fibroblast growth factor receptor 3. *Nat. Genet.* **16,** 260–264 (1997).
12. Seo, A. N. *et al.* FGFR1 amplification is associated with poor prognosis and smoking in non-small-cell lung cancer. *Virchows Arch.* **465,** 547–558 (2014).
13. Turner, N. *et al.* FGFR1 amplification drives endocrine therapy resistance and is a therapeutic target in breast cancer. *Cancer Res.* **70,** 2085–2094 (2010).

14. Xie, L. *et al.* FGFR2 gene amplification in gastric cancer predicts sensitivity to the selective FGFR inhibitor AZD4547. *Clin. Cancer Res.* **19,** 2572–2583 (2013).
15. Richelda, R. *et al.* A novel chromosomal translocation t(4;14)(p16.3;q32) in multiple myeloma involves the fibroblast growth-factor receptor 3 gene. *Blood* **90,** 4062–4070 (1997).
16. Fantl, W. J., Johnson, D. E. & Williams, L. T. Signaling by Receptor Tyrosine Kinases. *Annu. Rev. Biochem.* **62,** 453–481 (1993).
17. L'Horte, C. G. M. & Knowles, M. A. Cell responses to FGFR3 signaling: growth, differentiation and apoptosis. *Exp. Cell Res.* **304,** 417–431 (2005).
18. Linggi, B. & Carpenter, G. ErbB receptors: new insights on mechanisms and biology. *Trends Cell Biol.* **16,** 649–656 (2006).
19. Schlessinger, J. Common and distinct elements in cellular signaling via EGF and FGF receptors. *Science* **306,** 1506–1507 (2004).
20. Mohammadi, M. *et al.* Identification of six novel autophosphorylation sites on fibroblast growth factor receptor 1 and elucidation of their importance in receptor activation and signal transduction. *Mol. Cell. Biol.* **16,** 977–989 (1996).
21. Spivakkroizman, T. *et al.* How do heparin and heparan-sulfate activate Fgf-mitogenic activity. *Trends Glycosci. Glycotechnol.* **7,** 447–449 (1995).
22. Spivakkroizman, T. *et al.* Heparin-induced oligomerization of Fgf molecules is responsible for Fgf receptor dimerization, activation, and cell-proliferation. *Cell* **79,** 1015–1024 (1994).
23. Mohammadi, M., Olsen, S. K. & Ibrahimi, O. A. Structural basis for fibroblast growth factor receptor activation. *Cytokine Growth Factor Rev.* **16,** 107–137 (2005).
24. Goetz, R. & Mohammadi, M. Exploring mechanisms of FGF signalling through the lens of structural biology. *Nat. Rev. Mol. Cell Biol.* **14,** 166–180 (2013).
25. Olsen, S. K. *et al.* Insights into the molecular basis for fibroblast growth factor receptor autoinhibition and ligand-binding promiscuity. *Proc. Natl Acad. Sci. USA* **101,** 935–940 (2004).
26. Plotnikov, A. N., Schlessinger, J., Hubbard, S. R. & Mohammadi, M. Structural basis for FGF receptor dimerization and activation. *Cell* **98,** 641–650 (1999).
27. Schlessinger, J. *et al.* Crystal structure of a ternary FGF-FGFR-heparin complex reveals a dual role for heparin in FGFR binding and dimerization. *Mol. Cell* **6,** 743–750 (2000).
28. Hidai, C. *et al.* FGF-1 enhanced cardiogenesis in differentiating embryonal carcinoma cell cultures, which was opposite to the effect of FGF-2. *J. Mol. Cell. Cardiol.* **35,** 421–425 (2003).
29. Ahmed, Z. *et al.* Direct binding of Grb2 SH3 domain to FGFR2 regulates SHP2 function. *Cell. Signal.* **22,** 23–33 (2010).
30. Takeda, M. *et al.* AZD2171 shows potent antitumor activity against gastric cancer over-expressing fibroblast growth factor receptor 2/keratinocyte growth factor receptor. *Clin. Cancer Res.* **13,** 3051–3057 (2007).
31. Kunii, K. *et al.* FGFR2-amplified gastric cancer cell lines require FGFR2 and Erbb3 signaling for growth and survival. *Cancer Res.* **68,** 2340–2348 (2008).
32. Bryant, M. R., Marta, C. B., Kim, F. S. & Bansal, R. Phosphorylation and lipid raft association of fibroblast growth factor receptor-2 in oligodendrocytes. *Glia* **57,** 935–946 (2009).
33. Chen, L. R., Novicky, L., Merzlyakov, M., Hristov, T. & Hristova, K. Measuring the energetics of membrane protein dimerization in mammalian membranes. *J. Am. Chem. Soc.* **132,** 3628–3635 (2010).
34. Sarabipour, S., Chan, R. B., Zhou, B., Di Paolo, G. & Hristova, K. Analytical characterization of plasma membrane-derived vesicles produced via osmotic and chemical vesiculation. *Biochim. Biophys. Acta* **1848,** 1591–1598 (2015).
35. Del Piccolo, N., Placone, J. & Hristova, K. Effect of thanatophoric dysplasia type I mutations on FGFR3 dimerization. *Biophys. J.* **108,** 272–278 (2015).
36. Sarabipour, S. & Hristova, K. FGFR3 unliganded dimer stabilization by the juxtamembrane domain. *J. Mol. Biol.* **427,** 1705–1714 (2015).
37. Li, E., Placone, J., Merzlyakov, M. & Hristova, K. Quantitative measurements of protein interactions in a crowded cellular environment. *Anal. Chem.* **80,** 5976–5985 (2008).
38. Del Piccolo, N., Placone, J., He, L., Agudelo, S. C. & Hristova, K. Production of plasma membrane vesicles with chloride salts and their utility as a cell membrane mimetic for biophysical characterization of membrane protein interactions. *Anal. Chem.* **84,** 8650–8655 (2012).
39. Sarabipour, S. & Hristova, K. Glycophorin A transmembrane domain dimerization in plasma membrane vesicles derived from CHO, HEK 293 T, and A431 cells. *Biochim. Biophys. Acta* **1828,** 1829–1833 (2013).
40. Chen, L., Placone, J., Novicky, L. & Hristova, K. The extracellular domain of fibroblast growth factor receptor 3 inhibits ligand-independent dimerization. *Sci. Signal.* **3,** ra86 (2010).
41. Sarabipour, S., Del Piccolo, N. & Hristova, K. Characterization of membrane protein interactions in plasma membrane derived vesicles with quantitative imaging forster resonance energy transfer. *Acc. Chem. Res.* **48,** 2262–2269 (2015).
42. Moscatelli, D. High and low affinity binding-sites for basic fibroblast growth-factor on cultured-cells - absence of a role for low affinity binding in the stimulation of plasminogen-activator production by bovine capillary endothelial-cells. *J. Cell. Physiol.* **131,** 123–130 (1987).
43. Evers, T. H., van Dongen, E. M. W. M., Faesen, A. C., Meijer, E. W. & Merkx, M. Quantitative understanding of the energy transfer between fluorescent proteins connected via flexible peptide linkers. *Biochemistry* **45,** 13183–13192 (2006).
44. Lin, C. C. *et al.* Inhibition of basal FGF receptor signaling by dimeric Grb2. *Cell* **149,** 1514–1524 (2012).
45. Ahmed, Z. *et al.* Grb2 controls phosphorylation of FGFR2 by inhibiting receptor kinase and Shp2 phosphatase activity. *J. Cell Biol.* **200,** 493–504 (2013).
46. Bocharov, E. V. *et al.* Structure of FGFR3 transmembrane domain dimer: implications for signaling and human pathologies. *Structure* **21,** 2087–2093 (2013).
47. Li, E. & Hristova, K. Role of receptor tyrosine kinase transmembrane domains in cell signaling and human pathologies. *Biochemistry* **45,** 6241–6251 (2006).
48. He, L., Horton, W. A. & Hristova, K. The physical basis behind achondroplasia, the most common form of human dwarfism. *J. Biol. Chem.* **285,** 30103–30114 (2010).
49. Chen, F., Degnin, C., Laederich, M. B., Horton, A. W. & Hristova, K. The A391E mutation enhances FGFR3 activation in the absence of ligand. *Biochim. Biophys. Acta* **1808,** 2045–2050 (2011).
50. Chen, F. H. & Hristova, K. The physical basis of FGFR3 response to fgf1 and fgf2. *Biochemistry* **50,** 8576–8582 (2011).
51. Chen, F., Sarabipour, S. & Hristova, K. Multiple consequences of a single amino acid pathogenic RTK mutation: the A391E mutation in FGFR3. *PLoS One* **8,** e56521 (2013).
52. Lemmon, M. A., Treutlein, H. R., Adams, P. D., Brünger, A. T. & Engelman, D. M. A dimerization motif for transmembrane alpha-helices. *Nat. Struct. Biol.* **1,** 157–163 (1994).
53. Doura, A. K., Kobus, F. J., Dubrovsky, L., Hibbard, E. & Fleming, K. G. Sequence context modulates the stability of a GxxxG-mediated transmembrane helix-helix dimer. *J. Mol. Biol.* **341,** 991–998 (2004).
54. Cymer, F., Veerappan, A. & Schneider, D. Transmembrane helix-helix interactions are modulated by the sequence context and by lipid bilayer properties. *Biochim. Biophys. Acta* **1818,** 963–973 (2012).
55. Li, E., Wimley, W. C. & Hristova, K. Transmembrane helix dimerization: beyond the search for sequence motifs. *Biochim. Biophys. Acta* **1818,** 183–193 (2012).
56. Russ, W. P. & Engelman, D. M. The GxxxG motif: a framework for transmembrane helix-helix association. *J. Mol. Biol.* **296,** 911–919 (2000).
57. Meyers, G. A., Orlow, S. J., Munro, I. R., Przylepa, K. A. & Jabs, E. W. Fibroblast-growth-factor-receptor-3 (Fgfr3) transmembrane mutation in Crouzon-syndrome with acanthosis nigricans. *Nat. Genet.* **11,** 462–464 (1995).
58. Li, E., You, M. & Hristova, K. FGFR3 dimer stabilization due to a single amino acid pathogenic mutation. *J. Mol. Biol.* **356,** 600–612 (2006).
59. Burgess, A. W. *et al.* An open-and-shut case? Recent insights into the activation of EGF/ErbB receptors. *Mol. Cell* **12,** 541–552 (2003).
60. Brewer, M. R. *et al.* The juxtamembrane region of the EGF receptor functions as an activation domain. *Mol. Cell* **34,** 641–651 (2009).
61. Belov, A. A. & Mohammadi, M. Grb2, a double-edged sword of receptor tyrosine kinase signaling. *Sci. Signal.* **5,** pe49 (2012).
62. Bell, C. A. *et al.* Rotational coupling of the transmembrane and kinase domains of the Neu receptor tyrosine kinase. *Mol. Biol. Cell* **11,** 3589–3599 (2000).
63. Webster, M. K. & Donoghue, D. J. FGFR activation in skeletal disorders: too much of a good thing. *Trends Genet.* **13,** 178–182 (1997).
64. Robertson, S. C., Tynan, J. A. & Donoghue, D. J. RTK mutations and human syndromes - when good receptors turn bad. *Trends Genet.* **16,** 265–271 (2000).
65. Cohen, M. M. Some chondrodysplasias with short limbs: molecular perspectives. *Am. J. Med. Genet.* **112,** 304–313 (2002).
66. Sarabipour, S., King, C. & Hristova, K. Un-induced high-yield bacterial expression of fluorescent proteins. *Anal. Biochem.* **449,** 155–157 (2014).
67. King, C., Sarabipour, S., Byrne, P., Leahy, D. J. & Hristova, K. The FRET signatures of non-interacting proteins in membranes: simulations and experiments. *Biophys. J.* **106,** 1309–1317 (2014).
68. He, L. & Hristova, K. Pathogenic activation of receptor tyrosine kinases in mammalian membranes. *J. Mol. Biol.* **384,** 1130–1142 (2008).
69. Plotnikov, A. N., Hubbard, S. R., Schlessinger, J. & Mohammadi, M. Crystal structures of two FGF-FGFR complexes reveal the determinants of ligand-receptor specificity. *Cell* **101,** 413–424 (2000).

Acknowledgements

This work was supported by NIH GM68619, GM95930 and NSF MCB1157687. We thank Dr D. J. Donoghue for the full-length FGFR3 plasmid, and Dr Moosa Mohammadi for the full-length FGFR1 and FGFR2 plasmids. We are grateful to Drs Michael Edidin, Daniel Leahy, William C. Wimley, Eduard Bocharov and Pavel Krejci for critical reading of the manuscript prior to publication, and for insightful comments. We thank Christopher King for help with data analysis and Anna Frishman for help with cloning.

Author contributions

S.S. performed all the experiments. K.H. and S.S. wrote the paper.

Additional information

Competing financial interests: The authors declare no competing financial interests.

4

Using synthetic bacterial enhancers to reveal a looping-based mechanism for quenching-like repression

Michal Brunwasser-Meirom[1,*], Yaroslav Pollak[2,*], Sarah Goldberg[1], Lior Levy[1], Orna Atar[1] & Roee Amit[1,2]

We explore a model for 'quenching-like' repression by studying synthetic bacterial enhancers, each characterized by a different binding site architecture. To do so, we take a three-pronged approach: first, we compute the probability that a protein-bound dsDNA molecule will loop. Second, we use hundreds of synthetic enhancers to test the model's predictions in bacteria. Finally, we verify the mechanism bioinformatically in native genomes. Here we show that excluded volume effects generated by DNA-bound proteins can generate substantial quenching. Moreover, the type and extent of the regulatory effect depend strongly on the relative arrangement of the binding sites. The implications of these results are that enhancers should be insensitive to 10–11 bp insertions or deletions (INDELs) and sensitive to 5–6 bp INDELs. We test this prediction on 61 σ^{54}-regulated *qrr* genes from the *Vibrio* genus and confirm the tolerance of these enhancers' sequences to the DNA's helical repeat.

[1] Department of Biotechnology and Food Engineering, Technion—Israel Institute of Technology, Haifa 32000, Israel. [2] Russell Berrie Nanotechnology Institute, Technion—Israel Institute of Technology, Haifa 32000, Israel. * These authors contributed equally to this work. Correspondence and requests for materials should be addressed to R.A. (email: roeeamit@technion.ac.il)

Distal regulation by transcription factors, which are bound *in cis*, yet some distance away from the core promoters, is a regulatory phenomenon ubiquitous in all organisms[1–4]. Mechanistically, DNA looping has been implicated in distal regulation in eukaryotes[5–7] and has been shown directly to be involved in σ^{54} promoter expression in bacteria[8]. However, most structural features of distal regulatory regions such as the importance of having several binding sites for a given transcription factor, the genomic distance of these binding sites from the basal promoter and the functional significance of particular arrangements of the binding sites remain poorly understood.

In bacteria, distal regulatory regions called 'bacterial enhancers' are positioned within 200 bp of a σ^{54} promoter and seem to fall into one of two broad classes. The enhancers in the first class contain activator-binding sites and an additional site for the ubiquitous nucleoid-associated protein integration host factor (IHF)[9,10]. The enhancers in the second class do not contain IHF-binding sites, but either harbour additional sites for other transcription factors[11–13] or exhibit some conserved AT-rich sequence[14,15]. In eukaryotes, the diversity of distal regulatory regions is vastly richer. There are several well-known examples of 'promoter-proximal' distal regulatory regions, which are clusters of transcription factor-binding sites that are located within 100–300 bp away from a core PolII promoter and are thus similar in sequence length and regulatory content to bacterial enhancers. Examples include the *c-fos* promoter in mammalian cells[16,17], the gal1 promoter in *Saccharomyces cerevisiae*[18] and the *hb* promoter[19] in *Drosophila melanogaster*. In addition, in most higher eukaryotes, there is another class of distal regulatory elements, which are often called 'enhancers'. These regions may be located one kbp to several Mbp away from their regulated promoter and contain clusters of transcription factor-binding sites as well[19]. As most of the sequence for distal regulatory regions (independent of their proximity to the core promoter) is believed to be non-coding, the evolutionary pressure on conservation is small, resulting in highly divergent sequences for homologous promoter-proximal regions and enhancers[20]. Consequently, computationally predicting the regulatory output function for an unannotated distal regulatory region in all organisms has proven to be a difficult task[21,22].

To develop a deeper understanding of distal regulatory regions, not only is there a need for additional gene expression data sets but the underlying mechanistic models must be formulated as well. To address this problem, we present here a model-focused variation of the 'synthetic enhancer' approach (that is, libraries of bacterial enhancers that are engineered via high-capacity double-stranded DNA synthesis techniques)[8,23], to provide one possible mechanism for 'quenching-like' repression[20]. 'Quenching' is a form of repression originally observed in fly enhancers, where repressors such as Snail[24], Kruppel[25], Knirps[26] or Giant[27] downregulate expression not via a competition with an activator for binding, but rather through having its binding sites positioned several 10 to $\sim$100 bp away from the nearest activator. 'Quenching-like' repression effects have also been reported for eukaryotic promoter-proximal regulatory regions and in bacterial enhancers as well. Here, repressors are bound in between the activators and the core promoter in repressed complexes. Well-documented examples include the YY1 repressor in the *c-fos* and other promoters in mammalian cells[16,17]; the $\alpha 2$ repressor was found to be co-bound with the Gal4 activator in a tightly repressed complex in *S. cerevisiae*[28], the glnAp2 σ^{54} promoter in *Escherichia coli*[11] and the Nac σ^{54} promoter in *Klesbiella aerogenes*[13], and quenching-like effects have also been attributed to inadequate positioning of an IHF-binding site with respect to the activator or σ^{54} promoter sequences in other promoters as well[29]. However, despite the many observations of closely bound ensembles of proteins on distal regulatory elements, which interact in a repressive manner to regulate gene expression, establishing a broadly applicable mechanistic explanation has remained elusive.

Results

Simulating quenching repression. As quenching is a phenomenon that is associated with proteins that seem to bind DNA several tens of base pairs away from an activator, we hypothesized that the underlying mechanism for repression might be an excluded volume effect, where a bound protein alters the propensity of DNA to form a loop by its mere presence (Fig. 1a). We opted to explore this hypothesis by devising a numerical simulation using the worm-like chain (WLC) model[30] as a basis (see Supplementary Note 1). To do so, we modified the wormlike chain model to generate chains made of finite volume links (Supplementary Fig. 1a,b). Such 'thick' chains can be used to probe excluded volume effects, as only configurations where parts of the chain do not cross each other are considered. In addition, thick chains can be 'deformed' locally by additional volumes or protrusions (Supplementary Fig. 1c,d), and using numerical simulations the effects of these local protrusions on various chain properties can be estimated. Since these protrusions can be likened to proteins bound to DNA, the model's results can be used to estimate the likelihood that a particular protrusion-bound looped configuration will occur. We term this approach the self-avoiding WLC model[31].

To obtain an initial set of predictions, we generated ensembles containing 10^7–10^9 configurations of thick chains with protrusions (see Fig. 1b), up to a chain length of 300 links, with one link corresponding to 1 bp. Each chain in our ensemble was grown one link at a time, avoiding chain–chain collision events via a method termed weighted biased sampling[32] (see Supplementary Note 2). To model quenching effects, we generated a thick chain architecture containing three protrusions: one at each end of the chain, simulating the activator and holoenzyme complex, respectively, and an additional protrusion simulating a generic transcription factor (TF) positioned somewhere along the chains (Fig. 1b).

Using these ensembles, we computed chain observables. For the case of thick-chain looping, there is considerable freedom in choosing the particular end-to-end separation criterion, which differentiates between looped and non-looped configurations. For the predictions shown in this study, we chose a looping boundary condition (shown in Supplementary Fig. 1e) that mimics the actual geometry of the bacterial σ^{54} interaction with its upstream activator[33]. In our simulations, the TF binding site or protrusion is located some k links away from the 'activator' and $N-k$ links away from the promoter (Fig. 1c top). In Fig. 1c middle, we plot the results. The figure is plotted as a heat map representation of the looping probability ratio $R_1(N,k)$ (Fig. 1c bottom). Each point in the two-dimensional (N,k) map corresponds to the ratio of the probabilities of looping for an enhancer-mimicking thick chain of length N with a TF-like protrusion located k links away from the activator, to a thick chain without the TF-like protrusion. The figure shows a two-dimensional oscillatory pattern with alternating patches of upregulation (red, with looping ratio values $>100\%$) and downregulation (blue, looping ratio values $<100\%$) in the probability of looping, with periodicities that are consistent with the 10.5-bp helical periodicity that was preset for our simulation. The value of 100% (yellow–green) corresponds to the case where the looping ratio is one and thus no effect of excluded volume is predicted for those particular binding-site arrangements.

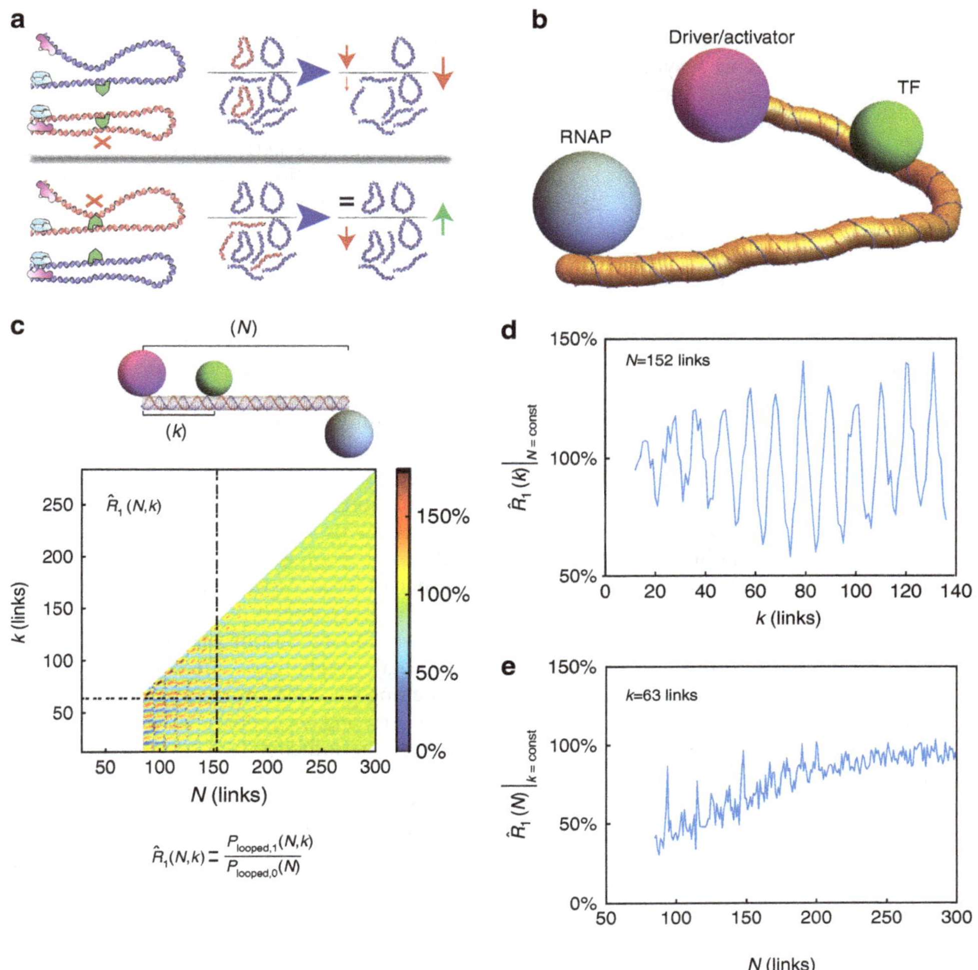

Figure 1 | Transcriptional regulation via an excluded volume effect. (**a**) Schematic representation of the looping probability ratio corresponding to the effects of a TF excluded volume on the probability of looping. We focus on two extreme binding configurations: first, the TF is positioned inside the loop 'in-phase'; thus, its binding site will be separated from the activator's by an integer multiple of the helical repeat ($\sim n \times 10.5$ bp). Second, the TF is positioned outside of the loop or 'out-of-phase' from the activator; thus, the binding site separation must be an odd integer multiple of half of the helical repeat ($\sim (n+1/2) \times 10.5$ bp). In the top schematic (**a**, top), the transcription factor bound 'in-phase' excludes many of the looped configurations that were available to the polymer not bound by the TF. As a result, the propensity of the bound complex to loop should be reduced as compared with a complex not bound by the TF. This should lead to a repression regulatory effect. Conversely, the 'out-of-phase' TF is positioned outside of the loop, which mainly results in the exclusion of previously available non-looped configurations. This results in a slight increase in the propensity of the TF bound complex to loop. Consequently, the out-of-phase configuration should yield a small upregulation or activation-like effect (**a**, bottom). Together, the 'in-phase' and 'out-of-phase' arrangements of the TF should generate a regulatory response that oscillates between down and upregulation as a function of the position of the TF within the loop. (**b**) Sample conformation of a thick chain with three protrusions generated using the self-avoiding WLC (SAWLC) algorithm. (**c**) Heat-map of the simulated looping probability ratio for a thick chain. Top: schematic with protrusions simulating a minimal enhancer with a single TF-binding site as a function of looping length (N) and TF protrusion position (k). The size of the TF protrusion was set to be 5.44 nm in diameter. Bottom: schematic showing the definition of the looping probability ratio. (**d,e**) One-dimensional cross-sections of the simulated probability ratio heat map at constant $N = 152$ bp and $k = 64$ bp, respectively.

A closer examination of various one-dimensional cross-sections of the heat map shows additional phenomena. For a cross-section at constant N (Fig. 1d), the amplitude of the oscillations between the up-regulation maxima and down-regulation minima increases the further the protrusion is positioned away from the chain origin, with a maximal amplitude reached at the position which bisects the chain ($k \approx N/2$). Furthermore, as the protrusion's location approaches the terminal end of the chain ($k \to N$), the effect is shifted towards upregulation. Thus, our model predicts that the strongest quenching is achieved not only when the TF is 'in-phase' (see Fig. 1) with the activator but also at locations inside the loop that are not immediately adjacent to either the activator or promoter, but rather equidistant from both of them. The origin of this effect is rooted in the geometry of the most probable looped configuration, which is shaped like a 'tear drop'. This shape is characterized by high curvature in the middle of the loop and low curvature in proximity to the chain termini (see Supplementary Fig. 2 and Supplementary Note 4). Finally, the heat map shows that for the cross-sections with constant k and varying N (Fig. 1e), the looping probability ratio increases initially for smaller looping lengths and saturates at values that are $< 100\%$ for large N.

Synthetic enhancers with a single binding site. To test the predictions of the self-avoiding WLC looping model, we constructed libraries of synthetic enhancers in bacteria. Previously, we showed that bacterial enhancers are a unique platform to test distal regulatory models in a synthetic context due to their simplicity of architecture and suitability for high-throughput *in vivo* work[8]. A minimal bacterial enhancer comprises a tandem of activator, or driver, binding sites (for example, NtrC, PspF and so on) located several tens to hundreds of base pairs upstream of a poised σ^{54} promoter (Fig. 2a). These enhancers can be made systematically more complex by adding cassettes of binding sites for other transcription factors in various configurations. We measure our synthetic enhancers' regulatory output via an expression-level assay, where a minimal NRI/NtrC-σ^{54} enhancer[34] drives the expression of an mCherry reporter, which is encoded downstream from the σ^{54} promoter.

Our first library design consisted of 81 regulatory sequences containing a single binding site for one of three different TFs: LacI[35], TetR[36] and TraR[37]. The choice of LacI, TetR and TraR was predicated on their ability to bind DNA dependent on the absence (LacI and TetR) or presence (TraR) of a ligand, whose concentration we controlled externally. In addition, we varied two additional control parameters: the looping length N, defined as the distance from the centre of the NtrC/NRI activator-binding site tandem to the centre of the σ^{54} promoter, and the distance k separating the centre of the transcription factor-binding site from the middle of the NtrC/NRI-binding site tandem. We limited our enhancers to looping lengths $N > 150$ bp but did not restrict the binding site address k, allowing its bound TF to be positioned on the same or the opposite DNA face with respect to the activator and the σ^{54}, or at some intermediate position.

In Fig. 2b–d, we plot expression-level ratio $R_1(N,k)$ for constant looping length (N) as a function of binding-site position (k). $R_1(N,k)$ is defined as the ratio of expression levels between a protein-bound enhancer and that of an unbound enhancer, for synthetic enhancers containing a single binding site for the three transcription factors: TraR (Fig. 2b), LacI (Fig. 2c) and TetR (Fig. 2d; see Methods and Supplementary Note 3 for a detailed definition). In addition, for all three data sets the position of the binding site (k) was varied across the looping region in 1–3 bp intervals. For the TetR and TraR cases, we observe a long-range oscillatory function in the expression-level ratio between quenching and upregulation with a period $\sim$10–11 bp, which is consistent with the accepted value for the DNA helical repeat ($\sim$10.5–10.9 bp (refs 38–41)). Here we define quenching as expression-level ratio values that are $<100\%$, as the protein-bound case yields a lower total mCherry reporter level than the unbound case. Conversely, we define upregulation as the case in which the expression-level ratio is $>100\%$. Interestingly, maximal quenching seems to occur at k-values that are roughly integer multiples of the DNA helical repeat, whereas binding-site positions that are displaced 5–6 bp away from the minimas resulted in either weaker quenching repression or slight upregulation ($>100\%$) of the expression level of the bound enhancer with respect to the unbound case.

A closer examination of the regulatory response curves shows distinct differences that are strongly dependent on the TF type. For TraR synthetic enhancers (Fig. 2b), the effect of the transcription factor on the probability of looping is small and the total regulatory effect observed varies between weak quenching to slight upregulation ($\sim$75–120%). For LacI synthetic enhancers (Fig. 2c), a barely detectible oscillatory behaviour with $\sim$11 bp periodicity is observed. Here, the small-amplitude oscillations vary between intermediate ($\sim$50%) to weak quenching ($\sim$70%). Moreover, the amplitude of the oscillations seems to diminish as k increases, settling on an intermediate quenching level of $\sim$60%. Finally, a third distinct regulatory response is observed for TetR in Fig. 2d. Here, regulatory effects persist for the entire segment of the loop tested and both significant

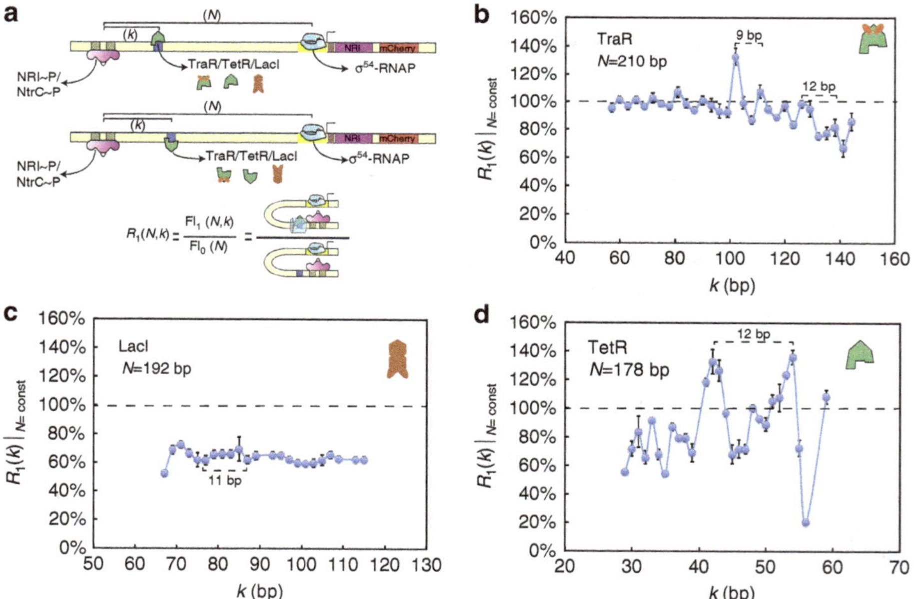

Figure 2 | Synthetic enhancers with a single TF-binding site. (**a**) Schematic for the minimal bacterial enhancer system used in our experiments, showing the poised holoenzyme complex at the σ^{54} promoter, NtrC activator and the additional binding site for either TraR, TetR or LacI. The schemas represent the two 'extreme' configurations. Top: binding sites are positioned 'out-of-phase' relative to the activator. Bottom: binding sites are positioned 'in-phase' relative to the activator. The schematic for TraR is drawn with two ovals corresponding to the 3OC8 ligand. (**b–d**) Expression level ratio results for synthetic enhancers with a single binding site for TraR (**b**), LacI (**c**) and TetR (**d**) at constant N. k was varied in 3, 2 and 1 bp steps for TraR, LacI and TetR, respectively. In addition, we show in Supplementary Note 3 that the expression-level ratio observable as defined here is approximately equal to the probability-of-looping ratio given known rates for the NtrC-σ^{54} system, thus allowing us to quantitatively compare experimental results to theoretical predictions. Error bars correspond to the s.d. from multiple measurements.

quenching and upregulation effects are observed (20–140%). Thus, although the oscillatory quenching/upregulation phenomenon observed clearly for two of the three proteins (TraR and TetR) supports the excluded volume looping-based regulatory mechanism, the differences in the expression-level ratio responses suggest that additional protein-specific aspects need to be added to the model, to better explain the data.

Additivity in synthetic enhancers. We reasoned that the simplest protein-specific parameter that can affect the regulatory response is the total volume in the loop, as both TetR (25 kDa) and TraR (26 kDa) are significantly smaller than LacI (38 kDa). To try to compensate for the mass and volume difference, we hypothesized that an increase in the number of binding sites for a smaller protein such as TraR should lead to a larger cumulative quenching effect, which will be comparable to the maximal affect achieved by LacI.

To do this, we constructed a second synthetic enhancer library with two TraR-binding sites. To determine the optimal binding-site arrangement for quenching, we first scanned the expression-level ratio values for a set of tandem TraR synthetic enhancers characterized by simultaneously varying values of the inter-site spacing s and the looping length N at 1 bp increments, while keeping k constant (Fig. 3a). We plot the results for the expression-level ratio as a function of the spacing s in Fig. 3a. The figure shows an oscillating function with a significantly stronger maximal quenching response ($\sim$30%) observed for $s=23$ than the one observed for the single TraR-binding site ($\sim$70%) and slightly larger maximal quenching than the response obtained for the single LacI-binding site. The figure also shows expression-level ratio minima at inter-site spacing values that are integer multiples of the helical repeat (that is, $s=23$–24, 34–35 and 44–45), whereas no quenching is observed for odd half-integer multiples of the spacing, as predicted by the model. Interestingly, the values of the expression-level ratio minima and maxima shift to higher values (from 30 to 60% and 80 to 100%, respectively) as the number of helical repeats between binding sites increases from two to three, to four. In Fig. 3b, we plot the model's predictions for the effects of inter-protrusion spacing (s) on the probability of looping. As in the experiment, the cross-section was taken for values of s and N that varied together by one-link increments for each successive point (Supplementary Fig. 3a dashed line), whereas the protrusion position (k) was kept constant at four helical repeats (42 links). Remarkably, the simulation exhibits not only oscillations in the probability ratio levels as expected but also an overall upward shift in the values of the probability ratio minima and maxima, in close agreement with the experimental data.

To further validate the volume additivity prediction, our second library also included synthetic enhancers with a tandem of TraR binding characterized by a fixed inter-site spacing ($s=23$ bp, in phase), while the placement of the proximal TraR-binding site from the (k) was varied. In Fig. 3c, we plot the expression-level ratio results (blue circles). The data stably oscillate from a strong quenching value of $\sim$40% to no-quenching or slight upregulation values of 100–110%. Although this behaviour persists for binding site positions that are spread over 100 bps, as k increases further so that the location of the tandem binding sites approaches the promoter, the amplitude of the oscillations diminishes and a clear bias towards upregulation emerges, with a maximal upregulation value of 160% observed for $k=189$ bp. The oscillatory pattern is highly repetitive with a periodicity of 10.5 ± 0.3 bp, an expression-level ratio amplitude that is approximately twice as large as for the synthetic enhancer with a single TraR-binding site (Fig. 3c red x's) and persists for nearly the entire looping length ($\sim$227 bp) with little dependence on the position of the first binding site.

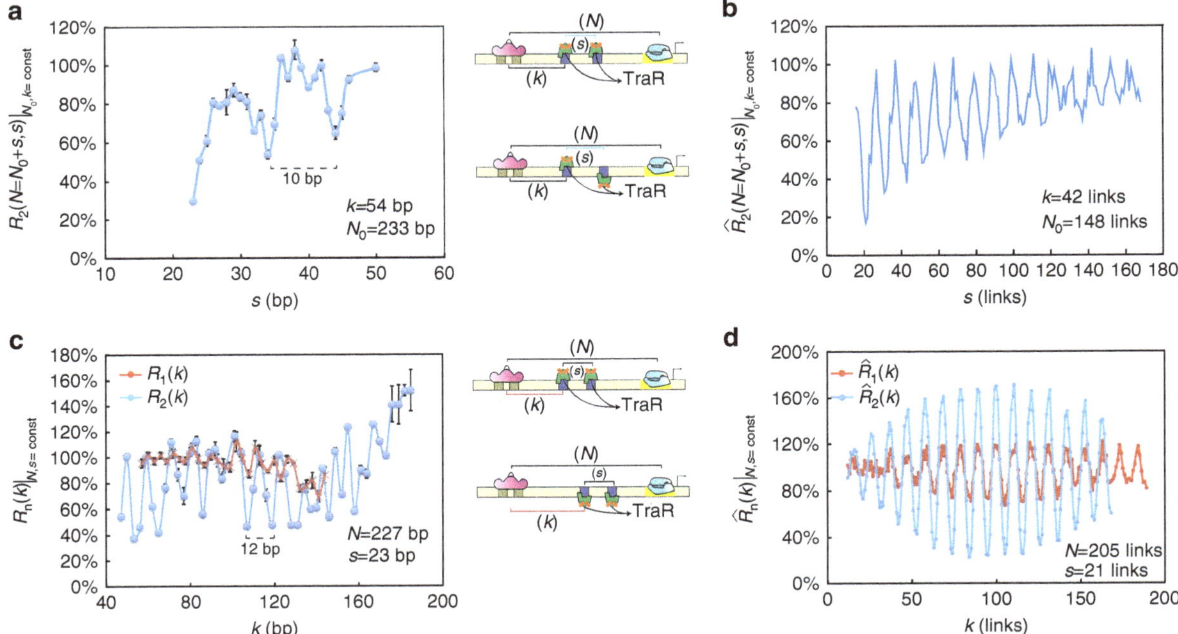

Figure 3 | Excluded volume is additive. (**a**) Expression level ratio for synthetic enhancers with a tandem of TraR-binding sites at various inter-TF spacing (s) and looping lengths (N); see schematic on the right. (**b**) Model predictions for looping probability ratio as a function of intra-binding site spacing s and looping length N. (**c**) Expression level ratio measurements for synthetic enhancers with a tandem of TraR-binding sites (blue) that are in-phase with fixed inter-TF spacing ($s=23$ bp); see schematic on the right. The single TraR-binding site synthetic enhancer data are overlayed as reference in red. (**d**) Model predictions for looping probability ratio as a function of distance to the chain origin (k) for the following configurations: blue, in phase tandem protrusions ($s=21$ bp); red, single protrusion. Error bars correspond to the s.d. from multiple measurements.

Comparing the experimental data with one-dimensional cross-sections of the modelling results (Fig. 3d and Supplementary Fig. 3b dashed line) for thick chains with a single protrusion (red line) and a tandem of in-phase protrusions ($s = 21$ links, blue line), the model captures the experimental trends nicely. Here, the in-phase tandems exhibit a probability ratio response, which is characterized by a significantly larger amplitude of oscillations, as compared with the chain containing a single protrusion. In addition, the in-phase tandems exhibit oscillations whose amplitude first increases as k varies from small values, reaches a maximum at $k \sim N/2$, decreases for $k > N/2$ and increases again at $k \sim N$, which is in agreement with similar trends observed experimentally.

Finally, we also varied the looping length N for in-phase ($s = 23$ bp) and out-of-phase ($s = 28$ bp) tandem TraR synthetic enhancers, while keeping the position inside the loop (k) constant, and once again found that experimental expression-level ratio measurements matched nicely with looping probability ratio predictions for the tandem TraR enhancer constructs (see Supplementary Fig. 4).

Stiffening and bending effects. An anomaly observed in the experimental data for the tandem-TraR and LacI synthetic enhancers as compared with model predictions is the lack of significant upregulation in the former (except near the promoter for large values of k) and a complete absence thereof in the latter. We hypothesized that a bias towards quenching can emerge if the transcription factors also 'stiffen' the DNA, making it slightly harder to bend locally. Based on our model, we expect such a bias to be dependent on the loop length (diminishing quickly for large N), the extent of the stiffened region (that is, the number of stiffening binding sites) and the binding sites' proximity to the centre of the loop (see Supplementary Notes 2 and 4 for additional discussion and Supplementary Fig. 5a,b).

To experimentally test the validity of the stiffening-excluded volume-looping regulatory model and provide further support for the additivity finding, we fused the 25-kDa glutathione S-transferase (GST) domain to the carboxy terminus of LacI to make a new 63 kDa TF: LacI-GST. This allowed us to generate a significantly larger LacI (see Fig. 4a top schema), while not affecting its capacity to bind DNA (see Supplementary Methods). As both LacI (38 kDa, Fig. 2c) and a tandem of TraR proteins bound to two binding sites (2 × 25 kDa, Fig. 3c) seem to stiffen DNA as compared with a single bound TraR (25 kDa), we reasoned that a larger transcription factor may also add to the stiffening effect. Thus, according to the excluded volume portion of the model, the larger protein should generate a larger amplitude of oscillations between the regulatory minima and maxima, while the stiffening effect should shift the mean regulatory levels of these oscillations towards quenching.

To quantify the regulatory effect induced by LacI-GST when bound to the synthetic enhancers and compare with the effect generated by the native LacI, we measured the expression-level ratio for LacI-GST on the same synthetic enhancer library as the one used for LacI. The data are plotted in Fig. 4a. The figure shows that the expression-level ratio for LacI-GST (blue) exhibits an oscillatory function that varies from very strong quenching values (25–30%) to intermediate quenching (40–50%). The oscillations exhibit the 10.5-bp periodicity observed for the TraR tandems and the overall extent of the expression-level ratio indicates that LacI-GST generates significantly stronger mean quenching response (38% ± 7%) than the native LacI (64% ± 4%,

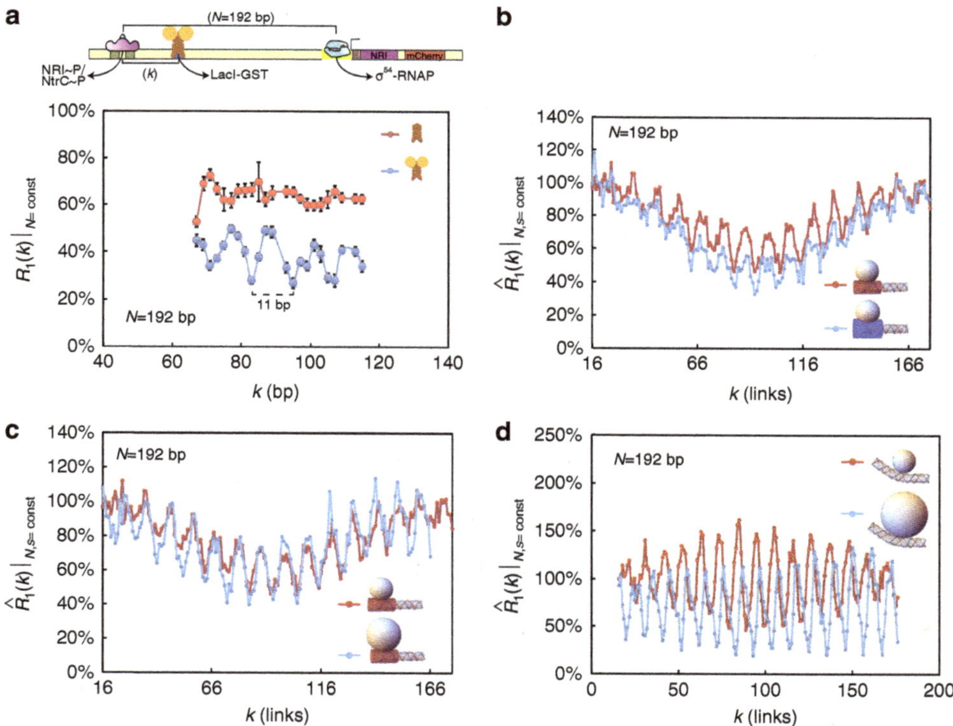

Figure 4 | Combined elastic and entropic effects on looping. (**a**) Expression-level ratio measurements for the synthetic enhancers with a single LacI-binding site. Blue and red circles correspond LacI-GST and LacI expression-level ratios, respectively. (**b–d**) One-dimensional cross-sections for $N = 192$ links comparing: (**b**) two values of stiffening with constant protrusion volume (blue) $2b$ and (red) $1.5b$, where b is the DNA's Kuhn length (106 nm). (**c**) Two protrusion volumes (blue, 8.16 nm; red, 5.44 nm) at constant stiffening ($1.5b$). (**d**) Two protrusion volumes that can also bend the thick chain by 10° (blue, 8.16 nm; red, 5.44 nm). Error bars correspond to the s.d. from multiple measurements.

red). Moreover, the amplitude of the oscillations that are observed for LacI-GST-bound synthetic enhancers (19% ± 4%) is approximately twice as large as the amplitude exhibited by the LacI synthetic enhancers (9% ± 5%). Finally, the LacI-GST expression-level ratio oscillatory function is phase flipped. Namely, the peaks of the LacI-GST data set appear at the minima of the LacI data set, and vice versa.

To compare the experimental results to the model, we plot in Fig. 4b–d three cross-sections. In Fig. 4b we show that by increasing the stiffness parameter (blue versus red line) the probability ratio gains an additional bias towards quenching. In Fig. 4c we show that for the same stiffness value, increasing the size of the protrusion by a small amount (blue versus red line) leads to an increase in the amplitude of the oscillations, as expected. However, to account for the phase flipping, another mechanism is needed. One possibility is bending, in which the transcription factor also 'bends' the DNA locally (see Supplementary Notes 2 and 4 for additional discussion). We find (see Supplementary Fig. 5c,d) that when the bending protrusion is inside the loop, the probability of looping is upregulated, whereas when the protrusion is outside of the loop the probability of looping is reduced. Thus, to account for the phase flipping observed for LacI-GST as compared with LacI, we plot (Fig. 4d) the looping probability ratio for two scenarios. In the first, we simulate a thick chain with a small protrusion inside the loop that also bends the chain by 10° (red line). In the second, we simulate a thick chain with a 3 × larger protrusion positioned inside the loop, which bends the DNA by the same amount (blue line). The data show that for the thick chain with a smaller bending protrusion, the oscillations are consistent with a dominant bending effect generating an upregulation prediction for in-phase locations k. However, for the thick chain with the larger protrusion, the oscillations are consistent with a dominant excluded volume and stiffening effect generating a phase-flipped signal, which is similar to the one observed when comparing the LacI and LacI-GST synthetic enhancers.

Deciphering higher-order TF-binding configurations. We next hypothesized that a pair of binding sites arranged in an out-of-phase configuration should generate a regulatory response with a periodicity that differs from the in-phase-tandems or single-binding-site cases. In Fig. 5a,b we plot model predictions for the regulatory output for an out-of-phase 'Z-shaped' binding configuration showing that 5–6 bp oscillations are generated with a pattern of alternating strong and weak maxima/minima. This periodicity is a result of the fact that a 180° rotation of the Z-shaped tandem around the thick chain axis yields a similar configuration. The deviation between these two configurations is responsible for the alternating extrema. Although the overall regulatory effect predicted by the model is relatively small for such a configuration, for larger TFs a detectable signature may be observed.

We constructed a final synthetic enhancer library to test the 5–6 bp periodicity and alternating weak/strong extrema predictions for synthetic enhancers bound by Z-shaped TF structures. First, we characterized 18 synthetic enhancers with tandems of LacI-binding sites whose centre-to-centre spacing was set at 38 bp. Such an arrangement not only places the LacI dimers in opposite orientation, but also strongly restricts their ability to tetramerize. The binding sites' positions inside the loop were shifted together in 2 bp increments, thus covering a range of 36 bp of intra-loop positions. In Fig. 5c (red) we plot the results. The

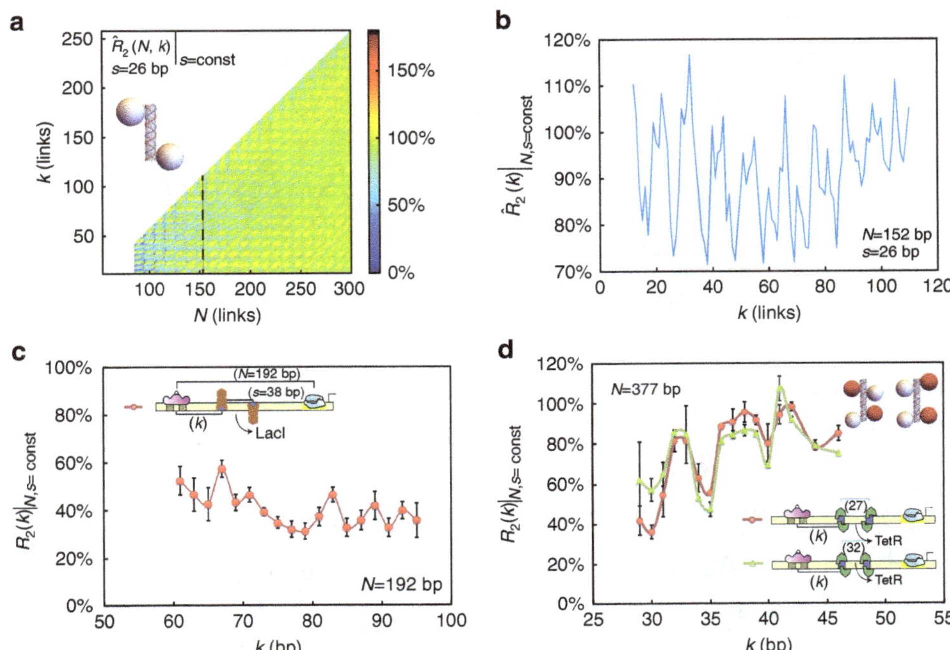

Figure 5 | Periodicity of a half-helical repeat. (**a**) Heat-map depiction of the probability-ratio level for a thick chain with two 'out-of-phase' protrusions taken at a fixed inter-site distance (s). Inset: chain schematic showing the Z-shaped binding architecture. (**b**) Cross-section of the heat map at $N = 152$ links, showing oscillations with a 5- to 6-bp periodicity. Protrusion size was taken to be 5.44 nm in diameter. (**c**) Expression level ratio data for LacI synthetic enhancers with tandem binding sites spaced at 38 bp. (**d**) Expression level ratio data for synthetic enhancers with tandems of TetR-binding sites that are spaced by half integer ($s = 27$ bp, red line) and integer multiples of the helical repeat ($s = 32$ bp, green line), showing identical regulatory function. Inset: the structural binding model showing a thick chain with two dumbbell-shaped protrusions positioned in-phase (right) and out-of-phase (left), respectively. Error bars correspond to the s.d. from multiple measurements.

data show that the tandem-LacI synthetic enhancer strains exhibit a fluctuating regulatory response with a distinct 4–6 bp periodicity for the majority of intra-loop positions of the tandems and a slight increase in the overall magnitude of quenching as the binding sites are moved towards the centre of the loop.

Given these results and Z-shape model predictions, we wondered whether there was something amiss with our interpretation of the single binding-site expression-level ratio results for the TetR synthetic enhancer shown in Fig. 2d. This data set shows a strong regulatory response with sharp fluctuations between quenching and upregulation despite the fact that TetR is a small protein (25 kDa[42]), which is nearly the size of TraR (26 kDa[43]). In addition, a closer look at the expression-level ratio scan (1 bp increments) reveals that the oscillations do not exhibit the ~11-bp periodicity expected from a single binding-site synthetic enhancer. Rather a complex pattern of strong peak/weak trough–weak peak/strong trough seems to emerge with a 5- to 6-bp periodicity between adjacent peaks and an 11-bp periodicity between 'strong' peaks or troughs is also apparent. These results are reminiscent of the model predictions shown in Fig. 5a,b for the out-of-phase tandems and match nicely with a slightly more complex Z-binding configuration as shown in Supplementary Fig. 6b,c (orange line).

Consequently, to account for the periodicity and size effect anomalies, we hypothesized that our form of TetR (TetR-B[44]) might bind its binding site not as a dimer but rather as a dimer-of-dimers oriented in dumbbell-like configuration. This binding architecture is known for a member of the TetR family QacR[36] and in this interpretation an additional cryptic binding site overlaps the major site, allowing a dumbbell-like bound TetR structure to form. To test our hypothesis, the second part of our final library was designed with synthetic enhancers containing tandems of TetR-binding sites. We designed two sets: (i) first, with the TetR-binding sites in-phase ($s = 32$ bp) and (ii) second, with the binding sites out-of-phase ($s = 27$ bp). If the dimer-of-dimer structural interpretation was correct, then the expression-level ratio for both binding site configurations should be nearly identical. This can be seen from a schematic of a thick chain with protrusions (Fig. 5d inset). In both configurations, two dimer-of-dimer protrusion structures are shown on the thick chain with an overall 'out-of-phase' arrangement of two dimers inside the loop and two outside for the chosen inter-site spacings s.

In Fig. 5d we plot the expression level ratio measured as a function of k for these synthetic enhancers ($N = 377$ bp), with green triangles and red circles for the in-phase and out-of-phase inter-site spacing configurations, respectively. The figure shows that the expression level ratio regulatory response generated by both tandems is nearly identical as predicted by the model, with a distinct 5–6 bp periodicity over a range of values for (k) that spans 20 bp at a single base-pair resolution. The regulatory pattern for both cases exhibits three distinct peaks and four troughs with a slightly increasing overall expression level ratio trend. In addition, the data sets lack the strong peak–weak peak pattern of the single binding-site synthetic enhancer. As a result, the dimer-of-dimer dumbbell binding structure for TetR may indeed be a possibility *in vivo*, despite not having been observed in classic *in vitro* experiments (see Supplementary Fig. 6d).

INDEL mutations in natural bacterial enhancers. Finally, we wondered whether we could find evidence for the excluded-

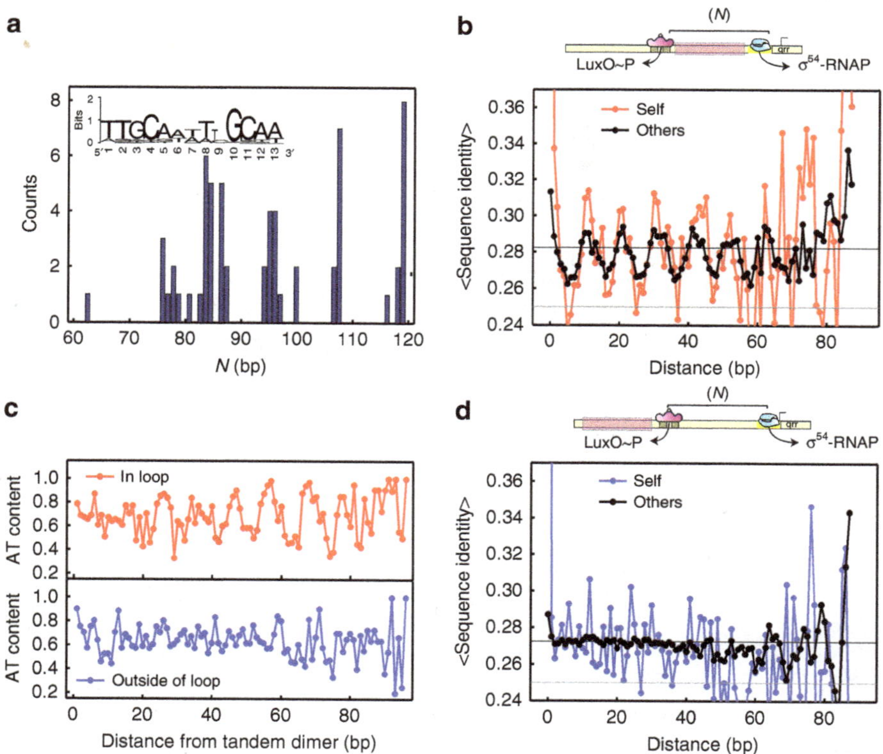

Figure 6 | Sensitivity to the DNA's helical repeat in the *Vibrio qrr* enhancers. (a) Histogram of loop lengths for all putative loop sequences. Inset, LuxO consensus sequence. (b) Relative identity of all putative loop sequences to themselves (red) and to all other putative loop sequences (black). Horizontal lines indicate the expected identity level for sequences with equal probability for all four nucleotides (grey) and for the putative loop sequences (black) (see the Supplementary Methods). (c) AT nucleotide content as function of position for the loop sequences (top, red) and for the non-looping upstream sequences (bottom, blue). (d) Relative identity for all non-looping upstream sequences (from the tandem LuxO binding sites, see top schematic) to themselves (blue) and to all other upstream sequences (black).

volume regulatory model in bacterial genomes. As this type of regulation depends on a fixed relative arrangement of the self-avoiding volumes, we speculated that naturally occurring enhancers should exhibit a conserved evolutionary signature for this mechanism if it does indeed play a biological role. Specifically, we speculated that bacterial enhancers with similar regulatory function should be insensitive to ~11 bp INDEL mutations that conserve both the regulatory TF and activator orientations relative to the promoter and sensitive to function-altering INDELs that are 5 or 6 bp long. As a result, we expected that looping sequences should reflect this tolerance to the DNA helical repeat.

To test this hypothesis, we analysed the *qrr* (quorum regulatory RNA) genes in the *Vibrio* genus. Some of the *qrr* genes in this genus are known to be regulated by LuxO, an NtrC-like activator, which drives σ^{54} promoters. This system was implicated in the quorum-sensing pathway and was characterized for *Vibrio cholerae*[45,46]. Using standard bioinformatic tools (see Supplementary Methods), we annotated 61 *qrr* enhancers (see Supplementary Data 2), which included a putative LuxO-binding site tandem (see Fig. 6a (inset) for consensus sequence and also Supplementary Table 1), a looping region and a single putative σ^{54} promoter (Supplementary Fig. 7). In Fig. 6a we plot the distribution of the looping lengths for all 61 putative *qrr* enhancers. The figure shows a set of clustered looping lengths (*N*) ranging from an average length of ~80 to ~120 bp, which are displaced from one another by ~11 bp, thus providing initial support for our hypothesis.

To further test the sensitivity of the *qrr* looping regions to integer multiples of the helical repeat, we checked the average identities of each loop sequence to itself and to all other loop sequences. To do this, we calculated the relative identity of each 9 bp window within a given loop sequence to all other 9 bp windows either on the same sequence, or on all other sequences, noting the distance between the positions of the first bases of compared windows (the relative identity is defined as the number of positions (that is, from 0 to 9) for which both windows contain the same base, divided by the window length). We then computed the mean relative identity for each window separation by averaging over all relative identities in a particular window-separation value. In Fig. 6b we plot the results. The figure shows that the mean relative identity for the annotated *qrr* enhancers exhibits an oscillatory behaviour, which persists for all possible values of the distance between the windows. Interestingly, the oscillatory pattern is detected not only for cross-correlated enhancers, but also within each enhancer to itself (self, red; other, black), with the first maxima appearing at ~0 bp displacement and with periodicity of 10.45 bp.

Next, we checked whether there was some underlying signature for a conserved sequence within the looping region. To that end, we computed the average AT/GC content of each position within the loop and plotted the results in Fig. 6c (top). The figure shows that the AT content is enriched at positions that are integer multiples of 10.6 bp with at least six distinct peaks visible in the data. In addition, the minima between the positions of AT enrichment converge on a content value of ~0.5, which is the value expected for a random allocation of AT or GC at those particular positions. Thus, loop sequences are similar either at the same relative position or alternatively at positions displaced by an integer multiple of the helical repeat from the position of the reference sequence, with a preference to AT segments.

Finally, in Fig. 6c (bottom) we plot the average AT/GC content and in Fig. 6d the average relative identities of the *qrr* enhancer 'upstream sequences' that are immediately adjacent to the annotated LuxO-binding site tandem (Fig. 6d schematic). Unlike the looping region, the analysis on the non-looping region results in no particular repetitive pattern of AT/GC content within the upstream sequence. In addition, a monotonic or slightly varying signal across all possible values of the window displacement is observed in the upstream region without any detectable characteristic oscillations. Thus, the striking difference between Fig. 6b,d and in Fig. 6c (red versus blue line) provides further support to the special sensitivity of *qrr* enhancer sequences to the helical periodicity as compared with non-looping sequences.

Discussion

We presented a new mechanism for quenching-like repression in bacterial enhancers using a combined thermodynamic modelling, synthetic biology and bioinformatic approach. First, we constructed a preliminary DNA looping-based mechanistic structure–function model. The model established a direct relationship between the values of enhancer structural control parameters (for example, number of binding sites for a TF, inter-site spacing, looping length and so on) to a predicted regulatory function that was based on elastic (bending and stiffening) and entropic (excluded volume) characteristics of a thick chain with protrusions. Second, based on the numerical results we characterized synthetic enhancer libraries focused on testing the predictions of the structure–function model, allowing us to improve the mechanistic model. Finally, we provided biological validation for the findings of the model and synthetic enhancer experiments using bioinformatic predictions on a multitude of naturally occurring enhancers. Altogether, we needed only 303 synthetic bacterial enhancers to mechanistically characterize the excluded-volume quenching mechanism, showing that our model-based design and experimentation approach can achieve insight into a natural phenomenon with a much smaller data set as compared with conventional high-throughput methodologies.

To fully account for all of our experimental observations, we had to incorporate small local bending and stiffening effects into our excluded-volume model. Interestingly, these additional elastic effects, which fine-tuned our model, have made both our experiments and model applicable to past observations made on bacterial enhancers. These include studies of IHF-dependent enhancers[9,10], where IHF was shown to both upregulate and downregulate expression depending on position[29], and non-IHF-dependent enhancers[11–13]. In these studies, the 'regulator' binding sites were typically located halfway between the promoter and activator (consistent with our prediction that the maximal regulatory effect should occur near the centre of the loop, see Supplementary Fig. 2 and associated discussion), and the regulatory effect depended strongly on whether the DNA-binding protein was positioned inside or outside the loop. As a result, a novelty of our experimental and modelling results is that the conservation of TF-binding orientation within bacterial enhancers seems to be a generic phenomenon for all transcription factors and is not limited to a handful of DNA-bending proteins (see also Supplementary Table 2 and Supplementary Note 4). Finally, although our bioinformatic observation is consistent with this conclusion, our model and computational analysis cannot decipher whether conserved sequences in natural bacterial enhancers correspond to TF-binding sites or to other sequences with special physical characteristics (for example, AT-rich curved segments).

An additional implication of our results is that complex binding arrangements of several transcription factors within an enhancer should generate unique regulatory signatures. Such a signature was predicted and observed for the Z-shaped binding arrangement for the anti-phase LacI synthetic enhancers, exhibiting a regulatory signature with a 4–6 bp periodicity.

Building on this observation, we used the results from our assay and model to propose that TetR-B may actually bind our synthetic enhancers *in vivo* as dumbbell-like dimer-of-dimer structure in a similar manner to its homologue QacR[47]. This interpretation of our data was based on two observations: the first was an anomalously large regulatory effect as compared with TetR's size (25 kDa ~ TraR) and the second was the 5- to 6-bp periodicity observed for the single TetR and for all the tandem TetR synthetic enhancer binding-site architectures. However, our inability to provide further support for this structural interpretation using our own or past[48] *in vitro* gel-shift experiments indicates that there are still many unresolved issues with our understanding of protein–DNA interactions *in vivo*.

Finally, there are broad genomic implications to our results. As excluded volume can affect the behaviour of polymers in the entropic regime (implying that it may be applicable to long-range interactions, unlike local bending and stiffening), the conservation of relative binding orientation between adjacently bound transcription factors may not only be relevant to bacterial enhancers or promoter-proximal regulatory regions in eukaryotes, but also to eukaryotic enhancers as well. Owing to the sensitivity of our model to boundary conditions, applying it to such systems will have to take into account molecular details to properly set the looping criterion, as was done here for the bacterial σ^{54} promoter. At the very least, our experimental, modelling and bioinformatic analysis suggest that sensitivity to INDELs that are integer multiples of the helical repeat could be an evolutionary fingerprint for enhancers, whereas INDELs of odd half-integer multiples of the helical repeat should be flagged as candidates for important regulatory variation.

Methods

Synthetic enhancer cassette design. Synthetic enhancer cassettes (see Supplementary Table 3) were ordered as double-stranded DNA minigenes from Gen9 Inc. Each minigene ordered was ~500 bp long and contained the following parts: BamHI restriction site, tandem NRI-binding sites from glnAp2 promoter (also containing the σ^{70} glnAp1 promoter), the σ^{54} glnAp2 promoter and a HindIII restriction site. In addition, each minigene contained a looping segment in between the NRI tandem binding sites and the σ^{54} promoter. The looping segment was of variable length (N) and contained either one or two binding sites for TraR, TetR or LacI. The binding sites were positioned in varying inter-site spacings (s) from one another and locations away from the NRI binding sites (k) within the looping region. For insertion into synthetic enhancer plasmids, minigene cassettes were first double-digested with BamHI/HindIII before being used as an insert in the cloning step. Cloning was then carried out into a basic template synthetic enhancer plasmid as previously described[8,49]. Briefly, synthetic enhancer sequences were computationally designed to have a minimal probability to bind DNA-binding proteins. This was done by constructing an algorithm that randomly generated a set of sequences, which were compared with the roughly 2,000 known specific DNA-binding sites for *E. coli* transcription factors obtained from RegulonDB (http://regulondb.ccg.unam.mx). Sequences were design with 40–50% AT/GC content (see Supplementary Data 1).

Strain construction. The synthetic enhancer strains were constructed as described by Amit *et al.*[8,49]. Briefly, *E. coli* strain 3.300LG[50] with deletions for *glnL* and *glnG* genes was transformed with sequence-verified pACT and synthetic enhancer plasmids. The pACT family of plasmids was constructed by modifying p3Y15 (ref. 50). We inserted a *lacI* gene and either a *tetR* or *traR* gene into the parent plasmid, under the control of the same glnL promoter controlling the NRII2302 mutant. The tetR sequence that we used is that of TetR-B[44], which we refer to as TetR. Selection was carried out via double Kan/Amp resistance (20 and 100 mg ml^{-1}, respectively). Candidate synthetic enhancer strains were tested for fluorescence in the presence and absence of the suitable inducer (see below) on a plate reader (Tecan, Infinite F200), to ensure that a proper strain was constructed. All synthetic enhancer sequences can be found in Supplementary Data 1.

Expression level ratio measurement assay. Expression level measurements for all synthetic enhancers without LacI-binding sites were carried out as follows: first, synthetic enhancer strains were grown in fresh Luria-Bertani with appropriate antibiotics (Kan/Amp) to midlog range (OD_{600} of ~0.6) as measured by a spectrophotometer (Novaspec III, Amersham Biosciences) and were resuspended in low-growth/low-autofluorescence bioassay buffer (for 1 litre: 0.5 g Tryptone (Bacto), 0.3 ml glycerol, 5.8 g NaCl, 50 ml of 1 M $MgSO_4$, 1 ml 10 × PBS buffer at pH7.4 and 950 ml double distilled water). Isopropyl β-D-1-thiogalactopyranoside (IPTG; 1 mM) was added at this point, to deactivate the LacI protein that represses the glnAp2 promoter in the pACT plasmid. Two millilitres of resuspended culture with IPTG were dispensed into each well of a 48-well plate. Appropriate concentrations of anhydrotetracycline (aTc, Cayman Chemical 10009542) or *N*-(3-Oxooctanoyl)-L-homoserine lactone (3OC8, Sigma-Aldrich O1764) were dispensed into each well, spanning four to six orders of magnitude. Up to 24 levels of aTc or 3OC8 concentration were used for each strain. The plates were then incubated in a 37 °C shaker until cultures reached steady-state growth. Measurements of fluorescence levels were taken by dispensing 200 µl of culture into each well of a 96-well plate and were carried out on a plate reader (Tecan F200). Two wells were used as IPTG controls. We carried out each measurement in duplicate. All fluorescence measurements were divided by optical density, to measure the normalized expression, and autofluorescence levels (cells with no plasmid) were subtracted from the normalized values. Some TetR and TraR synthetic enhancers were also tested in strains lacking the LacI protein and no distinguishable difference in the regulatory response was observed.

LacI experiments. Synthetic enhancer cassettes containing LacI-binding sites were cloned into a similar template plasmid as other cassettes, containing glnAp2 promoter and two NRI-binding sites, except the LacI-binding site was removed from the glnAp2 promoter. The removal of the exogenous LacI sites were done to ensure that no tetramerization could take place between the site on the enhancer and these sites, thus affecting the regulatory outcome. The cassettes were cloned into 3.300LG strain along with pACT plasmid (expressing LacI). The experiment was carried out as described above using IPTG as an inducer for LacI removal from the synthetic enhancer-binding site.

Expression level ratio and probability of looping ratio. To compute the expression level ratio $R_1(N,k)$ for a synthetic enhancer (see Fig. 1c bottom schematic and Supplementary Equation (39) for precise definition), we take the ratio in fluorescence expression levels between the protein-unbound regime to the protein-bound case for each measurement of a synthetic enhancer's regulatory response curve. Typical regulatory-response curves for individual synthetic enhancers are shown in Supplementary Fig. 8. For example, in Supplementary Fig. 8a–f we plot data sets obtained for two TetR, one LacI and three TraR synthetic enhancers, showing sigmoidal transfer functions when varying the concentration of aTc, IPTG and 3OC8 from low values to saturating levels, respectively. All expression level ratio measurements in our experiments were obtained using such a procedure with the protein-unbound case regimes set at low 3OC8, high aTc and high IPTG for TraR, TetR and LacI, respectively.

Error estimation. We employed two distinct error estimation methods, which both yielded a 3–7% estimated error in expression-level ratios. Variance in expression-level ratio measurements was calculated as follows: multiple (≥) measurements were taken at 30-min intervals for each of the 2 true duplicates of each strain at 24 inducer concentrations. For the raw data shown in Supplementary Fig. 8, the multiple measurements of duplicates were averaged and the s.d. of the measurements was used as the error bars.

For the expression-level ratio, two concentration ranges in which the fluorescence data levels remained constant were selected for each strain. For instance, in Supplementary Fig. 8a,b we chose the low aTc concentration range (in which the DNA is highly likely to be bound by a TF) and high aTc concentration range (in which the DNA is highly unlikely to be bound by a TF). For each of the two duplicates (1,2), the data in both ranges were averaged, yielding $\text{mean}_{i,l}$ and $\text{mean}_{i,h}$, respectively, where $i \in [1,2]$ denotes the duplicate index. The expression-level ratio for each duplicate was then obtained by $P_i = \frac{\text{mean}_{i,l}}{\text{mean}_{i,h}}$. The final expression-level ratio was obtained by averaging the two expression level ratios P_1, P_2. The expression-level ratio error was obtained using error propagation analysis on the variance in the different mean values as follows: $\sigma_R = \sqrt{\frac{\sigma^2_{\text{mean}_{i,l}}}{{\text{mean}_{i,l}}^2} + P_i^2 \frac{\sigma^2_{\text{mean}_{i,h}}}{{\text{mean}_{i,h}}^2}}$.

To verify the error estimate and expression-level ratio estimates obtained by the previous method, we employed a Hill function fit analysis. Here we took fluorescence level measurements as a function of inducer concentration (for example, Supplementary Fig. 8d,e) and fitted this data with a Hill function of order 1. This allowed us to determine the fluorescence expression levels for the occupied and unoccupied enhancer states from the Hill function fit parameters. The ratio of those two numbers was taken to be the expression level ratio. Error bars were determined by taking into account the errors in the hill function fitting parameters. The results from this analysis were indistinguishable within experimental error to the resulting from the alternative error analysis.

Simulation parameters. In all of our simulations for the synthetic-enhancer-based regulation, we used the following parameters for DNA: Kuhn length $b = 100$ nm (ref. 51), bending constant $a = b/2$, chain link length of $l = 0.34$ nm (corresponding to a single base pair of the DNA), width of $w = 4.6$ nm (ref. 52), $\Delta i \approx \frac{4}{3}\frac{w}{l}$

(see Supplementary Equation (13)), a helical repeat of $P = 10.5$ bp (ref. 53) and a twisting constant of $c = 294$ (corresponding to 100 nm)[51] (Supplementary Equation (2)). Our choice for the value for Δi was made, on the one hand, to be reasonably low, to allow the hard-wall potentials to interact at as close a distance as possible, whereas on the other hand we verified that it fell within a range of values where the simulation was not sensitive to the particular choice of this parameter (data not included). The radius of the corresponding protrusion is ≈ 3.5 nm (refs 3,54) and the polymerase protrusion radius is ≈ 7.2 nm (ref. 55). The TF protrusion radius is ≈ 2.72 nm (corresponding to a TetR transcription factor)[56] unless stated otherwise. We also used the following parameters for the looping boundary conditions (see Supplementary Note 2 section 2.3.1): $\varepsilon = 8.4$ nm, $\delta\omega = 90°$.

Robot measurements. High-resolution experiments were performed on a Tecan EVO 100 MCA 96 multichannel liquid handling system. Experiments were run as described above with slight changes: cells were grown to OD_{600} of ~ 0.1 and 200 µl volume in a 96-well plate, centrifuged and resuspended in bioassay buffer. Inducers were added manually to an inducer plate and then spread in different concentrations automatically to the 96 wells. Fluorescence measurements were taken using a plate reader (Tecan, F200) every 30 min.

Fusion proteins construction. Fusion proteins were designed by connecting the GST coding sequence to either the C or N terminus of TraR, N terminus of TetR and C terminus of LacI coding sequences. GST was PCR amplified from PGEX-4T1 and cloned into pACT, pACT-TetR or pAct-TraR plasmid using Gibson assembly[57]. His-GST-TetR fusion was created by cloning GST-TetR into PGEX-4T. Fusions were sequence verified and their expression was verified by Coomassie or western blot analysis.

Simulation and bioinformatic code. All computer code used in this work either for the numerical simulations or bioinformatic analysis can be made available on request.

References

1. Buck, M., Gallegos, M. -T., Studholme, D. J., Guo, Y. & Gralla, J. D. The bacterial enhancer-dependent sigma 54 (sigma N) transcription factor. *J. Bacteriol.* **182,** 4129–4136 (2000).
2. Ninfa, A. J. & Atkinson, M. R. PII signal transduction proteins. *Trends Microbiol.* **8,** 172–179 (2000).
3. Rappas, M., Bose, D. & Zhang, X. Bacterial enhancer-binding proteins: unlocking sigma54-dependent gene transcription. *Curr. Opin. Struct. Biol.* **17,** 110–116 (2007).
4. Levo, M. & Segal, E. In pursuit of design principles of regulatory sequences. *Nat. Rev. Genet.* **15,** 453–468 (2014).
5. Zhao, Z. *et al.* Circular chromosome conformation capture (4C) uncovers extensive networks of epigenetically regulated intra- and interchromosomal interactions. *Nat. Genet.* **38,** 1341–1347 (2006).
6. Fullwood, M. J. *et al.* An oestrogen-receptor-α-bound human chromatin interactome. *Nature* **462,** 58–64 (2009).
7. Li, G. *et al.* Extensive promoter-centered chromatin interactions provide a topological basis for transcription regulation. *Cell* **148,** 84–98 (2012).
8. Amit, R., Garcia, H. G., Phillips, R. & Fraser, S. E. Building enhancers from the ground up: a synthetic biology approach. *Cell* **146,** 105–118 (2011).
9. Hoover, T. R., Santero, E., Porter, S. & Kustu, S. The integration host factor stimulates interaction of RNA polymerase with NIFA, the transcriptional activator for nitrogen fixation operons. *Cell* **63,** 11–22 (1990).
10. Claverie-Martin, F. & Magasanik, B. Role of integration host factor in the regulation of the glnHp2 promoter of *Escherichia coli*. *Proc. Natl Acad. Sci. USA* **88,** 1631–1635 (1991).
11. Atkinson, M. R., Pattaramanon, N. & Ninfa, A. J. Governor of the glnAp2 promoter of *Escherichia coli*. *Mol. Microbiol.* **46,** 1247–1257 (2002).
12. Kiupakis, A. K. & Reitzer, L. ArgR-independent induction and ArgR-dependent superinduction of the astCADBE operon in *Escherichia coli*. *J. Bacteriol.* **184,** 2940–2950 (2002).
13. Feng, J., Goss, T. J., Bender, R. A. & Ninfa, A. J. Repression of the *Klebsiella aerogenes* nac promoter. *J. Bacteriol.* **177,** 5535–5538 (1995).
14. Carmona, M. & Magasanik, B. Activation of transcription at σ54-dependent promoters on linear templates requires intrinsic or induced bending of the DNA. *J. Mol. Biol.* **261,** 348–356 (1996).
15. Pérez-Martín, J., Rojo, F. & de Lorenzo, V. Promoters responsive to DNA bending: a common theme in prokaryotic gene expression. *Microbiol. Rev.* **58,** 268–290 (1994).
16. Natesan, S. & Gilman, M. Z. DNA bending and orientation-dependent function of YY1 in the c-fos promoter. *Genes Dev.* **7,** 2497–2509 (1993).
17. Kim, J. & Shapiro, D. J. In simple synthetic promoters YY1-induced DNA bending is important in transcription activation and repression. *Nucleic Acids Res.* **24,** 4341–4348 (1996).
18. Frolova, E., Majors, J. & Johnston, M. Binding of the glucose-dependent Mig1p repressor to the GAL1 and GAL4 promoters in vivo: regulation by glucose and chromatin structure. *Nucleic Acids Res.* **27,** 1350–1358 (1999).
19. Driever, W., Thoma, G. & Nüsslein-Volhard, C. Determination of spatial domains of zygotic gene expression in the *Drosophila* embryo by the affinity of binding sites for the bicoid morphogen. *Nature* **340,** 363–367 (1989).
20. Davidson, E. H. *The Regulatory Genome: Gene Regulatory Networks In Development and Evolution* (Academic Press, 2010).
21. Erceg, J. *et al.* Subtle changes in motif positioning cause tissue-specific effects on robustness of an enhancer's activity. *PLoS Genet.* **10,** e1004060 (2014).
22. Kulkarni, M. M. & Arnosti, D. N. cis-Regulatory logic of short-range transcriptional repression in *Drosophila melanogaster*. *Mol. Cell. Biol.* **25,** 3411–3420 (2005).
23. Sharon, E. *et al.* Inferring gene regulatory logic from high-throughput measurements of thousands of systematically designed promoters. *Nat. Biotechnol.* **30,** 521–530 (2012).
24. Chopra, V. S., Kong, N. & Levine, M. Transcriptional repression via antilooping in the *Drosophila* embryo. *Proc. Natl Acad. Sci. USA* **109,** 9460–9464 (2012).
25. Gray, S. & Levine, M. Short-range transcriptional repressors mediate both quenching and direct repression within complex loci in *Drosophila*. *Genes Dev.* **10,** 700–710 (1996).
26. Small, S., Blair, A. & Levine, M. Regulation of two pair-rule stripes by a single enhancer in the *Drosophila* embryo. *Dev. Biol.* **175,** 314–324 (1996).
27. Arnosti, D. N., Barolo, S., Levine, M. & Small, S. The eve stripe 2 enhancer employs multiple modes of transcriptional synergy. *Dev. Camb. Engl.* **122,** 205–214 (1996).
28. Redd, M. J., Stark, M. R. & Johnson, A. D. Accessibility of alpha 2-repressed promoters to the activator Gal4. *Mol. Cell. Biol.* **16,** 2865–2869 (1996).
29. Wassem, R., De Souza, E. M., Yates, M. G., de Pedrosa, F. O. & Buck, M. Two roles for integration host factor at an enhancer-dependent nifA promoter. *Mol. Microbiol.* **35,** 756–764 (2000).
30. Kratky, O. & Porod, G. Röntgenuntersuchung gelöster Fadenmoleküle. *Recl. Trav. Chim. Pays Bas.* **68,** 1106–1122 (1949).
31. Pollak, Y., Goldberg, S. & Amit, R. Self-avoiding wormlike chain model for double-stranded-DNA loop formation. *Phys. Rev. E* **90,** 052602 (2014).
32. De Pablo, J. J. & Escobedo, F. A. in *Advances in Chemical Physics*. (eds Prigogine, I. & Rice, S. A.) 337–367 (John Wiley & Sons, Inc., 1999; at <http://onlinelibrary.wiley.com/doi/10.1002/9780470141649.ch11/summary>.
33. Beck, L. L., Smith, T. G. & Hoover, T. R. Look, no hands! Unconventional transcriptional activators in bacteria. *Trends Microbiol.* **15,** 530–537 (2007).
34. Ninfa, A. J. *et al.* Using two-component systems and other bacterial regulatory factors for the fabrication of synthetic genetic devices. *Methods Enzymol.* **422,** 488–512 (2007).
35. Lewis, M. *et al.* Crystal structure of the lactose operon repressor and its complexes with DNA and inducer. *Science* **271,** 1247–1254 (1996).
36. Ramos, J. L. *et al.* The TetR family of transcriptional repressors. *Microbiol. Mol. Biol. Rev.* **69,** 326–356 (2005).
37. Qin, Y., Keenan, C. & Farrand, S. K. N- and C-terminal regions of the quorum-sensing activator TraR cooperate in interactions with the alpha and sigma-70 components of RNA polymerase. *Mol. Microbiol.* **74,** 330–346 (2009).
38. Becker, N. a., Kahn, J. D. & Maher, L. J. Bacterial repression loops require enhanced DNA flexibility. *J. Mol. Biol.* **349,** 716–730 (2005).
39. Law, S. M., Bellomy, G. R., Schlax, P. J. & Record, M. T. *In vivo* thermodynamic analysis of repression with and without looping in lac constructs. Estimates of free and local lac repressor concentrations and of physical properties of a region of supercoiled plasmid DNA *in vivo*. *J. Mol. Biol.* **230,** 161–173 (1993).
40. Lee, D. H. & Schleif, R. F. *In vivo* DNA loops in araCBAD: size limits and helical repeat. *Proc. Natl Acad. Sci. USA* **86,** 476–480 (1989).
41. Müller, J., Oehler, S. & Müller-Hill, B. Repression of lac promoter as a function of distance, phase and quality of an auxiliary lac operator. *J. Mol. Biol.* **257,** 21–29 (1996).
42. Wray, L. V. & Reznikoff, W. S. Identification of repressor binding sites controlling expression of tetracycline resistance encoded by Tn10. *J. Bacteriol.* **156,** 1188–1191 (1983).
43. Qin, Y. *et al.* Quorum-sensing signal binding results in dimerization of TraR and its release from membranes into the cytoplasm. *EMBO J.* **19,** 5212–5221 (2000).
44. Hillen, W. & Berens, C. Mechanisms underlying expression of Tn10 encoded tetracycline resistance. *Annu. Rev. Microbiol.* **48,** 345–369 (1994).
45. Svenningsen, S. L., Waters, C. M. & Bassler, B. L. A negative feedback loop involving small RNAs accelerates *Vibrio cholerae*'s transition out of quorum-sensing mode. *Genes Dev.* **22,** 226–238 (2008).
46. Tu, K. C. & Bassler, B. L. Multiple small RNAs act additively to integrate sensory information and control quorum sensing in *Vibrio harveyi*. *Genes Dev.* **21,** 221–233 (2007).

47. Schumacher, M. A. *et al.* Structural basis for cooperative DNA binding by two dimers of the multidrug-binding protein QacR. *EMBO J.* **21,** 1210–1218 (2002).
48. Scholz, O., Schubert, P., Kintrup, M. & Hillen, W. Tet repressor induction without Mg2 + †. *Biochemistry (Mosc.)* **39,** 10914–10920 (2000).
49. Amit, R. Anti-cooperative and cooperative protein-protein interactions between TetR isoforms on synthetic enhancers. *J. Comput. Biol.* **19,** 115–125 (2012).
50. Atkinson, M. R., Savageau, M. A., Myers, J. T. & Ninfa, A. J. Development of genetic circuitry exhibiting toggle switch or oscillatory behavior in *Escherichia coli. Cell* **113,** 597–607 (2003).
51. Moroz, J. D. & Nelson, P. Torsional directed walks, entropic elasticity, and DNA twist stiffness. *Proc. Natl Acad. Sci.* **94,** 14418–14422 (1997).
52. Rybenkov, V. V., Cozzarelli, N. R. & Vologodskii, A. V. Probability of DNA knotting and the effective diameter of the DNA double helix. *Proc. Natl Acad. Sci. USA* **90,** 5307–5311 (1993).
53. Allemand, J. -F., Cocco, S., Douarche, N. & Lia, G. Loops in DNA: an overview of experimental and theoretical approaches. *Eur. Phys. J. E Soft Matter* **19,** 293–302 (2006).
54. De Carlo, S. *et al.* The structural basis for regulated assembly and function of the transcriptional activator NtrC. *Genes Dev.* **20,** 1485–1495 (2006).
55. Darst, S. A., Kubalek, E. W. & Kornberg, R. D. Three-dimensional structure of *Escherichia coli* RNA polymerase holoenzyme determined by electron crystallography. *Nature* **340,** 730–732 (1989).
56. Orth, P., Schnappinger, D., Hillen, W., Saenger, W. & Hinrichs, W. Structural basis of gene regulation by the tetracycline inducible Tet repressor–operator system. *Nat. Struct. Mol. Biol.* **7,** 215–219 (2000).
57. Gibson, D. G. *et al.* Enzymatic assembly of DNA molecules up to several hundred kilobases. *Nat Methods* **6,** 12–16 (2009).

Acknowledgements

This project has received funding from the European Union's Horizon 2020 Research And Innovation Programme under grant agreement number 664918—MRG-GRammar, by the Israel Science Foundation through grant number 1677/12, by the I-CORE Program of the Planning and Budgeting Committee and the Israel Science Foundation (grant number 152/11), and the Marie Curie Reintegration grant number PCIG11-GA-2012-321675. Y.P. acknowledges support provided by the Russell Berrie Nanotechnology Institute, Technion. S.G. was supported in part at the Technion by an Aly Kaufman Fellowship. The authors thank Beate Kaufmann and Noa Katz for helpful discussions.

Author contributions

M.B.-M. designed most of the synthetic enhancer libraries and carried out most of the experiments both *in vivo* and *in vitro*. Y.P. designed and carried out all of the numerical simulations. Y.P. also carried out most of the TetR experiments. S.G. carried out the bioinformatic analysis. L.L. assisted with TetR experiments. R.A., M.B.-M., Y.P. and S.G. wrote the manuscript.

Additional information

Competing financial interests: The authors declare no competing financial interests.

5

4Pi-RESOLFT nanoscopy

Ulrike Böhm[1], Stefan W. Hell[1] & Roman Schmidt[1]

By enlarging the aperture along the optic axis, the coherent utilization of opposing objective lenses (4Pi arrangement) has the potential to offer the sharpest and most light-efficient point-spread-functions in three-dimensional (3D) far-field fluorescence nanoscopy. However, to obtain unambiguous images, the signal has to be discriminated against contributions from lobes above and below the focal plane, which has tentatively limited 4Pi arrangements to imaging samples with controllable optical conditions. Here we apply the 4Pi scheme to RESOLFT nanoscopy using two-photon absorption for the on-switching of fluorescent proteins. We show that in this combination, the lobes are so low that low-light level, 3D nanoscale imaging of living cells becomes possible. Our method thus offers robust access to densely packed, axially extended cellular regions that have been notoriously difficult to super-resolve. Our approach also entails a fluorescence read-out scheme that translates molecular sensitivity to local off-switching rates into improved signal-to-noise ratio and resolution.

[1] Department of NanoBiophotonics, Max Planck Institute for Biophysical Chemistry, Am Fassberg 11, Göttingen 37077, Germany. Correspondence and requests for materials should be addressed to S.W.H. (email: stefan.hell@mpibpc.mpg.de) or to R.S. (email: roman.schmidt@mpibpc.mpg.de).

The three to seven fold improved axial resolution provided by 4Pi microscopy[1–3] in the 1990s marked a first step in the quest for radically improving the resolution in far-field fluorescence microscopy. Yet the resolution provided by 4Pi microscopy remained diffraction-limited, because by jointly using two opposing lenses for focusing the excitation and/or the fluorescence light, this method just optimized the focusing conditions for feature separation. Modern far-field fluorescence nanoscopy[4], or superresolution microscopy, such as the methods called stimulated emission depletion (STED), reversible fluorescent saturable optical transition (RESOLFT) and later also photoactivated localization microscopy (PALM)/stochastical optical reconstruction microscopy (STORM) fundamentally departed from such early superresolution concepts by discerning features through a molecular state transition. The use of a state transition for feature separation, typically a transition between a fluorescent (ON) and a non-fluorescent (OFF) state, opened the road to lens-based fluorescence microscopy with resolution that is conceptually not limited by diffraction.

Yet diffraction plays a role in these 'diffraction-unlimited' techniques because the resolution of these 'nanoscopy' methods still benefits strongly from focusing the light as sharply as possible. While in STED and RESOLFT, it is the focusing of the illumination light in sample space that matters, in PALM/STORM it is the focusing of the emitted light at the detector. Therefore, the optimization of focusing remains very timely. 4Pi arrangements can also facilitate the doubling of the detected fluorescence without compromising the resolution in the focal plane (x,y), and offer significantly sharper axial (z) intensity gradients than single lenses for both the illumination and the detected light. Consequently, the combination of 4Pi with STED, RESOLFT and PALM/STORM approaches currently offers the most powerful optical setting for three-dimensional (3D) fluorescence nanoscopy[5–7].

Yet 4Pi-type super-resolution arrangements are scarcely reported for STED and PALM and entirely unexplored for RESOLFT, a STED-derivative that typically uses reversibly switchable fluorescent proteins (RSFP) for providing the mandatory ON and OFF states. RSFP-based RESOLFT is particularly attractive because it operates with low light levels, making it gentle to living cells[8].

The difficulties of realizing a 4Pi setup are commonly attributed to the counter-alignment of the two high numerical aperture (NA) lenses. In practice, however, the alignment can be controlled and stabilized over many hours. Instead, a far more general problem that is inherent to all fluorescent imaging modalities comes to the fore. The fluorescence signal (that emanates from each sub-diffraction pixel volume under investigation) needs to be discriminated against 'background' signal from outside of this volume. This 'background' largely stems from optical aberrations that preclude precise spatial control of the illumination or fluorescence beam positions and, in case of STED, RESOLFT or PALM/STORM, from imperfections of the ON/OFF-state transfer (switching) process. Discrimination against this 'background' signal is most challenging along the optical (z) axis, especially when the probed volume is of sub-diffraction dimensions. Lack of sufficient discrimination along the z-axis (optical sectioning) manifests itself as artifacts in the image, particularly as 'ghost features' above and below the real features.

When describing the imaging process in the spatial frequency domain, the appearance of axial lobes corresponds to local depressions in the amplitude of the optical transfer function, that is, the modulation transfer function (MTF) of the microscope. Structural information of the sample can only be retrieved in those spatial frequency bands where the MTF is strong enough to convey a signal that sufficiently exceeds the local noise level. In a 4Pi microscope, MTF depressions are typically restricted to sharp local minima at the so-called critical frequencies[9]. As their amplitude strongly depends on the aperture angle α of the objective lenses used[9], combinations of 4Pi with diffraction-unlimited super-resolution/nanoscopy methods such as isoSTED[5,10,11] and iPALM[6,12], have unfortunately been limited to imaging fixed samples that are more easily accessible with high angle lenses ($\alpha \geq 74°$, as for oil immersion lenses with NA $\geq$ 1.46). Furthermore, the imaged objects were rather thin and labelled very sparsely, as both properties alleviate the requirements on optical sectioning, that is, on suppressing ('background') signal from above and below the focal plane. Fortunately, in a coordinate-targeted nanoscopy method such as RESOLFT, the signal received from the targeted nanosized pixel volume scales with the average number of molecules located within this volume, allowing for tailoring of the pixel volume[11], and hence the resolution and the signal, to the actual imaging conditions to render the 'background' (mathematically) treatable.

Here we report the realization of 4Pi-RESOLFT nanoscopy, that is, of a conceptually diffraction-unlimited resolving method which, by virtue of 4Pi microscopy, provides spatially uniform 3D resolution for imaging (living) cells at the nanometre scale, offers strong optical sectioning due to multiple background suppressing mechanisms and operates at low light levels in 3D.

Results

On-switching order and optical sectioning. The effective point-spread-function (PSF) $h_{ef}(\mathbf{r})$ of a coordinate-targeted super-resolution imaging modality ultimately quantifies the 3D-coordinate range where the fluorescent molecules are allowed to yield measurable signal. If the fluorophores from a certain range are imaged onto a (confocal point) detector, $h_{ef}(\mathbf{r})$ is given by the normalized distribution $S^{ON}(\mathbf{r})$ showing where a molecule is allowed to be in the ON-state at the time of read-out, multiplied by a normalized function $h^{det}(\mathbf{r})$ that describes the detection probability:

$$h_{ef}(\mathbf{r}) = S^{ON}(\mathbf{r}) \cdot h^{det}(\mathbf{r}) \tag{1}$$

$S_{ON}(\mathbf{r})$ is proportional to a product of normalized terms h^{on} and $\tilde{h}^{off}$ that describe the generation of the ON-state by the use of on- and off-switching processes, respectively. h^{on} describes the spatial probability to assume the ON-state in the absence of off-switching light. It can typically be written as a product of terms h_i^{on} that each express the relative probability for absorption of a single photon that drives a transition to a (virtual) state, and therefore scales with the intensity of the light patterns used (for example, $h_{exc}^{on} \sim I_{exc}^{on}$ for single-photon excitation with intensity I_{exc}^{on}; in case of two-photon excitation: $h_{2phexc}^{on} = h_{exc}^{on} \cdot h_{exc}^{on}$). $\tilde{h}^{off}$ describes the effect of the off-switching light on a potential ON-state distribution; $\tilde{h}^{off} = 1$ where molecules are always allowed to assume the ON-state, and $\tilde{h}^{off} = 0$ where they are forced to stay in an OFF-state. Due to the forced assumption of an OFF-state by 'saturating' off-switching, $\tilde{h}^{off}$ is usually much sharper than the off-switching light intensity patterns.

We formally define h^{det} as the first on-switching term $h_1^{on} \equiv h^{det}$ (because of its similar effect on h_{ef}), drop any h_i^{on} that does not significantly sharpen h_{ef} (for example, widefield-detection or sample-wide switching) and obtain:

$$h_{ef}(\mathbf{r}) = \prod_{i=1}^{O^{on}} h_i^{on}(\mathbf{r}) \cdot \tilde{h}^{off}(\mathbf{r}) \tag{2}$$

Here the number of on-factors O^{on} denotes the on-(switching) order of h_{ef}, for example, $O^{on} = 2$ (excitation by single-photon absorption and confocal detection) for a typical confocal (STED) microscope. Optical sectioning can generally be improved by engineering $\tilde{h}^{\text{off}}$ such that molecules in out-of-focus areas are switched off more effectively (Supplementary Fig. 1), or by requiring the absorption of multiple photons for the occupation of the ON-state, which increases O^{on}. The latter can be realized directly through standard two-photon absorption[13–17], or by requiring the sequential occupation of multiple real states to reach the ON-state[18]. Such sequential state occupation is easily realized using the switching steps offered by RSFP (that are central to the RSFP-based RESOLFT concept[4,19]).

The 4Pi-RESOLFT modality. In this study, we devised a coordinate-targeted 4Pi-RESOLFT modality that utilizes negative-switching RSFP (that is, those that are switched off at a wavelength that is also used for generating the fluorescent signal, Fig. 1a) and that resorts to all the processes mentioned above for strong optical sectioning. Concretely, we opted for the RSFP Dronpa-M159T[20–22], which stands out by relatively fast switching kinetics with comparatively low background. At each scanning position, the local RSFP molecules were cycled through their ON- and OFF-states by consecutive light pulses that defined our RESOLFT imaging sequence (Fig. 1a). In the initial step ('activation' pulse), we applied a μs-long train of 170-fs pulses at 90 MHz/780 nm in a focal pattern h_{ac} to (partially) transfer ('activate') local RSFP to their meta-stable 'active'-state by two-photon absorption, as described by the activation distribution $S^{\text{B}}(\mathbf{r})$. Subsequently, we applied a μs–ms-long 'deactivation' pulse of continuous-wave (CW) irradiation at 488 or 491 nm, focused to form a hollow deactivation pattern (for example, a 'z-donut' h_{zd}, Fig. 1a, Supplementary Methods). This drove active RSFP outside the targeted pixel volume (for example, above and below the focal plane) back to their inactive state, which effectively denied them a further excitation to the fluorescent ON-state and thus improved the spatial ON/OFF-contrast during the following 'read-out' pulse. We finally probed the remaining active RSFP by a second μs–ms-long CW pulse at 491 nm with a focal pattern h_{ro}, which transferred them to their fluorescent (ON) state, and detected the fluorescence through a confocal pinhole.

Our scheme thus entails a number of advantages for live-cell 3D-imaging. First, RSFP are inherently live-cell compatible protein markers, and selection of sufficiently bright and stable RSFP is readily available[8,20,23]. Second, optical sectioning benefits from the additional switching step (activation) involved in the RSFP switching cycle with respect to modalities that do not make use of a meta-stable state. This additional switching becomes especially powerful if it is mediated by two-photon absorption in a 4Pi configuration, as O^{on} rises to 4 and the activation and read-out patterns (h_{ac}, h_{ro}) can be setup to a limited zone of overlap in the focal region (Fig. 1a). While overlapping several pattern h_i^{on} also forms the basis of 4Pi microscopy of type C using two-photon excitation[24], here we do not require coherent double-lens (4Pi) detection of the emitted fluorescence, and therefore do not need broad-band intra-cavity dispersion compensation. The scheme presented here thus acts to the same effect with much less technical complexity. Finally, activation by two-photon absorption entails much less photo damage than two-photon excitation, as it takes place at a time during the switching cycle when virtually no markers can assume their excited fluorescent state.

Under ideal conditions, the effective PSF h_{ef} of such a system is virtually free of axial lobes (Fig. 1a) even without a deactivation pulse. In practice, incomplete deactivation and optical aberrations

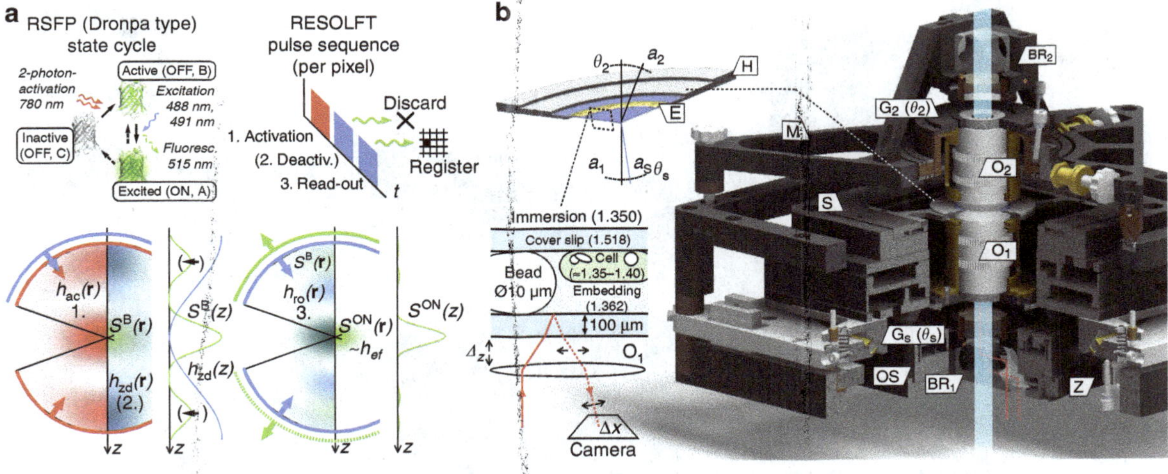

Figure 1 | 4Pi-RESOLFT principle and sample optics. (**a**) Coherent double-lens illumination cycles RSFP markers between dark (OFF) and bright (ON) states to generate spatial ON/OFF-contrast. For each pixel, an activation light pulse (focal pattern h_{ac}) induces two-photon activation of RSFP (state transition C->B) in a pattern $S^{\text{B}}(\mathbf{r})$ with axial side-maxima (lobes) that are optionally suppressed by a subsequently applied deactivation pulse (h_{zd}, B<->A->C). Fluorescence generated by the ON-state A is detected during read-out (B<->A->C) by a pattern h_{ro}. Its mutual overlap with $S^{\text{B}}(\mathbf{r})$ is constrained to the focal centre, resulting in an effective PSF $h_{\text{ef}} \sim S^{\text{ON}}$ that exhibits ≈100 nm axial FWHM and exceptionally low side-maxima. Profiles show on-axis values. (**b**) The upright 4Pi unit of the microscope. Cells are mounted on a ring-shaped sample holder (H), between two cover glasses fixed at 10 μm distance by spacer beads and epoxy resin (E). The set of refractive indices (in brackets) of the immersion and embedding medium, cover slip thickness and correction collar settings of the objective lenses (O_1, O_2) diminishes aberrations from the sample. The sample stage (S) is mounted on a vertically movable (Z) goniometer (G_S), accepts the sample holder (H) and provides five degrees-of-freedom for coarse *xyz*-positioning and *z*-scanning of the sample, as well as tip-/tilt-alignment (θ_S) of the cover slip normal (a_S) to the optic axis of O_1 (a_1). O_1 itself is mounted on a *xyz*-piezo stage (OS) that provides online fine control over the displacement of both foci. A triangular mount (M) allows for tip/tilt-(θ_2) and coarse *xyz*-alignment of O_2 (axis a_2) with respect to O_1, and can conveniently be detached to change the sample. Two polarizing beam splitter/quarter-wave retarder pairs ($BR_{1,2}$) clean up and tune the polarization of the incident beam pairs to opposing circular states. One beam splitter furthermore serves as a port for an alignment laser beam that provides optical feedback for online-stabilization of the axial sample position (Δz); the beam traverses the respective objective lens off-axis (solid red path), gets reflected at the embedding medium interface and is imaged onto a camera (dotted red path).

may give rise to lobe amplitudes that are still relevant. To counteract these effects, we applied dedicated lobe deactivation by h_{zd} and developed low-aberration[25], live-cell 4Pi optics (Fig. 1b, Methods). These measures enabled volume scans of over 5-μm-thick mammalian cells without noticeable bleaching at an axial (z) resolution in the 100 nm range and axial lobes of only $\sim 15\%$ of the main peak of the z-response, that is, measured on laterally (xy) integrated data (Fig. 2). The base acquisition time of 7–21 s μm^{-3} (depending on the brightness of the labelled structure) was short enough to capture the subtly moving cytoskeleton of a living cell (Fig. 2b,c, total acquisition times incl. drift correction overhead b: 115 min per 703 μm^3, c: 160 min per 400 μm^3). For highly mobile organelles, such as mitochondria (Fig. 2a), fixation of the sample by paraformaldehyde incubation (Supplementary Methods) offered a means to prevent motion blur. While this treatment irreversibly arrests the cell, its potential to introduce structural artifacts is very low with respect to staining/embedding protocols that involve membrane permeabilization.

To resolve smaller features, we implemented a second switching pattern for deactivation of RSFP around the focal centre: A hollow '3D-donut' h_{3d} (Fig. 3a), created by a single focused 4Pi beam pair[26] (Methods), allowed us to tune h_{ef} to a near-isotropic resolution below the diffraction limit (Fig. 3a). Calculations using a vectorial diffraction theory[27] predicted on-axis MTF values of over 40% of the MTF maximum within the MTF bandwidth up to a resolution of 30 nm. This feature keeps the signal well over the noise level in most applications and exemplifies the improvement brought about by higher order on-switching in comparison to modalities of second order such as those reported in isoSTED microscopy[5] (Fig. 3b). Furthermore, the confinement of the fluorescent on-state, that is, of h_{ef}, to sub-diffraction 3D volumes means that fewer fluorophores are interrogated at any point in time. This reduction in number of interrogated molecules (that are inherently co-localized) greatly facilitates the quantitative assessment of the properties of the fluorescent labels as they vary in the sample. We found that in time-resolved recordings, the on/off separation contrast decayed over time, hinting to the contribution of multiple deactivation rates. Thus, we introduced a 'rate-gated' RESOLFT detection scheme that improved both the signal-to-noise ratio in the image and the resolution by discriminating individual signal components (Fig. 3c,d, Methods).

Following this approach, we recorded images of Lifeact-Dronpa-M159T-expressing cells and adjusted rate-gating and the RESOLFT pulse sequence for target resolutions of 30-50 nm; the parameters were established by a PSF simulation using measured rate kinetics. Optical xz-sections taken perpendicular to the run of solitary actin fibre bundles confirmed the effectiveness of rate-gating (Fig. 3c,d) and the overall shape of the effective PSF (Fig. 3e). Illumination with the z-donut-shaped (h_{zd}) focus for 1 ms at an average light power of 1.8 μW (488 nm, CW) was sufficient to virtually eliminate lobe background from the image (Fig. 3e, $+h_{zd}$), while the low gradients around the central zero of h_{zd} with respect to h_{3d} facilitated the mutual alignment of these patterns (Supplementary Fig. 1). Turning to the finer structured actin network inside the cell body, we measured apparent feature sizes well below 40 nm (Fig. 4a–c). At a relaxed target resolution of 50 nm and an acquisition time of 3.3 min μm^{-2}, we observed the time evolution of the actin scaffold at a vertical contact region of two neighbouring cells (Fig. 4e–g).

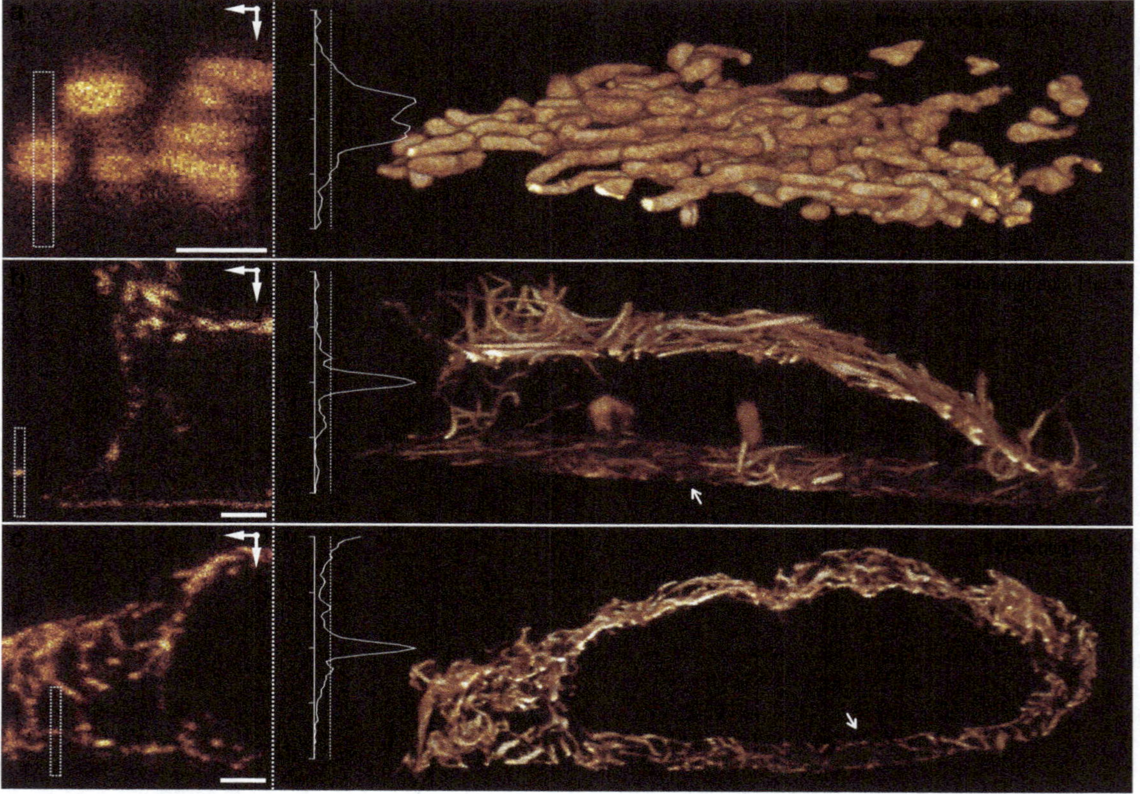

Figure 2 | 4Pi-RESOLFT imaging exhibits only minor axial lobes. 4Pi-RESOLFT raw data (left) and volume renderings (right) of Dronpa-M159T targeted to (**a**) the lumen of mitochondria, (**b**) actin microfilaments and (**c**) intermediate filaments of the cytoskeleton. The sample in **a** was subject to PFA fixation to freeze the motion of mitochondria; the filament networks in **b,c** were recorded from living cells, and exhibit regions of reduced density adjacent to the cover slip (arrows). Estimates of the z-response (insets), measured as box profiles over extended structures, exhibit only minor axial lobes in the 15 % range. Fast-to-slow order of scan axes, xzy. Pulse parameters, E_{ac}, E_{zd}, E_{ro} = 1.6 mW · 50 μs, 18 μW · 50 μs, 3.1 μW · 50 μs. Scale bars, 1 μm.

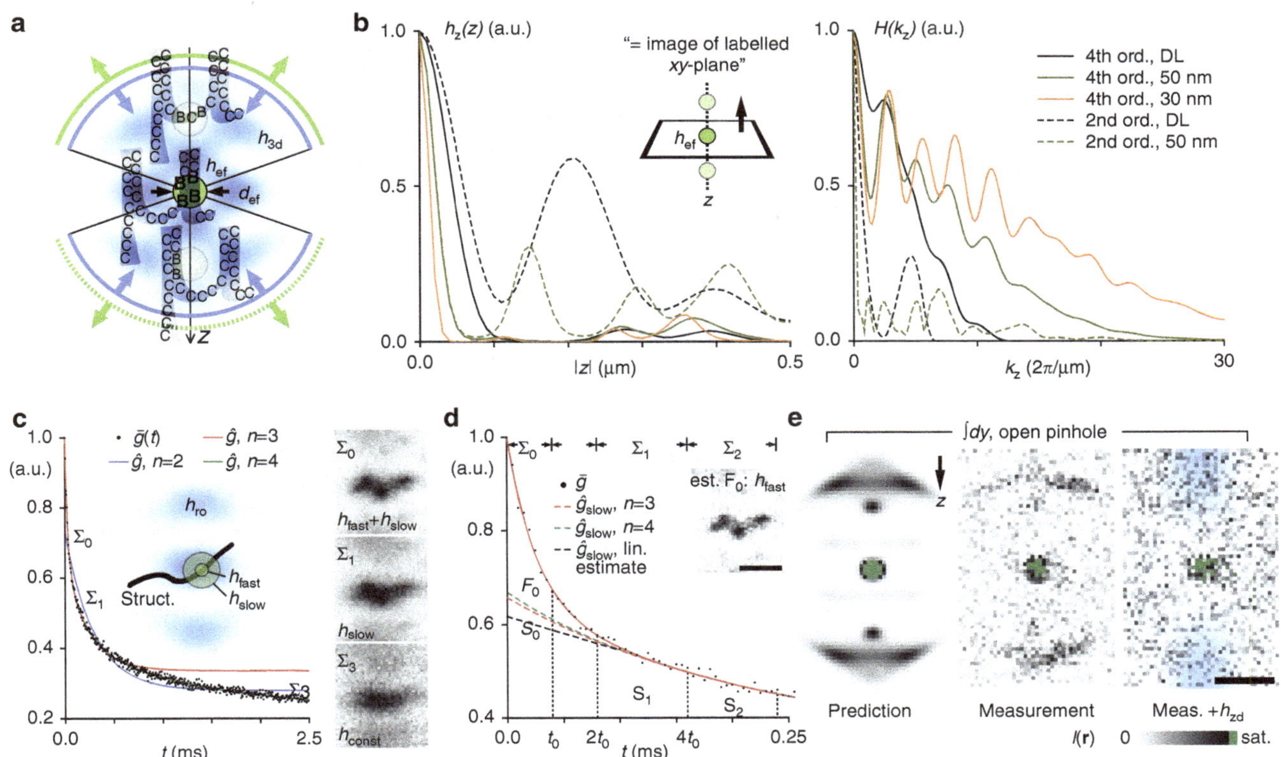

Figure 3 | 4Pi-RESOLFT image formation with <100 nm isotropic resolution. **(a)** A hollow switching pattern h_{3d} confines the central effective PSF to a spot with diameter d_{ef} by switching activated markers (B) back to their inactive state (C). Side-lobes due to inefficient switching at low off-centre h_{3d} amplitudes rise in relative strength as d_{ef} is reduced. μ, labelled structure. **(b)** Simulated z-response $h_z(z)$ (laterally integrated h_{ef}) and axial MTF profile $H(k_z)$ of the 4Pi-RESOLFT microscope (fourth on-order switching, solid lines) at different target resolutions d_{ef}. DL, diffraction limit. Graphs for an isoSTED microscope under similar conditions are included for reference (second on-order, dotted lines). **(c)** Normalized time-resolved, mean fluorescence signal $\bar{g}(t)$ collected from an xz-section through an actin fibre bundle (struct.) in a cell expressing Lifeact-Dronpa-M159T. Target resolution 50 nm, read-out pattern h_{ro} with a total power of $P_{ro} = 3.1\,\mu W$ incident on the sample. An n-component multi-exponential fit to the data corresponds to n apparent switching speeds $\hat{\lambda}_i$. A minimum of $n = 3$ is required to adequately represent the data from the beginning of the read-out pulse $t = 0$ up to 0.5 ms, $\hat{\lambda}_i = 40.5, 4.3, 0.0\,\mathrm{ms}^{-1}$ (for up to 2.5 ms: $n = 4$, $\hat{\lambda}_i = 41, 5.9, 0.85, 0.0\,\mathrm{ms}^{-1}$). Images $\Sigma_{0,1,3}$ integrated over time regimes that are dominated by fast (h_{fast}), slow (h_{slow}) and about constant PSF components (h_{const}) exhibit a declining resolution. **(d)** Rate-gated 4Pi-RESOLFT. Extrapolation of the initial contribution of h_{slow} ($= S_0$), based on integrated images Σ_1 ($\approx S_1$) and Σ_2 ($\approx S_2$), $t_0 = 40\,\mu s$, provides an estimate of the partial image generated by h_{fast} ($F_0 \approx \Sigma_0 - S_0$, inset), improving resolution and image fidelity over Σ_0. Details are provided in Methods. **(e)** Rate-gated xz-sections through actin fibres, recorded with open pinhole to boost out-of-focus signal. The measured (y-integrated) side-lobe structure closely resembles the numerical prediction and can be further suppressed (right) by an additional z-donut h_{zd} (overlay, $E_{zd} = 1.8\,\mu W \cdot 1.0\,ms$). Simulation parameters, numerical aperture 1.20, refractive index 1.362, pinhole diameter 0.5 airy units (**e**: open pinhole). Pulse parameters, E_{ac}, E_{3d}, $E_{ro} = 1.6\,mW \cdot 0.2\,ms$, $1.3\,\mu W \cdot 1.6\,ms$, $3.1\,\mu W \cdot 2.5\,ms$ (**e**: 0.5 ms). Scale bars, 250 nm.

Discussion

Using the current RSFP Dronpa-M159T, rate-gating allowed us to obtain images based on-switching speeds (switch-off half-time $T^{1/2} = 10$–$17\,\mu s$ at $11.5\,kW\,cm^{-2}$ illumination intensity, Fig. 3c, Supplementary Fig. 2, Supplementary Table 1) that were over an order of magnitude faster than the previously reported corresponding values for rapid switching RSFP ($T^{1/2}$, rsEGFP2: 250 μs (ref. 23), Dronpa-M159T: 230 μs (ref. 28)). Still the recording speed of our 4Pi-RESOLFT nanoscopy scheme can be made substantially faster by parallelization using a multi-spot scanning arrangement. In this case, the recording time of a certain sample area or volume would be cut down by the number of individual recording channels, that is, by the degree of parallelization.

In this study, we opted for cellular structures that are more demanding for 3D-superresolution imaging due to their high spatial density and wide axial extension. Under conditions exacerbated by the optical inhomogeneity of living cells, the signal from a (sub-diffractive) ensemble is easily buried in background (lobe) fluorescence beyond recovery. Nevertheless, owing to the consistently robust MTF of our 4Pi-RESOLFT scheme (Fig. 3b), we obtained raw (Fig. 2, insets) and rate-gated image data (Fig. 4e–g) that were conclusive without the mathematical post-processing (that is, deconvolution) dedicated to lobe-removal that is usually applied in 4Pi-based methods. The actin network seen in the exemplary time-lapse recording (Fig. 4e–g) appeared particularly crowded and extended over 8 μm along the optic axis, which forced the light to pass through several micrometres of cellular material from all angles. Still, the rearrangement of the entwined actin fibres could be traced in great detail, which was possible because the obtained images were practically devoid of axial lobes.

Notably, our scheme of reducing the global refractive index (RI) variance (Fig. 1b) turned out to sufficiently mitigate sample-induced aberrations without adding the complexity associated with adaptive optical elements. The most prominent residual aberration effect was a position-dependent 4Pi phase offset that stemmed from the uneven thickness of the cell layer; it has been accounted for during our recordings by the automated correction mechanism that also counteracted thermal drift (Supplementary Methods).

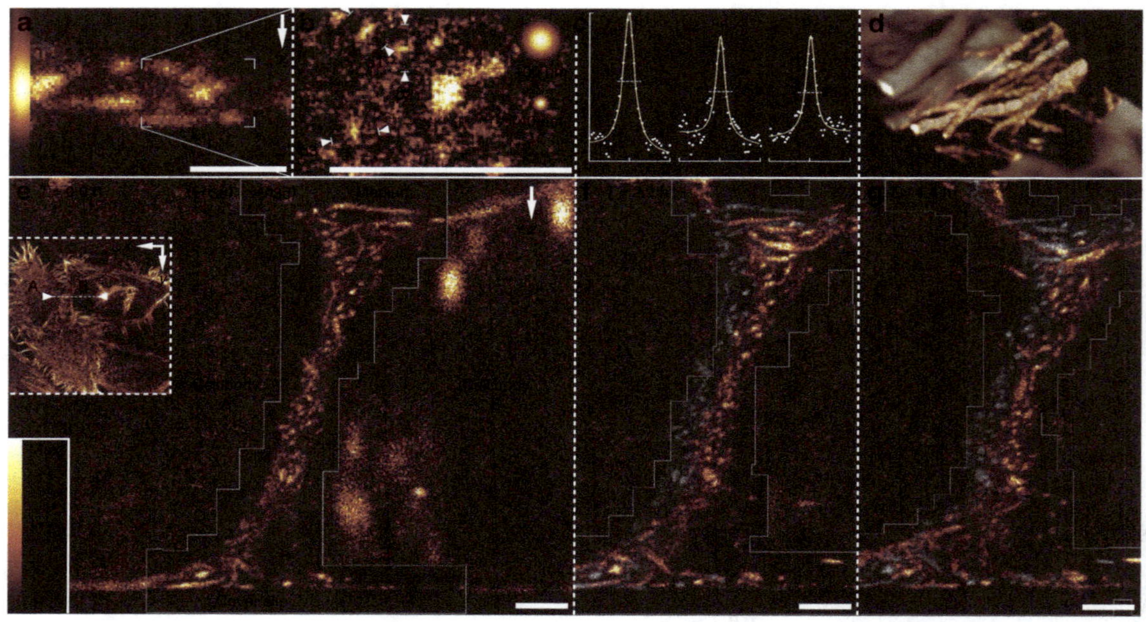

Figure 4 | 3D nanoscopy with strong optical sectioning. *xz*-sections of live HeLa cells expressing Lifeact-Dronpa-M159T. (**a**) Overview (optical *xz*-section) of actin fibre bundles at an axial base resolution in the 100 nm range. Left inset, confocal reference. (**b**) Addition of a 3D deactivation donut ($+h_{3d}$, $E_{3d} = 2.6\,\mu W \cdot 3.2\,ms$) to the RESOLFT pulse sequence reveals Dronpa patterns with apparent feature sizes well below 40 nm (inset, Gaussian reference spheres); (**c**) Lorentzian fits, plus a linear local background, to box profiles p1-3 over marked features in **b** along different directions. Numbers indicate full widths at half maximum (FWHM) over background. (**d**) Rendering of the volume surrounding **a**. (**e-g**) Time (*T*) evolution of an 8-μm-thick, densely labelled, vertical contact region between two adjacent cells (*xz*-section as marked in the *xy*-overview). Grayscale overlays of the preceding time step (**f,g**) aid in the tracking of individual features. A narrowed region of interest was generated online from initial overview scans ($-h_{3d}$) at each time frame and imaged at 50 nm target resolution ($+h_{3d}$, grey outline, E_{3d}, $E_{zd} = 1.3\,\mu W \cdot 1.6\,ms$, $1.8\,\mu W \cdot 0.5\,ms$). Despite the challenging imaging conditions, stacked actin structures are unambiguously resolved across the full axial extent of the cell layer. *xz*-panels depict rate-gated 4Pi-RESOLFT raw data, solely subjected to noise reduction. Fluorescence intensities *I*(*r*). Common pulse parameters (**b,e-g**), E_{ac}, $E_{ro} = 1.6\,mW \cdot 0.2\,ms$, $3.1\,\mu W \cdot 0.5\,ms$. Scale bars, 1 μm.

In conclusion, by realizing 4Pi-RESOLFT nanoscopy based on RSFPs, we have demonstrated exceptional optical sectioning in coordinate-targeted far-field fluorescence nanoscopy, which greatly facilitates nanometre scale 3D fluorescence imaging in living cells. Many accepted constraints to the sample can be lifted, which opens up an imaging regime that has so far been systematically avoided.

Methods

4Pi sample optics for live-cell imaging. In a 4Pi arrangement, the RI (n) difference between the material forming a living mammalian cell ($n \approx 1.35$–1.40) and the surrounding culture medium (typically $n \approx 1.33$) is a major source of optical aberrations. Aberrations generally reduce the attainable signal-to-noise ratio (*S/N*) by blurring the light intensity distributions in the focal region and by raising the intensity of the central minimum of the off-switching light patterns (Figs 1a and 3a). We therefore devised a mounting procedure that minimizes optical aberrations by raising the RI of a standard cell culture medium to $n = 1.362$ (Supplementary Fig. 3, Supplementary Methods) and by designing the optical setup accordingly (Fig. 1b, Supplementary Figs 4 and 5): The 4Pi foci are jointly created by two 1.20 NA water immersion objectives that are outfitted with individual tip/tilt-correction to prevent aberrations that arise from lens-coverslip misalignment[29]. The refractive indices of the embedding ($n = 1.362$) and immersion media ($n = 1.350$), the correction collar settings and the cover slip thickness were chosen to minimize spherical aberrations[25] over at least 10 μm of sample depth. The changes in the optical path lengths of the two 4Pi-interferometer arms due to *z*-scanning of the sample were compensated[30] by synchronous position adjustment of the main beam splitter cube (Supplementary Fig. 4).

A single-focus 3D light pattern for deactivation. To resolve features smaller than 100 nm, we added a RSFP deactivation beam to the microscope. It was imprinted with a circular phase ramp that was subsequently imaged into the back pupil planes of both objective lenses. In contrast to the configuration of a single-lens RESOLFT setup, the direction of rotation of the phase ramp was oriented in countersense with respect to the circular beam polarization at the back pupil planes, which produced a 4Pi off-switching pattern h_{3d} that completely surrounded a central zero[26] (Fig. 3a). Applying h_{3d} during the deactivation phase of the RESOLFT switching cycle squeezed the central full width at half maximum (FWHM) d_{ef} of h_{ef} and allowed us to tune h_{ef} to a resolution below the diffraction limit.

RESOLFT imaging with rate-gated detection. In the present case of RESOLFT imaging of negative-switching RSFP, deactivation of fluorophores during read-out gives rise to a time-dependent signal and hence a time-dependent effective PSF $h_{ef}(\mathbf{r},t)$. The time-resolved image $g(\mathbf{r},t)$ obtained by imaging a structure $s(\mathbf{r})$ is thus given by

$$g(\mathbf{r},t) = s(\mathbf{r}) \otimes h_{ef}(\mathbf{r},t). \quad (3)$$

We simplistically assume a deactivation rate $\lambda(\mathbf{r})$ that only depends on the read-out intensity $I_{ro}(\mathbf{r})$ and therefore obtain

$$h_{ef}(\mathbf{r},t) = h_{ef}(\mathbf{r}) \cdot e^{-\lambda(\mathbf{r})\cdot t} = h_{ef}(\mathbf{r}) \cdot e^{-\lambda \cdot I_{ro}(\mathbf{r})\cdot t} \quad (4)$$

with t denoting time relative to the start of the read-out pulse. The deactivation pattern $h_{3d}(\mathbf{r})$ typically confines the effective volume from which fluorescence is collected to a region of FWHM d_{ef} around the primary zero of the deactivation pattern. This region is much narrower than the FWHM d_{ro} of the diffraction pattern used for read-out:

$$d_{ef} \ll d_{ro} \quad (5)$$

The read-out intensity can then be considered constant, and $h_{ef}(\mathbf{r},t)$ follows a mono-exponential decay at a rate that only depends on the peak read-out intensity $I^0_{ro} := I_{ro}(0)$. Thus, $g(\mathbf{r},t)$ becomes separable and transforms into:

$$g(\mathbf{r},t) = g(\mathbf{r}) \cdot g(t) := (s(\mathbf{r}) \otimes h_{ef}(\mathbf{r})) \cdot e^{-\lambda \cdot I^0_{ro} \cdot t} \quad (6)$$

Data from test structures, however, exhibit a distinct multi-exponential behaviour that requires additional components for a proper fit (Fig. 3c):

$$\widehat{g}(\mathbf{r},t) = \sum_{i=0}^{n-1} \widehat{c}_i(\mathbf{r}) \cdot e^{-\widehat{\lambda}_i(\mathbf{r})\cdot t} \quad (7)$$

where n components with ordered switching rates $\widehat{\lambda}_i$, $\widehat{\lambda}_i > \widehat{\lambda}_{i+1}$ and coefficients $\widehat{c}_i$ are fitted to the data ('^' marks fit results). While we attribute the fastest rate $\widehat{\lambda}_0$ to signal from unimpaired RSFP at the focal centre, the presence of additional rates suggests the co-existence of RSFP species that exhibit significantly slower switching kinetics. A slowed-down switching observed after fixation supports this notion.

Unintended processes during image recording also potentially contribute to the observed signal behaviour, for example, the re-activation of RSFP by the read-out light which generates a constant background.

Without loss of generality, we assume a position independent mixture of n species with discrete switching rates λ_j. Since the deactivation pattern h_{3d} shrinks the effective PSF by a λ-dependent factor, $h_{ef}(\mathbf{r},t)$ has to be generalized to a superposition of n individual h^j_{ef} that each correspond to a λ_j. Furthermore, we assume our experimental parameters are chosen such that the in-focus part h_{fast} of h^0_{ef} obeys the analogue to equation (5), and pool the remaining contributions in h_{slow}:

$$h_{ef}(\mathbf{r},t) = h_{fast}(\mathbf{r},t) + h_{slow}(\mathbf{r},t) := h_{fast}(\mathbf{r}) \cdot e^{-\lambda_0 \cdot I^0_{ro} \cdot t} + \sum_{j=1}^{n-1} h^j_{slow}(\mathbf{r}) \cdot e^{-\lambda_j \cdot I_{ro}(\mathbf{r}) \cdot t} \tag{8}$$

Consequently, the apparent resolution of the acquired image is less than the potential resolution provided by h_{fast} and declines over time, as faster components vanish first (Fig. 3c).

To access the full image information that is mediated by h_{fast}, we implemented an unmixing scheme that isolates the fastest switching signal component (rate λ_0), and that we hence termed 'rate-gating': according to equation (3), the image generated by a PSF equation (8) takes on the form

$$g(\mathbf{r},t) = g_{fast}(\mathbf{r},t) + g_{slow}(\mathbf{r},t) := g_0(\mathbf{r}) \cdot e^{-\lambda_0 \cdot I^0_{ro} \cdot t} + \int_0^{\lambda_1} c(\mathbf{r},\Lambda) \cdot e^{-\Lambda t} d\Lambda \tag{9}$$

whereby g_{slow} is represented by a continuum of exponentials with switching speeds $\Lambda \in [0,\lambda_1]$ and coefficients $c(\mathbf{r},\Lambda)$ as the result of $s \otimes h_{slow}$.

Hence, a fit $\hat{g}(\mathbf{r},t)$ to an imaged structure according to equation (7), in principle, provides a position invariant estimate for $\lambda_0 = \hat{\lambda}_0 / I^0_{ro}$ by $\hat{g}_{fast}$, on top of a local approximation of $g_{slow}(\mathbf{r},t)$. In practice however, an insufficient photon count often prohibits local fitting of equation (9). We therefore implemented a robust approximation scheme for $g_{fast}(\mathbf{r})$ that only relies on parameters that can be extracted from a fit $\hat{g}(t)$ to the global (that is, from a region much larger than the corresponding diffraction limit) spatial average $\bar{g}(t)$ of the measured data:

First, we estimated the time t_0 at which the integrated signal exhibits the maximum S/N with respect to g_{fast}:

$$t_0 := \operatorname{argmax}\left(\frac{\int_0^T \bar{g}_{fast}(t) dt}{\sqrt{\int_0^T \bar{g}_{fast}(t) + \bar{g}_{slow}(t) dt}} \right) \tag{10}$$

Locally calculated values for t_0 would slightly differ, but as this only affects the statistical error of the result, equation (10) is usually sufficiently precise. Second, we determined a cut-off time t_1 such that

$$g_{fast}(t) \ll g_{slow}(t) \quad (t > t_1) \tag{11}$$

which is usually the case for $t_1 = 2t_0$. By further choosing t_2 and t_3 such that $t_i > t_{i+1} (i = 0..3)$, we partition the measured signal into time bins $\sum_{0,1,2}$ (Fig. 3d),

$$\sum\nolimits_{0,1,2}(\mathbf{r}) := \int_{0,t_1,t_2}^{t_0,t_2,t_3} g(\mathbf{r},t) dt \approx F(\mathbf{r}) + S_0(\mathbf{r}), S_1(\mathbf{r}), S_2(\mathbf{r}) \tag{12}$$

with $F(\mathbf{r})$ and $S(\mathbf{r})$ denoting the time integrals over the fast and slow components:

$$F(\mathbf{r}) := \int_0^{t_0} g_{fast}(\mathbf{r},t) dt, \quad S_{0,1,2} := \int_{0,t_1,t_2}^{t_0,t_2,t_3} g_{slow}(\mathbf{r},t) dt \tag{13}$$

Finally, we estimated $F(\mathbf{r})$, and thereby $g_0(\mathbf{r})$, by linear extrapolation in either zeroth or first order:

$$F(\mathbf{r})^{0th} := \sum\nolimits_0(\mathbf{r}) - u \cdot \sum\nolimits_1(\mathbf{r}) \tag{14}$$

$$F(\mathbf{r})^{1st} := F(\mathbf{r})^{0th} - v \cdot \left(\sum\nolimits_1(\mathbf{r}) - \sum\nolimits_2(\mathbf{r}) \right) \tag{15}$$

with $u,v = u,v(t_{0..3})$ denoting geometrical factors that account for the particular choice of the $t_{0..3}$. Narrowing the integration intervals defined by $t_{0..3}$ and moving them closer to $t = 0$, just as the inclusion of the first extrapolation order, reduces the systematic error, but also raises the statistic error due to a reduced photon count. To mitigate this effect, and owing to equation (11), we substituted $\sum_{1,2}$ with their respective resolution-neutral local averages, for example, by applying a Gaussian filter with a FWHM sufficiently far below the FWHM of $h^{1,2}_{slow}$.

References

1. Hell, S. W. Double-confocal scanning microscope. European Patent 0491289 (1992).
2. Hell, S. W. & Stelzer, E. H. K. Properties of a 4pi confocal fluorescence microscope. *J. Opt. Soc. Am. A Opt. Image Sci. Vis.* **9,** 2159–2166 (1992).
3. Gustafsson, M. G. L., Agard, D. A. & Sedat, J. W. Sevenfold improvement of axial resolution in 3D widefield microscopy using two objective lenses. *Proc. Soc. Photo-Opt. Instrum. Eng.* **2412,** 147–156 (1995).
4. Eggeling, C., Willig, K. I., Sahl, S. J. & Hell, S. W. Lens-based fluorescence nanoscopy. *Q. Rev. Biophys.* **48,** 178–243 (2015).
5. Schmidt, R. *et al.* Spherical nanosized focal spot unravels the interior of cells. *Nat. Methods* **5,** 539–544 (2008).
6. Shtengel, G. *et al.* Interferometric fluorescent super-resolution microscopy resolves 3D cellular ultrastructure. *Proc. Natl Acad. Sci. USA* **106,** 3125–3130 (2009).
7. Hell, S. W., Schmidt, R. & Egner, A. Diffraction-unlimited three-dimensional optical nanoscopy with opposing lenses. *Nat. Photon.* **3,** 381–387 (2009).
8. Grotjohann, T. *et al.* Diffraction-unlimited all-optical imaging and writing with a photochromic GFP. *Nature* **478,** 204–208 (2011).
9. Lang, M. C., Engelhardt, J. & Hell, S. W. 4Pi microscopy with linear fluorescence excitation. *Opt. Lett.* **32,** 259–261 (2007).
10. Schmidt, R. *et al.* Mitochondrial cristae revealed with focused light. *Nano Lett.* **9,** 2508–2510 (2009).
11. Ullal, C. K., Schmidt, R., Hell, S. W. & Egner, A. Block copolymer nanostructures mapped by far-field optics. *Nano Lett.* **9,** 2497–2500 (2009).
12. Kanchanawong, P. *et al.* Nanoscale architecture of integrin-based cell adhesions. *Nature* **468,** 580–584 (2010).
13. Denk, W., Strickler, J. H. & Webb, W. W. 2-photon laser scanning fluorescence microscopy. *Science* **248,** 73–76 (1990).
14. Schneider, M., Barozzi, S., Testa, I., Faretta, M. & Diaspro, A. Two-photon activation and excitation properties of PA-GFP in the 720-920-nm region. *Biophys. J.* **89,** 1346–1352 (2005).
15. Glaschick, S. *et al.* Axial resolution enhancement by 4Pi confocal fluorescence microscopy with two-photon excitation. *J. Biol. Phys.* **33,** 433–443 (2007).
16. Moneron, G. & Hell, S. W. Two-photon excitation STED microscopy. *Opt. Express* **17,** 14567–14573 (2009).
17. York, A. G., Ghitani, A., Vaziri, A., Davidson, M. W. & Shroff, H. Confined activation and subdiffractive localization enables whole-cell PALM with genetically expressed probes. *Nat. Methods* **8,** 327–333 (2011).
18. Hell, S. W. Improvement of lateral resolution in far-field light microscopy using two-photon excitation with offset beams. *Opt. Commun.* **106,** 19–24 (1994).
19. Hell, S. W. Toward fluorescence nanoscopy. *Nat. Biotechnol.* **21,** 1347–1355 (2003).
20. Stiel, A. C. *et al.* 1.8 angstrom bright-state structure of the reversibly switchable fluorescent protein dronpa guides the generation of fast switching variants. *Biochem. J.* **402,** 35–42 (2007).
21. Willig, K. I., Stiel, A. C., Brakemann, T., Jakobs, S. & Hell, S. W. Dual-label STED nanoscopy of living cells using photochromism. *Nano Lett.* **11,** 3970–3973 (2011).
22. Testa, I. *et al.* Nanoscopy of living brain slices with low light levels. *Neuron* **75,** 992–1000 (2012).
23. Grotjohann, T. *et al.* rsEGFP2 enables fast RESOLFT nanoscopy of living cells. *eLife* **1,** e00248 00241-00214 (2012).
24. Gugel, H. *et al.* Cooperative 4pi excitation and detection yields sevenfold sharper optical sections in live-cell microscopy. *Biophys. J.* **87,** 4146–4152 (2004).
25. Wan, D. S., Rajadhyaksha, M. & Webb, R. H. Analysis of spherical aberration of a water immersion objective: application to specimens with refractive indices 1.33-1.40. *J. Microsc.* **197,** 274–284 (2000).
26. Schmidt, R. *3D Fluorescence Microscopy with Isotropic Resolution on the Nanoscale* (PhD thesis, Univ. Heidelberg, 2008).
27. Richards, B. & Wolf, E. Electromagnetic diffraction in optical systems. II. structure of the image field in an aplanatic system. *Proc. R. Soc. Lond. A* **253,** 358–379 (1959).
28. Testa, I., D'Este, E., Urban, N. T., Balzarotti, F. & Hell, S. W. Dual channel RESOLFT nanoscopy by using fluorescent state kinetics. *Nano Lett.* **15,** 103–106 (2015).
29. Arimoto, R. & Murray, J. M. A common aberration with water-immersion objective lenses. *J. Microsc.* **216,** 49–51 (2004).
30. Hell, S. W., Schrader, M. & VanderVoort, H. T. M. Far-field fluorescence microscopy with three-dimensional resolution in the 100-nm range. *J. Microsc.* **187,** 1–7 (1997).

Acknowledgements

We thank C. Gregor for providing the plasmid Lifeact-Dronpa-M159T, T. Gilat and E. Rothermel for cloning support and assistance with cell culture preparation, B. Thiel for Imspector software support and A. Pucher-Diehl for precision mechanic support for custom microscope parts. S.W.H. and R.S. acknowledge a grant by the Deutsche Forschungsgemeinschaft within SFB 775.

Author contributions

U.B. prepared the samples and carried out the measurements. U.B. and R.S. designed and built the RESOLFT microscope and evaluated the data. R.S. designed and implemented the 4Pi unit and the data acquisition and rate-gating algorithms. The project was defined and supervised by R.S. and S.W.H. All authors discussed the project as it evolved and edited the final version of the paper. The paper was written by R.S. and S.W.H. with contributions by U.B.

Additional information

Competing financial interests: The authors declare no competing financial interests.

6

Nanopore sensing at ultra-low concentrations using single-molecule dielectrophoretic trapping

Kevin J. Freedman[1], Lauren M. Otto[2], Aleksandar P. Ivanov[1], Avijit Barik[2,3], Sang-Hyun Oh[2,3] & Joshua B. Edel[1]

Single-molecule techniques are being developed with the exciting prospect of revolutionizing the healthcare industry by generating vast amounts of genetic and proteomic data. One exceptionally promising route is in the use of nanopore sensors. However, a well-known complexity is that detection and capture is predominantly diffusion limited. This problem is compounded when taking into account the capture volume of a nanopore, typically 10^8–10^{10} times smaller than the sample volume. To rectify this disproportionate ratio, we demonstrate a simple, yet powerful, method based on coupling single-molecule dielectrophoretic trapping to nanopore sensing. We show that DNA can be captured from a controllable, but typically much larger, volume and concentrated at the tip of a metallic nanopore. This enables the detection of single molecules at concentrations as low as 5 fM, which is approximately a 10^3 reduction in the limit of detection compared with existing methods, while still maintaining efficient throughput.

[1] Department of Chemistry, Imperial College London, South Kensington, London SW7 2AZ, UK. [2] Department of Electrical and Computer Engineering, University of Minnesota, Minneapolis, Minnesota 55455, USA. [3] Department of Biomedical Engineering, University of Minnesota, Minneapolis, Minnesota 55455, USA. Correspondence and requests for materials should be addressed to J.B.E. (email: joshua.edel@imperial.ac.uk).

One of the greatest challenges facing physical and biological scientists is the accurate detection and identification of single molecules. The drive to performing such experiments is a result of experimental observations typically being an 'ensemble average' whereas a single-molecule measurement enables the possibility of detecting kinetics and dynamics from individual molecules in real time. Since the mid 1990s, nanopores have been a rapidly growing technology for use in applications such as single-molecule DNA sequencing[1–3], protein detection[3–6] and the study of RNA-drug targets[7] largely owing to the method being label-free while still maintaining a high temporal resolution. As a consequence, nanopore sensors have the potential to revolutionize the healthcare industry with numerous clinical applications already in the pipeline. Although promising, nanopore sensing has several drawbacks with one of the most problematic being the lack of efficiency when trying to detect individual molecules from the bulk solution. In fact, this problem extends further and is fundamental to many surface-based biosensors[8]. This stems from the fact that the dominant mechanism of capture and detection (for example, the process of DNA translocating through the pore) is diffusion-limited resulting in only a small fraction of the total sample volume being accessed. For example, a 10 µl solution at a concentration of 1 nM will have on average 34 molecules in the capture volume assuming a capture radius of 3 µm. This is roughly 10^8 times lower than the actual number of molecules in the solution. At sub-picomolar concentrations, the average number of molecules inside the capture volume would be well below 0.03 leading to long measurement timescales (>1 h). It is clear that new strategies are needed to enable 'rare event' and 'needle-in-a-haystack' detection experiments.

Importantly, molecular transport through nanopores has been studied extensively[9–11], however, there have been much fewer documented techniques which have demonstrated capture rate enhancement. In 2010, the limit of detection was reduced from several hundred picomolar to 3.8 pM by utilizing high salt gradients to manipulate the voltage drop outside the nanopore[12]. Using this technique, an event rate of ~ 60 molecules per minute was achieved. More recently, a subclass of nanopores called nanopipettes was shown to offer similar levels of sensitivity by loading DNA inside the pipette and applying voltage pulses to improve the efficiency of delivery from the nanopore[13]. Protein pores also offer unique biological methods of increasing the translocation rate including modifying the internal charge of the pore[14,15]. Both studies made use of the electrophoretic properties of DNA to enhance the capture rate, while others have attempted to use pressure gradients to add a level of control to the translocation process[16]. However, all these methods have limited ability to concentrate and perform high-throughput detection of ultra-dilute samples. Although these studies have focused on DNA, it should be noted that the goal to increase capture rate has far-reaching applications related to rare event detection and is not exclusively directed towards DNA sequencing; especially if DNA translocations are successfully slowed down such that the inter-event time is not a limiting factor.

In this manuscript, we developed a novel method to improve the detection efficiency by 1,000-fold by incorporating a dielectrophoretic (DEP) trap at the nanopore opening. The DEP trap not only provides on-demand control of the capture volume (that is, the volume in which molecules become drawn towards the pore) but can also significantly increase the number of molecules being detected per unit time even at concentrations of a few femtomolar. The reported technique pushes the envelope of high-sensitivity and amplification-free single-molecule detection and paves the way for high-speed and high-throughput detection of ultra-dilute samples and rare events.

Results

Nanopipettes incorporating dielectrophoretic traps. Typically, nanopore experiments use a constant DC bias to prompt the translocation of an analyte molecule. In this study, an AC voltage (10–20 V and 0.5–4 MHz) is applied to the metallized layer surrounding the nanopipette followed by a DC voltage which translocates the molecules, Fig. 1a. The AC and DC voltages are applied to the system using two Au electrodes and two Ag/AgCl electrodes, respectively (Fig. 1a,b). Metallic tips and apertures, used for DEP experiments and biosensing[17], offers several key advantages over planar electrodes which include (i) a three-dimensional trapping volume, (ii) being able to control the electrode gap distance and therefore the field gradient forces and (iii) added enhancement due to the sharpness of the metallic tip[18].

Nanopipettes are glass or quartz capillaries that are heated and pulled until a nanometer-sized aperture is formed at the tip. When biological molecules are passed through the nanopore, the exclusion of ions causes a decrease in bulk ion flow thereby allowing single molecules to be detected. Although it is in principle possible to measure the ionic current in the presence of a superimposed AC field to enable molecular trapping, calibration of the competing DEP and electrophoretic forces are far from trivial. For simplicity, experiments were split into a DEP (capture) phase (Fig. 1b, (i)) and an electro-kinetic translocation (detection) phase (Fig. 1b, (ii)). To adapt the nanopipettes (25 nm diameter) for use in dielectrophoretic trapping, a thin 5 nm layer of Au was deposited on the barrel (Fig. 1c) and were characterized by filling with 1 mM KCl and measuring the ionic current when a DC voltage bias is applied between the Ag/AgCl electrodes (Fig. 1a). Further details of the experimental setup and nanopipette positioning are available in Supplementary Note 1 and Supplementary Figs 1 and 2. Using uncoated quartz nanopipettes, the current–voltage (I–V) curve showed rectification (that is, unequal conduction depending on voltage polarity) consistent with that expected by negatively charged conical glass nanopores[19,20]. On coating with gold, the rectification ratio is reduced but is still present ($|I_{-600\,\mathrm{mV}}/I_{600\,\mathrm{mV}}| = 1.3$) compared with the bare pipette ($|I_{-600\,\mathrm{mV}}/I_{600\,\mathrm{mV}}| = 4.8$; Fig. 1d), indicating a reduction in the surface charge on the pipette[13,21–24]. The noise characteristics after gold deposition showed more than an order of magnitude reduction in low frequency $1/f$ noise typically associated with conductance fluctuations (flicker noise) and a mild increase in higher frequency noise attributed to capacitance (representative plots of >50 nanopipettes shown in Fig. 1e,f)[25]. Importantly, while measuring ionic current, relatively small changes in the noise demonstrated that metallized nanopipettes are suitable for the envisioned application of DEP-enhanced DNA detection.

Modelling of dielectrophoretic fields. Incorporating a DEP trap to the nanopore was achieved by adding gold to the area surrounding the nanopore (that is, the nanopore entrance), as demonstrated by the change in ionic conductance. With the gold electrode in proximity to the pore, high gradient forces can be generated. However, theoretical approaches are needed to find the distribution of the electric fields around the tip of the nanopipette. The counterion fluctuation (CIF) model is proposed as the best possible method for describing the true nature of the polarizability of DNA, which is based on the redistribution of the counterions from the solution which surround the molecule near its charged sites, see Supplementary Note 2 and Supplementary Figs 3–5. The polarizability of the 10 kbp DNA used in experiments was calculated to be $1.59 \times 10^{-30}\,\mathrm{F\,m^2}$ and was

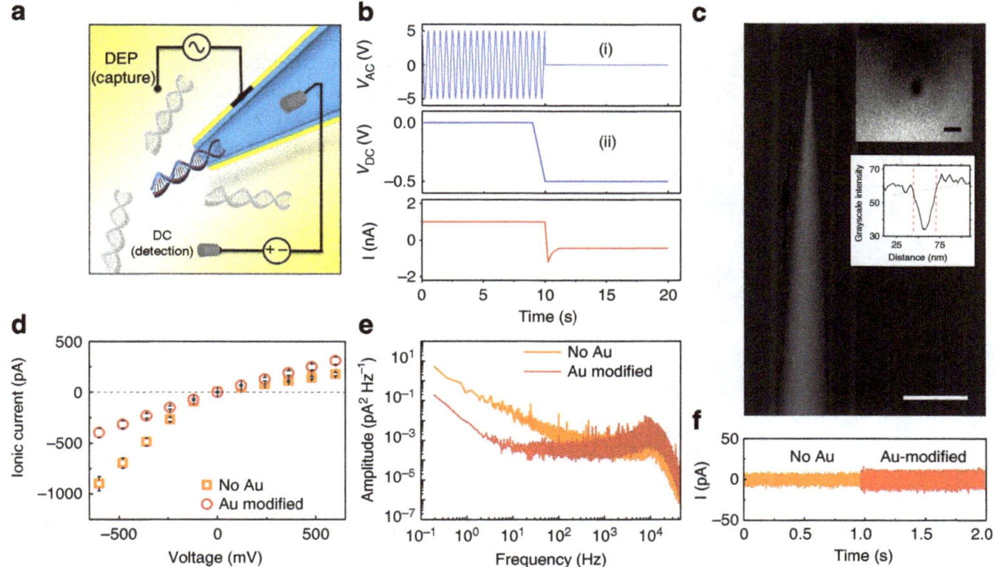

Figure 1 | Experimental setup and characterization of gold-coated nanopipettes. (**a**) Schematic of DNA being threaded through the tip of a gold-coated nanopipette. (**b**) Schematic of the voltage protocol used for DEP pre-concentration phase via AC voltage (i) and subsequent nanopore translocations via DC voltage (ii). All three channels (V_{AC}, V_{DC}, I) are simultaneously recorded. (**c**) SEM of a gold-coated nanopipette; scale bar, 5 μm (insets: SEM and intensity line plot of the tip visualized parallel to the barrel; scale bar, 50 nm). (**d**) Current-voltage curves for glass nanopipettes before and after gold coating. Gold coating thickness was ~5 nm. (**e**) Power spectral density of pipettes under a negative 500 mV voltage bias. (**f**) Baseline-subtracted time traces of the pipettes before and after gold coating at a negative 500 mV voltage bias and 1 mM KCl.

determined by normalizing the subunit polarization, α_s, to the entire length of the DNA (see Supplementary Information for further details). The polarizability of the DNA from the modified CIF model was multiplied with the electric field intensity gradient extracted from finite-element method (FEM) models to determine the force on the DNA molecules.

FEM modelling was performed using COMSOL Multiphysics 4.3a to determine the strength of the electric field intensity gradient, $|\nabla|E|^2|$. Plots of $\log_{10}|\nabla|E|^2|$ surrounding the nanopipette were extracted from the model to show the strength of the electric field intensity gradient, and therefore the force, near the nanopipette (Fig. 2a) and several micrometers from the nanopipette (Fig. 2b). Critical to DEP trapping, it was discovered that DEP forces are strongest at the edges of the nanopipette tip and therefore DNA will be drawn directly to the nanopore. The threshold force (that is, the force required to overcome Brownian motion: 9.92 fN) extends ~4–5 μm from the nanopipette tip in our models (marked with a black contour line at the threshold $|\nabla|E|^2| = 10^{16.4}\,V^2\,m^{-3}$). It should also be noted that the threshold force and therefore the trapping volume is adjustable based on the parameters of the AC voltage applied to the pipette.

The DEP force acting on the particles under fixed conditions (10 V, 1 MHz) is shown by plotting the $\log_{10}|\nabla|E|^2|$ (Fig. 2c) along the z axis extending out from the nanopipette tip toward the flat electrode which is 50 μm from the nanopipette tip. One of the key parameters that the simulations probed was the distance between the nanopipette tip and the planar electrode, d_{gap}, which was experimentally set using a micromanipulator. In Fig. 2d, the dependence of $-\nabla|E|^2$ along the z axis is shown for both changing distances from the nanpopipette tip (30 nm, 300 nm and 3 μm), as well as changing electrode gap distances (d_{gap} = 20–100 μm). Function fitting shows a logarithmically decreasing trend in the force as the electrode gap distance increases, which matches our expectations based on previous literature[26].

Discussion

Single-molecule fluorescence imaging of dielectrophoretic traps. Quantification of the DEP trapping kinetics was accomplished using YOYO-1 labelled 10 kbp DNA in conjunction with fluorescence microscopy. The key parameters which were controlled include the electrode gap distance (d_{gap}), the AC frequency (f_{AC}) and the AC peak-to-peak voltage (V_{pp}). Qualitatively, once the AC field was turned on, the DNA would localize and surround the region of the pipette with the sharpest geometric features (that is, the tip of the nanopipette, Fig. 3a,b), as expected from simulations[17]. As the trapping field was kept on for longer time periods, the fluorescent region around the tip would grow in size owing to the accumulation of labelled DNA (Fig. 3c). At d_{gap} = 50 μm and V_{pp} = 20 V, various frequencies were applied to the two gold electrodes. Using the CIF model, the frequency dependence is owing to the relaxation time constant of the ions surrounding the DNA; specifically the time duration for charges to relax after being perturbed by an electric field. For a 12 kbp DNA molecule and a similar dielectric decrement, the relaxation frequency was ~2 MHz (ref. 27), which explains the sudden loss of DEP trapping efficiency at the higher AC frequencies (2, 3 and 4 MHz). Indeed the highest fluorescence intensity was obtained for the 1 MHz condition, which maximizes at about 4.5 s after applying the AC field. Interestingly, a decrease in the intensity past this point is observed which is likely due to the DNA closest to the tip (trapped within the focal plane for the most time) being photobleached (exponential decay in intensity shown in Fig. 3e)[28,29].

To characterize the effects of DEP acting on the DNA, the spatial position of the DNA relative to the tip was tracked using image processing. As the DNA diffuses close to the tip (16–19 μm), the velocity of the DNA increased as the molecules enter a region of AC field-induced motion (DNA trajectories plotted in Fig. 3f). The mean squared displacement (MSD) of DNA was calculated using the two-dimensional (2D) diffusion equation ($\langle x^2 \rangle = 4Dt$), as the optical images represent a

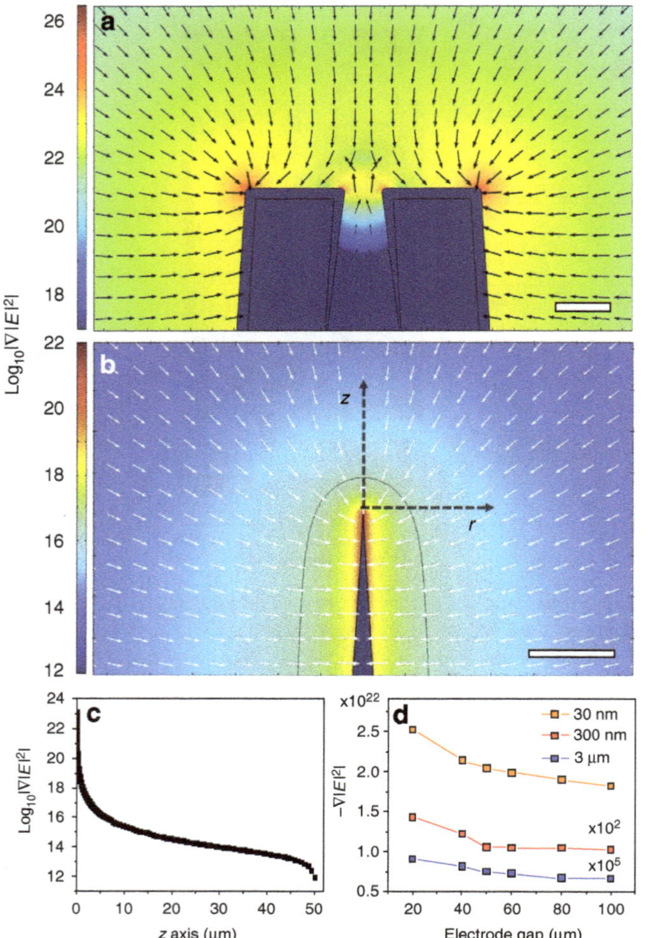

Figure 2 | Finite-element method modelling of DEP with the nanopipette. Two-dimensional-axisymmetric electrostatic modelling of a 50 μm electrode gap between the nanopipette tip and a conductive flat electrode was simulated with a 10 V DC signal applied. Plots of the magnitude of the field intensity gradient, which is proportional to the force on a particle, are shown for regions (**a**) near the end of the pipette (scale bar, 25 nm) and (**b**) the DEP trapping volume and surrounding area (scale bar, 10 μm). The logarithmically scaled arrows show the direction of the force. The black contour line in **b** is along $|\nabla|E|^2| = 10^{16.4}\,V^2\,m^{-3}$, which corresponds approximately to the threshold field intensity gradient for trapping 10 kbp DNA against Brownian motion. (**c**) The magnitude of the field intensity gradient along the *z* axis (black dashed arrow) in **b**. (**d**) The strength of $|\nabla|E|^2|$ was also tracked at different distances from the tip of the nanopipette along the *z* axis for electrode gaps ranging from 20 to 100 μm with a 10 V DC applied signal. The field intensity gradient strength decreases logarithmically with increasing gap size.

projection of the three-dimensional molecular movement into a 2D image. The MSD was averaged over two frames (15 ms per frame) and plotted against distance away from the nanopipette tip (Fig. 3g; $D_{10\,kbp\ DNA} = 1.05 \times 10^{-8}\,cm^2\,s^{-1}$; ref. 30). The point where DNA crosses over from a diffusion-limited regime to an AC field-enhanced regime occurred between 16 and 19 μm from the tip (marked by arrows in Fig. 3f).

Interestingly, DEP is not the only force experienced by DNA. Electrothermal flow (ETF), stemming from temperature inhomogeneity in the fluid medium surrounding the pipette, became apparent at elevated voltages, see Supplementary Notes 3 and 4 and Supplementary Fig. 6. To demonstrate the distinct forces observed for both DEP and ETF, polystyrene beads were tracked at the conditions used for DNA trapping ($V_{pp} = 12$ V, $f_{AC} = 1$ MHz, $d_{gap} = 50$ μm) as well as at higher voltages and a larger electrode gap distance (Fig. 3h,i; $V_{pp} = 20$ V, $f_{AC} = 1$ MHz, $d_{gap} = 200$ μm). As DNA trapping experiments were performed at frequencies of 1 MHz and higher, we do not expect electro-osmotic flow to play a significant role. Although Fig. 3h,i show rather distinct trajectories, in reality, both DEP and ETF exist around the nanopipette and play a role in the trapping of DNA. For example, instead of DNA diffusing into the DEP-trapping volume, the ETF plays a role in enhancing the capture efficiency by delivering DNA into the trapping volume. An approximation of the trapping volume size for the above conditions is supplied in the Supplementary Note 5 and Supplementary Fig. 7, however, it should be kept in mind that the trapping volume is controllable depending on the applied AC potential. Therefore, the capture rate of DNA can be actively controlled by varying the DEP parameters.

Single-molecule translocations and capture rate enhancement. The DEP phase of the experiment not only results in efficient molecule capture close to the pore entrance, but also enables the subsequent translocation. Using a DNA (10 kbp) concentration of 500 pM, DEP trapping and subsequent DNA translocations were performed. The dwell time of the DNA was measured as the full width half maximum of the ionic current signature for each translocation event. For reference, single-molecule translocations without DEP trapping showed consistent ionic current blockade amplitudes, dwell times and charge distributions to what is expected for conventional nanopore experiments (see Supplementary Note 6 and Supplementary Figs 8 and 9). These values were then plotted as a histogram along with the data obtained without DEP trapping (Fig. 4b). The data were time-normalized (recording time = 180 s) and plotted on a log scale to show both populations (linearly scaled plots are shown in Supplementary Fig. 10. Typical enhancement factors (EF = number of events with DEP/number of events without DEP) range between 80 and 100 for the conditions tested (DEP duration = 10 s, $V_{pp} = 12$ V, $f_{AC} = 1$ MHz and $d_{gap} = 50$ μm). EF values (calculated using the same DNA concentration) differ from our previously mentioned definition of detection efficiency which strictly refers to the ability to detect reduced concentrations of DNA (that is, the limit of detection). Using one particular experiment with EF ≈ 80, with and without DEP, the number of events were 2,715 and 34, respectively. Importantly, translocation characteristics are not affected by DEP trapping as shown by a count-normalized histogram (Fig. 4d). Similarly, the maximum current drop obtained for each event was plotted as a log-scale histogram (time-normalized) and a linearly scaled histogram (count-normalized; Fig. 4b,d). Importantly, pipette to pipette reproducibility was good, see Supplementary Fig. 11.

The inter-event time (δt) was extracted from the data by taking the difference between the start times of two consecutive events. The δt parameter has an exponential distribution which can be fitted by a linear curve on a log-scaled axis (as shown in Fig. 4d). Importantly, the δt values during experiments where DEP trapping was used show a marked decrease. This is expected as there are more events per unit time due to the pre-concentration effects of DEP. Alternatively, a cumulative histogram can be used to show the percentage of events which occur below a certain inter-event time (Supplementary Fig. 12). Comparing between experiments with and without DEP trapping, it was observed that 95% of inter-event times were below 8,750 ms when trapping was not used while the same number of events occurred below 400 ms with trapping forces being used to pre-concentrate the DNA.

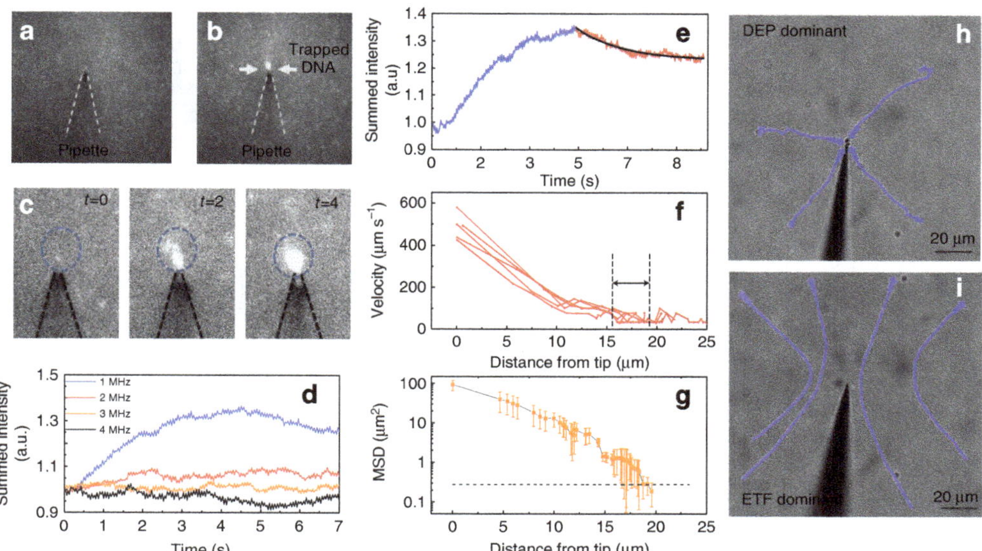

Figure 3 | Fluorescence images and analysis of DNA trapped at the nanopipette tip. (**a**) YOYO-labelled 10 kbp DNA sample with the gold-coated nanopipette positioned 50 μm above the surface of the planar counter electrode without applied AC voltage. (**b**) YOYO-labelled 10 kbp DNA sample with an AC voltage ($V_{pp}=20$ V, $f_{AC}=1$ MHz) being applied between the gold layer on the nanopipette and the counter electrode. (**c**) Images showing the progression of DNA trapping which occurs at the tip of the nanopipette. (**d**) Fluorescent intensity plotted over time for video recordings conducted at $V_{pp}=20$ V and $f_{AC}=1, 2, 3, 4$ MHz. (**e**) Fitting the reduction of fluorescent intensity to an exponential ($\tau_{decay}=1.54$ s) for the 1 MHz condition. (**f**) DNA velocity profiles for five 10 kbp DNA molecules being trapped using the following conditions: $V_{pp}=12$ V, $f_{AC}=1$ MHz. (**g**) Average mean squared displacement (MSD) of 10 kbp DNA as they transition into the trapping volume. Black dotted line: MSD expected for diffusion alone. (**h,i**) Trajectories for 1 μm polystyrene beads showing DEP-dominant ($f_{AC}=1$ MHz, $V_{pp}=12$ V, $d_{gap}=50$ μm) conditions and ETF-dominant conditions ($f_{AC}=1$ MHz, $V_{pp}=20$ V, $d_{gap}=200$ μm).

Further than simply increasing the capture rate before DNA translocations, multiple DEP cycles are shown to have a compounded effect on the capture rate of DNA. Typically the first cycle of trapping/translocation (1 cycle = 10 s trapping + 10 s translocations) yielded an increase in capture rate followed by further enhancement with each subsequent cycle (cycles with only 5 s of trapping were also tested; see Supplementary Note 8 and Supplementary Fig. 15). The trend shows an exponential increase in capture rate as a function of cycle number (Fig. 5a) up until equilibrium is reached, see also Supplementary Note 9. If it is assumed that each cycle is modelled as a rate balance ($R_{in}-R_{out}=R_{acc}$) where the input of mass (that is, DNA) is governed by DEP and the output of mass is governed by the electrophoretic transport of DNA through the pipette, the increase in capture rate across cycles, R_{acc}, is justified as being the result of an imbalance between these two rates. Losses due to diffusion are also present and make up a portion of the R_{out} term of the equation. The accumulation of DNA at the tip therefore seems to be a key contributor to the enhancement observed in our experiments.

By reducing the peak-to-peak voltage used for DEP trapping (20 to 12 V), the timescale for DNA accumulation, and post-DEP enhancement, was effectively reduced so that within two cycles, the capture rate fell to near-baseline values. The ability to reduce the post-DEP enhancement signifies that the DNA accumulation term was reduced and most of the DNA which was loosely bound to the pipette surface (that is, capable of desorbing) was removed by translocations. It is likely that the functionality of the nanopipette can be tuned to minimize surface adsorption to increase the collection efficiency. Cycles where DNA pre-concentration was used can clearly be identified by plotting the capture rate across cycles (Fig. 5b) where various AC frequencies were used in combination with a DC voltage of $|\Delta V|=500$ mV.

The V_{pp} parameter used for DEP trapping is critical to the enhancement of DNA sensing as it determines the relative strengths of both DEP as well as ETF. As V_{pp} was increased from 10 to 20 V, a linear increase in the capture rate was observed. Surprisingly, while V_{pp} was only doubled at the extreme ends of the values tested (10 and 20 V), a 10-fold increase in the capture rate was discovered. Due to the presence of ETF, DNA is not delivered into the DEP capture volume by simple diffusion mechanisms but rather delivered at an enhanced rate via ETF. Furthermore, as ETF scales with voltage to the fourth power (V^4; ref. 31), the strong dependence with V_{pp} is reasonable and provides further evidence for the existence of ETF in the nanopipette system[32]. In analysing the data presented in Fig. 5c–e, 10 consecutive DEP pre-concentrating cycles were used and each data point represents the average of the last five cycles.

To establish how the capture rate depends on the electrode gap distance, the pipette tip location was changed in 20 μm steps and DEP/translocation cycles were recorded for each position. As the gap between the two gold electrodes became smaller, the capture rate increased for gap distances between 100 and 40 μm (Fig. 5d), as expected from the FEM models (Fig. 2d). Interestingly, at 20 μm, the capture rate was reduced slightly. On the basis of the trapping volume data obtained using YOYO-1-labelled DNA described earlier, 20 μm is on the same scale as the distance away from the tip where DNA begins to respond to DEP forces. It is likely that the decrease in capture rate was due to the trapping volume (or ETF) becoming geometrically confined by the surface of the planar electrode.

High sensitivity detection in concentrations as low as 5 fM was achieved using a tip-to-surface gap distance of 20 μm and optimized DEP trapping conditions ($V_{pp}=20$ V and 1 MHz). The capture rate logarithmically decreases as a function of

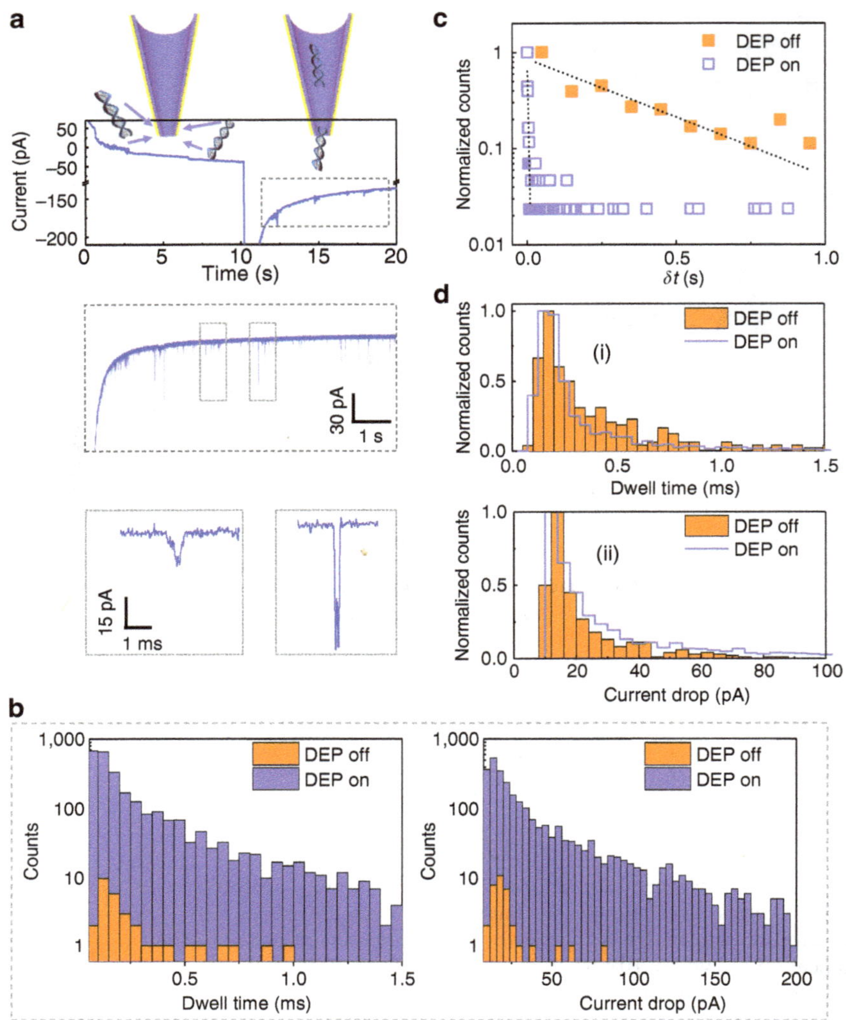

Figure 4 | Single-molecule detection and analysis with and without DEP. (**a**) Current traces of a typical DEP pre-concentration/recording cycle with various time scaling. The first 10 s is the pre-concentration phase and the later 10 s is the translocation/detection phase. The lower panel shows typical events representing a single DNA molecule translocation (~96% of events), as well as a DNA aggregate translocating the pore (~4% of events; see Supplementary Note 7 and Supplementary Figs 13 and 14). (**b**) Time-normalized histogram of the dwell time and current drop comparing translocations with and without DEP pre-concentration. (**c**) Normalized histogram of the inter-event time (δt) with and without DEP pre-concentration. (**d**) Count-normalized distributions with and without DEP pre-concentration for the dwell time (i) and current drop (ii). All DEP data were obtained at $V_{pp} = 12$ V, $f_{AC} = 1$ MHz and $d_{gap} = 50$ μm.

bulk concentration (Fig. 5e). Although sub-picomolar DNA concentrations are typically not capable of being sensed using nanopores, DEP pre-concentration allowed the full range of fM concentrations to be detected (Fig. 5f,g, Table 1). Previous studies which used salt gradients, and later using controlled DNA delivery methods, showed the ability to enhance the capture rate of DNA down to ~3 pM (refs 12,13). Both of the previous methods rely on the electrophoretic properties of DNA to enhance the capture radius and the local concentration, respectively. The DEP-based method, which uses the polarizability of DNA, is shown here to be a much more powerful method for enhancing the capture rate of DNA. At 5 fM DNA, a capture rate of 315 ± 147 events per minute is achieved (Fig. 5f). Before the experiments, a DC bias was applied to the same pipette for 2 min with a total of four events being detected. To obtain accurate event statistics, an arbitrary threshold of 1,000 events was deemed reasonable. If true, the recording time for the 5 fM sample would require a duration of 8.3 h. Using DEP, the recording time is 3.2 min.

We have demonstrated the use of metallized nanopipettes, and more generally nanopores, for DEP trapping and DNA pre-concentration. The foreseeable benefits of a hybrid DEP-nanopore device on genetic analyses are enormous owing to the need for ultra-sensitive methods that can analyse low concentrations of genomic DNA with high throughput. This report has demonstrated the ability to sense DNA at a concentration as low as 5 fM at an event rate of 315 events per minute. The capture rate enhancement is shown to stem from a larger DEP trapping volume in combination with enhanced delivery into the trapping volume by electrothermal flow. Importantly, the trapping volume size is a controlled and easily manipulated parameter which allows the capture rate to be changed on command. No longer is nanopore sensing dictated by the diffusion and electrophoretic properties of the analytes, but by their polarizability and the controlled AC voltage applied. Future work will be directed towards combining AC and DC voltages in an effort to apply two independent forces to the translocating

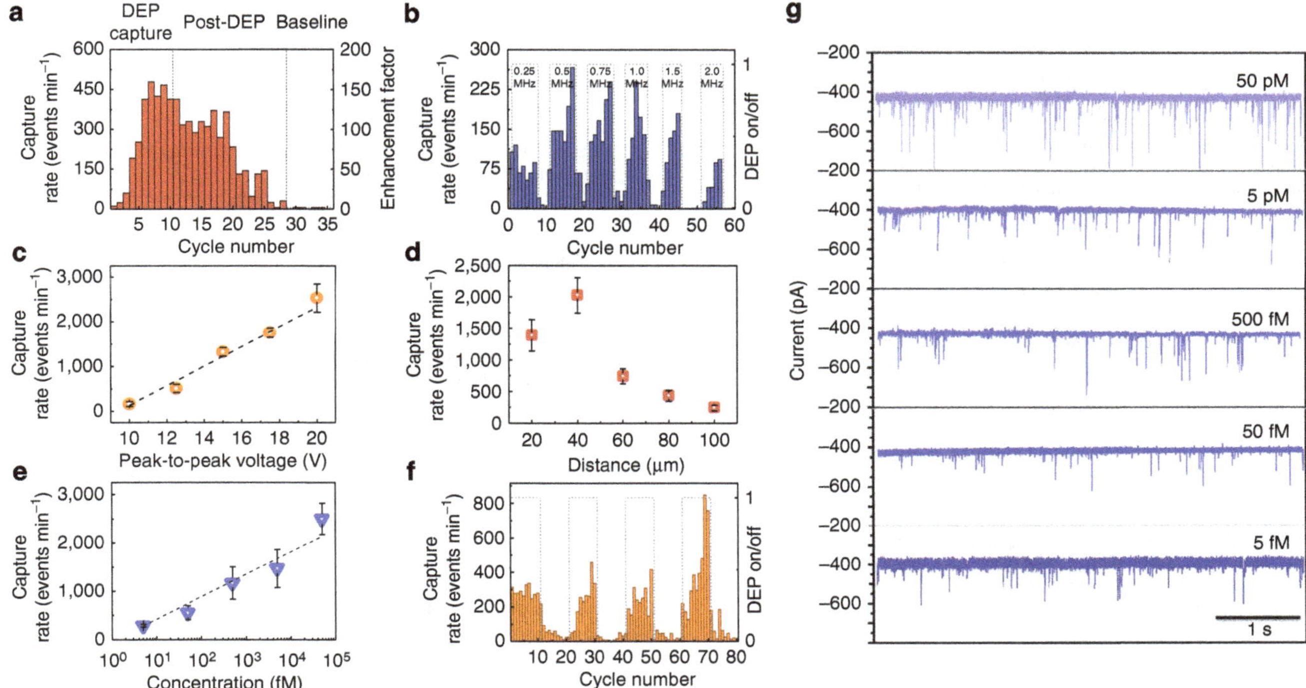

Figure 5 | Analysis of enhanced capture rates. (**a**) Capture rate per recording cycle where each recording cycle was 10 s in duration. The first 10 cycles were performed with 10 s of DEP pre-concentration, whereas the last 25 cycles were proceeded by 10 s of no trapping forces. (**b**) Capture rate per recording cycle where DEP pre-concentration was turned on/off. The DEP trapping frequency was also increased with each subsequent DEP pre-concentration phase (V_{pp} = 12 V). (**c**) Capture rate as a function of the peak-to-peak voltage used for the DEP trapping (f_{AC} = 1 MHz). (**d**) Capture rate as a function of the nanopipette tip-counter electrode gap distance (V_{pp} = 12 V, f_{AC} = 1 MHz). (**e**) Capture rate as a function of the 10 kbp DNA concentration. (**f**) Capture rate per recording cycle for a 5 fM DNA sample where DEP pre-concentration was turned on and off for four consecutive cycles (V_{pp} = 20 V, f_{AC} = 1 MHz and d_{gap} = 20 μm). (**g**) Current traces obtained using DEP as a pre-concentration step for five different concentrations (50 pM, 5 pM, 500 fM, 50 fM and 5 fM). A bias of $|\Delta V|$ = 500 mV was applied across the nanopore for all the experiments.

Table 1 | Comparison between methods to control capture rate.

	Nanopore capture rate enhancement technologies			
Pore type	**Mechanism**	**Factor increase**	**Event rate**	**References**
α-Hemolysin	Electrostatics at pore	10	0.1 s^{-1}nM^{-1}	14,15
Solid-state	Electric field enhancement	40	263 s^{-1}nM^{-1}	12
Nanopipette	Dielectrophoresis	80	1,040,000 s^{-1}nM^{-1}	This study

DNA. Achieving this would enable the slowing down of the molecule providing access to novel sub-molecular information. The proposed DEP-based method of pre-concentrating an analyte using a nanopipette could also be extended to other analytes (proteins, RNA) as well as other spectroscopies including SERS where concentration is a critical parameter[33–36]. Most importantly, this device can bridge a major technological gap in science where currently there are few rare event detection strategies. DEP electrodes can be integrated into other nanopore systems such as solid-state pores using an additional photolithography step to deposit gold or any other conductive material. Protein pores either suspended in a polymer membrane or integrated into a silicon-based membrane[37] could also benefit from DEP preconcentration. Finally, this platform opens up the door to single-cell experimentation whereby the sharp nanopipette tip can be used to efficiently extract low copy numbers of nucleic acids.

Methods

DNA labelling and imaging. Double stranded DNA with a length of 10 kbp and with a stock concentration of 500 μg ml^{-1} were obtained from New England Biolabs. DNA solutions (500 pM, 50 pM, 5 pM, 500 fM, 50 fM and 5 fM) were prepared by serial dilution. For fluorescence measurements, DNA was incubated with YOYO-1 (Molecular Probes) at a ratio of five base pairs per molecule. Images and video were collected by a × 60 water-immersion objective and directed to an electron multiplying CCD (emCCD) camera (Cascade II, Photometrics). The CCD camera has a pixel size of 16 μm, however, when used in conjunction with the × 60 objective, the final effective pixel size was 266 nm. In the fluorescence-based measurement, a constant AC field was applied to the outer gold layer of the nanopipette (no DC phase). The pipette was filled with the same buffer used in translocation recordings; however, it should be noted that the filling of the pipette with buffer does not affect the DEP forces as the gold electrode and the maximum field gradients exist just outside the nanopipette's tip.

Nanopipette fabrication. Nanopipettes were fabricated using a P-2000 laser puller (Sutter Instrument Co.) from quartz capillaries with an outer diameter of 1.0 mm and an inner diameter of 0.5 mm (QF100-50-7.5; Sutter Instrument Co). Nanopipettes were fabricated using a two-line protocol: (1) HEAT: 575; FIL: 3; VEL: 35; DEL: 145; PUL: 75, followed by (2) HEAT: 900; FIL: 2; VEL: 15; DEL: 128;

PUL: 200. It should be noted that the above parameters are instrument specific and were optimized to yield 25 nm openings at the tip of the nanopipette. Pipettes were then sputter coated for 60 s to produce a 5 nm coating of gold (Quorum Technologies; Q150R S) and used within several weeks of coating. In rare cases (approximately one in twenty pipettes) gold would delaminate from the pipette and this was observed optically by the DNA being attracted upstream from the tip to where the gold layer was still intact. It was more likely to see the delamination of gold from the second gold electrode: a glass slide coated with 5–10 nm of gold. The conical geometry of the pipette may have attributed to the stability of the gold, which we also observed to increase over time. In cases where pipettes had to be used immediately after gold coating, a thin layer of chromium (2 nm) could be used to increase the level of gold adhesion. Although thicker gold layers were initially tested and proved successful for DEP trapping, thinner gold layers were preferred since longer deposition times had a higher probability of blocking the pore. As for the lower limit of gold deposition, sub-5-nm gold layers had a lower success rate which was due to the lack of conformal coating and/or higher electrical resistance.

Single channel recordings. The ionic current was measured using an AxoPatch 200B patch-clamp amplifier (Molecular Devices, USA) in voltage clamp mode. The signal was filtered using a low-pass filter at 10 kHz and digitized with a Digidata 1,440 at a rate of 111 kHz and recorded using WINWCP software. WINWCP was used instead of pClamp because it allowed for synchronized triggering of both the AC and DC components. Data analysis was carried out using a custom-written MATLAB analysis routine. The baseline current was calculated via moving window for every data point. Event widths (dwell time) were obtained by measuring the full width half maximum of the current reduction. Current drop was calculated as current peak maximum after subtraction of the baseline current.

Numerical simulations. FEM computational models of the DEP experiments were created using COMSOL Multiphysics 4.3a along with the AC/DC module. An electrostatic DC approximation was used since the feature sizes of the nanopipette were significantly smaller than the wavelength of the applied AC field. In addition, a 2D-axisymmetric model was used to approximate the nanopipette at normal incidence to the planar electrode (in contrast to the actual 60° angle in experiments). COMSOL Multiphysics 5.0 along with the CFD module was used for ETF simulations, which are further discussed in the Supplementary Information. The threshold $|\nabla|E|^2|$ was determined from calculating a threshold force for trapping a particle of radius R required to overcome Brownian motion, which is given as $F_{\mathrm{th}} = \frac{k_B T}{2R}$ where k_B is the Boltzmann constant and T is the temperature of the particle/solution (298.15 K for all calculations). For the particle radius R, the hydrodynamic radius R_H for 10 kbp DNA was calculated to be 207 nm from the Einstein–Stokes equation[12,38]: $D = \frac{k_B T}{6\pi\eta_s R_H}$, where D is the diffusion constant for 10 kbp DNA (1.05×10^{-8} cm^{-2} s) and η_s is the viscosity of the solvent (water, 1.002 mPa s).

References

1. Kasianowicz, J. J., Brandin, E., Branton, D. & Deamer, D. W. Characterization of individual polynucleotide molecules using a membrane channel. *Proc. Natl Acad. Sci. USA* **93,** 13770–13773 (1996).
2. Clarke, J. *et al.* Continuous base identification for single-molecule nanopore DNA sequencing. *Nat. Nanotechnol.* **4,** 265–270 (2009).
3. Miles, B. N. *et al.* Single molecule sensing with solid-state nanopores: novel materials, methods, and applications. *Chem. Soc. Rev.* **42,** 15–28 (2013).
4. Rosen, C. B., Rodriguez-Larrea, D. & Bayley, H. Single-molecule site-specific detection of protein phosphorylation with a nanopore. *Nat. Biotechnol.* **32,** 179–181 (2014).
5. Freedman, K. J., Haq, S. R., Edel, J. B., Jemth, P. & Kim, M. J. Single molecule unfolding and stretching of protein domains inside a solid-state nanopore by electric field. *Sci. Rep.* **3,** 1638 (2013).
6. Japrung, D. *et al.* Single-molecule studies of intrinsically disordered proteins using solid-state nanopores. *Anal. Chem.* **85,** 2449–2456 (2013).
7. Shasha, C. *et al.* Nanopore-based conformational analysis of a viral RNA drug target. *ACS Nano* **8,** 6425–6430 (2014).
8. Squires, T. M., Messinger, R. J. & Manalis, S. R. Making it stick: convection, reaction and diffusion in surface-based biosensors. *Nat. Biotechnol.* **26,** 417–426 (2008).
9. Plesa, C., van Loo, N., Ketterer, P., Dietz, H. & Dekker, C. Velocity of DNA during translocation through a solid-state nanopore. *Nano Lett.* **15,** 732–737 (2014).
10. Cressiot, B. *et al.* Protein transport through a narrow solid-state nanopore at high voltage: experiments and theory. *ACS Nano* **6,** 6236–6243 (2012).
11. Kasianowicz, J. J. Nanopores: flossing with DNA. *Nat. Mater.* **3,** 355–356 (2004).
12. Wanunu, M., Morrison, W., Rabin, Y., Grosberg, A. Y. & Meller, A. Electrostatic focusing of unlabelled DNA into nanoscale pores using a salt gradient. *Nat. Nanotechnol.* **5,** 160–165 (2010).
13. Ivanov, A. P. *et al.* On-demand delivery of single DNA molecules using nanopipets. *ACS Nano* **9,** 3587–3595 (2015).
14. Maglia, G., Restrepo, M. R., Mikhailova, E. & Bayley, H. Enhanced translocation of single DNA molecules through α-hemolysin nanopores by manipulation of internal charge. *Proc. Natl Acad. Sci. USA* **105,** 19720–19725 (2008).
15. Laszlo, A. H. *et al.* Decoding long nanopore sequencing reads of natural DNA. *Nat. Biotechnol.* **32,** 829–833 (2014).
16. Lu, B. *et al.* Pressure-controlled motion of single polymers through solid-state nanopores. *Nano Lett.* **13,** 3048–3052 (2013).
17. Jose, J. *et al.* Individual template-stripped conductive gold pyramids for tip-enhanced dielectrophoresis. *ACS Photonics* **1,** 464–470 (2014).
18. Yeo, W.-H. *et al.* Dielectrophoretic concentration of low-abundance nanoparticles using a nanostructured tip. *Nanotechnology* **23,** 485707 (2012).
19. Siwy, Z. *et al.* Protein biosensors based on biofunctionalized conical gold nanotubes. *J. Am. Chem. Soc.* **127,** 5000–5001 (2005).
20. Siwy, Z., Heins, E., Harrell, C. C., Kohli, P. & Martin, C. R. Conical-nanotube ion-current rectifiers: the role of surface charge. *J. Am. Chem. Soc.* **126,** 10850–10851 (2004).
21. Lan, W.-J., Holden, D. A. & White, H. S. Pressure-dependent ion current rectification in conical-shaped glass nanopores. *J. Am. Chem. Soc.* **133,** 13300–13303 (2011).
22. Siwy, Z. & Fuliński, A. Fabrication of a synthetic nanopore ion pump. *Phys. Rev. Lett.* **89,** 198103 (2002).
23. Wei, C., Bard, A. J. & Feldberg, S. W. Current rectification at quartz nanopipet electrodes. *Anal. Chem.* **69,** 4627–4633 (1997).
24. Sze, J. Y., Kumar, S., Ivanov, A. P., Oh, S.-H. & Edel, J. B. Fine tuning of nanopipettes using atomic layer deposition for single molecule sensing. *Analyst* **140,** 4828–4834 (2015).
25. Tabard-Cossa, V., Trivedi, D., Wiggin, M., Jetha, N. N. & Marziali, A. Noise analysis and reduction in solid-state nanopores. *Nanotechnology* **18,** 305505 (2007).
26. Belaidi, S., Girard, P. & Leveque, G. Electrostatic forces acting on the tip in atomic force microscopy: modelization and comparison with analytic expressions. *J. Appl. Phys.* **81,** 1023–1030 (1997).
27. Bakewell, D. J., Ermolina, I., Morgan, H., Milner, J. & Feldman, Y. Dielectric relaxation measurements of 12 kbp plasmid DNA. *Biochim. Biophys. Acta* **1493,** 151–158 (2000).
28. Eggeling, C., Widengren, J. & Rigler, R. Seidel CAM. Photobleaching of fluorescent dyes under conditions used for single-molecule detection: evidence of two-step photolysis. *Anal. Chem.* **70,** 2651–2659 (1998).
29. Giloh, H. & Sedat, J. W. Fluorescence microscopy: reduced photobleaching of rhodamine and fluorescein protein conjugates by n-propyl gallate. *Science* **217,** 1252–1255 (1982).
30. Regtmeier, J., Eichhorn, R., Bogunovic, L., Ros, A. & Anselmetti, D. Dielectrophoretic trapping and polarizability of DNA: the role of spatial conformation. *Anal. Chem.* **82,** 7141–7149 (2010).
31. Sasaki, N., Kitamori, T. & Kim, H. B. Fluid mixing using AC electrothermal flow on meandering electrodes in a microchannel. *Electrophoresis* **33,** 2668–2673 (2012).
32. Morgan, H. & Green, N. G. *AC Electrokinetics: Colloids and Nanoparticles* (Research Studies Press, 2003).
33. Cecchini, M. P. *et al.* Rapid ultrasensitive single particle surface-enhanced raman spectroscopy using metallic nanopores. *Nano Lett.* **13,** 4602–4609 (2013).
34. Ivanov, A. P., Freedman, K. J., Kim, M. J., Albrecht, T. & Edel, J. B. High precision fabrication and positioning of nanoelectrodes in a nanopore. *ACS Nano* **8,** 1940–1948 (2014).
35. Barik, A. *et al.* Dielectrophoresis-enhanced plasmonic sensing with gold nanohole arrays. *Nano Lett.* **14,** 2006–2012 (2014).
36. Brolo, A. G., Arctander, E., Gordon, R., Leathem, B. & Kavanagh, K. L. Nanohole-enhanced raman scattering. *Nano Lett.* **4,** 2015–2018 (2004).
37. Hall, A. R. *et al.* Hybrid pore formation by directed insertion of α-haemolysin into solid-state nanopores. *Nat. Nanotechnol.* **5,** 874–877 (2010).
38. Hirano, K. *et al.* Plasmonic imaging of Brownian motion of single DNA molecules spontaneously binding to Ag nanoparticles. *Nano Lett.* **13,** 1877–1882 (2013).

Acknowledgements

K.J.F. acknowledges support of the Whitaker International Program by the IIE. J.B.E. acknowledges BBSRC and the receipt of an ERC starting investigator grant. L.M.O., A.B. and S.-H.O. acknowledge support from the U.S. National Science Foundation (NSF CAREER Award) and the MnDRIVE Initiative from the State of Minnesota. L.M.O. acknowledges support from the National Institutes of Health biotechnology training grant (T32 GM008347) and the NSF Graduate Research Fellowship Program. Computational modelling using COMSOL Multiphysics 4.3a was performed through the

University of Minnesota Supercomputing Institute. Finally, we thank Ulrich Keyser for useful discussions.

Author contributions

J.B.E., S.-H.O., K.J.F. and A.P.I. designed and supervised the research. K.J.F., L.M.O., A.P.I. and A.B. conducted the experiments. All the authors contributed to the analysis of the data and writing of the manuscript.

Additional information

Competing financial interests: The authors declare no competing financial interests.

7

Manipulation of prenylation reactions by structure-based engineering of bacterial indolactam prenyltransferases

Takahiro Mori[1], Lihan Zhang[1], Takayoshi Awakawa[1], Shotaro Hoshino[1], Masahiro Okada[1], Hiroyuki Morita[2] & Ikuro Abe[1]

Prenylation reactions play crucial roles in controlling the activities of biomolecules. Bacterial prenyltransferases, TleC from *Streptomyces blastmyceticus* and MpnD from *Marinactinospora thermotolerans*, catalyse the 'reverse' prenylation of (−)-indolactam V at the C-7 position of the indole ring with geranyl pyrophosphate or dimethylallyl pyrophosphate, to produce lyngbyatoxin or pendolmycin, respectively. Using *in vitro* analyses, here we show that both TleC and MpnD exhibit relaxed substrate specificities and accept various chain lengths (C_5-C_{25}) of the prenyl donors. Comparisons of the crystal structures and their ternary complexes with (−)-indolactam V and dimethylallyl *S*-thiophosphate revealed the intimate structural details of the enzyme-catalysed 'reverse' prenylation reactions and identified the active-site residues governing the selection of the substrates. Furthermore, structure-based enzyme engineering successfully altered the preference for the prenyl chain length of the substrates, as well as the regio- and stereo-selectivities of the prenylation reactions, to produce a series of unnatural novel indolactams.

[1] Graduate School of Pharmaceutical Sciences, The University of Tokyo, 7-3-1 Hongo, Bunkyo-ku, Tokyo 113-0033, Japan. [2] Department of Medicinal Resources, Institute of Natural Medicine, University of Toyama, 2630-Sugitani, Toyama 930-0194, Japan. Correspondence and requests for materials should be addressed to H.M. (email: hmorita@inm.u-toyama.ac.jp) or to I.A. (email: abei@mol.f.u-tokyo.ac.jp.).

As prenylated molecules usually exhibit improved interactions with proteins and biological membranes, prenylation reactions catalysed by prenyltransferases (PTs) play crucial roles in controlling the activities of biomolecules, as exemplified by the biosyntheses of Vitamin K and menaquinones in primary metabolism[1], the posttranslational modifications of peptides[2,3] and proteins[4], as well as the production of various biologically active secondary metabolites in plants, fungi and bacteria[5–9]. This is also the case for the indole alkaloids, which include the specific cell cycle inhibitors tryprostatins[5], the ergot alkaloids anti-migraine drugs[6] and the potent protein kinase C activator teleocidins[7] (Fig. 1). In their biosyntheses, soluble indole PT superfamily enzymes, which are the most common aromatic PTs in bacteria and fungi, usually select the dimethylallyl pyrophosphate (DMAPP) as the prenyl donor to catalyse the Friedel–Crafts alkylation of the indole ring, thereby contributing to the structural diversification and exertion of their prominent biological activities[10–13]. In nature, enzymes that regioselectively catalyse the prenylation of all of the possible positions of the indole ring have been identified[10–13] (Supplementary Figs 1 and 2). There are two types of the prenyltransfer reactions: 'normal' prenylation (attack at the C-1 primary centre of the prenyl carbocation) and 'reverse' prenylation (attack at the C-3 tertiary centre of the prenyl carbocation). These indole PT enzymes share the αββα (ABBA) barrel overall structural fold; however, the structure–function relationships of the enzymes still remained to be elucidated and the structure-based rational control of the enzyme reactions has been awaited.

We recently reported the biosynthetic gene cluster for teleocidin B, a unique prenylated indolactam with a monoterpenoid moiety fused with C-6 and C-7 of the indole ring[14] (Fig. 1a). The gene cluster from *S. blastmyceticus* encodes an indolactam PT, TleC, which is a 42-kDa protein consisting of 391 amino acids that shares 40% and 38% identities with LtxC from the cyanobacterium *Moorea producens*[15] and MpnD from the deep sea bacterium *M. thermotolerans*[16], respectively (Supplementary Fig. 3). Heterologous expression in *Streptomyces lividans* TK21 and an *in vivo* analysis revealed that TleC catalyses the 'reverse' prenylation of the cyclic dipeptide (−)-indolactam V (**1**) at the C-7 position of the indole ring with geranyl pyrophosphate (GPP) (C_{10}) to produce lyngbyatoxin A (**2**), with the C19*R* configuration, as in the case of LtxC[15] (Fig. 1a). MpnD also catalyses the 'reverse' prenylation of **1** at the C-7 position, with DMAPP (C_5) to yield pendolmycin[16] (**3**) (Fig. 1b). The two bacterial PTs thus yield products with different prenyl chain lengths from the common indolactam. Furthermore, notably, the 'reverse' prenylation at the C-7 position of the indole ring is limited only to the enzyme reactions catalysed by TleC, LtxC and MpnD, although several 'reverse' prenylation enzymes have been reported[17–19].

Here we report the biochemical characterization, crystal structures and structure-based engineering of *S. blastmyceticus* TleC and *M. thermotolerans* MpnD. Interestingly, the indolactam-specific TleC and MpnD both accepted different chain lengths (from C_5 to C_{25}) of prenyl pyrophosphate *in vitro*. Comparisons of the crystal structures revealed the intimate structural details of the enzyme-catalysed reverse prenylation reactions. Furthermore, we succeeded in structure-based enzyme engineering, to rationally control the preference for the chain lengths of the prenyl donors, as well as the regio- and stereo-selectivities of the enzyme reactions, to generate unnatural novel indolactams.

Results

***In vitro* characterizations of TleC and MpnD**. *S. blastmyceticus* TleC and *M. thermotolerans* MpnD were heterologously expressed in *Escherichia coli* as His-tagged fusion proteins and purified to homogeneity by Ni-chelate affinity chromatography. Although most of the aromatic PTs and soluble indole PTs are metal ion-independent enzymes, divalent cations such as Mg^{2+}, Ca^{2+} and Co^{2+} are known to affect the enzyme activities[10–13]. Notably, a recent study on an insect isoprenyl diphosphate synthase demonstrated that the prenyl chain length of the enzyme reaction products is controlled by metal ions[20]. Therefore, we first investigated the metal ion dependency of the enzyme reactions, by using (−)-indolactam V and GPP/DMAPP as substrates in the presence of various metal ions (Mg^{2+}, Ca^{2+}, Co^{2+}, Ni^{2+}, Mn^{2+}, Fe^{2+}, Zn^{2+} or Cu^{2+}). The results demonstrated that the presence of metal ions did not change the product profile of the enzyme reactions but the presence of Mg^{2+} and Ca^{2+} slightly enhanced (<10%) their activities. Furthermore, the enzyme activities were not influenced by the addition of the chelating agent EDTA (Supplementary Fig. 4). These results clearly indicated that both TleC and MpnD are also metal ion-independent enzymes, with activities slightly enhanced by the addition of Mg^{2+} and Ca^{2+}, as in the cases of other indole PTs[10–13].

Next, to investigate the substrate specificities of TleC and MpnD, the purified recombinant enzymes were incubated with

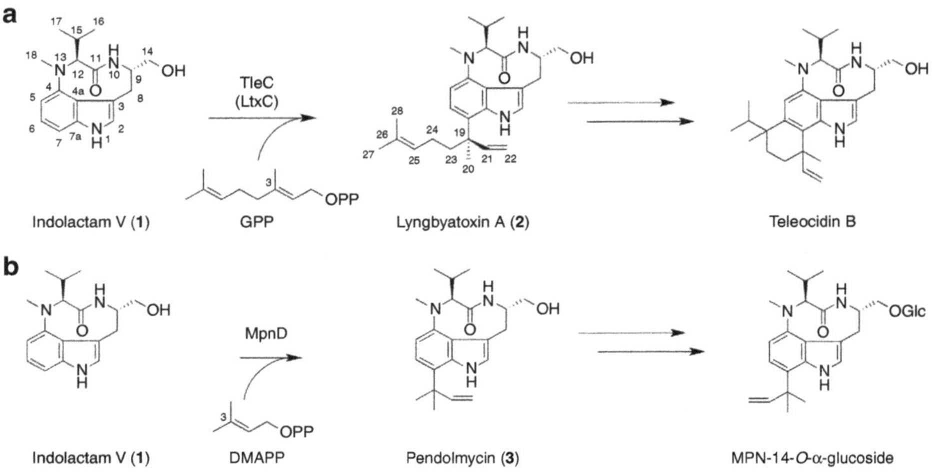

Figure 1 | Reaction schemes of indole prenyltransferases. Reaction schemes for the conversion of (−)-indolactam V (**1**) into (**a**) lyngbyatoxin A (**2**) by *S. blastmyceticus* TleC and *M. producens* LtxC, and (**b**) pendolmycin (**3**) by *M. thermotolerans* MpnD.

several indole derivatives and prenyl pyrophosphates with various chain lengths as substrates (Supplementary Fig. 5). The liquid chromatography–mass spectrometry (LC–MS) analysis of the enzyme reaction products demonstrated that both enzymes exhibit strict substrate specificity towards the prenyl acceptors. Thus, they are indolactam-specific and do not accept indole, 4-aminoindole, 4-methoxyindole, indole acetic acid, L-tryptophan, harman and brevianamide F as a prenyl acceptor. In contrast, both TleC and MpnD exhibited relaxed substrate specificities towards the prenyl donors. First, the indolactam-specific TleC and MpnD both accepted DMAPP (C_5) and GPP (C_{10}) as substrates to produce pendolmycin (**3**), and lyngbyatoxin A (**2**) and 5-geranylindolactam V (**4**) (**2**:**4** = 26:1 for TleC; **2**:**4** = 1:3 for MpnD), respectively (Fig. 2). The structure of **3** and **2** were determined by comparisons of the NMR data with the reported data. TleC exhibited 23-fold higher k_{cat}/K_M value for the C_{10} prenylation activity over the C_5 prenylation activity. In contrast, MpnD showed 109-fold higher activity for the C_5 over the C_{10} (Table 1). Thus, TleC and MpnD clearly prefer to accept GPP and DMAPP as the prenyl donors, respectively. Meanwhile, the high-resolution-mass spectrometry (HR-MS) of **4** revealed a 136-Da larger molecular weight than that of **1**, suggesting the presence of a geranyl group in compound **4**. The ^{1}H-NMR signals at δ_H 5.15 (1H, *t*, J = 6.8 Hz) and 5.30 (1H, *t*, J = 6.4 Hz) of the geranyl moiety and δ_H 7.04 (1H, *d*, J = 2.0 Hz), 7.12 (1H, *d*, J = 8.0 Hz), 7.24 (1H, *d*, J = 8.0 Hz) and 8.26 (1H, *brs*) of the indole ring indicated that GPP is attached to the C-5 or C-7 position of the indolactam in the 'normal' manner (the key signals due to the methylidene group in the 'reverse' prenylated structure are missing). Further ^{1}H-detected multi-bond heteronuclear multiple quantum coherence spectra revealed that H-19a and H-19b (δ_H 3.73 and δ 3.55, 1H, *dd*, J = 6.5 Hz) correlated with C-4 (δ_c 141.8), indicating prenylation at the C-5 position (Supplementary Fig. 6a). The structure of **4** was thus determined to be 5-geranylindolactam V (Fig. 2).

The relaxed substrate specificity of the enzymes prompted us to test longer chain prenyl donors as substrates. Interestingly, both TleC and MpnD also accepted farnesyl pyrophosphate (FPP) (C_{15}), geranylgeranyl pyrophosphate (GGPP) (C_{20}) and geranylfarnesyl pyrophosphate (GFPP) (C_{25}) as prenyl donors, to produce **5**, **6** and **7**, respectively (Fig. 2 and Supplementary Fig. 7). HR-MS indicated that **5** was $C_{32}H_{46}N_3O_2$, corresponding to the farnesylated **1**. The ultraviolet and NMR spectra of **5** were quite similar to those of **4**, except for the presence of an additional isoprenyl unit (δ_H 1.59, 1.97, 2.06, and 5.09; δ_C 17.7, 26.8, 39.8, 124.4, and 131.3) (Supplementary Figs 6b and 8). Thus, the structure of **5** was determined to be 5-farnesylindolactam V. In a similar manner, **6** and **7** are postulated to be 5-geranylgeranylindolactam V and 5-geranylfarnesylindolactam V, respectively (Fig. 2). In contrast to typical indole PTs that exhibit broad substrate specificities towards the prenyl acceptors but with strict specificities towards the prenyl donors[10–13], it is remarkable that the indolactam-specific TleC and MpnD exhibited unusually relaxed substrate specificities towards the prenyl donors and accepted different chain lengths (from C_5 to C_{25}) of prenyl pyrophosphate *in vitro*. Although TleC and MpnD predominantly catalyse the 'reverse' prenylation of (−)-indolactam V at the C-7 position of the indole ring with GPP (C_{10}) or DMAPP (C_5), respectively, these enzymes also catalysed the C-5 'normal' prenylation *in vitro* when longer chain prenyl donors (C_{15}, C_{20} and C_{25} in TleC and C_{10}, C_{15}, C_{20} and C_{25} in MpnD) were used as substrates. These results suggested that the binding modes of the longer chain substrates are different from those of the physiological substrates. An indole PT BAE61387 from *Aspergillus oryzae* DSM1147 reportedly catalyses the prenylation of up to the C_{15} FPP[21]. However, to the best our knowledge, this is the first demonstration that indole PT enzymes catalyse the C_{20} and C_{25} prenylation reactions.

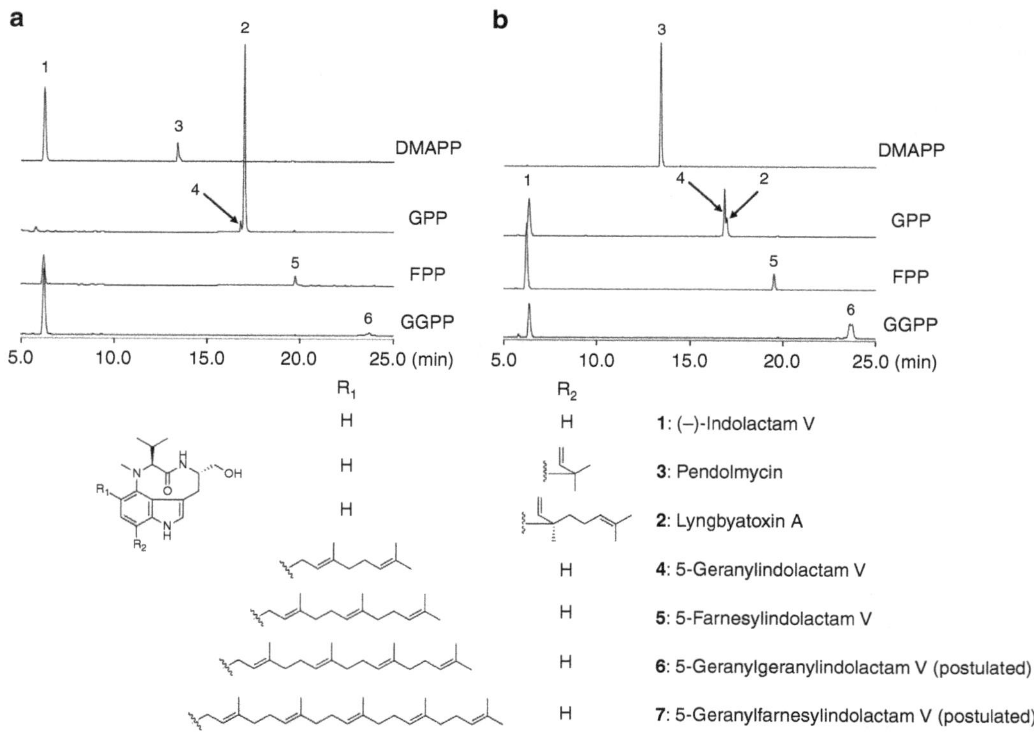

Figure 2 | The *in vitro* enzyme reactions of *S. blastmyceticus* TleC and *M. thermotolerans* MpnD. HPLC elution profiles of the enzyme reaction products of (**a**) TleC and (**b**) MpnD from (−)-indolactam V and prenyl pyrophosphates with various chain lengths as substrates.

Table 1 | The steady-state enzyme kinetics values of the wild-type and mutants of (A) *S. blastmyceticus* TleC and (B) *M. thermotolerans* MpnD.

(A) TleC	k_{cat} (min)	K_M (μM)	k_{cat}/K_M ($s^{-1}mM^{-1}$)	Product
TleC WT				
DMAPP	1.5 ± 0.1	21.7 ± 4.3	1.1	**3**
GPP	15.0 ± 1.5	9.5 ± 3.3	26.4	**2**:**4** = 26:1
TleC W97Y				
DMAPP	24.1 ± 0.8	11.3 ± 1.6	35.5	**3**
GPP	10.1 ± 0.4	7.7 ± 1.0	21.8	**2**:**4** = 24:1
TleC A173M				
DMAPP	3.9 ± 0.1	24.6 ± 3.0	2.6	**3**
GPP	n.d.	n.d.	ND	**2**:**4**:**8** = 1:0.4:0.6
TleC W97Y/A173M				
DMAPP	2.0 ± 0.1	41.2 ± 5.6	0.8	**3**
GPP	0.7 ± 0.1	34.4 ± 9.3	0.3	**2**:**4**:**8** = 1:0.6:2.3
TleC W97Y/F170W/A173M				
DMAPP	2.6 ± 0.2	46.6 ± 8.2	0.9	**3**
GPP	0.7 ± 0.0	46.8 ± 6.7	0.3	**2**:**4**:**8** = 1:1:2.9
(B)				
MpnD WT				
DMAPP	52.9 ± 2.6	4.8 ± 1.0	183.0	**3**
GPP	2.6 ± 0.2	26.0 ± 4.9	1.7	**2**:**4** = 1:3
FPP	5.2 ± 0.5	95.5 ± 17.1	0.9	**5**
MpnD Y80W				
DMAPP	7.1 ± 0.3	15.3 ± 2.5	7.7	**3**
GPP	5.4 ± 0.1	9.7 ± 1.1	9.3	**2**:**4** = 1:2.3
FPP	4.8 ± 0.7	148.3 ± 46.8	0.5	**5**
MpnD M159A				
DMAPP	30.3 ± 2.0	24.2 ± 4.2	20.9	**3**
GPP	4.3 ± 0.2	9.1 ± 1.2	7.9	**2**:**4** = 15.6:1
FPP	10.0 ± 0.9	48.7 ± 14.6	3.4	**5**
MpnD Y80W/M159A				
DMAPP	4.5 ± 0.2	4.3 ± 0.7	17.0	**3**
GPP	5.8 ± 0.7	33.0 ± 11.0	2.9	**2**:**4** = 1:1.9
FPP	5.6 ± 0.4	96.5 ± 14.3	1.0	**5**

DMAPP, dimethylallyl pyrophosphate; FPP, farnesyl pyrophosphate; GPP, geranyl pyrophosphate.

Overall structures of TleC and MpnD. To understand the intimate structural details of the enzyme reactions, we solved the X-ray crystal structures of TleC and MpnD as the apo structures (at 2.0 and 1.6 Å resolutions, respectively) and the ternary complexes with (−)-indolactam V and a DMAPP analogue dimethylallyl *S*-thiophosphate (DMSPP) (at 2.1 and 1.4 Å resolutions, respectively) (Table 2). The overall structures of TleC and MpnD adopt the ABBA-fold, which consists of ten antiparallel β-strands assembled into a circular β-barrel, surrounded by a ring of solvent-exposed α-helices, as observed in the structures of indole PTs and aromatic PTs[22–27] (Supplementary Fig. 9). Furthermore, the overall structures of TleC and MpnD are highly homologous to each other, with a root mean square deviation (RMSD) of 1.9 Å for the Cα-atoms. The structural comparison using the Dali programme revealed that the TleC structure shares high similarities with those of the known indole PTs, including *Neosartorya fischeri* AnaPT[24], *Aspergillus fumigatus* FgaPT2 (ref. 25) and *A. fumigatus* FtmPT1 (ref. 26), with RMSD values of 2.7, 2.8 and 2.8 Å for the Cα-atoms, respectively. MpnD is also quite similar to these enzymes, with RMSD values of 2.9, 3.0 and 2.8 Å, respectively. A structure-based sequence comparison indicated that TleC and MpnD share low sequence identity/similarity to FgaPT2 (TleC: 14/27%, MpnD: 12/24%), FtmPT1 (TleC: 12/24%, MpnD: 15/25%), CdpNPT (TleC: 12/25%, MpnD: 13/25%) and AnaPT (TleC: 16/29%, MpnD: 15/26%) (Supplementary Fig. 10). The comparison of overall structures of TleC, MpnD and the structurally characterized other indole PTs revealed that the indole PT's conserved amino-terminal α-helices, locating outside of β-barrel, are converted into loops and cover the centre of circular β-barrel in TleC and MpnD. Furthermore, TleC and MpnD also lack the two helices between α4 and β3, and β4 and α5, corresponding to α6 and α7 in the other indole PTs.

Substrate binding. The ternary complex structures revealed the binding modes of their substrates. TleC binds the (−)-indolactam V (**1**) and DMSPP in the centre of the β-barrel and the catalytic Glu106 protrudes towards the N-1 of the indole ring, as in the cases of other indole PTs[24–27] (Fig. 3a and Supplementary Figs 11a and 12a). A hydrogen bond network is formed among Glu28, Glu106, Lys297 and Asn367 neighbouring **1**, in which Glu106 and Asn367 hydrogen bond with the N-1 of **1**, while Lys297 and Glu28 hydrogen bond with the oxygen atom of the 14-OH of **1**. Furthermore, **1** is also fixed by hydrogen bond networks among the C-11 carbonyl oxygen, the N-10 of **1**, the main chain amine of Ile26, the N-1 of Trp97 and the main chain carbonyl of Pro298 via two water molecules W1 and W2.

Table 2 | Data collection, phasing and refinement statistics.

	TleC SeMet	TleC apo	TleC complex	MpnD apo	MpnD complex
Data collection					
Space group	$P2_12_12_1$	$P2_12_12_1$	$P2_12_12_1$	$P2_12_12_1$	$P2_12_12_1$
Cell dimensions					
a, b, c (Å)	41.9, 73.9, 110.4	42.0, 74.1, 111.0	42,4, 74.1, 112.2	46.7, 53.7, 132.7	46.8, 54.1, 132.2
Resolution (Å)	50.0-2.1 (2.23-2.11)*	50.0-1.95 (2.07-1.95)*	50.0-2.1 (2.22-2.10)*	50.0-1.6 (1.70-1.60)*	50.0-1.4 (1.48-1.40)*
R_{merge} (%)	10.7 (70.4)	11.4 (73.5)	8.4 (51.9)	4.9 (31.5)	4.9 (45.2)
$I/\sigma I$	15.4 (3.3)	13.4 (3.1)	22.7 (4.4)	25.0 (5.8)	22.8 (4.0)
Completeness (%)	99.4 (96.5)	99.5 (97.1)	99.8 (98.6)	99.6 (98.3)	99.7 (99.2)
Redundancy	7.7 (7.4)	7.2 (7.2)	7.2 (7.1)	7.2 (7.0)	6.4 (6.3)
Refinement					
Resolution (Å)	44.2-2.1	44.4-2.0	44.7-2.1	34.1-1.6	41.9-1.4
No. reflections	20415	25878	21401	44640	66871
R_{work}/R_{free} (%)	18.8/23.5	18.6/23.2	18.9/23.7	18.0/20.4	17.8/20.3
No. atoms					
Protein	2606	2652	2766	2791	2816
Ligand/ion	—	—	36	—	51
Water	125	224	142	363	429
B-factors					
Protein	30.5	26.6	28.6	18.5	15.6
Ligand/ion	—	—	25.1	—	20.8
Water	34.3	32.4	30.7	27.3	25.6
Root mean square deviations					
Bond lengths (Å)	0.009	0.007	0.009	0.006	0.009
Bond angles (°)	1.211	1.032	1.182	1.127	1.247

*One crystal was used for data collection. Values in parentheses are for highest-resolution shell.

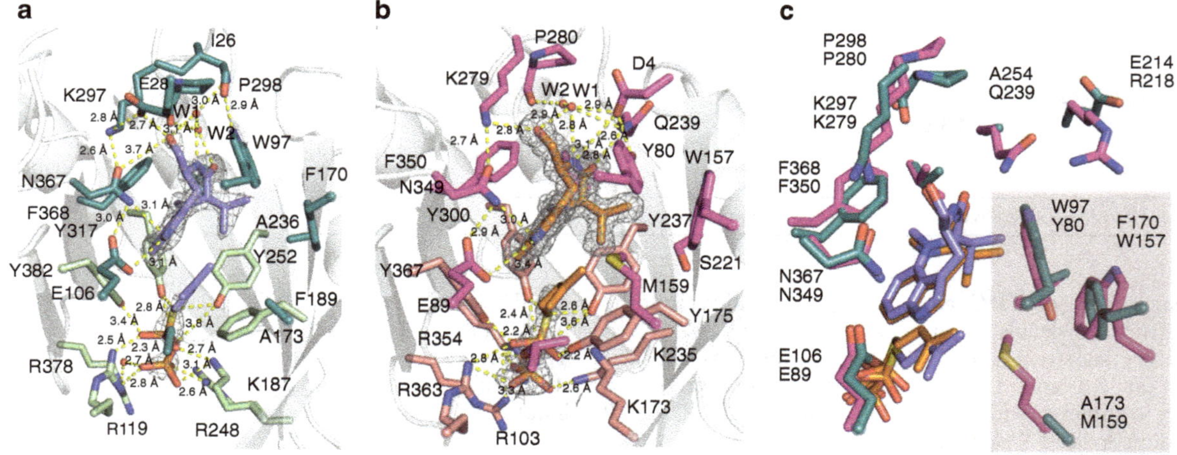

Figure 3 | Comparison of the substrate binding modes in *S. blastmyceticus* TleC and *M. thermotolerans* MpnD. Closeup views of the active site architectures in (**a**) the TleC-**1**-DMSPP ternary complex and (**b**) the MpnD-**1**-DMSPP ternary complex. (**c**) Superimposition of the active site residues of TleC and MpnD. The key three residues discussed here are highlighted by red squares. (−)-Indolactam V (**1**) and DMSPP are depicted by blue and orange stick models, respectively. The amino acid residues located in the pyrophosphate and indolactam-binding sites are represented by limegreen and green stick models in TleC, and salmon and magenta stick models in MpnD, respectively. The $F_o - F_c$ electron density maps of the **1** and DMSPP are represented as a black mesh, contoured at $+2.5\sigma$. Dashed yellow lines represent hydrogen bonds. Red spheres depict the water molecules.

In contrast, the α and β-phosphates of DMSPP are anchored by positively charged residues, Arg119, Lys187, Arg248 and Arg378. In addition, the α-phosphate is held by the so-called 'tyrosine shield' composed of Tyr252, Tyr317 and Tyr382. These tyrosines have been proposed to play an important role in stabilizing the dimethylallyl cation during the enzyme reaction[24–27]. Finally, the dimethylallyl moiety is sandwiched between the indole ring of **1** and the benzene ring of Phe189, and is stabilized by π-π interactions.

The active site architecture of the ternary structure revealed that MpnD binds **1** and DMSPP in a similar manner to that of TleC, even though these enzymes share only moderate amino acid identities (Fig. 3b and Supplementary Figs 11b and 12b). In addition to Glu89, Lys279, Pro280, Asn349 and the two structurally conserved water molecules (corresponding to Glu106, Lys297, Pro298, Asn367, W1 and W2 in TleC, respectively), Asp4, Tyr80 and Gln239 participate in the tight hydrogen bond network with **1**. In MpnD, the two water molecules, along

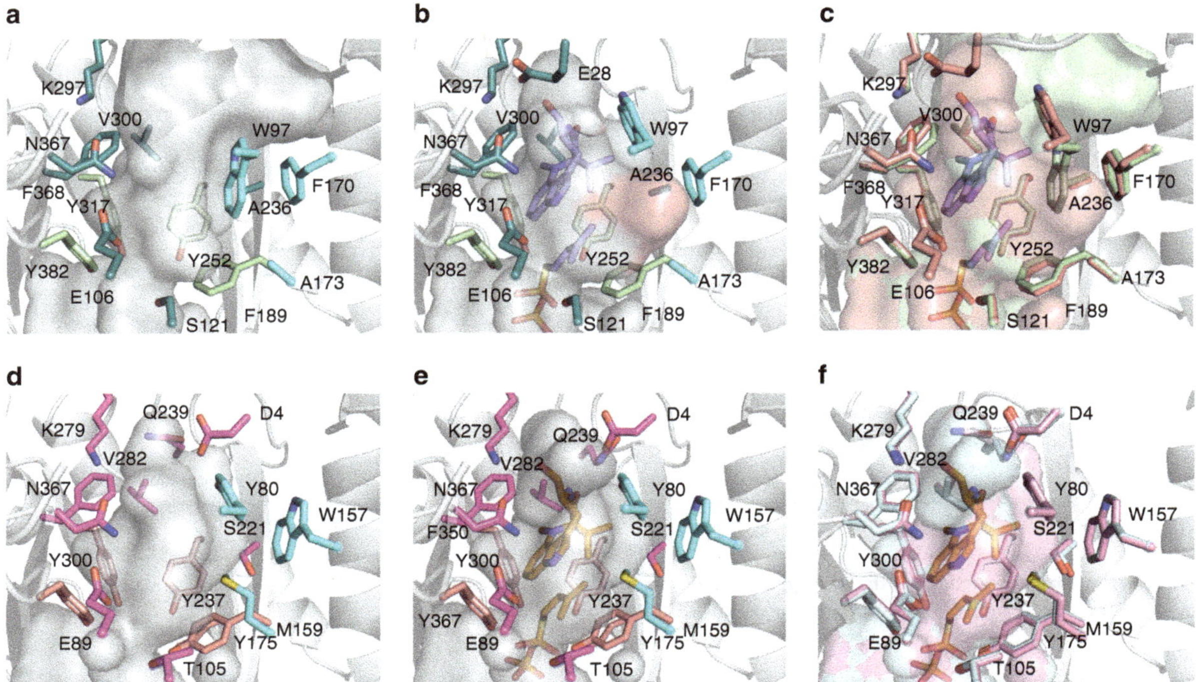

Figure 4 | Comparison of the active site structures of *S. blastmyceticus* TleC and *M. thermotolerans* MpnD. Closeup views of the active site cavities of (**a**) the TleC apo structure and (**b**) the TleC complex structure with **1** and DMSPP, (**c**) superimposed view of apo and complex structure of TleC, (**d**) the MpnD apo structure, (**e**) the MpnD complex structure with **1** and DMSPP, and (**f**) superimposed view of apo and complex structure of MpnD. (−)-Indolactam V (**1**) and DMSPP are depicted by a cyan and blue stick models, respectively. The amino acid residues located in the pyrophosphate and indolactam-binding sites are represented by limegreen and green stick models in TleC, and magenta and salmon stick models in MpnD, respectively. The three important residues discussed in the main text are depicted by cyan stick models. The 'prenyl binding pocket' in the TleC-**1**-DMSPP ternary complex structure is highlighted by a red surface. The residues and surface of apo structure and complex structure of TleC are represented by palegreen and salmon, respectively. The residues and surface of apo structure and complex structure of MpnD are represented by pink and pale cyan, respectively.

with Asp4, Tyr80 and Gln239, form further hydrogen bonds with the N-10, C-11 carbonyl and 14-OH of **1**, to tightly accommodate the indolactam. In contrast, in the previously reported crystal structures of indole PTs, which exhibited relatively broad substrate specificity for prenyl acceptors, the indole substrate is only fixed by a few hydrogen bonds, in addition to a hydrogen bond with highly conserved glutamic acid, while the pyrophosphate is anchored tightly by numerous salt bridges, electrostatic interactions and hydrogen bond networks[25,27]. For example, the structure of CdpNPT complexed with (*S*)-benzodiazepinedione revealed that only hydroxyl group of Thr108 construct a hydrogen bond with O-13′ (ref. 27). Furthermore, the structure of FgaPT2 complexed with substrate L-tryptophan also indicated that the indole substrate is only fixed by hydrogen bonds with side chain of Arg244, main chain of Ile80 and Leu81 (ref. 25). Thus, the presence of the tight hydrogen bond networks in TleC and MpnD would explain why they are both indolactam specific and do not accept other indole derivatives as the prenyl acceptor substrate. In contrast, as in the case of TleC, the dimethylallyl moiety is sandwiched between the indole ring of **1** and the benzene ring of Tyr175 in MpnD. The binding mode of DMAPP is thus quite similar; however, the amino acid residues lining the prenyl binding pocket Trp97, Phe170 and Ala173 in TleC are uniquely substituted with Tyr80, Trp157 and Met159, respectively, in MpnD (Figs 3c and 4, and Supplementary Figs 11–13). Notably, Trp97 in TleC and the corresponding Tyr80 in MpnD are located on the so-called 'substrate-binding loop' between α2 and β1, which was proposed to be important for substrate binding by forming a hydrogen bond network between the substrate and the backbone oxygens of Leu94 and Met94 in FgaPT2 and FtmPT1, respectively[25,26].

Remarkably, a comparison of the active site architectures of the TleC apo structure and the ternary complex structure with substrates revealed that the TleC-specific Trp97 rotated by about 70° when the substrates bound to the active site (Fig. 4a–c). This rotation generated a new large pocket to accommodate the prenyl side chain of the C_{10} GPP substrate in TleC, but this pocket is absent in the active site of MpnD. The pocket is mainly composed of hydrophobic residues Trp97, Glu123, Phe170, Ala173 and Phe189 at the superior region of the isoprenyl moiety of DMSPP. In contrast, the rotation of the corresponding Tyr80 was not observed in the MpnD structures, because of the steric hindrance between Tyr80 and Met159 (Fig. 4d–f). This is due to the small-to-large substitutions of Phe170 and Ala173 in TleC with Trp157 and Met159 in MpnD and the different orientation of the side chain of Tyr80, corresponding to Trp97 in TleC, which leans slightly towards Met159 to form a hydrogen bond network with the carboxyl group of Asp4. Moreover, the significant conformational differences of the loop between residues 166–174 in TleC and 152–160 in MpnD (RMSD of 1.9 Å for the Cα atoms), and the lack of a residue between 157 and 159 in MpnD (170 and 173 in TleC), result in the displacement of the Cα atom of the active site forming residue Met159 in MpnD by 1.3 Å towards the inside of the active site, as compared with Ala173 in TleC. These observations along with the similar binding modes of substrates between TleC and MpnD suggested that the presence of the 'prenyl binding pocket' controls the preference of the chain length

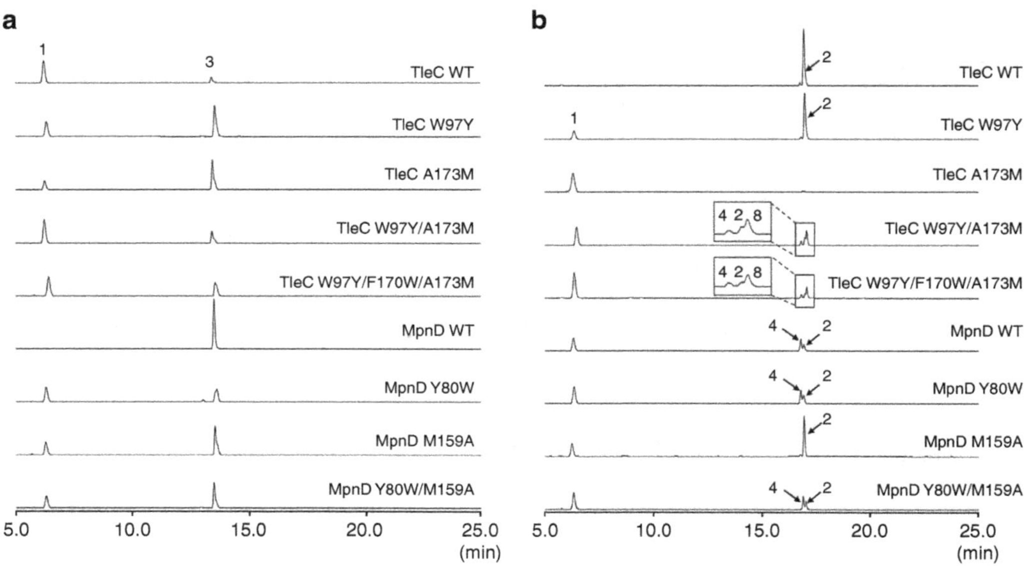

Figure 5 | The *in vitro* enzyme reactions of the wild-type and mutants of *S. blastmyceticus* TleC and *M. thermotolerans* MpnD. HPLC elution profiles of the enzyme reaction products from (−)-indolactam V and (**a**) DMAPP or (**b**) GPP as substrates.

of the prenyl donors, as well as the regio- and stereo-selectivities of the enzyme-catalysed prenylation reactions (Fig. 3c).

Structure-based engineering of TleC and MpnD. On the basis of the structural analyses, we envisaged that the modulation of the 'prenyl binding pocket' should alter the substrate and product specificities of the enzyme reactions. We therefore focused on the three uniquely substituted key residues lining the pocket: Trp97, Phe170 and Ala173 in *S. blastmyceticus* TleC and the corresponding Tyr80, Trp157 and Met159 in *M. thermotolerans* MpnD. To evaluate the importance of these residues, we first constructed the W97Y and A173M mutants for TleC and the Y80W and M159A mutants for MpnD, and investigated the effects of the mutagenesis on the enzyme activities with (−)-indolactam V and prenyl pyrophosphate (DMAPP, GPP or FPP) as substrates. Indeed, as anticipated, both the W97Y and A173M mutants of TleC significantly increased the activity for the C_5 DMAPP substrate, to produce pendolmycin (**3**) (Fig. 5 and Table 1). Furthermore, remarkably, the A173M mutant no longer accepted the C_{10} GPP to yield lyngbyatoxin A (**2**). The kinetic analyses revealed that the TleC A173M mutant exhibits a 2.3-fold increase in the k_{cat}/K_M value for the C_5 prenylation activity, whereas the C_{10} prenylation activity was almost abolished, as compared with the wild-type enzyme. Thus, TleC was functionally converted to the C_5 PT by the small-to-large A173M single amino acid substitution. In contrast, notably, the TleC W97Y mutant exhibited a 31-fold increase for the C_5 prenylation activity, although it still retained the C_{10} prenylation activity comparable to that of the wild-type enzyme.

The Y80W and M159A mutants of *M. thermotolerans* MpnD both decreased the activity for the C_5 substrate, but instead increased the C_{10} prenylation activity (Fig. 5 and Table 1). Furthermore, the large-to-small M159A substitution even increased the C_{15} prenylation activity. The kinetic analyses revealed that the M159A mutant exhibited an 8.8-fold decrease in the k_{cat}/K_M value for the C_5 prenylation activity, whereas it exhibited 4.7- and 3.8-fold increases in the k_{cat}/K_M for the C_{10} and C_{15} prenylation activities, respectively. Moreover, surprisingly, the product ratio of lyngbyatoxin A (**2**) (the 'reverse' C_{10} prenylation at the C-7 position of **1**) and 5-geranylindolactam V (**4**) (the 'normal' C_{10} prenylation at the C-5 position of **1**) was significantly altered in the M159A mutant, which now produced **2** as the major product. In contrast, the Y80W mutant exhibited a 24-fold decrease in the C_5 prenylation activity, but a 5.5-fold increase in the k_{cat}/K_M value for the C_{10} activity and a 2.7-fold higher affinity to GPP. This would be due to the loss of the hydrogen bond network from the hydroxyl group of Tyr80 and the narrowed active site entrance. These results suggested that the hydrophobicity and the shape of the active site are important for the substrate binding.

To further modulate the 'prenyl binding pocket', we constructed double and triple mutants of the three key residues: the W97Y/A173M and W97Y/F170W/A173M mutants of TleC and the Y80W/M159A and Y80W/W157F/M159A mutants of MpnD. As a result, although the double and triple mutants of TleC significantly decreased the activities by 33- and 28-fold, respectively, they still preferred to accept DMAPP as the prenyl donor (Fig. 5 and Table 1). Interestingly, a careful examination of the enzyme reaction products of these mutants revealed the newly generated novel product **8**, as the major product, in addition to **2** and **4** (**2**:**4**:**8** = 1:0.6:2.3 for the W97Y/A173M mutant of TleC; **2**:**4**:**8** = 1:1: 2.9 for the W97Y/F170W/A173M mutant of TleC) from GPP as the prenyl donor (Fig. 5 and Supplementary Fig. 14). The structure of **8** was established to be teleocidin A-2, which is the 19*S*-epimer of **2**, by comparisons of the NMR data. The ^{1}H-NMR of **8** showed good accordance with **2**, except the signals due to the prenyl moiety (that is, slight differences in the chemical shifts of H-20, H-21 and H-23, which are completely identical to those of teleocidin A-2 in the literature[28]). Thus, the mutations affected the stereochemistry of the enzyme-catalysed prenylation reaction to produce 19*S*-**8** as the major product along with 19*R*-**2** and its regioisomer 5-gerany-**4**. In contrast, the MpnD double mutant exhibited C_{10} and C_{15} prenylation activities comparable to those of the wild-type enzyme, whereas the C_5 activity was decreased by 11-fold. Unfortunately, as the MpnD triple mutant was expressed as an insoluble inclusion body, we could not evaluate its activities. These results clearly demonstrated that the structure-based site-directed mutagenesis successfully altered not only the preference for the chain length of the prenyl donors but also the regio- and stereo-selectivities of the enzyme reactions.

Enzyme reaction mechanisms of TleC and MpnD. The mechanism for the formation of lyngbyatoxin A (**2**) was proposed to proceed through *N*-geranylindolactam V as an intermediate, suggesting the involvement of the initial 'normal' prenylation at the N-1 position of (−)-indolactam V (**1**) and the subsequent aza-Claisen rearrangement in the installation of the geranyl moiety on the C-7 position of **1** (Supplementary Fig. 15)[29]. This hypothesis is based on the presence of blastmycetin E, which is easily converted into des-*O*-methyl olivoretin E, with the terpenoid moiety fused with C-6 and C-7 of the indole ring by the aza-Claisen rearrangement under mild acidic conditions. To test this hypothesis, we synthesized *N*-geranylindolactam V, according to the published method[30], and incubated it with TleC. As a result, no conversion was observed after an 18-h incubation, even in the presence of methanol containing 1% acetic acid (Supplementary Fig. 16). This result suggested that *N*-geranylindolactam V is not the intermediate and the 'reverse' prenylation at the C-7 position of the indolactam does not proceed through the aza-Claisen rearrangement. Furthermore, the crystal structures of TleC and MpnD complexed with the substrates suggested that the distance between the C-3 position of DMSPP and the C-7 position of **1** (3.3 Å in TleC and 3.1 Å in MpnD) is closer than that between the C-1 position of the prenyl substrates and the N-1 position of **1** (6.0 Å in TleC and MpnD) (Supplementary Fig. 17a). Moreover, the docking model of TleC with GPP also suggested that GPP binds in the 'prenyl binding pocket' in the active site and the distances between the C-3 position of GPP and the C-7 position of **1**, and the C-1 position of the prenyl substrates and the N-1 position of **1**, are almost the same as those with DMSPP (3.3 and 5.8 Å, respectively) (Supplementary Fig. 17b). These results strongly suggested that both TleC and MpnD catalyse the 'reverse' prenylation directly onto the C-7 position of **1** to produce **2** and **3**, respectively. Interestingly, the docking model of MpnD with GPP also suggested that the loss of the 'prenyl binding pocket' causes the alteration of the GPP binding mode in the active site of MpnD. In this model, GPP is predicted to bind between **1** and the aromatic ring of Tyr175, and the distance between the C-5 of **1** and the C-1 of GPP (3.2 Å) is almost the same as that between the C-7 of **1** and the C-3 of GPP (3.1 Å) (Supplementary Fig. 17d). This would be the reason why MpnD produces not only the C-7 'reverse' prenylation product **2** but also the C-5 'normal' prenylation product **4**.

Discussion

Our *in vitro* analyses of *S. blastmyceticus* TleC and *M. thermotolerans* MpnD revealed that these enzymes strictly recognize the cyclic dipeptide (−)-indolactam V as the prenyl acceptor. Remarkably, the indolactam-specific PTs accept different chain lengths of prenyl pyrophosphate, to generate a series of unnatural indolactam derivatives. Furthermore, although previous *in vivo* analyses of TleC and MpnD demonstrated that these enzyme catalyse C-7 'reverse' prenylation.

To the best our knowledge, this is the first demonstration that a single PT enzyme catalyses from the C_5 to C_{25} prenylation reactions. In contrast, most of the soluble indole PTs from fungi and bacteria show broad substrate specificities towards the prenyl acceptors, but not for the prenyl donors, except for several recently reported indole PTs, including AnaPT from *N. fischeri*[31], 6-DMAT$_{sa}$ from *S. ambofaciens* and 6-DMAT$_{sv}$ from *S. violaceusniger*, which accept both DMAPP and GPP as prenyl donors[32], and BAE61387 from *A. oryzae* DSM1147, which accepts DMAPP, GPP and FPP as prenyl donors[21]. These characteristic features of TleC and MpnD make the enzymes excellent targets for the manipulation of the prenylation reactions.

The structure-based engineering of TleC and MpnD demonstrated that the modulation of the 'prenyl binding pocket' successfully altered the substrate and product specificities of the enzyme reactions. In particular, the three uniquely substituted key residues lining the pocket, Trp97, Phe170 and Ala173 in *S. blastmyceticus* TleC, and the corresponding residues Tyr80, Trp157 and Met159 in *M. thermotolerans* MpnD, are important for determining the preference of the chain length of the prenyl donors, as well as the regio- and stereo-selectivities of the enzyme reactions. The large-to-small M159A substitution in MpnD newly create the 'prenyl binding pocket', which is absent in the active site of the wild-type MpnD, to significantly enhance the C_{10} and C_{15} prenylation activities. In contrast, the small-to-large A173M substitution in the TleC mutant leads to the loss of the pocket, which results in a dramatic decrease in the C_{10} prenylation activity. Furthermore, the additional W97Y substitution in TleC decreases the steric hindrance between the enzyme and GPP, as the double and triple mutants were able to accept GPP as a prenyl substrate. However, the different configuration of the active site architecture results in the alteration of the binding mode of GPP. Although the C-3 methyl group of GPP should face towards Ser121 in the wild-type TleC, to produce **2** with the C19*R* configuration, the C-3 methyl group may turn in the opposite direction in the mutants, leading to the production of **8** with the C19*R* configuration (Supplementary Fig. 17). These results strongly support our proposal that both TleC and MpnD catalyse the 'reverse' prenylation directly onto the C-7 position of **1**, to produce **2** and **3**, respectively.

A comparison of the structures of the indolactam-specific TleC (and MpnD), with other indole PTs revealed that these enzymes share the ABBA barrel overall fold, but the modes of the substrate binding in the active site are totally different (Fig. 6 and Supplementary Fig. 18). The indole ring of the indolactam is positioned proximally to the side chains of Tyr317 and Tyr382 in TleC (Tyr300 and Tyr367 in MpnD) and the prenyl substrate is sandwiched between the indole ring and the benzene ring of Phe189, proximally to the side chain of Phe170 in TleC (Tyr175 and Met159 in MpnD, respectively). In contrast, in the crystal structures of FgaPT2 (ref. 25), FtmPT1 (ref. 26) and CdpNPT[27] from *A. fumigatus*, their respective indole substrates, L-tryptophan, brevianamide F and (*S*)-benzodiazepinedione, bind in the opposite direction and are positioned proximally to the side chains of Lys174 and Tyr189 in FgaPT2, Leu187 and Tyr203 in FtmPT1 and in Val206 and Tyr221 in CdpNPT, respectively (the residues corresponding to Ala173 and Phe189 in TleC). In these cases, the dimethylallyl group of DMAPP is sandwiched by π–π interactions between the aromatic rings of the substrate and Tyr345 in FgaPT2 (Tyr382 in FtmPT1). Furthermore, Met328 and Tyr398 in FgaPT2 (Met364 and Tyr435 in FtmPT2, and Met349 and Trp419 in CdpNPT) protrude towards the active site centre to anchor the superior region of the dimethyl moiety of DMAPP (Supplementary Fig. 18d–f). In contrast, the corresponding residues are replaced with the slightly smaller amino acids Val and Trp in AnaPT[24], Leu and Tyr in BAE61387 (ref. 21), and Val and Phe in TleC and MpnD, respectively. As a result, their active site pockets are relatively larger than those of the DMAPP-specific indole PTs (Supplementary Figs. 18a–c). Furthermore, the crystal structure of AnaPT and the model structure of LtxC from *M. producens*, which catalyses the same reaction as TleC, suggested that these enzymes also have the 'prenyl binding pocket', and the three key residues discussed above are uniquely replaced with Gly102, Phe212 and Val195 in AnaPT and with Asn89, Ile169 and Gly173 in LtxC, respectively (Supplementary Fig. 19). These observations would explain why these enzymes can accept not only DMAPP but also longer prenyl pyrophosphates as substrates. In particular,

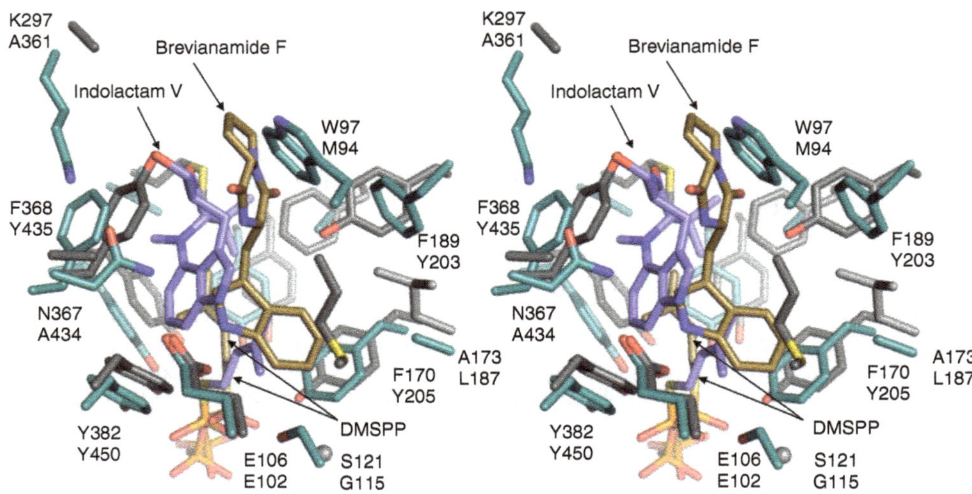

Figure 6 | The stereo view of the active site structures of TleC and FtmPT1. The active site residues of TleC and FtmPT1 from *A. fumigatus* are shown by green and grey stick models, respectively. The respective substrates **1** and DMSPP in TleC, and brevianamide F and DMSPP in FtmPT1 are represented by blue and brown stick models, respectively. Structures were superposed based on the Cα-atom.

it is noteworthy that both TleC and MpnD have larger active site cavities than these GPP- or FPP-accepting enzymes, which would be the reason why they accept the longer GGPP and GFPP substrates. Most importantly, our results expanded the substrate scope of the enzyme reactions, which will set the stage for further development of new enzymes with novel catalytic functions.

In conclusion, a comparison of the crystal structures of the indolactam-specific TleC and MpnD revealed the intimate structural details of the enzyme-catalysed 'reverse' prenylation reactions and identified the active site residues governing the selection of the prenyl donors. Furthermore, structure-based enzyme engineering successfully altered the preference for the chain lengths of the prenyl donors, as well as the regio- and stereo-selectivities of the prenylation reactions. These findings provided insights into the catalytic machinery and the strategies towards expanding the catalytic repertoire of the enzymes, to generate designed novel molecular scaffolds for drug discovery.

Methods

Materials. (−)-Indolactam V was prepared by using *S. blastmyceticus* according to the published method[14]. DMSPP was purchased from Echelon Bioscience, Inc. Oligonucleotides were obtained from Eurofins Genomics, Inc.

Compound characterization. High-resolution electrospray ionization mass spectrometry (HRESIMS) spectra were measured with an Agilent 1100 series HPLC-microTOF mass spectrometer (JEOL), using electrospray ionization (ESI). NMR spectra were obtained at 500 MHz (^{1}H) and 125 MHz (^{13}C) with a JEOL ECX-500 spectrometer.

5-Geranylindolactam V (**4**): ultraviolet: λ_{max} 291 nm. HR-MS (ESI): found for $[C_{27}H_{38}N_3O_2]^-$ 436.29567: calcd 436.29640. ^{1}H NMR (500 MHz, $CDCl_3$): δ 8.26 (*s*, 1H, NH-1), 7.24 (*d*, 1H, *J* = 8.0 Hz, H-7), 7.12 (*d*, 1H, *J* = 8.0 Hz, H-6), 7.04 (*d*, 1H, *J* = 2.0 Hz, H-2), 5.30 (*t*, 1H, *J* = 6.5 Hz, H-2′), 5.15 (*t*, 1H, *J* = 7.0 Hz, H-6′), 4.96 (*d*, 1H, *J* = 10.5 Hz, NH-10), 4.48 (*m*, 1H, H-9), 3.73 (dd, 1H, *J* = 6.5, 15.5 Hz, H-1′), 3.55 (dd, 1H, *J* = 6.5, 15.5 Hz, H-1′), 3.43 (dd, 1H, *J* = 6.5, 11.0 Hz, H-14), 3.38 (dd, 1H, *J* = 6.5, 11.0 Hz, H-14), 3.25 (dd, 1H, *J* = 4.5, 14.5 Hz, H-8), 3.05 (*d*, 1H, *J* = 10.5 Hz, H-12), 2.81 (dd, 1H, *J* = 1.5, 14.5 Hz, H-8), 2.77 (*s*, 3H, H-18), 2.69 (*m*, 1H, H-15), 2.15 (*q*, 2H, *J* = 7.0 Hz, H-5′), 2.09 (*t*, 2H, *J* = 7.0 Hz, H-4′), 1.75 (*s*, 3H, H-9′), 1.71 (*s*, 3H, H-8′), 1.63 (*s*, 3H, H-10′), 1.31 (*d*, 3H, *J* = 7.0 Hz, H-16), 0.99 (*d*, 3H, *J* = 6.5 Hz, H-17). ^{13}C NMR (125 MHz, $CDCl_3$): δ 173.4 (C-11), 141.8 (C-4), 138.3 (C-7a), 135.3 (C-3′), 134.2 (C-5), 131.4 (C-7′), 127.9 (C-4a), 124.8 (C-6), 124.7 (C-2 and C-2′), 124.4 (C-6′), 110.4 (C-7), 109.8 (C-3), 73.5 (C-12), 63.1 (C-14), 55.4 (C-9), 39.8 (C-4′), 36.2 (C-18), 30.8 (C-1′), 29.6 (C-8), 26.6 (C-5′), 26.4 (C-15), 25.7 (C-8′), 20.9 (C-17), 19.8 (C-16), 17.7 (C-10′), 16.2 (C-9′). For full NMR spectra of compound **4**, see Supplementary Fig. 20.

5-Farnesylindolactam V (**5**): ultraviolet: λ_{max} 291 nm. HR-MS (ESI): found for $[C_{32}H_{46}N_3O_2]^-$ 504.35752: calcd 504.35900. ^{1}H NMR (500 MHz, $CDCl_3$): δ 8.17 (*s*, 1H, NH-1), 7.21 (*d*, 1H, *J* = 8.5 Hz, H-7), 7.10 (*d*, 1H, *J* = 8.5, H-6), 7.02 (*s*, 1H, H-2), 5.29 (*t*, 1H, *J* = 7.0 Hz, H-2′), 5.14 (*t*, 1H, *J* = 6.5 Hz, H-6′), 5.09 (*m*, 1H, H-10′), 4.93 (*d*, 1H, *J* = 10.5 Hz, NH-10), 4.46 (*m*, 1H, H-9), 3.71 (dd, 1H, *J* = 7.0, 15.5 Hz, H-1′), 3.53 (dd, 1H, *J* = 7.0, 15.5 Hz, H-1′), 3.41 (dd, 1H, *J* = 6.5, 12.0 Hz, H-14), 3.36 (dd, 1H, *J* = 6.5, 12.0 Hz, H-14), 3.23 (dd, 1H, *J* = 4.5, 14.5 Hz, H-8), 3.02 (d, 1H, *J* = 11.0 Hz, H-12), 2.79 (d, 1H, *J* = 14.5, H-8), 2.74 (*s*, 3H, H-18), 2.66 (*m*, 1H, H-15), 2.14 (*m*, 2H, H-5′), 2.07 (*m*, 2H, H-4′), 2.06 (*m*, 2H, H-9′), 1.97 (*m*, 2H, H-8′), 1.73 (*s*, 3H, H-13′), 1.67 (*s*, 3H, H-12′), 1.60 (*s*, 3H, H-14′), 1.59 (*s*, 3H, H-15′), 1.29 (*d*, 3H, *J* = 7.0 Hz, H-16), 0.97 (*d*, 3H, *J* = 6.5 Hz, H-17). ^{13}C NMR (125 MHz, $CDCl_3$): δ 173.4 (C-11), 141.9 (C-4), 138.3 (C-7a), 135.5 (C-3′), 135.1 (C-7′), 134.3 (C-5), 131.3 (C-11′), 127.9 (C-4a), 124.8 (C-6), 124.7 (C-2 and C-2′), 124.4 (C-10′), 124.2 (C-6′), 110.3 (C-7), 109.9 (C-3), 73.5 (C-12), 63.1 (C-14), 55.4 (C-9), 39.8 (C-4′ and C-8′), 36.2 (C-18), 30.8 (C-1′), 29.6 (C-8), 26.8 (C-9′), 26.6 (C-5′), 26.4 (C-15), 25.7 (C-12′), 20.9 (C-17), 19.8 (C-16), 17.7 (C-15′), 16.3 (C-13′), 16.1 (C-14′). For full NMR spectra of compound **5**, see Supplementary Fig. 21.

Enzyme expression and purification of the TleC and MpnD enzymes. The DNA encoding full-length TleC was amplified from the genome of *S. blastmyceticus* by PCR, using 5′-CATATGGAGTCCGCAGGTCCCGGAACC-3′ as the sense primer, which introduces an NdeI restriction site, and 5′-AAGCTT CGCCTCCCGGTAGACCCCCGGC-3′ as the antisense primer, which introduces a HindIII restriction site. The DNA encoding full-length MpnD from *M. thermotolerans* was purchased from Eurofins Genomics, Inc., and has NdeI and XhoI restriction sites at the 5′- and 3′-termini, respectively. The DNA fragments were digested with NdeI/HindIII (TleC) and NdeI/XhoI (MpnD), and ligated into the NdeI/HindIII (TleC) and NdeI/XhoI (MpnD) sites of the digested pET22b vector (Merck Millipore), for expression as fusion proteins with a His_6-tag at the carboxy terminus. After confirmation of the sequence, the pET22b vectors encoding full-length TleC and MpnD were transformed into *E. coli* BLR (DE3). The cells harbouring the plasmids were cultured to an OD_{600} of 0.6 in Luria-Bertani medium containing 100 μg ml^{-1} ampicillin at 37 °C. Isopropyl β-D-1-thiogalactopyranoside was then added to a final concentration of 0.2 mM, to induce gene expression, and the cultures were incubated further for 16 h at 20 °C.

All of the following procedures were performed at 4 °C. The *E. coli* cells were harvested by centrifugation at 5,000 *g* and resuspended in 50 mM Tris-HCl buffer pH 7.5, containing 200 mM NaCl, 5% (v/v) glycerol and 5 mM imidazole (buffer A). The cells were disrupted by sonication and the lysate was centrifuged at 12,000 *g* for 30 min. The supernatant was loaded onto a COSMOGEL His-Accept (Nacalai Tesque) column, equilibrated with buffer A. After washing the resin with buffer A containing 10 mM imidazole, the recombinant TleC protein was subsequently eluted with buffer A containing 300 mM imidazole. The protein solution was concentrated to 5 ml, purified to homogeneity by gel-filtration chromatography on a HiLoad 16/60 Superdex 200 pg column (GE Healthcare), and concentrated to 10 mg ml^{-1} in 20 mM Tris-HCl buffer pH 7.5, containing 150 mM NaCl and 1 mM dithiothreitol. The protein concentration was calculated by measuring ultraviolet absorption at A_{280} (ref. 33). MpnD was purified by the same method as for TleC, except Tris-HCl buffer pH 8.5 instead of Tris-HCl buffer pH 7.5.

The selenomethionine-labelled TleC protein was overexpressed in *E. coli* strain B834 (DE3). The cells were grown at 37 °C in 8 l of LeMaster medium, containing 25 mg l^{-1} L-selenomethionine, Sigma Kao and Michayluk Vitamin Solution (Sigma-Aldrich) and 100 μg ml^{-1} ampicillin, until the cell density reached an OD_{600} of 1.2. At this point, 0.2 mM isopropyl β-D-1-thiogalactopyranoside was added to induce the expression of the selenomethionine-labelled and C-terminally His_6-fused TleC, and the growth was continued for 18 h at 20 °C. The TleC (SeMet) protein was purified as described above.

Enzyme assays. TleC or MpnD (10 μg) was incubated with 100 μM of indolactam V and 200 μM of prenyl pyrophosphates (DMAPP, GPP or FPP), in a final volume of 100 μl of 50 mM Tris–HCl buffer containing 150 mM NaCl pH 7.0, at 42 °C for 10 min. When FPP (TleC reaction), GGPP and GFPP were used as substrates, the enzyme reactions were performed at 37 °C for 18 h. The reactions were stopped by the addition of 100 μl of MeOH. After the aggregated protein was removed by centrifugation, an aliquot of the sample (100 μl) was subjected to HPLC on a Mightysil RP-18 GP column (4.6 × 250 mm) (KANTO KAGAKU), at a flow rate of 1.0 ml min^{-1}. Gradient elution was performed with H_2O and CH_3CN, both containing 0.1% trifluoroacetic acid: 0–6 min, 50–65% CH_3CN; 6–10 min, 65–100% CH_3CN; and 10–25 min, 100% CH_3CN. The absorbance of the enzyme reaction products was monitored at 303 nm. For the LC–MS analysis, the samples were prepared in the same manner and analysed by LC–MS on a COSMOSIL C18-MS-II column (2 × 75 mm) (Nacalai Tesque), at a flow rate of 0.2 ml min^{-1}. Gradient elution was performed with H_2O and CH_3CN, both containing 0.1% acetic acid: 0–3 min, 30% CH_3CN; 3–15 min, 30–100% CH_3CN; and 15–20 min, 100% CH_3CN. The enzyme reaction products was purified by preparative HPLC on a COSMOSIL 5C18-MS-II column (10 × 250 mm) (Nacalai Tesque), at a flow rate of 3.0 ml min^{-1}. Isocratic elution was performed with 60% CH_3CN/H_2O containing 0.1% acetic acid.

To determine the kinetic parameters of wild-type TleC and MpnD with **1**, the assays contained prenyl pyrophosphate (TleC: GPP, MpnD: DMAPP, 200 μM), **1** (0.1–20 μM) and purified TleC and MpnD. The reactions for the determination of the kinetic parameters were incubated for 3 min at 42 °C. The kinetic parameters of wild-type TleC and MpnD, and their mutants with DMAPP, GPP and FPP, were determined with **1** (150 μM), DMAPP (1–150 μM) and either GPP (1–150 μM) or FPP (5–300 μM). The reaction mixtures were incubated for 5 min at 42 °C. The enzymatic reactions were stopped by the addition of MeOH (100 μl) and analysed by HPLC. The authentic concentrations of the products were calculated by the total peak areas of the products. The experiments were performed in triplicate. The K_M and k_{cat} values were calculated using GraphPad Prism 6 for Mac OS X (GraphPad Software, Inc., La Jolla, CA).

Crystallization and structure determination. Well-diffracting selenomethionine-labelled TleC and wild-type TleC crystals were obtained at 20 °C, in 100 mM MES-NaOH (pH 6.5) containing 1,700–1,785 mM $(NH_4)_2SO_4$, with 10 mg ml^{-1} of the SeMet-labelled purified TleC solution, by using the sitting-drop vapour-diffusion method. The crystals of wild-type of MpnD were obtained at 20 °C, in 100 mM Tris–HCl pH 8.5 containing 1,610 mM $(NH_4)_2SO_4$, with 15 mg ml^{-1} of purified MpnD solution, by using the sitting-drop vapour-diffusion method. The crystals were transferred into the reservoir solution with 18-22% (v/v) glycerol as a cryo-protectant and then flash cooled at − 173 °C in a nitrogen-gas stream. X-ray diffraction data sets were collected at BL-17A, BL-5A and NW12A at the Photon Factory, Tsukuba, Japan. The ternary complex containing both substrates was prepared by incubating the TleC and MpnD crystals for 1 min at 20 °C with 5 mM **1** and 5 mM DMSPP. We used wavelengths of 0.9788 Å for the SeMet-labelled TleC, based on the fluorescence spectrum of the Se K absorption edge[34], 0.9800 Å for TleC and 1.0000 Å for MpnD, for data collection.

The diffraction data for TleC SeMet were processed and scaled using the *XDS* programme package[35]. The Se sites were determined with *AutoSol* in *PHENIX*[36,37]. The sites were refined and the initial phases were calculated with *AutoBuild* in *PHENIX*[36,38]. The structure was modified manually with *Coot*[39] and refined with *PHENIX*[36,40]. The initial phases of the TleC ternary complex structure and the MpnD ternary complex structure were determined by molecular replacement, using the structure of TleC SeMet as the search model. Molecular replacement was performed with *Phaser* in *PHENIX*[36,41]. The refinements were performed in the same manner as for TleC SeMet. The final crystal data and intensity statistics are summarized in Table 2. A structural similarity search was performed by using the Dali programme[42]. The cavity volumes were calculated with the programme CASTP (http://cast.engr.uic.edu/cast/). All crystallographic figures were prepared with PyMOL (DeLano Scientific, http://www.pymol.org).

Site-directed mutagenesis. The plasmids expressing the mutants of TleC (A173M, W97Y/A173M and W97Y/F170W/A173M) and MpnD (M159A, Y80W/M159A and Y80W/W157F/M159A) were constructed with a QuikChange Site-Directed Mutagenesis Kit (Stratagene), according to the manufacturer's protocol. The mutant proteins were expressed and purified in the same manner as for the wild-type enzymes and the enzyme reactions were performed with the same methods as for the wild-type enzymes.

Molecular modelling. The three-dimensional model of GPP was generated by the Chem3D Ultra 13 programme (CambridgeSoft). Docking calculations were performed using AutoDock4 (ref. 43). The mutant models were constructed by manually swapping of the GPP in the active site of the TleC mutant model structure by using *Coot* and the energy minimization calculation by simulated annealing with *PHENIX* was then performed. The parameters of the intermediates for the energy minimization calculation were obtained from the PRODRG server.

References

1. Nakagawa, K. *et al.* Identification of UBIAD1 as a novel human menaquinone-4 biosynthetic enzyme. *Nature* **468,** 117–121 (2010).
2. Sakagami, Y., Yoshida, M., Isogai, A. & Suzuki, A. Peptidal sex hormones inducing conjugation tube formation in compatible mating-type cells of *Tremella mesenterica*. *Science* **212,** 1525–1527 (1981).
3. Okada, M. *et al.* Structure of the *Bacillus subtilis* quorum-sensing peptide pheromone ComX. *Nat. Chem. Biol.* **1,** 23–24 (2005).
4. Brown, M. S. & Goldstein, J. L. Protein prenylation. Mad bet for Rab. *Nature* **366,** 14–15 (1993).
5. Cui, C. B. *et al.* Tryprostatins A and B, novel mammalian cell cycle inhibitors produced by *Aspergillus fumigatus*. *J. Antibiot.* **48,** 1382–1384 (1995).
6. Haarmann, T., Rolke, Y., Giesbert, S. & Tudzynski, P. Ergot: from witchcraft to biotechnology. *Mol. Plant Pathol.* **10,** 563–577 (2009).
7. Harada, H., Nakata, H. & Hirata, Y. The structure of teleocidin B. *Tet. Lett.* **23,** 2515–2522 (1966).
8. Heide, L. Prenyl transfer to aromatic substrates: genetics and enzymology. *Curr. Opin. Chem. Biol.* **13,** 171–179 (2009).
9. Alkhalaf, L. M. & Ryan, K. S. Biosynthetic manipulation of tryptophan in bacteria: pathways and mechanisms. *Chem. Biol.* **22,** 317–328 (2015).
10. Li, S. M. Prenylated indole derivatives from fungi: structure diversity, biological activities, biosynthesis and chemoenzymatic synthesis. *Nat. Prod. Rep.* **27,** 57–78 (2010).
11. Tanner, M. E. Mechanistic studies on the indole prenyltransferases. *Nat. Prod. Rep.* **32,** 88–101 (2015).
12. Yu, X., Liu, Y., Xie, X., Zheng, X. D. & Li, S. M. Biochemical characterization of indole prenyltransferases: filling the last gap of prenylation positions by a 5-dimethylallyltryptophan synthase from *Aspergillus clavatus*. *J. Biol. Chem.* **287,** 1371–1380 (2012).
13. Winkelblech, J., Fan, A. & Li, S. M. Prenyltransferases as key enzymes in primary and secondary metabolism. *Appl. Microbiol. Biotechnol.* **99,** 7379–7397 (2015).
14. Awakawa, T. *et al.* A methyltransferase initiates terpene cyclization in teleocidin B biosynthesis. *J. Am. Chem. Soc.* **136,** 9910–9913 (2014).
15. Edwards, D. J. & Gerwick, W. H. Lyngbyatoxin biosynthesis: sequence of biosynthetic gene cluster and identification of a novel aromatic prenyltransferase. *J. Am. Chem. Soc.* **126,** 11432–11433 (2004).
16. Ma, J. *et al.* Characterization of a single gene cluster responsible for methylpendolmycin and pendolmycin biosynthesis in the deep sea bacterium *Marinactinospora thermotolerans*. *Chembiochem* **13,** 547–552 (2012).
17. Yin, W. B., Yu, X., Xie, X. L. & Li, S. M. Preparation of pyrrolo[2,3-b]indoles carrying a beta-configured reverse C3-dimethylallyl moiety by using a recombinant prenyltransferase CdpC3PT. *Org. Biomol. Chem.* **8,** 2430–2438 (2010).
18. Yin, S., Yu, X., Wang, Q., Liu, X. Q. & Li, S. M. Identification of a brevianamide F reverse prenyltransferase BrePT from *Aspergillus versicolor* with a broad substrate specificity towards tryptophan-containing cyclic dipeptides. *Appl. Microbiol. Biotechnol.* **97,** 1649–1660 (2013).
19. Qian, Q., Schultz, A. W., Moore, B. S. & Tanner, M. E. Mechanistic studies on CymD: a tryptophan reverse *N*-prenyltransferase. *Biochemistry* **51,** 7733–7739 (2012).
20. Frick, S. *et al.* Metal ions control product specificity of isoprenyl diphosphate synthases in the insect terpenoid pathway. *Proc. Natl Acad. Sci. USA* **110,** 4194–4199 (2013).
21. Pockrandt, D., Sack, C., Kosiol, T. & Li, S. M. A promiscuous prenyltransferase from *Aspergillus oryzae* catalyses C-prenylations of hydroxynaphthalenes in the presence of different prenyl donors. *Appl. Microbiol. Biotechnol.* **98,** 4987–4994 (2014).
22. Kuzuyama, T., Noel, J. P. & Richard, S. B. Structural basis for the promiscuous biosynthetic prenylation of aromatic natural products. *Nature* **435,** 983–987 (2005).
23. Metzger, U., Keller, S., Stevenson, C. E., Heide, L. & Lawson, D. M. Structure and mechanism of the magnesium-independent aromatic prenyltransferase CloQ from the clorobiocin biosynthetic pathway. *J. Mol. Biol.* **404,** 611–626 (2010).
24. Yu, X. *et al.* Catalytic mechanism of stereospecific formation of cis-configured prenylated pyrroloindoline diketopiperazines by indole prenyltransferases. *Chem. Biol.* **20,** 1492–1501 (2013).
25. Metzger, U. *et al.* The structure of dimethylallyl tryptophan synthase reveals a common architecture of aromatic prenyltransferases in fungi and bacteria. *Proc. Natl Acad. Sci. USA* **106,** 14309–14314 (2009).
26. Jost, M. *et al.* Structure-function analysis of an enzymatic prenyl transfer reaction identifies a reaction chamber with modifiable specificity. *J. Am. Chem. Soc.* **132,** 17849–17858 (2010).
27. Schuller, J. M. *et al.* Structure and catalytic mechanism of a cyclic dipeptide prenyltransferase with broad substrate promiscuity. *J. Mol. Biol.* **422,** 87–99 (2012).
28. Sakai, S. *et al.* Absolute configuration of lyngbyatoxin A (teleocidin A-1) and teleocidin A-2. *Tet. Lett.* **27,** 5219–5220 (1986).

29. Irie, K. *et al.* Biosynthesis of indole alkaloid tumor promoters teleocidins (I) possible biosynthetic pathway of the monoterpenoid moieties of teleocidins. *Tetrahedron* **46,** 2773–2788 (1990).
30. Endo, Y., Shudo, K., Itai, A., Hasegawa, M. & Sakai, S. Synthesis and stereochemistry of indolactam-V, an active fragment of teleocidins. Structural requirements for tumor-promoting activity. *Tetrahedron* **42,** 5905–5924 (1986).
31. Pockrandt, D. & Li, S. M. Geranylation of cyclic dipeptides by the dimethylallyl transferase AnaPT resulting in a shift of prenylation position on the indole ring. *Chembiochem* **14,** 2023–2028 (2013).
32. Winkelblech, J. & Li, S. M. Biochemical investigations of two 6-DMATS enzymes from *Streptomyces* reveal new features of L-tryptophan prenyltransferases. *Chembiochem* **15,** 1030–1039 (2014).
33. Pace, C. N., Vajdos, F., Fee, L., Grimsley, G. & Gray, T. How to measure and predict the molar absorption coefficient of a protein. *Protein Sci.* **4,** 2411–2423 (1995).
34. Rice, L. M., Earnest, T. N. & Brunger, A. T. Single-wavelength anomalous diffraction phasing revisited. *Acta Crystallogr. D Biol. Crystallogr.* **56,** 1413–1420 (2000).
35. Kabsch, W. Xds. *Acta Crystallogr. D Biol. Crystallogr.* **66,** 125–132 (2010).
36. Adams, P. D. *et al.* PHENIX: a comprehensive Python-based system for macromolecular structure solution. *Acta. Crystallogr. D Biol. Crystallogr.* **66,** 213–221 (2010).
37. Terwilliger, T. C. *et al.* Decision-making in structure solution using Bayesian estimates of map quality: the PHENIX AutoSol wizard. *Acta Crystallogr. D Biol. Crystallogr.* **65,** 582–601 (2009).
38. Terwilliger, T. C. *et al.* Iterative model building, structure refinement and density modification with the PHENIX AutoBuild wizard. *Acta Crystallogr. D Biol. Crystallogr.* **64,** 61–69 (2008).
39. Emsley, P. & Cowtan, K. Coot: model-building tools for molecular graphics. *Acta Crystallogr. D Biol. Crystallogr.* **60,** 2126–2132 (2004).
40. Afonine, P. V. *et al.* Towards automated crystallographic structure refinement with phenix.refine. *Acta Crystallogr. D Biol. Crystallogr.* **68,** 352–367 (2012).
41. McCoy, A. J. *et al.* Phaser crystallographic software. *J. Appl. Crystallogr.* **40,** 658–674 (2007).
42. Holm, L. & Sander, C. Dali: a network tool for protein structure comparison. *Trends Biochem. Sci.* **20,** 478–480 (1995).
43. Seeliger, D. & VGroot, B. L. Ligand docking and binding site analysis with PyMOL and Autodock/Vina. *J. Comput. Aided Mol. Des.* **24,** 417–422 (2010).

Acknowledgements

This work was supported in part by a Grant-in-Aid for Scientific Research from the Ministry of Education, Culture, Sports, Science and Technology, Japan (to T.M., H.M. and I.A.). We thank Yuya Takeshige for his help in collecting the HR-MS data. We also thank beamline staff of Photon Factory for their help in collecting X-ray diffraction data.

Author contributions

T.M., H.M. and I.A. designed the experiments. T.M., S.H., L.Z. and T.A. performed the experiments. T.M., S.H., M.O., H.M. and I.A. analysed the data. T.M., H.M. and I.A. wrote the paper.

Additional information

Accession codes: Protein Data Bank (PDB): The coordinates and the structure factor amplitudes for the selenomethionine-labelled TleC, the TleC apo structure, the TleC complexed with (−)-indolactam V and DMSPP, the MpnD apo structure and the MpnD complexed with (−)-indolactam V and DMSPP were deposited under accession codes 4YZJ, 4YZK, 4YZL, 4YL7 and 4YLA, respectively.

Competing financial interests: The authors declare no competing financial interests.

8

Structures of the *E. coli* translating ribosome with SRP and its receptor and with the translocon

Ahmad Jomaa[1], Daniel Boehringer[1], Marc Leibundgut[1] & Nenad Ban[1]

Co-translational protein targeting to membranes is a universally conserved process. Central steps include cargo recognition by the signal recognition particle and handover to the Sec translocon. Here we present snapshots of key co-translational-targeting complexes solved by cryo-electron microscopy at near-atomic resolution, establishing the molecular contacts between the *Escherichia coli* translating ribosome, the signal recognition particle and the translocon. Our results reveal the conformational changes that regulate the latching of the signal sequence, the release of the heterodimeric domains of the signal recognition particle and its receptor, and the handover of the signal sequence to the translocon. We also observe that the signal recognition particle and the translocon insert-specific structural elements into the ribosomal tunnel to remodel it, possibly to sense nascent chains. Our work provides structural evidence for a conformational state of the signal recognition particle and its receptor primed for translocon binding to the ribosome-nascent chain complex.

[1] Department of Biology, Institute of Molecular Biology and Biophysics, Otto-Stern-Weg 5, ETH Zurich CH-8093, Switzerland. Correspondence and requests for materials should be addressed to N.B. (email: ban@mol.biol.ethz.ch).

Synthesis of membrane proteins requires that they are co-translationally targeted to the endoplasmic reticulum in eukaryotes and to the cell membrane in bacteria[1,2]. This process involves the recruitment of intricate cellular machineries that are conserved throughout all forms of life. This includes a ribonucleoprotein complex termed the signal recognition particle (SRP), the SRP receptor (SR) and the translocation machinery or Sec translocon[3–6]. Co-translational protein targeting is initiated by the emergence of a hydrophobic N-terminal signal sequence (SS) from a translating ribosome (termed the ribosome–nascent chain complex, RNC). The SRP, consisting of a hairpin-shaped 4.5S RNA (SRP RNA) and Ffh protein (SRP54 homologue in eukaryotes), then binds the RNC[7,8]. Ffh is composed of an N-terminal helical and a Ras-like GTPase domain (NG domain)[9] and methionine-rich M domain that binds to SRP RNA with pico-molar affinity[10–12]. The SRP interaction with the RNC is mediated through both its NG domain and the M domain, which interacts with the SS cargo via a hydrophobic groove sealed by a flanking loop, termed the 'fingerloop'[10,11]. The RNC–SRP complex is then recruited to the membrane by the SR (FtsY in bacteria), which is composed of an Ffh-homologous NG domain[13] and a largely unstructured A domain that serves to anchor the SR on the membrane and the translocon[14]. Membrane recruitment of the RNC–SRP depends on a guanidine triphosphate (GTP)-mediated heterodimerization of the NG domains of SRP and SR[15,16], resulting in an initial 'early' state of the SRP–SR complex, in which the NG domain of SR binds to the tetraloop of the SRP RNA hairpin[17,18]. Conformational rearrangements in the SRP–SR heterodimer[15,16,19] then lead to detachment of the SRP–SR NG dimer from the RNC[8,20,21] to form the 'closed' state[22]. At this stage, the Sec translocon binds to the RNC–SRP–SR complex, and the heterodimerized NG domains of SRP and SR dock at the distal end of the SRP RNA hairpin[23–25], forming an 'activated' state where GTP hydrolysis is triggered[26]. Concurrently, the SS is handed over to the Sec translocon, promoting the dissociation of the SRP and the SR once GTP is hydrolyzed[27]. Nascent proteins are then inserted into or translocated across the membrane by the Sec translocon[1].

Isolated components of the targeting and membrane insertion machinery were previously visualized by X-ray crystallography in different conformational states, including isolated SRP, SRP in complex with SRP RNA and the translocon[10,24,28–34]. However, structural information about the binding of the SRP, SR and the translocon to the ribosome in bacteria is only available at lower resolution[21,35–38], and a comprehensive view on how the SS is recognized by the SRP and then delivered to the translocon is still lacking. Using cryo-electron microscopy, we present structures of the co-translational-targeting pathway intermediates and a complex with Sec translocon at resolutions from 3.4 to 4.3 Å. Our results provide the structural basis of the interaction of the SRP M domain with the RNC, and we present a newly observed state of the SRP–SR complex in the presence of non-hydrolysable GTP analogue that maintains contacts with both the RNC and SRP RNA. In addition, we observe remodelling of the exit tunnel of the RNC and a direct interaction with the nascent chain by the cytoplasmic loops of the Sec translocon. Our results provide evidence for an active role played by the SRP and the Sec translocon on the RNC during cargo recognition and handover.

Results

Overview of the co-translational protein-targeting complexes. To obtain high-resolution structures of co-translational-targeting complexes using cryo-electron microscopy, we utilized a well-characterized system to produce SecM-stalled RNC[39] bearing a modified SS of the PhoA protein, which contains nine leucines and one alanine (1A9L)[40–42] (Methods). This SS construct exhibits strong binding affinity for both SRP and SRP–SR complexes and has been demonstrated to direct efficient co-translational protein targeting[40]. This approach in combination with the latest cryo-electron microscopy technology enabled us to reconstruct structures of several complexes involved in co-translational targeting and membrane insertion at near-atomic resolution and perform refinement of the atomic coordinates of our models.

Molecular basis of the RNC–SRP interactions. Our electron microscopy reconstructions reveal two states of the RNC–SRP complex (Supplementary Fig. 1). In one state, SRP RNA and the M domain of the Ffh are ordered, whereas in a second state, we additionally observe the NG domain bound to the RNC. These structures were refined to 3.8- and 4.3 Å resolution, respectively (Fig. 1; Supplementary Fig. 2). The SRP forms three contacts with the ribosomal surface in the vicinity of the nascent polypeptide exit site (Fig. 2a; Supplementary Fig. 3), in agreement with low-resolution studies[36,43]. The NG domain binds close to ribosomal protein uL29, positioning the G domain in close proximity to the tetraloop of the SRP RNA. The M domain is located in the vicinity of the tunnel exit, while the SRP RNA contacts the ribosomal protein bL32.

The resolution of the M domain approaches the mean resolution of the ribosome (3.8 Å), which allowed us to build and refine the coordinates of four alpha helices (MH1, MH2, MH3 and MH4) plus the SS (Supplementary Fig. 2c–d; Supplementary Table 1). The M domain interacts almost exclusively with the 23S ribosomal RNA (rRNA; Supplementary Fig. 3a–b), which spans several conserved regions of MH3 and MH4 (Supplementary Fig. 3c–d). The clearly visible density of the SS helix is sandwiched between a hairpin loop of uL24 and the tip of H59 and stacks against MH1 and MH4 (Fig. 2b), which indicates a role for these two ribosomal structural elements in cargo recognition and possibly handover. Although we could not assign the side chains of the SS, which reflects the promiscuity of SS recognition by the M domain, we consider it more likely that the positively charged N terminus of the SS points towards negatively charged H59 rRNA. This orientation is also consistent with the visible path of the nascent chain in the tunnel and the placement of the SS in previous studies (Fig. 2c)[28,29]. The density for the C terminus of the M domain, which is disordered in isolated complexes[10,12], can now be visualized. This region forms a short helix (MH5) that seals the SS-binding hydrophobic groove from the bottom. The C terminus further extends into the ribosomal tunnel, where it is in contact with the 23S rRNA, as evidenced by the electron microscopy density contoured at lower threshold (Supplementary Fig. 3a). This location would also allow the C terminus to interact with SSs before their emergence. Consistently, contacts between the C-terminal region of the M domain with the ribosomal tunnel were also observed in the eukaryotic SRP complex[44].

The interaction areas of the NG domain with the RNC approach the mean resolution of the complex (4.3 Å) and are established through NG loops 1 and 2, which clamp the N-terminal helix of uL29 and contact the C-terminal tail of uL23 (Fig. 2d–e). The interactions include highly conserved residues (Supplementary Figs 4 and 5) mediating both hydrophobic and electrostatic interactions in agreement with previous cross-linking and mutational experiments[7,8,27,45]. Taken together, the interactions of the M and NG domain with the RNC and the positioning of the NG domain in close proximity to the SRP RNA tetraloop indicates that the NG domain is in a state that primes it for the SR binding.

Architecture of the RNC–SRP–SR complexes. Recognition of the RNC–SRP complex by SR on the membrane initiates GTP-dependent conformational rearrangements that lead to a detachment of the SRP–SR NG domain dimer from the RNC. We assembled the RNC–SRP–SR complex in the presence of 5′-guanylyl imidodiphosphate and could reconstruct the structures of two different states, which were refined to 3.8- and 3.7 Å resolution, respectively (Supplementary Figs 6 and 7). In one state, we observed a detachment of the SRP–SR NG dimer, which was identified as the 'closed' state of the SRP–SR dimer in a recent cryo-electron microscopy study[21]. Concomitantly with the detachment of this dimer, we can now observe a release of the SRP RNA from its contact with the RNC (Fig. 1; Supplementary Fig. 7e–f). The detachment of the distal region of the SRP RNA may occur to accommodate the 'activated' state of the NG dimers at the RNA distal site[23,24].

In the second state, we identified a conformation of the SRP–SR-targeting complex in which the SR G domain is bound to the SRP RNA tetraloop, whereas the N domain of Ffh is still attached to the RNC at uL23 and uL29 (Fig. 1; Fig. 3a–b). This conformation of the SRP–SR complex was never observed before indicating that it is a short-lived intermediate, and thus we defined it as the 'early' state of the SRP–SR complex in accordance with previous biochemical data[17,18]. The G domains of SRP and SR form a tight complex and when compared with the RNC–SRP complex, the Ffh G domain is rotated relative to the N domain (Fig. 3c; Supplementary Fig. 7c–d). This conformational change could be required to accommodate the SR on the tetraloop. Intriguingly, this state of the Ffh NG domain is distinct from a conformation previously observed in the structures of the isolated Ffh NG domain[9], the SR–SRP NG dimer[15,16,46] and the activated NG dimer bound to the distal site of the SRP RNA[23,24]. Furthermore, the GM linker, connecting the NG to M domain, can be observed stacking against the fingerloop and MH1, extending the hydrophobic groove within which the SS is bound (Fig. 3a; Supplementary Fig. 8b). This linker and the preceding sequence exhibit a high degree of conservation (Supplementary Fig. 4), which is consistent with our observation of this region interacting with the SS. This would imply a role for the GM linker in communicating the presence of the SS at the M domain to the attached NG dimer before NG dimer detachment and relocation to the distal site of the SRP RNA.

In addition to the conformational changes in the NG dimer in our visualized complexes, we observe that the M domain of SRP adopts distinct conformations. Notably, in the RNC–SRP complex, the hydrophobic pocket is less well ordered and only weak density for the SS is visible, together with a flexibly disposed fingerloop and GM linker (Fig. 4a; Supplementary Fig. 8a). In contrast, the SS in the 'early' state of the RNC–SRP–SR-targeting complex exhibits a pronounced α-helical density, packs against MH4 and interacts with MH1 (Fig. 4b), along with a now ordered fingerloop and GM linker (Supplementary Fig. 8b). Fluorescence resonance energy transfer studies have also reported tighter packing of the SS against MH4 in the RNC–SRP–SR compared with the RNC–SRP complex[27]. In the 'closed' state of the SRP–SR complex, the fingerloop further clamps down onto the SS, compacting hydrophobic groove of the M domain, while the density for the GM linker is now absent (Fig. 4c; Supplementary Fig. 8c). Taken together, these results present further evidence that the conformational states of the M and NG domains are

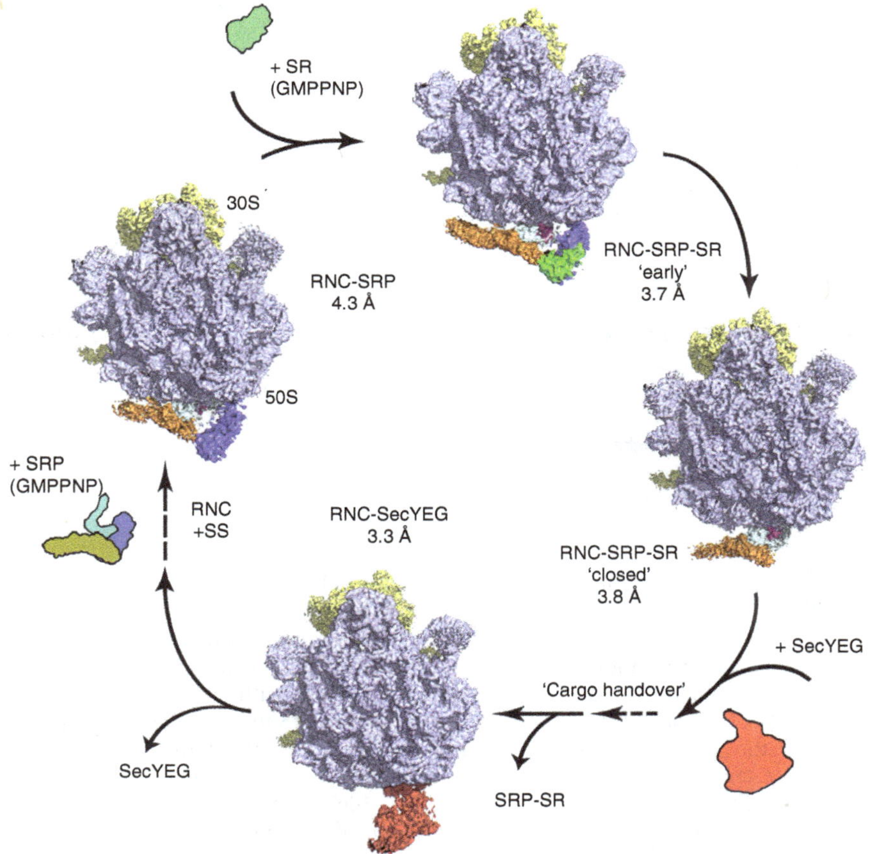

Figure 1 | Overview of the visualized co-translational protein-targeting complexes. Structures of the RNC-SRP, RNC-SRP-SR 'early' and 'closed' states, and the RNC-SecYEG protein-targeting complexes resolved to near-atomic resolution and placed in sequential order. SRP and SR electron microscopy densities were locally filtered to 6 Å.

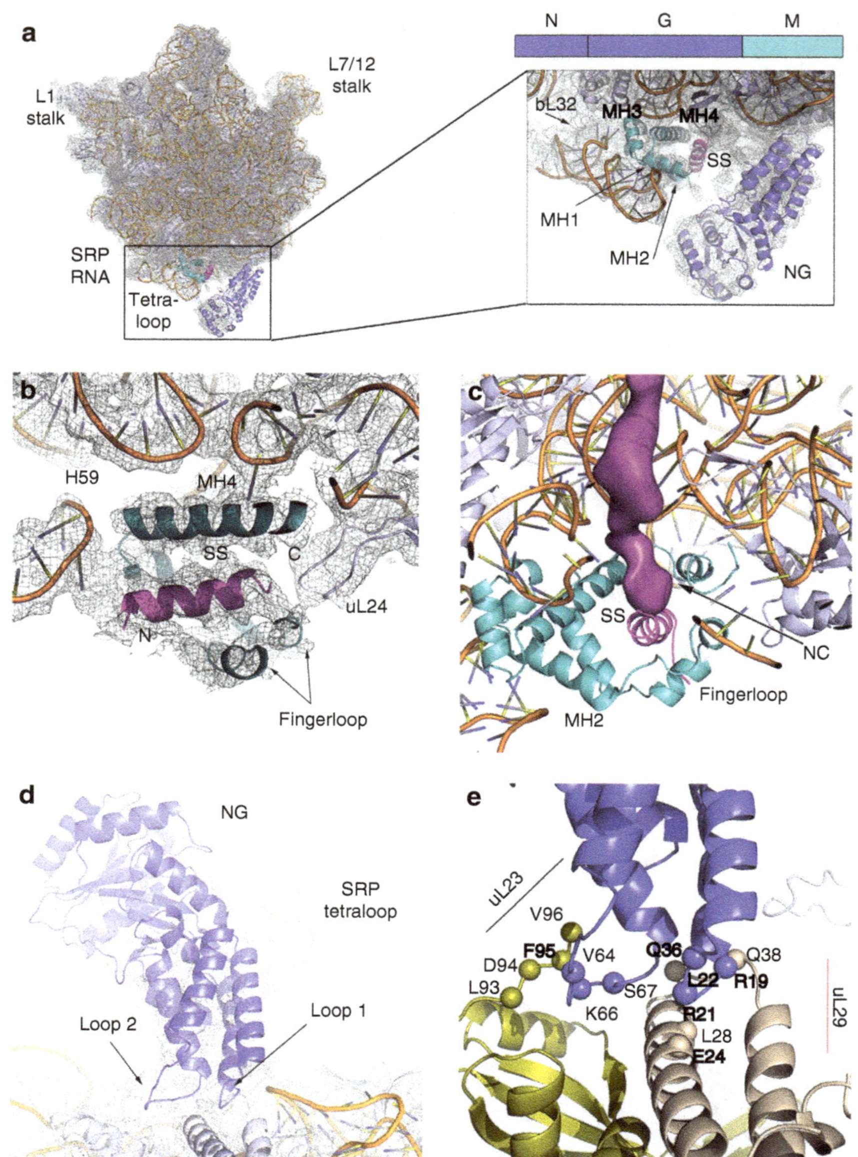

Figure 2 | Atomic model of the RNC-bound SRP M and NG domains. (**a**) Cartoon representation of the RNC-bound SRP M and NG domains with overlaid electron microscopy density filtered to 4.3 Å resolution. The electron microscopy density in the zoomed snapshot was locally filtered to 6 Å. NG and M domain are coloured blue and teal; RNA, ribosomal proteins and SS are coloured orange, light blue and magenta, respectively. (**b**) Cross-section of the SS-bound M domain hydrophobic groove with overlaid electron microscopy density filtered to 4.3 Å. (**c**) A view of the polypeptide exit site showing the M domain in the 'closed' SRP-SR state. The nascent chain (NC) electron microscopy density is displayed as surface (filtered to 6 Å) and coloured in magenta. (**d**) View of the NG domain bound to the RNC, fitted into electron microscopy density. (**e**) Conserved residues, spanning the contact regions, are displayed as spheres. Ribosomal proteins uL23 and uL29 are coloured olive and wheat, respectively.

connected and depend on the presence of the SS and the nucleotide state of the SRP–SR GTPase core.

Atomic model of SecYEG bound to a translating ribosome. The detachment of the SRP–SR from the tetraloop and docking to the distal SRP RNA site exposes the Sec translocon-binding site on the RNC. We have determined the structure of the RNC in complex with the 'translocating' *Escherichia coli* Sec translocon (SecYEG) at 3.3 Å (Supplementary Figs 9 and 10a), allowing us to directly visualize the next step in the pathway during co-translational targeting of proteins to the membranes (Fig. 1). Our structure provides a substantial improvement from previous low-resolution cryo-electron microscopy reconstructions[35,37] allowing us to visualize atomic details of the contact points with the RNC and resolve transmembrane helices within the less well-ordered micelle (Fig. 5a–c; Supplementary Fig. 10c). In particular, our structure of the translocating SecYEG reveals a displacement of the plug helix of the aqueous channel in comparison with its position in the structure of the idle translocon (Supplementary Fig. 10d)[31]. In addition, we now observe widening of the lateral gate indicating that the RNC-bound translocon adopts an open conformation (Supplementary Fig. 10e). The contacts with the RNC, including the cytosolic loops of the translocon (loops 6/7 and loop 8/9), can now be visualized at atomic level (~3.4 Å) and were therefore built *de novo* (Supplementary Fig. 10a–b; Supplementary Table 1). In contrast to the eukaryotic Sec translocon[47], loop 6/7 of SecY is further inserted into the exit tunnel of the ribosome and sandwiches the nascent chain together with a hairpin loop of uL23 (Fig. 5d; Supplementary Fig. 11a). In particular, this hairpin loop undergoes a shift compared with its position in a recent high-resolution X-ray structure of a non-translating ribosome[48] and in our RNC–SRP complexes. Loop 8/9 is inserted between rRNA helices H50, and H53, where a contact is established through stacking of residue R357 onto base A1392 (Fig. 5c), in a position similar to the one recently described for the eukaryotic translocon[47]. In contrast to the eukaryotic translocon, which contacts mainly 28S rRNA, loop 6/7 in the bacterial system interacts almost exclusively with ribosomal protein uL29 (Fig. 5c).

Interestingly, the contact residues on uL29 overlap with the SRP NG domain-binding site on the ribosome (Supplementary Fig. 11b). Previous crosslinking[7,8,45] and low-resolution cryo-electron microscopy studies[49,50] suggested that both the translocon and SRP interact with both uL23 and uL29. Our data now show that these two factors interact predominantly with uL29, while crosslinks to uL23 occur due to proximity. Moreover, the nascent chain density in the RNC–SecYEG complex is visualized to the end of the ribosomal exit tunnel, where it contacts hairpin loop 6/7. In addition, a tubular electron microscopy density between TM2, TM7 and TM8 at the lateral gate of the translocon is observed, which could represent a density for the SS (Supplementary Fig. 11a).

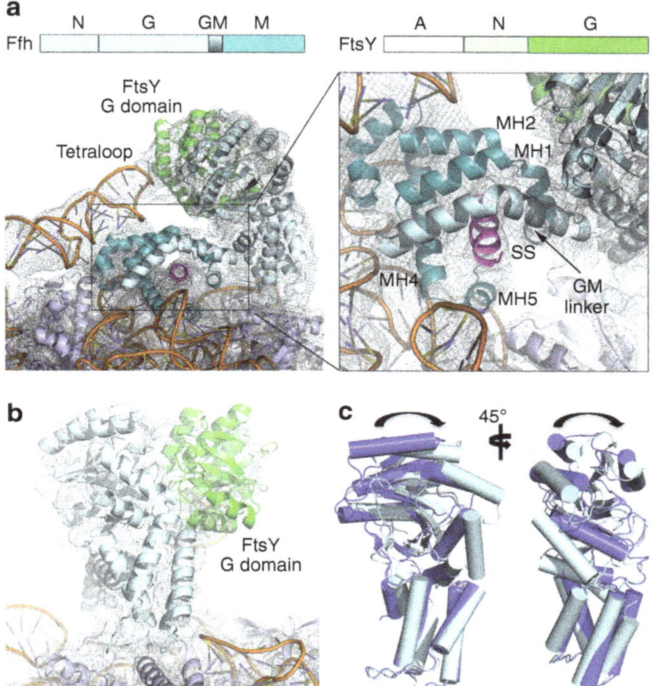

Figure 3 | Atomic model of the SRP-SR complex in the 'early' state. (**a**) Domain structure of the RNC-SRP-SR complex. Representative view of the 'early' RNC-SRP-SR-targeting complex, depicting the docked SRP-SR NG dimer on the RNC with a magnified snapshot of the M domain. Flexibly disposed A and N domains of FtsY were omitted from the final atomic model. (**b**) Snapshot of the SRP-SR NG domains. The electron microscopy densities of the NG domains and SRP RNA was locally filtered to 6-Å resolution. (**c**) Overlay of the Ffh NG domain in the RNC-SR complex (blue) and the RNC-SRP-SR 'early' complex (cyan).

Discussion

Our RNC–SR- and RNC–SRP–SR-targeting complexes imply an interplay between the SS, the fingerloop of the M domain and the GM linker: binding of the SRP to the ribosome involves the formation of interactions between the SRP NG domain, uL23 and uL29, and positioning of the M domain together with the SRP RNA next to the tunnel exit (Fig. 6a). Our data show that the C terminus of the M domain becomes inserted into the ribosomal tunnel, where it is able to interact with the nascent chain. This would imply that the SRP has increased affinity for the translating ribosome even before the SS exits the tunnel, which is consistent with previous biochemical observations[51,52]. Higher affinities of SRP binding to the RNC can also be established once the SS emerges from the ribosome tunnel[53]. Recognition of the SS by SRP on the ribosome positions the NG domain and the SRP tetraloop to promote dimerization of the SRP–SR NG domains[38,40]. Our results now reveal that the fingerloop remains disordered, presumably due to the conformation of the docked SRP NG domain, which would sterically clash with an ordered fingerloop. Furthermore, our results show that the formation of the 'early' SRP–SR complex leads to a rotation of the Ffh NG domain and structuring of the GM linker and the fingerloop (Fig. 6b). The ordered GM linker, in the newly observed 'early' targeting complex, appears to play an important role in binding and positioning the SS. This is in addition to its role in repositioning the NG dimer to the distal region of the SRP RNA at a later stage, where GTP hydrolysis is triggered[24,25]. In the 'closed' SRP–SR complex, we now observe that the fingerloop moves closer to the SS further compacting the hydrophobic groove, whereas the NG domains and the distal region of the SRP RNA detach from the ribosomal surface (Fig. 6c). These results provide evidence for communicating the presence of a SS between the M domain with the NG domain via the fingerloop, which is also corroborated by previous biochemical data[54].

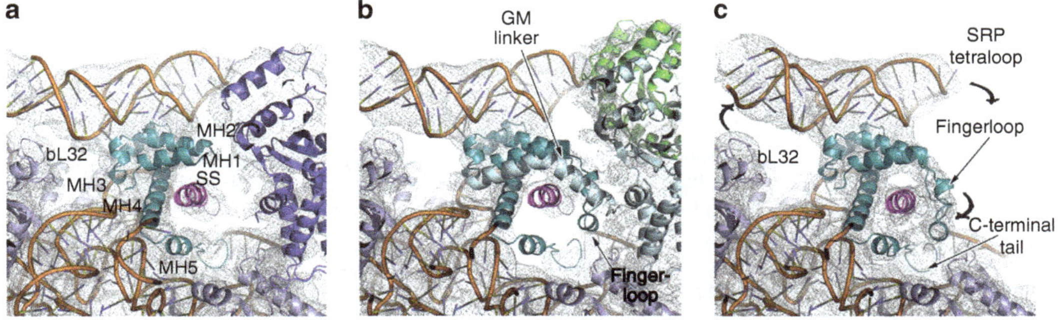

Figure 4 | Conformational rearrangements of the M domain during co-translational targeting. (**a–c**) Structures of the M domain depicting distinct conformational states in RNC-SRP and RNC-SRP-SR in the 'early' and 'closed' states, respectively, with overlaid electron microscopy densities. Curved arrows indicate detachment of the SRP RNA and compaction of the SS hydrophobic groove. The colour scheme is identical to that in Figs 2 and 3.

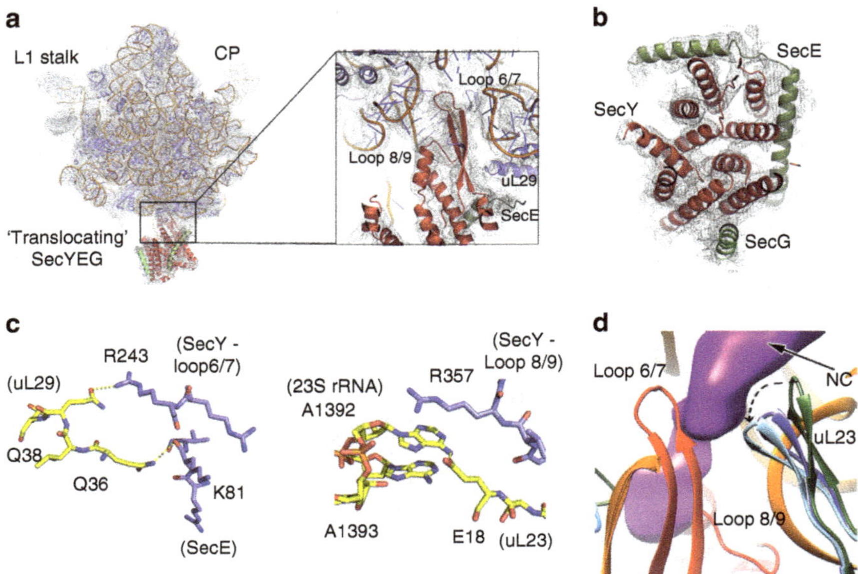

Figure 5 | Structure of the 'translocating' bacterial Sec translocon. (**a**) Atomic model of the RNC-SecYEG complex with overlaid electron microscopy density of the Sec translocon locally filtered to 4.8 Å. (**b**) Representative density of the TM helices of the translocon. (**c**) Contacts between the Sec translocon and the RNC (uL23 and uL29). (**d**) Close-up view of loops 6/7 and 8/9 (red) and the conformation of the hairpin loop of uL23 in the 70S (green), RNC-1A9L and RNC-SRP (orange), and RNC-SecYEG (cyan).

Our data indicate that the 'closed' state of the SRP–SR-targeting complex is in a conformation that would allow the translocon to approach the M domain where the SS is bound (Fig. 6d; Supplementary Fig. 11c). Previous Fluorescence resonance energy transfer studies also proposed a simultaneous binding mode of the SRP–SR complex and the Sec translocon at the polypeptide exit tunnel[26]. Intriguingly, the only structural elements from our ribosome-bound SRP that would clash with the translocon are the fingerloop and the C terminus of the M domain. Their position on the ribosome can now be seen occupied by the cytosolic loop 6/7 of the translocon in the RNC–SecYEG complex, suggesting an interplay mechanism between these elements during cargo handover. This is further corroborated by biochemical studies reporting that the deletion of the fingerloop renders the SRP unable to stably engage with the translocation machinery[54]. Once SS handover is complete, the translocon may employ a similar mechanism as the SRP to sense the presence of the nascent chain in the tunnel. In particular, our results imply that the translocon would maintain affinity for the RNC even after the fully synthetized protein detaches from the transfer RNA, by sandwiching the nascent chain between loop 6/7 of the translocon and the hairpin loop of uL23 as long as it is in the tunnel (Fig. 6e).

The structural snapshots of several intermediates of the targeting process presented here provide insights into the molecular mechanism of co-translational protein targeting and membrane protein insertion. Particularly, the complexes resolved in this study track the series of interactions that a particular SS forms during this process, and underscore important roles for the GM linker and the fingerloop of the M domain in coordinating cargo recognition and cargo handover. We also reveal the molecular basis for the interactions between the ribosome, the SRP and the translocon, which allow a better understanding of the spatial rearrangements of these factors on the surface of the ribosome. Finally, our data depict remodelling of the exit site of the ribosomal tunnel, induced by the binding of the Sec translocon and of the SRP, possibly as a nascent chain-sensing mechanism.

Methods

Protein expression and purification. The plasmids pET24aFfh and puC19Ffs[38,43] were co-transformed into *E. coli* strain BL21Star(DE3) (Invitrogen) and the cells were grown in LB media at 37 °C. Cultures were induced at an $OD_{600\,nm}$ of 0.6 with 1 mM isopropyl-β-D-thiogalactoside for 3 h and broken using a French press in buffer A (50 mM HEPES-KOH, 100 mM KCl, 10 mM $MgCl_2$, 1 mM TCEP, pH 8.0). The cleared lysate was purified on His-Trap, MonoQ and S200 columns (GE Healthcare). The presence of the protein component, Ffh, and the SRP RNA, Ffs, was verified using double stained 12% SDS–polyacrylamide gel electrophoresisgels with two drops of 0.1% ethidium bromide added, followed by Comassie brilliant blue. The buffer of purified SRP was exchanged to buffer B (50 mM HEPES-KOH, 100 mM KCl, 10 mM $MgCl_2$, 5% glycerol, pH 7.3), and flash frozen and stored at − 80 °C until further use. FtsY was expressed from a pET24aFtsY vector, purified and stored in buffer B in a similar manner as described above. The plasmid pTrc99a_SecYEG was transformed in BL21C43(DE3) and the cells were grown at 30 °C in TB media to an OD_{600nm} of 2.0, followed by induction with 1 mM isopropyl-β-D-thiogalactoside at 18 °C overnight. Cells were passed once through a French Press and the lysate was cleared twice in a 70 Ti rotor (Beckman Coulter) at 15,000 r.p.m. for 5 min. Membranes were solubilized for 1 h in buffer A using a final concentration of 1% n-dodecyl β-D-maltopyranoside (DDM) per 10 mg ml^{-1} of protein. Subsequently, solubilized membranes were clarified at 40,000 r.p.m. in a 70 Ti rotor (Beckman Coulter) for 1 h. The supernatant was transferred to a Dounce homogenizer, resuspended in buffer A (+ 0.02% DDM) and homogenized using 10 strokes. Purification of the resuspended membranes was carried out using His-Trap, SP Sepharose and S200 columns (GE Healthcare). Purified SecYEG sample was dialysed in buffer B (+ 0.02% DDM) and flash frozen at − 80 °C. All purification procedures were performed at 4 °C unless otherwise stipulated.

Preparation of RNC complexes. RNCs were generated using an *in vitro* translation system using a membrane-free extract from *E. coli* BL21(DE3)[39]. Messenger RNA containing an N-terminal 3 × Strep-tag followed by an engineered SS (1A9L) based on the first 85 amino acids of the PhoA[40] protein sequence and a SecM-stalling sequence was produced using an cell-free translation system for 30 min at 37 °C. Stalled RNCs were applied on a 10–40% sucrose gradient and subjected to centrifugation for 18 h at 19,000 r.p.m. and 4 °C in a SW 32 Ti rotor (Beckman Coulter) to remove polysomes. Monosomes were then loaded onto a Strep-Tactin Sepharose column (GE Healthcare) and eluted with 2.5 mM D-desthiobiotin (IBA). Eluted fractions were concentrated by pelleting for 3 h in an MLA80 rotor (Beckman Coulter) at 65,000 r.p.m. and 4 °C. Pellets were resuspended in buffer C (50 mM HEPES-KOH, 100 mM KOAc, 10 mM $Mg(OAc)_2$, pH 7.3) and flash frozen at − 80 °C.

Cryo-electron microscopy data acquisition. RNC complexes were incubated at 37 °C for 30 min followed by incubation at 4 °C for 30 min using a 1:3 molar excess

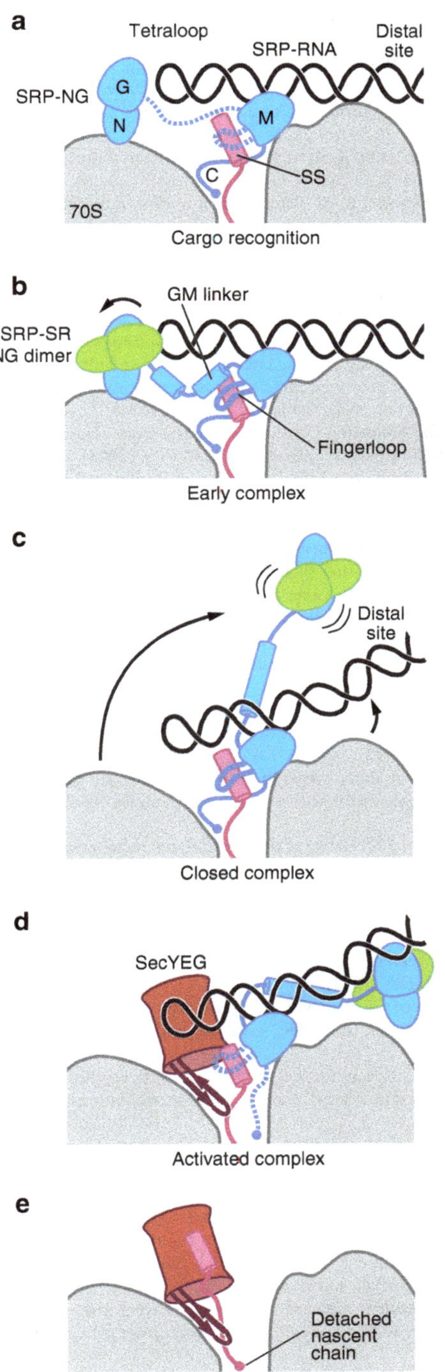

Figure 6 | Schematic of the proposed regulatory mechanism of co-translational protein targeting. (**a–e**) SRP, SR and SecYEG binding to the RNC based on the structures of the co-translational protein-targeting complexes reported in this study. For details see main text.

of factors (either SRP, SRP and SR or SecYEG) to yield a final concentration of 250 nM in reaction buffer (50 mM HEPES-KOH, 50 mM KOAc, 25 mM $Mg(OAc)_2$, 0.02% DDM, 2 mM GMPNP, 5 mM spermidine, 0.5 mM spermine). Samples were applied to Quantifoil grids upon which an additional thin layer of carbon had been previously deposited. Samples were incubated for 60 s, blotted for 10 s using filter paper at 6 °C and 100% relative humidity, and then plunged directly into liquid ethane cooled with liquid nitrogen temperature using a Vitrobot (FEI Company). Data were collected on a Titan Krios cryo-transmission electron microscope (FEI Company) operated at 300 KeV and equipped with a Falcon II direct electron detector. EPU software was used for data collection within a defocus range (− 0.8 to − 3.2 µm) and at × 101,083 magnification. A total of seven frames were collected for each image with a total dosage of 20 electrons per $Å^2$. The movie frames were aligned using DOSEFGPU DRIFTCORR[55] to correct for beam-induced movement.

Structures calculations. Three different data sets were collected for RNC–SRP, RNC–SRP–SR and RNC–SecYEG complexes, respectively. Overview images of holes were inspected for poor quality ice and cracks in the carbon. The power spectra of individual frames were then carefully inspected and only images exhibiting Thon rings that extended beyond 5 Å were retained. Contrast transfer function (CTF) was estimated using CTFFIND3 (ref. 56). Particles were selected using the batchboxer from EMAN[57]. An initial round of two-dimensional classification was performed on fourfold binned images using the maximum-likelihood refinement algorithm implemented in RELION[58] to select for two-dimensional averages exhibiting high-resolution features (Supplementary Figs 1, 6 and 9). A three-dimensional classification approach was the preformed using as a reference an empty 70S ribosome low-pass filtered to 50 Å. This step was used to remove ratcheted ribosomes or ribosomes with weakly bound factors. Twofold binned particle images were then refined against an empty 70S ribosome filtered to 50 Å using RELION. An additional classification step, applying a mask over the bound factor area, was performed using local search and skipping alignments, and limiting the resolution to 25 Å. Masks were calculated using SPIDER[59]. Finally, three-dimensional classes exhibiting strong density corresponding to SRP, SRP–SR or translocon were refined by applying a mask on the 50S subunit and using information to Nyquist frequency in RELION. These final maps were used for model building, validation and refinement.

Model building, refinement and validation. Models of the RNC–SRP and RNC–SecYEG complexes was built using O[60,61] and COOT[62]. The coordinates were refined using PHENIX[63] as described previously[64]. The atomic coordinates of the 70S[48], SRP RNA (domain IV) and the Ffh M domain[65] from *E. coli* were initially docked as a rigid bodies into the cryo-electron microscopy maps. Homology models for the *E. coli* SecY, E and G were obtained using Phyre2 (ref. 66) using the structures of *T. thermophilus* SecY[33] and *Methanocaldococcus jannaschii* Sec E and G[31] as templates. Regions of RNA and proteins not fitting into the electron microscopy densities were manually readjusted, including the SecYEG helices. Due to limited local resolution, the fingerloop and the C-terminal extension of SRP were modelled as unassigned (UNK) residues, and residues in loops 1 and 2 were stripped to the backbone.

The resulting models were subjected to nine cycles of individual B-factor and coordinate refinement against figure of merit (FOM) weighted experimental electron microscopy phases and back-calculated structure factors[64] using the phased maximum-likelihood (MLHL) target. Information during refinement was limited to the FSC gold standard 0.143 cutoff criteria for each cryo-electron microscopy map (Supplementary Table 1). Good main chain geometry, especially for the less well-resolved areas of electron microscopy density, was maintained by imposing base pair, Ramachandran and secondary structural restraints during coordinate refinement[64]. To prevent over-fitting of the coordinates, the weighting between model geometry and structure factors was screened to produce a model with good geometry and low R-values, resulting in fix_wxc values of 1.0 (RNC–SRP–SR, closed), 1.1 (RNC–SRP, class 3) and 1.2 (RNC–SecYEG). Both RNC–SRP (class 3) and 'closed' RNC–SRP displayed near-identical conformation of the M domain, and thus were used to produce the atomic model for this region. For validation, the refined coordinates were randomly shifted by 0.5 Å and the B-values were reset. Subsequently, both structures were refined into one data half-set of the data following a similar procedure as described above, and the resulting FSCs of the models were compared with the FSCs calculated for the other half-set (Supplementary Figs 2d and 9a).

To interpret the electron microscopy maps of the RNC–SRP with stable NG domain density and the 'early' RNC–SRP–SR complex, the refined atomic model of the 50S subunit from the RNC–SRP (class 3) complex was fitted as a rigid body into the electron microscopy densities. To account for the rotation of the SRP on the 50S, the Ffh M domain together with SRP RNA and the SS were fitted as a rigid body, and minor adjustments were applied to the distal end of the SRP RNA helix. Although weak electron microscopy density for the distal end of the RNA was visible, this region was not modelled due to its flexibility. Homology models of the NG domain of Ffh (PDB:1JPJ)[67] and the NG dimers of the Ffh–FtsY (PDB:2J7P)[46], derived from *Thermus aquaticus* in the GPPNP state, were docked as rigid bodies, followed by manual readjustment of several helices and loops. These included loop 1 and loop 2 of the Ffh N domain contacting the ribosome, the contact area with the SRP RNA of the FtsY G domain, the Ffh N domain helical bundle, which was in a different orientation relative to the G-domain, and the C-terminal helix of the Ffh G domain. Due to limited local resolution, the fingerloop, the GM linker helix and the C-terminal extension of SRP were modelled as unassigned (UNK) residues, and residues in loops 1 and 2 were stripped to the backbone. The N domain of FtsY was only visible at low resolution, which means that this region was flexibly disposed. To correct the final models for geometry errors and to remove sterical clashes, the rebuilt SRP and ribosomal contact areas were subjected to 100 iterations of geometry minimization using PHENIX, while secondary structure, Ramachandran and base pair restraints were applied.

Making figures and plots. All figures were produced either using UCSF CHIMERA[68] or Pymol (The Pymol Molecular Graphics System Version 1.7 Schrödinger, LLC.). Local resolution maps were produced using ResMap[69].

References

1. Driessen, A. J. & Nouwen, N. Protein translocation across the bacterial cytoplasmic membrane. *Annu. Rev. Biochem.* **77,** 643–667 (2008).
2. Kudva, R. *et al.* Protein translocation across the inner membrane of Gram-negative bacteria: the Sec and Tat dependent protein transport pathways. *Res. Microbiol.* **164,** 505–534 (2013).
3. Doudna, J. A. & Batey, R. T. Structural insights into the signal recognition particle. *Annu. Rev. Biochem.* **73,** 539–557 (2004).
4. Keenan, R. J., Freymann, D. M., Stroud, R. M. & Walter, P. The signal recognition particle. *Annu. Rev. Biochem.* **70,** 755–775 (2001).
5. Denks, K. *et al.* The Sec translocon mediated protein transport in prokaryotes and eukaryotes. *Mol. Membr. Biol.* **31,** 58–84 (2014).
6. Osborne, A. R., Rapoport, T. A. & van den Berg, B. Protein translocation by the Sec61/SecY channel. *Annu. Rev. Cell Dev. Biol.* **21,** 529–550 (2005).
7. Gu, S. Q., Peske, F., Wieden, H. J., Rodnina, M. V. & Wintermeyer, W. The signal recognition particle binds to protein L23 at the peptide exit of the Escherichia coli ribosome. *RNA* **9,** 566–573 (2003).
8. Pool, M. R., Stumm, J., Fulga, T. A., Sinning, I. & Dobberstein, B. Distinct modes of signal recognition particle interaction with the ribosome. *Science* **297,** 1345–1348 (2002).
9. Freymann, D. M., Keenan, R. J., Stroud, R. M. & Walter, P. Structure of the conserved GTPase domain of the signal recognition particle. *Nature* **385,** 361–364 (1997).
10. Keenan, R. J., Freymann, D. M., Walter, P. & Stroud, R. M. Crystal structure of the signal sequence binding subunit of the signal recognition particle. *Cell* **94,** 181–191 (1998).
11. Zopf, D., Bernstein, H. D., Johnson, A. E. & Walter, P. The methionine-rich domain of the 54 kd protein subunit of the signal recognition particle contains an RNA binding site and can be crosslinked to a signal sequence. *EMBO J.* **9,** 4511–4517 (1990).
12. Batey, R. T., Rambo, R. P., Lucast, L., Rha, B. & Doudna, J. A. Crystal structure of the ribonucleoprotein core of the signal recognition particle. *Science* **287,** 1232–1239 (2000).
13. Montoya, G., Svensson, C., Luirink, J. & Sinning, I. Crystal structure of the NG domain from the signal-recognition particle receptor FtsY. *Nature* **385,** 365–368 (1997).
14. Weiche, B. *et al.* A cleavable N-terminal membrane anchor is involved in membrane binding of the Escherichia coli SRP receptor. *J. Mol. Biol.* **377,** 761–773 (2008).
15. Egea, P. F. *et al.* Substrate twinning activates the signal recognition particle and its receptor. *Nature* **427,** 215–221 (2004).
16. Focia, P. J., Shepotinovskaya, I. V., Seidler, J. A. & Freymann, D. M. Heterodimeric GTPase core of the SRP targeting complex. *Science* **303,** 373–377 (2004).
17. Jagath, J. R. *et al.* Important role of the tetraloop region of 4.5S RNA in SRP binding to its receptor FtsY. *RNA* **7,** 293–301 (2001).
18. Siu, F. Y., Spanggord, R. J. & Doudna, J. A. SRP RNA provides the physiologically essential GTPase activation function in cotranslational protein targeting. *RNA* **13,** 240–250 (2007).
19. Shan, S. O., Chandrasekar, S. & Walter, P. Conformational changes in the GTPase modules of the signal reception particle and its receptor drive initiation of protein translocation. *J. Cell Biol.* **178,** 611–620 (2007).
20. Halic, M. *et al.* Signal recognition particle receptor exposes the ribosomal translocon binding site. *Science* **312,** 745–747 (2006).
21. von Loeffelholz, O. *et al.* Ribosome-SRP-FtsY cotranslational targeting complex in the closed state. *Proc. Natl Acad. Sci. USA* **112,** 3943–3948 (2015).
22. Zhang, X., Schaffitzel, C., Ban, N. & Shan, S. O. Multiple conformational switches in a GTPase complex control co-translational protein targeting. *Proc. Natl Acad. Sci. USA* **106,** 1754–1759 (2009).
23. Voigts-Hoffmann, F. *et al.* The structural basis of FtsY recruitment and GTPase activation by SRP RNA. *Mol. Cell* **52,** 643–654 (2013).
24. Ataide, S. F. *et al.* The crystal structure of the signal recognition particle in complex with its receptor. *Science* **331,** 881–886 (2011).
25. Shen, K., Arslan, S., Akopian, D., Ha, T. & Shan, S. O. Activated GTPase movement on an RNA scaffold drives co-translational protein targeting. *Nature* **492,** 271–275 (2012).
26. Akopian, D., Dalal, K., Shen, K., Duong, F. & Shan, S. O. SecYEG activates GTPases to drive the completion of cotranslational protein targeting. *J. Cell Biol.* **200,** 397–405 (2013).
27. Saraogi, I., Akopian, D. & Shan, S. O. Regulation of cargo recognition, commitment, and unloading drives cotranslational protein targeting. *J. Cell Biol.* **205,** 693–706 (2014).
28. Hainzl, T. & Sauer-Eriksson, A. E. Signal-sequence induced conformational changes in the signal recognition particle. *Nat. Commun.* **6,** 7163 (2015).
29. Janda, C. Y. *et al.* Recognition of a signal peptide by the signal recognition particle. *Nature* **465,** 507–510 (2010).
30. Rosendal, K. R., Wild, K., Montoya, G. & Sinning, I. Crystal structure of the complete core of archaeal signal recognition particle and implications for interdomain communication. *Proc. Natl Acad. Sci. USA* **100,** 14701–14706 (2003).
31. Van den Berg, B. *et al.* X-ray structure of a protein-conducting channel. *Nature* **427,** 36–44 (2004).
32. Egea, P. F. & Stroud, R. M. Lateral opening of a translocon upon entry of protein suggests the mechanism of insertion into membranes. *Proc. Natl Acad. Sci. USA* **107,** 17182–17187 (2010).
33. Tsukazaki, T. *et al.* Conformational transition of Sec machinery inferred from bacterial SecYE structures. *Nature* **455,** 988–991 (2008).
34. Zimmer, J., Nam, Y. & Rapoport, T. A. Structure of a complex of the ATPase SecA and the protein-translocation channel. *Nature* **455,** 936–943 (2008).
35. Bischoff, L., Wickles, S., Berninghausen, O., van der Sluis, E. O. & Beckmann, R. Visualization of a polytopic membrane protein during SecY-mediated membrane insertion. *Nat. Commun.* **5,** 4103 (2014).
36. Halic, M. *et al.* Following the signal sequence from ribosomal tunnel exit to signal recognition particle. *Nature* **444,** 507–511 (2006).
37. Park, E. *et al.* Structure of the SecY channel during initiation of protein translocation. *Nature* **506,** 102–106 (2014).
38. Estrozi, L. F., Boehringer, D., Shan, S. O., Ban, N. & Schaffitzel, C. Cryo-EM structure of the E. coli translating ribosome in complex with SRP and its receptor. *Nat. Struct. Mol. Biol.* **18,** 88–90 (2011).
39. Schaffitzel, C. & Ban, N. Generation of ribosome nascent chain complexes for structural and functional studies. *J. Struct. Biol.* **158,** 463–471 (2007).
40. Zhang, X., Rashid, R., Wang, K. & Shan, S. O. Sequential checkpoints govern substrate selection during cotranslational protein targeting. *Science* **328,** 757–760 (2010).
41. Doud, S. K., Chou, M. M. & Kendall, D. A. Titration of protein transport activity by incremental changes in signal peptide hydrophobicity. *Biochemistry* **32,** 1251–1256 (1993).
42. Valent, Q. A. *et al.* Early events in preprotein recognition in E. coli: interaction of SRP and trigger factor with nascent polypeptides. *EMBO J.* **14,** 5494–5505 (1995).
43. Schaffitzel, C. *et al.* Structure of the *E. coli* signal recognition particle bound to a translating ribosome. *Nature* **444,** 503–506 (2006).
44. Voorhees, R. M. & Hegde, R. S. Structures of the scanning and engaged states of the mammalian SRP-ribosome complex. *eLife* **4,** e07975 (2015).
45. Ullers, R. S. *et al.* Interplay of signal recognition particle and trigger factor at L23 near the nascent chain exit site on the Escherichia coli ribosome. *J. Cell Biol.* **161,** 679–684 (2003).
46. Gawronski-Salerno, J. & Freymann, D. M. Structure of the GMPPNP-stabilized NG domain complex of the SRP GTPases Ffh and FtsY. *J. Struct. Biol.* **158,** 122–128 (2007).
47. Voorhees, R. M., Fernandez, I. S., Scheres, S. H. & Hegde, R. S. Structure of the mammalian ribosome-Sec61 complex to 3.4A resolution. *Cell* **157,** 1632–1643 (2014).
48. Noeske, J. *et al.* High-resolution structure of the Escherichia coli ribosome. *Nat. Struct. Mol. Biol.* **22,** 336–341 (2015).
49. Menetret, J. F. *et al.* Ribosome binding of a single copy of the SecY complex: implications for protein translocation. *Mol. Cell* **28,** 1083–1092 (2007).
50. Frauenfeld, J. *et al.* Cryo-EM structure of the ribosome-SecYE complex in the membrane environment. *Nat. Struct. Mol. Biol.* **18,** 614–621 (2011).
51. Bornemann, T., Jockel, J., Rodnina, M. V. & Wintermeyer, W. Signal sequence-independent membrane targeting of ribosomes containing short nascent peptides within the exit tunnel. *Nat. Struct. Mol. Biol.* **15,** 494–499 (2008).
52. Holtkamp, W. *et al.* Dynamic switch of the signal recognition particle from scanning to targeting. *Nat. Struct. Mol. Biol.* **19,** 1332–1337 (2012).
53. Noriega, T. R., Chen, J., Walter, P. & Puglisi, J. D. Real-time observation of signal recognition particle binding to actively translating ribosomes. *eLife* **3** **(2014).**
54. Ariosa, A. R. *et al.* Fingerloop activates cargo delivery and unloading during cotranslational protein targeting. *Mol. Biol. Cell.* **24,** 63–73 (2013).
55. Li, X., Zheng, S. Q., Egami, K., Agard, D. A. & Cheng, Y. Influence of electron dose rate on electron counting images recorded with the K2 camera. *J. Struct. Biol.* **184,** 251–260 (2013).
56. Mindell, J. A. & Grigorieff, N. Accurate determination of local defocus and specimen tilt in electron microscopy. *J. Struct. Biol.* **142,** 334–347 (2003).
57. Ludtke, S. J., Baldwin, P. R. & Chiu, W. EMAN: semiautomated software for high-resolution single-particle reconstructions. *J. Struct. Biol.* **128,** 82–97 (1999).
58. Scheres, S. H. RELION: implementation of a Bayesian approach to cryo-EM structure determination. *J. Struct. Biol.* **180,** 519–530 (2012).
59. Frank, J. *et al.* SPIDER and WEB: processing and visualization of images in 3D electron microscopy and related fields. *J. Struct. Biol.* **116,** 190–199 (1996).

60. Jones, T. A., Zou, J. Y., Cowan, S. W. & Kjeldgaard, M. Improved methods for building protein models in electron density maps and the location of errors in these models. *Acta. Crystallogr. A.* **47**(Pt 2): 110–119 (1991).
61. Jones, T. A. Interactive electron-density map interpretation: from INTER to O. *Acta. Crystallogr. D. Biol. Crystallogr.* **60,** 2115–2125 (2004).
62. Emsley, P., Lohkamp, B., Scott, W. G. & Cowtan, K. Features and development of Coot. *Acta. Crystallogr. D. Biol. Crystallogr.* **66,** 486–501 (2010).
63. Adams, P. D. *et al.* PHENIX: a comprehensive Python-based system for macromolecular structure solution. *Acta. Crystallogr. D. Biol. Crystallogr.* **66,** 213–221 (2010).
64. Greber, B. J. *et al.* The complete structure of the large subunit of the mammalian mitochondrial ribosome. *Nature* **515,** 283–286 (2014).
65. Batey, R. T., Sagar, M. B. & Doudna, J. A. Structural and energetic analysis of RNA recognition by a universally conserved protein from the signal recognition particle. *J. Mol. Biol.* **307,** 229–246 (2001).
66. Kelley, L. A. & Sternberg, M. J. Protein structure prediction on the Web: a case study using the Phyre server. *Nat. Protoc.* **4,** 363–371 (2009).
67. Padmanabhan, S. & Freymann, D. M. The conformation of bound GMPPNP suggests a mechanism for gating the active site of the SRP GTPase. *Structure* **9,** 859–867 (2001).
68. Pettersen, E. F. *et al.* UCSF Chimera--a visualization system for exploratory research and analysis. *J. Comput. Chem.* **25,** 1605–1612 (2004).
69. Kucukelbir, A., Sigworth, F. J. & Tagare, H. D. Quantifying the local resolution of cryo-EM density maps. *Nat. Methods* **11,** 63–65 (2014).

Acknowledgements

We are grateful for Professor Shu-ou Shan (Caltech) for providing the 1A9L SS construct and critical comments. Cryo-electron microscopy data were collected at the scientific center for optical and electron microscopy at the ETH Zurich (ScopeM). We are thankful to Peter Tittman and members of the ScopeM facility for technical support. We thank Basil Greber and Nikolaus Schmitz for general discussions. We thank Christopher Aylett for critically reading the manuscript. We acknowledge the use of computing infrastructure provided by the Central Information Technology Services of ETH Zurich. This work was supported by the Swiss National Science Foundation, the National Center of Excellence in Research Structural Biology and RNA & Disease programs of the SNSF and European Research Council grant 250071 under the European Community's Seventh Framework Programme (to N.B.).

Author contributions

A.J. and N.B. initiated the project; A.J. prepared all samples; A.J. and D.B. acquired electron microscopy data; A.J. and D.B. performed cryo-electron microscopy reconstructions. M.L. and A.J. performed model building. M.L. performed model refinements. All authors interpreted the structures and contributed to the final version of the manuscript.

Additional information

Accession codes: Cryo-EM maps have been deposited in the Electron Microscopy Databank with accession codes EMD-8000, EMD-8001, EMD-8002, EMD-8003 and EMD-8004. The coordinates of the atomic structures of the 50S ribosomal subunit in complex SRP–SR, Sec translocon and SRP have been deposited in the protein databank with PDB codes 5GAD, 5GAE, 5GAF, 5GAG and 5GAH.

Competing financial interests: The authors declare no competing financial interests.

9

Optical micromanipulation of nanoparticles and cells inside living zebrafish

Patrick Lie Johansen[1,*], Federico Fenaroli[1,*], Lasse Evensen[1], Gareth Griffiths[1] & Gerbrand Koster[1,†]

Regulation of biological processes is often based on physical interactions between cells and their microenvironment. To unravel how and where interactions occur, micromanipulation methods can be used that offer high-precision control over the duration, position and magnitude of interactions. However, lacking an *in vivo* system, micromanipulation has generally been done with cells *in vitro*, which may not reflect the complex *in vivo* situation inside multicellular organisms. Here using optical tweezers we demonstrate micromanipulation throughout the transparent zebrafish embryo. We show that different cells, as well as injected nanoparticles and bacteria can be trapped and that adhesion properties and membrane deformation of endothelium and macrophages can be analysed. This non-invasive micromanipulation inside a whole-organism gives direct insights into cell interactions that are not accessible using existing approaches. Potential applications include screening of nanoparticle-cell interactions for cancer therapy or tissue invasion studies in cancer and infection biology.

[1] Department of Biosciences, University of Oslo, Blindernveien 31, 0371 Oslo, Norway. * These authors contributed equally to this work. † Present address: Inven2 AS, Forskningsparken, Gaustadalléen 21, 0349 Oslo, Norway. Correspondence and requests for materials should be addressed to G.K. (email: gerbrand.koster@inven2.com).

Many cellular mechanisms depend on regulation of the physical contact between biological structures. A prominent example of structures rich in cellular interactions are the blood vessels where the lumen is surrounded by a layer of endothelial cells; these form a crucial barrier between the blood vessel lumen and the surrounding tissue. This barrier is important for many functions, such as preventing blood clotting, in inflammation, in formation of new blood vessels and in the control of blood pressure[1]. Endothelial cells can interact with different immune cells in the blood, such as neutrophils and macrophages; they can also interact with cancer cells that can pass through the endothelium in the process of extravasation, as well as with blood-borne pathogens[1].

As *in vivo* interactions are especially hard to characterize in detail since they are difficult to image and often occur at unpredictable positions and in short time windows, novel approaches are needed to increase the efficiency of 'catching' these interactions and to study them in a controlled manner. In addition, to properly understand why and how cellular interactions are established, we need analytical methods operating at the sub-micrometre scale of cells and on the time scale of bond formation.

Optical tweezers (OT)[2,3] are a highly sensitive and flexible micromanipulation tool that uses the force of light for non-invasive manipulation of nano- to micrometre sized particles. Moreover, by controlling the strength of the OT, interaction forces can be studied. Because of their versatility, OT have become a central tool for soft matter and life sciences in the last two decades[3]. Using this approach allows one to decipher when and where an interaction occurs, while parallel sensitive imaging and detection methods can be applied for detailed structural analysis. In order to achieve maximum control of the surrounding environment, and to simplify interpretation, a major part of OT experiments have been done *in vitro*, either in minimal reconstituted systems or with cells in culture. More recently, there have been developments towards optical trapping in the, technically more challenging, complex active environments inside living cells[4–7]. Ultimately, however, one would want to study interactions in a multicellular organism, where the interactions of interest occur deeper inside the organism, beneath layers of tissue. Towards this goal there has been one recent report describing optical trapping of erythrocytes in mouse[8]; however, these experiments were necessarily limited to red blood cells in thinner, superficial blood vessels in the ear.

Here we introduce the use of the zebrafish (ZF) larva for applying OT inside a living vertebrate. In the last years, the use of the ZF larva model has exploded in popularity for studying processes such as development and disease models[9]. An obvious advantage of using the ZF larva for OT is that it is optically transparent; moreover, many transgenic zebrafish lines are available with fluorescent cell types, such as macrophages and endothelial cells[10]. In a recent analysis of the biodistribution of different nanoparticles in the zebrafish we found that many of the nanoparticles bound tightly to the endothelium[11]. Here we take advantage of OT to follow and quantify these interactions. In another earlier project we have also followed the process of infection of *Mycobacterium marinum*, the organism of fish tuberculosis in the zebrafish and developed nanoparticle-based therapies against the disease[12].

Using a flexible optical tweezers-imaging system in combination with zebrafish lines, we demonstrate here that many different types of structures, such as microinjected nanoparticles or bacteria, or different cell types such as macrophages (fluorescent) or erythrocytes can be identified and manipulated; this allows their interactions to be determined inside the living zebrafish. We analyse the details of particle–endothelium interactions and quantify 'stealth' properties of particles; such stealth particles are designed to avoid interactions with phagocytic cells, important for using nanoparticles against cancer[13]. Using multiplexing, we show that a number of traps can be activated to 'fish' out multiple nanoparticles from the blood stream simultaneously. We also demonstrate a procedure using OT where a region was first 'cleaned' of erythrocytes after which the interaction details of nanoparticles with the endothelium could be studied in detail in a cell-free environment, thereby simplifying the interpretation. This experiment reveals the formation and detection of tethering nanotubes that are formed when adhered particles are pulled away from endothelial cells, providing convincing evidence that the particles were indeed tightly bound to the cell plasma membrane. Collectively, these data establish the zebrafish larva as a powerful model for optical trap micromanipulation and for analysis of *in vivo* interactions under controlled experimental conditions.

Results and Discussion

Optical trapping of nanoparticles inside living zebrafish. The optical transparency of the thin zebrafish larvae (Supplementary Movie 1; Supplementary Fig. 1) is unique among *in vivo* vertebrate models and one of the main reasons for its immense popularity as a model system. With this in mind, we tested whether the zebrafish was also 'optically see-through' for the infrared laser beam of the optical tweezers and whether this approach could be used to manipulate different structures inside the living larva.

One recent emerging application of zebrafish as a model system is in the nanomedicine field[12]. For this, nanoparticles loaded with drugs and decorated on their surface with (potentially) targeting factors can be injected, enabling their biodistribution and blood circulation to be monitored using live imaging. To determine whether such injected particles can be trapped inside the complex environment of the fish, we injected latex particles into 2-day-old fish larvae (according to a recently developed protocol[12], see also Methods, Supplementary Movie 2 and Supplementary Fig. 2). The particles readily distributed throughout the circulation of the blood stream and with time an increasing portion of the nanoparticles either adhered to the endothelium lining of the blood vessels or were taken up by macrophages.

After mounting the fish in the optical tweezers microscope (Supplementary Fig 3), we used transmission microscopy to select particles of interest that had adhered to the endothelium lining the caudal vein (Fig. 1a). Next, the OT were turned on and we were able to carefully move away some particles off the endothelium (Fig. 1a at 3.5 s, see also Supplementary Movie 3). Subsequently, the particles could be displaced and also moved against the direction of fast blood flow (200 $\mu m\,s^{-1}$ in the vein and $\sim$700 $\mu m\,s^{-1}$ in the artery[14]) indicating strong trapping. In the example shown (Fig. 1a), at time point 5.4 s an erythrocyte hits the optical trap, thereby dislodging the particle from the trap. Intriguingly, the dislodged particle was subsequently pulled back, spring-like, to the original adhesion point. This indicated that the particle was connected through a nanotube (as described previously[15]), which exerted a pulling force on the particle, that re-incorporated the tether into the endothelium after retraction. These manipulations could be done at a trapping laser power settings of 500 mW, about 10% of the maximum force available (corresponding to about max 75 mW in the sample). This demonstrates that statically adhering particles can be trapped in the zebrafish, making possible the investigating of details of interaction that are not possible with other methods.

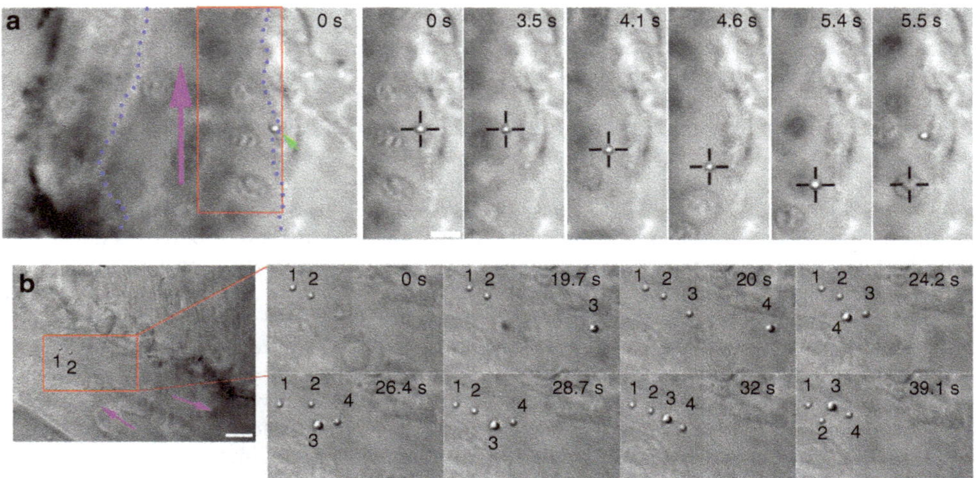

Figure 1 | *In vivo* optical micromanipulation of microinjected particles. (**a**) A particle (green arrowhead) adhered to the endothelium of the caudal vein (indicated with blue dotted lines) is pulled away from the endothelium into the fast blood flow (purple arrow) using optical tweezers (black crosshairs). At time 5.5 s an erythrocyte is drawn into the trap. This replaces the particle in the trap which is subsequently pulled back towards the original adhesion point of the endothelium, presumably due to a connecting nanotube. Experiment repeated at least 80 times. (**b**) Four separate particles (numbered) are fished out of the blood flow and moved towards a sheltered region at the tip of the tail. Purple arrows indicate flow direction. Experiment was repeated at least 10 times. Scale bar, 5 µm.

After having established that an already adhered particle could be trapped, we next investigated whether it was possible to catch particles that were injected into the blood stream and moving with the flow at high speed[14]. To do this, we made use of the available time-sharing multiplexing option of the optical tweezers system; by scanning the trapping laser at high speed over several positions, multiple traps can be created that can be used to trap in parallel a high number of objects[16]. We thus distributed several 'fishing' traps throughout the blood stream. In Fig. 1b (Supplementary Movie 4) using two traps, two particles (marked with '1' and '2') had already been moved towards the tail to a region with lower flow velocity outside the main blood flow (purple arrows). Next, another particle ('3') was caught (19.7 s) and moved together with the two other particles while another particle ('4') was trapped in the flow at $t = 20$ s. To demonstrate the positioning control that the multiplexed tweezers allow *in vivo*, the particles were positioned first in a straight line (32 s) and next repositioned into a square shape (39.1 s). The trapping of multiple particles within the blood flow and the subsequent reorganization of their relative positions demonstrates the robustness of the trapping, and the versatility of experiments that are possible; for example, by establishing simultaneous contact between several particles and specific cells (Supplementary Movie 5; Supplementary Fig. 4).

Trapping was possible throughout the fish and up to depths of 100 µm away from the bottom cover glass (including zebrafish and medium, Supplementary Fig. 3) and we for example trapped cells and particles inside the beating heart. However, we found that most straightforward trapping could be done in the caudal vein and artery in the (thinner) tail region of the fish, at a depth of about 50 µm, where the arterial blood flow turns from retrograde transport towards the tail to anterograde transport towards the head of the fish. The experiments demonstrate that the adhesive properties of particles can be tested inside the living fish using micromanipulation, even in the fast flowing blood in the caudal artery. We successfully trapped and moved polystyrene particles of 200 nm to 1 µm diameter, but conducted most experiments with particles of 840 nm diameter (see Methods section, fluorescence emission peak at ~700 nm) or 1 µm (non-fluorescent) because of their clear visibility and higher trapping stiffness[17]. Slowing down the blood flow using the anaesthetic tricaine[18] facilitates optical trapping, and allows also smaller particles down to 200 nm to be moved, although this was more difficult due to the lower trapping stiffness and the fact that smaller particles were difficult to discriminate using bright field transmission microscopy. However, using confocal fluorescence microscopy the particles could be identified. The larger particles could be trapped without obvious heating damage for extended periods of time inside the vasculature with laser powers 100 mW to 3 W (about 15 mW to ~450 mW in the sample). However, a few scattered darker areas (possibly pigments that had developed despite use of phenylthiourea (PTU), see Methods) interacted strongly with the OT, and when the OT was moved transiently over these areas they became visibly damaged, presumably due to heating. However, by avoiding these areas no discernable damage was observed over the course of the experiments.

Trapping of cells in living zebrafish. Traditionally for experiments with optical tweezers spherical particles with high refractive index are used because they can be trapped with a high stiffness and they are especially suitable for trap calibration for quantitative experiments. However, for other applications, spatial micromanipulation is the crucial feature, as objects such as cells, particles or other nano–micron scale objects can be brought into contact with each other at well-defined time points and positions, opening up the possibility for analysing dynamics not accessible through passive observation.

As we found that particles can be easily and robustly trapped throughout the whole zebrafish, we next asked whether cells could also be micromanipulated inside the fish. In the absence of particles and using higher laser powers (about 2 W system setting, ~250 mW in the sample), we activated the optical trap in the middle of the caudal vein, which almost instantly resulted in the immobilization of an erythrocyte in the trap. These cells could be stably held in position against the full flow of the blood stream (Fig. 2a; Supplementary Movie 6) and could even be moved in the direction against the blood flow. We found that the nucleus of the zebrafish erythrocytes[19] was trapped most strongly. In Fig. 2a, a membrane in a 'bag'-like shape can be observed behind the

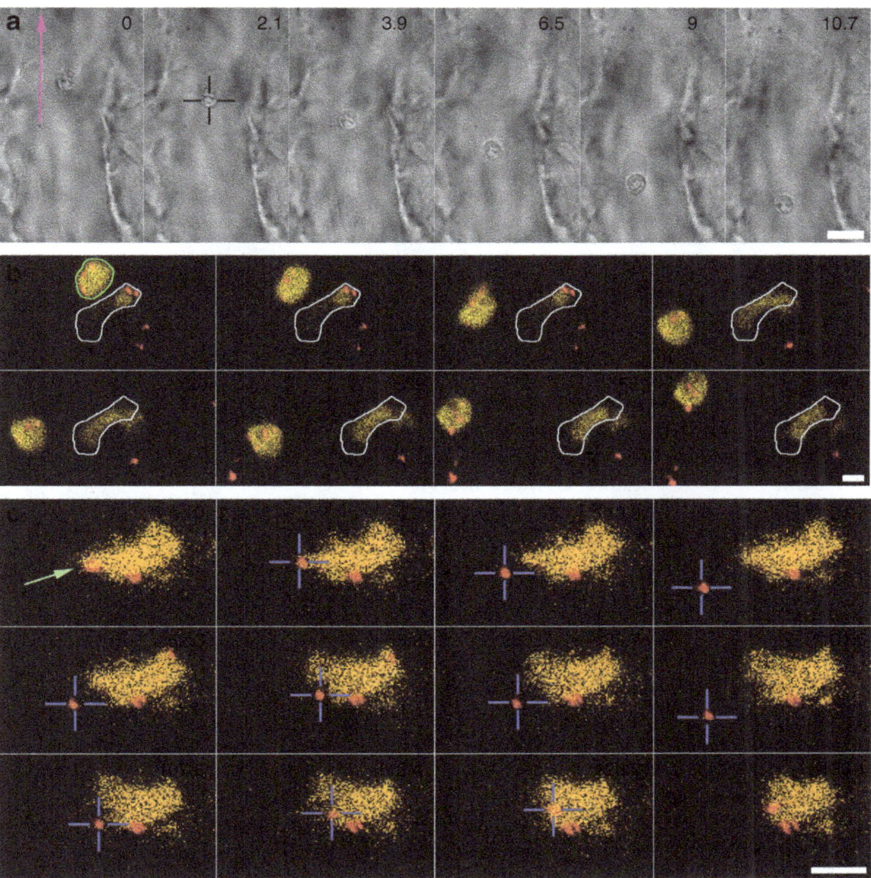

Figure 2 | Trapping of erythrocytes and macrophages. (**a**) An erythrocyte is trapped and moved in the blood flow. Scale bar, 10 μm. Experiment repeated at least 10 times. (**b**) A blood-resident fluorescent macrophage (yellow, green outline in $t = 0$ s) was micromanipulated and moved in 3D in a blood vessel. The white outline indicates another, non-mobile macrophage. The red dots are injected particles. Scale bar, 5 μm. Experiment repeated at least 5 times. (**c**) An injected particle (red colour) that was associated with a macrophage was tested for adhesion. First the particle was moved away ($t = 0$–21.7 s) after which the OT was briefly shut off. This did not result in the particle flowing away with the blood, suggesting that a nanotube (not visible) was tethering the particle; next the particle was carefully brought into contact and moved away again (52.3–101.6 s), indicating that no strong binding was established. Finally, the particle was moved further into the macrophage with a higher pushing force after which the particle could not be detached anymore ($t = 107.6$–136.1 s). Scale bar, 10 μm. Experiment was repeated at least 5 times.

stably trapped nucleus, this deformation is caused by the drag force that the blood exerts and resembles the blebs that form due to local heating of cells[20]. This experiment shows that in the zebrafish optical trapping of erythrocytes is strong and seems more robust than that achieved in mouse ear[8], where trapping had to be done gradually while making crucial use of the wall of the blood vessel. A possible explanation for this may be that erythrocytes in mice lack a nucleus.

Next, we tested whether other cell types could also be trapped in the zebrafish larva. However, it is not always trivial to identify different cell types based on solely using transmission imaging in the crowded and dynamic environment of a living organism. We therefore made use of a feature we implemented on the optical tweezers imaging system used here (adapted Nanotracker JPK Instruments AS, Berlin, see Methods), namely the possibility to do parallel trapping and confocal and transmission imaging (see also ref. 21). Using a transgenic zebrafish line with fluorescent macrophages, (Tg(mpeg1:mcherry))[22] we thus set out to test the trapping potential of these cells. In this fish line, a relatively high number of macrophages are resident in the tissue and these cells cannot be manipulated using OT. However, several macrophages were also present freely in the lumen of blood vessels. In Fig. 2b (see also Supplementary Movie 7) an area with two macrophages in the caudal vein was selected; these macrophages could be identified using fluorescence imaging, one that was immobile (white contour) and another that could be moved using the OT. First, the macrophage was moved relative to the fish (using acousto-optic deflection-based repositioning; $t = 0$–18 s and 39–81 s), while at $t = 20$ s the whole stage, including the fish and the non-mobile macrophage, was moved (horizontally and axially), while the trap was held in a stable position. The combined trapping and confocal imaging in the zebrafish demonstrates that specific cell types can be identified and trapped using this versatile system. A plethora of transgenic zebrafish lines are available[10] and different cell types and physiological questions can thereby be addressed. As another example, in Supplementary Movie 8 (and Supplementary Fig. 5) a particle (red colour, emission 640 nm) was moved in zebrafish with (green) fluorescent endothelial cells (Tg(fli1:EGFP)[23].

We then asked whether particles that were associated with a macrophage could be moved using the tweezers; specifically, we wondered whether they were adhering to the macrophage surface or had been internalized by the cell. For this, we again used tricaine to slow the blood flow. In Fig. 2c (Supplementary Movie 9), a fluorescent macrophage was identified in the caudal vein of the zebrafish larva using confocal imaging. Next, a particle (red colour, indicated with a green arrow) was trapped with the tweezers and moved away, which resulted in its detachment from the macrophage ($t = 0$–21.7 s), thus indicating that it was not inside the macrophage. Shutting the OT off briefly did not result in the particle flowing away with the blood as other particles (the red dots) can be observed to do while they move past the stationary macrophage and stationary red nanoparticles. This suggests that the particle remained tethered to the macrophage through a nanotube (not visible either because of it being in a different plane than the confocal imaging plane or because it was very thin and not very bright). Next, the particle was moved towards the macrophage again; however, no interaction could be observed (52.3–101.6 s). Finally, the particle was moved with a higher pushing force against the macrophage after which it adhered robustly, and it was not possible to detach the particle anymore (even at 5 W, the highest power available on our system). Macrophages play a central role in clearing of larger objects[24], and it is crucial to understand this interaction for nanoparticle-based nanomedicine and delivery of drugs. The zebrafish-optical tweezers system makes it possible to study the dynamics of the interactions, make contact with a specific area of the cell, and may also be used to investigate the role of the contact forces with which particle interact when they are pushed against a macrophage in a controlled manner.

Multiple traps for *in vivo* nanotube formation. To study in more detail the particle adhesion to endothelial cells *in vivo*, a more complex system was tested involving multiple cells. In a zebrafish where the blood flow was slowed down (using tricaine), we first located a particle that adhered to the endothelium. Next, we applied several optical traps to clear the operation area of all cells: four erythrocytes that were present close to the adhered particle were removed (Fig. 3a; Supplementary Movie 10). First, cell '1' was moved away (behaving like a billiard ball pushing the other cells forward) and after moving it away it was kept trapped at a distance, allowing the assembly of a 'fence' together with several other traps placed in adjacent positions. Next, cell '2' was moved and placed behind cell '1'; the 'fence' formed by several optical traps prevented the cells from flowing and diffusing back into the operation region. In addition, cells that were out of focus experienced a scattering force that 'blew' them away from the area of the fence. At $t = 36.4$ s the earlier identified particle was next moved away from the endothelium. However, this particle did not fully detach, as a thin membrane protrusion was pulled out from the endothelial cell. Such membrane tethers or nanotubes have been studied in great detail *in vitro* since they can reveal information about, for example, membrane-cytoskeleton interactions, continuity of the membrane and its biophysical properties[25–27]. Our experiments using OT inside the live zebrafish larva show for the first time that nanotubes can be formed in a multicellular organism *in vivo* using active micromanipulation and that these nanotubes could function as a flexible adhesion, where adhered objects can be moved away while remaining tethered.

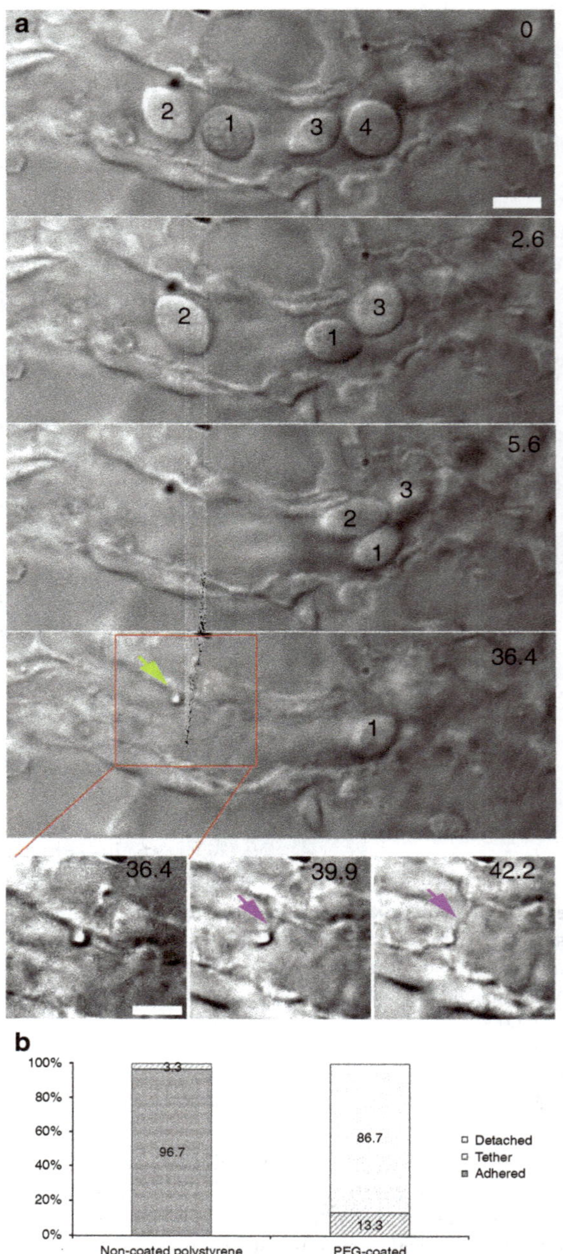

Figure 3 | Multiple traps for *in vivo* nanotube formation in cell-free blood vessels (**a**) In a smaller blood vessel several erythrocytes are cleared (0–36.4 s) and fenced off, after which an adhered particle was moved away from the endothelium, and a tethering nanotube was formed (36.4–42.2 s). Scale bar, 5 μm. (**b**) Quantification of the adhesion probability of naked and PEG-coated particles, classified as: detachable (white), strongly adhered (solid) and tethered (lines). Experiment was repeated at least 60 times.

Quantification of adhesion and 'stealth' properties. Understanding how nanoparticles adhere to different cell types and how the dynamics of adhesion is regulated is crucial for nanoparticle-based drug delivery. We therefore wanted to test systematically the adhesion of particles to the endothelium cells lining the vessels. To do this we experimented with two types of particles: 'naked'—and polyethylene glycol (PEG)-coated polystyrene particles (1 μm diameter). The PEG should provide a coat that lowers the non-specific affinity of the particles for cells such as macrophages. PEG has been widely used to provide 'stealth' properties to nanoparticles designed for, for example, cancer therapy, facilitating a longer circulation time in the blood stream. In another project we used PEG-coated particles in ZF and effectively monitored their biodistribution and characterized their targeting and 'stealth' properties to cancer cells[11]. Injected 1 μm particles adhering to the endothelium were detected and attempts were made to pull them away at a fixed laser power of 500 mW. For the naked polystyrene particles we found that the majority of the particles adhered strongly to the endothelium (Fig. 3b) and it was not possible to move them away at the laser power used (often also not at higher powers as they were strongly adhered). Out of 30 particles tested in the larvae, 29 could not be moved and only 1 could be pulled away but nevertheless remained connected through a tether and was pulled back after the trap was shut off (Fig. 3b, left bar). Next, we tested injected particles with a PEG-coating. These particles were also found to be adhered to the endothelium, and we determined the binding of these PEGylated particles (Fig. 3b, right). Out of the 30 PEG-coated particles that were found immobilized to the endothelium 26 could be detached, whereas four formed a tether. None of these particles were attached so strongly that it was impossible to move them away from the endothelium. This demonstrates that PEG lowers the binding affinity of the polystyrene particles for endothelial cells *in vivo*, even though they still adhered to some extent to the endothelium.

The fact that the PEG-coated particles more often form tethers indicates that the particles do adhere, but with a significantly lower strength than non-coated particles. The initial force required for tether formation depends strongly on the size of the adhesion zone and more frequent tether formation implies that the area of adhesion to endothelial cells is smaller in the PEGylated-particle-case. Naked-particles likely adhere through larger areas resulting in a higher threshold force for nanotube formation[15,28] than the tweezers can provide.

We then investigated whether heparin[29] would affect the binding of nanoparticles to the endothelial lining, given that proteins and drugs with heparin binding sites that bind the endothelium can be detached in the presence of heparin. We evaluated this first by comparing the circulation of fluorescent particles in ZF injected with mixtures (3 nl) of particles (non-coated polystyrene) together with or without heparin (concentrations 40 and 100 mg ml^{-1}). Confocal fluorescence imaging revealed no difference in adhesion properties in the presence or absence of heparin even at the highest drug concentration. We then used optical tweezers to evaluate whether the adhesion properties of particles had changed on a 'more subtle level' in the presence of heparin. As with the PEG-coated particles (Fig. 3b) we used the OT to pull on nanoparticles adhered to the endothelium (500 mW laser power). These experiments show that there was no significant difference in the adhesion properties in the presence or absence of heparin (Supplementary Fig. 6).

Optical trapping and manipulation of injected bacteria. After having demonstrated trapping of injected particles and naturally occurring zebrafish cells inside the living zebrafish larva, we tested whether this approach can also be applied to study injected bacteria. Bacteria were amongst the first biological particles to be optically trapped *in vitro* in aqueous solutions[2], but as far as we know there are no reports of trapping bacteria inside living vertebrates. Rod-shaped bacteria are more difficult to trap than spherical (polystyrene) particles and they are smaller than macrophages and erythrocytes. As a proof of principle of optical manipulation of bacteria in zebrafish, we injected the fish bacterium *M. marinum*, which causes fish tuberculosis, and which has been widely and effectively used as a model for human tuberculosis in the zebrafish larva[12,30].

For this experiment we selected a region where the blood flow was very slow (again using tricaine), and using either fluorescence or transmission imaging, we could detect the presence of (red fluorescent) bacteria that had been injected (Fig. 4). Next, a bacterium was trapped and moved against the endothelium using the OT (Supplementary Movie 11). Before the trap was activated, the bacterium was seen moving in a Brownian manner through the blood vessel, rotating and diffusing. It could be imaged in a snapshot while oriented in the imaging plane (Fig. 4a, see also Supplementary Movie 11). Activating the optical tweezers (Fig. 4b) resulted in the bacterium becoming orientated in a direction perpendicular to the imaging plane (in the direction of the tweezer beam). This reorientation is expected for elongated objects. Next, the bacterium was pushed against the endothelium (Fig. 4c). Intriguingly, in Fig. 4d a cell (presumably a

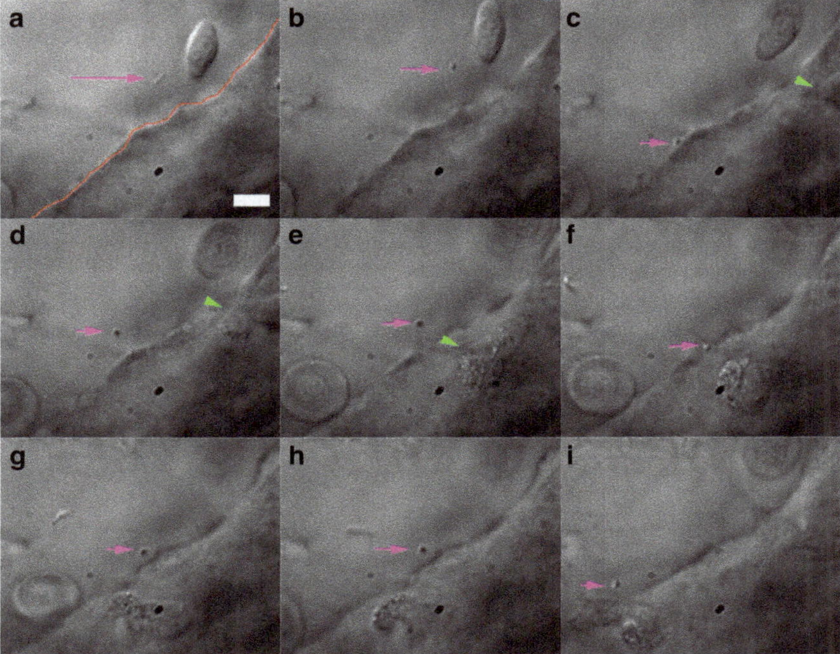

Figure 4 | Trapping of injected bacterium. (**a**) A diffusing bacterium (purple arrow) is (**b-d**) trapped, pushed against and moved away from the endothelium (red line in **a**). This seems to activate an immune cell (green arrowheads), which moves towards the contact point (**c-f**). Repeated contact with the endothelium seems to again attract the attention of the crawling cell, which (**g-i**) finally moves into the vein. The total duration of experiment is ~2 min. Scale bar, 10 μm. Experiment was repeated at least 4 times.

macrophage[31]) could be observed crawling towards the contact point. After the bacterium was repeatedly moved against the endothelium, the immune cell was seen to arrest its movement and finally seemed to 'decide' to move across the endothelial barrier (Fig. 4g-i). Macrophages are known to collect bacteria[31], and controlling the position and timing of bacteria and interactions with different cells makes it possible to study this phenomenon in a controlled manner. The OT can for example be used to determine how long a bacterium needs to stay in contact to adhere to a cell or to evoke a response, or whether multiple bacteria will increase the recruitment of macrophages or influence phagocytosis dynamics.

We also tested the effects of anti-inflammatory drugs on the migration of immune cells to bacteria-invaded areas. Our preliminary experiments using fish lines with either fluorescent neutrophils or fluorescent macrophages confirmed that the anti-inflammatory drugs diclofenac and indomethacin[32] inhibited the recruitment of neutrophils (and macrophages to a lesser degree, unpublished results) to sites where bacteria had been injected[12]. However, these experiments need a systematic follow-up for their significance to be verified.

Collectively, the experiments we have described demonstrate for the first time active micromanipulation of a full scale of nano- to micron-sized structures inside a living vertebrate using the transparent zebrafish larva. The manipulated structures ranged from injected nanoparticles and bacteria to naturally occurring zebrafish cells as erythrocytes and macrophages.

We foresee many uses of this approach such as (but not limited to), the characterization of interaction properties of nanoparticles with specific cells for nanomedicine applications. In particular the properties of nanoparticles could be studied, for example, by functionalizing them with ligands for targeting to specific cells or with coats such as PEG to prevent interactions with other cells. Alternatively, optically manipulated nanoparticles releasing a specific compound could be brought in the proximity of the organismal structure of interest for testing of local cellular responses to chemicals (as has been already demonstrated elegantly with cells *in vitro*[33]) such as for studies of vascular function and endothelial integrity.

Controlled investigations of recruitment and activation of immune cells, by micromanipulating bacteria, other microorganisms or antigen-coated particles to specific regions in the organism will make possible to investigate adhesion to and activation/recruitment of immune cells, for example in the presence of anti-inflammatory drugs, especially in combination with imaging. Finally, quantitative optical tweezers have been instrumental to understanding cellular biomechanical properties and their regulatory role in function. However, this has been mostly done *in vitro* with cells in culture, and the work presented here opens up many possibilities to perform such experiments throughout a living vertebrate.

Methods

Zebrafish care and treatment. Two lines of transgenic ZF larvae were used, Tg(fli1:EGFP) and Tg(mpeg1:mcherry) with green fluorescent endothelial cells and red fluorescent macrophages, respectively. The ZF larvae were kept in Petri dishes containing salt-containing water[18] with 0.003% w/v phenylthiourea (PTU, Sigma-Aldrich, St. Louis, USA) to keep the fish transparent by preventing pigmentation in retinal epithelium and melanophores[34]. All experiments were done at 28.5 °C. Experiments were conducted in agreement with the ethical provisions enforced by the Norwegian national animal research authority (NARA).

Microinjections of zebrafish larvae. Injections were done using a glass micropipette (Harvard apparatus, Holliston, USA), with an outer diameter of 1.0 mm and inner diameter of 0.78 mm. The glass micropipettes were made using a micropipette puller Model P-97 (Sutter Instruments Co., Novato, USA). Manipulation of the micropipette was done using a Narishige MN-153 micromanipulator (Narishige, London, UK), and injection time and pressure was controlled using a FemtoJet Express micro injector (Eppendorf, Hamburg Germany). Visualization of the ZF larvae during injections was done with a stereomicroscope (Leica DFC365FX with a × 1.0 Planapo lens).

For all injections, the ZF larvae were anaesthetized using (0.5–2 mg ml^{-1}) tricaine (Finquel, Argent Laboratories, Redmond, USA) in appropriate solution[18]. The ZF embryo was then placed on a gel made of hardened 2% agarose (Sigma-Aldrich) in water, and excess fluid was removed from around the ZF embryo using a pipette. This was done to immobilize the fish before injections.

Nanoparticles of five different sizes were used; four Fluoresbrite Microparticles (Polysciences Inc., Warrington, USA) with diameters of 100, 200, 500 and 1000 nm, and SPHERO particles (Spherotech Inc., Lake Forest, USA) with 840 nm diameter. The 200 nm Fluoresbrite particles contained fluorescein dye (yellow-green) with excitation maximum at 441 nm, while the 500 nm and 1 μm nanoparticles contained coumarin dye (bright blue) with excitation maxima 360 nm. The 840 nm SPHERO nanoparticles (Spherotech Inc.) contained sky blue, with excitation at 640 nm. For the experiments on the 'stealth' effect of PEG coating, the Fluoresbrite nanoparticles were modified with MPEG5000-NH2 on the surface[11]. The nanoparticles were diluted in PBS to a concentration of 2×10^8 nanoparticles per ml and loaded into a glass micropipette. Subsequently, 3–6 nl was injected into the posterior cardinal vein of the ZF larvae. *M. marinum* carrying the fluorescent reporter construct DsRed was injected (250 c.f.u.) in the posterior cardinal vein at 48 h post fertilization, as in ref. 12.

Sample preparation for optical tweezers experiments. Following injection, the ZF larvae were moved to a Petri dish containing a tricaine solution. The concentration of tricaine was between 0.1 and 0.4 mg ml^{-1} in salt water[18] (depending on how slow blood flow was desired for the experiment).

Two parallel lines of silicone grease were applied to a 22 × 60 mm cover glass using a hypodermic needle and a syringe. The length of the grease lines was ~30 mm in length and the distance between the two lines was ~15 mm. Between these silicone lines the ZF embryo was placed in ~100 μl tricaine solution using a pipette. Using a small paint brush or hair loop, the embryo can be manipulated into a suitable position so that its body and tail are close to the cover glass (Supplementary Fig. 3, right), which facilitates optical trapping. A 22 × 22 mm coverslip (Karl Roth nr 1, thickness 0.13–0.16 μm) was carefully placed on top, resting on the two lines of silicone. The coverslip was pushed down onto the silicone, carefully, in order not to damage the embryo. It is important that the 22 × 22 mm coverslip is pushed far enough down to make sure the ZF embryo does not float around or move during the experiment. Excess fluid was removed from the edges using filter paper, or fill up the remaining space between the coverslips was filled with an embryo water-tricaine solution if necessary. The remaining openings between the cover glasses were sealed using clear nail polish. The sample was next mounted (with the 22 × 22 mm cover glass downwards) on the sample stage in the optical tweezers microscope.

Optical tweezers and imaging microscope. An adapted version of the NanoTracker2 system (JPK Instruments AG, Berlin) was used. This custom-built system was developed in collaboration with JPK Instruments for parallel confocal, transmission and optical trapping (Supplementary Fig. 3, left). A 1,064-nm trapping laser (5 W) was split into two polarizations for independent trapping. One of these beams is controlled through a piezo-mirror and the other passes through Acousto-Optic Deflectors for position control and multiplexing. The trap position relative to the fish could also be controlled through movement of the whole sample with a piezo stage.

The optical trapping system was merged on a NIKON C2 confocal microscope with a × 60 (numerical aperture (NA) 1.2, WD 0.27 mm) water immersion objective for imaging and trapping, and we used Zeiss' 'Immersol' immersion fluid for water objectives ($n = 1.334$). Transmission light was focused in the sample through a × 60 water dipping condenser (NA 1, WD 2.5 mm). To be able to image in parallel in confocal mode, the ~700 to 900 nm band was used for transmission, which does not interfere with the confocal imaging.

Quantification of adhesion with and without PEG coating. For the quantification of the binding affinities of the 1 μm polystyrene nanoparticles with and without PEG, we used three individual fishes for both PEG- and non-coated nanoparticles. In each of these 6 individuals 10 particles were trapped and manipulated using 500-mW laser power. Three possible outcomes were considered in the experiment: (1) the particle remained adhered to the endothelium; (2) the nanoparticle detached from the endothelial cell; or (3) a tether was pulled from the cell, allowing the particle to be moved away but maintaining the connection to the endothelium.

References

1. Pober, J. S. & Sessa, W. C. Evolving functions of endothelial cells in inflammation. *Nat. Rev. Immunol.* **7,** 803–815 (2007).
2. Ashkin, A. & Dziedzic, J. M. Optical trapping and manipulation of viruses and bacteria. *Science* **235,** 1517–1520 (1987).
3. Bendix, P. M., Jauffred, L., Norregaard, K. & Oddershede, L. B. Optical trapping of nanoparticles and quantum dots. *IEEE J. Sel. Top. Quantum Electron.* **20,** 15–26 (2014).

4. Blehm, B. H., Schroer, T. A., Trybus, K. M., Chemla, Y. R. & Selvin, P. R. *In vivo* optical trapping indicates kinesin's stall force is reduced by dynein during intracellular transport. *Proc. Natl Acad. Sci. USA* **110,** 3381–3386 (2013).
5. Hendricks, A. G., Holzbaur, E. L. & Goldman, Y. E. Force measurements on cargoes in living cells reveal collective dynamics of microtubule motors. *Proc. Natl Acad. Sci. USA* **109,** 18447–18452 (2012).
6. Oddershede, L. B. Force probing of individual molecules inside the living cell is now a reality. *Nat. Chem. Biol.* **8,** 879–886 (2012).
7. Hansen, P. M. & Oddershede, L. B. Optical trapping inside living organisms. *Proc. SPIE* **5930,** 1–9 (2005).
8. Zhong, M. C., Wei, X. B., Zhou, J. H., Wang, Z. Q. & Li, Y. M. Trapping red blood cells in living animals using optical tweezers. *Nat. Commun.* **4,** 1768 (2013).
9. Liu, S. & Leach, S. D. Zebrafish models for cancer. *Annu. Rev. Pathol. Mech. Dis.* **6,** 71–93 (2011).
10. Lieschke, G. J. & Currie, P. D. Animal models of human disease: zebrafish swim into view. *Nat. Rev. Genet.* **8,** 353–367 (2007).
11. Evensen, L. *et al.* Zebrafish as a model system for characterization of nanoparticles against cancer. *Nanoscale* **8,** 862–877 (2015).
12. Fenaroli, F. *et al.* Nanoparticles as drug delivery system against tuberculosis in zebrafish embryos: direct visualization and treatment. *ACS Nano* **8,** 7014–7026 (2014).
13. Kamaly, N., Xiao, Z., Valencia, P. M., Radovic-Moreno, A. F. & Farokhzad, O. C. Targeted polymeric therapeutic nanoparticles: design, development and clinical translation. *Chem. Soc. Rev.* **41,** 2971–3010 (2012).
14. Fieramonti, L. *et al.* Quantitative measurement of blood velocity in zebrafish with optical vector field tomography. *J Biophoton.* **8,** 52–59 (2015).
15. Koster, G., Cacciuto, A., Derenyi, I., Frenkel, D. & Dogterom, M. Force barriers for membrane tube formation. *Phys. Rev. Lett.* **94,** 068101 (2005).
16. Visscher, K., Gross, S. P. & Block, S. M. Construction of multiple-beam optical traps with nanometer-resolution position sensing. *IEEE J. Sel. Top. Quantum Electron.* **2,** 1066–1076 (1996).
17. Rohrbach, A. Stiffness of optical traps: quantitative agreement between experiment and electromagnetic theory. *Phys. Rev. Lett.* **95,** 168102 (2005).
18. Cosma, C. L., Swaim, L. E., Volkman, H., Ramakrishnan, L. & Davis, J. M. Zebrafish and frog models of Mycobacterium marinum infection. *Curr. Protoc. Microbiol.* Chapter 10, Unit 10B.2 (2006).
19. Zon, L. I. Developmental biology of hematopoiesis. *Blood* **86,** 2876–2891 (1995).
20. Oyama, K. *et al.* Directional bleb formation in spherical cells under temperature gradient. *Biophys. J.* **109,** 355–364 (2015).
21. Sorre, B. *et al.* Curvature-driven lipid sorting needs proximity to a demixing point and is aided by proteins. *Proc. Natl Acad. Sci. USA* **106,** 5622–5626 (2009).
22. Ellett, F., Pase, L., Hayman, J. W., Andrianopoulos, A. & Lieschke, G. J. mpegl promoter transgenes direct macrophage-lineage expression in zebrafish. *Blood* **117,** e49–e56 (2011).
23. Lawson, N. D. & Weinstein, B. M. In vivo imaging of embryonic vascular development using transgenic zebrafish. *Dev. Biol.* **248,** 307–318 (2002).
24. Desjardins, M. & Griffiths, G. Phagocytosis: latex leads the way. *Curr. Opin. Cell Biol.* **15,** 498–503 (2003).
25. Chaudhuri, O., Parekh, S. H., Lam, W. A. & Fletcher, D. A. Combined atomic force microscopy and side-view optical imaging for mechanical studies of cells. *Nat. Methods* **6,** 383–387 (2009).
26. Davis, D. M. & Sowinski, S. Membrane nanotubes: dynamic long-distance connections between animal cells. *Nat. Rev. Mol. Cell Biol.* **9,** 431–436 (2008).
27. Derényi, I. *et al.* in Controlled nanoscale motion Vol. 711 *Lecture Notes in Physics*. (eds Linke, H. & Månsson, A.) Ch. 7 141–159 (Springer Berlin Heidelberg, 2007).
28. Pontes, B. *et al.* Membrane elastic properties and cell function. *PLoS ONE* **8,** e67708 (2013).
29. Xu, D. & Esko, J. D. Demystifying heparan sulfate-protein interactions. *Annu. Rev. Biochem.* **83,** 129–157 (2014).
30. Ramakrishnan, L. Looking within the zebrafish to understand the tuberculous granuloma. *Adv. Exp. Med. Biol.* **783,** 251–266 (2013).
31. Herbomel, P., Thisse, B. & Thisse, C. Ontogeny and behaviour of early macrophages in the zebrafish embryo. *Development* **126,** 3735–3745 (1999).
32. Cordero-Maldonado, M. L. *et al.* Optimization and pharmacological validation of a leukocyte migration assay in zebrafish larvae for the rapid in vivo bioactivity analysis of anti-inflammatory secondary metabolites. *PLoS ONE* **8,** e75404 (2013).
33. Kress, H. *et al.* Cell stimulation with optically manipulated microsources. *Nat. Methods* **6,** 905–909 (2009).
34. Karlsson, J., von Hofsten, J. & Olsson, P.-E. Generating transparent zebrafish: a refined method to improve detection of gene expression during embryonic development. *Mar. Biotechnol.* **3,** 522–527 (2001).

Acknowledgements

JPK Instruments is acknowledged for excellent support and development of the optical tweezers imaging station, which is part of the NorMIC imaging platform at IBV (University of Oslo). We thank Dirk Linke for critical reading of the manuscript. G.K. acknowledges the Research Council of Norway for funding. We are grateful to the Norwegian Cancer Society for their generous funding (to F.F. and L.E.).

Author contributions

G.G. and G.K. conceived the project: P.L.J., F.F., G.G. and G.K. designed research; P.L.J., F.F., L.E., and G.K. conducted experiments: G.G. and G.K. supervised the project, P.L.J., G.G., and G.K. wrote the manuscript.

Additional information

Competing financial interests: The authors declare no competing financial interests.

10

The Atg1-kinase complex tethers Atg9-vesicles to initiate autophagy

Yijian Rao[1,*], Marco G. Perna[1,*], Benjamin Hofmann[1], Viola Beier[1] & Thomas Wollert[1]

Autophagosomes are double-membrane vesicles that sequester cytoplasmic material for lysosomal degradation. Their biogenesis is initiated by recruitment of Atg9-vesicles to the phagophore assembly site. This process depends on the regulated activation of the Atg1-kinase complex. However, the underlying molecular mechanism remains unclear. Here we reconstitute this early step in autophagy from purified components *in vitro*. We find that on assembly from its cytoplasmic subcomplexes, the Atg1-kinase complex becomes activated, enabling it to recruit and tether Atg9-vesicles. The scaffolding protein Atg17 targets the Atg1-kinase complex to autophagic membranes by specifically recognizing the membrane protein Atg9. This interaction is inhibited by the two regulatory subunits Atg31 and Atg29. Engagement of the Atg1-Atg13 subcomplex restores the Atg9-binding and membrane-tethering activity of Atg17. Our data help to unravel the mechanism that controls Atg17-mediated tethering of Atg9-vesicles, providing the molecular basis to understand initiation of autophagosome-biogenesis.

[1] Molecular Membrane and Organelle Biology, Max Planck Institute of Biochemistry, Am Klopferspitz 18, Martinsried 82152, Germany. * These authors contributed equally to this work. Correspondence and requests for materials should be addressed to T.W. (email: wollert@biochem.mpg.de).

The homeostasis of cells depends critically on recycling pathways to remove superfluous or damaged cytoplasmic material[1]. During macroautophagy, to which we will refer to as autophagy in the following, cytoplasmic cargo is engulfed by a cup-shaped membrane, termed phagophore. Formation of such phagophores requires, in addition to canonical membrane remodelling and fusion machines, a dedicated set of autophagy-related (Atg) proteins[2]. Most Atg-proteins have first been identified in yeast[3–5], but many of them possess mammalian homologues, emphasizing the high degree of conservation among eukaryotes[6].

Autophagy is initiated at the phagophore assembly site (PAS) in yeast, to which Atg-proteins and donor-membranes are recruited in a spatiotemporally coordinated manner[7,8]. Sequestration of cargo from the cytoplasm involves expansion of the phagophore. Sealing of the membrane generates the double-membrane surrounded autophagosome. Enclosed autophagic cargo is finally delivered to the vacuole for degradation[9].

Although Atg-proteins have been studied extensively[9], a central question in autophagy research concerning the nucleation of autophagosomes remained unclear. A recent study suggests that an average of three Atg9-vesicles coalesce at the PAS[10]. The recruitment of these vesicles depends on the scaffolding protein Atg17, which is the earliest protein to arrive at the PAS[7]. Atg17 adapts a highly elongated crescent shape[11,12] and constitutively interacts with Atg29 and Atg31 *in vivo*, forming a trimeric complex (Atg17^{TC} in the following)[13,14]. On autophagy induction, Atg1 and Atg13 assemble with Atg17^{TC} into the pentameric Atg1–kinase complex (Atg1^{PC} in the following)[13,15]. Interestingly, mammalian homologues of Atg1 (ULK1/2), Atg13 (ATG13) and potentially Atg17 (FIP200) have been identified[16–18]. Moreover, the ULK1-complex regulates mATG9 localization and is involved in early steps of autophagy, suggesting that yeast Atg1- and mammalian ULK1/2-complexes possess similar activities in autophagy[19,20]. Initiation of mammalian autophagy occurs, however, at specialized ER-domains, termed omegasomes[6]. Thus, although mammalian and yeast autophagy pathways share conserved elements, both are driven by distinct yet to be characterized mechanisms.

Atg1^{PC} integrates nutrient dependent signals from other kinases such as the protein kinase A (PKA) and the target of rapamycin complex (TORC) 1 (refs 21,22). Under vegetative conditions, TORC1 phosphorylates Atg13, preventing its interaction with Atg17 (ref. 23) and thus the assembly of Atg1^{PC} from its subunits. On deactivation of TORC1 Atg13 becomes partially dephosphorylated and functional Atg1^{PC} assembles at the PAS[24].

Significant progress has been made in the past to decipher the function of Atg1^{PC} as well as its interaction with Atg9-vesicles[12,23,25,26]. The molecular mechanism of Atg1^{PC}-activation and how this translates into its regulated interaction with Atg9 remained, however, elusive. Here we analysed for the first time the Atg1^{PC}-Atg9-interaction in a fully reconstituted *in vitro* system using purified components. Our results show that the Atg17-dimer binds two Atg9-vesicles independently and functions as vesicle-tether without contacting the membrane directly. The Atg31–Atg29 subcomplex sterically blocks the Atg9-binding site in Atg17 such that Atg17^{TC} becomes inactive. On autophagy induction, the pentameric complex Atg1^{PC} assembles and restores the full Atg9-binding and vesicle-tethering activity of Atg17.

Results

Reconstitution of Atg1^{PC} *in vitro*. To study the intricate interplay of Atg1^{PC} and Atg9-vesicles in a fully reconstituted system, we purified individual subunits and subcomplexes to incrementally assemble full-length Atg1^{PC} from its components (Fig. 1a, Supplementary Fig. 1a). Recombinant Atg1 was phosphorylated at T226 (Supplementary Fig. 1b, Supplementary Table 1), representing the active enzyme that functions in autophagy[27,28]. In contrast to the recombinant Atg1-MIT-domain, which dimerized in solution[12,25], full-length Atg1 was monomeric (Supplementary Fig. 1c). Atg17^{TC} was produced by co-expressing Atg17, Atg31 and Atg29 or by combining Atg17 with co-expressed Atg31–Atg29. Atg17 and Atg17^{TC} form well-characterized S-shaped dimers with dimerization being mediated by the C-terminus of Atg17 (refs 11,12). To confirm that our recombinant Atg17 and its related complex form similar dimers, we determined their Stokes radii (R_H) using dynamic light scattering (DLS). Atg17 and Atg17^{TC} possessed very similar R_H of 10.4 ± 0.1 nm and 10.3 ± 0.1 nm ($\pm$ refers to s.d., $n = 3$), respectively. Thus, Atg17 and Atg17^{TC} form extended dimers comparable to those observed previously[11]. Correspondingly, the C-terminally truncated variants Atg17^{mono} (Atg17$^{1-354}$) and Atg17^{monoTC}, which lack the dimerization domain, are monomeric ($R_H = 5.3 \pm 0.1$ nm and 5.6 ± 0.2 nm, $\pm$ refers to s.d., $n = 5$).

Atg1 and Atg13 formed a stable subcomplex and incubating both with Atg17^{TC} resulted in the formation of Atg1^{PC} (Fig. 1a,b, Supplementary Fig. 1d). Multi-angle light scattering (MALS) revealed that Atg1^{PC} is a dimer of pentamers (Supplementary Fig. 1e). Although the Atg1–Atg13 subcomplex contributes two-thirds of the total mass of Atg1^{PC}, the R_H ($\sim$13 nm) of Atg1^{PC} is only slightly larger than that of Atg17^{TC}. Atg1, Atg13 and Atg29 have been reported to contain intrinsically disordered regions that are involved in intermolecular interactions that stabilize Atg1^{PC} (refs 29,30). Moreover, previous electron microscopy studies showed that the overall S-shaped structure of Atg17^{TC} was not perturbed upon binding of the C-terminal domains of Atg1 and Atg13 (ref. 30). These observations are in agreement with our data demonstrating that Atg1^{PC} has similar overall hydrodynamic properties as Atg17^{TC}. Moreover, our results show that N-terminal domains of Atg1 and Atg13 do not significantly increase the R_H of the complex, suggesting that Atg1 and Atg13 form a compact subcomplex that partially occupies the Atg17-crescent. The monomeric variant Atg17^{monoTC} assembled with Atg1–Atg13 into monomeric Atg1^{PC} (Atg1^{monoPC}) as demonstrated by a shift in the retention volume (V_R) from 10.2 ml (Atg1^{PC}) to 12.6 ml (Atg1^{monoPC}) in size exclusion chromatography (Supplementary Fig. 1f). Thus, complex formation and Atg17-mediated dimerization are independent of each other.

Atg1 and Atg13 are membrane-binding proteins. The C-terminal tandem MIT-domain of yeast Atg1 has been found to bind highly curved membranes *in vitro*[12]. To what extend full-length Atg1 targets fully assembled Atg1^{PC} to membranes remained, however, elusive.

To address this question, we analysed membrane binding of individual Atg1^{PC}-components. While Atg17 and Atg17^{TC} did not interact with membranes of any curvature, both Atg1 and Atg13 were efficiently recruited to highly curved membranes composed of yeast polar lipid (YPL) extracts (Supplementary Fig. 1g).

To test whether specific lipids are recognized by Atg1 and Atg13, we compared binding to small unilamellar vesicles (SUVs) made of YPL-extracts with that to SUVs containing synthetic lipid mixtures. Acidic SUVs containing 10 mol% or 40 mol% phosphatidylserine (PS) did not facilitate Atg1 or Atg13 recruitment. Interestingly, however, Atg1 strongly bound

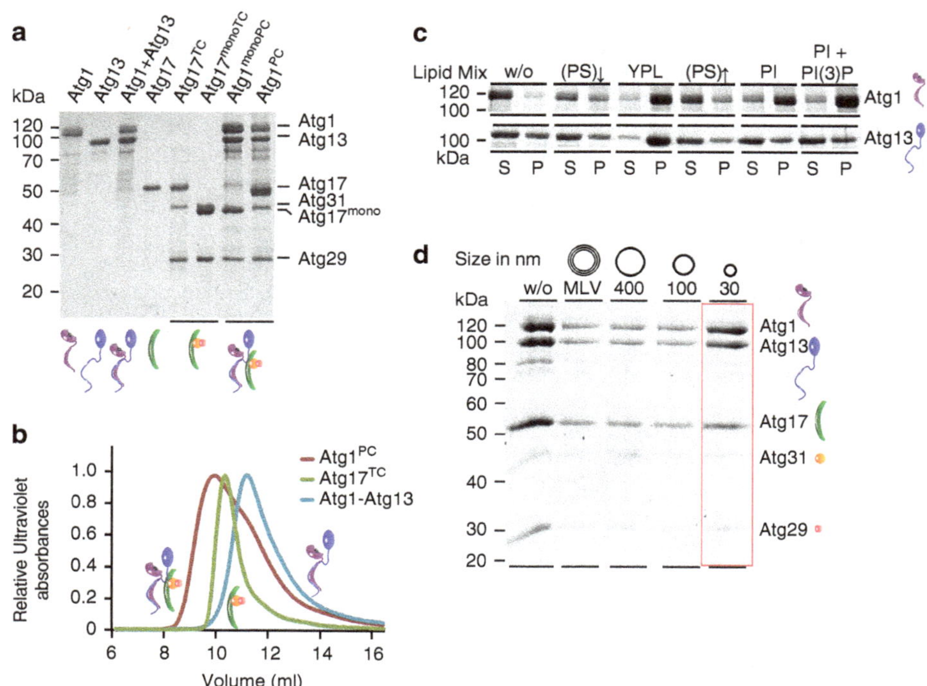

Figure 1 | Reconstitution of Atg1^{PC} from purified components. (**a**) SDS-PAGE of recombinantly expressed and SEC-purified Atg1-kinase complex subunits. Atg17^{TC} and Atg17^{monoTC} were co-expressed; Atg1–Atg13, Atg1^{PC} and Atg1^{monoPC} were reconstituted from single components. (**b**) SEC elution profiles of Atg1^{PC} (red) compared with that of Atg17^{TC} (green) and Atg1–Atg13 subcomplexes (blue). (**c**) Liposome sedimentation assays of Atg1 and Atg13. Lipid specificity of Atg1 and Atg13 was assessed by comparing their interaction with SUV containing 10 mol% PS ([PS]↓), YPL-extracts, 40 mol% PS ([PS]↑), 30 mol% PI, or 25 mol% PI supplemented with 2.5 mol% PI(3)P. S, supernatant; P, pellet. (**d**) Floatation assay of YPL-large unilamellar vesicles (LUVs) and MLVs of different sizes with Atg1^{PC}. Atg1^{PC} is efficiently recruited to membranes with high curvature (red box). Cartoons represent subunits and complexes.

phosphatidylinositol (PI)-containing SUVs. Replacing 5 mol% PI by PI(3)-phosphate (PI(3)P) enhanced Atg1 recruitment further (Fig. 1c). By contrast, Atg13 was only recruited to SUVs prepared from YPL-extracts, suggesting that other lipids than those present in tested synthetic mixtures are required for Atg13 binding (Fig. 1c). We next investigated whether full-length Atg1^{PC} is targeted to membranes by Atg1 and Atg13. Using floatation assays we found similar membrane-binding characteristics of Atg1^{PC} (Fig. 1d) compared with those of Atg1 (Supplementary Fig. 1g). The specificity of Atg1 for PI-containing membranes of high curvature is consistent with the localization of Atg1^{PC} *in vivo*. Atg1^{PC} is one of the earliest complexes to assemble at the PAS followed by Atg13-mediated recruitment of the autophagy specific PI(3)-kinase complex[31]. The conversion of PI to PI(3)P within autophagic membranes occurs, thus, downstream of Atg1 (ref. 32).

Atg17 binds the conserved Atg9-core to promote autophagy. During autophagy, Atg1^{PC} assembles at the PAS and recruits Atg9-vesicles in order to initiate the formation of autophagosomes[10,33]. Thus lipid-binding as observed for Atg1 and Atg13 needs to be complemented by a specific recognition of Atg9. Atg17 is the first Atg1^{PC}-subunit to arrive at the PAS on induction of autophagy[7], which is followed by Atg17-dependent recruitment of Atg9-vesicles[34]. We therefore aimed to compare binding of Atg17, Atg17^{TC} and Atg1^{PC} to Atg9-vesicles in a fully reconstituted system using purified components. Atg9 possesses in addition to weakly conserved and mostly disordered N- and C-terminal regions a highly conserved core region (Atg9^{core} in the following), which corresponds to the ubiquitously expressed metazoan Atg9A-orthologue. We therefore focused on Atg9^{core} (Fig. 2a) and studied its interaction with Atg1^{PC}.

We first confirmed that Atg9^{core} retained its capacity to bind Atg17 *in vivo* by co-immunoprecipitating HA-tagged Atg9^{core} with Atg17^{GFP} (Supplementary Fig. 2a). Next, we reconstituted recombinant Atg9^{core} in proteoliposomes (Atg9–PLs). Circular dichroism (CD) spectroscopy of detergent-solubilized and purified Atg9^{core} revealed the protein to be correctly folded (secondary structure content of 77.5%), possessing an α-helix content of 46.2% (Supplementary Fig. 2b), which is comparable to that of similar membrane proteins[35]. Assessing the orientation of reconstituted Atg9^{core} in PLs by protease protection revealed that ~50% exposed their cytoplasmic domain to the exterior, adapting the native membrane topology (Supplementary Fig. 2c).

We next tested whether Atg17 directly interacts with Atg9–PLs and found that Atg17 was indeed recruited to Atg9–PLs, but not to liposomes lacking Atg9 in co-floatation assays (Fig. 2b, Supplementary Fig. 2d). To determine which region in Atg9 contributes to Atg17-recognition, we further truncated Atg9^{core} to produce deletion variants in which the entire N-terminal domain (Atg9$^{\Delta N}$), the cytoplasmic domain between transmembrane helix two and three (Atg9$^{\Delta cD}$), or both domains (Atg9$^{\Delta N\Delta cD}$) were missing (Fig. 2a). CD-spectroscopy confirmed that all recombinant Atg9-variants were correctly folded (Supplementary Fig. 2b). In contrast to Atg9^{core}, which did not prefer a specific orientation, Atg9-variants were incorporated such that their cytoplasmic domains were facing the exterior. The orientation of Atg9-variants in PLs is thus similar to the native topology of Atg9 *in vivo*. (Supplementary Fig. 2c). As a consequence, the amount of accessible binding sites is larger than that of equal amounts of liposomes containing Atg9^{core}. Even

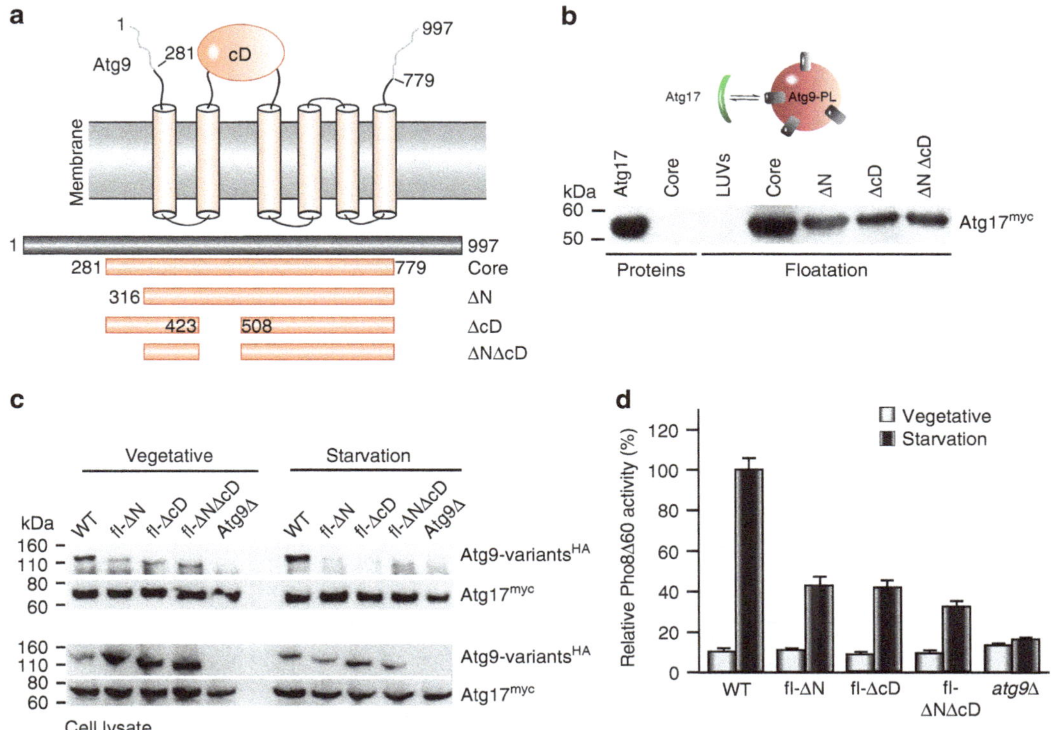

Figure 2 | Cytoplasmic domains of Atg9 interact with Atg17 to promote autophagy. (**a**) The scheme shows the topology of Atg9 in membranes (transmembrane domains, pink cylinders; membrane bilayer in grey; cD, cytoplasmic domain). The grey bar represents $Atg9^{WT}$, the pink bars illustrate Atg9-variants with corresponding domain borders indicated by numbers. (**b**) α-myc immunoblot of co-floatation assays of Atg9-PL, containing $Atg9^{core}$ or its variants, with Atg17. Large unilamellar vesicles (LUVs) lacking Atg9 served as a control for unspecific membrane binding. Proteins, recombinant Atg17 and Atg9-PLs used for experiments. (**c**) Immunoprecipitation (IP) of $Atg17^{myc}$ from $ATG9^{WT}$, $ATG9^{\Delta N}$, $ATG9^{\Delta cD}$, $ATG9^{\Delta N \Delta cD}$ and *ATG9Δ* lysates under vegetative conditions and after 2 h of starvation. *Atg9*-variants contained the very N- (1–280) and C-terminal (779–997) unstructured regions, which were not present in $Atg9^{core}$ (indicated by the prefix fl), because these regions are essential for proper trafficking of Atg9 *in vivo*[10,49,50]. Atg17 and co-immunoprecipitated Atg9-variants were detected by α-myc and α-HA immunoblots as indicated. Atg17 strongly interacts with wild-type Atg9, but not with Atg9 lacking N-, cD- or both domains. Bottom panel: total protein amounts in cell lysates. (**d**) Pho8Δ60 assay of vegetative (white bars) and starved (black bars) $ATG9^{WT}$, $ATG9^{\Delta N}$, $ATG9^{\Delta cD}$, $ATG9^{\Delta N \Delta cD}$ and *ATG9Δ*. The Pho8Δ60 activity of starved $ATG9^{WT}$ was set to 100% and used for normalization. Mean values ± s.d. of $N = 3$ independent experiments are shown.

though more binding sites are available in PLs containing Atg9-variants, a strong reduction in co-floating Atg17 was observed (Supplementary Fig. 2d). To eliminate the interference of Atg9 on the intensity of Atg17-signals, we quantified co-floating myc-tagged Atg17 ($Atg17^{myc}$) by immunoblotting (Fig. 2b, Supplementary Fig. 2e). All Atg9-variants were strongly impaired in recruiting Atg17 by >60%, indicating that both N-terminal and cytoplasmic domains of Atg9 are crucial for Atg17-binding.

To investigate binding of Atg17 to Atg9 and its variants *in vivo*, we complemented yeast *atg9Δ* cells with $atg9^{WT}$, $atg9^{\Delta N}$ (Δ281–315), $atg9^{\Delta cD}$ (Δ424–507), and $atg9^{\Delta N \Delta cD}$ in trans. We first analysed whether Atg9-variants are forming peripheral Atg9-pools using subcellular fractionation as previously reported[33] and found distribution-profiles of Atg9-variants under autophagy-induced conditions to be comparable to that of $Atg9^{WT}$ under vegetative conditions (Supplementary Fig. 2f). We next studied binding of Atg17 to Atg9. As reported previously[34], Atg17 co-immunoprecipitated Atg9 even under vegetative conditions and inducing autophagy by starvation strongly enhanced their interaction (Fig. 2c). Consistent with our *in vitro* data, we observed a strongly decreased interaction of Atg17 with $Atg9^{\Delta N}$, $Atg9^{\Delta cD}$ and $Atg9^{\Delta N \Delta cD}$ under vegetative conditions, which was entirely abolished on starvation (Fig. 2c). We inferred from these results that due to the missing interaction of Atg17 with Atg9-variants, Atg9-vesicles are not recruited from peripheral pools to the PAS, which is consistent with previous observations[7,34]. Consequently, autophagic activity which we quantified using Pho8Δ60-assays was strongly decreased in $atg9^{\Delta N}$, $atg9^{\Delta cD}$, and $atg9^{\Delta N \Delta cD}$ cells (Fig. 2d). Our data thus demonstrate that both N- and cD-domains are required for Atg17-binding to facilitate efficient autophagy.

Competition between Atg9 and Atg31 regulates Atg17 activity. We identified a stable C-terminal degradation product of Atg17, which apparently retained the capacity to interact with Atg9. This fragment contained the entire helix α4 of Atg17 and its C-terminal dimerization-domain (Fig. 3a). The corresponding recombinantly expressed Atg17-fragment ($Atg17^{235-417}$) was dimeric as expected. Moreover, co-floatation experiments demonstrated that $Atg17^{235-417}$ indeed strongly interacted with Atg9–PLs (Fig. 3b). To further narrow down the potential Atg9-binding site in Atg17, we expressed $Atg17^{235-354}$ (lacking the dimerization domain, thus being monomeric) and $Atg17^{354-417}$ (dimerization domain) separately. Although both constructs bind Atg9–PLs to a certain extent, binding of $Atg17^{235-354}$ was much stronger than that of $Atg17^{354-417}$, suggesting that the N-terminal part of α4 harbours the primary binding site for Atg9 (Fig. 3b). Interestingly, Atg31 occupies a similar region in Atg17 (ref. 36; Fig. 3a). To investigate whether Atg31 and Atg9 compete with each other, we performed co-floatation experiments of Atg17 and Atg9–PLs adding

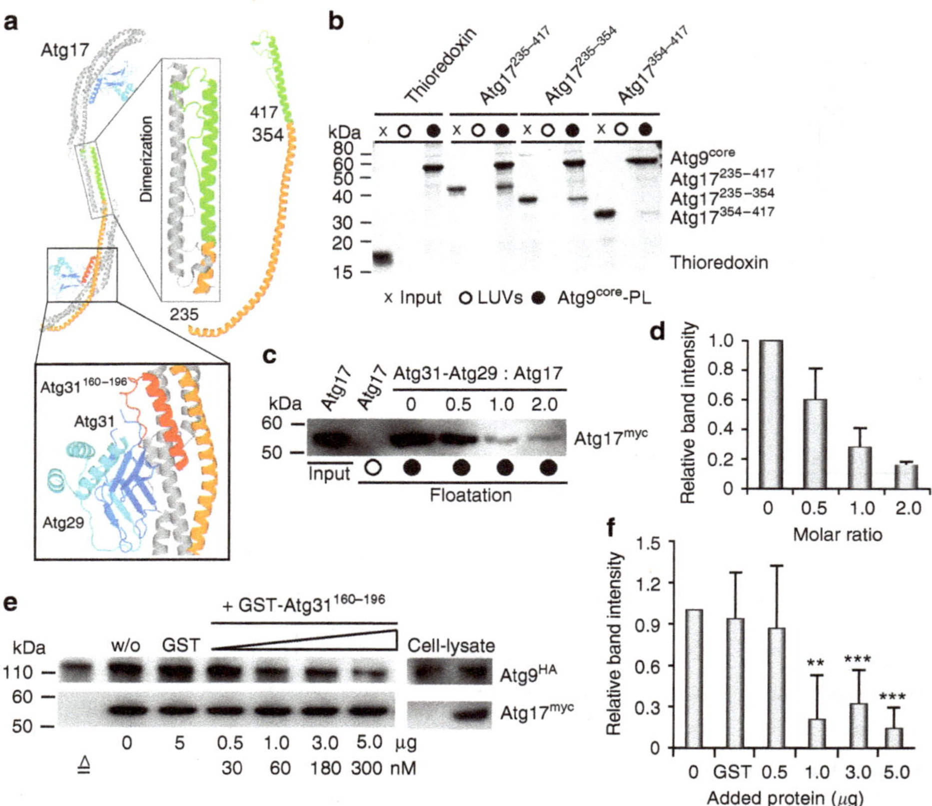

Figure 3 | Atg9 binds the central crescent of Atg17. (**a**) Cartoon representation of the crystal structure of Atg17^{TC} (pdb 4HPQ). Helix α4 (orange and green) spans the entire Atg17-crescent and forms part of the dimerization domain (green). Domain borders that have been used to design Atg17-variants are indicated by numbers. The minimal Atg17-binding region of Atg31 is highlighted in red with numbers indicating corresponding domain borders. Images were prepared in PYMOL (The PyMOL Molecular Graphics System, Version 1.2r3pre, Schrödinger, LLC). (**b**) SDS-PAGE gel of co-floatation experiments of Atg9-PLs with recombinant thioredoxin-tagged Atg17-fragments. Thioredoxin-tag lacking Atg17 (w/o Atg17) and co-floatation with large unilamellar vesicles (LUVs) lacking Atg9^{core} served as controls. Input corresponds to 10% of total protein used for co-floatation. (**c**) α-myc immunoblot of co-floatation experiments of Atg17 with Atg9-PLs and increasing stoichiometric ratios of Atg31-Atg29. After incubating Atg9-PLs with Atg17, purified Atg31-Atg29 was added. Atg31-Atg29 inhibits binding of Atg17 to Atg9 *in vitro*. (**d**) Quantification of Atg17 retrieved from the floating fraction in co-floatation experiments with Atg9-PLs as shown in **c**. The intensity of the Atg17-band in the absence of Atg31-Atg29 was set to one and used for normalization. Mean values ± s.d. of $N=3$ independent experiments are shown. (**e**) Immunoprecipitation (IP) of Atg17^{myc} from cell lysates after 2 h of starvation. Increasing amount of GST-Atg31$^{160-196}$ as indicated or 5 µg GST were added to cell lysates prior IP. Atg17 and co-immunoprecipitated Atg9 were detected by α-myc and α-HA immunoblots. (**f**) Quantification of Atg9^{HA} as shown in **e**. The intensity of the Atg9^{HA}-band (w/o GST-Atg31$^{160-196}$) was set to one and used for normalization. Mean values ± s.d. of $N=3$ independent experiments are shown. *P* values were calculated using two-tailed Student's *T*-test (*$P<0.05$; **$P<0.01$; ***$P<0.005$).

increasing stoichiometric amount of purified Atg31–Atg29 subcomplex (Fig. 3c). Remarkably, Atg31–Atg29 strongly inhibited Atg17 for Atg9-binding. In the presence of equimolar amounts of Atg17 and Atg31–Atg29, only ~30% of Atg17 was recovered from the Atg9-PL fraction (Fig. 3d).

The interaction interface between Atg17 and its regulatory Atg31–Atg29 subcomplex is composed of two C-terminal α-helices of Atg31 (residues 160–196), which form a helical bundle with Atg17 (Fig. 3a). We therefore predicted that Atg31$^{160-196}$ efficiently inhibits binding of Atg17 to Atg9 *in vivo* given that a similar molecular mechanism as observed in our reconstituted system regulates Atg17 activity in yeast. To test this hypothesis, we co-immunoprecipitated Atg9^{HA} with Atg17^{myc} from cell lysates which were incubated with increasing amounts of recombinant, purified GST-Atg31$^{160-196}$ (Fig. 3e). We observed a strong and concentration-dependent decrease in co-immunoprecipitated Atg9 (Fig. 3e), resulting in ~20% residual binding on incubation with one to five µg GST-Atg31$^{160-196}$, which amounts to concentrations between 60 and 300 nM GST-Atg31$^{160-196}$ in cell lysates (Fig. 3f). The estimated concentration of Atg17 in cell lysates (based on a Atg17-copy number of ~500 per cell[26] and 50 OD yeast cells in 500 µl lysate) is ~15 nM. Thus Atg31$^{160-196}$ is a strong competitive inhibitor of the Atg17–Atg9 interaction. The residual ~20% binding activity might be attributed to an Atg17-independent interaction of Atg1^{PC} with Atg9, which has recently been found to be mediated by Atg13 (ref. 14). We next confirmed that the inhibitory effect of GST-Atg31$^{160-196}$ is caused by direct and competitive interaction with Atg17 by co-immunoprecipitating Atg17 with GST-Atg31$^{160-196}$ (Supplementary Fig. 3). In summary, our *in vitro* and *in vivo* data show that the Atg31–Atg29 subcomplex regulates binding of Atg17 to Atg9. Moreover, the direct physical interaction between Atg17 and Atg9 appears to dominate other interactions of Atg1^{PC}-subunits with Atg9 *in vivo*, since Atg31$^{160-196}$ strongly inhibits binding of Atg17 to Atg9. Thus, the interaction of Atg1^{PC} with Atg9-vesicles appears to be controlled by a direct and tightly regulated binding of the scaffolding subunit Atg17 to Atg9 *in vitro* and *in vivo*.

The Atg1–Atg13 subcomplex activates Atg17^{TC}. Under vegetative conditions, most of Atg17^{TC} resides in the cytoplasm in its inactive state. On starvation, however, Atg1–Atg13 associates with Atg17^{TC} at the PAS to initiate autophagy. In context of the observed competition between Atg9 and Atg31–Atg29, we predicted that Atg1–Atg13 activates Atg17^{TC} to restore Atg9-binding. We thus compared binding of Atg17, Atg17^{TC} and Atg1^{PC} with Atg9–PLs (Fig. 4a).

Atg9–PLs used in this study were prepared from synthetic lipid mixtures lacking PI. Consequently, Atg1 and Atg13 were not recruited to Atg9–PLs in the absence of Atg17 or Atg17^{TC} (Supplementary Fig. 4a). Consistent with our competition assay, co-purified Atg17^{TC} was strongly impaired in Atg9 binding. However, combining Atg1–Atg13 with Atg17^{TC} fully restored the binding capacity of Atg17 (Fig. 4b,c, Supplementary Fig. 4b). In agreement with our and other data that only Atg13 directly contacts Atg17, Atg1 alone did not and Atg13 alone partially activated Atg17^{TC} for Atg9 binding (Fig. 4b,c).

To confirm that this mechanism of activation regulates Atg17 activity *in vivo* and thus autophagy initiation, we co-immunoprecipitated Atg9^{HA} with Atg17^{myc} from cell lysates of starved wildtype, *atg1Δ*, and *atg13Δ* cells (Fig. 4d). Consistent with our observations *in vitro*, deletion of *atg1* had a minor impact on the interaction of Atg17 and Atg9. In the absence of Atg13, however, a reduction in co-immunoprecipitated Atg9 by ~70% was observed (Fig. 4e). In conclusion our experiments suggest that the constitutive Atg17-Atg31–Atg29 complex[7,15] is inactivated by occlusion of the Atg9-binding site. On autophagy induction, Atg17^{TC} recruits Atg1–Atg13 to the PAS[7,15] and their direct interaction releases the inhibition of Atg17 to restore its full Atg9-binding capacity.

The recruitment of Atg9-vesicles by Atg17 to the PAS is one of the earliest steps in autophagy. We thus investigate whether regulation impacts on the co-localization of the two proteins

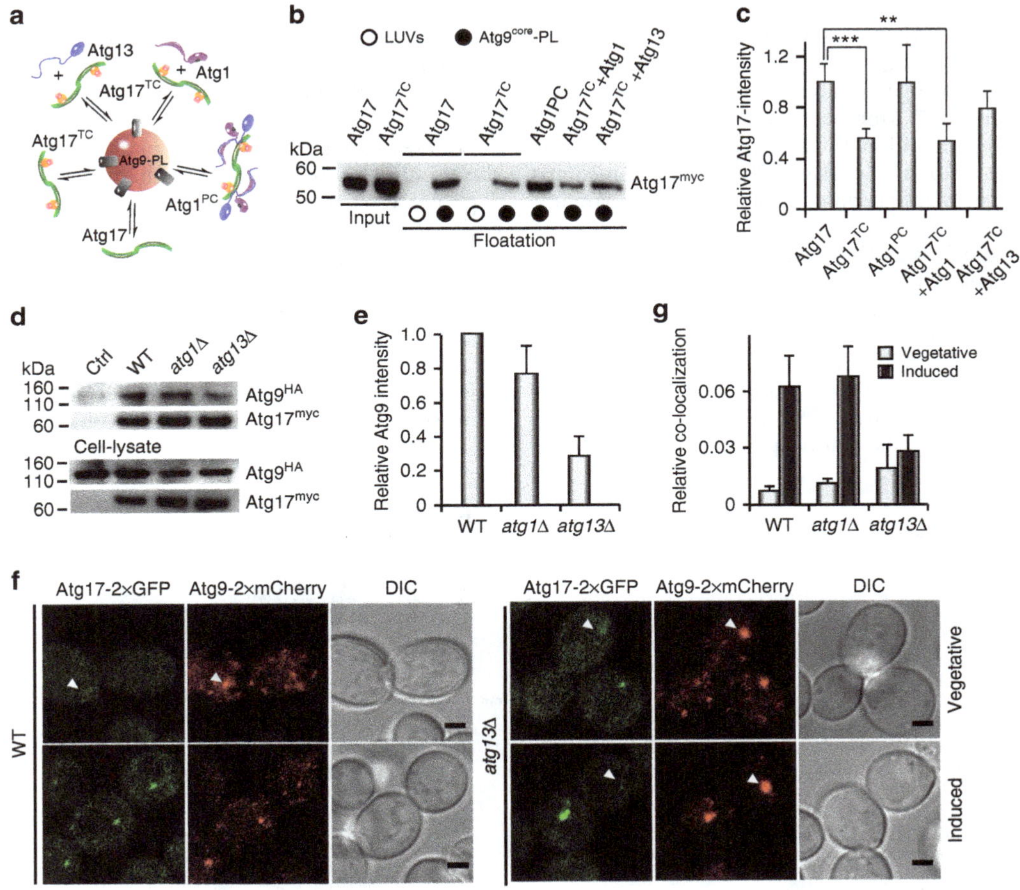

Figure 4 | The activity of Atg17 is regulated by Atg31-Atg29 and Atg1-Atg13 subcomplexes *in vitro* and *in vivo*. (**a**) The schematic drawing illustrates the experimental set-up of floatation assays. (**b**) α-myc immunoblot from co-floatation experiments of Atg9-PLs with recombinant myc-tagged Atg17 with and without other Atg1^{PC}-subunits as indicated. Similar experiments with large unilamellar vesicles (LUVs) lacking Atg9^{core} served as controls for unspecific membrane binding. Input corresponds to 10% of total protein used for co-floatation. (**c**) Semiquantitative analysis of immunoblots as shown in **b** from co-floatation experiments of Atg17^{myc} with Atg9-PLs in the presence of other subunits of Atg1^{PC} as indicated. All Atg17^{myc}-intensities were corrected for Atg9-intensities. The band intensity of Atg17^{myc} (w/o other Atg1^{PC}-subunits) was set to one and used for normalization. Mean values ± s.d. of $N=3$ independent experiments are shown. *P* values were calculated using two-tailed Student's *T*-test (**P*<0.05; ***P*<0.01; ****P*<0.005). (**d**) Immunoprecipitation (IP) of Atg17^{myc} from lysates of wildtype (WT), *atg1Δ* and *atg13Δ*cells after two hours of starvation. Atg17 and co-immunoprecipitated Atg9 were detected by α-myc and α-HA immunoblots as indicated. (**e**) Quantification of Atg9^{HA} as shown in **d**. The intensity of the Atg9^{HA}-band in wildtype (WT), *atg1Δ*, and *atg13Δ* cells were corrected for unspecific binding of Atg9^{HA} in the control (Ctrl, cells w/o myc-tag). The WT Atg17-band intensity was set to one and used for normalization. Mean values ± s.d. of $N=3$ independent experiments are shown. (**f**) Co-localization of Atg9-tandem-mCherry with Atg17-tandem-GFP in wildtype (WT) and *atg13Δ* cells under vegetative and autophagy-induced conditions. Arrowheads indicate positions of non-colocalizing Atg9-puncta. Scale bar, 2 μm. (**g**) Quantification of relative co-localization of Atg17-puncta with Atg9-puncta (number of co-localizing puncta per cell, normalized by number of Atg17 puncta per cell) in WT, *atg1Δ*, and *atg13Δ* cells under vegetative (white bars) and autophagy-induced (black bars) conditions. Mean values ± s.d. of $N=3$ independent experiments are shown (>200 cells per strain were examined).

using wildtype, *atg13Δ* and *atg1Δ* cells. In the wild-type background, multiple Atg17-negative Atg9-puncta were detected under vegetative conditions and inducing autophagy increased the relative co-localization of both proteins drastically (Fig. 4f,g). A similar co-localization was observed in *atg1Δ* cells, arguing that Atg13 itself sufficiently activates Atg17^{TC} (Fig. 4g, Supplementary Fig. 4c) By contrast, autophagy induction did not promote co-localization of Atg17 and Atg9 in *atg13Δ* cells, demonstrating that inhibited Atg17^{TC} is strongly impaired in recruiting Atg9-vesicles to the PAS (Fig. 4f,g).

The dimerization domain of Atg17 is involved in regulation. Our observation that the dimerization domain of Atg17 is dispensable for Atg9-binding (Fig. 3b) prompted us to investigate, whether monomeric Atg17 is regulated in a similar manner as observed for full-length Atg17. We thus compared binding of Atg17^{mono}, Atg17^{monoTC} and Atg17^{monoPC} with Atg9-PL using floatation assays. Consistent with our data that helix α4 of Atg17 harbours the primary Atg9-binding site, Atg17^{mono} was efficiently recruited to Atg9-PLs (Fig. 5a). Remarkably, Atg17^{mono} was much stronger inhibited by Atg31–Atg29 as it has been observed for Atg17. Moreover, Atg1–Atg13 was not able to release this strong inhibition (Fig. 5a), although both sub-complexes assemble into stable Atg1^{monoPC} (Supplementary Fig. 1f). Thus, the dimerization domain of Atg17 is apparently involved in regulating Atg17-function. The strong inhibition of Atg1^{monoPC} also demonstrates that membrane binding of Atg1 and Atg13 *per se* does not contribute to enhanced binding of full-length Atg1^{PC} to Atg9-PLs.

We next tested whether the residual Atg9-binding activity of Atg17^{monoTC} and Atg1^{monoPC} suffices to induce autophagy. Consistent with previous findings[12], autophagy is strongly impaired in cells expressing Atg17^{mono} (Fig. 5b). This earlier study also demonstrated that the autophagy-defect was caused by an impaired recruitment of Atg8 to the PAS[12]. Since Atg17 functions upstream of Atg8, we investigated whether Atg17^{mono} localizes to the PAS by analysing mCherry-Atg17- and GFP-Atg8-puncta formation in cells expressing Atg17 or Atg17^{mono} (Fig. 5c). We performed these experiments in *atg11Δ* cells to prevent Cvt-mediated GFP-Atg8-puncta formation. As expected, both Atg17 and Atg8 were recruited to the PAS in cells expressing full-length Atg17 under autophagy-induced conditions. Interestingly, however, neither Atg8- nor Atg17-puncta were observed in Atg17^{mono} expressing cells. (Fig. 5c). Thus, expression of Atg17^{mono} inhibits early steps in autophagy, arguing that dimerization of Atg17 itself is important for its proper localization and function *in vivo*.

Atg1^{PC} is a membrane-tethering complex. Our data revealed the mechanism that regulates Atg17 activity, implying that dimerization of Atg17 is important for its targeting to and activation at the PAS. Moreover, the S-shaped structure[11,12] of Atg17^{TC} suggests that one Atg17-dimer binds two Atg9-vesicles independently. We thus tested whether Atg17 has membrane-tethering activity by measuring the R_H of Atg9-PLs in the presence and absence of Atg17 using DLS.

Atg9-PLs are polydisperse in solution, indicating that they are clustering. This is consistent with previous data showing that Atg9 possesses a highly conserved self-interacting motif that is required for Atg9-trafficking and function *in vivo*[37]. To circumvent Atg9-mediated self-interaction, we prepared Atg9-PLs in the presence of either Atg17 (Atg9–Atg17–PLs) or Atg17^{mono}. In contrast to SUVs lacking Atg9 ($R_H = 60 \pm 2$ nm, Fig. 6a), Atg9–Atg17–PLs possessed an R_H of 135 ± 2 nm (± refers to s.d., $n = 5$), indicating two Atg9-PLs to be tethered together by dimeric Atg17. To exclude that other effects than specific tethering cause Atg9-PLs to appear larger by DLS, we analysed Atg17^{mono} for its tethering activity. The size of

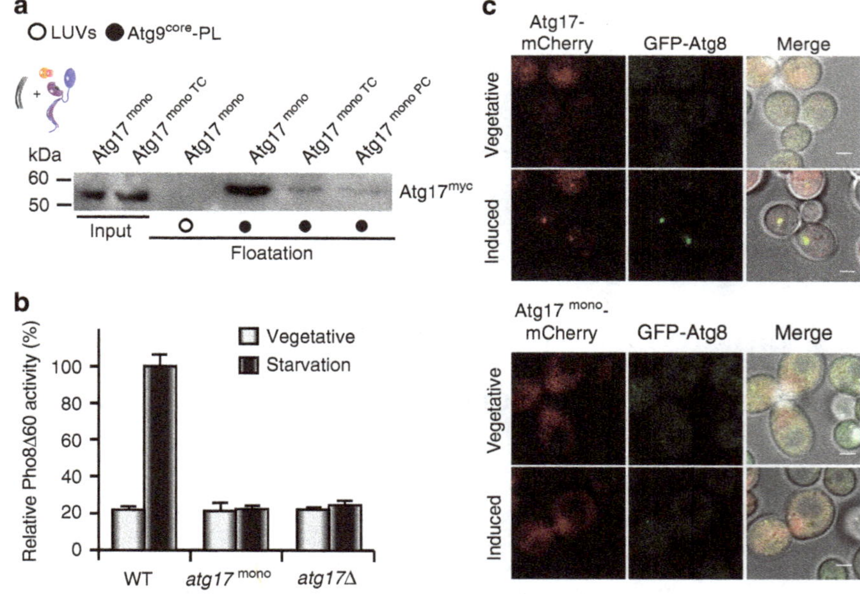

Figure 5 | The dimerization domain of Atg17 regulates Atg17-function. **(a)** α-myc immunoblots from co-floatation experiments of Atg9-PLs with recombinant myc-tagged, monomeric Atg17 (Atg17^{mono}), Atg17^{monoTC} and Atg1^{monoPC}. Atg17^{mono} is strongly inhibited by Atg31-Atg29, but Atg17^{monoTC} is not activated by Atg1-Atg13. Similar experiments with large unilamellar vesicles (LUVs) lacking Atg9^{core} served as controls. Input corresponds to 10% of total protein used for co-floatation. **(b)** Pho8Δ60 assay of wildtype (WT), *atg17*^{mono}, and *atg17Δ* cell lysates to assess autophagic activity under vegetative (grey bars) and starvation (black bars) conditions. Mean values ± s.d. of $N = 3$ independent experiments are shown. **(c)** Co-localization of GFP-Atg8 with mCherry-tagged Atg17 or Atg17^{mono} under vegetative or autophagy-induced conditions. Atg17^{mono} is not forming puncta under autophagy-induced conditions. Consequently, autophagosomes are not being formed as apparent from the lack of Atg8-puncta. Scale bar, 2 μm.

Atg9–Atg17^{mono}–PLs ($R_H = 65 \pm 1$ nm, (refers to s.d., $n = 5$) was similar to that of corresponding SUVs lacking Atg9^{core} (Fig. 6a). Thus, dimerization of Atg17 is required to tether Atg9–PLs.

Atg17 constitutively interacts with Atg31–Atg29 *in vivo* and we demonstrated that Atg17 is inhibited by Atg31–Atg29 for Atg9-binding *in vitro* and *in vivo* (Fig. 3c,d). We therefore tested whether this inhibition impacts on tethering. Atg9–PLs, prepared in the presence of Atg17^{TC} and Atg17^{monoTC} were monodisperse with a R_H of 72 ± 2 nm and 82 ± 2 nm ($\pm$ refers to s.d., $n = 5$), respectively (Fig. 6b,c). Thus Atg31–Atg29 does not only interfere with Atg9-binding, it also regulates tethering of Atg9–PLs by Atg17. Consequently, Atg1–Atg13-mediated activation of Atg17^{TC} should promote membrane-tethering by Atg1^{PC}. In fact, Atg9–Atg1^{PC}–PLs were found to be twice the size ($R_H = 180 \pm 4$ nm) of corresponding SUVs lacking Atg9^{core} ($R_H = 87 \pm 2$ nm, $\pm$ refers to s.d., $n = 5$) (Fig. 6b,c). Consistent with our data on Atg17^{TC}-regulation (Fig. 4c), Atg1 alone did not promote tethering of Atg9–PLs by Atg17^{TC}. In the presence of Atg13, however, Atg9–Atg17^{TC}–PLs were significantly larger ($R_H = 105 \pm 5$ nm) and polydisperse ($22 \pm 3\%$ compared to 12–16% in other samples, $\pm$ refers to s.d., $n = 5$), indicating that a mixture of tethered and dispersed vesicles is present in these samples (Supplementary Fig. 5a). Thus, Atg13 partially activates Atg17^{TC}, which allows the resulting complex to tether Atg9–PLs, but with lower efficiency compared with Atg17 or Atg1^{PC}. Furthermore, Atg1^{monoPC} was deficient in tethering Atg9–PLs (Fig. 6b), demonstrating that dimerization and activation of Atg17 are required to tether Atg9-vesicles.

Since DLS is an indirect measurement for vesicle tethering, we set out to visualize this process by cryoelectron microscopy. We found that Atg9–Atg17–PLs are 20–50 nm in diameter (Fig. 6d, Supplementary Fig. 5b,c), closely resembling the size of Atg9-vesicles that are recruited to the PAS to nucleate the phagophore *in vivo*[10]. Most importantly, however, the majority of such vesicles was found to be tethered, resulting in two vesicles contacting each other (Fig. 6d, Supplementary Fig. 5b,d). Consistent with our DLS measurements, Atg9–Atg17^{mono}–PLs and Atg9–Atg17^{TC}–PLs were homogeneously dispersed with diameters similar to that of Atg9–Atg17–PLs (Supplementary Fig. 5b,c). In contrast, multiple Atg9–PLs were contacting each other in the presence of Atg1^{PC} (Fig. 6d, Supplementary Fig. 5b, d). Moreover, Atg1^{PC}-mediated tethering progressed from two vesicles to multiple vesicles, resulting in a steady increase in R_H and polydispersity in DLS measurements over time. Interestingly, an average of three Atg9-vesicles have been shown to coalesce at the PAS to nucleate the phagophore[10], which is consistent with our finding that Atg1^{PC} tethers multiple vesicles.

Phosphorylated Atg13 is deficient in activating Atg17^{TC}. Under vegetative conditions, Atg13 is phosphorylated by TORC1 (ref. 24), preventing its interaction with Atg17 and thus the assembly of Atg1^{PC} at the PAS[13,23]. In order to recapitulate phosphoregulation *in vitro*, we made use of the kinase activity of recombinant Atg1. By adding ATP to fully assembled Atg1^{PC}, the electrophoretic mobility of Atg13, Atg29 and Atg1 shifted on SDS–PAGE gels, indicating these proteins to be phosphorylated by Atg1 (Supplementary Fig. 6a). We thus determined the phosphorylation sites by mass spectrometry. Surprisingly, many previously described regulatory sites, which are phosphorylated in

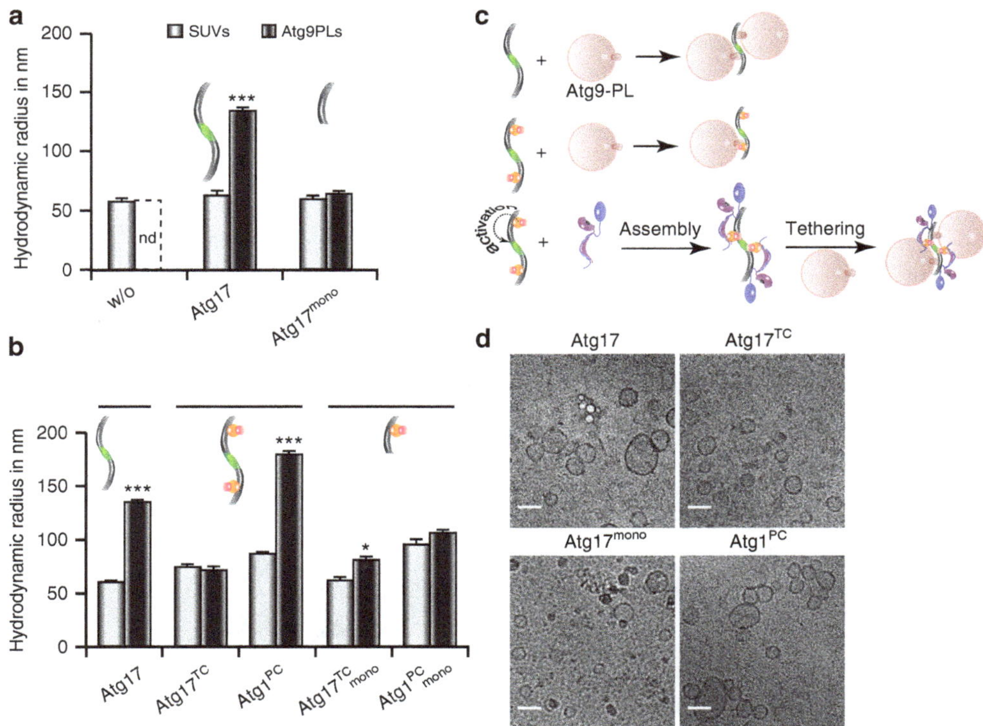

Figure 6 | Atg1^{PC} tethers Atg9-PLs. (**a**) The hydrodynamic radius of SUVs (white bar) and Atg9-PLs (black bar) obtained from DLS-experiments in the absence or presence of Atg17 and corresponding variants as illustrated by cartoons are shown. ND, not determinable (polydisperse). (**b**) Similar experiments shown in **a** were performed for Atg17 and Atg17-variants and that of corresponding Atg17^{TC} and Atg1^{PC} (sub)complexes. (**a**,**b**). Mean values $\pm$ s.d. of $N = 5$ experiments are shown for both **a** and **b**. *P* values were calculated using two-tailed Student's *T*-test (*$P < 0.05$; **$P < 0.01$; ***$P < 0.001$). (**c**) The schematic drawing illustrates the experimental setup and summarizes the results obtained from DLS-experiments in **a** and **b**. (**d**) Cyro-electron micrographs of Atg9-PLs from samples used for DLS-experiments containing Atg17, Atg17^{TC}, Atg17^{mono}, and Atg1^{PC}. Atg17 tethers two and Atg1^{PC} tethers multiple vesicles together. Scale bar, 50 nm.

response to nutrient conditions *in vivo*, are phosphorylated by Atg1-kinase *in vitro* (Supplementary Table 1). The observed phosphorylation patterns *in vitro* did, however, not recapitulate starvation-induced regulation *in vivo* since Atg29 is phosphorylated, but Atg13 is dephosphorylated on autophagy induction[24,29].

We next tested whether $Atg1^{PC}$ is disassembled on phosphorylation. In fact, SEC and SDS–PAGE of SEC-fractions demonstrated that $Atg1^{PC}$ dissociated on phosphorylation into $Atg17^{TC}$ and Atg1–Atg13 (Fig. 7a, Supplementary Fig. 6b,c). Interestingly, although Atg13 was phosphorylated at several positions including serine (S)496 within the C-terminal MIM, which contacts Atg1, the Atg1–Atg13(P) subcomplex remained tightly associated. Consistent with previous findings that phosphorylation of S428 and S429 in Atg13 prevents its interaction with Atg17 (ref. 23), both residues were phosphorylated on ATP-addition to $Atg1^{PC}$.

As indicated above, phosphorylation of Atg1, Atg13 and Atg29 is caused by promiscuous Atg1-kinase activity *in vitro*. To investigate the specific impact of Atg13-phosphorylation on Atg17-regulation, we therefore made use of the recently described phosphomimic mutant $Atg13^{S428D,S429D}$ ($Atg13^{SD}$)[23]. Consistent with this study, SEC-profile and SDS–PAGE of SEC-fractions demonstrate that the Atg1–$Atg13^{SD}$ subcomplex does not interact with $Atg17^{TC}$ (Fig. 7a, Supplementary Fig. 6d). Moreover, Atg1–$Atg13^{SD}$ did not activate $Atg17^{TC}$ for Atg9-binding (Fig. 7b). Consequently and consistent with all data in our study, Atg1–$Atg13^{SD}$ did not promote tethering of Atg9–PLs by $Atg17^{TC}$ (Fig. 7c).

In summary, our *in vitro* reconstitution study revealed important aspects of $Atg1^{PC}$-function, including its assembly, regulation, disassembly, as well as its interaction with and tethering of Atg9-vesicles.

Discussion

The Atg1-kinase complex assembles at the PAS upon dephosphorylation of Atg13 (refs 13,24), where it extensively regulates autophagosome biogenesis[38]. However, the molecular function of $Atg1^{PC}$ during phagophore biogenesis, its regulation as well as its intricate interplay with Atg9 remained controversial.

In this study, we revealed the molecular mechanism of $Atg1^{PC}$-activation and how this impacts on its interaction with Atg9-vesicles. The key elements of this study comprise: (1) reconstitution of full-length $Atg1^{PC}$ from purified subunits to investigate assembly, disassembly and regulation of $Atg1^{PC}$ as well as (2) reconstituting $Atg9^{core}$ in proteoliposomes to biochemically dissect interactions of both components *in vitro* and *in vivo*. Our results led to a comprehensive model for $Atg1^{PC}$-function, revealing insights into the molecular basis to initiate autophagy. We found that Atg17 has membrane-tethering activity, which is inhibited by the regulatory Atg31–Atg29 subcomplex. On dephosphorylation of Atg13, the Atg1–Atg13 subcomplex engages $Atg17^{TC}$ to assemble $Atg1^{PC}$. This releases the steric inhibition of Atg17, allowing Atg9 to bind the central crescent of Atg17. Consequently, each of the two Atg17-subunits in $Atg1^{PC}$ binds an Atg9-vesicle, thereby tethering these vesicles together (Fig. 7d).

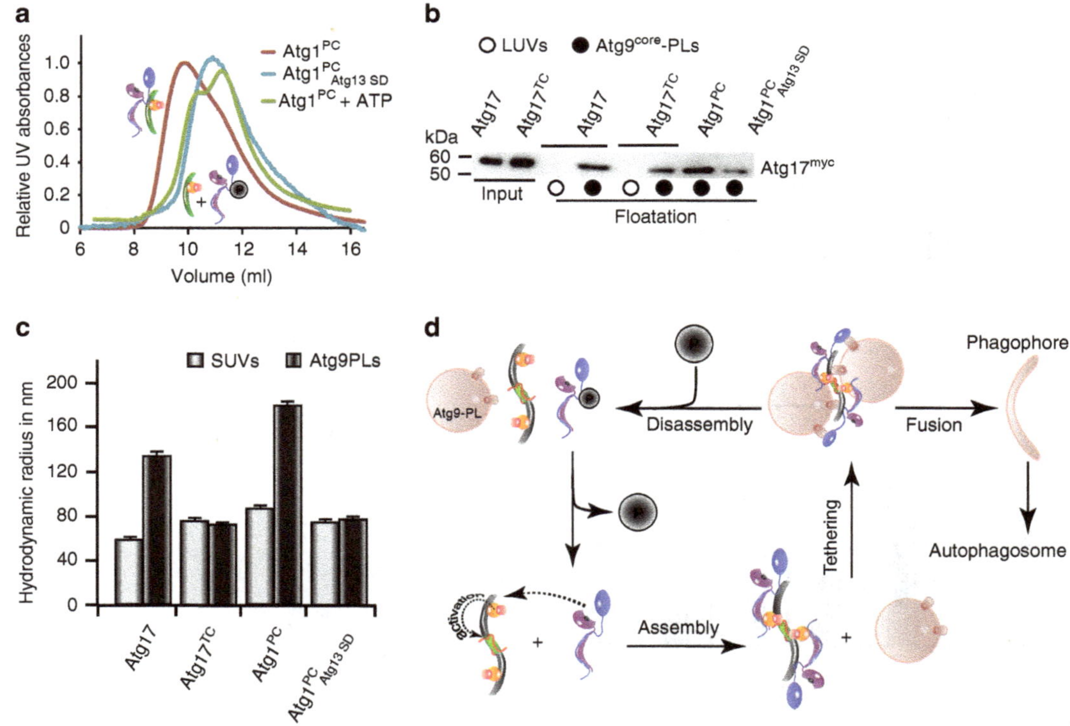

Figure 7 | Phosphorylation of Atg13 disassembles $Atg1^{PC}$ and prevents tethering. (**a**) SEC elution profiles of $Atg17^{TC}$ in the presence of Atg1–Atg13 (red), Atg1–$Atg13^{SD}$ (blue), and Atg1–Atg13 + ATP (green). $Atg13^{SD}$ and addition of ATP prevents assembly of $Atg1^{PC}$. (**b**) α-myc immunoblot of co-floatation experiments of Atg9-PLs with recombinant myc-tagged Atg17 and $Atg17^{TC}$ without and with Atg1–Atg13 ($Atg1^{PC}$) or Atg1–$Atg13^{SD}$ (ref. 23). Similar experiments with large unilamellar vesicles (LUVs) lacking $Atg9^{core}$ served as controls. Input corresponds to 10% of total protein used for co-floatation. (**c**) The hydrodynamic radius of SUVs (white bar) and Atg9-PLs (black bar) with similar samples as shown in (**b**) obtained from DLS-experiments. Mean values ± s.d. of $N = 5$ experiments. (**d**) Model for $Atg1^{PC}$ function, tethering of Atg9-vesicles, and its regulated assembly from and disassembly into subcomplexes. P indicates phosphorylation of Atg13 under vegetative conditions.

The most interesting insight of our study concerns the molecular function of Atg1^{PC} and its subunits at the PAS. A previous study established that an average of three Atg9-vesicles, 30–60 nm in size, coalesce at the PAS to nucleate the phagophore[10]. Here we demonstrate that Atg1^{PC} is a membrane-tethering factor, which specifically recognizes Atg9-vesicles of high curvature. The principle membrane-tethering subunit of Atg1^{PC} is Atg17, which has been shown to form elongated, S-shaped dimers in solution[11,12]. We could show that Atg31–Atg29 occludes the Atg9-binding site in Atg17, preventing Atg9-vesicles from being recruited under non-induced conditions by constitutive Atg17^{TC}. Molecular basis for this competition are at least overlapping binding sites of Atg31$^{160-196}$ and cytoplasmic domains of Atg9 in Atg17. Atg31$^{160-196}$ forms a four-helical bundle with Atg17 by directly contacting helices α1 and α4[12]. Atg17-helix α4 also harbours the Atg9-binding site, thus preventing recruitment of Atg9 when Atg31 occupies the Atg17-crescent. The interaction of Atg17 with Atg9–PLs was depleted by 70–80% when Atg31–Atg29 was added to preassembled Atg17–Atg9–PLs *in vitro* or when Atg31$^{160-196}$ was added to cell lysates *in vivo*. This suggests that a direct inhibition of Atg17 by Atg31–Atg29 is a key determinant in regulating the recruitment of Atg9-vesicles to the PAS in response to starvation. Moreover, the minimal Atg17-binding region of Atg31 is a very potent inhibitor of Atg17. Adding increasing amounts of Atg31$^{160-196}$ to cell lysates which resulted in concentrations well above the estimated total concentration of Atg17 did, however, not result in full inhibition of the Atg17–Atg9 interaction. Residual binding might be mediated by Atg13, which has recently been found to bind Atg9 *in vitro* and *in vivo*[26].

Assembly of Atg1^{PC} from its Atg17^{TC} and Atg1–Atg13-subcomplexes restores binding of Atg17 to Atg9 and promotes tethering of Atg9–PLs *in vitro*. The direct contact between both subcomplexes is mediated by Atg13, which directly interacts with Atg17 (refs 15,39). This implies that the activation of Atg17 requires at least Atg13. In agreement with this notion, we found that not only the interaction between Atg17 and Atg9 is significantly impaired in *atg13Δ* cells, but also targeting of Atg9 to the PAS. Moreover, the phosphomimicking mutant Atg13^{S428,429D} does not form a pentameric complex with Atg17^{TC}, thereby preventing Atg17 from interacting with and tethering of Atg9–PLs. A recent structural study identified the Atg13-binding site in Atg17, comprising helices α3 and α4 and being in close proximity to Atg31 (ref. 23). Moreover, the Atg31–Atg29 complex was suggested to be flexibly connected to Atg17 (ref. 12), thereby potentially adopting alternative conformations. Altogether with our observation that Atg13 activates Atg17^{TC} for Atg9-binding *in vitro* and *in vivo*, it appears plausible that Atg13 induces a conformational rearrangement in Atg17^{TC}, such that Atg31–Atg29 move away from the central crescent to expose the Atg9-binding site of Atg17. This model is supported by our observation that although Atg13 did not directly interact with Atg9^{core}, because Atg9^{core} lacks the previously identified Atg13-binding site[26], Atg13 restores binding of Atg17^{TC} to Atg9–PLs to promote tethering of Atg9-vesicles. We therefore propose a model in which Atg13 exerts two different activities. First, Atg13 cooperates with Atg17 in recruiting Atg9-vesicles to the PAS by direct recognition of Atg9. Furthermore, Atg13 activates Atg17^{TC} such that fully assembled Atg1^{PC} binds and tethers Atg9-vesicles through its scaffolding subunit Atg17 (Fig. 7d).

Interestingly, we observed a significant difference in Atg1–Atg13-mediated activation of monomeric or dimeric Atg17^{TC}. Even though Atg17^{monoTC} assembled together with Atg1–Atg13 into stable Atg1^{monoPC} *in vitro*, binding of Atg17 to Atg9 was not restored. Consistently, the low residual binding activity of Atg17^{monoTC} and Atg1^{monoPC} for Atg9 *in vitro* correlates with a lack of Atg9-recuitment to the PAS in Atg17^{mono}-expressing cells. Atg17^{mono} was, however, also not recruited to the PAS on starvation, arguing that dimerization of Atg17 is not only important for regulating Atg17 activity, but also for its recruitment to the PAS.

Another interesting aspect of our study regards the membrane-binding activity of Atg1 and Atg13. Although the recombinantly expressed MIT-domain of Atg1 has been found to recognize membranes previously[12], it was not clear how this affects Atg1^{PC}. According to this study, we showed that full-length Atg1 also binds highly curved membranes. Moreover, we found that Atg1 possesses specificity for PI, thereby targeting Atg1^{PC} to PI-containing membranes with high curvature. While the MIT-domain of Atg1 has no structural similarity to canonical membrane binding domains[23], a recent study showed that homologous MIT-domains in Vps4 and SNX15a bind PI and PI-phosphates as well[40]. Whether MIT-domains represent a new class of PI-specific membrane binding modules remains to be addressed by future studies. In contrast to our findings that Atg17 and Atg9 are not sensing membrane curvature, both proteins localize to the highly bent membrane edges of the phagophore[41]. Given that Atg1 and Atg13 sense curvature and that the Atg1–Atg13 subcomplex interacts with Atg17 at the PAS, it is tempting to speculate that Atg1–Atg13 restricts the localization of Atg17 and consequently that of Atg9 at the phagophore. This implies, on the other hand, that Atg1^{PC} might not only tether vesicles for phagophore nucleation, but also for phagophore expansion.

Methods

Reagents. Synthetic lipids 1-palmitoyl-2-oleoyl-sn-glycerol-3-phosphocholine (POPC), 1-palmitoyl-2-oleoyl-sn-glycerol-3-phosphoethanolamine (POPE), 1-palmitoyl-2-oleoyl-sn-glycerol-3-phosphoserine (POPS), 1-palmitoyl-2-oleoyl-sn-glycerol-3-phosphoinositol (POPI), cholesterol, 1,2-dioleoyl-*sn*-glycero-3-phospho-1′-myo-inositol-3′-phosphate (PI(3)-phosphate) as well as yeast polar lipid (YPL) extracts were purchased from Avanti Polar Lipids. The detergent *n*-Dodecyl-*N*,*N*-Dimethylamine-*N*-Oxide (LDAO) was from Anatrace.

The following antibodies were used in this study: αPep12p (Invitrogen, Catalog No: 710037, 1:1,000 dilution), αPgk1p (Invitrogen, Catalog No: 459250, 1:10,000 dilution), αHA (Santa Cruz, sc-7392, 1:200 dilution,), and αMyc (Santa Cruz, sc-789 and sc-40, 1:1,000 dilution), and αGFP antibodies (Roche, Catalog No: 11814460001).

Yeast strains and growth conditions. Derivatives of yeast strain *Saccharomyces cerevisiae* BY4741 (Euroscarf) used for this study, are listed in Supplementary Table 2. Genomic gene deletion and tagging was performed as described[42]. Atg9-variants were ectopically expressed in *atg9Δ* cells. Therefore, the coding sequences of the respective Atg9-variants were sub-cloned into the pTL58 vector (pAtg9::Leu2) and transformed in *atg17^{myc}atg9Δ* and *pho8Δ60atg9Δ* strains. To adjust the expression levels of Atg9 to that of its variants, the Pma1-promoter in pTL58-Atg9 was exchanged against the endogenous Atg9-promoter. To investigate whether Atg9^{core} is able to interact with Atg17 *in vivo*, pRS316-Atg9^{core}-3xHA (pPma1::URA3) was transformed into *atg17^{GFP}atg9Δ* cells. To analyse the regulation of Atg17 by Atg1 and Atg13 *in vivo*, pTL58-Atg9–3xHA (under control of the Atg9-promoter, pAtg9::Leu2) was transformed into *atg17^{myc9}atg9Δatg1Δ* and *atg17^{myc9}atg9Δatg13Δ* cells.

Co-localization of Atg17-variants and Atg8 was analysed by transforming plasmids pTL58-Atg17(variants)-mCherry (pPma1::Leu2)) in *atg8-GFPΔatg11 Δatg17Δ* strains. To study the co-localization of Atg17 and Atg9, pTL58-Atg17-2 × GFP (under control of its native promoter, pAtg17::Leu2) and pRS316-Atg9-2xmCherry (under control of its native promoter, pAtg9::URA3) were co-transformed into *atg17Δatg9Δ*, *atg17Δatg9Δatg1Δ*, *and atg17Δatg9Δatg13Δ* cells.

Cells were grown in synthetic media (0.67% yeast nitrogen base, 0.5% ammonium sulphate, 2% glucose, amino acids) at 30 °C to mid-log phase. For co-immunoprecipitation experiments and Pho8Δ60 assays, autophagy was induced by starvation, cells were collected by centrifugation, washed twice and resuspended in synthetic medium without amino acids (0.17% yeast nitrogen base, 2% glucose). For fluorescence imaging, autophagy was induced by addition of rapamycin (0.2 μg ml^{-1}, Sigma) for 1 h.

Preparation of yeast cell extracts and immunoprecipitation. Yeast cells corresponding to 50 OD_{600} (optical density at 600 nm) were collected by centrifugation (3 min, 1,500*g*), resuspended in ice-cold lysis buffer (25 mM Tris-HCl pH 7.2, 150 mM NaCl, 0.2% NP-40 supplemented with protease inhibitor cocktail (Sigma) and 3 mM PMSF) and lysed by vortexing with acid washed glass beads. Lysates were cleared by centrifugation (20 min, 17,000*g*).

Immunoprecipitation was performed by incubating the cell lysates with an α-myc antibody coupled to magnetic protein A beads (raised from murine B-cell-hybridomas) or GFP-Trap-M beads (Chromoteck) for 2h at 4 °C. For competition assays, different amounts of recombinant, purified GST-$Atg31^{160-196}$ were added to lysate. For GST pull-down assay, 10 μg of GST or GST-$Atg31^{160-196}$ was incubated with cell lysate, prepared as described above. Lysates were incubated with Glutathione Sepharose 4B beads, extensively washed with lysing-buffer containing an additional 100 mM NaCl and resuspended in SDS-sample buffer. Samples were separated on NuPage Bis-Tris gels (Life Technologies) prior to immunoblotting. Corresponding uncropped blots are shown in Supplementary Fig. 7.

Subcellular fraction assay. Yeast cells were grown to mid-logarithmic phase (OD_{600} of 0.8–1.0) and autophagy was induced by treatment with 0.2 μg ml^{-1} rapamycin for 1 h at 30 °C. Cells corresponding to 500 OD_{600} were collected by centrifugation (3 min, 1,500*g*) and resuspended in 50-ml buffer containing 100 mM Tris-H_2SO_4, pH 9.4 and 10 mM DTT. After incubation at 30 °C for 10 min, cells were spheroblasted by resuspending them in 50 ml synthetic medium containing 20 mM Tris-HCl pH 7.5 and 1.0 M sorbitol supplemented with 2.5 μg per OD_{600} zymolyase-100T (Biomol), and incubation for 30 min at 30 °C. Spheroblasts were washed twice in 50 ml synthetic medium containing 1.0 M sorbitol and lysed in 5-ml ice-cold hypoosmotic buffer (50 mM HEPES-KOH, pH 7.6, 200 mM sorbitol, 1 mM EDTA) containing complete protease inhibitors (Sigma) and 2 mM PMSF using 50 strokes in a Dounce homogenizer. The cell lysate was centrifuged sequentially at 300*g* for 5 min and 13,000*g* for 10 min and 1.9 ml of the supernatant was loaded on top of a 22–60% sucrose step gradient. The gradient was prepared by overlaying 0.5 ml 60% sucrose with 1.0 ml 37%, 1.5 ml 34%, 2 ml 32%, 2.0 ml 29%, 1 ml 27%, and 1.5 ml 22% sucrose solution. After ultra-centrifugation at 170,000 g for 17–18 h using a SW40Ti rotor (Beckman Coulter), 0.8 ml fractions were collected from top to bottom and analysed by western blotting using αPep12p (Invitrogen), αPgk1p (Invitrogen), αHA, and αMyc (Santa Cruz) antibodies.

Pho8Δ60 assay. Autophagic activity was assessed as described[43]. Briefly, yeast cells corresponding to 4 OD_{600} were collected by centrifugation (3 min, 1,500*g*) and resuspended in ice-cold assay buffer (250 mM Tris-HCl, pH 9.0, 10 mM $MgSO_4$ and 10 μM $ZnSO_4$). Cells were lysed by vortexing with acid washed glass beads and cleared by centrifugation (20 min, 17,000*g*). 1-naphthyl-phosphate was added to final concentration of 5 mM and the enzymatic reaction was stopped after 20 min incubation at 30 °C by adding one volume of stopping buffer (2 M Glycin-NaOH, pH 11.0). The enzymatic conversion of 1-naphthyl-phosphate by alkaline phosphatases was determined by measuring the fluorescence with a TECAN infinite M1000Pro using an excitation wavelength of 345 nm and detecting emission at 472 nm. The fluorescence signal was corrected for the total protein content in the respective sample as determined using the Pierce BCA protein assay kit (Thermo Scientific).

Recombinant expression and purification of the $Atg1^{PC}$-subunits. The open reading frames encoding Atg1, Atg13, Atg17, $Atg17^{1-354}$, $Atg17^{235-417}$, $Atg17^{235-354}$, $Atg17^{354-417}$, Atg29, $Atg31^{160-196}$ and Atg31 were amplified from yeast *S. cerevisiae* BY4741 (Euroscarf) genomic DNA and cloned into pCoofy-vectors with N- or C-terminal affinity-tags for expression in insect cells (Atg1) or *Escherichia coli* (all others) by homologous recombination[44]. Oligonucleotides used are shown in Supplementary Table 3. Atg17 and the Atg17-Atg31–Atg29 complex as well as its monomeric variant $Atg17^{1-354}$-Atg31–Atg29 were (co)expressed from the polycystronic vector pST39 (ref. 45) with a His_6-tag fused to the N-terminus of Atg17 or a myc-tag fused to the C-terminus of Atg17.

Atg1 was expressed in High Five (BTI-TN-5B1-4) cells applying the titerless infected-cells preservation and scale-up (TIPS) method as described[46]. Cells were resuspended in lysis buffer (100 mM Tris-HCl, pH 7.4, 300 mM NaCl, 20 mM imidazole, 5% glycerol, 5 mM β-mercaptoethanol, protease inhibitor cocktail (Sigma), benzonase (Sigma), 10 U ml^{-1}) supplemented with 1% Triton-X and disrupted by high pressure homogenization. All other proteins were expressed in *E.coli* Rosetta cells. Cultures were grown in LB-medium at 37 °C until they reached an OD_{600} of 1.0 and shifted to 18 °C. Expression was induced by adding Isopropyl-β-D-thiogalactopyranosid (0.1 mM). The cells were collected by centrifugation after 16 h and resuspended in lysis-buffer. Cells were disrupted by sonication. All proteins were purified by Ni-nitrilotriacetate (Ni-NTA) affinity chromatography, applying a linear imidazole gradient in running buffer (25 mM Tris-HCl pH 7.4, 300 mM NaCl, 20 mM or 1 M imidazole, 5% glycerol, 5 mM β-mercaptoethanol). For Atg1 purification, a running buffer supplemented with 0.1% Triton-X was used. The N- and C-terminal tags were cleaved by PreScission protease digest for 30 min at room temperature. Digested proteins were immediately subjected to anion-exchange chromatography (Atg13) equilibrated with 25 mM Tris-HCl pH 7.4, 125 mM NaCl, 5% glycerol, 2 mM DTT or size exclusion chromatography (SEC) on a HiLoad Superdex 200 prep grade column (GE Healthcare) equilibrated with 25 mM Tris HCl pH 7.4, 300 mM NaCl, 5% glycerol, 5 mM β-mercaptoethanol. Fractions containing the protein of interest were pooled, concentrated and further purified by Ni-NTA rechromatography. Aliquots of the purified proteins were flash frozen in liquid nitrogen and stored at − 80 °C until use. The Atg1–Atg13 subcomplex or $Atg1^{PC}$ and its monomeric variant were assembled from purified subunits by incubating respective mixtures at room temperature for 30 min followed by gelfitration using a Superose 6 column.

Expression and purification of Atg9. The coding sequence of $Atg9^{core}$ ($Atg9^{281-779}$) was amplified from *S. cerevisiae* BY4741 and sub-cloned into a modified pET28a(+)vector (Novagen). Truncated constructs $Atg9^{\Delta N}$ (residues 316–779), $Atg9^{\Delta cD}$ (residues 316–423 and 508–779) and $Atg9^{\Delta N \Delta cD}$ were generated using the In-Fusion HD Cloning Kit (Clontech). $Atg9^{core}$ and its variants were expressed in *E. coli* Rosetta cells as described above, but resuspended in buffer containing 25 mM Tris-HCl, pH 8.0 and 150 mM NaCl. After lysing the cells with a high-pressure microfluidizer, cell debris was removed by centrifugation (10 min, 24,000*g*). The supernatant was subjected to a second centrifugation step to separate the membrane fraction (1 h, 150,000*g*). The membrane fraction was resuspended in buffer containing 25 mM Tris-HCl pH 8.0, 150 mM NaCl, and 40 mM LDAO (critical micelle concentration = 1 mM) and incubated for 1 h at 4 °C to solubilize the membrane proteins. Not solubilized material was removed by ultracentrifugation (1 h, 150,000*g*) and the supernatant was purified by Ni-NTA-chromatography as described above, with Atg9-buffer (25 mM Tris-HCl pH 8.0, 150 mM NaCl, 4 mM LDAO), supplemented with imidazole. Atg9-containing fractions were pooled, concentrated and subjected to SEC using a Superdex-200 10/300GL column (GE Healthcare) for further purification using Atg9-buffer. Purified proteins were stored at − 80 °C until used as described above.

Circular dichroism spectroscopy. The quality of $Atg9^{core}$ preparations was assessed using CD spectroscopy. CD-spectra of detergent-solubilized $Atg9^{core}$ (-variants) were recorded using a JASCO J-715 spectrophotometer, covering a 190–250 nm spectral range, at 4 °C and a path length of 0.1 mm. The protein concentration was determined by measuring the OD at 280 nm and the secondary structure content of Atg9 was estimated by fitting experimental data with standard curves using the contin-algorithm[47], provided by the Jasco Standard Analysis Program.

Liposome preparation and sedimentation. Liposome sedimentation assays were performed to investigate protein–membrane interactions in the absence of Atg9. Liposomes were prepared from mixtures of synthetic lipids containing 20 mol% cholesterol, 10 mol% POPE, 60 mol% POPC, 10 mol% POPS or from YPL-extracts. For liposomes containing 30 mol% POPI, 25 mol% POPI + 2.5 mol% PI(3)-phosphate, or 40 mol% POPS, the amount of POPC was adjusted accordingly. A thin lipid film was produced in a glass vial by evaporating the chloroform of the respective lipid-mixtures under a nitrogen-stream and subsequent incubation under vacuum. Multilamellar vesicles (MLVs) were generated by hydrating the lipid film with hydration-buffer containing 25 mM Tris-HCl pH 8.0, 150 mM NaCl, adjusted to a total lipid concentration of 1 mg ml^{-1}. Large unilamellar vesicles were produced by extruding the resulting MLV-suspension through filters with appropriate pore size as indicated. SUVs were prepared by sonicating the MLV-suspension on ice until the solution became clear followed by extrusion through 30-nm pore-size filters.

Liposome sedimentation assays were performed by incubating 100-μl large unilamellar vesicles or SUVs with the protein of interest (2 μM final concentration) for 30 min at room temperature. The liposomes were collected by centrifugation (30 min, 140 000 g, 4 °C) for subsequent analysis using SDS–PAGE.

Reconstitution of Atg9 in liposomes and floatation assay. $Atg9^{core}$ or its variants were reconstituted in liposomes using the rapid dilution approach. Liposomes were prepared from mixtures of synthetic lipids containing 20 mol% cholesterol, 10 mol% POPE, 60 mol% POPC and 10 mol% POPS, yielding a total lipid concentration of 1 mg ml^{-1}. After rehydration of liposomes, lipids were resolubilized with LDAO yielding a final concentration of 6 mM LDAO before Atg9-variants were added. $Atg9^{core}$-(variants) solubilized in 4 mM LDAO were added in a protein:lipid ratio of 1:200 and the mixture was 30-fold diluted with hydration buffer and incubated for 30 min at room temperature. Atg9 containing proteoliposomes (Atg9-PL) were collected by centrifugation (30 min, 150 000 g, 4 °C), resuspended in 25 mM HEPES pH 7.0, 100 mM NaCl (interaction buffer) and extruded using a filter pore size of 200 nm.

The orientation of Atg9 was assessed by incubating Atg9-PL with PreScission protease (50 U ml^{-1}) at room temperature for 1 h. Proteolytical cleavage of the C-terminal His-tag resulted in a significant band shift of the corresponding protein on SDS–PAGE gels.

The interaction of $Atg1^{PC}$, its subunits, or its subcomplexes with Atg9-PL was assessed by floatation experiments. Therefore, 250 μl of Atg9–PLs were mixed with an equivalent volume of 80% Nycodenz in interaction buffer containing the respective $Atg1^{PC}$-subunits in a stoichiometric ratio of 1:1. A step gradient was generated by overlaying this mixture with 250 μl 30% Nycodenz and 100 μl interaction buffer. After centrifugation at 650,000*g* for 2 h (at 4 °C), the top 100 μl co-floated fraction contained accumulated Atg9–PLs (at the Nycodenz–buffer interface) and bound interaction partners. This fraction was subsequently analysed by SDS–PAGE and immunoblotting.

Dynamic light scattering. The protein amount used for tethering assays was calculated according to the amount of Atg9–PLs with native orientation in the membrane (50% of total protein). For each assay, 200 μl Atg9–PLs were mixed with 100 μl protein of interest, yielding molar ratios of 1:1, diluted in 25 mM HEPES pH 7.0 and 100 mM NaCl buffer and incubated at room temperature for 15 min. The mixture was co-sonicated at 4 °C until the solution appeared clear (15 min). Dynamic light scattering was carried out at 4 °C on a DynaPro NanoStar Instrument (Wyatt) using 50-μl cuvettes and a total sample volume of 20 μl. Raw data were analysed by the DYNAMICS software package. The dispersity of the solution was assessed, and the average hydrodynamic radius (R_H) was calculated.

Multi-angle light scattering. Size exclusion chromatography–MALS (SEC–MALS) measurements were performed on a HP1100 HPLC system (Agilent) using a Superdex-200 Increase 10/300GL (GE-Healthcare, Atg1) or a Superose-6 10/300GL (GE-Healthcare, Atg1^{PC}) column. The MALS signal was recorded on-line using a miniDAWN Treos MALS detector (Wyatt) and the corresponding protein concentration was determined with an Optilab t-Rex RI detector (Wyatt). Atg1 was injected at a concentration of 5 mg ml^{-1} and Atg1^{PC} at a concentration of 3.4 mg ml^{-1} in a volume of 50 μl and 100 μl, respectively. SEC–MALS data were analysed using the Astra software package (Wyatt) applying the Zimm light scattering model. The molar mass at the concentration peak apex (MW max, refractive index peak apex) and the weight-average molar mass of the respective peak (MW peak, ultraviolet peak borders selected by hand) were determined.

Tandem mass spectrometry. Proteins for tandem mass spectrometry were used at a 1 μM concentration. Proteolytic digestion was performed over night with LysC at RT and subsequently with Trypsin for 2 h at 37 °C. For injection to the LTQ Orbitrap (Thermo Fisher) tandem mass spectrometer, the resulting peptides were separated by C18 reversed phase nanoscale liquid chromatography. Data were analysed using the MaxQuant software package[48].

Confocal microscopy. Images were acquired on a Zeiss LSM780 and a Leica TSC SP8 laser scanning confocal microscope with × 63/1.4 numerical aperture (NA) objectives using the Zeiss ZEN 2011 and LAS AF software packages, respectively. GFP and mCherry were excited with 488-nm and 561-nm laser light, respectively. Multi-tracking mode was used for image acquisition on the Zeiss LSM780 microscope. Pinholes were set for acquiring 1.5–2.0 μm optical slices. All images were analysed using ImageJ (Schneider *et al.*, 2012) and Zeiss ZEN 2011 software packages. For yeast cell imaging, 5 ml yeast cells were grown to log phase and treated with rapamycin for 1 h. Two hundred-microlitre samples were added to an observation chamber (Lab-Tek #1.0 Borosilicate), which had been coated with 1 mg ml^{-1} concanavalin A (Sigma).

Cryo-EM analysis. Samples (5 μl co-sonicated Atg9–PLs) were applied to lacey-carbon-grids (Quantifoil), blotted and frozen in liquid ethane using a Vitrobot (FEI). Images were recorded on a CM120 FEG under low-dose conditions with a 4k × 4k CCD camera (TVIPS) and a defocus of − 2 μm.

References

1. Choi, A. M. K., Ryter, S. W. & Levine, B. Autophagy in human health and disease. *N. Engl. J. Med.* **368,** 651–662 (2013).
2. Lamb, C. A., Dooley, H. C. & Tooze, S. A. Endocytosis and autophagy: shared machinery for degradation. *Bioessays* **35,** 34–45 (2013).
3. Tsukada, M. & Ohsumi, Y. Isolation and characterization of autophagy-defective mutants of *Saccharomyces cerevisiae*. *FEBS Lett.* **333,** 169–174 (1993).
4. Harding, T. M., Hefner-Gravink, A., Thumm, M. & Klionsky, D. J. Genetic and phenotypic overlap between autophagy and the cytoplasm to vacuole protein targeting pathway. *J. Biol. Chem.* **271,** 17621–17624 (1996).
5. Thumm, M. *et al.* Isolation of autophagocytosis mutants of *Saccharomyces cerevisiae*. *FEBS Lett.* **349,** 275–280 (1994).
6. Yang, Z. & Klionsky, D. J. Mammalian autophagy: core molecular machinery and signaling regulation. *Curr. Opin. Cell Biol.* **22,** 124–131 (2010).
7. Suzuki, K., Kubota, Y., Sekito, T. & Ohsumi, Y. Hierarchy of Atg proteins in pre-autophagosomal structure organization. *Genes Cells* **12,** 209–218 (2007).
8. Chan, S. N. & Tang, B. L. Location and membrane sources for autophagosome formation—from ER-mitochondria contact sites to Golgi-endosome-derived carriers. *Mol. Membr. Biol.* **30,** 394–402 (2013).
9. Reggiori, F. & Klionsky, D. J. Autophagic processes in yeast: mechanism, machinery and regulation. *Genetics* **194,** 341–361 (2013).
10. Yamamoto, H. *et al.* Atg9 vesicles are an important membrane source during early steps of autophagosome formation. *J. Cell Biol.* **198,** 219–233 (2012).
11. Chew, L. H., Setiaputra, D., Klionsky, D. J. & Yip, C. K. Structural characterization of the Saccharomyces cerevisiae autophagy regulatory complex Atg17-Atg31-Atg29. *Autophagy* **9,** 1467–1474 (2013).
12. Ragusa, M. J., Stanley, R. E. & Hurley, J. H. Architecture of the Atg17 complex as a scaffold for autophagosome biogenesis. *Cell* **151,** 1501–1512 (2012).
13. Kawamata, T., Kamada, Y., Kabeya, Y., Sekito, T. & Ohsumi, Y. Organization of the pre-autophagosomal structure responsible for autophagosome formation. *Mol. Biol. Cell* **19,** 2039–2050 (2008).
14. Kabeya, Y. *et al.* Characterization of the Atg17-Atg29-Atg31 complex specifically required for starvation-induced autophagy in Saccharomyces cerevisiae. *Biochem. Biophys. Res. Commun.* **389,** 612–615 (2009).
15. Kabeya, Y. *et al.* Atg17 functions in cooperation with Atg1 and Atg13 in yeast autophagy. *Mol. Biol. Cell* **16,** 2544–2553 (2005).
16. Jung, C. H. *et al.* ULK-Atg13-FIP200 complexes mediate mTOR signaling to the autophagy machinery. *Mol. Biol. Cell* **20,** 1992–2003 (2009).
17. Hosokawa, N. *et al.* Nutrient-dependent mTORC1 association with the ULK1-Atg13-FIP200 complex required for autophagy. *Mol. Biol. Cell* **20,** 1981–1991 (2009).
18. Hara, T. *et al.* FIP200, a ULK-interacting protein, is required for autophagosome formation in mammalian cells. *J. Cell Biol.* **181,** 497–510 (2008).
19. Koyama-Honda, I., Itakura, E., Fujiwara, T. K. & Mizushima, N. Temporal analysis of recruitment of mammalian ATG proteins to the autophagosome formation site. *Autophagy* **9,** 1491–1499 (2013).
20. Young, A. R. J. *et al.* Starvation and ULK1-dependent cycling of mammalian Atg9 between the TGN and endosomes. *J. Cell Sci.* **119,** 3888–3900 (2006).
21. Kamada, Y. *et al.* Tor-mediated induction of autophagy via an Apg1 protein kinase complex. *J. Cell Biol.* **150,** 1507–1513 (2000).
22. Stephan, J. S., Yeh, Y.-Y., Ramachandran, V., Deminoff, S. J. & Herman, P. K. The Tor and PKA signaling pathways independently target the Atg1/Atg13 protein kinase complex to control autophagy. *Proc. Natl Acad. Sci. USA* **106,** 17049–17054 (2009).
23. Fujioka, Y. *et al.* Structural basis of starvation-induced assembly of the autophagy initiation complex. *Nat. Struct. Mol. Biol.* **21,** 513–521 (2014).
24. Kamada, Y. *et al.* Tor directly controls the Atg1 kinase complex to regulate autophagy. *Mol. Cell Biol.* **30,** 1049–1058 (2010).
25. Stjepanovic, G. *et al.* Assembly and dynamics of the autophagy-initiating Atg1 complex. *Proc. Natl Acad. Sci. USA* **111,** 12793–12798 (2014).
26. Suzuki, S. W. *et al.* Atg13 HORMA domain recruits Atg9 vesicles during autophagosome formation. *Proc. Natl Acad. Sci. USA* **112,** 3350–3355 (2015).
27. Kijanska, M. *et al.* Activation of Atg1 kinase in autophagy by regulated phosphorylation. *Autophagy* **6,** 1168–1178 (2010).
28. Yeh, Y.-Y., Wrasman, K. & Herman, P. K. Autophosphorylation within the Atg1 activation loop is required for both kinase activity and the induction of autophagy in Saccharomyces cerevisiae. *Genetics* **185,** 871–882 (2010).
29. Mao, K. *et al.* Atg29 phosphorylation regulates coordination of the Atg17-Atg31-Atg29 complex with the Atg11 scaffold during autophagy initiation. *Proc. Natl Acad. Sci. USA* **110,** E2875–E2884 (2013).
30. Chew, L. H. *et al.* Molecular interactions of the Saccharomyces cerevisiae Atg1 complex provide insights into assembly and regulatory mechanisms. *Autophagy* **11,** 891–905 (2015).
31. Obara, K., Noda, T., Niimi, K. & Ohsumi, Y. Transport of phosphatidylinositol 3-phosphate into the vacuole via autophagic membranes in Saccharomyces cerevisiae. *Genes Cells* **13,** 537–547 (2008).
32. Backer, J. M. The regulation and function of Class III PI3Ks: novel roles for Vps34. *Biochem. J.* **410,** 1–17 (2008).
33. Mari, M. *et al.* An Atg9-containing compartment that functions in the early steps of autophagosome biogenesis. *J. Cell Biol.* **190,** 1005–1022 (2010).
34. Sekito, T., Kawamata, T., Ichikawa, R., Suzuki, K. & Ohsumi, Y. Atg17 recruits Atg9 to organize the pre-autophagosomal structure. *Genes Cells* **14,** 525–538 (2009).
35. Sreerama, N. & Woody, R. W. On the analysis of membrane protein circular dichroism spectra. *Protein Sci.* **13,** 100–112 (2004).
36. Geng, J. F., Baba, M., Nair, U. & Klionsky, D. J. Quantitative analysis of autophagy-related protein stoichiometry by fluorescence microscopy. *J. Cell Biol.* **182,** 129–140 (2008).
37. He, C., Baba, M., Cao, Y. & Klionsky, D. J. Self-interaction is critical for Atg9 transport and function at the phagophore assembly site during autophagy. *Mol. Biol. Cell* **19,** 5506–5516 (2008).
38. Lamb, C. A., Yoshimori, T. & Tooze, S. A. The autophagosome: origins unknown, biogenesis complex. *Nat. Rev. Mol. Cell Biol.* **14,** 759–774 (2013).
39. Cheong, H. *et al.* Atg17 regulates the magnitude of the autophagic response. *Mol. Biol. Cell* **16,** 3438–3453 (2005).
40. Iwaya, N. *et al.* MIT domain of Vps4 is a Ca2 + -dependent phosphoinositide-binding domain. *J. Biochem.* **153,** 473–481 (2013).
41. Suzuki, K., Akioka, M., Kondo-Kakuta, C., Yamamoto, H. & Ohsumi, Y. Fine mapping of autophagy-related proteins during autophagosome formation in Saccharomyces cerevisiae. *J. Cell Sci.* **126,** 2534–2544 (2013).

42. Janke, C. *et al.* A versatile toolbox for PCR-based tagging of yeast genes: new fluorescent proteins, more markers and promoter substitution cassettes. *Yeast* **21,** 947–962 (2004).
43. Noda, T., Matsuura, A., Wada, Y. & Ohsumi, Y. Novel system for monitoring autophagy in the yeast Saccharomyces cerevisiae. *Biochem. Biophys. Res. Commun.* **210,** 126–132 (1995).
44. Scholz, J., Besir, H., Strasser, C. & Suppmann, S. A new method to customize protein expression vectors for fast, efficient and background free parallel cloning. *BMC Biotechnol.* **13,** 12 (2013).
45. Selleck, W. & Tan, S. Recombinant protein complex expression in *E. coli. Curr. Protoc. Protein Sci.* Chapter 5, Unit 5 21 (2008).
46. Wasilko, D. J. *et al.* The titerless infected-cells preservation and scale-up (TIPS) method for large-scale production of NO-sensitive human soluble guanylate cyclase (sGC) from insect cells infected with recombinant baculovirus. *Protein Expr. Purif.* **65,** 122–132 (2009).
47. Provencher, S. W. & Glöckner, J. Estimation of globular protein secondary structure from circular dichroism. *Biochemistry* **20,** 33–37 (1981).
48. Cox, J. & Mann, M. MaxQuant enables high peptide identification rates, individualized p.p.b.-range mass accuracies and proteome-wide protein quantification. *Nat. Biotechnol.* **26,** 1367–1372 (2008).
49. Yen, W. L., Legakis, J. E., Nair, U. & Klionsky, D. J. Atg27 is required for autophagy-dependent cycling of Atg9. *Mol. Biol. Cell* **18,** 581–593 (2007).
50. Legakis, J. E., Yen, W. L. & Klionsky, D. J. A cycling protein complex required for selective autophagy. *Autophagy* **3,** 422–432 (2007).

Acknowledgements

We thank Roland Wedlich-Söldner for yeast strains, vectors and technical advice, Michael Thumm for providing reagents, Monica Zobawa for SEC–MALS measurements and Judith Scholz for expressing Atg1. The work was funded by the Max Planck Society (T.W.).

Author contributions

Y.R. performed all experiments and analysed the data. Y.R. purified and reconstituted Atg9, M.P. expressed and purified Atg1–KC and its related subunits, M.P. and B.H. characterization Atg1–KC *in vitro*, Y.R. and V.B. performed co-IPs. T.W. conceived the experiments, analysed the data and wrote the manuscript.

Additional information

Competing financial interests: The authors declare no competing financial interests.

11

Local viscoelasticity of living cells measured by rotational magnetic spectroscopy

J.-F. Berret[1]

When submitted to a magnetic field, micron-size wires with superparamagnetic properties behave as embedded rheometers and represent interesting sensors for microrheology. Here we use rotational magnetic spectroscopy to measure the shear viscosity of the cytoplasm of living cells. We address the question of whether the cytoplasm is a viscoelastic liquid or an elastic gel. The main result of the study is the observation of a rotational instability between a synchronous and an asynchronous regime of rotation, found for murine fibroblasts and human cancer cells. For wires of susceptibility 3.6, the transition occurs in the range 0.01–1 rad s^{-1}. The determination of the shear viscosity (10–100 Pa s) and elastic modulus (5–20 Pa) confirms the viscoelastic character of the cytoplasm. In contrast to earlier studies, it is concluded that the interior of living cells can be described as a viscoelastic liquid, and not as an elastic gel.

[1] Matière et Systèmes Complexes, UMR 7057 CNRS Université Denis Diderot Paris-VII, Bâtiment Condorcet, 10 rue Alice Domon et Léonie Duquet, 75205 Paris, France. Correspondence and requests for materials should be addressed to J.-F.B. (email: jean-francois.berret@univ-paris-diderot.fr).

In rheology, viscoelastic liquids and viscoelastic solids differ from each other by their stress relaxation function $G(t)$. In viscoelastic liquids, $G(t)$ is a decreasing function of the time which tends to zero as $t \to \infty$, whereas in viscoelastic solids $G(t)$ tends to a finite elastic modulus, indicating the presence of residual unrelaxed stresses even at very long time durations[1]. Hence, the static shear viscosity, which is defined as $\eta_0 = \int_0^\infty G(t)dt$ takes a finite value for liquids and it is generally expressed as a product of a relaxation time and of an elastic modulus. For viscoelastic solids, the static viscosity is not defined, and the rheological properties are described in terms of a low frequency elastic modulus $G'(\omega \to 0)$ and of a yield stress, that is, the critical value of the stress above which the sample does flow. In the following, viscoelastic solids will be termed 'elastic gels' for convenience. Here we address the question of the rheological properties of the intracellular medium of mammalian cells, and in particular the question to know whether the cytoplasm of such cells is a viscoelastic liquid or an elastic gel.

Rheometers are used to determine the relationship between the strain and stress on material sample of volume of the order of millilitres. Microrheology in contrast uses micron-size probes embedded in the material and needs much less sample, of the order of 1 picolitre. The past 20 years have seen increasingly rapid advances in this field, specifically in cell and tissue biomechanics[2,3]. For living cells, techniques including atomic force microscopy (AFM)[4,5], optical and magnetic tweezers[6–10], parallel microplates[11] and active tracking of internalized particles[7,10,12–14] were developed and provide consistent evaluation of the time or frequency dependencies of rheological parameters, such as the elastic compliance $J(t)$ or of the complex modulus $G^*(\omega)$. Numerous studies performed on a wide variety of cell lines have shown that $J(t)$[9,14,15] and $G^*(\omega)$[8,9,12,14,16] obey scaling laws as a function of time or frequency, with exponents comprised between 0.1 and 0.5. As for $G^*(\omega)$ measurements, the elastic modulus $G'(\omega)$ was found to be larger than the loss modulus $G''(\omega)$ on broad frequency ranges. These results were interpreted as an indication that the mechanical response of the cell or of the cytoplasm is that of a weak elastic solid[8–10,12,15,17], or in the terminology adopted previously that of an elastic gel. Out-of-equilibrium modelling suggests that the cell interior is a kind of soft glassy material with an effective temperature higher than that of the physiological temperature[17]. Alternative approaches of biomechanics were also attempted and provided different view points. On the basis of AFM micro-indentation and force-relaxation tests, Moeendarbary *et al.*[5] proposed that the cytoplasm behaves as a poroelastic material, where the water redistribution within the cytoplasm plays a fundamental role in setting the internal rheology. Other studies, in particular the earlier work of the Sackmann group[6,7,18] suggest that the intracellular medium behaves like a liquid and is characterized by a finite value of the shear viscosity[13]. More recently, Kuimova *et al.*[19,20] developed porphyrin-dimer-based molecular rotors in which fluorescence emitting frequencies scale with the viscosity of the surrounding medium. Evaluated on immortalized cervical cancer cells (HeLa cells), these authors found intracellular viscosities in the range 0.05–0.2 Pa s. Chevry *et al.* show that micron-size wires internalized in mouse fibroblasts and MDCK epithelial cells exhibit Brownian-like orientational fluctuations, associated to finite effective viscosity of 0.1–1 Pa s, in relative agreement with the determinations of Kuimova *et al.*[21].

Regarding the complexity of the dynamics in living systems, and the contrasting results obtained so far, there is a need to develop new types of probes and protocols for testing the cell biomechanics. Following the pioneer work by Crick and Hughes some 60 years ago[22], recent studies have shown that microrheology based on the tracking and monitoring of anisotropic probes could bring significant advances to the field[23–34]. It has been proposed, for example, that the shear viscosity of a fluid could be determined from the motion of a micro-actuator submitted to a rotating electric[28] or magnetic field[27,32–36]. These techniques are described as electric or magnetic rotational spectrometry[29,30]. With micro-actuators rotating at increasing frequency, a transition between a steady and a hindered motion is predicted for viscous fluids, and it is found experimentally[25,27,29,30,37,38]. A recent quantitative analysis shows that for viscoelastic liquids, the elasticity does not affect the onset of the instability and that the expression for the critical frequency remains the same as in a viscous fluid[25]. For an elastic gel of infinite static viscosity, in contrast, extrapolation yields a critical frequency going to zero, suggesting that the technique of the rotating wire is well adapted to differentiate liquid- from gel-like behaviours, as defined in rheology.

In the present paper, we exploit this concept to study the mechanical response of the cytoplasm of mouse fibroblast NIH/3T3 and human cancer HeLa cells. It is first demonstrated that the micron-size wires specifically fabricated for the study are nontoxic, enter spontaneously inside living cells and that they are not comprised in membrane-bound compartments. Actuated by a low-frequency external field, the wires rotate at the same angular speed than the field, whereas at higher frequency their motion exhibits back-and-forth oscillations. Values for the static viscosity, elastic modulus and relaxation time of the cytoplasm are derived from a complete analysis of the wire temporal trajectories. It is concluded that the interior of fibroblasts and HeLa cells can be appropriately described as viscoelastic liquids.

Results

Magnetic wire synthesis and characterization. Crystalline iron oxide nanoparticles were synthesized by co-precipitation of iron(II) and iron(III) salts in alkaline aqueous media and by further oxidation of the magnetite (Fe_3O_4) into maghemite (γ-Fe_2O_3). The size and dispersity (ratio between standard deviation and average diameter) of the particles prepared were determined from transmission electron microscopy ($D_{TEM} = 13.2$ nm, $s = 0.23$, Fig. 1 and Supplementary Fig. 1). Their crystallinity and structure was assessed by electron beam microdiffraction (Supplementary Fig. 2). The magnetization curves at different volume fractions were obtained by vibrating sample magnetometry and adjusted using a paramagnetic model[39,40]. The adjustment provided a specific magnetization of 3.5×10^5 A m^{-1} and a magnetic diameter of 10.7 nm (Supplementary Fig. 3). Light scattering was used to measure the weight-average molecular weight ($M_w = 12 \times 10^6$ g mol^{-1}) and the hydrodynamic diameter ($D_H = 27$ nm) for the bare particles[40]. Values of the electrophoretic mobility and zeta-potential were also derived ($\mu_E = -3.8 \times 10^{-4}$ cm^2 V^{-1}, $\zeta = -48$ mV, Supplementary Table 1) and indicated that the dispersions were stabilized by electrostatic forces[41].

For the wire synthesis, the particles were coated with poly(sodium acrylate)[39,41] (Aldrich) of molecular weight $M_w = 5{,}100$ g mol^{-1}. The wires were made according to a bottom-up co-assembly process using γ-Fe_2O_3-coated particles and oppositely charged polyelectrolytes. The polymer used was poly(trimethylammoniumethylacrylate)-b-poly(acrylamide) with molecular weights 11,000 g mol^{-1} for the cationic block and 30,000 g mol^{-1} for the neutral block[42,43]. Supplementary Fig. 4 illustrates the protocol for the fabrication of the wires. Figure 1b displays a TEM image of a single wire illustrating its nanostructure, whereas Fig. 1c shows a collection of wires

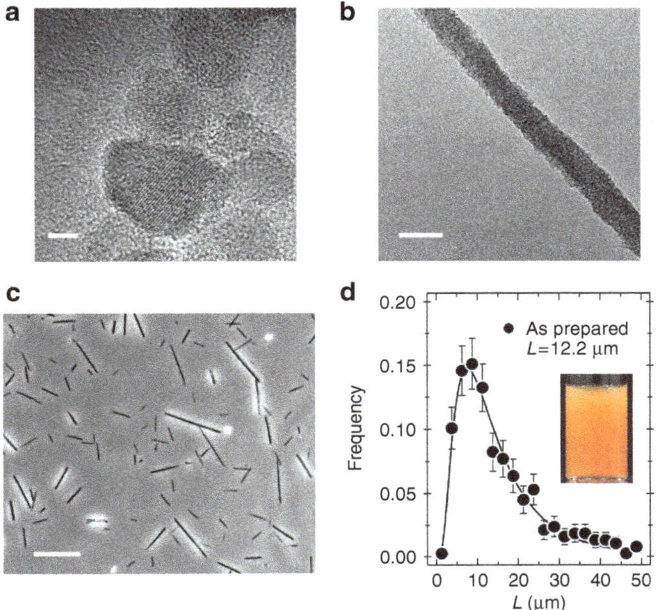

Figure 1 | Characterization of particles and wires. (**a**) High-resolution transmission electron microscopy (TEM) of 13.2 nm iron oxide nanoparticles (scale bar, 2 nm). (**b**) TEM of magnetic wires made by co-assembly using oppositely charged particles and polyelectrolytes (scale bar, 200 nm). (**c**) Phase-contrast image of magnetic wires deposited on a glass substrate and observed by optical microscopy (× 60, scale bar, 10 μm). (**d**) Size distribution of the wires adjusted using a log-normal distribution function with length of 12.2 μm and dispersity of 0.70. Error bars are defined as s.e.m. Inset: image of a vial containing a dispersion of wires at the concentration of 10^6 wires μl^{-1}.

deposited on a glass substrate and observed by optical microscopy (40 ×). The as-prepared samples contained anisotropic objects of median length 12.2 μm and of diameter 0.49 μm (Fig. 1d). Before incubation with cells, the wires were sonicated for 120 s to shorten their lengths and sorted by magnetic separation. The sonicated wires had an average length of 2.4 μm and a dispersity of 0.35. The wires dispersion was then autoclaved (Tuttnauer Steam Sterilizer 2340M) at 120 °C and atmospheric pressure for 20 minutes to prevent bacterial contamination for cell culture, concentrated by magnetic sedimentation at 10^6 wires μl^{-1} and stored at 4 °C in a secure environment before use.

Wire microrheology on model fluids and gels. First we derive the equation of motion of a wire immersed in well-known rheological models[1] (Supplementary Fig. 5) and submitted to a rotating magnetic field, and then assess the validity of the technique by comparing the predictions to observations obtained on a wormlike surfactant solution and on a polysaccharide gel.

For a viscous liquid of viscosity η_0, a wire submitted to a rotating field experiences restoring a viscous torque that slows down its rotation. As a result, its motion undergoes a transition between a synchronous and an asynchronous rotation. The critical frequency ω_C between these two regimes reads:

$$\omega_C = \frac{3}{8}\frac{\mu_0 \Delta\chi}{\eta_0} g\left(\frac{L}{D}\right)\frac{D^2}{L^2}H^2 \quad (1)$$

where μ_0 is the permeability in vacuum, L and D the length and diameter of the wire, H the amplitude of the magnetic excitation and $g(\frac{L}{D}) = Ln(\frac{L}{D}) - 0.662 + 0.917\frac{D}{L} - 0.050(\frac{D}{L})^2$ is a dimensionless function of the anisotropy ratio[44,45]. In equation (1), $\Delta\chi = \chi^2/(2+\chi)$ where χ denotes the magnetic susceptibility[27]. For data treatment, the geometrical characteristics are combined into the dimensionless parameter $L^* = L/D\sqrt{g(L/D)}$. Under these conditions, equation (1) becomes:

$$\omega_C = \frac{3\mu_0 \Delta\chi}{8\eta_0}\frac{H^2}{L^{*2}} \quad (2)$$

The average angular velocity $\Omega(\omega)$ in the two regimes expresses as[27,35]:

$$\begin{array}{ll} \omega \leq \omega_C & \Omega(\omega) = \omega \\ \omega \geq \omega_C & \Omega(\omega) = \omega - \sqrt{\omega^2 - \omega_C^2} \end{array} \quad (3)$$

With increasing frequency, the average velocity increases linearly, passes through a cusp-like maximum at the critical frequency and then decreases. The transition between the synchronous and asynchronous regimes was used to calibrate the wire rheometer and determine the susceptibility parameter $\Delta\chi$. Experiments were performed at $T = 26$ °C on a 85 wt. % water–glycerol mixture of static viscosity $\eta_0 = 0.062$ Pa s^{-1}. In a purely viscous fluid, the critical frequency is found to decrease as L^{*-2} in agreement with equation (2). For wires made from 13.2 nm particles and block copolymers, we found $\Delta\chi = 2.3 \pm 0.7$, and a magnetic susceptibility $\chi = 3.6 \pm 0.9$ (Supplementary Fig. 6). Here and in the following, the uncertainty is defined as s.e.m.

A Maxwell fluid is described by a spring and dashpot in series[1]. An actuated wire immersed in such a medium experiences a viscous and an elastic torque that both oppose the applied magnetic torque. The differential equation describing the wire motion has been derived and solved, leading to the following predictions (Supplementary Note 1). With increasing ω, the wire undergoes the same type of transition as the one described previously and the critical frequency ω_C expresses as in equation (1). The static viscosity η_0 in equation (1) is however replaced by the product $G\tau$, where G and τ denote the shear elastic modulus and the relaxation time of the fluid. The set of equations in equation (3) is also identical to that of a Newtonian fluid. From the amplitude of the oscillations $\theta_B(\omega)$ (more precisely the angle by which the wire returns back after a period of increase) in the asynchronous regime, it is possible to determine the shear elastic modulus G using[25]:

$$\lim_{\omega\to\infty} \theta_B(\omega) = \frac{3}{4}\frac{\mu_0 \Delta\chi}{G} g\left(\frac{L}{D}\right)\frac{D^2}{L^2}H^2 \quad (4)$$

The above predictions were tested by monitoring the wire motion in a wormlike micellar solution made from cetylpyridinium chloride and sodium salicylate ([NaCl] = 0.5 M) at 2 wt. % (ref. 46). The rheological parameters of the surfactant solution as determined by cone-and-plate rheometry were $\eta_0 = 1.0 \pm 0.1$ Pa s, $G = 7.1 \pm 0.1$ Pa and $\tau = 0.14 \pm 0.01$ s (Supplementary Note 2). From the evolution of the average angular velocity $\Omega(\omega)$ of actuated wires and from the position of the maximum (equation (3)), a static viscosity of 1.3 ± 0.3 Pa s was obtained. From the oscillation amplitudes in the regime $\omega\tau \gg 1$ (equation (4)), a modulus of 9.4 ± 2 Pa was derived (Supplementary Fig. 7). Experiments performed with wires of different lengths and in various magnetic field conditions confirmed the good agreement with cone-and-plate rheometry, and demonstrates the ability of the technique to measure the linear viscoelasticity of a Maxwell fluid[25].

A Kelvin–Voigt element aims to describe a viscoelastic solid, and it is represented by a spring and dashpot in parallel[1]. The resolution of the differential equation of motion of a wire leads to the following predictions (Supplementary Note 1). At all frequencies, the wire rotation exhibits back-and-forth oscillations at a frequency twice that of the field, and the average angular velocity $\Omega(\omega) = 0$. In addition, the amplitude of the oscillations $\theta_B(\omega)$ varies inversely with the shear elastic modulus according to:

$$\theta_B(\omega) = \frac{3}{4}\frac{\mu_0 \Delta\chi}{G'(\omega)} g\left(\frac{L}{D}\right)\frac{D^2}{L^2}H^2 \quad (5)$$

Gellan gum (phytagel, Sigma-Aldrich), a linear anionic polysaccharide comprising glucose, glucuronic acid and rhamnose building units[47] was added slowly to a 1 mM calcium chloride solution at room temperature with rapid stirring before heating up to 50 °C. Samples prepared at concentrations 0.3, 0.5, 0.75, 1 and 2 wt. % and studied by cone-and-plate rheometry exhibited a gel-like behaviour: the elastic modulus exhibits scaling of the form $G'(\omega) \sim \omega^{0.2}$ and $G'(\omega) > G''(\omega)$ (Supplementary Note 3). At frequencies 5×10^{-3} to $10\ \mathrm{rad\ s^{-1}}$, the rotation angle of the wires reveal steady oscillations, and an average angular speed $\Omega(\omega)$ equal to zero, in agreement with the Kelvin–Voigt predictions (Supplementary Fig. 8). Using equation (5), we found an elastic modulus $G'(\omega)$ of 2.5 ± 0.8 Pa, 9.5 ± 3 Pa and 95 ± 30 Pa for the 0.3, 0.5 and 1 wt. % samples, respectively, again in good agreement with the cone-and-plate data ($G'(\omega) = 3 \pm 0.5$ Pa, 13 ± 2 Pa and 130 ± 10 Pa). In conclusion of this part, we have found that theory correctly predicts the motion of rotating magnetic wires immersed in viscoelastic model systems and that the viscoelastic parameters retrieved, such as the viscosity or the elastic modulus are those of the linear shear rheology.

Localization of the wires inside living cells. Here we establish the experimental conditions under which the wires interact with the cells, and in which part of the cytoplasm they are located. The localization of the wires inside the cells was determined by transmission electron microscopy. The incubation time was 24 h and the amount of wires per cell was 10 for the two cell lines. Under such conditions, no perturbations of the cell morphology or cell cycle were observed (Supplementary Fig. 9). In a previous report, it was shown that the capture rate for neutral wires was ~10% at 24 h, leading to an average number of one wire per cell in the present experiment[48]. Figure 2a,b display TEM images of NIH/3T3 and HeLa cell interiors, respectively. The elongated objects appearing in the TEM images are in the range 200 nm–1 µm and they are made of densely packed particles of high electronic contrast. These threads are pieces of wires that were taken up by the cells. Shorter than their initial lengths, the threads also have sharp and diffuse extremities. These observations are interpreted by assuming that elongated structures not in the plane of the microtome section were shortened during the sample preparation, an outcome that was already noted in previous studies[48,49]. In addition to wires, clusters in the range of 200 nm are also visible. These clusters result from the wire degradation occurring during incubation or the sample preparation. Close-up views of the TEM images in Fig. 2c,d indicate that the wires are not in membrane-bound compartments[50]. This result was already pointed out in our previous study on fibroblasts. Here however, the TEM outcomes are similar for the cancerous HeLa cells and for human lymphoblasts, indicating that the absence of endosomal membrane around internalized wires may be specific to anisotropic objects. Concerning the entry mechanism, several scenarios were suggested, including the perforation of the cellular membrane[51,52], macropinocytosis[53] or multiple combined processes[48,54]. At this stage, no definite conclusion can be made concerning the entry mechanism into NIH/3T3 or HeLa cells. The absence of cytoplasmic membrane around the internalized probes ensures, however, that the wires will probe the mechanical properties of the cytosol.

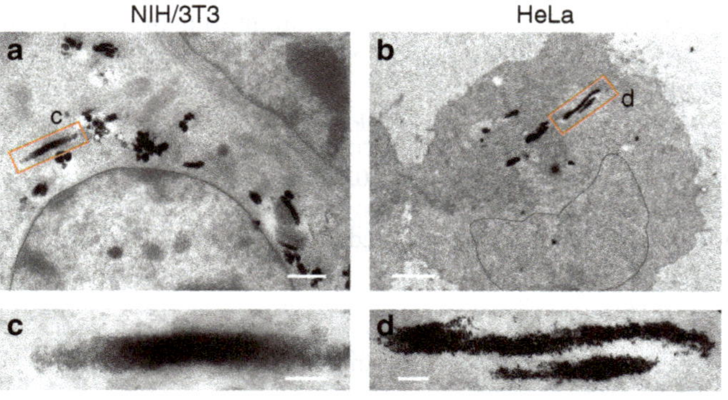

Figure 2 | Localization of wires inside living cells. Representative transmission electron microscopy image of (**a**) NIH/3T3 fibroblasts and (**b**) HeLa cancer cells incubated with 2.4 µm wires for 24 h at the concentration of 10 wires per cell. For the fibroblasts, clusters in the range of 200 nm are also visible. Scale bars in **a**,**b** are 1 and 2 µm, respectively. (**c**,**d**) Close-up views of regions marked by an orange rectangle in **a**,**b** respectively, indicating that the wires are not enclosed in membrane-bound compartments. Scale bars are 200 nm.

Evidence of wire rotational instability in living cells. We then examined the motions of wires internalized into cells and submitted to a 14 mT rotating field. In this part, emphasis is put on the time-dependent behaviour. Wires of lengths 2 to 6 µm were studied as a function of the frequencies ω. For each condition, a 100 s movie was recorded and digitalized, from which the position of the centre-of-mass and orientation angle of the wire were retrieved and plotted as a function of time. Figure 3a shows the rotation of a 2.8 µm wire internalized into a fibroblast at the angular frequency of $0.14\ \mathrm{rad\ s^{-1}}$ and Fig. 3b illustrates the time dependence of the angle $\theta(t)$ at different frequencies (see Supplementary Movie 1). For the conditions tested, $\theta(t)$ increases linearly with time, and the slope $\Omega = \langle d\theta(t)/dt \rangle_t$ corresponds exactly to the actuating frequency. Here, $\langle \ldots \rangle_t$ denotes the time average in the steady regime. In this first regime, the wire rotates with the field, and $\theta(t) = \omega t$.

Figure 3c,d illustrate the change of regime as the angular frequency is increased above a critical value, here $\omega_C = 0.15\ \mathrm{rad\ s^{-1}}$. The microscopy images on the left hand side show that after a clockwise steady rotation (between 23.8 and 26.4 s), the wire comes back rapidly by 50 degrees in an anticlockwise motion, indicating that the wire rotation is hindered. On a longer period, the wires are animated with back-and-forth motion characteristic of the asynchronous regime (Fig. 3d). The traces recorded at $\omega = 0.41$, 0.76 and $1.87\ \mathrm{rad\ s^{-1}}$ display steady oscillations, with a rotation angle that continues to increase with time (see Supplementary Movies 2 and 3). In this second regime, however, it is observed that $\Omega(\omega) \ll \omega$. The data shown in the figure are representative of the overall behaviour found for wires in living fibroblast and HeLa cells. In these time-resolved experiments, it was verified that the centre-of-mass of the wires remained punctual during the steady or hindered rotations, insuring that the same volume of the cytosol is probed during the measurements. In conclusion to this part, with increasing frequency, wires dispersed in the intracellular medium of fibroblasts or of Hela cells undergo a transition between a synchronous and asynchronous regime. The critical frequency is found in the range $0.01–1\ \mathrm{rad\ s^{-1}}$.

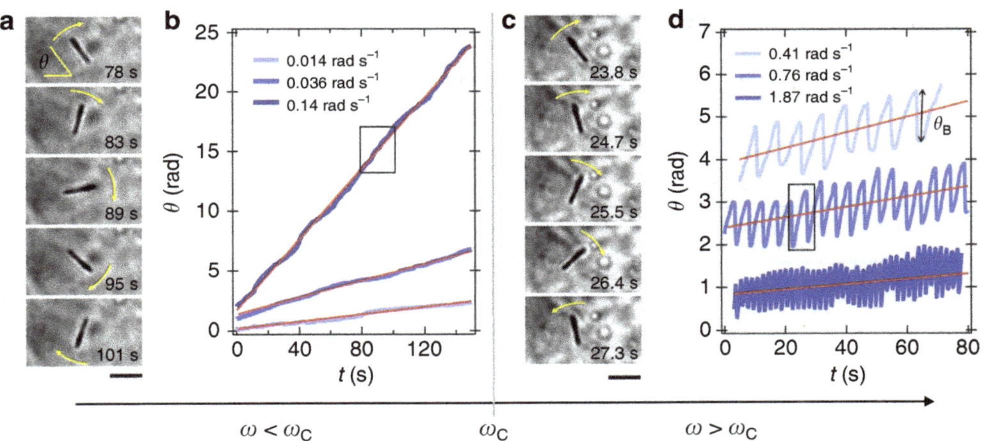

Figure 3 | Evidence of wire rotational instability in living cells. (**a**) Optical microscopy images of a 2.8 μm internalized wire subjected to a rotating field of 14 mT and the frequency of 0.14 rad s^{-1}. (**b**) Time dependences of the angle $\theta(t)$ at varying frequencies $\omega = 0.014$, 0.036 and 0.14 rad s^{-1}. The rectangle around 80 s provides the time range for the images on the left hand side. The straight lines in red were calculated from the expression $\theta(t) = \omega t$, indicating that the wire rotates synchronously with the field, and at the same angular frequency. The synchronous rotation of the wire is illustrated in Supplementary Movie 1. (**c**,**d**) Same representation as in **a**,**b** but for angular frequencies (0.41, 0.76 and 1.87 rad s^{-1}) above the critical frequency $\omega_C = 0.15$ rad s^{-1}. The microscopy images on the left-hand side shows that after a clockwise rotation, the wire comes back rapidly in an anti-clockwise motion, indicating that the wire rotation is hindered (see Supplementary Movies 2 and 3. Scale bars in **a**,**c** are 2 μm.

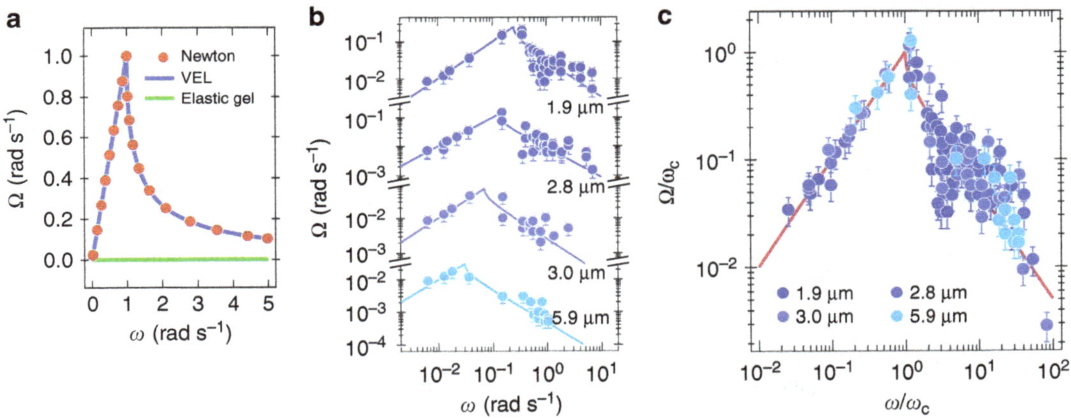

Figure 4 | Master curve for wire average rotation velocity in cells. (**a**) Average rotation velocity $\Omega(\omega)$ of a magnetic wire calculated for a Newtonian fluid, for a viscoelastic liquid (indicated as VEL) and for an elastic gel according to a set of models listed in Supplementary Note 1. For the sake of convenience, the viscous and the viscoelastic liquids have here the same static viscosity and $\omega_C = 1$ rad s^{-1}. (**b**) $\Omega(\omega)$-evolution for wires of lengths between 1.9 and 5.9 μm internalized in the cytoplasm of mouse fibroblasts. The continuous lines (from equation (1)) indicate the existence of a critical frequency in the rotation dynamics. (**c**) Data from Fig. 4b plotted in reduced units, Ω/ω_C versus ω/ω_C. The continuous lines are $\Omega(X) = X$ and $\Omega(X) = X - \sqrt{X^2 - 1^2}$ for $X \leq 1$ and $X \geq 1$, respectively. Error bars in **b**,**c** are defined as s.e.m.

Master curve for the wire average rotation velocity. To elucidate the rheological profile of NIH/3T3 and HeLa intracellular medium, the $\theta(t)$-traces of internalized wires were analysed and translated into a set of two parameters: the average rotation velocity $\Omega(\omega)$ and the amplitude $\theta_B(\omega) = \langle\theta_B(t,\omega)\rangle_t$ of the oscillations in the unstable regime. As mentioned in the section on the validity of the magnetic wire spectroscopy, it was found that for Newton, Maxwell and Kelvin–Voigt models, $\Omega(\omega)$ and $\theta_B(\omega)$ display specific asymptotic behaviours as a function of the frequency[25]. For the viscous and viscoelastic liquids, the average frequency $\Omega(\omega)$ is superimposed and show a cusp-like maximum at ω_C, whereas for an elastic gel the average rotation velocity is constant and equal to 0. These different behaviours are illustrated in Fig. 4a.

Figure 4b displays the $\Omega(\omega)$-evolution for wires of lengths between 1.9 and 5.9 μm. With increasing frequency, the average velocity increases linearly, passes through a maximum at ω_C and then decreases. As already mentioned, the transition corresponds to the change of rotation regime, from synchronous rotation to back-and-forth oscillations. Note that the cut-off frequency increases as the length of the wire decreases (Fig. 4b). The data in Fig. 4 were adjusted using equation (3)[27,35]. In both regimes, the agreement between the data and the model is excellent. At high frequency, the $\Omega(\omega)$-data exhibit some scattering that could come from time-dependent effects of the viscosity, and be related to the cellular activity. Table 1 provides the values of the critical frequency and static viscosity for the experiments in Fig. 4b. There, ω_C varies from 0.03 to 0.2 rad s^{-1}, and is associated to static viscosities between 20 and 80 Pa s. Figure 4c displays the data of Fig. 4b in reduced units, Ω/ω_C versus ω/ω_C. In this representation, the Ω/ω_C-data are found to collapse onto the same master curve, now observed over four decades in ω. The

Table 1 | Viscoelastic parameters of NIH/3T3 fibroblast.

Wire length	ω_C (rad s^{-1})	η_0 (Pa s)	$\lim_{\omega\to\infty} \theta_B(\omega)$ (rad)	G (Pa)
$L = 1.9\ \mu m$	0.20 ± 0.05	32 ± 8	0.8 ± 0.1	16 ± 2
$L = 2.8\ \mu m$	0.15 ± 0.02	26 ± 5	0.8 ± 0.1	9 ± 2
$L = 3.0\ \mu m$	0.07 ± 0.02	59 ± 15	0.8 ± 0.1	11 ± 2
$L = 5.9\ \mu m$	0.03 ± 0.01	78 ± 20	0.6 ± 0.1	8 ± 2

ω_C denotes the critical cut-off frequency between the synchronous and asynchronous regimes for the data in Fig. 4b. η_0 is the static viscosity derived from equation (1), $\lim_{\omega\to\infty} \theta_B(\omega)$ is the angle by which the wire returns back at high frequency and G is the elastic modulus obtained from equation (4). The error bars are defined as s.e.m.

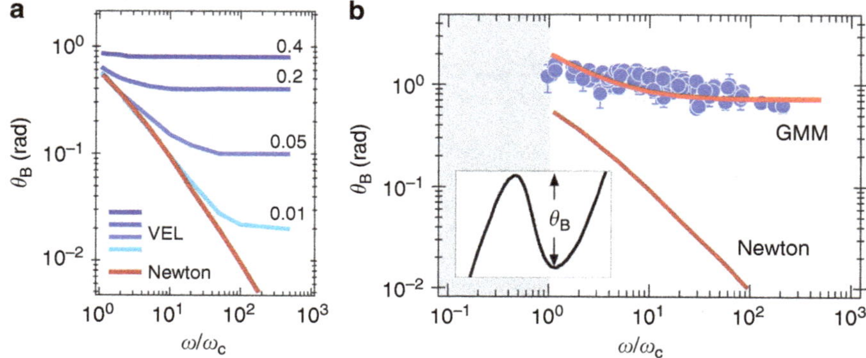

Figure 5 | Master curve for wire oscillation amplitude in cells. (**a**) Average rotation velocity $\Omega(\omega)$ of a magnetic wire calculated for a Newtonian fluid and for a viscoelastic liquid (indicated as VEL) of relaxation times ranging from $0.01/\omega_C$ (light blue) to $0.4/\omega_C$ (dark blue). (**b**) Variation of θ_B as a function of the reduced frequency ω/ω_C for wires of lengths 1.9 to 5.9 μm. The continuous lines were obtained from the Newtonian and from the Generalized Maxwell (GMM) models. Inset: definition of θ_B. Error bars in **b** are defined as s.e.m.

agreement obtained over a large frequency window attests to the reliability of the model. A comparison with the predictions shown in Fig. 4a suggests that the rheology of the intracellular medium is that of a viscous or of a viscoelastic liquid. The data in Fig. 4c also rule out the hypothesis of a gel-like rheology.

Master curve for the wire oscillation amplitude. Figure 5a displays the behaviours of the oscillation amplitudes for Newtonian and Maxwell fluids as a function of the reduced frequency ω/ω_C. θ_B being related to the asynchronous regime, it is defined only for $\omega/\omega_C \geq 1$. For the viscoelastic predictions (labelled VEL), the curves were calculated for different relaxation times ranging from $0.01/\omega_C$ to $0.4/\omega_C$, as indicated in the margin. For the Newtonian fluid, the amplitude of the oscillations decreases with increasing frequency. In contrast, the Maxwell fluid displays a crossover between a viscous and an elastic regime that occurs at a fixed value of the reduced frequency, here $\omega\tau = 1/2$. The value of 1/2 is explained by the fact that the frequency of the oscillations in this regime is twice that of the field. The existence of a plateau at high frequency is related to the elastic response of the fluid, and its height is given by equation (4) (ref. 21).

Figure 5b shows the dependence of θ_B in NIH/3T3 fibroblasts as a function of the reduced frequency for different wires of lengths between 1.9 and 5.9 μm. The major result that emerges from the figure is that the angles θ_B collected over many experimental conditions (that is, changing ω and L) are well superimposed when plotted against the reduced frequency. Starting at $\theta_B = 1.5$ rad slightly above ω_C, the angle exhibits a moderate but steady decrease with increasing ω. Also displayed in the figure is the prediction for a Newtonian fluid, in which decrease is much stronger and cannot account for the data found in cells. Attempts to use a single-mode Maxwell model (Fig. 5a) were also inconclusive, and the θ_B-data could not be reproduced. The adjustment was noticeably improved using a Generalized Maxwell Model (noted GMM on the figure) and assuming a constant distribution of relaxation times. The spread of the distribution was found to go from $1/\omega_C$ to $100/\omega_C$. For ω_C-values of 0.1 rad s^{-1}, the time distribution extends hence from 0.1 to 10 s. From this fitting, the elastic modulus G of the cell interior was estimated. We found an elastic modulus $G = 12$ Pa (with a s.d. of 4 Pa, $n = 9$) for the fibroblasts and $G = 14$ Pa (with a s.d. of 9 Pa, $n = 10$) for the HeLa cells. Such values for the elasticity are in agreement with the results obtained by tracking techniques, specifically those probing the interior of the cell[13]. In contrast, G-values found here are one to two decades smaller than that determined by techniques probing the mechanical response of the entire cell, or that of the cell surface[4,5,9]. Such differences were already discussed in the literature, and for the latter experiments they were attributed to the cortical actin network[12].

Shear viscosity of the cytoplasm. The model of rotating wires in a viscoelastic liquid contains an important prediction: the critical frequency ω_C varies inversely with the square of the wire length. This relationship was verified for fluids of known viscosity[25,27] and tested again here for a 85% glycerol/water mixture (Supplementary Fig. 6). Figure 6a,b display the dependences of the critical frequency ω_C as a function of the reduced wire length $L^* = L/D\sqrt{g(L/D)}$ for the two cell lines studied. Least square calculations using a power law dependence of the form $\omega_C(L^*) \sim L^{*\alpha}$ provide exponents $\alpha = -3.0 \pm 0.5$ and $\alpha = -6.5 \pm 1.0$ for NIH/3T3 and HeLa, respectively (straight lines in red in Fig. 6). Two related effects could account for the discrepancy with respect to equation (2): the heterogeneities of the intracellular medium, which also contribute to a relatively broad scattering of the data points, and the possible dependence

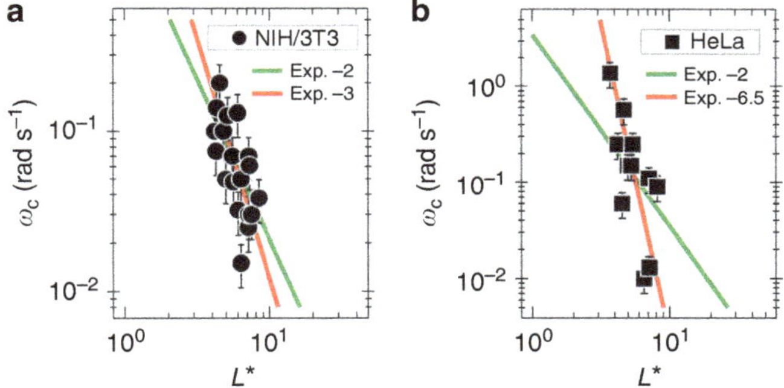

Figure 6 | Scaling critical frequency and wire length. Variation of the critical frequency ω_C as a function of the reduced wire length $L^* = L/D\sqrt{g(L/D)}$ for **(a)** NIH/3T3 mouse fibroblasts and **(b)** for HeLa cervical cancer cells. Straight lines in red are least-square fits using power laws with exponent (exp.) -3 and -6.5, respectively, whereas straight lines in green are obtained from equation (2). Error bars are defined as s.e.m.

of the mechanical response on the length scale. This later feature was observed in active gels of F-actin using two-particle microrheology experiments[55]. By forcing the exponent to be -2 and repeating the fitting procedure, the static shear viscosity of fibroblasts could be estimated, providing $\eta_0 = 47(+26/-14)$ Pa s. This value is larger than those reported using passive microrheology in different culture and probe conditions[13,21]. For the Hela cells, the deviation between the observed and predicted scaling is significant, and η_0 could not be determined. This result confirms the high sampling variability of rheological parameters associated with this cell line.

Discussion

In this work, we used magnetic wires between 2 and 6 µm to measure the static viscosity of living cells. The technique, known as rotational magnetic spectroscopy, is based on the experimental determination of a critical frequency ω_C between two different rotation regimes, one where the wire and the field are synchronous, and one where the wire performs back-and-forth oscillations. The measure of the viscosity is obtained directly from the critical frequency ω_C, through the relationship given by equation (1). This wire-based technique is known for more than 20 years but was applied only recently to complex fluids composed of colloids of synthetic and natural origin[25,27,29,30,37,38,56]. To the best of our knowledge, it was not applied yet to living cells. In a first step, we assess the validity of the wire-based microrheology technique using known rheological models. We have found that theory and experiments agree well with each other, and that the viscoelastic parameters retrieved (including the static viscosity η_0 and the elastic modulus G) are those of the linear shear rheology. With living cells, one of the key results here is that the rotational instability as in the range 10^{-2}–1 rad s^{-1} for the wires of magnetic susceptibility $\chi = 3.6$. It is, therefore, accessible to standard rotating field devices mounted on an optical microscopy platform. In this paper, the analysis has been extended to the amplitude of the oscillations in the asynchronous regime and to its frequency dependence and scaling. The high frequency limit was used to estimate the shear elastic modulus G of the intracellular medium for the NIH/3T3 fibroblasts and for the HeLa cervical cancer cells. The modulus is in the range 5–20 Pa. The determinations of the viscosity and elasticity confirm the viscoelastic character of the cytoplasm. However here, and in stark contrast to several studies[9,10,56], we conclude that the interior of living cells is best described in terms of a viscoelastic liquid, and not of an elastic gel. The conclusion was made possible by gaining access to a frequency range not explored before (down to 6×10^{-3} rad s^{-1}), and for which the cytoplasm was shown to flow. The present magnetic rotational spectroscopy method does not allow to determine the distribution of relaxation times with accuracy, in part because of the restricted range of frequencies explored. An estimate is nevertheless provided and it is found that the time distribution is constant in the range 0.1–10 s (ref. 9). The rheological model used to fit the data is in agreement with a Generalized Maxwell model combined with a distribution of relaxation times. In conclusion, the present work shows the potential of the wire-based magnetic rotation spectroscopy as an accurate rheological technique to distinguish between flow and yield stress behaviours in highly confined environment.

Methods

Microrheology set-up. Bright field microscopy was used to monitor the actuation of the wires as a function of time. Stacks of images were acquired on an IX73 inverted microscope (Olympus) equipped with a × 100 objective. For the measure of the magnetic properties of the wires, 65 µl of a dispersion of a know viscosity were deposited on a glass plate and sealed to a Gene Frame (Abgene/Advanced Biotech) dual adhesive system. The glass plate was introduced into a homemade device (Supplementary Fig. 10) generating a rotational magnetic field, thanks to two pairs of coils working with a 90° phase shift. An electronic set-up allowed measurements in the frequency range 10^{-2}–100 rad s^{-1} and at magnetic fields $B = 0$–20 mT. A stream of air directed toward the measuring cell was used to thermalize the sample at $T = 26$ or 37 °C. The image acquisition system consisted of an EXi Blue CCD camera (QImaging) working with Metaview (Universal Imaging Inc.). Images of wires were digitized and treated by the ImageJ software and plugins (http://rsbweb.nih.gov/ij/). The three-dimensional motion was extracted from their two-dimensional projection according a procedure described previously[26,44]. In brief, the out-of-plane motion of the wires is deduced from their projection on the focal plane of an optical microscope objective. An angular variable reflecting the out-of-plane motion is computed from the apparent wire length and its projected angular displacement.

Calibration of the wire-based rheometer. To determine the susceptibility parameter $\Delta\chi$ in equations (1) and (2), steady rotation experiments were carried out on a fluid of known viscosity. An 85 wt. % glycerol–water mixture of static viscosity $\eta_0 = 0.062$ Pa s^{-1} (T = 26 °C) was used as suspending medium. In a typical optical microscopy experiment, a wire was first selected, its length and diameter measured with the objective × 100 and it was then submitted to rotation frequency sweep from 0.1 to 100 rad s^{-1}. For a statistically relevant sample, the protocol was applied on 44 wires of lengths 2–20 µm and at a magnetic field of 7 mT. In Supplementary Fig. 10, ω_C is shown as a function of L^*. The critical frequency was found to decrease as $\omega_C \sim L^{*-2}$, in agreement with the prediction of equation (2). From the prefactor ($3\mu_0\Delta\chi H^2/8\eta_0 = 590 \pm 180$ rad s^{-1}), we infer $\Delta\chi = 2.3 \pm 0.7$, and $\chi = 3.6 \pm 0.9$. Knowing $\Delta\chi$, a wire submitted to a rotating field will be subjected to a known magnetic torque, a condition that is a prerequisite for quantitative rotational microrheology.

Cell culture. NIH/3T3 fibroblast (ATCC-CRL-1658) and HeLa (ATCC-CCL-2) cells were grown in T25-flasks as a monolayer in DMEM with high glucose (4.5 g l^{-1}) and stable glutamine (PAA Laboratories GmbH, Austria). The medium was supplemented with 10% fetal bovine serum and 1% penicillin/streptomycin (PAA Laboratories GmbH), referred to as cell culture medium. Exponentially growing cultures were maintained in a humidified atmosphere of 5% CO_2 and 95% air at 37 °C, and in these conditions, the plating efficiency was 70–90% and the cell duplication time was 12–14 h. Cell cultures were passaged twice weekly using trypsin–EDTA (PAA Laboratories GmbH) to detach the cells from their culture flasks and wells. The cells were pelleted by centrifugation at 260 g for 5 min. The supernatant was removed and cell pellets were re-suspended in assay medium and counted using a Malassez counting chamber.

Transmission electron microscopy. NIH/3T3 fibroblast cells were seeded onto the six-well plate. After a 24-h incubation with 4 µm wires, the excess medium was removed. The cells were washed in 0.2 M phosphate buffer (phosphate-buffered saline, PBS) and fixed in 2% glutaraldehyde-phosphate buffer (0.1 M) for 1 h at room temperature. Fixed cells were further washed in 0.2 M PBS. The cells were then postfixed in a 1% osmium-phosphate buffer for 45 min at room temperature in dark conditions. After several washes with 0.2 M PBS, the samples were dehydrated by addition of ethanol. Samples were infiltrated in 1:1 ethanol:epon resin for 1 h and finally in 100% epon resin for 48 h at 60 °C for polymerization. Ninety-nanometre-thick sections were cut with an ultramicrotome (LEICA, Ultracut UCT) and picked up on copper–rhodium grids. They were then stained for 7 min in 2%-uranyl acetate and for 7 min in 0.2%-lead citrate. Grids were analysed with a transmission electron microscope (ZEISS, EM 912 OMEGA) equipped with a LaB_6 filament, at 80 kV. Images were recorded with a digital camera (SS-CCD, Proscan 1024 × 1024), and the iTEM software.

References

1. Larson, R. G. *The Structure and Rheology of Complex Fluids* (Oxford Univ. Press, 1998).
2. Bao, G & Suresh, S. Cell and molecular mechanics of biological materials. *Nat. Mater.* **2,** 715–725 (2003).
3. Squires, T. M. & Mason, T. G. Fluid mechanics of microrheology. *Annu. Rev. Fluid Mech.* **42,** 413–438 (2010).
4. Park, S., Koch, D., Cardenas, R., Kas, J & Shih, C. K. Cell motility and local viscoelasticity of fibroblasts. *Biophys. J.* **89,** 4330–4342 (2005).
5. Moeendarbary, E. *et al.* The cytoplasm of living cells behaves as a poroelastic material. *Nat. Mater.* **12,** 253–261 (2013).
6. Bausch, A. R., Ziemann, F., Boulbitch, A. A., Jacobson, K. & Sackmann, E. Local measurements of viscoelastic parameters of adherent cell surfaces by magnetic bead microrheometry. *Biophys. J.* **75,** 2038–2049 (1998).
7. Bausch, A. R., Moller, W. & Sackmann, E. Measurement of local viscoelasticity and forces in living cells by magnetic tweezers. *Biophys. J.* **76,** 573–579 (1999).
8. Fabry, B. *et al.* Scaling the microrheology of living cells. *Phys. Rev. Lett.* **87,** 148102 (2001).
9. Balland, M. *et al.* Power laws in microrheology experiments on living cells: comparative analysis and modeling. *Phys. Rev. E* **74,** 17 (2006).
10. Guo, M. *et al.* Probing the stochastic, motor-driven properties of the cytoplasm using force spectrum microscopy. *Cell* **158,** 822–832 (2014).
11. Mitrossilis, D. *et al.* Real-time single-cell response to stiffness. *Proc. Natl Acad. Sci. USA* **107,** 16518–16523 (2010).
12. Hoffman, B. D., Massiera, G., Van Citters, K. M. & Crocker, J. C. The consensus mechanics of cultured mammalian cells. *Proc. Natl Acad. Sci. USA* **103,** 10259–10264 (2006).
13. Panorchan, P. *et al.* in *Cell Mechanics* (eds Wang, Y. L. & Discher, D. E.) (Elsevier Academic Press Inc, 2007).
14. Weihs, D., Mason, T. G. & Teitell, M. A. Bio-microrheology: a frontier in microrheology. *Biophys. J.* **91,** 4296–4305 (2006).
15. Lenormand, G., Millet, E., Fabry, B., Butler, J. P. & Fredberg, J. J. Linearity and time-scale invariance of the creep function in living cells. *J. R. Soc. Interface* **1,** 91–97 (2004).
16. Deng, L. H. *et al.* Fast and slow dynamics of the cytoskeleton. *Nat. Mater.* **5,** 636–640 (2006).
17. Gallet, F., Arcizet, D., Bohec, P. & Richert, A. Power spectrum of out-of-equilibrium forces in living cells: amplitude and frequency dependence. *Soft Matter* **5,** 2947–2953 (2009).
18. Feneberg, W., Westphal, M. & Sackmann, E. Dictyostelium cells' cytoplasm as an active viscoplastic body. *Eur. Biophys. J. Biophys.* **30,** 284–294 (2001).
19. Kuimova, M. K., Yahioglu, G., Levitt, J. A. & Suhling, K. Molecular rotor measures viscosity of live cells via fluorescence lifetime imaging. *J. Am. Chem. Soc.* **130,** 6672–6673 (2008).
20. Kuimova, M. K. *et al.* Imaging intracellular viscosity of a single cell during photoinduced cell death. *Nat. Chem.* **1,** 69–73 (2009).
21. Chevry, L., Colin, R., Abou, B. & Berret, J. F. Intracellular micro-rheology probed by micron-sized wires. *Biomaterials* **34,** 6299–6305 (2013).
22. Crick, F. H. C. & Hughes, A. F. W. The physical properties of cytoplasm—a study by means of the magnetic particle method .1. Experimental. *Exp. Cell Res.* **1,** 37–80 (1950).
23. Celedon, A., Hale, C. M. & Wirtz, D. Magnetic manipulation of nanorods in the nucleus of living cells. *Biophys. J.* **101,** 1880–1886 (2011).
24. Chang, W. S., Ha, J. W., Slaughter, L. S. & Link, S. Plasmonic nanorod absorbers as orientation sensors. *Proc. Natl Acad. Sci. USA* **107,** 2781–2786 (2010).
25. Chevry, L., Sampathkumar, N. K., Cebers, A. & Berret, J. F. Magnetic wire-based sensors for the microrheology of complex fluids. *Phys. Rev. E Stat. Nonlin. Soft Matter Phys.* **88,** 062306 (2013).
26. Colin, R., Chevry, L., Berret, J.-F. & Abou, B. Rotational microrheology of Maxwell fluids using micron-sized wires. *Soft Matter* **10,** 1167–1173 (2014).
27. Frka-Petesic, B. *et al.* Dynamics of paramagnetic nanostructured rods under rotating field. *J. Magn. Magn. Mater.* **323,** 1309–1313 (2011).
28. Pedaci, F., Huang, Z. X., van Oene, M., Barland, S. & Dekker, N. H. Excitable particles in an optical torque wrench. *Nat. Phys.* **7,** 259–264 (2011).
29. Tokarev, A. *et al.* Probing viscosity of nanoliter droplets of butterfly saliva by magnetic rotational spectroscopy. *Appl. Phys. Lett.* **102,** 033701 (2013).
30. Tokarev, A., Luzinov, I., Owens, J. R. & Kornev, K. G. Magnetic rotational spectroscopy with nanorods to probe time-dependent rheology of microdroplets. *Langmuir* **28,** 10064–10071 (2012).
31. Zhang, L., Petit, T., Peyer, K. E. & Nelson, B. J. Targeted cargo delivery using a rotating nickel nanowire. *Nanomedicine* **8,** 1074–1080 (2012).
32. Tierno, P. Recent advances in anisotropic magnetic colloids: realization, assembly and applications. *Phys. Chem. Chem. Phys.* **16,** 23515–23528 (2014).
33. Cappallo, N., Lapointe, C., Reich, D. H. & Leheny, R. L. Nonlinear microrheology of wormlike micelle solutions using ferromagnetic nanowire probes. *Phys. Rev. E Stat. Nonlin. Soft Matter Phys.* **76,** 031505 (2007).
34. Rovner, J. B., Lapointe, C. P., Reich, D. H. & Leheny, R. L. Anisotropic stokes drag and dynamic lift on cylindrical colloids in a nematic liquid crystal. *Phys. Rev. Lett.* **105,** 228301 (2010).
35. Helgesen, G., Pieranski, P. & Skjeltorp, A. T. Nonlinear phenomena in systems of magnetic holes. *Phys. Rev. Lett.* **64,** 1425–1428 (1990).
36. Helgesen, G., Pieranski, P. & Skjeltorp, A. T. Dynamic behavior of simple magnetic hole systems. *Phys. Rev. A* **42,** 7271–7280 (1990).
37. Keshoju, K., Xing, H. & Sun, L. Magnetic field driven nanowire rotation in suspension. *Appl. Phys. Lett.* **91,** 123114 (2007).
38. McNaughton, B. H. *et al.* Compact sensor for measuring nonlinear rotational dynamics of driven magnetic microspheres with biomedical applications. *J. Magn. Magn. Mater.* **321,** 1648–1652 (2009).
39. Berret, J.-F., Sandre, O. & Mauger, A. Size distribution of superparamagnetic particles determined by magnetic sedimentation. *Langmuir* **23,** 2993–2999 (2007).
40. Fresnais, J. *et al.* Poly(acrylic acid)-coated iron oxide nanoparticles: Quantitative evaluation of the coating properties and applications for the removal of a pollutant dye. *J. Colloid Interface Sci.* **395,** 24–30 (2013).
41. Chanteau, B., Fresnais, J. & Berret, J.-F. Electrosteric enhanced stability of functional sub-10 nm cerium and iron oxide particles in cell culture medium. *Langmuir* **25,** 9064–9070 (2009).
42. Berret, J.-F. *et al.* Stable oxide nanoparticle clusters obtained by complexation. *J. Colloid Interface Sci.* **303,** 315–318 (2006).
43. Fresnais, J., Berret, J.-F., Frka-Petesic, B., Sandre, O. & Perzynski, R. Electrostatic co-assembly of iron oxide nanoparticles and polymers: towards the generation of highly persistent superparamagnetic nanorods. *Adv. Mater.* **20,** 3877–3881 (2008).
44. Colin, R., Yan, M., Chevry, L., Berret, J.-F. & Abou, B. 3D rotational diffusion of micrometric wires using 2D video microscopy. *Europhys. Lett.* **97,** 30008 (2012).
45. Tirado, M. M., Martinez, C. L. & Delatorre, J. G. Comparison of theories for the translational and rotational diffusion-coefficients of rod-like macromolecules - application to short DNA fragments. *J. Chem. Phys.* **81,** 2047–2052 (1984).
46. Walker, L. M., Moldenaers, P. & Berret, J.-F. Macroscopic response of wormlike micelles to elongational flow. *Langmuir* **12,** 6309–6314 (1996).
47. Nussinovitch, A. *Hydrocolloid Applications: Gum Technology in the Food and Other Industries* (Springer Science + Business Media Dordrecht, 1997).
48. Safi, M. *et al.* Interactions between magnetic nanowires and living cells: uptake, toxicity, and degradation. *ACS Nano* **5,** 5354–5364 (2011).
49. Leung, K. C. F. *et al.* Biological and magnetic contrast evaluation of shape-selective Mn-Fe nanowires. *IEEE Trans. Nanobiosci.* **8,** 192–198 (2009).
50. Hillaireau, H. & Couvreur, P. Nanocarriers' entry into the cell: relevance to drug delivery. *Cell. Mol. Life Sci.* **66,** 2873–2896 (2009).
51. Doshi, N. & Mitragotri, S. Needle-shaped polymeric particles induce transient disruption of cell membranes. *J. R. Soc. Interface* **7,** S403–S410 (2010).
52. Shi, X., von dem Bussche, A., Hurt, R. H., Kane, A. B. & Gao, H. Cell entry of one-dimensional nanomaterials occurs by tip recognition and rotation. *Nat. Nanotechnol.* **6,** 714–719 (2011).
53. Cacchioli, A. *et al.* Cytocompatibility and cellular internalization mechanisms of SiC/SiO2 nanowires. *Nano Lett.* **14,** 4368–4375 (2014).
54. Champion, J. A. & Mitragotri, S. Role of target geometry in phagocytosis. *Proc. Natl Acad. Sci. USA* **103,** 4930–4934 (2006).
55. Liu, J. *et al.* Microrheology probes length scale dependent rheology. *Phys. Rev. Lett.* **96,** 118104 (2006).
56. Castillo, M. *et al.* Local mechanical response of cells to the controlled rotation of magnetic nanorods. *J. Biomed. Mater. Res. B Appl. Biomater.* **102,** 1779–1785 (2014).

Acknowledgements

K. Anselme, A. Baeza, A. Cebers, L. Chevry, E. Fodor, M.-A. Fardin, A. Hallou, F. Montel and P. Visco are acknowledged for fruitful discussions. Interns who participated in the research, C. Leverrier, A. Çonte-Daban, C. Lixi, L. Carvhalo and R. Chan are also acknowledged. The author is grateful to the ImagoSeine facility (Jacques Monod Institute, Paris, France), and to France BioImaging infrastructure supported by the French National Research Agency (ANR-10-INSB-04, « Investments for the future »), and to L. Vitorazi, F. Mousseau and R. Le Borgne for the TEM images of the HeLa cells. The Laboratoire de Physico-chimie des Electrolytes et Nanosystémes Interfaciaux (UMR Université Pierre et Marie Curie-CNRS n° 7612) is acknowledged for providing us with the magnetic nanoparticles.

Author contributions

J.-F.B. conceived the project, developed, performed and supervised the experiments, carried out the data analysis and calculations, and wrote the paper.

Additional information

Competing financial interests: The author declares no competing financial interests.

12

Multiplexed single-molecule force spectroscopy using a centrifuge

Darren Yang[1,*], Andrew Ward[2,3,*], Ken Halvorsen[2,3,4] & Wesley P. Wong[2,3,5]

We present a miniature centrifuge force microscope (CFM) that repurposes a benchtop centrifuge for high-throughput single-molecule experiments with high-resolution particle tracking, a large force range, temperature control and simple push-button operation. Incorporating DNA nanoswitches to enable repeated interrogation by force of single molecular pairs, we demonstrate increased throughput, reliability and the ability to characterize population heterogeneity. We perform spatiotemporally multiplexed experiments to collect 1,863 bond rupture statistics from 538 traceable molecular pairs in a single experiment, and show that 2 populations of DNA zippers can be distinguished using per-molecule statistics to reduce noise.

[1] School of engineering and applied sciences, Harvard University, Cambridge, Massachusetts 02138, USA. [2] Program in Cellular and Molecular Medicine, Boston Children's Hospital, Boston, Massachusetts 02115, USA. [3] Department of Biological Chemistry and Molecular Pharmacology, Harvard Medical School, Boston, Massachusetts 02115, USA. [4] The RNA Institute, University at Albany, Albany, New York 12222, USA. [5] Wyss Institute for Biologically Inspired Engineering, Harvard University, Boston, Massachusetts 02115, USA. * These authors contributed equally to this work. Correspondence and requests for materials should be addressed to K.H. (email: khalvorsen@albany.edu) or to W.P.W. (email: wesley.wong@childrens.harvard.edu).

The ability to mechanically manipulate single molecules is leading to insights throughout biomedical research, from the action of molecular motors in replication and transcription to the role of mechanical forces in development[1–5]. While in principle these approaches enable the full characterization of individual molecular complexes and the study of population heterogeneity at the single-molecule level, in practice key challenges exist. The first challenge for force spectroscopy studies is the low throughput of most single-molecule approaches, which is just starting to be addressed with recently developed multiplexed methods[6–13]. Furthermore, sufficient statistics must be collected not only for the population[14], but also for each individual molecule, which can be a challenge for studying catastrophic transitions such as bond rupture[15–17]. Another challenge is the positive identification of the single-molecule interactions of interest over non-specific and multiple interactions. Finally, there is the subtle challenge of noise, both thermal and experimental, that makes distinguishing different populations of molecules with similar force properties difficult.

We have met all of these challenges with spatiotemporally multiplexed force spectroscopy, a combination of massively parallel spatial multiplexing with repeated interrogation enabled by self-assembled nanoscale devices. First, we introduce a miniature centrifuge force microscope (CFM) for high-throughput single-molecule experimentation that utilizes a commercial benchtop centrifuge. Functionally similar to our first prototype device[18], an entire microscope imaging system is rotated to observe microscopic objects subjected to uniform centrifugal force (unlike earlier 'spinning disk' centrifuge microscopes[19,20]). This inexpensive design brings new features including temperature control and high-resolution particle tracking (~2 nm). Second, we introduce a high-throughput CFM assay that integrates mechanical nanoswitches[17,21,22] to provide important new functionality. The nanoswitches serve two roles—one as a molecular signature to facilitate reliable and automated analysis of large data sets, and the second to enable the repeated interrogation of each single-molecule pair, increasing throughput and enabling new measurements of heterogeneity in single-molecule experiments. By making repeated force measurements on hundreds of single-molecule complexes, we can collect multiple statistics on each molecule that comprises the population. We additionally show that by averaging multiple rupture forces on a per-molecule basis, we can reduce noise to enable super-resolved force spectroscopy—the identification of different populations of molecules below the thermal force-resolution limit. Averaging allows us to reduce the spread in force distributions (averaging reduces noise by a factor of $\sim\sqrt{(N)}$) without losing information about differences between molecules. Furthermore, the rich and relatively large data sets provided by our technique could also complement other analysis techniques for statistical deconvolution[23,24].

Results

Benchtop CFM with nanoscale tracking. The redesigned CFM is a miniaturized microscope that fits into a commercial centrifuge so that force can be applied to samples while their micro- to nano-scale motions are tracked (Fig. 1 and Supplementary Fig. 1). To accommodate the microscope components in the 400 ml bucket volume, we selected components with size and weight in mind, and induced two right-angle bends in the optical path (Supplementary Fig. 1b). To ensure accessibility of the instrument, most components are commercially available and can be easily assembled (Supplementary Fig. 1d,e and Supplementary Table 1). A three-dimensional (3D)-printed housing was made to encapsulate the device and to ensure a tight fit within the bucket (Solidworks drawings provided as Supplemental Files). Transmission of camera data out of the CFM during centrifugation is accomplished by converting the camera's gigabit Ethernet signal to a fibre-optic signal, then passing this data out of the centrifuge through a fibre rotary joint (Supplementary Fig. 1a,f). Details of the design are presented in the Methods.

We implemented a unique solution to measure nanometer-level extensions of tethers in the CFM by projecting tether length changes onto the x–y plane of the coverslip, enabling these measurements to be made in a relatively simple and computationally efficient way. Taking advantage of the fact that we can precisely control the direction of force application by mechanically constraining the angle of the centrifuge bucket, we intentionally misaligned the force and imaging axes (Supplementary Fig. 2a and Supplementary Note 1). Using this technique, we can 'tune' our tether length resolution based on the bucket angle with a range of ~2.5 nm (at 20°) to 12 nm (at 80°) based on our lateral tracking resolution of ~2 nm (Supplementary Fig. 2b,c). To validate and demonstrate this approach, we measured DNA force extension for over 100 molecules simultaneously over a span of <1 min (Supplementary Fig. 3a), and fit each with the standard worm-like chain model. The most likely contour length and persistence length was 8.2 ± 0.2 μm and 46 ± 1 nm, respectively, in agreement with expected values[25]. We additionally performed multiplexed overstretching measurements of lambda DNA (Supplementary Fig. 3c), yielding an overstretching force of 63.5 ± 1.7 pN (mean ± s.d.), consistent with previous measurements at these conditions[25].

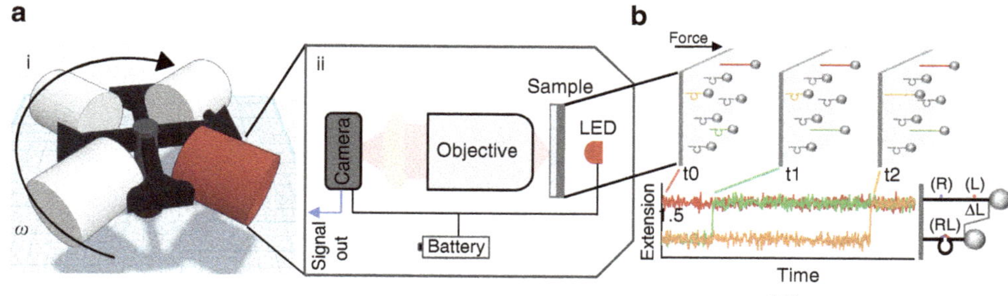

Figure 1 | Overview of the benchtop CFM. (**a**,i) Illustration of a spinning centrifuge rotor containing the CFM in the red swing bucket; the rotation induces forces on molecules tethered between beads and the coverslip surface. The direction of force application can be set by constraining the angle of the centrifuge bucket, as described in Supplementary Fig. 5. (**a**,ii) Schematic of the CFM. (**b**) Tether extension is monitored as a function of time while force is applied. When the DNA nanoswitch is integrated, molecular transitions such as bond rupture between a receptor (R)-ligand (L) pair causes a well-defined change in tether extension, providing a distinct signature for detecting interactions between two molecules of interest.

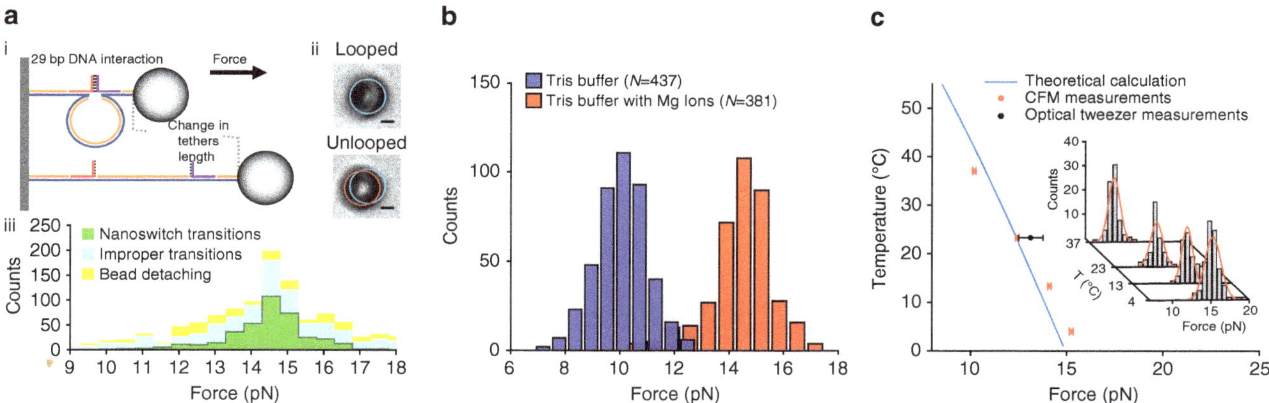

Figure 2 | DNA unzipping force measured using DNA nanoswitches on the CFM. (**a**,i) Schematic of the unzipping construct. Two complementary oligos (red and purple strands) hybridize to form a looped DNA nanoswitch. Force can unzip the two complementary strands, resulting in a measurable increase in tether length, providing a signature of DNA unzipping. (**a**,ii) Images of a bead tethered to the surface via a DNA nanoswitch showing the looped and unlooped states. The scale bar, 1-μm long. (**a**,iii) In one example of rupture-force measurement, we identified 381 tethers with the DNA nanoswitch transitions signature to collect rupture forces while the remaining 673 transitions that corresponded to bead detachment and improper transitions were omitted (see also Supplementary Fig. 5 for detail). (**b**) Unzipping force histograms of 29 bp dsDNA measured with the DNA nanoswitch under two different buffer conditions. (**c**) Average unzipping force of 29 bp dsDNA under different temperatures with PBS buffer (Total $n = 306$), with histograms of rupture forces shown as an inset. The theoretical line is calculated using a previously described thermodynamic model[30].

DNA nanoswitches for authenticating single-molecule data. To enable robust and repeatable rupture experiments, we integrated DNA nanoswitches into the CFM. These molecular switches are designed to adopt a looped structure when the molecules of interest are interacting and a linear structure when they are not (Fig. 2a), providing a distinct 'signature' (that is, increase in tether length) for rupture events (Supplementary Fig. 4b). We performed DNA unzipping experiments on a 29 bp DNA interaction, and used the 'signature' unlooping of the nanoswitches to positively identify and discriminate valid single-molecule data from multiple tethers and non-specific interactions (Fig. 2a and Supplementary Fig. 5). Following this approach, unzipping force measurements were carried out in the presence and absence of magnesium ions, with hundreds of rupture statistics for each condition collected in under 30 s of centrifuge run time. We found that magnesium stabilizes the duplex as observed previously[26], with the average unzipping force ($\pm$ s.d.) increasing from 10.1 ± 0.9 to 14.6 ± 1.1 pN with the addition of magnesium (Fig. 2b).

Temperature-dependent DNA unzipping. We additionally performed DNA unzipping experiments at four temperatures, 4, 13, 23 and 37 °C (Fig. 2c). Seminal works introduced temperature control into optical tweezers over 10 years ago[27,28], but even today the vast majority of single-molecule pulling experiments are carried out at room temperature, and temperature control remains an active area of development[29]. The benchtop CFM has the advantages of built-in temperature control and portability to move into cold (4 °C) or warm (37 °C) rooms available to most biologists. Accurate, real-time measurements of the sample temperature during experiments were made with a wireless thermocouple embedded within the centrifuge bucket (Methods for full details). We found an increase in the unzipping force with decreasing temperature, in good quantitative agreement with the thermodynamic models of DNA unzipping and previous observations[30] (see also Supplementary Note 2).

Repeated interrogation and super-resolved force spectroscopy. Finally, we demonstrate spatiotemporally multiplexed force spectroscopy using the DNA nanoswitch assay on the CFM to repeatedly interrogate a population of molecules at the single-molecule level (Fig. 3a). We measured 1,863 statistics of DNA unzipping from repeated pulls of 538 molecules in a single sample to demonstrate the large amounts of single-molecule force data that can be accumulated with this approach (Fig. 3b). Even more significantly, the nanoswitches enable the unique properties of each molecule in a sample to be characterized from repeated measurements, as we demonstrate by determining a rupture-force histogram for each molecule in a sample, illuminating population heterogeneity at the single-molecule level (Fig. 3c). Furthermore, we show that by averaging data from multiple pulls of the same molecular pair the spread in force is reduced without losing the unique characteristics of each molecule. When applied to data from a single population, this per-molecule force averaging generates a super-resolved histogram with the expected narrowing when compared with the raw histogram of all the data (Fig. 3d). When applied to combined statistics from two populations of DNA zippers (introducing another G–C-rich zipper with a higher unzipping force[31]), we show that the super-resolved histogram generated from per-molecule averaging can separate out two populations that are unresolveable from the raw histograms due to the intrinsic broadening of force that results from thermal noise and instrumental noise (Fig. 3e).

Discussion

As we have shown, together the CFM and the DNA nanoswitch assay provide powerful single-molecule capabilities, and give non-specialists unprecedented access to force spectroscopy. And even separately these technologies provide unique benefits that can make single-molecule experiments better, cheaper and faster. Our miniature CFM presented here, which integrates into standard benchtop centrifuges, has important new features including an expanded biologically relevant dynamic force range (Supplementary Fig. 6), easily implemented temperature control between 4 and 37 °C, improved safety, lower cost (Supplementary Table 1 and Supplementary Note 3), and greater ease of use compared with the first-generation CFM. These last two points are important strides towards increasing the accessibility of single-molecule experiments for non-specialists. The total reduction in cost is particularly significant when considering the additional infrastructure typically needed for single-molecule

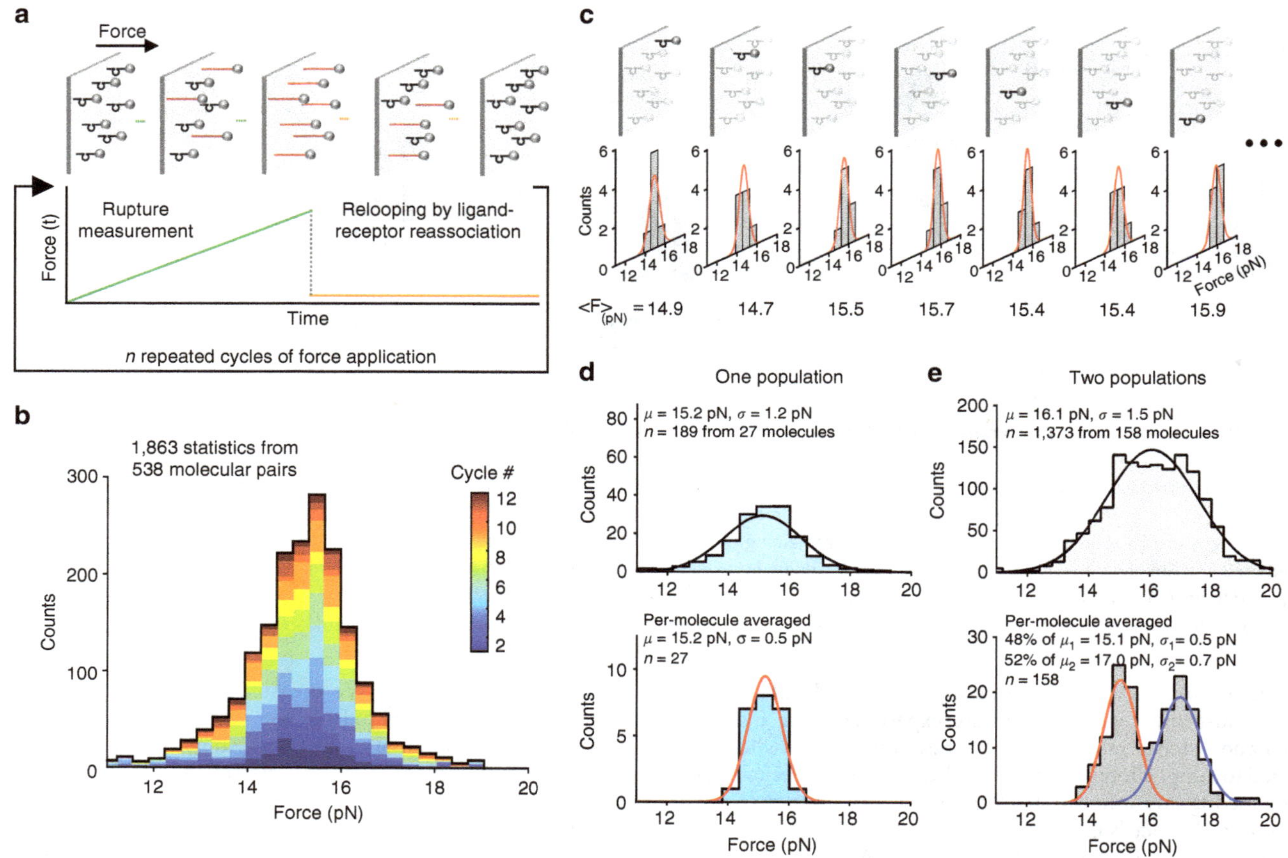

Figure 3 | Repeated rupture-force measurement of single molecular pairs. (**a**) Protocol for repeated cycles of force application, with each cycle consisting of a linear force ramp to induce rupture and DNA nanoswitch unlooping, followed by a low force reassociation period allowing the molecular pairs to rebind. (**b**) DNA unzipping force histogram of 1,863 rupture events collected from a total of 538 molecular pairs with 12 cycles of force application. The colour from blue to red corresponds to statistics collected from each cycle. (**c**) Example rupture-force histograms generated for individual molecular pairs. The calculated average rupture force is shown below in red. (**d**) Combined histogram of rupture forces from 27 molecules with 7 cycles of force rupture each (top), and histogram of the per-molecule averaged rupture force (bottom), showing a reduced width. (**e**) Combined histogram for two populations of DNA unzipping experiments (top), and the per-molecule averaged super-resolved histogram (bottom) that recovers the two separate populations from the mixed data.

measurements that are obviated by our approach, as the centrifuge itself provides an isolated, temperature-regulated enclosure. In addition, our instrument can be operated without significant training, even by undergraduate researchers. Similarly, the DNA nanoswitch-based single-molecule assay has its own benefits, significantly reducing the effort to obtain reliable and accurate measurements by standardizing sample preparation, and by supplying a distinct molecular signature that enables reliable data verification and automated analysis. Furthermore, it enables repeated interrogation of each molecular complex for building up statistics of both the population and the individual molecules, revealing population heterogeneity at the single-molecule level.

There is difficulty in generalized comparisons of single-molecule instruments due to the wide variety of designs, but the miniature CFM presented here (using the 2.8 μm dynabeads) is comparable to our own optical tweezers set-up[17,32] in force range (~0.1 to 100 pN), force resolution (~3–5%) and spatial resolution (~2 nm). It is worth noting, however, that the most advanced optical tweezers push spatial resolution into the sub-Angstrom range, have force resolution of <0.1 pN and have temporal resolutions >10 kHz (this CFM is 15 Hz)[2]. Regarding multiplexing, magnetic tweezers have achieved up to 357 statistics in a single experiment[8], acoustic manipulation has achieved 145 statistics[6] and other multiplexing methods show great promise as well[12,13]. However, our demonstration of 1,863 validated single-molecule statistics in one experiment is to our knowledge more than in any previous work.

As the potential for single-molecule approaches to address critical problems in biology becomes increasingly clear, the availability of such simple and powerful tools that do not require months of training or hundreds of thousands of dollars becomes a key. We believe the technologies presented here are an important step in that direction, enabling both a more detailed view of single molecules as part of a population, and a more streamlined way to amass large numbers of statistics. It is not unreasonable to imagine that this approach, paired with sample scanning and/or fast force cycling could eventually enable the collection of millions of single-molecule statistics from a single sample. With these advances, previously unrealistic applications such as high-throughput single-molecule screening and diagnostics become more viable, and single-molecule analysis may start to become as ubiquitous as standard bulk assays.

Methods

CFM instrumentation. The CFM consists of an optical microscope and digital image acquisition system, which is integrated into a centrifuge (Fig. 1a and Supplementary Fig. 1). We used a refrigerated benchtop centrifuge (Heraeus X1R, Thermo Scientific) with two modifications. First, the TX-400 rotor was modified to mount a fibre-optic rotary joint along the central axis (Princetel MJX). This was accomplished by removing the rotor's central push-release mechanism (by loosening the screw on the side of the button), and threading the four existing holes to accept 10–32 screws. An adapter was designed and installed on the rotor to hold the slip ring (Supplementary Fig. 1a,f). To enable the fibre optic to pass through the lid, the central plastic viewing window was removed. Second, the centrifuge's control module was upgraded to enable computer control. An upgraded module was kindly provided by Thermo Fisher Scientific to support this project, and was a simple drop-in replacement.

The light microscope of the benchtop CFM was constructed using mainly Thorlabs SM1 compatible components (Supplementary Fig. 1b,d,e and Supplementary Table 1). A red LED (LED630E, Thorlabs) threaded to a tube mount (S1LEDM, Thorlabs) served as the illumination source. A glass diffuser positioned between the LED source and sample cell provided uniform illumination across the field of view. The 25 mm diameter of the sample cell was designed to be compatible with the SM1 lens tube (see below CFM sample cell). The sample was magnified and imaged onto a CCD camera (GC 2,450, Prosilica, AVT) with a 40 × Olympus Plan Achromat objective (infinity corrected, 0.65 numerical aperture and 0.6 mm working distance) and Ø1″ 100 mm tube lens (AC254-100-A, Thorlabs). Due to the limited depth of the centrifuge TX-400 bucket (~16 cm), we shortened the length of the microscope by bending the light path 180° using a pair of turning mirrors glued to a custom turning cube (part #17 in Supplementary Fig. 1e). Moreover, a 3D-printed enclosure made of acrylonitrile butadiene styrene was used to secure and integrate the imaging and acquisition system inside of the centrifuge bucket. The enclosure also included an open slot for a battery (PRT-00339, SparkFun), and a connected direct current (DC)-to-DC step-up circuit (PRT-08290, SparkFun) that served as the power source for the LED, camera and media converter. The camera used the standard GigE Vision interface, outputting the data as a gigabit Ethernet signal. To enable live imaging, we utilized a fibre-optic rotary joint installed at the centre of the centrifuge rotor (Supplementary Fig. 1f). The camera signal was converted from twisted-pair Ethernet to a fibre-optic signal by a small media converter inside of the centrifuge, transferred through the rotary joint, then converted back to a standard Ethernet signal by a second media converter connected to the acquisition computer. (855–10,734 and 855–10,735, IMC Networks) (Supplementary Fig. 1a). The images collected from the camera were recorded with custom LabView software developed with the Vision Module.

We embedded a portable wireless thermocouple connector (MWTC-D-K-915, Omega Engineering) with a surface adhesive thermocouple (SA1XL-K, Omega Engineering) within the bucket that contains the CFM to measure the sample temperature. A wireless receiver (WTC-REC1-915, Omega Engineering) was used to acquire the temperature from the thermocouple connector to record the temperature in real time.

Software and technical drawings are available as Supplemental Files and on request.

Molecular constructs. The looped DNA nanoswitch construct was made using our previously published DNA self-assembly protocol[17,21]. Circular M13mp18 single-stranded DNA (ssDNA) (N4040S, New England Biolabs) was linearized by hybridizing a 40 bp oligo that created a double-stranded restriction site for the BtsCI enzyme (R0647S, New England Biolabs). Subsequently, a set of complementary oligos (Integrated DNA Technologies) was hybridized onto the linear ssDNA. Functionalized oligos (biotinylated and digoxigenin-modified) were hybridized onto the 3′ and 5′ ends of the ssDNA, respectively. The hybridization was carried out with 15 nM of linearized ssDNA and 10 M excess of the complementary oligos in 1 × NEBuffer 2 with a temperature ramp from 90 to 20 °C (−1 °C min^{-1}) in a thermocycler. After this initial hybridization two specific single-stranded regions remained, which were bridged by two partially complimentary oligos to form the final looped construct (Fig. 2a). The sequence of the complimentary bridge oligo that formed the loop was: 3′-CTCAAATA TCAAACCCTCAATCAATATCT-5′. This secondary hybridization step was carried out at a final construct concentration of 250 pM with a 1.25 molar excess of the bridge oligos in 1 × NE buffer 2 at room temperature for 1 h. We verified looping of the construct using gel-shift assays, single-molecule optical trap measurements and atomic force microscopy imaging (Supplementary Fig. 4). In the optical trap measurement, we applied force on the looped DNA construct via tethering between laser-trapped streptavidin and anti-digoxigenin-functionalized silica beads.

For the DNA overstretching measurements in the CFM, we functionalized both ends of lambda DNA with biotin to provide strong anchorage to the streptavidin-functionalized glass coverslip and bead surfaces. First, 20 µl of lambda DNA (0.28 µg ml^{-1}, 10745782001, Roche) was incubated for 20 min at 65 °C to remove the hybridized overhangs. Subsequently a nucleotide mixture that consists of Biotin-14-dATP, Biotin-14-dCTP, dTTP and dGTP, each at 100 µM final concentration, was added to the lambda DNA solution with 0.25 U ml^{-1} Klenow Fragment (M0212S, New England Biolabs). This mixture was incubated for 1 h at 37 °C. The dual-end biotin lambda DNA was purified from the excess nucleotides and enzyme using the Qiagen PCR Purification Kit.

For the parallel force extension measurements, we made the half-length lambda DNA functionalized with digoxigenin and biotin. First, the biotin-labelled full-lambda DNA construct was cut near the middle using the XbaI restriction enzyme (R0145S, New England Biolab). The resulting overhangs were functionalized with digoxigenin to produce a heterobifunctional 24-kbp construct labelled with digoxigenin on one side and biotin on the other.

CFM sample cell. The mini-CFM sample cell was constructed using double-sided Kapton tape sandwiched between a 25 mm diameter support glass and a 19 mm diameter cover glass (3346, Gold Seal). The 0.7-mm-thick support glass was ordered from S.I. Howard Glass (D263) and two 1 mm diameter ports were drilled that served as the solution inlet and outlet. The cover and support glasses were cleaned by immersing in 100 ml a 1% (v/v) Hellmanex III solution, microwaving for 1 min, then sonicating for 30 min. Subsequently, the slides were rinsed thoroughly with Millipore water then dried with nitrogen flow. A 1 × 7 mm rectangular flow channel was cut on the double-sided Kapton tape using a cut plotter (Graphtec). To form tethers with digoxigenin-functionalized construct, we functionalized the cover glass with anti-digoxigenin using a modified version of a previously developed protocol[33]. First the cover glass was coated with a nitrocellulose solution by depositing 2 µl of amyl acetate solution with 0.2% (m/v) dissolved nitrocellulose. We then incubated the channel with phosphate buffered saline (PBS, 137 mM NaCl, 2.7 mM KCl, 10 mM phosphate buffer, pH 7.4) solution containing 100 µg ml^{-1} anti-digoxigenin (11333089001, Roche) for 15 min. The channel was then washed and further incubated with a surface passivation solution (10 mg ml^{-1} Roche Blocking Reagent in PBS) for 1 h. After the passivation step, the channel was flushed with experimental buffer then incubated with 5 pM of construct for 15 min. At 5 pM construct concentration, the construct was limited to an average spacing of roughly 2 µm on the surface, making formation of double tethers a rare event. After tethering, the construct to the surface, the flow channel was washed with 20 µl of the experimental buffer then incubated with 15 mg ml^{-1} streptavidin beads (M-270, Invitrogen). For each experiment, beads were washed excessively with the experimental buffer before loading them to the sample cell. Before loading the sample cell to the mini-CFM, the solution inlet and outlet ports were sealed with vacuum grease. The tris experimental buffer in Fig. 2 consisted of 10 mM Tris, 30 mM NaCl at pH 7.5 with or without 10 mM $MgCl_2$.

For the overstretching experiment, the surface tethering was strengthened by replacing digoxigenin–anti-digoxigenin with biotin–streptavidin—in other words, biotin–streptavidin interactions were used to anchor both ends of each tether. The nitrocellulose surface was functionalized by incubating it with 1 mg ml^{-1} streptavidin in PBS solution that contained 1 mg ml^{-1} of Roche Blocking Reagent for 12 h, followed by incubation with passivation solution (10 mg ml^{-1} Roche Blocking Reagent in PBS buffer) for 1 h. The channel was flushed with PBS and incubated with 5 pM of dual-biotin λ-DNA for 15 min before loading in the streptavidin-coated beads. Under such conditions, the density of streptavidin on the surface was sparse enough that only one end of the biotin-labelled λ-DNA bound to the surface, leaving the other biotinylated end free to bind to the streptavidin-coated bead.

Swinging bucket angle measurement. To measure the angle of the swinging bucket relatives to the axis of rotation, we attached an oil-based marker to the bottom of the bucket, which marks the height of the bucket on the wall as the centrifuge spins (Supplementary Fig. 2d). The uncertainty of the angle measurement was based on the distance measurement error estimate of 1 mm. At the rotational speed of 300 r.p.m. the bucket swings out to an angle of (81.4 ± 0.8)°. At a much higher speed of 1,800 r.p.m. the angle increases by 2.3%. Further increase of the rotational speed does not increase the angle beyond the error of the measurement.

Data analysis. The loop opening signature was identified by tracking the beads' x and y positions. As the angle of the centrifuge swing bucket was not 90° with respect to the axis of rotation, a component of the centrifugal force was directed in the x–y plane (Supplementary Fig. 5). As the rotational speed of the CFM increased, the looped DNA tethers were continuously extended. The opening of the loop was identified as a discontinuous change in extension. Each movie contained ~1,000 beads. To track these beads, they were first identified using the Matlab function imfindcircles. A template image for each bead was stored. To identify the bead in the subsequent frame, the template image was scanned in the x–y plane to find the position of maximum correlation. First the image was scanned in the x direction over a 25 pixel search region centred on the bead position from the previous frame. A 2nd order parabola was then fit to the correlation coefficient as a function of position. The position of maximum correlation was identified as the new bead position. The template image was than centred on the new x position, and the same procedure was done in the y direction. During the course of each experiment, there was some drift in the x–y plane. This was corrected for by taking the median change in x and y for all beads being tracked from frame to frame. We found this drift correction to be sufficient for identifying the looped to unlooped transition. Transitions were identified by filtering out all bead trajectories except those that contained a discontinuous change in extension of both the correct magnitude and direction (Supplementary Fig. 5b,c). This removed false transitions that may have resulted from non-specific interactions or the formation of multiple bonds between the bead and the surface. Following this automated filtering procedure, the transition events were visually inspected, in random order, to reject any remaining erroneous transitions, which may have occurred due to particle tracking artifacts that manifest as discontinuous changes in position (that is, particle mislabeling, overlapping beads and so on).

For DNA unzipping experiments, the number of transitions as a function of time was converted to the number of transitions as a function of force as follows:

The rotational speed of the centrifuge was recorded during each movie using WinMess software (provided by Thermo Fisher Scientific, R&D) which enables communication and control with the computer. The rotational speed was then converted to force (F) as a function of time using the following equation:

$$F = m_{\text{bead}} R_{\text{CFM}} \omega^2$$

where R_{CFM} and ω are the rotational radius and speed, respectively. To determine the effective mass of the beads (m_{bead}) in solution, we measured the Invitrogen M-270 bead density through sink-float analysis using aqueous sodium polytungstate solution (71913, Sigma-Aldrich) and obtained a density of $1.61 \pm 0.02\,\text{g cm}^{-3}$ that confirmed the manufacturer's (Life Technologies) reported value of $1.6\,\text{g cm}^{-3}$. The manufacturer also provided the bead diameter of our specific lot giving a mean of 2.80 μm with <1.6% CV (lot #144315600). The density of the silica beads used in the DNA force extension and overstretching experiment was measured similarly using the sodium polytungstate solution, yielding a density of $1.50 \pm 0.03\,\text{g cm}^{-3}$. The diameter of the silica beads was measured using transmission electron microscopy, yielding an average and s.d. of 4.27 ± 0.16 μm.

Tethered particle motion. We carried out tethered particle motion analysis for our tethered beads prior to the rupture-force measurement. We analysed the lateral fluctuations of each bead, calculating the root mean square of the drift-subtracted displacement and the symmetry ratio, following previously established methods[34]. Here the symmetry ratio was calculated as the square root of the ratio between the minimum and maximum eigen values of the covariance matrix for the in-plane displacement. The in-plane position of each bead was recorded for 10 s at an acquisition rate of 10 Hz.

Repeated cycles of rupture-force measurement. Multiple rupture events were collected for each molecule in a set by repeatedly spinning the same sample multiple times. For the data presented in Fig. 3, the sample was spun up with an effective force-loading rate of $1\,\text{pN s}^{-1}$ to a maximum force of 22 pN. The speed was then ramped down to 0 r.p.m. for ~1 h to allow rebinding between each pair of molecules. Beads were identified from cycle to cycle by their positions relative to a set of fiducial beads, which were common to each cycle. For the measurement of the two populations of molecules, we made two different DNA nanoswitch unzipping constructs, one with 48% GC content (5′-CACGAATTCTCTGCCTC CCTTTTAACCCTAG-3′) and one with 31% GC content (5′-CTCAAATATCAAA CCCTCAATCAATATCT-3′).

References

1. Bustamante, C., Cheng, W. & Meija, Y. X. Revisiting the central dogma one molecule at a time. *Cell* **144,** 480–497 (2011).
2. Neuman, K. C. & Nagy, A. Single-molecule force spectroscopy: optical tweezers, magnetic tweezers and atomic force microscopy. *Nat. Methods* **5,** 491–505 (2008).
3. Ritort, F. Single-molecule experiments in biological physics: methods and applications. *J. Phys. Condens. Matter* **18,** R531–R583 (2006).
4. Bustamante, C. J., Kaiser, C. M., Maillard, R. A., Goldman, D. H. & Wilson, C. A. M. Mechanisms of cellular proteostasis: insights from single-molecule approaches. *Annu. Rev. Biophys.* **43,** 119–140 (2014).
5. Greenleaf, W. J., Woodside, M. T. & Block, S. M. High-resolution single-molecule measurements of biomolecular motion. *Annu. Rev. Biophys. Biomol. Struct.* **36,** 171–190 (2007).
6. Sitters, G. *et al.* Acoustic force spectroscopy. *Nat. Methods* **12,** 47–50 (2015).
7. Soltani, M. *et al.* Nanophotonic trapping for precise manipulation of biomolecular arrays. *Nat. Nanotechnol.* **9,** 448–452 (2014).
8. De Vlaminck, I. *et al.* Highly parallel magnetic tweezers by targeted DNA tethering. *Nano Lett.* **11,** 5489–5493 (2011).
9. Fazio, T., Visnapuu, M. L., Wind, S. & Greene, E. C. DNA curtains and nanoscale curtain rods: High-throughput tools for single molecule imaging. *Langmuir* **24,** 10524–10531 (2008).
10. Ribeck, N. & Saleh, O. A. Multiplexed single-molecule measurements with magnetic tweezers. *Rev. Sci. Instrum.* **79,** 094301 (2008).
11. Kim, S. J., Blainey, P. C., Schroeder, C. M. & Xie, X. S. Multiplexed single-molecule assay for enzymatic activity on flow-stretched DNA. *Nat. Methods* **4,** 397–399 (2007).
12. Otten, M. *et al.* From genes to protein mechanics on a chip. *Nat. Methods* **11,** 1127–1130 (2014).
13. Chiou, P. Y., Ohta, A. T. & Wu, M. C. Massively parallel manipulation of single cells and microparticles using optical images. *Nature* **436,** 370–372 (2005).
14. Evans, E. Probing the relation between force-lifetime-and chemistry in single molecular bonds. *Annu. Rev. Biophys. Biomol. Struct.* **30,** 105–128 (2001).
15. Kim, J., Zhang, C.-Z., Zhang, X. & Springer, T. A. A mechanically stabilized receptor-ligand flex-bond important in the vasculature. *Nature* **466,** 992–995 (2010).
16. Li, P. T., Bustamante, C. & Tinoco, I. Unusual mechanical stability of a minimal RNA kissing complex. *Proc. Natl Acad. Sci. USA* **103,** 15847–15852 (2006).
17. Halvorsen, K., Schaak, D. & Wong, W. P. Nanoengineering a single-molecule mechanical switch using DNA self-assembly. *Nanotechnology* **22,** 494005 (2011).
18. Halvorsen, K. & Wong, W. P. Massively parallel single-molecule manipulation using centrifugal force. *Biophys. J.* **98,** L53–L55 (2010).
19. Harvey, E. N. & Loomis, A. L. A microscope-centrifuge. *Science* **72,** 42–44 (1930).
20. Oiwa, K., Chaen, S., Kamitsubo, E., Shimmen, T. & Sugi, H. Steady-state force-velocity relation in the ATP-dependent sliding movement of myosin-coated beads on actin cables in vitro studied with a centrifuge microscope. *Proc. Natl Acad. Sci. USA* **87,** 7893–7897 (1990).
21. Koussa, M. A., Halvorsen, K., Ward, A. & Wong, W. P. DNA nanoswitches: a quantitative platform for gel-based biomolecular interaction analysis. *Nat. Methods* **12,** 123–U148 (2015).
22. Chandrasekaran, A. R., Zavala, J. & Halvorsen, K. Programmable DNA nanoswitches for detection of nucleic acid sequences. *ACS Sens.* **1,** 120–123 (2016).
23. Cheng, W., Arunajadai, S. G., Moffitt, J. R., Tinoco, I. & Bustamante, C. Single–base pair unwinding and asynchronous RNA release by the hepatitis C virus NS3 helicase. *Science* **333,** 1746–1749 (2011).
24. Yu, Z. *et al.* Tertiary DNA structure in the single-stranded hTERT promoter fragment unfolds and refolds by parallel pathways via cooperative or sequential events. *J. Am Chem. Soc.* **134,** 5157–5164 (2012).
25. Baumann, C. G., Smith, S. B., Bloomfield, V. A. & Bustamante, C. Ionic effects on the elasticity of single DNA molecules. *Proc. Natl Acad. Sci. USA* **94,** 6185–6190 (1997).
26. Lee, C. H., Danilowicz, C., Conroy, R. S., Coljee, V. W. & Prentiss, M. Impacts of magnesium ions on the unzipping of gimel-phage DNA. *J. Phys. Condens. Matter* **18,** S205–S213 (2006).
27. Mao, H., Arias-Gonzalez, J. R., Smith, S. B., Tinoco, I. & Bustamante, C. Temperature control methods in a laser tweezers system. *Biophys. J.* **89,** 1308–1316 (2005).
28. Williams, M. C., Wenner, J. R., Rouzina, I. & Bloomfield, V. A. Entropy and heat capacity of DNA melting from temperature dependence of single molecule stretching. *Biophys. J.* **80,** 1932–1939 (2001).
29. Stephenson, W. *et al.* Combining temperature and force to study folding of an RNA hairpin. *Phys. Chem. Chem. Phys.* **16,** 906–917 (2014).
30. Danilowicz, C. *et al.* Measurement of the phase diagram of DNA unzipping in the temperature-force plane. *Phys. Rev. Lett.* **93,** 078101 (2004).
31. Rief, M., Clausen-Schaumann, H. & Gaub, H. E. Sequence-dependent mechanics of single DNA molecules. *Nat. Struct. Mol. Biol.* **6,** 346–349 (1999).
32. Zhang, X., Halvorsen, K., Zhang, C.-Z., Wong, W. P. & Springer, T. A. Mechanoenzymatic cleavage of the ultralarge vascular protein von Willebrand factor. *Science* **324,** 1330–1334 (2009).
33. Lipfert, J., Kerssemakers, J. W. J., Jager, T. & Dekker, N. H. Magnetic torque tweezers: measuring torsional stiffness in DNA and RecA-DNA filaments. *Nat. Methods* **7,** 977–U954 (2010).
34. Nelson, P. C. *et al.* Tethered particle motion as a diagnostic of DNA tether length. *J. Phys. Chem.* **110,** 17260–17267 (2006).

Acknowledgements

We thank Andreas Karl and Thermo Fisher Scientific for providing materials and support for centrifuge automation, Chris Stokes for technical assistance with early prototypes, Marco Catipovic for assistance with experimental protocols and preparations. W.P.W. acknowledges support from the Arnold and Mabel Beckman Foundation through the Beckman Young Investigator Award, US National Institutes of Health R21 RGM107907A, the Wyss Institute at Harvard University and Boston Children's Hospital startup funds. D.Y. was supported by the National Science Foundation Graduate Research Fellowship DGE-1144152.

Author contributions

The concept was initially conceived by W.P.W. and K.H. and expanded by D.Y. and A.W. Experiments were designed by all authors and were carried out by D.Y. and A.W. All authors participated in data analysis, critical discussions and writing of the manuscript.

Additional information

Competing financial interests: Patent applications have been filed for various aspects of this work by D.Y., A.W., K.H. and W.P.W.

13

Removal of the mechanoprotective influence of the cytoskeleton reveals PIEZO1 is gated by bilayer tension

Charles D. Cox[1,*], Chilman Bae[2,*], Lynn Ziegler[2], Silas Hartley[2], Vesna Nikolova-Krstevski[1], Paul R. Rohde[1], Chai-Ann Ng[1,3], Frederick Sachs[2,4], Philip A. Gottlieb[2,4,**] & Boris Martinac[1,3,**]

Mechanosensitive ion channels are force-transducing enzymes that couple mechanical stimuli to ion flux. Understanding the gating mechanism of mechanosensitive channels is challenging because the stimulus seen by the channel reflects forces shared between the membrane, cytoskeleton and extracellular matrix. Here we examine whether the mechanosensitive channel PIEZO1 is activated by force-transmission through the bilayer. To achieve this, we generate HEK293 cell membrane blebs largely free of cytoskeleton. Using the bacterial channel MscL, we calibrate the bilayer tension demonstrating that activation of MscL in blebs is identical to that in reconstituted bilayers. Utilizing a novel PIEZO1-GFP fusion, we then show PIEZO1 is activated by bilayer tension in bleb membranes, gating at lower pressures indicative of removal of the cortical cytoskeleton and the mechanoprotection it provides. Thus, PIEZO1 channels must sense force directly transmitted through the bilayer.

[1] Victor Chang Cardiac Research Institute, Darlinghurst, New South Wales 2010, Australia. [2] Department of Physiology and Biophysics, State University of New York at Buffalo, Buffalo, New York 14214, USA. [3] St Vincent's Clinical School, University of New South Wales, Darlinghurst, New South Wales 2010, Australia. [4] The Centre for Single Molecule Biophysics, State University of New York at Buffalo, Buffalo, New York 14214, USA. * These authors contributed equally to this work. ** These authors jointly supervised this work. Correspondence and requests for materials should be addressed to P.A.G. (email: philgott@buffalo.edu) or to B.M. (email: B.martinac@victorchang.edu.au).

PIEZO channels are a recently cloned class of eukaryotic mechanosensitive channels identified initially in a murine neuroblastoma cell line (neuro-2a cells)[1,2]. The two subtypes, PIEZO1 and PIEZO2, are currently the subject of intense research and have been shown to be involved in numerous physiological and pathophysiological processes. PIEZO2 has been linked to light touch via expression in Merkel cells in the skin[3,4]. PIEZO1 has been linked to the maintenance of cell volume in erythrocytes and its dysfunction has been linked to specific hereditary anemias (for example, xerocytosis)[5–7]. In addition, it plays a role in vascular development, vascular physiology and cell differentiation[8–10].

Identifying force transduction mechanisms in mechanosensitive channels has been difficult owing to the complex environment of a cell membrane patch and partly owing to the multiple pathways leading to mechanosensitive channel activation[11–15]. For example, mechanosensitive channels might be gated (i) directly by bilayer tension as seen with the MscS and MscL bacterial channel families as well as the two-pore domain K^+ channels TREK-1 and TRAAK[16,17] (This is referred to as the 'Force-from-lipids' concept[18,19]), or (ii) by tethering to the cytoskeleton and/or the extracellular matrix[20–22]. It is important to note that even if a channel is gated directly by bilayer tension, the tension in the bilayer can be modified by cytoskeletal proteins and linkages to the extracellular matrix. Thus the sensitivity of a channel to the mean stress in the cortex can be modified by alterations in an array of scaffold proteins[23]. One way to probe the role of the cytoskeleton in the activation of a mechanosensitive channel is to use cytoskeletal deficient membrane 'blebs'.

Blebs are rounded protrusions of the bilayer where connections between the cytoskeleton and the bilayer are largely broken[24–26]. They are found in numerous processes including cytokinesis, cell motility and programmed cell death[25,27,28]. The blebbing process itself is dynamic and tightly regulated by numerous factors. In a physiological setting, the lifetime of a bleb ranges from tens of seconds to minutes[25,26]. Their formation may be induced by numerous stimuli including changes in external osmolarity. Bleb formation usually starts with a breakdown of cortical f-actin (Fig. 1a)[25,29]. Subsequent bleb growth is modulated by hydrostatic pressure and actomyosin contractility. During the expansion phase, the bleb membrane is still amenable to electrophysiological study[30–33]. The challenge for electrophysiology is to maintain the decoupling of the cortical cytoskeleton long enough to measure channel activity since the cytoskeleton can be rebuilt[26]. Using this research paradigm, we asked whether the PIEZO1 channel is gated by bilayer tension.

We first generated blebs deficient in f-actin and β-tubulin long enough for electrophysiological recording. We then characterized the mechanical environment of the blebbed membranes using as a probe MscL, a mechanosensitive channel gated directly by bilayer tension[34]. The pipette suction required to gate MscL in blebs was three times less than in cell-attached patches, reflecting the fact that much of the cortical tension is supported by the cytoskeleton[23]. Interestingly, the gating of MscL in blebs was almost identical to that observed in liposomes. To apply the same experimental paradigm to PIEZO1 channels, we created a green fluorescent protein (GFP) fusion construct of PIEZO1 (with the fluorophore inserted in the centre of the protein at position 1591), and showed that not only do these channels behave like wild-type (WT) PIEZO1 channels but they are present in the membrane blebs and can be activated by applied negative pressure to the patch pipette. Like MscL, PIEZO1 channels in blebs were more sensitive to applied tension than in cell-attached patches. Thus PIEZO1 appears to sense tension in the bilayer and is gated according to the "force-from lipids" principle, an evolutionarily conserved gating mechanism.

Results

Blebbing HEK293 cells. We first explored the best way to make stable blebs[30,31,33]. We focused on three solutions: (1) Hypoosmotic sodium gluconate solution (~140 mOsm), (2) Hyperosmotic sodium gluconate (~440 mOsm), and (3) KCl Ca^{2+} free solution[35]. The results were summarized by the number of cells that blebbed at various time points up to 12 h (Fig. 1b–e). The membrane integrity of cells after blebbing was monitored using a trypan blue assay. While the KCl Ca^{2+} free solution generated blebs, a high percentage of cells (>50%) stained with trypan blue after ~4 h. The most effective solution was hypoosmotic sodium gluconate with >60% cells exhibiting at least one bleb after 6 h. Furthermore, this blebbing seemed to be dependent on the activity of myosin II based on the fact that blebbistatin greatly reduced the number of blebs (Fig. 1e).

Cytoskeletal deficiency in HEK293 membrane blebs. Next we investigated whether the resulting blebs were deficient in cortical cytoskeleton. We assayed f-actin using an Alexa Fluor 568 phalloidin conjugate. We also visualized three GFP containing constructs expressed in HEK293 cells: a pIRES2–EGFP PIEZO1 where GFP is expressed separately from the PIEZO1 channel, and two fusion proteins, one of PIEZO1 and one of MscL (Fig. 2a and Supplementary Fig. 1). All the images show large rounded protrusions that grew up to 15 μm in diameter. It was simple to identify the GFP fluorescence of the fusion proteins at the membrane boundary. Figure 2b shows blebs from cells expressing pIRES2–EGFP PIEZO1 where free GFP was present in the cytoplasm and the cell was also stained with phalloidin. The arrows show blebs with GFP fluorescence but f-actin is not visible (Fig. 2b,c). In many cases, the blebs detached from the membrane surface and resembled liposomes as previously observed with Xenopus oocytes[33]. Using a β-tubulin-GFP, we showed that β-tubulin is also absent (Supplementary Fig. 2)[36].

Blebs present a functionally different membrane environment. To assess whether the blebs had a different mechanical environment for the channel than a cell-attached patch, we used the well-characterized prokaryotic mechanosensitive channel MscL. This channel has been extensively studied and is known to gate in response to bilayer tension[34]. We generated a carboxy (C)-terminal MscL fusion protein and characterized its electrophysiological properties and found that in cell-attached patches it activated with a pressure threshold of 144 ± 5 mm Hg ($n = 12$; ± s.e.m.; Fig. 3). The pressure threshold is defined as the pressure at which the first channel current appears. The threshold was higher in cell-attached than in inside-out patches where the cytoskeleton is more disrupted (120 ± 8 mm Hg; $n = 17$; Fig. 3a,c). The pressure threshold was also 30 ± 6% ($n = 12$) lower in excised patches where cell-attached activity as a reference was available before excision. The average number of channels per patch in excised patches was 13 ± 3 ($n = 17$). The pressure that gated the MscL–cGFP fusion protein was higher than reported for MscL in HEK cells (~90 mm Hg)[37]. This is likely due to modulation by the fusion of GFP to the C-terminal since the same effect was observed in bacterial spheroplasts[38]. In blebbed membranes, MscL activated at 56 ± 5 mm Hg ($n = 7$), close to the values reported for gating in liposomes[39] (Fig. 3b,c). Thus, the blebbed membrane is a different mechanical environment than the cell cortex and is more similar to liposomes rather than cell membranes. This emphasizes the mechanical influence of the

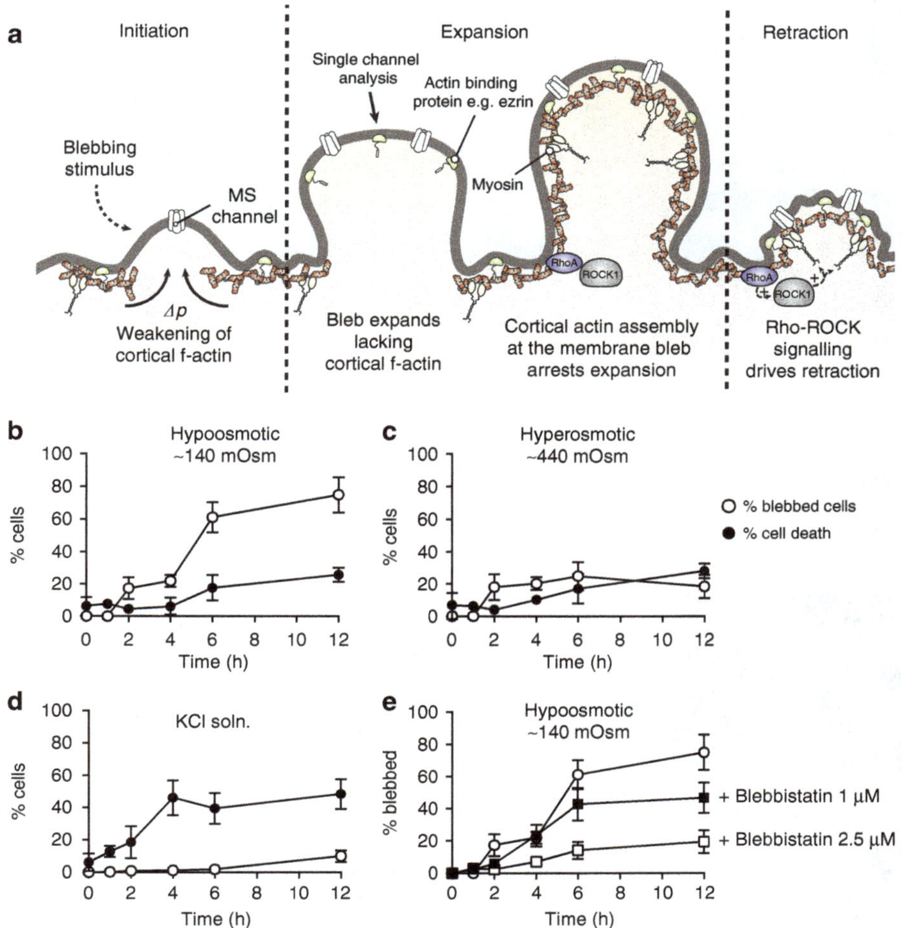

Figure 1 | Overview of membrane bleb formation and the efficiency of blebbing solutions on HEK293 cells including the effect on cell viability. (**a**) Formation of membrane blebs consists of three phases: initiation, expansion and retraction. Initiation of blebs can be instigated by a variety of stimuli, and certain cell lines continually bleb (for example, M12 cell line[68]). The initiation phase usually involves a focal weakening or rupture of the cortical cytoskeleton and bleb expansion continues largely devoid of cortical f-actin driven by hydrostatic pressure[69] and actomyosin contractility. Once polymerization of f-actin begins at the bleb membrane, expansion is halted. In the physiological setting, Rho-ROCK signalling then drives bleb retraction again via actomyosin contractility. Illustration of the number of cells blebbing and the corresponding % cells stained with trypan blue in response to treatment with (**b**) a hypoosmotic NaGluconate solution (~140 mOsm), (**c**) a hyperosmotic NaGluconate solution (~440 mOsm) and (**d**) a Ca^{2+} free KCl-based solution (first described for use in myocytes[35,70]). (**e**) Illustration of the dependence of blebbing on the activity of myosin II with almost complete abolition of bleb formation in the presence of 2.5 μM blebbistatin (data points represent mean ± s.e.m.; $n = 4$ with each replicate including >250 cells).

cytoskeleton on plasma membrane stress and reaffirms the functional presence of the cytoskeleton in excised patches[13,23,40].

We further probed the environment using the more-sensitive MscL mutant, MscL–G22S–cGFP (Fig. 4) where a hydrophilic residue in the hydrophobic pore causes a reduction in the energy required for gating (Fig. 4a,b)[41]. This channel showed a similar reduction in the pressure threshold for activation in the different membrane environments (Fig. 4b). The G22S mutant enabled us to activate multichannel currents in the cell-attached configuration without rupturing the patch (Fig. 4b–i). Interestingly, the MscL–GFP fusion constructs were not only confined to the plasma membrane but were also present in the membranes of organelles (Fig. 4e).

We measured the gating curve in three patch configurations (Fig. 4c) and found a consistent leftward shift indicative of increased tension in the bilayer of blebbed membranes for a given amount of pipette suction. A similar shift was observed with ramp pressure protocols (Fig. 4d,f,g). Ramp protocols allow more time for a redistribution of force than a pulse and this produced a larger shift in the $P_{1/2}$ for MscL–G22S–cGFP since stress in the cytoskeleton had more time to relax (Fig. 4h,i). This is depicted diagrammatically in Fig. 4c.

We further probed how mechanoprotection by the cytoskeleton affected the activity of MscL–G22S–cGFP by applying cytochalasin D (10 μM) and colchicine (10 μM) to HEK293 cells. In cells pre-treated with these cytoskeletal-disrupting agents, the pressure threshold was reduced in cell-attached and excised patches (Supplementary Fig. 3). As expected, the leftward shift (lower stress) in the gating curve elicited by excision was lost. Thus membranes treated with these compounds behave more like membrane blebs.

PIEZO1–1591–GFP fusion protein behaves as WT. After characterizing MscL in blebbed membranes, we repeated the experiments using a fluorescently labelled PIEZO1 fusion construct. The necessity for a fluorescent construct in this situation stems from the fact that some proteins can be excluded from blebbing membranes[42]. With GFP at either the amino (N)- or

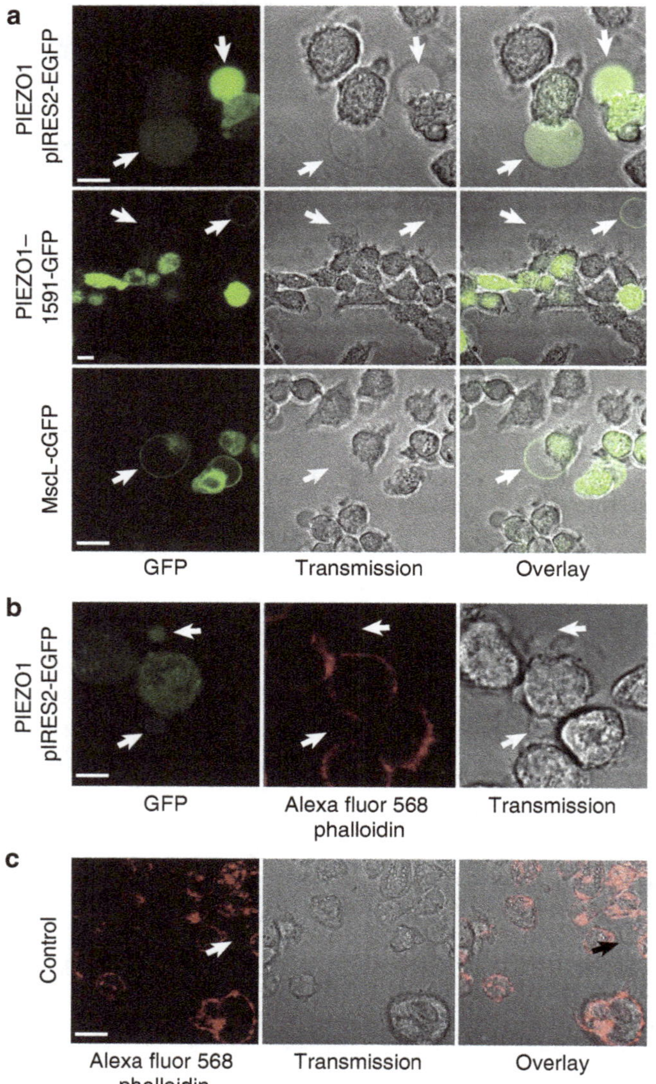

Figure 2 | Fluorescence imaging of HEK293 cell membrane blebs generated using a hypo-osmotic solution. (a) Example of pIRES2-EGFP PIEZO1 expression in HEK blebs. The blebs can be clearly seen with free GFP inside the cytoplasm. Blebs sometimes grew to diameters greater than 15 μM (upper row). GFP fusion proteins of PIEZO1 (middle row) and MscL (lower row) can also be seen in the bleb membrane (scale bar, 10 μm). **(b)** Illustration of staining of cells expressing a pIRES2-EGFP PIEZO1 construct. Arrows represent blebs that contain GFP but are devoid of f-actin staining (red; scale bar, 7 μm). **(c)** As a control, we also stained cells blebbed in the same manner and stained with phalloidin. Again, the arrow represents a bleb that is not stained and hence deficient in f-actin (scale bar, 20 μm).

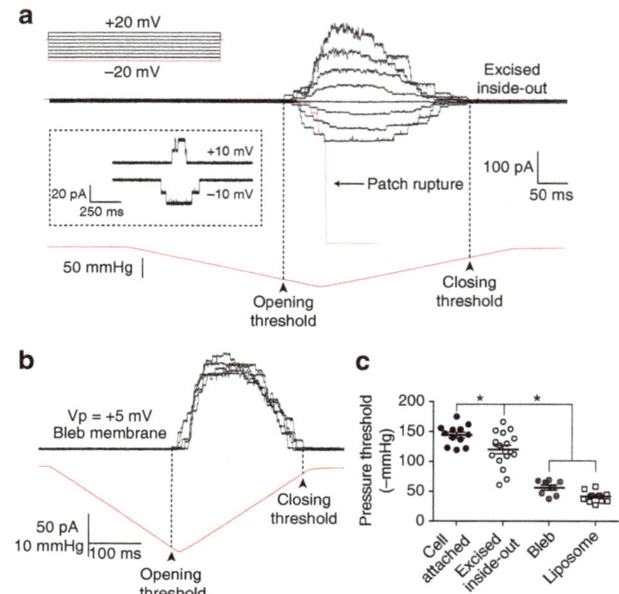

Figure 3 | Activation of MscL-GFP in bleb-attached patches clearly illustrates the different mechanical environments. (a) A family of MscL current activity in an excised inside-out patch in response to pressure ramps (300 s to peak) at voltages ranging from +20 to −20 mV. (Inset shows single channel activity elicited by manual pressure application at two opposing voltages). In symmetrical NaCl, 145 mM MscL gives a conductance of 1.57 nS in excised patches. The conductance is somewhat lower than in previous reports owing to a reduction in the bulk conductivity of this recording solution in comparison with the widely used KCl 200 mM and $MgCl_2$ 40 mM buffer. **(b)** Illustration of MscL activity in membrane blebs; traces show six sweeps of identical pressures. **(c)** Pressure thresholds of activation for WT MscL in four configurations; cell-attached, excised, bleb-attached and excised azolectin liposome (data points represent individual experiments with mean ± s.e.m. shown for comparison). Pressure thresholds in cell-attached and excised were some three times higher than those seen in bleb-attached patches (*P value <0.01; one-way analysis of variance, Tukey's *post hoc* test).

C-terminal of PIEZO1, we found that in the cell-attached configuration, the channel kinetics showed delayed inactivation relative to controls. In addition, the channel number per patch for the N-GFP fusion was much lower than WT, despite the presence of comparable amounts of protein. This may be a result of abrogated trafficking (Fig. 5a–c).

We then introduced mCherry or GFP into an internal site in an effort to maintain activity. We tested five internal sites that are putative loop structures (Fig. 5a,b). Surprisingly when we introduced mCherry to position 1591 (called PIEZO1-1591-mCherry), the kinetics were similar to the WT channel (Fig. 5b) with an inactivation time constant of ~40 ms in cell-attached mode. The other mutants displayed abnormal activity, for example mCherry inserted at position 1851 provided a response similar to that observed for N- or C-terminal modification as it did not inactivate and had small currents ($n=4$, Fig. 5b). Three mutants (at positions 160, 724 and 855) produced no current ($n=4$). We confirmed that three of these constructs (N-terminal, C-terminal and 1591 insertion) expressed full-length protein by probing with a GFP antibody (Fig. 5c). The response of PIEZO1-1591-mCherry using whole-cell recording is shown in Fig. 6, and the response was similar to the WT channel with a reversal potential near 0 mV and robust mechanically induced currents.

The PIEZO1-1591-GFP construct also behaved similarly to wild-type channels including a reduction in unitary conductance with increasing external Ca^{2+} and voltage-dependent inactivation (Fig. 7a–d)[5,43,44]. The kinetics of inactivation gave time constants similar to those previously reported for whole-cell mode ($\Delta V_{patch} = -55$ mV: $\tau = 55 \pm 11$ ms; $n=6$). The pressure sensitivity in cell-attached patches gave a $P_{1/2}$ of 44 ± 2 mm Hg ($n=6$), almost identical to the WT PIEZO1 channel $P_{1/2}$ of 45 ± 3 mm Hg ($n=5$). The slope sensitivity (the slope of the P_{open} curve versus pressure at the midpoint, $1/\alpha$) was also similar to the WT (0.10 ± 0.02 mm Hg^{-1} and PIEZO1-1591-GFP fusion; 0.15 ± 0.04 mm Hg^{-1}). The channel could also be gated by positive pressure in the cell-attached configuration as seen

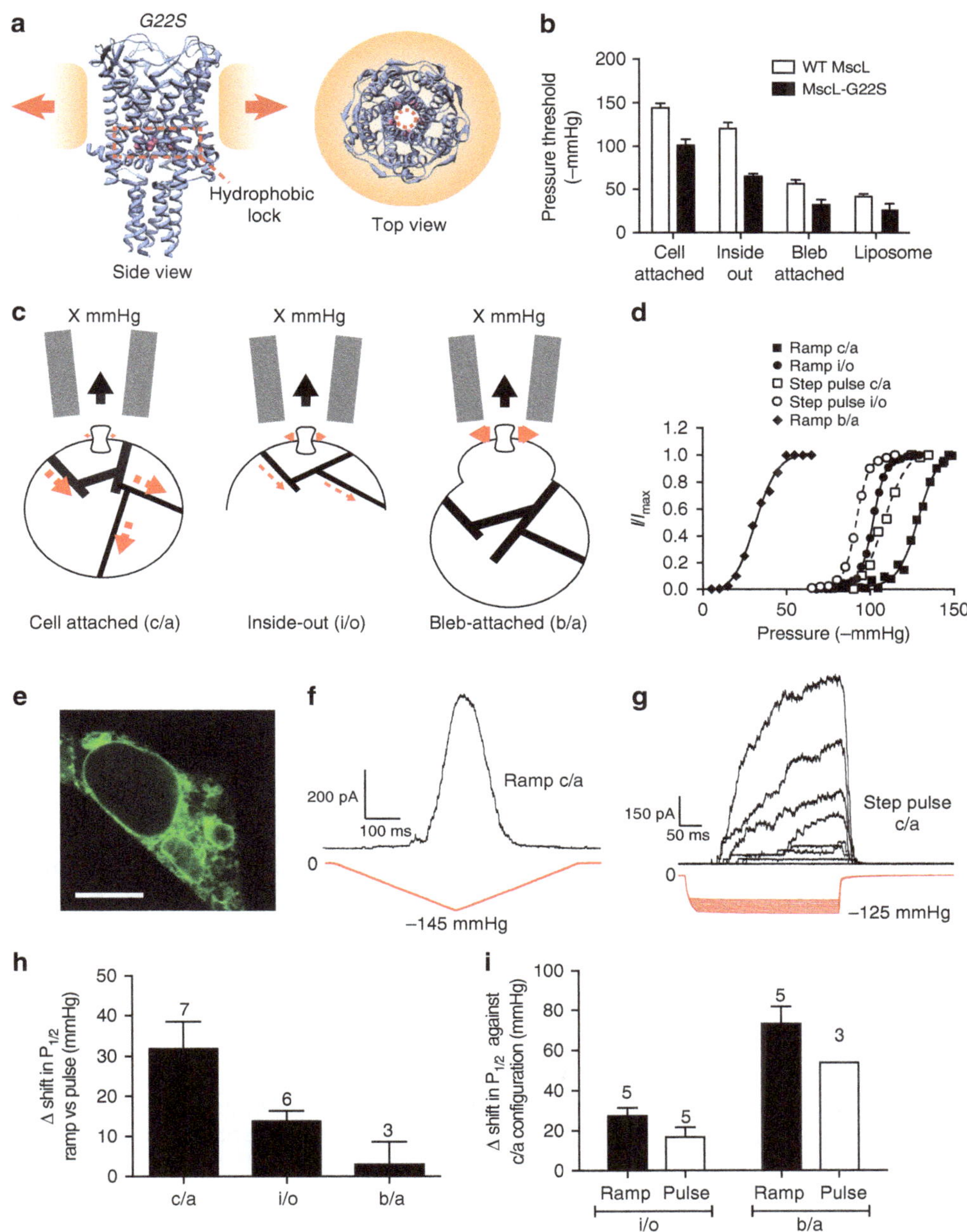

Figure 4 | Activity of MscL-G22S-cGFP construct in cells provides further evidence for the lack of cortical cytoskeleton in bleb membranes. (**a**) Illustration of MscL structure with reference to the hydrophobic lock where the G22S mutation is housed. (**b**) Comparison of the activation pressures of the G22S mutation with the WT MscL channel in various configurations. (**c**) Graphic illustration indicating that there is a progressive loss of cortical cytoskeletal involvement from cell-attached to bleb-attached configuration with a corresponding reduction in the ability of the cytoskeleton to redistribute the applied force. (**d**) Activity of G22S-MscL-cGFP in the three configurations in response to either a ramp (350 ms to peak) or a square wave pressure pulse. A leftward shift in the Boltzmann distribution is obvious, not only between configurations but also between ramps and pulses, both applied via a high-speed pressure servo. (**e**) Confocal image of MscL-G22S-cGFP expressed in HEK293 cells. The channel is not confined to the plasma membrane and seems to be also incorporated in many organelle membranes (scale bar, 10 μm). (**f**) Representative trace showing G22S-MscL-cGFP activity in response to a pressure ramp up to a peak of −145 mm Hg (350 ms to peak). (**g**) Representative trace showing G22S-MscL-cGFP activity in response to a pressure pulse of 350 ms duration. (**h**) Quantification of leftward shift of $P_{1/2}$ between ramp and pulse responses of G22S-MscL-cGFP. (**i**) Quantification of leftward shift of $P_{1/2}$ between ramp and pulse responses of G22S-MscL-cGFP in comparison with the cell-attached patches (c/a, cell-attached; b/a, bleb-attached; i/o, inside-out; data points throughout represent mean ± s.e.m.).

for other mechanosensitive channels gated by membrane tension[17,45,46] (Supplementary Fig. 4). In the following experiments, we used channels with the non-perturbing insertions of mCherry or GFP.

Recently the cryo-EM structure of murine PIEZO1 was solved and shown to be a trimer and along with biochemical data it provides structural clues to the functional effects that we saw upon fluorophore insertion (Fig. 5)[47,48]. For example, positions 724 and 855 where insertions give rise to non-functional channels appear to be in an extracellular region termed the 'blades', which have been suggested to be important in the sensing of mechanical stimuli. Both biochemical data and cryo-EM data suggest that a

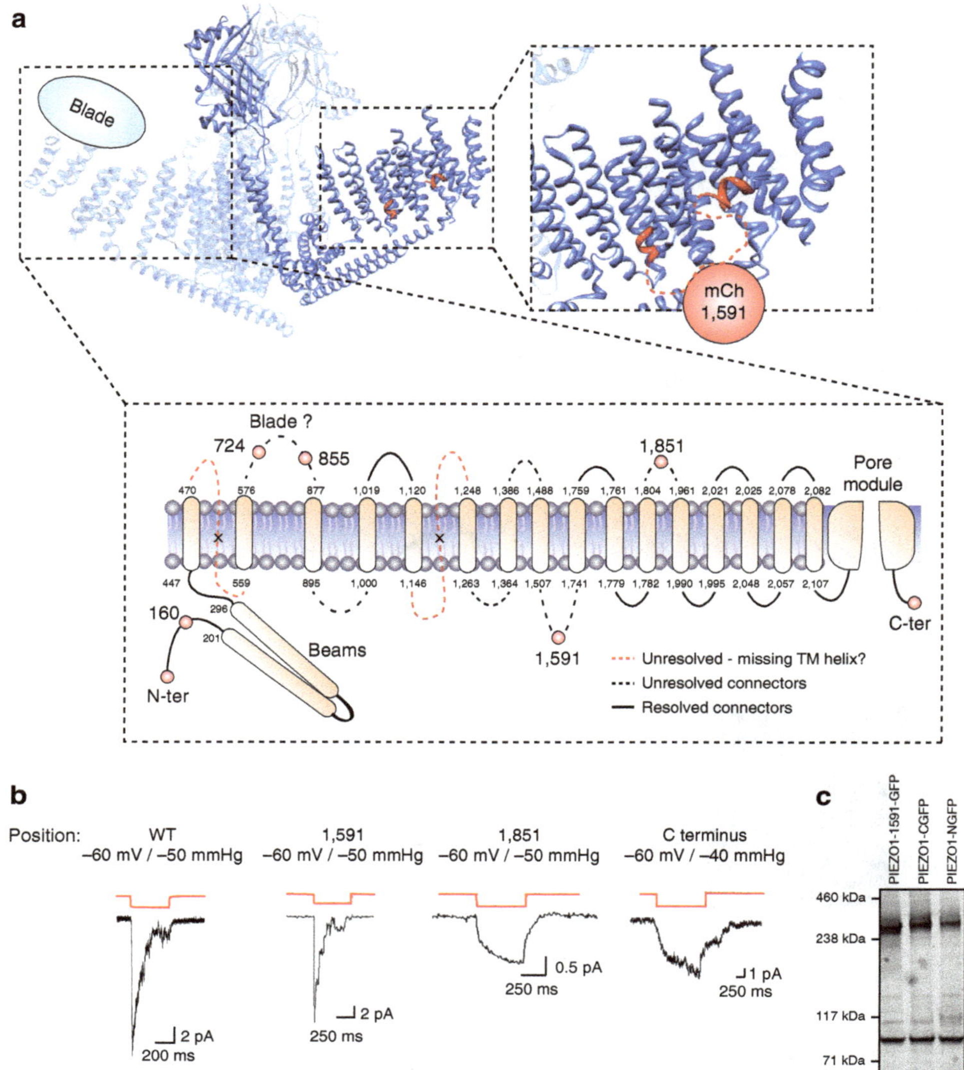

Figure 5 | Activity of different PIEZO1 fluorescent constructs using GFP and mCherry. (**a**) Cryo-EM structure of mouse PIEZO1 with an inset showing the likely position of the 1591 insertion (mCherry or GFP)[47]. In addition, we show the putative topology of the transmembrane segments and the positioning of our fluorophore insertions using biochemical data[48]. (**b**) To create a fluorescent GFP-PIEZO1 protein, we first attached GFP to the N- and C-terminal end of the channel protein. The response of the channel was disrupted as it did not inactivate, and gave rise to small currents (trace labelled C terminus). We then introduced mCherry fluorescent protein into the predicted loop positions indicated in **a**. We were unable to observe any channel response from channels with mCherry insertions at positions 160, 724 and 855. (**b**) However, the addition of mCherry into position 1591 (called PIEZO1-1591-mCherry) was similar to WT with an inactivation time constant of 40 ms. The introduction of the fluorescent protein at position 1851 produced low currents and the channel did not inactivate. (**c**) A representative western blot showing three PIEZO1-GFP fusion constructs (N- and C-linked PIEZO-GFP and PIEZO1-1591-GFP) express full-length protein ~300 kDa. Despite the lower number of channels encountered in the N- and C-terminal fusion proteins, total expression levels were similar. Lower band (~100 kDa) shows α-actinin for comparison ($n=3$).

fluorophore inserted at position 1591, is likely to be intracellular and arranged close to the periphery of each 'wing'. This gives a plausible reason as to why such large insertions can result in channel activity almost indistinguishable from WT.

Visualizing the fluorescent PIEZO1 constructs in HEK293 cells. With PIEZO1-1591-mCherry (250 ng) and TREK-1-GFP (50 ng) co-transfected into HEK-293 cells, images made with Structured Illumination Microscopy (SIM-API) (Supplementary Fig. 5) show that the two mechanical channels are in separate domains of differing sizes. In some regions of the cell, TREK-1 appears in bead-like chains suggesting that they are organized along cytoskeletal fibres. The association of TREK-1 to the underlying actin cytoskeleton has been noted previously[46].

PIEZO1 in bleb membranes. We activated PIEZO1-1591-GFP in bleb-attached patches (Fig. 8a) and quantitative analysis showed that the basal activity was higher than that in cell-attached patches indicative of pretension resulting from gigaseal formation[15]. While 'gating at rest' is often encountered in cell-attached patches, it is not as pronounced as in bleb-attached patches (Fig. 8a–c). The adhesion energy of membranes to the glass pipette creates baseline tension and this tension background is common to all patches[13]. As seen with MscL, channel activation in blebs showed a leftward shift in the gating curve

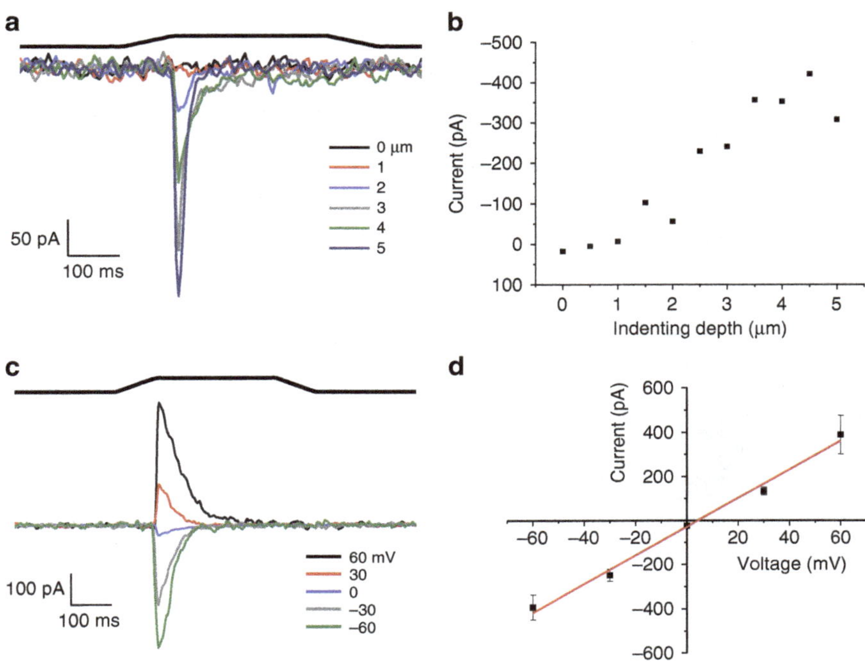

Figure 6 | Whole-cell recording for PIEZO1-1591-mCherry on HEK293 cells produce robust currents. (**a**) Whole-cell currents elicited by pressing on the cell with a glass probe to the indicated depths. The membrane voltage was set to −60 mV. Above the current trace is the stimulus waveform. (**b**) Current plotted as a function of depth, which was incrementally increased at 0.5 micron intervals. Increased depth increases the current. (**c**) Currents recorded when the probe is set to single depth at varying membrane potentials. (**d**) Current as a function of voltage (from **c**) showing that the reversal potential is near 0 mV as in the wild-type PIEZO1 channel (data points represent mean ± s.e.m.; $n = 5$).

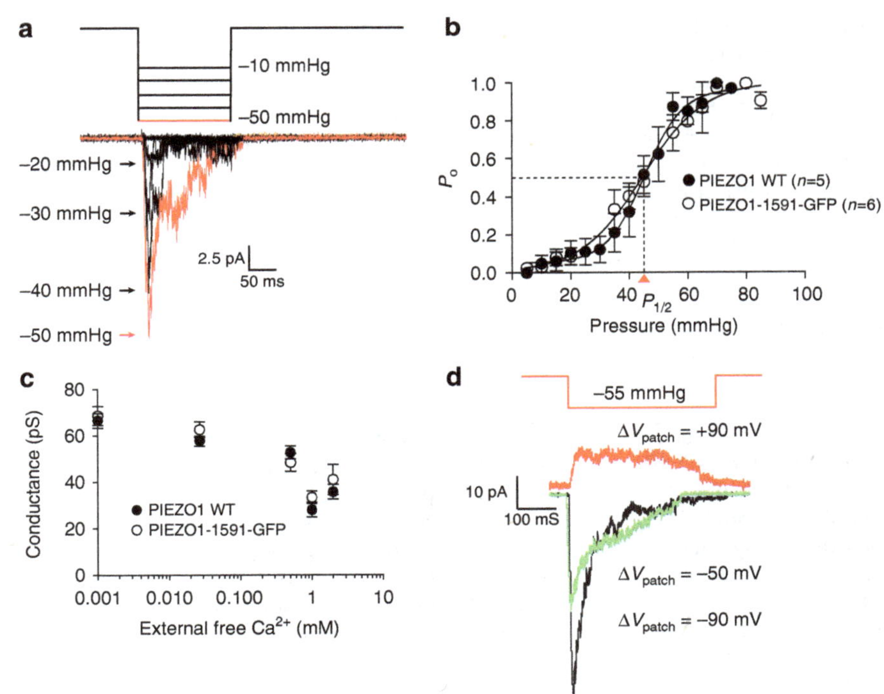

Figure 7 | PIEZO1-1591-GFP fusion protein behaves similar to WT PIEZO1. (**a**) PIEZO1-1591-GFP currents elicited from a stepwise increase (−10 to −50 mm Hg) in pressure of 150 ms duration. The peak current generated from each pulse is labelled for clarity. (**b**) Open probability (P_o) plotted against pressure pulse magnitude for PIEZO1-1591-GFP and WT PIEZO1. Here P_o is estimated as I/I_{max} from peak currents with increasing pressure pulses. (**c**) Effect of raising external Ca^{2+} concentration on single-channel conductance of PIEZO1-1591-GFP. Initial conductance in CsCl with ~1 μM free Ca^{2+} is ~62 pS (buffered using EGTA and calculated using Ca-EGTA Calculator v1.3, an online EGTA Ca^{2+} chelator calculator), and is reduced by almost half by 1 mM external Ca^{2+} (data points represent mean ± s.e.m.; $n = 4$). (**d**) Example of PIEZO1-1591-GFP voltage-dependent inactivation at three voltages. Inactivation markedly slows with depolarization as seen in WT PIEZO1 (for comparison with previous results: $\Delta V_{patch} = -55$ mV; $\tau = 55 \pm 11$ ms, $n = 6$).

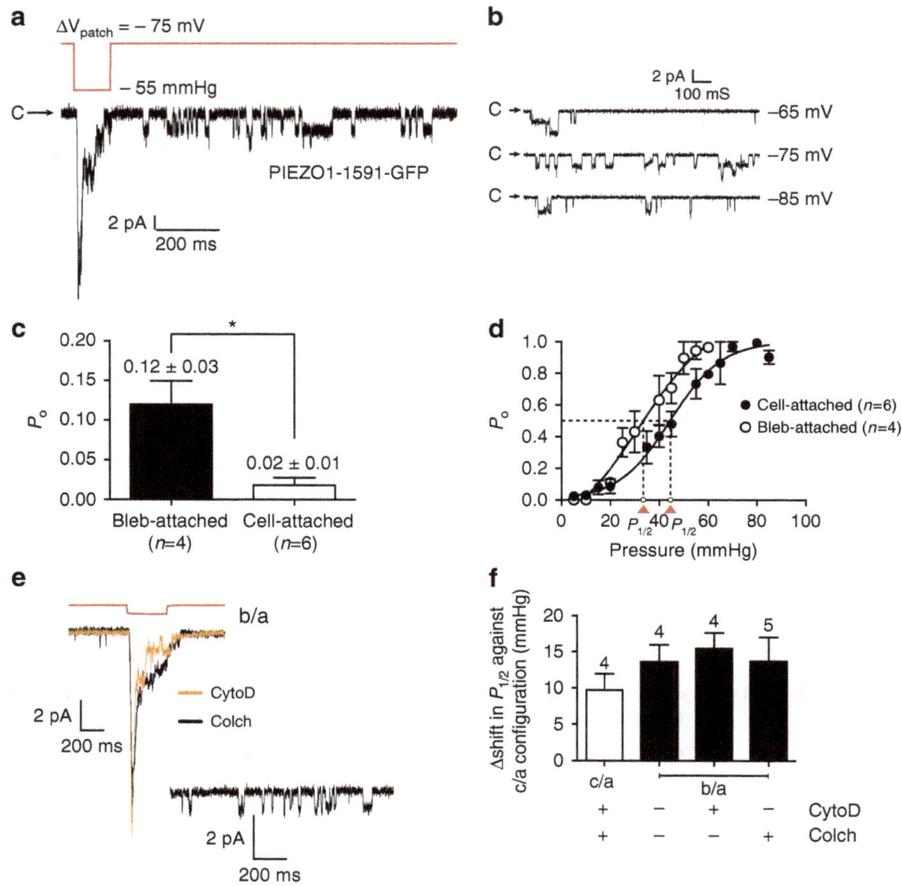

Figure 8 | Activation of the PIEZO1-1591-GFP fusion protein in bleb-attached patches in comparison with cell-attached. (**a**) Mechanosensitivity of PIEZO1-1591-GFP in bleb-attached patches in response to a 150 ms square wave pressure pulse of – 55 mm Hg (pipette: CsCl). (**b**) Single-channel activity at rest as seen post pressure at three different voltages. (**c**) The increase in P_o at zero pressure is calculated using 5 s records. There is at least a fivefold increase in basal activity (*$P<0.01$; Student's *t*-test). (**d**) Leftward shift in Boltzmann distribution function in bleb-attached patches (cell-attached $P_{1/2}=44 \pm 2$ mm Hg; $n=6$: bleb-attached $P_{1/2}=33 \pm 3$ mm Hg; $n=4$). (**e**) Example traces of PIEZO1-1591-GFP activated in blebbed membranes pre-treated with Cytochalasin D (CytoD) 10 µM and colchicine (Colch) 10 µM. Cells were treated for 1 h pre-blebbing, then blebbed in the presence of these agents for 2 h using a hypo-osmotic solution. Final patch-clamp electrophysiology was carried out immediately with the bath solution again containing the same pharmacological agents. (**f**) Quantification of leftward shift of $P_{1/2}$ between cell-attached patches and bleb-attached patches with the identified cytoskeletal interfering agents (c/a, cell-attached; b/a, bleb-attached; data represents mean ± s.e.m.; $n=4/5$).

($P_{1/2}=33 \pm 3$ mm Hg, $n=4$) relative to cell-attached patches. The slope sensitivity (0.15 ± 0.04 mm Hg^{-1}) was the same as the WT suggesting the closed–open conformational change was the same (0.18 ± 0.05 mm Hg^{1}; Fig. 8d).

To further compound the lack of cortical cytoskeleton in the blebbed membranes, we pre-treated cells with cytochalasin D (10 µM, $n=4$) and colchicine (10 µM, $n=5$) to see whether the activity of PIEZO1-1591-GFP was modified (Fig. 8e). Treatment with neither agent modified the gating curve implying that bleb formation disrupts most of the structural cytoskeleton (Fig. 8f). Consistent again with MscL, and as further evidence of mechanoprotection in the cell-attached patches co-treatment of cell-attached patches with both cytochalasin D (10 µM) and colchicine (10 µM) reduced the activation pressures for PIEZO1-1591-GFP (Fig. 8f).

Estimating the tension sensitivity of gating PIEZO1. Thus far, we have shown that PIEZO1 can be gated in bleb-attached patches in the absence of the cytoskeleton from which we can strongly suggest that the channel is gated by bilayer tension. The next question is what is the absolute tension sensitivity of PIEZO1? We co-transfected MscL-G22S-cGFP and PIEZO1 and measured the $P_{1/2}$ for gating ($P_{1/2}$ is the pressure at which $P_{open}=1/2$ of saturation). The ratio of $P_{1/2}$ was ~0.4, that is, the $P_{1/2}$ of WT PIEZO1 was 40% of the $P_{1/2}$ for MscL-G22S-cGFP in cell-attached patches (Supplementary Fig. 6). We examined the tension required to gate MscL-G22S-cGFP in excised inside-out patches that are easily imaged using confocal microscopy. We activated MscL-G22S-cGFP with 2 s pulses of negative pressure of increasing magnitude and monitored the corresponding patch deformation (Fig. 9a,b) allowing us to use Laplace's law to determine mean tension in the patch dome[13,23]. Under these conditions, the channels began to gate at ~8.5–10 mN m^{-1} with a $T_{1/2}$ (tension where $P_o=0.5$) of 11.3 ± 0.6 mN m^{-1} ($n=3$, Fig. 9c). This corresponds to a free energy of gating, $\Delta E=40 \pm 4$ kT ($n=3$, the energy required to open the channels with no applied tension) which corresponds to an in-plane area change (ΔA) of 15 ± 2 nm^2. This large expansion is characteristic of mechanosensitive channels. From the co-expression data, we estimate that the tension at $P_{1/2}$ is ~4.5 mN m^{-1} for PIEZO1.

To validate this estimate, we expressed PIEZO1 channels in HEK293 cells with a bicistronic plasmid also expressing GFP to sharpen the cell boundaries (Fig. 9c–e). In the cell-attached

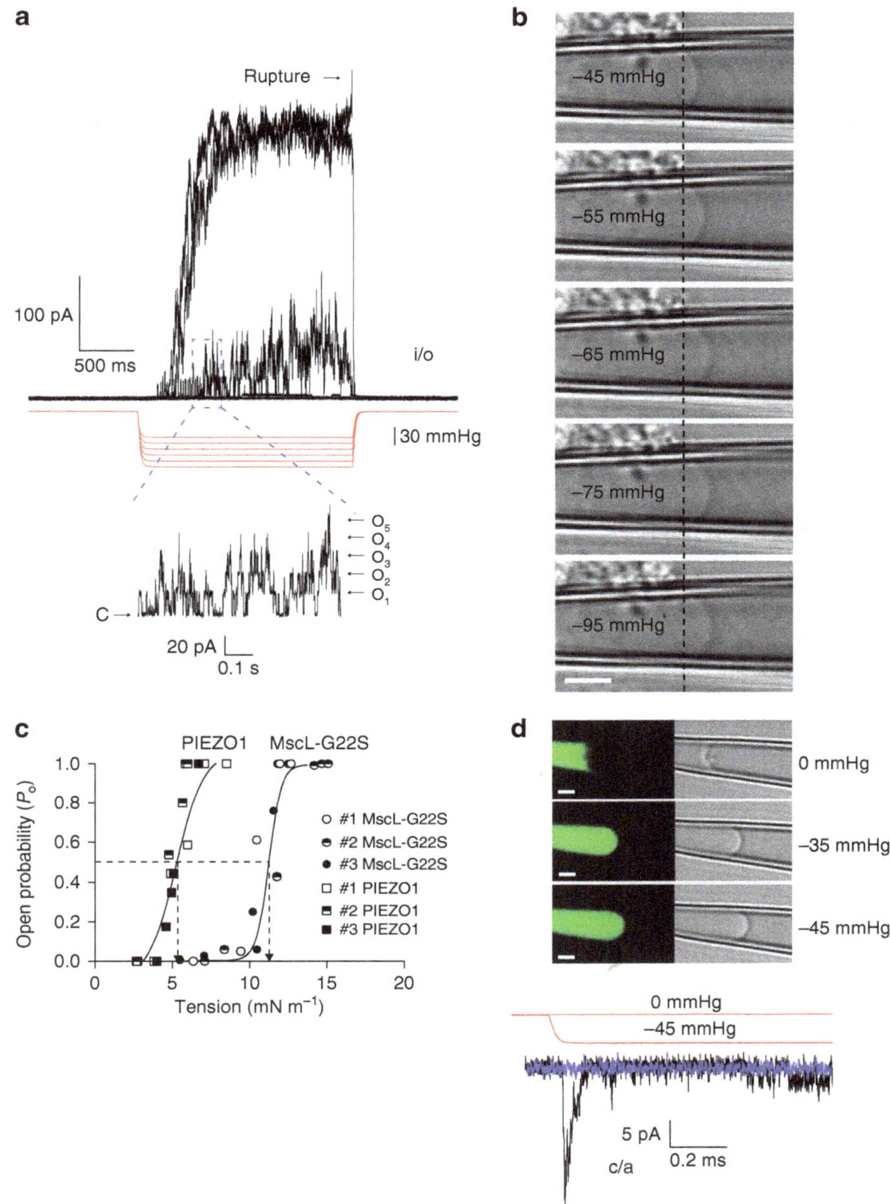

Figure 9 | Estimating the tension required to gate PIEZO1 channels. (**a**) Family of MscL-G22S-cGFP channel currents elicited in an excised inside-out patch at +10 mV pipette potential. Inset shows enlargement of a segment of one sweep clearly documenting single MscL channel transitions. (**b**) Confocal images of the corresponding patch membrane and its deformation over time under the negative pressures. The images corresponding to the pressure steps seen in **a** allow calculation of membrane tension using Laplace's law ($T = Pr/2$; where T is tension, P is the applied pressure and r is the radius of patch curvature). Laplace's law provides an upper limit for the tension sensitivity. (**c**) A Boltzmann distribution is shown for three independent experiments illustrating how membrane tension is linked to channel open probability. The same analysis is shown for PIEZO1 using data accrued from the cell-attached patches as shown in **d**. The fluorescence is that of GFP being expressed on the same plasmid as used to deliver the WT PIEZO1 channels. The patch-clamp recording below the visualized patch membranes shows the corresponding currents elicited by membrane deformation (blue, 0 mm Hg; black, −45 mm Hg; $\Delta V_{patch} = -55$ mV; scale bar, 2 μm in all the images).

configuration, $T_{1/2}$ of PIEZO1 was 5.1 ± 0.2 mN m^{-1} ($n = 3$, Fig. 9c–e), and $\Delta E = 9.7 \pm 1.5$ kT ($n = 3$) with $\Delta A = 8 \pm 1$ nm^2.

Discussion

Our experimental strategy of generating membrane blebs has followed the study of mechanosensitive channels by Zhang *et al.*[33], using Xenopus oocytes[30,31]. The blebs are clearly deficient in cytoskeletal components including f-actin and β-tubulin[25,26]. As an internal calibrator of tension, we used MscL that has been calibrated in liposomes. The pressure sensitivity in bleb membranes was threefold less than cell-attached patches and is close to that seen in liposomes[39]. This demonstrates the protective effect of the cytoskeleton on membrane-embedded proteins, that is, 'mechanoprotection' and emphasizes that any intervention that alters cytoskeletal structure is likely to affect channel gating[12,13,49,50]. MscL channels in excised inside-out patches from cells required more pressure (2.5 times the pressure) to gate than in blebs. Excision interferes with, but does not completely disrupt tension sharing by the cytoskeleton[51–53]. This is supported by pretreatment with colchicine and cytochalasin D, which also reduced the force required to gate MscL in both cell-attached and excised configurations. The imaging of fluorescent PIEZO1 and TREK-1 showed that the two channel types are segregated to separate domains. A critical consequence is that the tension within each domain may be different since the line tension of the domains requires that the internal tension is different from external tension[54]. An interesting feature of the TREK-1 distribution was an apparent alignment of some channels to underlying linear cytoskeletal elements. This interaction is supported by the observation that a non-conducting point mutation in TREK-1 has a dramatic effect on the cytoskeletal reorganization[46]. Although these data do not unequivocally support the idea that PIEZO1 resides in a 'cytoskeleton free' domain, they do illustrate that PIEZO1 resides in a different cellular domain to TREK-1. This is interesting and fits with the proposed functional crosstalk between these two types of channels[17]. However, ultimately TREK-1 as well as the related protein TRAAK, sense force transmitted directly through the lipid bilayer despite their previously affirmed association with the cytoskeleton[16,17]. If PIEZO1 channels were gated by bilayer tension, like TREK-1, TRAAK[16,17], MscL and MscS-like channels[55], we would expect to see a similar trend in its activity when compared with MscL in bleb-attached patches, and this is what is found. PIEZO1 can be activated in bleb membranes with less applied pressure than in cells. Furthermore, this is unaffected by interfering with either actin polymerization (CytoD) or microtubule polymerization (colchicine) providing further support for an absence of cytoskeletal linkages in blebbed membranes. However, if PIEZO1 is gated by bilayer tension, an obvious question that arises is: 'does purified PIEZO1 protein exhibit mechanosensitive gating when reconstituted?'[2]. The answer thus far seems to be no and there are a number of possible explanations for this. The data from planar bilayers do not exhibit inactivation kinetics[2]. This could arise because the data actually came from occasional channels returning from the inactivated (tension insensitive) state since the resting tension in bilayers is on the order of 4–6 $\mathrm{mN\,m^{-1}}$. Alternatively, as others have reported, we saw rapid rundown of PIEZO1 channel currents in the excised inside-out configuration, and recent work has shown that this rundown is correlated with a loss of phosphoinositides, in particular PIP2 (ref. 56). The lack of PIP_2 in reconstituted bilayers may explain the lack of response to applied membrane tension documented in this environment[2]. The fact that PIEZO1 channels can be activated in blebs virtually eliminates the possibility that activation requires binding to the cytoskeleton and biochemical data have shown no specific association of PIEZO1 with other proteins[2]. Future work should aim to address the question as to why does purified PIEZO protein not exhibit mechanosensitive gating when reconstituted; answering such a central question will no doubt uncover important functional information.

It is important to note that our work does not preclude modulation by structural proteins of the cytoskeleton or extracellular matrix. For example, PIEZO1 is sensitized by the expression of STOML3, a cytoskeleton-associated protein[57], and this may well alter the force distribution to the bilayer[49]. In fact we have already shown that PIEZO1 currents elicited by cell indentation are markedly reduced by pre-treatment with cytochalasin D. This, however, is a result of force transmission in this configuration and not a direct link to the channel as when we pre-stressed the bilayer with hypo-osmotic shock in the presence of cytochalasin D, the currents were boosted[43].

Here we have provided the first direct assessment of the tension sensitivity of PIEZO1. The first method utilized the gating of MscL–G22S–cGFP channels in inside-out patches ($T_{1/2} = 11.3 \pm 0.6\ \mathrm{mN\,m^{-1}}$; $n = 3$). We then used this value to estimate the tension required to gate PIEZO1 channels from the pressures required to gate PIEZO1 when co-expressed with MscL–G22S–cGFP. This gives a $T_{1/2}$ of $\sim 4.5\ \mathrm{mN\,m^{-1}}$. We confirmed this by imaging the patch geometry and estimated $T_{1/2} = 5.1 \pm 0.2\ \mathrm{mN\,m^{-1}}$ ($n = 3$). Applying Laplace's law to heterogeneous three-dimensional structures is a simplification of the complex material properties of a cell[13,58], but it does provide a useful upper limit of the magnitude of the forces required to gate the channel. Furthermore, an inherent characteristic of mechanosensitive channels is that they undergo large in-plane area expansion (ΔA) during gating allowing the tension field to do work on the channel. From the tension and the free energy difference between the closed and open states, we estimated $\Delta A \sim 15\ \mathrm{nm}^2$ for MscL–G22S–cGFP fitting well with previous estimates[59,60]. PIEZO1 also appears to undergo a large in-plane expansion of $\sim 8\ \mathrm{nm}^2$ (ref. 5). These calculations assume that the free energy of gating is driven purely by a change of in-plane area and that may not be completely accurate[61].

The quantification of PIEZO1 gating may be used as an internal calibrator of how efficiently a stimulus can reach the channel. Our data also emphasize that drug screens of PIEZO channels are predicted to be sensitive to cytoskeletal modulation and lipid modulation that can change the distribution of stresses between the bilayer, the cytoskeleton and the extracellular matrix. Collectively, these results illustrate the utility of prokaryotic mechanosensitive channels as models and tools for studying mechanosensory transduction, and the wide applicability of basic biophysical principles in the gating of mechanosensitive channels irrespective of their evolutionary provenance.

Methods

PIEZO1 and MscL GFP fusion constructs. *Construction of C- and N-terminally labelled PIEZO1.* Two vectors, pNGFP-EU and pCGFP-EU (gift of Eric Gouaux), were used to introduce an EGFP-fusion label at either the N- or C-terminal of *PIEZO1*, respectively. For pNGFP-EU, *PIEZO1* was inserted between the HindIII and BamHI sites. To achieve this, we amplified the gene with a forward primer having a HindIII restriction site and a reverse primer having a BamHI site with Prime star GXL DNA polymerase (Clontech/Takara). The DNA was purified by ZymoResearch Clean Kit according to manufacturer's specifications. The vector pNGFP-EU was treated with BamHI and HindIII restriction endonuclease and purified by ZymoResearch Clean Kit. *PIEZO1* DNA (100 ng) and treated vector (30 ng) were assembled by InFusion (Clontech/Takara). Stellar cells were transformed after 15 min. Colonies were analysed by restriction analysis and then by sequencing.

A similar protocol was used to covalently attach EGFP to the C-terminal of PIEZO1. We amplified *PIEZO1* with primers carrying HindIII and EcoRI restriction sites as described above. pCGFP-EU was prepared by treatment with HindII and EcoRI restriction endonucleases. The vector and *PIEZO1* was assembled by InFusion as described above.

Construction of internally labelled PIEZO1. We inserted the fluorescent protein, mCherry, into five internal sites at amino acid positions 160, 724, 855, 1591 and 1851. This was achieved by first introducing AgeI and SpeI sites into DNA encoding PIEZO1 at the indicated positions using QuikChange II XL Mutagenesis kit (Agilent).

All clones were analysed by first isolating plasmid DNA using the ZymoResearch mini prep kit according to the manufacturer's specification and restriction endonuclease digestion. To insert DNA coding for mCherry fluorescent protein into each AgeI/SpeI site, we amplified the DNA for the mCherry protein with primers that added AgeI (forward) and SpeI (Reverse) restriction sites at the ends.

Plasmid and mCherry PCR products were cut with AgeI and SpeI and then purified. Ligation of the mCherry DNA and each plasmid (3:1 or 10:1 concentration) was achieved using Quick Ligase, (New England Biolabs). Each reaction was transformed into Stellar cells with indicated antibiotic for selection. The plasmid DNA was purified and sequenced.

The expression of EGFP from the pIRES vector was silenced by a frameshift mutation using Agilent's QuikChange II XL kit with 100 ng of template DNA and with an extension time of 21 min.

To replace the mCherry protein with the GFP fluorescent protein at position 1591, we used the above primers and amplified the DNA from the pIRES-EGFP vectors. The Plasmid PIEZO1–1591–mCherry was cut with SpeI and AgeI to remove the DNA that expresses the mCherry protein and purified by gel electrophoresis. The amplified DNA for GFP was treated with SpeI and AgeI and purified. The final products were ligated and transformed (3–4 µl) into Stellar cells and the plasmid DNA isolated and sequenced.

MscL with C-terminal-fused EGFP constructs were cloned by PCR, using PfuUltra (Agilent), for insertion into pTRE-Tight (Clontech) EcoRI and HindIII restriction sites. The MscL forward primer contained the following Kozak sequence 5′-gccAccATGGcg-3′ after the EcoRI site, which consequently inserted an alanine following the MscL methionine 1 ATG start site. An EGFP with an N-terminal thrombin cleavage site construct contained within the laboratory was used as the PCR template for EGFP. The reverse MscL primer, and the forward thrombin-site EGFP primer both contained an in-frame NheI restriction site that was utilized to ligate MscL to EGFP while both inserts were ligated together into pTRE-Tight. MscL mutants were cloned by the same method using a previously made mutant (MscL-G22S) contained within the laboratory as MscL PCR template. The final pTRE-Tight MscL–T–eGFP construct insertions were sequence verified. Plasmid maxipreps were prepared using a PureLink kit (Life Technologies), with a NucleoBond Finalizer (Macherey–Nagel) using 1 ml TE for elution.

Transient transfection. HEK293 cells were transiently transfected with 250–650 ng of cDNA using Lipofectamine 3000 in OptiMEM. Electrophysiological analysis was undertaken 24–72 h post transfection.

Bleb formation. Blebs of HEK293 cells were formed using a modified Ca^{2+} free KCl solution (in mM: 140 KCl, 10 EGTA, 1 $MgCl_2$, 15 HEPES pH 7.2 adjusted with KOH)[35], hyperosmotic sodium gluconate solution ($\sim$440 mOsm) or a hypo-osmotic sodium gluconate solution ($\sim$140 mOsm). Sodium gluconate solutions were diluted with ddH_2O from stock containing in mM: 350 sodium gluconate, 2 $CaCl_2$, 3 KCl, 15 glucose and 20 HEPES pH 7.2, adjusted with NaOH. Blebs took between 45 min and 2 h to appear depending on the solution used. For electrophysiological recordings, cells were plated out onto 35 mm circular dishes (FluoroDish, World Precision Instruments, Inc) coated with poly-L-lysine and were treated for 3 h using a hypo-osmotic solution to induce blebbing. After this, blebs were directly patched using the recording solutions mentioned for up to 45 min. No bleb retraction was seen during this period of time.

Liposome reconstitution. MscL proteoliposomes were prepared using the dehydration/rehydration method. This methodology is a modified procedure from[39,45,62]. WT MscL for reconstitution was purified using immobilized metal affinity chromatography[63]. MscL-G22S protein for reconstitution was expressed by cell-free expression using an automated system (Exiprogen, Bioneer, Daejeon, Korea) that also performed the protein purification steps after expression. Ten micrograms of expression plasmid was added to the kit used (EC-1, Bioneer), with a previously used detergent combination of 6 mM Brij-58 detergent (Sigma) added to the expression well, and 1% n-Octyl-β-D-glucopyranoside (OG) (Anatrace) added to the binding/wash buffer wells, as well as the elution buffer well[64]. Protein concentration was estimated by SDS–polyacrylamide gel electrophoresis (SDS–PAGE) with staining using SimplyBlue SafeStain (Life Technologies).

Patch-clamp recording. Transiently transfected HEK293 cells were plated on coverslips for patch-clamp analysis at a density of $\sim$3,000 cells per coverslip. Coverslips were placed in a recording chamber containing 145 mM NaCl, 3 mM KCl, 1 mM $MgCl_2$ and 10 mM HEPES (pH 7.2) adjusted using NaOH. In cell-attached and bleb-attached recordings, the pipette solution contained either 145 mM CsCl or 145 mM NaCl with 10 mM HEPES (pH 7.2) adjusted using the respective hydroxide. EGTA was added to control levels of free pipette (extracellular) Ca^{2+} using an available online EGTA calculator—Ca-EGTA Calculator TS v1.3—Maxchelator. Negative pressure was applied to patch pipettes using a High Speed Pressure Clamp-1 (ALA Scientific Instruments) or recorded in millimeters of mercury (mmHg) using a piezoelectric pressure transducer (WPI, Sarasota, FL, USA). Borosilicate glass pipettes (Sigma, St Louis, MO, USA) were pulled using a vertical pipette puller (PP-83, Narashige, Japan) to produce electrodes with a resistance of 3.5–5.0 MΩ and coated with Sylgard to within $\sim$100 μm of the tip. Single-channel PIEZO1 currents were amplified using an AxoPatch 200B amplifier (Axon Instruments), and data were acquired at a sampling rate of 10 kHz with 1–2 kHz filtration and analysed using pCLAMP10 software (Axon Instruments). Boltzmann distribution functions shown in Figs 4, 6 and 9 describe dependence of single MscL and PIEZO1 channel currents and open probability, respectively, on the negative pressure applied to patch pipettes. The Boltzmann plots were obtained by fitting open probability $P_o \sim I/I_{max}$ versus negative pressure using the expression $P_o/(1-P_o) = \exp[\alpha(P-P_{1/2})]$, where P is the negative pressure (suction) [mmHg], $P_{1/2}$ is the suction at which $P_o = 0.5$, and α [$mmHg^{-1}$] is the slope of the plot $\ln[P_o/(1-P_o) = [\alpha(P-P_{1/2})]$ reflecting the channel mechanosensitivity[65].

The membrane tension T was estimated using Laplace's law ($T = Pr/2$) by measuring the radius of membrane patches r at applied pressures P (For a more detailed explanation, see supplement of ref. 66).

Whole-cell recording. In whole-cell recording mode, cells were mechanically stimulated by pressing on the cell with a fire-polished glass pipette (diameter of 2–4 μm) positioned at an angle of 30° with respect to the cover glass[5]. The probe was coarsely positioned $\sim$20 μm above the cell with a MP-285 manipulator (Sutter Instruments Co.), and from that position, the probe was moved up and down with a trapezoidal waveform by a piezoelectric stage (P-280.20 XYZ NanoPositioner, Physik Instrumente). Control of the depth was controlled by LabVIEW software with 40 nm resolution. The probe velocity was constant at 0.15 $\mu m\, ms^{-1}$ during transitions, and the stimulus was held constant for 300 ms. All currents were recorded at room temperature. The experiments were performed using an Axopatch 200B amplifier (Axon Instruments) sampled at 10 kHz and filtered at 1 kHz. Data acquisition and stimulation were all controlled by QUBIO software.

Confocal microscopy. Cells were grown on 35-mm-diameter cell culture dishes (FluoroDish, World Precision Instruments, Inc) coated with poly-L-lysine (Sigma, 5 $\mu g\, cm^{-2}$) and subjected to blebbing solutions identical to those used for electrophysiological study. Cells were then fixed for 15 min using 4% paraformaldehyde at room temperature. Subsequently, control cells or those expressing PIEZO1 or MscL GFP fusion proteins, were permeabilized using 0.5% Triton X-100 (Sigma) for 2 min and stained for 20 min at room temperature with Alexa Fluor 568 conjugated phalloidin (Sigma). Confocal images were made using an inverted confocal microscope (LSM 700; Carl Zeiss) equipped with a water immersion long working distance objective ($\times$63; NA 1.15; Carl Zeiss). Both 488 and 555 nm lasers were used to excite the respective fluorophores (GFP, Alexa Fluor 568).

SIM microscopy. Plasmids for PIEZO1–1591–mCherry and TREK-1–EGFP were transfected into HEK293 cells, and allowed to incubate for 24 h. Cells were then washed twice with phosphate-buffered saline (PBS) and fixed with a 4% solution of paraformaldehyde in PBS for 30 min. After removal of the fixing solution, cells were washed and mounting media, Prolong Gold (Life Technologies), was applied to the cells. These were subsequently inverted and sealed on a glass coverslip. Data were collected by SIM (API, Seattle) with a $\times$60 objective NA 1.4 oil. Data were collected (24 z-stack) and reconstructed into a three-dimensional model using Vaad3D software[67].

Western blotting. HEK293 cells were transiently transfected in 24-well plates using Liposome-based transfection reagent Lipofectamine 3000 (Life Technologies, Carlsbad, CA, USA) to deliver PIEZO1 cDNA into the cells. Cells were lysed in a lysis buffer containing 1% NP-40 (Sigma-Aldrich, Castle Hill, NSW, Australia) and protease inhibitors (Roche, Cromer, NSW, Australia) after 48 h and equal volume of lysate loaded for SDS–PAGE and western blot analysis. The nitrocellulose membranes were probed simultaneously with a rabbit polyclonal anti-GFP antibody at a 1:5,000 dilution (Abcam, Cambridge, UK; Cat no—ab290) and a mouse monoclonal anti α-actinin antibody at 1:1,000 dilution (Santa Cruz Biotechnology, Dallas, TX, USA Cat no—sc17829) overnight. Both anti-mouse IRDye800 and anti-rabbit IRDye680 (Li-Cor Biotechnology, Lincoln, NE, USA) at a 1:20,000 dilution were incubated with the membrane and the PIEZO1 proteins were detected using the Li-Cor Odyssey system (Li-Cor Biotechnology). Western blot images were produced using ImageStudioLite (Li-Cor Biotechnology).

References

1. Coste, B. *et al.* Piezo1 and Piezo2 are essential components of distinct mechanically activated cation channels. *Science* **330,** 55–60 (2010).
2. Coste, B. *et al.* Piezo proteins are pore-forming subunits of mechanically activated channels. *Nature* **483,** 176–181 (2012).
3. Ikeda, R. & Gu, J. G. Piezo2 channel conductance and localization domains in Merkel cells of rat whisker hair follicles. *Neurosci. Lett.* **7,** 210–215 (2014).
4. Woo, S. H. *et al.* Piezo2 is required for Merkel-cell mechanotransduction. *Nature* **509,** 622–626 (2014).
5. Bae, C., Gnanasambandam, R., Nicolai, C., Sachs, F. & Gottlieb, P. A. Xerocytosis is caused by mutations that alter the kinetics of the mechanosensitive channel PIEZO1. *Proc. Natl Acad. Sci. USA* **110,** E1162–E1168 (2013).
6. Albuisson, J. *et al.* Dehydrated hereditary stomatocytosis linked to gain-of-function mutations in mechanically activated PIEZO1 ion channels. *Nat. Commun.* **4,** 1884 (2013).
7. Demolombe, S., Duprat, F., Honore, E. & Patel, A. Slower Piezo1 inactivation in dehydrated hereditary stomatocytosis (xerocytosis). *Biophys. J.* **105,** 833–834 (2013).
8. Pathak, M. M. *et al.* Stretch-activated ion channel Piezo1 directs lineage choice in human neural stem cells. *Proc. Natl Acad. Sci. USA* **111,** 16148–16153 (2014).
9. Ranade, S. S. *et al.* Piezo1, a mechanically activated ion channel, is required for vascular development in mice. *Proc. Natl Acad. Sci. USA* **111,** 10347–10352 (2014).
10. Retailleau, K. *et al.* Piezo1 in smooth muscle cells is involved in hypertension-dependent arterial remodeling. *Cell Rep.* **13,** 1161–1171 (2015).
11. Bavi, N. *et al.* Biophysical implications of lipid bilayer rheometry for mechanosensitive channels. *Proc. Natl Acad. Sci. USA* **111,** 13864–13869 (2014).

12. Suchyna, T. M. & Sachs, F. Mechanosensitive channel properties and membrane mechanics in mouse dystrophic myotubes. *J. Physiol.* **581,** 369–387 (2007).
13. Suchyna, T. M., Markin, V. S. & Sachs, F. Biophysics and structure of the patch and the gigaseal. *Biophys. J.* **97,** 738–747 (2009).
14. Hamill, O. P. & McBride, Jr D. W. Induced membrane hypo/hyper-mechanosensitivity: a limitation of patch-clamp recording. *Annu. Rev. Physiol.* **59,** 621–631 (1997).
15. Slavchov, R. I., Nomura, T., Martinac, B., Sokabe, M. & Sachs, F. Gigaseal mechanics: creep of the gigaseal under the action of pressure, adhesion, and voltage. *J. Phys. Chem. B.* **118,** 12660–12672 (2014).
16. Brohawn, S. G., Campbell, E. B. & MacKinnon, R. Physical mechanism for gating and mechanosensitivity of the human TRAAK K^+ channel. *Nature* **516,** 126–130 (2014).
17. Brohawn, S. G., Su, Z. & MacKinnon, R. Mechanosensitivity is mediated directly by the lipid membrane in TRAAK and TREK1 K^+ channels. *Proc. Natl Acad. Sci. USA* **111,** 3614–3619 (2014).
18. Teng, J., Loukin, S., Anishkin, A. & Kung, C. The force-from-lipid (FFL) principle of mechanosensitivity, at large and in elements. *Pflugers Arch.* **467,** 27–37 (2015).
19. Martinac, B., Adler, J. & Kung, C. Mechanosensitive ion channels of Escherichia coli activated by amphipaths. *Nature* **348,** 261–263 (1990).
20. Prager-Khoutorsky, M., Khoutorsky, A. & Bourque, C. W. Unique interweaved microtubule scaffold mediates osmosensory transduction via physical interaction with TRPV1. *Neuron* **83,** 866–878 (2014).
21. Poole, K., Moroni, M. & Lewin, G. R. Sensory mechanotransduction at membrane-matrix interfaces. *Pflugers Arch.* **467,** 121–132 (2015).
22. Cueva, J. G., Mulholland, A. & Goodman, M. B. Nanoscale organization of the MEC-4 DEG/ENaC sensory mechanotransduction channel in Caenorhabditis elegans touch receptor neurons. *J. Neurosci.* **27,** 14089–14098 (2007).
23. Akinlaja, J. & Sachs, F. The breakdown of cell membranes by electrical and mechanical stress. *Biophys. J.* **75,** 247–254 (1998).
24. Beyder, A. & Sachs, F. Electromechanical coupling in the membranes of Shaker-transfected HEK cells. *Proc. Natl Acad. Sci. USA* **106,** 6626–6631 (2009).
25. Charras, G. T., Coughlin, M., Mitchison, T. J. & Mahadevan, L. Life and times of a cellular bleb. *Biophys. J.* **94,** 1836–1853 (2008).
26. Charras, G. T. A short history of blebbing. *J. Microsc.* **231,** 466–478 (2008).
27. Goudarzi, M. *et al.* Identification and regulation of a molecular module for bleb-based cell motility. *Dev. Cell.* **23,** 210–218 (2012).
28. Shiratsuchi, A., Mori, T. & Nakanishi, Y. Independence of plasma membrane blebbing from other biochemical and biological characteristics of apoptotic cells. *J. Biochem.* **132,** 381–386 (2002).
29. Fackler, O. T. & Grosse, R. Cell motility through plasma membrane blebbing. *J. Cell Biol.* **181,** 879–884 (2008).
30. Maksaev, G., Milac, A., Anishkin, A., Guy, H. R. & Sukharev, S. Analyses of gating thermodynamics and effects of deletions in the mechanosensitive channel TREK-1: comparisons with structural models. *Channels* **5,** 34–42 (2011).
31. Liu, X. *et al.* Stretch-activated potassium channels in hypotonically induced blebs of atrial myocytes. *J. Membr. Biol.* **226,** 17–25 (2008).
32. Honoré, E., Patel, A. J., Chemin, J., Suchyna, T. & Sachs, F. Desensitization of mechano-gated K2P channels. *Proc. Natl Acad. Sci. USA* **103,** 6859–6864 (2006).
33. Zhang, Y., Gao, F., Popov, V. L., Wen, J. W. & Hamill, O. P. Mechanically gated channel activity in cytoskeleton-deficient plasma membrane blebs and vesicles from Xenopus oocytes. *J. Physiol.* **523,** 117–130 (2000).
34. Moe, P. & Blount, P. Assessment of potential stimuli for mechano-dependent gating of MscL: effects of pressure, tension, and lipid headgroups. *Biochemistry (Mosc)* **44,** 12239–12244 (2005).
35. Hilgemann, D. W. Giant excised cardiac sarcolemmal membrane patches: sodium and sodium-calcium exchange currents. *Pflugers Arch.* **415,** 247–249 (1989).
36. Gan, P. P. *et al.* Microtubule dynamics, mitotic arrest, and apoptosis: drug-induced differential effects of betaIII-tubulin. *Mol. Cancer. Ther.* **9,** 1339–1348 (2010).
37. Doerner, J. F., Febvay, S. & Clapham, D. E. Controlled delivery of bioactive molecules into live cells using the bacterial mechanosensitive channel MscL. *Nat. Commun.* **3,** 990 (2012).
38. Norman, C. *et al.* Visualisation of the mechanosensitive channel of large conductance in bacteria using confocal microscopy. *Eur. Biophys. J.* **34,** 396–402 (2005).
39. Nomura, T. *et al.* Differential effects of lipids and lyso-lipids on the mechanosensitivity of the mechanosensitive channels MscL and MscS. *Proc. Natl Acad. Sci. USA* **109,** 8770–8775 (2012).
40. Sachs, F. Stretch activated ion channels: what are they? *Physiology* **25,** 50–56 (2010).
41. Sawada, Y., Murase, M. & Sokabe, M. The gating mechanism of the bacterial mechanosensitive channel MscL revealed by molecular dynamics simulations: from tension sensing to channel opening. *Channels* **6,** 317–331 (2012).
42. Tank, D. W., Wu, E. S. & Webb, W. W. Enhanced molecular diffusibility in muscle membrane blebs: release of lateral constraints. *J. Cell Biol.* **92,** 207–212 (1982).
43. Gottlieb, P. A., Bae, C. & Sachs, F. Gating the mechanical channel Piezo1: a comparison between whole-cell and patch recording. *Channels* **6,** 282–289 (2012).
44. Bae, C., Gottlieb, P. A. & Sachs, F. Human PIEZO1: removing inactivation. *Biophys. J.* **105,** 880–886 (2013).
45. Nomura, T., Cox, C. D., Bavi, N., Sokabe, M. & Martinac, B. Unidirectional incorporation of a bacterial mechanosensitive channel into liposomal membranes. *FASEB J.* **29,** 4334–4345 (2015).
46. Lauritzen, I. *et al.* Cross-talk between the mechano-gated K2P channel TREK-1 and the actin cytoskeleton. *EMBO Rep.* **6,** 642–648 (2005).
47. Ge, J. *et al.* Architecture of the mammalian mechanosensitive Piezo1 channel. *Nature* **527,** 64–69 (2015).
48. Coste, B. *et al.* Piezo1 ion channel pore properties are dictated by C-terminal region. *Nat. Commun.* **6,** 7223 (2015).
49. Gnanasambandam, R., Bae, C., Gottlieb, P. A. & Sachs, F. Ionic selectivity and permeation properties of human PIEZO1 channels. *PLoS ONE* **10,** e0125503 (2015).
50. Morris, C. E. Mechanoprotection of the plasma membrane in neurons and other non-erythroid cells by the spectrin-based membrane skeleton. *Cell. Mol. Biol. Lett.* **6,** 703–720 (2001).
51. Dzhura, I. *et al.* Cytoskeletal disrupting agents prevent calmodulin kinase, IQ domain and voltage-dependent facilitation of L-type Ca^{2+} channels. *J. Physiol.* **545,** 399–406 (2002).
52. Sokabe, M. & Sachs, F. The structure and dynamics of patch-clamped membranes: a study using differential interference contrast light microscopy. *J. Cell Biol.* **111,** 599–606 (1990).
53. Ruknudin, A., Song, M. J. & Sachs, F. The ultrastructure of patch-clamped membranes: a study using high voltage electron microscopy. *J. Cell Biol.* **112,** 125–134 (1991).
54. Markin, V. S. & Sachs, F. Thermodynamics of mechanosensitivity. *Curr. Top. Membr.* **58,** 87–119 (2007).
55. Cox, C. D., Nakayama, Y., Nomura, T. & Martinac, B. The evolutionary 'tinkering' of MscS-like channels: generation of structural and functional diversity. *Pflugers Arch.* **467,** 3–13 (2015).
56. Borbiro, I., Badheka, D. & Rohacs, T. Activation of TRPV1 channels inhibits mechanosensitive Piezo channel activity by depleting membrane phosphoinositides. *Sci. Signal.* **8,** ra15 (2015).
57. Poole, K., Herget, R., Lapatsina, L., Ngo, H. D. & Lewin, G. R. Tuning Piezo ion channels to detect molecular-scale movements relevant for fine touch. *Nat. Commun.* **5,** 3520 (2014).
58. Lim, C. T., Zhou, E. H. & Quek, S. T. Mechanical models for living cells—a review. *J. Biomech.* **39,** 195–216 (2006).
59. Sukharev, S. I., Sigurdson, W. J., Kung, C. & Sachs, F. Energetic and spatial parameters for gating of the bacterial large conductance mechanosensitive channel, MscL. *J. Gen. Physiol.* **113,** 525–540 (1999).
60. Wang, Y. *et al.* Single molecule FRET reveals pore size and opening mechanism of a mechano-sensitive ion channel. *Elife* **3,** e01834 (2014).
61. Wiggins, P. A. & Phillips, R. Analytic models for mechanotransduction: gating the mechanosensitive channel (MscL). *Biophys. J.* **86,** 467a–467a (2004).
62. Cox, C. D. *et al.* Selectivity mechanism of the mechanosensitive channel MscS revealed by probing channel subconducting states. *Nat. Commun.* **4,** 2137 (2013).
63. Martinac, B. *et al.* Studying mechanosensitive ion channels using liposomes. *Methods Mol. Biol.* **606,** 31–53 (2010).
64. Yoshimura, K., Usukura, J. & Sokabe, M. Gating-associated conformational changes in the mechanosensitive channel MscL. *Proc. Natl Acad. Sci. USA* **105,** 4033–4038 (2008).
65. Hamill, O. P. & Martinac, B. Molecular basis of mechanotransduction in living cells. *Physiol Rev.* **81,** 685–740 (2001).
66. Shaikh, S., Cox, C. D., Nomura, T. & Martinac, B. Energetics of gating MscS by membrane tension in azolectin liposomes and giant spheroplasts. *Channels* **8,** 321–326 (2014).
67. Peng, H., Bria, A., Zhou, Z., Iannello, G. & Long, F. Extensible visualization and analysis for multidimensional images using Vaa3D. *Nat. Protoc.* **9,** 193–208 (2014).
68. Charras, G. T., Hu, C. K., Coughlin, M. & Mitchison, T. J. Reassembly of contractile actin cortex in cell blebs. *J. Cell Biol.* **175,** 477–490 (2006).
69. Charras, G. T., Yarrow, J. C., Horton, M. A., Mahadevan, L. & Mitchison, T. J. Non-equilibration of hydrostatic pressure in blebbing cells. *Nature* **435,** 365–369 (2005).
70. Standen, N. B., Stanfield, P. R., Ward, T. A. & Wilson, S. W. A new preparation for recording single-channel currents from skeletal muscle. *Proc. R Soc. Lond. B Biol. Sci.* **221,** 455–464 (1984).

Acknowledgements

First we thank Professor Jamie I Vandenberg, Dr Ruhma Syeda and Mr Navid Bavi for helpful, robust discussions. This work was supported by the NHMRC grant APP1047980 and NHMRC Principal Research Fellowship to B.M. This work was also supported in part by funds from the Office of Health and Medical Research, NSW State Government, Australia and NIH 5R01HL054887 to F.S. and P.A.G. We are grateful to Robert Hard (University at Buffalo, New York) for assistance with SIM microscopy.

Author contributions

C.D.C., C.B. and P.A.G. performed the electrophysiological experiments. C.D.C. and V.N.-K. designed and carried out confocal microscopy imaging. C.D.C. carried out the patch fluorometry and blebbing experiments. P.R.R., L.Z. and S.H. generated mutations and carried out cloning. P.A.G. carried out SIM microscopy. C.-A.N. and C.D.C. carried out western blotting. P.R.R. and L.Z. provided technical help. C.D.C., F.S., P.A.G. and B.M. designed the experiments and wrote the manuscript. All the authors discussed and contributed to the final version of the manuscript.

Additional information

Competing financial interests: The authors declare no competing financial interests.

14

Transition state theory demonstrated at the micron scale with out-of-equilibrium transport in a confined environment

Christian L. Vestergaard[1,†], Morten Bo Mikkelsen[1], Walter Reisner[1,†], Anders Kristensen[1] & Henrik Flyvbjerg[1]

Transition state theory (TST) provides a simple interpretation of many thermally activated processes. It applies successfully on timescales and length scales that differ several orders of magnitude: to chemical reactions, breaking of chemical bonds, unfolding of proteins and RNA structures and polymers crossing entropic barriers. Here we apply TST to out-of-equilibrium transport through confined environments: the thermally activated translocation of single DNA molecules over an entropic barrier helped by an external force field. Reaction pathways are effectively one dimensional and so long that they are observable in a microscope. Reaction rates are so slow that transitions are recorded on video. We find sharp transition states that are independent of the applied force, similar to chemical bond rupture, as well as transition states that change location on the reaction pathway with the strength of the applied force. The states of equilibrium and transition are separated by micrometres as compared with angstroms/nanometres for chemical bonds.

[1] Department of Micro- and Nanotechnology, Technical University of Denmark, DK-2800 Kgs. Lyngby, Denmark. † Present addresses: Aix Marseille Université, Université de Toulon, CNRS, CPT, UMR 7332, 13288 Marseille, France (C.L.V.); Department of Physics, McGill University, Montreal, Quebec, Canada H3A 2T8 (W.R.). Correspondence and requests for materials should be addressed to C.L.V. (email: cvestergaard@gmail.com).

Transition state theory (TST), with its scenario of a reaction pathway through a free-energy landscape (Fig. 1), provides concepts for understanding how thermally activated processes take place. Its development can be traced back to the second half of the nineteenth century[1], notably to 1889 when Arrhenius proposed his famous empirical relation between the reaction rate r of an irreversible chemical reaction and temperature T:

$$r \propto e^{-\frac{\Delta\mathcal{F}^{\ddagger}}{k_B T}}. \quad (1)$$

Here $\Delta\mathcal{F}^{\ddagger}$ is the height of the free-energy barrier separating the initial state (reactants) and the end state (product), and k_B is the Boltzmann constant. Theoretical efforts to describe such reactions led to the development of TST in the second half of the 1930s, notably by Eyring, Polanyi, Evans and Wigner[1–3]. TST for elementary chemical reactions assumes, as Wigner summarized it, statistical mechanics, classical motion of atomic nuclei, adiabatically changing electronic states and what has become known as TST's fundamental assumption, fundamental dynamical assumption or no-recrossing assumption. When the reaction process is described by a single reaction coordinate x, as in Fig. 1, the no-recrossing assumption states that if x crosses the point of maximal free energy—the 'transition state' $x^{\ddagger}$—from left to right, it does not recross from right to left. This is plausible if motion in x is inertial, as in chemical reactions between colliding gas molecules.

To investigate the validity of TST, Kramers introduced in 1940 a model, which has become known as 'Kramers' problem'[4,5]. This model relaxes TST's no-recrossing assumption. It considers the reaction to be described by a fictive particle undergoing Brownian motion with more or less friction in the free-energy landscape along the reaction coordinate (Fig. 1). The case of large friction does not model colliding gas molecules, but reactants diffusing in a liquid. It differs from Wigner's TST by having x diffuse across $x^{\ddagger}$ with multiple recrossing expected from its trajectory of Brownian motion. In many cases Kramers' formalism allows calculation of the proportionality factor between the reaction rate in equation (1) and the Boltzmann factor, but it does not change the exponential dependence on the barrier height in equation (1).

Despite their simplicity, TST and Kramers' model are surprisingly successful at predicting chemical reaction rates, and they are unrivalled at providing conceptual insight into how such reactions occur. Though devised to describe chemical reactions (Fig. 1a), where reaction pathways are measured in fractions of angstroms and reaction times in femtoseconds[6], their formalism has been extended to processes taking place at timescales and length scales that are orders of magnitude longer. At the nanometre scale the formalism has been applied to rupture of chemical bonds[7–11] (Fig. 1b), protein (un)folding[12–14] and RNA unzipping[15] under both constant and time-dependent loads[16–18]; at the micron scale it has been applied to polymers crossing entropic barriers[19–22]. Han and Craighead notably showed that

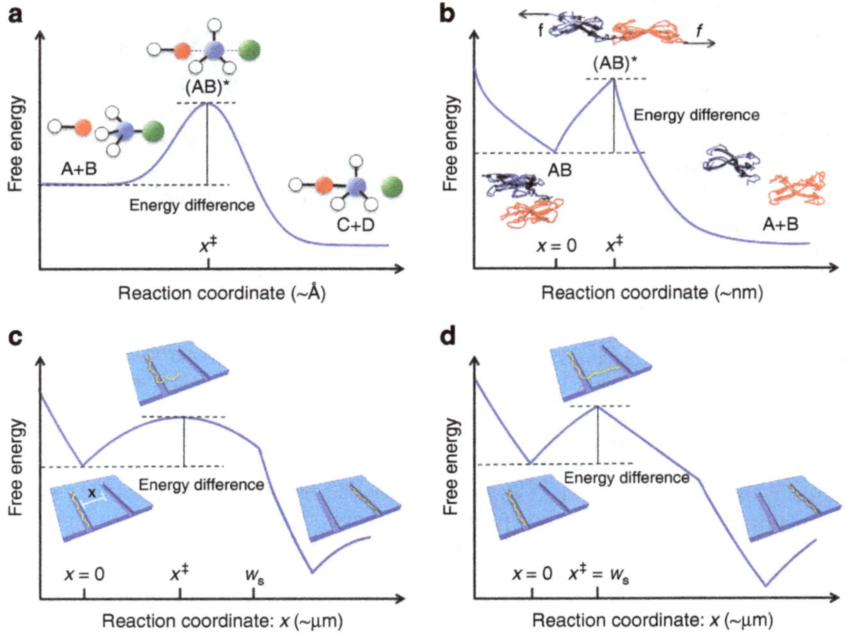

Figure 1 | The scenario of TST. A free-energy landscape is traversed by a reaction pathway that is parameterized by a reaction coordinate;[1,2] typical length scales of reaction pathways are given in parentheses. Insets portray physical situations corresponding to (quasi) equilibrium and transition states (at $x^{\ddagger}$). (**a**,**c**) Transition states that will change location on the reaction pathway with the strength of an applied force, exemplified by (**a**) a chemical reaction (with transition state (AB)*), and in the present study (**c**) a polymer crossing an entropic barrier in the form of nanoslit separating two nanogrooves, where the transition state lies inside the nanoslit. (**b**,**d**) Sharp transition states that are independent of an applied force, similar to the situation in chemical bond rupture; here exemplified by (**b**) the separation of two binding proteins under an external force (with transition state (AB)*), and in the present study (**d**) a polymer crossing through a nanoslit where the transition state is located at the end of the nanoslit. In **c**,**d** the reaction coordinate x parameterizes the continuous shifting of DNA in a transition from the upstream to the downstream nanogroove. Specifically, it measures the extension of the leading end of the DNA into the nanoslit, until this end of the DNA enters the next nanogroove, which happens at $x = w_s$. For $0 < x < w_s$, x is approximately proportional to the contour length ℓ of the DNA that has left the upstream nanogroove. After the leading end of the DNA has entered the downstream nanogroove, we let x denote a fixed fraction of the contour length ℓ of the DNA that has left the upstream nanogroove, the same fraction as x denoted for $0 < x < w_s$. Note that we need not know the value of this fraction, and its existence can be an approximation. The qualitative picture described here still captures the essence of Fig. 2's experimental observations of transitions. The trailing end of the DNA leaves the upstream nanogroove when ℓ equals the full contour length of the DNA molecule, L_{DNA}. We denote the value of x at that point by $x = x_{DNA}$. After this point, we let x denote x_{DNA} plus the distance that the DNA molecule's trailing end has moved into the nanoslit. After the DNA has completely entered the next nanogroove, the landscape repeats itself as from $x = 0$ (see also Fig. 3).

TST describes the mean waiting time before translocation of a randomly coiled DNA molecule from one micro-groove to another through a nanoslit, driven by an external electric field. The transition state occurs where the external force that squeezes the coil into the slit balances entropic recoil forces[19,20].

We have here replaced the microgrooves of Han and Craighead, in which the DNA assumes a bulk coiled conformation, with nanogrooves that force the molecule to extend linearly and transversely to the axis of propagation (Figs 1c,d and 2). This ensures that the molecule can escape via clearly defined excursions of its end points into the nanoslit (Fig. 2c), leading to a well-defined single reaction coordinate, the DNA strand's extension into the nanoslit (Fig. 2d). The corresponding free-energy landscape for the DNA is (a) one dimensional; (b) tuneable; (c) so large that we can see the DNA moving through it, from one quasi-equilibrium state, over a free-energy barrier, into another quasi-equilibrium state, and so on repeatedly (Fig. 2e); (d) periodic, so each escape is an independent repetition of the same process, which allows us to accumulate good statistics; and (e) so simple that we find a closed formula for transition rates.

We study the translocation of DNA strands between nanogrooves driven by an external flow (Figs 1c,d and 2). For sufficiently weak flow, Brownian motion dominates the dynamics of the DNA to such a degree that its translocation between neighbouring nanogrooves is diffusive[23]. As one increases the flow, and hence its force on the DNA, the system transitions smoothly from the 'diffusion-dominated regime' to a 'force-dominated regime' in which translocation is irreversible and described by TST; see below. In the force-dominated regime, Brownian motion still plays a pivotal role by providing the fluctuations that let the DNA cross the entropic barrier separating neighbouring grooves. However, after the DNA has crossed the transition state, translocation is effectively deterministic and dictated by the external force field of the imposed flow.

We show theoretically, using TST, and verify experimentally by measuring the waiting time in each nanogroove that in the force-dominated regime two distinct sub-regimes exist for the transition of the DNA molecule from one groove to the next: (i) for large separation between grooves and high flow speeds, the transition state lies inside the nanoslit (Fig. 1c). Its location is determined by the balance between entropic and drag forces[19,20,22] and thus changes with the applied flow speed. (ii) As we decrease the external field (the flow), the location of the transition state moves in the field's direction (downstream), until it reaches the end of the nanoslit. Below the critical field strength at which this happens, the transition state does not move further downstream. It remains fixed at the width of the nanoslit (Fig. 1d). In this previously unobserved low-force (yet force dominated) regime, the transition state is independent of the external field. Both the initial and transition states are here 'sharp'—that is, the derivative of the free energy with respect to the reaction coordinate is not continuous at these points; it changes value abruptly. This is why the initial and transition states do not move along the reaction coordinate when we alter the external field. The dynamics in the low-force regime is consequently described by the Bell–Evans[7–9] model for chemical bond breaking under external load.

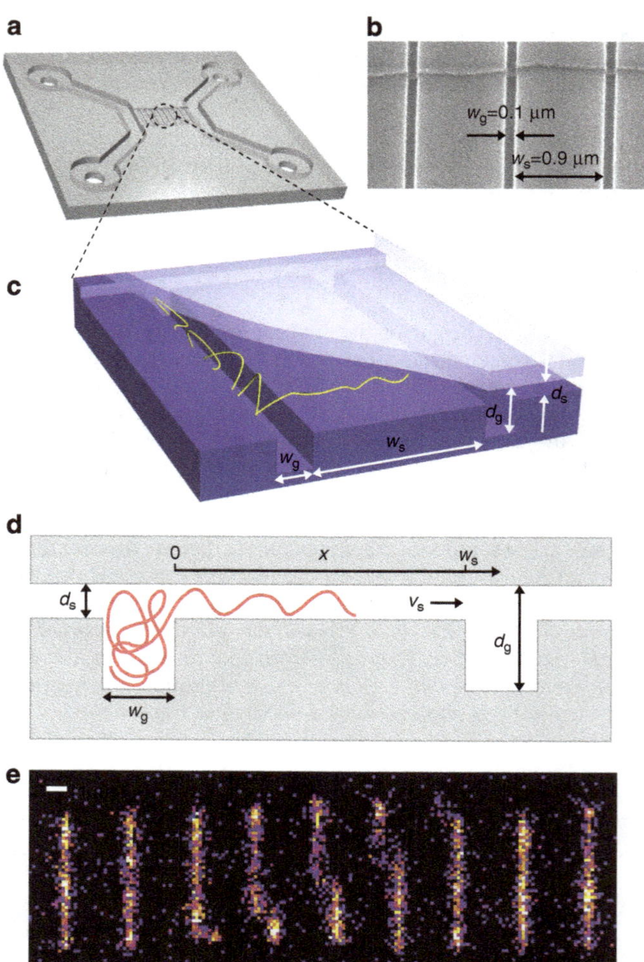

Figure 2 | Experimental set-up. (**a**) Schematic drawing of the microfluidic device containing the nanogroove array. A hydrodynamic flow (from left to right) is induced in the chip by imposing a pressure difference ΔP over the chip. Fluorescently labelled λ-DNA molecules (48.5 kb, $L_{DNA} = 21\,\mu m$) were introduced into the nanogroove array by the flow. (**b**) Electron micrograph of a section of a nanogroove array. (**c**) Schematic representation of a DNA strand trapped in a nanogroove and attempting to cross the nanoslit separating two grooves. The nanogroove geometry extends the DNA molecule transversally to the flow direction. Consequently, escape into the nanoslit is initiated by an end of the DNA. This vastly simplifies the dynamics compared with other entropic trapping geometries where polymers tend to form herniations inside the nanoslit. (**d**) Same as **c**, but showing a cross-section perpendicular to the nanogrooves. The extent of the DNA molecule's end inside the slit in the direction of the flow is called x. The hydrodynamic drag force on the DNA is proportional to x and υ_s, where υ_s is the mean speed of the buffer flow inside the nanoslit. The relevant dimensions of the nanogroove array are the height of the nanoslit, $d_s \approx 50$ nm; the total height of a nanogroove plus the nanoslit, $d_g \approx 150$ nm; the width of a nanogroove, $w_g \approx 100$ nm; and the width of a nanoslit separating two grooves, $w_s = 0.4$, 0.9, 1.9, 3.9 µm. (**e**) Montage of fluorescence images of a DNA molecule performing a sidewinder transition from one groove to the next[21]. The timelapse between consecutive images is 0.1 s. The fluorescence intensity is indicated with false colours. Uneven fluorescence of DNA in nanochannels is due to thermal fluctuations in the density of DNA in channels, where it coils a little as indicated in **c**. The lower fluorescence of DNA where it connects two channels in frames 3–6 (counting left to right) is due to the DNA being stretched in the slit, as indicated in **c**. Frame 7 shows the very last part of the transition between channels/barrier crossing. **a**,**b** and **e** are adapted from ref. 21.

Results

Transition state theory for DNA translocation via a nanoslit. We consider a DNA strand trapped in a nanogroove. Thermal fluctuations will now and then move one of its ends into the nanoslit (Fig. 2c,d). Let x denote the position of this end inside the nanoslit, measured in the direction parallel to the flow

(Fig. 2d). This x is our reaction coordinate. Thus, $x=0$ denotes the equilibrium state of a DNA strand trapped in a nanogroove. Similarly, $x^{\ddagger}$ denotes the transition state for crossing into the next groove downstream. $x=0$ and $x^{\ddagger}$ correspond to local minima and maxima, respectively, of the free-energy landscape experienced by a DNA strand moving through the chip (Fig. 3a).

The drag force f pulling at the DNA in the slit is proportional to the length ℓ of this DNA, $f=\gamma \upsilon_s \ell$, where υ_s is the mean flow speed and γ is the effective drag coefficient of the DNA inside the slit (Fig. 2d). Since the flow in the chip is laminar, υ_s is proportional to the pressure drop over the microchip, ΔP, which we control experimentally. Assuming that ℓ is approximately proportional to the DNA's extension parallel to the flow, x, we find that the drag force on the part of the DNA strand inside the nanoslit is proportional to $x\Delta P$. Thus, the decrease in free energy associated with the hydrodynamic drag force on the DNA is proportional to $x^2\Delta P$.

Note that the part of the DNA that rests in the nanogroove also experiences a drag force. It is proportional to the flow speed inside the nanogroove, $\upsilon_g \approx \upsilon_s/(1+d_g/d_s)=\upsilon_s/3$. This force, however, does not contribute to the free-energy difference along the reaction coordinate, since the part of the DNA inside the nanogrove does not move downstream with the drag force it experiences in the groove.

The decrease in entropy caused by the introduction of the DNA into a nanoslit, where it is more confined than in a nanogroove, gives rise to an entropic recoil force that tends to pull the DNA back out of the slit. This decrease in entropy is proportional to x (ref. 24).

Introduction of an end of the DNA strand into a nanoslit thus changes its free energy by the amount $\Delta\mathcal{F} = k_B T(bx - ax^2\Delta P/2)$ compared with its equilibrium state, where the whole strand resides in the nanogroove, $x=0$ (Fig. 3b)[19,20,24,25]. Here a and b are constants of proportionality that depend, respectively, on the mean drag coefficient on the DNA inside the nanoslit and the increase in entropy per unit length of DNA introduced into the nanoslit.

TST then predicts that the waiting times of a DNA strand in a groove are exponentially distributed (Fig. 4 and Supplementary Fig. 1) with a mean value that is given by $\tau=\tau_0 \exp[bx^{\ddagger} - a(x^{\ddagger})^2\Delta P/2]$. Two regimes exist, separated by a critical pressure difference ΔP_{crit} (Fig. 3c): (i) a high-force regime, characterized by $\Delta P > \Delta P_{crit}$, with the transition state inside the nanoslit at $x^{\ddagger}=b/(a\Delta P)$, and the dynamics of barrier crossing independent of w_s; and (ii) a low-force regime, characterized by $\Delta P < \Delta P_{crit} = b/(aw_s)$, with the transition state given by $x^{\ddagger}=w_s$, where w_s is the width of the nanoslit.

The mean trapping time is thus given by

$$\tau = \begin{cases} \tau_0 \exp(bw_s - aw_s^2\Delta P/2) & \text{for } \Delta P \leq \Delta P_{crit}, \\ \tau_0 \exp[b^2/(2a\Delta P)] & \text{for } \Delta P \geq \Delta P_{crit}, \end{cases} \quad (2)$$

where the prefactor τ_0 is related to the effective timescale of the motion along the reaction coordinate[1-4]. Equation (2) shows that for $\Delta P < \Delta P_{crit}$, the trapping time is described by the Bell–Evans model and $\log(\tau/\tau_0)$ is a first-degree polynomial in ΔP (Fig. 5a,c,e). For $\Delta P > \Delta P_{crit}$, equation (2) shows that $\log(\tau/\tau_0)$ is proportional to $1/\Delta P$ (Fig. 5b,d,f), as also observed in refs 19,20. At $\Delta P = \Delta P_{crit}$ we have a continuous transition between the two distinct regimes (Fig. 5a,b). The values found for the parameters of equation (2) (Table 1) are connected with microscopic physical

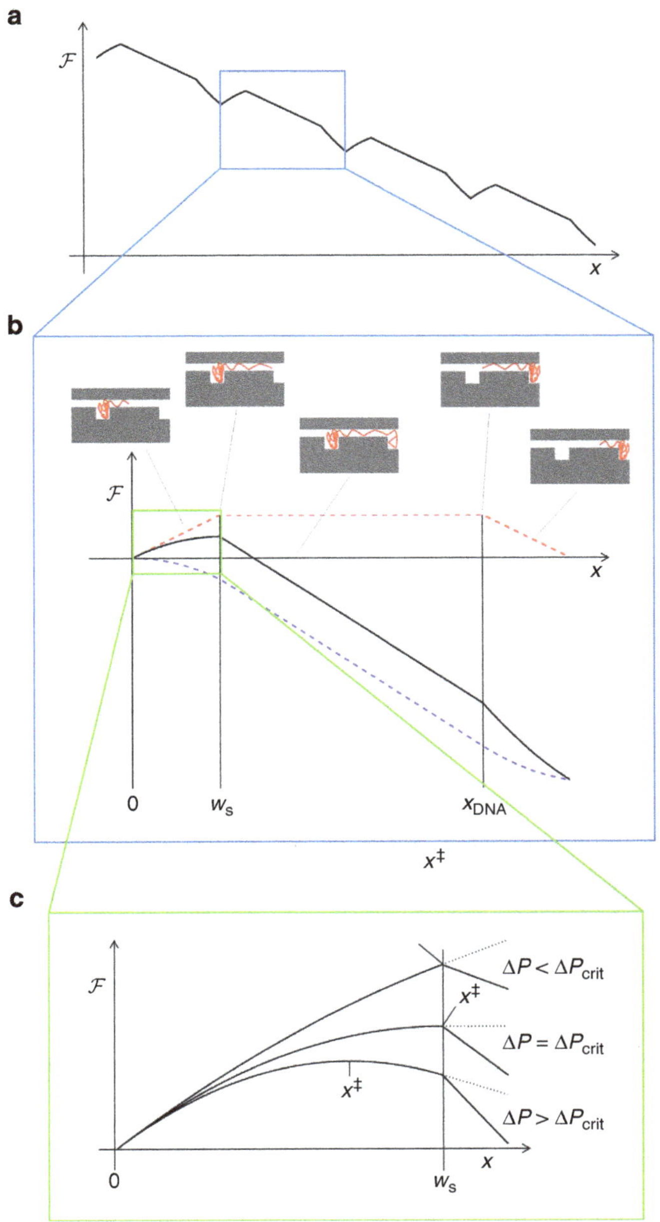

Figure 3 | Energy landscape experienced by a DNA molecule in the nanogroove array. (**a**) Free energy $\mathcal{F}(x)$ experienced by the molecule when driven by the force field from the buffer flow through a series of entropic traps. (**b**) Free energy $\mathcal{F}(x)$ experienced by a the molecule during a single transition between adjacent grooves. Insets show schematic drawings of the physical situation corresponding to five values of x. Here x is the x coordinate inside the nanoslit of the leading end of the DNA strand until that end descends into the next nanogroove. From that point and until the next state of quasi-equilibrium has been reached, further increase in x describes the length of DNA that has entered the next nanogroove. Thus, $x=0$ corresponds to the equilibrium state in which the whole strand resides in the left nanogroove. At $x=w_s$ the leading end enters the next nanogroove. At $x=x_{DNA}$, the trailing end of the DNA leaves the upstream nanogroove. At $x=x_{DNA}+w_s$, the DNA is again in quasi-equilibrium in the next groove, and the energy landscape repeats itself downstream from there as shown in **a**. Red dashed line: contribution to the free energy due to loss of entropy; blue dashed line: loss of potential energy of the DNA due to higher hydrodynamic drag on the part of the strand inside the nanoslit; black full line: total free-energy difference $\mathcal{F}(x)$. (**c**) Zoom on the energy landscape for $0<x\leq w_s$. The transition state at $x^{\ddagger}\leq w_s$ is the point with maximal free energy. For $\Delta P \leq \Delta P_{crit}$, $x^{\ddagger}=w_s$; for $\Delta P > \Delta P_{crit}$, $x^{\ddagger}<w_s$. (Note that we have assumed that $x_{DNA}\geq w_s$, which is always true in the present study. For DNA too short to span the width of the slit, $x_{DNA}<w_s$, and TST predicts $x^{\ddagger}\leq x_{DNA}$.)

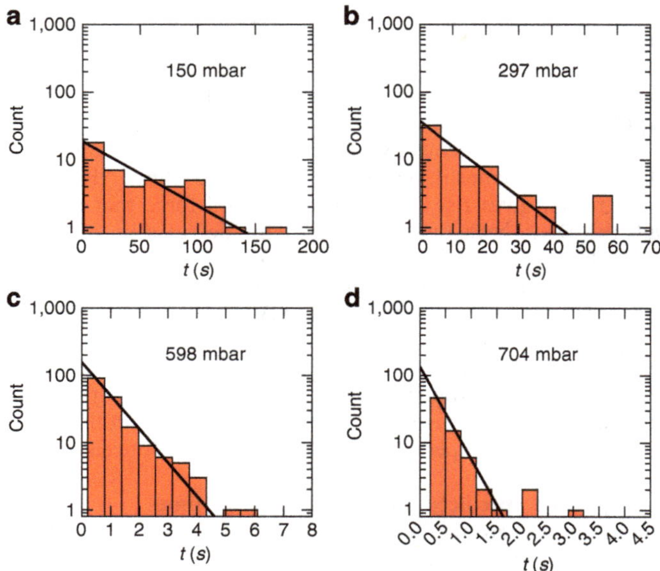

Figure 4 | Example distributions of waiting times in a groove. Measured waiting times in nanogrooves separated by nanoslits of width $w_s = 0.4\,\mu m$ (chip 1 below) for different values of the pressure difference imposed over the chip, spanning most of the parameter range explored in experiments here. Histograms shown agree well with single exponential fits (solid lines—obtained from maximum likelihood estimation, see Methods). Apparent 'outliers' arise from finite statistics in the tails and should be there, as their numbers agree with the expected numbers given by the areas under the tails of the theoretical distributions. Numbers of measured transition events are (**a**) 47, (**b**) 72, (**c**) 180 and (**d**) 73.

quantities. The value for τ_0 suggest that the timescale of relaxation of the DNA inside the slit is of the order of milliseconds, while a and b are determined by the DNA's effective drag coefficient and persistence length, the degree of stretching and the effective confinement energy of the DNA inside the slit. The latter four quantities cannot be found from our values for a and b alone. But by using that the drag coefficient and persistence length for DNA under similar conditions were found to be $\gamma \approx 1\text{–}2$ fN s μm^{-2} and $\ell_p \approx 40$ nm, respectively, we can give rough estimates of the degree of stretching and effective confinement energy (see the section 'Microscopic interpretation of parameter values' in Methods). We find that the degree of stretching of the DNA inside the slit is 30–50%, and the effective confinement energy of the DNA inside the slit is 0.4–0.6k_BT per persistence length ($\ell_p \approx 40$ nm) of DNA introduced into the slit.

Finally, by renormalizing τ and ΔP as $u = (\tau/\tau_0)^{1/(bw_s)}$ and $\xi = \Delta P/\Delta P_{crit}$, we find that all data fall on the same curve (Fig. 5g) given by

$$u = \begin{cases} \exp(1-\xi/2) & \text{for } \xi \le 1, \\ \exp[1/(2\xi)] & \text{for } \xi \ge 1. \end{cases} \quad (3)$$

Why TST works here. We made several simplifying assumptions to derive equations (2) and (3). These assumptions hold for our experiments for the following reasons.

(i) We assume that the DNA strand is in a state of thermal quasi-equilibrium when trapped in a nanogroove with its ends occasionally, randomly entering the nanoslit—more specifically, each point of the free-energy landscape inside the trap should be visited with a probability given by its Boltzmann factor. The validity of this assumption depends on the timescale of relaxation

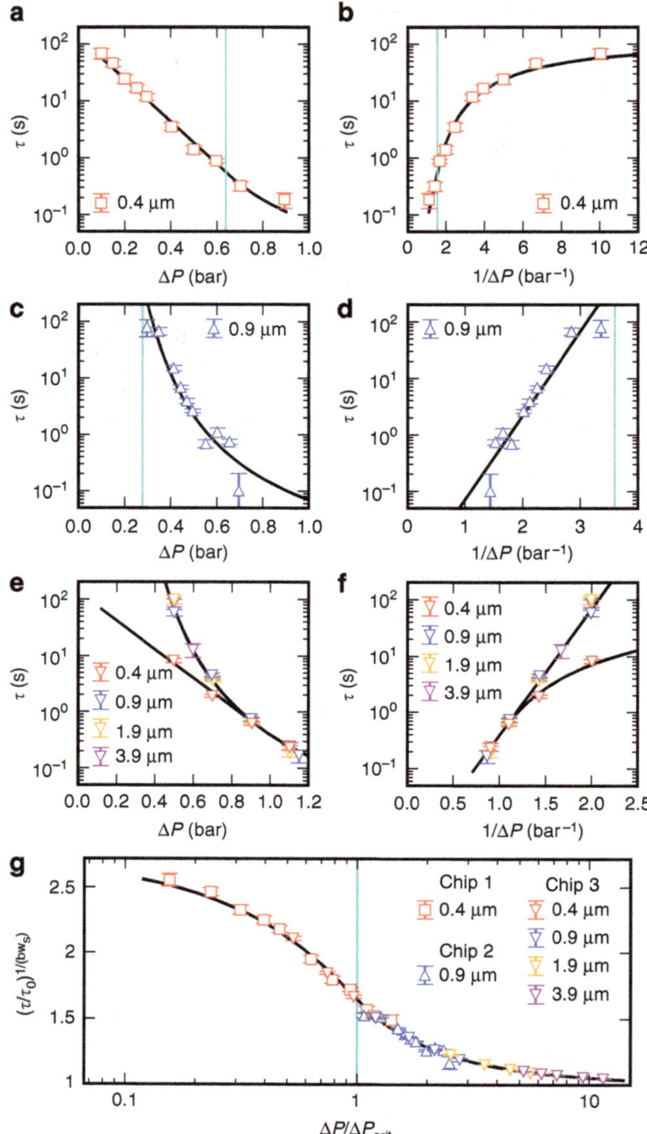

Figure 5 | Average lifetimes of quasi-stationary states in various force fields. Experimentally measured average residence times τ of DNA strands in a nanogroove as function of the pressure drop ΔP over the microfluidic chip. Different symbols correspond to different chips, different colors correspond to different nanoslit widths, see legends; chip 3 has different nanoslit widths at different places and hence does several distinct experiments all on the same chip. Data were collected for three different chips: (**a**,**b**) chip 1 (642 transition events), (**c**,**d**) chip 2 (1,604 events) and (**e**,**f**) chip 3 (2,873 events). (**a**,**c**,**e**) τ as function of ΔP—data follow a straight line for $\Delta P < \Delta P_{crit}$ and a hyperbolic curve for $\Delta P > \Delta P_{crit}$; (**b**,**d**,**f**) τ as function of $1/\Delta P$—data follow a straight line for $1/\Delta P < 1/\Delta P_{crit}$ and a hyperbolic curve for $1/\Delta P > 1/\Delta P_{crit}$. Symbols: experimental data, mean ± s.e.m. (examples of full distributions are shown in Fig. 4 and Supplementary Fig. 1); black lines: single fit of the theory (equation (2)) to all data in **a**–**f** (see Methods section); blue vertical lines: ΔP_{crit} for **a**,**c**,**e**, $1/\Delta P_{crit}$ for **b**,**d**,**f**. (**g**) All data from the three microfluidic chips fall on a single curve given by equation (3), independent of chip geometry when expressed in dimensionless variables $\Delta P/\Delta P_{crit}$ and $(\tau/\tau_0)^{1/(bw_s)}$.

of the DNA strand in a trap, τ_{relax}, being much shorter than the average time to escape from the trap.

For λ-DNA in a nanogroove with cross-section $\approx 100 \times 150\,nm^2$ (similar to here), $\tau_{relax} \sim 1$ s (ref. 26). Inside the slit, the relaxation

Table 1 | Estimated parameter values (top) and estimated values of related microscopic quantities (bottom).

τ_0 (ms)	a_1 ($\mu m^{-2} mbar^{-1}$)	a_2 ($\mu m^{-2} mbar^{-1}$)	a_3 ($\mu m^{-2} mbar^{-1}$)	b (μm^{-1})
2.2 (68% CI (0.6, 7.6))	0.11 ± 0.03	0.11 ± 0.01	0.07 ± 0.01	27 ± 4
	$\langle x/\ell \rangle$	$\langle x/\ell \rangle$	$\langle x/\ell \rangle$	$\delta\mathcal{F}_{conf}$ ($k_B T/\ell_p$)
	0.4-0.5	0.3-0.5	0.4-0.6	0.4-0.6

CI, confidence interval.
Microscopic quantities were estimated using $\gamma = 1$-$2\,fN\,s\,\mu m^{-2}$ and $\ell_p = 40$ nm as described in the section 'Microscopic interpretation of parameter values' in Methods.

time is much smaller than this due to additional confinement[26], the tension on the strand due to drag and the short length of DNA that is inside the slit (the relaxation timescales as $\sim 1/L$)[26]. Confinement reduces the relaxation time by a factor ~ 5 (ref. 26). Less than one-tenth of the DNA is in the slit before the transition state is traversed, as the longest distance separating the initial and transition states is 0.8 μm, further reducing τ_{relax} in the slit by a factor ~ 10.

These two effects alone then reduce the relaxation time to ~ 10 ms, while drag reduces it further, in agreement with the fitted value of τ_0 being of the order of milliseconds. This is fast enough compared with typical waiting times and experimental resolution that we may consider the DNA to be in quasi-equilibrium in the energy landscape before crossing the barrier; the exponential distribution of recorded waiting times to escape confirms this (Fig. 4 and Supplementary Fig. 1), while any 'inertial' effects making a second transition more probable immediately after a transition has occurred, say, due to incomplete relaxation in the trap, would result in an excess of counts for low waiting times.

(ii) We do not need the no-recrossing assumption, since Kramers' problem with large friction covers our case. It is approximately valid, however, if the free-energy landscape is steep and drops far, starting just past the transition state, and this condition is satisfied in our experiment.

Consider the case of lowest force ($w_s = 0.4$ μm and $\Delta P = 100$ mbar), where the free-energy landscape is the least steep. The height of the free-energy barrier here is $\Delta\mathcal{F} \approx 10 k_B T$, and the slope of the energy landscape to the right of the transition state is $\approx -5 k_B T$ per μm. Since recrossing happens with a probability that is proportional to the Boltzmann factor, x needs not be much larger than $x^{\ddagger}$ for recrossings to become highly unlikely, and translocation is hence effectively irreversible in our experiments.

Our data confirm this understanding. Recrossings over the transition state leads to 'dynamical corrections' of τ_0: it is expected to scale with the width of the free-energy barrier[4]. For $\Delta P \gg \Delta P_{crit}$, the barrier is essentially a parabolic potential, and one finds $\tau_0 \sim \Delta P^{-1/2}$, while for $\Delta P \ll \Delta P_{crit}$, the landscape is essentially rectilinear around the barrier, and one finds $\tau_0 \sim \Delta P^{-1}$ (ref. 4). For our experiments, this dependence on ΔP is so weak compared with the exponential factor in equation (2) that Fig. 5 shows agreement between data and theory without these dynamical corrections.

Further assumptions used here are the following: the shape of the free-energy landscape of the DNA in the nanoslit was derived assuming (iii) that nonlinear effects of hydrodynamic self-screening of the DNA in the slit is negligible; (iv) that the degree of stretching of the DNA in the slit does not depend on the amount of DNA in the slit; (v) that the increase in confinement energy caused by introduction of an end of the DNA into the slit scales linearly with the amount of DNA contour introduced; and (vi) that escape over the barrier through formation of a hernia, that is, a hairpin-like protrusion, of DNA in the slit happens so rarely that it can be ignored.

The effects of screening (iii) and uneven stretching (iv) tend to cancel each other, while both are diminished by the high degree of stretching of the DNA inside the nanoslit.

Assumption (v) that the free-energy cost per unit length of escaping contour is constant can be justified by noting that the free energy of confinement $\delta\mathcal{F}^{(slit)}_{conf}$ and $\delta\mathcal{F}^{(groove)}_{conf}$ both scale linearly with contour present in the slit and groove, respectively. Thus, the cost in free energy per unit length of contour, $\left(\delta\mathcal{F}^{(slit)}_{conf} - \delta\mathcal{F}^{(groove)}_{conf}\right)/\Delta\ell$, of driving contour length $\Delta\ell$ from the groove to the slit is constant. This linear scaling is fundamental and will hold regardless of the specific confinement regime as long as the size of the slit- or groove-confined polymer is much larger than the size of a 'statistical blob' in the slit or groove (which is true for the λ-DNA used here). To see this, simply note that for a confined polymer of contour length L, with $k_B T$ stored per blob, $\mathcal{F} = k_B T(L/L_{blob}) \sim L$. For a semiflexible chain, this linear scaling of confinement free energy with contour has been explicitly demonstrated in ref. 27 (for a slit) and ref. 28 (for a nanochannel, which approximates a nanogroove geometry well in as much as both force DNA into linearly extended configurations).

Finally, we assumed (vi) that escape over the entropic barrier through formation of a hernia inside the slit is an event sufficiently rare to be ignored. This escape event would lead to a non-exponential distribution of waiting times since the timescale of escape via this mechanism differs from that of end-induced escape. The reasons we do not observe escape via hernias here are threefold. First, as a hernia may form anywhere along a DNA strand, the rate of escape via hernias scales linearly with the length of the DNA[19,21,29,30]. Since λ-DNA is relatively short, this suppresses escape via herniation. Second, the free energy of a herniation inside the slit is more than twice that of an end (for an ideal flexible chain, it is exactly twice as high)[29,30]. So the probability of finding a hernia (as opposed to its multiplicity) is the square of the probability of finding a given end extending equally far into the slit. The latter being small, the former is very small. Third, since the DNA in the groove is stretched and the timescale of herniation is much faster than DNA relaxation in the groove (compare $\tau_0 \sim 1$ ms with $\tau_{relax} \sim 1$ s), introduction of a hernia into the nanoslit must stretch the DNA in the groove close to the hernia, thus increasing the free-energy barrier against escape via herniation even more. In contrast, introduction of an end into the nanoslit does not decrease the entropy per unit length of the DNA remaining in the groove.

Discussion

The use of an external pressure gradient to control and understand translocation of molecules by nanofluidic flows is poorly represented in the literature. So are simple models of such processes and their experimental verification.

Here we have shown that TST describes translocation of DNA driven by a hydrodynamic flow through a nano-confined environment that forms a series of entropic traps. We observed two distinct regimes: (i) a high-force regime in which the free-energy barrier is parabolic around the transition state. The transition state consequently moves along the reaction coordinate when the external force is altered; and (ii) a low-force regime in which the transition state is sharp and thus does not move when the force is altered. Observation of this low-force regime was made possible by reducing the barrier width considerably compared with earlier experiments[19,20,22]. A simple order-of-magnitude calculation shows that one would have to wait on the order of 100 years to see a single translocation in the geometries used in refs 19,20,22 (see the section 'Size matters' in Methods).

The applicability of TST to DNA translocation over entropic barriers relies on two conditions on the energy landscape describing the barriers: (i) the barrier separating two traps must be sufficiently high for quasi-equilibrium to exist before translocation; and (ii) the barrier must be steep enough beyond the transition state for recrossing not to occur, effectively.

If (i) is not satisfied, for example, for very high force, the motion becomes (partially) ballistic[21], leading to a non-exponential distribution of waiting times with an excess of short waits. If (ii) is not satisfied, for example, for very low force, Brownian motion dominates over drift, so the escape process is no longer irreversible[23] and hence ill defined as an 'escape.'

Microscopic 'bottom-up' models for barrier crossing of ideal Rouse polymers (polymers without bending rigidity and excluded volume effects)[29–33] yield an expression (equation (25) of ref. 30) for the rate of barrier crossing that is similar to our simple TST result (equation (1)). In these models the polymer crosses the barrier by stretching out in a 'kink' configuration, if it is long enough, since this lowers its free energy. An experimental demonstration of such stretching is provided in ref. 34, which shows DNA stretching where it crosses a potential barrier created by a conservative thermophoretic force field.

Note, however, that the physical mechanism responsible for the stretching of DNA in our nanoslit is entirely different. In refs 29–33, and in essence also in ref. 34, each monomer experiences the same potential energy barrier. Stretching lowers the energy barrier for the whole polymer by placing fewer monomers on top of the barrier, while stretching also costs a decrease in entropy, since some degrees of freedom are suppressed. The polymer stretches across the barrier when this decreases its potential energy by more than the ensuing cost in entropic free energy. However, this stretching is just a side effect of having the potential energy barrier; it is an adaption of the polymer to it. The barrier remains a potential energy barrier, while it affects the configurational entropy of the polymer.

In our experiments, on the other hand, DNA entering the nanoslit is less free to move thermally than it is in the nanogroove. The ensuing cost in entropic free energy constitutes the whole free-energy barrier. Thus, this barrier is entropic in nature. (Note that Figs 4 and 5 in ref. 30 resemble ours, but are artefacts resulting from displaying the DNA in an extra, non-existing dimension.) The final results are similar, because it is only the total free energy (containing both the energetic and entropic contributions) that matters. It should be interesting to extend the models of refs 29–33 to a non-smooth energy landscape to see if these microscopic models can predict the crossover between the high- and low-force regimes observed here.

The methods developed here might be useful also in the study of translocation of other biomolecules with more complex topologies (RNA, proteins, circular DNA or branched polymers) and in biological phenomena such as chromatin translocation in the cell nucleus and nuclear export. In particular, the quadratic shape of the energy landscape seen here is predicted also for more complex polymers, by scaling arguments[25,35]. It would be interesting to investigate experimentally whether in consequence the statistics of barrier crossing for such more complex polymers also is described by the simple formulas derived here.

The study presented here demonstrates the wide applicability of TST and the Bell–Evans model, in particular to out-of-equilibrium transport in confined environments. It is, to the best of our knowledge, the first time that the Bell–Evans model for barrier crossing under external load has been demonstrated on the micron scale and for polymers crossing entropic barriers. The fact that the process can be monitored with video microscopy should appeal to anyone who teaches or has been taught TST.

In general, our study may serve as a reminder that TST applies where thermal activation is possible, irrespective of length scales. Its rates are dominated by the factor given in equation (1) above, so the Boltzmann energy sets the scale, while length scales are irrelevant in a first approximation. This physics insight can be used deliberately in engineering, in microfluidic handling of polymers, particles or cells. Or, if ignored, it might cause problems.

This fact that TST applies on any length scale where thermal activation is possible, is a small demonstration of the universality of many physical theories. TST is more universal than that, however: The Brownian motion at its core needs not be thermal in origin[36], so TST can describe other random processes as well.

Methods

Device fabrication and experimental set-up. The nanofluidic devices were fabricated from fused silica wafers (JINSOL)[37]. Electron beam lithography in zep520A resist was used to define the nanogrooves, and photolithography in AZ5214E resist was used to define the nanoslit and the inlet channels. The structures were transferred to the silica through CF_4:CHF_3 reactive ion etching, and the channels were closed by fusion bonding of a 157-μm-thick fused silica coverslip to the wafer surface. Experiments were performed using λ-phage DNA (48.5 kb, New England Biolabs) stained with the fluorescent dye YOYO-1 (Molecular Probes) at a ratio of 1 dye molecule per 5 base pairs. A buffer solution of 0.5 × TBE (0.445 M Tris-base, 0.445 M boric acid and 10 mM EDTA) with 3% beta-mercaptoethanol was used. The buffer was driven through the nanogroove channel by applying air pressure, controlled to a precision of 1 mbar, to the inlets of the device. The DNA molecules were observed using an epi-fluorescence microscope (Nikon Eclipse TE2000-U) with × 60 and × 100 oil immersion 1.4 NA objectives. Movies of the DNA were recorded at up to 10 fps with an electron multiplying charge-coupled device camera (Cascade II, Photometrics). Identification of DNA molecules in the sidewinder state was performed using two distinct criteria[21]: (i) the DNA rested at least two frames in a groove between two transitions; and (ii) during a transition, both the DNA contour that connects two grooves and the DNA inside each groove were stretched, along the flow direction and the nanogroove, respectively. Waiting times were defined as the durations between the time at which the DNA found its equilibrium configuration inside a nanogroove after a transition and the time at which the leading end of the DNA had crossed over into the next nanogroove. In total, 5,119 sidewinder transitions were observed (chip 1: 642; chip 2: 1,604; and chip 3: 2,873).

Length and persistence length of YOYO-1-stained DNA. Intercalating YOYO dye affects both the length L and the persistence length ℓ_p of DNA. The dye increases L by approximately a factor 1.3 for a concentration of 1 YOYO-1 molecule per 5 base pairs[38]. For λ-DNA this leads to $L \approx 21$ μm. There is controversy in the literature as to its effect on ℓ_p, however, with some studies reporting an increase[39,40] in ℓ_p and others[38,41,42] reporting a decrease in ℓ_p. We here use the most recent results of ref. 38, which, for a dye concentration of 1 YOYO-1 molecule per 5 base pairs, gives $\ell_p \approx 40$ nm.

Note that using the results of refs 39,40, which give $\ell_p = 65$ nm, does not change the conclusions presented in the sections 'Microscopic interpretation of parameter values' and 'Size matters' below. The following argument shows this.

For $\ell_p \approx 65$ nm, we find from our estimated value of b an effective confinement energy per persistence length inside the nanoslit of $\delta\mathcal{F}_{\mathrm{conf}} \approx 0.6 - 1.0 k_B T$, which should be compared with $\delta\mathcal{F}_{\mathrm{conf}} \approx 0.33 k_B T$ from equations (6) and (7). The change of entropy due to the introduction of one persistence length of DNA into the nanoslit in the microarray of refs 19,20 is found to be $\delta\mathcal{F}_{\mathrm{conf}} \approx 0.57 k_B T$ and $\delta\mathcal{F}_{\mathrm{conf}} \approx 0.36 k_B T$ for ref. 22. For $\ell_p \approx 65$ nm, the confinement energy in the geometries of refs 19,20,22 would thus be roughly twice as high as for $\ell_p = 40$ nm, and the argument used below would then give that the expected waiting time for a single DNA strand to cross the barrier at critical field strength is $\sim 10^{10}$ years or more.

Calculating the mean of exponentially distributed data. As described in ref. 21, the DNA may at random switch from the 'sidewinder state', in which it is trapped for a time t_i in thermal equilibrium in a nanogroove, before it escapes to the next nanogroove, to a 'tumbleweed state', in which the DNA strand moves through the array without getting trapped in the grooves. Thus, there are no waiting times in the tumbleweed state; motion is continuous. The interaction with a groove may, however, slow the speed of the DNA in the tumbleweed state. This may falsely be detected as a waiting time by our movie-analysis software. As a filter against such false positives, we discard all waiting times that are shorter than twice the time Δt between frames in the movie. This means, however, that we also reject some true positives: DNA strands that are in the sidewinder state, but escape sooner than $2\Delta t$. Thus, the average $\bar{t} = \sum_i t_i/n$ of the measured dwell times is a biased estimate of $\tau = \langle t_i \rangle$. Instead we use the unbiased (maximum likelihood) estimator $\hat{\tau} = \bar{t} - 2\Delta t$.

Parameter estimation. Equation (2) is our theory for the observed waiting times. We fit the parameters $\theta = (a_1, a_2, a_3, b, \tau_0)$ of this theory to data $(\hat{\tau}_n \pm \sigma_n, w_{s,n}, \Delta P_n)_{n=1}^N$ using weighted least squares with weights $1/\hat{\sigma}_n^2$. Here $\hat{\sigma}_n$ is the empirically estimated s.e.m. of $\hat{\tau}_n$. We fit simultaneously to the data from all three microchips. In this fit, the parameters (a_1, a_2, a_3) are allowed individual values for each chip, since they depend on the hydraulic resistance, which differs between chips. The parameters b and τ_0 depend only on the DNA, solvent and temperature. They should not differ between chips, so we fit values shared by all chips. Fitting to data presented in Fig. 5, we obtain the estimates given in Table 1.

The variances of errors on fitted parameter values were estimated as

$$\operatorname{var}(\hat{\theta}_i) = \hat{\chi}^2_{\text{norm}} \left(\hat{J}\hat{\Sigma}^{-1}\hat{J}^{\top}\right)^{-1}_{ii}, \tag{4}$$

where $\hat{\chi}^2_{\text{norm}} = \hat{\chi}^2/(N-5)$ is $\hat{\chi}^2$ per degree of freedom, $\hat{\Sigma}$ is the covariance matrix of $\hat{\tau}_n$ with entries given by $\hat{\Sigma}_{mn} = \hat{\sigma}_n^2 \delta_{m,n}$ and $\hat{J}$ is the Jacobian of the vector $\tau(\hat{\theta}) = (\tau_n(\hat{\theta}))_{n=1}^N$, with entries given by $\hat{J}_{in} = \partial\tau_n(\hat{\theta})/\partial\theta_i$.

Microscopic interpretation of parameter values. The parameter values returned by the fit described above can be interpreted at the microscopic level. Thus, the value obtained for τ_0 means that the effective timescale of DNA motion in the free-energy potential inside the slit is of the order of milliseconds.

The values found for a is connected with the degree of stretching of the DNA inside the nanoslit. From our definition of a and x, we have $\gamma v_s \ell^2/(2k_B T) \approx a\Delta P x^2/2$. The mean degree of stretching of the DNA inside the slit is then given by

$$\langle x/\ell \rangle \approx \sqrt{\gamma v_s/(a\Delta P k_B T)}. \tag{5}$$

Here $v_s \approx \Delta P/(d_s w_{ch} R_{hyd})$, where $w_{ch} = 50\,\mu\text{m}$ is the chip width perpendicular to the flow, and $R_{hyd} \approx 12\eta N w_s/(d_s^3 w_{ch})$ is the hydraulic resistance over the chip[43], where N is the number of slits in the chip and η is the dynamic viscosity of water. Since equation (5) depends on the value of γ, our estimate for a does not directly give us the degree of stretching of the DNA inside the nanoslit. However, the drag coefficient for a flow parallel to the DNA backbone has been found previously for similar conditions to be $\gamma_{||} \approx 1\,\text{fN s}\,\mu\text{m}^{-1}$ (J.N. Pedersen, personal communication). For DNA segments aligned perpendicular to the flow, $\gamma_{\perp} = 2\gamma_{||}$, so we expect the effective drag coefficient on DNA inside the slit to lie between these two values.

We use this range of values for γ in the following and that $\eta = 10^{-3}\,\text{Pa}\cdot\text{s}$, $k_B T = 4.1\,\text{fN}\,\mu\text{m}$ and $d_s = 0.05\,\mu\text{m}$. For chip 1 we have $N = 900$, $w_s = 0.4$ and $a = 0.11$, yielding $\langle x/\ell \rangle \approx 0.4 - 0.5$. For chip 2 we have $N = 450$, $w_s = 0.9$ and $a = 0.11$, yielding $\langle x/\ell \rangle \approx 0.3 - 0.5$. For chip 3 we have $N = 375$, average slit width $\overline{w_s} = 1.1$ and $a = 0.07$, yielding $\langle x/\ell \rangle \approx 0.4 - 0.6$.

Finally, we may use the above estimate of the degree of stretching inside the nanoslit and the value of b to give a rough estimate of the effective confinement energy of the DNA inside the slit compared with the groove. Since b gives the effective confinement energy in the slit per μm along the reaction coordinate, we may find the confinement energy per persistence length ℓ_p as $\delta\mathcal{F}_{\text{conf}} \approx b\langle x/\ell \rangle \ell_p \approx 0.4 - 0.6 k_B T$ by using $\ell \approx 40$ nm and the values for $\langle x/\ell \rangle$ found above.

We may compare this result with the expected difference in confinement energies between the slit and groove, $\delta\mathcal{F}_{\text{conf}} = \delta\mathcal{F}^{(\text{slit})}_{\text{conf}} - \delta\mathcal{F}^{(\text{groove})}_{\text{conf}}$. The DNA in the groove and in the slit is in a crossover regime between the De Gennes and Odjik regimes. Here interpolation formulas for the confinement energy were recently determined from a combination of high precision simulations and experiments. The free energy of confinement per persistence length of DNA in a channel (which approximates the nanogroove geometry) is[28]

$$\delta\mathcal{F}^{(\text{groove})}_{\text{conf}} = \frac{k_B T(\pi^2/3)\left(\ell_p/d_s\right)^2}{\left[5.147\left(\ell_p/d_s\right)^2 + 3.343\ell_p/d_s + 1\right]^{2/3}}, \tag{6}$$

while the confinement energy in the nanoslit is[27,44]

$$\delta\mathcal{F}^{(\text{slit})}_{\text{conf}} = \frac{k_B T(\pi^2/3)\left(\ell_p/d_s\right)^2}{\left[5.146\left(\ell_p/d_s\right)^2 + 1.984\ell_p/d_s + 1\right]^{2/3}}. \tag{7}$$

The expected confinement energy per ℓ_p is thus $\delta\mathcal{F}_{\text{conf}} \approx 0.3 k_B T$, which is somewhat smaller than the energy estimated from b. This difference may be explained by the flow stretching the DNA inside the slit, such that b reflects not only the confinement energy due the walls of the slit but also contains a term from the stretching of the DNA due to the flow. Table 1 collects our estimates for microscopic quantities.

Size matters. In the present experiment, the barrier width is significantly smaller than in earlier similar experiments[19,20,22]. This makes all the difference for our ability to observe the low-force regime ($\Delta P < \Delta P_{\text{crit}}$).

In refs 19,20, the potential wells of the microarray with the smallest dimensions were separated by $w_s \approx 2\,\mu\text{m}$ of nanoslit, while (ref. 22) studied DNA translocation through nanopores of 6 μm or more. This made observation of the low-force regime impossible according to the following order-of-magnitude estimate of the time—100 years or more—that one would have to wait for a single DNA molecule to traverse the nanoslit/nanopore in these geometries. Our argument hinges on the fact that the mean trapping time of a DNA strand at critical pressure difference ΔP_{crit} depends exponentially on bw_s, that is, $\tau_{\text{crit}} = \tau_0 \exp(bw_s/2)$. Thus, a linear increase in b or w_s leads to an exponential increase in the waiting time.

The entropic traps (microgrooves) of refs 19,20 had dimensions $d_g \approx 1\,\mu\text{m}$ by $w_g \approx 2\,\mu\text{m}$. Since the radius of gyration, $R_g \approx 0.5\,\mu\text{m}$, of YOYO-stained λ-DNA is smaller than both d_g and w_g, the DNA was essentially in bulk conformation there. The increase in free energy due to confinement caused by the introduction of a unit length ℓ of DNA from bulk into the nanoslit was thus simply equal to its confinement free energy in the slit, given by equation (7).

For a height of $d_s \approx 0.09\,\mu\text{m}$ of the nanoslits of refs 19,20, this gives $\delta\mathcal{F}_{\text{conf}} \approx 0.32 k_B T$, which is comparable to the confinement energy in our set-up of $\delta\mathcal{F}_{\text{conf}} \approx 0.28 k_B T$ found from equations (6) and (7). So while the DNA strand is more confined in the nanoslit in our set-up than in the one of refs 19,20 ($d_s = 50$ nm in our case versus $d_s \approx 90$ nm in refs 19,20), the preconfinement in the nanogroove lowers the relative confinement energy here such that the entropic recoil force is similar in magnitude to, or lower than, the one in refs 19,20.

In the present experiment, we have $\tau_{\text{crit}} \approx 0.5$ s from $\tau_0 = 0.002$ s and $b = 27\,\mu\text{m}^{-1}$, found by fitting equation (2) to data (see 'Parameter estimation' above). We assume that τ_0 does not change much for different geometries, and since $b \propto \delta\mathcal{F}_{\text{conf}}$, we may assume that b was roughly the same in the set-up of refs 19,20 as here. Thus, since their narrowest nanoslit measured 2 μm, the expected mean waiting time at critical field strength would in their geometry be $\tau_{\text{crit}} \sim 100$ years. Similarly, for the least constricted geometry of ref. 22, we have $\delta\mathcal{F}_{\text{conf}} = 0.17 k_B T$ (approximating their nanopore as a nanochannel), that is, around half of the confinement energy of the present study. However, since their pore measured 10 μm, we find here $\tau_{\text{crit}} \sim 10^{20}$ years.

References

1. Laidler, K. J. & King, M. C. The development of transition-state theory. *J. Phys. Chem.* **87**, 2657–2664 (1983).
2. Laidler, K. J. A lifetime of transition-state theory. *Chem. Intell.* **4**, 39–47 (1998).
3. Garrett, B. C. Perspective on 'The transition state method". *Theor. Chem. Acc.* **103**, 200–204 (2000).
4. Kramers, H. Brownian motion in a field of force and the diffusion model of chemical reactions. *Physica* **7**, 284–304 (1940).
5. Mel, V. The Kramers problem: fifty years of development. *Phys. Rep.* **209**, 1–71 (1991).
6. Zewail, A. H. Laser femtochemistry. *Science* **242**, 1645–1653 (1988).
7. Zhurkov, S. Kinetic concept of strength of solids. *Int. J. Fract. Mech.* **1**, 311–323 (1965).
8. Bell, G. I. Models for the specific adhesion of cells to cells. *Science* **200**, 618–627 (1978).
9. Evans, E. Probing the relation between force-lifetime-and chemistry in single molecular bonds. *Annu. Rev. Biophys. Biomol. Struct.* **30**, 105–128 (2001).
10. Freund, L. Characterizing the resistance generated by a molecular bond as it is forcibly separated. *Proc. Natl Acad. Sci. USA* **106**, 8818–8823 (2009).
11. Tian, Y. & Boulatov, R. Comparison of the predictive performance of the Bell-Evans, Taylor-expansion and statistical-mechanics models of mechanochemistry. *Chem. Commun.* **49**, 4187–4189 (2013).
12. Schuler, B. & Clarke, J. Biophysics: rough passage across a barrier. *Nature* **502**, 632–633 (2013).
13. Chung, H. S. & Eaton, W. A. Single-molecule fluorescence probes dynamics of barrier crossing. *Nature* **502**, 685–688 (2013).
14. Woodside, M. T. & Block, S. M. Folding in single riboswitch aptamers. *Science* **180**, 2006–2009 (2008).
15. Dudko, O. K., Mathé, J., Szabo, A., Meller, A. & Hummer, G. Extracting kinetics from single-molecule force spectroscopy: nanopore unzipping of DNA hairpins. *Biophys J.* **92**, 4188–4195 (2007).
16. Dudko, O., Hummer, G. & Szabo, A. Intrinsic rates and activation free energies from single-molecule pulling experiments. *Phys. Rev. Lett.* **96**, 1–4 (2006).
17. Dudko, O. K., Hummer, G. & Szabo, A. Theory, analysis, and interpretation of single-molecule force spectroscopy experiments. *Proc. Natl Acad. Sci. USA* **105**, 15755–15760 (2008).

18. Dudko, O. K. Single-molecule mechanics: new insights from the escape-over-a-barrier problem. *Proc. Natl Acad. Sci. USA* **106,** 8795–8796 (2009).
19. Han, J., Turner, S. & Craighead, H. Entropic trapping and escape of long DNA molecules at submicron size constriction. *Phys. Rev. Lett.* **83,** 1688–1691 (1999).
20. Han, J., Turner, S. & Craighead, H. Erratum: entropic trapping and escape of long DNA molecules at submicron size constriction. *Phys. Rev. Lett.* **86,** 9007 (2001).
21. Mikkelsen, M. B. *et al.* Pressure-driven DNA in nanogroove arrays: complex dynamics leads to length-and topology-dependent separation. *Nano Lett.* **11,** 1598–1602 (2011).
22. Auger, T. *et al.* Zero-mode waveguide detection of flow-driven DNA translocation through nanopores. *Phys. Rev. Lett.* **113,** 028302 (2014).
23. Hoogerheide, D. P., Albertorio, F. & Golovchenko, J. A. Escape of DNA from a weakly biased thin nanopore: experimental evidence for a universal diffusive behavior. *Phys. Rev. Lett.* **111,** 248301 (2013).
24. De Gennes, P.-G. *Scaling Concepts in Polymer Physics* (Cornell Univ. Press, 1979).
25. Sakaue, T., Raphaël, E., de Gennes, P.-G. & Brochard-Wyart, F. Flow injection of branched polymers inside nanopores. *Europhys. Lett.* **72,** 83 (2005).
26. Reisner, W. *et al.* Statistics and dynamics of single DNA molecules confined in nanochannels. *Phys. Rev. Lett.* **94,** 196101 (2005).
27. Chen, J. & Sullivan, D. Free energy of a wormlike polymer chain confined in a slit: crossover between two scaling regimes. *Macromolecules* **39,** 7769–7773 (2006).
28. Tree, D. R., Wang, Y. & Dorfman, K. D. Extension of DNA in a nanochannel as a rod-to-coil transition. *Phys. Rev. Lett.* **110,** 208103 (2013).
29. Sebastian, K. L. Kink motion in the barrier crossing of a chain molecule. *Phys. Rev. E* **61,** 3245 (2000).
30. Sebastian, K. L. & Paul, A. K. R. Kramers problem for a polymer in a double well. *Phys. Rev. E* **62,** 927 (2000).
31. Lee, S. K. & Sung, W. Coil-to-stretch transition, kink formation, and efficient barrier crossing of a flexible chain. *Phys. Rev. E* **63,** 021115 (2001).
32. Sebastian, K. L. & Debnath, A. Polymer in a double well: dynamics of translocation of short chains over a barrier. *J. Phys. Condens. Matter* **18,** S283 (2006).
33. Debnath, A., Paul, A. K. R. & Sebastian, K. L. Barrier crossing in one and three dimensions by a long chain. *J. Stat. Mech. Theor. Exp.* **11,** P11024 (2010).
34. Pedersen, J. N. *et al.* Thermophoretic forces on DNA measured with a single-molecule spring balance. *Phys. Rev. Lett.* **113,** 268301 (2014).
35. De Gennes, P. G. Flexible polymers in nanopores. *Adv. Polym. Sci.* **138,** 92 (1999).
36. D'Anna, G. *et al.* Observing brownian motion in vibration-fluidized granular matter. *Nature* **424,** 909–912 (2003).
37. Reisner, W. *et al.* Nanoconfinement-enhanced conformational response of single DNA molecules to changes in ionic environment. *Phys. Rev. Lett.* **99,** 058302 (2007).
38. Maaloum, M., Muller, P. & Harlepp, S. DNA-intercalator interactions: structural and physical analysis using atomic force microscopy in solution. *Soft Matter* **9,** 11233–11240 (2013).
39. Quake, S. R., Babcock, H. & Chu, S. The dynamics of partially extended single molecules of DNA. *Nature* **388,** 151–154 (1997).
40. Makita, N., Ullner, M. & Yoshikawa, K. Conformational change of giant DNA with added salt as revealed by single molecular observation. *Macromolecules* **39,** 6200–6206 (2006).
41. Sischka, A. *et al.* Molecular mechanisms and kinetics between DNA and DNA binding ligands. *Biophys J.* **88,** 404–411 (2005).
42. Murade, C. U., Subramaniam, V., Otto, C. & Bennink, M. L. Force spectroscopy and fluorescence microscopy of dsDNA-YOYO-1 complexes: implications for the structure of dsDNA in the overstretching region. *Nucleic Acids Res.* **38,** 3423–3431 (2010).
43. Bruus, H. *Theoretical Microfluidics* (Oxford Univ. Press, 2007).
44. Klotz, A. R. *et al.* Measuring the confinement free energy and effective width of single polymer chains via single-molecule tetris. *Macromolecules* **48,** 5028–5033 (2015).

Acknowledgements

The work was partly funded by the Danish Council for Strategic Research Grant 10-092322 (PolyNano), by the Danish Council for Technology and Innovation through the Innovation Consortium OCTOPUS, by the Danish Research Council for Technology and Production, FTP Grant 274-05-0375 and by the European Commission-funded project READNA (Contract HEALTH-F4-2008-201418).

Author contributions

C.L.V., M.B.M., W.R., A.K. and H.F. designed the study; M.B.M. performed the experiments; C.L.V. and M.B.M. analysed the experimental data; C.L.V. derived theoretical results; C.L.V., M.B.M. and W.R. produced figures; C.L.V. and H.F. wrote the paper.

Additional information

Competing financial interests: The authors declare no competing financial interests.

15

Cryo-EM structures of the eukaryotic replicative helicase bound to a translocation substrate

Ferdos Abid Ali[1,*], Ludovic Renault[1,2,*], Julian Gannon[1,†], Hailey L. Gahlon[3], Abhay Kotecha[4], Jin Chuan Zhou[1], David Rueda[3] & Alessandro Costa[1]

The Cdc45-MCM-GINS (CMG) helicase unwinds DNA during the elongation step of eukaryotic genome duplication and this process depends on the MCM ATPase function. Whether CMG translocation occurs on single- or double-stranded DNA and how ATP hydrolysis drives DNA unwinding remain open questions. Here we use cryo-electron microscopy to describe two subnanometre resolution structures of the CMG helicase trapped on a DNA fork. In the predominant state, the ring-shaped C-terminal ATPase of MCM is compact and contacts single-stranded DNA, via a set of pre-sensor 1 hairpins that spiral around the translocation substrate. In the second state, the ATPase module is relaxed and apparently substrate free, while DNA intimately contacts the downstream amino-terminal tier of the MCM motor ring. These results, supported by single-molecule FRET measurements, lead us to suggest a replication fork unwinding mechanism whereby the N-terminal and AAA+ tiers of the MCM work in concert to translocate on single-stranded DNA.

[1] Macromolecular Machines, Clare Hall Laboratory, The Francis Crick Institute, Blanche Lane, South Mimms EN6 3LD, UK. [2] National Institute for Biological Standards and Control, Microscopy and Imaging, Blanche Lane, South Mimms EN6 3QG, UK. [3] Section of Virology and Single Molecule Imaging Group, Department of Medicine, MRC Clinical Centre, Imperial College London, London W12 0NN, UK. [4] Division of Structural Biology, Wellcome Trust Centre for Human Genetics, University of Oxford, Roosevelt Drive, Oxford OX3 7BN, UK. * These authors contributed equally to this work. † Present address: Synthetic and Systems Biochemistry of the Microtubule Cytoskeleton, Lincoln's Inn Fields Laboratory, The Francis Crick Institute, 44 Lincoln's Inn Fields, London WC2A 3LY, UK. Correspondence and requests for materials should be addressed to A.C. (email: alessandro.costa@crick.ac.uk).

DNA replication onset requires an initiator that loads a set of two helicases for double-helix unwinding. This provides the single-stranded DNA template for the replicative polymerases. In eukaryotic cells, helicase recruitment and origin activation are temporally separated[1]. The origin recognition complex partakes in loading an inactive dimer of ring-shaped MCM helicase motors that encircle double-stranded DNA[2–4]. Origin firing depends on the recruitment of a set of replication factors[5], including the GINS and Cdc45 activators that bind to each MCM ring in the dimer, forming a pair of multisubunit Cdc45-MCM-GINS (CMG) holo-helicases[6,7]. Upon origin activation, the two CMG particles are believed to separate and move in opposite directions to unwind DNA[8], however, the molecular basis of MCM double-ring uncoupling is unknown. The MCM helicase motor is a ring-shaped hetero-hexamer containing six homologous polypeptides belonging to the superfamily of AAA+ ATPases. The N-terminal domain (NTD) of the MCM forms a DNA-binding collar and a co-axial carboxy-terminal ATPase motor powers substrate translocation through the ring central channel[9]. Whether DNA unwinding involves MCM motor translocation on duplex- or single-stranded DNA remains unclear[4,10–13]. DNA fork progression depends on the ATPase function of the MCM motor[5,14], however, it is unknown how the energy derived from ATP hydrolysis is converted into motion and fork unwinding[15]. To start to address these outstanding questions, we have determined two cryo-electron microscopy (cryo-EM) structures of the CMG helicase trapped on a model DNA fork (by incubation with the slowly hydrolysable ATP analogue, ATPγS). We have also obtained two similar basic structures of the CMG helicase in the absence of DNA, imaged in conditions that allow for ATP turnover. Combined with single-molecule FRET analysis of DNA deformation by the CMG, our data provide important novel insights into the mechanism of replication fork progression in eukaryotic cells.

Subnanometre resolution structure of the CMG

Catalytically active, baculovirus-expressed *Drosophila melanogaster* CMG was incubated with a model replication fork in the presence of ATPγS, required for stable DNA binding[6]. Particles embedded in vitrified ice were imaged on a FEI Polara electron microscope equipped with an energy filter and a K2 Summit direct electron detector (Gatan, Inc.; Supplementary Fig. 1). Following two-dimensional (2D) and three-dimensional (3D) classification, a first structure was refined to 7.4 Å resolution (Supplementary Fig. 2). Atomic docking was employed to interpret the cryo-EM map, using the coordinates of known holo-helicase components. These efforts provide an exhaustive description of the CMG intersubunit interaction network. The structure contains a closed hexameric ring face that matches the N-terminal DNA-interacting collar of yeast MCM[4] (Fig. 1a,b, PDB entry 3JA8), albeit with significant inter-domain rearrangements (Supplementary Fig. 3 and Supplementary Movie 1). Combined with previous subunit mapping studies[16,17], our data confirm that GINS components Psf2 and Psf3 (PDB entry 2Q9Q) interact with the outer perimeter of MCM subunits 5 and 3 (Fig. 1a–c). Remarkably, Psf2 α-helices 3 and 5 (as defined in the human GINS structure[18]) contact a region of the Mcm5 N-terminal 'A domain' that is protected by the N-terminal extension of the MCM subunit 7 from the opposing ring in the double hexamer, as described in the atomic resolution yeast structure[4] (Supplementary Fig. 4).

As previously proposed[16,17,19], unoccupied density mapping next to GINS is assigned to Cdc45 and indeed matches the secondary structure elements of RecJ[20] (PDB entry 1IR6), a distant prokaryotic exonuclease homologue (Fig. 1d and Supplementary Fig. 5)[21,22]. In addition, a C-terminal protrusion projects from the RecJ-like catalytically defunct core of Cdc45 and wedges in between the Mcm5 and Mcm2 A domains[16], as previously observed at low resolution[19] (Figs 1a and 2a,b).

AAA+ ATPase in the activated MCM helicase motor

DNA unwinding by hexameric helicases involves the ATP hydrolysis-driven reconfiguration of the ATPase centres, causing allosteric changes that promote substrate translocation[23–25]. Seeking to identify various intermediates in the DNA unwinding reaction, we refined all intact CMG 3D classes and identified two similar, higher-resolution CMG forms that differ only in their ATPase module. The first form has been introduced

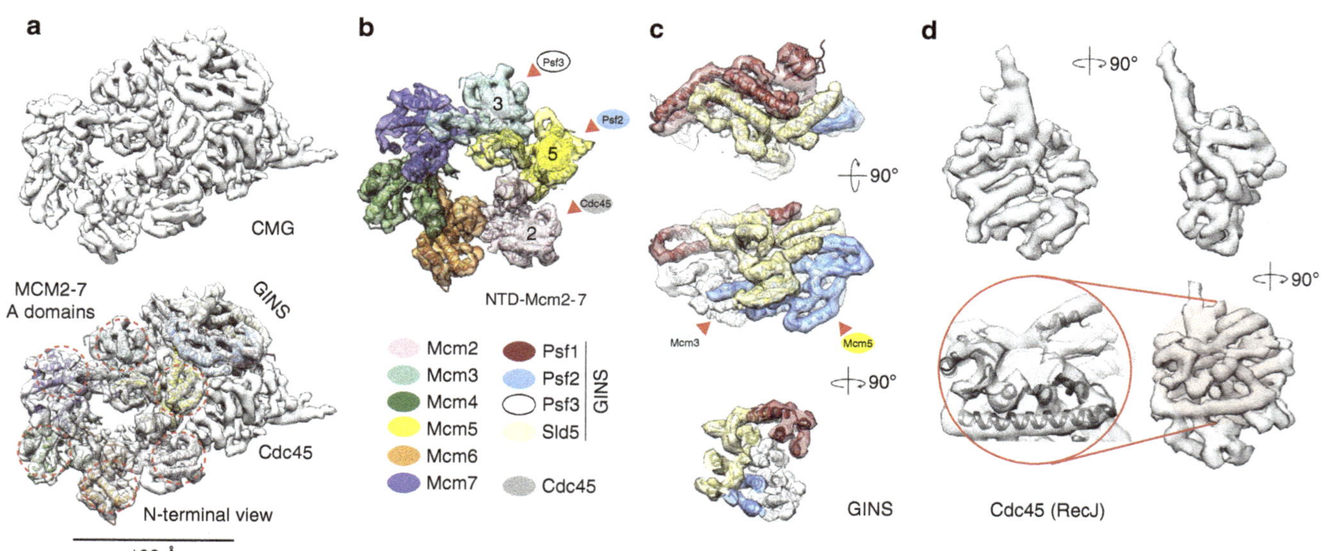

Figure 1 | CMG helicase structure at subnanometre resolution. (**a**) Resolution density (7.4 Å) map of the CMG viewed from the MCM N-terminal face, without or with docked MCM and GINS atomic structures. (**b**) Detailed view of the MCM N-terminal DNA-interacting collar. Psf3 contacts Mcm3, Psf2 contacts Mcm5 and Cdc45 contacts Mcm2. (**c**) Detailed view of the GINS assembly with docked human atomic structure (PDB entry 2Q9Q). (**d**) Density assigned to Cdc45. The Cdc45 core matches the secondary structure elements of the bacterial RecJ exonuclease (PDB entry 1IR6).

before, while a second form was refined to 9.8 Å resolution (Fig. 2a,b, Supplementary Figs 2,6 and 7, and Supplementary Movie 2). Inspection of the two helicase structures reveals that the ATPase tier is found as either topologically closed ('compact ATPase') or notched ('relaxed ATPase'). To define the interdomain arrangement in the helicase motor, we compared the recently established atomic model of yeast (double-hexameric) MCM[4] with our newly determined CMG structures. The compact ATPase configuration is highly reminiscent of the yeast MCM, with a closed ATPase ring and nearly parallel NTD and AAA+ tiers. In this configuration, MCM protomers are aligned and NTD interacts with AAA+ in *cis*, with a slight left-handed twist[4], confirming a previous model proposed by us[16,19] and others[17] (Fig. 2b). Conversely, in the relaxed ATPase form, the GINS-Cdc45 elements wedge in between the NTD and AAA+ tiers, a configuration previously observed at low resolution (Fig. 2a) (ref. 19). In this form, individual AAA+ domains become rearranged with respect to the cognate NTDs, approximating a conformer captured in a recent crystallographic study on an archaeal MCM hexamer[26]. DNA density can be recognized in both CMG structures and the nature of the DNA contacts will be discussed below.

Remodelling of the ATPase motor

To accurately describe ATPase motor dynamics, the yeast Mcm2-7 AAA+ hexamer was fitted into the CMG structure and further energy minimization was allowed for the individual docked ATPase modules into the two refined 3D classes. This docking exercise highlights a remarkable plasticity of the Mcm2-7 protomers, with relative tilting and rocking of the AAA+ domains, occurring both within the six Mcm2-7 ring subunits and between the two CMG conformers (Figs 2 and 3a). As previously proposed[16,19], we can assign a discontinuity in the AAA+ domain to the Mcm5-2 ATPase interface (Fig. 2a,b). By comparing the compact and relaxed structures, a conformational transition can be modelled in the ATPase tier. This reveals two types of ATPase interfaces that behave differently. Movement mainly depends on the loosening and tightening of the Mcm5-2 interface (root mean squared deviation (r.m.s.d.) 8.2 Å between the open and closed configurations; Fig. 3b,c). Mcm5-2 loosening is chaperoned by compensatory tightening of the neighbouring Mcm3-5 and 2-6 interfaces (r.m.s.d.: 5.3 and 8.3 Å, respectively; Fig. 3b,c), while the rest of the subunits undergo more subtle reconfigurations (with Mcm4-7 and Mcm7-3 r.m.s.d.: 3.1 and 3.3 Å, respectively; Fig. 3b,c). Coherent with these asymmetric rearrangements, *in vitro* studies of the *Drosophila* CMG have shown that certain Mcm2-7 ATPase sites (found at the static hinge in our structures) can be inactivated with minimal effects on DNA unwinding by the CMG, while other sites (the active Mcm5-2 ATPase and its immediate neighbors) are required for translocation[6] (Fig. 3c).

We have previously observed a compact Mcm5-2 interface at low resolution for a CMG in the presence of a non-hydrolysable ATP analogue (with or without DNA), while the nucleotide-free CMG contained a notched 5-2 interface[16,19]. These data support the notion that the relaxed ATPase form observed in our new, subnanometre resolution structure might contain an empty (post hydrolysis) ATPase site at the 5-2 interface. In line with this notion, the compact ATPase state is the predominant form (74% of particles that reach subnanometre resolution) in the ATPγS–CMG–DNA preparation, while the relaxed structure is underrepresented (the remaining 26%), as would be expected when using a slowly hydrolysable ATP analogue. To test whether ATP hydrolysis affects the particle distribution of the two conformers, we have solved the cryo-EM structure of the CMG incubated with ATP (hydrolysis rate is unaffected by the presence or absence of DNA[6]; imaging performed on a Titan Krios electron microscope, equipped with a Falcon II direct electron detector, FEI). Three-dimensional classification led to the identification of two CMG forms highly similar to the ATPγS/DNA-bound structures (although DNA free; Supplementary Figs 8–10 and Supplementary Movies 2 and 3). Remarkably, when the ATP-treated CMG is imaged, the notched Mcm5-2 structure is more represented than the compact state (72% versus 28%), supporting the notion that ATP hydrolysis by the Mcm5-2 site could convert the compact ATPase into a relaxed configuration (Supplementary Fig. 11). Although we cannot rule out the possibility that one of the two forms might be an inactive state of the helicase[12], it is possible that the two structures described here represent two intermediates of the DNA translocation reaction.

The CMG helicase is a single-stranded DNA translocase

Comparison of the compact ATPase DNA–CMG structure with the DNA-free counterpart reveals the presence of apparently bent, rod-shaped density features contacting the C-terminal face

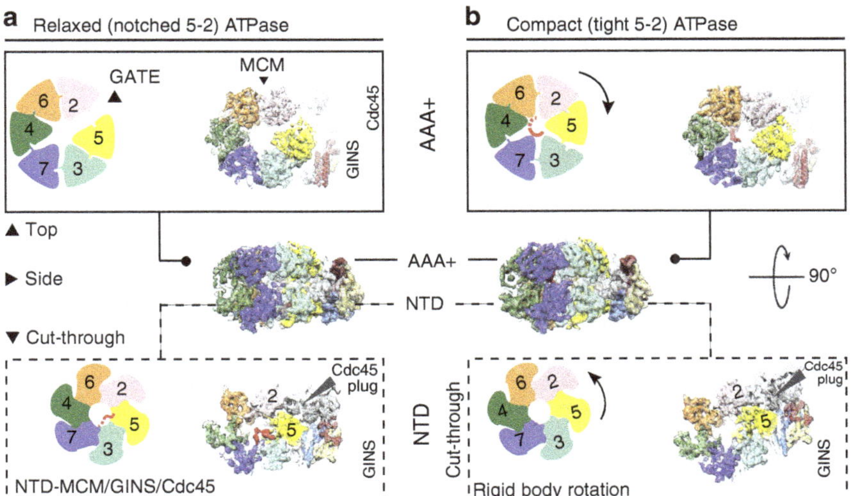

Figure 2 | Two configurations in the ATPγS-CMG-DNA complex. (**a**) AAA+, side and cut-through view of the CMG in a relaxed ATPase configuration. Cdc45 topologically locks the Mcm5-2 gate. (**b**) AAA+, side and cut-through N-terminal view of the CMG in a compact ATPase configuration. DNA density surmounting the AAA+ domain has been removed for visualization purposes (also see Fig. 4).

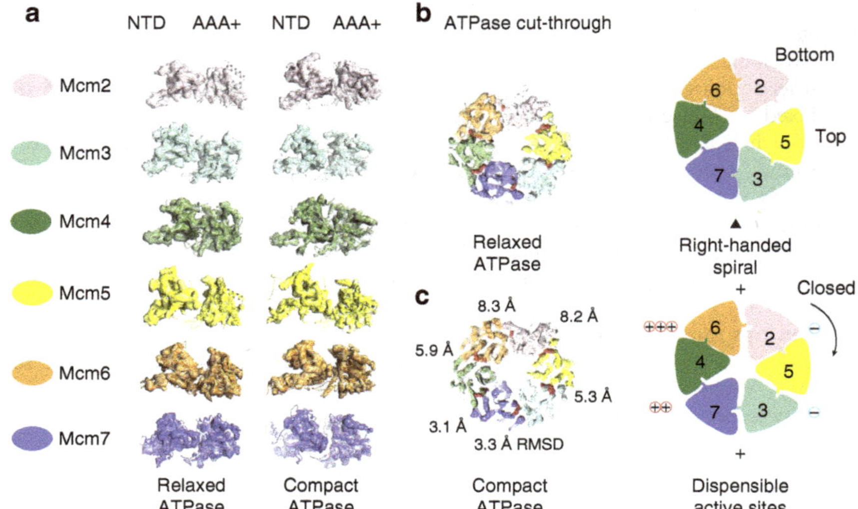

Figure 3 | The MCM ATPase centres. (a) Segmented density and docked atomic structures of the six Mcm2-7 protomers in the two conformers. The NTD and AAA+ domains tilt and rock with respect to one another. **(b)** ATPase sites in the relaxed ATPase conformer. **(c)** ATPase sites in the compact ATPase conformer. One side of the MCM ring is more static and the corresponding active sites are dispensable for viability (− means Walker A mutation kills DNA unwinding, + means tolerated mutation).

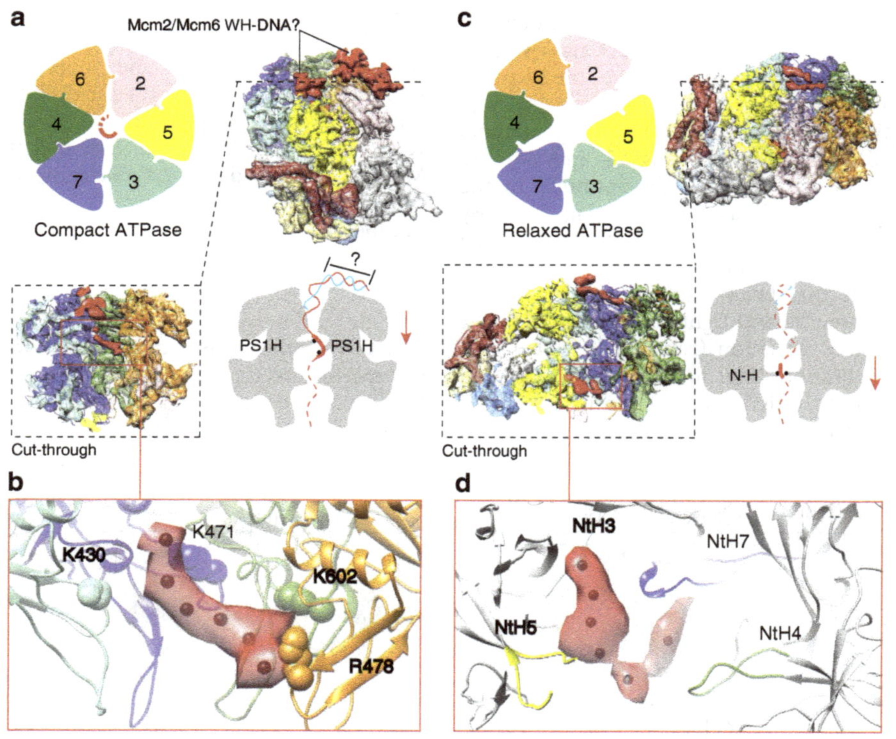

Figure 4 | DNA-bound form of the CMG helicase. (a) The compact ATPase form contains rod-shaped, bent density features surmounting the ATPase face tentatively assigned to duplex DNA engaged by the MCM winged helix (WH) C-terminal extensions. An MCM slice through the side view reveals an extended density feature, which we assign to single-stranded DNA, traversing the AAA+ pore. **(b)** Single-stranded DNA contacts conserved positively charged residues on the AAA+ PS1 hairpins that have a key role in DNA unwinding. **(c)** The relaxed ATPase form contains a thin density feature surmounting the AAA+ ring, which we assign to the flexible C-terminal MCM WH extensions. Although the AAA+ tier appears substrate free, a well-resolved elongated density threads through the MCM N-terminal collar and **(d)** contacts a set of N-terminal hairpins important for DNA binding and helicase activity.

of Mcm2-6 and surmounting the pore of the MCM motor (Fig. 4a and Supplementary Fig. 11). The local resolution is low in this region of the cryo-EM map (compared with the core protein components) and not suitable for atomic docking (Supplementary Figs 6 and 7). Nonetheless, the dimensions and overall shape of the additional density feature matches that of duplex DNA, possibly engaged by one or more C-terminal MCM extensions (containing a winged helix fold)[4]. The notion that duplex DNA enters through the AAA+, and not the N-terminal tier of MCM, is in agreement with our earlier streptavidin labelling of the DNA fork[19]. The inner core of the AAA+ domain is better defined and shows continuous density spanning three MCM subunits (Mcm7-4-6) and contacting the pre-sensor 1 (PS1) and helix-2 insert pore loops (Fig. 4b). This density can accommodate single- but not double-stranded DNA (Fig. 5), indicating that the AAA+ domain of the CMG helicase is a single-stranded DNA translocase (Supplementary Movie 3). A set of positively charged residues on the PS1 hairpins (Mcm7 K471, Mcm4 K602 and Mcm6 R478) contact and spiral around the nucleic acid (Fig. 4b). These DNA-binding elements are highly conserved from archaea through eukarya and their mutation affects DNA binding and abrogate DNA unwinding[11,12,27]. Noticeably, Mcm3 K430 is also poised in close proximity to the DNA density (Fig. 4b). This is a known DNA translocation element that is essential for viability in yeast[27].

Comparison of the relaxed ATPase structures in the presence or absence of DNA shows more subtle differences (Supplementary Fig. 11). In both structures, thin density features

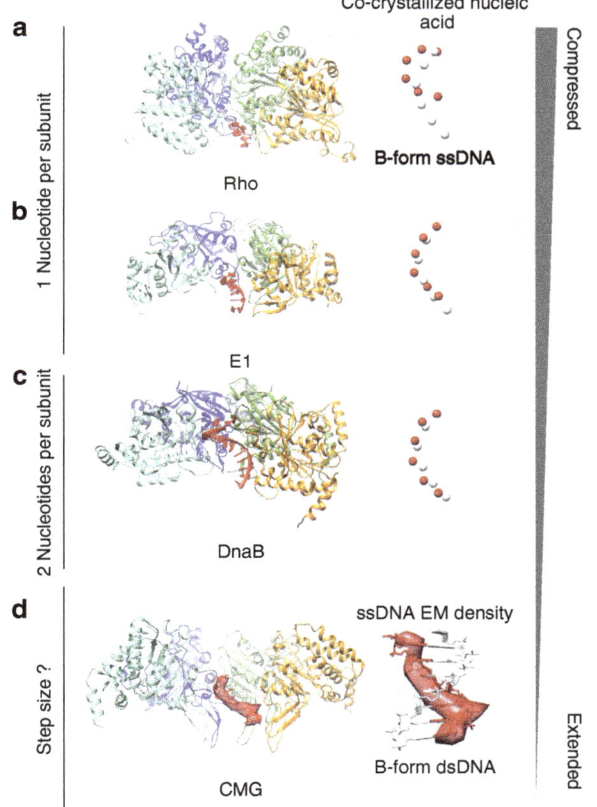

Figure 5 | Comparison of the ATPγS-CMG-DNA structure with available helicase-nucleic acid assemblies. **(a)** The Rho termination factor (PDB entry 3ICE) contacts and compresses single-stranded RNA. The RNA structure is compared with one single DNA strand extracted from a B-form double helix. Spheres represent the centre of mass of each nucleotide. **(b)** The Papillomavirus E1 replicative helicase (PDB entry 2GXA) contacts and compresses single-stranded DNA. **(c)** The bacterial DnaB replicative helicase contacts and slightly compresses single-stranded DNA (approximating A-form DNA, PDB entry 4ESV). **(d)** The MCM motor of the CMG harbours a more extended form of single-stranded DNA, which appears to match B-form DNA. The DNA density matches the structure of one single strand in the double helix. This supports the notion that the CMG is a single-stranded DNA translocase.

surmount the C-terminal face of the AAA+ pore and probably correspond to DNA-free winged helix MCM appendices[4], raising the question of whether DNA occupancy in this state is lower than in the compact ATPase state. Remarkably, although the AAA+ pore loops appear DNA free, a second set of N-terminal pore loops encircle an elongated density feature visible in the nucleic-acid-treated CMG but not in the DNA-free CMG. Thus, we assign the N-terminal contacting density to DNA. Similar to the PS1 hairpins, these 'N-terminal hairpins' are also conserved and are essential for DNA binding and have a role in unwinding[28–30] (Fig. 4d). It remains to be established whether the compact and relaxed DNA-binding modes in the ring channel correspond to a translocation and a paused state[12], respectively, or represent two intermediates in the DNA translocation reaction.

Comparison of the DNA-engaged MCM ATPase tier with other structures of nucleic acid-bound hexameric helicases highlights a mixture of similarities and differences (Fig. 5). A universal feature appears to be the presence of a right-handed spiral formed by the ATPase pore loops, which follows the helical character of the nucleic acid substrate[15]. Whereas in previously reported helicase structures (Rho, E1 and, to a lesser extent, DnaB) the translocation substrate is more compressed than that of canonical B-form DNA[24,31,32] (Fig. 5a–c), the ATPγS–CMG–DNA complex matches the structure of one single strand extracted from a B-form double helix (Fig. 5d). Noticeably, the Rho and E1 helicases, which significantly compress the translocation substrate as they encircle it, contact one base per ATPase protomer, while DnaB contacts two bases per protomer. These observations are compatible with an unwinding step size of 1- (Rho and E1) and 2-bp (DnaB) per ATP hydrolysed. Although the resolution of our compact ATPγS–CMG–DNA complex is not high enough to count bases in the single-stranded nucleic acid, the more extended state of the DNA substrate suggests that the CMG might unwind at least two base pairs per ATP molecule hydrolysed. To confirm the notion that single-stranded DNA is stabilized in an extended configuration on CMG binding, we performed single-molecule experiments, in the presence of either ATPγS or ATP. CMG–DNA binding was monitored by measuring FRET efficiency between a Cy3 donor and a Cy5 acceptor fluorophore, separated by seven nucleotides in surface-immobilized single-stranded DNA. This is an established method to measure DNA stretching by multi-subunit, DNA-encircling ATPases[33] (Fig. 6a,b). CMG binding to DNA in the presence of ATPγS results in a dramatic decrease from 0.94 ± 0.01 (naked DNA; Fig. 6a) to 0.44 ± 0.01 (DNA+CMG; Fig. 6b and Supplementary Fig. 12) in mean FRET (x_0), due to both a reduction in molecular flexibility relative to naked single-stranded DNA and CMG-mediated single-stranded DNA stretching. Similar results were obtained with ATP ($x_0 = 0.96 \pm 0.01$ for naked DNA and $x_0 = 0.48 \pm 0.01$ for CMG), although binding/stretching events occurred with lower frequency (Supplementary Fig. 12). These data agree with the observation that use of a slowly hydrolysable ATP analogue is required to stabilize the interaction between the MCM AAA+ domain and single-stranded DNA[6,12], and support the notion that CMG binding stabilizes and stretches single-stranded DNA. We note that the lower frequency of DNA binding/stretching observed when CMG is ATP incubated correlates with the predominance of the relaxed ATPase form observed in the ATP–CMG cryo-EM data set.

ATPase controlled rotation of the CMG DNA-binding domain

How does the ATPase state of the MCM motor affect the rest of the CMG structure? To address this question we have aligned the AAA+ rings of the two ATP–CMG conformers and used

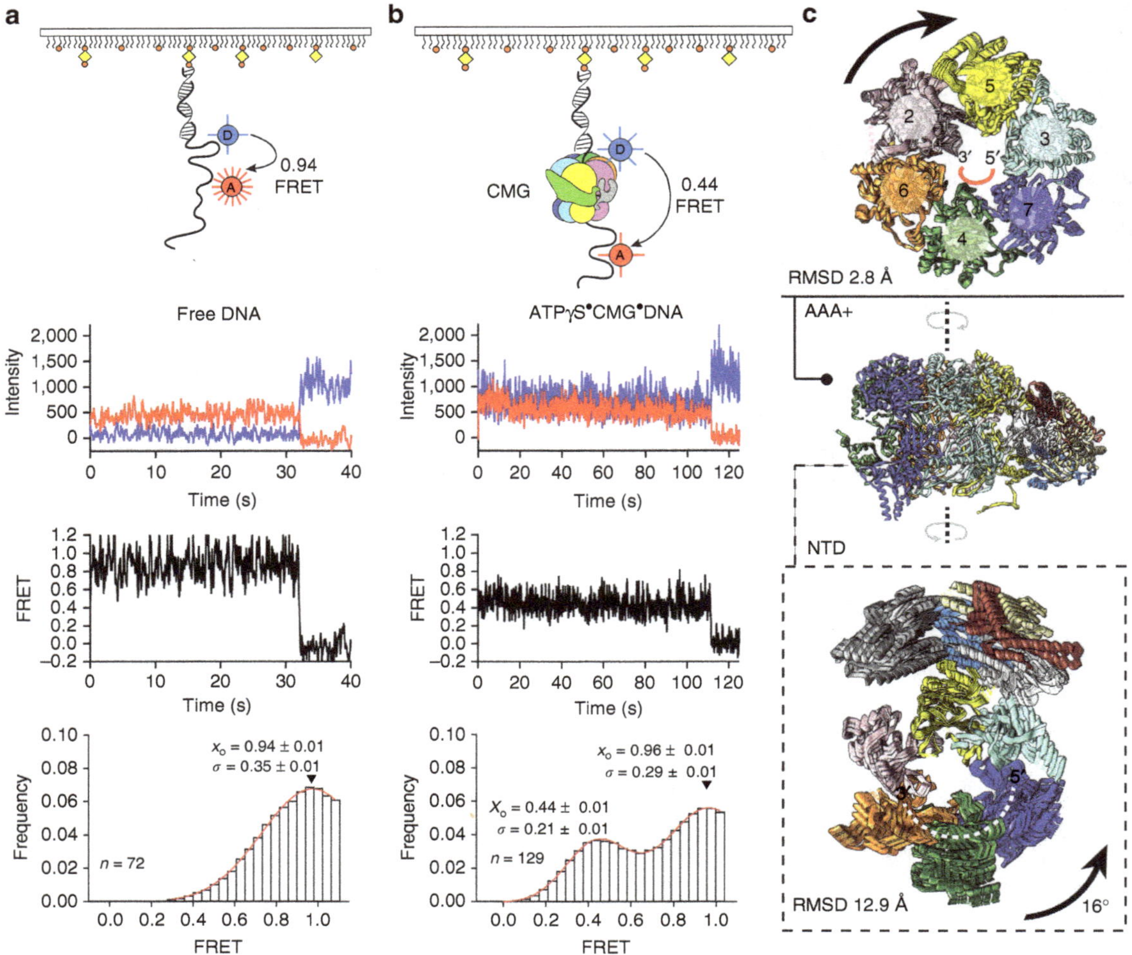

Figure 6 | DNA engagement and deformation by the CMG helicase. Single-molecule FRET analysis of (**a**) isolated DNA and (**b**) DNA + ATPγS CMG. Top to bottom: cartoon schematic of the single-molecule FRET experiment indicating FRET between Cy3 (Donor, blue sphere) and Cy5 (Acceptor, red sphere, 7-nt separation) attached to biotinylated 3′-tailed duplex DNA, immobilized on a neutravidin-coated biotin-PEG surface; donor (blue) and acceptor (red) intensity trajectories are anti-correlated until single-step photobleaching of the acceptor. FRET trajectories (black) between Cy3 and Cy5 exhibit a sharp drop to zero FRET when the acceptor photobleaches; histogram of FRET values collected from all molecules with a Gaussian fit are shown in red, where x_0 is the mean FRET and σ is the distribution width. ATPase controlled ring rotation and DNA stretching by the CMG helicase. (**c**) The AAA + tier of the Mcm2-7 motor rotates clockwise as Mcm2 moves towards the Mcm5 protomer, to close a gap in the motor domain. The DNA-interacting NTD rotates anticlockwise as a rigid body, together with GINS and Cdc45.

molecular morphing to model a transition from a relaxed to a compact ATPase conformation. In our model, the motor undergoes minor subunit rearrangements, with an overall r.m.s.d. of 2.8 Å, mainly due to the breathing of the Mcm5-2 interface (Fig. 6c). Conversely, the non-catalytic portion of the CMG complex, formed by Cdc45, NTD-Mcm2-7 and GINS, rotates by 16° as a rigid body with respect to the motor (overall r.m.s.d. of 12.9 Å between the two forms; Figs 2a,b and 6c, and Supplementary Movie 4). Remarkably, the NTD and AAA + tiers of Mcm2-7 rotate in opposite directions, as they transition from a relaxed to a compact ATPase form (Fig. 6c). This rotation is compatible with a nucleotide-state-controlled inter-subunit movement, previously observed using double electron–electron resonance in the archaeal MCM motor[29].

Previous crystallographic studies in archaea have shown that single-stranded DNA can line the NTD-MCM pore. This interaction (preserved in the eukaryotic Mcm6-4-7) is essential for DNA replication initiation in yeast and probably important for the initial melting of the origin DNA duplex (Fig. 6c). According to our model, as the ATPase motor tightens to engage the translocation strand, the AAA+ domain rotates clockwise, while the NTD DNA-interacting collar of the MCM rotates anticlockwise. Considering that DNA interacts with the inner perimeter of NTD-MCM running 3′→5′ in an anticlockwise manner[30], we note that a further anticlockwise NTD rotation would result in DNA duplex underwinding, indeed compatible with origin DNA opening.

Discussion

DNA replication start sites need to be licensed for initiation and this process involves the deposition of a pair of MCM helicases that form a two-fold symmetric, double-hexameric assembly[2,3]. GINS/Cdc45 recruitment is necessary for origin firing and multiple lines of evidence suggest that two CMG holo-helicases separate and move in opposite directions during replication. For example, single-molecule studies have demonstrated that two replisomes need not be physically linked for efficient DNA replication to occur[8]. Further evidence of helicase separation is derived from EM studies on the CMG, which primarily form monomeric assemblies[16] (although loosely tethered dimeric CMGs have also been observed)[19]. A recent high-resolution study describes the dimerization interface of the yeast MCM double ring, which is in part formed by an N-terminal Mcm7 helical insert that latches onto the Mcm5 A domain in the opposing hexamer[4]. Our subnanometre resolution CMG structure reveals that the A domain of Mcm5 undergoes a reconfiguration in the CMG (Supplementary Movie 1), which would be incompatible with Mcm7 engagement (at least in yeast). Remarkably, the highly conserved Mcm5 site that is generally protected by Mcm7 in the double hexamer is found GINS associated in the CMG (Supplementary Fig. 4). GINS recruitment depends on the DDK phosphorylation of Mcm4, 6 and 7 (refs 5,34); however, DDK requirement in yeast can be bypassed by a mutation of Mcm5 (ref. 35) that promotes a reconfiguration of the Mcm5 A domain[28]. DDK phosphorylation probably promotes a similar conformational change in Mcm5 that could disrupt the (yeast specific) Mcm5-7 double hexamer interface and enable GINS binding. These events, we suggest, might help destabilize the double hexamer and contribute to replication intiation (Fig. 7).

Our new CMG structures better explain the role of Cdc45, containing a globular extension emanating from the RecJ-like core. This extension inserts in between the A domains of Mcm5 and 2 (refs 16,19,36), plugging a gate that is generally found open in the isolated MCM complex[16,36] and used for DNA loading[37]. During elongation, ring opening is thought to allow for the bypass of a roadblock on the translocation strand in certain viral replicative helicases[38], while the eukaryotic replisome stalls on meeting an obstruction on the translocation strand[10,39]. A topologically sealed Mcm2-7 ring (the CMG structure described here) provides a rationale to explain this observation. With Cdc45 plugging the Mcm5-2 gate, full ring opening is energetically disfavoured in the holo-helicase (Fig. 2a,b), possibly explaining helicase stalling.

Whether the CMG helicase is a single-stranded or a duplex DNA translocase is still debated[4,9,10,40]. As the Mcm2-7 is initially loaded onto duplex DNA, replication fork establishment has been suggested to occur via two alternative mechanisms. According to the 'strand extrusion' model duplex, DNA enters the AAA+ ATPase tier of MCM and the two strands become spatially segregated before exiting the helicase toroid[9,13]. A second model, called 'strand exclusion', envisages translocation on one single strand and steric exclusion of the other strand[9–11] (Supplementary Fig. 13). Recent structural characterization of the yeast MCM double hexamer provides some evidence in support of the strand extrusion model[4]. In the double-ring structure, a unique lateral channel exists between the Mcm6 and 2 subunits, which might be used for extrusion of the lagging strand. Given the position of this side pore, mapping in between the AAA+ and NTD tiers, strand extrusion would occur after the double strand has been spooled through the AAA+ motor (Supplementary Fig. 13). This hypothesis is supported by the notion that the CMG helicase contains a central pore that is large enough to accommodate double-stranded DNA[16,17,19]. Our new ATPγS-CMG-DNA structure, however, provides robust evidence against a duplex translocase model and favours the steric exclusion mechanism. In fact, in the compact ATPase state we observe thin continuous density that spirals inside the AAA+ MCM hexamer and contacts a set of positively charged residues known to be important for DNA binding and translocation. This density can accommodate single- but not double-stranded DNA (Fig. 5d), indicating that the MCM motor unwinds a replication fork by translocating on the leading strand template, whereas the lagging strand template is sterically excluded (Fig. 7). Support for our model derives from studies on *Xenopus* egg extracts, according to which the replisome can bypass a road block on the lagging strand but not on the translocation (leading) strand[10]. Future, high-resolution studies on higher-order assemblies will be needed to define the replication fork trajectory in the eukaryotic replisome[17].

Three main mechanisms have been proposed for coupling nucleotide turnover with substrate translocation in hexameric NTPases, involving either stochastic, concerted or rotary firing[15]. The stochastic model, mainly supported by studies on protein unfoldases, suggests that the NTPase firing does not need to follow an ordered sequence of events for substrate translocation

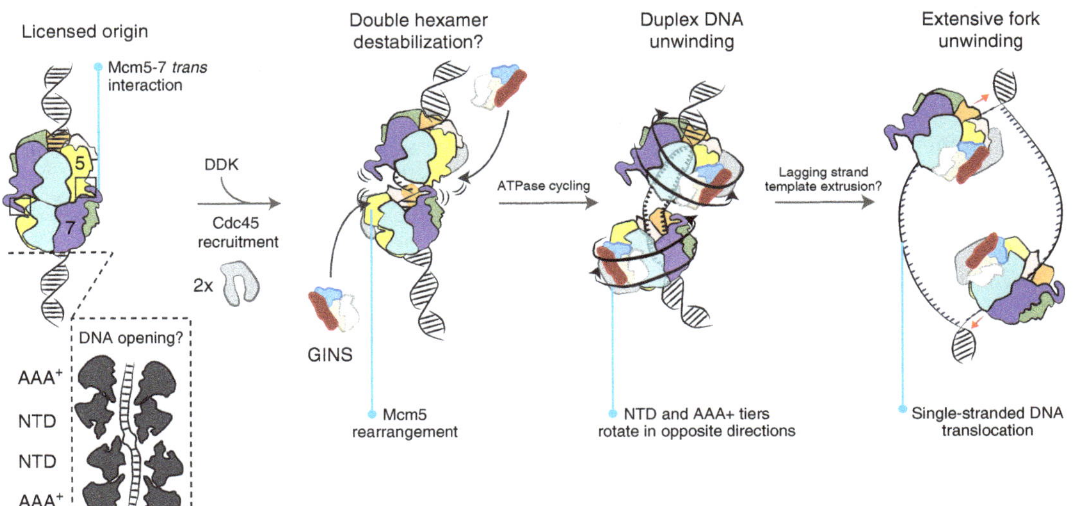

Figure 7 | Origin activation and replication fork unwinding by the CMG helicase. Origin licensing involves the loading of a head-to-head double hexameric ring that encircles duplex DNA, which might become partially deformed. Cdc45 is loaded onto the double hexamer, in a process that requires DDK phosphorylation of MCM. DDK phosphorylation might cause a rearrangement in Mcm5 and disrupt the (yeast specific) Mcm5-7 *trans* interaction. This rearrangement would expose a GINS interacting element in Mcm5. ATP hydrolysis by the MCM promotes the relative rotation of the NTD and AAA+ tiers of the helicase, in a movement that might promote duplex DNA underwinding. Following a poorly understood lagging-strand extrusion process, the CMG helicase extensively unwinds the replication fork, by translocating on single-stranded DNA.

to occur[41,42]. The concerted model envisions six NTPase subunits that simultaneously bind, hydrolyse or release the nucleotide. This mechanism is supported by symmetrized crystallographic structures of the SV40 large T-antigen replicative helicase[23]. According to the rotary mechanism, nucleotide hydrolysis occurs sequentially from one hydrolase centre to the next, similar to a Mexican wave around a stadium[15,24,31]. The rotary cycling model is supported by biochemical and crystallographic studies of the T7 gp4 helicase[43] and by three helicase structures, the Rho transcription termination[24] factor, the Papillomavirus E1 (ref. 31) and the bacterial DnaB replicative helicases[32], which have all been imaged bound to their translocation substrate (single-stranded nucleic acid). Importantly, in the E1 and Rho translocation–substrate complexes, the hydrolase centres are radially distributed in an order that sequentially visits the steps of the NTPase cycle[24,31]. In both structures, a staircase formed by the ATPase pore loops follows the right-handed spiral of a single-stranded nucleic acid and the position of the pore loops in the assembly correlates with the nucleotide occupancy state of the ATPase centres. An ATPase-controlled rotary pore-loop movement would therefore promote substrate translocation through the ATPase channel. Owing to the homo-oligomeric nature of the E1 helicase complex however, a rotary cycling movement in the helicase hexamer can only be modelled and not visually proven. Because of its inherently asymmetric nature, we reasoned that the Mcm2-7 hetero-hexamer would be a useful tool to test the rotary cycling model. The asymmetric distribution of ATPase centres and the ordered spiral organization of the DNA-interacting PS1 hairpins indeed favour a sequential rotary over a concerted firing or stochastic mechanism for substrate translocation by the CMG. However, to our surprise, we were unable to characterize multiple rotational states of the ATPase, with DNA contacting different ATPase protomers. Rather, the ATPase exists in two states: compact (DNA gripping) and relaxed (DNA ungripping; Supplementary Fig. 14). Our results lend themselves to two possible explanations. A first possibility is that one of six rotational states of the MCM is more stable than the other, short-lived ATPase ring permutations. However, we favour a second possibility. The ATPase motor could fluctuate between the compact state (where DNA is stabilized/extended by Mcm7-4-6) and the relaxed state where the AAA+ motor releases the substrate. In this second configuration, DNA could be handed off to the N-terminal collar and our observation of stable NTD–DNA association in the relaxed ATPase structure supports this model. Alternatively, the relaxed ATPase configuration could represent a stalled form of the helicase, as could be found, for example, at a pausing replication fork (Supplementary Fig. 14).

Notably, not all ATPase sites in the MCM motor of the CMG equally contribute to DNA translocation[6,44] and this feature is shared by other hetero-hexameric AAA+ ATPases such as the Rpt1-6 assembly[45] of the proteasome or the dynein motor (which in some organisms contains ATPase centres that have become inactivated during evolution)[46]. Both the proteasome and dynein have been imaged in a compact and a relaxed ATPase form, with movement dependent on the breathing of one lone hydrolase interface and chaperoning activity of the neighbouring interfaces[47,48] (such as the Mcm5-2 site and its immediate neighbours in our CMG structure). Although these ATPase motors are functionally very distinct (a DNA, a polypeptide and a microtubule translocase, respectively), phylogenetic[49,50] and mechanochemical[19] kinship has been noticed before for the three systems, suggesting a shared mechanism of substrate translocation for hetero-hexameric AAA+ ATPases. Taken together, our results suggest that the CMG helicase might translocate on single-stranded DNA via a mechanism distinct from strictly sequential rotary cycling[6,44] and probably involves the interplay between the N-terminal and AAA+ tiers of the MCM motor[29].

Methods

Purification of the CMG helicase. pFastBac1 plasmids containing genes that encode the individual subunits of the CMG complex (a gift from Dr Michael R. Botchan) were used to produce bacmids, which were subsequently transfected into Sf9 cells using the Invitrogen Bac-to-Bac Baculovirus Expression System methods[6]. P3 baculoviruses used in protein expression experiments were freshly amplified from the P2 stocks for 5 days in 100 ml of Sf9 cells grown in 250 ml flasks in Graces medium supplemented with 10% FCS. The infections for protein purification were carried out by inoculating 4 l of Hi5 cells at $10^6\,ml^{-1}$ with a multiplicity of infection of 5 by adding 200 ml of each virus stock at a titre 10^8 pfu ml^{-1}. Infected cells incubated for 72 h at 27 °C were harvested by centrifugation and washed with PBS + 5 mM $MgCl_2$. The following steps were performed either on ice or at 4 °C, unless indicated otherwise. The collected cells were resuspended in 200 ml of buffer C (25 mM Hepes pH 7.6, 0.02% Tween-20, 10% glycerol, 1 mM EDTA, 1 mM EGTA) supplemented with 15 mM KCl, 2 mM $MgCl_2$, 0.4 mM phenylmethylsulfonyl fluoride, 2 mM 2-mercaptoethanol and the complete protease inhibitors cocktail from Roche Diagnostics. The cell suspension was snap frozen in 10 ml aliquots and stored at − 80 °C. To purify CMG complexes from the extract, the infected cell suspension was thawed and cells were broken in a Dounce homogenizer. KCl was added to 100 mM and the extract was cleared by centrifugation at 14,000 r.p.m. in an Avanti J-26S XP centrifuge for 10 min. The cleared extract was incubated with 2 ml of anti-FLAG M2-agarose beads (Sigma-Aldrich) for 2–3 h with continuous end-over-end mixing. The beads were then collected in a 20-ml Poly-Prep disposable chromatography column (BioRad) and washed in 30 ml C-100 buffer (buffer C with 100 mM KCl and 1 mM dithiothreitol (DTT)) before transferring to a 10 ml column (BioRad). The column was washed twice with 5 ml C-100 buffer. Bound complexes were eluted with 200 $\mu g\,ml^{-1}$ flag peptide in C-100 supplemented with complete protease inhibitor cocktail, the first elution step was performed in 5 ml of elution buffer for 15 min at room temperature with end-over-end mixing, then collecting the flowthrough and repeating the step with 4 ml of elution buffer for 10 min. Both fractions were pooled, cooled on ice and pumped through a Mono S HiTrap SP FF column equilibrated in C-100 buffer. The flowthrough and a further 4 ml of wash was collected and injected onto a MonoQ HR 5/5 column equilibrated in C-100 buffer[6]. The column was washed in 15 ml of C-100 and bound complexes were eluted with 20-ml 100–550 mM KCl gradient in buffer C supplemented with 1 mM DTT. Fractions (0.5 ml) were collected, the CMG peak fractions (eluted at 410–440 mM KCl) were diluted to 150 mM KCl in buffer C and injected onto a Mono Q PC 1.6/5 column connected to AKTAmicro purification system equilibrated in buffer D (25 mM Hepes pH 7.6, 1 mM EDTA, 1 mM EGTA, 1 mM DTT) supplemented with 150 mM KCl. Elution was performed using a 2-ml 150–550 mM KCl linear gradient in buffer D and 75 μl fractions were collected. The CMG peak fractions were pooled and dialysed for 16 h into buffer A (25 mM Hepes pH 7.6, 50 mM sodium acetate, 10 mM magnesium acetate, 1 mM DTT). The CMG preparation was supplemented with 1 mM ATP or treated as described in the 'CMG–DNA complex reconstitution' section. Protein concentration was measured by using known MCM3 protein standards serving as a reference on a silver stain SDS–PAGE gel. A yield of 150 μl with a concentration of 1 μM was achieved.

Helicase assay. Assays were performed in 20 μl reactions contacting 25 mM Hepes pH 7.5, 75 mM NaCl, 0.5 mM ATP, 10 mM magnesium acetate, 1 mM DTT, 0.1 mg ml^{-1} BSA. Purified CMG (123 to 410 fmol) was incubated in the presence of 27 μM ATPγS and 1.8 pmol of [γ-^{32}P]-ATP-labelled fork DNA substrate for 3 h at 30 °C, to allow for fork loading. Unwinding was induced by adding ATP to 7 mM for 5 min at 30 °C and stopped with 6 × stop buffer (150 mM Tris pH 8.0, 3% SDS, 120 mM EDTA). The unwinding product was separated using 8% PAGE in 0.1% SDS 1 × TBE.

CMG-DNA complex reconstitution. The oligonucleotides used for CMG–DNA reconstitution were synthesized by Integrated DNA Technology. Sequences of the two partially complementary oligos used are shown below:

Leading strand template: 5′-CACTCGGGCTCGTTTTACAACGTCGTGACT GGGCACTTGATCGGCCAACCTTTTTTTTTTTTTTTTTTTTTTTTTTTTTTTT TT TTTTTTT-3′

Lagging strand template: 5′-CTGGCGTCGGGTCGGCGGTTGGCCGAT CAAGTGCCCAGTCACGACGTTGTAA AACGAGCCCGAGTG-3′

When annealed together, this model replication fork contains a 50-bp double-stranded region with two single-stranded fork overhangs (16-nt on the 5′-end and 40-nt poly-T on the 3′-end, based on Petojevic *et al.*[12]).

Annealing of the two strands was performed by mixing each of the oligos in equimolar amounts, heating at 95 °C for 3 min and slow cooling to room temperature for 60 min. To reconstitute the nucleoprotein complex, forked DNA substrate was added to dialysed ~300 nM CMG in a 2:1 molar ratio in the presence of 0.1 mM ATPγS (Sigma). The CMG–DNA–ATPγS mixture was left to incubate for 2 h at 30 °C, to allow for complex association before cryo-grid preparation.

Cryo-grid preparation and data collection for CMG–DNA. Four microlitres of reconstituted CMG–DNA assembly at 300 nM concentration were applied onto freshly glow-discharged Quantifoil 1.2/1.3 or C-flat 1/1 grids. After a 30-s incubation in 100% humidity, the Quantifoil grid was double-side blotted for 4 s using a Vitrobot (FEI) and plunged into liquid ethane. For the C-flat grid, the sample was incubated for 2 min and double-side blotted for 3 s in a Cp3 (Gatan, Inc.), operating at 90% humidity and plunged into liquid ethane. Cryo-grids were screened for ice quality using on a JEOL-2100 or a FEI Spirit LaB6 operated at 120 kV and equipped with an 4 k × 4 k or 2 k × 2 k Ultrascan charge-coupled device camera (Gatan, Inc.), respectively. Data were collected on a Tecnai F30 Polara electron microscope operated at 300 kV and equipped with a K2 Summit direct electron detector (Gatan, Inc.) and an energy filter in zero-loss mode (GIF Quantum, Gatan, Inc.). Twenty-five-frame movies (2,098) were manually collected, with a single frame duration of 0.4 ms. Movies were acquired using SerialEM, in single-electron counting mode with a total dose of 48 e^{-} Å^{-2} and a − 3.5 to 1.7 defocus range, at a constant nominal magnification of × 37,037, yielding a 1.35-Å pixel size.

Cryo-grid preparation and data collection for ATP–CMG. The ATP–CMG sample at 330 nM concentration was applied onto freshly glow-discharged open-holes Quantifoil 1.2/1.3. After 30 s incubation, grids were double-side blotted for 5 s in a Vitrobot (FEI) at 100% humidity and plunge frozen into liquefied ethane. Grids were loaded onto the LMB Cambridge Titan Krios electron microscope (FEI) operated at 300 kV for automated data collection with the EPU software (FEI). Images were recorded on a FEI Falcon II detector at a nominal magnification of × 47,000 (yielding a pixel size of 1.77 Å). An in-house built system described in ref. 51 was used to collect 17 frames per second. Five hundred and thirty-six movies of 51 frames were recorded using a − 2 to − 4 µm defocus range with an electron dose of 51 e^{-} Å^{-2}.

CMG–DNA image processing. To correct for beam-induced drift, whole frame alignment and averaging was performed for each movie using MotionCorr (http://cryoem.ucsf.edu/software/driftcorr.html)[52]. Particles (340,573) were picked semi-automatically in EMAN2 (ref. 53). Contrast transfer function parameters were estimated using CTFFIND4 (ref. 54) and low-quality integrated movies were excluded. All further processing was performed in RELION 1.4 (ref. 55). Extracted particles were binned by 2, yielding a pixel size of 2.7 Å per pixel. Two-dimensional classification allowed for the isolation of a set of 60,287 high-quality CMG particles (Supplementary Fig. 2). A first 3D refinement was performed using a 50-Å low-pass filtered compact ATP–CMG structure as a starting model, resulting in an initial 8.1 Å structure. Postprocessing (using automatically estimated B factor and an arbitrarily chosen mask) improved the resolution to 7.2 Å but resulted in a poorly defined ATPase tier (Supplementary Fig. 2). To separate various ATPase conformers in the data set, the refined 3D volume was filtered to 50 Å and used as a model for 3D classification without particle alignment (ten classes). After excluding poor-quality classes, 2 structures were isolated, referred to as 'compact ATPase' (38,792 particles; Supplementary Fig. 6) and 'relaxed ATPase' (13,692 particles; Supplementary Fig. 7, also refer to Figs 2 and 4, and Supplementary Fig. 2 for comparisons). Each particle subset was separately refined and postprocessed, resulting in a 7.4-Å resolution structure (compact ATPase) and 9.8 Å (relaxed ATPase).

ATP–CMG image processing. Similar to the CMG–DNA data set, beam-induced drift was corrected and averaging was performed for each movie as described in ref. 52. Contrast transfer function parameters were estimated using CTFFIND3 (ref. 56) and the best 471 out of 536 integrated movies were selected for particle picking in XMIPP3 (ref. 57) and subsequent processing in RELION 1.3 (ref. 55). Semi-manually picked particles (160,401) were 2D-classified, to get rid of bad particles, resulting in a clean data set containing 78,601 particles. A first ATP–CMG 3D volume was determined to 9.5 Å, using a 50-Å-filtered version of EMDB entry 2,772 as an initial model (Supplementary Fig. 8). Using the 9.5-Å structure filtered to 50 Å, 3D classification was performed to identify four recognizable CMG classes[58] as the new, '3D-cleaned' data set. Using statistical movie processing[59], an 8.3-Å reconstruction (Supplementary Fig. 8) was then obtained from these merged (29,772) particles. As the resulting structure showed a disordered AAA+ motor domain, further focused 3D classification[55,58] was performed to identify different motor conformers. This resulted in two classes (compact ATPase, 5,111 particles and relaxed ATPase, 13,182 particles), which were independently refined and postprocessed as described above (Supplementary Figs 8–10).

Resolution estimation and model building. Resolution of the obtained cryo-EM maps was estimated using the 'gold-standard' Fourier Shell Correlation (FSC) method, using the 0.143 FSC criterion. Local resolution was estimated using ResMap[60] and visualized using UCSF Chimera[61]. Automated atomic docking was performed using UCSF Chimera and further model manipulation performed using The PyMOL Molecular Graphics System. The A and B–C domains of the NTD tier of the yeast Mcm2-7 structure (PDB entry 3JA8 (ref. 4)) were docked as isolated rigid bodies. GINS (PDB entry 2Q9Q[18]) and RecJ (PDB entry 1IR6 (ref. 20)) were docked as rigid bodies. The C-terminal Psf1 domain GINS was modelled as described in ref. 19. The AAA+-MCM domains were docked as six individual rigid bodies after docking of the cryo-EM MCM AAA+ tier (PDB entry 3JA8 (ref. 4)). FSC to compare the CMG atomic models and the cryo-EM maps were computed using RELION 1.4 and visualized with the PDBe FSC server. Figures and Movies were generated using UCSF Chimera[61].

Single-molecule FRET. DNA oligonucleotides were purchased from Operon and labelled with fluorescent dyes Cy3 and Cy5. Reactions were carried out using an amino-modified C6-dT oligonucleotide and a mono-reactive Cy3 or Cy5 dye (GE Healthcare). Labelling reactions were performed with 1 nmol of DNA in 44 µl of 100 mM sodium carbonate buffer pH 8.5 and 10 nmol of Cy3 or Cy5 dye dissolved in 7 µl of dimethyl sulfoxide. Reactions were performed overnight at room temperature. Reverse-phase HPLC purification was performed on an analytical C8-column (Sigma-Aldrich Supelco Discovery BIO wide pore C8, 25 cm × 4.6 mm × 5 µm), to separate labelled and unlabelled DNA; fractions containing labelled DNA were collected and stored at − 20 °C in 10 mM Tris-HCl pH 8.0. DNA sequences are as follows; for the Cy3 strand 5′-Cy3-CGCGAGGA ATGGATGTAGGG-biotin-3′ and for the Cy5 strand 5′-CCCTACATCCA TTCCTCGCGTTTTTT (Cy5-T)$(T)_{65}$-3′. Quartz slides and cover slips were prepared following established protocols[62]. Briefly, quartz slides and coverslips were passivated with methoxy-PEG-SVA (M_r = 5,000, Laysan Bio, Inc.) containing 10% biotin-PEG-SVA (M_r = 3,400, Laysan Bio, Inc.) in 100 mM sodium bicarbonate. Reaction chambers were first incubated with 0.2 mg ml^{-1} BSA (Sigma-Aldrich) in T50 buffer (10 mM Tris-HCl pH 7.0 and 50 mM NaCl) for 10 min. Next, BSA was washed with T50 buffer and neutravidin (0.2 mg ml^{-1} in T50 buffer) was injected and incubated for 10 min. Excess neutravidin was removed by washing with buffer E (25 mM Hepes pH 7.6, 50 mM sodium acetate, 10 mM magnesium acetate, 10% glycerol, 1 mM DTT and 2 mM Trolox). DNA was surface-immobilized by incubation for 10 min with an annealed biotinylated DNA duplex (25 pM). Excess DNA was then washed with imaging buffer containing the CMG protein (Buffer E supplemented with 50 µM ATPγS (Sigma-Aldrich) or ATP (Sigma-Aldrich), 7 nM CMG protein and an oxygen scavenging system containing 10 mM 3,4-dihydroxybenzoic acid (Sigma-Aldrich) and 120 nM protocatechuate dioxygenase (Sigma-Aldrich; the final concentration of protocatechuate dioxygenase is corrected for the presence of 40% stabilizer), to minimize dye photobleaching). DNA molecules were imaged on a home-built, prism-based total internal reflection fluorescence microscope. All single-molecule measurements were recorded at room temperature using continuous green excitation (532 nM laser) at ~1.0 mW and 30 ms time resolution. Apparent FRET efficiencies were calculated as FRET $= I_A/(I_D + I_A)$, where I_A and I_D are the acceptor and donor intensities, respectively. Acceptor intensity (I_A) was corrected for donor emission in the acceptor channel (11%), no direct excitation of the acceptor was observed.

References

1. Remus, D. & Diffley, J. F. Eukaryotic DNA replication control: lock and load, then fire. *Curr. Opin. Cell Biol.* **21,** 771–777 (2009).
2. Remus, D. *et al.* Concerted loading of Mcm2-7 double hexamers around DNA during DNA replication origin licensing. *Cell* **139,** 719–730 (2009).
3. Evrin, C. *et al.* A double-hexameric MCM2-7 complex is loaded onto origin DNA during licensing of eukaryotic DNA replication. *Proc. Natl Acad. Sci. USA* **106,** 20240–20245 (2009).
4. Li, N. *et al.* Structure of the eukaryotic MCM complex at 3.8A. *Nature* **524,** 186–191 (2009 (2015).
5. Yeeles, J. T., Deegan, T. D., Janska, A., Early, A. & Diffley, J. F. Regulated eukaryotic DNA replication origin firing with purified proteins. *Nature* **519,** 431–435 (2015).
6. Ilves, I., Petojevic, T., Pesavento, J. J. & Botchan, M. R. Activation of the MCM2-7 helicase by association with Cdc45 and GINS proteins. *Mol. Cell* **37,** 247–258 (2010).
7. Moyer, S. E., Lewis, P. W. & Botchan, M. R. Isolation of the Cdc45/Mcm2-7/GINS (CMG) complex, a candidate for the eukaryotic DNA replication fork helicase. *Proc. Natl Acad. Sci. USA* **103,** 10236–10241 (2006).
8. Yardimci, H., Loveland, A. B., Habuchi, S., van Oijen, A. M. & Walter, J. C. Uncoupling of sister replisomes during eukaryotic DNA replication. *Mol. Cell* **40,** 834–840 (2010).
9. Costa, A. & Onesti, S. Structural biology of MCM helicases. *Crit. Rev. Biochem. Mol. Biol.* **44,** 326–342 (2009).
10. Fu, Y. V. *et al.* Selective bypass of a lagging strand roadblock by the eukaryotic replicative DNA helicase. *Cell* **146,** 931–941 (2011).
11. McGeoch, A. T., Trakselis, M. A., Laskey, R. A. & Bell, S. D. Organization of the archaeal MCM complex on DNA and implications for the helicase mechanism. *Nat. Struct. Mol. Biol.* **12,** 756–762 (2005).
12. Petojevic, T. *et al.* Cdc45 (cell division cycle protein 45) guards the gate of the eukaryote replisome helicase stabilizing leading strand engagement. *Proc. Natl Acad. Sci. USA* **112,** E249–E258 (2015).
13. Takahashi, T. S., Wigley, D. B. & Walter, J. C. Pumps, paradoxes and ploughshares: mechanism of the MCM2-7 DNA helicase. *Trends Biochem. Sci.* **30,** 437–444 (2005).

14. Heller, R. C. *et al.* Eukaryotic origin-dependent DNA replication *in vitro* reveals sequential action of DDK and S-CDK kinases. *Cell* **146,** 80–91 (2011).
15. Lyubimov, A. Y., Strycharska, M. & Berger, J. M. The nuts and bolts of ring-translocase structure and mechanism. *Curr. Opin. Struct. Biol.* **21,** 240–248 (2011).
16. Costa, A. *et al.* The structural basis for MCM2-7 helicase activation by GINS and Cdc45. *Nat. Struct. Mol. Biol.* **18,** 471–477 (2011).
17. Sun, J. *et al.* The architecture of a eukaryotic replisome. *Nat. Struct. Mol. Biol.* **22,** 976–982 (2015).
18. Chang, Y. P., Wang, G., Bermudez, V., Hurwitz, J. & Chen, X. S. Crystal structure of the GINS complex and functional insights into its role in DNA replication. *Proc. Natl Acad. Sci. USA* **104,** 12685–12690 (2007).
19. Costa, A. *et al.* DNA binding polarity, dimerization, and ATPase ring remodeling in the CMG helicase of the eukaryotic replisome. *Elife* **3,** e03273 (2014).
20. Yamagata, A., Kakuta, Y., Masui, R. & Fukuyama, K. The crystal structure of exonuclease RecJ bound to Mn2 + ion suggests how its characteristic motifs are involved in exonuclease activity. *Proc. Natl Acad. Sci. USA* **99,** 5908–5912 (2002).
21. Krastanova, I. *et al.* Structural and functional insights into the DNA replication factor Cdc45 reveal an evolutionary relationship to the DHH family of phosphoesterases. *J. Biol. Chem.* **287,** 4121–4128 (2012).
22. Sanchez-Pulido, L., Diffley, J. F. & Ponting, C. P. Homology explains the functional similarities of Treslin/Ticrr and Sld3. *Curr. Biol.* **20,** R509–R510 (2010).
23. Gai, D., Zhao, R., Li, D., Finkielstein, C. V. & Chen, X. S. Mechanisms of conformational change for a replicative hexameric helicase of SV40 large tumor antigen. *Cell* **119,** 47–60 (2004).
24. Thomsen, N. D. & Berger, J. M. Running in reverse: the structural basis for translocation polarity in hexameric helicases. *Cell* **139,** 523–534 (2009).
25. Enemark, E. J., Chen, G., Vaughn, D. E., Stenlund, A. & Joshua-Tor, L. Crystal structure of the DNA binding domain of the replication initiation protein E1 from papillomavirus. *Mol. Cell* **6,** 149–158 (2000).
26. Miller, J. M., Arachea, B. T., Epling, L. B. & Enemark, E. J. Analysis of the crystal structure of an active MCM hexamer. *Elife* **3,** e03433 (2014).
27. Lam, S. K. *et al.* The PS1 hairpin of Mcm3 is essential for viability and for DNA unwinding in vitro. *PLoS ONE* **8,** e82177 (2013).
28. Fletcher, R. J. *et al.* The structure and function of MCM from archaeal *M. thermoautotrophicum*. *Nat. Struct. Biol.* **10,** 160–167 (2003).
29. Barry, E. R., Lovett, J. E., Costa, A., Lea, S. M. & Bell, S. D. Intersubunit allosteric communication mediated by a conserved loop in the MCM helicase. *Proc. Natl Acad. Sci. USA* **106,** 1051–1056 (2009).
30. Froelich, C. A., Kang, S., Epling, L. B., Bell, S. P. & Enemark, E. J. A conserved MCM single-stranded DNA binding element is essential for replication initiation. *Elife* **3,** e01993 (2014).
31. Enemark, E. J. & Joshua-Tor, L. Mechanism of DNA translocation in a replicative hexameric helicase. *Nature* **442,** 270–275 (2006).
32. Itsathitphaisarn, O., Wing, R. A., Eliason, W. K., Wang, J. & Steitz, T. A. The hexameric helicase DnaB adopts a nonplanar conformation during translocation. *Cell* **151,** 267–277 (2012).
33. Taylor, M. R. *et al.* Rad51 paralogs remodel pre-synaptic Rad51 filaments to stimulate homologous recombination. *Cell* **162,** 271–286 (2015).
34. Labib, K. How do Cdc7 and cyclin-dependent kinases trigger the initiation of chromosome replication in eukaryotic cells? *Genes Dev.* **24,** 1208–1219 (2010).
35. Hardy, C. F., Dryga, O., Seematter, S., Pahl, P. M. & Sclafani, R. A. mcm5/cdc46-bob1 bypasses the requirement for the S phase activator Cdc7p. *Proc. Natl Acad. Sci. USA* **94,** 3151–3155 (1997).
36. Lyubimov, A. Y., Costa, A., Bleichert, F., Botchan, M. R. & Berger, J. M. ATP-dependent conformational dynamics underlie the functional asymmetry of the replicative helicase from a minimalist eukaryote. *Proc. Natl Acad. Sci. USA* **109,** 11999–12004 (2012).
37. Samel, S. A. *et al.* A unique DNA entry gate serves for regulated loading of the eukaryotic replicative helicase MCM2-7 onto DNA. *Genes Dev.* **28,** 1653–1666 (2014).
38. Yardimci, H. *et al.* Bypass of a protein barrier by a replicative DNA helicase. *Nature* **492,** 205–209 (2012).
39. Duxin, J. P., Dewar, J. M., Yardimci, H. & Walter, J. C. Repair of a DNA-protein crosslink by replication-coupled proteolysis. *Cell* **159,** 346–357 (2014).
40. Takahashi, K., Yamada, H. & Yanagida, M. Fission yeast minichromosome loss mutants mis cause lethal aneuploidy and replication abnormality. *Mol. Biol. Cell* **5,** 1145–1158 (1994).
41. Martin, A., Baker, T. A. & Sauer, R. T. Rebuilt AAA + motors reveal operating principles for ATP-fuelled machines. *Nature* **437,** 1115–1120 (2005).
42. Yakamavich, J. A., Baker, T. A. & Sauer, R. T. Asymmetric nucleotide transactions of the HslUV protease. *J. Mol. Biol.* **380,** 946–957 (2008).
43. Singleton, M. R., Sawaya, M. R., Ellenberger, T. & Wigley, D. B. Crystal structure of T7 gene 4 ring helicase indicates a mechanism for sequential hydrolysis of nucleotides. *Cell* **101,** 589–600 (2000).
44. Moreau, M. J., McGeoch, A. T., Lowe, A. R., Itzhaki, L. S. & Bell, S. D. ATPase site architecture and helicase mechanism of an archaeal MCM. *Mol. Cell* **28,** 304–314 (2007).
45. Beckwith, R., Estrin, E., Worden, E. J. & Martin, A. Reconstitution of the 26S proteasome reveals functional asymmetries in its AAA + unfoldase. *Nat. Struct. Mol. Biol.* **20,** 1164–1172 (2013).
46. Roberts, A. J., Kon, T., Knight, P. J., Sutoh, K. & Burgess, S. A. Functions and mechanics of dynein motor proteins. *Nat. Rev. Mol. Cell. Biol.* **14,** 713–726 (2013).
47. Schmidt, H., Gleave, E. S. & Carter, A. P. Insights into dynein motor domain function from a 3.3-A crystal structure. *Nat. Struct. Mol. Biol.* **19,** 492–497, S1 (2012).
48. Matyskiela, M. E., Lander, G. C. & Martin, A. Conformational switching of the 26S proteasome enables substrate degradation. *Nat. Struct. Mol. Biol.* **20,** 781–788 (2013).
49. Erzberger, J. P. & Berger, J. M. Evolutionary relationships and structural mechanisms of AAA + proteins. *Annu. Rev. Biophys. Biomol. Struct.* **35,** 93–114 (2006).
50. Gleave, E. S., Schmidt, H. & Carter, A. P. A structural analysis of the AAA + domains in *Saccharomyces cerevisiae* cytoplasmic dynein. *J. Struct. Biol.* **186,** 367–375 (2014).
51. Bai, X. C., Fernandez, I. S., McMullan, G. & Scheres, S. H. Ribosome structures to near-atomic resolution from thirty thousand cryo-EM particles. *Elife* **2,** e00461 (2013).
52. Li, X. *et al.* Electron counting and beam-induced motion correction enable near-atomic-resolution single-particle cryo-EM. *Nat. Methods* **10,** 584–590 (2013).
53. Tang, G. *et al.* EMAN2: an extensible image processing suite for electron microscopy. *J. Struct. Biol.* **157,** 38–46 (2007).
54. Rohou, A. & Grigorieff, N. CTFFIND4: Fast and accurate defocus estimation from electron micrographs. *J. Struct. Biol.* **192,** 216–221 (2015).
55. Scheres, S. H. RELION: implementation of a Bayesian approach to cryo-EM structure determination. *J. Struct. Biol.* **180,** 519–530 (2012).
56. Mindell, J. A. & Grigorieff, N. Accurate determination of local defocus and specimen tilt in electron microscopy. *J. Struct. Biol.* **142,** 334–347 (2003).
57. Sorzano, C. O. *et al.* XMIPP: a new generation of an open-source image processing package for electron microscopy. *J. Struct. Biol.* **148,** 194–204 (2004).
58. Amunts, A., Brown, A., Toots, J., Scheres, S. H. & Ramakrishnan, V. Ribosome. The structure of the human mitochondrial ribosome. *Science* **348,** 95–98 (2015).
59. Scheres, S. H. Beam-induced motion correction for sub-megadalton cryo-EM particles. *Elife* **3,** e03665 (2014).
60. Kucukelbir, A., Sigworth, F. J. & Tagare, H. D. Quantifying the local resolution of cryo-EM density maps. *Nat. Methods* **11,** 63–65 (2014).
61. Pettersen, E. F. *et al.* UCSF Chimera--a visualization system for exploratory research and analysis. *J. Comput. Chem.* **25,** 1605–1612 (2004).
62. Lamichhane, R., Solem, A., Black, W. & Rueda, D. Single-molecule FRET of protein-nucleic acid and protein-protein complexes: surface passivation and immobilization. *Methods* **52,** 192–200 (2010).

Acknowledgements

We thank Shaoxia Chen and Christos Savva (LMB Cambridge) for assistance with automatic data collection on the Titan Krios. We also thank Michael Botchan for the gift of the CMG pFastBacs and for useful discussion; Kristy MacLellan-Gibson (NIBSC), Raffaella Carzaniga and Lucy Collinson (The Crick) for technical support with cryo screening; and Bik Tye and Ning Gao for sharing EMDB and PDB data before release. We thank Ivar Ilves and James Berger for interminable conversations; Hasan Yardimci for helpful discussion, advise and support with the DNA unwinding reaction; Peter Cherepanov, Giulia Zanetti and Adelina Davies for critical reading of the manuscript; and the members of the Costa lab for useful comments. This work was supported by The Francis Crick Institute (which receives its core funding from Cancer Research UK, the UK Medical Research Council and the Wellcome Trust, A.C.) and a PhD fellowship from the Boehringer Ingelheim Fonds (to F.A.A.). A.K. is supported by the Wellcome Trust. The Division of Structural Biology—Particle Imaging Center Electron Microscopy Facility at University of Oxford was founded by The Wellcome Trust JIF award (060208/Z/00/Z) and is supported by a WT equipment grant (093305/Z/10/Z). The Wellcome Trust, MRC and BBSRC also supports the National EM facility, which enabled provision of the K2 detector at Oxford. The D.R. laboratory is funded by the Clinical Sciences Center of the Medical Research Council (RCUK MC-A658-5TY10) and a startup grant from Imperial College London. H.L.G. is funded by a postdoctoral fellowship from the Swiss National Science Foundation.

Author contributions

F.A.A. and J.G. purified the CMG complex. F.A.A. reconstituted the CMG–DNA complex. L.R., F.A.A. and A.C. prepared the cryo-EM grids. F.A.A., L.R., A.K. and A.C. performed cryo-EM data collection. F.A.A. and L.R. performed the cryo-EM reconstruction. H.G. and F.A.A. performed single-molecule FRET experiments supervised by D.R. F.A.A. and A.C. wrote the manuscript with inputs from all authors.

Additional information

Accession codes: ATP–CMG: compact ATPase EMDB entry EMD-3320, relaxed ATPase EMDB entry EMD-3321. ATPγS–CMG–DNA: compact ATPase EMDB entry EMD-3318, relaxed ATPase EMDB entry EMD-3319.

Competing financial interests: The authors declare no competing financial interests.

16

F-actin mechanics control spindle centring in the mouse zygote

Agathe Chaigne[1,†], Clément Campillo[2], Raphaël Voituriez[3,*], Nir S. Gov[4,*], Cécile Sykes[5,6,7], Marie-Hélène Verlhac[1] & Marie-Emilie Terret[1]

Mitotic spindle position relies on interactions between astral microtubules nucleated by centrosomes and a rigid cortex. Some cells, such as mouse oocytes, do not possess centrosomes and astral microtubules. These cells rely only on actin and on a soft cortex to position their spindle off-centre and undergo asymmetric divisions. While the first mouse embryonic division also occurs in the absence of centrosomes, it is symmetric and not much is known on how the spindle is positioned at the exact cell centre. Using interdisciplinary approaches, we demonstrate that zygotic spindle positioning follows a three-step process: (1) coarse centring of pronuclei relying on the dynamics of an F-actin/Myosin-Vb meshwork; (2) fine centring of the metaphase plate depending on a high cortical tension; (3) passive maintenance at the cell centre. Altogether, we show that F-actin-dependent mechanics operate the switch between asymmetric to symmetric division required at the oocyte to embryo transition.

[1] CIRB, Collège de France, and CNRS-UMR7241 and INSERM-U1050, Equipe labellisée Ligue contre le Cancer, Paris F-75005, France. [2] Université Evry Val d'Essonne, LAMBE, Boulevard F Mitterrand, Evry 91025, France. [3] UMR7600-CNRS/UPMC, 4 Place Jussieu, Paris F-75005, France. [4] Department of Chemical Physics, Weizmann Institute of Science, Rehovot 76100, Israel. [5] CNRS-UMR168, Paris F-75248, France. [6] UPMC, 4 Place Jussieu, Paris F-75248, France. [7] Institut Curie, Centre de Recherche, Laboratoire Physico-Chimie, Paris F-75248, France. † Present address: Laboratory for Molecular Cell Biology, UCL, London WC1E 6BT, UK. * These authors contributed equally to this work. Correspondence and requests for materials should be addressed to M.-H.V. (email: marie-helene.verlhac@college-de-france.fr) or to M.-E.T. (email: marie-emilie.terret@college-de-france.fr).

Mouse oocytes undergo a very asymmetric division in size during meiosis I. This asymmetry is a consequence of the migration of the microtubule spindle from the cell centre towards the closest cortex[1]. Oocytes are devoid of centrioles and astral microtubules[2]. As such, spindle positioning does not depend on microtubules[3] as in most mitotic cells[4], but on two actin networks. One is an F-actin cytoplasmic meshwork, nucleated by the cooperation between two types of actin nucleators, Spire1/2 and Formin-2 (refs 5–10). It is present in Prophase I and dismantled at nuclear envelope breakdown (NEBD), maybe favouring meiotic spindle assembly in the absence of mechanical constraints[11]. This meshwork reforms progressively as meiosis I progresses. It is then composed of a dense and dynamic cytoplasmic F-actin meshwork and of an actin cage surrounding the microtubule spindle, connecting it to the cortex[5,9,12,13]. The other actin network shown to function during meiotic spindle positioning is a cortical F-actin thickening nucleated progressively by the Arp2/3 complex, which triggers the removal of Myosin-II from the cortex, promoting a drop of cortical tension[12,14]. The Arp2/3 complex lies downstream of a signalling cascade activated by Mos, a MAPKKK specific to the female gamete[15,16]. Myosin-II, localized at the poles of the actin cage[9,17], pulls the spindle towards the cortex. However, since the spindle does not form at the exact cell centre[18,19], one pole is closer to the cortex and thus pulled harder. The decrease in cortical tension allows the cortex to deform and to constitute a large overlap on which Myosin-II can further exert pulling forces, accelerating spindle migration[14].

The first embryonic mitosis following fertilization in the mouse also occurs in the absence of true centrosomes and astral microtubules since the centrioles brought by the sperm are rapidly degraded[20] and rebuilt *de novo* at the 64-cell stage[21]. This division is peculiar because the male and female DNAs are present in two different nuclei called pronuclei (we will refer to the period where both pronuclei are present as Pronucleus stage). They migrate from the periphery towards the centre of the embryo for several hours[22,23] and are located roughly in the central region of the embryo at mitosis onset. They further undergo NEBD simultaneously, and then the two sets of chromosomes migrate and merge at the exact embryo centre, forming the metaphase plate. This division is symmetric, and requires F-actin since depolymerizing microfilaments before metaphase leads to an asymmetric division[24,25]. However, it is not known how F-actin is organized and controls the geometry of the division, how central positioning of the spindle is achieved and specifically whether spindle positioning also depends on the mechanical properties of the cortex as in oocytes. Yet, any deviation in spindle centring could perturb the symmetry of the division and compromise further embryo development and viability. We show here that F-actin-dependent mechanics are crucial for zygotic spindle positioning, ensuring first the coarse centring of pronuclei and second the fine centring of the metaphase plate, which is then maintained at the exact cell centre by a passive mechanism.

Results

Coarse centring of pronuclei depends on an F-actin meshwork. To investigate the mechanism ensuring proper central position of the spindle, we performed *in vitro* fertilization and followed the first embryonic division in live confocal microscopy (Supplementary Movie 1, Fig. 1a). Following fertilization, the two pronuclei assemble at the periphery of the embryo (Supplementary Movie 1, Fig. 1a) and migrate towards the embryo centre in 12–15 h (Supplementary Movie 1, Fig. 1a). At the end of their migration, the two pronuclei are rarely at the perfect centre of the zygote, represented by the theoretical central position of the female and male pronuclei (Fig. 1b right panel, pink- and purple-dotted circles, respectively, as measured in ref. 26). They can be far apart as observed, for example, in the 30 μm distance separating the two orange spots (Fig. 1b right panel). Although it was hypothesized previously that the coarse centring could be F-actin dependent[22], the precise mechanism for pronuclei centration has never been addressed. We first confirmed that the migration of the two pronuclei was indeed dependent on microfilaments by depolymerizing actin with 1 μg ml^{-1} Cytochalasin D after pronuclei formation but during their migration (Fig. 1c, right panel). In these embryos, NEBD occurred normally as seen on Fig. 1c (left panel) but pronuclei migration was abolished, leading to embryos with off-centred sets of chromosomes (Fig. 1c). Both male and female pronuclei end up mis-positioned with respect to the embryo centre and with respect to each other (Fig. 1d). Mis-positioning the pronuclei is sufficient to perturb the symmetry of the first zygotic division since when Cytochalasin D is washed out before NEBD there is a higher occurrence of asymmetric divisions (Supplementary Fig. 1a, quantification of the proportion of asymmetric divisions on Supplementary Fig. 1b).

Since the migration of the two pronuclei is dependent on actin, we followed the dynamics of microfilaments in Pronucleus stage embryos (Supplementary Movie 2, Fig. 2a) using a green fluorescent protein–UtrCH (GFP–UtrCH) probe that specifically binds to F-actin[27]. At the Pronucleus stage, the embryo is filled with a very dense and dynamic F-actin mesh that seems to emanate from discrete foci, which could correspond to vesicles (Supplementary Movie 2) as described in oocytes[10,28,29]. These dots move at a mean speed of 10.55 ± 3.70 μm min^{-1}, a speed similar to the motion of actin-positive vesicles present in Prophase I oocytes (~13 μm min^{-1})[28,29]. In Prophase I, the movement of the nucleus towards the cell centre is promoted by the active diffusion of actin-positive vesicles that establish a pressure gradient and is favoured by non-directed global cytoplasmic flows fluidizing the cytoplasm[29]. We thus checked for the presence of similar cytoplasmic flows in the zygote (Supplementary Movie 3) using spatiotemporal image correlation spectroscopy (STICs) analysis. As in Prophase I oocytes, we could detect the presence of non-organized cytoplasmic flows presenting a maximum speed of 2 μm min^{-1} (Fig. 2b). Consistent with the presence of cytoplasmic flows, inert fluorescent beads injected in early Pronucleus stage were found to move at a mean velocity of 18.2 ± 7.8 μm min^{-1} from 2 to 3 h after pronuclei assembly (Supplementary Movie 4), close to the speed of the actin-positive foci.

In Prophase I, the active diffusion mechanism responsible for nucleus centring depends on the F-actin motor Myosin-Vb[29]. Indeed the movement of actin-positive vesicles induced by the activity of Myosin-Vb has a global effect, indirectly putting the cytoplasm in motion by dragging the fluid around moving vesicles[30,31]. To test if the same mechanism could be at play here, we expressed a Myosin-Vb tail which acts like a dominant negative construct for Myosin-Vb[10]. Myosin-Vb tail expressing embryos no longer display cytoplasmic flows (Supplementary Movie 5, Fig. 2c right panel), indicating that their cytoplasmic activity depends on Myosin-Vb activity. These embryos display massive cortical deformations (Supplementary Movies 6 and 7, Fig. 2d), but their pronuclei do not move from their cortical positions (Fig. 2d). Thus, when they undergo NEBD, they are mis-positioned with respect to the embryo centre and with respect to each other (Fig. 2d,e). Depending on the entry site of the spermatozoid, the two sets of chromosomes will start far or close from one another. When the two sets of chromosomes are close (Supplementary Movie 6, Fig. 2d, upper panel), they do gather after Myosin-Vb tail expression, but are improperly

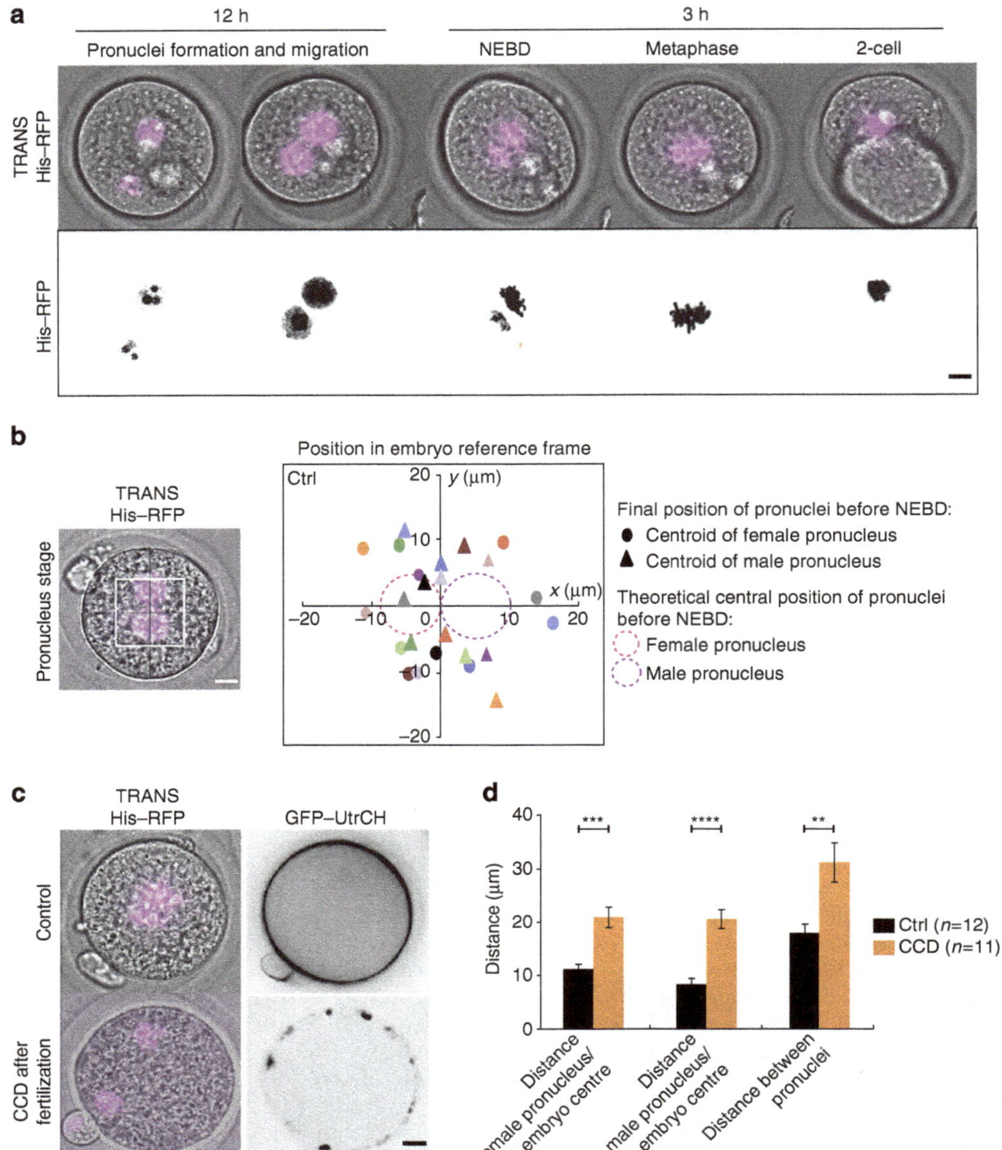

Figure 1 | The coarse centring of the two pronuclei is F-actin dependent. (**a**) Control embryo expressing His-RFP (upper panel: purple, overlay on transmitted light, Z-projection over 20 μm; lower panel: black). The first two frames are 12 h apart then one picture is shown every 1 h (Scale bar: 10 μm). (**b**) Left panel: control embryo expressing His-RFP (purple, overlay on transmitted light, Z-projection over 20 μm) at the Pronucleus stage (scale bar, 10 μm). Right panel: dot plot showing the position of the centroid of the two pronuclei in the referential of the embryo (0 being the centre) 1 h before NEBD for 10 control embryos. The scale is represented by the white box on the left panel. Each colour represents an embryo, the circle is the female pronucleus and the triangle is the male pronucleus. The mean size of the maternal (pink) and paternal (purple) pronuclei is represented by the two dotted circles at the embryo centre. (**c**) Embryos just after NEBD expressing His-RFP (left panel, purple, overlay on transmitted light, Z-projection over 20 μm) and GFP-utrCH (right panel, black, one Z-plane is shown). Upper panel: control embryo; lower panel: embryo treated during pronuclei migration with 1 μg ml^{-1} Cytochalasin D (CCD). Scale bar, 10 μm. (**d**) Bar graph showing the male and female pronuclei distance to the embryo centre and to each other at NEBD for control embryos (black bars) or embryos treated during pronuclei migration with 1 μg ml^{-1} CCD (orange bars). Mean is shown of 12 controls and 11 CCD-treated embryos from 6 independent experiments. s.e.m. is plotted on each bar. Statistical significance of differences is assessed with *t*-test or *t*-test with Welch correction (*P* values: female pronuclei/embryo centre: 0.0004; male pronuclei/embryo centre <0.0001; between pronuclei: 0.0035).

positioned since the metaphase plate ends up 10 μm away from the embryo centre, instead of 2 μm in controls (Fig. 2f). Strikingly, when the two sets of chromosomes are far away from each other (Supplementary Movie 7, Fig. 2d, lower panel), they are not able to merge and each pronucleus undergoes anaphase independently, leading to the formation of four haploid nuclei (Fig. 2d, lower panel). This phenotype recapitulates the Cytochalasin D phenotype (Fig. 1c,d) but with cortical deformations, which suggests that these cortical deformations are not responsible for the defects in pronuclei positioning observed after Myosin-Vb tail expression.

Altogether, these results argue that F-actin- and Myosin-Vb-driven cytoplasmic flows induce the motion of the two pronuclei during the Pronucleus stage, as it is the case in Prophase I oocytes that centre their nucleus. We then searched for what gives the directionality towards the centre to the movement.

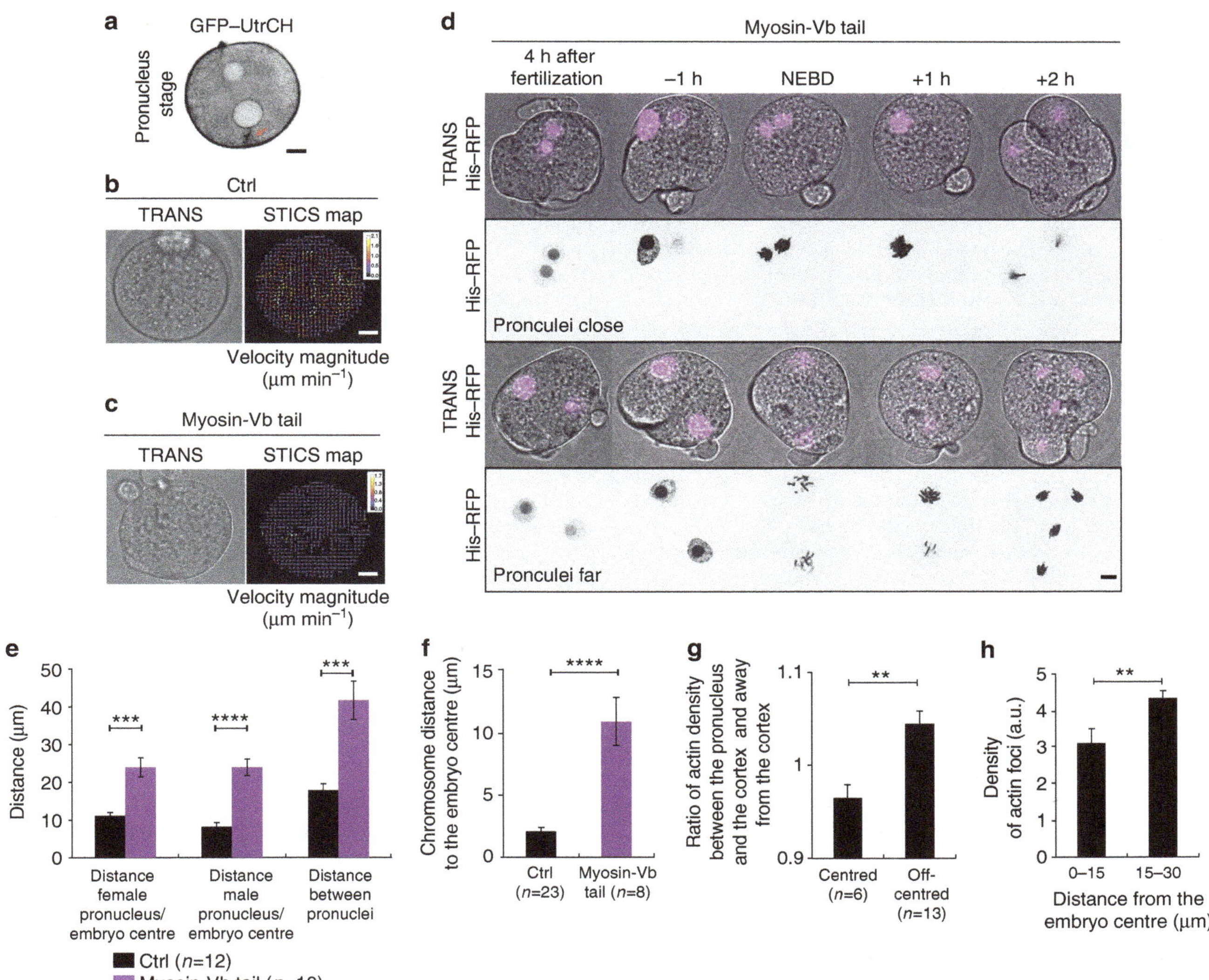

Figure 2 | F-actin and Myosin-Vb-driven cytoplasmic flows position the pronuclei in the central region of the embryo. (**a**) Control embryo expressing GFP-UtrCH (black, one Z-plane is shown). The red arrow points at actin enrichment. Scale bar, 10 μm. (**b**,**c**) Left panels: embryos expressing (**c**) or not (**b**) Myosin-Vb tail. Right panels: associated vector maps of cytoplasmic flows obtained using STICS analysis. Heat bar unit in micrometres per minute. Scale bars, 10 μm. One example is shown out of 11 controls and 13 Myosin-Vb tail embryos. (**d**) Embryos expressing His-RFP (purple or black, Z-projection over 20 μm). Upper panels: pronuclei close; lower panels: pronuclei far. Scale bar, 10 μm. (**e**) Bar graph showing the pronuclei distance to the embryo centre and to each other 1 h before NEBD for controls (black bars) or Myosin-Vb tail-expressing embryos (purple bars). Mean is shown of 12 controls and 10 Myosin-Vb tail embryos. s.e.m. is plotted on each bar. Statistical significance of differences is assessed with a *t*-test or a *t*-test with Welch correction (*P* values: female pronuclei/embryo centre: 0.0005; male pronuclei/embryo centre <0.0001; between pronuclei: 0.0009). (**f**) Bar graph showing the chromosome mass distance to the embryo centre in metaphase for controls (black bar) and Myosin-Vb tail-expressing embryos (purple bar). Mean is shown for 23 controls and 8 Myosin-Vb tail embryos. s.e.m. is plotted on each bar. Statistical significance of differences is assessed with a Mann-Whitney test (*P* value <0.0001). (**g**) Graph bar showing the ratio of GFP-UtrCH fluorescent intensity between the pronucleus and the cortex and on the other side of the pronucleus. Mean is shown for 6 centred and 13 off-centred pronuclei embryos. s.e.m. is plotted on each bar. Statistical significance of the difference is calculated with a Mann-Whitney test (*P* value 0.0014). (**h**) Graph bar showing the mean percentage of actin-positive foci in each region of the embryo normalized by the effective surface of the region. Mean is shown from 9 embryos out of 2 independent experiments. s.e.m. is plotted on each bar. Statistical significance of the difference is calculated with a Mann-Whitney test (*P* values 0.007).

Mesh heterogeneities could drive coarse pronuclei centring. What could be the mechanism for coarse pronuclei centring? To address this question, we first searched for an anisotropy that could drive directional motion. Interestingly, the cytoplasmic actin meshwork does not seem homogeneous. Specifically, once the pronuclei have assembled under the cortex (Fig. 2a and referred in Fig. 2g as off-centred), the meshwork between the pronuclei and the adjacent cortex seems to be denser than the meshwork on the other side of the pronuclei (Fig. 2a red arrow). We then measured the fluorescence intensity of the F-actin probe (GFP–UtrCH) as a proxy for filament density between the pronuclei and the adjacent cortex (Fig. 2a red arrow) and on the other side of the pronuclei (see Methods). Before pronuclei migration, more actin is indeed present between pronuclei and cortex (Fig. 2g, the ratio is >1 for off-centred pronuclei). This is not the case once the pronuclei have migrated in the central region of the embryo, (Fig. 2g, the ratio is <1 for centred pronuclei). Close observation of the actin-positive foci revealed that there are no heterogeneities in their speed, but in their density, with more foci close to the cortex and less in the central region of the embryo (Fig. 2h). We can model the actin-positive foci as active particles propelled by Myosin-Vb, like for Prophase I oocytes[29]. Heterogeneities in actin density could be responsible for pronuclei centring. Indeed, similarly to a perfect gas, an assembly of n self-propelled particles induces a pressure that can be written: $P = nk_BT_e$, with k_B the Bolzmann constant and T_e the effective temperature of the particles. k_BT_e is a measure of the agitation energy of the particles, and can be inferred from their mean squared speed ($<v^2>$). The mean square speed of the actin-positive foci in embryos is uniform, however, their density is not (Fig. 2h). Such a gradient in the number of particles n would then result in a pressure gradient (∇P).

What would be the characteristics of such a pressure gradient? Here the resulting active force on a test object would be $F_a = \nabla P.V$, with V the volume of the object to be moved. V is proportional to R^3 with R the radius of the object. The force is counteracted by a friction force F_f proportional to $R\eta v_p$ (with η the viscosity and v_p the speed of the object). Hence, $v_p \propto <v^2> \nabla n \frac{R^2}{\eta}$, which means that the speed of an object moving in the density gradient would be proportional to the square radius of this object: the bigger the object the higher the speed. This strong size dependence of the velocity predicted by the model may explain why pronuclei are effectively centred, while fluorescent beads are not (small fluorescent latex beads of 0.1 μm in diameter are not centred 10 to 12 h after pronuclei assembly, Supplementary Movie 8). Altogether, this suggests that a gradient of density of the actin-positive foci could be responsible for the coarse centring of big objects such as the two pronuclei.

Chromosome merging determines the metaphase plate position. At mitosis onset, the two pronuclei are apposed more or less in the central region of the embryo (Fig. 1b). At NEBD, the two sets of chromosomes condense separately and are brought together in a slow but directed motion (roughly 1 h) (Supplementary Movie 9, Fig. 3a). The one-cell zygote is unique in this respect, having to achieve the fusion of maternal and paternal chromosome complements on a single metaphase plate[32]. Then the chromosome mass stays aligned in metaphase for roughly another hour before anaphase is triggered (Supplementary Movie 9, Fig. 3a). We followed precisely the position of the chromosomes during the first of these two phases, namely from NEBD to metaphase (Fig. 3b). During the migration of both sets of chromosomes (Fig. 3b), the two masses are brought almost towards the exact embryo centre (0 on the graph in Fig. 3b) with a mean maximum distance travelled of ~10 μm both for the male and the female chromosomes (Supplementary Fig. 1c). At the end of this phase, the metaphase plate is positioned in close proximity with the exact embryo centre (Fig. 3b, almost all the curves converge around 0).

Merging depends on microtubules but its position on F-actin. Since in mouse oocytes, spindle positioning depends on F-actin[5–10,12], we investigated its role during mitosis in the zygote. We followed actin organization throughout the first mitosis (Supplementary Movie 10, Fig. 3c). At the Pronucleus stage, an F-actin meshwork fills the cytoplasm of the embryo (Figs 2a and 3c). The meshwork is dismantled around NEBD (Figs 1c and 3c, Supplementary Fig. 1d,e). Interestingly, such a dismantlement of F-actin meshwork has also been observed at meiosis resumption in oocytes[11]. The cytoplasmic meshwork then reforms and includes an F-actin cage, probably around the microtubule spindle (Fig. 3c), that is anchored at the cortex in metaphase (Fig. 3d). To test if F-actin plays any role in the merging of parental chromosomes, we acutely treated embryos just after NEBD with 1 μg ml^{-1} Cytochalasin D, sufficient to induce massive depolymerization of cytoplasmic F-actin (Fig. 3e, upper panel). Interestingly, F-actin depolymerization alters the location of the merging of chromosome (Supplementary Movie 11, Fig. 3e,f). Indeed, tracking of chromosomes indicates that their merging occurs on time (Fig. 3e) but rarely at the embryo centre (Fig. 3f). To assess the efficiency of the centring process, we measured the distance between the merged chromosomes and the embryo centre. In controls, the chromosomes merged around 2 μm away from the centre of the embryo (Fig. 3g, black bar). However, in embryos treated with CCD around NEBD, the chromosomes merge rather 4 μm away from the embryo centre (Fig. 3g), suggesting that F-actin is important for the proper positioning of the chromosome merge. Also, we measured the amplitude of chromosome displacement and observed that paternal chromosomes undergo a movement of smaller amplitude than controls (Fig. 3h). Altogether, these results show that F-actin participates in precise metaphase plate centring.

Microtubules play a crucial role in spindle morphogenesis and positioning in most cells[33], thus we decided to investigate their contribution to the process of chromosome merging. We therefore treated embryos after pronuclei migration, around NEBD, with 1 μM Nocodazole and monitored the position of chromosome complements (Supplementary Movie 12, Fig. 4a). The Nocodazole treatment induced extensive microtubule spindle depolymerization after 30 min (Fig. 4a). Strikingly, parental chromosomes never merge onto a unique metaphase plate in the absence of microtubules (Supplementary Movie 12, Fig. 4a,b). Precise tracking of the two sets of chromosomes indicates that they undergo a non-oriented motion (Fig. 4b), with a maximum distance travelled smaller than in controls (Fig. 4c). Altogether, these results suggest that maternal and paternal chromosomes are brought together in a microtubule-dependent manner. Since Myosin-Vb tail expressing embryos harbour pronuclei that can be close or far apart (Fig. 2d), following chromosome merging in these embryos allows to gain insight into the reach of microtubules. Indeed, when the two sets of chromosomes are separated by <40 μm they are able to merge, whereas when they are >40 μm apart their merging fails (Fig. 2d, quantification on Fig. 4d). Therefore the microtubules are not capable of bridging a longer distance than 40 μm, possibly explaining why the pronuclei have first to be coarsely centred via an F-actin-dependent mechanism in order to merge.

Myosin-II drives cortical tension increase at NEBD. Cortical tension being essential for spindle positioning in mouse oocytes[12,14], we decided to investigate whether it also regulates spindle centring in embryos. Like other cell types, mouse zygotes round up at mitosis onset (Supplementary Movie 13 and Fig. 5a,b). Opposite to the measurements made in oocytes where cortical tension decreases during the division[12,34], cortical tension does increase during mitosis (Table 1) as in other mitotic cells[35]. Interestingly, even if the increase in cortical tension is modest (2-fold, Table 1) compared with the important (30-fold) drop experienced by oocytes during meiosis I (Table 1)[12] it is sufficient to induce the rounding up of extremely deformed embryos. This is the case, for example, in embryos expressing the Myosin-Vb tail (Fig. 2d and Supplementary Movies 6 and 7). To decipher the molecular basis for this change in cortex tension, we followed the localization of Myosin-II[36], a major regulator of tension in cells[37–39], in particular in oocytes[12,14,34]. Myosin-II is present in the cytoplasm both at the Pronucleus stage and throughout the first mitotic division (Supplementary Movie 14, Fig. 5c). However, during mitosis it is slightly enriched at the cell cortex (Supplementary Movie 14, Fig. 5c,d, a ratio closer to 1 corresponds to cortex enrichment). The cortical recruitment of PhosphoMyosin-II at mitosis onset was also observed on fixed embryos (Fig. 5e). The cortical recruitment of Myosin-II is modest compared with its enrichment in Prophase I oocytes[12].

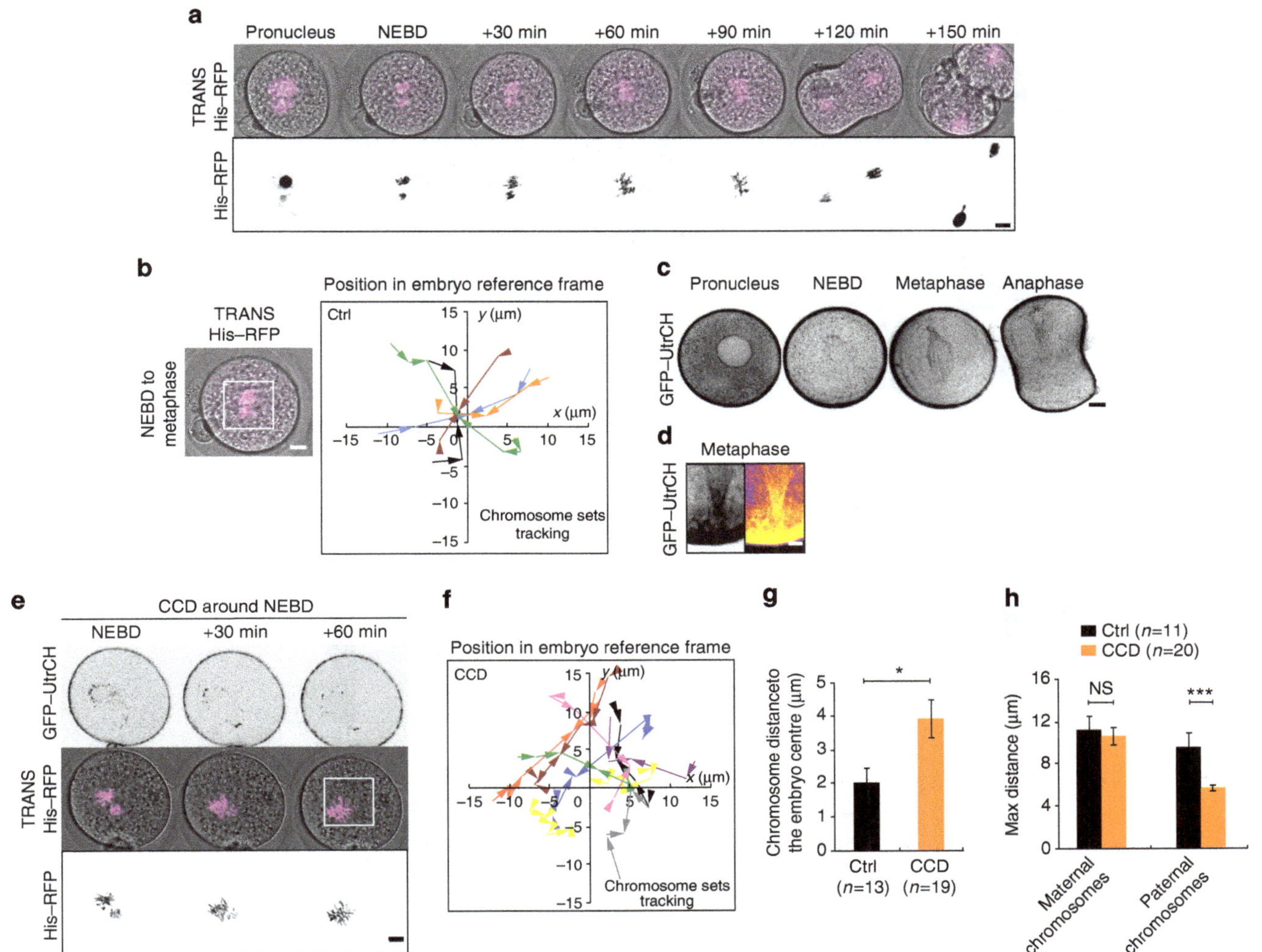

Figure 3 | Fine chromosome centring in mitosis is actin dependent. (**a**) Control embryo expressing His-RFP (purple or black, Z-projection over 20 μm). Scale bar, 10 μm. (**b**) Left panel: Control embryo expressing His-RFP (purple, Z-projection over 20 μm) during chromosome migration. Scale bar, 10 μm. Right panel: graph showing the position of the centroid of the two sets of chromosomes in the referential of the embryo (0 being the centre) from NEBD to metaphase for controls. The scale is represented by the white box on the left panel. Each colour represents an embryo. The arrows point to the position where the chromosomes merge. One time point every 15 or 20 min over 1 h. (**c**) Embryo expressing GFP-UtrCH (black). One Z-plane is shown every 40 min. Scale bar, 10 μm. (**d**) Metaphase embryo expressing GFP-UtrCH (left: black. Right: white, higher intensity and black, lower intensity). Zoom on the connection between the actin cage and the cortex (cortex saturated to allow visualization of the actin cage filaments). Scale bar, 5 μm. (**e**) Embryo treated with 1 μg ml^{-1} Cytochalasin D (CCD) around NEBD expressing GFP-UtrCH (black, one Z-plane is shown) and His-RFP (purple or black, Z-projection over 20 μm). Scale bar, 10 μm. (**f**) Same as **b** for embryos treated with CCD around NEBD. (**g**) Bar graph showing the position of the centroid of the merged chromosomes for controls (black bar) and embryos treated with CCD around NEBD (orange bar). Mean is shown of 13 controls and 19 CCD embryos. s.e.m. is plotted on each bar. Statistical significance of differences is assessed with a Mann-Whitney test (P value 0.0175). (**h**) Graph showing the maximum distance (that is, the distance between the points furthest apart on the trajectories) of chromosome motion in controls (black bars, from Supplementary Fig. 1c) and embryos treated CCD around NEBD (orange bars) during the migration of the two sets of chromosomes towards the embryo centre. Mean of 11 controls and 20 CCD embryos are shown over 3 independent experiments. s.e.m. is plotted on each bar. Statistical significance of differences is assessed with t-test or Mann-Whitney tests (P values: female 0.6776; male 0.0004).

This is consistent with the amplitude of cortical tension variations observed during meiosis I in oocytes versus the one observed here in the zygote (Table 1). Altogether, our results suggest that differently from oocytes[12,14] and similarly to other cell types[39], Myosin-II is recruited at the cortex at mitosis onset in one-cell embryos, leading to an increase in cortical tension.

Central merging relies on a high cortical tension. To investigate the role of the increase in cortical tension, we decided to artificially decrease it using a membrane-targeted cVCA, a construct promoting F-actin nucleation exclusively at the cortex and triggering cortical Myosin-II removal, thus lowering cortical tension, as observed in oocytes[14]. The cVCA efficiently decreased cortical tension during zygote's mitosis (Fig. 6a). The amplitude of the decrease in cortical tension due to cVCA expression was, however, smaller compared with the one measured during meiosis (3-fold decrease compared with 100-fold[14]). Interestingly, the mechanism responsible for the drop in cortical tension was comparable to the one present in oocytes. First, cVCA-expressing embryos harboured a cortical F-actin thickening (Supplementary Fig. 2a) of 2.5 μm compared with a thickness of 1 μm in controls (Supplementary Fig. 2b). The size and width of the actin cage were unchanged arguing that here, as for oocytes, the membrane-targeted cVCA construct acted mostly at the cortex

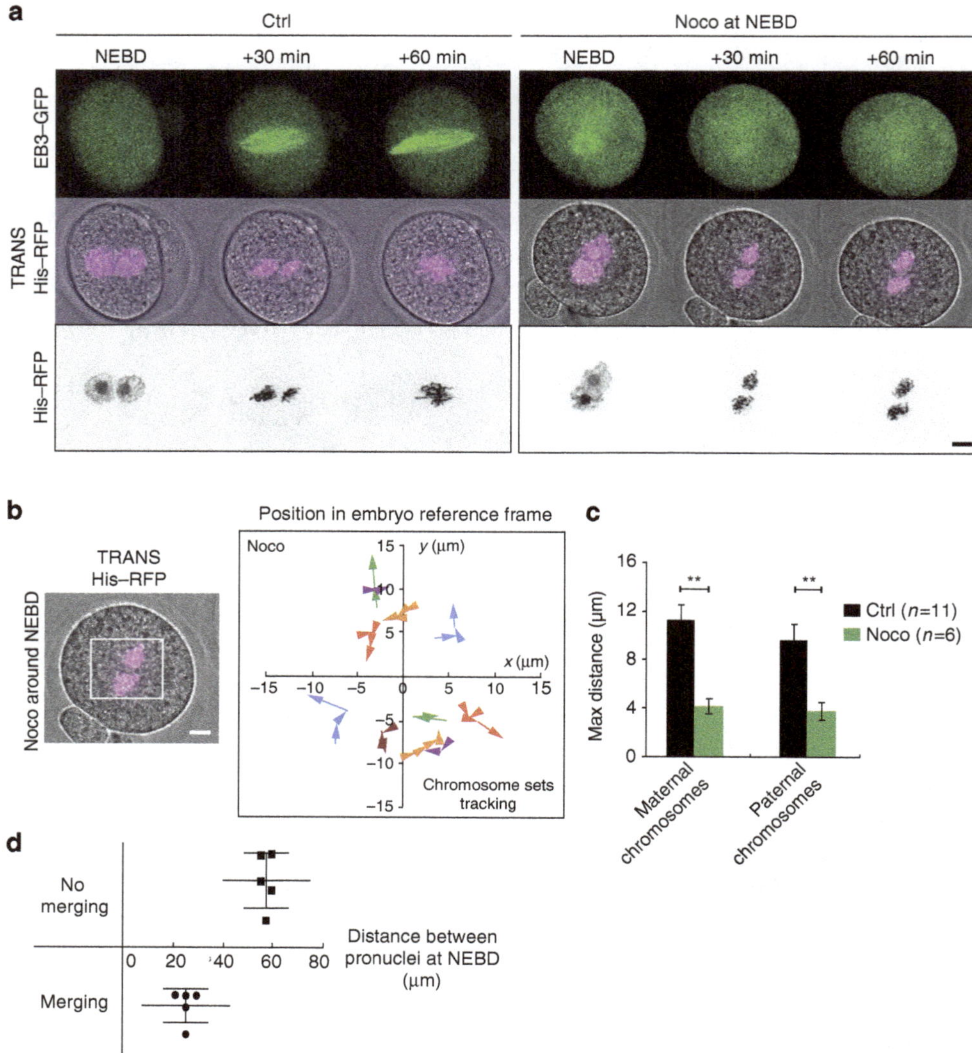

Figure 4 | Chromosome merge is dependent on microtubules. (**a**) Control embryo (left panels) or embryo treated with 1 µM Nocodazole around NEBD (right panels) expressing EB3-GFP (upper panel, green, Z-projection over 20 µm) and His-RFP (purple or black, Z-projection over 20 µm). One image is shown every 30 min. Scale bar, 10 µm. (**b**) Left panel: Embryo treated with 1 µM Nocodazole around NEBD expressing His-RFP (purple, Z-projection over 20 µm) during the migration of the two sets of chromosomes. Scale bar, 10 µm. Right panel: graph showing the position of the centroid of the two sets of chromosomes in the referential of the embryo (0 being the centre) from NEBD for embryos treated with 1 µM Nocodazole around NEBD. The scale is represented by the white box on the left panel. Each colour represents an embryo. The arrows point to the last position recorded. Points are 30 min apart. (**c**) Graph showing the maximum distance (that is, the distance between the points furthest apart on the trajectories) of chromosome motion in controls (black bars, from Supplementary Fig. 1c) and embryos treated with 1 µM Nocodazole around NEBD (green bars) during the migration of the two sets of chromosomes towards the embryo centre (female: left bars; male: right bars). Mean of 11 controls and 6 Nocodazole-treated embryos are shown over 2 independent experiments. s.e.m. is plotted on each bar. Statistical significance of differences is assessed with Mann-Whitney tests (P values: female 0.0011; male 0.0011). (**d**) Dot plot showing if the two sets of chromosome merge as a function of initial chromosome distance at NEBD for 10 embryos expressing Myosin-Vb. Mean distance and s.e.m. are plotted for each condition.

and not indirectly on the cytoplasmic actin mesh (Supplementary Fig. 2c). Second, cVCA expression was sufficient to impair Myosin-II recruitment at the cortex (Supplementary Fig. 2d,e: the level of Myosin-II at the cortex is constant in embryos expressing cVCA between Pronucleus stage and mitosis, whereas it increases in mitosis in control embryos). cVCA-expressing embryos harbour properly positioned pronuclei (Supplementary Fig. 2f), reinforcing the idea that pronuclei centring could depend on F-actin-driven active diffusion, this process being also tension independent in oocytes[29]. Importantly, none of the defects are due to the presence of Ezrin in the cVCA construct, since embryos solely overexpressing Ezrin behave like controls (Supplementary Figs 2f and 4h).

In cVCA-expressing embryos, the two chromosome complements merge with timings comparable to controls (Supplementary Movie 15, Fig. 6c, quantification on Supplementary Fig. 2g). However, the position of the merge is not at the exact cell centre (Fig. 6b–d). Indeed, precise quantification of the trajectories shows that parental chromosomes merge up to 10 µm away from the embryo centre (Fig. 6b, purple curve, for example; almost no curves converge at 0). This was very different to what was observed in controls (Fig. 3b). Interestingly, the lack of precise centring is not due to a reduction in chromosome motion since the mean maximal distance travelled by chromosomes is comparable to controls (Supplementary Fig. 2h). These results demonstrate that a high

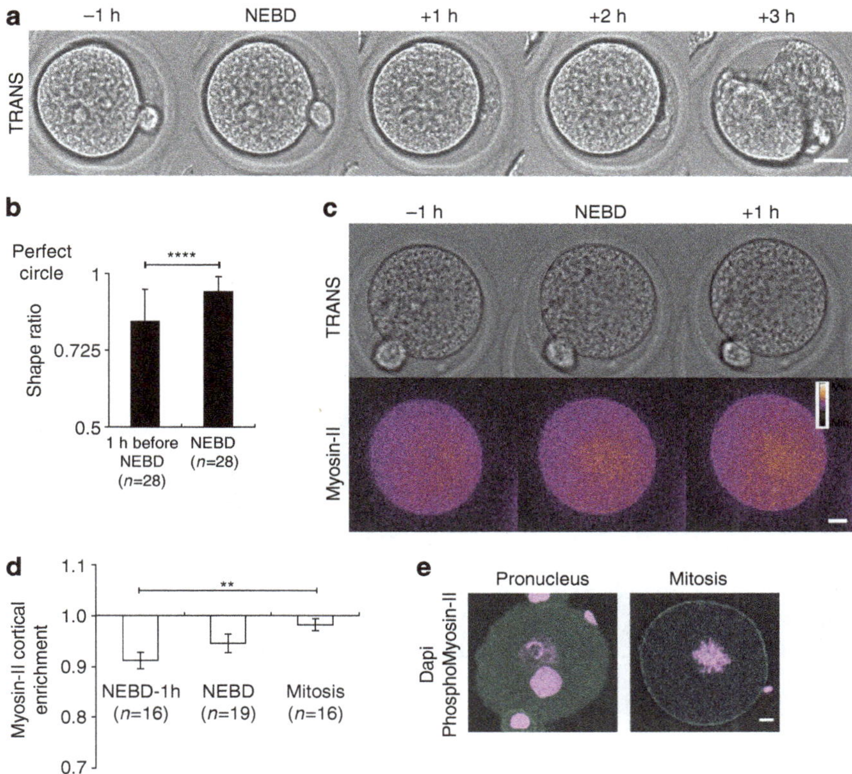

Figure 5 | Embryos round up at NEBD following a Myosin-II-dependent cortical tension increase. (**a**) Embryo before and during the first mitosis. One Z-plane is shown every 1 h. Scale bar, 10 μm. (**b**) Graph bar showing the shape ratio (see Methods) of embryos before and during mitosis. Mean is shown of 28 embryos from 10 independent experiments. s.d. is plotted on each bar. Statistical significance of differences is assessed with a Mann-Whitney test (*P* value <0.0001). (**c**) Embryo observed by transmitted light (upper panel) expressing SF9-GFP (Myosin-II intrabody, blue: lower intensity, orange: higher intensity). One Z-plane is shown every 1 h. Scale bar, 10 μm. (**d**) Bar graph showing the ratio between the average intensities of cortical versus cytoplasmic Myosin-II in embryos expressing SF9-GFP 1 h before NEBD, at NEBD or during mitosis. For each embryo, six measurements were taken in the cortex and six in the cytoplasm (see Methods). Mean of 16 embryos 1 h before NEBD, 19 at NEBD and 16 mitotic embryos is shown assessed over 6 independent experiments. S.e.m. is plotted on each bar. Statistical significance of differences is assessed with a *t*-test (*P* values: 1 h before NEBD/NEBD 0.1854; NEBD/mitotic 0.1066; 1 h before NEBD/mitotic 0.0013). (**e**) Control embryo at the Pronucleus stage (left panel) or during mitosis (right panel) stained for chromosomes (magenta, Z-projection is shown over 25 μm) and PhosphoMyosin-II (green, one Z-plane is shown). Scale bar, 5 μm.

Table 1 | Cortical tension and viscosity in oocytes and embryos.

	Early meiosis I oocyte	Late meiosis I oocyte	Pronucleus stage embryo	Mitotic stage embryo
Cortical tension (nN μm^{-1})	0.92 ± 0.2 (from ref. 12)	0.034 ± 0.1 (from ref. 12)	0.35 ± 0.04	0.61 ± 0.05
Cytoplasmic viscosity (Pa · s)			$4.4 \times 10^2 \pm 40$	$12 \times 10^2 \pm 300$

For cortical tension in embryos, mean of 13 Pronucleus stage and 9 mitotic embryos is shown, measured over 8 independent experiments. The statistical significance of differences is assessed with a *t*-test (*P* value 0.0011). The values of cortical tension for early and late oocytes are extracted from ref. 12. For cytoplasmic viscosity, mean of 16 Pronucleus stage and 15 mitotic embryos is shown, measured over 6 independent experiments. Statistical significance of differences is assessed with a Mann-Whitney test (*P* value <0.0001).

cortical tension defines the exact embryo centre where maternal and paternal chromosomes will eventually merge.

Both decreasing cortical tension as well as depolymerizing F-actin lead to merging of the chromosomes away from the perfect embryo centre. This suggests that cortical tension is transmitted to the chromosomes via microfilaments. The F-actin cage surrounds the chromosomes and is connected to the cortex, it could thus transmit forces to the spindle. What is the molecular mechanism ensuring this relay of tension from the cortex to the spindle? To address this question, we observed Myosin-II localization, which is responsible for F-actin-based force generation, in particular in mouse oocytes[9,12,14]. A closer inspection of the active form of Myosin-II staining in mitosis, PhosphoMyosin-II, showed that it is present at spindle poles, colocalizing with pericentrin (Supplementary Fig. 3a, yellow foci at the poles of the spindle) as in meiosis I where it is responsible for spindle traction to the cortex[9,12,17]. We thus wondered whether Myosin-II could put this F-actin cage under tension. To address this question, we treated one-cell zygotes in metaphase with ML-7, which reduces Myosin-II activity[40]. Less than 20% of embryos treated with 30 μM ML-7 undergo normal cytokinesis (Supplementary Fig. 3b), consistent with a role for Myosin-II in cytokinesis[41]. Embryos treated with ML-7 have shorter actin cages (Fig. 6e,f), suggesting that Myosin-II puts the actin cage under tension. ML-7 does not act on the actin cage by modulating the amount of microfilaments connected to the cage, since the density of F-actin (relative to the size of the cage), quantified using the fluorescence level of GFP–UtrCH, is comparable in controls versus ML-7-treated embryos (Supplementary Fig. 3c). Therefore, Myosin-II may put the actin cage under tension in metaphase.

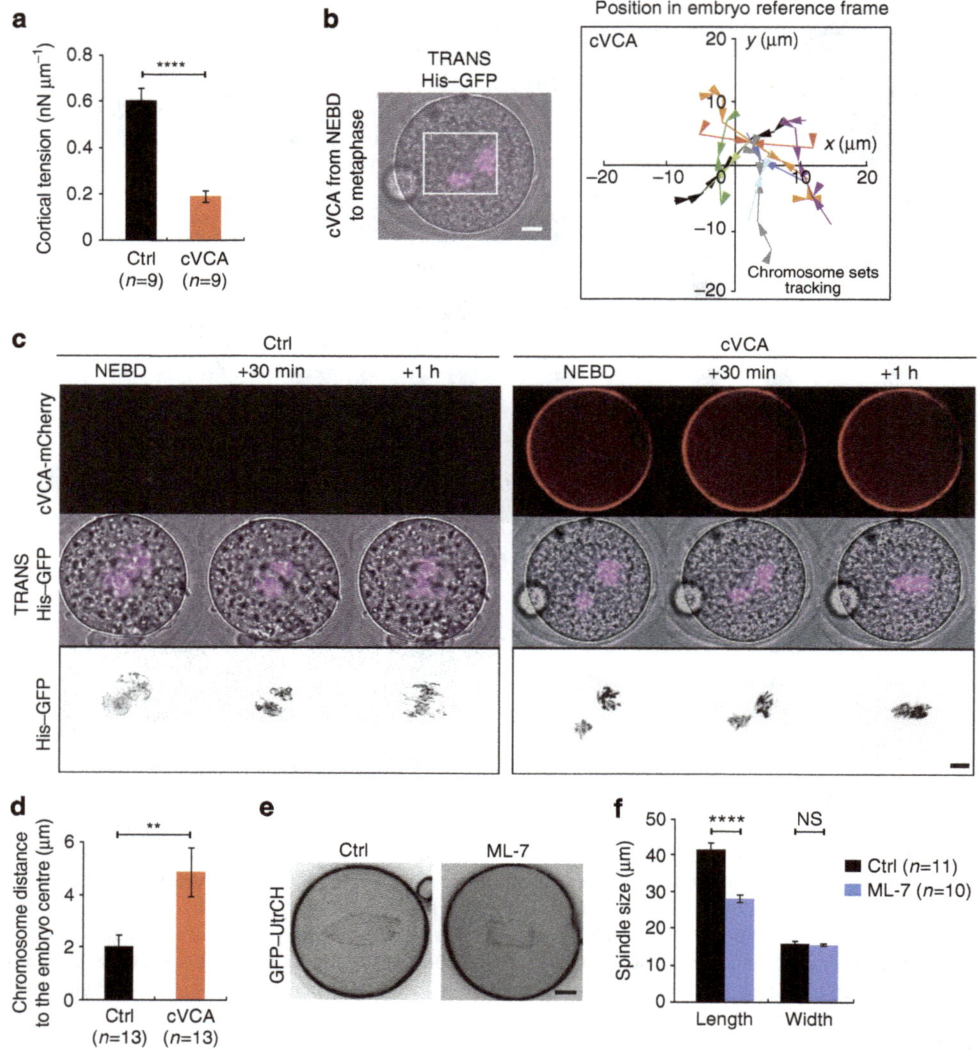

Figure 6 | Decreasing cortical tension impairs the fine centring of the chromosomes. (**a**) Bar graph showing the values of cortical tension of a mitotic embryo expressing cVCA (red bar) or not (black bar). Mean is shown over two independent experiments. s.e.m. is plotted on each bar. Statistical significance of differences is assessed with a *t*-test (*P* value <0.0001). (**b**) Left panel: Embryo expressing His-RFP (purple, Z-projection over 20 µm) together with cVCA during the migration of the two sets of chromosomes. Scale bar, 10 µm. Right panel: graph showing the position of the centroid of the two sets of chromosomes in the referential of the embryo (0 being the centre) from NEBD to metaphase for representative embryos expressing cVCA. The scale is represented by the white box on the left panel. Each colour represents an embryo. The arrows point to the position where the two sets of chromosomes merge. One time point is shown every 30 min. (**c**) Embryos expressing His-RFP alone (left panels) or together with cVCA (right panels). Upper panel: cVCA (red), one Z-plane is shown; middle and lower panels: His-RFP, purple or black, Z-projection over 20 µm. One frame is shown every 30 min. Scale bar, 10 µm. (**d**) Bar graph showing the position of the centroid of the merged chromosomes for controls (black bar) and embryos expressing cVCA (red bar). Mean is shown of 13 controls and 13 embryos expressing cVCA. s.e.m. is plotted on each bar. Statistical significance of differences is assessed with a Mann-Whitney test (*P* value 0.0015). (**e**) Embryos expressing GFP-UtrCH (black) treated (right panel) or not (left panel) in metaphase with 30 µM ML-7. Scale bar, 10 µm. (**f**) Bar graph showing the mean length and width of the actin cage in metaphase for 11 controls (black bars) and 10 ML-7 (blue bars) treated embryos from two independent experiments. s.e.m. is plotted on each bar. Statistical significance of differences is assessed with a *t*-test (*P* value length <0.0001) or a Mann-Whitney test (*P* value width 0.3167).

Altogether, these results suggest that Myosin-II could transmit the forces exerted by the increase in cortical tension to the microtubule spindle via the F-actin cage, placing it at the exact embryo centre.

Maintenance of the chromosomes at the cell centre is passive. Once the chromosomes have merged and are positioned at the exact embryo centre, they stay at this position until anaphase (Figs 1a and 3a). We precisely monitored and tracked the position of the chromosomes during metaphase in control embryos (Fig. 7a,b) and showed that they undergo very little motion until anaphase (Fig. 7a,b). Thus, the position where the chromosomes merge specifies the division plane. To test whether F-actin is involved in maintaining the metaphase plate at the embryo centre, we acutely depolymerized cytoplasmic F-actin in metaphase with 1 µg ml^{-1} Cytochalasin D (Fig. 7c, lower panel). This treatment has little incidence on the motion of chromosomes (Supplementary Movie 16; Fig. 7c, right panel; Supplementary Fig. 4a), despite more variability in the amplitude of chromosomes movements in CCD-treated embryos (Supplementary Fig. 4a). However, the position of chromosomes before anaphase is comparable to controls

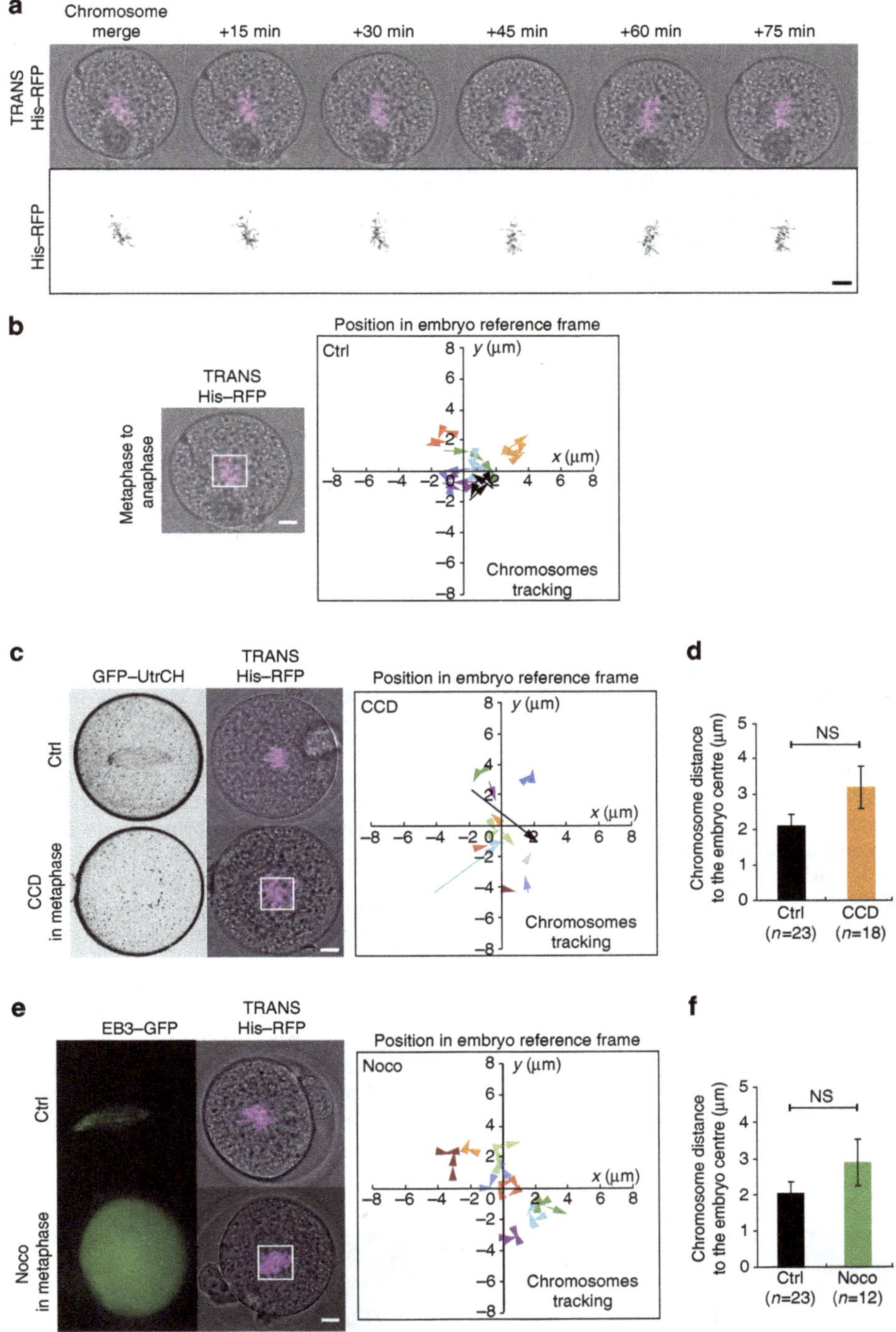

Figure 7 | Maintenance of the central position of the metaphase plate is actin and microtubule independent. (**a**) Control embryo expressing His-RFP (purple or black, Z-projection over 20 μm). One picture is shown every 15 min. Scale bar, 10 μm). (**b**) Left panel: Control embryo expressing His-RFP (purple, Z-projection over 20 μm) after chromosome merging. Scale bar, 10 μm. Right panel: graph showing the position of the centroid of the chromosomes in the referential of the embryo (0 being the centre) from metaphase to anaphase for representative control embryos. The scale is represented by the white box on the left panel. Each colour represents an embryo. The arrows point to the last position recorded before anaphase. One time point is shown every 15 or 20 min. (**c**) Left panel: control metaphase embryo (upper panel) and embryo treated with 1 μg ml^{-1} Cytochalasin D (CCD) in metaphase (lower panel) expressing GFP-UtrCH (black, one Z-plane is shown) and His-RFP (purple, Z-projection over 20 μm). Scale bar, 10 μm. Right panel: Same as **b** for embryos treated with CCD in metaphase. (**d**) Bar graph showing the chromosome mass distance to the embryo centre in metaphase for controls (black bar) and embryos treated with CCD in metaphase (orange bar). Mean is shown of 23 controls and 18 CCD-treated embryos from 2 independent experiments. s.e.m. is plotted on each bar. Statistical significance of differences is assessed with a Mann-Whitney test (*P* value 0.204). (**e**) Left panel: control metaphase embryo (upper panel, same as in Fig. 4a) and embryo treated with 1 μM Nocodazole in metaphase (lower panel) expressing EB3-GFP (green, one Z-plane is shown) and His-RFP (purple, Z-projection over 20 μm). Scale bar, 10 μm. Right panel: Same as **b** for embryos treated with 1 μM Nocodazole in metaphase. (**f**) Bar graph showing the chromosome mass distance to the embryo centre in metaphase for controls (black bar) and embryos treated with Nocodazole in metaphase (green bar). Mean is shown of 23 controls and 12 Nocodazole-treated embryos from 2 independent experiments. s.e.m. is plotted on each bar. Statistical significance of differences is assessed with a Mann-Whitney test (*P* value 0.2203).

(Fig. 7d). Thus the overall maintenance of the spindle at the cell centre is actin independent. Consistent with this result, when Myosin-II is inhibited, the chromosomes stay centrally located (Supplementary Fig. 4b), indicating that the tension imposed by Myosin-II to the cage is not necessary for the maintenance of the spindle at the centre of the cell.

We then tested the contribution of microtubules in this process by adding 1 μM Nocodazole in metaphase. Similarly, we could not detect any effect on the position of the metaphase plate (Supplementary Movie 17, Fig. 7e,f, Supplementary Fig. 4c). These results show that the global spindle maintenance at the embryo centre is microtubule independent. Eventually, when both microfilaments and microtubules were depolymerized in metaphase, the central position of the chromosomes was unaffected (Supplementary Movie 18, Supplementary Fig. 4d,e). In this condition, however, the chromosomes moved slightly less than in controls (Supplementary Fig. 4f), probably due to the fact that the microtubule spindle promotes slight chromosomes oscillations as in mitotic cells[42].

We also followed metaphasic chromosomes in embryos with artificial reduction in cortical tension. As shown above, their chromosomes do merge, but not at the embryo centre (Supplementary Movie 15, Figs 6b,c,8a). Following their gathering, during metaphase the chromosomes undergo small movements as in controls (Supplementary Movie 15, Fig. 8b), with a maximal amplitude similar to controls (Supplementary Fig. 4g). Thus in metaphase, the chromosomes stay at the position where they merged, that is, slightly off-centre (Supplementary Movie 15, Fig. 8b). Altogether, these results suggest that a high cortical tension is essential for accurate merging of maternal and paternal chromosomes at the exact embryo centre, but is not necessary for spindle maintenance there.

Since spindle maintenance at the cell centre does not seem to be achieved via active cytoskeletal mechanics, we wondered whether viscosity could be impeding on spindle motion. We thus measured viscosity using a micropipette aspiration technique (see Methods) and observed that it increases in metaphase up to threefold its value at the Pronucleus stage (Table 1). Accordingly, when beads were injected in embryos they were found to move fast at the early and late Pronucleus stages (Supplementary Movies 4–8; mean bead velocity of $18.5 \pm 7.8\,\mu m\,min^{-1}$ for early and $17.2 \pm 9.4\,\mu m\,min^{-1}$ for late Pronucleus stage) but significantly slower during mitosis (Supplementary Movie 19; mean bead velocity of $13.8 \pm 9.9\,\mu m\,min^{-1}$; Mann–Whitney test P values: early versus mitosis <0.0001; late versus mitosis 0.0013). Therefore maintenance of spindle position during metaphase could be due to a passive mechanism, a consequence of an increased viscosity in mitosis.

Discussion

Mouse zygotes are not polarized, and the positioning of the first cleavage plane depends only on the position of the two pronuclei[43]. However, how the two pronuclei are positioned has never been precisely addressed. Some studies suggested that it was actin dependent[22] whereas studies in other model systems pointed to a microtubule-dependent mechanism[44]. Here we unambiguously show that coarse pronuclei centring in mouse embryos depends on a dynamic F-actin/Myosin-Vb meshwork but not on cortical tension, suggesting that pronuclei centring could occur via an actin-based mechanism similar to the one described in Prophase I mouse oocytes (Supplementary Fig. 5)[29]. Interestingly, in *Arabidopsis thaliana*, also lacking centrosomes, the male nucleus migrates toward the female nucleus in an actin-myosin dependent manner[45]. This suggests that mechanisms of nucleus positioning that rely only on F-actin could be common to all cells that are devoid of centrioles.

Importantly, previous works using *in vitro* systems established that in metaphase, due to the dynamic instability of microtubules, spindle size scales with cell size, suggesting an upper limit for spindle reach around 40 μm in cells of a size comparable to oocytes[46–48], which is exactly the maximal distance for proper chromosome merging *in vivo* here. Altogether, this strongly argues that the F-actin-dependent coarse centring of pronuclei is absolutely required to bring the two chromosome complements within range of microtubule reach. Indeed, even in the huge starfish oocyte that does possess centrioles, the microtubules are too short to capture the chromosomes and an F-actin fishnet is responsible for their gathering[49,50].

However, this is a rather coarse centring and how the perfect symmetry is achieved has never been investigated. We propose that the main event controlling acute centring of the spindle is a centrally oriented gathering of the two sets of chromosomes, which requires F-actin and a Myosin-II-dependent increase in cortical tension (Supplementary Fig. 5). The gathering itself relies on microtubules, but the position of the gathering depends on an

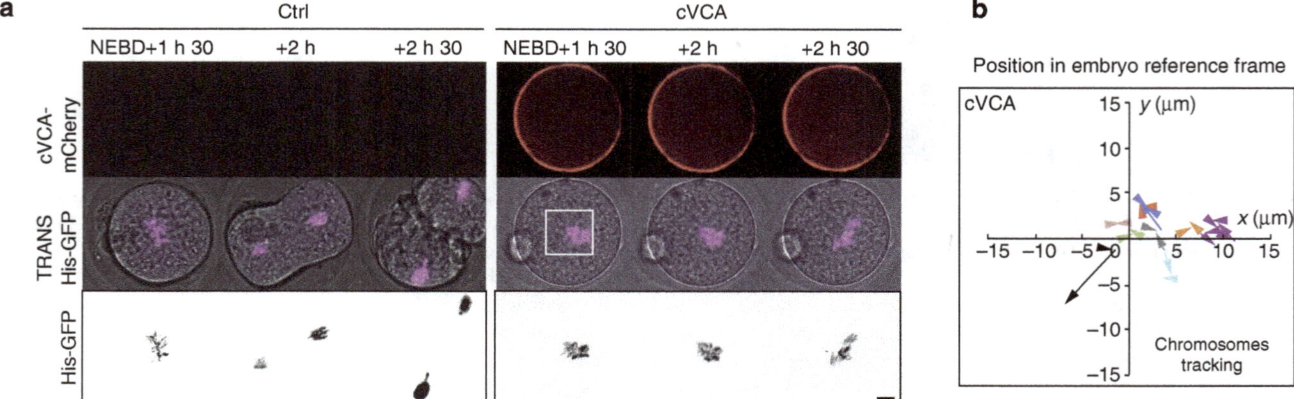

Figure 8 | Maintenance of the central position of the metaphase plate does not rely on tension but could depend on a viscous cytoplasm. (**a**) Embryos expressing His-RFP alone (left panels) or together with cVCA (right panels). Upper panel: cVCA (red), one Z-plane is shown; middle and lower panels: His-RFP, purple or black, Z-projection over 20 μm. One frame is shown every 30 min. Scale bar, 10 μm. (**b**) Graph showing the position of the centroid of the chromosomes in the referential of the embryo (0 being the centre) from metaphase to anaphase for representative embryos expressing cVCA. The scale is represented by the white box in Fig. 8a. Each colour represents an embryo. The arrows point to the last position recorded before anaphase. One time point is shown every 20 or 30 min.

increase of cortical tension and the presence of F-actin. The formation of an F-actin cage connected to the cortex (Fig. 3c,d), probably put under tension by Myosin-II at its poles (Supplementary Fig. 3a, Fig. 6e,f) and thus relaying cortical tension, could therefore allow the chromosomes to be centrally positioned when they merge, acting like a spring. The precise mechanism remains to be determined. Interestingly, chromosome mis-positioning is never drastic (Figs 3f and 6b), which likely results from the inherent geometry of the system: a long microtubule spindle (Fig 4a) and a long F-actin cage (Figs 3c and 6e) anchored to the cortex that cannot move from more than a few microns.

We also hypothesize here how, after chromosome merge, they are maintained at their central position. Neither F-actin nor microtubules or cortical tension play any role in this process. We show that a passive mechanism relying on an increase in cytoplasmic viscosity could be at play. Interestingly, this is very different from meiosis I where the spindle is dragged from the centre towards the cortex of the oocyte in an ~ 10 times less viscous cytoplasm[14]. Thus the viscosity of the cytoplasm could be differently modulated in these two systems to allow opposite functions regarding spindle positioning. In the future, it will be interesting to determine how actin meshes organized during M-phase can end up increasing or reducing cytoplasmic viscosity to influence organelle positioning.

Strikingly, similar cellular processes involving F-actin and cortical tension are used to position the spindle at the cell periphery in oocytes and at the cell centre in one-cell embryos. One difference is the presence or not of the upstream regulator, Mos, which is absent in embryos[15,16]. Thus, it could imply that Mos, a specificity of the female gamete from jelly fish to mammals[1,51], degraded after fertilization[52–54], could participate in the switch from an oocyte to an embryo mode of division.

Methods

In vitro fertilization microinjection and live imaging. *In vitro* fertilization was performed by releasing the sperm in home-made Fraser medium[55] and letting it capacitate for 1.5–2 h, then releasing 5–10 µl of capacitated sperm on Metaphase II oocytes from 5–8-week-old OF1 females stimulated around 19:00 with 50 IU of pregnant mare serum and 50 IU human chorionic gonadotropin at 48-h interval. Fertilization was allowed for 2–4 h typically from 14:00 to 16:00 and then embryos were transferred and cultured in M2 medium. Embryos were micro-injected at the Pronucleus stage with complementary RNAs (cRNAs) using an Eppendorf Femtojet micro-injector[56]. Culture and imaging were carried out under oil at 37 °C. Due to asynchrony of fertilization, embryos initiated mitosis during a 4-h period. The chromosome state was observed to infer mitotic progression. Two different females minimum were used for each experiment, and experiments were replicated at least twice to confirm the reproducibility of the results. The animal facility scientific council of the CIRB granted permission for the animal experiments described here.

Measurements of cortical tension in embryos were performed as previously described for oocytes[12]. Briefly, the zona pellucida of embryos was removed using acid tyrod (pH = 2.3), and embryos were analysed at desired stages. Embryos were loaded onto a chamber equilibrated with M2 + BSA medium. A glass micropipette of a diameter five times smaller than the embryo diameter was connected to a water reservoir of adjustable height to apply a defined aspiration pressure. Zero aspiration pressure was set prior to each experiment by checking the absence of visible flow inside the pipette. Observations were made through an inverted microscope (Axiovert 200, Zeiss) equipped with a × 40 immersion oil objective (Neofluar 1.3 numerical aperture (NA)) and connected to a charge-coupled device (CCD) camera (XC-ST70CE, Sony, Japan). For every applied pressure, we monitored the length L of the embryo portion aspirated in the pipette as a function of time and derived the speed dl/dt at which the embryo cortex enters the pipette. Then, we extrapolated the critical aspiration pressure ΔP_c at which $dl/dt = 0$ from the plot dl/dt versus the aspiration pressure. To obtain the cortical tension T_c, we used the 'viscous drop' model previously used for cells[12] that gives,

$$T_c = \frac{R\Delta P_c}{2\left(1 - \frac{R}{R_c}\right)}$$

where R is the pipette radius and R_c is the cell radius.

Spinning disk images were acquired using a Plan-APO × 40/1.25 NA objective on a Leica DMI6000B microscope enclosed in a thermostatic chamber (Life Imaging Service) equipped with a CoolSnap HQ2/CCD camera (Princeton Instruments) coupled to a Sutter filter wheel (Roper Scientific) and a Yokogawa CSU-X1-M1spinning disc[57]. The Metamorph software (Universal Imaging) was used to collect data and ImageJ (NIH) to analyse and process data. Since the embryo is a round cell with two objects of interest, the two sets of parental DNA, all analysis were performed in two dimensions in the medial plane formed by these two objects.

Plasmids and *in vitro* transcription of cRNAs. We used the following constructs: pspe3-GFP–UtrCH[5], pRN3-histone–red fluorescent protein (pRN3-histone–RFP)[58], pRN3-SF9–GFP[36], pRN3-EB3–GFP[59], pRN3-EzTD-mCherry-VCA[14], pRN3-histone–GFP[14] and pGEMHE-mCherry-Myosin-Vb (gift from Melina Schuh, MRC Cambridge, UK)[10,29].

In vitro synthesis of capped cRNAs was performed using the mMessage mMachine kit (Ambion) and the RNeasy kit (Qiagen) following the manufacturer's instructions[60]. cRNAs were centrifuged at 4 °C for 45 min at 20,000*g* before microinjection.

Trajectories, STICS analysis and trackings. Embryos were followed from NEBD until anaphase in two dimensions and only the ones with chromosomes moving in the focal plane and that divided in the focal plane were quantified. The position of the centroid of the two sets of chromosomes (male and female) was tracked using Metamorph until their merging, then the position of the centroid of the chromosome mass was recorded. The position of the centroid of the embryo was recorded at each time point. The trajectories were drawn by calculation of the coordinates of the centroid of chromosomes in the embryo frame of reference. The tracking of actin-positive foci and STICS analysis were performed as previously described[29]. Briefly, GFP–UtrCH-expressing embryos were denoised using the Safir plugin for Metamorph, background was subtracted, the images were realigned using the StackReg plugin of Fiji for embryos expressing GFP–UtrCH and Linear Stack Alignment with SIFT for transmitted images, and corrected for bleaching if required on Fiji. Tracking was performed using the TrackMate plugin of Fiji. Heat maps were generated using the ImageJ plugin 'ICS tools: STICS map jru v2' using regions avoiding the cortex. Controls are taken from different experiments (the black bars on Figs 3h and 4c, and Supplementary Fig. 2h come from Supplementary Fig. 1c; the controls on Figs 3g and 6d are the same; the controls on Fig. 7d,f and Supplementary Fig. 4b are the same; the controls on Supplementary Fig. 4a,c,f,g are the same).

Actin vesicles density analysis. To assess the difference in actin intensity between cortex and cytoplasm close to the pronuclei, GFP–UtrCH fluorescence intensity was measured in embryos in a defined round region. For each pronucleus, four measures were taken between the cortex and the pronucleus, and four between the pronucleus and the adjacent cytoplasm, all around the pronucleus. The ratio between the sum of fluorescence intensities on the cortex side and the sum of fluorescence intensities on the cytoplasm side is then calculated.

To assess the density of actin-positive foci on each region, the number of foci in each region was extracted using the Trackmate analysis described above. The percentage of foci in each region over the total number of foci in each embryo was calculated. The effective surface of the region was calculated by measuring the surface occupied by each pronucleus. The actin-positive foci density is the ratio of these two values.

Drug treatments. Embryos were treated with 1 µM Nocodazole[1,5] (Sigma, Ref. M1404), 1 µg ml^{-1} Cytochalasin D (Life Technologies, Ref PHZ1063)[1,5] or 30 µM ML-7 (refs 9,12) (Sigma, Ref. I 2764), diluted in M2 medium. Treatments were performed during pronuclei migration, around NEBD or in early metaphase. The controls presented in all panels represent control embryos extracted from different experiments.

Latex fluorescent beads (0.1 µm; Life Technologies, F8803) were rinsed several times in nuclease-free water before use to remove traces of sodium azide and diluted 500 times before injection to avoid aggregates.

Myosin-II and F-actin measurements. The overall thickness of both the cortical outer bright layer and the cortical inner dimmer layer (which increases in cVCA-expressing embryos) was measured manually in Metamorph. All measures were taken on one focal plane corresponding to the embryo's biggest diameter to avoid projection artifacts as done in ref. 9.

To assess the cortical enrichment of Myosin-II, SF9-GFP fluorescence intensity was measured in embryos, after background subtraction, in a defined square region smaller than cortex width. For each embryo, six measures were taken in the cortex and six in the cytoplasm, all around the embryo. The cortical enrichment is then defined as the ratio between the sum of fluorescence intensities in the cortex and the sum of fluorescence intensities in the cytoplasm as done in ref. 9.

Volume measurements. Pictures were taken 2 µm apart along 100 µm to cover all embryo volume on GFP–UtrCH-expressing embryos as a marker for cortex borders. The volume of the two daughter cells was calculated using the Measure stack plugin for ImageJ (Copyright (c) 2002, 2005, OptiNav, Inc.).

Shape ratio measurements. For each embryo, the inner and outer radius were measured. The shape ratio is defined as the ratio between these inner and outer radius, such as the closer the ratio to 1, the rounder the cell.

Immunofluorescence. The zona pellucida of Pronucleus stage or mitotic embryos was removed using Acid Tyrod (pH = 2.3) and embryos were fixed on coated slides[61] in 3.7% formaldehyde and stained with PhosphoMyosin-II (Cell Signaling TECHNOLOGY, Ref #3671) diluted 1:200, pericentrin (BD Biosciences, Ref 611814) diluted 1:500 and mounted in Prolongold with DAPI[7]. Images were acquired using a Leica SP5 confocal inverted microscope using a Plan-APO × 63/1.25 NA objective with Z-steps every 1 μm.

Viscosity measurements. The zona pellucida of embryos was removed using Acid Tyrod (pH = 2.3). Cells were aspirated in a 50-μm diameter micropipette until they were completely inside the pipette. Great care was taken that all cells stay inside the pipette for the same duration (~100 s). They were then released and the shape recovery was recorded. The curve representing the evolution of the reduced diameter $\frac{L}{l} - 1$ with L the bigger diameter and l the smaller diameter was plotted as a function of time and fitted as an exponential. The characteristic time τ was extracted and the viscosity η was deduced from the formula $\tau = \frac{\eta R_c}{k}$ with R_c as the cell radius and k as the cortical tension.

Statistical analysis. The statistical analysis was performed using GraphPad Prism version 6.00 for MacOS, GraphPad Software, La Jolla, CA, USA (www.graphpad.com). For comparison of several means, the normality of the variables was checked and parametric (Student's t-tests with Welch correction where indicated or one-way analysis of variance) or non-parametric comparison tests were performed with a confidence interval of 95%. For contingency analysis, χ^2 or Fischer exact tests were used with a confidence interval of 95%. In all figures, '*' corresponds to a P value <0.05, '**' to a P value <0.005, '***' to a P value <0.0005 and '****' to a P value <0.0001; NS, not statistically significant. In all figure legends, s.e.m. stands for standard error of the mean and s.d. for standard deviation.

References

1. Verlhac, M.-H., Lefebvre, C., Guillaud, P., Rassinier, P. & Maro, B. Asymmetric division in mouse oocytes: with or without Mos. *Curr. Biol.* **10,** 1303–1306 (2000).
2. Szollosi, D., Calarco, P. & Donahue, R. P. Absence of centrioles in the first and second meiotic spindles of mouse oocytes. *J. Cell Sci.* **11,** 521–541 (1972).
3. Longo, F. & Chen, D. Development of cortical polarity in mouse eggs: Involvement of the meiotic apparatus. *Dev. Biol.* **107,** 382–394 (1985).
4. Kozlowski, C., Srayko, M. & Nedelec, F. Cortical microtubule contacts position the spindle in *C. elegans* embryos. *Cell* **129,** 499–510 (2007).
5. Azoury, J. *et al.* Spindle positioning in mouse oocytes relies on a dynamic meshwork of actin filaments. *Curr. Biol.* **18,** 1514–1519 (2008).
6. Leader, B. *et al.* Formin-2, polyploidy, hypofertility and positioning of the meiotic spindle in mouse oocytes. *Nat. Cell Biol.* **4,** 921–928 (2002).
7. Dumont, J. *et al.* Formin-2 is required for spindle migration and for the late steps of cytokinesis in mouse oocytes. *Dev. Biol.* **301,** 254–265 (2007).
8. Pfender, S., Kuznetsov, V., Pleiser, S., Kerkhoff, E. & Schuh, M. Spire-type actin nucleators cooperate with formin-2 to drive asymmetric oocyte division. *Curr. Biol.* **21,** 955–960 (2011).
9. Schuh, M. & Ellenberg, J. A new model for asymmetric spindle positioning in mouse oocytes. *Curr. Biol.* **18,** 1986–1992 (2008).
10. Holubcová, Z., Howard, G. & Schuh, M. Vesicles modulate an actin network for asymmetric spindle positioning. *Nat. Cell Biol.* **15,** 937–947 (2013).
11. Azoury, J., Lee, K. W., Georget, V., Hikal, P. & Verlhac, M.-H. Symmetry breaking in mouse oocytes requires transient F-actin meshwork destabilization. *Development* **138,** 2903–2908 (2011).
12. Chaigne, A. *et al.* A soft cortex is essential for asymmetric spindle positioning in mouse oocytes. *Nat. Cell Biol.* **15,** 958–966 (2013).
13. Chaigne, A., Verlhac, M.-H. & Terret, M.-E. Spindle positioning in mammalian oocytes. *Exp. Cell Res.* **318,** 1442–1447 (2012).
14. Chaigne, A. *et al.* A narrow window of cortical tension guides asymmetric spindle positioning in the mouse oocyte. *Nat. Commun.* **6,** 6027 (2015).
15. Verlhac, M. H. *et al.* Mos is required for MAP kinase activation and is involved in microtubule organization during meiotic maturation in the mouse. *Development* **122,** 815–822 (1996).
16. Kalab, P., Kubiak, J. Z., Verlhac, M. H., Colledge, W. H. & Maro, B. Activation of p90rsk during meiotic maturation and first mitosis in mouse oocytes and eggs: MAP kinase-independent and -dependent activation. *Development* **122,** 1957–1964 (1996).
17. Simerly, C., Nowak, G., de Lanerolle, P. & Schatten, G. Differential expression and functions of cortical myosin IIA and IIB isotypes during meiotic maturation, fertilization, and mitosis in mouse oocytes and embryos. *Mol. Biol. Cell* **9,** 2509–2525 (1998).
18. Brunet, S. & Verlhac, M. H. Positioning to get out of meiosis: the asymmetry of division. *Hum. Reprod. Update* **17,** 68–75 (2010).
19. Brunet, S. & Maro, B. Germinal vesicle position and meiotic maturation in mouse oocyte. *Reproduction* **133,** 1069–1072 (2007).
20. Manandhar, G., Schatten, H. & Sutovsky, P. Centrosome reduction during gametogenesis and its significance. *Biol. Reprod.* **72,** 2–13 (2005).
21. Gueth-Hallonet, C. *et al.* gamma-Tubulin is present in acentriolar MTOCs during early mouse development. *J. Cell Sci.* **105,** 157–166 (1993).
22. Maro, B., Johnson, M. H., Pickering, S. J. & Flach, G. Changes in actin distribution during fertilization of the mouse egg. *J. Embryol. Exp. Morphol.* **81,** 211–237 (1984).
23. Schatten, G., Simerly, C. & Schatten, H. Microtubule configurations during fertilization, mitosis, and early development in the mouse and the requirement for egg microtubule-mediated motility during mammalian fertilization. *Proc. Natl Acad. Sci. USA* **82,** 4152–4156 (1985).
24. Chew, T. G., Lorthongpanich, C., Ang, W. X., Knowles, B. B. & Solter, D. Symmetric cell division of the mouse zygote requires an actin network. *Cytoskeleton* **69,** 1040–1046 (2012).
25. Yu, X.-J. *et al.* The subcortical maternal complex controls symmetric division of mouse zygotes by regulating F-actin dynamics. *Nat. Commun.* **5,** 4887 (2014).
26. Adenot, P. G., Mercier, Y., Renard, J. P. & Thompson, E. M. Differential H4 acetylation of paternal and maternal chromatin precedes DNA replication and differential transcriptional activity in pronuclei of 1-cell mouse embryos. *Development* **124,** 4615–4625 (1997).
27. Burkel, B. M., von Dassow, G. & Bement, W. M. Versatile fluorescent probes for actin filaments based on the actin-binding domain of utrophin. *Cell Motil. Cytoskeleton* **64,** 822–832 (2007).
28. Schuh, M. An actin-dependent mechanism for long-range vesicle transport. *Nat. Cell Biol.* **13,** 1431–1436 (2011).
29. Almonacid, M. *et al.* Active diffusion positions the nucleus in mouse oocytes. *Nat. Cell Biol.* **17,** 470–479 (2015).
30. Gutzeit, H. O. The role of microfilaments in cytoplasmic streaming in Drosophila follicles. *J. Cell Sci.* **80,** 159–169 (1986).
31. Brangwynne, C. P., Koenderink, G. H., MacKintosh, F. C. & Weitz, D. A. Intracellular transport by active diffusion. *Trends Cell Biol.* **19,** 423–427 (2009).
32. Courtois, A., Schuh, M., Ellenberg, J. & Hiiragi, T. The transition from meiotic to mitotic spindle assembly is gradual during early mammalian development. *J. Cell Biol.* **198,** 357–370 (2012).
33. Betschinger, J. & Knoblich, J. A. Dare to be different: asymmetric cell division in drosophila, *C. elegans* and vertebrates. *Curr. Biol.* **14,** R674–R685 (2004).
34. Larson, S. M. *et al.* Cortical mechanics and meiosis II completion in mammalian oocytes are mediated by myosin-II and ezrin-radixin-moesin (ERM) proteins. *Mol. Biol. Cell* **21,** 3182–3192 (2010).
35. Stewart, M. P. *et al.* Hydrostatic pressure and the actomyosin cortex drive mitotic cell rounding. *Nature* **469,** 226–230 (2011).
36. Nizak, C. *et al.* Recombinant antibodies against subcellular fractions used to track endogenous golgi protein dynamics *in vivo*: recombinant antibodies for cell biology. *Traffic* **4,** 739–753 (2003).
37. Matthews, H. K. *et al.* Changes in Ect2 localization couple actomyosin-dependent cell shape changes to mitotic progression. *Dev. Cell* **23,** 371–383 (2012).
38. Kunda, P., Pelling, A. E., Liu, T. & Baum, B. Moesin controls cortical rigidity, cell rounding, and spindle morphogenesis during mitosis. *Curr. Biol.* **18,** 91–101 (2008).
39. Ramanathan, S. P. *et al.* Cdk1-dependent mitotic enrichment of cortical myosin II promotes cell rounding against confinement. *Nat. Cell Biol.* **17,** 148–159 (2015).
40. Saitoh, M., Ishikawa, T., Matsushima, S., Naka, M. & Hidaka, H. Selective inhibition of catalytic activity of smooth muscle myosin light chain kinase. *J. Biol. Chem.* **262,** 7796–7801 (1987).
41. West-Foyle, H. & Robinson, D. N. Cytokinesis mechanics and mechanosensing. *Cytoskeleton* **69,** 700–709 (2012).
42. Kiyomitsu, T. & Cheeseman, I. M. Chromosome- and spindle-pole-derived signals generate an intrinsic code for spindle position and orientation. *Nat. Cell Biol.* **14,** 311–317 (2012).
43. Hiiragi, T. & Solter, D. First cleavage plane of the mouse egg is not predetermined but defined by the topology of the two apposing pronuclei. *Nature* **430,** 360–364 (2004).
44. Wühr, M., Tan, E. S., Parker, S. K., Detrich, H. W. & Mitchison, T. J. A model for cleavage plane determination in early amphibian and fish embryos. *Curr. Biol.* **20,** 2040–2045 (2010).
45. Kawashima, T. *et al.* Dynamic F-actin movement is essential for fertilization in *Arabidopsis thaliana*. *eLife* **3,** e04501 (2014).
46. Good, M. C., Vahey, M. D., Skandarajah, A., Fletcher, D. A. & Heald, R. Cytoplasmic volume modulates spindle size during embryogenesis. *Science* **342,** 856–860 (2013).
47. Hazel, J. *et al.* Changes in cytoplasmic volume are sufficient to drive spindle scaling. *Science* **342,** 853–856 (2013).

48. Wühr, M. *et al.* Evidence for an upper limit to mitotic spindle length. *Curr. Biol.* **18,** 1256–1261 (2008).
49. Lénárt, P. *et al.* A contractile nuclear actin network drives chromosome congression in oocytes. *Nature* **436,** 812–818 (2005).
50. Mori, M. *et al.* Intracellular transport by an anchored homogeneously contracting F-actin meshwork. *Curr. Biol.* **21,** 606–611 (2011).
51. Amiel, A., Leclère, L., Robert, L., Chevalier, S. & Houliston, E. Conserved functions for Mos in eumetazoan oocyte maturation revealed by studies in a cnidarian. *Curr. Biol.* **19,** 305–311 (2009).
52. Lorca, T. *et al.* Degradation of the proto-oncogene product p39mos is not necessary for cyclin proteolysis and exit from meiotic metaphase: requirement for a Ca(2 +)-calmodulin dependent event. *EMBO J.* **10,** 2087–2093 (1991).
53. Verlhac, M. H., Kubiak, J. Z., Clarke, H. J. & Maro, B. Microtubule and chromatin behavior follow MAP kinase activity but not MPF activity during meiosis in mouse oocytes. *Development* **120,** 1017–1025 (1994).
54. Weber, M., Kubiak, J. Z., Arlinghaus, R. B., Pines, J. & Maro, B. c-mos proto-oncogene product is partly degraded after release from meiotic arrest and persists during interphase in mouse zygotes. *Dev. Biol.* **148,** 393–397 (1991).
55. Ajduk, A., Ciemerych, M. A., Nixon, V., Swann, K. & Maleszewski, M. Fertilization differently affects the levels of cyclin B1 and M-phase promoting factor activity in maturing and metaphase II mouse oocytes. *Reproduction* **136,** 741–752 (2008).
56. Verlhac, M.-H. *et al.* Mos activates MAP kinase in mouse oocytes through two opposite pathways. *EMBO J.* **19,** 6065–6074 (2000).
57. Breuer, M. *et al.* HURP permits MTOC sorting for robust meiotic spindle bipolarity, similar to extra centrosome clustering in cancer cells. *J. Cell Biol.* **191,** 1251–1260 (2010).
58. Tsurumi, C. *et al.* The spindle assembly checkpoint is not essential for CSF arrest of mouse oocytes. *J. Cell Biol.* **167,** 1037–1050 (2004).
59. Kolano, A., Brunet, S., Silk, A. D., Cleveland, D. W. & Verlhac, M.-H. Error-prone mammalian female meiosis from silencing the spindle assembly checkpoint without normal interkinetochore tension. *Proc. Natl Acad. Sci USA* **109,** E1858–E1867 (2012).
60. Terret, M. E. *et al.* DOC1R: a MAP kinase substrate that control microtubule organization of metaphase II mouse oocytes. *Development* **130,** 5169–5177 (2003).
61. Luksza, M., Queguigner, I., Verlhac, M.-H. & Brunet, S. Rebuilding MTOCs upon centriole loss during mouse oogenesis. *Dev. Biol.* **382,** 48–56 (2013).

Acknowledgements

We wish to thank the members of the Verlhac team, Sophie Louvet (UPMC, Paris, France), Keith Jones (the University of Southampton, UK) for helpful discussions and Melina Schuh (MRC Cambridge, UK) for the gift of the Myosin-Vb plasmid.
This work was supported by grants from the Ligue Nationale Contre le Cancer (EL2012/LNCC/MHV) and from the Fondation ARC (PJA20131200412 to M.-E.T.). This work has received support under the program Investissements d'Avenir launched by the French Government and implemented by the ANR, with the references: ANR-10-LABX-54 MEMO LIFE, ANR-11-IDEX-0001-02 PSL* Research University.

Author contributions

A.C., M.-H.V. and M.-E.T. designed the experiments, interpreted the results and wrote the manuscript. A.C. carried out most of the experiments. A.C. and C.C. carried out the micropipette experiments. A.C., C.C. and C.S. designed the micropipette experiments and interpreted the results. N.S.G. and R.V. designed the physical model. M.-E.T. and M.-H.V. supervised the project.

Additional information

Competing financial interests: The authors declare no competing financial interests.

17

Connecting two proteins using a fusion alpha helix stabilized by a chemical cross linker

Woo Hyeon Jeong[1,*], Haerim Lee[2,*], Dong Hyun Song[3], Jae-Hoon Eom[1], Sun Chang Kim[2], Hee-Seung Lee[1], Hayyoung Lee[4] & Jie-Oh Lee[1]

Building a sophisticated protein nano-assembly requires a method for linking protein components in a predictable and stable structure. Most of the cross linkers available have flexible spacers. Because of this, the linked hybrids have significant structural flexibility and the relative structure between their two components is largely unpredictable. Here we describe a method of connecting two proteins via a 'fusion α helix' formed by joining two pre-existing helices into a single extended helix. Because simple ligation of two helices does not guarantee the formation of a continuous helix, we used EY-CBS, a synthetic cross linker that has been shown to react selectively with cysteines in α-helices, to stabilize the connecting helix. Formation and stabilization of the fusion helix was confirmed by determining the crystal structures of the fusion proteins with and without bound EY-CBS. Our method should be widely applicable for linking protein building blocks to generate predictable structures.

[1] Department of Chemistry, KAIST, Daejeon 34141, Korea. [2] Department of Biological Sciences, KAIST, Daejeon 34141, Korea. [3] Agency for Defense Development, Daejeon 34186, Korea. [4] Institute of Biotechnology, Chungnam National University, Daejeon 34134, Korea. * These authors contributed equally to this work. Correspondence and requests for materials should be addressed to J.-O.L. (email: jieoh@kaist.ac.kr).

Linking two protein components to form a predictable and rigid structure is a prerequisite for generating complex protein assemblies in a pre-designed fashion[1]. Most of the chemical cross linkers available have long and flexible spacers to help them approach the reactive side chains of the target proteins. Because of this, the resulting hybrids have significant structural flexibility and the relative orientation and distance between their two components is largely unpredictable. This is the case even when the chemical cross linkers themselves have rigid structures since they are attached to flexible side chains such as cysteines or lysines. Recently several new methods have been proposed to assemble proteins in a pre-designed fashion.

Radford *et al.*[2–4] have shown that two proteins can be linked in a rigid and predictable way using a metal coordination method. Since the bound metal ion has additional coordination sites that are available for binding of another protein molecule, it can induce homodimerization of the protein. Because the metal coordination requires precise positioning of the coordinating amino acid side chains, the resulting protein dimer will have a predictable structure. The authors used a similar metal coordination method to produce a variety of protein assemblies[5–12]. King *et al.*[13] reported a computational method for designing cage-like structures by connecting two protein components. One of them was chosen from among natural proteins that can form stable homodimers and the other from among proteins that can form stable homotrimers. By optimizing the interface between these two protein components in the fusion proteins using computer calculations, they were able to produce several fusion proteins that could assemble into large cage-like structures.

Short alpha-helical linkers joining two protein components have been used to build large and highly symmetric assemblies with rigid and predictable structures. Because an α-helix is under strict structural restraint, the structure of the fusion protein ought to be predictable so long as a connecting helix is formed. Diverse architectures are possible, depending on the specific geometric arrangement between the protein components[14]. Using this method, Lai *et al.*[1,15,16] connected M1 matrix protein, which can form homodimers, and BPO protein, which can form stable homotrimers, with a short linker. By testing various linker sequences they were able to find one that could form an α-helix and supported the formation of cage-like structures. Recently, they applied a similar method to generate large and porous cube-like closed structures with high accuracy[17]. Rigidity of the helical linker is essential for the success of the design because apparently minor deviation of the structure from the intended ideal helix can lead to failure of protein assembly[18].

Here we propose a novel method for connecting the pre-existing α-helices of two proteins into a single extended helix using a chemical cross linker. To test our method, we chose as a model system an ankyrin family protein containing a C-terminal helix and a protein A fragment containing an N-terminal helix. We fused the two proteins by connecting the two terminal α-helices into a single continuous helix. In general, simple genetic ligation of the helices in the proteins does not lead to fusion of the helices. Therefore, we used a synthetic chemical that selectively reacts with α helices and stabilizes them. EY-CBS (3,3′-ethyne-1,2-diylbis-{6-[(chloroacetyl)amino]benzenesulfonic acid}) is a chemical cross linker that can react simultaneously with two cysteines in the i and i + 11 positions of the α-helix[19]. To test whether EY-CBS detects the formation of a helix, we chose two amino acid residues separated by 11 amino acids in the proposed fusion helix and mutated them to cysteines. The successful fusion helices showed high reactivity with EY-CBS, and cross linking with EY-CBS forced the structure of the fusion α-helix to resemble that of an ideal helix. Our method does not require metal ions in the buffer, nor complicated computer simulation in the design process. It can be used to assemble protein components into asymmetrical and non-repeating structures.

Results

Design of the fusion helices. We chose an artificial ankyrin protein, named 'mbp3-16', that has been engineered to bind maltose-binding protein (MBP)[20], and the B4 domain of protein A as a model system to test our fusion helix method. We chose these proteins because both of them are composed entirely of α-helices and lack cysteine residues. In addition to this, they have been proven to be useful as building blocks of artificial protein assemblies because they can be engineered to bind other proteins[21,22]. We produced 17 fusion proteins by connecting the C-terminal helix of the ankyrin domain and the N-terminal helix of the protein A domain (Fig. 1 and Table 1). While designing connections between the two helices, obvious steric collision was avoided by using a molecular graphics program. To generate the EY-CBS reaction site, two amino acid residues separated by 11 amino acids, one from the ankyrin domain and the other from the protein A domain, were chosen and mutated to cysteines. Also, two additional amino acids, in positions i + 4 and i + 7, were mutated to alanines because long side chains may lead to steric collision with the EY-CBS molecules[19]. Using a similar

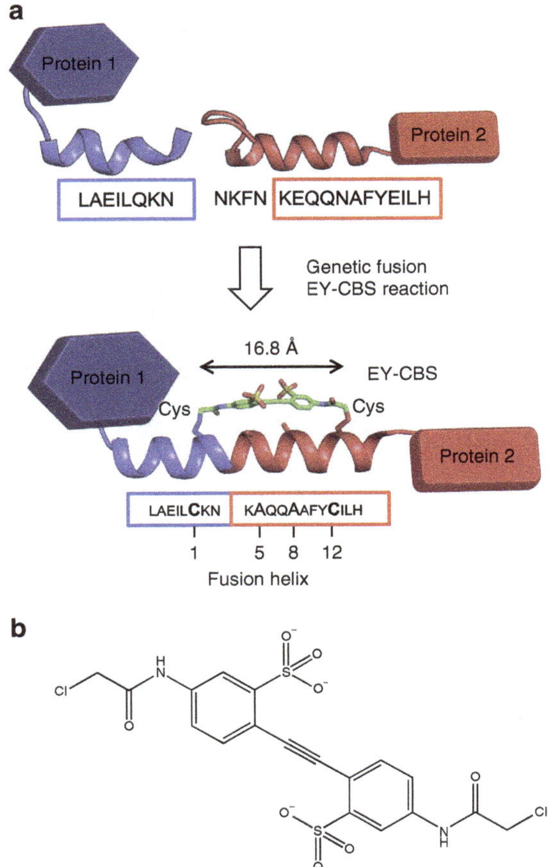

Figure 1 | Design of the fusion proteins. (**a**) The C-terminal helix of protein 1 and the N-terminal helix of protein 2 were genetically connected. Two amino acids separated by 11 amino acids were mutated to cysteines for reaction with EY-CBS. (**b**) Chemical structure of EY-CBS. EY-CBS has a rigid chemical structure with two α-chlorocarbonyl groups reactive with the thiol groups of cysteines.

Table 1 | Design of the fusion proteins.

Name	First protein	Second protein		Sequence of the fusion helix	
3,305	Protein A	mbp3-16	266*	QCAREAAREAAACD	13*
3,306	mbp3-16	Protein A	133	QCAREAAAREAACN	217
3,307	Protein A	mbp3-16	263	NCAQEAAAREAACD	13
3,308	mbp3-16	Protein A	129	ACILQAAAREAACN	217
3,309	Protein A	mbp3-16	266	QCAREAAARDLGCK	17
3,310	mbp3-16	Protein A	133	QCAREAAAREANCE	219
3,311	mbp3-16	Protein A	132	LCKNKAQQAAFYCI	227
3,312	mbp3-16	Protein A	125	NCRLAAILAKNKCQ	220
3,313	Protein A	mbp3-16	259	ACKLNAAQALGRCL	18
3,315	Protein A	mbp3-16	260	ACLNAAQAALGRCL	18
4,254	Protein A	mbp3-16	260	KCLNDALGAKLLCA	21
4,256	Protein A	mbp3-16	260	KCLNDAQAAKLLCA	21
4,257	Protein A	mbp3-16	259	ACKLNAAQALGRCL	18
4,258	Protein A	mbp3-16	259	ACKLNAAQAAGRCL	18
4,260	Protein A	mbp3-16	261	KCNDAAARALLECA	22
4,261	Protein A	mbp3-16	261	KCNDAALGAKLLCA	21
4,262	Protein A	mbp3-16	261	KCNDAAAGAKLLCA	21
6,758	Protein A	Calmodulin	260	KCLNDAEQIAEFCE	15
6,759	Protein A	Calmodulin	260	KCLNDEQIAEFKCA	16
6,760	Protein A	Calmodulin	259	ACKLNAQIAEFKCA	16
6,761	Protein A	Calmodulin	260	KCLNDAQAAAEECI	10

*Residue numbers are taken from the UniProt (protein A and calmodulin code number P38507 and P62158, respectively) and the GenBank (mbp3-16, AY326426) databases.

method, we designed four more fusion proteins with the protein A domain and the calmodulin N-terminal domain[23]. We intentionally positioned positively charged residues such as lysines and arginines in some of the fusion proteins to test whether the negatively charged EY-CBS had higher reactivity with positively charged targets (Fig. 1b). However, we found no obvious correlation between the location of the positively charged residues and the reactivity of the protein towards EY-CBS, and we did not artificially include positively charged amino acids in the later design. When necessary, we inserted alanine residues between the two proteins to control the relative orientation of the two protein components (Table 1). Note that insertion of one amino acid residue in an α-helix will rotate the helix by ~100 degrees. Therefore, one can control the relative orientation of the connected proteins by inserting precise number of amino acids into the fusion helix. Amino acid residues that may have roles in stabilizing the native helices were not changed.

Reaction with EY-CBS. The fusion proteins produced were purified and incubated with 1 mM EY-CBS (Fig. 2 and Supplementary Figs 1–3). Five of the 21 fusion proteins shifted substantially in the SDS–PAGE analysis. The molecular weight of EY-CBS is 521 daltons and if the two cysteines in the fusion helix react with EY-CBS, the shift in molecular weight should be by 521 daltons. In fact the change of molecular weight estimated from the shift of the protein bands was more than ~2 kDa (Fig. 3a). Therefore, we assumed the band shift was not only due to the additional molecular weight of the bound EY-CBS but to a change in the local structure of the fusion helix induced by circularization of the two cysteine side chains by EY-CBS. To analyse the situation further, we chose the 3,311 fusion protein because it had the highest reaction efficiency with EY-CBS (Fig. 3 and Supplementary Figs 4–6). To confirm that the 3,311 fusion indeed reacted with EY-CBS, the absorption spectrum of the reacted protein was measured after removing unreacted EY-CBS by gel filtration chromatography. The isolated EY-CBS molecule has an absorption peak at 340 nm due to the conjugated phenyl rings. As expected, the reacted fusion protein had a strong absorption peak at 340 nm, demonstrating that

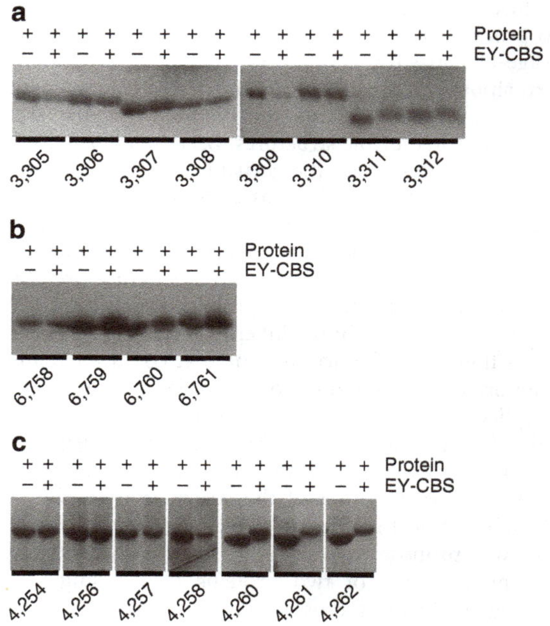

Figure 2 | Reactivity of the designed fusion proteins. The protein A-ankyrin (**a**, **c**) and the protein A-caldmodulin (**b**) fusion proteins were treated with 1mM EY-CBS. Constructs 3,311, 3,312, 4,260, 4,261 and 4,262 are substantially upshifted after the reaction.

EY-CBS was covalently linked to the protein (Fig. 3b). The ratio of the absorption peaks at 280 and 340 nm suggests that one EY-CBS molecule is bound per protein molecule.

To maximize the reaction efficiency of EY-CBS, we measured the reactivity of the 3,311 protein as a function of pH, anticipating that the deprotonated thiol groups of the cysteines would have higher reactivity at higher pH. As expected, the reaction was strongly dependent on pH and proceeded to near completion at pH 8 (Fig. 3c). At pH >9, the protein band and reaction

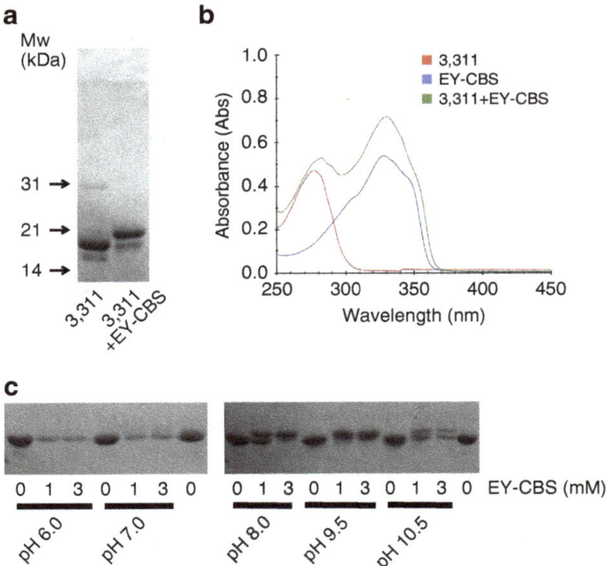

Figure 3 | Reactivity of the 3,311 fusion protein toward EY-CBS. (**a**) SDS-PAGE analysis of the 3,311 protein. Nearly 100% of the 3,311 protein is upshifted by reaction with EY-CBS. (**b**) Absorption spectrum of the 3,311 protein reacted with EY-CBS. Excess EY-CBS after reaction was removed by gel filtration chromatography. (**c**) pH dependence of the EY-CBS reaction. The 3,311 protein was reacted with EY-CBS at different pH and analysed by SDS-PAGE.

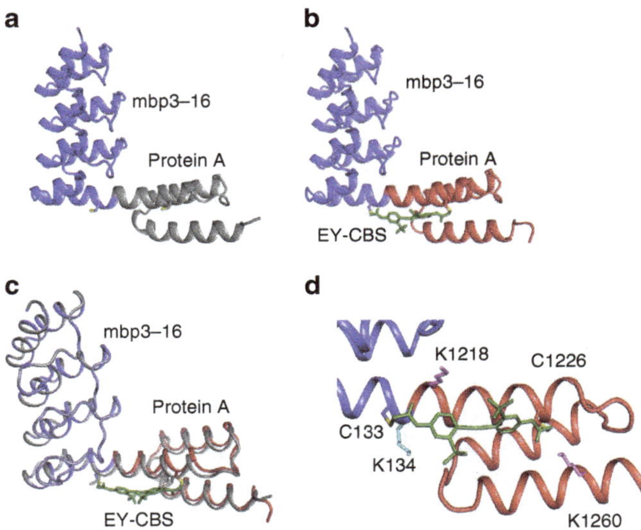

Figure 4 | Crystal structure of the 3,311 fusion protein after reaction with EY-CBS. (**a**) The intended structure of the 3,311 fusion protein. The ankyrin domain of mbp3-16 and the protein A domain are connected by an α-helix. (**b**) Crystal structure of 3,311 EY-CBS. EY-CBS is attached to the fusion helix and shown in green. For clarity the structure of the MBP bound to the ankyrin domain is not shown. (**c**) Superimposition of the intended and crystal structures of the 3,311 fusion. The intended and crystal structures are coloured in grey and blue/red, respectively. (**d**) Close-up view of the fusion helix. Positively charged amino acid side chains that interact with the sulfate groups of EY-CBS are shown.

efficiency declined presumably due to denaturation and precipitation of the fusion protein.

Crystal structure of the 3,311 fusion protein. To verify fusion of the helices, we crystallized the 3,311 fusion protein before and after reaction with EY-CBS and determined the structures of the two products (Figs 4 and 5). Since the ankyrin domain, mbp3–16, had been engineered to bind MBP, we were able to use MBP to facilitate crystallization[20]. In the crystal structure, one MBP molecule interacted exclusively with the mbp3–16 component of the hybrid protein and there was no interaction with either protein A or the fusion helix. The electron density map calculated using the refined protein structure showed a clear density corresponding to the EY-CBS, and one EY-CBS molecule could be modelled into the density (Supplementary Fig. 7). Both reactive ends of the EY-CBS are covalently linked to the cysteines in the fusion helix. In the crystal structure, the fusion helix adopts a nearly ideal α-helical structure and matches closely to the intended structure (Fig. 4c). The Cα atoms of the cysteines are separated by 16.8 angstrom and the mutated alanines are located under the phenyl rings of EY-CBS, as designed. The structures of the ankyrin and protein A domains are easily superimposable on those of the isolated proteins, which demonstrates that the helix fusion has no impact on the structure of the protein other than near the fusion site (Fig. 4b and c). The sulfate moieties of the EY-CBS groups form nonspecific ionic interactions with three lysine groups of the protein A and ankyrin domains (Fig. 4d). Other than that, EY-CBS does not have any noticeable interaction with the protein.

We determined the crystal structure of the unreacted 3,311 fusion protein at 2.8 angstrom resolution to understand why it reacts strongly with EY-CBS (Fig. 5). MBP was not included in the crystallized protein solution. The crystal structure showed that the protein A and ankyrin domains were connected by a slightly bent helix (Fig. 5a). When the structures of the fusion helix with and without the bound EY-CBS were superimposed, the fusion helix was seen to be bent by ~20 degrees with respect to that of the EY-CBS-bound helix (Fig. 5b). The asymmetric unit of the crystal contains 12 fusion proteins that can be divided into two groups, each group containing 6 proteins (Supplementary Fig. 8). The proteins belonging to the same group are in a similar crystal environment and, as expected, have superimposable structures. However, the proteins belonging to different groups are in completely different crystal environments. Despite this, the proteins in the different groups have almost identical structures, with a Cα root mean square difference value of 0.32 angstrom. Furthermore, the simulated annealing omit map calculated using the refined protein structure showed a clear density near the fusion helix (Supplementary Fig. 9). Because of these, we conclude that formation of the fusion helix is unlikely to be a crystallization artifact.

Crystal structure of the unreacted 3,311 clearly explains why the 3,311 fusion protein is strongly reactive with EY-CBS: the fusion helix is already present before EY-CBS is bound; after the covalent reaction with EY-CBS the fusion helix resembles more closely the ideal helix because the distance between the two cysteines is forced to be 16.8 angstrom. In the crystal structure, Asn107 and Lys1218 are separated by 2.9 angstrom and are hydrogen bonded. In addition, Asn135 is connected to Gln1220 via a hydrogen bond and contributes to stability of the fusion helix (Fig. 5c). However, it is not clear whether these relatively weak interactions are enough to stabilize the helical connection between the two protein components. More sophisticated calculations are required to analyse this stabilization.

Crystal structure of a fusion protein that failed the EY-CBS test. Of the 21 fusion proteins, 16 failed to display any significant

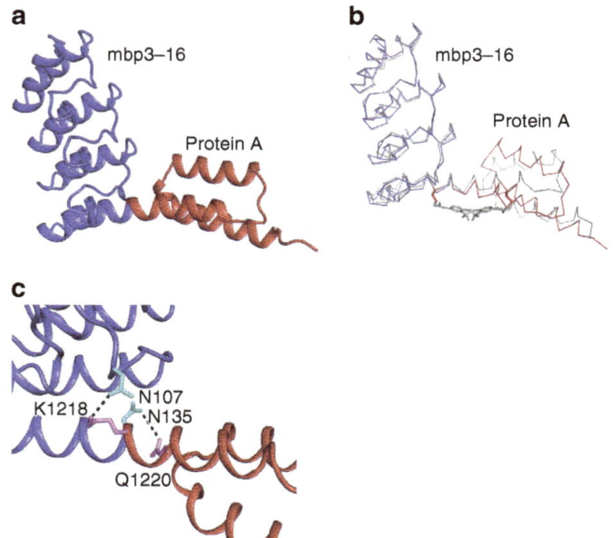

Figure 5 | Crystal structure of the 3,311 fusion protein not reacted with EY-CBS. (**a**) Crystal structure of the fusion protein. The fusion helix is bent by ~20 degrees. (**b**) Superposition of the EY-CBS-bound and -unbound structures of 3,311 fusion protein. The 3,311 EY-CBS and 3,311 structures are coloured in grey and blue/red, respectively. (**c**) Close-up view of the fusion helix. Amino acid residues that may stabilize the fusion helix are labeled.

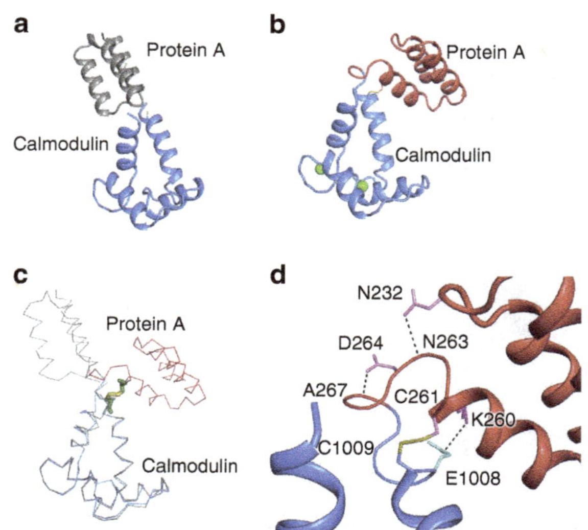

Figure 6 | Crystal structure of the 6,761 fusion protein containing the protein A domain and the calmodulin domain. (**a**) Intended structure of the 6,761 fusion protein. The two protein domains are connected by a continuous α-helix. (**b**) Crystal structure of the 6,761 fusion protein. An unexpected disulfide bridge that connects the mutated cysteines is shown. (**c**) Superposition of the intended and crystal structures of the 6,761 fusion protein. The intended structure is coloured in grey and the crystal structure is coloured in blue and red. (**d**) Close-up view of the fusion site. The cysteine 261 of the protein A domain and cysteine 1,009 of the calmodulin domain form a disulfide bridge. Amino acid residues that stabilize the interface of the two helices are drawn and labeled.

reactivity towards EY-CBS (Fig. 2). We were unable to crystallize any of these proteins except 6,761, presumably due to the structural instability of the interface between the fused protein components.

The crystallized 6,761 fusion protein is composed of an N-terminal protein A and a C-terminal calmodulin domain (Supplementary Fig. 10). Calmodulin is a calcium sensor composed of two homologous calcium-binding domains. Each calmodulin domain contains two EF hand motifs that bind calcium ions[24]. The crystal structure of the 6,761 fusion protein was determined at 2.7 angstrom resolution (Fig. 6). The structures of the two domains of the 6,761 fusion protein were easily superimposable with those of the separate protein A and calmodulin domains, and there were no noticeable structural changes except around the fusion site. In the crystal structure, it could be seen that the two helices designed to fuse had not formed a continuous helix, and the cysteines introduced to bind EY-CBS were close to one another and formed an unexpected disulfide bridge (Fig. 6c). The last and the first turns of the helices of the protein A and calmodulin domains, respectively, formed irregular loops and made sharp ~120 degree turns. The crystal structure of the connecting loop and the disulfide bridge formed a distinct electron density demonstrating that the fusion site was structurally stable (Supplementary Fig. 11). It appears that the 6,761 fusion protein can be crystallized because the interface between the two protein components of the fusion protein is stabilized by the opportunistic disulfide bridge. Several strong interactions can be seen at the fusion site and these account for why the disulfide bridge is formed (Fig. 6d). Lys260 of the protein A domain and Glu1008 of the calmodulin domain are separated by only 3.8 angstrom and form a strong ionic interaction. In addition, hydrogen bonding of the backbone nitrogen of Ala267 and the carboxylic oxygen of Asp264 should also contribute to the stabilization of the disulfide bridge. Furthermore, the side chain of Asn232 and the backbone carbonyl oxygen of Asn263 are separated by 2.9 angstrom and form a strong hydrogen bond.

We conclude that that simple-minded ligation of two helices does not guarantee the formation of a single extended helix, and that high reactivity of the fusion helix with EY-CBS is a reliable indicator of helix formation.

Insertion of a binding adaptor protein into an internal loop. To confirm the validity of our EY-CBS method, we designed several fusion proteins containing the protein A domain inserted into loops preceding internal rather than terminal α-helices of T4 lysozyme. To identify the best insertion site, we tested four loop regions and inserted an artificial protein A analogue, Ztaq (Supplementary Fig. 12)[25]. Three of the four fusion proteins were produced as efficiently as wild-type lysozyme when expressed in *Escherichia coli*. The exception was the 8,133 fusion. All three fusion proteins eluted as monomers in size-exclusion chromatography and were resistant to limited subtilisin digestion, confirming their structural integrity. Of the three possible insertion sites, we chose the loop containing residue 37 and designed three helix-fusion proteins, 8,155, 8,157 and 8,158 containing the EY-CBS-binding sites (Supplementary Figs 13–15). Fusion protein 8,157 has two more amino acids in the fusion helix than protein 8,155 and therefore its protein A domain should be rotated by ~200 degrees and translated by ~7 angstroms. Fusion proteins 8,157 and 8,158 have the same EY-CBS sites except that the positions of the reactive cysteines are shifted by four amino acids.

All three helix-fusion proteins were produced in *E. coli* and purified to homogeneity. Unexpectedly, reaction with EY-CBS did not result in clearly visible upshifts of the protein bands on SDS–PAGE presumably because size of the fusion proteins are too big and resolution of the SDS–PAGE analysis is not good enough to detect small changes in the structure (Fig. 7b and Supplementary Figs 16 and 17). However, we believe that all three

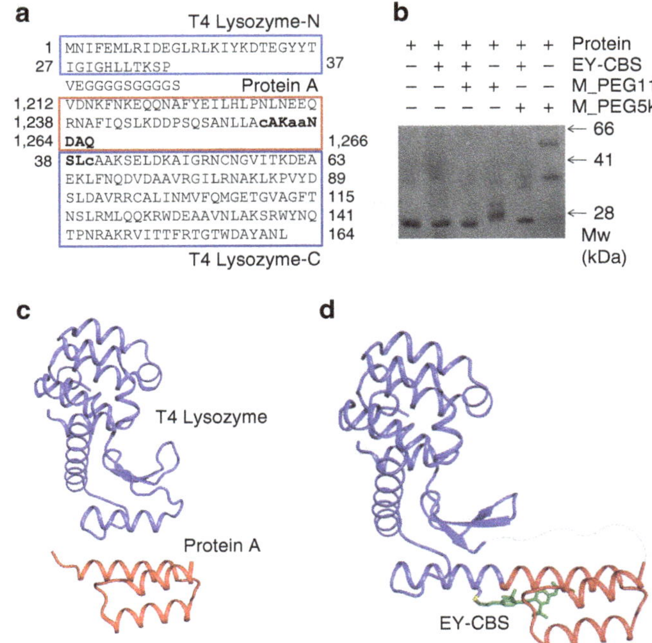

Figure 7 | Insertion of the protein A domain into an internal loop of T4 lysozyme. (**a**) The amino acid sequence of fusion protein 8,155. (**b**) Reactivity of fusion protein 8,155 with EY-CBS. After one hour of the reaction, PEG-Maleimide reagent Maleimide-PEG$_{11}$-Biotin (M_PEG11) or PEG-Maleimide 5000 (M_PEG5k) was added where appropriate. (**c**) Previously reported crystal structures of T4 lysozyme and protein A. (**d**) Crystal structure of fusion protein 8,155. The T4 lysozyme and the protein A parts of the fusion protein are coloured in blue and red, respectively. The 'GGGGS' linker and the N-terminal seven amino acids of protein A that are disordered in the crystal structure are drawn as a broken line.

proteins had reacted with EY-CBS with high efficiency because their cysteines became resistant to two maleimide-containing reagents, Maleimide-PEG$_{11}$-Biotin and PEG-Maleimide 5000. These reagents are highly reactive with the free thiol groups of cysteines, and their reactivity is easier to detect after SDS–PAGE because they have high-molecular weights, 1.1 and 5 kDa, respectively. As shown in Fig. 7b and Supplementary Figs 16 and 17, the fusion proteins were resistant to the PEG-maleimide reagents after EY-CBS treatment, presumably because they had already formed covalent bonds with EY-CBS, whereas the SDS–PAGE bands formed by the same fusion proteins not reacted with EY-CBS were clearly shifted upwards, showing that their cysteines were free to react with the PEG-maleimide reagents.

To confirm fusion of the α-helices connecting protein A and lysozyme, we crystallized 8,155 after reaction with EY-CBS and determined its crystal structure. The purified and reacted 8,157 and 8,158 fusions were also crystallized, but we have not tried to optimize the crystallization conditions nor determined their structures. The 8,155 crystals diffracted X-rays towards 2.7 angstrom resolution. In the crystal structure, one EY-CBS molecule is covalently connected to the two cysteines in the fusion helix as expected (Supplementary Fig. 18). The distance between the Cα atoms of the reacted cysteines is 16.7 angstrom, which is shorter by only 0.1 angstrom than that of 3,311 treated with EY-CBS. The fusion helix thus adopts a nearly ideal α-helical structure and closely matches the intended structure. The lysozyme and protein A parts of the structure can be superimposed with the structures of the individual proteins, demonstrating that fusion of the two helices had little impact on the overall structure of the individual protein components (Supplementary Fig. 19).

We chose the protein A domain as the insertion partner because it can be mutated to bind a variety of target proteins, as shown previously[26]. Because of this, it can be used as a universal adaptor protein mediating dimerization of pairs of target proteins. Provided we identify a suitable connecting helix, we can use the same helix to connect all other mutant protein A's for the following reasons. First, all the mutant protein A proteins adopt an essentially identical conformation, as shown by many crystal and NMR structures. Second, the C-terminal helix where our EY-CBS site is located is not changed in the mutant proteins because the mutations are limited to the first two α-helices. Among the known mutants, the Ztaq and anti-Ztaq proteins were selected for our study because they can form stable heterodimers[25]. To confirm that the mutations in the Ztaq and anti-Ztaq proteins do not affect the EY-CBS reaction, we replaced the protein A regions of fusions 8,155, 8,157 and 8,158 with the Ztaq or anti-Ztaq domain. The substituted fusion proteins retained similar reactivity with EY-CBS, as shown in Supplementary Fig. 16. Because the structure of the Ztaq-anti-Ztaq heterodimer was already known[25], we could predict the structures of the homo- and heterocomplexes 8,155–8,155, 8,157–8,157 and 8,155–8,157 with considerable confidence (Supplementary Fig. 20).

Thus, we have shown that universal adaptor proteins such as protein A or a coiled-coil domain can be inserted into a target protein to generate a predictable homo- or heterodimeric structure using our EY-CBS method.

Discussion

In this research, we have developed a novel method for connecting two proteins using a rigid fusion helix that is stabilized by a chemical cross linker. Using this method, we successfully fused the C-terminal helix of an ankyrin family protein to the N-terminal helix of a protein A fragment. We also succeeded inserting a protein A domain inside T4 lysozyme with predictable structure. Subsequent reaction with EY-CBS stabilized the structure of the fusion helix to that of an ideal helix. Our method, in principle, can be applied to majority of proteins as long as they contain at least one helix in their structure. In our method, EY-CBS reacts with thiol groups of cysteines. Therefore, proteins with natural cysteines may have unwanted reactions with EY-CBS. In such cases all the cysteines exposed to the surface need to be mutated before applying the method. Fortunately, most exposed cysteines found in proteins are not important functionally or structurally and can be converted into other amino acids without deleterious effects. In the PDB database of more than 100,000 protein structures, the majority have at least one α-helix and could therefore be used as building blocks for assembly of complex nanostructures by the helix-fusion method. Furthermore, application area of the helix-fusion method can be expanded by combining it with other methods like the metal chelation or the interface design method.

Several chemicals besides EY-CBS have been reported to stabilize the α-helical content of peptides[27]. Here are some notable examples of these helix stabilizers that can be applied to full size proteins. Two bisarylmethylene bromides named Bph and Bpy can selectively cross-link cysteines at the i and i+7 positions in the α-helices[28]; the cross-linked peptides show increased helicity and cell permeability. Also spermine contains positively charged amines at 4.6 angstrom and 6.0 angstrom spacings, and has been shown to stabilize a peptide with four negatively charged glutamate or aspartate residues separated by

3–4 residues[29]. The stabilizing effect is highly dependent on pH, suggesting that ionic interactions play the major role in the helix stabilization. Similarly a diguanidium-containing compound has been shown to enhance the helical content of a peptide containing aspartate groups in the i and i + 3 positions[30]. These compounds may be able to replace EY-CBS in the helix-fusion method. Unlike EY-CBS, some of these helix stabilizing compounds do not use cysteines for cross linking. Therefore, they may be useful for connecting proteins with reactive cysteines.

Crystallization chaperones are proteins that can be easily produced and crystallized[31,32]. Fusing them to a target protein has proven successful in improving the crystallization property of the target protein. For example, T4 lysozyme has been inserted into a flexible loop of the β2-adrenergic receptor[33,34]. The authors generated a panel of fusion proteins and chose one where the linker between lysozyme and the receptor was rigid enough for crystallization. A similar method using cytochrome cb562 instead of T4 lysozyme has been essential for crystallizing several G-protein-coupled receptor receptors and determining their structures[35–38]. The success of this method relies on structural rigidity of the fusion site, because proteins with flexible structures cannot readily be crystallized. Since it is generally impossible to predict the structure of fusion sites, a large number of fusion proteins have to be screened and the linker sequence optimized for crystallization. Our method can be used to test whether a fusion protein has a rigid fusion site before extensive and time-consuming crystallization trials. For example, one can insert a helical chaperone protein like cytochrome cb562 into a loop preceding or following an α-helix in the target protein and connect the α-helices of the target and the chaperone protein using EY-CBS as shown in the crystal structure of 8,155 (Fig. 7). The resulting fusion protein will have a rigid helical connection and there will be more chance of successful crystallization.

In this study, we showed that reaction with EY-CBS could force the structure of a fusion helix into that of an ideal helix. The crystal structure demonstrated that the fusion helix was slightly bent at the fusion site before reaction with EY-CBS and was converted to a near-ideal helix after reaction with EY-CBS. For applications that involve crystallization of the fusion protein, the rigidity of the fusion protein is critical and reaction with EY-CBS would be desirable. However, in other applications small deviations from the ideal helical structure and some structural flexibility may be allowed. In such cases the EY-CBS reaction could be useful as a probe to test whether the two helices are indeed fused and a hybrid helix has been formed.

In conclusion, we have developed a new method that can convert two pre-existing α-helices into one long extended helix. This method can be used to link proteins to form a desired structure and may be useful in constructing complex protein nanoassemblies for a variety of purposes.

Methods

Design of fusion proteins. The fusion helices between pairs of proteins were designed using the molecular graphics program, COOT[39], assuming that their structures would not be significantly perturbed by the fusion. The coordinate files of the individual protein components were obtained from the PDB database[20,23,40,41]. The fusion helix was modelled by aligning the helices from the individual proteins with that of an ideal α-helix. Possible clashes between the two proteins were visually checked and avoided by addition or deletion of amino acid residues. Amino acid positions that were open to the solvent for easy approach of EY-CBS were chosen visually and mutated to cysteines for the EY-CBS reaction. Computer programs predicting protein structure were not used in our modelling.

Protein expression and purification. The genes of the fusion proteins were cloned into pET28a vector and over-expressed using *E. coli* strain BL21(DE3). Protein production was induced by adding 0.5 mM IPTG when OD 600 of the culture reached 0.7 and incubating the cells for an additional 20 h at 25 °C.

For proteins 3,305–3,315, glutathione-*S*-transferase was used as the purification tag. The cells were harvested by centrifugation, resuspended in lysis buffer containing 50 mM Tris pH 8.0, 200 mM NaCl and 10 mM β-mercaptoethanol and homogenized using a microfluidizer, model M-110L (Microfluidics). The supernatant after centrifugation was injected into a Glutathione-Sepharose (GE Healthcare) column equilibrated with a buffer containing 20 mM Tris pH 8.0, 200 mM NaCl and 1 mM dithiothreitol (DTT). The glutathione-*S*-transferase-tagged protein was eluted in a buffer containing 50 mM Tris pH 8.0, 100 mM NaCl, 10 mM glutathione and 1 mM DTT. The protein was cleaved overnight by thrombin to remove the purification tag and purified by a Q-Sepharose (GE Healthcare) anion exchange column equilibrated with a buffer containing 20 mM Tris 8.0 and 1 mM DTT. The bound protein was eluted by linear gradient of 1 M NaCl. The fractions containing the target protein were pooled and purified using a Supderdex 200 (GE Healthcare) gel filtration column equilibrated with a buffer containing 20 mM Tris pH 8.0, 200 mM NaCl and 1 mM TCEP.

For proteins 4,254–4,262, chitin-binding domain was used as the purification tag[42]. The harvested cells were homogenized in lysis buffer containing 50 mM Tris pH 8.0, 200 mM NaCl and 1 mM DTT. The supernatant after centrifugation was loaded onto a chitin affinity column (New England Biolab) equilibrated with 20 mM Tris 8.0, 200 mM NaCl and 1 mM DTT. The protein was eluted by thrombin cleavage and purified by a Q-Sepharose anion exchange column equilibrated with a buffer containing 20 mM Tris 8.0 and 1 mM DTT. The bound protein was eluted by linear gradient of 1 M NaCl. The fractions containing the target protein were pooled and purified using a Supderdex 200 gel filtration column equilibrated with a buffer containing 20 mM Tris pH 8.0, 200 mM NaCl and 1 mM TCEP.

Hexa-histidine was used as the purification tag for constructs 6,758–6,761. The harvested cells were homogenized in lysis buffer containing 50 mM Tris pH 8.0, 200 mM NaCl and 1 mM DTT. The supernatant was loaded onto a Ni-NTA affinity column equilibrated with 20 mM Tris 8.0, 200 mM NaCl and 1 mM $CaCl_2$. One millimolar DTT was added to the protein immediately after elution using 500 mM imidazole. The protein was cleaved overnight by thrombin to remove the purification tag. The cleaved protein was purified by Q-Sepharose anion exchange chromatography equilibrated with a buffer containing 20 mM Tris 8.0, 1 mM $CaCl_2$ and 1 mM DTT. The bound protein was eluted by linear gradient of 1 M NaCl. The fractions containing the target protein were pooled and purified using a Supderdex 200 gel filtration column equilibrated with a buffer containing 20 mM Tris pH 8.0, 200 mM NaCl, 1 mM $CaCl_2$ and 1 mM TCEP. The purified proteins were concentrated with an ultrafiltration kit (Amicon) and used for the EY-CBS reaction.

For proteins 8,132–8,136 and 8,155–8,158, hexa-histidine was used as the purification tag. The harvested cells were homogenized in lysis buffer containing 20 mM Tris pH 8.0, 200 mM NaCl, 0.1 mM phenylmethyl sulphonyl fluorideand 10 mM β-mercaptoethanol. The supernatant after centrifugation was loaded onto a Ni-NTA affinity column equilibrated with 20 mM Tris pH 8.0 and 200 mM NaCl. One millimolar TCEP was added immediately to the protein eluted using 300 mM imidazole. The eluted protein was reacted with EY-CBS according to the EY-CBS reaction protocol. The EY-CBS reacted protein was purified using a SP-Sepharose (GE Healthcare) cation exchange column equilibrated with a buffer containing 20 mM MES pH 5.5. The bound protein was eluted by linear gradient of 1 M NaCl. The fractions containing the target protein were pooled and cleaved overnight by thrombin to remove the purification tag. The cleaved protein was purified using a Supderdex 200 gel filtration column equilibrated with a buffer containing 20 mM Tris pH 8.0 and 200 mM NaCl.

EY-CBS reaction. EY-CBS was synthesized as described by Zhang *et al.*[19] The EY-CBS reaction, typically with 1 mg ml^{-1} fusion protein, was carried out at 23 °C for 1 h in 20 mM Tris HCl buffer at pH 8.0 and 200 mM NaCl. One millimolar TCEP was added to the reaction buffer to prevent oxidation of the cysteine during the reaction. Progress of the reaction was monitored by SDS–PAGE. The reaction was stopped and unreacted EY-CBS was removed by Superdex 200 gel filtration or SP-Sepharose ion exchange chromatography. The reacted and purified proteins were concentrated with an ultrafiltration kit and used for crystallization if necessary. The absorption spectrum of the protein after the EY-CBS reaction was measured with a JASCO UV-530 spectrophotometer.

The PEG-Maleimide reaction. For the PEG-Maleimide reaction, 10 mM Maleimide-PEG_{11}-Biotin (Thermo Scientific) or PEG-Maleimide 5000 (Nanocs) was added to fusion proteins that either had, or had not, been reacted with EY-CBS. The progress of the reaction was monitored by SDS–PAGE and the reaction was stopped by adding SDS–PAGE sample buffer containing 570 mM β-mercaptoethanol.

Crystallization. Purified and concentrated proteins were used for crystallization. Initial crystallization conditions were screened using a crystallization robot (Mosquito, TTP Labtech). Unreacted 3,311 fusion protein was crystallized in a buffer containing 0.1 M Bis-Tris pH 6.5, 45% v/v PEG 400 and 10 mM hexamine cobalt (III) chloride. The crystals were frozen in liquid nitrogen. Addition of cryoprotectant was not necessary. After EY-CBS reaction and addition of excess amount of MBP, the 3,311-EY-CBS–MBP complex protein was purified by Superdex 200 gel filtration chromatography and crystallized in a buffer containing 0.1 M MOPS pH 6.5, 0.2 M magnesium chloride and 36% w/v PEG 2000. Crystals were frozen in a buffer

Table 2 | Data collection and refinement statistics.

	3,311 + EY-CBS	3,311	6,761	8,155 + EY-CBS
Data collection				
Space group	$P2_12_12_1$	$P2_1$	$I4_1$	C2
Resolution (Å)	2.30-50	2.80-50	2.70-50	2.60-50
Coordinations::				
a, b, c (Å)	64.3, 75.8, 111.4	85.4, 220.0, 85.4	98.8, 98.8, 31.7	89.5, 149.9, 55.8
α, β, γ (°)	90.0, 90.0, 90.0	90.0, 60.0, 90.0	90.0, 90.0, 90.0	90.0, 122.7, 90.0
R_{sym}	0.063 (0.513)	0.051 (0.152)	0.114 (0.762)	0.079 (0.352)
$I/\sigma I$	31.9	25.3	17.2	18.3
Completeness (%)	99.7 (99.2)	92.9 (80.0)	98.7 (97.0)	99.1 (98.5)
Redundancy	6.8	3.2	6.5	3.5
Search probes (PDB code)	1SVX, 1DEE	1SVX, 1DEE	2SPZ, 1CLL	1LYD, 1DEE
Refinement				
No. of reflections	24,741	62,287	4,460	35,397
R_{work}/R_{free}	0.194/0.262	0.207/0.250	0.206/0.259	0.200/0.259
Twin				
Operator	NA	h, -k, h-l	NA	NA
Fraction	NA	0.46	NA	NA
No. of protein molecules	1 (MBP) + 1 (3311)	12	1	2
No. of atoms	4,330	16,233	978	3,451
Protein	4,188	16,116	966	3,408
Ligands	30	0	0	60
Water	112	117	12	43
Average B factor (Å^2)	51.2	29.6	65.4	57.4
R.m.s. deviations				
Bond length (Å)	0.008	0.010	0.009	0.009
Bond angles (°)	1.080	1.081	1.105	1.451

EY-CBS, 3,3′-ethyne-1,2-diylbis-{6-[(chloroacetyl)amino]benzenesulfonic acid}; NA, not applicable; r.m.s., root mean square.
Highest resolution shell is shown in parenthesis.

containing the crystallization solution plus 14% PEG 2000 as the cryoprotectant. The 6,761 fusion protein was crystallized in a solution containing 10% w/v PEG 1000 and 10% w/v PEG 8000. Crystals were frozen in a buffer containing the crystallization solution plus 30% ethylene glycol as the cryoprotectant. The 8,155 protein reacted with EY-CBS was crystallized in a solution containing 1.84 M NaK phosphate pH 7.5. Crystals harvested with the crystallization buffer were flash frozen in a drop of paraffin oil (Hampton Research) using liquid nitrogen.

X-ray diffraction data and structure refinement. Diffraction data were collected at Pohang Accelerator Laboratory, beam lines 5C and 7A. Diffraction images were integrated and scaled with HKL2000. All the structures were solved by the molecular replacement method using the PHASER software[43]. Initial phasing data were then refined using PHENIX[44]. The program COOT was used for manual correction of the structure. The crystallographic data are summarized in Table 2. The 3,311 crystals were merohedrally twinned with a twin fraction of 0.46. Initially we could not solve the structure by the molecular replacement technique because the diffraction data were incorrectly indexed for the $P6_322$ space group. Later we found that the crystals were almost perfectly twinned (Supplementary Fig. 21). Therefore, the data were re-indexed for space groups with lower symmetry. Promising solutions could be found for both the $P6_3$ and $P2_1$ space groups in molecular replacement calculations. However, refinement could not reduce the free R factor below 40% when the data were indexed for the $P6_3$ space group. Therefore, they were finally indexed for the $P2_1$ space group, and the structure was refined. The free R factor dropped from 0.369 to 0.261 after incorporation of the twin operator in the refinement protocol.

References

1. Lai, Y. T., King, N. P. & Yeates, T. O. Principles for designing ordered protein assemblies. *Trends Cell Biol.* **22,** 653–661 (2012).
2. Radford, R. J., Nguyen, P. C., Ditri, T. B., Figueroa, J. S. & Tezcan, F. A. Controlled protein dimerization through hybrid coordination motifs. *Inorg. Chem.* **49,** 4362–4369 (2010).
3. Radford, R. J., Nguyen, P. C. & Tezcan, F. A. Modular and versatile hybrid coordination motifs on alpha-helical protein surfaces. *Inorg. Chem.* **49,** 7106–7115 (2010).
4. Radford, R. J. & Tezcan, F. A. A superprotein triangle driven by nickel(II) coordination: exploiting non-natural metal ligands in protein self-assembly. *J. Am. Chem. Soc.* **131,** 9136–9137 (2009).
5. Brodin, J. D. *et al.* Metal-directed, chemically tunable assembly of one-, two- and three-dimensional crystalline protein arrays. *Nat. Chem.* **4,** 375–382 (2012).
6. Brodin, J. D. *et al.* Evolution of metal selectivity in templated protein interfaces. *J. Am. Chem. Soc.* **132,** 8610–8617 (2010).
7. Medina-Morales, A., Perez, A., Brodin, J. D. & Tezcan, F. A. *In vitro* and cellular self-assembly of a Zn-binding protein cryptand via templated disulfide bonds. *J. Am. Chem. Soc.* **135,** 12013–12022 (2013).
8. Salgado, E. N. *et al.* Metal templated design of protein interfaces. *Proc. Natl Acad. Sci. USA* **107,** 1827–1832 (2010).
9. Salgado, E. N., Faraone-Mennella, J. & Tezcan, F. A. Controlling protein-protein interactions through metal coordination: assembly of a 16-helix bundle protein. *J. Am. Chem. Soc.* **129,** 13374–13375 (2007).
10. Salgado, E. N., Lewis, R. A., Faraone-Mennella, J. & Tezcan, F. A. Metal-mediated self-assembly of protein superstructures: influence of secondary interactions on protein oligomerization and aggregation. *J. Am. Chem. Soc.* **130,** 6082–6084 (2008).
11. Song, W. J., Sontz, P. A., Ambroggio, X. I. & Tezcan, F. A. Metals in protein-protein interfaces. *Annu. Rev. Biophys.* **43,** 409–431 (2014).
12. Song, W. J. & Tezcan, F. A. A designed supramolecular protein assembly with in vivo enzymatic activity. *Science* **346,** 1525–1528 (2014).
13. King, N. P. *et al.* Computational design of self-assembling protein nanomaterials with atomic level accuracy. *Science* **336,** 1171–1174 (2012).
14. Padilla, J. E., Colovos, C. & Yeates, T. O. Nanohedra: using symmetry to design self assembling protein cages, layers, crystals, and filaments. *Proc. Natl Acad. Sci. USA* **98,** 2217–2221 (2001).
15. Lai, Y. T., Tsai, K. L., Sawaya, M. R., Asturias, F. J. & Yeates, T. O. Structure and flexibility of nanoscale protein cages designed by symmetric self-assembly. *J. Am. Chem. Soc.* **135,** 7738–7743 (2013).
16. Lai, Y. T., Cascio, D. & Yeates, T. O. Structure of a 16-nm cage designed by using protein oligomers. *Science* **336,** 1129 (2012).
17. Lai, Y. T. *et al.* Structure of a designed protein cage that self-assembles into a highly porous cube. *Nat. Chem.* **6,** 1065–1071 (2014).
18. Lai, Y. T., Jiang, L., Chen, W. & Yeates, T. O. On the predictability of the orientation of protein domains joined by a spanning alpha-helical linker. *Protein Eng. Des. Sel.* **28,** 491–500 (2015).
19. Zhang, F., Sadovski, O., Xin, S. J. & Woolley, G. A. Stabilization of folded peptide and protein structures via distance matching with a long, rigid cross-linker. *J. Am. Chem. Soc.* **129,** 14154–14155 (2007).

20. Binz, H. K. *et al.* High-affinity binders selected from designed ankyrin repeat protein libraries. *Nat. Biotech.* **22,** 575–582 (2004).
21. Nygren, P. A. Alternative binding proteins: affibody binding proteins developed from a small three-helix bundle scaffold. *FEBS J.* **275,** 2668–2676 (2008).
22. Pluckthun, A. Designed ankyrin repeat proteins (DARPins): binding proteins for research, diagnostics, and therapy. *Annu. Rev. Pharmacol. Toxicol.* **55,** 489–511 (2015).
23. Chattopadhyaya, R., Meador, W. E., Means, A. R. & Quiocho, F. A. Calmodulin structure refined at 1.7A resolution. *J. Mol. Biol.* **228,** 1177–1192 (1992).
24. Meador, W. E., Means, A. R. & Quiocho, F. A. Target enzyme recognition by calmodulin: 2.4A structure of a calmodulin-peptide complex. *Science* **257,** 1251–1255 (1992).
25. Lendel, C., Dogan, J. & Hard, T. Structural basis for molecular recognition in an affibody:affibody complex. *J. Mol. Biol.* **359,** 1293–1304 (2006).
26. Feldwisch, J. & Tolmachev, V. Engineering of affibody molecules for therapy and diagnostics. *Methods Mol. Biol.* **899,** 103–126 (2012).
27. Henchey, L. K., Jochim, A. L. & Arora, P. S. Contemporary strategies for the stabilization of peptides in the alpha-helical conformation. *Curr. Opin. Chem. Biol.* **12,** 692–697 (2008).
28. Muppidi, A., Wang, Z., Li, X., Chen, J. & Lin, Q. Achieving cell penetration with distance-matching cysteine cross-linkers: a facile route to cell-permeable peptide dual inhibitors of Mdm2/Mdmx. *Chem. Commun.* **47,** 9396–9398 (2011).
29. Tabet, M., Labroo, V., Sheppard, P. & Sasaki, T. Spermine-Induced Conformational-Changes of a Synthetic Peptide. *J. Am. Chem. Soc.* **115,** 3866–3868 (1993).
30. Albert, J. S., Goodman, M. S. & Hamilton, A. D. Molecular recognition of proteins—sequence-selective binding of aspartate pairs in helical peptides. *J. Am. Chem. Soc.* **117,** 1143–1144 (1995).
31. Koide, S. Engineering of recombinant crystallization chaperones. *Curr. Opin. Struct. Biol.* **19,** 449–457 (2009).
32. Lieberman, R. L., Culver, J. A., Entzminger, K. C., Pai, J. C. & Maynard, J. A. Crystallization chaperone strategies for membrane proteins. *Methods* **55,** 293–302 (2011).
33. Cherezov, V. *et al.* High-resolution crystal structure of an engineered human beta2-adrenergic G protein-coupled receptor. *Science* **318,** 1258–1265 (2007).
34. Rosenbaum, D. M. *et al.* GPCR engineering yields high-resolution structural insights into beta2-adrenergic receptor function. *Science* **318,** 1266–1273 (2007).
35. Liu, W. *et al.* Structural basis for allosteric regulation of GPCRs by sodium ions. *Science* **337,** 232–236 (2012).
36. Thompson, A. A. *et al.* Structure of the nociceptin/orphanin FQ receptor in complex with a peptide mimetic. *Nature* **485,** 395–399 (2012).
37. Fenalti, G. *et al.* Molecular control of delta-opioid receptor signalling. *Nature* **506,** 191–196 (2014).
38. Wu, H. *et al.* Structure of a class C GPCR metabotropic glutamate receptor 1 bound to an allosteric modulator. *Science* **344,** 58–64 (2014).
39. Emsley, P., Lohkamp, B., Scott, W. G. & Cowtan, K. Features and development of Coot. *Acta Crystallogr. D Biol. Crystallogr.* **66,** 486–501 (2010).
40. Graille, M. *et al.* Crystal structure of a Staphylococcus aureus protein A domain complexed with the Fab fragment of a human IgM antibody: Structural basis for recognition of B-cell receptors and superantigen activity. *Proc. Natl Acad. Sci. USA* **97,** 5399–5404 (2000).
41. Mooers, B. H. *et al.* Repacking the Core of T4 lysozyme by automated design. *J. Mol. Biol.* **332,** 741–756 (2003).
42. Mitchell, S. F. & Lorsch, J. R. Protein affinity purification using intein/chitin binding protein tags. *Methods Enzymol.* **559,** 111–125 (2015).
43. McCoy, A. J. *et al.* Phaser crystallographic software. *J. Appl. Crystallogr.* **40,** 658–674 (2007).
44. Adams, P. D. *et al.* PHENIX: a comprehensive Python-based system for macromolecular structure solution. *Acta Crystallogr. D Biol. Crystallogr.* **66,** 213–221 (2010).

Acknowledgements

We thank the staff of the beam lines 5C and 7A, Pohang Accelerator Laboratory for help with data collection, and Dr Julian Gross for critical reading of the manuscript. This research was supported by the Intelligent Synthetic Biology Center of Global Frontier Project (2011-0031955) and the National Research Foundation (NRF-2014R1A2A1A10050436) funded by the Ministry of Science, ICT and Future Planning of Korea.

Author contributions

W.H.J., H.L., D.H.S. and J.-H.E. performed experiments and data analysis. W.H.J., H.L., D.H.S., S.C.K., H.-S.L., H.L. and J.-O.L. designed experiments. W.H.J., H.L., S.C.K., H.-S. L., H.L. and J.-O.L. contributed to writing the paper. J.-O.L. supervised the project.

Additional information

Accession codes: Atomic coordinates and diffraction data have been deposited in the Protein Data Bank with code numbers 5CBN, 5CBO, 5COC and 5EWX for the 3311 + EY-CBS, 3311, 6761 and 8155 + EY-CBS structures, respectively.

Competing financial interests: The authors declare no competing financial interests.

18

Phosphorylation modifies the molecular stability of β-amyloid deposits

Nasrollah Rezaei-Ghaleh[1], Mehriar Amininasab[2], Sathish Kumar[3], Jochen Walter[3] & Markus Zweckstetter[1,4,5]

Protein aggregation plays a crucial role in neurodegenerative diseases. A key feature of protein aggregates is their ubiquitous modification by phosphorylation. Little is known, however, about the molecular consequences of phosphorylation of protein aggregates. Here we show that phosphorylation of β-amyloid at serine 8 increases the stability of its pathogenic aggregates against high-pressure and SDS-induced dissociation. We further demonstrate that phosphorylation results in an elevated number of hydrogen bonds at the N terminus of β-amyloid, the region that is critically regulated by a variety of post-translational modifications. Because of the increased lifetime of phosphorylated β-amyloid aggregates, phosphorylation can promote the spreading of β-amyloid in Alzheimer pathogenesis. Our study suggests that regulation of the molecular stability of protein aggregates by post-translational modifications is a crucial factor for disease progression in the brain.

[1]German Center for Neurodegenerative Diseases (DZNE), 37075 Göttingen, Germany. [2]Department of Cell and Molecular Biology, School of Biology, College of Science, University of Tehran, 1417466191 Tehran, Iran. [3]Department of Neurology, University of Bonn, 53127 Bonn, Germany. [4]Department for NMR-based Structural Biology, Max Planck Institute for Biophysical Chemistry, 37077 Göttingen, Germany. [5]Department of Neurology, University Medical Center Göttingen, University of Göttingen, 37073 Göttingen, Germany. Correspondence and requests for materials should be addressed to N.R.-G. (email: Nasrollah.Rezaei-Ghaleh@dzne.de) or to J.W. (email: Jochen.Walter@ukb.uni-bonn.de) or to M.Z. (email: Markus.Zweckstetter@dzne.de).

Protein aggregation plays a crucial role in several neurodegenerative diseases such as Alzheimer (AD) and Parkinson disease[1]. A remarkable feature of neurotoxic protein aggregation is the ubiquitous presence of post-translational modifications of aggregated proteins[2,3]. A particular abundant modification of amyloid deposits is phosphorylation[4]. Phosphorylation influences protein–protein and protein–lipid interactions, protein turnover and subcellular localization of neurodegeneration-related proteins[4–8]. In addition, phosphorylation plays a significant role in regulating the clearance mechanisms of proteinaceous deposits[4,7,9,10]. Despite its relevance in neurodegenerative diseases, the effect of phosphorylation on the molecular properties of amyloid deposits is largely unknown.

Protein amyloid fibrils exhibit remarkable stability against physical perturbations suggesting that they may represent the lowest free-energy state in the conformational energy landscape of proteins[11,12]. Amyloid fibrils also possess intriguing mechanical properties with an elastic Young's modulus comparable to that of the most rigid proteinaceous materials in nature[13]. The nanomechanical features of amyloid fibrils influence the kinetics of fibril growth by modulating their inherent fragmentation tendency[14]. Fragmentation of protein fibrils is therefore thought to be crucial for propagation and spreading of disease pathology[15]. Less ordered but more toxic aggregate species such as protofibrils and early oligomeric assemblies often have lower physical and mechanical stability than fully formed fibrils[13,16,17], further supporting a strong link between the stability of protein aggregates and their pathological function[14].

Early-onset familial forms of AD are generally caused by mutations, which increase amyloid β (Aβ) production, elevate Aβ42/Aβ40 ratio or enhance Aβ propensity for formation of toxic aggregates[18]. In contrast, causes of the much more common late-onset sporadic AD are poorly understood. It has been suggested that post-translational modifications such as phosphorylation, nitration and pyroglutamination may play an important role in the initiation and progression of sporadic AD[4,19–21]. Indeed, extracellular Aβ can undergo phosphorylation by a cell surface-localized or secreted form of protein kinase A and Aβ phosphorylated at serine 8 is detectable in human AD brain[22,23]. Phosphorylation of Aβ at serine 8 markedly accelerates aggregation of Aβ *in vitro* and *in vivo* and increases Aβ-mediated toxicity in *Drosophila* models of AD[22]. In addition, it attenuates proteolytic degradation of Aβ by certain proteases[24].

Here we investigate how phosphorylation at serine 8 alters the kinetic and thermodynamic stability of Aβ aggregates. Using high-pressure nuclear magnetic resonance (NMR) and molecular dynamics (MD) simulation, we demonstrate that phosphorylation modifies the stability of amyloid aggregates of Aβ and propose a structural basis for the higher stability of phosphorylated Aβ aggregates. Our study supports an important role of phosphorylation and other post-translational modifications of amyloid deposits in neurodegenerative diseases.

Results

Aβ aggregates are sensitive to pressure. To investigate the effect of phosphorylation on the stability of Aβ aggregates, we employed high pressure as a tool to induce protein disaggregation. In general, the susceptibility of protein aggregates to high pressures is largely governed by their degree of compaction[25–27]. Mature amyloid fibrils are largely resistant to pressure, because tight packing of their backbone and side-chain groups minimizes void volumes inside the fibrils. In contrast, aggregates, which appear early during amyloid aggregation, are often rich of internal water-excluded cavities[28], rendering them sensitive to pressure-induced dissociation[11,16]. In addition, ionic interactions often experience significant volume decrease on disruption and are therefore sensitive to high pressure[29]. To assist pressure-induced dissociation of Aβ aggregates, we performed the experiments at 0 °C.

Wild-type Aβ (npAβ) and Aβ phosphorylated at serine 8 (pS8Aβ) were incubated in aggregation-prone conditions for 4 days, which led to the formation of mature amyloid fibrils (Fig. 1a). Subsequently the aggregates were exposed to increasing pressures up to 2,000 bar and the release of Aβ monomers was monitored by one-dimensional proton NMR spectroscopy. Figure 1b demonstrates that Aβ monomers were released from the preformed aggregates of both npAβ and pS8Aβ in

a

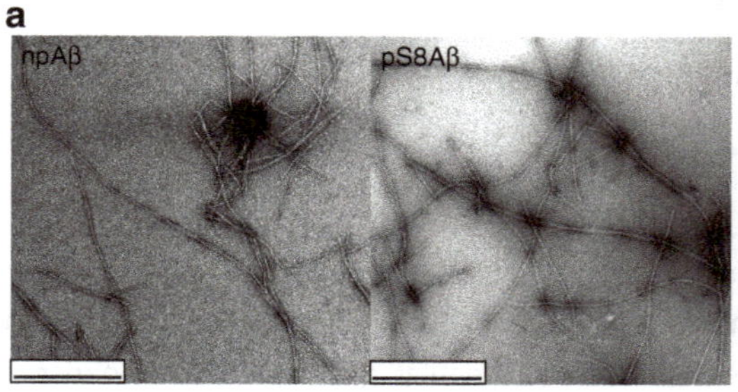

b

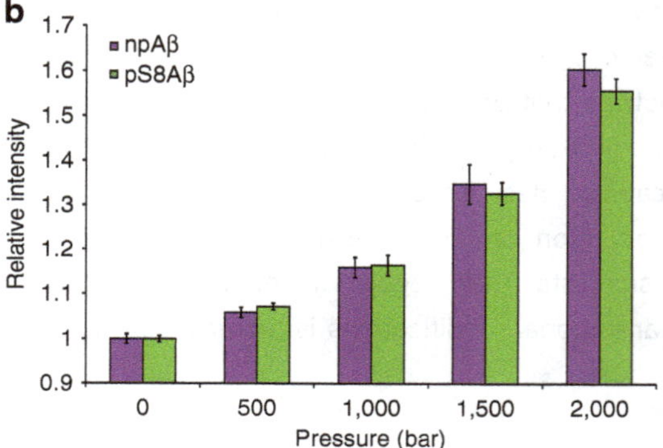

c

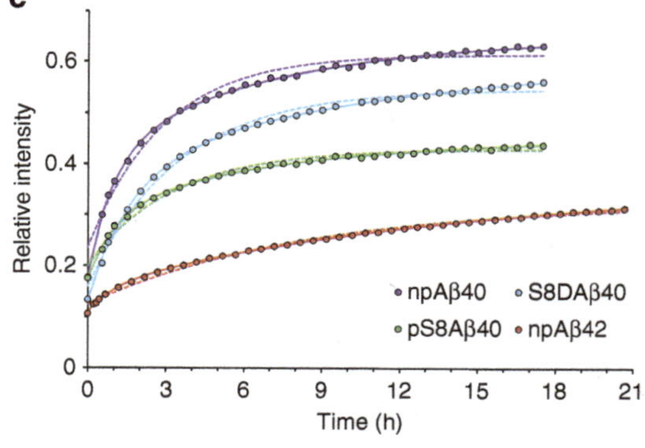

Figure 1 | Aβ aggregates have distinct pressure stability. (**a**) Electron micrographs of aggregated Aβ40 in its non-phosphorylated (npAβ) and S8-phosphorylated (pS8Aβ) state reveal amyloid fibrils of similar morphology. Scale bar, 500 nm. (**b**) Pressure dependence of Aβ monomer release from npAβ (magenta) and pS8Aβ (green) aggregates. Error bars represent the s.d. between three consecutive measurements each taking 10 min. (**c**) Time dependence of Aβ monomer release, as followed by NMR signal intensities, on incubation of aggregated Aβ at a hydrostatic pressure of 2,000 bar. Data fit to mono- and bi-exponential functions are presented in dotted and solid lines, respectively. Phosphorylation at S8 attenuates the monomer release from Aβ40 aggregates. In addition, Aβ42 aggregates (red) are more stable than Aβ40 aggregates.

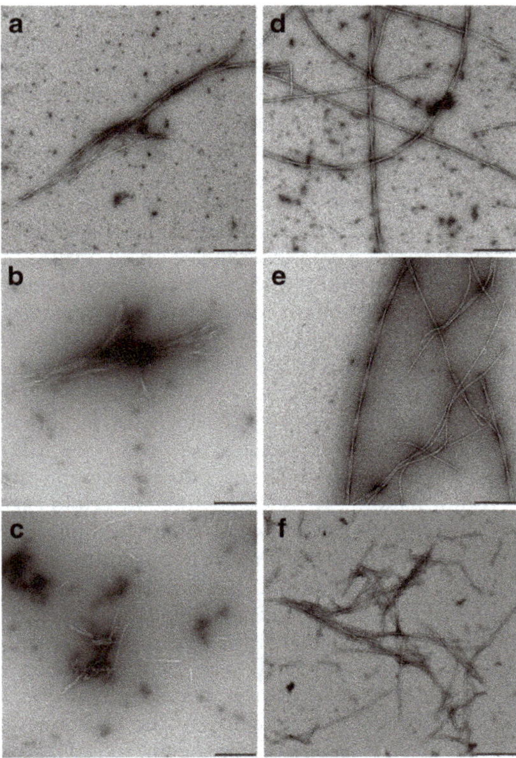

Figure 2 | Effect of high pressure on Aβ fibrils. Electron micrographs of npAβ (**a–c**) and pS8Aβ (**d–f**) aggregates, before (**a,d**) and after (**b–c,e–f**) application of high hydrostatic pressure (2,000 bar, 18 h). Many fibrils were unaffected by the high pressure (**b,e**), but some broken fibrils were also observed (**c,f**). Scale bar, 200 nm.

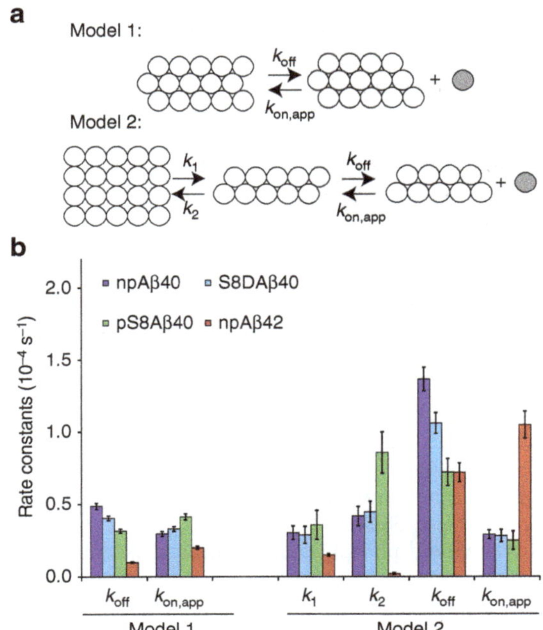

Figure 3 | Phosphorylation modulates the kinetic and thermodynamic stability of Aβ aggregates. (**a**) Schematic representation of two kinetic models for aggregate dissociation. Model 1 consists of a single step of reversible Aβ monomer (grey) dissociation from aggregates. Model 2 involves a reversible step in which Aβ aggregates are converted to an intermediate state under rate constants k_1 and k_2, followed by a second reversible step of monomer release from the intermediate state. (**b**) Rate constants obtained from the analysis of monomer release kinetics (Fig. 1c) according to the two kinetic models shown in **a**. Error bars represent s.e.s of fitting parameters.

dependence of the applied pressure. Consistent with a pressure-induced perturbation of the aggregates, a slight reduction in the ThT fluorescence signal intensities was observed after pressure application (Supplementary Fig. 1). The drop in ThT intensity was lower than the level of concomitant Aβ monomer release suggesting that the released Aβ monomers originated partly from non-ThT reactive aggregates. Application of pressure up to 2,000 bar did not have a large effect on the morphology Aβ fibrils, although a number of broken fibrils were detected after pressure application (Fig. 2).

Phosphorylation modifies the stability of Aβ aggregates. For both npAβ and pS8Aβ, incubation in aggregation-prone conditions for 96 h decreased the NMR signal intensity to ~18% of the initial value, indicating conversion of >80% of Aβ molecules to NMR-invisible aggregates. We then followed the kinetics of Aβ monomer release at 2,000 bar: after ~18 h of incubation at 2,000 bar, the NMR intensity increased from 18 to ~63% for npAβ and to ~44% for pS8Aβ (Fig. 1c). To quantify the release of Aβ monomers from the aggregates, we fitted the gradual rise in NMR signal intensities to a monoexponential function. From the analysis an apparent rate R of $0.79 \pm 0.04 \times 10^{-4}\,s^{-1}$ and limiting intensity value of 0.62 were obtained for npAβ40. In case of pS8Aβ40, values of $0.73 \pm 0.04 \times 10^{-4}\,s^{-1}$ and 0.43 were calculated. Notably, the quality of fit was significantly improved when a bi-exponential function was used (Fig. 1c; Supplementary Table 1).

To interpret the fitting parameters in terms of physical rates, we first assumed the simplest model involving only a single step of reversible Aβ monomer dissociation from aggregates (Fig. 3a). Assuming that the association reaction obeys (pseudo) first-order kinetics in terms of monomer concentration, the rates of monomer dissociation (k_{off}) of $0.49 \pm 0.02 \times 10^{-4}\,s^{-1}$ for npAβ and $0.32 \pm 0.01 \times 10^{-4}\,s^{-1}$ for pS8Aβ were obtained. The corresponding association rates ($k_{on,app}$) were $0.30 \pm 0.02 \times 10^{-4}\,s^{-1}$ (npAβ) and $0.42 \pm 0.05 \times 10^{-4}\,s^{-1}$ (pS8Aβ). The obtained k_{off} rate of $0.49 \pm 0.02 \times 10^{-4}\,s^{-1}$ for npAβ is comparable to the off-rate value of $1.2 \times 10^{-4}\,s^{-1}$ previously reported for the protofibrils of wild-type Aβ40 (25 °C, ambient pressure) on the basis of fluorescence measurements[30]. In case of amyloid fibrils formed by an SH3 domain (28 °C, ambient pressure), a dissociation rate of $1.4 \times 10^{-4}\,s^{-1}$ had been estimated[31]. Compared with the extrapolated k_{off} value of $\sim 0.06 \times 10^{-4}\,s^{-1}$ at 0 °C in ref. 30, the use of high pressure in our study has resulted in an about eightfold increase in the monomer dissociation rate of npAβ.

In the next step, we considered a more complex model involving two steps: first, a reversible step in which Aβ aggregates are converted to an intermediate state under rate constants k_1 and k_2, and second, the reversible monomer dissociation from these intermediate aggregates governed by the rate constants k_{off} and $k_{on,app}$ as described above (Fig. 3a). This model implies bi-exponential behaviour of Aβ monomer level in response to increased pressure, which was shown above to better fit to the experimental data than the monoexponential model (Fig. 1c; Supplementary Tables 1 and 2). Serine 8 phosphorylation did not have a significant effect on the forward kinetics of the first step, but diminished k_{off} rate in the second step (Fig. 3b; Supplementary Table 2). The lower level of pressure-induced monomer release from pS8Aβ than from npAβ aggregates provides substantial support for an enhanced stability of Aβ aggregates in the

S8-phosphorylated state. In addition, the decrease in the monomer dissociation rate (k_{off}) for both models demonstrates the increased kinetic stability of phosphorylated Aβ aggregates.

Aspartate variants are widely used to model defined phosphorylation states of neurotoxic proteins *in vivo*. Indeed, replacement of serine 8 by aspartate in Aβ promotes its aggregation similar to pS8Aβ (Supplementary Fig. 2)[22]. We therefore asked whether the S8D substitution could also mimic the effect of S8-phosphorylation on the stability of Aβ aggregates. To this end, we exposed S8DAβ aggregates to high pressure. Figure 1c shows that the level of pressure-induced monomer release from S8DAβ aggregates was in between that of npAβ and pS8Aβ (Supplementary Table 1). Comparison of the kinetic parameters of npAβ, S8DAβ and pS8Aβ monomer release demonstrated that the off-rate k_{off} was similarly reduced in S8DAβ albeit at a smaller level than pS8Aβ (Fig. 3b; Supplementary Table 2). In contrast, no significant changes in the apparent on-rates $k_{on,app}$ were observed (Fig. 3b). This suggests differences between phosphomimetic variants and naturally occurring phosphorylated proteins even when, as it is the case for the pS8Aβ and S8DAβ, the general aggregation behaviour of phosphorylated and phosphomimetic protein is similar.

Volume change on Aβ monomer release from aggregates. To gain insight on the impact of S8 phosphorylation on volume changes during Aβ monomer release, we monitored how the equilibrium between aggregated and monomeric Aβ is affected by pressure. After 2 days of pressure application to aggregated Aβ samples (at 500, 1,000, 1,500 and 2,000 bar, 0 °C), the Aβ monomer levels were quantified by NMR and the equilibrium constants between Aβ aggregates and monomers derived (equation (3)). As shown in Fig. 4, the pressure dependence of free-energy change on monomer release from npAβ40 aggregates ($\Delta G_{A\rightarrow M}$) resulted in an apparent volume change of 23 ± 1 μl mol^{-1}. Since the free-energy change at ambient pressure was quite close to the intercept of the fitted line, the volume change on monomer release seemed to be effectively constant over the studied pressure range. On the other hand, the volume change on monomer release from pS8Aβ40 aggregates was only 8 ± 1 μl mol^{-1}, three times lower than that of npAβ40. Notably, the free-energy change at ambient pressure was significantly above the fitted line (green data point at ambient pressure in Fig. 4), indicating that the volume change on monomer release was decreased by pressure, that is, the compressibility of pS8Aβ40 aggregates in water was higher than its monomers.

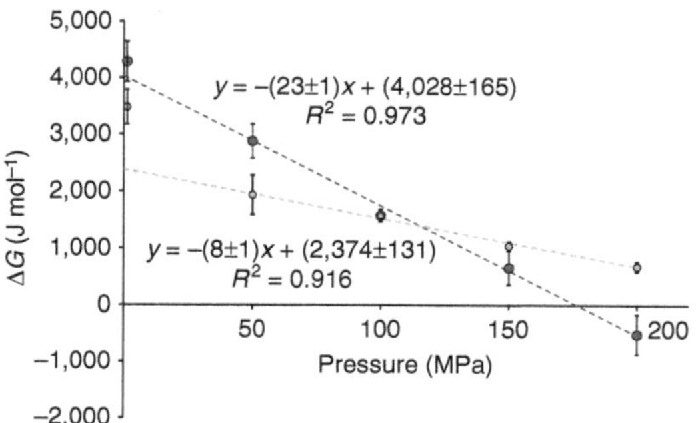

Figure 4 | Volume change on Aβ monomer release from aggregates. The pressure dependence of free-energy change on monomer dissociation from npAβ40 (magenta) and pS8Aβ40 (green) aggregates is shown. Slopes of fitted lines represent volume changes on Aβ monomer release from aggregates, which is approximately three times smaller in pS8Aβ40. Error bars indicate 1 s.d. of 2-5 repeated measurements.

Mechanistic basis of altered stability of amyloid fibrils. Despite the drastically higher rate and amount of amyloid aggregation of pS8Aβ, the morphology of fibrils differed only subtly from that of npAβ (Fig. 1a). In the negatively stained transmission electron micrographs, the fibrils formed by the two peptides were similarly twisted and their apparent diameter (maximum width) was 13.7 ± 3.6 nm for npAβ and 11.5 ± 1.9 nm for pS8Aβ. The size and morphology of Aβ40 fibrils resembled those generated in a previous study under quiescent condition and lower temperature of 24 °C (ref. 32). The size distribution of npAβ fibrils exhibited a weakly bimodal behaviour, with the major and minor modes centered at ~12 and 24 nm, respectively, (Supplementary Fig. 3). Fibrils of pS8Aβ were more strongly centered at the 12-nm mode, suggesting that phosphorylation of serine 8 affected the higher order assembly of Aβ40 fibrils.

To gain insight into the molecular basis of the increased structural stability of Aβ aggregates, we performed MD simulations of Aβ40 fibrils in different phosphorylation states. As starting conformation we selected the threefold symmetric structure of Aβ40 fibrils prepared after seeding by *in vivo* plaques[33]. This structure was chosen because it is currently the only available experimental model of Aβ fibrils, which contains detailed coordination information about the structure of the N-terminal residues of Aβ. The high level of similarity in fibrillar morphology of npAβ and pS8Aβ (Fig. 1a; Supplementary Fig. 3) supports the use of same structural model for both species.

Figure 5a shows how the mobility of the backbone atoms of Aβ40 amyloid fibrils is influenced by the phosphorylation state of serine 8. The most flexible part of the wild-type Aβ40 sequence in the MD simulation is the N-terminal region: the backbone atoms of D1-Y10 experienced high root-mean-square fluctuations (r.m.s.f.) when compared with other parts of Aβ40. When serine 8 was phosphorylated, an overall decrease in the r.m.s.f. of the backbone atoms was observed, but the strongest reduction was found for residues F4-Y10. In addition, V24-N27 and G38-V40 showed a slight drop in r.m.s.f. (Fig. 5a). Mapping of r.m.s.f. values onto a three-dimensional structure of Aβ40 fibrils[33] showed that the two patches of residues, F4-Y10 and V24-N27 that partially lose their mobility on S8-phosphorylation, are in spatial proximity to the phosphorylation site (Supplementary Fig. 4). Comparison of the average conformation of the last 25 ns of the MD trajectory further highlights the distinct mobility of Aβ40 fibrils when phosphorylated at serine 8 (Supplementary Fig. 5). In case of the phosphomimetic mutation, the reduction in backbone mobility was less pronounced and was restricted more to the N-terminal region (Fig. 5a; Supplementary Fig. 5).

Phosphoserines have a higher capacity for hydrogen bond formation than serine residues, because the phosphate group can act both as donor and acceptor in hydrogen bonds. To see whether the higher rigidity of the N-terminal region in S8-phosphorylated Aβ40 fibrils is caused by the stronger hydrogen bond capacity of phosphoserine, we calculated the existence probability of several hydrogen bonds involving the phosphorylated residue. While the total number of hydrogen bonds in npAβ and pS8Aβ fibrils was highly similar (~23 for Aβ molecule), phosphorylation of serine 8 enhanced hydrogen bond formation between S8 and the nearby residues R5-Y10 within the same or adjacent Aβ molecules (Fig. 5b). In addition, a long-range hydrogen bond between S8 and S26 was slightly more populated when serine 8 was phosphorylated. Notably, some hydrogen bonds not involving the phosphoserine were also more frequently populated, for example, between D7 and S26, and to a lesser extent between Y10 and N27.

To investigate how S8 phosphorylation influences backbone conformation in the proximity of the phosphorylation site, we compared the intramolecular distance between CA atoms of

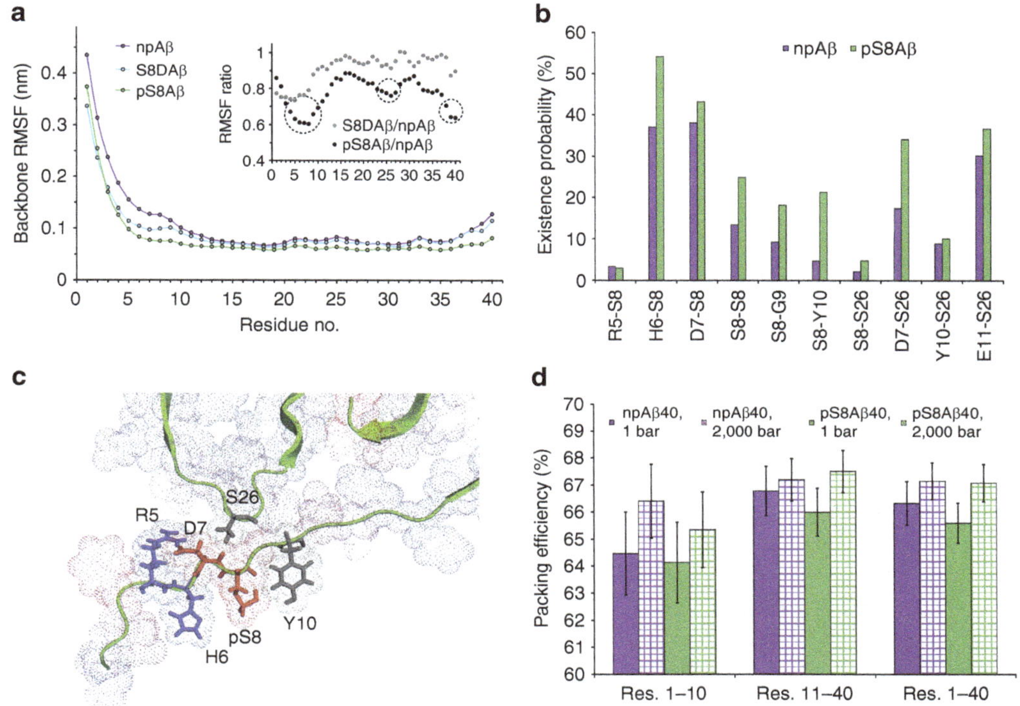

Figure 5 | Structural basis for the increased rigidity of Aβ aggregates phosphorylated at serine 8. (**a**) Residue-specific r.m.s. of fluctuations (RMSF) in backbone atom positions averaged over the last 25 ns of MD trajectories of npAβ (magenta), S8DAβ (blue) and pS8Aβ (green) amyloid fibrils. The inset shows the respective ratios with strongly affected regions circled. (**b**) Probability of hydrogen bond formation between the N-terminal residues of Aβ, as well as interstrand hydrogen bond formation with residues S26-N27. Phosphorylation at serine 8 increases the frequency of hydrogen bond formation. (**c**) Favourable electrostatic interactions in the N-terminal region of Aβ40 fibrils phosphorylated at serine 8. (**d**) Packing efficiency of npAβ40 and pS8Aβ40 fibrils at low (1 bar) and high (2,000 bar) pressure, reported separately for residues 1–10 in the N-terminal region, residues 11–40 and the whole Aβ molecule. Compared with npAβ40, the pS8Aβ40 fibrils experience larger pressure-induced increase in packing efficiency, particularly in residues 11–40. Error bars represent one s.d. over the last 25 ns of MD simulation trajectories.

residues E3-Y10 in npAβ and pS8Aβ fibrils. The average E3CA-Y10CA distance increased from 1.94 ± 0.02 nm in npAβ fibrils to 2.06 ± 0.01 nm in pS8Aβ fibrils, an increase of ~6% which is probably caused by additional steric demand in the phosphorylated fibrils (change to a fully extended conformation would increase this distance by 30%). In addition, the average structures of the Aβ fibrils over the MD trajectory offer a potential basis for the local conformation (Fig. 5c): a favourable electrostatic interaction between H6 and the phosphorylated S8 leads to the rearrangement of residues R5-Y10. Due to this rearrangement, the electrostatic attraction between the side chains of R5 and D7 is stabilized and the side chains of pS8 and Y10 form a hydrogen bond. In addition, hydrogen bond formation between the backbone atoms of D7 and the side-chain OH of S26 is favoured, contributing to the partial rigidification of S26 and its nearby residues.

Phosphorylation increases compressibility of Aβ fibrils. To address how S8 phosphorylation influences the volume of npAβ40 and pS8Aβ40 fibrillar state in dependence of pressure, we performed MD simulations at the elevated pressure of 2,000 bar. The packing efficiency of Aβ fibrils was then calculated as the ratio of the van der Waals to Voronoi volumes and compared between low and high pressures. For npAβ40 fibrils, the average packing efficiency increased from 66.3 ± 0.8% at 1 bar to 67.2 ± 0.7% at 2,000 bar, a small rise of ~1.2% (Fig. 5d). On the other hand, the packing efficiency of pS8Aβ40 fibrils increased from 65.6 ± 0.7% at 1 bar to 67.1 ± 0.7% at 2,000 bar, an increase of 2.3%, which was twice the corresponding value for npAβ40 fibrils. Unlike npAβ40 fibrils, in which the increase in packing efficiency was more prominent in the N-terminal region, the packing efficiency of pS8Aβ40 was enhanced nearly uniformly over the Aβ molecule (Fig. 5d).

As an added mechanism, the ionic interaction between the side chains of H6 and pS8 in phosphorylated Aβ fibrils may contribute to the higher compressibility of pS8Aβ40 fibrils, since breakage of intramolecular electrostatic interactions and resultant electrostriction of solvent molecules by separated charges could decrease the volume of the system without monomer release[29]. We therefore investigated how high pressure altered the distance between neighbouring charges around the phosphorylation site. In line with the increase in packing efficiency at high pressure, the distance between R5 and D7 side chains decreased in both npAβ40 and pS8Aβ40 fibrils. On the other hand, while the distance between H6 and (p)S8 side chains remained intact in npAβ40, it was increased by approximately 8–9% in pS8Aβ40. The relative charge separation induced in pS8Aβ40 fibrils therefore enables the solvent electrostriction effect to further reduce the system volume at high pressure.

Stability of Aβ aggregates against SDS-induced dissociation. To further investigate the stability of phosphorylated Aβ assemblies, soluble and insoluble assemblies of pS8Aβ and npAβ were separated by ultracentrifugation after different time points of aggregation and subjected to denaturing SDS-PAGE followed by western blotting. In the soluble fractions, only monomeric and dimeric Aβ40 was observed indicating that soluble oligomers of both npAβ and pS8Aβ were dissociated under denaturing SDS-

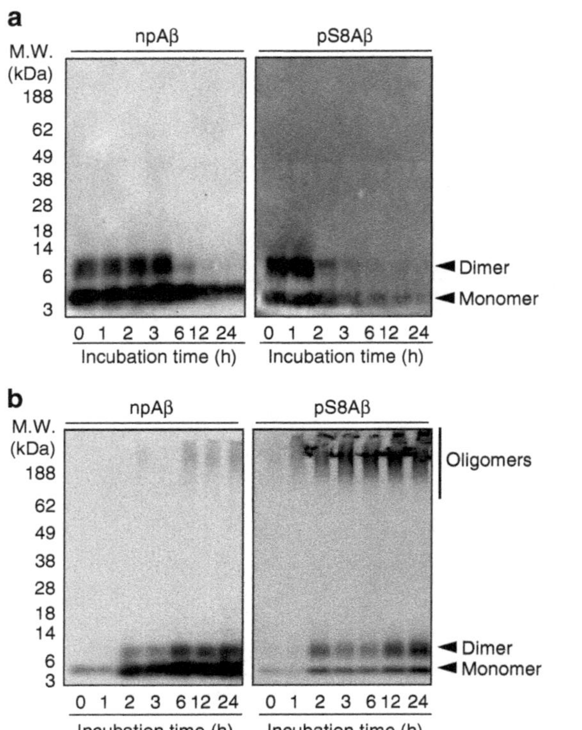

Figure 6 | Phosphorylation of Aβ at serine 8 enhances the stability of Aβ oligomers. (**a**) SDS-PAGE of the soluble fraction of aggregated npAβ and pS8Aβ. All soluble oligomers were dissociated under denaturing SDS-PAGE condition and only monomeric and dimeric Aβ was detected. The disappearance of monomeric/dimeric bands occurred more quickly for pS8Aβ than for npAβ, consistent with the higher rate of insoluble Aβ aggregation in case of pS8Aβ. (**b**) SDS-PAGE of the insoluble fraction of aggregated npAβ and pS8Aβ samples demonstrates that high-molecular-weight oligomers of pS8Aβ are more resistant against SDS-induced dissociation.

PAGE conditions (Fig. 6a). In contrast, the insoluble aggregates of pS8Aβ and npAβ showed different stability against SDS-induced dissociation (Fig. 6b; Supplementary Fig. 6). Only very little SDS-resistant oligomeric forms were detected for npAβ, even in samples taken after 12 and 24 h of incubation when robust amounts of ThT-reactive aggregates were present (Supplementary Fig. 7). In contrast, the insoluble assemblies of pS8Aβ were more stable in SDS–PAGE. High-molecular-weight oligomeric bands were observed and the intensity of the monomeric and dimeric bands was decreased in S8-phosphorylated Aβ40. Whether these SDS-stable assemblies of pS8Aβ represent the intermediate oligomeric species populated over the course of aggregation remains, however, to be elucidated.

Aβ42 aggregates are more stable than Aβ40 aggregates. Between the two abundant variants of Aβ found in human brains, Aβ42 has a stronger aggregation propensity than Aβ40 *in vitro* and *in vivo*, and is more toxic in a variety of animal models of AD[34]. To examine whether the distinct aggregation and toxicity behaviour of Aβ42 is accompanied by differences in the stability of Aβ42 aggregates, we performed high-pressure NMR measurements for Aβ42 aggregates. After 21 h of incubation at 2,000 bar, the NMR signal intensity reached ~31% of the initial (monomeric) signal intensity, which is by a factor of two smaller than the value of 63% observed for Aβ40 (Fig. 1c; Supplementary Table 1). Consistent with the kinetic data for Aβ40 fibrils, the kinetics of monomer release from Aβ42 aggregates was more suitably described by the two-step model. Analysis of the calculated kinetic parameters revealed that the first step, which involves the reversible conversion of the preformed Aβ aggregates to an intermediate aggregate state was slowed down in Aβ42 when compared with Aβ40. In addition, the off-rate of monomer release from the intermediate Aβ aggregates was reduced, while the rate of monomer association to the intermediate aggregates was increased (Fig. 3b). Taken together the data reveal that Aβ42 aggregates are thermodynamically and kinetically more stable than Aβ40 aggregates.

Discussion

Phosphorylation of aggregated proteins is the pathological hallmark of a wide variety of neurodegenerative diseases[4]. It affects conformational stability of globular proteins and influences the self- and hetero-association of both globular and intrinsically disordered proteins[4,35]. Phosphorylation of Aβ at S26 blocks amyloid fibrillation[36]. In contrast, phosphorylation of Aβ at serine 8 drastically increases the rate and amount of Aβ oligomerization and amyloid fibril formation[22]. Here we demonstrated that Aβ aggregates phosphorylated at serine 8 are more resistant against high-pressure and SDS-induced dissociation than non-phosphorylated aggregates (Figs 1 and 6). A quantitative analysis of pressure-induced monomer release showed that Aβ40 aggregates phosphorylated at serine 8 have higher thermodynamic ($k_{on,app}/k_{off}$) and kinetic ($1/k_{off}$) stability than their non-phosphorylated counterparts (Fig. 3). The combination of faster aggregation and higher concentration of oligomeric aggregates results in a higher toxicity of Aβ phosphorylated at serine 8 (ref. 22).

Several lines of evidence attribute an important role for aggregation of Aβ to its N-terminal region[37–40]. The results from our MD simulation suggested that the high capacity of phosphate groups to act both as the donor and acceptor of hydrogen bonds can lead to the formation of a rich network of hydrogen bonds in the N-terminal region and significantly diminishes backbone dynamics in phosphorylated Aβ aggregates (Fig. 5). To accommodate the bulky phosphate group and maximize the number of stabilizing hydrogen bonds and electrostatic interactions around the phosphoserine, a more extended conformation is induced in the vicinity of the phosphorylation site. We propose that this more extended conformation of the N-terminal region of Aβ may form the structural basis for the enhanced stability of phosphorylated Aβ oligomers. This is in agreement with a recent report showing that the disruption of local extended conformation through S8P mutagenesis leads to selective destabilization of Aβ oligomers[37].

Pressure-induced dissociation of protein aggregates is governed by volume change on monomer release. Our experimental data demonstrated a volume change of ~23 μl mol^{-1} for npAβ40, which corresponds roughly to ~40 Å^3 per each Aβ molecule (Fig. 4). The corresponding value for pS8Aβ40 was only 8 μl mol^{-1}. MD simulations suggested that the lower volume change on pS8Aβ40 monomer release was caused by the higher compressibility of pS8Aβ40 fibrils, which might be attributed to its lower degree of compaction. Another contribution to the higher compressibility of pS8Aβ40 fibrils could potentially arise from the intramolecular ionic interaction between the side chain of pS8 and H6, which is partially disrupted at high pressure and, because of the electrostriction of solvent molecules by separated charges, could lead to a volume decrease without Aβ monomer release[29].

Despite only two additional residues at the C terminus, Aβ42 displays aggregation and toxicity properties that strongly exceed those of Aβ40 (ref. 34). In addition, Aβ42 and Aβ40 follow different aggregation pathways[41] and populate distinct oligomeric and fibrillar states[42,43]. Using high-pressure NMR we here showed that aggregates of Aβ42 are kinetically and thermodynamically more stable than those of Aβ40 (Fig. 3b). In particular, the smaller off-rate of Aβ42 assemblies indicates that the lifetime of these potentially toxic Aβ aggregates is significantly prolonged and might be an important factor that contributes to the toxic role of Aβ42 in AD. Further studies are required to explain the structural basis of the enhanced stability of Aβ42 assemblies.

A key step in the prion-like spreading of neurodegenerative pathology is the seeding of soluble protein by previously formed and transmitted protein aggregates[44]. As shown for the yeast prion protein, seed generation is controlled by the rate at which amyloid fibrils fragment, which is in turn governed by their stability[15]. Because aggregation seeds have to be transmitted to the next cell and different brain areas, the efficiency of spreading relies on the stability of aggregation seeds. In case of Aβ, it is known that multiple states of Aβ assemblies are capable of inducing amyloid aggregation *in vivo*, suggesting that small soluble Aβ seeds are potent mediators of the spreading of extracellular β-amyloidosis in AD[45]. Our data demonstrate that phosphorylation enhances the stability of Aβ assemblies (Figs 1, 3 and 6). In addition, the dissociation rate of monomeric Aβ from phosphorylated Aβ aggregates is reduced, which prolongs the lifetime of phosphorylated Aβ assemblies. Because pS8Aβ assemblies can cross-seed the aggregation of non-phosphorylated Aβ (ref. 22), the prolonged lifetime of S8-phosphorylated Aβ aggregates might increase spreading of AD pathology.

In summary, our study demonstrates that phosphorylation of Aβ at serine 8 enhances the stability and prolongs the lifetime of Aβ aggregates. Serine 8 is located in the N-terminal region of Aβ and therefore likely to be exposed in amyloid deposits and thus available to the enzymatic machinery of phosphorylation. Importantly, our work indicates that phosphorylation changes the molecular properties of protein aggregates even after they are formed. Phosphorylation and potentially other post-translational modifications can thus modulate aggregate-induced toxicity in the human brain.

Methods

Materials. Aβ40 and Aβ42, as well as Aβ40 phosphorylated at serine 8 (pS8Aβ) were obtained from Peptide Specialty Laboratory (Germany) and used without further purification. The Aβ40 variant S8D (S8DAβ) was purchased from EZBiolab (USA). The monoclonal Aβ antibody 82E1 was obtained from IBL Corporation (Japan). Unless stated otherwise, Aβ stock solutions were prepared by dissolving peptide powders in 10 mM NaOH at a concentration of 2 mg ml^{-1} (~460 μM), then flash freezing the aliquots of Aβ stock solutions in liquid nitrogen and storing them at −80 °C until further use. Stock solutions of Aβ42 were prepared by a two-step solubilization method[46].

PAGE and western blotting. npAβ and pS8Aβ (50 μM, pH 7.4 buffered with 20 mM sodium phosphate containing 50 mM sodium chloride) were incubated at 37 °C with gentle stirring. Aliquots of samples collected at different time points were centrifuged at 100,000*g* for 30 min to separate the soluble Aβ species in the supernatant from insoluble Aβ assemblies in the pellet. The supernatant and pellets were subjected to native-PAGE or denaturing SDS–PAGE and Aβ was detected by western blotting. The samples were mixed with an equal volume of 4 × SDS sample buffer and then boiled at 100 °C for 5 min. The samples were electrophoresed on 4–12% Bis-Tris gels (Novex, San Diego, CA, USA) and then transferred onto 0.2-μM nitrocellulose membrane (Whatmann GmbH, Germany) at 400 mA for 2 h at 4 °C. After 2 h, membranes were washed in TBS-T and incubated overnight with the primary antibody 82E1 in 0.5 μg ml^{-1} TBS-T solution at 4 °C. The bound antibody was detected using horseradish peroxidase-conjugated anti-mouse IgG (1:50,000; Sigma) using enhanced chemiluminescence (ECL reagent, GE Healthcare) with an ECL imager (BioRad Inc.).

Electron microscopy. npAβ, pS8Aβ and S8DAβ samples of 0.2 mg ml^{-1} (~50 μM) concentration in HEPES buffer (25 mM, pH 7.4, 50 mM NaCl) were incubated at 37 °C with gentle stirring. After 48 h of incubation, samples were diluted, deposited onto carbon-coated copper mesh grids and negatively stained with 2% (w/v) uranyl acetate. Excess stain was washed away, and the sample grids were allowed to air dry. Samples were examined using a Philips CM 120 BioTwin transmission electron microscope (Philips Inc., Eindhoven, The Netherlands). The apparent diameter (largest width) of negatively stained Aβ fibrils were measured using the ITEM software (Olympus Soft Imaging Solutions, Münster, Germany) and the size distribution of npAβ and pS8Aβ fibrils was determined after manual measurement of >300 fibrils per peptide variant.

To investigate the effect of high pressure on Aβ fibrils, npAβ and pS8Aβ samples of 0.25 mg ml^{-1} (~60 μM) concentration in phosphate buffer (50 mM, pH 7.4, 50 mM NaCl) were incubated in aggregation-prone condition (37 °C, with gentle stirring) for 96 h. Before and after application of high hydrostatic pressure (2,000 bar, 18 h) during which the pressure-induced Aβ monomer release was monitored, Aβ samples were taken for the EM examination. The EM images were obtained as described above.

High-pressure NMR experiments. npAβ40, pS8Aβ40, S8DAβ40 and Aβ42 samples of 0.25 mg ml^{-1} (~60 μM) concentration in phosphate buffer (50 mM, pH 7.4, 50 mM NaCl, 100 μM DSS, 10% D_2O) were incubated in aggregation-prone condition (37 °C, with gentle stirring). After the 96 h of incubation, the aggregated Aβ samples were transferred without further fractionation to high-pressure NMR tubes and subjected to hydrostatic pressures up to 2,000 bar (Daedalus Innovations LLC, PA) and low temperature (273 K). NMR experiments were performed on a 700 MHz Bruker spectrometer (Germany). At each pressure, Aβ samples were first incubated for 20 min, then three consecutive 1D proton experiments each taking 10 min were recorded, before raising the pressure to the next higher level. The amount of Aβ monomer released as a function of the applied pressure was estimated through proton signal intensities in the methyl region, normalized by the intensity of an internal DSS signal. For the kinetics of pressure-induced Aβ monomer release from aggregates, the aggregated Aβ samples were directly subjected to the highest pressure, that is, 2,000 bar, and 1D proton NMR spectra were measured for 18 h at 30-min intervals. The time-dependent relative intensity data (*I*) were fitted to mono- and bi-exponential equations:

$$I = I_{\infty} + A.e^{-Rt} \tag{1}$$

$$I = I_{\infty} + A_1.e^{-R_1 t} + A_2.e^{-R_2 t} \tag{2}$$

Assuming a simple model for the conversion between NMR-invisible Aβ assemblies (*A*) and monomers (*M*):

$$A \rightleftarrows M \tag{3}$$

in which both the forward dissociation and backward association reactions obey first-order kinetics governed by the rate constants k_{off} and k_{on}. The fitting parameters derived from the monoexponential equation (equation (1)) were interpreted as:

$$I_{\infty} = \frac{k_{\text{off}}}{k_{\text{off}} + k_{\text{on}}} \tag{4}$$

$$R = k_{\text{off}} + k_{\text{on}} \tag{5}$$

and apparent values of k_{off} and k_{on} were derived accordingly. The assumption of (pseudo) first-order kinetics for the reverse reaction in equation (3) is reasonably justified noting that the number of growing ends in Aβ aggregates does not change during dissociation–association reaction. The fitting parameters, which were derived from the bi-exponential model (equation (2)), were interpreted according to the following model:

$$A \rightleftarrows I \rightleftarrows M \tag{6}$$

which involves reversible conversion of Aβ assemblies to an intermediate state (I) before the reversible monomer dissociation–association reaction occurs. An in-house Mathematica 8.0 (Wolfram Research, Inc.) script was used for data fitting.

To obtain the change in volume on Aβ monomer release from aggregates, the npAβ40 and pS8Aβ40 samples (0.25 mg ml^{-1} in 50 mM phosphate, pH 7.4, 50 mM NaCl, 100 μM DSS, 10% D_2O) were incubated in aggregation-prone conditions (37 °C, with gentle stirring) for 96 h. Subsequently, they were transferred to high-pressure NMR tubes and incubated for 48 h at different pressure levels (500, 1,000, 1,500 and 2,000 bar) and 273 K. The 1D NMR proton spectra were recorded before aggregation (day 0), after aggregation (day 4) and after 48-h application of high pressure to Aβ aggregates. Aβ monomer levels were quantified as described above, the equilibrium constants between Aβ aggregates and monomers ($K \equiv \frac{k_{\text{off}}}{k_{\text{on}}}$) were calculated according to equation (3) and then converted to free-energy change on Aβ monomer release from aggregates (ΔG):

$$\Delta G = -RT \ln K \tag{7}$$

in which *R* is the ideal gas constant and *T* is (absolute) temperature. Assuming that the pressure dependence of free-energy change is governed by the following linear relation,

$$\Delta G_P = \Delta G_{P_0} + \Delta V.(P - P_0) \tag{8}$$

the (apparent) volume change on Aβ monomer release from aggregates (ΔV) was estimated from the slope of the fitted ΔG versus pressure line.

MD simulation. MD simulations were performed using the GROMACS simulation package, version 4.6.5 (ref. 47). The Amber99sb force field parameters were used as implemented in GROMACS[48]. The initial protein model was based on a solid-state NMR structure of Aβ40 fibrils (PDB code: 2M4J (ref. 33)). This protein database (PDB) structure contains three layers of a threefold symmetric repeat unit each constituted by three Aβ molecules. Taking the lowest-energy model of this PDB entry, three models were built for the unmodified, S8-phosphorylated and S8D-Aβ40 fibrils. To reduce the contribution of end effects in the final results, each model was elongated by adding new layers in the direction of the fibrillar axis. The final models were 21-layered structures of Aβ40 fibrils, composed of a total of 63 polypeptide chains. Each fibril model was solvated in a cubic box of dimension $11 \times 11 \times 11\,nm^3$ with TIP4P-Ew water molecules and neutralized by adding monovalent ions. To remove atomic clashes, each solvated and neutralized model was then subjected to energy minimization using the steepest descent algorithm until the maximum force on atoms was reduced to values smaller than $1{,}000\,kJ\,mol^{-1}\,nm^{-1}$. After energy minimization, short-length position-restrained NVT and position-restrained NPT simulations (1 ns) at 300 K were applied to set atomic velocities and adjust densities. After equilibration, the production MD simulations were performed for each model at constant temperature (300 K) and pressure (1 bar) over a 75-ns period. The LINCS algorithm was used to constrain the bonds. Lennard-Jones and short-range electrostatic interaction cutoff radii of 1 nm were used. Coulomb interactions at longer distances were calculated by the Particle Mesh Ewald algorithm. The protein fibril and water molecules plus ions were coupled separately to a thermal bath using the velocity rescale algorithm (modified Berendsen thermostat) with $\tau_T = 0.1$ ps at 300 K (refs 49,50). The isotropic pressure coupling was used to keep the pressure at desired values by applying the Berendsen algorithm with $\tau_P = 2.0$ ps and a compressibility of $4.5 \times 10^{-5}\,bar^{-1}$ (ref. 50). The neighbour lists were updated every five steps and a time step of 2 fs was chosen for integration. Hydrogen bonds were calculated using the angle and distance cutoff values of 30° and 0.35 nm, respectively.

To evaluate the impact of high pressure on Aβ fibrils, two additional 75-ns MD simulations were performed on unmodified and S8-phosphorylated Aβ40 fibrils using the same set-up as mentioned above except of pressure and temperature which were set to 2,000 bar and 273 K in accordance with the experimental condition of our high-pressure NMR experiments. Packing efficiencies were then calculated as the ratio of the van der Waals volume to the Voronoi volume over the last 25 ns of MD trajectories, using the computational tool of *trjVoronoi* (ref. 51).

Details on ThT binding assays are provided in Supplementary Methods.

References

1. Knowles, T. P., Vendruscolo, M. & Dobson, C. M. The amyloid state and its association with protein misfolding diseases. *Nat. Rev. Mol. Cell Biol.* **15**, 384–396 (2014).
2. Schmid, A. W., Fauvet, B., Moniatte, M. & Lashuel, H. A. Alpha-synuclein post-translational modifications as potential biomarkers for Parkinson disease and other synucleinopathies. *Mol. Cell. Proteomics* **12**, 3543–3558 (2013).
3. Russell, C. L., Koncarevic, S. & Ward, M. A. Post-translational modifications in Alzheimer's disease and the potential for new biomarkers. *J. Alzheimers Dis.* **41**, 345–364 (2014).
4. Tenreiro, S., Eckermann, K. & Outeiro, T. F. Protein phosphorylation in neurodegeneration: friend or foe? *Front. Mol. Neurosci.* **7**, 42 (2014).
5. McFarland, M. A., Ellis, C. E., Markey, S. P. & Nussbaum, R. L. Proteomics analysis identifies phosphorylation-dependent alpha-synuclein protein interactions. *Mol. Cell. Proteomics* **7**, 2123–2137 (2008).
6. Paleologou, K. E. *et al.* Phosphorylation at S87 is enhanced in synucleinopathies, inhibits alpha-synuclein oligomerization, and influences synuclein-membrane interactions. *J. Neurosci.* **30**, 3184–3198 (2010).
7. Oueslati, A., Schneider, B. L., Aebischer, P. & Lashuel, H. A. Polo-like kinase 2 regulates selective autophagic alpha-synuclein clearance and suppresses its toxicity *in vivo*. *Proc. Natl Acad. Sci. USA* **110**, E3945–E3954 (2013).
8. Goncalves, S. & Outeiro, T. F. Assessing the subcellular dynamics of alpha-synuclein using photoactivation microscopy. *Mol. Neurobiol.* **47**, 1081–1092 (2013).
9. Tenreiro, S. *et al.* Phosphorylation modulates clearance of alpha-synuclein inclusions in a yeast model of Parkinson's disease. *PLoS Genet.* **10**, e1004302 (2014).
10. Mahul-Mellier, A. L. *et al.* c-Abl phosphorylates alpha-synuclein and regulates its degradation: implication for alpha-synuclein clearance and contribution to the pathogenesis of Parkinson's disease. *Hum. Mol. Genet.* **23**, 2858–2879 (2014).
11. Meersman, F. & Dobson, C. M. Probing the pressure-temperature stability of amyloid fibrils provides new insights into their molecular properties. *Biochim. Biophys. Acta* **1764**, 452–460 (2006).
12. Gazit, E. The "Correctly Folded" state of proteins: is it a metastable state? *Angew. Chem. Int. Ed.* **41**, 257–259 (2002).
13. Knowles, T. P. *et al.* Role of intermolecular forces in defining material properties of protein nanofibrils. *Science* **318**, 1900–1903 (2007).
14. Knowles, T. P. & Buehler, M. J. Nanomechanics of functional and pathological amyloid materials. *Nat. Nanotechnol.* **6**, 469–479 (2011).
15. Tanaka, M., Collins, S. R., Toyama, B. H. & Weissman, J. S. The physical basis of how prion conformations determine strain phenotypes. *Nature* **442**, 585–589 (2006).
16. Dirix, C., Meersman, F., MacPhee, C. E., Dobson, C. M. & Heremans, K. High hydrostatic pressure dissociates early aggregates of TTR105-115, but not the mature amyloid fibrils. *J. Mol. Biol.* **347**, 903–909 (2005).
17. Relini, A. *et al.* Detection of populations of amyloid-like protofibrils with different physical properties. *Biophys. J.* **98**, 1277–1284 (2010).
18. Tanzi, R. E. The genetics of Alzheimer disease. *Cold Spring Harb. Perspect. Med.* **2**, a006296 (2012).
19. Kumar, S. & Walter, J. Phosphorylation of amyloid beta (Abeta) peptides—a trigger for formation of toxic aggregates in Alzheimer's disease. *Aging* **3**, 803–812 (2011).
20. Kummer, M. P. *et al.* Nitration of tyrosine 10 critically enhances amyloid beta aggregation and plaque formation. *Neuron* **71**, 833–844 (2011).
21. Schilling, S. *et al.* Glutaminyl cyclase inhibition attenuates pyroglutamate Abeta and Alzheimer's disease-like pathology. *Nat. Med.* **14**, 1106–1111 (2008).
22. Kumar, S. *et al.* Extracellular phosphorylation of the amyloid beta-peptide promotes formation of toxic aggregates during the pathogenesis of Alzheimer's disease. *EMBO J.* **30**, 2255–2265 (2011).
23. Rijal Upadhaya, A. *et al.* Biochemical stages of amyloid-beta peptide aggregation and accumulation in the human brain and their association with symptomatic and pathologically preclinical Alzheimer's disease. *Brain* **137**, 887–903 (2014).
24. Kumar, S. *et al.* Phosphorylation of amyloid-beta peptide at serine 8 attenuates its clearance via insulin-degrading and angiotensin-converting enzymes. *J. Biol. Chem.* **287**, 8641–8651 (2012).
25. Mishra, R. & Winter, R. Cold- and pressure-induced dissociation of protein aggregates and amyloid fibrils. *Angew. Chem. Int. Ed.* **47**, 6518–6521 (2008).
26. Silva, J. L., Foguel, D. & Royer, C. A. Pressure provides new insights into protein folding, dynamics and structure. *Trends Biochem. Sci.* **26**, 612–618 (2001).
27. Nucci, N. V., Fuglestad, B., Athanasoula, E. A. & Wand, A. J. Role of cavities and hydration in the pressure unfolding of T4 lysozyme. *Proc. Natl Acad. Sci. USA* **111**, 13846–13851 (2014).
28. Akasaka, K. *et al.* Amyloid protofibril is highly voluminous and compressible. *Biochemistry* **46**, 10444–10450 (2007).
29. Urbauer, J. L., Ehrhardt, M. R., Bieber, R. J., Flynn, P. F. & Wand, A. J. High-resolution triple-resonance NMR spectroscopy of a novel calmodulin peptide complex at kilobar pressures. *J. Am. Chem. Soc.* **118**, 11329–11330 (1996).
30. Gruning, C. S. *et al.* The off-rate of monomers dissociating from amyloid-beta protofibrils. *J. Biol. Chem.* **288**, 37104–37111 (2013).
31. Carulla, N. *et al.* Molecular recycling within amyloid fibrils. *Nature* **436**, 554–558 (2005).
32. Petkova, A. T. *et al.* Self-propagating, molecular-level polymorphism in Alzheimer's beta-amyloid fibrils. *Science* **307**, 262–265 (2005).
33. Lu, J. X. *et al.* Molecular structure of beta-amyloid fibrils in Alzheimer's disease brain tissue. *Cell* **154**, 1257–1268 (2013).
34. Younkin, S. G. Evidence that A beta 42 is the real culprit in Alzheimer's disease. *Ann. Neurol.* **37**, 287–288 (1995).
35. Nishi, H., Hashimoto, K. & Panchenko, A. R. Phosphorylation in protein-protein binding: effect on stability and function. *Structure* **19**, 1807–1815 (2011).
36. Rezaei-Ghaleh, N. *et al.* Turn plasticity distinguishes different modes of amyloid-beta aggregation. *J. Am. Chem. Soc.* **136**, 4913–4919 (2014).
37. Haupt, C. *et al.* Structural basis of beta-amyloid-dependent synaptic dysfunctions. *Angew. Chem. Int. Ed.* **51**, 1576–1579 (2012).
38. Gardberg, A. S. *et al.* Molecular basis for passive immunotherapy of Alzheimer's disease. *Proc. Natl Acad. Sci. USA* **104**, 15659–15664 (2007).
39. Di Fede, G. *et al.* A recessive mutation in the APP gene with dominant-negative effect on amyloidogenesis. *Science* **323**, 1473–1477 (2009).
40. Ono, K., Condron, M. M. & Teplow, D. B. Effects of the English (H6R) and Tottori (D7N) familial Alzheimer disease mutations on amyloid beta-protein assembly and toxicity. *J. Biol. Chem.* **285**, 23186–23197 (2010).
41. Bitan, G. *et al.* Amyloid beta -protein (Abeta) assembly: Abeta 40 and Abeta 42 oligomerize through distinct pathways. *Proc. Natl Acad. Sci. USA* **100**, 330–335 (2003).
42. Parthasarathy, S. *et al.* Structural Insight into an Alzheimer's Brain-Derived Spherical Assembly of Amyloid beta by Solid-State NMR. *J Am. Chem. Soc.* **137**, 6480–6483 (2015).
43. Xiao, Y. *et al.* Abeta(1-42) fibril structure illuminates self-recognition and replication of amyloid in Alzheimer's disease. *Nat. Struct. Mol. Biol.* **22**, 499–505 (2015).
44. Brettschneider, J., Del Tredici, K., Lee, V. M. & Trojanowski, J. Q. Spreading of pathology in neurodegenerative diseases: a focus on human studies. *Nat. Rev. Neurosci.* **16**, 109–120 (2015).
45. Langer, F. *et al.* Soluble Abeta seeds are potent inducers of cerebral beta-amyloid deposition. *J. Neurosci.* **31**, 14488–14495 (2011).
46. Bouter, Y. *et al.* N-truncated amyloid beta (Abeta) 4-42 forms stable aggregates and induces acute and long-lasting behavioral deficits. *Acta Neuropathol.* **126**, 189–205 (2013).

47. Berendsen, H. J. C., Vanderspoel, D. & Vandrunen, R. Gromacs—a Message-Passing Parallel Molecular-Dynamics Implementation. *Comput. Phys. Commun.* **91,** 43–56 (1995).
48. Hornak, V. *et al.* Comparison of multiple amber force fields and development of improved protein backbone parameters. *Proteins* **65,** 712–725 (2006).
49. Bussi, G., Donadio, D. & Parrinello, M. Canonical sampling through velocity rescaling. *J. Chem. Phys.* **126,** 014101 (2007).
50. Berendsen, H. J. C., Postma, J. P. M., Vangunsteren, W. F., Dinola, A. & Haak, J. R. Molecular-dynamics with coupling to an external bath. *J. Chem. Phys.* **81,** 3684–3690 (1984).
51. Abel, S., Dupradeau, F. Y. & Marchi, M. Molecular dynamics simulations of a characteristic DPC micelle in water. *J. Chem. Theory Comput.* **8,** 4610–4623 (2012).

Acknowledgements

We thank Dr Dietmar Riedel and Gudrun Heim for electron microscopy and for help with the analysis of EM images.

Author contributions

N.R.-G. designed, performed and analysed high-pressure NMR experiments, contributed to MD simulation and analysis, performed size distribution analysis in TEM and wrote the manuscript; M.A. performed MD simulation and analysis; S.K. and J.W. performed PAGE experiments; J.W. and M.Z. conceived the project; M.Z. supervised the project and wrote the manuscript.

Additional information

Competing financial interests: The authors declare no competing financial interests.

19

Direct evidence for sequence-dependent attraction between double-stranded DNA controlled by methylation

Jejoong Yoo[1,*], Hajin Kim[2,3,*], Aleksei Aksimentiev[1,4] & Taekjip Ha[1,5,6,7,8]

Although proteins mediate highly ordered DNA organization *in vivo*, theoretical studies suggest that homologous DNA duplexes can preferentially associate with one another even in the absence of proteins. Here we combine molecular dynamics simulations with single-molecule fluorescence resonance energy transfer experiments to examine the interactions between duplex DNA in the presence of spermine, a biological polycation. We find that AT-rich DNA duplexes associate more strongly than GC-rich duplexes, regardless of the sequence homology. Methyl groups of thymine acts as a steric block, relocating spermine from major grooves to interhelical regions, thereby increasing DNA-DNA attraction. Indeed, methylation of cytosines makes attraction between GC-rich DNA as strong as that between AT-rich DNA. Recent genome-wide chromosome organization studies showed that remote contact frequencies are higher for AT-rich and methylated DNA, suggesting that direct DNA-DNA interactions that we report here may play a role in the chromosome organization and gene regulation.

[1] Department of Physics and the Center for the Physics of Living Cells, University of Illinois at Urbana-Champaign, Urbana, Illinois 61801, USA. [2] School of Life Sciences, Ulsan National Institute of Science and Technology, Ulsan, Korea. [3] Center for Soft and Living Matter, Institute for Basic Science, Ulsan, Korea. [4] Beckman Institute for Advanced Science and Technology, University of Illinois at Urbana-Champaign, Urbana, Illinois 61801, USA. [5] Howard Hughes Medical Institute, Baltimore, Maryland 21205, USA. [6] Department of Biophysics and Biophysical Chemistry, Johns Hopkins University, Baltimore, Maryland 21205, USA. [7] Department of Biophysics, Johns Hopkins University, Baltimore, Maryland 21205, USA. [8] Department of Biomedical Engineering, Johns Hopkins University, Baltimore, Maryland 21205, USA. * These authors contributed equally to this work. Correspondence and requests for materials should be addressed to A.A. (email: aksiment@illinois.edu) or to T.H. (email: tjha@jhu.edu).

Formation of a DNA double helix occurs through Watson–Crick pairing mediated by the complementary hydrogen bond patterns of the two DNA strands and base stacking. Interactions between double-stranded (ds)DNA molecules in typical experimental conditions containing mono- and divalent cations are repulsive[1], but can turn attractive in the presence of high-valence cations[2]. Theoretical studies have identified the ion–ion correlation effect as a possible microscopic mechanism of the DNA condensation phenomena[3-5]. Theoretical investigations have also suggested that sequence-specific attractive forces might exist between two homologous fragments of dsDNA[6], and this 'homology recognition' hypothesis was supported by *in vitro* atomic force microscopy[7] and *in vivo* point mutation assays[8]. However, the systems used in these measurements were too complex to rule out other possible causes such as Watson–Crick strand exchange between partially melted DNA or protein-mediated association of DNA.

Here we present direct evidence for sequence-dependent attractive interactions between dsDNA molecules that neither involve intermolecular strand exchange nor are mediated by proteins. Further, we find that the sequence-dependent attraction is controlled not by homology—contradictory to the 'homology recognition' hypothesis[6]—but by a methylation pattern. Unlike the previous *in vitro* study that used monovalent (Na^+) or divalent (Mg^{2+}) cations[7], we presumed that for the sequence-dependent attractive interactions to operate polyamines would have to be present. Polyamine is a biological polycation present at a millimolar concentration in most eukaryotic cells and essential for cell growth and proliferation[9,10]. Polyamines are also known to condense DNA in a concentration-dependent manner[2,11]. In this study, we use spermine^{4+} (Sm^{4+}) that contains four positively charged amine groups per molecule.

Results

Sequence dependence of DNA–DNA forces. To characterize the molecular mechanisms of DNA–DNA attraction mediated by polyamines, we performed molecular dynamics (MD) simulations where two effectively infinite parallel dsDNA molecules, 20 base pairs (bp) each in a periodic unit cell, were restrained to maintain a prescribed inter-DNA distance; the DNA molecules were free to rotate about their axes. The two DNA molecules were submerged in 100 mM aqueous solution of NaCl that also contained 20 Sm^{4+} molecules; thus, the total charge of Sm^{4+}, 80 e, was equal in magnitude to the total charge of DNA ($2 \times 2 \times 20\ e$, two unit charges per base pair; Fig. 1a). Repeating such simulations at various inter-DNA distances and applying weighted histogram analysis[12] yielded the change in the interaction free energy (ΔG) as a function of the DNA–DNA distance (Fig. 1b,c). In a broad agreement with previous experimental findings[13], ΔG had a minimum, ΔG_{min}, at the inter-DNA distance of $25-30$ Å for all sequences examined, indeed showing that two duplex DNA molecules can attract each other. The free energy of inter-duplex attraction was at least an order of magnitude smaller than the Watson–Crick interaction free energy of the same length DNA duplex. A minimum of ΔG was not observed in the absence of polyamines, for example, when divalent or monovalent ions were used instead[14,15].

Unexpectedly, we found that DNA sequence has a profound impact on the strength of attractive interaction. The absolute value of ΔG at minimum relative to the value at maximum separation, $|\Delta G_{min}|$, showed a clearly rank-ordered dependence on the DNA sequence: $|\Delta G_{min}|$ of $(A)_{20} > |\Delta G_{min}|$ of $(AT)_{10} > |\Delta G_{min}|$ of $(GC)_{10} > |\Delta G_{min}|$ of $(G)_{20}$. Two trends can be noted. First, AT-rich sequences attract each other more strongly than GC-rich sequences[16]. For example, $|\Delta G_{min}|$ of $(AT)_{10}$ (1.5 kcal mol^{-1} per turn) is about twice $|\Delta G_{min}|$ of $(GC)_{10}$ (0.8 kcal mol^{-1} per turn) (Fig. 1b). Second, duplexes having identical AT content but different partitioning of the nucleotides between the strands (that is, $(A)_{20}$ versus $(AT)_{10}$ or $(G)_{20}$ versus $(GC)_{10}$) exhibit statistically significant differences ($\sim$0.3 kcal mol^{-1} per turn) in the value of $|\Delta G_{min}|$.

To validate the findings of MD simulations, we performed single-molecule fluorescence resonance energy transfer (smFRET)[17] experiments of vesicle-encapsulated DNA molecules. Equimolar mixture of donor- and acceptor-labelled 120-bp dsDNA molecules was encapsulated in sub-micron size, porous lipid vesicles[18] so that we could observe and quantitate rare binding events between a pair of dsDNA molecules without triggering large-scale DNA condensation[2]. Our DNA constructs were long enough to ensure dsDNA–dsDNA binding that is stable on the timescale of an smFRET measurement, but shorter than the DNA's persistence length ($\sim$150 bp (ref. 19)) to avoid intramolecular condensation[20]. The vesicles were immobilized on a polymer-passivated surface, and fluorescence signals from individual vesicles containing one donor and one acceptor were selectively analysed (Fig. 1d). Binding of two dsDNA molecules brings their fluorescent labels in close proximity, increasing the FRET efficiency (Fig. 1e).

FRET signals from individual vesicles were diverse. Sporadic binding events were observed in some vesicles, while others exhibited stable binding; traces indicative of frequent conformational transitions were also observed (Supplementary Fig. 1A). Such diverse behaviours could be expected from non-specific interactions of two large biomolecules having structural degrees of freedom. No binding events were observed in the absence of Sm^{4+} (Supplementary Fig. 1B) or when no DNA molecules were present. To quantitatively assess the propensity of forming a bound state, we chose to use the fraction of single-molecule traces that showed any binding events within the observation time of 2 min (Methods). This binding fraction for the pair of AT-rich dsDNAs (AT1, 100% AT in the middle 80-bp section of the 120-bp construct) reached a maximum at $\sim$2 mM Sm^{4+} (Fig. 1f), which is consistent with the results of previous experimental studies[2,3]. In accordance with the prediction of our MD simulations, GC-rich dsDNAs (GC1, 75% GC in the middle 80 bp) showed much lower binding fraction at all Sm^{4+} concentrations (Fig. 1b,c). Regardless of the DNA sequence, the binding fraction reduced back to zero at high Sm^{4+} concentrations, likely due to the resolubilization of now positively charged DNA–Sm^{4+} complexes[2,3,13].

Because the donor and acceptor fluorophores were attached to the same sequence of DNA, it remained possible that the sequence homology between the donor-labelled DNA and the acceptor-labelled DNA was necessary for their interaction[6]. To test this possibility, we designed another AT-rich DNA construct AT2 by scrambling the central 80-bp section of AT1 to remove the sequence homology (Supplementary Table 1). The fraction of binding traces for this nonhomologous pair of donor-labelled AT1 and acceptor-labelled AT2 was comparable to that for the homologous AT-rich pair (donor-labelled AT1 and acceptor-labelled AT1) at all Sm^{4+} concentrations tested (Fig. 1f). Furthermore, this data set rules out the possibility that the higher binding fraction observed experimentally for the AT-rich constructs was caused by inter-duplex Watson–Crick base pairing of the partially melted constructs.

Next, we designed a DNA construct named ATGC, containing, in its middle section, a 40-bp AT-rich segment followed by a 40-bp GC-rich segment (Fig. 1g). By attaching the acceptor to the end of either the AT-rich or GC-rich segments, we could compare the likelihood of observing the parallel binding mode that brings the two AT-rich segments together and

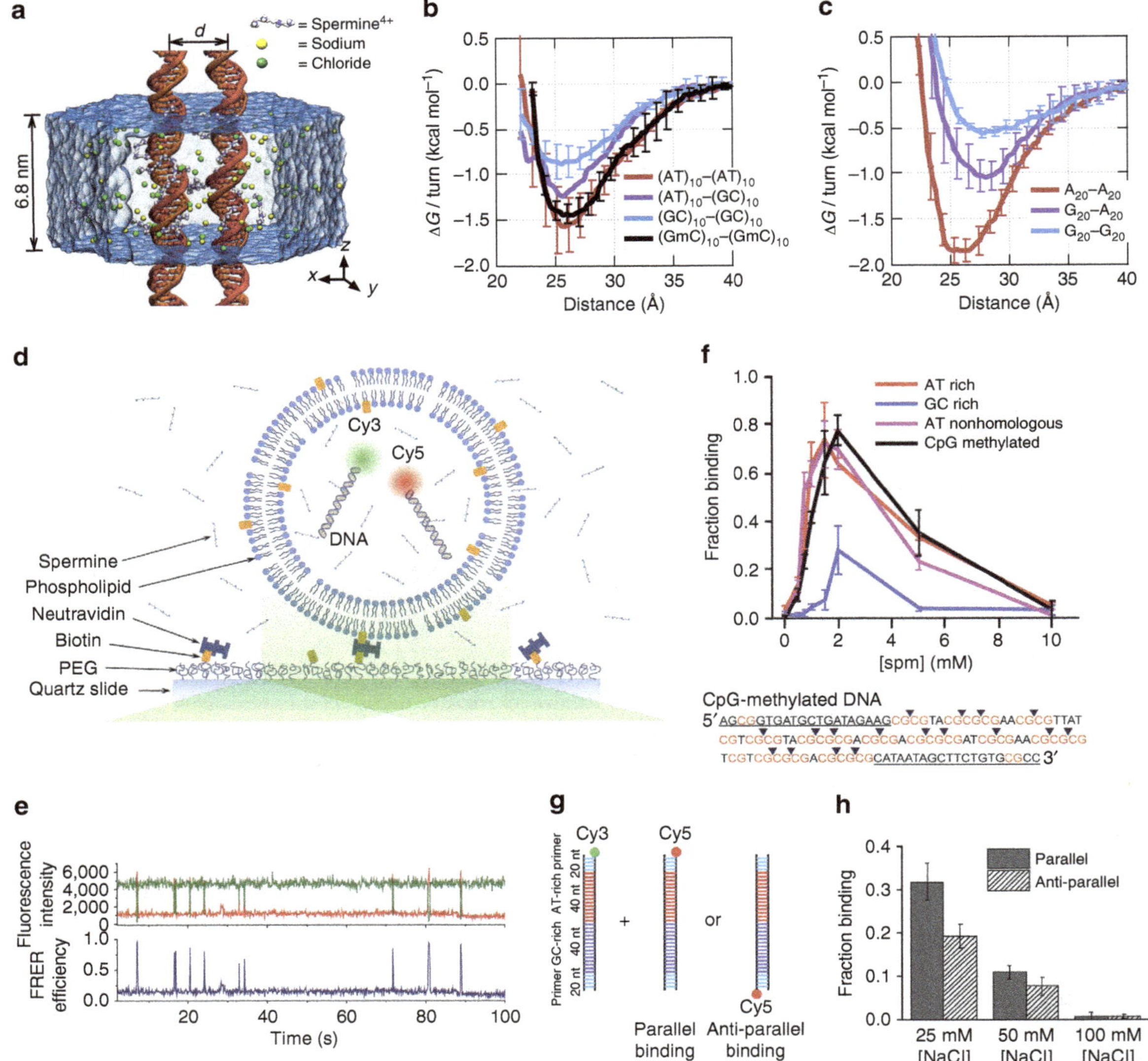

Figure 1 | Polyamine-mediated DNA sequence recognition observed in MD simulations and smFRET experiments. (**a**) Set-up of MD simulations. A pair of parallel 20-bp dsDNA duplexes is surrounded by aqueous solution (semi-transparent surface) containing 20 Sm^{4+} molecules (which compensates exactly the charge of DNA) and 100 mM NaCl. Under periodic boundary conditions, the DNA molecules are effectively infinite. A harmonic potential (not shown) is applied to maintain the prescribed distance between the dsDNA molecules. (**b,c**) Interaction free energy of the two DNA helices as a function of the DNA–DNA distance for repeat-sequence DNA fragments (**b**) and DNA homopolymers (**c**). (**d**) Schematic of experimental design. A pair of 120-bp dsDNA labelled with a Cy3/Cy5 FRET pair was encapsulated in a ~200-nm diameter lipid vesicle; the vesicles were immobilized on a quartz slide through biotin-neutravidin binding. Sm^{4+} molecules added after immobilization penetrated into the porous vesicles. The fluorescence signals were measured using a total internal reflection microscope. (**e**) Typical fluorescence signals indicative of DNA–DNA binding. Brief jumps in the FRET signal indicate binding events. (**f**) The fraction of traces exhibiting binding events at different Sm^{4+} concentrations for AT-rich, GC-rich, AT nonhomologous and CpG-methylated DNA pairs. The sequence of the CpG-methylated DNA specifies the methylation sites (CG sequence, orange), restriction sites (BstUI, triangle) and primer region (underlined). The degree of attractive interaction for the AT nonhomologous and CpG-methylated DNA pairs was similar to that of the AT-rich pair. All measurements were done at [NaCl] = 50 mM and T = 25 °C. (**g**) Design of the hybrid DNA constructs: 40-bp AT-rich and 40-bp GC-rich regions were flanked by 20-bp common primers. The two labelling configurations permit distinguishing parallel from anti-parallel orientation of the DNA. (**h**) The fraction of traces exhibiting binding events as a function of NaCl concentration at fixed concentration of Sm^{4+} (1 mM). The fraction is significantly higher for parallel orientation of the DNA fragments.

the anti-parallel binding mode. Measurements at 1 mM Sm^{4+} and 25 or 50 mM NaCl indicated a preference for the parallel binding mode by ~30% (Fig. 1h). Therefore, AT content can modulate DNA–DNA interactions even in a complex sequence context. Note that increasing the concentration of NaCl while keeping the concentration of Sm^{4+} constant enhances competition between Na^{+} and Sm^{4+} counterions, which reduces the concentration of Sm^{4+} near DNA and hence the frequency of dsDNA–dsDNA binding events (Supplementary Fig. 2).

Methylation determines the strength of DNA–DNA attraction. Analysis of the MD simulations revealed the molecular mechanism of the polyamine-mediated sequence-dependent attraction (Fig. 2). In the case of the AT-rich fragments, the bulky

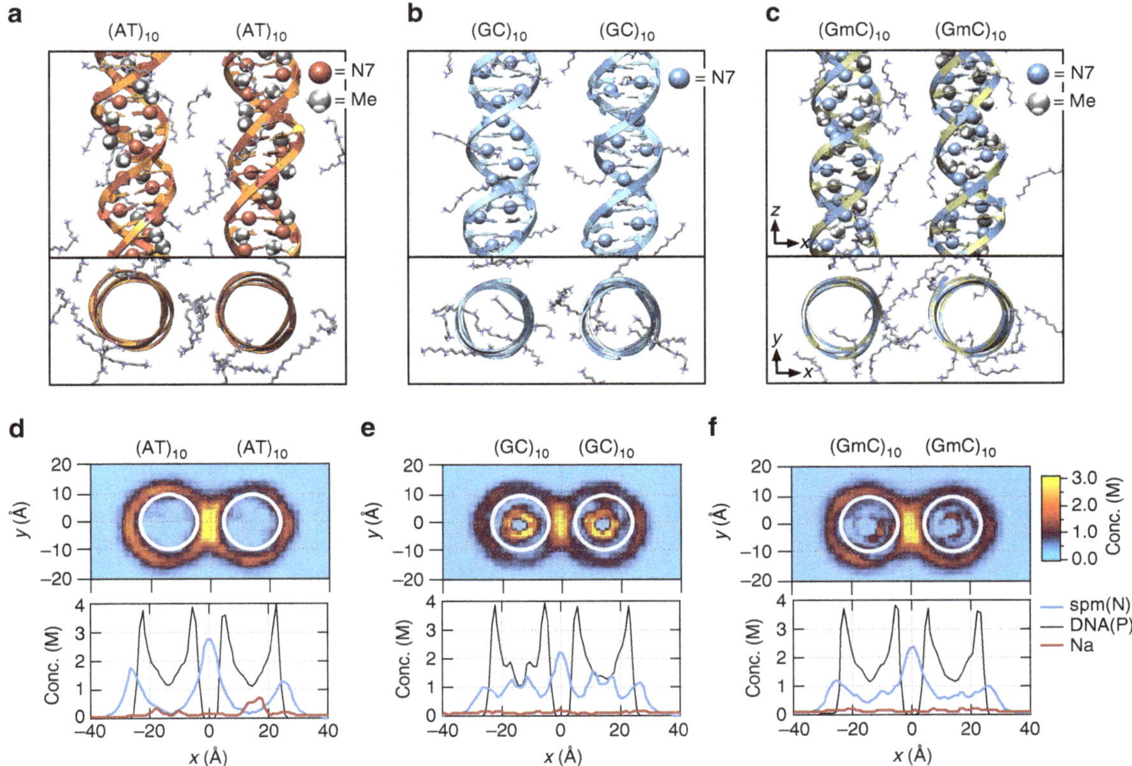

Figure 2 | Molecular mechanism of polyamine-mediated DNA sequence recognition. (**a-c**) Representative configurations of Sm^{4+} molecules at the DNA-DNA distance of 28 Å for the $(AT)_{10}$-$(AT)_{10}$ (**a**), $(GC)_{10}$-$(GC)_{10}$ (**b**) and $(GmC)_{10}$-$(GmC)_{10}$ (**c**) DNA pairs. The backbone and bases of DNA are shown as ribbon and molecular bond, respectively; Sm^{4+} molecules are shown as molecular bonds. Spheres indicate the location of the N7 atoms and the methyl groups. (**d-f**) The average distributions of cations for the three sequence pairs featured in **a-c**. Top: density of Sm^{4+} nitrogen atoms ($d = 28$ Å) averaged over the corresponding MD trajectory and the z axis. White circles (20 Å in diameter) indicate the location of the DNA helices. Bottom: the average density of Sm^{4+} nitrogen (blue), DNA phosphate (black) and sodium (red) atoms projected onto the DNA-DNA distance axis (x axis). The plot was obtained by averaging the corresponding heat map data over $y = [-10, 10]$ Å. See Supplementary Figs 4 and 5 for the cation distributions at $d = 30$, 32, 34 and 36 Å.

methyl group of thymine base blocks Sm^{4+} binding to the N7 nitrogen atom of adenine, which is the cation-binding hotspot[21,22]. As a result, Sm^{4+} is not found in the major grooves of the AT-rich duplexes and resides mostly near the DNA backbone (Fig. 2a,d). Such relocated Sm^{4+} molecules bridge the two DNA duplexes better, accounting for the stronger attraction[16,23-25]. In contrast, significant amount of Sm^{4+} is adsorbed to the major groove of the GC-rich helices that lacks cation-blocking methyl group (Fig. 2b,e).

If indeed the extra methyl group in thymine, which is not found in cytosine, is responsible for stronger DNA-DNA interactions, we can predict that cytosine methylation, which occurs naturally in many eukaryotic organisms and is an essential epigenetic regulation mechanism[26], would also increase the strength of DNA-DNA attraction. MD simulations showed that the GC-rich helices containing methylated cytosines (mC) lose the adsorbed Sm^{4+} (Fig. 2c,f) and that $|\Delta G_{min}|$ of $(GC)_{10}$ increases on methylation of cytosines to become similar to $|\Delta G_{min}|$ of $(AT)_{10}$ (Fig. 1b).

To experimentally assess the effect of cytosine methylation, we designed another GC-rich construct GC2 that had the same GC content as GC1 but a higher density of CpG sites (Supplementary Table 1). The CpG sites were then fully methylated using M. SssI methyltransferase (Supplementary Fig. 3; Methods). As predicted from the MD simulations, methylation of the GC-rich constructs increased the binding fraction to the level of the AT-rich constructs (Fig. 1f).

The sequence dependence of $|\Delta G_{min}|$ and its relation to the Sm^{4+} adsorption patterns can be rationalized by examining the number of Sm^{4+} molecules shared by the dsDNA molecules (Fig. 3a). An Sm^{4+} cation adsorbed to the major groove of one dsDNA is separated from the other dsDNA by at least 10 Å, contributing much less to the effective DNA–DNA attractive force than a cation positioned between the helices, that is, the 'bridging' Sm^{4+} (ref. 23). An adsorbed Sm^{4+} also repels other Sm^{4+} molecules due to like-charge repulsion, lowering the concentration of bridging Sm^{4+}. To demonstrate that the concentration of bridging Sm^{4+} controls the strength of DNA–DNA attraction, we computed the number of bridging Sm^{4+} molecules, N_{spm} (Fig. 3b). Indeed, the number of bridging Sm^{4+} molecules ranks in the same order as $|\Delta G_{min}|$: N_{spm} of $(A)_{20} > N_{spm}$ of $(AT)_{10} \approx N_{spm}$ of $(GmC)_{10} > N_{spm}$ of $(GC)_{10} > N_{spm}$ of $(G)_{20}$. Thus, the number density of nucleotides carrying a methyl group (T and mC) is the primary determinant of the strength of attractive interaction between two dsDNA molecules. At the same time, the spatial arrangement of the methyl group carrying nucleotides can affect the interaction strength as well (Fig. 3c). The number of methyl groups and their distribution in the $(AT)_{10}$ and $(GmC)_{10}$ duplex DNA are identical, and so are their interaction free energies, $|\Delta G_{min}|$ of $(AT)_{10} \approx |\Delta G_{min}|$ of $(GmC)_{10}$. For AT-rich DNA sequences, clustering of the methyl groups repels Sm^{4+} from the major groove more efficiently than when the same number of methyl groups is distributed along the DNA (Fig. 3b). Hence, $|\Delta G_{min}|$ of

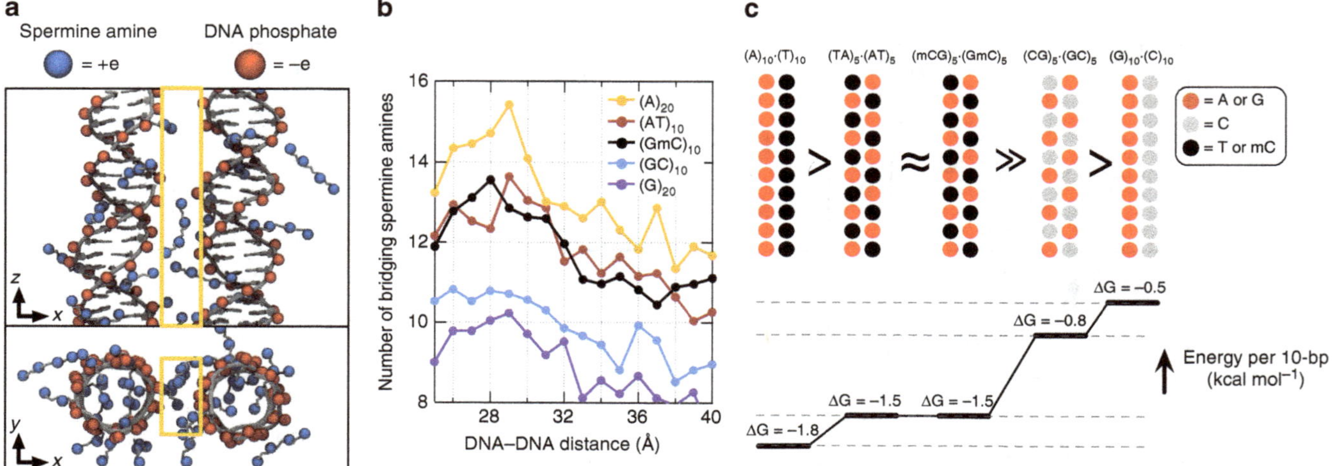

Figure 3 | Methylation modulates the interaction free energy of two dsDNA molecules by altering the number of bridging Sm^{4+}. (**a**) Typical spatial arrangement of Sm^{4+} molecules around a pair of DNA helices. The phosphates groups of DNA and the amine groups of Sm^{4+} are shown as red and blue spheres, respectively. 'Bridging' Sm^{4+} molecules reside between the DNA helices. Orange rectangles illustrate the volume used for counting the number of bridging Sm^{4+} molecules. (**b**) The number of bridging amine groups as a function of the inter-DNA distance. The total number of Sm^{4+} nitrogen atoms was computed by averaging over the corresponding MD trajectory and the 10 Å (*x* axis) by 20 Å (*y* axis) rectangle prism volume (**a**) centred between the DNA molecules. (**c**) Schematic representation of the dependence of the interaction free energy of two DNA molecules on their nucleotide sequence. The number and spatial arrangement of nucleotides carrying a methyl group (T or mC) determine the interaction free energy of two dsDNA molecules.

$(A)_{20} > |\Delta G_{min}|$ of $(AT)_{10}$. For GC-rich DNA sequences, clustering of the cation-binding sites (N7 nitrogen) attracts more Sm^{4+} than when such sites are distributed along the DNA (Fig. 3b), hence $|\Delta G_{min}|$ is larger for $(GC)_{10}$ than for $(G)_{20}$.

Discussion

Genome-wide investigations of chromosome conformations using the Hi-C technique revealed that AT-rich loci form tight clusters in human nucleus[27,28]. Gene or chromosome inactivation is often accompanied by increased methylation of DNA[29] and compaction of facultative heterochromatin regions[30]. The consistency between those phenomena and our findings suggest the possibility that the polyamine-mediated sequence-dependent DNA–DNA interaction might play a role in chromosome folding and epigenetic regulation of gene expression.

Methods

MD protocols. All MD simulations were carried out in a constant-temperature/constant-pressure ensemble using the Gromacs 4.5.5 package[31]. Integration time step was 2 fs. The temperature was set to 300 K using the Nosé–Hoover scheme[32,33]. The pressure in the *xy* plane (normal to DNA) was kept constant at 1 bar using the Parrinello–Rahman scheme[34]; the length of the box in the *z* direction was kept constant at 68 Å. Van der Waals forces were evaluated using a 7–10-Å switching scheme. Long-range electrostatic forces were computed using the particle-Mesh Ewald summation scheme[35], a 1.5-Å Fourier-space grid and a 12-Å cutoff for the real-space Coulomb interaction. Covalent bonds to hydrogen in water, and in non-water molecules were constrained using SETTLE[36] and LINCS[37] algorithms, respectively. All simulations were carried out using the AMBER99bsc0 force field for DNA[38,39], NaCl parameters of Joung *et al.*[40] and the TIP3P water model[41]. Parameters describing spermine ($NH_2(CH_2)_3NH(CH_2)_4NH(CH_2)_3NH_2$) were based on the AMBER99 force field[24]. Custom Van der Waals parameters (CUFIX) were used to describe non-bonded interactions between spermine amine and DNA phosphate, between sodium ion and DNA phosphate, and between sodium and chloride ions[14,15,25,42].

Potential of mean force calculations. Initially, a pair of 20-bp duplexes DNA was placed in a hexagonal water box parallel to the *z* axis. The water box measured ~130 Å within the *xy* plane and 68 Å along the *z* axis. Each DNA strand was effectively infinite under the periodic boundary conditions. The relatively large lateral size of the simulation box was chosen to avoid finite size artefacts. Twenty Sm^{4+} molecules were randomly placed in the box to neutralize the charge of the DNA molecules; sodium and chloride ions were added corresponding to a 100 mM concentration. Five variants of the system were built, different only by the nucleotide sequence of the DNA molecules. Each system was equilibrated for at least 50 ns; the DNA molecules were free to move about the simulation system during the equilibration. The last frame of each equilibration trajectory was used to initiate umbrella sampling simulations that determined the free energy (ΔG) of the pair of parallel dsDNA molecules as a function of the inter-DNA distance. The reaction coordinate was defined as the distance between the centres of mass of the two DNA molecules projected onto the *xy* plane. The harmonic restraints used for umbrella sampling simulations had a force constant of 2,000 kJ mol^{-1} nm^{-2}; the inter-DNA distance varied from 23 to 42 Å with a 1-Å window spacing. The umbrella sampling simulations were ~200 ns in duration in each sampling window; the inter-DNA distance was recorded every 2 ps. Except for the umbrella restraining potential, no additional constraints were applied to DNA. The weighted histogram analysis method implemented in the Gromacs package was used for the reconstruction of the free energy from the recorded inter-DNA distance data[12].

DNA design and synthesis. Supplementary Table 1 specifies the design of the 120-bp long DNA template and other molecules used in this study. All DNA constructs had the same 20-bp primer regions at both the ends of the constructs (primer A and B, designed by Clone Manager Suite 7). Primer B was labelled at the 5′ end with either Cy3 or Cy5 dye with the efficiency of 90% or higher. The nucleotide sequence of the middle 80-bp section of the constructs varied among the constructs. Note that we did not use 100% GC constructs because they are known to contain quadruplexes[43]. The 120-bp DNA templates were made by Integrated DNA Technologies. The dsDNA constructs were synthesized from these templates and two primers by PCR using Phusion High-Fidelity PCR Master Mix kit (New England BioLabs) and following the standard protocol of the kit. The PCR products were purified using the QIAquick PCR purification kit (Qiagen) and their concentrations were measured by ultraviolet-visible absorption. The CpG-methylated constructs were obtained by performing an 8-h methylation reaction on the dsDNA constructs using the CpG methyltransferase M. SssI (New England BioLabs, M0226L) following the company's standard protocol. The product was purified by the PCR purification kit. The methylation efficiency was estimated by digesting the dsDNA products with the BstUI restriction enzyme (New England BioLabs, R0518L), which can cut only unmethylated CGCG sequence, and subsequent electrophoresis of the digested fragments through a polyacrylamide gel (Supplementary Fig. 3).

DNA encapsulation. We encapsulated the purified dsDNA in lipid vesicles by modifying the protocol previously developed for single-stranded DNA[18]. In short, we mixed biotinyl cap phosphoethanolamine with 1,2-dimyristoyl-sn-glycero-3-phosphocholine in 1:100 molar ratio, dried and hydrated with buffer solution of 100 mM NaCl, and 25 mM Tris (pH 8.0). After hydration, the mixture was flash-frozen in liquid nitrogen and thawed seven times to create large unilamellar vesicles. The desired pair of dsDNAs was added to the solution, each at 400 nM concentration (corresponding to 1 molecule in a spherical volume of 200 nm diameter) and then the solution was extruded through a membrane filter with

200 nm pores to create uniformly sized unilamellar vesicles. Typical acceptor co-encapsulation yield (defined as the fraction of vesicles detected with a pair of donor and acceptor among all vesicles with any acceptor signal) was 10–20%.

Single-molecule measurements. The lipid vesicles were immobilized on a PEG-coated surface through the biotin–neutravidin interaction (Fig. 2a). Fluorescence signals from individual vesicles were collected by total internal reflection microscopy as previously described[44]. Imaging solution contained 1 mg ml^{-1} glucose oxidase, 0.04 mg ml^{-1} catalase, 0.8% dextrose, saturated Trolox (~3 mM) and 25 mM Tris in addition to the desired amounts of NaCl and Sm^{4+}. The gel–liquid transition temperature of 1,2-dimyristoyl-sn-glycero-3-phosphocholine is 24 °C, which results in a bistable membrane structure at room temperature, allowing the exchange of ions and small molecules (Sm^{4+}, Trolox and so on) through the membrane. All measurements and solution exchanges were carried out at 25 °C. Fluorescence movies were taken with the rate of 100 ms per frame. For Cy3 and Cy5 dyes, 532- and 647-nm solid-state lasers were used as the excitation sources, respectively.

Analysis of single-molecule traces. We quantified the strength of binding by measuring the fraction of traces that showed any binding events among all traces containing a single pair of donor and acceptor dyes. Binding traces exhibited a variety of behaviours (Supplementary Fig. 1A), which in part can be attributed to the variation in the vesicle size. First, we selected the traces containing single pair of Cy3 and Cy5 by examining their bleaching steps and signal intensities from either excitation. Among these, we selected the traces showing clear binding behaviours. The criteria we used as the binding behaviours was either the FRET efficiency jumping over 0.5 or showing no clear jumps but maintaining a FRET level of 0.25 or higher. The number of these binding traces over the number of all single pair traces was measured for triplicate data sets at each Sm^{4+} concentrations for each DNA sample. The error bars in Fig. 1f represents the s.e. of mean between these triplicate measurements.

References

1. Rau, D. C., Lee, B. & Parsegian, V. A. Measurement of the repulsive force between polyelectrolyte molecules in ionic solution: hydration forces between parallel DNA double helices. *Proc. Natl Acad. Sci. USA* **81,** 2621–2625 (1984).
2. Raspaud, E., Olvera de la Cruz, M., Sikorav, J. L. & Livolant, F. Precipitation of DNA by polyamines: a polyelectrolyte behavior. *Biophys. J.* **74,** 381–393 (1998).
3. Besteman, K., Van Eijk, K. & Lemay, S. G. Charge inversion accompanies DNA condensation by multivalent ions. *Nat. Phys.* **3,** 641–644 (2007).
4. Lipfert, J., Doniach, S., Das, R. & Herschlag, D. Understanding nucleic acid-ion interactions. *Annu. Rev. Biochem.* **83,** 813–841 (2014).
5. Grosberg, A. Y., Nguyen, T. T. & Shklovskii, B. I. The physics of charge inversion in chemical and biological systems. *Rev. Mod. Phys.* **74,** 329–345 (2002).
6. Kornyshev, A. A. & Leikin, S. Sequence recognition in the pairing of DNA duplexes. *Phys. Rev. Lett.* **86,** 3666–3669 (2001).
7. Danilowicz, C. *et al.* Single molecule detection of direct, homologous, DNA/DNA pairing. *Proc. Natl Acad. Sci. USA* **106,** 19824–19829 (2009).
8. Gladyshev, E. & Kleckner, N. Direct recognition of homology between double helices of DNA in Neurospora crassa. *Nat. Commun.* **5,** 3509 (2014).
9. Tabor, C. W. & Tabor, H. Polyamines. *Annu. Rev. Biochem.* **53,** 749–790 (1984).
10. Thomas, T. & Thomas, T. J. Polyamines in cell growth and cell death: molecular mechanisms and therapeutic applications. *Cell. Mol. Life Sci.* **58,** 244–258 (2001).
11. DeRouchey, J. E. & Rau, D. C. Role of amino acid insertions on intermolecular forces between arginine peptide condensed DNA helices: implications for protamine-DNA packaging in sperm. *J. Biol. Chem.* **286,** 41985–41992 (2011).
12. Kumar, S., Rosenberg, J. M., Bouzida, D., Swendsen, R. H. & Kollman, P. A. The weighted histogram analysis method for free-energy calculations on biomolecules. I. The method. *J. Comput. Chem.* **13,** 1011–1021 (1992).
13. Todd, B. A., Parsegian, V. A., Shirahata, A., Thomas, T. J. & Rau, D. C. Attractive forces between cation condensed DNA double helices. *Biophys. J.* **94,** 4775–4782 (2008).
14. Yoo, J. & Aksimentiev, A. Improved parametrization of Li^+, Na^+, K^+, and $Mg2^+$ ions for all-atom molecular dynamics simulations of nucleic acid systems. *J. Phys. Chem. Lett.* **3,** 45–50 (2012).
15. Yoo, J. & Aksimentiev, A. Improved parameterization of amine-carboxylate and amine-phosphate interactions for molecular dynamics simulations using the CHARMM and AMBER force fields. *J. Chem. Theory Comput.* **12,** 430–443 (2016).
16. Tolokh, I. S. *et al.* Why double-stranded RNA resists condensation. *Nucleic Acids Res.* **42,** 10823–10831 (2014).
17. Ha, T. *et al.* Probing the interaction between two single molecules: fluorescence resonance energy transfer between a single donor and a single acceptor. *Proc. Natl Acad. Sci. USA* **93,** 6264–6268 (1996).
18. Cisse, I. I., Kim, H. & Ha, T. A rule of seven in Watson-Crick base-pairing of mismatched sequences. *Nat. Struct. Mol. Biol.* **19,** 623–627 (2012).
19. Baumann, C. G., Smith, S. B., Bloomfield, V. A. & Bustamante, C. Ionic effects on the elasticity of single DNA molecules. *Proc. Natl Acad. Sci. USA* **94,** 6185–6190 (1997).
20. Hud, N. V. & Vilfan, I. D. Toroidal DNA condensates: unraveling the fine structure and the role of nucleation in determining size. *Annu. Rev. Biophys. Biomol. Struct.* **34,** 295–318 (2005).
21. Ruiz-Chica, J., Medina, M. A., Sánchez-Jiménez, F. & Ramírez, F. J. Fourier transform Raman study of the structural specificities on the interaction between DNA and biogenic polyamines. *Biophys. J.* **80,** 443–454 (2001).
22. Ouameur, A. A. & Tajmir-Riahi, H. A. Structural analysis of DNA interactions with biogenic polyamines and cobalt(III)hexamine studied by Fourier transform infrared and capillary electrophoresis. *J. Biol. Chem.* **279,** 42041–42054 (2004).
23. de la Cruz, M. O. *et al.* Precipitation of highly charged polyelectrolyte solutions in the presence of multivalent salts. *J. Chem. Phys.* **103,** 5781–5791 (1995).
24. Dai, L., Mu, Y., Nordenskiöld, L. & van der Maarel, J. R. Molecular dynamics simulation of multivalent-ion mediated attraction between DNA molecules. *Phys. Rev. Lett.* **100,** 118301 (2008).
25. Yoo, J. & Aksimentiev, A. The structure and intermolecular forces of DNA condensates. *Nucleic Acids Res.* doi: 10.1093/nar/gkw081.
26. Jaenisch, R. & Bird, A. Epigenetic regulation of gene expression: how the genome integrates intrinsic and environmental signals. *Nat. Genet.* **33,** 245–254 (2003).
27. Lieberman-Aiden, E. *et al.* Comprehensive mapping of long-range interactions reveals folding principles of the human genome. *Science* **326,** 289–293 (2009).
28. Imakaev, M. *et al.* Iterative correction of Hi-C data reveals hallmarks of chromosome organization. *Nat. Methods* **9,** 999–1003 (2012).
29. Cremer, T. & Cremer, C. Chromosome territories, nuclear architecture and gene regulation in mammalian cells. *Nat. Rev. Genet.* **2,** 292–301 (2001).
30. Rego, A., Sinclair, P. B., Tao, W., Kireev, I. & Belmont, A. S. The facultative heterochromatin of the inactive X chromosome has a distinctive condensed ultrastructure. *J. Cell Sci.* **121,** 1119–1127 (2008).
31. Van Der Spoel, D. *et al.* GROMACS: fast, flexible and free. *J. Comput. Chem.* **26,** 1701–1718 (2005).
32. Nose, S. & Klein, M. L. Constant pressure molecular dynamics for molecular systems. *Mol. Phys.* **50,** 1055–1076 (1983).
33. Hoover, W. G. Canonical dynamics: equilibrium phase-space distributions. *Phys. Rev. A* **31,** 1695–1695 (1985).
34. Parrinello, M. & Rahman, A. Polymorphic transitions in single crystals: a new molecular dynamics method. *J. Appl. Phys.* **52,** 7182–7190 (1981).
35. Darden, T., York, D. & Pedersen, L. Particle mesh Ewald: an N log(N) method for Ewald sums in large systems. *J. Chem. Phys.* **98,** 10089–10092 (1993).
36. Miyamoto, S. & Kollman, P. A. SETTLE: An analytical version of the SHAKE and RATTLE algorithms for rigid water models. *J. Comput. Chem.* **13,** 952–962 (1992).
37. Hess, B., Bekker, H., Berendesen, H. J. C. & Fraaije, J. G. E. M. LINCS: a linear constraint solver for molecular simulations. *J. Comput. Chem.* **18,** 1463–1472 (1997).
38. Cornell, W. D. *et al.* A second generation force field for the simulation of proteins, nucleic acids, and organic molecules. *J. Am. Chem. Soc.* **117,** 5179–5197 (1995).
39. Pérez, A. *et al.* Refinement of the AMBER force field for nucleic acids: improving the description of alpha/gamma conformers. *Biophys. J.* **92,** 3817–3829 (2007).
40. Joung, I. S. & Cheatham, T. E. Determination of alkali and halide monovalent ion parameters for use in explicitly solvated biomolecular simulations. *J. Phys. Chem. B* **112,** 9020–9041 (2008).
41. Jorgensen, W. L., Chandrasekhar, J., Madura, J. D., Impey, R. W. & Klein, M. L. Comparison of simple potential functions for simulating liquid water. *J. Chem. Phys.* **79,** 926–935 (1983).
42. Yoo, J. & Aksimentiev, A. Competitive binding of cations to duplex DNA revealed through molecular dynamics simulations. *J. Phys. Chem. B* **116,** 12946–12954 (2012).
43. Sen, D. & Gilbert, W. Formation of parallel four-stranded complexes by guanine-rich motifs in DNA and its implications for meiosis. *Nature* **334,** 364–366 (1988).
44. Kim, H., Tang, G. Q., Patel, S. S. & Ha, T. Opening-closing dynamics of the mitochondrial transcription pre-initiation complex. *Nucleic Acids Res.* **40,** 371–380 (2012).

Acknowledgements

This work was supported by the National Science Foundation (USA) through the PHY-1430124 award. J.Y. and A.A. gladly acknowledge supercomputer time provided through XSEDE Allocation Grant MCA05S028 and the Blue Waters Sustained Petascale Computer System (UIUC). H.K. was supported by the National Research Foundation of

Korea grants 2014R1A1A1003949, IBS-R020-D1 and the 2014 Research Fund (1.130091.01) of UNIST. T.H. was funded in part by a grant from National Institutes of Health (GM065367), and is an investigator with the Howard Hughes Medical Institute. We thank Dr Sang Hak Lee for the discussion of the DNA methylation protocol.

Author contributions

J.Y., H.K., A.A. and T.H. designed all computational and experimental assays. J.Y. and H.K. performed the simulations and experiments, respectively. All authors contributed to the analysis and interpretation of the data, development of the theory and writing of the manuscript.

Additional information

Competing financial interests: The authors declare no competing financial interests.

20

Regulation of cell-to-cell variability in divergent gene expression

Chao Yan[1,2], Shuyang Wu[1], Christopher Pocetti[3] & Lu Bai[1,2,3]

Cell-to-cell variability (noise) is an important feature of gene expression that impacts cell fitness and development. The regulatory mechanism of this variability is not fully understood. Here we investigate the effect on gene expression noise in divergent gene pairs (DGPs). We generated reporters driven by divergent promoters, rearranged their gene order, and probed their expressions using time-lapse fluorescence microscopy and single-molecule fluorescence *in situ* hybridization (smFISH). We show that two genes in a co-regulated DGP have higher expression covariance compared with the separate, tandem and convergent configurations, and this higher covariance is caused by more synchronized firing of the divergent transcriptions. For differentially regulated DGPs, the regulatory signal of one gene can stochastically 'leak' to the other, causing increased gene expression noise. We propose that the DGPs' function in limiting or promoting gene expression noise may enhance or compromise cell fitness, providing an explanation for the conservation pattern of DGPs.

[1] Department of Biochemistry and Molecular Biology, University Park, Pennsylvania 16802, USA. [2] Center for Eukaryotic Gene Regulation, University Park, Pennsylvania 16802, USA. [3] Department of Physics, The Pennsylvania State University, University Park, Pennsylvania 16802, USA. Correspondence and requests for materials should be addressed to L.B. (email: lub15@psu.edu).

Cell proliferation and differentiation depend on rigorously controlled gene activities, yet gene expression is intrinsically variable. In recent years, large effort has been dedicated to characterizing the cell-to-cell variability, or 'noise', of gene expression[1–4]. Large variations in the expression of housekeeping genes are likely to compromise cell fitness[5–7]; variations in some other genes can be beneficial by generating heterogeneous phenotypes in a population of genetically identical cells[8–15]. Given the physiological relevance, it is conceivable that cells have developed mechanisms to regulate gene expression noise. Indeed, recent studies have shown that noise can be modulated by promoter architecture, including the configuration of transcription factor binding sites, the sequence of TATA box and positioning of nucleosomes[6,8,16–18]. Beyond these *cis*-regulatory elements, a gene is embedded in a chromosomal context. The arrangement of neighbouring genes, also known as gene order, is thought to regulate the average level of gene expression[19]. However, the relationship between gene order and gene expression noise is unknown.

In this study, we focused on the noise regulation by divergent gene pairs (DGPs), that is, two neighbouring genes transcribed on opposite stands in a 'head-to-head' configuration. About half of the whole yeast genome is organized in DGPs, most of which have short intergenic regions (between 200 and 800 bp)[20]. In fruit fly and human genomes, 32 and 10% of all genes form DGPs with intergenic distances <1 kb (refs 21,22). Given the close proximity, a transcription factor associated with a DGP promoter may activate the two genes simultaneously, causing synchronized fluctuations in their expressions in individual cells (Fig. 1a,b). In other words, DGP may suppress the 'uncorrelated expression noise' between the two genes. Such effect can be beneficial if the two gene products form a complex or function in the same pathway. In contrast, if the two genes in a DGP are differentially regulated by different transcription factors, there may be 'crosstalk' between the two regulatory signals (Fig. 1c). If the 'crosstalk' occurs stochastically and infrequently, it may result in enhanced variability of gene expression among genetically identical cells as well as in a single cell over time. The function of DGPs in regulating gene expression noise may also shape its conservation pattern through evolution. In particular, it may provide explanation to previous observations that the conservation of DGPs is positively correlated with co-expression[23,24].

Here we tested these ideas by combining yeast genetics, single-cell gene expression assay and bioinformatics. We showed that two genes in a co-regulated DGP have higher covariance in their expression compared with other spatial configurations, and this higher covariance is caused by more synchronized firing of the divergent transcriptions. We also examined two differentially regulated DGPs and found that the regulatory signal of one gene can stochastically 'leak' to the other, and in these two cases, the leakage causes increased cell-to-cell variability of gene expression. Finally, we proposed that the DGPs' function in limiting/promoting gene expression noise may enhance/compromise cell fitness, which is consistent with the observation that highly co-regulated DGPs are more conserved through evolution, and differentially regulated DGPs are less conserved.

Results

Probe *HTA2–HTB2* divergent promoter activity in single cells. To study the effect of co-regulated DGPs on gene expression covariance, we chose the yeast *HTA2–HTB2* gene pair as our model. *HTA2* and *HTB2* encode the highly conserved histones H2A and H2B. H2A and H2B function as a dimer *in vivo*, making it important to maintain their stoichiometry[25,26]. *HTA2–HTB2* promoter is cell cycle regulated[27] and thus presents a stringent system for quantitative measurement of gene expression.

HTA2–HTB2 promoter fires during a short window of time between G1/S and G2 (ref. 27). To evaluate its activity in single cells, we fused unstable Venus (a yellow fluorescent protein)[28] to either side of the entire *HTA2–HTB2* intergenic region to generate *HTB2pr*-Venus and *HTA2pr*-Venus (same divergent promoter with different orientations) (Fig. 2a,b). We then integrated the reporters into yeast chromosome and monitored Venus intensity in live cells using time-lapse fluorescence microscopy over multiple cell generations (Fig. 2a,b; Methods)[6,29–31]. We also used Myo1-mCherry in these strains as a cell cycle marker[6]. These strains have intact endogenous *HTA2–HTB2* genes, and their growth rates are the same as wild-type cells.

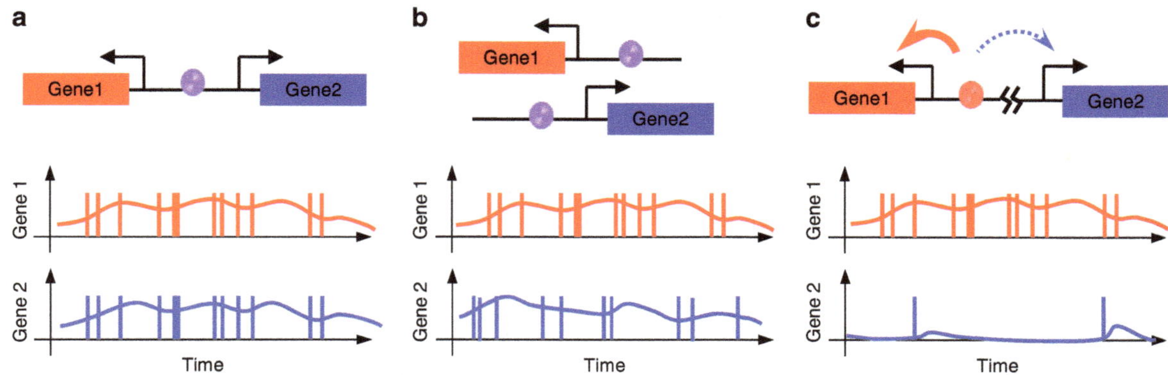

Figure 1 | Models of the stochastic DGP transcription. (**a**) Two genes in a DGP may have enhanced covariance in their expression. Stochastic binding of transcription factors contributes to gene expression noise. In the case of a co-regulated DGP, where the two genes are under the control of the same transcription factor binding events, the transcription initiation of the two genes may occur simultaneously (vertical bars in the lower panel). Although the synchrony level of gene expression is likely to be reduced by subsequent stochastic events such as transcription elongation and termination, we still expect more synchronized fluctuations in the protein level of the two genes (curves in the lower panel). (**b**) Two separate genes driven by the same divergent promoter as in **a** may experience independent factor binding and asynchronized transcription firing. (**c**) In a differentially regulated DGP where the two genes respond to different transcription factors, the regulatory signal may 'leak' stochastically to the wrong side. For every activation event of gene 1, if the leakage occurs sporadically in a fraction of cells, it may result in enhanced expression noise of gene 2. The notations in the diagrams are: black arrows, TSS; purple and red ovals, transcription factors; red and blue arrow in **c**, major and minor activation. The same notation applies to the following figures if not specified.

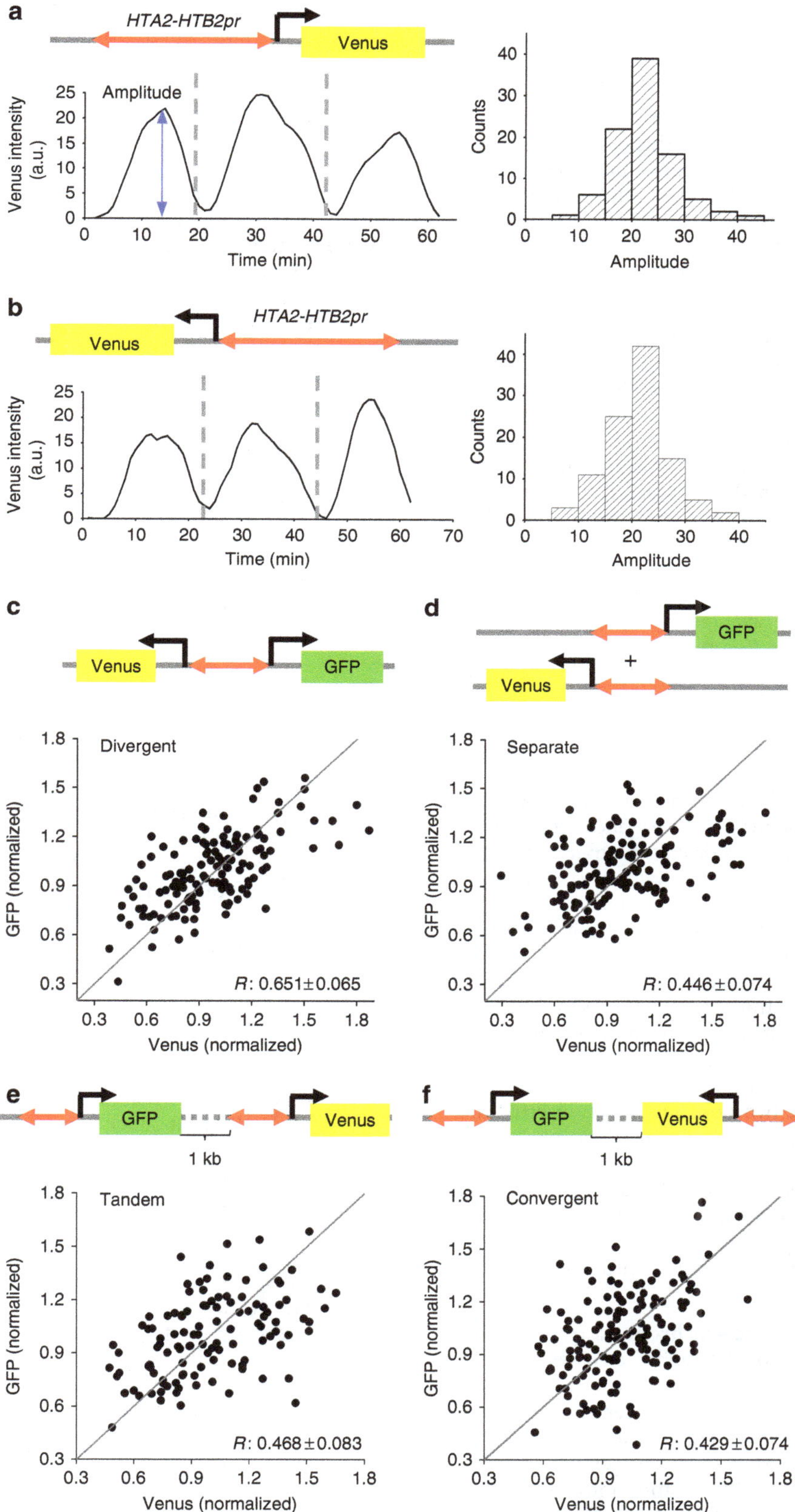

Figure 2 | *HTA2-HTB2* DGP enhances expression covariance. (**a**,**b**) Characterization of the cell cycle regulated activity of the *HTA2-HTB2* promoter. The traces represent the Venus intensity as a function of time driven by this promoter from either orientation. The grey vertical lines indicate the time of cell division. The histograms of the cell cycle amplitudes, calculated as the peak-to-trough difference in the Venus signal per cell cycle, are also shown. a.u.: arbitrary unit. (**c**–**f**) Covariance between the GFP and Venus expression driven by the *HTA2-HTB2* promoter in divergent (**c**), separate (**d**), tandem (**e**) or convergent (**f**) configurations. The *x* and *y* axis for each dot in these plots correspond to the normalized Venus and GFP amplitudes in the same cell cycle. The *R* values are Pearson correlation coefficient (mean ± s.e.m.; see Methods). Number of cell cycles: divergent: 140; separate 149; tandem: 115; convergent: 150.

Consistent with previous bulk measurements, Venus driven from either side of this promoter showed strong periodic changes in its intensity once every cell cycle (Fig. 2a,b). On average, *HTB2pr*-Venus and *HTA2pr*-Venus have identical cell cycle amplitudes, indicating that this divergent promoter has the same firing strength in both orientations (histograms in Fig. 2a,b; Methods). Note that in both cases, this amplitude varies up to four-fold among individual cell cycles. Such variation mostly reflects real changes in Venus expression since the measurement error accounts for $<9\%$ of the total fluctuations in the fluorescent intensity (Methods; Supplementary Fig. 1).

Co-regulated DGPs have higher expression covariance. The experiment above monitored the *HTA2* and *HTB2* promoter activity individually. We next investigated the covariance of their activities in single cells by constructing a strain containing this promoter driving GFP and Venus as a DGP (Fig. 2c, Methods). For comparison, we separated *HTA2pr*-Venus and *HTB2pr*-GFP by keeping *HTB2pr*-GFP at the original locus and inserting *HTA2pr*-Venus into another chromosome (each with full-length *HTA2–HTB2* promoter; Fig. 2d). GFP and Venus have similar brightness and maturation rate in yeast, allowing us to directly compare their intensity after crosstalk elimination (Methods; Supplementary Fig. 2). The average Venus expression in these two strains are very similar (Supplementary Fig. 3), indicating that there is no position-dependent gene expression bias between these two chromosomal loci.

In both strains, the amplitudes of GFP and Venus expression in individual cell cycles are positively correlated (Fig. 2c,d). Such positive correlation is expected because several global factors, such as cell size and factor concentration, would have similar impact on both reporter genes in the same cell. Importantly, the covariance between GFP and Venus is significantly higher when divergently transcribed (Pearson correlation: $R=0.65$ versus 0.45, P value $=0.0062$; Fig. 2c,d; Methods), supporting a role of DGP in coordinating the expression of two genes in single cells. These R values are highly reproducible in three independent measurements.

The higher covariance of divergent GFP and Venus expression may be due to synchronized transcription firings (as hypothesized in Fig. 1) or simply due to their chromosomal proximity. In the latter scenario, two genes at the same locus have similar nuclear surroundings, which may contribute to their co-regulation. To differentiate between these two possibilities, we inserted the two reporters into neighbouring loci in tandem or convergent configurations (Fig. 2e,f). GFP and Venus expression in these strains showed low covariance ($R=0.47$ and 0.43), comparable to the case where the two reporters are separated on different chromosomes (Fig. 2d). Therefore, co-regulation was not enhanced by the spatial proximity *per se* in our system.

Co-regulated DGPs show more synchronized transcription. To probe the molecular origin of the higher DGP expression covariance, we used single molecule fluorescence *in situ* hybridization (smFISH) to directly visualize the nascent transcripts driven by *HTA2–HTB2* promoter in divergent and convergent configurations (Methods). We started from the genetic constructs in Fig. 2c,f, introduced a frame-shift mutation into Venus (Venus*) to avoid the interference from its fluorescent signal, and replaced GFP with a frame-shifted mCherry gene (mCherry*) so that its sequence can be differentiated from that of Venus* (Fig. 3a). After hybridization with a mixture of two differentially labelled probe sets targeting Venus* and mCherry* transcripts, we detected individual fluorescent 'particles' that were only visible in the presence of target mRNAs (Supplementary Fig. 4; Methods). More than 90% of the particle-containing cells were either unbudded or small-budded, consistent with the G1/S activation of the *HTA2-HTB2* promoter (Supplementary Fig. 5; Methods). These observations confirmed the specificity of our smFISH signals.

For cells containing both Venus* and mCherry* reporters, the brightest red and green fluorescent particles often co-localize inside the nucleus (Fig. 3a,b). These particles represent the clusters of nascent Venus* and mCherry* mRNAs at the transcription sites (TSs)[32,33]. For every cell with detected TS (Venus* and/or mCherry*), we evaluated the fluorescent intensity at the TS in both red and green channels (Methods). In both divergent and convergent strains, the two fluorescent intensities fluctuate significantly relative to each other (Fig. 3c). These fluctuations largely reflect real differences in the transcriptional activities since the measurement error is much smaller (Supplementary Fig. 6). The covariance between the divergent transcription signals ($R=0.59$) is again significantly higher than the convergent ones ($R=0.43$; P value $=0.0116$; Methods) (Fig. 3c), indicating that the divergent genes are fired in a more synchronized fashion. These observations are consistent with a previous FISH experiment showing that the expression correlation between *GAL1* and *GAL10* (a DGP) is higher than that between *GAL1* and *GAL7* (non-divergent genes closely located on the same chromosome). Overall, our data in Figs 2 and 3 support the model in Fig. 1a,b.

Leakage expression in differentially regulated DGPs. Unlike *HTA2–HTB2*, most DGPs in the genome are differentially regulated with two genes under the control of different transcription factors[34]. Our previous work showed that the regulation of two divergent genes in yeast can be 'decoupled' by sequence-specific DNA-binding factors (blockage factors) that shield the proximal promoter from the action of more distant transcription regulators[34]. If this blockage mechanism is not 100% robust, there will be 'leakage' of the regulatory signal from the upstream promoter, leading to undesired gene expression pattern (Fig. 1c).

To test this idea, we first examined a differentially regulated DGP, *PFK26–MOB1*. Under the heat-shock condition, the master stress regulator Msn2/4 binds and activates *PFK26pr* (Fig. 4a)[35,36]. The heat-shock activation of *MOB1*, however, is prevented by a Mcm1 protein associated downstream the Msn2/4 (ref. 34). To probe the potential leakage of the Msn2/4 signal, we measured *MOB1* heat-shock response in single cells (Methods). Interestingly, while *MOB1pr* has little heat-shock activity in most cells, mild activation of *MOB1pr*-Venus was detected in 15% of the population (Fig. 4b). Correspondingly, the slopes of these traces have a bimodal distribution that can be fit by two Gaussians (Fig. 4c; Methods). As a result, there is an increased cell-to-cell variability of *MOB1pr*-Venus expression during heat shock (Fig. 4b,c).

To understand the bimodality of *MOB1pr* heat-shock activity, we truncated the promoter sequence on the *PFK26* side (ΔP), eliminating all the potential Msn2/4 sites (Fig. 4d). Venus driven by the ΔP promoter exhibited close-to-zero heat-shock responses in all cells (Fig. 4d), indicating that Msn2/4 was responsible for the stochastic activation. In contrast, when we mutated the binding site of the blockage factor Mcm1 (mcm1*) in the full-length promoter, all cells showed positive heat-shock responses (Fig. 4e). The corresponding histogram of the heat-shock slopes had a unimodal distribution centered at ~ 0.49, similar to the higher peak in that of the WT *MOB1pr* (centred at 0.5). These observations indicate that the heat-shock response of the WT *MOB1pr* is either fully 'off' or fully 'on' in individual cells, and the 'on' cases may be due to the loss of Mcm1 blockage for extended period of time (see Discussion).

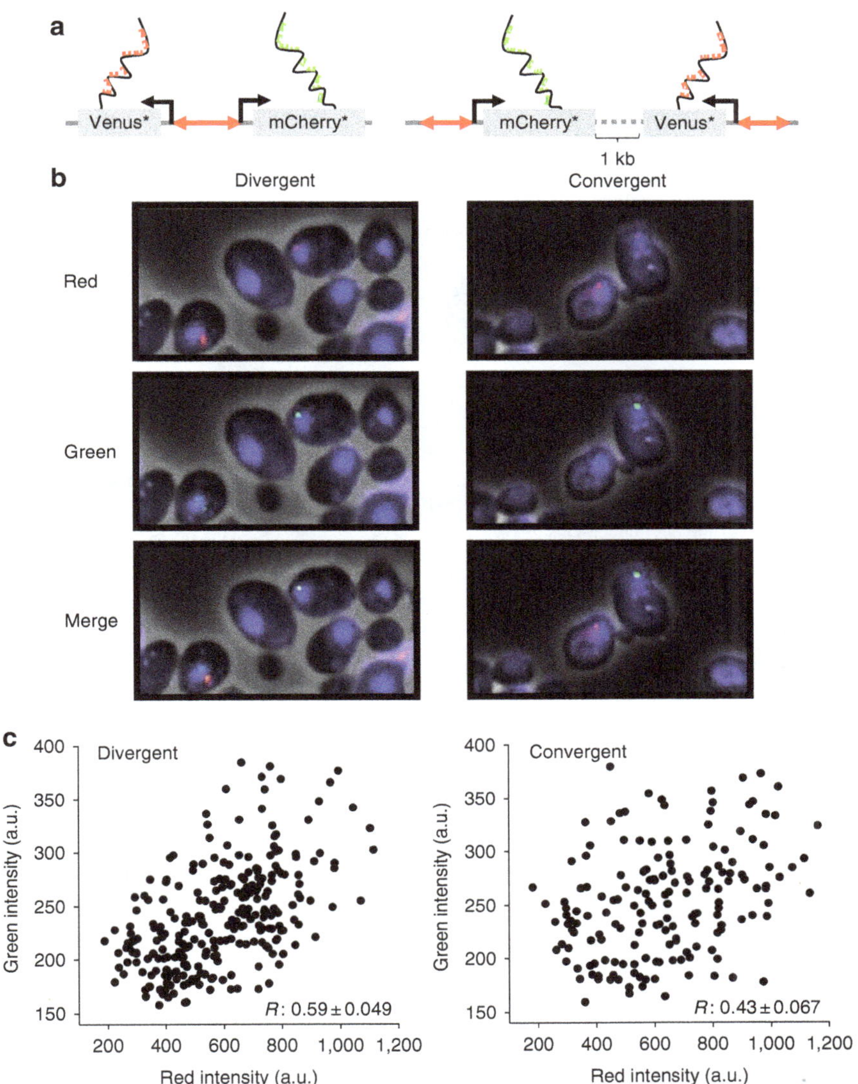

Figure 3 | smFISH detected higher correlation in divergent transcription. (**a**) Cartoons illustrating the smFISH method for probing the divergently (left) or convergently transcribed (right) Venus* and mCherry* genes. The black curves represent nascent mRNAs at the transcription site (TS). The red and green blocks represent the CAL Fluor Red 610-labelled probes for Venus* and FAM-labelled probes for mCherry* transcript, respectively. (**b**) Typical FISH images of Venus* and mCherry* transcripts driven by *HTA2-HTB2* promoter in divergent (left) or convergent (right) configurations. Phase and DAPI channels merged with red, green or both fluorescent images were shown in the upper, middle or lower panels, respectively. (**c**) Correlation of the Venus* and mCherry* transcriptional activities. Each dot represents the fluorescent intensity of one TS in the red (*x*) and green (*y*) channels. The *R* values are Pearson correlation coefficient (mean ± s.e.m.; see Methods). Number of TSs: divergent, 272; convergent, 185.

Similar stochastic crosstalk was observed on another differentially regulated DGP, *PRX1–KIP1* (Fig. 5a). *PRX1–KIP1* promoter has two separated nucleosome-depleted regions (Supplementary Fig. 7a)[37], each containing the respective regulatory elements for *PRX1* or *KIP1*. A late S-phase transcription factor Hcm1 resides closer to the *KIP1* gene and drives the cell cycle regulation of *KIP1*, but not *PRX1* (refs 38,39). Consistent with literature, we observed robust cell cycle regulation of *KIP1pr*-Venus with an average amplitude much higher than that of *PRX1pr*-Venus (Fig. 5b). Nevertheless, *PRX1pr*-Venus expression showed strong oscillations in a fraction of cell cycles (Fig. 5c), and the corresponding cell cycle amplitude had a bimodal distribution (Fig. 5d). The higher peak in this distribution disappeared when we deleted the *KIP1*-proximal sequences that contain the Hcm1 binding site (ΔK; Fig. 5e), indicating that the stochastic cell cycle regulation of *PRX1* originates from the leakage of Hcm1 activation. The mechanism underlying *PRX1* and *KIP1* differential regulation is still unknown. Deletion of an ~300-bp nucleosomal sequence in the middle of the promoter joined the two nucleosome-depleted regions together (Supplementary Fig. 7b) and mildly increased the probability of high *PRX1* cell cycle regulation from 20 to 29% (ΔNuc; Fig. 5f). This result shows that the nucleosome separation (or simply the distance) only plays a minor role in decoupling the two genes.

Highly co-regulated DGPs are more conserved in evolution. The effect on gene expression noise through divergent transcription likely plays a physiological role. For co-regulated DGPs, the maintenance of the stoichiometry of two subunits in a protein complex such as H2A and H2B may help to avoid the toxicity of the unpaired subunits. It may also facilitate the coordination of multiple proteins in the same reaction pathway. In contrast, differentially regulated DGPs cause leaky expression. Although

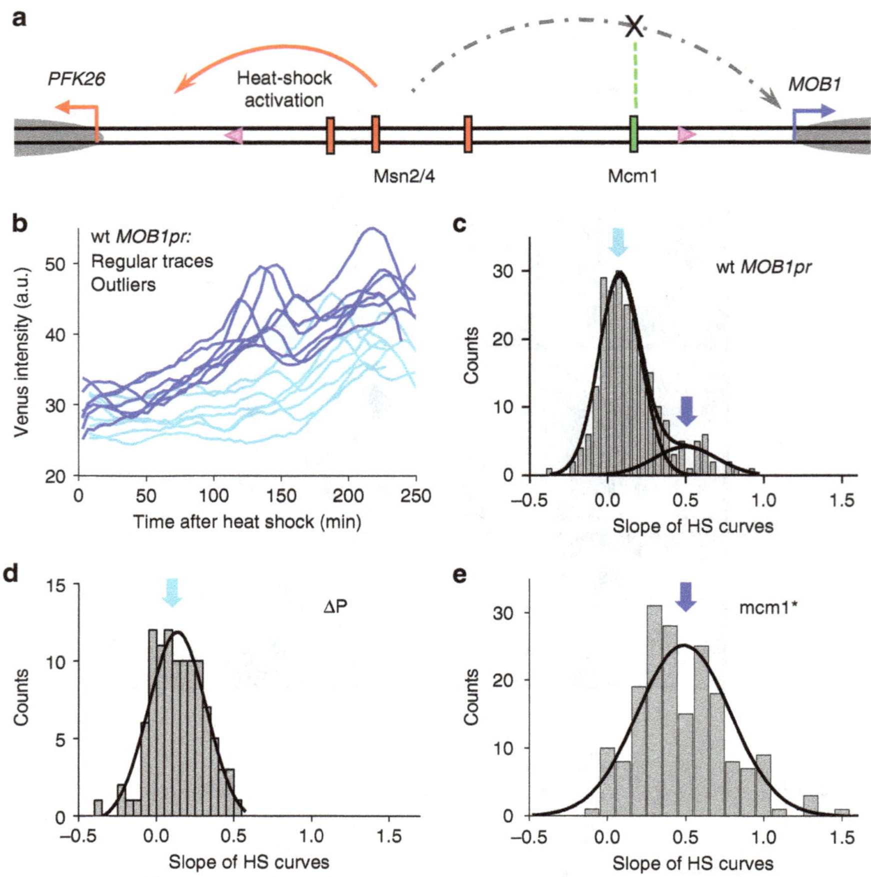

Figure 4 | Stochastic *MOB1* heat-shock activation from the *PFK26*UAS. (**a**) Genomic structure of *PFK26–MOB1* divergent promoter. Red rectangles, putative Msn2/4 binding sites; green rectangle, Mcm1 binding site; pink triangles, two non-consensus TATA elements; red and blue arrows, TSSs and grey ovals, nucleosomes. (**b**) Representative heat-shock (HS) traces of Venus driven by *PFK26–MOB1* promoter from the *MOB1* side in single yeast cells. In most cells, *MOB1pr*-Venus showed no activation during the heat-shock phase (first 2 h), but started to oscillate when cells adapted and resumed growth (cyan traces). However, *MOB1pr* showed significant heat-shock response in a small fraction of cells (blue traces). (**c–e**) Histograms of heat-shock intensity driven by WT *MOB1* promoter (**c**) and its variants (**d**,**e**) among individual cells. This intensity was quantified by the slopes of the fluorescent traces during heat shock. ΔP, deletion of the *PFK26* side of the promoter including all the Msn2/4 binding sites; mcm1*, mutation of the Mcm1 binding site that leads to Mcm1 depletion from the promoter[34]. The WT *MOB1* promoter showed bimodal heat-shock response (**c**); in contrast, the two mutant promoters showed unimodal response corresponding to the lower (**d**) and higher peaks (**e**) respectively. Number of traces: WT, 236; ΔP, 105; mcm1*, 185.

large cell-to-cell variability in gene expression was shown to be beneficial for several environmental-responsive genes[8–11], we suspect that in most cases (especially for house-keeping genes), sporadic gene expression triggered by unscheduled regulatory signal would have either neutral or detrimental effect on cell fitness. This idea is consistent with previous reports that the conservation of DGPs is positively correlated with their co-regulation[23,24].

To confirm the conservation data, and in particular, to compare the conservation of DGP with the genome-wide baseline, we examined the change of neighbouring gene arrangement across the whole-genome duplication (WGD). WGD dramatically altered genome organization through duplication and deletion, which provided a unique opportunity to evaluate the evolution of gene order[40]. We expect gene orders that enhance/reduce cell fitness to be more/less conserved than the neutral selection.

Based on the gene mapping in Kellis *et al.*[40], we found 1,151 DGPs in *K. waltii* (pre-WGD) where both genes have orthologues in *S. cerevisiae* (post WGD). About 40% of these DGPs are conserved, and the rest are 'lost' during evolution, that is, their orthologues are no longer in the DGP form in *S. cerevisiae* (Fig. 6a). There are also 700 newly formed DGPs in *S. cerevisiae*. This level of conservation is comparable to that of the tandem and convergent gene pairs (Fig. 6b). To understand how the conservation level is related to co-expression, we collected the co-expression score (CES) between two *S. cerevisiae* genes in the 'conserved', 'lost' and 'newly formed' categories (Methods). Higher CES represents higher level of co-regulation. In agreement with previous reports[23,24], highly co-regulated DGPs (CES > 3) are more conserved than average (P value $< 10^{-2}$; Pearson's Chi-square test), and this trend was not observed for tandem and convergent gene pairs (Fig. 6b). Importantly, the average CES of the conserved DGPs (1.66 ± 0.03; mean ± s.e.m.) is significantly higher than that of the newly formed DGPs (new: 1.28 ± 0.02) (Fig. 6c), indicating that the high CES is not merely a consequence of the DGP configuration. Out of the 12 conserved DGPs with CES > 3, 8 encode subunits of histone and ribosome (Supplementary Table 1), indicating that there is a motivation to maintain the stoichiometry of these proteins. In contrast, the differentially regulated DGPs (CES < 1.5) are significantly less conserved than average (P value $< 10^{-3}$; Pearson's Chi-square test) (Fig. 6b). These observations support the idea that the high covariance of co-regulated DGPs is beneficial for cell fitness and

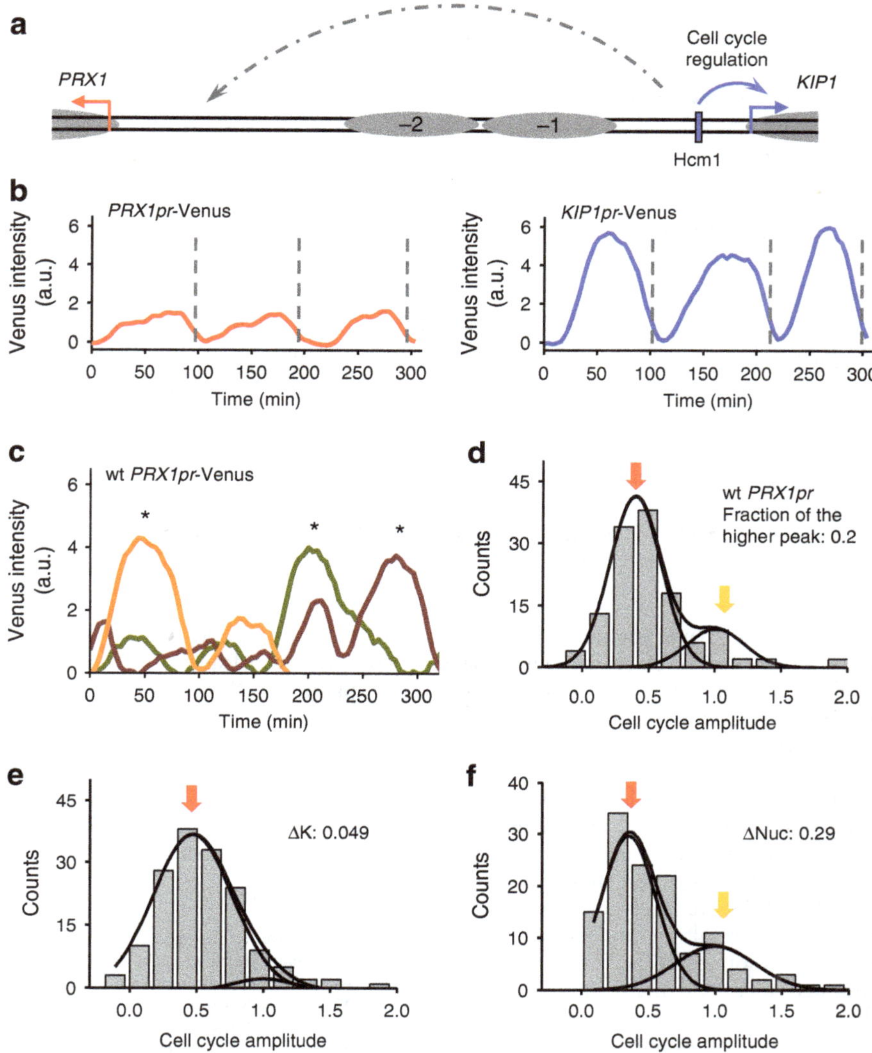

Figure 5 | Stochastic *PRX1* cell cycle activation from the *KIP1* UAS. (**a**) Genomic structure of *PRX1-KIP1* divergent promoter. Blue rectangles, putative Hcm1 binding sites and grey ovals, nucleosomes. (**b**) Typical traces of Venus expression driven by WT *PRX1-KIPpr* from the *PRX1* orientation (left) or *KIP1* orientation (right) during vegetative growth. Each trace represents the Venus intensity as a function of time in a single yeast cell over multiple cell cycles. The vertical dash lines indicate the cell division time. In most cells, *KIP1pr*, but not *PRX1pr*, shows strong cell cycle-regulated activity. (**c**) Selected traces of *PRX1pr*-Venus with high cell cycle oscillation (marked by '*'). These traces occur in a small fraction of cells. (**d-f**) Histograms of the cell cycle amplitude of Venus driven by WT or mutant *PRX1* promoter. ΔK, deletion of the *KIP1* side of the promoter including nucleosome −1; ΔNuc, deletion of the nucleosomal sequences in the middle of the promoter. The WT *PRX1* promoter and ΔNuc showed bimodal cell cycle amplitude, and the higher peak is eliminated in the ΔK promoter. Number of cell cycles: WT, 129; ΔK, 155; ΔNuc, 125.

therefore these DGPs tend to be conserved. In the opposite, crosstalk between differentially regulated DGPs may be harmful and therefore minimized by natural selection.

Discussion

Some divergent gene pairs are highly conserved among eukaryotic species but their biological function is not clear. Since the two genes in a DGP potentially share the same upstream regulatory sequences, it is commonly thought that DGP leads to co-regulation of functionally related genes[41]. However, two genes do not need to be divergently transcribed to achieve co-regulation; they simply need to be regulated by the same transcription factors. Indeed, in the budding yeast genome, only 34 out of the 44,489 gene pairs with CESs greater than or equal to 3 are DGPs. Even though the probability of finding high CES among DGPs (0.026) is still one order of magnitude higher than the genome-wide average (0.0022), these numbers clearly show that divergent transcription is not necessary for achieving high correlation between two genes. Therefore, the functional advantage of maintaining co-regulated DGPs in the genome is not clear.

The CESs were calculated based on microarray data, which measured the average gene expression among millions of cells. The cell-to-cell variability of gene expression is therefore not captured by these data. For some genes that function in multi-subunit complexes, it is conceivable that their balance needs to be stringently maintained in individual cells at all times. By measuring gene expression in single cells, we found that organizing two genes into a DGP serves as a mechanism to suppress the uncorrelated expression variation. Notably, for ribosome biosynthesis genes, it was proposed that adjacent gene pairs in all orientations (tandem, divergent and convergent) contribute to their coordination[41]. This was not found in our study: the tandem or convergent dual reporter genes driven by

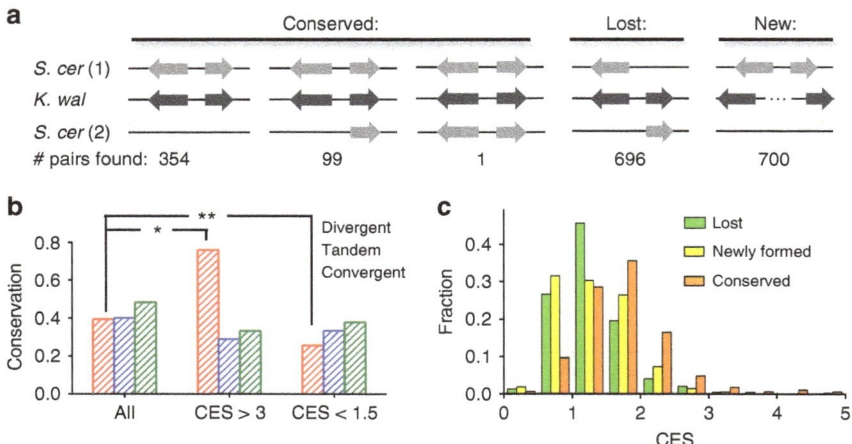

Figure 6 | Evolution of DGPs across whole-genome duplication. (**a**) Definition of 'conserved', 'lost' and 'new' DGPs by comparing the *S. cerevisiae* (post WGD) and *K. waltii* (pre-WGD) genome. (**b**) The conservation fraction for all divergent, convergent or tandem gene pairs, or the ones with CES > 3 (highly co-regulated), or CES < 1.5 (differentially regulated). Comparing with average, DGPs with high CESs are more likely to be conserved, and the ones with low CESs are less likely to be conserved. *P value < 10^{-2}; **P value < 10^{-3}. (**c**) The histogram of the CES for the conserved, lost and newly formed DGPs.

the *HTA2–HTB2* promoter do not have higher expression correlation than the separate ones. Therefore in our case, physical proximity alone does not play a major role in co-regulation. Instead, the smFISH measurement supported the idea that the suppression of uncorrelated noise is caused by more synchronized firing of the divergent transcriptions. When combined with literature on histone gene regulation, these data are consistent with the following picture: *HTA2* and *HTB2* are both activated by a transcription factor, Spt10, and the divergent promoter contains at least four Spt10 recognition sites[42]. Because of its strong binding cooperativity[43], Spt10 tends to occupy these sites in an all-or-none fashion, resulting in simultaneous activation and coordinated expression of *HTA2* and *HTB2*. It should be noted that, although there are many multi-subunit complexes in yeast, highly correlated DGPs are limited to a few complexes, including histone and ribosome (Supplementary Table 1). We suspect that this is because these complexes play such fundamental roles, and the expressions of their subunits need to be better balanced.

Besides coordinating the expression of two genes in single cells, co-regulated DGPs may also allow the co-evolution of their regulation. Mutation in the divergent promoter sequence, especially in the upstream activating sequence (UAS), can have a similar impact on the two genes. These changes can fine-tune the expression level of the two genes without affecting their stoichiometry. This potential advantage may also contribute to the conservation of co-regulated DGPs.

Most DGPs in the genome are differentially regulated. So far we have characterized gene expression noise in two such DGPs and found transcription leakage in both cases, suggesting this is a wide-spread phenomenon. The decoupling between *PFK26* and *MOB1* is due to sequence-specific 'blockage factors'[34], and we suspect that the differential regulation of *PRX1* and *KIP1* is based on a similar mechanism. The leakages of the *PFK26* and *KIP1* signals occur sporadically in a small fraction of cells, causing bimodal gene activation in WT *MOB1* and *PRX1* promoters (Fig. 4c,d). This may be due to infrequent association and dissociation of the blockage factors. Alternatively, it can be caused by a double negative feedback loop between the binding of blockage factors and the leakage transcription[44,45]: the factors prevent the transcription initiation in the wrong orientation, but once such transcription starts, the blockage factors can no longer bind. In fact, in our previous experiments with synthetic promoters, we found some evidence showing that a strong transcription can lead to the eviction of blockage factors[34]. This can explain why the leakage transcription is either fully 'on' or fully 'off' among individual cells (Fig. 4c,e). Cells can prevent these unintended transcripts through more blockage factors, longer intergenic regions or other decoupling mechanisms. The current configuration of the differentially regulated divergent genes may reflect a balance between the fitness cost of the leakage transcripts and the cost to achieve substantial noise suppression.

Methods

Strains and plasmids. Standard methods were used to construct the strains and plasmids. All strains are based on W303. All the 'wild-type' promoters were constructed as the entire intergenic region between the two divergent open reading frames (ORFs). The divergent *HTA2–HTB2* promoter driving GFP and Venus were integrated into *CLN2* locus at chrXVI: 65106 (Fig. 2c). The strains in Fig. 2d–f were derived from a common strain that contains *HTB2pr* driving GFP inserted into the *CLN2* locus at chrXVI: 65106. *HTA2pr* driving Venus was then integrated with different configurations relative to *HTB2pr*: (1) at chrXVI: 64037 in a tail to tail orientation (convergent); (2) at chrXVI: 64037 in a tail to head orientation (tandem); (3) at the *TRP1* locus on ChrIV (separate). Construction of the strains containing *PFK26–MOB1* promoter driving reporters were described in our previous work[34], and the strains with *PRX1–TRP1* promoters were constructed with similar methods.

Time-lapse fluorescence microscopy and data analysis. For the time-lapse measurements, we placed live yeast cells under an agar pad made with growth media and recorded fluorescent images every 4 min over 5–10 h (refs 6,34). The cell cycle and heat-shock expression data was collected under 30 °C and 37 °C, respectively. An objective heater was used to keep the sample surface temperature as desired (measured by Omega surface thermocouple). The cell cycle curves were smoothed and corrected by subtracting a baseline connecting flanking troughs, and the peak height in each cell cycle represents the cell cycle amplitude. The heat-shock response was quantified by the slopes of fluorescence trace during the first 2 h of heat shock before cells adapted to the higher temperature and resumed growth. S.e.'s of the Pearson correlation coefficient in Figs 2 and 3 were calculated as $\text{s.e.}_R = \sqrt{\frac{1-R^2}{n-2}}$. To evaluate the significance between two independent correlation coefficients, each R value was first transformed to Z scores through Fisher z-transformation, and the pooled s.e. was calculated as $\text{s.e.}_z = \sqrt{\frac{1}{n_1-3} + \frac{1}{n_2-3}}$. Z score between the two R values was then calculated as $Z = \frac{Z_1 - Z_2}{\text{s.e.}_Z}$ and then transformed to P value.

Estimation of measurement error in single-cell microscopy. The measurement uncertainty most likely originates from the variations in the focusing position and the opening duration of the fluorescence shutter. To estimate the magnitude of this uncertainty, we repeated the time-lapse fluorescence measurement using the same cells with stage repositioning and re-focusing (Supplementary Fig. 1). The measurement error was calculated by the squared deviation between measurement 1 and 2.

Two-colour data analysis. GFP and Venus have significantly overlapped spectrum, and for two-colour experiment, we eliminated the crosstalk by the following linear equation:

$$\begin{cases} S_g = [\mathrm{GFP}] + a[\mathrm{Venus}] \\ S_y = b[\mathrm{GFP}] + [\mathrm{Venus}] \end{cases} \Rightarrow \begin{cases} [\mathrm{GFP}] = \left(S_g - aS_y\right)/(1-ab) \\ [\mathrm{Venus}] = \left(S_y - bS_g\right)/(1-ab) \end{cases}$$

Where S_g and S_y are the signals we acquired in the GFP and YFP channel; [GFP] and [Venus] present the GFP and Venus concentration. The crosstalk parameters a and b were obtained using strains containing only GFP or Venus. The values of a and b depend on the filter, light source spectrum and the exposure settings in the GFP and YFP channel. With our current experimental condition (GFP filter: Chroma 49002, YFP filter: 49003, light source: OSRAM ARC Lamp, GFP exposure with 30% intensity for 0.1 s, YFP exposure with 55% intensity for 0.2 s), a and b equal to 0.409 and 1.036, respectively. See Supplementary Fig. 2 for fluorescent traces after crosstalk elimination. Importantly, the GFP peak value of the Venus-only strain was close to 0 after crosstalk elimination (0.056 ± 0.052; mean ± s.d. normalized by the average intensity of *HTB2pr*-GFP), and so is the Venus peak value in GFP-only strain (0.064 ± 0.047) (normalized by the average intensity of *HTA2pr*-Venus). These numbers show that our method can effectively remove crosstalk.

smFISH assay. We slightly modified a previously developed protocol of single molecule FISH[46,47]. We targeted the Venus* transcript with 28 CAL Fluor Red 610-labelled probes and mCherry* with 31 FAM-labelled probes (LGC Biosearch Technologies). After fixing the yeast cells with formaldehyde for 45 min at room temperature and permeating the membranes with 70% ethanol overnight at 4 °C, we performed hybridization in 50 µl solution overnight at 30 °C. The final concentration of each set of the probes is 150 nM. After DAPI staining, the cells were placed between a coverslip and a 1.5% agar pad made with 2 × SSC buffer, and imaged under a Leica DMI6000b fluorescent microscope. To detect three-dimensional FISH signals, the images were taken at seven focal planes (z-stack), 0.4 µm apart. The images were then analysed using MATLAB programs developed in our lab. After processing, the programs generated the cell and nucleus boundaries, as well as the locations of FISH dots with intensities above a threshold. The brightest dot inside the nucleus was assigned as the TS. For all detected TSs (either in the red or green channel, or both), we quantified the TS intensity within a 3 by 3 pixel area at the TS location for all seven z positions, and used the maximum across the z-stack as the final intensity.

Nucleosome mapping. Nucleosome occupancy was measured by MNase digestion followed by stacking quantitative PCR. We first grew 10 ml cell to OD ~0.15, collected the cell, and washed in 0.5 ml water. Then we re-suspended the cells in 0.5 ml of sphaeroplasting solution (1 M sorbitol, 0.5 mM 2-mercaptoethanol, 0.18 mg ml^{-1} zymolyase), incubated at room temperature for ~5 min with gentle stir. We then collected the cell, wash in 1 ml of 1 M sorbitol, then re-suspend the pellet in 200 µl of digestion buffer (1 M sorbitol, 50 mM NaCl, 100 mM Tris-Cl (pH 7.4), 5 mM $MgCl_2$, 1 mM $CaCl_2$, 1 mM 2-mercaptoethanol, 0.5 mM spermidine, 0.075% NP-40, micrococcal nuclease with final concentration 1–10 unit ml^{-1}) for ~8 min in 37 °C. After terminating the MNase digestion by adding 20 µl quench buffer (250 mM EDTA, 5% SDS), we purified the ~150 bp DNA, and proceeded with the qPCR analysis with stacking PCR primer pairs[48]. The PCR products were all ~100 bp in length, and the distances between adjacent primers were typically ~50 bp. We used the nucleosome −1 on the *PHO5pr* as the standard to scale the occupancy from 0 to 1.

Bioinformatics analysis of DGP conservation. We developed MATLAB software for the statistical analysis. In both *S. cerevisiae* and *K. waltii*, we defined 'DGPs' as head-to-head gene pairs with distance between the two ORFs <1,000 bp. We used the data in ref. 40 to map the homologous genes in these two species. Lacking CES data in *K. waltii*, we retrieved the CES in *S. cerevisiae* from Serial Pattern of Expression Levels Locator (SPELL)[49].

References

1. Raser, J. M. & O'Shea, E. K. Noise in gene expression: origins, consequences, and control. *Science* **309,** 2010–2013 (2005).
2. Munsky, B., Neuert, G. & van Oudenaarden, A. Using gene expression noise to understand gene regulation. *Science* **336,** 183–187 (2012).
3. Raj, A. & van Oudenaarden, A. Nature, nurture, or chance: stochastic gene expression and its consequences. *Cell* **135,** 216–226 (2008).
4. Elowitz, M. B., Levine, A. J., Siggia, E. D. & Swain, P. S. Stochastic gene expression in a single cell. *Science* **297,** 1183–1186 (2002).
5. Fraser, H. B., Hirsh, A. E., Giaever, G., Kumm, J. & Eisen, M. B. Noise minimization in eukaryotic gene expression. *PLoS Biol.* *2, e137 (2004).*
6. Bai, L., Charvin, G., Siggia, E. D. & Cross, F. R. Nucleosome-depleted regions in cell-cycle-regulated promoters ensure reliable gene expression in every cell cycle. *Dev. Cell* **18,** 544–555 (2010).
7. Newman, J. R. *et al.* Single-cell proteomic analysis of S. cerevisiae reveals the architecture of biological noise. *Nature* **441,** 840–846 (2006).
8. Blake, W. J. *et al.* Phenotypic consequences of promoter-mediated transcriptional noise. *Mol. Cell* **24,** 853–865 (2006).
9. Lopez-Maury, L., Marguerat, S. & Bahler, J. Tuning gene expression to changing environments: from rapid responses to evolutionary adaptation. *Nat. Rev. Genet.* **9,** 583–593 (2008).
10. Thattai, M. & van Oudenaarden, A. Stochastic gene expression in fluctuating environments. *Genetics* **167,** 523–530 (2004).
11. Kussell, E. & Leibler, S. Phenotypic diversity, population growth, and information in fluctuating environments. *Science* **309,** 2075–2078 (2005).
12. Rister, J., Desplan, C. & Vasiliauskas, D. Establishing and maintaining gene expression patterns: insights from sensory receptor patterning. *Development* **140,** 493–503 (2013).
13. Charlebois, D. A., Abdennur, N. & Kaern, M. Gene expression noise facilitates adaptation and drug resistance independently of mutation. *Phys. Rev. Lett.* **107,** 218101 (2011).
14. Sharma, S. V. *et al.* A chromatin-mediated reversible drug-tolerant state in cancer cell subpopulations. *Cell* **141,** 69–80 (2010).
15. Brock, A., Chang, H. & Huang, S. Non-genetic heterogeneity--a mutation-independent driving force for the somatic evolution of tumours. *Nat. Rev. Genet.* **10,** 336–342 (2009).
16. Jones, D. L., Brewster, R. C. & Phillips, R. Promoter architecture dictates cell-to-cell variability in gene expression. *Science* **346,** 1533–1536 (2014).
17. Sanchez, A., Garcia, H. G., Jones, D., Phillips, R. & Kondev, J. Effect of promoter architecture on the cell-to-cell variability in gene expression. *PLoS Comput. Biol.* **7,** e1001100 (2011).
18. Carey, L. B., van Dijk, D., Sloot, P. M., Kaandorp, J. A. & Segal, E. Promoter sequence determines the relationship between expression level and noise. *PLoS Biol.* **11,** e1001528 (2013).
19. Hurst, L. D., Pal, C. & Lercher, M. J. The evolutionary dynamics of eukaryotic gene order. *Nat. Rev. Genet.* **5,** 299–310 (2004).
20. Chang, D. T., Wu, C. Y. & Fan, C. Y. A study on promoter characteristics of head-to-head genes in Saccharomyces cerevisiae. *BMC Genomics* **13,** S11 (2012).
21. Yang, L. & Yu, J. A comparative analysis of divergently-paired genes (DPGs) among Drosophila and vertebrate genomes. *BMC Evol. Biol.* **9,** 55 (2009).
22. Trinklein, N. D. *et al.* An abundance of bidirectional promoters in the human genome. *Genome Res.* **14,** 62–66 (2004).
23. Kensche, P. R., Oti, M., Dutilh, B. E. & Huynen, M. A. Conservation of divergent transcription in fungi. *Trends Genet.* **24,** 207–211 (2008).
24. Sugino, R. P. & Innan, H. Natural selection on gene order in the genome reorganization process after whole-genome duplication of yeast. *Mol. Biol. Evol.* **29,** 71–79 (2012).
25. Meeks-Wagner, D. & Hartwell, L. H. Normal stoichiometry of histone dimer sets is necessary for high fidelity of mitotic chromosome transmission. *Cell* **44,** 43–52 (1986).
26. Gunjan, A. & Verreault, A. A Rad53 kinase-dependent surveillance mechanism that regulates histone protein levels in S. cerevisiae. *Cell* **115,** 537–549 (2003).
27. Heintz, N. The regulation of histone gene expression during the cell cycle. *Biochim. Biophys. Acta* **1088,** 327–339 (1991).
28. Mateus, C. & Avery, S. V. Destabilized green fluorescent protein for monitoring dynamic changes in yeast gene expression with flow cytometry. *Yeast* **16,** 1313–1323 (2000).
29. Zhang, Q. *et al.* Stochastic expression and epigenetic memory at the yeast HO promoter. *Proc. Natl Acad. Sci. USA* **110,** 14012–14017 (2013).
30. Charvin, G., Cross, F. R. & Siggia, E. D. A microfluidic device for temporally controlled gene expression and long-term fluorescent imaging in unperturbed dividing yeast cells. *PloS ONE* **3,** e1468 (2008).
31. Bean, J. M., Siggia, E. D. & Cross, F. R. Coherence and timing of cell cycle start examined at single-cell resolution. *Mol. Cell* **21,** 3–14 (2006).
32. Trcek, T., Larson, D. R., Moldon, A., Query, C. C. & Singer, R. H. Single-molecule mRNA decay measurements reveal promoter- regulated mRNA stability in yeast. *Cell* **147,** 1484–1497 (2011).
33. Raj, A., Peskin, C. S., Tranchina, D., Vargas, D. Y. & Tyagi, S. Stochastic mRNA synthesis in mammalian cells. *PLoS Biol.* **4,** e309 (2006).
34. Yan, C., Zhang, D., Raygoza Garay, J. A., Mwangi, M. M. & Bai, L. Decoupling of divergent gene regulation by sequence-specific DNA binding factors. *Nucleic Acids Res.* **43,** 7292–7305 (2015).
35. Gasch, A. P. *et al.* Genomic expression programs in the response of yeast cells to environmental changes. *Mol. Biol. Cell* **11,** 4241–4257 (2000).
36. Elfving, N. *et al.* A dynamic interplay of nucleosome and Msn2 binding regulates kinetics of gene activation and repression following stress. *Nucleic Acids Res.* **42,** 5468–5482 (2014).
37. Lee, W. *et al.* A high-resolution atlas of nucleosome occupancy in yeast. *Nat. Genet.* **39,** 1235–1244 (2007).

38. Pramila, T., Wu, W., Miles, S., Noble, W. S. & Breeden, L. L. The Forkhead transcription factor Hcm1 regulates chromosome segregation genes and fills the S-phase gap in the transcriptional circuitry of the cell cycle. *Genes Dev.* **20,** 2266–2278 (2006).
39. Spellman, P. T. *et al.* Comprehensive identification of cell cycle-regulated genes of the yeast Saccharomyces cerevisiae by microarray hybridization. *Mol. Biol. Cell* **9,** 3273–3297 (1998).
40. Kellis, M., Birren, B. W. & Lander, E. S. Proof and evolutionary analysis of ancient genome duplication in the yeast Saccharomyces cerevisiae. *Nature* **428,** 617–624 (2004).
41. Arnone, J. T., Robbins-Pianka, A., Arace, J. R., Kass-Gergi, S. & McAlear, M. A. The adjacent positioning of co-regulated gene pairs is widely conserved across eukaryotes. *BMC Genomics* **13,** 546 (2012).
42. Eriksson, P. R., Ganguli, D. & Clark, D. J. Spt10 and Swi4 control the timing of histone H2A/H2B gene activation in budding yeast. *Mol. Cell. Biol.* **31,** 557–572 (2011).
43. Mendiratta, G., Eriksson, P. R. & Clark, D. J. Cooperative binding of the yeast Spt10p activator to the histone upstream activating sequences is mediated through an N-terminal dimerization domain. *Nucleic Acids Res.* **35,** 812–821 (2007).
44. Ingolia, N. T. & Murray, A. W. Positive-feedback loops as a flexible biological module. *Curr. Biol.* **17,** 668–677 (2007).
45. Ferrell, Jr J. E. *et al.* Simple, realistic models of complex biological processes: positive feedback and bistability in a cell fate switch and a cell cycle oscillator. *FEBS Lett.* **583,** 3999–4005 (2009).
46. Raj, A., van den Bogaard, P., Rifkin, S. A., van Oudenaarden, A. & Tyagi, S. Imaging individual mRNA molecules using multiple singly labeled probes. *Nat. Methods* **5,** 877–879 (2008).
47. Hocine, S., Raymond, P., Zenklusen, D., Chao, J. A. & Singer, R. H. Single-molecule analysis of gene expression using two-color RNA labeling in live yeast. *Nat. Methods* **10,** 119–121 (2013).
48. Sekinger, E. A., Moqtaderi, Z. & Struhl, K. Intrinsic histone-DNA interactions and low nucleosome density are important for preferential accessibility of promoter regions in yeast. *Mol. Cell* **18,** 735–748 (2005).
49. Hibbs, M. A. *et al.* Exploring the functional landscape of gene expression: directed search of large microarray compendia. *Bioinformatics* **23,** 2692–2699 (2007).

Acknowledgements

We acknowledge Dr Robert Singer and Bin Wu for scientific advice, Qian Zhang for technical support, and all the members in Bai lab for insightful comments on the manuscript. This work was supported by the Penn State start-up funding to L. Bai.

Author contributions

C.Y. and L.B. designed and performed most of the experiments; C.Y. and L.B. performed all the bioinformatics analysis; S.W. participated in the strain construction, time-lapse measurements and data analysis; C.P. participated in the software development for FISH data analysis; C.Y. and L.B. wrote the manuscript.

Additional information

Competing financial interests: The authors declare no competing financial interests.

21

Identification of a new electron-transfer relaxation pathway in photoexcited pyrrole dimers

Simon P. Neville[1,†], Oliver M. Kirkby[2], Nikolas Kaltsoyannis[2,†], Graham A. Worth[1] & Helen H. Fielding[2]

Photoinduced electron transfer is central to many biological processes and technological applications, such as the harvesting of solar energy and molecular electronics. The electron donor and acceptor units involved in electron transfer are often held in place by covalent bonds, π–π interactions or hydrogen bonds. Here, using time-resolved photoelectron spectroscopy and *ab initio* calculations, we reveal the existence of a new, low-energy, photoinduced electron-transfer mechanism in molecules held together by an NH$\cdots\pi$ bond. Specifically, we capture the electron-transfer process in a pyrrole dimer, from the excited π-system of the donor pyrrole to a Rydberg orbital localized on the N-atom of the acceptor pyrrole, mediated by an N–H stretch on the acceptor molecule. The resulting charge-transfer state is surprisingly long lived and leads to efficient electronic relaxation. We propose that this relaxation pathway plays an important role in biological and technological systems containing the pyrrole building block.

[1] School of Chemistry, University of Birmingham, Edgbaston, Birmingham B15 2TT, UK. [2] Department of Chemistry, University College London, 20 Gordon Street, London WC1H 0AJ, UK. † Present addresses: Department of Chemistry, University of Ottawa, 10 Marie Curie, Ottawa, Ontario, Canada K1N 6N5 (S.P.N.); School of Chemistry, The University of Manchester, Oxford Road, Manchester M13 9PL, UK (N.K.). Correspondence and requests for materials should be addressed to G.A.W. (email: g.a.worth@bham.ac.uk) or to H.H.F. (email: h.h.fielding@ucl.ac.uk).

Electron-transfer (ET) reactions in which an electron moves from a donor atom or molecule to an acceptor atom or molecule are ubiquitous in nature and technology[1,2]; important examples include the ET chains that drive photosystems I and II (ref. 3) and nanoscale electronic devices[4]. ET processes are often mediated by covalently bound molecular bridges that hold the donor and acceptor in place; for example, in bridged zinc porphyrin-quinone complexes, which are structurally related to the photosynthetic reaction centre, excitation of the porphyrin results in rapid ET to the quinone accompanied by large-scale vibrations of the bridges[5,6]. Photoinduced ET can also take place between non-covalently bonded donors and acceptors; for example, between π-stacked nucleobases[7]. Recently, there has been a great deal of interest in ET reactions occurring between hydrogen-bonded donors and acceptors accompanied by proton transfer. This class of ET reaction has been shown to provide an efficient mechanism for excited state deactivation following absorption of UV photons. Such proton-coupled ET reactions are important in the photostability of proteins, DNA and other essential bio-macromolecules[8] and in the photorepair cycles of UV-damaged DNA[9]. Proton-coupled ET has also been found to play an important role in the photostability of hydrogen-bonded heterocycles, such as the pyrrole dimer[10,11].

The pyrrole molecule is a common motif in biology; for example, in tryptophan and porphyrins. It is the key to the photoactivation of the phytochrome enzyme[12]. Pyrrole is also a basic building block in many biologically and technologically important systems, such as polyamide DNA-binding agents in chemical biology[13,14], dye-sensitized solar cells[15,16], the polypyrrole conducting polymer[17] and self-assembled architectures containing donor–acceptor units[18]. Moreover, it is a prototypical heteroaromatic molecule that undergoes highly efficient non-radiative relaxation on a $\pi\sigma^*$ potential surface[19–21]. Dimerization of pyrrole has been found to open up additional relaxation pathways, such as a high-energy channel involving proton-coupled ET[10,11].

In this article, we report the observation of a new, low-energy relaxation channel in pyrrole dimers in the gas-phase, involving electron transfer between the two pyrrole molecules. Using time-resolved photoelectron spectroscopy, we find that at high gas pressures, when dimers are present, a spectroscopic feature appears on a sub-picosecond timescale that is not observed at low pressures when only monomers are present. High level *ab initio* calculations show that the lowest excited state of the dimer corresponds to a charge-transfer (CT) state and that the ionization energy of this state is consistent with the spectral feature associated with the dimers. Analysis of the orbitals and molecular structures shows that ET takes place from a π-orbital on the donor pyrrole molecule to a Rydberg orbital localized on the N-atom of the acceptor pyrrole molecule. Time-resolved photoelectron spectra recorded following excitation at different wavelengths, together with calculations of potential energy surfaces, suggest that the N–H bond of the acceptor pyrrole must stretch for ET to take place; in the framework of Marcus theory, the N–H bond of the acceptor pyrrole stretches to bring the dimer to a geometry in which there is a strong vibronic coupling between neutral and CT states. As with the high-energy, proton-coupled ET process identified by Slavíček and co-workers[10,11], this new channel competes with dynamics on the $\pi\sigma^*$ state. A surprising feature of this CT complex is its lifetime. One might assume that, since the dimer is composed of identical units, the back transfer is fast and the CT complex has a negligible lifetime; however, we do not find any evidence for back transfer, which suggests that the CT complex is sufficiently long lived to potentially form a pyrrole ion pair. The elucidation of this new ET mechanism in the pyrrole dimer is significant and may provide a basis for practical applications, such as the rational design of new photoactive materials.

Results

Photoelectron spectra of pyrrole and pyrrole clusters. He carrier gas was passed through a reservoir of pyrrole held at 25 °C and the mixture was expanded through a 50 μm diameter, 1-mm-long nozzle. The distribution of pyrrole cluster ions formed by 1 + 1 ionization of the clusters was 100% monomer when using 0.4 bar He and ~89% monomer, 10% dimer and 1% trimer when using 1.8 bar He.

The spectral signatures of the pyrrole dimer were identifed by comparing photoelectron spectra obtained with different backing pressures of He gas. Figure 1 shows the 240 nm (5.17 eV) 1 + 1 multiphoton photoelectron spectra, recorded with helium carrier gas pressures of 0.4 and 1.8 bar. At 0.4 bar, the spectrum is identical to the one-photon, vacuum UV photoelectron spectrum of pyrrole[22], indicating that it is predominantly a 1 + 1 multiphoton photoelectron spectrum of the monomer. The maximum electron kinetic energy (eKE) possible following 1 + 1 photoionization of pyrrole, calculated using the central wavelength of the laser pulse and the experimental adiabatic ionization energy (8.207 eV) (refs 23,24), is 2.13 eV and is marked on Fig. 1. The photoelectron spectrum at 1.8 bar has an additional feature ~2.45 eV, which lies above the maximum eKE possible following 1 + 1 photoionization of the monomer. The intensity of this feature is independent of laser power but increases with

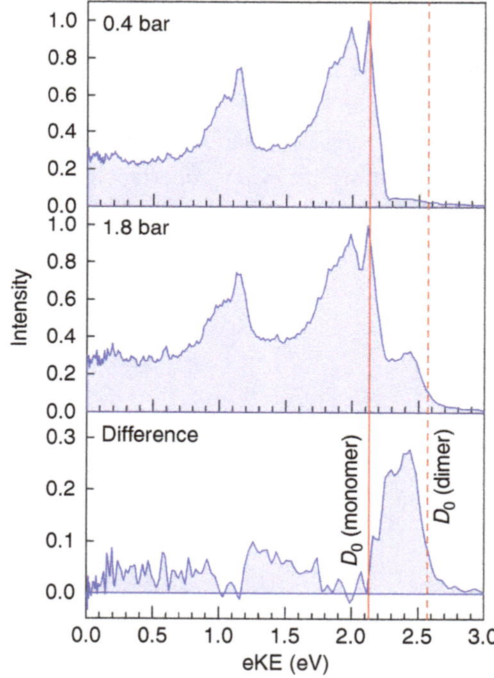

Figure 1 | Photoelectron spectra of pyrrole at different backing pressures. 240 nm (5.17 eV) 1+1 photoelectron spectra of pyrrole with He backing pressures of 0.4 bar (top) and 1.8 bar (middle) together with the difference spectrum (bottom) obtained by subtracting the 0.4 bar spectrum from the 1.8 bar spectrum. The 0.4 and 1.8 bar spectra have been normalized to the maximum intensity peak at 2.13 eV. The solid and dashed red lines mark the maximum eKEs possible from 1+1 ionization of the pyrrole monomer and dimer, respectively, calculated using the central wavelength of the laser pulses; the experimental adiabatic ionization potential of the monomer[23] and the EOM-IP-CCSD/aug-cc-pVDZ calculated vertical ionization energy of the dimer (Table 2).

increasing He backing pressure. At higher pressures, pyrrole clusters are formed and this feature can be attributed to photoionization of these clusters. Since our molecular beam at 1.8 bar is dominated by monomers and the concentration of clusters decreases exponentially with size, it seems most likely that this feature arises from photoionization of pyrrole dimers. To test this suggestion, we have calculated the vertical ionization energies (VIEs) of the monomer, dimer and trimer at the EOM-IP-CCSD/aug-cc-pVDZ level of theory. For the monomer, the lowest VIE is 8.15 eV (ref. 21), in good agreement with the experimental adiabatic ionization energy of 8.207 eV (refs 23,24). For the dimer, the lowest VIE from the T-shaped minimum energy geometry is 7.77 eV. The pyrrole trimer has three stable conformers and the minimum energy form is predicted by theory and experiment to have a cyclic triangular shape with C_{3h} symmetry[25,26]. From this conformer, the lowest VIE of the trimer is 8.10 eV, very close to that calculated for the monomer. The next most stable trimer structure is 0.16 eV (15 kJ mol^{-1}) higher in energy and will not contribute significantly to our experiments. The same trend in VIEs has been calculated at the MP2 level[27]. The optimized geometries for the dimer and trimer at the MP2 level are given in Supplementary Tables 2 and 3.

Using these calculated VIEs, the maximum eKEs for electrons generated by 240 nm 1 + 1 photoionization of the dimer and trimer are 2.57 and 2.24 eV, respectively, confirming our suggestion that the feature observed around 2.45 eV in the spectrum recorded with 1.8 bar He backing pressure is attributed to 1 + 1 photoionization of the pyrrole dimer. The 240 nm 1 + 1 photoelectron spectrum of the dimer can be obtained by subtracting the 0.4 bar spectrum (mostly monomer) from the 1.8 bar spectrum (monomer and clusters) and this is presented in the lower panel of Fig. 1. It should be noted that this is only approximate since the normalization of the two individual spectra is rather arbitrary and there will be some contributions from larger clusters. Nonetheless, it is clear that the photoelectron spectrum of the dimer is dominated by a relatively sharp and symmetric feature centred ~2.4 eV and that the broad feature that lies under the monomer photoelectron spectrum is significantly weaker than the monomer photoelectron spectrum.

Electronic structure of the pyrrole dimer. The pyrrole dimer is known to have a T-shaped structure with C_s symmetry[25,26], in which the two pyrrole molecules are held together by an NH···π bond (Fig. 2a), a motif found in protein structures[28,29]. This non-typical hydrogen bond is also responsible for the self-assembly of pyrrole dimers in apolar solvents[30]. The monomer whose N–H bond points into vacuum is labelled A and the monomer whose N–H bond is directed towards the π electron system of A is labelled B. The centres of mass of the monomers are calculated to be separated by $R = 4.06$ Å, the planes of the monomers are at angle of $\phi = 51.5°$, and the centre of mass vector has an angle of $\theta = 13.6°$ from the norm of the plane of monomer A, in agreement with the results of gas-phase experiments[31,32] and other *ab initio* calculations[11,33], including recent dispersion-corrected density functional theory (DFT) calculations[34]. The latter study also employed molecular dynamics simulations to show that the NH···π bond persists in liquid pyrrole.

The results of electronic structure calculations for the pyrrole dimer are summarized in Tables 1 and 2 and the energy level diagram in Fig. 2a. At all levels of theory, both the S_1 and S_2 states of the neutral dimer are found to be of A″ symmetry and have 3s character at the Franck–Condon point. The $S_1(\pi_B 3s_A)$ state has CT character, with the dominant configuration corresponding to excitation from a π-orbital localized on monomer B to the 3s orbital localized on the N-atom of monomer A (Fig. 2b). In contrast, the $S_2(\pi_A 3s_A)$ state is dominated by a configuration corresponding to excitation from a π-orbital localized on monomer A to the 3s orbital localized on the same monomer (Fig. 2b), reminiscent of the $A_2(\pi 3s)$ state of the pyrrole monomer[35].

The $D_1(\pi_A)$ and $D_2(\pi_A)$ states of the pyrrole dimer cation have A′ and A″ symmetries, respectively, and are dominated by configurations corresponding to removal of an electron from π-orbitals localized on monomer A. From a consideration of the dominant configurations and the symmetries of these states, the dimer cation states $D_1(\pi_A)$ and $D_2(\pi_A)$ correlate with the $D_1(\pi)$ and $D_0(\pi)$ states of the monomer cation, which have symmetries B_1 and A_2 respectively, and correspond to removing an electron from the HOMO-1 π-orbitals. The $D_0(\pi_B)$ dimer cation state is dominated by a configuration corresponding to removal of an electron from a π-orbital localized on monomer B and lies below the lowest two monomer-like states with a calculated VIE of 7.77 eV.

The photoionization cross-sections corresponding to ionization to the $D_0(\pi_B)$ and $D_2(\pi_A)$ cation states from the $S_1(\pi_B 3s_A)$ and $S_2(\pi_A 3s_A)$ dimer neutral states are plotted in Fig. 2c. The $S_1(\pi_B 3s_A)$ state is predicted to ionize preferentially to the $D_0(\pi_B)$ state in the region of the probe photon energy (4.13 eV), while the $S_2(\pi_A 3s_A)$ state is predicted to ionize preferentially to the higher lying $D_2(\pi_A)$ state. The cross-section for ionization to the $D_1(\pi_A)$ state of the cation is close to zero for ionization from either neutral excited state, similar to the negligible photoionization cross-section found for the $S_1(\pi 3s)$ state of the pyrrole monomer to the monomer cation $D_1(\pi)$ state[21].

Time-resolved photoelectron spectra. Figure 3a–d shows the time-resolved 1 + 1′ photoelectron spectra recorded using 1.8 bar He carrier gas (monomers and clusters) with pump wavelengths of 249.5 nm (4.97 eV), 245 nm (5.06 eV), 240 nm (5.17 eV) and 200 nm (6.20 eV) and a probe wavelength of 300 nm (4.13 eV). It is worth noting that the eKE axes are not directly comparable with those in Fig. 1 because different wavelengths are employed in the 1 + 1 and 1 + 1′ photoionization schemes.

For each photoexcitation wavelength, the integrated areas of the photoelectron spectra have been scaled to the total integrated photoelectron signals at the corresponding pump–probe delays and plotted as a contour map. No obvious variation in anisotropy parameter with pump–probe delay was observed. The lifetimes determined from the time-resolved photoelectron spectra are presented in Table 3. To gain insight into the flow of excited-state population, the time-resolved photoelectron spectra were fit to sums of the exponential decays listed in Table 3 and convoluted with the pump–probe cross-correlation $g(t)$,

$$S(\mathrm{eKE}, t) = \sum_i C_i(\mathrm{eKE}) e^{-t/\tau_i} \otimes g(t). \quad (1)$$

The coefficients C_i(eKE) represent the contribution of the ith decay at a given eKE. The spectra of the fit coefficients C_i(eKE) are plotted in Fig. 3e–h. Positive values of C_i(eKE) represent exponential decay on a timescale τ_i and negative values represent exponential growth on a timescale τ_i.

Only a single lifetime, $\tau_1 = 47$ fs, was required to fit the 249.5 nm data. The spectrum associated with this lifetime (Fig. 3e) has positive amplitudes everywhere, indicating a rapid decay of population out of the photoionization window. Following excitation at 245 and 240 nm, the spectra associated with the τ_1 timescale have positive amplitudes in the range 0–1.3 eV but negative amplitudes ~1.45 eV (Fig. 3f,g). This suggests an evolution along the excited potential energy surface from a region with a photoelectron spectrum with lower eKE to a

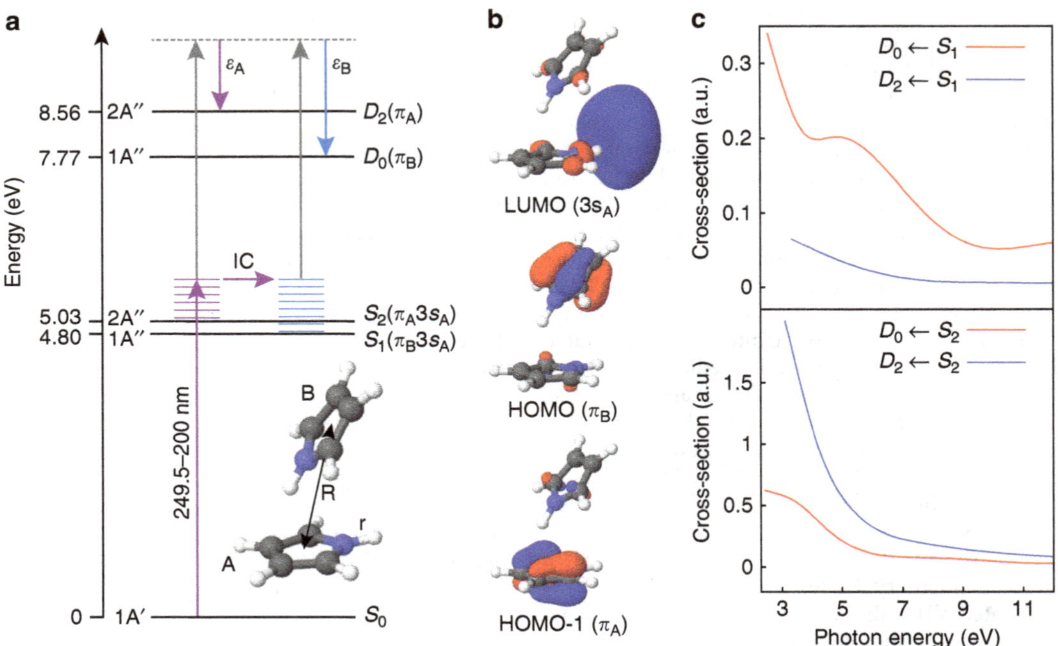

Figure 2 | Electronic structure of the pyrrole dimer. (**a**) Energy level diagram showing the ordering of the lowest electronic states of the pyrrole dimer calculated at the CASPT2(8,8) level and the cation calculated at the EOM-IP-CCSD/aug-cc-pVDZ level. Upward arrows represent pump (purple) and probe (grey) laser pulses and the horizontal lines represent vibrational energy in the $S_1(\pi_B 3s_A)$ and $S_2(\pi_A 3s_A)$ states. The horizontal arrow represents IC from the $S_2(\pi_A 3s_A)$ state to the $S_1(\pi_B 3s_A)$ state. Downward arrows represent the eKE of the emitted photoelectrons following $S_1(\pi_B 3s_A) \rightarrow D_0(\pi_B)$ and $S_2(\pi_A 3s_A) \rightarrow D_2(\pi_A)$ ionization. Inset: ground-state minimum energy geometry of the pyrrole dimer calculated at the SCS-MP2/aug-cc-pVDZ level. The monomer whose N-H bond points into vacuum is labelled A and the other monomer is labelled B. (**b**) Molecular orbitals involved in transitions from S_0 to the $S_1(\pi_B 3s_A)$ and $S_2(\pi_A 3s_A)$ states of the neutral dimer. (**c**) Photoionization cross-sections, as a function of photon energy, for ionization from the $S_1(\pi_B 3s_A)$ and $S_2(\pi_A 3s_A)$ states to the $D_0(\pi_B)$ and $D_2(\pi_A)$ cation states, calculated at the EOM-IP-CCSD/aug-cc-pVDZ level. IC, internal conversion.

Table 1 | VEEs of the pyrrole dimer.

State	Configuration	ΔE(CASSCF)	ΔE(CASPT2)	ΔE(EOM-CCSD)	ΔE(DFT/MRCI)
S_1, 1A″	$\pi_B \rightarrow 3s_A$	4.84	4.80	5.19	4.79
S_2, 2A″	$\pi_A \rightarrow 3s_A$	5.18	5.03	5.56	5.24

DFT, density functional theory; MRCI, multi-reference configuration interaction; VEE, vertical excitation energy.
VEEs and dominant configurations of the first two excited states of the pyrrole dimer calculated at the CASSCF(8,8), CASPT2(8,8), EOM-CCSD and DFT/MRCI levels using the aug-cc-pVDZ basis. VEEs are given in units of eV.

Table 2 | VIEs of the pyrrole dimer.

State	Configuration	VIE
D_0, 1A″	$\pi_B \rightarrow \psi_k$	7.77
D_1, 1A″	$\pi_A \rightarrow \psi_k$	8.52
D_2, 2A″	$\pi_A \rightarrow \psi_k$	8.56

VIE, vertical ionization energy.
VIEs and dominant configurations of the first three cationic states of the pyrrole dimer, calculated at the EOM-IP-CCSD level using the aug-cc-pVDZ basis set. VIEs are given in units of eV.

region with a photoelectron spectrum with higher eKE. At 245 and 240 nm, spectra associated with a second ultrafast timescale, $\tau_2 \approx 190$–270 fs, are centred around 1.45 eV with positive amplitude, suggesting that once populated, the region of the excited potential energy surface with a photoelectron spectrum with higher eKE decays out of the photoionization window with a slightly longer timescale τ_2. This sequential population and decay is clearly visible as a delayed rise and fall of photoelectron signal ~1.45 eV in Fig. 3b,c as well as in plots showing the integrated photoelectron signals corresponding to τ_1 and τ_2 (Fig. 3j,k). At 240 nm, a third lifetime is required to fit the data, $\tau_3 \approx 2.5$ ps. The spectrum associated with this timescale has very low amplitude, which is either the signature of larger clusters or may be attributed to a small fraction of the population remains trapped on the excited potential energy surface of the dimer within the photoionization window and decaying on a much slower timescale. At 200 nm, the situation is similar to that for 240 nm. The spectrum associated with the τ_1 timescale has positive amplitudes across the range 0–2.5 eV, but there is a dip in the amplitude ~1.45 eV that coincides with the spectrum associated with the τ_2 timescale that has positive amplitude. This is consistent with a flow of population from a region of the excited potential energy surface that has a broad photoelectron spectrum with maximum eKE ~2.5 eV to a region of the photoelectron spectrum that has a maximum eKE ~1.45 eV. A third lifetime, $\tau_3 \approx 1.1$ ps, is also required to fit the 200 nm photoelectron spectra. Again, the spectrum associated with this timescale has low amplitude; however, it is larger than the equivalent spectrum for 240 nm and has a small dip ~1.5 eV. Again, this is either a signature of larger clusters or it may be attributed to population that remains trapped on the excited potential energy surface of

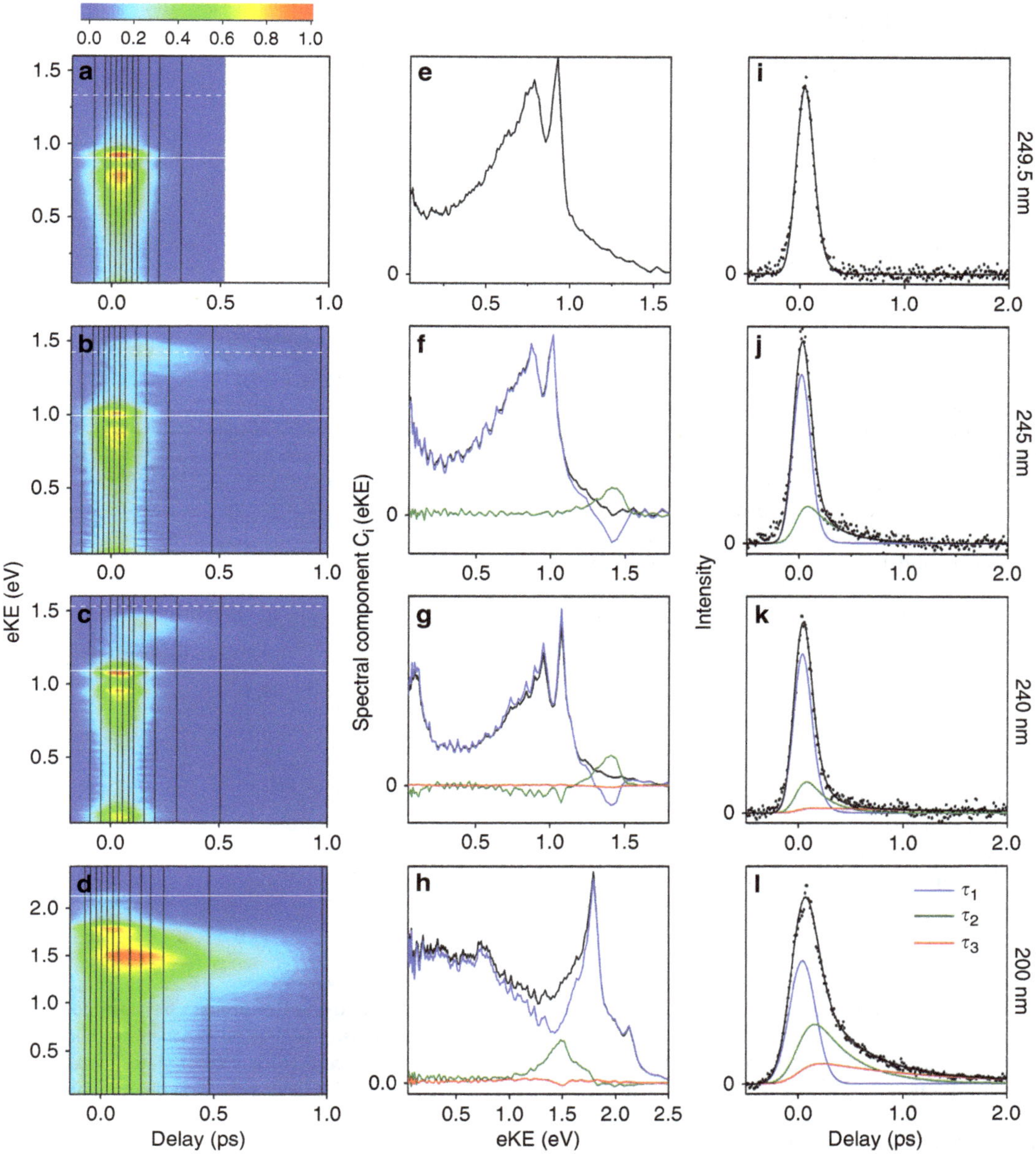

Figure 3 | Excited state dynamics of pyrrole at 1.8 bar. (a-d) Contour plots showing experimental time-resolved photoelectron spectra following excitation at 249.5 nm (4.97 eV), 245 nm (5.06 eV), 240 nm (5.17 eV) and 200 nm (6.20 eV). Individual plots were normalized to their maximum photoelectron signals. The vertical black lines mark the pump-probe delays at which photoelectron spectra were recorded and the shading was smoothed using linear interpolation. The horizontal white lines mark the maximum eKEs possible from 1 + 1′ ionization, calculated using the central wavelengths of the pump and probe laser pulses and the experimental adiabatic ionization potentials of the pyrrole monomer, 8.207 eV (solid lines)[23], and EOM-IP-CCSD/aug-cc-pVDZ calculated vertical ionization energies of the pyrrole dimer (dashed lines), 7.77 eV (this work). **(e-h)** Spectral components of $C_i(eKE)$ extracted from the time-resolved photoelectron spectra using the decay times listed in Table 3 and equation (1). The sum of $C_i(eKE)$ (black lines) represents the combined photoelectron spectrum of the initially excited states of the pyrrole monomer and dimer. **(i-l)** Integrated photoelectron spectra of the initially populated momomer and $S_2(\pi_A 3s_A)$ state of the dimer (blue) that decay on a timescale <60 fs, the $S_1(\pi_B 3s_A)$ state of the dimer (green) whose population rises on a timescale <60 fs and subsequently decays on a timescale 190–360 fs, the longer-lived excited state population (red) and the total integrated photoelectron spectrum (black), together with experimental data (scattered points).

the dimer and flows into the region of the photoelectron spectrum with maximum eKE ~1.45 eV.

Discussion

The time-resolved photoelectron spectra are dominated by the excited-state dynamics of the pyrrole monomer. Recently, Wu *et al.* used time-resolved photoelectron spectroscopy to probe the relaxation dynamics of the pyrrole monomer following excitation in the range 242–217 nm (5.1–5.7 eV) (ref. 21). They found that photoexcitation in this range resulted in ultrafast decay of the system from the ionization window on a timescale <30 fs, in agreement with other femtosecond studies[36–39]. Our time-resolved photoelectron spectra of the pyrrole monomer using 0.4 bar He carrier gas and the lifetimes (Supplementary Table 1) are in agreement with those of Wu *et al.* The main difference between the time-resolved photoelectron spectra recorded at 0.4 and 1.8 bar is the feature around 1.45 eV, which has already been assigned to the dimer and is observed to

Table 3 | Summary of pyrrole lifetimes and pump-probe cross-correlation measurements.

λ_{pump}/nm	τ_1/fs	τ_2/fs	τ_3/ps	g(t)/fs
249.5	47 ± 1			180 ± 1
245	46 ± 8	270 ± 40		179 ± 8
240	57 ± 8	190 ± 30	2.5 ± 1.5	179 ± 9
200	31 ± 1	360 ± 120	1.1 ± 0.2	289 ± 3

The 1/e lifetimes and pump-probe cross-correlation, g(t), extracted from time-resolved photoelectron spectra of pyrrole using 1.8 bar He carrier gas with different pump wavelengths. The errors quoted represent 2 s.d.

have a delayed rise on an ultrafast timescale (<60 fs) and subsequent decay on a slightly longer timescale (190–360 fs).

To a first approximation, the photoelectron spectrum of the dimer can be understood using conservation of energy. Assuming the excess vibrational energy for a given electronic state, $E_{vib} = h\nu_{pump} - E(S_n)$, is conserved during photoionization, $eKE \approx h\nu_{probe} - [E(D_n) - E(S_n)]$. Thus, from the vertical excitation energies (VEEs) of the neutral dimer calculated at the CASPT2 level (Table 1) and the VIEs (Table 2), ionization from the $S_2(\pi_A 3s_A)$ state of the dimer to the $D_2(\pi_A)$ state is predicted to produce photoelectrons with eKE ~ 0.6 eV, coinciding with the photoelectron spectrum of the monomer. The $S_1(\pi_B 3s_A)$ state is predicted to ionize preferentially to the $D_0(\pi_B)$ cation and is predicted to produce photoelectrons with eKE ~ 1.2 eV. Thus, our photoelectron spectra show that the $S_2(\pi_A 3s_A)$ state of the dimer, which has a similar photoelectron spectrum to the $S_1(\pi 3s)$ state of the monomer, is populated directly and has a lifetime that is indistinguishable from that of the $S_1(\pi 3s)$ state of the monomer. The $S_1(\pi_B 3s_A)$ CT state of the dimer is populated indirectly, from the $S_2(\pi_A 3s_A)$ state on a timescale of <60 fs and subsequently decays on a slightly longer timescale (190–360 fs).

For the dimer, the calculated VEEs (Table 1), confirm that excitation at 249.5, 245 and 240 nm will result predominantly in population of the $S_2(\pi_A 3s_A)$ state, which has an oscillator strength an order of magnitude higher than the $S_1(\pi_B 3s_A)$ state (Supplementary Table 4). Excitation at 200 nm is likely to result in transitions to higher lying optically bright $\pi \to \pi^*$ or $\pi \to 3p$ states, predominantly localized on monomer A (Supplementary Table 4). The potential energy surfaces of the S_0, $S_1(\pi_B 3s_A)$ and $S_2(\pi_A 3s_A)$ states along the N–H dissociation coordinate $\boldsymbol{r}$ of monomer A and the vector $\boldsymbol{R}$ connecting the centres of mass of the two monomers are presented in Fig. 4. In the N–H dissociation limit, the adiabatic S_1 state correlates with the diabatic $S_2(\pi_A 3s_A)$ state, while the adiabatic S_2 state correlates with the diabatic $S_1(\pi_B 3s_A)$state. This change in character is represented on the plot by a change in colour on the adiabatic curves along the N–H stretching coordinate $\boldsymbol{r}$. In a diabatic picture, the $S_1(\pi_B 3s_A)$ CT state is essentially bound with respect to N–H dissociation and the $S_2(\pi_A 3s_A)$ state has only a shallow barrier to dissociation. However, in the adiabatic picture, the N–H bond becomes bound in the $S_2(\pi_A 3s_A)$ state due to the avoided crossing with the lower $S_1(\pi_B 3s_A)$ CT state. The $S_1(\pi_B 3s_A)$ state is found to be weakly bound with respect to the monomerization coordinate $\boldsymbol{R}$, with a barrier to monomerization of around 0.1 eV, whereas the $S_2(\pi_A 3s_A)$ state potential is found to be repulsive along this $\boldsymbol{R}$ coordinate. Motions along the $\boldsymbol{r}$ and $\boldsymbol{R}$ coordinates will be the key initial motions because the N–H bond in pyrrole is photolabile and the cluster is only weakly bound; however, motions along other degrees of freedom may also be involved in the electronic relaxation pathway. For example, motion along the N out-of-plane coordinate and some ring stretches are known be involved in the relaxation dynamics of pyrrole[35] so may also play a role in the relaxation of the pyrrole dimer. Motion along a tilting coordinate that brings the monomers into co-planarity may play a role at longer timescales since it is known that in the benzene dimer the monomers are co-planar[40] and such a geometry might stabilize the CT state in the pyrrole dimer. Nonetheless, these motions will not change the major character of the initial dynamics which will be dominated by motions along the N–H stretch and monomerization coordinates.

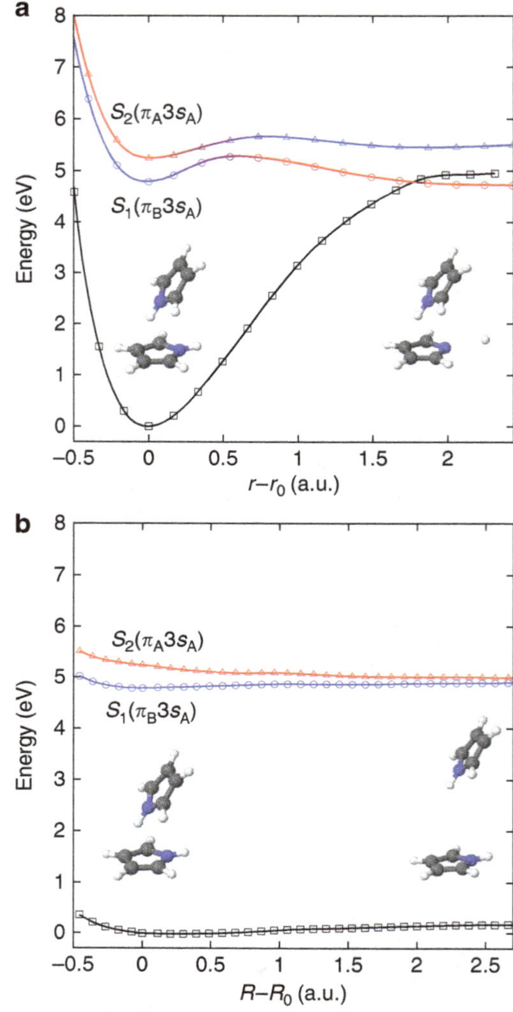

Figure 4 | Electron-transfer process between non-covalently bonded pyrrole molecules. Adiabatic potential energy curves for the S_0 (black), $S_1(\pi_B 3s_A)$ (blue) and $S_2(\pi_A 3s_A)$ (red) states calculated at the DFT/MRCI/aug-cc-pVDZ level, (**a**) along the N-H dissociation coordinate of monomer A and (**b**) along the vector $\boldsymbol{R}$ connecting the centres of masses of the two monomers.

At 249.5 nm, a single ultrafast decay out of the photoionization window of the predominantly populated $S_2(\pi_A 3s_A)$ state is observed. Since there is only one timescale, which is indistinguishable from that of the monomer, it seems that monomer separation together with ultrafast N–H dissociation of monomer A is likely to be the predominant decay channel of the dimer at this excitation energy. At this wavelength, excitation is to the bottom of the $S_2(\pi_A 3s_A)$ potential well plotted as a function of the N–H bond length; however, as soon as the dimers start to separate along the dissociative R monomerization coordinate, the barrier to N–H dissociation will approach that of the monomer and thus the dimer will display only monomer-like behaviour.

At 245 and 240 nm, ultrafast decay out of the photoionization window is still observed, but there is a delayed rise and subsequent decay of the higher-energy dimer band around 1.45 eV (Fig. 3b,c), which is consistent with a competing process populating the $S_1(\pi_B 3s_A)$ state. At 200 nm, a similar ultrafast population and subsequent decay of the $S_1(\pi_B 3s_A)$ state is observed, also $\sim$1.45 eV because of the propensity for conservation of excess vibrational energy during photoionization. These observations suggest that a crossing point between the $S_1(\pi_B 3s_A)$ and $S_2(\pi_A 3s_A)$ states becomes accessible between 249.5–245 nm (4.97–5.06 eV) and crossing occurs to form the CT complex on a timescale that is faster than that of monomer separation in the $S_2(\pi_A 3s_A)$ state. This population is likely to occur at the avoided crossing between the $S_1(\pi_B 3s_A)$ and $S_2(\pi_A 3s_A)$ states along the N–H bond stretch (Fig. 4). However, as discussed above, motions along other coordinates are also expected to play a role in the dynamics, possibly turning this avoided crossing into a conical intersection (although as the states have the same symmetry this would be an accidental intersection) and moving towards co-planarity. Nonetheless, these motions will not change the basic picture of initial N–H stretching followed by crossing to form the CT state that must occur.

It can also be inferred that the decay time $\tau_2 \sim 190$–360 fs (Table 3) is the lifetime of the dimer CT complex. However, how the CT complex relaxes cannot be ascertained from our experiments and calculations and various pathways are possible. Significantly, there is no experimental evidence for re-crossing from the CT complex back to the neutral dimer excited state. This is surprising given the equivalent electronegativities of the monomers. Another possibility might be N–H dissociation in the $S_1(\pi_B 3s_A)$ state, but it should be noted that the barrier to H-atom loss in the $S_1(\pi_B 3s_A)$ state (0.5 eV, calculated at the DFT/multi-reference configuration interaction (MRCI) level), is significantly larger than that of the monomer (0.25 eV (ref. 35)). The most plausible pathways are: direct crossing to the ground-state, as suggested by Poterya *et al.*[10]; slow separation to form a monomer pair as the barrier to monomerization calculated to exist in the $S_1(\pi_B 3s_A)$ state at the DFT/MRCI level is very small (Fig. 4); or formation of a co-planar excimer ion pair as seen in benzene[40] if the tilt motion becomes important during separation.

In conclusion, the dynamics of the excited states of the pyrrole dimer have been studied using a combination of time-resolved photoelectron spectroscopy and *ab initio* calculations. It has been revealed that the formation of pyrrole dimers opens up a new relaxation pathway involving ET from the excited π-system of one pyrrole molecule to a Rydberg orbital localized on the N-atom of the other pyrrole molecule and that the ET is mediated by stretching of the N–H bond on the acceptor molecule leading to efficient curve crossing. Furthermore, our results reveal that the resulting CT complex has a lifetime of a few 100 fs and that it may form a pyrrole ion pair. We propose that this relaxation pathway will play a role in the photochemistry and photophysics of biologically and technologically important systems containing the pyrrole building block or indeed other NH-π bonded molecular units.

Methods

Cluster source. Helium carrier gas (0.4–1.8 bar) was passed through a reservoir of liquid pyrrole held at room temperature (25 °C) outside the vacuum chamber and carried into the vacuum chamber through 1/16″ tubing to a 1-mm-long nozzle of 50 µm diameter. The vapour pressure of pyrrole at 25 °C is 8.35 mm Hg and the concentration of pyrrole in 1.8 bar He is estimated to be 0.6%. The molecular beam passed through a 1 mm skimmer, located $\sim$40 mm from the nozzle, into the interaction region of a photoelectron velocity map imaging spectrometer[41–43]. To confirm the formation of clusters, we recorded ion images following two-photon ionization at 249.5 nm. The time-of-flight from the interaction region to the detector depends on the mass of the ion, $t_{\mathrm{ToF}} \propto \sqrt{m}$, and we observe images of dimer ions displaced by $x_{\mathrm{py}}\sqrt{2}$ and trimer ions displaced by $x_{\mathrm{py}}\sqrt{3}$, where x_{py} is the displacement along the molecular beam axis of the pyrrole monomer ions, which is perpendicular to the time-of-flight axis. With a carrier gas pressure of 0.4 bar, 100% of the ion signal corresponded to $m/z \equiv C_4H_5N^+$. At 1.8 bar, the cluster ion distribution at the detector is exponential with around 89% monomer, 10% dimer and 1% trimer (Supplementary Fig. 1 and Supplementary Note 1) and the average cluster ion size at the detector is $\bar{N} \approx 1$. We note that these measurements are lower limits for the fractions of clusters in the interaction region because they do not account for fragmentation of cluster ions on the way to the detector. Nonetheless, our observations are consistent with those of Poterya *et al.*[10] and Profant *et al.*[27] who measured average cluster sizes $\bar{N} = 3$ following supersonic expansion of pyrrole through a 2-mm-long conical nozzle of 60 µm diameter, with a He backing pressure of 1.5 bar and pyrrole reservoir and nozzle temperatures 8 and 9 °C, respectively, substantially lower than those in our experiments.

Time-resolved photoelectron spectra. The time-resolved photoelectron spectra were recorded with helium gas pressures of 1.8 bar. After collimation by a 1 mm skimmer, the molecular beam was intersected by femtosecond pump (249.5–200 nm) and probe (300 nm) laser pulses. Pump–probe cross-correlation full-width half-maximum measurements were in the range 180–290 fs. The pump and probe pulses were focussed to a spot size of diameter $\sim$50 µm after being attenuated to <1 µJ per pulse ($<10^{11}$ W cm^2) to minimize multiphoton processes and space-charge effects. For each excitation wavelength, a set of $\sim$15 photoelectron images with pump–probe delays in the range -0.25 to 1 ps were recorded, together with the total integrated photoelectron signal. Photoelectron velocity distributions were recovered from the raw photoelectron images using the pBasex image inversion algorithm[44] and the energy scale was calibrated by recording the 2 + 1 resonance-enhanced multiphoton ionization spectrum of Xe at 249.6 nm (ref. 45). The resolution of this instrument is $\sim$3.5%. To extract decay times from a set of time-resolved photoelectron spectra, the total integrated areas of the photoelectron spectra recorded at each pump–probe delay were scaled to the total integrated photoelectron signal intensity. A least-squares fitting procedure using the Levenberg–Marquardt optimization algorithm was used to fit integrated portions of the set of scaled spectra to sums of exponentially decaying profiles convoluted with a Gaussian cross-correlation function, representing the cross-correlation of the pump and probe laser pulses $g(t)$ (refs 43,46),

$$S(t) = \sum_i c_i e^{-t/\tau_i} \otimes g(t). \quad (2)$$

c_i is the intensity of the ith decay with $1/e$ decay time τ_i. The residuals from the fits are shown in Supplementary Fig. 2.

Computational methodology. Ground-state minimum energy geometries of the pyrrole monomer, dimer and trimer were optimized at the spin-component-scaled MP2 (SCS-MP2)[47] level using the aug-cc-pVDZ basis. Ionization potentials for the lowest cationic states were calculated at the EOM-IP-CCSD/aug-cc-pVDZ level. For the dimer, vertical excitation energies of the low-lying excited states at the Franck–Condon point geometry were calculated using the EOM-CCSD, CASSCF, CASPT2 and DFT/MRCI methods and the aug-cc-pVDZ basis set. The DFT/MRCI calculations[48] were performed using the BH-LYP functional. The CASSCF and CASPT2 calculations were carried out using eight electrons in eight orbitals and the aug-cc-pVDZ basis set, that is, CAS(8,8)/aug-cc-pVDZ and CASPT2(8,8)/aug-cc-pVDZ. The active space was the four occupied orbitals $21a'(\pi_A)$, $22a'(\pi_A)$, $13a''(\pi_B)$ and $14a''(\pi_A)$, and four virtual orbitals $23a'(3s_A)$, $24a'(3p_{Z,A})$, $25a'(\pi_A^*)$ and $15a''(\pi_A^*)$. The subscripts A and B denote the monomer on which a given orbital is localized. The CAS calculations were performed using MOLPRO[49], the EOM-CCSD calculations using QChem[50] and the MP2 and DFT/MRCI calculations using Turbomole[51].

Photoionization cross-sections, $P_I^{\alpha k}(E)$, corresponding to ionization from the low-lying excited neutral states were calculated under the assumption of vertical excitation and the applicability of Fermi's golden rule,

$$\begin{aligned} P_I^{\alpha k}(E) &\propto \left|\left\langle \Phi_I^{(N)} |\hat{\mu}| \Phi_\alpha^{(N-1)} \psi_k \right\rangle\right|^2 \delta(E - E_I^\alpha - E_k) \\ &\approx \left|\left\langle \phi_I^\alpha |\hat{\mu}| \psi_k \right\rangle\right|^2 \delta(E - E_I^\alpha - E_k) \end{aligned} \quad (3)$$

Here, the $\Phi_I^{(N)}$ and Φ_α^{N-1} correspond to the N-electron neutral and (N-1)-electron cationic states, respectively, while ψ_k denotes the photoelectron orbital (taken as a Coulomb partial wave with kinetic energy $E_k = k^2/2$). The energy of the applied laser pulse is denoted by E, whilst E_I^α denotes the vertical energy difference between the netural and cationic states, and $\hat{\mu}$ is the molecular dipole operator. The one-electron quantity ϕ_I^α is a so-called Dyson orbital, obtained from the overlap of the neutral and cationic states[52].

Dyson orbitals corresponding to ionization from both the S_1 and S_2 states to each of the D_0, D_1 and D_2 states were calculated at the EOM-CCSD/EOM-IP-CCSD level using the aug-cc-pVDZ basis set. Evaluation of the photoionization matrix elements $\langle \phi_I^\alpha |\hat{\mu}| \psi_k \rangle$ was performed numerically on a grid and with isotropic averaging over molecular orientations using the ezDyson program[53].

References

1. Delor, M., Sazanovich, I. V., Towrie, M. & Weinstein, J. A. Probing and exploiting the interplay between nuclear and electronic motion in charge transfer processes. *Acc. Chem. Res.* **48,** 1131–1139 (2015).
2. Zhang, W. & Gaffney, K. J. Mechanistic studies of photoinduced spin crossover and electron transfer in inorganic complexes. *Acc. Chem. Res.* **48,** 1140–1148 (2015).
3. Romero, E. *et al.* Quantum coherence in photosynthesis for efficient solar-energy conversion. *Nat. Phys.* **10,** 676–682 (2014).
4. Sukegawa, J. *et al.* Electron transfer through rigid organic molecular wires enhanced by electronic and electronvibration coupling. *Nat. Chem.* **6,** 899–905 (2014).
5. Dreuw, A., Worth, G. A., Cederbaum, L. S. & Head-Gordon, M. Ultrafast photoinitiated long-range electron transfer in cyclophane-bridged zincporphyrin—quinone complexes via conical intersections. *J. Phys. Chem. B* **108,** 19049–19055 (2004).
6. Delor, M. *et al.* Toward control of electron transfer in donor-acceptor molecules by bond-specific infrared excitation. *Science* **346,** 1492–1495 (2014).
7. Zhang, Y. *et al.* Photoinduced electron transfer in DNA: charge shift dynamics between 8-oxo-guanine anion and adenine. *J. Phys. Chem. B* **119,** 7491–7502 (2015).
8. Domcke, W. & Sobolewski, A. L. Spectroscopy meets theory. *Nat. Chem.* **5,** 257 (2013).
9. Zhong, D. Electron transfer mechanisms of DNA repair by photolyase. *Annu. Rev. Phys. Chem.* **66,** 691–715 (2015).
10. Poterya, V., Profant, V., Fárnk, M., Slavíček, P. & Buck, U. Experimental and theoretical study of the pyrrole cluster photochemistry: closing the $\pi\sigma^*$ dissociation pathway by complexation. *J. Chem. Phys.* **127,** 064307 (2007).
11. Slavíček, P. & Fárnk, M. Photochemistry of hydrogen bonded heterocycles probed by photodissociation experiments and *ab initio* methods. *Phys. Chem. Chem. Phys.* **13,** 12123–12137 (2011).
12. Ulijasz, A. T. *et al.* Structural basis for the photoconversion of a phytochrome to the activated Pfr form. *Nature* **463,** 250–254 (2010).
13. White, S., Szewczyk, J. W., Turner, J. M., Baird, E. E. & Dervan, P. B. Recognition of the four Watson-Crick base pairs in the DNA minor groove by synthetic ligands. *Nature* **391,** 468–471 (1998).
14. Pazos, E., Mosquera, J., Vázquez, M. E. & Mascareñas, J. L. DNA recognition by synthetic constructs. *ChemBioChem.* **12,** 1958–1973 (2011).
15. Rasmussen, S. C. & Evenson, S. J. Dithieno[3,2-b:2,3-d]pyrrole-based materials: synthesis and application to organic electronics. *Prog. Polym. Sci.* **38,** 1773–1804 (2013).
16. Wang, Z. *et al.* Organic dyes containing dithieno[2,3-d:2,3-d]thieno[3,2-b:3, 2-b]dipyrrole core for efficient dye-sensitized solar cells. *J. Mater. Chem. A* **3,** 4865–4874 (2015).
17. MacDiarmid, a. G. Synthetic metals: a novel role for organic polymers (Nobel Lecture). *Angew Chem. Int. Ed. Engl.* **40,** 2581–2590 (2001).
18. Souto, M. *et al.* Self-assembled architectures with segregated donor and acceptor units of a dyad based on a monopyrrolo-annulated TTF-PTM radical. *Chemistry* **21,** 8816–8825 (2015).
19. Sobolewski, A. L., Domcke, W., Dedonder-Lardeux, C. & Jouvet, C. Excited-state hydrogen detachment and hydrogen transfer driven by repulsive $\pi\sigma^*$ states: a new paradigm for nonradiative decay in aromatic biomolecules. *Phys. Chem. Chem. Phys.* **4,** 1093–1110 (2002).
20. Ashfold, M. N. R., Cronin, B., Devine, A. L., Dixon, R. N. & Nix, M. G. D. The role of $\pi\sigma^*$ excited states in the photodissociation of heteroaromatic molecules. *Science* **312,** 1637–1640 (2006).
21. Wu, G. *et al.* Excited state non-adiabatic dynamics of pyrrole: a time-resolved photoelectron spectroscopy and quantum dynamics study. *J. Chem. Phys.* **142,** 074302 (2015).
22. Klasinc, L., Sabljić, A., Kluge, G., Rieger, J. & Scholz, M. Chemistry of excited States. Part 13. Assignment of lowest π-ionizations in photoelectron spectra of thiophen, furan and pyrrole. *J. Chem. Soc. Perkin Trans. 2* **2,** 539–542 (1982).
23. Derrick, P., Åsbrink, L., Edqvist, O. & Lindholm, E. Photoelectron-spectroscopical study of the vibrations of furan, thiophene, pyrrole and cyclopentadiene. *Spectrochim. Acta* **27,** 2525–2537 (1971).
24. van den Brom, A. J. *et al.* Photodissociation and photoionization of pyrrole following the multiphoton excitation at 243 and 364.7 nm. *Phys. Chem. Chem. Phys.* **7,** 892–899 (2005).
25. Gómez-Zavaglia, A. & Fausto, R. Self-aggregation in pyrrole: matrix isolation, solid state infrared spectroscopy, and DFT study. *J. Phys. Chem. A* **108,** 6953–6967 (2004).
26. Dauster, I., Rice, C. A., Zielke, P. & Suhm, M. A. N-H $\cdots\pi$ interactions in pyrroles: systematic trends from vibrational spectroscopy in clusters. *Phys. Chem. Chem. Phys.* **10,** 2827–2835 (2008).
27. Profant, V., Poterya, V., Fárnk, M., Slavíček, P. & Buck, U. Fragmentation dynamics of size-selected pyrrole clusters prepared by electron impact ionization: forming a solvated dimer ion core. *J. Phys. Chem. A* **111,** 12477–12486 (2007).
28. Worth, G. & Wade, R. The aromatic—(i + 2) amine interaction in peptides. *J. Phys. Chem.* **99,** 17473–17482 (1995).
29. Chakrabarti, P. & Bhattacharyya, R. Geometry of nonbonded interactions involving planar groups in proteins. *Prog. Biophys. Mol. Biol.* **95,** 83137 (2007).
30. Stefov, V., Pejov, L. & Soptrajanov, B. Experimental and quantum chemical study of pyrrole self-association through N H p hydrogen bonding. *J. Mol. Struct.* **649,** 231–243 (2003).
31. Columberg, G. & Bauder, A. Pure rotational spectrum, quadrupole coupling constants and structure of the dimer of pyrrole. *J. Chem. Phys.* **106,** 504–510 (1997).
32. Lukeš, V., Breza, M. & Biskupic, S. Regular article interaction energy anisotropy of the pyrrole dimer : *ab initio* theoretical study. *Theor. Chem. Acc.* **101,** 319–324 (1999).
33. Kabeláč, M., Hobza, P. & Špirko, V. The structure and vibrational dynamics of the pyrrole dimer. *Phys. Chem. Chem. Phys.* **11,** 3885–3891 (2009).
34. Gao, W., Feng, H., Xuan, X. & Chen, L. A theoretical study of N—H $\cdots\pi$ h-bond interaction of pyrrole: from clusters to the liquid. *Mol. Phys.* **110,** 2151–2161 (2012).
35. Neville, S. P. & Worth, G. A. A reinterpretation of the electronic spectrum of pyrrole: a quantum dynamics study. *J. Chem. Phys.* **140,** 034317 (2014).
36. Roberts, G. M. *et al.* Probing ultrafast dynamics in photoexcited pyrrole: timescales for 1 ps* mediated H-atom elimination. *Faraday Discuss* **163,** 95–116 (2013).
37. Montero, R. *et al.* Femtosecond evolution of the pyrrole molecule excited in the near part of its UV spectrum femtosecond evolution of the pyrrole molecule excited in the near part of its UV spectrum. *J. Chem. Phys.* **2012,** 064317 (2012).
38. Montero, R., Ovejas, V., Fernández-Fernández, M., Peralta Conde, A. & Longarte, A. Revisiting the relaxation dynamics of isolated pyrrole. *J. Chem. Phys.* **141,** 014303 (2014).
39. Lippert, H., Ritze, H.-H., Hertel, I. V. & Radloff, W. Femtosecond time-resolved hydrogen-atom elimination from photoexcited pyrrole molecules. *ChemPhysChem.* **5,** 1423–1427 (2004).
40. Diri, K. & Krylov, A. I. Electronic states of the benzene dimer: a simple case of complexity. *J. Phys. Chem. A* **116,** 653–662 (2012).
41. Minns, R. S., Parker, D. S. N., Penfold, T. J., Worth, G. A. & Fielding, H. H. Competing ultrafast intersystem crossing and internal conversion in the channel 3 region of benzene. *Phys. Chem. Chem. Phys.* **12,** 15607–15615 (2010).
42. Spesyvtsev, R., Kirkby, O. M., Vacher, M. & Fielding, H. H. Shedding new light on the role of the Rydberg state in the photochemistry of aniline. *Phys. Chem. Chem. Phys.* **14,** 9942–9947 (2012).
43. Kirkby, O. M. *et al.* Comparing the electronic relaxation dynamics of aniline and d_7-aniline following excitation at 272–238 nm. *Phys. Chem. Chem. Phys.* **17,** 16270–16276 (2015).
44. Garcia, G. A., Nahon, L. & Powis, I. Two-dimensional charged particle image inversion using a polar basis function expansion. *Rev. Sci. Instrum.* **75,** 4989–4996 (2004).
45. Hansen, J. E. & Persson, W. Revised analysis of singly ionized xenon, Xe II. *Phys. Scripta* **36,** 602–643 (1987).
46. Spesyvtsev, R., Kirkby, O. M. & Fielding, H. H. Ultrafast dynamics of aniline following 269–238 nm excitation and the role of the $S_2(\pi 3s/\pi\sigma^*)$ state. *Faraday Discuss* **157,** 165–179 (2012).
47. Grimme, S. Improved second-order MøllerPlesset perturbation theory by separate scaling of parallel- and antiparallel-spin pair correlation energies. *J. Chem. Phys.* **118,** 9095–9102 (2003).
48. Grimme, S. & Waletzke, M. A combination of Kohn Sham density functional theory and multi-reference configuration interaction methods. *J. Chem. Phys.* **111,** 5645–5655 (1999).
49. Werner, H.-J. *et al.* MOLPRO, v.2012.1. https://www.molpro.net (2012).
50. Shao, Y. *et al.* Advances in molecular quantum chemistry contained in the Q-Chem 4 program package. *Mol. Phys.* **113,** 184–215 (2015).
51. TURBOMOLE v.6.2. TURBOMOLE GmbH. http://www.turbomole.com (2010).
52. Oana, C. M. & Krylov, A. I. Dyson orbitals for ionization from the ground and electronically excited states within equation-of-motion coupled-cluster formalism: theory, implementation, and examples. *J. Chem. Phys.* **127,** 234106 (2007).
53. Gozem, S. & Krylov, A. I. ezDyson v.3. http://iopenshell.usc.edu/downloads/ezdyson (2005).

Acknowledgements

We are grateful to Michael Parkes for his help in the preparation of Fig. 3. This work was enabled by support from EPSRC grants EP/G055270/1 and GR/T20304/01.

Author contributions

H.H.F. conceived and designed the experiments; O.M.K. performed the experiments; N.K. performed preliminary calculations; G.A.W. and S.P.N. designed the calculations presented in the paper; and S.P.N. performed the calculations. All authors discussed the results of the experiments and calculations. H.H.F., G.A.W. and S.P.N. wrote the manuscript.

Additional information

Competing financial interests: The authors declare no competing financial interests.

22

CETSA screening identifies known and novel thymidylate synthase inhibitors and slow intracellular activation of 5-fluorouracil

Helena Almqvist[1,*], Hanna Axelsson[1,*], Rozbeh Jafari[2,†], Chen Dan[3], André Mateus[4], Martin Haraldsson[1], Andreas Larsson[5], Daniel Martinez Molina[2], Per Artursson[4,6,7], Thomas Lundbäck[1] & Pär Nordlund[2,3,8]

Target engagement is a critical factor for therapeutic efficacy. Assessment of compound binding to native target proteins in live cells is therefore highly desirable in all stages of drug discovery. We report here the first compound library screen based on biophysical measurements of intracellular target binding, exemplified by human thymidylate synthase (TS). The screen selected accurately for all the tested known drugs acting on TS. We also identified TS inhibitors with novel chemistry and marketed drugs that were not previously known to target TS, including the DNA methyltransferase inhibitor decitabine. By following the cellular uptake and enzymatic conversion of known drugs we correlated the appearance of active metabolites over time with intracellular target engagement. These data distinguished a much slower activation of 5-fluorouracil when compared with nucleoside-based drugs. The approach establishes efficient means to associate drug uptake and activation with target binding during drug discovery.

[1] Laboratories for Chemical Biology, Karolinska Institutet, Science for Life Laboratory Stockholm, Division of Translational Medicine & Chemical Biology, Department of Medical Biochemistry & Biophysics, Karolinska Institutet, Tomtebodavägen 23A, Solna 171 65, Sweden. [2] Department of Medical Biochemistry & Biophysics, Division of Biophysics, Karolinska Institutet, Scheeles väg 2, Stockholm 171 77, Sweden. [3] School of Biological Sciences, Nanyang Technological University, 61 Biopolis Drive (Proteos), Singapore 138673, Singapore. [4] Department of Pharmacy, Uppsala University, BMC, Box 580, Uppsala SE-751 23, Sweden. [5] School of Biological Sciences, Nanyang Technological University, SBS-04s-45, 60 Nanyang Drive, Singapore 639798, Singapore. [6] Uppsala University Drug Optimization and Pharmaceutical Profiling Platform (UDOPP), Department of Pharmacy, Uppsala University, BMC, Box 580, Uppsala SE-751 23, Sweden. [7] Science for Life Laboratory Drug Discovery and Development platform, Uppsala University, Uppsala SE-751 23, Sweden. [8] Institute of Cellular and Molecular Biology, ASTAR, 61 Biopolis Drive (Proteos), Singapore 138673, Singapore. * These authors contributed equally to this work. † Present address: Clinical Proteomics Mass Spectrometry, Department of Oncology-Pathology, Science for Life Laboratory and Karolinska Institutet, Stockholm, Sweden. Correspondence and requests for materials should be addressed to T.L. (email: thomas.lundback@ki.se) or to P.N. (email: par.nordlund@ki.se).

Therapeutic efficacy is achieved when drugs bind their relevant molecular targets in the physiologically relevant setting. Despite this known fact, insufficient control of target engagement is surprisingly common and contributes to high failure rates in clinical trials[1–3]. Methods that allow for robust measurements of drug target engagement in primary cells, tissues and patient biopsies are thus urgently needed, but have been hard to establish[4,5].

Ligand-induced changes in protein thermal stability are frequently used to monitor binding to isolated proteins in thermal shift assays[6–9]. The recently developed cellular thermal shift assay (CETSA; see Supplementary Note 1 for a list of abbreviations) builds on the discovery that ligand induced thermal shifts can also be measured in the context of cell lysates, whole cells or tissues[10]. This finding effectively allows for biophysical binding studies in native environments—preserving expression levels, posttranslational modifications and the local environment for the endogenous protein. Whereas the original CETSA study included multiple case studies, recent work extends this method to include melting transitions for a significant portion of the proteome, thus expanding the putative use of the methodology to a large number of protein families[11–13]. Of practical importance is that the melting transitions are established for individual proteins by the use of protein affinity reagents[10,14] or quantitative mass spectrometry (MS)[11–13]. As a consequence these measurements are amenable to either high-throughput measurements or proteome-wide multiplexing.

To improve current strategies for drug development, stringent control of target engagement should ideally be established from initial hit identification, through preclinical and clinical development. The same demands apply to the validation of chemical probes discovered in academic settings[2,4,15]. To probe the value of CETSA in earlier stages of the discovery process we applied it for primary screening of thymidylate synthase (TS) in live human myelogenous leukemia cells. TS is a pivotal enzyme in production of thymidine monophosphate and a well validated cancer target[16,17]. Inhibition of TS leads to thymineless death characterized by DNA-damage, chromosomal fragmentation and concomitant induction of apoptosis. Novel classes of TS inhibitors with improved efficacy and resistance profiles could provide important complements to current TS directed drugs, for which there are reports of resistance[18,19].

Here, we show for the first time that a CETSA-based screen for direct physical target engagement constitutes an attractive high throughput screening (HTS) strategy, which allows for the detection of known and novel TS inhibitors with cellular activity. Furthermore, we establish a hit validation strategy, in which time-dependent target engagement is explored in parallel with measurement of intracellular compound concentration. Taken together this provides a sound and efficient strategy to establish control of target engagement from an early stage of the drug discovery process, and which is likely to minimize problems in subsequent stages.

Results

Microplate-based CETSA measuring target engagement of TS. CETSA is based on measurements of remaining soluble target protein against a background of thermally denatured and precipitated proteins following a heat challenge[10,14]. To enable large-scale screening and automation we developed a no-wash immunoassay for TS using AlphaScreen technology in 384-well plates (see Supplementary Figs 1–6 and Supplementary Table 1). As outlined in Fig. 1a the assay workflow starts with a pre-incubation of K562 cells with library compounds or controls to allow cellular uptake, potential compound metabolism and binding to TS. The treated samples in the plates are next transiently heated in a PCR machine, resulting in denaturation and precipitation of intracellular TS unless stabilized by ligand. After cooling to room temperature the cells are lyzed and the remaining (stabilized) levels of TS are measured.

We validated the assay by investigating the response to two drugs of structurally different classes, that is, floxuridine and raltitrexed. Both drugs require intracellular enzymatic conversion prior to high-affinity TS binding[17]. Pyrimidine-based inhibitors, such as floxuridine, bind to TS as the corresponding monophosphate, whereas folate-based drugs, such as raltitrexed, are polyglutamylated and bind TS in a ternary complex with 2′-deoxyuridine 5′-monophosphate (dUMP). Two assay formats were employed for validation. First, the heating was done at a series of different temperatures at a fixed compound concentration to establish aggregation temperature (T_{agg}) curves (Fig. 1b). As expected both drugs resulted in substantial shifts of the thermal stability of TS, thus confirming cellular uptake and intracellular enzymatic conversion to the active forms that bind TS. Based on these curves, 50 °C was selected for further characterization in isothermal dose-response fingerprint ($ITDRF_{CETSA}$) experiments. In these experiments the compound concentration is titrated during the pre-incubation, after which all samples are heated to the same temperature (Fig. 1c,d). Both drugs showed dose-dependent stabilization of TS with half maximal effective concentration (EC_{50}) values in the sub-nM range. Data from parallel experiments using quantitative western blots for assessment of stabilized TS confirmed a significant shift in T_{agg} in the presence of 5 μM of either of these drugs as well as potent dose-dependent stabilization (Supplementary Fig. 7). No change in total TS levels was observable at 5 μM concentration following a 2 h pre-incubation time in the K562 cells, demonstrating that the thermal stabilization data were not influenced by drug-induced changes in total protein levels under these conditions.

Small molecule library screening and hit confirmation. We screened a library of 10,928 compounds at Chemical Biology Consortium Sweden (CBCS; www.cbcs.se) using the TS assay described above. The library includes a structurally-diverse selection of lead-like compounds[20], nucleosides and known drugs (see Methods section for details on the library). These latter subsets include folate and nucleoside-based drugs known to act on TS, suppress thymidine incorporation into DNA and reduce cell proliferation[16,21]. A schematic outline of the screen logistics is available in Supplementary Fig. 8. Screening was performed at a compound concentration of 50 μM and resulted in a reproducible response to the controls and the appearance of several stabilizing compounds (Fig. 2a). Additional graphs illustrating the screen performance and statistics are available in Supplementary Fig. 9 and Supplementary Table 2, respectively. The campaign involved one day of screening, with the AlphaScreen readings done the following morning to ensure equilibration of the antibody recognition.

The threshold for active solutions was calculated at 11.7% stabilization and resulted in 65 hits (Supplementary Table 3). Solutions for 63 of these were available for cherry-picking from our vial-based compound stores, that is, a different source intended for long-term storage. The activities of these solutions were examined in $ITDRF_{CETSA}$ experiments to confirm the screen results (Supplementary Table 3). The majority of compounds with an apparent stabilization above 30% in the primary screen confirmed activity. We also found that 12 out of the 15 vial-based solutions that failed to reproduce activity (highlighted at the bottom of Supplementary Table 3) had been contaminated with

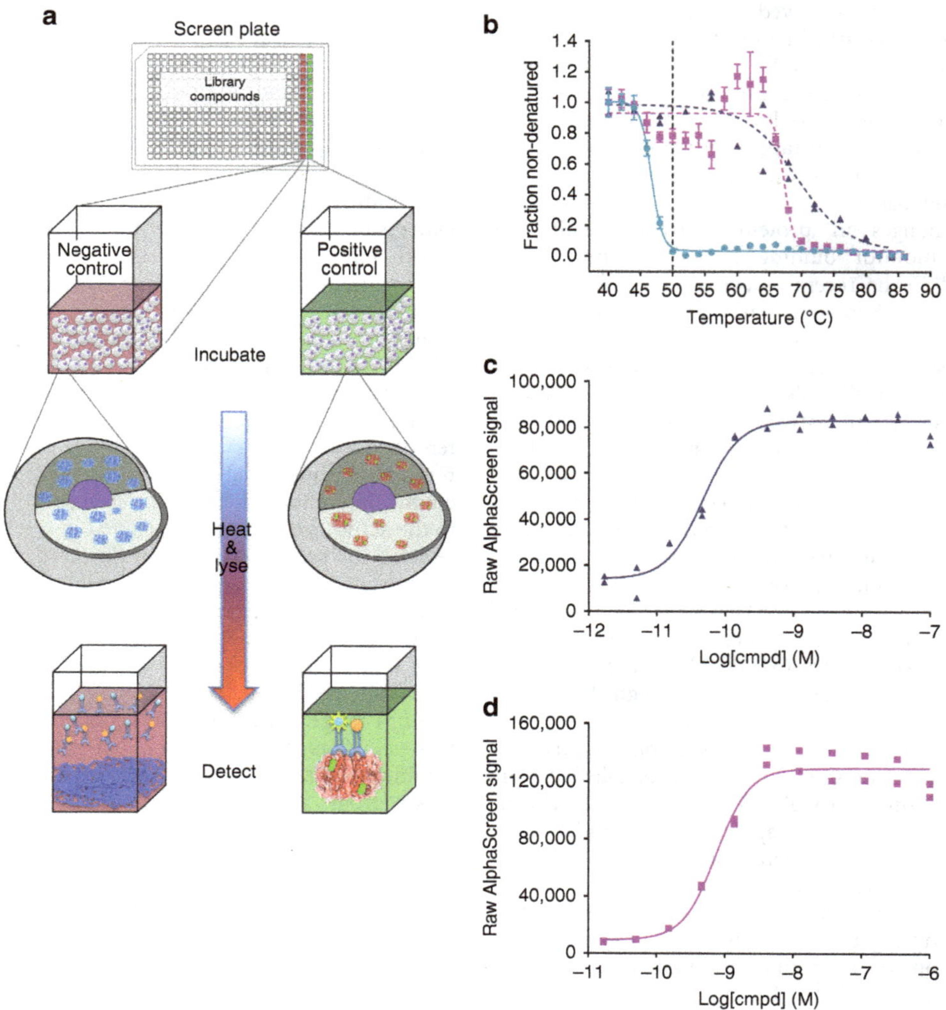

Figure 1 | Development of a no-wash CETSA for human TS. (**a**) Overview of the assay principle with live K562 cells seeded into a 384-well PCR plate. The plate contains controls or library compounds that are taken up by the cells. Following a pre-incubation period the plate is transiently heated for 3 min followed by cooling and cell lysis. Part of the cell lysate is transferred to a detection plate, to which antibodies and AlphaScreen beads are added to allow measurements of remaining soluble TS. (**b**) CETSA derived T_{agg} curves for TS in K562 cells in the presence of DMSO (0.5%) (green circle), 15 μM floxuridine (blue triangle) or 1 μM raltitrexed (magenta square). All data were normalized to the response observed for each treatment condition at the lowest test temperature. The solid line represents the best fit to the Boltzmann sigmoid equation resulting in an apparent T_{agg} of 46.7 ± 0.2 °C for the DMSO control, whereas both floxuridine and raltitrexed stabilized TS above 65 °C (we do not consider higher T_{agg} values reliable as these temperatures influence cell membrane integrity[10]). The vertical dotted line is at 50 °C, the temperature selected for the isothermal screen. Data are provided as the average and standard error of mean (s.e.m.) from two independent experiments performed in duplicate for raltitrexed and as individual data points from one experiment in duplicate for floxuridine. (**c**) $ITDRF_{CETSA}$ of floxuridine (blue triangle) at 50 °C based on raw data from the AlphaScreen readings. The solid line represents the best fit to a saturation binding curve resulting in an EC_{50} of 47 ± 16 pM. Data are provided as two individual data points from one test occasion. (**d**) The corresponding $ITDRF_{CETSA}$ for raltitrexed (magenta square) at 50 °C resulting in an EC_{50} of 0.75 ± 0.2 nM. Data are provided as two individual data points from one test occasion.

highly active compounds because of insufficient tip washing during the transfer from vials to screen library plates (Supplementary Fig. 10). Consequently these hits reproducibly confirmed activity when tested from the contaminated source plates, while the original solutions were inactive. Taken together the confirmation rate was 90% for hits yielding more than 30% apparent thermal stabilization of TS in the original screen.

Fluoropyrimidines, anti-folates and their analogs. The majority of confirmed hits were pyrimidine-based nucleosides and analogs thereof. At the top of the list were substituted 2′-deoxyuridines, including floxuridine (three independent occurrences in the library), 5-trifluoro-2′-deoxythymidine (TFT), and 5-ethynyl-2′-deoxyuridine (EdU) (Fig. 2b and Supplementary Table 3). All of these are known to be taken up and metabolized intracellularly to the active monophosphate forms that interact with TS[16,17]. Novel findings among the nucleosides included the two drugs azacitidine and its deoxyribose analog decitabine, as well as two purine nucleosides (8-bromoadenosine and 8-allyloxy-N2-isobutyryl-2′-deoxyguanosine).

With regards to folate analogs, methotrexate was present at two instances in the library, both as a racemate and as its L form. As expected both appeared as strong hits. The CETSA screen also identified two other marketed drugs, triamterene, a sodium channel inhibitor used to treat hypertension, and pyrimethamine, an inhibitor of dihydrofolate reductase from *Plasmodium falciparum* that is used to treat malaria. They have related

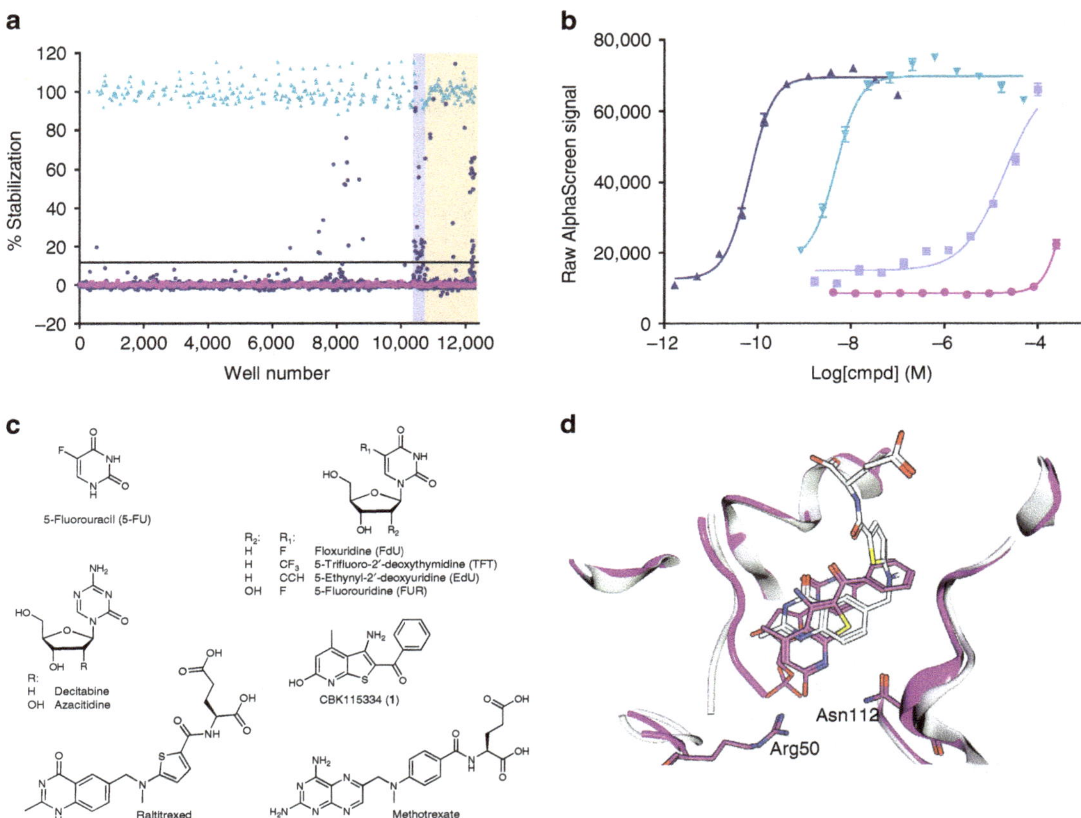

Figure 2 | Primary screen using CETSA to measure target engagement of human thymidylate synthase. (**a**) Scatter plot illustrating normalized screen data, where 0% corresponds to the TS signal observed in the presence of DMSO only (magenta square) and 100% corresponds to the TS signal observed in the presence of 100 nM raltitrexed (green triangle). Data for library compounds at a concentration of 50 μM are shown in blue (blue circle). The hit limit was calculated based on the average plus three standard deviations for the library compounds and is illustrated as a black solid line at 11.7%. The locations of the Prestwick drug set (yellow) and a nucleoside subset (purple) are highlighted. (**b**) $ITDRF_{CETSA}$ data illustrating the ranking of floxuridine (blue upwards triangle), 5-fluorouridine (FUR) (green downwards triangle), and 5-FU (lavender blue square) after 2 h of preincubation time. Data are also included for CBK115334 (magenta circle).The solid lines represent best fits to a saturation binding curve resulting in an apparent EC_{50} of TS at a concentration of 65 ± 9 pM, 47 ± 15 nM, 19 ± 4 μM and 0.46 ± 0.08 mM, respectively. Data are provided as the average and s.e.m. from one independent hit confirmation experiment done in quadruplicate. (**c**) Structures of known drugs and hit compounds discussed in the main text. (**d**) Structure of CBK115334 (magenta) and dUMP bound to TS, shown overlayed on the structure of the complex of raltitrexed (white) and dUMP (PDB 1HVY).

structures and can potentially act as folic acid antagonists[22], but they have not been previously shown to bind TS. Given the scarcity of anti-folates in the hit list we also looked whether there were any obvious false negatives in the screen and confirmed this was not the case (Supplementary Tables 4 and 5).

CBK115334 as a novel TS inhibitor. Besides the pyrimidine- and folate-based inhibitors there were 17 additional weak hits of different chemical classes (Supplementary Table 3). We investigated one of these, CBK115334 or 3-amino-2-benzoyl-4-methylthieno(2,3-b)pyridin-6-ol (**1**), which was chemically distinct from known TS inhibitors (Fig. 2c and Supplementary Fig. 11). It also appeared for the first time as a hit in our screens (Supplementary Table 6). Confirmatory data were obtained using CETSA on K562 cell lysates, which demonstrated a 3.7 °C shift at 200 μM (Supplementary Fig. 12). When applied to isolated recombinant human TS, **1** showed a 2.6 °C shift at 25 μM and a 5.2 °C shift at 100 μM and confirmed binding in the low μM range using surface plasmon resonance (Supplementary Fig. 12). We tested whether binding affected enzymatic activity of TS *in vitro* and observed 60% inhibition at 10 μM concentration and near complete inhibition at 100 μM (Supplementary Fig. 12). A crystal structure of TS with **1** revealed that the compound binds the active site of TS occupying the folate-binding pocket (Fig. 2d). The binding involves π–π stacking interactions with the substrate dUMP and polar interactions with residues lining the catalytic cavity (Asn112 and Arg50 in particular). This constitutes a novel mode of binding as compared with other anti-folates occupying this space[23,24]. Finally, we investigated the impact on cell proliferation in K562 cells. A clear impact was seen with a half-maximal inhibitory concentration (IC_{50}) value just below 100 mM (Supplementary Fig. 12), in line with the weak CETSA response.

Addressing kinetics of compound transport and metabolism. We performed time-traces of the $ITDRF_{CETSA}$ experiments, that is, by varying the time during which cells were exposed to compound prior to the heating step. $ITDRF_{CETSA}$ data obtained after various pre-incubation times are shown in Fig. 3a,b for 5-fluorouracil (5-FU) and floxuridine, demonstrating several orders of magnitude lower potency for 5-FU. The corresponding data on additional nucleosides are available in Supplementary Fig. 13. To examine whether the observed target engagement coincides with appearance of the active forms of these compounds, we monitored levels of compounds and their anticipated active metabolites using liquid chromatography

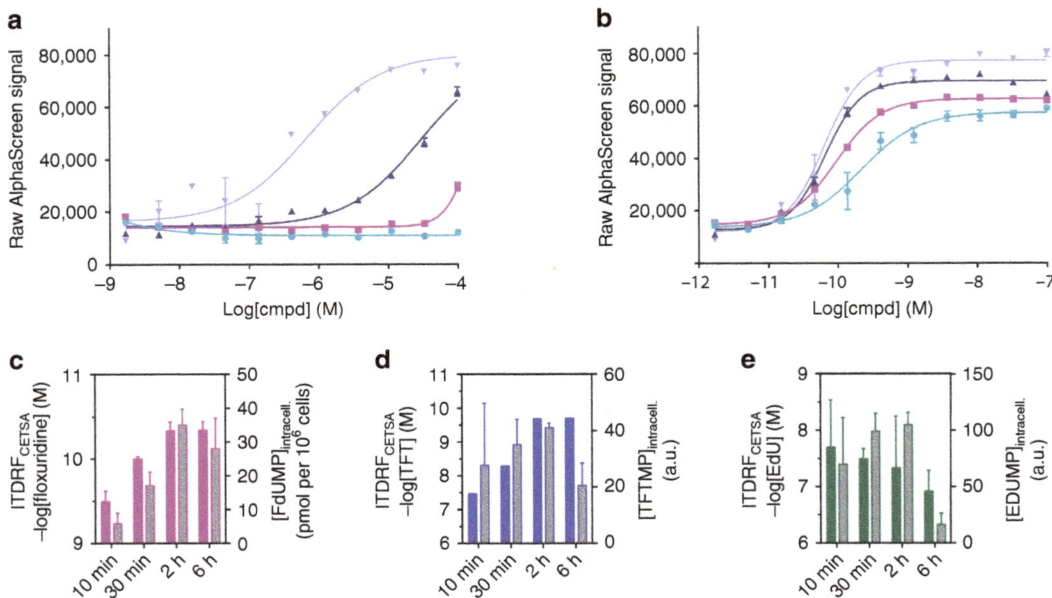

Figure 3 | Time dependence of target engagement and correlation with the appearance of intracellular active metabolites. (**a**) Representative ITDRF$_{CETSA}$ curves for 5-fluorouracil as a function of preincubation time in K562 cells; 10 min (green circle), 30 min (magenta square), 2 h (blue upwards triangle) and 6 h (lavender blue downwards triangle). The solid lines represent best fits to a saturation binding curve function to yield ITDRF$_{CETSA}$ values for half-maximal stabilization of TS. Data are provided as the average and s.e.m. from experiments done in quadruplicate at a single test occasion. (**b**) The corresponding ITDRF$_{CETSA}$ data for floxuridine. (**c**) Half-maximal stabilization of TS (magenta) and intracellular concentration of FdUMP (grey) as a function of preincubation time with floxuridine. The CETSA data are presented as the average and range from two independent experiments. The LC-MS/MS data are provided as the average and s.e.m. from experiments done at three different occasions. (**d**) The corresponding data for TFT (blue) and TFTMP (grey). (**e**) The corresponding data for EdU (green) and EdUMP (grey).

coupled to tandem mass spectrometry (LC–MS/MS)[25,26]. Intracellular and extracellular concentrations as a function of incubation time are shown in Supplementary Fig. 14 for selected nucleosides and their corresponding monophosphate species.

The cellular import and metabolic activation of floxuridine and 5-FU to generate the common active species 5-fluoro-2′-deoxyuridine 5′-monophosphate (FdUMP) require different enzymatic pathways[16,17,27]. CETSA data for floxuridine showed stabilization at low nM concentrations after only 10 min of pre-incubation (Fig. 3b). The potency improved during the first 2 h and persisted throughout the experiment. This time trace was consistent with the intracellular appearance of FdUMP, which was measureable already after 10 min and increased during the first hours of incubation (Fig. 3c). However, for 5-FU the CETSA response increased slowly in the first 6 h (Fig. 3a), with undetectable intracellular levels of FdUMP (Supplementary Fig. 14). Meanwhile the concentration of 5-FU in cells and media remained relatively constant at all time points, thus demonstrating a fast cellular uptake (Supplementary Fig. 14). We hence conclude that the enzymatic conversion of 5-FU to FdUMP is much slower than for floxuridine in K562 cells under these conditions and that the enzymatic conversion to the active species is mirrored by the CETSA responses.

EdU and TFT are structurally related to floxuridine differing only at position 5 of the uracil moiety (Fig. 2c). Although their primary activity on cell viability is believed to result from their misincorporation into DNA, they are also known to inhibit TS following intracellular phosphorylation[28,29]. The uptake and metabolism of TFT, as well as its CETSA time trace, was similar to that observed for floxuridine (Fig. 3d). This was consistent with a build-up of 5-trifluoro-2′-deoxythymidine 5′-monophosphate (TFTMP) and binding to TS in the first hours, generating full target engagement after 2 h of incubation. TFTMP is known to be a tight-binding inhibitor that forms covalent complexes with TS also in absence of the folate-based cofactor[30–32], in line with the observation of a persistent target engagement as measured by CETSA. EdU behaved differently with a more rapid uptake and faster decay of both the extracellular nucleoside and the active form 5-ethynyl-2′-deoxyuridine 5′-monophosphate (EdUMP) (Fig. 3e). The CETSA response was consistent with the fast uptake and activation with an early maximal response that then decayed slightly after the first 2 h, presumably due to the disappearance of the active species.

Phosphorylation and deamination of decitabine. An unexpected hit in the screen was decitabine, which primarily acts as an inhibitor of DNA methyltransferase[33]. The identification of a 2′-deoxycytidine analog as a hit was surprising, but reinforced by the concurrent appearance of the corresponding ribose azacitidine (Fig. 2c). To shed further light on the generation of the active compound, TS stabilization by decitabine itself was first investigated in K562 cell lysates, where activating metabolism is lower because of significant dilution of intracellular enzymes and their substrates. As shown in Supplementary Fig. 15 TS was not stabilized by decitabine in treated lysates. Likewise decitabine did not stabilize recombinant TS in a thermal shift assay, in line with observations for other nucleosides including deoxyuridine, floxuridine and TFT (Supplementary Fig. 15).

The structural analogy to the known nucleoside-based inhibitors of TS triggered the question as to whether decitabine is also phosphorylated, and potentially also deaminated, to generate a TS ligand (Fig. 4a). To investigate the importance of phosphorylation we performed ITDRF$_{CETSA}$ experiments in the presence of DI-82 (ref. 34). This compound is a potent inhibitor of deoxycytidine kinase (DCK) (Fig. 4b), which is required for formation of decitabine monophosphate[35,36]. Dose-dependent stabilization of TS was confirmed in the absence of DI-82,

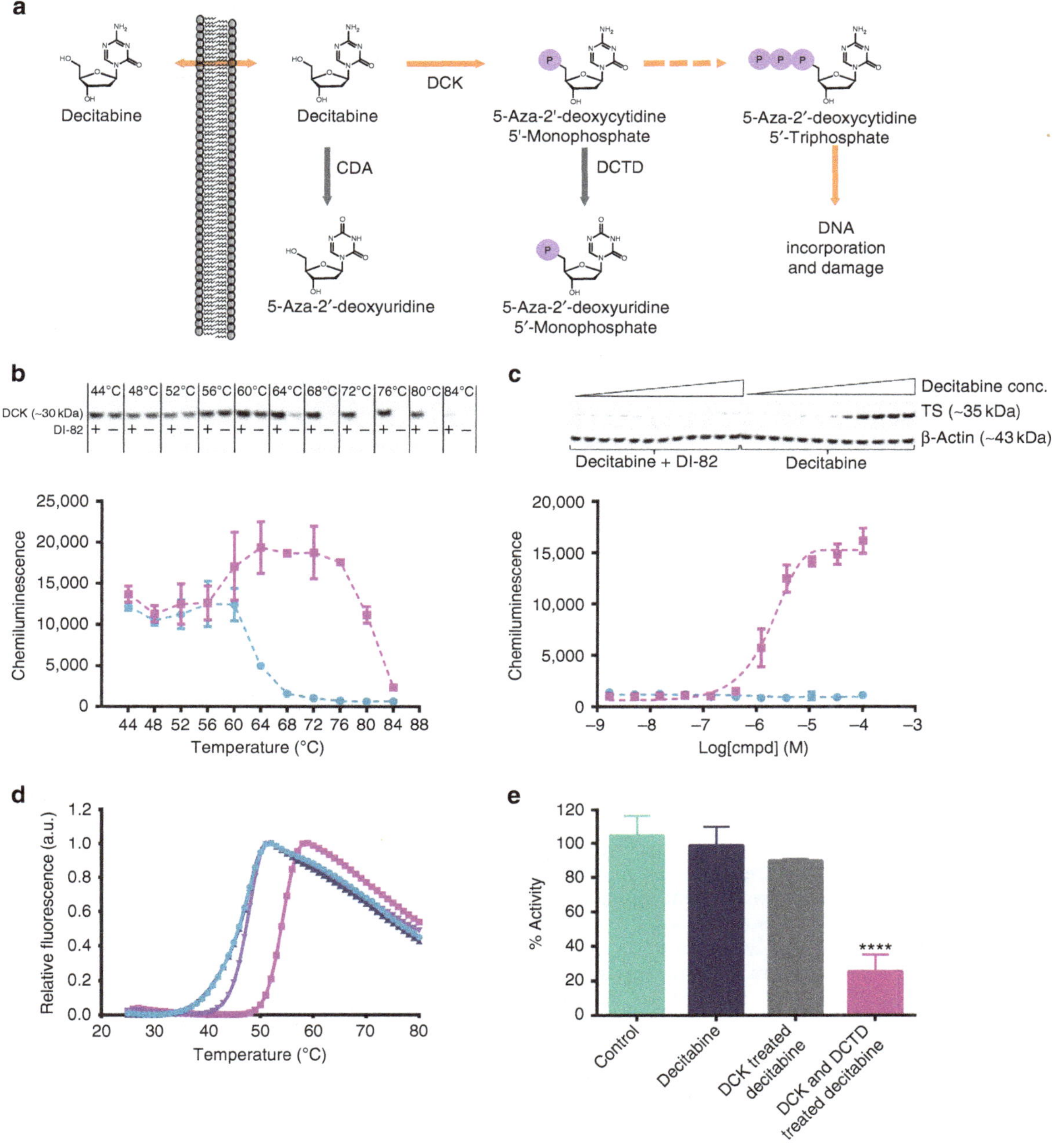

Figure 4 | Target engagement by decitabine is dependent on its metabolic activation. (**a**) Schematic overview of decitabine treatment, cellular uptake and intracellular metabolic conversion. After uptake decitabine is phosphorylated to form 5-aza-2′-deoxycytidine 5′-monophosphate by DCK. This compound is further phosphorylated in two steps to yield the triphosphate that is incorporated into DNA. Cytidine deaminase (CDA) and DCTD are known to be involved in the metabolism and clearance of decitabine[39]. (**b**) T_{agg} experiments for CDK in the absence (green circle) and presence of 200 μM of the DCK inhibitor DI-82 (magenta square). Above the graphs are the chemiluminescence data (full blots are available in Supplementary Fig. 16). The experiments were performed in K562 cells at two independent occasions. (**c**) $ITDRF_{CETSA}$ data for decitabine in the absence (magenta square) and presence of 200 μM of the DCK inhibitor DI-82 (green circle). Full blots are available in Supplementary Fig. 16. The experiments were performed in K562 cells at two independent occasions. (**d**) Normalized thermal shift assay response for recombinant human TS in the absence (magenta square) and presence of 1 mM decitabine without prior enzyme treatment (blue upwards triangle), following DCK treatment (lavender blue downwards triangle) and following treatment with both DCK and DCTD (magenta square). The data are shown as the average and s.e.m. from triplicate samples at one test occasion. (**e**) Enzyme inhibition data for TS in the presence of control and enzymatically treated decitabine samples. The data are shown as the average and s.e.m. from triplicate samples at one test occasion.

whereas its presence at 200 μM completely blunted the ability of decitabine to bind TS (Fig. 4c). However, studies using recombinant DCK to generate decitabine monophosphate resulted in only a marginal stabilization of TS (Fig. 4d). The sample was therefore additionally treated with deoxycytidylate deaminase (DCTD), which is known to deaminate decitabine monophosphate[37], resulting in a thermal shift of nearly 10 °C (Fig. 4d). The inhibitory capacity of these samples mirrored these data, that is, minor inhibition was observed after DCK treatment in the TS enzymatic assay, whereas near full inhibition appeared after treatment with both DCK and DCTD (Fig. 4e). Taken together with the structural analogy to the natural substrate of TS these data strongly infer that the TS ligand is 5-aza-2′-deoxyuridine 5′-monophosphate, that is, the expected product of phosphorylation and deamination of decitabine.

Discussion

Target engagement is essential for efficacy of targeted therapies and validation of new chemical probes[2,4,15]. These validating experiments are ideally performed for many representatives within a chemical series to allow comparisons of structure–activity relationships. To push towards the goal of having a procedure amenable to automation and screening we applied CETSA for assessment of intracellular target engagement at the stage of primary screening. Prior to this work the methodology had been applied on a growing number of drugs and chemical probes[10–14], but it remained challenging to apply to large chemical libraries.

To achieve this evaluation, we developed a homogeneous CETSA and applied it to screening in live, non-engineered cells expressing thymidylate synthase, which is targeted by several different chemical classes of drugs in clinical use[16,17]. Importantly, the screen identified all drugs within the test library that act on TS, as well as novel compounds capable of binding and inhibiting this enzyme. Collectively the known drugs and new inhibitors span over a broad range of affinities. Amongst the new hits was **1**, a μM inhibitor of the purified enzyme with sufficient cell penetration to result in intracellular target engagement and anti-proliferative effects in the high μM range. The binding mode of **1** to TS is partly new and it thus provides a potential starting point for further chemistry optimization.

Several other marketed drugs not previously known to inhibit TS emerged as hits, including triamterene, pyrimethamine and decitabine. As these are clinically-used compounds it is of interest to understand whether there are instances where the interaction with TS plays a role in either efficacy or toxicity. The identification of decitabine illustrates the relevance of monitoring target binding in live cells as this finding was dependent on active cellular metabolism. Decitabine itself was largely inactive on the protein such that, in analogy to the already known uridine-based inhibitors, enzymatic conversion to an active species is a prerequisite for observation of binding. We showed that this conversion does not take place to a significant extent in cell lysates, but can be reproduced with the *in vitro* application of enzymes for phosphorylation and deamination to yield the substrate analog 5-aza-2′-deoxyuridine 5′-monophosphate. Although our data give strong support that a metabolite of decitabine yield significant cellular inhibition of TS, further studies are required to determine whether this is relevant for polypharmacology or toxicity at typical therapeutic doses.

Time-traces of $ITDRF_{CETSA}$ were used to analyze different scenarios for cellular target engagement, thereby integrating aspects of drug transport and metabolism. Combination of these results with measurements of intracellular drug and metabolite concentrations allowed for a comprehensive dissection of cellular drug kinetics. Overall, the intracellular concentrations of the active species of the drugs correlated with the observed target engagement, consistent with the notion that CETSA directly reports on target binding. The combination of high-throughput target engagement studies with LC–MS/MS measurements of intracellular concentrations of drugs and metabolites constitutes a new paradigm for hit validation and optimization in the discovery of chemical probes and drugs. Importantly, CETSA is applicable to studies in native cells and tissue samples[10,14]. Thus the basic scenarios for compound metabolism and target engagement derived from these cell culture studies should be possible to translate towards studies of activities and resistance development to drugs in man. Of particular interest in this regard was the observation that we nearly missed the identification of 5-FU in the screen because of the relatively slow appearance of target engagement. It will be interesting to extend these experiments to patient cells.

The present work demonstrates that CETSA constitutes a robust high-throughput screening strategy that allows for target proteins to be approached in their natural cellular environment. This is in contrast to the majority of targeted cellular HTS assays as these rely on overexpressed and tagged proteins. Since CETSA does not require engineered cells or compounds, it could be particularly attractive for screening in primary cells, tissues or patient-derived material. Our approach can be applied to a large number of different proteins, with the generic assay development path being established in this work. The work also introduces the combination of time-dependent $ITDRF_{CETSA}$ and measurement of intracellular concentrations of metabolites as a stringent approach for hit validation, where the same assay format can be utilized and provide value throughout the drug discovery process.

Methods

Cell culture conditions in AlphaScreen-based experiments. Human myelogenous leukemia cell line K562 (ATCC no. CCL-243) were cultured in RPMI-1640 (SH30027.01, HyClone) supplemented with 10% fetal bovine serum (SV30160.03, HyClone), 0.3 g l^{-1} L-glutamine (G7513, Sigma-Aldrich) and 100 units ml^{-1} Penicillin-Streptomycin (P4333, Sigma-Aldrich). The same cell medium composition was used for all experiments unless otherwise stated.

Development of an AlphaScreen-based assay for TS. Measurements of remaining levels of soluble TS in cell lysates were achieved based on an AlphaScreen-based assay. Establishment of this assay required the identification of a pair of antibodies that simultaneously recognize TS (Supplementary Fig. 1). Combinations of four mouse-derived and three rabbit-derived antibodies directed towards different epitopes of TS (see Supplementary Table 1) were tested for this ability. Four different conditions were tested for each pair, with two of those being the absence and presence of target protein. Given our previous experience of ligand induced quenching of protein target recognition by the antibody pair[14] we also included a control containing an excess of dUMP with and without the additional presence of raltitrexed, which are known binders to the active sites of TS. Recombinant TS, diluted in 1 × AlphaLISA buffer (AL000F, PerkinElmer), was preincubated at room temperature in the presence of buffer only, 100 μM dUMP or 100 μM dUMP and 10 μM raltitrexed in a total volume of 4 μl in a ProxiPlate (#6008280, PerkinElmer). After this preincubation all 12 possible combinations of antibody pairs were added to the sample in a volume of 4 μl followed by incubation for 30 min at room temperature. A mix of AlphaScreen acceptor and donor beads was finally added in a volume of 4 μl under subdued light and allowed to incubate at room temperature for 2 h before reading in an Envision plate reader (PerkinElmer). Final concentrations of the reagents in the detection step were 2 nM recombinant TS, 2 nM of each antibody, 40 μg ml^{-1} AlphaScreen anti-mouse donor beads (#AS104D, PerkinElmer) and 10 μg ml^{-1} AlphaScreen anti-rabbit acceptor beads (#AL104C, PerkinElmer). The plates were sealed with TopSeal-A PLUS (6050185, PerkinElmer). The data were analyzed using microsoft excel and GraphPad Prism 6.

Four different antibody pair combinations based on sc-376161, WH0007298M1, 15047-1-AP and D5B3 (Supplementary Table 1) were selected from the antibody screen to study the kinetics of their recognition of TS in cell lysate. Two batches of 7.5 million K562 cells per ml in supplemented cell culture medium were prepared to serve as max and min controls. One culture was left at room temperature and the other was heated to 52 °C for 3 min in a PCR machine (TECHNE TC-PLUS thermal cycler). Both batches of cells were then lysed by the addition of an equal volume of 2X AlphaScreen SureFire Lysis Buffer (TGRLB100ML, PerkinElmer). After thorough mixing, 4 μl aliquots of the lysates were transferred to a ProxiPlate and detected and analyzed as described above except that the bead incubation was performed at 2 h, 6 h and overnight.

The sc-376161 and 15047-1-AP antibodies were titrated to match their concentrations with the AlphaScreen bead concentrations, that is, to ensure they do not exceed concentrations where hook effects are observed. Fifteen million K562 cells per ml were prepared in supplemented cell culture medium and lysed as described above. Aliquots of 4 μl were transferred to a ProxiPlate followed by the addition of 4 μl of a mix of different concentrations of the two antibodies (final concentrations of each antibody in the detection varied between 0 and 10 nM). Detection and analysis was done as described above, except the bead incubation was performed overnight.

Optimization of cell numbers was achieved by serial dilution of a cell suspension of K562 cells in supplemented cell culture medium. Each sample was then split into two aliquots, which were either kept at room temperature or heated to 52 °C as described above. Both aliquots were then lysed as described above. After thorough mixing, 3 µl of the lysates were transferred to a ProxiPlate followed by the addition of 6 µl of a mix of antibodies and AlphaScreen acceptor and donor beads in AlphaLISA buffer under subdued light. Detection and analysis was achieved as described above except the antibody concentrations were modified to 1 nM 15047-1-AP and 0.4 nM sc-376161.

The control experiment, in which recombinant TS was seeded to cell lysates was prepared based on a serial dilution of recombinant human TS. Dilutions were done in equal volumes of supplemented cell medium and 2 × AlphaScreen SureFire Lysis Buffer. K562 cells at a cell density of 2 million cells per ml were lysed as described above and split in two samples, which were either kept at room temperature or heated to 52 °C as described above. A 10 µl aliquot of each TS dilution was then added to the same volume of each of the two lysates as well as to the mixture of cell medium and lysis buffer. A 3 µl aliquot of each sample was then transferred to a ProxiPlate and detected and analyzed as described above.

Thermal aggregation experiments using AlphaScreen. Floxuridine and raltitrexed were diluted from dimethyl sulfoxide (DMSO) stock solutions to concentrations of 30 µM and 20 µM respectively in supplemented cell culture medium (final DMSO content 1%). These solutions were transferred in a volume of 10 µl to a skirted Twin.tec PCR 96-wellplate (0030 128 672, Eppendorf). A suspension of K562 cells in a volume of 10 µl and a density of 10 million cells per ml were then added to all wells. The PCR plates containing the compounds and cells were sealed with a breathable plate seal (3345, Corning) and incubated for 2 h in a humidified incubator at 37 °C and 5% CO_2. The cells were then transiently heated to different temperatures ranging from 40 °C to 86 °C for 3 min, followed by a controlled cooling to 20 °C for 1 min using a real-time PCR machine (ProFlex, Applied Biosystems). After the heating step the plate was centrifuged briefly (1,000 × *g* for 1 min) followed by lysis of the heated cells by the addition of 20 µl of 2 × AlphaScreen SureFire Lysis Buffer using a Flexdrop IV (PerkinElmer). To ensure sufficient lysis the cell lysates were mixed by 10 repetitive aspiration and dispensing cycles using a Bravo liquid handling platform (Agilent). The lysates (3 µl) were then transferred to 384-well ProxiPlates followed by the addition of 6 µl of a mix of antibodies and AlphaScreen acceptor and donor beads in AlphaLISA buffer under subdued light. Final concentrations of the assay reagents in the detection step were 1 nM rabbit polyclonal anti-TS IgG (15047-1-AP, Proteintech), 0.4 nM mouse monoclonal anti-TS IgG (sc-376161, Santa Cruz), 40 µg ml^{-1} AlphaScreen anti-mouse donor beads and 10 µg ml^{-1} AlphaScreen anti-rabbit acceptor beads. The plates were sealed with TopSeal-A PLUS and incubated over night at room temperature prior to detection in an Envision plate reader. The data were analyzed using microsoft excel and GraphPad Prism 6.

Composition and storage of the primary screening set. The library of compounds applied in this screening campaign consists of 10,928 compounds and is part of the primary screening set at CBCS. The majority of these compounds was donated by Biovitrum AB and originates from both in-house and commercial sources. Compounds included in the primary screening set were selected to represent a diverse selection of a larger set of 65,000 compounds, while keeping a certain depth to allow crude structure–activity relationship studies. The selection was also biased towards lead-like and drug-like profiles with regards to molecular weight, hydrogen bond donors/acceptors and LogP[20]. The library also includes a nucleoside set from Berry & Associates and a set of approved drugs from Prestwick. Compound stock solutions at 10 mM in DMSO are stored frozen at approximately − 20 °C in individual capped tubes in REMP 96 Storage Tube Racks. The racks are stored in a REMP Small-Size Store, which allows cherrypicking while the solutions are still frozen to minimize repetitive freeze-thaw cycles. For screening purposes the compound solutions have been replicated from the REMP racks to Labcyte 384 LDV plates (LP-0200) and then further into Labcyte 1536 HighBase plates (LP-03730) to enable dispensing using acoustic liquid handling equipment.

Compound handling. Assay ready plates were prepared by transferring 200 nl of the 10 mM DMSO solutions of compounds and controls by means of acoustic dispensing (Echo 550, Labcyte) to 384-well polypropylene plates (784201, Greiner). Compounds were placed in columns 1–22. DMSO controls were placed in column 23 and raltitrexed controls were placed in column 24. The assay ready plates were heat sealed with a Peelable Aluminium seal (24210-001 Agilent) using a thermal microplate sealer (PlateLoc, Agilent) and stored at − 20 °C until use. At the day of the experiment the plates were allowed to thaw for 30 min followed by a brief centrifugation step (1,000 × *g* for 1 min) prior to removal of the seal. The compounds were then diluted with 20 µl supplemented cell culture medium using a Multidrop Combi reagent dispenser (Thermo Scientific). Finally 5 µl of the diluted compounds were transferred to a 384 well hardshell PCR plate (HSR480, BIORAD) using a Bravo liquid handling platform equipped with a 384-well head (Agilent). The final concentrations in the incubation with cells (see below) were 50 µM of test compounds and 100 nM of the positive control raltitrexed. The final concentration of DMSO in the assay was 0.5% in all samples.

For the $ITDRF_{CETSA}$ experiments 11-point dose-response curves with three-fold difference in concentration between wells were generated using the Bravo liquid handling system (all serial dilutions were done in 100% DMSO). The final highest concentrations of the test compounds in the incubation with cells were ranging from 100 nM–50 µM depending on estimated potency. The concentration of DMSO was 0.5% in all samples. The final concentration of the positive control raltitrexed was 100 nM. The assay ready plates were prepared as outlined above for the screen.

Screening and dose-response characterization by AlphaScreen. The screen procedure started with the addition of 5 µl of a suspension of K562 cells at a density of 10 million cells per ml to all wells of a 384 well hardshell PCR plate (HSR480, BIORAD) using an electronic multichannel pipette (Biohit). The plates were then heat sealed with Peelable Aluminium seal in a thermal microplate sealer (PlateLoc, Agilent) and allowed to incubate for 2 h in an incubator at 37 °C and 5% CO_2. For the time-course experiments the incubation times were altered to include also 10 min, 30 min and 6 h. To allow gas exchange during the longer incubation times these plates were instead sealed with at breathable plate seal (3345, Corning). After the incubation step the plates were transiently heated at 50 °C for 3 min followed by a controlled cooling to 20 °C for 1 min using a real-time PCR machine (Light-Cycler480 system, Roche). Plate handling was then as described above for the thermal aggregation experiments. The data were analyzed using microsoft excel and GraphPad Prism 6.

Screen and $ITDRF_{CETSA}$ data analysis. Screen data were imported into microsoft excel and normalized for each compound based on the negative and positive controls on each plate, that is, with the response in the presence of DMSO defining 0% stabilization and the response in the presence of 100 nM raltitrexed defining 100% stabilization. A calculation of the average and standard error of means for each set of controls also allowed an illustration of how these responses varied over the 32 screening plates. The Z' factor[38] is commonly used as a measure of how well the assay separates between the controls and this was calculated as described based on the calculated averages and standard deviations of the controls on a per plate basis. For the T_{agg} shift and the $ITDRF_{CETSA}$ experiments the data were analyzed in GraphPad Prism using the Boltzmann sigmoid equation and the saturation binding curve (rectangular hyperbola; binding isotherm) function, respectively. As already discussed[14] these methods make use of equilibrium models for data analysis although the methodology depends on the irreversible aggregation of denatured material. For this reason we refer to the observed responses as apparent and isothermal dose-response fingerprints and are careful with any quantitative interpretations, being well aware of their dependency on experimental conditions. Experiments are on-going to address the quantitative interpretation of CETSA data.

Measurements of identity and purity of test compound solutions. Assessments of identity and purity of the test solutions that were used for hit confirmation purposes, that is, those being stored in REMP vials, was done by means of high-pressure liquid chromatography coupled to mass spectrometry (HPLC–MS). A small aliquot of each test solution (2 µl of a 10 mM solution) was placed in a 96-well plate (267245, Nunc) and diluted with 20 µl of methanol. The plate was then placed in an Agilent 1,100 HPLC UV/MS with electrospray ionization (ESI +). The HPLC method was based on an ACE C8 3 µm column (3.0 × 50 mm) and a mobile phase (CH_3CN)/(0.1% TFA/H_2O). All solvents were HPLC grade and absorbance was monitored at 220 nm. Compounds that did not give satisfactory data were re-analyzed using a method based on a Waters XBridge C18 3.5 µm column (3.0 × 50 mm), 3.5 min gradient mobile phase (CH_3CN)/(10 mM NH_4HCO_3/H_2O). The instrument software was used to integrate the UV response for each peak and provided a list of the peaks and their associated masses. The estimated purity was calculated based on the integrated area for the expected mass compared with the areas of all other peaks. The result was manually controlled and if there were deviations from the expected outcome a meticulous investigation of the UV-response and MS was performed.

Cell viability assay. A concentration–response curve of CBK115334 was generated using the Bravo liquid handling system (serial dilution of a 50 mM stock solution was done in 100% DMSO). A total of 150 nl of the serially diluted solutions and controls (positive control 0.67 mM staurosporine and negative control DMSO) were transferred to a white 384-well assay plate (3570, Corning) by means of acoustic dispensing (Echo 550, Labcyte). A Multidrop reagent dispenser (Thermo Scientific) was used to dispense 30 µl of a K562 cell suspension at a density of 33 × 10^3 cells per ml in supplemented cell culture medium. The cells were incubated at 37 °C in the presence of 5% CO_2 for 72 h before addition of 30 µl CellTiter-Glo Luminescent Cell Viability Assay reagent (Promega) using a Multidrop Combi reagent dispenser (Thermo Scientific). The plate was placed on a plate shaker for 15 min prior to detection of the luminescence signal in an Envision plate reader (PerkinElmer). The final highest concentration of CBK115334 in the incubation with cells was 250 µM and the final concentration of the positive control staurosporine was 3 µM. All samples contained 0.5% DMSO. The data were analyzed using microsoft excel and GraphPad Prism 6.

Chemicals and buffers in western blot-based experiments. The cell lysis buffer contained 100 mM 4-(2-hydroxyethyl)-1-piperazineethanesulfonic acid (HEPES, pH 7.5), 1 mM Tris-(2-carboxyethyl)phosphine hydrochloride (TCEP) and 10 mM magnesium chloride (Sigma-Aldrich) supplemented with complete (EDTA-free) protease inhibitor cocktail from Roche (Switzerland). Tris-buffered saline with tween (TBST) buffer (150 mM NaCl, 0.05% (v/v) Tween-20, 50 mM Tris-HCl buffer at pH 7.6) was prepared by dissolving TBS-TWEEN tablets obtained from Merck KGaA (Darmstadt, Germany) in ddH_2O. The blocking buffer consisted of 5% (w/v) non-fat milk (Semper AB, Sundbyberg, Sweden) diluted in tris-buffered saline with tween. Hank's Balanced Salt Solution (HBSS) was from Gibco/Life Technologies. Raltitrexed monohydrate, dUMP and decitabine was purchased from Sigma-Aldrich and Selleckchem, respectively. DI-82 was kindly provided by Prof. Caius G. Radu and Raymond M. Gipson at the Department of Molecular and Medical Pharmacology, University of California, Los Angeles.

Cell lines and cultures in western blot-based experiments. Human cell line K562 (ATCC no. CCL-243) was cultured in RPMI-1640 medium (Sigma-Aldrich) supplemented with 0.3 $g\,l^{-1}$ L-glutamine and 10% fetal bovine serum (FBS, Gibco/Life Technologies, Carlsbad, CA, USA), 100 units per ml penicillin and 100 units per ml streptomycin (Gibco/Life Technologies). Short-term passages (<20) were used for experiments.

Cell lysate thermal shift experiments. For the cell lysate thermal shift experiments, cultured K562 cells were harvested and washed with Hank's Balanced Salt Solution. The cells were diluted in lysis buffer supplemented with complete protease inhibitor cocktail. The cell suspensions were freeze-thawed three times using liquid nitrogen and passed through a 27″ gauge needle five times. The soluble fraction (lysate) was separated from the cell debris by centrifugation at 20,000 × *g* for 20 min at 4 °C. For the thermal aggregation curve experiments cell lysates were diluted with lysis buffer supplemented with 200 μM dUMP and divided into two aliquots, with one aliquot being treated with ligand and the other aliquot with vehicle (control). After 10 min incubation at room temperature the respective lysates were divided into smaller (50 μl) aliquots and heated individually at different temperatures for 3 min in a Veriti thermal cycler (Applied Biosystems/Life Technologies) followed by cooling for 3 min at room temperature. The heated lysates were centrifuged at 20,000 × *g* for 20 min at 4 °C in order to separate the soluble fractions from precipitates. The supernatants containing the remaining soluble proteins were transferred to new 0.2 ml microtubes and analyzed by sodium dodecyl sulfate polyacrylamide gel electrophoresis (SDS-PAGE) followed by western blot analysis.

For the in-cell experiments K562 cells were harvested and resuspended with culture medium to a cell density of 5 million cells per ml and seeded into T25 flasks. Cells were treated with either raltitrexed, floxuridine, DI-82 or vehicle (DMSO) for 2 h in an incubator at 37 °C and 5% CO_2. The cell suspensions were then divided into 100 μl aliquots in 0.2 ml tubes and heated at designated temperatures ranging from 40 to 84 °C for 3 min in a Veriti thermal cycler (Life Technologies) followed by 3 min of cooling at room temperature. The heat-treated cell suspensions were freeze-thawed three times using liquid nitrogen and a heating block set at 25 °C. Tubes were gently vortexed between the freeze-thaw cycles. The resulting cell lysates were centrifuged at 20,000 × *g* for 20 min at 4 °C. The supernatant was removed from the cell debris and aggregates and the remaining soluble TS was analyzed using western blot.

For the $ITDRF_{CETSA}$ in cell experiments, raltitrexed and floxuridine were serially diluted to generate an 11 point dose–response curve with three-fold difference in concentration between each point. K562 cells were treated with each respective compound concentrations and one vehicle as control in 100 μl aliquots in 0.2 ml tubes for 2 h in an incubator at 37 °C and 5% CO_2. The cell aliquots were heated at 50 °C and analyzed with western blot following the procedure described above.

For the decitabine metabolism, decitabine was serially diluted to generate an 11 point dose–response curve and an $ITDRF_{CETSA}$ experiment was performed as described above in absence and presence of 200 μM of the DCK inhibitor DI-82 (referred to as 12R in the main text of the original publication and DI-82 in the Supplementary Material)[34].

SDS-PAGE and western blot. NuPage Novex Bis-Tris 4–12% polyacrylamide gels with NuPAGE MES SDS running buffer (Life Technologies) were used for separation of proteins in the samples. Proteins were transferred to nitrocellulose membranes using the iBlot2 blotting system (Life Technologies). Primary antibodies anti-TS (D5B3) XP (Cell Signaling), anti-dCK (sc-393099), anti-β-actin (sc-69879); secondary goat anti-mouse HRP-IgG (sc-2055) and bovine anti-rabbit HRP-IgG (sc-2374) antibodies (Santa Cruz Biotechnology, Santa Cruz, CA, USA) were used for immunoblotting. All membranes were blocked with blocking buffer; standard transfer and western blot protocols recommended by the manufacturers (listed above) were used. All antibodies were diluted in blocking buffer. The membranes were developed using Clarity Western ECL substrate HRP-Substrate (Bio-Rad) according to the manufacturer's recommendations. Chemiluminescence intensities were detected and quantified using a ChemiDoc XRS+ imaging system (Bio-Rad) with Image Lab software (Bio-Rad).

Expression and purification of human thymidylate synthase. The gene encoding human TS (NM_001071.2) was subcloned into the pNIC28-Bsa4 vector and expressed in Rosetta BL21-DE3 Escherichia coli (Novagen) in Terrific Broth media supplemented with 50 $\mu g\,ml^{-1}$ of kanamycin and 34 $\mu g\,ml^{-1}$ chloramphenicol. Cells were grown at 37 °C until $OD_{600\,nm}$ reached about 2.0 and induced with 0.5 mM isopropyl-beta-D-1-thiogalactopyranoside (IPTG) at 18 °C overnight. The cells were harvested by centrifugation at 4,500 × *g* for 15 min at 15 °C. The cell pellet was re-suspended in lysis buffer (100 mM HEPES, 500 mM NaCl, 10 mM imidazole, 10% (v/v) glycerol and 1 mM TCEP at pH 8.0) supplemented with 1:1,000 (v/v) EDTA-free protease inhibitor cocktail (Calbiochem) and 125 $U\,ml^{-1}$ of Benzonase (Merck). Cells were lysed by sonication on ice at 70% amplitude, 3 s on/off for 3 min. The lysate was clarified by centrifugation at 47,000 × *g* for 25 min at 4 °C, and the supernatant was filtered through a 1.3 μm syringe filter to remove cell debris. The cell-free extract was loaded on a pre-equilibrated HisTrapTM HP column (GE Healthcare) in IMAC wash buffer 1 (20 mM HEPES, 500 mM NaCl, 10 mM imidazole, 10% (v/v) glycerol and 1 mM TCEP at pH 7.5) and subsequently washed with 20 column volumes (CVs) of IMAC wash buffer 1 and 15 CVs of IMAC wash buffer 2 (20 mM HEPES, 500 mM NaCl, 25 mM imidazole, 10% (v/v) glycerol and 1 mM TCEP at pH 7.5). Bound protein was eluted with 5 CVs of elution buffer (20 mM HEPES, 500 mM NaCl, 500 mM imidazole, 10% (v/v) glycerol and 1 mM TCEP at pH 7.5) and loaded onto a HiLoad 16/60 Superdex-200 column (GE Healthcare) pre-equilibrated with buffer (20 mM HEPES, 300 mM NaCl, 10% (v/v) glycerol, and 1 mM TCEP at pH 7.5). Based on Nu-PAGE gel results pure protein fractions were pooled and concentrated using a 50 kDa cutoff centrifugal driven filter concentrator (Sartorius Stedium Biotech). The protein concentration was determined by the absorbance at 280 nm using a Nanodrop spectrophotometer (Thermo Scientific).

Thermal shift *in vitro* assay on recombinant protein. The assay was performed on the iCycler iQ Real Time PCR Detection System (Bio-Rad), using the 96-well thin-wall PCR plate (Bio-Rad). The experiment was conducted in a buffer containing 20 mM HEPES at pH 7.5 and 150 mM NaCl. A total volume of 25 μl solution containing 0.2 $mg\,ml^{-1}$ protein, compounds and ×5 Sypro Orange dye (Invitrogen) diluted from 5,000× stock was dispensed into the 96-well plate. The same amount of DMSO was added in the control wells. The plates were sealed with Microseal B adhesive sealer (Bio-Rad) and heated in iCycler from 25 to 80 °C (56 heating cycles in 28 min). Fluorescent filter used for Sypro Orange measurements was $\lambda_{excitation} = 492$ nm and $\lambda_{emission} = 610$ nm. The calculation of

Table 1 | X-ray diffraction data and refinement statistics.

	TS-dUMP-CBK115334
Data collection	
Space group	$P4_32_12$
Cell dimensions	
a, *b*, *c* (Å)	108.2, 108.2, 313.9
α, β, γ (°)	90, 90, 90
Resolution (Å)	30-3.1 (3.2-3.1)*
R_{merge}	0.147 (0.619)
$I/\sigma I$	15.56 (4.98)
Completeness (%)	99.9 (99.9)
Redundancy	11.9 (10.9)
Refinement	
Resolution (Å)	30-3.1 (3.2-3.1)
No. reflections	34,453
R_{work}/R_{free}	18.0/26.8
No. atoms	
Protein	13,596
Ligand dUMP	186
Ligand CBK115334	96
Water	11
B-factors	
Protein	45.3
Ligand dUMP	45.1
Ligand CBK115334	51.9
Water	30.9
R.m.s. deviations	
Bond lengths (Å)	0.014
Bond angles (°)	1.096

*Highest resolution shell is shown in parenthesis.

the midpoint of the curves (T_m) was performed using the software package XLfit from IDBS within microsoft excel.

TS enzyme inhibition assay. Enzymatic activity of recombinant human TS was measured spectrophotometrically at 340 nm by monitoring the absorbance change during the conversion of 5,10-methylenetetrahydrofolate to dihydrofolate using an Infinite M200 spectrometer (Tecan). Measurements were carried out at room temperature and in a buffer of 50 mM Tris at pH 7.5 and 150 mM NaCl. Initial velocities were measured with 250 nM of purified protein, 100 μM 5,10-methylenetetrahydrofolate and 100 μM dUMP in the presence of compounds using the same amount of DMSO as control. Initial rates and activity were analyzed with the software package Prism (GraphPad Software).

Surface plasmon resonance. Recombinant human TS protein (25 μg ml^{-1} in a buffer of 10 mM sodium acetate at pH 5.0, 1 mM dUMP and 200 μM methotrexate) were captured on Sensor Chip S-CM5 via amine coupling to a level of ~5,000 resonance units using Biacore T-200. Raltitrexed was used as a positive control to ensure that the protein remained active after immobilization and during the run. A concentration series (20 nM to 10 μM) of CBK115334 was injected over the prepared surface for 60 s and allowed to dissociate for 60 s with a flow rate of 70 μl min^{-1} at 25 °C. The assay buffer was 20 mM HEPES, 150 mM NaCl, pH 7.5 and 0.005% Tween-20 supplemented with 1% DMSO and 200 μM dUMP. Response data was processed using the BIAevaluation software. Responses were double referenced and solvent-corrected. The data sets were fitted to 1:1 steady-state model for determination of binding constants.

Decitabine treatment with DCK and DCTD. Decitabine at a concentration of 1 mM was incubated with 1 mg ml^{-1} of recombinant DCK and 2.5 mM ATP in a buffer of 50 mM Tris at pH 7.5, 150 mM NaCl and 0.5 mM $MgCl_2$ at room temperature for 60 min to generate the corresponding monophosphates. To probe for nucleotide deamination the samples were further treated with 1 mg ml^{-1} of recombinant DCTD in the presence of 1 mM ATP and 10 μM $ZnCl_2$ for 20 min at room temperature. After incubation, the samples were heated at 95 °C for 10 min and centrifuged for 10 min at the highest speed at 4 °C using a benchtop centrifuge. The supernatant was tested without dilution in thermal shift *in vitro* assays and the TS enzyme inhibition assay. The supernatants from samples without nucleoside were used as the treatment controls.

Crystallization and structure determination. TS protein crystallized in sitting drops comprising equal volume of protein (about 24 mg ml^{-1}) and reservoir solution at 20 °C. The crystallization condition was composed of 0.1 M sodium cacodylate at pH 6.5 and 15% PEG 4,000. Crystals were soaked with 1 mM compound and 2 mM dUMP in cryo-protectant buffer containing 0.1 M sodium cacodylate at pH 6.5 and 25% PEG 4,000 and 10% DMSO for 15 min, followed by flash frozen in liquid nitrogen. Data collection was performed on beamline MX1 at Australian Synchrotron. X-ray diffraction data was collected at 100 K with a wavelength of 0.9537 Å. It should be noted that CBK115334 is subject to keto–enol isomerization. While the resolution is not sufficient to distinguish between these, the included illustrations are based on the enol form (both forms make interactions with Asn112 and Arg50, but with different donor–acceptor pairs). The structure was solved by molecular replacement using Phaser with the hTS-dUMP-raltitrexed structure (PDB code 1HVY) as the search model. Structure was refined with phenix refine. Ligand structures and restraints files were generated using eLBOW. In the TS-dUMP-CBK115334 complex structure, 90.43% of residues were in favoured regions and 7.92% of residues were in allowed regions. The data collection parameters and refinement statistics are summarized in Table 1. An image of the electron density map of the active site in the co-crystal structure is available in Supplementary Fig. 17.

Table 2 | Monitored LC-MS/MS transitions.

Compound/Metabolite	Parent and daughter ions
5-FU	128.8 > 41.9
FdU	244.9 > 155.0
FdUMP	324.9 > 195.0
FUR	260.9 > 171.0
EdU	250.9 > 135.8
EdUMP	330.9 > 195.0
TFT	249.9 > 179.7
TFTMP	374.9 > 179.4
Decitabine	227.0 > 93.8
Decitabine-MP	307.0 > 195.0
Pyrimethamine*	248.8 > 176.9
CBK115334*	284.9 > 188.9

*Analyzed with electrospray ionization in positive mode.

Intracellular compound and metabolite concentrations. Intracellular concentrations of compounds were measured as previously described[25,26]. Briefly, K562 cells were incubated with the compounds for a predefined time at 37 °C in a 5% CO_2 atmosphere. After incubation, the cells were centrifuged (300 × *g* for 5 min) and a medium sample (supernatant) was collected and diluted 1:10 in a 50 nM warfarin solution (internal standard). The cells were washed with phosphate buffered salt solution and lysed with methanol. Methanol was evaporated and the cell samples were reconstituted in 50 nM warfarin. Samples were analyzed with LC–MS/MS with electrospray ionization in negative mode with transitions monitored as listed in Table 2.

References

1. Morgan, P. *et al.* Can the flow of medicines be improved? Fundamental pharmacokinetic and pharmacological principles toward improving Phase II survival. *Drug Discov. Today* **17,** 419–424 (2012).
2. Bunnage, M. E., Chekler, E. L. P. & Jones, L. H. Target validation using chemical probes. *Nat. Chem. Biol.* **9,** 195–199 (2013).
3. Cook, D. *et al.* Lessons learned from the fate of AstraZeneca's drug pipeline: a five-dimensional framework. *Nat. Rev. Drug Discov.* **13,** 419–431 (2014).
4. Simon, G. M., Niphakis, M. J. & Cravatt, B. F. Determining target engagement in living systems. *Nat. Chem. Biol.* **9,** 200–205 (2013).
5. Durham, T. B. & Blanco, M.-J. Target engagement in lead generation. *Bioorg. Med. Chem. Lett.* **25,** 998–1008 (2015).
6. Pantoliano, M. W. *et al.* High-density miniaturized thermal shift assays as a general strategy for drug discovery. *J. Biomol. Screen.* **6,** 429–440 (2001).
7. Ericsson, U. B., Hallberg, B. M., DeTitta, G. T., Dekker, N. & Nordlund, P. Thermofluor-based high-throughput stability optimization of proteins for structural studies. *Anal. Biochem.* **357,** 289–298 (2006).
8. Senisterra, G. A. *et al.* Screening for ligands using a generic and high-throughput light-scattering-based assay. *J. Biomol. Screen* **11,** 940–948 (2006).
9. Niesen, F. H., Berglund, H. & Vedadi, M. The use of differential scanning fluorimetry to detect ligand interactions that promote protein stability. *Nat. Protoc.* **2,** 2212–2221 (2007).
10. Martinez Molina, D. *et al.* Monitoring drug target engagement in cells and tissues using the cellular thermal shift assay. *Science* **341,** 84–87 (2013).
11. Savitski, M. M. *et al.* Tracking cancer drugs in living cells by thermal profiling of the proteome. *Science* **346,** 1255784 (2014).
12. Franken, H. *et al.* Thermal proteome profiling for unbiased identification of direct and indirect drug targets using multiplexed quantitative mass spectrometry. *Nat. Protoc.* **10,** 1567–1593 (2015).
13. Huber, K. V. M. *et al.* Proteome-wide drug and metabolite interaction mapping by thermal-stability profiling. *Nat. Methods* **12,** 1055–1057 (2015).
14. Jafari, R. *et al.* The cellular thermal shift assay for evaluating drug target interactions in cells. *Nat. Protoc.* **9,** 2100–2122 (2014).
15. Arrowsmith, C. H. *et al.* The promise and peril of chemical probes. *Nat. Chem. Biol.* **11,** 536–541 (2015).
16. Longley, D. B., Harkin, D. P. & Johnston, P. G. 5-fluorouracil: mechanisms of action and clinical strategies. *Nat. Rev. Cancer* **3,** 330–338 (2003).
17. Wilson, P. M., Danenberg, P. V, Johnston, P. G., Lenz, H.-J. & Ladner, R. D. Standing the test of time: targeting thymidylate biosynthesis in cancer therapy. *Nat. Rev. Clin. Oncol.* **11,** 282–298 (2014).
18. Assaraf, Y. G. Molecular basis of antifolate resistance. *Cancer Metastasis Rev.* **26,** 153–181 (2007).
19. Gonen, N. & Assaraf, Y. G. Antifolates in cancer therapy: Structure, activity and mechanisms of drug resistance. *Drug Resist Updat.* **15,** 183–210 (2012).
20. Lipinski, C. A. Lead- and drug-like compounds: the rule-of-five revolution. *Drug Discov. Today Technol.* **1,** 337–341 (2004).
21. Houghton, J. A., Tillman, D. M. & Harwood, F. G. Ratio of 2′-deoxyadenosine-5′-triphosphate/thymidine-5′-triphosphate influences the commitment of human colon carcinoma cells to thymineless death. *Clin. Cancer Res.* **1,** 723–730 (1995).
22. Lambie, D. G. & Johnson, R. H. Drugs and folate metabolism. *Drugs* **30,** 145–155 (1985).
23. Almog, R., Waddling, C. A., Maley, F., Maley, G. F. & Van Roey, P. Crystal structure of a deletion mutant of human thymidylate synthase Δ (7-29) and its ternary complex with Tomudex and dUMP. *Protein Sci.* **10,** 988–996 (2001).
24. Sayre, P. H. *et al.* Multi-targeted antifolates aimed at avoiding drug resistance form covalent closed inhibitory complexes with human and Escherichia coli thymidylate synthases. *J. Mol. Biol.* **313,** 813–829 (2001).
25. Mateus, A., Matsson, P. & Artursson, P. Rapid measurement of intracellular unbound drug concentrations. *Mol. Pharm.* **10,** 2467–2478 (2013).
26. Mateus, A., Matsson, P. & Artursson, P. A high-throughput cell-based method to predict the unbound drug fraction in the brain. *J. Med. Chem.* **57,** 3005–3010 (2014).

27. O'Connor, O. A. in *Cancer Drug Discovery and Development: Combination Cancer Therapy: Modulators and Potentiators* (ed. Schwartz, G. K.) 133–174 (Humana Press Inc., 2005).
28. De Clercq, E. *et al.* Thymidylate synthetase as target enzyme for the inhibitory activity of 5-substituted 2′-deoxyuridines on mouse leukemia L1210 cell growth. *Mol. Pharmacol.* **19,** 321–330 (1981).
29. Temmink, O. H., Comijn, E. M., Fukushima, M. & Peters, G. J. Intracellular thymidylate synthase inhibition by trifluorothymidine in FM3A cells. *Nucleosides Nucleotides Nucleic Acids* **23,** 1491–1494 (2004).
30. Danenberg, P. V, Langenbach, R. J. & Heidelberger, C. Structures of reversible and irreversible complexes of thymidylate synthetase and fluorinated pyrimidine nucleotides. *Biochemistry* **13,** 926–933 (1974).
31. Danenberg, P. V & Lockshin, A. Fluorinated pyrimidines as tight-binding inhibitors of thymidylate synthetase. *Pharmac. Ther.* **13,** 69–90 (1981).
32. Eckstein, J. W., Foster, P. G., Finer-Moore, J., Wataya, Y. & Santi, D. V. Mechanism-based inhibition of thymidylate synthase by 5-(trifluoromethyl)-2′-deoxyuridine 5′-monophosphate. *Biochemistry* **33,** 15086–15094 (1994).
33. Stresemann, C. & Lyko, F. Modes of action of the DNA methyltransferase inhibitors azacytidine and decitabine. *Int. J. Cancer* **123,** 8–13 (2008).
34. Nomme, J. *et al.* Structure-Guided Development of Deoxycytidine Kinase Inhibitors with Nanomolar Affinity and Improved Metabolic Stability. *J. Med. Chem.* **57,** 9480–9494 (2014).
35. Momparler, R. L. Molecular, cellular and animal pharmacology of 5-aza-2′-deoxycytidine. *Pharmacol. Ther.* **30,** 287–299 (1985).
36. Momparler, R. L. Pharmacology of 5-aza-2′-deoxycytidine (decitabine). *Semin. Hematol.* **42,** S9–S16 (2005).
37. Momparler, R. *et al.* Kinetic interaction of 5-AZA-2′-deoxycytidine-5′-monophosphate and its 5′-triphosphate with deoxycytidylate deaminase. *Mol. Pharmacol.* **25,** 436–440 (1984).
38. Zhang, J.-H., Chung, T. D. Y. & Oldenburg, K. R. A simple statistical parameter for use in evaluation and validation of high throughput screening assays. *J. Biomol. Screen.* **4,** 67–73 (1999).
39. Chabot, G. G., Bouchard, J. & Momparler, R. L. Kinetics of deamination of 5-aza-2′-deoxycytidine and cytosine arabinoside by human liver cytidine deaminase and its inhibition by 3-deazauridine, thymidine or uracil arabinoside. *Biochem. Pharmacol.* **32,** 1327–1328 (1983).

Acknowledgements

T.L., H.Al., M.H. and H.Ax. acknowledge Karolinska Institutet, SciLifeLab and the Swedish Research Council (Vetenskapsrådet), which funds Chemical Biology Consortium Sweden. P.N. acknowledges funding from Karolinska Institutet (DPA), the Swedish Research Council, the Swedish Cancer Society, the Knut and Alice Wallenberg foundation, as well as a startup grant from Nanyang Technological University. P.A. acknowledges the Swedish Research Council and SciLifeLab. A.M. acknowledges the Portuguese Research Council (Fundação para a Ciência e Tecnologia). R.J. acknowledges Magnus Bergvalls Stiftelse, The Lars Hierta Memorial Foundation and Helge Ax:son Johnsons stiftelse. We are very grateful to Caius G. Radu and Raymond M. Gipson for providing the DCK inhibitor DI-82. We would also like to thank Protein Science Facility (PSF) at Karolinska Institutet for providing DCTD and Staffan Eriksson from the Swedish University of Agricultural Sciences for providing the plasmid for DCK. We are also grateful to Carlo Zaniol at PerkinElmer for valuable support in development of the AlphaScreen assays. Brinton Seashore-Ludlow is acknowledged for careful reviewing and commenting on the manuscript.

Author contributions

H.Al., H.Ax., R.J., D.M.M., T.L. and P.N. conceived the study. H.Al., H.Ax. and T.L. developed the homogenous TS CETSA assay supported by R.J. and D.M.M. and ran the screen. R.J. generated the western blot CETSA data. A.M. and P.A. planned and generated the intracellular concentration data. C.D. generated the *in vitro* thermal shift and surface plasmon resonance data, and solved the crystal structure of CBK115334 together with A.L. M.H. performed the cheminformatics and metabolic pathway analysis. P.N., P.A. and T.L. lead the project. T.L. and P.N. wrote the first draft of the manuscript. All co-authors contributed to the final draft of the manuscript.

Additional information

Accession codes: The crystal structure of CBK115334 and dUMP in complex with TS has been deposited with the RCSB Protein Data Bank under the accession code 5HS3.

Competing financial interests: D.M.M. and P.N. are founders of the company Pelago Bioscience AB. D.C., A.L., A.M., P.A, R.J., T.L., H.Al., M.H. and H.Ax. declare no competing financial interests.

23

Structure of the intact ATM/Tel1 kinase

Xuejuan Wang[1,2], Huanyu Chu[2], Mengjuan Lv[2], Zhihui Zhang[2], Shuwan Qiu[2], Haiyan Liu[2], Xuetong Shen[3], Weiwu Wang[1] & Gang Cai[2,4,5]

The ataxia-telangiectasia mutated (ATM) protein is an apical kinase that orchestrates the multifaceted DNA-damage response. Normally, ATM kinase is in an inactive, homodimer form and is transformed into monomers upon activation. Besides a conserved kinase domain at the C terminus, ATM contains three other structural modules, referred to as FAT, FATC and N-terminal helical solenoid. Here we report the first cryo-EM structure of ATM kinase, which is an intact homodimeric ATM/Tel1 from *Schizosaccharomyces pombe*. We show that two monomers directly contact head-to-head through the FAT and kinase domains. The tandem N-terminal helical solenoid tightly packs against the FAT and kinase domains. The structure suggests that ATM/Tel1 dimer interface and the consecutive HEAT repeats inhibit the binding of kinase substrates and regulators by steric hindrance. Our study provides a structural framework for understanding the mechanisms of ATM/Tel1 regulation as well as the development of new therapeutic agents.

[1] Key Laboratory of Agricultural and Environmental Microbiology, Ministry of Agriculture, College of Life Sciences, Nanjing Agricultural University, Nanjing 210095, China. [2] School of Life Sciences, University of Science and Technology of China, Anhui 230027, China. [3] Department of Epigenetics and Molecular Carcinogenesis, University of Texas M.D. Anderson Cancer Center, Smithville, Texas 78957, USA. [4] Hefei National Laboratory for Physical Sciences at the Microscale, Center for Integrative Imaging, Anhui 230027, China. [5] Center for Biomedical Engineering, University of Science and Technology of China, Anhui 230027, China. Correspondence and requests for materials should be addressed to W.W. (email: wwwang@njau.edu.cn) or to G.C. (email: gcai@ustc.edu.cn).

The genome is constantly under assault by environmental exposure to irradiation, chemical agents, ultraviolet light, as well as endogenous agents such as free radicals generated during normal metabolic processes. DNA double-strand breaks (DSBs) are considered to be the most deleterious DNA damage, which can directly lead to cell death if not repaired. When repaired incorrectly, DSBs can give rise to mutations and chromosomal rearrangements, leading to cancer in multicellular organisms[1]. The ataxia-telangiectasia mutated (ATM) protein is the apical kinase to trigger DNA-damage signalling required for the maintenance of genomic stability on DSB formations. Once activated, ATM phosphorylates a multitude (more than 700) of substrates[2] such as p53 and checkpoint kinase Chk2, which are involved in cell-cycle control, DNA repair, cell survival and other cellular processes[2,3]. These ATM-dependent phosphorylation events are essential to arrest cell-cycle progression to allow DNA repair or to induce apoptotic cell death[4]. Consistently, ATM deficiency correlates with hypersensitivity to DNA-damaging agents, and in human ATM deficiency leads to ataxia telangiectasia (AT), a genetic disorder that is characterized by premature aging, cerebellar neuropathy, immunodeficiency and predisposition to cancer. Besides the role as an apical activator of the DNA-damage response for DSBs, ATM has also been suggested to be a versatile kinase involved in response to various genotoxic stresses and in diverse aspects of cellular homeostasis[3].

The ATM protein, a large serine–threonine protein kinase (>300 kDa), belongs to the evolutionarily conserved phosphatidylinositol-3-kinase-related protein kinase (PIKK) family. There are six members in the family; each plays pivotal roles in controlling cellular homeostasis, including DNA-damage response (ATM, ATR and DNA-PKcs), cell growth (mTOR), mRNA decay (SMG1) and transcriptional regulation (TRRAP)[5]. The PIKKs contain a highly conserved C-terminal FAT/kinase/FATC domain architecture[6,7] and an extended N-terminal helical solenoid of lower sequence similarity[8,9]. Owing to their important and diverse functions, sophisticated regulatory mechanisms are required to control the kinase activities of the PIKKs to adapt to particular pathways[10]. Under normal conditions, ATM kinase in cells is inactive and in the form of homodimers. Upon DSB formation, ATM homodimers quickly undergo intermolecular autophosphorylation and are transformed into monomers to fully activate the kinase activity[11]. Given the central roles of ATM in genome integrity and human diseases, it is essential to understand the mechanisms of ATM regulation. In particular, molecular and structural insights into ATM are critical to facilitate the design of therapeutic agents targeting the ATM/Tel1 kinase[4,5]. However, because of the large size (300 ∼ 500 kDa) and structural complexity, obtaining the high-resolution structure of PIKKs has remained a challenge. At present, except for the crystal structure of mTOR catalytic domain[12], a 6.6-Å crystal structure of DNA-PKcs is the only model of a near-full-length PIKK[8]. Several EM structures were published, but the majority was determined under negative staining, which suffer from deformation artefacts induced by particle dehydration and distortion. Furthermore, the resolution of the structure is too low to even allow reliable docking of the crystal structures of the catalytic domain[13].

In the fission yeast *Schizosaccharomyces pombe*, ATM is encoded by *Tel1* (telomere maintenance 1) gene[14]. The ATM/Tel1 kinase from *S. pombe* is one of the largest proteins, composed of 2,812 amino-acid residues with a molecular weight of 327 kDa. Cryo-EM is uniquely suited for structural analysis for such large and complex protein assembly. Here we report the first cryo-EM structure of the intact ATM kinase, which is in a non-covalent homodimer state. The 8.7-Å structure illuminated the intricate interface of ATM/Tel1 kinase homodimer and the unusual winding tertiary structure of the consecutively stacked N-terminal helical solenoid, as well as the distinct conformation of the N-terminal helical solenoid tightly packing against the FAT and kinase domains. These novel structural insights suggest that the dimer interface and the N-terminal helical solenoid could potentially allow tight regulation of the kinase activity by redundantly regulating the bindings of substrates and regulators. The detailed structural analysis on ATM/Tel1 also improves our understanding of structural organization and functional properties of other PIKKs. Moreover, this information could be used to develop specific inhibitors to lock the ATM in the inactive dimer state as potential radiosensitizers for cancer radiotherapy.

Results

Biochemical purification of ATM/Tel1 kinase. We have established an efficient purification procedure to acquire endogenous ATM/Tel1 directly from the yeast cells, which involves ammonium sulfate precipitation to enrich the ATM/Tel1-containing fraction and one-step affinity chromatography purification using IgG column. To further remove minor contaminants, an ion exchange Mono S column was employed. The procedure yielded highly homogeneous ATM/Tel1 kinase that is uniform in composition using SDS–PAGE analysis (Fig. 1a) and is satisfying for cryo-EM analysis. The particles observed with EM appeared well-preserved under negative stain and were similar to each other in size and overall shape (Supplementary Fig. 1a).

Cryo-EM structure of homodimeric ATM/Tel1 kinase. We first performed negative-staining EM analysis to obtain the initial model for cryo-EM refinement using Random Conical Tilt (RCT) method (Supplementary Fig. 1b). Then, we determined the cryo-EM structure of the homodimeric ATM/Tel1 kinase. Consistent with the biochemical homogeneity of our ATM preparation as demonstrated by SDS–PAGE, cryo-EM images of the ATM/Tel1 particles preserved in vitrified ice (Supplementary Fig. 2a) are mostly homogeneous and are readily picked from the EM micrographs (Supplementary Fig. 2b). Through rigorous two-dimensional (2D; Fig. 1b and Supplementary Fig. 3a) and three-dimensional (3D) classifications using the RCT reconstruction of ATM/Tel1 under stain (low-pass-filtered to 60 Å) as an initial model, ∼32% particles are stably classified into a highly homogeneous class (Supplementary Fig. 3b). To refine this subset to high resolution in a fully automated manner, the 3D auto-refine procedure in RELION was used, resulting in a cryo-EM reconstruction of ATM/Tel1 kinase (Fig. 1c). The resolution was estimated objectively at 8.7 Å (Supplementary Figs 3c and 4b), using the gold-standard Fourier Shell Correlation (FSC) calculations.

The intact ATM/Tel1 kinase directly purified from the yeast cells is a homodimer, and the dimeric architecture resembles a butterfly stretching out its wings (Fig. 1c). Two ATM/Tel1 molecules are juxtaposed in a side-by-side manner, with their top region interacting with each other. This high-resolution structure has allowed us to dissect the structural mechanisms of ATM/Tel1 at an unprecedented level. Particularly noteworthy is that the majority of the ATM/Tel1 domains are composed of helix-turn-helix repeats, with the α-helix being the dominant secondary structure (Supplementary Fig. 4c).

ATM/Tel1 structural modules. The unprecedented details apparent in the ATM/Tel1 structure prompted us to delineate its modular architecture. All the PIKKs contain a highly conserved C-terminal FAT/kinase/FATC domain architecture[6,7]. Consistently, the crystal structure of mTOR catalytic core (PDB ID: 4JSV)[12] could be unambiguously fitted in the Head region of

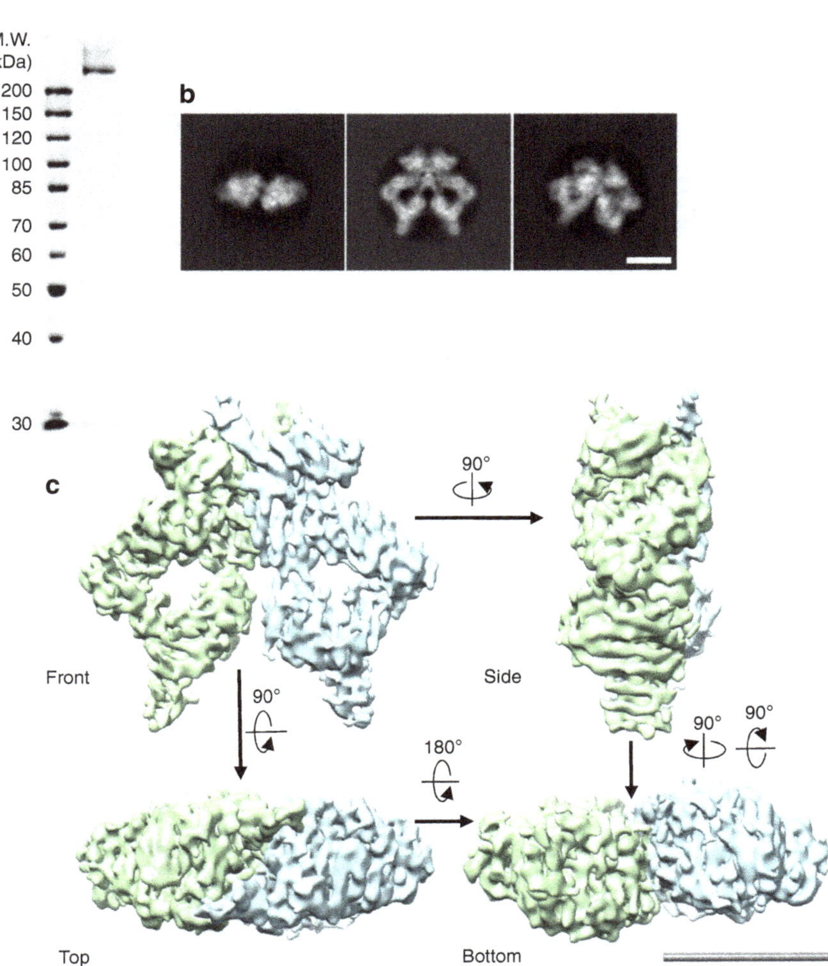

Figure 1 | Purification and cryo-EM structure of intact ATM/Tel1 homodimer. (**a**) SDS-PAGE analysis of the intact ATM/Tel1 kinase endogenously purified from the *S. pombe*. (**b**) Three representative 2D class averages of ATM/Tel1 kinase. Scale bar, 100 Å. (**c**) The structure of ATM/Tel1 homodimer at 8.7 Å resolution, colour-coded by monomer assignment. Scale bar, 100 Å.

the ATM/Tel1 kinase cryo-EM map without flexible fitting (Fig. 2a and Supplementary Fig. 5). The only exception is that the unique FRB domain insertion of mTOR could not be fitted and sticks out from the N-lobe of the kinase domain. A general correspondence between the crystal structure of mTOR catalytic core and the cryo-EM structure of the Head domain of the ATM/Tel1 corroborates the high conservation of C terminus of PIKKs.

On the basis of the crystal structure of mTOR kinase (PDB ID: 4JSV), we modelled the C-terminal 1,027 residues of ATM/Tel1 through homology modelling followed by Molecular Dynamics Flexible Fitting (MDFF)[15] to optimize the fitting into the EM map. The kinase domain of ATM/Tel1 kinase has a typical bi-lobed structure, and the active site is deeply recessed inside a cleft, which is generally in the intrinsically active conformation[5,12,13]. The FAT domain intimately wraps around ~50% of the kinase domain, which has been suggested to regulate the binding of kinase substrates or regulators[5] (Fig. 2b and Supplementary Fig. 6). In addition, adopting a highly sinuous superhelical structure, the N-terminal helical solenoid formed two arms, cradling the catalytic core on top. These two observations are consistent with the model that the intrinsically active conformation of the kinase domain and restricted access of the active site are general features of the PIKKs[10,12].

ATM/Tel1 dimerization interface. The dimer architecture of PIKKs is conserved and dimerization has been implicated in the proper functions of several PIKKs, such as ATM, DNA-PKcs and mTOR[10,13]. Similar to PI3K in its auto-inhibited basal state, each subunit of the dimer likely participates in blocking the activity of the other subunit[4]. It has been suggested that inactive ATM exists as a non-covalent homodimer in cells, but is rapidly activated and converted into monomer on sensing the DNA DSBs[11,14,16]. Owing to the lack of the structural information, the detailed mechanisms for PIKK dimerization and activation remain largely unknown[4].

The structure of the ATM/Tel1 homodimer provides unprecedented structural insight into the dimer architecture of PIKKs and the mechanisms of ATM's auto-inhibition in the dimeric state (Fig. 3a).

The cryo-EM map fitted with the homology model of the C-terminal of ATM/Tel1 illuminated that the dimer interface is mainly centred on the TRD2 regions of the FAT domains between the two monomers, which constitutes the largest part of this interface (Fig. 3b). The arrangement of only helices close to the dimer interface generates a large curved surface across the TRD2 domains (Supplementary Fig. 7). It is tempting to propose that the dimer interface observed in the ATM/Tel1 kinase is conserved among the PIKKs. In this case, several mutations in

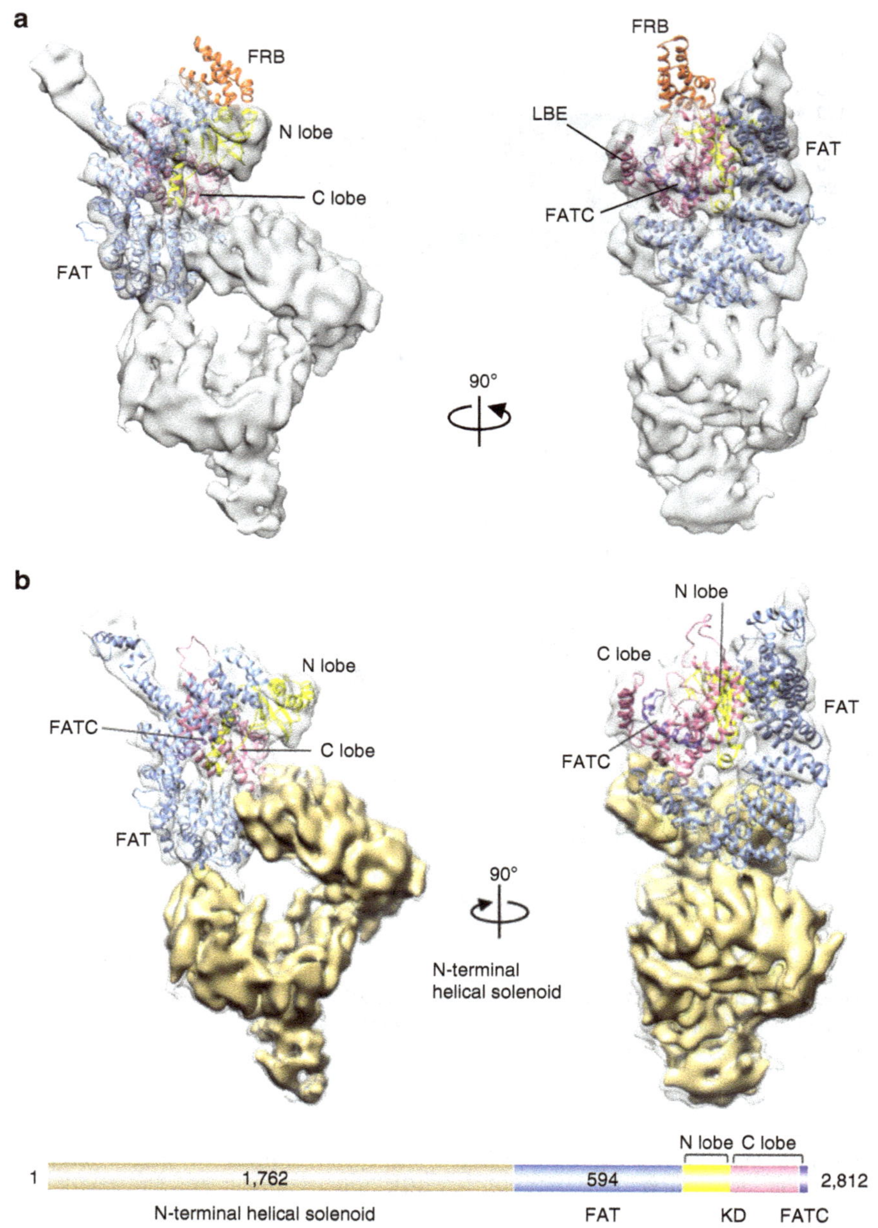

Figure 2 | Modular architecture of ATM/Tel1 kinase. (**a**) The structure of ATM/Tel1 monomer fitted with crystal structure of mTOR catalytic core (PDB ID: 4JSV)[12] as a rigid body. (**b**) Model of the ATM/Tel1 monomer, colour-coded by domain assignment (top). The EM structure of ATM/Tel1 monomer is fitted with the C-terminal ATM/Tel1 homology modelling model (residues: 1,786–2,812) based on the crystal structure of mTOR catalytic core (PDB ID: 4JSV). Domain organization of ATM/Tel1 kinase (bottom). The number of residues of N-terminal helical solenoid and FAT domains are indicated.

mTOR that lead its activation (W1449, A1459, L1460, C1483, and A1577) are also localized in the TRD2 domain[12], which is consistent with the notion that dimerization interface of ATM/Tel1 across the TRD2 domains seems critical to stabilize the dimeric state. Thus, the cryo-EM structure of ATM/Tel1 suggests that these mutations in the TRD2 domain likely destabilize the dimer interface of mTOR.

All PIKKs also contain a LST8-binding element (LBE)-like insertion in the C-lobe of the kinase domain, which is potentially involved in interactions with PIKK regulators[10,12,13]. For example, the LBE domain of mTOR directly interacts with its regulator mLST8 (ref. 5). In the dimer structure of ATM/Tel1, a rod-shaped protrusion density leans on the LBE domain, which corresponds to the linker region of the α21 and α22 of the TRD3 domain (Fig. 3c). Sequence analysis suggests that the secondary structure of the linker region is mainly α-helix, which is consistent with the observed rod-shaped density of the protrusion. Intriguingly, the region of ATM/Tel1 interacting with the conserved LBE-like domain is also highly conserved (Supplementary Fig. 8a). Therefore, we name the protrusion as LBE-interacting domain (LID). In the dimeric state, the LID from one ATM monomer contacts the LBE of the other molecule, possibly blocking a putative ATM regulator binding to LBE, and thus could inhibit the activation of ATM/Tel1 kinase. Consistently, in the mTOR dimer model derived through fitting the crystal structure of mTOR catalytic core into the cryo-EM structure of ATM/Tel1 dimer, the LID domain would directly block mLST8 binding to the LBE domain[12] (Supplementary Fig. 9). Such an intermolecular LID–LBE interaction of ATM/Tel1 constitutes the second dimer interface, which could

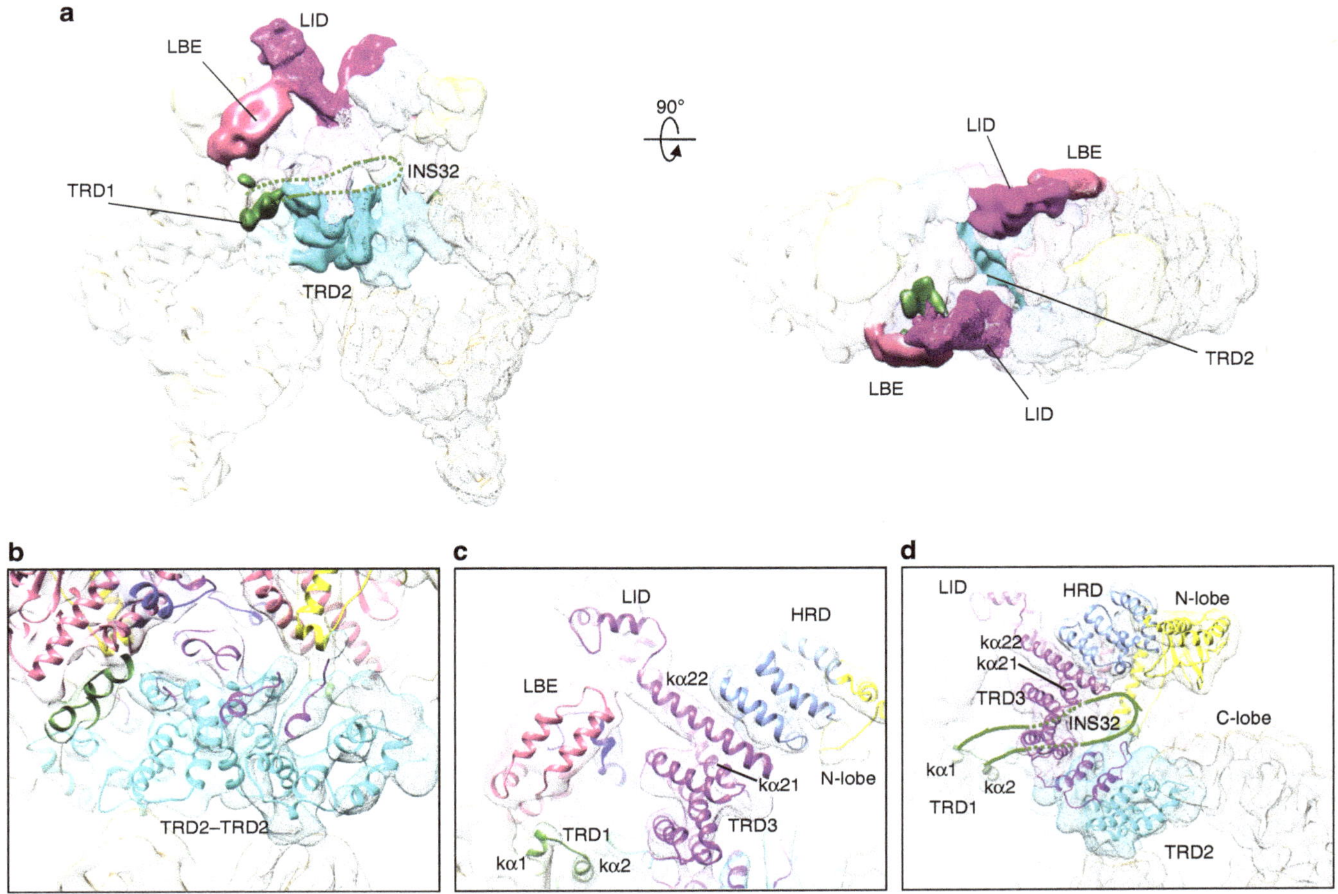

Figure 3 | Three prominent dimer interfaces of ATM/Tel1 kinase. (**a**) EM models of ATM/Tel1 dimer highlighting the following three dimer interfaces. Two monomers of ATM/Tel1 are displayed as surface and mesh. Domains are colour-coded as in Supplementary Fig. 6. (**b**) Enlarged view of the TRD2–TRD2 interface. (**c**) Enlarged view of the LID–LBE interface. (**d**) Enlarged view of the putative INS32-AS (active site) interface in the ATM kinase from higher eukaryotes.

play a critical role in inhibiting the activation of ATM/Tel1 (probably in other PIKKs as well) by blocking the binding of regulators (Fig. 3c).

In higher eukaryotes, intermolecular autophosphorylation of Ser1981 has been proposed to induce dimer dissociation and it initiates ATM kinase activity[11]. Ser1981 lies in the middle of a 32-amino-acid insertion between the α1 and α2 of the TRD1 region of the FAT domain. We named the 32-amino-acid insertion INS32 (Supplementary Fig. 8b). Insertion INS32 and the transphosphorylation of Ser1981 within the insertion are highly conserved in the ATM of higher eukaryotes, but are absent in yeast ATM/Tel1 (ref. 4). This part of the FAT domain plays important roles not only in ATM (S1981) but also in ATR (T1989)[17] and DNA-PKcs (2,023–2,056 and 2,609–2,647; ref. 18). The disordered INS32 probably intertwines with the active site of another molecule, which enables the kinase domain to bind to the region surrounding Ser1981 for its transphosphorylation[11]. In addition, it could potentially restrict the substrate access (Fig. 3d). In this regard, one effect of the Ser1981 autophosphorylation in higher eukaryotes could be to induce the conformational changes of the INS32 to release its binding and inhibition of the active site.

Structure of the N-terminal helical solenoid. The N terminus of all the PIKKs is composed entirely of HEAT repeats, forming a superhelix or α-solenoid, for which limited structural information is available[10]. The consecutively stacked N-terminal helical solenoid of ATM/Tel1 kinase yields an unusually winding tertiary structure with two arms. Therefore, we morphologically segment the N-terminal helical solenoid as C-pincer and N-spiral. In a minor population of the ATM/Tel1 particles, the bottom part of the N-terminal helical solenoid, corresponding to the N-spiral, could sway far from the dimer interface (data not shown). We thus deduce that the C-pincer of the N-terminal helical solenoid is covalently connected to the FAT domain of the ATM/Tel1 kinase (Fig. 4a).

Similar to the α-solenoids of DNA-PKcs[8], the two arms of the N-terminal helical solenoid of ATM/Tel1 directly interact with both the FAT and kinase domains (Fig. 4a,b). In particular, one tip of the C-pincer binds to the TRD2 region of the FAT domain; another tightly interacts with both the C-lobe of the kinase domain and the TRD1 region of the FAT domain (Fig. 4c,d). The N-spiral packs against another region within the TRD2 domain (Fig. 4a,c). Therefore, the entire catalytic core of PIKKs appears tightly coupled to the N-terminal helical solenoid, consistent with the previous structural and functional finding that the N-terminal helical solenoid may serve as a scaffold[10]. Furthermore, the N-terminal helical solenoid also has been reported to be essential for binding proteins that are associated with the PIKKs to regulate their activities and cellular localizations[10]. In summary, our ATM/Tel1 structure has revealed the architecture of this highly important kinase, and provided novel mechanistic insights

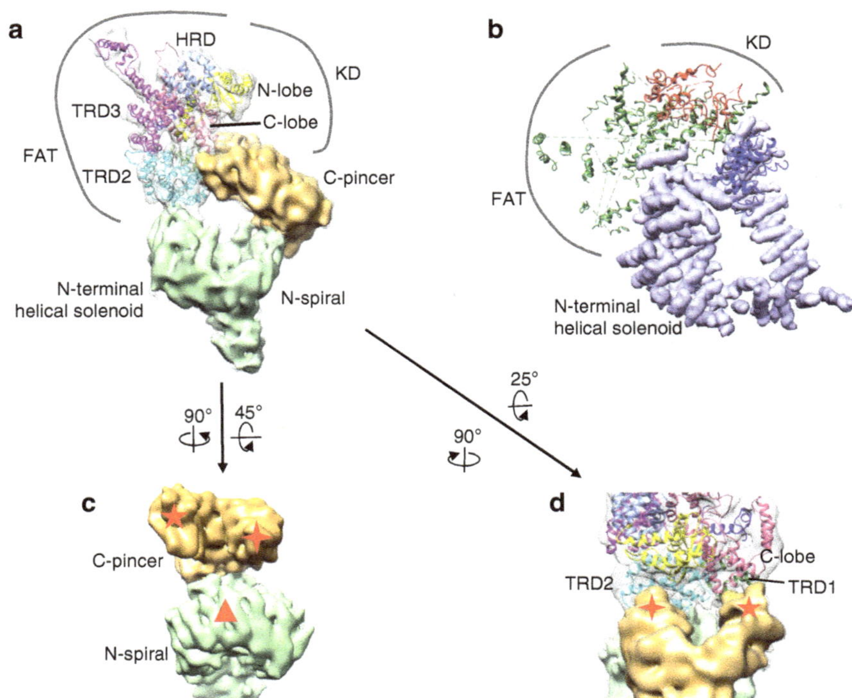

Figure 4 | Overall architecture of the N-terminal helical solenoid. (**a**) Model of ATM/Tel1 monomer highlighting the two arms of the N-terminal helical solenoid: C-pincer and N-spiral are coloured as orange and green, respectively. (**b**) A model for DNA-PK (PDB ID: 3KGV)[8] coloured by the FAT/kinase/N-terminal helical solenoid domains. (**c**) Three interfaces of the C-pincer and N-spiral interacting with the FAT and kinase domains. (**d**) Enlarged view of two interactions of the C-pincer with the FAT domain and C-lobe of kinase domain. The triangle shows the interface of N-spiral contacting with the TRD2 domain of the FAT region. The four-pointed star shows one tip of the C-pincer interacting with the TRD2 domain of the FAT region. The five-pointed star shows another tip of the C-pincer binding with TRD1 domain of the FAT region and C-lobe of the kinase domain.

into how the N-terminal helical solenoid interact with FAT and kinase domains.

Discussion

Since ATM is a central and versatile kinase involved in diverse aspects of cellular homeostasis[3], it is critical to tightly regulate its kinase activity. Accordingly, it has been proposed that ATM is held inactive in cells as a homodimer and quickly transits from dimer into active monomer in response to DNA damage[11]. The intact ATM/Tel1 kinase directly purified from *S. pombe* grown under normal conditions is an inactive homodimer, which suggests that each molecule likely participates in blocking the activity of the other, locking the dimer in an auto-inhibited basal state[11]. Importantly, the dimeric architecture of PIKKs appears to be conserved[5,13], which has been implicated in the proper functions of ATM[11], DNA-PKcs[19] and mTOR[20].

The dimeric ATM/Tel1 structure suggests that each ATM monomer likely participates in inhibiting the activity of the other via at least three interfaces: TRD2–TRD2, LBE–LID and INS32–active site. The TRD2–TRD2 interaction surface constitutes the main scaffold of the dimer interface. Furthermore, the LBE–LID and INS32–active site interfaces could directly inhibit the binding of ATM regulators and block the substrate-binding site, respectively. It is conceivable that the dual blockage mechanism may also be utilized by other PIKK dimers. Moreover, not only with a scaffolding function, the N-terminal helical solenoid could also contribute to control the recruitment and delivery of specific substrates and regulators[10]. Through these multiple and sophisticated mechanisms, the kinase activity of ATM/Tel1 is strongly and redundantly auto-inhibited in the homodimeric state. The dimer–monomer transition of ATM/Tel1 may therefore function to uncover the substrate and regulator-binding domains and allow the formation of active complexes. Therefore, the homodimeric ATM/Tel1 structure offers a structural framework to understand the mechanisms of ATM kinase activation and its regulation.

ATM/Tel1 is the apical kinase to initiate the multifaceted DNA-damage responses to DNA DSBs. Owing to its pivotal roles in the regulation of genomic integrity, ATM has been a potentially viable therapeutic target[4,5,21]. On the basis of the structure of ATM/Tel1 dimer, specific ATM inhibitors could be developed to lock the ATM kinase into the inactive dimeric state. Moreover, conjugating such specific ATM inhibitors with coupling agents harboring high affinity to cancer cells would result in new therapeutic agents as radiosensitizers. By targeting tumour cells during the radiation therapy, radiosensitizers may selectively induce catastrophic genomic instability and cause lethality in cancer cells. Our dimeric ATM/Tel1 structure provides an initial molecular blueprint for the development of potential radiosensitizers for cancer radiotherapy. However, the resolution of the current EM structure of ATM is still too low to directly guide the design of such radiosensitizers. The major limiting factor in increasing the resolution of the ATM/Tel1 cryo-EM structure is the strong preferred orientations of the particles in the vitrified ice (Supplementary Fig. 4a). Further improving the cryo-EM reconstruction of ATM/Tel1 kinase through optimizing the coverage of spatial orientation will critically contribute to providing critical structural information required for the development of drugs targeting ATM.

Methods

Purification of Yeast ATM/Tel1. *S. pombe* stain CC5060 (*h-leu1-32 ura4-D18 kanMX6:nmt1:TAP-tel1*)[14] was grown in YPD medium to the stationary phase.

About 100 g cells were harvested, washed and resuspended in extraction buffer (50 mM HEPES (pH 7.6), 300 mM KOAc, 0.5 mM EDTA, 5 mM β-ME, 10% (v/v) glycerol, 0.1% (v/v) NP-40 and protease inhibitors), and a whole-cell extraction was prepared as previously described[22]. This whole-cell extract was selectively precipitated in 30–55% ammonium sulfate and resuspended using 1 × TEZ buffer (50 mM Tris-HCl (pH 7.5), 1 mM EDTA, 10 mM $ZnCl_2$, 5 mM β-ME and protease inhibitors). After the suspension was clarified using centrifugation, the supernatant was incubated for 2 h at 4 °C with 1 ml of a 50% slurry of IgG-sepharose resin beads (GE Healthcare) that had been pre-equilibrated with 1 × TEZ plus 250 mM ammonium sulfate. After incubation, the beads were washed with 50 ml of 1 × TEZ plus 500 mM ammonium sulfate, followed by a second wash with 50 ml of 1 × TEZ plus 50 mM ammonium sulfate. After equilibration of the column with 1 × TEZ plus 100 mM ammonium sulfate (without protease inhibitors), 100 units of AcTEV protease (Invitrogen) was added to the resin beads and incubated overnight at 4 °C. The ATM/Tel1 fraction was then eluted with three column volumes of 1 × TEZ plus 100 mM ammonium sulfate and 10% glycerol was added, and the resulting aliquot was snap-frozen in liquid nitrogen and temporarily stored at − 80 °C. For the next purification step, the IgG eluate fractions were thawed in ice and applied on a Mono S column (GE Healthcare) in Q100 buffer (100 mM ammonium sulfate, 50 mM Tris, pH 7.6, 10% v/v glycerol, 1 mM EDTA, 10 µM $ZnSO_4$, 0.02% NP-40 and 10 mM β-ME) and was resolved over a 100–1,000-mM ammonium sulfate gradient. The ATM/Tel1 elution was flash-frozen in liquid nitrogen, and was analysed using SDS–PAGE and EM examination.

EM sample preparation and RCT reconstruction. The Mono S peak fraction was diluted eight times (20 mM HEPES (pH 8.0), 40 mM KOAc, 5 mM $MgCl_2$, 0.1% trehalose, 2 mM dithiothreitol (DTT) and 0.01% NP-40) and was applied to a freshly glow-discharged carbon-coated 400-mesh Cu EM specimen grid, and then preserved by staining with 0.75% (w/w) uranyl formate solution. Images were recorded at a magnification of × 62,000 on a 4,096 × 4,096 CCD (charge-coupled device) detector (FEI Eagle) with a Tecnai F20 electron microscope (FEI) operating at an acceleration voltage of 200 kV. Images were recorded by using low-dose procedures at ~0.6–0.8 µm under focus. Twofold pixel binning of the original CCD images resulted in a final pixel size of 3.54 Å per pixel.

The particles of the ATM/Tel1 kinase show a strongly preferred orientation on adsorption to amorphous carbon support films. To obtain the initial model for cryo-EM refinement, 3D reconstructions were calculated by using the RCT method[23]. Tilted (− 55°) and untilted image pairs were obtained under low-dose conditions, and particles were selected using the TiltPicker programme[24] and were montaged for interactive screening, yielding ~11,000 tilt-pair images of the ATM/Tel1. We run iterative alternating rounds of supervised multireference alignment and classification as well as reference-free alignment to improve the homogeneity of the image classes. All the 3D reconstructions were calculated with SPIDER[25] and SPARX[26].

Sample vitrification and cryo-EM data collection. Samples were diluted to a final concentration of 20–50 µg ml^{-1} (20 mM HEPES (pH 8.0), 40 mM KOAc, 5 mM $MgCl_2$, 0.1% trehalose, 2 mM DTT and 0.01% NP-40), and 3 µl of aliquots was applied to freshly glow-discharged Quantifoil R2/1 grids coated with a second layer of thin carbon film. The grids were blotted for 3–4 s at 4 °C in 100% humility, and then plunged into liquid ethane using an FEI Vitrobot (FEI Company). Frozen grids were stored in liquid nitrogen. The grids were first loaded into a Gatan 626 cryo-holder and transferred to an FEI Tecnai TF20 electron microscope to check the quality of the sample vitrification. Then, the grids were transferred to Titan Krios equipped with a field emission source and were operated at 300 kV. Images were recorded on a Falcon 2 direct electron detector at a nominal magnification of × 59,000 with a defocus range of 2–4 µm, resulting in a calibrated sampling of 1.42 Å per pixel. The total accumulated dose rate was set to be 35 e^2 per $Å^2$ on the specimen, and the exposure time was 1.5 s. Each image was fractionated into 23 frames.

Image processing. Frames were summed to a single micrograph for subsequent processing using motion-correction procedure described by Li *et al.*[27]. Contrast transfer function (CTF) parameters and defocus values for each micrograph were determined by using CTFFIND3 (ref. 28). A semi-automated procedure in RELION[29] was used to pick particles. Overall, 57,435 particles were picked after carefully sorting and cleaning. 2D and 3D classification and auto-refinement were performed using RELION. The initial RCT model was low-pass-filtered to 60 Å and was used as the starting model for the 3D classification. 3D refinement used gold-standard FSC calculations to avoid overfitting, and reported resolutions are based on the FSC 0.143 criterion. The 17,894 'good' particles' group was used to polish the refinement. A soft mask was used during the auto-refinement procedure to improve the resolution, and postprocessing was carried out by 'postprocess procedure' in Relion using a provided B-factor − 400 and a soft spherical mask (with a 5-pixel fall-off). The final resolution reported was estimated from the masking-effect-corrected FSC curve using the FSC 0.143 criterion. All the 3D structures were displayed using Chimera[30].

Model building and simulation. A homology model of the C-terminal 1,027 residues of ATM/Tel1 was built with Modeller[31] using the crystal structure of mTOR catalytic core (PBD ID: 4JSV)[12] as a template. Missing and unaligned residues, including residues 2,055D–2,080N, 2,185F–2,231I, 2,393L–2,433K and 2,721L–2,754P, were rebuilt with I-TASSER[32]. This model was used as input for MDFF[15] to fit the density map. All the simulations were prepared using VMD[33] and were performed using NAMD[34]. The mdff package was used for the density-fitting-related preparations. The griddx command was used to convert the density map to MDFF potential. The CHARMM27 force field was used for the molecular dynamics. The ssrestrains command was used to enforce restraints on secondary structure elements. MDFF modelling was carried out in two consecutive stages, as outlined in the MDFF tutorial. In the first stage, a short 200-step minimization was carried out to optimize bond geometries and remove clashes. Then, a 1-ns simulation was performed using a low-density scaling factor of 0.3 (gscale = 0.3). In the second stage, a 200-ps simulation using a higher-density scaling factor of 5 (gscale = 5) followed the simulation in the first stage.

Data availability statement. The 8.7-Å cryo-EM structure of ATM/Tel1 kinase from *S. pombe* is available from the EMDB with accession code EMD-6399. The authors declare that all other data supporting the findings of this study are available within the article and its Supplementary Information Files.

References

1. Aplan, P. D. Causes of oncogenic chromosomal translocation. *Trends Genet.* **22,** 46–55 (2006).
2. Matsuoka, S. *et al.* ATM and ATR substrate analysis reveals extensive protein networks responsive to DNA damage. *Science* **316,** 1160–1166 (2007).
3. Shiloh, Y. & Ziv, Y. The ATM protein kinase: regulating the cellular response to genotoxic stress, and more. *Nat. Rev. Mol. Cell Biol.* **14,** 197–210 (2013).
4. Paull, T. T. Mechanisms of ATM Activation. *Annu. Rev. Biochem.* **84,** 711–738 (2015).
5. Lovejoy, C. A. & Cortez, D. Common mechanisms of PIKK regulation. *DNA Repair (Amst).* **8,** 1004–1008 (2009).
6. Bosotti, R., Isacchi, A. & Sonnhammer, E. L. FAT: a novel domain in PIK-related kinases. *Trends Biochem. Sci.* **25,** 225–227 (2000).
7. Jiang, X., Sun, Y., Chen, S., Roy, K. & Price, B. D. The FATC domains of PIKK proteins are functionally equivalent and participate in the Tip60-dependent activation of DNA-PKcs and ATM. *J. Biol. Chem.* **281,** 15741–15746 (2006).
8. Sibanda, B. L., Chirgadze, D. Y. & Blundell, T. L. Crystal structure of DNA-PKcs reveals a large open-ring cradle comprised of HEAT repeats. *Nature* **463,** 118–121 (2010).
9. Perry, J. & Kleckner, N. The ATRs, ATMs, and TORs are giant HEAT repeat proteins. *Cell* **112,** 151–155 (2003).
10. Baretic, D. & Williams, R. L. PIKKs--the solenoid nest where partners and kinases meet. *Curr. Opin. Struct. Biol.* **29,** 134–142 (2014).
11. Bakkenist, C. J. & Kastan, M. B. DNA damage activates ATM through intermolecular autophosphorylation and dimer dissociation. *Nature* **421,** 499–506 (2003).
12. Yang, H. *et al.* mTOR kinase structure, mechanism and regulation. *Nature* **497,** 217–223 (2013).
13. Angel Rivera-Calzada, A. L.-P. *et al.* Structure and assembly of the PI3K-like protein kinases (PIKKs) revealed by electron microscopy. *AIMS Biophys.* **2,** 21 (2015).
14. You, Z., Chahwan, C., Bailis, J., Hunter, T. & Russell, P. ATM activation and its recruitment to damaged DNA require binding to the C terminus of Nbs1. *Mol. Cell Biol.* **25,** 5363–5379 (2005).
15. Trabuco, L. G., Villa, E., Schreiner, E., Harrison, C. B. & Schulten, K. Molecular dynamics flexible fitting: a practical guide to combine cryo-electron microscopy and X-ray crystallography. *Methods* **49,** 174–180 (2009).
16. Lee, J. H. & Paull, T. T. ATM activation by DNA double-strand breaks through the Mre11-Rad50-Nbs1 complex. *Science* **308,** 551–554 (2005).
17. Liu, S. *et al.* ATR autophosphorylation as a molecular switch for checkpoint activation. *Mol. Cell* **43,** 192–202 (2011).
18. Douglas, P. *et al.* Identification of *in vitro* and *in vivo* phosphorylation sites in the catalytic subunit of the DNA-dependent protein kinase. *Biochem. J.* **368,** 243–251 (2002).
19. Spagnolo, L., Barbeau, J., Curtin, N. J., Morris, E. P. & Pearl, L. H. Visualization of a DNA-PK/PARP1 complex. *Nucleic Acids Res.* **40,** 4168–4177 (2012).
20. Yip, C. K., Murata, K., Walz, T., Sabatini, D. M. & Kang, S. A. Structure of the human mTOR complex I and its implications for rapamycin inhibition. *Mol. Cell* **38,** 768–774 (2010).
21. Chaudhary, M. W. & Al-Baradie, R. S. Ataxia-telangiectasia: future prospects. *Appl. Clin. Genet.* **7,** 159–167 (2014).
22. Takagi, Y., Chadick, J. Z., Davis, J. A. & Asturias, F. J. Preponderance of free mediator in the yeast *Saccharomyces cerevisiae*. *J. Biol. Chem.* **280,** 31200–31207 (2005).

23. Radermacher, M. Three-dimensional reconstruction of single particles from random and nonrandom tilt series. *J. Electron. Microsc. Tech.* **9,** 359–394 (1988).
24. Voss, N. R., Yoshioka, C. K., Radermacher, M., Potter, C. S. & Carragher, B. DoG Picker and TiltPicker: software tools to facilitate particle selection in single particle electron microscopy. *J. Struct. Biol.* **166,** 205–213 (2009).
25. Frank, J. *et al.* SPIDER and WEB: processing and visualization of images in 3D electron microscopy and related fields. *J. Struct. Biol.* **116,** 190–199 (1996).
26. Hohn, M. *et al.* SPARX, a new environment for Cryo-EM image processing. *J. Struct. Biol.* **157,** 47–55 (2007).
27. Li, X. *et al.* Electron counting and beam-induced motion correction enable near-atomic-resolution single-particle cryo-EM. *Nat. Methods* **10,** 584–590 (2013).
28. Mindell, J. A. & Grigorieff, N. Accurate determination of local defocus and specimen tilt in electron microscopy. *J. Struct. Biol.* **142,** 334–347 (2003).
29. Scheres, S. H. RELION: implementation of a Bayesian approach to cryo-EM structure determination. *J. Struct. Biol.* **180,** 519–530 (2012).
30. Pettersen, E. F. *et al.* UCSF Chimera--a visualization system for exploratory research and analysis. *J. Comput. Chem.* **25,** 1605–1612 (2004).
31. Webb, B. & Sali, A. Comparative protein structure modeling using MODELLER. *Curr. Protoc. Bioinformatics* **47,** 5 6 1–5 6 32 (2014).
32. Yang, J. *et al.* The I-TASSER Suite: protein structure and function prediction. *Nat. Methods* **12,** 7–8 (2015).
33. Yahyavi, M., Falsafi-Zadeh, S., Karimi, Z., Kalatarian, G. & Galehdari, H. VMD-SS: a graphical user interface plug-in to calculate the protein secondary structure in VMD program. *Bioinformation* **10,** 548–550 (2014).
34. Phillips, J. C. *et al.* Scalable molecular dynamics with NAMD. *J. Comput. Chem.* **26,** 1781–1802 (2005).

Acknowledgements

We thank Professor Paul Russell (The Scripps Research Institute) for providing the CC5060 strain for expression of *S. pombe* ATM/Tel1. EM data were collected at the Center for Bio-imaging, Institute of Biophysics, Chinese Academy of Sciences. We thank G. Ji and X. Huang for their technical help and support with electron microscopy. EM sample quality check was performed at the Center for Integrative Imaging, Hefei National Laboratory for Physical Science at Microscale. We thank P. Tang for her support with electron microscopy facility. We also thank Professor Xiangwei He for revising the manuscript. This work was supported by the National Basic Research Program (2014CB910700 and 2013CB910200), the National Natural Science Foundation of China (31222017) and the Ministry of Education of PRC (NCET-11-0874).

Author contributions

M.L. initiated the project and optimized the purification procedure; X.W. purified the protein samples, freezed the grids, processed and analysed data; H.C. and H.L. performed the model building; Z.Z. collected cryo-EM data; S.Q. helped grow the yeast cells; X.S. critically revised the manuscript; G.C. and W.W. designed experiments, analysed data and wrote the manuscript with X.W.

Additional information

Competing financial interests: The authors declare no competing financial interests.

Permissions

All chapters in this book were first published in NC, by Nature Publishing Group; hereby published with permission under the Creative Commons Attribution License or equivalent. Every chapter published in this book has been scrutinized by our experts. Their significance has been extensively debated. The topics covered herein carry significant findings which will fuel the growth of the discipline. They may even be implemented as practical applications or may be referred to as a beginning point for another development.

The contributors of this book come from diverse backgrounds, making this book a truly international effort. This book will bring forth new frontiers with its revolutionizing research information and detailed analysis of the nascent developments around the world.

We would like to thank all the contributing authors for lending their expertise to make the book truly unique. They have played a crucial role in the development of this book. Without their invaluable contributions this book wouldn't have been possible. They have made vital efforts to compile up to date information on the varied aspects of this subject to make this book a valuable addition to the collection of many professionals and students.

This book was conceptualized with the vision of imparting up-to-date information and advanced data in this field. To ensure the same, a matchless editorial board was set up. Every individual on the board went through rigorous rounds of assessment to prove their worth. After which they invested a large part of their time researching and compiling the most relevant data for our readers.

The editorial board has been involved in producing this book since its inception. They have spent rigorous hours researching and exploring the diverse topics which have resulted in the successful publishing of this book. They have passed on their knowledge of decades through this book. To expedite this challenging task, the publisher supported the team at every step. A small team of assistant editors was also appointed to further simplify the editing procedure and attain best results for the readers.

Apart from the editorial board, the designing team has also invested a significant amount of their time in understanding the subject and creating the most relevant covers. They scrutinized every image to scout for the most suitable representation of the subject and create an appropriate cover for the book.

The publishing team has been an ardent support to the editorial, designing and production team. Their endless efforts to recruit the best for this project, has resulted in the accomplishment of this book. They are a veteran in the field of academics and their pool of knowledge is as vast as their experience in printing. Their expertise and guidance has proved useful at every step. Their uncompromising quality standards have made this book an exceptional effort. Their encouragement from time to time has been an inspiration for everyone.

The publisher and the editorial board hope that this book will prove to be a valuable piece of knowledge for researchers, students, practitioners and scholars across the globe.

List of Contributors

Arlie J. Rinaldi
Single Molecule Analysis Group, Department of Chemistry, University of Michigan, Ann Arbor, Michigan 48109, USA

Paul E. Lund
Single Molecule Analysis Group, Department of Chemistry, University of Michigan, Ann Arbor, Michigan 48109, USA
Program in Chemical Biology, University of Michigan, Ann Arbor, Michigan 48109, USA

Mario R. Blanco and Nils G. Walter
Single Molecule Analysis Group, Department of Chemistry, University of Michigan, Ann Arbor, Michigan 48109, USA

Chuang Liu, Jiying Ning, Gongpu Zhao, Angela M. Gronenborn, Jinwoo Ahn, In-Ja Byeon, and Peijun Zhang
Department of Structural Biology, University of Pittsburgh School of Medicine, Pittsburgh, Pennsylvania 15260, USA
Pittsburgh Center for HIV Protein Interactions, University of Pittsburgh School of Medicine, Pittsburgh, Pennsylvania 15260, USA

Juan R. Perilla and Klaus Schulten
Department of Physics and Beckman Institute, University of Illinois at Urbana-Champaign, Urbana, Illinois 61801, USA

Manman Lu, Guangjin Hou and Tatyana Polenova
Pittsburgh Center for HIV Protein Interactions, University of Pittsburgh School of Medicine, Pittsburgh, Pennsylvania 15260, USA
Department of Chemistry and Biochemistry, University of Delaware, Newark, Delaware 19716, USA

Ruben Ramalho and Itay Rousso
Department of Physiology and Cell Biology, Ben-Gurion University of the Negev, Be'er-Sheva, 84105, Israel

Benjamin A. Himes
Department of Structural Biology, University of Pittsburgh School of Medicine, Pittsburgh, Pennsylvania 15260, USA

Gregory J. Bedwell
Department of Microbiology, University of Alabama at Birmingham, Birmingham Alabama 35294, USA

Peter E. Prevelige
Pittsburgh Center for HIV Protein Interactions, University of Pittsburgh School of Medicine, Pittsburgh, Pennsylvania 15260, USA
Department of Microbiology, University of Alabama at Birmingham, Birmingham Alabama 35294, USA

Christopher Aiken
Pittsburgh Center for HIV Protein Interactions, University of Pittsburgh School of Medicine, Pittsburgh, Pennsylvania 15260, USA
Department of Pathology, Microbiology and Immunology, Vanderbilt University School of Medicine, Nashville, Tennessee 37232, USA

Sarvenaz Sarabipour and Kalina Hristova
Department of Materials Science and Engineering, Johns Hopkins University, Baltimore, Maryland 21218, USA

Michal Brunwasser-Meirom, Sarah Goldberg, Lior Levy andOrna Atar1
Department of Biotechnology and Food Engineering, Technion—Israel Institute of Technology, Haifa 32000, Israel

Yaroslav Pollak
Russell Berrie Nanotechnology Institute, Technion—Israel Institute of Technology, Haifa 32000, Israel

Roee Amit
Russell Berrie Nanotechnology Institute, Technion—Israel Institute of Technology, Haifa 32000, Israel
Department of Biotechnology and Food Engineering, Technion—Israel Institute of Technology, Haifa 32000, Israel

Ulrike Bo¨hm, Stefan W. Hell and Roman Schmidt
Department of NanoBiophotonics, Max Planck Institute for Biophysical Chemistry, Am Fassberg 11, Go¨ttingen 37077, Germany

Kevin J. Freedman, Aleksandar P. Ivanov and Joshua B. Edel
Department of Chemistry, Imperial College London, South Kensington, London SW7 2AZ, UK

Lauren M. Otto
Department of Electrical and Computer Engineering, University of Minnesota, Minneapolis, Minnesota 55455, USA

Avijit Barik and Sang-Hyun Oh
Department of Electrical and Computer Engineering, University of Minnesota, Minneapolis, Minnesota 55455, USA
Department of Biomedical Engineering, University of Minnesota, Minneapolis, Minnesota 55455, USA

Takahiro Mori, Lihan Zhang, Takayoshi Awakawa, Shotaro Hoshino, Masahiro Okada and Ikuro Abe
Graduate School of Pharmaceutical Sciences, The University of Tokyo, 7-3-1 Hongo, Bunkyo-ku, Tokyo 113-0033, Japan

Hiroyuki Morita
Department of Medicinal Resources, Institute of Natural Medicine, University of Toyama, 2630-Sugitani, Toyama 930-0194, Japan

Ahmad Jomaa, Daniel Boehringer, Marc Leibundgut and Nenad Ban
Department of Biology, Institute of Molecular Biology and Biophysics, Otto-Stern-Weg 5, ETH Zurich CH-8093, Switzerland

Patrick Lie Johansen, Federico Fenaroli, Lasse Evensen, Gareth Griffiths and Gerbrand Koster
Department of Biosciences, University of Oslo, Blindernveien 31, 0371 Oslo, Norway

Yijian Rao, Marco G. Perna, Benjamin Hofmann, Viola Beier and Thomas Wollert
Molecular Membrane and Organelle Biology, Max Planck Institute of Biochemistry, Am Klopferspitz 18, Martinsried 82152, Germany

J.-F. Berret
Matière et Systèmes Complexes, UMR 7057 CNRS Universitè Denis Diderot Paris-VII, Bâtiment Condorcet, 10 rue Alice Domon et Léonie Duquet, 75205
Paris, France

Darren Yang
School of engineering and applied sciences, Harvard University, Cambridge, Massachusetts 02138, USA

Andrew Ward
Program in Cellular and Molecular Medicine, Boston Children's Hospital, Boston, Massachusetts 02115, USA
Department of Biological Chemistry and Molecular Pharmacology, Harvard Medical School,Boston, Massachusetts 02115, USA

Ken Halvorsen
Program in Cellular and Molecular Medicine, Boston Children's Hospital, Boston, Massachusetts 02115, USA
Department of Biological Chemistry and Molecular Pharmacology, Harvard Medical School,Boston, Massachusetts 02115, USA
The RNA Institute, University at Albany, Albany, New York 12222, USA

Wesley P. Wong
Program in Cellular and Molecular Medicine, Boston Children's Hospital, Boston, Massachusetts 02115, USA
Department of Biological Chemistry and Molecular Pharmacology, Harvard Medical School,Boston, Massachusetts 02115, USA
Wyss Institute for Biologically Inspired Engineering, Harvard University, Boston, Massachusetts 02115, USA

Christian L. Vestergaard, Morten Bo Mikkelsen, Walter Reisner, Anders Kristensen and Henrik Flyvbjerg
Department of Micro- and Nanotechnology, Technical University of Denmark, DK-2800 Kgs. Lyngby, Denmark

Alessandro Costa, Ferdos Abid Ali,Julian Gannon and Jin Chuan Zhou
Macromolecular Machines, Clare Hall Laboratory, The Francis Crick Institute, Blanche Lane, South Mimms EN6 3LD, UK

Ludovic Renault
Macromolecular Machines, Clare Hall Laboratory, The Francis Crick Institute, Blanche Lane, South Mimms EN6 3LD, UK
National Institute for Biological Standards and Control, Microscopy and Imaging, Blanche Lane, South Mimms EN6 3QG, UK

Hailey L. Gahlon and David Rueda
Section of Virology and Single Molecule Imaging Group
Department of Medicine, MRC Clinical Centre, Imperial College London, London W12 0NN, UK

Abhay Kotecha
Division of Structural Biology, Wellcome Trust Centre for Human Genetics, University of Oxford, Roosevelt Drive, Oxford OX3 7BN, UK

Woo Hyeon Jeong, Jae-Hoon Eom, Hee-Seung Lee and Jie-Oh Lee
Department of Chemistry, KAIST, Daejeon 34141, Korea

Haerim Lee and Sun Chang Kim
Department of Biological Sciences, KAIST, Daejeon 34141, Korea

Dong Hyun Song
Agency for Defense Development, Daejeon 34186, Korea

Hayyoung Lee
Institute of Biotechnology, Chungnam National University, Daejeon 34134, Korea

Nasrollah Rezaei-Ghaleh
German Center for Neurodegenerative Diseases (DZNE), 37075 Go¨ttingen, Germany

Mehriar Amininasab
Department of Cell and Molecular Biology, School of Biology, College
of Science, University of Tehran, 1417466191 Tehran, Iran

Sathish Kumar and Jochen Walter
Department of Neurology, University of Bonn, 53127 Bonn, Germany

Markus Zweckstetter
German Center for Neurodegenerative Diseases (DZNE), 37075 Go¨ttingen, Germany
Department for NMRbased Structural Biology, Max Planck Institute for Biophysical Chemistry, 37077 Göttingen, Germany
Department of Neurology, University Medical Center Göttingen, University of Göttingen, 37073 Göttingen, Germany

Jejoong Yoo
Department of Physics and the Center for the Physics of Living Cells, University of Illinois at Urbana-Champaign, Urbana, Illinois 61801, USA

Hajin Kim
School of Life Sciences, Ulsan National Institute of Science and Technology, Ulsan, Korea
Center for Soft and Living Matter, Institute for Basic Science, Ulsan, Korea

Aleksei Aksimentiev
Department of Physics and the Center for the Physics of Living Cells, University of Illinois at Urbana-Champaign, Urbana, Illinois 61801, USA
Beckman Institute for Advanced Science and Technology, University of Illinois at Urbana-Champaign, Urbana, Illinois 61801, USA

Taekjip Ha
Department of Physics and the Center for the Physics of Living Cells, University of Illinois at Urbana-Champaign, Urbana, Illinois 61801, USA
Howard Hughes Medical Institute, Baltimore, Maryland 21205, USA
Department of Biophysics and Biophysical Chemistry, Johns Hopkins University, Baltimore, Maryland 21205, USA
Department of Biophysics, Johns Hopkins University, Baltimore, Maryland 21205, USA. 8 Department of Biomedical Engineering, Johns Hopkins
University, Baltimore, Maryland 21205, USA

Chao Yan
Department of Biochemistry and Molecular Biology, University Park, Pennsylvania 16802, USA
Center for Eukaryotic Gene Regulation, University Park, Pennsylvania 16802, USA

Shuyang Wu
Department of Biochemistry and Molecular Biology, University Park, Pennsylvania 16802, USA

Christopher Pocetti
Department of Physics, The Pennsylvania State University, University Park, Pennsylvania 16802, USA

Lu Bai
Department of Biochemistry and Molecular Biology, University Park, Pennsylvania 16802, USA
Center for Eukaryotic Gene Regulation, University Park, Pennsylvania 16802, USA
Department of Physics, The Pennsylvania State University, University Park, Pennsylvania 16802, USA

Simon P. Neville and Graham A. Worth
School of Chemistry, University of Birmingham, Edgbaston, Birmingham B15 2TT, UK

Oliver M. Kirkby, Nikolas Kaltsoyannis and Helen H. Fielding
Department of Chemistry, University College London, 20 Gordon Street, London WC1H 0AJ, UK

Helena Almqvist, Hanna Axelsson, Martin Haraldsson, Thomas Lundbäck
Laboratories for Chemical Biology, Karolinska Institutet, Science for Life Laboratory Stockholm, Division of Translational Medicine & Chemical Biology, Department of Medical Biochemistry & Biophysics, Karolinska Institutet, Tomtebodavägen 23A, Solna 171 65, Sweden

Rozbeh Jafari and Daniel Martinez Molina
Department of Medical Biochemistry & Biophysics, Division of Biophysics, Karolinska Institutet, Scheeles väg 2, Stockholm 171 77, Sweden

Chen Dan
School of Biological Sciences, Nanyang Technological University, 61 Biopolis Drive (Proteos), Singapore 138673, Singapore

André Mateus
Department of Pharmacy, Uppsala University, BMC, Box 580, Uppsala SE-751 23, Sweden

Andreas Larsson
School of Biological Sciences, Nanyang Technological University, SBS-04s-45, 60 Nanyang Drive, Singapore 639798, Singapore

Per Artursson
Department of Pharmacy, Uppsala University, BMC, Box 580, Uppsala SE-751 23, Sweden
Uppsala University Drug Optimization and Pharmaceutical Profiling Platform (UDOPP), Department of Pharmacy, Uppsala University, BMC, Box 580, Uppsala SE-751 23, Sweden
Science for Life Laboratory Drug Discovery and Development platform, Uppsala University, Uppsala SE-751 23, Sweden

Pär Nordlund
Department of Medical Biochemistry & Biophysics, Division of Biophysics, Karolinska Institutet, Scheeles väg 2, Stockholm 171 77, Sweden
School of Biological Sciences, Nanyang Technological University, 61 Biopolis Drive (Proteos), Singapore 138673, Singapore
Institute of Cellular and Molecular Biology, ASTAR, 61 Biopolis Drive (Proteos), Singapore 138673, Singapore

Xuejuan Wang
Key Laboratory of Agricultural and Environmental Microbiology, Ministry of Agriculture, College of Life Sciences, Nanjing Agricultural University, Nanjing 210095, China
School of Life Sciences, University of Science and Technology of China, Anhui 230027, China

Huanyu Chu, Mengjuan Lv, Zhihui Zhang, Shuwan Qiu and Haiyan Liu
School of Life Sciences, University of Science and Technology of China, Anhui 230027, China

Xuetong Shen
Department of Epigenetics and Molecular Carcinogenesis, University of Texas M.D. Anderson Cancer Center, Smithville, Texas 78957, USA

Weiwu Wang
Key Laboratory of Agricultural and Environmental Microbiology, Ministry of Agriculture, College of Life Sciences, Nanjing Agricultural University, Nanjing 210095, China

Gang Cai
School of Life Sciences, University of Science and Technology of China, Anhui 230027, China
Department of Epigenetics and Molecular Carcinogenesis, University of Texas M.D. Anderson Cancer Center, Smithville, Texas 78957, USA
Hefei National Laboratory for Physical Sciences at the Microscale, Center for Integrative Imaging, Anhui 230027, China. 5Center for Biomedical Engineering, University of Science and Technology of China, Anhui 230027, China

Index

www.ingramcontent.com/pod-product-compliance
Lightning Source LLC
Chambersburg PA
CBHW080250230326
41458CB00097B/4260

* 9 7 8 1 6 3 2 3 8 5 4 4 4 *

Gastroenterology and Hepatology

Edited by **Greg Callister**

New Jersey

Published by Foster Academics,
61 Van Reypen Street,
Jersey City, NJ 07306, USA
www.fosteracademics.com

Gastroenterology and Hepatology
Edited by Greg Callister

International Standard Book Number: 978-1-63242-465-5 (Hardback)

Printed in the United States of America.

Contents

Preface

The main aim of this book is to educate learners and enhance their research focus by presenting diverse topics covering this vast field. This is an advanced book which compiles significant studies by distinguished experts. This book addresses successive solutions to the challenges arising in the area of application, along with it; the book provides scope for future developments.

Gastroenterology is a part of medicinal studies which deals with the disorders related to the digestive system. It studies all diseases related to mouth till anus, usually referred to as gastrointestinal tract. Hepatology is a study of liver, pancreas and biliary tree and is generally considered a sub-field of gastroenterology. It uses techniques like endoscopic retrograde cholangiopancreatography (ERCP), transhepatic pancreato-cholangiography (TPC), transjugular intrahepatic portosystemic shunt (TIPSS), etc. to treat the diseases. This book outlines the processes and applications of various diagnostic methods and treatment strategies in detail. It traces the progress made in this field and highlights its key concepts. Those in search of information to further their knowledge will be greatly assisted by this book. It will serve as a valuable source of reference to gastroenterologists, hepatologists, researchers and students associated with these fields at various levels.

It was a great honour to edit this book, though there were challenges, as it involved a lot of communication and networking between me and the editorial team. However, the end result was this all-inclusive book covering diverse themes in the field.

Finally, it is important to acknowledge the efforts of the contributors for their excellent chapters, through which a wide variety of issues have been addressed. I would also like to thank my colleagues for their valuable feedback during the making of this book.

Editor

Transperineal ultrasonography in perianal Crohn disease: A valuable imaging modality

Emily K Wright MD[1,2], Kerri L Novak MD[1], Cathy Lu MD[1], Remo Panaccione MD[1], Subrata Ghosh MD FCAHS[1], Stephanie R Wilson MD[1,3]

EK Wright, KL Novak, C Lu, R Panaccione, S Ghosh, SR Wilson. Transperineal ultrasonography in perianal Crohn disease: A valuable imaging modality. Can J Gastroenterol Hepatol 2015;29(8):445-447.

Aims of treatment for Crohn disease have moved beyond the resolution of clinical symptoms to objective end points including endoscopic and radiological normality. Regular re-evaluation of disease status to safely, readily and reliably detect the presence of inflammation and complications is paramount. Improvements in sonographic technology over recent years have facilitated a growing enthusiasm among radiologists and gastroenterologists in the use of ultrasound for the assessment of inflammatory bowel disease. Transabdominal intestinal ultrasound is accurate, affordable and safe for the assessment of luminal inflammation and complications in Crohn disease, and can be performed with or without the use of intravenous contrast enhancement. Perianal fistulizing disease is a common, complex and often treatment-refractory complication of Crohn disease, which requires regular radiological monitoring. Endoanal ultrasound is invasive, uncomfortable and yields limited assessment of the perineal region. Although magnetic resonance imaging of the pelvis is established, timely access may be a problem. Transperineal ultrasound has been described in small studies, and is an accurate, painless and cost-effective method for documenting perianal fluid collections, fistulas and sinus tracts. In the present article, the authors review the literature regarding perineal ultrasound for the assessment of perianal Crohn disease and use case examples to illustrate its clinical utility.

Key Words: *Crohn disease; Inflammatory bowel disease; Perianal fistula; Ultrasound*

L'échographie transpérinéale en cas de maladie de Crohn périanale : un précieux mode d'imagerie

Les objectifs du traitement de la maladie de Crohn ne se limitent plus à la résolution des symptômes cliniques, mais visent désormais des paramètres objectifs, y compris la normalité endoscopique et radiologique. Il est essentiel de réévaluer régulièrement l'état de la maladie pour déceler rapidement, de manière sécuritaire et fiable, la présence d'inflammation et de complications. Ces dernières années, les améliorations à la technologie échographique ont favorisé un enthousiasme croissant de la part des radiologues et des gastroentérologues envers l'utilisation de l'échographie pour évaluer les maladies inflammatoires de l'intestin. L'échographie intestinale transabdominale, qui est précise, peu coûteuse et sécuritaire pour évaluer l'inflammation et les complications intracavitaires en cas de maladie de Crohn, peut être effectuée avec ou sans agent de contraste. Les fistules périanales sont une complication courante, complexe et souvent réfractaire au traitement, qui exigent une surveillance radiologique régulière. L'échographie endoanale est invasive, désagréable et permet peu d'évaluer la région périnéale. Même si l'imagerie par résonance magnétique du bassin est établie, son accès rapide peut être problématique. L'échographie transpérinéale, décrite dans de petites études, est précise, indolore et rentable pour vérifier l'accumulation de liquides périnéaux, les fistules et les voies sinusales. Dans le présent article, les auteurs examinent les publications scientifiques sur l'échographie périnéale pour évaluer la maladie de Crohn périnéale. Ils utilisent des exemples de cas pour en illustrer l'utilité clinique.

Perianal Crohn disease is a common, important, complex and distressing complication of Crohn disease (1,2). Even with regular anti-tumour necrosis factor therapy, long-term healing rates are poor (3). Despite closure of draining external orifices, deep fistula tracks may persist, placing patients at risk for further complications (4). Regular assessment with imaging is, therefore, important as part of comprehensive patient care. Magnetic resonance imaging (MRI) of the pelvis has been described extensively for the assessment of perianal Crohn disease. It offers good soft tissue discrimination and enables multiplanar image acquisition without radiation hazard; however, cost and timely access to this resource have limited frequent use for some patients. Endoanal ultrasound has also been described for the assessment of perianal Crohn disease, but is invasive, often poorly tolerated in the context of active perianal sepsis and impossible to perform if anal stenosis is present. With this technique, perineal views are incomplete and pathological changes expanding to the gluteal region cannot be assessed.

Transperineal ultrasound has been described as a simple and painless method to examine perianal pathology and provides imaging quality comparable with both endoanal ultrasound and MRI (5,6). For the assessment of perianal Crohn disease, including abscess collections, fistulas and sinus tracts, transperineal ultrasound has been validated by several groups (7-10).

Using case examples, we illustrate the utility of transperineal ultrasound in the assessment of patients with perianal Crohn disease. We hope to advocate for its wider use in Canadian practice. We summarize the clinical studies to date evaluating this imaging modality in this population.

CASE 1

A 26-year-old woman with ileocolonic and perianal Crohn disease undergoing therapy with adalimumab and methotrexate was referred for investigation of increasing perianal pain. Transperineal ultrasound was performed. Figure 1 shows the level of the low anal canal. There are two low internal openings at one o'clock and six o'clock. The tract at one o'clock (Figure 1 [A]) was observed on the left anterolateral wall; from this, there is a transphincteric tract that runs laterally to the

[1]*Department of Medicine, Division of Gastroenterology, University of Calgary, Foothills Medical Centre, Calgary, Alberta;* [2]*Department of Gastroenterology, St Vincent's Hospital and The University of Melbourne, Melbourne, Victoria, Australia;* [3]*Department of Diagnostic Imaging, University of Calgary, Foothills Medical Centre, Calgary, Alberta*

Correspondence: Dr Stephanie R Wilson, Diagnostic Imaging, Foothills Medical Centre, Calgary, Alberta T2N 4J8. e-mail emily.wright@svha.org.au

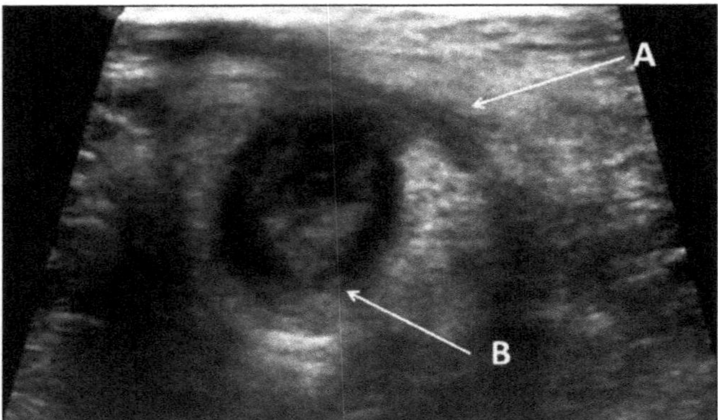

Figure 1) *Transverse view of the anal canal showing a transphincteric hook-shaped fistula* (**A**), *which has an internal opening on the left anterolateral wall of the anal canal at one o'clock. Its external opening on the skin of the left anterior perineum is not shown on this image. Arrow* (**B**) *indicates a small second internal opening directly posterior at six o'clock showing as a small defect in the sphincter*

left and comes to an external opening on the left lateral aspect of the anal canal. The tract at six o'clock (Figure 1 [B]) runs transphincteric before turning anteriorly and joining with the other tract.

CASE 2

A 49-year-old woman with colonic Crohn disease treated with corticosteroid only was referred for investigation of increased perianal pain and discharge. Transperineal ultrasound (Figure 2, top and bottom panels) was performed and demonstrated a fistula arising from the anterior wall of the anal canal between 12 o'clock and one o'clock (Figure 2, top panel). The internal opening is broad; from this, there is a hypoechoic tract-like area that runs directly anterior into the vagina.

DISCUSSION

Transperineal ultrasound for the assessment of perianal Crohn disease was first described by Stewart et al (7) in a cohort of 54 patients being investigated for perianal inflammatory disease. Regular convex, high-frequency, high-resolution linear probes were used. All patients had perianal pathology identified at ultrasound, and 49 had perianal fistulas (intersphincteric, transphincteric and extrasphincteric) or sinuses, of whom 30 underwent surgical intervention and validation of ultrasonographic findings. Of these, 26 patients had perianal fistulas and/or sinus tracts confirmed in the exact anatomical location and course accurately identified on preoperative sonography. Although the location and course of the fistula was confirmed surgically in three patients, the internal opening could not be seen. In one patient, a high perirectal internal opening was not detected on ultrasound but observed at the time of surgery, although the course of the fistula tract was correctly identified on ultrasound (7). The study found transperineal ultrasound to be an accurate and painless technique for the assessment of perianal inflammatory disease and recommended implementing transperineal sonography as a routine sonographic procedure for the evaluation of these patients.

These findings were reproduced in a larger cohort by Mallouhi et al (10). In this study, involving 87 patients with suspected perianal inflammatory disease, transperineal ultrasound performed well in the diagnosis (area under the curve [r] = 0.85; $P<0.001$) and characterization ($r=0.65$; $P<0.001$) of disease when compared with surgical findings. For the detection of perianal fistulas and abscesses, sensitivity and specificity of transperineal ultrasound ranged from 94% to 100% (10). A smaller study involving 25 patients by Wedemeyer et al (11) confirmed high sensitivity and specificity of transperineal ultrasound, when compared with pelvic MRI, for the diagnosis and characterization perianal fistulas and/or abscesses with excellent agreement between the two imaging methods (kappa >0.83). Comparison of transperineal ultrasound and endoanal ultrasound was performed in a study involving 46 patients with perianal Crohn disease by Maconi et al (9). In this study, transperineal ultrasound was accurate for diagnosis and classification of perianal fistulas and/or abscesses, with a sensitivity of 84.9%.

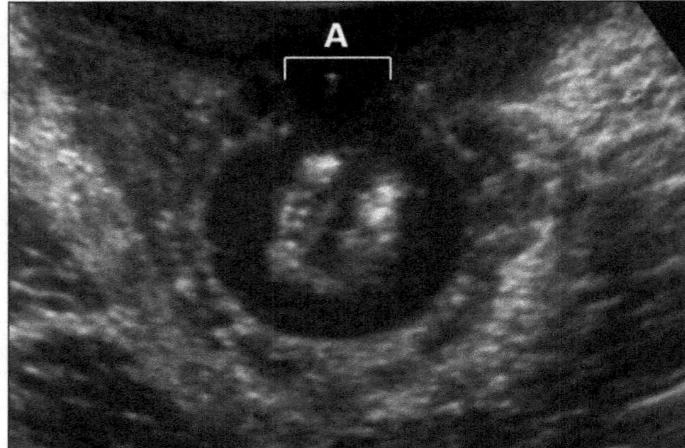

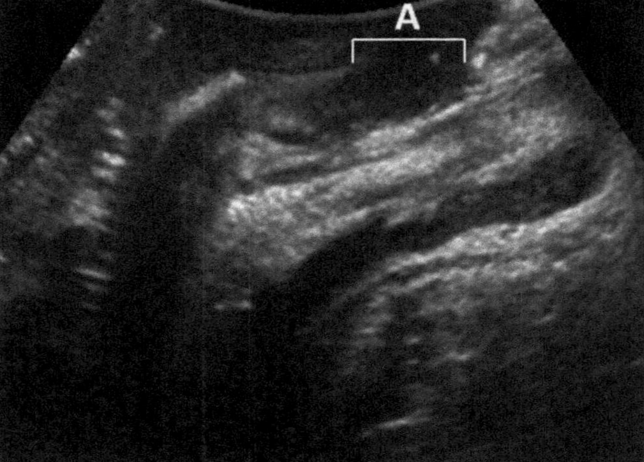

Figure 2) Top panel *Transverse view of the anal canal demonstrating the rectovaginal fistula in the 12 o'clock to one o'clock position* (**A**). **Bottom panel** *Sagittal view of the anal canal and vagina. Bracket A shows gas-filled fistula extending from the lumen of the anal canal to the posterior vaginal wall. In both images, the fistula is broad and appears black. A tiny punctate echogenicity within the fistula is a gas bubble*

The benefits of transperineal ultrasound are clear. The technique is simple, painless, inexpensive and can be performed in real-time in an outpatient clinic setting or in a radiology department. It is accurate, correlates well with surgical findings and performs well against established methods for perianal imaging including endoanal ultrasound and pelvic MRI. The relative simplicity of the technique and lack of patient preparation required lends itself to point-of-care testing as part of a complete physical assessment in the outpatient clinic setting, enabling prompt clinical decision making. However, despite this, transperineal ultrasound is performed routinely at only a small number of centres worldwide. A possible explanation for this may be the steep learning curve needed to establish proficiency in scanning and interpretation, and the limited opportunities for training. While there are no published guidelines regarding the number of scans or time required to establish proficiency, it has been suggested that competent ultrasonographers will achieve competency in 12 months (7).

DISCLOSURES: The authors have no financial disclosures or conflicts of interest to declare.

CONCLUSION

Transperineal ultrasound yields high detection rates, and comprehensive and accurate assessment of perianal fisulas and their complications in patients with Crohn disease. Complex perianal disease requires urgent, even emergent assessment. Although pelvic MRI is currently considered to be the modality of choice for evaluating patients with perianal Crohn disease, transperineal ultrasound has many advantages including lower cost, high patient tolerance, timely access and easy repeatability. Performed by appropriately skilled operators, point-of-care transperineal ultrasound in an outpatient clinic setting offers the possibility of prompt and informed clinical decision making. This technique is currently underappreciated and underused by gastroenterologists and radiologists, despite being first described >10 years ago as an efficient and accurate diagnostic tool in the investigation and characterization of perianal Crohn disease.

REFERENCES

1. Sandborn WJ, Fazio VW, Feagan BG, et al. AGA technical review on perianal Crohn's disease. Gastroenterology 2003;125:1508-30.
2. Mahadev S, Young JM, Selby W, et al. Quality of life in perianal Crohn's disease: What do patients consider important? Dis Colon Rectum 2011;54:579-85.
3. Sands BE, Anderson FH, Bernstein CN, et al. Infliximab maintenance therapy for fistulizing Crohn's disease. N Engl J Med 2004;350:876-85.
4. Van Assche G, Vanbeckevoort D, Bielen D, et al. Magnetic resonance imaging of the effects of infliximab on perianal fistulizing Crohn's disease. Am J Gastroenterol 2003;98:332-9.
5. Kleinubing H Jr, Jannini JF, Malafaia O, et al. Transperineal ultrasonography: New method to image the anorectal region. Dis Colon Rectum 2000;43:1572-4.
6. Rubens DJ, Strang JG, Bogineni-Misra S, et al. Transperineal sonography of the rectum: Anatomy and pathology revealed by sonography compared with CT and MR imaging. AJR Am J Roentgenol 1998;170:637-42.
7. Stewart LK, McGee J, Wilson SR. Transperineal and transvaginal sonography of perianal inflammatory disease. AJR Am J Roentgenol 2001;177:627-32.
8. Rasul I, Wilson SR, MacRae H, et al. Clinical and radiological responses after infliximab treatment for perianal fistulizing Crohn's disease. Am J Gastroenterol 2004;99:82-8.
9. Maconi G, Ardizzone S, Greco S, et al. Transperineal ultrasound in the detection of perianal and rectovaginal fistulae in Crohn's disease. Am J Gastroenterol 2007;102:2214-9.
10. Mallouhi A, Bonatti H, Peer S, et al. Detection and characterization of perianal inflammatory disease: Accuracy of transperineal combined gray scale and color Doppler sonography. J Ultrasound Med 2004;23:19-27.
11. Wedemeyer J, Kirchhoff T, Sellge G, et al. Transcutaneous perianal sonography: A sensitive method for the detection of perianal inflammatory lesions in Crohn's disease. World J Gastroenterol 2004;10:2859-63.

2

Noninvasive methods, including transient elastography, for the detection of liver disease in adults with cystic fibrosis

Matthew D Sadler MD FRCPC[1], Pam Crotty MSc[1], Linda Fatovich MN[2], Stephanie Wilson MD FRCPC[3], Harvey R Rabin MD FRCPC[2], Robert P Myers MD MSc FRCPC[1,4]

MD Sadler, P Crotty, L Fatovich, S Wilson, HR Rabin, RP Myers. Noninvasive methods, including transient elastography, for the detection of liver disease in adults with cystic fibrosis. Can J Gastroenterol Hepatol 2015;29(3):139-144.

BACKGROUND: Liver disease is the third leading cause of mortality in patients with cystic fibrosis (CF). However, detection of CF-associated liver disease (CFLD) is challenging.
OBJECTIVE: To evaluate the diagnostic performance of noninvasive methods for the detection of CFLD with a focus on transient elastography (TE).
METHODS: Patients at the Adult CF Clinic of Calgary and Southern Alberta (n=127) underwent liver stiffness measurement (LSM) by TE using the FibroScan (FS, Ecosens, France) M probe; aspartate aminotransferase to platelet ratio index (APRI) and FibroTest (FT) scores were also calculated. The diagnostic performance of these tools for the detection of CFLD (defined as two or more the following criteria: abnormal liver biochemistry, hepatomegaly or sonographic abnormalities other than steatosis) were compared using the area under ROC curves.
RESULTS: Forty-seven percent of the cohort was male. The median age was 27 years (interquartile range [IQR] 22 to 37 years) and body mass index 21 kg/m^2 (IQR 19 kg/m^2 to 23 kg/m^2); 25% of patients were on ursodeoxycholic acid and 12% had undergone lung transplantation. The prevalence of CFLD was 14% (n=18). FS was successful in all patients; one (0.8%) patient had poorly reliable results (IQR/M >30% and LSM ≥7.1kPa). Compared with patients without CFLD (n=109), individuals with CFLD had higher median LSM according to FS (3.9 kPa [IQR 3.4 to 4.9 kPa] versus 6.4 kPa [IQR 4.4 to 8.0 kPa]), APRI (0.24 [IQR 0.17 to 0.31] versus 0.50 [IQR 0.22 to 1.18]) and FT scores (0.08 [IQR 0.05 to 1.5] versus 0.18 [IQR 0.11 to 0.35]; all P<0.05). Area under ROC curve for FS, APRI and FT for the detection of CFLD were 0.78 (95% CI 0.65 to 0.92), 0.72 (95% CI 0.56 to 0.87) and 0.76 (95% CI 0.62 to 0.90) (P not significant). At a threshold of >5.2 kPa, the sensitivity, specificity, positive and negative predictive values of LSM according to FS for detecting CFLD were 67%, 83%, 40% and 94%, respectively.
CONCLUSIONS: FS, APRI and FT were useful noninvasive methods for detecting CFLD in adults.

Key Words: *Cystic fibrosis; FibroTest; Transient elastography*

Les méthodes non invasives, y compris l'élastographie transitoire, pour déceler les maladies hépatiques chez les adultes atteints de fibrose kystique

HISTORIQUE : Les maladies hépatiques sont la troisième cause de mortalité en importance chez les patients atteints de fibrose kystique (FK). Il est toutefois difficile de dépister les maladies hépatiques qui s'associent à la FK (MHFK).
OBJECTIF : Évaluer le rendement diagnostique de méthodes non invasives pour déceler les MHFK, en s'attardant sur l'élastographie transitoire (ÉT).
MÉTHODOLOGIE : Les patients de la clinique de FK pour adultes de Calgary et du sud de l'Alberta (n=127) ont subi une mesure de dureté du foie (MDF) par ÉT au moyen de la sonde M de FibroScan (FS, Ecosens, France). Les chercheurs ont également calculé le score de l'indice du ratio entre l'aspartate-aminotransférase et les plaquettes (APRI) et le score du FibroTest (FT). Ils en ont comparé le rendement diagnostique pour déceler la MHFK (définie comme au moins deux des critères suivants : biochimie hépatique anormale, hépatomégalie ou autres anomalies sonographiques qu'une stéatose) à l'aide des surfaces sous la courbe ROC.
RÉSULTATS : La cohorte, dont 47 % étaient de sexe masculin, avait un âge médian de 27 ans (plage interquartile [PIQ] de 22 à 37 ans) et un indice de masse corporelle de 21 kg/m^2 (PIQ de 19 kg/m^2 à 23 kg/m^2). De plus, 25 % des patients prenaient de l'acide ursodésoxycholique et 12 % avaient subi une transplantation pulmonaire. La prévalence de MHFK était de 14 % (n=18). La FS a pu être exécutée chez tous les patients, mais les résultats d'un patient (0,8 %) étaient peu fiables (PIQ/M>30 % et MDF ≥7,1 kPa). Par rapport aux patients sans MHFK (n=109), ceux ayant une MHFK présentaient une MDF médiane plus élevée selon la FS (3,9 kPa [PIQ 3,4 à 4,9 kPa] par rapport à 6,4 kPa [PIQ 4,4 à 8,0 kPa]), et les scores d'APRI (0,24 [PIQ 0,17 à 0,31] par rapport à 0,50 [PIQ 0,22 à 1,18]) et de FT (0,08 [PIQ 0,05 à 1,5] par rapport à 0,18 [PIQ 0,11 à 0,35]; tous les P<0,05). La surface sous la courbe ROC de la FS, de l'APRI et du FT pour déceler la MHFK correspondait à 0,78 (95 % IC 0,65 à 0,92), 0,72 (95 % IC 0,56 à 0,87) et 0,76 (95 % IC 0,62 à 0,90) (P non significatif). À un seuil de plus de 5,2 kPa, la sensibilité, la spécificité et les valeurs prédictives positives et négatives de la MDF par FS pour déceler la MHFK s'élevaient à 67 %, 83 %, 40 % et 94 %, respectivement.
CONCLUSIONS : La FS, l'APRI et le FT se sont révélés des méthodes non invasives utiles pour déceler la MHFK chez les adultes.

Hepatobiliary manifestations of cystic fibrosis (CF) are variable, but include alterations attributable to the underlying genetic defect, malnutrition, lesions of iatrogenic origin and changes that reflect the effects of a systemic disease. Focal biliary cirrhosis due to biliary obstruction with progressive periportal fibrosis is the most clinically relevant problem because multilobular biliary cirrhosis, portal hypertension and, eventually, liver failure may develop (1). In fact, successes in the management of CF-related respiratory and nutritional complications have resulted in an increased recognition of liver disease, which now ranks as the third leading cause of mortality in patients with CF (1).

Unfortunately, several challenges exist in the identification of CF-associated liver disease (CFLD) given the wide spectrum of hepatic complications associated with this condition. Standard screening using physical examination, liver biochemistry and abdominal ultrasound imaging lacks sensitivity for early hepatic involvement in CF and for predicting who is at risk for progression to cirrhosis (2). As such, liver biopsy remains the gold standard for diagnosing and staging hepatic fibrosis in patients with CF. However, there are several limitations with the routine use of liver biopsy, including the focal nature of early CF-related lesions, which may lead to sampling error and subjectivity in histological interpretation. Moreover, the invasiveness of

[1]*Liver Unit, Division of Gastroenterology and Hepatology;* [2]*Adult Cystic Fibrosis Clinic of Calgary and Southern Alberta, Department of Medicine;* [3]*Department of Radiology and Diagnostic Imaging;* [4]*Department of Community Health Sciences, University of Calgary, Calgary, Alberta*
Correspondence: Dr Robert P Myers, Liver Unit, University of Calgary, 6D22, Teaching, Research and Wellness Building, 3280 Hospital Drive Northwest, Calgary, Alberta T2N 4Z6. e-mail rpmyers@ucalgary.ca

liver biopsy makes it difficult to repeat for monitoring of changes in fibrosis over time (3). Importantly, the procedure carries a small – but real – risk for complications including hemorrhage and death (4,5).

Given these limitations, alternative means of identifying CFLD and monitoring for disease progression are needed. Noninvasive serum and imaging-based methods of liver fibrosis evaluation may have a role in this patient population. Preliminary data suggest that liver stiffness measurement (LSM) using transient elastography (TE) (FibroScan [FS], Echosens, France) and serum biomarkers (eg, the aspartate aminotransferase [AST] to platelet ratio index [APRI] and FibroTest [FT; Biopredictive, France]) hold promise (6-10). A novel technique known as acoustic radiation force impulse (ARFI) imaging (Virtual Touch Tissue Quantification, Acuson S2000, Siemens, USA) has also been studied (7,11,12). This elastographic method, which is incorporated in a conventional ultrasound device, uses acoustic pulses to generate a shear wave in the liver. The velocity of shear wave propagation (measured in m/s) is proportional to liver fibrosis determined histologically.

Because data regarding the use of these noninvasive fibrosis methods in patients with CF are limited, our objective was to assess their feasibility and diagnostic performance for the detection of CFLD, with an emphasis on LSM using TE. In addition, our unique clinical setting enabled us to determine the prevalence of liver disease in a population-based cohort of adults with CF.

METHODS

Study population

All adult patients (≥18 years of age) with CF followed at the Adult CF Clinic of Calgary and Southern Alberta were recruited for the present study between April 2010 and April 2011. The Adult CF Clinic is a multidisciplinary clinic that cares for all adults with CF in southern Alberta and a small number of patients from southeastern British Columbia and Saskatchewan. The catchment population of Calgary is approximately 1.2 million, with a reported CF prevalence of 11 to 12 cases per 100,000 population (13). Patients with serological evidence of chronic hepatitis B or hepatitis C were excluded. The Conjoint Health Research Ethics Board at the University of Calgary (Calgary, Alberta) approved the research protocol.

Clinical data

At enrollment, patient demographics, anthropometric measurements (weight, height, body mass index [BMI]), CF transmembrane regulator (CFTR) genetic mutations (ΔF508), ursodeoxycholic acid (UDCA) use and history of CF-related complications, including pancreatic exocrine insufficiency, diabetes mellitus and lung transplantation, were recorded. Pancreatic exocrine insufficiency was defined as severe malabsorption responding to pancreatic enzyme supplementation. All patients underwent a standard physical examination and routine laboratory investigations. Individuals with examination findings suggestive of liver disease or abnormal liver biochemistry underwent abdominal ultrasonography (n=78). The following laboratory investigations were performed: complete blood count, alanine aminotransferase, AST, gamma-glutamyl transferase (GGT), alkaline phosphatase (ALP), total bilirubin, albumin, creatinine, random glucose, international normalized ratio, haptoglobin, apolipoprotein A1 and alpha-2-macroglobulin. Spirometry values from pulmonary function testing on the day of enrollment were also recorded.

The primary outcome measure was CFLD defined according to previously published criteria (1,2). Specifically, CFLD was based on the presence of at least two of the following: clinical hepatomegaly and/or splenomegaly confirmed by ultrasonography; abnormal liver biochemistry consisting of elevated levels of any two of AST, ALT or GGT; and ultrasound abnormalities other than hepatomegaly (ie, increased heterogeneous echogenicity, nodularity of the liver surface, irregular margins, splenomegaly, presence of portosystemic collateral veins and ascites) (1,2,6). Ultrasonographic evidence of steatosis was not a diagnostic criterion. Liver biopsies are not routinely performed in the authors' clinic.

LSM

Two experienced operators performed all LSMs with TE using the FS, as previously described (14). The FS M probe was used in all patients except one, in whom the XL probe was used. Briefly, with the patient lying in the dorsal decubitus position, the tip of the transducer was placed on the skin between the ribs over the right lobe of the liver. An attempt was made to collect at least 10 valid LSMs, and the median LSM (in kPa) was considered to be representative of the elastic modulus of the liver. FS failure was defined as no valid LSMs after at least 10 attempts. LSMs ≥7.1 kPa with an interquartile range (IQR) to median LSM ratio (IQR/M) >30% were considered to be poorly reliable as defined by Boursier et al (15). In a subset of patients (n=18), ARFI imaging was used to determine hepatic shear wave velocity (SWV) based on the median of 12 measurements taken from the right lobe of the liver (in m/s; range 0.5 m/s to 5.5 m/s). A single experienced radiologist performed all ARFI measurements as previously described (11).

Serum markers of liver fibrosis

The APRI was calculated as follows (9,10):

$$\text{APRI} = (\text{AST/upper limit of normal for AST}) \times (100/\text{platelets } [\times 10^9/\text{L}])$$

The FT was calculated based on age, sex, GGT, total bilirubin, alpha-2-macroglobulin, apolipoprotein A1 and haptoglobin (16-19).

Both the APRI and FT were developed in patients with chronic hepatitis C, but have been validated in other liver disorders including CFLD (20-22).

Statistical analyses

Between-group comparisons were made using Fisher's exact and Mann-Whitney tests, and Spearman correlation coefficients were calculated between results of the noninvasive measures. The diagnostic performance of LSM according to TE, ARFI, APRI and FT for CFLD were determined using areas under ROC curves (AUROC). Comparisons between diagnostic methods (restricted to patients in whom all of the relevant tests had been performed) were made using the nonparametric method described by DeLong et al (23). Finally, the sensitivity, specificity, accuracy, positive predictive value (PPV) and negative predictive value (NPV) values of these tools were calculated with exact binomial 95% CIs. For these analyses, optimal thresholds defined by the maximal sum of sensitivity and specificity were calculated. Stata version 11.0 (StataCorp, USA) was used for all statistical analyses. A two-sided $P<0.05$ were considered to be statistically significant.

RESULTS

Patient characteristics and prevalence and predictors of CFLD

The characteristics of the study cohort (n=127) are summarized in Table 1. Forty-seven percent of the patients were male and the median age and BMI were 27 years (IQR 22 to 37 years) and 21 kg/m^2 (IQR 19 kg/m^2 to 23 kg/m^2) , respectively. Only three (2.5%) patients were obese (BMI ≥30 kg/m^2). The majority (84%) of patients had pancreatic insufficiency, 12% had undergone lung transplantation and 25% were prescribed UDCA (median daily dosage, 1000 mg [IQR 1000 mg to 1375 mg]). Most (90%) patients had the ΔF508 CF mutation (53% were homozygous).

CFLD was present in 18 (14%) patients. More patients with CFLD than those without CFLD were prescribed UDCA (83% versus 14%; $P<0.00005$). However, there were no significant differences between groups with respect to the prevalence of diabetes mellitus, pancreatic insufficiency, previous lung transplantation, pulmonary spirometry or the ΔF508 mutation (heterozygous or homozygous). As expected, patients with CFLD had higher median serum ALT, AST, GGT, alkaline phosphatase and bilirubin levels, and international normalized ratio values (Table 1).

Detection of CFLD using TE, ARFI and serum markers of fibrosis

All patients underwent a successful LSM with TE using the FibroScan (ie, 0% failure rate. A poorly reliable result was obtained in only one

TABLE 1
Characteristics of the cohort according to the presence of cystic fibosis-related liver disease (CFLD)

Characteristic	All patients (n=127)	With CFLD (n=18)	Without CFLD (n=109)	P
Demographics				
Age, years	27 (22–37)	28 (18–32)	27 (22–37)	0.22
Male sex	60 (47)	10 (56)	50 (42)	0.46
Body mass index†, kg/m^2	21 (19–23)	21 (19–23)	21 (19–23)	0.96
CF-related variables				
Diabetes mellitus	24 (19)	4 (22)	20 (17)	0.75
Pancreatic insufficiency	107 (84)	17 (94)	90 (76)	0.31
Lung transplantation	15 (12)	3 (17)	12 (10)	0.45
Ursodeoxycholic acid use	32 (25)	15 (83)	17 (14)	<0.00005
CFTR-genotype ΔF508 mutation	114 (90)	16 (89)	98 (90)	1.00
CFTR-genotype ΔF508 homozygous	67 (53)	12 (67)	55 (51)	0.31
Spirometry				
FEV_1, L	2.55 (1.71–3.28)	2.98 (1.86–3.26)	2.46 (1.70–3.28)	0.42
FEV_1, % predicted (%)	70 (52.5–88)	74 (58–80)	69 (50–89)	0.92
FVC, L	3.86 (2.97–4.73)	4.06 (3.01–4.58)	3.82 (3.02–4.75)	0.85
FVC, % predicted (%)	92 (78–105)	95 (77–103)	92 (78–106)	0.74
FEV_1/FVC (%)	73 (64–84)	74 (73–85)	73 (62–84)	0.28
Biochemistry				
Albumin†, g/L	38 (35–41)	38 (36–40)	39 (35–41)	0.48
Total bilirubin†, μmol/L	6 (4–8)	8 (6–10)	6 (4–8)	0.03
Alanine aminotransferase (ALT)†, U/L	21 (17–29)	31 (24–48)	21 (16–27)	0.001
Aspartate aminotransferase (AST)†, U/L	23 (19–28)	28 (22-45)	22 (19–26)	0.004
Gamma-glutamyl transferase (GGT)†, U/L	18 (13–28)	41 (26–68)	17 (13–25)	0.0001
Alkaline phosphatase (ALP)†, U/L	94 (78–123)	157 (125–205)	89 (77–108)	<0.00005
Platelets, ×10^9/L	272 (224–332)	218 (112–332)	274 (236–330)	0.053
International normalized ratio†	1.0 (1.0–1.1)	1.1 (1.0–1.1)	1.0 (1.0–1.1)	0.01
Creatinine, μmol/L†	66 (55–77)	60 (52–70)	67 (56–77)	0.23
Glucose, mmol/L‡	4.9 (4.3–6.6)	4.9 (4.4–6.9)	4.9 (4.2–6.4)	0.74

Data presented as n (%) or median (interquartile range). †Data missing for five patients (four without CFLD and one with CFLD). ‡Albumin, total bilirubin, ALT, AST, GGT, ALP, international normalized ratio and creatinine missing in 1, 1, 1, 5, 1, 1, 4 and 1 patients (all in the group without CFLD), respectively; ‡Glucose data missing in 11 patients (one with CFLD). FEV_1 Forced expiratory volume in 1 s; FVC Forced vital capacity

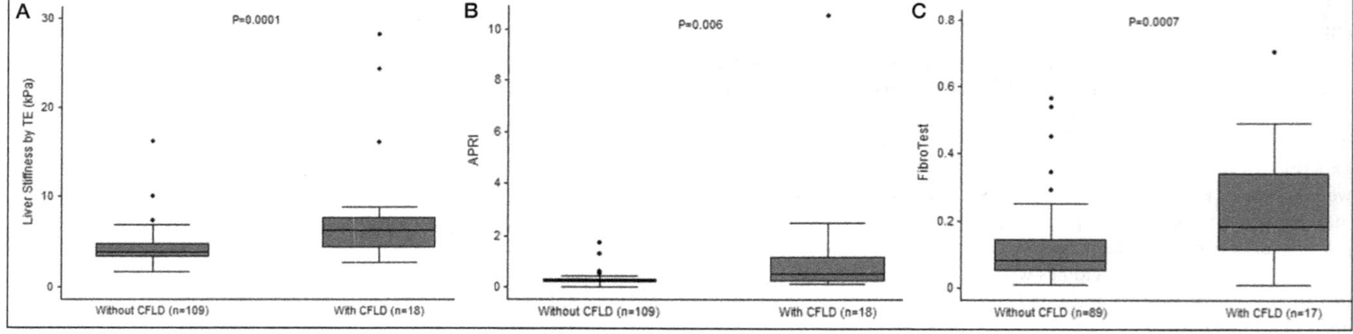

Figure 1) *Liver stiffness measurement using transient elastography (TE; FibroScan, Ecosens, France)* (**A**)*, aspartate aminotransferase to platelet ratio index (APRI)* (**B**) *and FibroTest (Biopredictive, France)* (**C**) *according to the presence or absence of cystic fibrosis-related liver disease (CFLD)*

patient (0.8%) who was severely obese (BMI 35 kg/m^2). In this patient – in whom CFLD was not diagnosed – the median LSM was 16.1 kPa using the FibroScan M probe and the IQR/M was 41%, indicating high measurement variability. Overall, 118 (93%) patients had at least 10 valid LSMs using TE (success rate ≥60%) and IQR/M ≤30%.

As shown in Figure 1 and Table 2, median LSM according to TE was significantly higher in patients with than those without CFLD (6.4 kPa [IQR 4.4 to 7.7 kPa] versus 3.9 kPa [IQR 3.4 to 4.9 kPa]; P=0.0001). Hepatic SWV measured using ARFI was available for 18 patients, three of whom had CFLD. Median SWV was not significantly different between patients with (1.50 m/s [IQR 0.99 to 1.50 m/s]) and without CFLD (1.07 m/s [IQR 1.01 to 1.10 m/s]) (P=0.24). In contrast, median APRI (0.50 [IQR 0.22 to 1.17] versus 0.24 [IQR 0.17 to 0.31]; P=0.006) and FT (0.18 [IQR 0.11 to 0.34] versus 0.08 [IQR 0.05 to 0.14]; P=0.0007) values were significantly higher in patients with CFLD.

Liver stiffness according to TE was moderately correlated with FT (ρ=0.32; P=0.001), but not APRI (ρ=0.02; P=0.83) values. APRI and FT were weakly correlated (ρ=0.20; P=0.04). SWV according to ARFI was moderately correlated with LSM according to TE (ρ=0.31; P=0.21) and FT (ρ=0.42; P=0.10); however, these correlations were not statistically significant.

Diagnostic accuracy of LSM according TE, ARFI and serum markers for CFLD

The AUROCs for LSM according TE, APRI and FT for the diagnosis of CFLD were 0.78 (95% CI 0.65 to 0.92 [n=127]), 0.70 (95% CI

TABLE 2
Noninvasive markers of liver fibrosis in adults with cystic fibrosis-related liver disease (CFLD)

Test	All patients	With CFLD	Without CFLD	P
Liver stiffness measurement using transient elastography (n=127)	4.1 (3.5–5.2)	6.4 (4.4–7.7)	3.9 (3.4–4.9)	0.0001
Aspartate aminotransferase to platelet ratio index (n=122)*	0.24 (0.17–0.33)	0.50 (0.22–1.17)	0.24 (0.17–0.31)	0.006
FibroTest† (n=106)	0.09 (0.06–0.18)	0.18 (0.11–0.34)	0.08 (0.05–0.14)	0.0007
Acoustic radiation force impulse, m/s (n=18)‡	1.07 (1.01–1.13)	1.50 (0.99–1.50)	1.07 (1.01–1.10)	0.24

*Data presented as median (interquartile range) unless otherwise indicated. *Data missing for five patients (all without CFLD); †Biopredictive, France (data missing for 21 patients [one with CFLD and 20 without CFLD]); ‡Data available for 18 patients (three with CFLD and 15 without CFLD)*

TABLE 3
Liver stiffness measurement (LSM) usung transient elastography (TE), aspartate aminotransferase to platelet ratio index (APRI) and FibroTest* for the diagnosis of cystic fibrosis-related liver disease

	AUROC	Sensitivity	Specificity	Predictive value: Negative	Predictive value: Positive	Accuracy
LSM using TE (n=127)						
≥3.7 kPa		89 (65–99)	37 (28–46)	95 (84–99)	19 (11–29)	44 (35–53)
≥5.3 kPa†	0.78 (0.65–0.92)	67 (41–87)	83 (75–90)	94 (87–98)	40 (23–59)	81 (73–88)
>6.0 kPa		56 (31–78)	91 (86–97)	93 (86–97)	50 (27–73)	86 (76–91)
APRI (n=122)						
>0.4	0.70 (0.54–0.86)	50 (27–73)	92 (85–97)	90 (83–95)	56 (31–78)	85 (78–91)
>0.5†		50 (26–74)	94 (88–98)	92 (85–96)	60 (32–84)	88 (81–93)
FibroTest (n=106)						
>0.10†	0.76 (0.62–0.90)	82 (57–96)	57 (46–68)	94 (85–99)	27 (16–41)	61 (51–71)
>0.20		38 (18–62)	89 (81–95)	85 (76–92)	47 (23–72)	79 (70–87)

*Data presented as % (95% CI). *Biopredictive, France; †Optimal cut-offs of tests defined by the maximal sum of sensitivity and specificity. AUROC Area under the ROC curve*

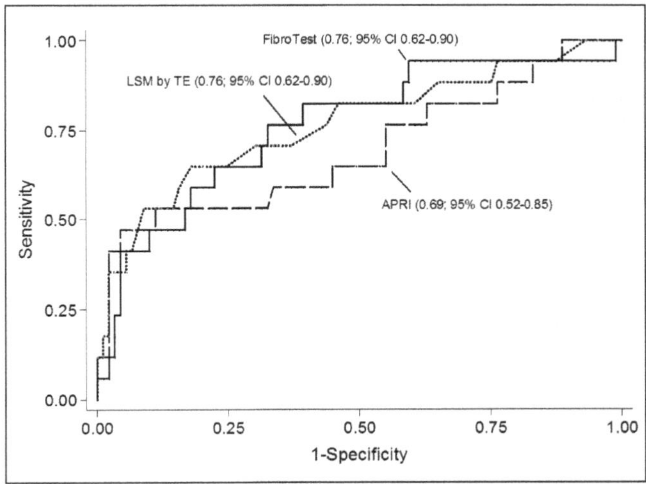

Figure 2) *Areas under the ROC curve of liver stiffness measurement (LSM) using transient elastography, aspartate aminotransferase to platelet ratio index (APRI) and FibroTest (Biopredictive, France) for the detection of cystic fibrosis-related liver disease. There were no significant differences among measures among 105 patients who underwent all three tests (P=0.64)*

0.54 to 0.86 [n=122]) and 0.76 (95% CI 0.62 to 0.90 [n=106]), respectively. In the 105 patients who underwent all three tests, there were no significant differences in AUROCs for the detection of CFLD (P=0.64; Figure 2). Among 18 patients who underwent ARFI imaging, the AUROC of SWV for detecting CLFD was 0.72 (95% CI 0.17 to 1.00).

Table 3 summarizes the operating characteristics of LSM using TE, APRI and FT for the detection of CFLD. ARFI is excluded due to the small number of patients. At an optimal cut-off value ≥5.3 kPa, LSM according to TE was 67% sensitive (95% CI 41% to 87%) and 83% specific (95% CI 75% to 90%). At a CFLD prevalence of 14%, FS values below this threshold excluded CFLD (NPV) with 94% (95% CI 87% to 98%) certainty. However, LSM values ≥5.3 kPa were poorly predictive of CFLD (PPV 40% [95% CI 23% to 59%]). An LSM cut-off of >6.0 kPa was more specific (91% [95% CI 86% to 97%]), but less sensitive (56% [95% CI 31% to 78%]). At an optimal cut-off of >0.10, the FT was 82% (95% CI 57% to 96%) sensitive and 57% (95% CI 46% to 68%) specific for CFLD. The NPV, PPV and accuracy were 94% (95% CI 85% to 99%), 27% (95% CI 16% to 41%) and 61% (95% CI 51% to 71%), respectively. At an optimal cut-off of >0.5, the APRI was 50% (95% CI 26% to 74%) sensitive and 94% (95% CI 88% to 98%) specific. The NPV, PPV and accuracy were 92% (95% CI 85% to 96%), 60% (95% CI 32% to 84%) and 88% (95% CI 81% to 93%), respectively.

DISCUSSION

Identifying patients who are at risk for CFLD and potential hepatic complications is an ongoing challenge. Previous studies have estimated that as many as 30% to 40% of CF patients have some form of liver involvement, while severe CFLD with portal hypertension is believed to occur in <10% of patients (1,24,25). The first step remains identifying which CF patients have liver disease to better determine those at risk for developing serious complications. In the present population-based cohort of adults with CF, we identified a CFLD prevalence of 14%. This figure is lower than previously reported, perhaps due to the population-based nature of our data. Moreover, by the time CF patients reach adulthood, some will have already succumbed to the complications of liver disease. Our data demonstrate that patients with CFLD have higher liver stiffness according to FS, APRI and FT values than those without CFLD suggesting that these noninvasive markers of liver fibrosis have a role in screening CF patients for hepatobiliary involvement. These data validate many other studies that have demonstrated the utility of these tools in liver disorders including other cholestatic conditions, nonalcoholic fatty liver disease and viral hepatitis (9,14,20-22,26-30).

Our findings illustrate that LSM according to TE, APRI and FT have acceptable diagnostic performance for the detection of CFLD (AUROCs 0.70 to 0.78) and that none clearly outperformed the others. Measurement of SWV according to ARFI imaging had similar utility, although the small sample size limits any firm conclusions. Our data indicate that the major role of these tests in the clinical management of patients with CF is in ruling out CFLD. For example, LSMs according to TE <5.3 kPa, APRI values <0.5, and FT values <0.10 excluded CFLD with 92% to 94% certainty. Given these data, there is likely little benefit of further investigation when measurements are below these thresholds. Unfortunately, the PPV of higher values was suboptimal (27% to 60% at these diagnostic cut-offs), suggesting that additional investigations (eg, liver biopsy) may be considered in these patients to confirm significant hepatic involvement.

LSM values in patients with CFLD were significantly higher than in those without CFLD despite both groups having relatively low median TE values (6.4 kPa versus 3.9 kPa [Table 2]). Other investigators have noted these small, but significant, differences. For example, Kitson et al (6) reported median LSMs of 8.1 kPa (IQR 6.8 to 9.5 kPa) in adults with CFLD compared with 5.0 kPa (IQR 4.1 to 5.6 kPa) in those without CFLD. Median LSM values in two other studies were 6.7 kPa and 9.2 kPa in patients with CFLD compared with 4.3 kPa and 4.5 kPa in patients without CFLD, respectively (7,31). In another study, Karlas et al (12) reported a trend toward higher LSMs in patients with CFLD; however, this difference was not statistically significant, likely due to the small number of patients with CFLD (n=14). In this study, TE measurements in patients with CF-related cirrhosis were found to be lower than those observed in cirrhotic patients without CF. It is, therefore, important to recognize that patients with CF may be at risk for CFLD despite having FS readings not considered to be grossly abnormal in patients with other forms of liver disease. This is not surprising considering the fact that CFLD includes a constellation of histological abnormalities above and beyond fibrosis.

Our study confirms the feasibility of LSM using TE in patients with CF. Only one patient had an unreliable LSM result, likely due to use of the FS M probe rather than the XL probe despite severe obesity (BMI 35 kg/m^2) (32). We have previously shown that patients with chronic lung disease (predominantly chronic obstructive pulmonary disease) have an increased risk for unreliable TE readings. We hypothesized that this was likely due to structural alterations in the chest wall (eg, hyperinflation) or technical difficulties with the FS procedure due to deep respirations in these patients (14). This does not appear to be a factor in patients with CF.

We have yet to complete follow-up TE measurements in our cohort, but one small study of patients with CFLD (8) reported that LSMs remained stable over a median follow-up period of nearly two years. This finding supports previous studies reporting that the development of CFLD generally occurs before 20 years of age and follows a relatively benign course in adults (1,25,33,34). However, in other cholestatic liver disorders, such as primary biliary cirrhosis and primary sclerosing cholangitis, liver stiffness increases more rapidly at greater fibrosis stages (28,30). It is, therefore, possible that progression of CFLD in patients with more advanced disease may occur faster than in those with milder disease. Large prospective studies examining serial TE readings will help clarify whether repeat FS assessment will be useful in the management of adults with CF.

We did not find an association between the development of CFLD and any CF-related complications (eg, pancreatic insufficiency or lung transplantation) or *CFTR* mutations. The underlying mechanism explaining why certain individuals with CF develop more significant liver disease than others is not clearly understood; however, one study suggests a genetic polymorphism in the *SERPINA1* gene may play a role (33). A study that examined the use of these noninvasive fibrosis measurements in young patients with the *SERPINA1* Z allele would possibly help clarify how to identify patients with CF who are at risk for progressing to severe liver disease.

Our study had several limitations. First, the definition of CFLD relied on noninvasive criteria (ie, abnormal liver biochemistry and ultrasonography), rather than liver biopsy. Nevertheless, even with histological assessment, there is no universally accepted definition of what constitutes CFLD. Moreover, many liver lesions of CF begin focally and may be missed by liver biopsy (3). Furthermore, the current consensus opinion has relied on the same noninvasive markers to define CFLD used in our study (2,35). On a related note, 25% of our cohort was prescribed UDCA, including 83% of those with CFLD. This finding is relevant because UDCA has been shown to improve **liver** biochemistry test results and the progression to severe liver disease in patients with CF (34,36,37). Therefore, the values of the noninvasive markers obtained in our study may have been affected. To date, the effect of UDCA on liver stiffness and serum fibrosis markers in patients with CF has not been studied. In addition, our reference standard for the definition of CFLD included abnormal liver biochemistry as one criterion. Therefore, biochemical improvement due to UDCA may have led to misclassification of CFLD in some patients. In fact, because 14% of our patients 'without' CFLD were prescribed UDCA, the prevalence of CFLD could be as high as 28%. The presence of extrahepatic cholestasis in patients with CF may have also led to falsely elevated LSM using TE (38). Finally, only a small subset of our cohort underwent ARFI imaging due to limited availability during the study period. This likely explains why we failed to show a significant difference in SWV between those with and without CFLD, particularly considering the correlation between LSM measured by TE and SWV. Previous studies involving a pediatric population have shown ARFI to be useful in detecting CFLD (39-41); however, a small study failed to confirm this finding in an adult population (12). Larger studies assessing the feasibility and performance characteristics of ARFI in the detection of CFLD in adults are needed.

SUMMARY

Noninvasive markers of liver fibrosis including LSM using FS, FT and APRI appear to be useful in patients with CF, particularly in the exclusion of CFLD. Additional studies are necessary to validate measurement of SWV using ARFI, and to confirm the prognostic significance of these tools for predicting CFLD-related complications.

ACKNOWLEDGEMENTS: This study was supported in part by grants to Dr Myers from Alberta Innovates – Health Solutions (AI-HS), the Canadian Institutes for Health Research (CIHR), and the Canadian Liver Foundation.

DISCLOSURES: The authors have no financial disclosures or conflicts of interest to declare.

REFERENCES

1. Colombo C, Battezzati PM, Crosignani A, et al. Liver disease in cystic fibrosis: A prospective study on incidence, risk factors, and outcome. Hepatology 2002;36:1374-82.
2. Debray D, Kelly D, Houwen R, Strandvik B, Colombo C. Best practice guidance for the diagnosis and management of cystic fibrosis-associated liver disease. J Cyst Fibros 2011;(10 Suppl 2):S29-36.
3. Lewindon PJ, Shepherd RW, Walsh MJ, et al. Importance of hepatic fibrosis in cystic fibrosis and the predictive value of liver biopsy. Hepatology 2011;53:193-201.
4. Bravo AA, Sheth SG, Chopra S. Liver biopsy. N Engl J Med 2001;344:495-500.
5. Myers RP, Fong A, Shaheen AAM. Utilization rates, complications and costs of percutaneous liver biopsy: A population-based study including 4275 biopsies. Liver Int 2008;28:705-12.
6. Kitson MT, Kemp WW, Iser DM, Paul E, Wilson JW, Roberts SK. Utility of transient elastography in the non-invasive evaluation of cystic fibrosis liver disease. Liver Int 2013;33:698-705.
7. Friedrich-Rust M, Schlueter N, Smaczny C, et al. Non-invasive measurement of liver and pancreas fibrosis in patients with cystic fibrosis. J Cyst Fibros 2013;12:431-9.
8. Karlas T, Neuschulz M, Oltmanns A, Wirtz H, Keim V, Wiegand J. ARFI and transient elastography for characterization of cystic fibrosis related liver disease: First longitudinal follow-up data in adult patients. J Cyst Fibros 2013;12:826-7.
9. Shaheen AAM, Myers RP. Diagnostic accuracy of the aspartate aminotransferase-to-platelet ratio index for the prediction of hepatitis C-related fibrosis: A systematic review. Hepatology 2007;46:912-21.

10. Wai C-T, Greenson JK, Fontana RJ, et al. A simple noninvasive index can predict both significant fibrosis and cirrhosis in patients with chronic hepatitis C. Hepatology 2003;38:518-26.
11. Friedrich-Rust M, Wunder K, Kriener S, et al. Liver fibrosis in viral hepatitis: Noninvasive assessment with acoustic radiation force impulse imaging versus transient elastography. Radiology 2009;252:595-604.
12. Karlas T, Neuschulz M, Oltmanns A, et al. Non-invasive evaluation of cystic fibrosis related liver disease in adults with ARFI, transient elastography and different fibrosis scores. PLoS One 2012;7:e42139.
13. Canadian Cystic Fibrosis Patient Data Registry Report 2010 <cysticfibrosisvictoria.ca/wp-content/uploads/2013/06/CPDR_ReportE.pdf> (Accessed November 1, 2014).
14. Pang JX, Pradhan F, Zimmer S, et al. The feasibility and reliability of transient elastography using Fibroscan: A practice audit of 2335 examinations. Can J Gastroenterol Hepatol 2014;28:143-9.
15. Boursier J, Zarski J-P, de Lédinghen V, et al. Determination of reliability criteria for liver stiffness evaluation by transient elastography. Hepatology 2013;57:1182-91.
16. Imbert-Bismut F, Ratziu V, Pieroni L, Charlotte F. Biochemical markers of liver fibrosis in patients with hepatitis C virus infection: A prospective study. Lancet 2001;357:1069-75.
17. Poynard T, Imbert-Bismut F, Ratziu V, et al. Biochemical markers of liver fibrosis in patients infected by hepatitis C virus: Longitudinal validation in a randomized trial. J Viral Hepat 2002;9:128-33.
18. Poynard Patent <http://patft.uspto.gov/netacgi/nph-Parser?Sect1=PTO1&Sect2=HITOFF&d=PALL&p=1&u=%2Fnetahtml%2FPTO%2Fsrchnum.htm&r=1&f=G&l=50&s1=6631330.PN.&OS=PN/6631330&RS=PN/6631330> (Accessed November 1, 2014).
19. Myers RP, De Torres M, Imbert-Bismut F, et al. Biochemical markers of fibrosis in patients with chronic hepatitis C: A comparison with prothrombin time, platelet count, and age-platelet index. Dig Dis Sci 2003;48:146-53.
20. Myers RP, Tainturier MH, Ratziu V, et al. Prediction of liver histological lesions with biochemical markers in patients with chronic hepatitis B. J Hepatol 2003;39:222-30.
21. Lieber CS, Weiss DG, Morgan TR, Paronetto F. Aspartate aminotransferase to platelet ratio index in patients with alcoholic liver fibrosis. Am J Gastroenterol 2006;101:1500-8.
22. Kim D, Kim WR, Kim HJ, Therneau TM. Association between noninvasive fibrosis markers and mortality among adults with nonalcoholic fatty liver disease in the United States. Hepatology 2013;57:1357-65.
23. DeLong ER, DeLong DM, Clarke-Pearson DL. Comparing the areas under two or more correlated receiver operating characteristic curves: A nonparametric approach. Biometrics 1988;44:837-45.
24. Chryssostalis A, Hubert D, Coste J, et al. Liver disease in adult patients with cystic fibrosis: A frequent and independent prognostic factor associated with death or lung transplantation. J Hepatol 2011;55:1377-82.
25. Lamireau T, Monnereau S, Martin S, Marcotte J-E, Winnock M, Alvarez F. Epidemiology of liver disease in cystic fibrosis: A longitudinal study. J Hepatol 2004;41:920-5.
26. Shaheen AAM, Wan AF, Myers RP. FibroTest and FibroScan for the prediction of hepatitis C-related fibrosis: A systematic review of diagnostic test accuracy. Am J Gastroenterol 2007;102:2589-600.
27. Myers RP, Elkashab M, Ma M, Crotty P, Pomier-Layrargues G. Transient elastography for the noninvasive assessment of liver fibrosis: A multicentre Canadian study. Can J Gastroenterol 2010;24:661-70.
28. Corpechot C, Gaouar F, Naggar El A, et al. Baseline values and changes in liver stiffness measured by transient elastography are associated with severity of fibrosis and outcomes of patients with primary sclerosing cholangitis. Gastroenterology 2014;146:970-6.
29. Corpechot C, Naggar El A, Poujol-Robert A, et al. Assessment of biliary fibrosis by transient elastography in patients with PBC and PSC. Hepatology (Baltimore, Md) 2006;43:1118-24.
30. Corpechot C, Carrat F, Poujol-Robert A, et al. Noninvasive elastography-based assessment of liver fibrosis progression and prognosis in primary biliary cirrhosis. Hepatology (Baltimore, Md) 2012;56:198-208.
31. Rath T, Menendez KM, Kügler M, et al. TIMP-1/-2 and transient elastography allow non invasive diagnosis of cystic fibrosis associated liver disease. Dig Liver Dis 2012;44:780-7.
32. Myers RP, Pomier-Layrargues G, Kirsch R, et al. Feasibility and diagnostic performance of the FibroScan XL probe for liver stiffness measurement in overweight and obese patients. Hepatology (Baltimore, Md) 2012;55:199-208.
33. Bartlett JR, Friedman KJ, Ling SC, et al. Genetic modifiers of liver disease in cystic fibrosis. JAMA 2009;302:1076-83.
34. Desmond CP, Wilson J, Bailey M, Clark D, Roberts SK. The benign course of liver disease in adults with cystic fibrosis and the effect of ursodeoxycholic acid. Liver Int 2007;27:1402-8.
35. Colombo C. Liver disease in cystic fibrosis. Curr Opin Pulm Med 2007;13:529-36.
36. Nousia-Arvanitakis S, Fotoulaki M, Economou H, Xefteri M, Galli-Tsinopoulou A. Long-term prospective study of the effect of ursodeoxycholic acid on cystic fibrosis-related liver disease. J Clin Gastroenterol 2001;32:324-8.
37. Kappler M, Espach C, Schweiger-Kabesch A, et al. Ursodeoxycholic acid therapy in cystic fibrosis liver disease – a retrospective long-term follow-up case-control study. Aliment Pharmacol Ther 2012;36:266-73.
38. Millonig G, Reimann FM, Friedrich S, et al. Extrahepatic cholestasis increases liver stiffness (FibroScan) irrespective of fibrosis. Hepatology (Baltimore, Md) 2008;48:1718-23.
39. Behrens CB, Langholz JH, Eiler J, et al. A pilot study of the characterization of hepatic tissue strain in children with cystic-fibrosis-associated liver disease (CFLD) by acoustic radiation force impulse imaging. Pediatr Radiol 2013;43:552-7.
40. Monti L, Manco M, Zupone Lo C, et al. Acoustic radiation force impulse (ARFI) imaging with Virtual Touch Tissue Quantification in liver disease associated with cystic fibrosis in children. Radiol Med 2012;117:1408-18.
41. Manco M, Zupone CL, Alghisi F, D'Andrea ML, Lucidi V, Monti L. Pilot study on the use of acoustic radiation force impulse imaging in the staging of cystic fibrosis associated liver disease. J Cyst Fibros 2012;11:427-32.

Gaps in the hepatitis C continuum of care among sex workers in Vancouver, British Columbia: Implications for voluntary hepatitis C virus testing, treatment and care

M Eugenia Socías MD[1,2], Kate Shannon PhD MPH[1,3,4], Julio S Montaner MD[1,3], Silvia Guillemi MD[1,3], Sabina Dobrer MA[1], Paul Nguyen PhD[1], Shira Goldenberg PhD[1,5], Kathleen Deering PhD[1,3]

ME Socías, K Shannon, JS Montaner, et al. Gaps in the hepatitis C continuum of care among sex workers in Vancouver, British Columbia: Implications for voluntary hepatitis C virus testing, treatment and care. Can J Gastroenterol Hepatol 2015;29(8):411-416.

BACKGROUND: Hepatitis C virus (HCV) eradication leads to reduced morbidity, mortality and transmission. Despite the disproportionate burden of HCV among sex workers, data regarding the HCV care continuum in this population remain negligible.

METHODS: Using baseline data from an ongoing cohort of women sex workers in Vancouver (An Evaluation of Sex Workers' Health Access, January 2010 to August 2013), the authors assessed HCV prevalence and engagement in the HCV care continuum within the past year. Multivariable logistic regression analyses were used to evaluate associations with recent (ie, in the past year) HCV testing.

RESULTS: Among 705 sex workers, 302 (42.8%) were HCV seropositive. Of these, 22.5% were previously unaware of their HCV status, 41.7% had accessed HCV-related care, 13.9% were offered treatment and only 1.0% received treatment. Among 552 HCV-seronegative sex workers, only one-half (52.9%) reported a recent HCV test. In multivariable analysis, women who self-identified as a sexual/gender minority (adjusted OR [aOR] 1.89 [95% CI 1.11 to 3.24]), resided in the inner city drug use epicentre (aOR 3.19 [95%CI 1.78 to 5.73]) and used injection (aOR 2.00 [95% CI 1.19 to 3.34]) or noninjection drugs (aOR 1.95 [95% CI 1.00 to 3.78]) had increased odds of undergoing a recent HCV test, while immigrant participants (aOR 0.24 [95% CI 0.12 to 0.48]) had decreased odds.

CONCLUSIONS: Despite a high burden of HCV among sex workers, large gaps in the HCV care continuum remain. Particularly concerning are the low access to HCV testing, with one-fifth of women living with HCV being previously unaware of their status, and the exceptionally low prevalence of HCV treatment. There is a critical need for further research to better understand and address barriers to engage in the HCV continuum for sex workers.

Key Words: *Hepatitis C; HCV testing; HCV treatment; People who use drugs; Sex workers; Women*

Les lacunes dans le continuum des soins de l'hépatite C chez les travailleuses du sexe de Vancouver, en Colombie-Britannique : les conséquences sur les tests volontaires de dépistage, le traitement et les soins de l'hépatite C

HISTORIQUE : L'éradication du virus de l'hépatite C (VHC) entraîne une diminution de la morbidité, de la mortalité et de la transmission de la maladie. Malgré le fardeau disproportionné du VHC chez les travailleuses du sexe, les données relatives au continuum des soins du VHC demeurent négligeables au sein de cette population.

MÉTHODOLOGIE : Au moyen de données initiales provenant d'une cohorte de travailleuses du sexe de Vancouver (une évaluation de l'accès des travailleuses du sexe à la santé, de janvier 2010 à août 2013), les auteurs ont évalué la prévalence du VHC et la participation au continuum des soins du VHC au cours de l'année précédente. Les analyses de régression logistique multivariées ont permis d'évaluer les associations avec les récents tests du VHC (depuis un an).

RÉSULTATS : Des 705 travailleuses du sexe, 302 (42,8 %) étaient séropositives au VHC. De ce nombre, 22,5 % ne le savaient pas auparavant, 41,7 % avaient eu accès à des soins liés au VHC, 13,9 % s'étaient fait offrir un traitement et seulement 1,0 % avait reçu un traitement. Chez les 552 travailleuses du sexe séronégatives au VHC, seulement la moitié (52,9 %) déclarait avoir récemment subi un test de dépistage. À l'analyse multivariée, celles qui affirmaient faire partie d'une minorité sexuelle ou de genre (RC rajusté [RCr] 1,89 [95 % IC 1,11 à 3,24]), qui habitaient dans l'épicentre de la consommation de drogue des quartiers pauvres (RCr 3,19 [95 % IC 1,78 à 5,73]) ou qui consommaient des drogues injectables (RCr 2,00 [95 % IC 1,19 à 3,34]) ou non injectables (RCr 1,95 [95 % IC 1,00 à 3,78]) étaient plus susceptibles d'avoir récemment subi un test de dépistage du VHC, tandis que les participantes immigrantes (RCr 0,24 [95 % IC 0,12 à 0,48]) l'étaient moins.

CONCLUSIONS : Malgré le fardeau élevé du VHC chez les travailleuses du sexe, d'importantes lacunes persistent dans le continuum des soins du VHC. Le faible accès au test de dépistage du VHC est particulièrement préoccupant, puisque le cinquième des femmes atteintes ne se savait pas infectées. La prévalence exceptionnellement faible de traitement du VHC est tout aussi préoccupante. Il est urgent de poursuivre les recherches pour mieux comprendre et vaincre les obstacles à la participation des travailleuses du sexe au continuum des soins du VHC.

Chronic hepatitis C virus (HCV) infection is an emerging global health problem. According to recent estimates >185 million individuals worldwide are living with chronic HCV infection (1). Without treatment, between 15% and 30% of individuals with HCV will progress to cirrhosis, end-stage liver disease, hepatocellular carcinoma and, eventually, death. In fact, HCV is believed to be responsible for 350,000 annual deaths globally (1). HCV infection accounts not only for an important health and social burden (2), but also for a significant economic burden, both for the health system (3-5) and for the affected individual (6). Importantly, the economic burden of HCV is heavily associated with the treatment of complications associated with end-stage liver disease (eg, cirrhosis, hepatocellular carcinoma), while viral clearance (ie, sustained virological response [SVR]) has been associated with lower costs (5,7,8).

The burden of chronic HCV infection is also substantial in Canada, with approximately 0.8% of the Canadian population – nearly 250,000 individuals – living with HCV in 2007 (9,10). However, this figure is likely to be an underestimate of the real prevalence, which could be up to 2.8% (11,12). Chronic HCV is the leading indication for liver transplantation (13), and modelling studies have projected that, under the current status quo scenario, the incidence of HCV-related complications will continue to rise until at least 2035 (5,10). In fact, a recent

[1]*British Columbia Centre for Excellence in HIV/AIDS, St Paul's Hospital;* [2]*Interdisciplinary Studies Graduate Program;* [3]*Department of Medicine;* [4]*School of Population and Public Health, University of British Columbia, Vancouver;* [5]*Faculty of Health Sciences, Simon Fraser University, Burnaby, British Columbia*

Correspondence: Dr Kate Shannon, Department of Medicine, University of British Columbia, St Paul's Hospital, 608-1081 Burrard Street, Vancouver, British Columbia V6Z 1Y6. e-mail gshi@cfenet.ubc.ca

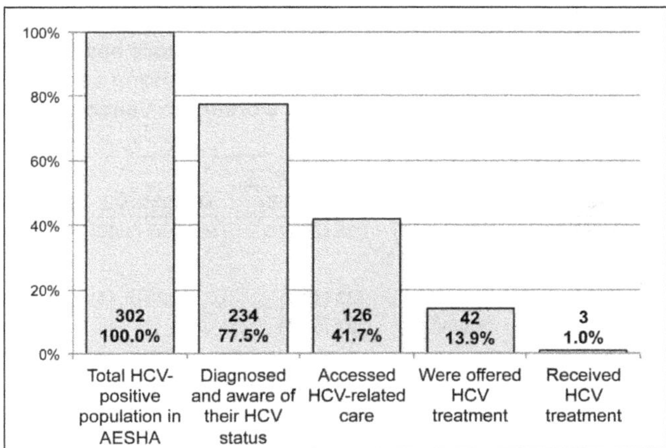

Figure 1) *Hepatitis C virus (HCV) continuum of care for women sex workers living with HCV, Vancouver, British Columbia (n=302). AESHA An Evaluation of Sex Workers' Health Access*

study from Ontario (2) found HCV to be the infectious disease accounting for the greatest loss of health-adjusted life years in the province.

Recent advances in HCV therapeutics have resulted in high success rates for all genotypes (>90% to 95% in most scenarios). Newer treatments are markedly more effective, safer, shorter and better tolerated (1). Importantly, HCV eradication is associated with up to 90% reductions in all-cause mortality, risk for developing hepatocellular carcinoma and need for liver transplantation (14). Additionally, achieving an SVR has a secondary prevention benefit, by eliminating the risk of onward transmission of HCV (15,16). In fact, given increasing availability of highly curative HCV drugs, and building on lessons from the HIV field (17), there have been increasing global calls to consider the implementation of HCV treatment as a prevention strategy (15,16). However, this needs to be tempered by the potential for HCV reinfection.

A comprehensive HCV treatment and prevention strategy would need to meet several criteria to achieve optimal individual- and population-level health outcomes (18). Knowledge of HCV status (ie, being tested for and diagnosed with HCV) represents a critical first step in the HCV cascade of care. However, a substantial proportion of individuals living with HCV (at least 21% in Canada) remain unaware of their HCV status (10,19) and are not connected to care and treatment; consequently, they are at risk for poor health outcomes. Collectively, these issues highlight the critical importance of access to and uptake of HCV treatment, as well as other prevention strategies, particularly among key affected populations, to maximize the impact of HCV efforts.

HCV disproportionally affects groups who are under-represented in health surveillance and who are underserved by the health system. Sex workers, particularly those who are street involved and use injection drugs, may be particularly vulnerable to HCV infection (1,20). In addition, multiple studies show that sex workers face many unique challenges that can severely reduce their access to and uptake of health services. These include fear of disclosing sex work or drug use due to criminalization and stigmatization, disrespect from health care providers, poor quality of care, and lack of women- and sex worker-tailored services, among others (21,22). However, and despite the increased vulnerability to HCV through dual sex/drug use routes and barriers to health services among sex workers, research examining access to HCV testing and care among this population remains negligible. Therefore, the aim of the current study was to characterize engagement in the HCV continuum of care and identify correlates of recent HCV testing among a cohort of women sex workers in metropolitan Vancouver, British Columbia.

METHODS

Study design

Data were drawn from An Evaluation of Sex Workers' Health Access (AESHA), an ongoing open prospective cohort of >720 women sex workers in metropolitan Vancouver, British Columbia, initiated in 2010. This study was developed based on community collaborations with sex work agencies since 2005 (23) and is monitored by a Community Advisory Board encompassing >15 agencies. Eligibility criteria for this cohort include age ≥14 years, self-identification as women (including male-to-female transgender individuals), and having exchanged money for sex in the previous 30 days. Participants are recruited through a combination of outreach to outdoor/public (eg, streets, alleys), indoor (eg, massage parlours, microbrothels, in-call locations) and off-street (eg, online and newspapers advertisements) sex work venues across metropolitan Vancouver, using time-location sampling (24,25). Sampling and recruitment procedures have been described in detail previously (23).

After providing written informed consent, participants complete an interview-administered questionnaire at baseline and every six months thereafter, and are offered HIV, HCV and other sexually transmitted infections testing, alongside pre- and post-test counselling. The questionnaire collects sociodemographic information, sex work patterns, sexual health and intimate partners, violence and trauma, drug use patterns, health care services access and utilization (including HIV and HCV care), and physical, social and structural characteristics of the working and living environment. A trained nurse provides basic treatment for sexually transmitted infections onsite, as well as referrals to health care services when needed. Participants receive an honorarium of $40 at each visit for their time and expertise. The study has been approved by the Providence Health Care/University of British Columbia Research Ethics Board.

The current study used cross-sectional baseline data from participants who enrolled between January 2010 and August 2013.

HCV continuum of care

Study population: For the analysis characterizing the HCV continuum of care, only participants with serologically confirmed HCV infection (ie, HCV antibody positive) at baseline were included.

Statistical analysis: Descriptive statistics were used to assess engagement in each step of the HCV continuum of care, including the proportion of: women living with HCV; diagnosed and aware of their infection; access to HCV related-care; offered treatment; and received treatment. Access to HCV care was evaluated through several questions regarding HCV care in the six months before the baseline visit, including whether women had received regular blood tests for HCV, had seen an HCV specialist, had been offered HCV treatment (including if declined) and had been receiving HCV treatment. The total number of HCV antibody-positive participants at baseline was used as the denominator for all calculations.

Recent HCV testing

Study population: The analysis of correlates of recent HCV testing was restricted to sex workers who self-reported being HCV-seronegative at enrollment or who acquired HCV <1 year before the date of their baseline visit.

Primary outcome: The primary outcome of interest for this particular analysis was recent HCV testing, defined as self-reporting having tested for HCV within the 12 months before the baseline visit. Participants who had never tested for HCV or those whose last test was >12 months were considered to not have a recent HCV test.

Explanatory variables: Based on previous studies examining correlates of uptake of HCV and HIV testing among sex workers and other vulnerable populations (26-32), a range of individual, contextual and social-structural explanatory variables were selected that were hypothesized to influence utilization of HCV testing in this particular setting. Individual-level factors examined included sociodemographic characteristics, and substance use behaviours including age (per year older), Aboriginal/Indigenous ancestry (including First Nations, Métis and Inuit peoples, yes versus no), international migration status (immigrant/migrant versus Canadian born), self-identification as sexual/gender minority (lesbian, gay, bisexual, transgender or two-spirit versus heterosexuals), baseline HIV status (positive versus negative), heavy alcohol drinking (>4 drinks per day, yes versus no), and any noninjection or injection drug use (yes versus no). Contextual factors evaluated included: current place of residence (residing in the inner city drug use epicentre [downtown eastside [DTES] versus elsewhere in Vancouver versus being homeless), physical and social features of the work environment such as primary place of soliciting clients (street/public place versus others, including off-street independent/ self-advertising or off-street indoor venue-based solicitation), servicing a higher number of clients (≥10 versus <10 clients/week), inconsistent condom use with clients (defined as 'usually', 'sometimes', 'occasionally' or 'never' versus 'always'), client-perpetrated physical or sexual violence (yes versus no), involvement in peer support (yes versus no) and recent incarceration, defined as having been in detention, prison or jail overnight or longer (yes versus no). Also considered were social-structural barriers to health services, including limited availability (defined as limited hours of operation or long wait times) and language barriers (yes versus no). Except for the sociodemographic variables, all other variables refer to the six-month period before the baseline visit. **Statistical analysis:** First, individual-, contextual- and social-structural-level factors were compared among participants who self-reported undergoing a recent HCV test with those who did not. In the bivariate analyses, logistic regression was used to analyze the categorical and continuous variables. Then, to examine the independent factors associated with recent HCV testing, a multivariable logistic regression model was fitted using an a priori defined model-building approach based on the examination of the Akaike information criterion (AIC) and P values. A backward stepwise technique was used in the selection of covariates. This approach has been used successfully in several studies (26,33). Starting with a full model containing all variables associated with the outcome at P<0.10 level in bivariate analyses, the AIC of the model is noted and the variable with the largest P value is dropped to build a reduced model. This iterative process is continued until no variables are left. The model with the lowest AIC value is selected as the final model. Two-sided P values, and unadjusted and adjusted ORs (aOR) with 95% CIs are reported. All statistical analyses were performed using SAS version 9.4 (SAS Institute, USA).

RESULTS

HCV continuum of care

Among 705 sex workers who received HCV testing at baseline, 302 were seropositive for HCV, corresponding to an HCV prevalence of 42.8%. Among them, the median age was 36 years (interquartile range 29 to 43 years), slightly more than one-half (51.7%) were of Indigenous/Aboriginal ancestry, and one-third identified as a sexual/ gender minority (33.1%). Recent use of noninjection or injection drugs was self-reported by 93.1% and 74.5%, respectively. With regard to the work environment, the majority (77.8%) were women who primarily solicited clients in public spaces. Overall, at baseline, 77.5% of participants were aware of their HCV infection, 41.7% reported having accessed HCV-related care, 13.9% having been offered treatment and only 1.0% reported receiving treatment (Figure 1).

Recent HCV testing

The present analysis included 552 women sex workers, of whom 420 (76.1%) self-reported being HCV-seronegative at baseline and an additional 132 (23.9%) being newly diagnosed with HCV within one year before enrollment. The median age was 34 years (interquartile range 27 to 42 years), 183 (33.2%) were of Indigenous/Aboriginal ancestry, and 46 (8.3%) were also living with HIV. More than one-quarter (27.0%) lived in the inner city drug use epicentre, and more than one-half primarily solicited clients in public/outdoor settings (51.6%). Recent use of noninjection or injection drugs was self-reported by 65.6% and 34.4% participants, respectively. In total, 292 (52.9%) reported undergoing a recent HCV test (ie, within the past year).

As shown in Table 1, factors positively and statistically significantly (ie, P<0.05) associated with recent HCV testing in bivariate analyses included being of Aboriginal/Indigenous ancestry, identifying as a sexual/ gender minority, living with HIV, use of injection and noninjection

TABLE 1

Bivariate logistic regression analyses of factors associated with recent hepatitis C virus (HCV) testing among self-reported HCV-negative female sex workers in Vancouver, British Columbia (n=552)

	Recent HCV test, n (%)			
Characteristic	Yes (n=292)	No (n=260)	OR (95% CI)	P
Age, median (IQR)	32.5 (26–41)	35 (29–42)	0.98 (0.97–1.00)	0.079
Sexual/gender minority				
Yes	99 (33.9)	33 (12.7)	3.53 (2.28–5.47)	<0.001
No	193 (66.1)	227 (87.3)		
Aboriginal ancestry				
Yes	129 (44.2)	54 (20.8)	3.02 (2.07–4.41)	<0.001
No	163 (55.8)	206 (79.2)		
Living with HIV*				
Yes	34 (11.6)	12 (4.6)	2.67 (1.35–5.27)	0.005
No	257 (88.0)	242 (93.1)		
Heavy drinking*				
Yes	9 (3.1)	4 (1.5)	2.02 (0.61–6.63)	0.248
No	282 (96.6)	253 (97.3)		
Non-injection drug use*				
Yes	254 (87.0)	108 (41.5)	9.41 (6.18–14.33)	<0.001
No	38 (13.0)	152 (58.5)		
Injection drug use*				
Yes	149 (51.0)	41 (15.8)	5.57 (3.71–8.34)	<0.001
No	143 (49.0)	219 (84.2)		
Immigrant to Canada				
Yes	25 (8.6)	142 (54.6)	0.08 (0.05–0.13)	<0.001
No	266 (91.1)	118 (45.4)		
Current place of residence* (Vancouver, British Columbia)				
Outside DTES	134 (45.9)	210 (80.8)	Reference	
DTES	125 (42.8)	24 (9.2)	8.16 (5.01–13.29)	<0.001
Homelessness	29 (9.9)	21 (8.1)	2.16 (1.19–3.95)	0.012
Primary place of soliciting clients*				
Outdoor/public	199 (68.2)	86 (33.1)	4.33 (3.03–6.19)	<0.001
Indoor/ off-street/ independent	93 (31.8)	174 (66.9)		
Higher number of clients*				
Yes (≥10/week)	140 (48.0)	148 (56.9)	0.69 (0.49–0.97)	0.035
No (<10/week)	146 (50.0)	107 (41.2)		
Inconsistent condom use with clients*				
Yes	65 (22.3)	31 (11.9)	2.16 (1.36–3.45)	0.001
No	219 (75.0)	226 (86.9)		
Involvement in peer support*				
Yes	53 (18.2)	22 (8.5)	2.47 (1.46–4.20)	0.001
No	231 (79.1)	237 (91.2)		
Physical/sexual violence by client*				
Yes	84 (28.8)	46 (17.7)	1.88 (1.25–2.82)	0.002
No	208 (71.2)	214 (82.3)		
Recent incarceration*				
Yes	57 (19.5)	19 (7.3)	3.11 (1.79–5.38)	<0.001
No	229 (78.4)	237 (91.2)		
Language barrier to health services*				
Yes	3 (1.0)	29 (11.2)	0.08 (0.03–0.28)	<0.001
No	289 (99.0)	231 (88.8)		
Availability barrier to health services*				
Yes	112 (38.4)	106 (40.8)	0.90 (0.64–1.27)	0.563
No	180 (61.6)	154 (59.2)		

**Refers to the six-month period before the baseline interview. Column percentages may not necessarily sum to 100% due to missing data or rounding error. DTES Downtown eastside; IQR Interquartile range*

TABLE 2
Multivariable logistic regression model of factors associated with recent hepatitis C virus (HCV) testing among self-reported HCV-negative female sex workers in Vancouver, British Columbia (n=552)*

Variable	Adjusted OR (95% CI)	P
Sexual/gender minority		
Yes versus no	1.89 (1.1–3.24)	0.020
Non-injection drug use†		
Yes versus no	1.95 (1.00–3.78)	0.049
Injection drug use*		
Yes versus no	2.00 (1.19–3.34)	0.008
Immigrant to Canada		
Yes versus no	0.24 (0.12–0.48)	<0.001
Current place of residence† (Reference: outside downtown eastside)		
Downtown eastside, Vancouver	3.19 (1.78–5.73)	<0.001
Homelessness	1.03 (0.50–2.12)	0.928
Higher number of clients†		
Yes versus no	0.67 (0.43–1.06)	0.088
Recent incarceration†		
Yes versus no	1.87 (0.95–3.67)	0.069

**Only the final list of variables included in the model with the best overall fit (ie, lowest Akaike information criterion value) are shown. Age, Aboriginal ancestry, HIV status, primary place of soliciting clients, client-perpetrated violence, inconsistent condom use by clients, peer support and language barriers to health services were included in the full model, but removed in the backwards selection approach; †Refers to the six-month period before the baseline interview*

drugs, currently residing in the inner city drug use epicentre, or being homeless (versus living outside the DTES), primarily soliciting clients in public/outdoor spaces, having been exposed to client-perpetrated violence, inconsistent condom use with clients, involvement in peer support and recent incarceration. In contrast, immigrant women, participants with higher number of clients and those reporting language barriers to health services had reduced odds of undergoing a recent HCV test.

As indicated in Table 2, in multivariable analyses, identifying as a sexual/gender minority (aAOR 1.89 [95% CI 1.11 to 3.24]), use of injection (aOR 2.00 [95% CI 1.19 to 3.34]) or noninjection drugs (aOR 1.95 [95% CI 1.00 to 3.78]) and living in the inner city drug use epicentre (versus elsewhere in Vancouver) (aOR 3.19 [95% CI 1.78 to 5.73]) remained independently and positively statistically significantly associated with undergoing a recent HCV test, while immigrant participants (aOR 0.24 [95% CI 0.12 to 0.48]) were less likely to have had a recent HCV test.

DISCUSSION

To our knowledge, the present study was the first to characterize access and uptake along the HCV care continuum among sex workers, a population with disproportionate burden of HCV (1,20). In the present analysis, only one-half of the women who self-reported being HCV seronegative or previously unaware of their HCV status had been tested for HCV in the past year. In addition, among women living with HCV, approximately one-quarter were unaware of their HCV status before undergoing an HCV test at their baseline visit. Importantly, and similar to a previous study of recent HIV testing among women sex workers in this setting (26), we found that a number of markers of vulnerability were associated with recent HCV testing, including use of noninjection and injection drugs, self-identification as a gender/sexual minority, and living in the inner city drug use epicentre (ie, the DTES). In contrast, immigrant participants had reduced odds of undergoing a recent HCV test.

A possible explanation to our findings is that to date, voluntary HCV screening programs in Canada have mostly used a risk-based approach, targeting individuals perceived to be at higher risk for HCV infection, including people who use injection drugs, or with clinical or laboratory evidence of liver disease (9). It is likely that the provincial government-sponsored "HIV Treatment as Prevention" efforts (ie, STOP HIV/AIDS®) to scale-up access to HIV testing and care, with a particular focus on the inner city DTES community, have contributed to increase uptake of HCV testing among these vulnerable populations (34,35). However, the high prevalence of undiagnosed HCV infection among sex workers found in the present analysis suggests that the effectiveness of risk-based testing strategies for HCV may be limited. Studies suggest that years of targeted screening programs for HCV might have contributed to community or health provider perceptions that only certain vulnerable populations are at-risk of acquiring HCV (eg, people who inject drugs) (36,37). Thus, given that HCV testing is only performed nominally, this stigma, as well as fear of disclosure of positive results, might pose barriers to HCV testing particularly in more underserved and criminalized populations such as women sex workers.

Of particular concern is that in our study, immigrant women were less likely to report having a recent HCV test. These results are consistent with previous research from different settings showing that migrant and new immigrant populations frequently experience enhanced barriers to health services (33,38-40). These barriers include limited knowledge on how to navigate the local health system, stigma, cultural and language barriers, as well as woman's immigration legal status, which could result in exclusion from publicly funded health insurance. Altogether, these findings highlight the need for tailored interventions to facilitate migrant sex workers' access to voluntary testing and other health services, such as peer/sex worker-led outreach to connect migrant sex workers to culturally appropriate and low-threshold health services. In addition, as lessons from successful global initiatives demonstrate, community empowerment through sex workers collectivization can play a key role in increasing access to health and social services for all sex workers, including migrant women (41-43).

Our results also showed that >40% of women sex workers in Vancouver had evidence of current/past HCV infection, likely due to dual parenteral and sexual routes (44-47). However, and despite this high HCV prevalence, fewer than one-half of sex workers living with HCV in our study reported being connected to HCV care and almost none were receiving treatment. These results are consistent with previous studies showing extremely low rates of HCV treatment uptake in British Columbia, particularly among women (4) and people who use illicit drugs (48).

Although not formally evaluated in the current study, determinants of low treatment uptake among this cohort of sex workers may reflect a combination of structural and individual barriers to health care. Misperceptions of HCV as relatively benign disease, absence of symptoms associated with chronic HCV until advanced stages, fear of side effects and associated treatment costs have been suggested as important individual-level barriers to HCV treatment uptake (49,50). For example, a recent study indicated that individuals with chronic HCV in British Columbia face substantial out-of-pocket expenses on HCV-related health care, even within a publicly funded universal health care system. The burden of these costs mostly falls on individuals with lower income, which could play a key role in patients forgoing or delaying treatment until advances stages of liver diseases (6). Collectively, these findings suggest an urgent need to explore alternative mechanisms to facilitate and support engagement in care and to ensure equitable access to HCV treatment to vulnerable populations including sex workers (4,6).

At the provider and macrostructural level, studies have demonstrated that stigma and misconceptions among health care providers regarding eligibility for HCV treatment of people who use drugs, as well as their perception of the clients potential to adhere to the regimen and eventual risk of re-infection represent major barriers to access to HCV treatment in this setting (49,51,52). Additionally, Canada's criminalized sex work and drug use environment poses a further challenge in this regard. In fact, previous studies have shown that policing practices can play an important role in shaping health care access among marginalized populations such as sex workers, people who use drugs or transwomen (22,53-55). Accordingly, decriminalization of sex work and drug use could offer an important first step toward increasing access to HCV testing and care to these vulnerable populations and a critical step for the success of any HCV 'treatment as prevention' strategy (16,56). For example, decriminalization of sex work could facilitate connections to health services of hidden migrant sex workers, who primarily work in indoor settings, do not disclose their involvement in sex work to friends, families and providers, and who are often disengaged from care. In addition, improving access to HCV care will require a combination intervention approach, including mobile and culturally competent services, as well as harm reduction and addiction management among other social-structural interventions (52).

The present study had several limitations. First, due its cross-sectional design, we were unable to determine temporal and causal relationships between the explanatory variables and outcome. Second, our sample was not randomly selected and, therefore, the results of the present study may not be generalizable to all sex workers in Vancouver or to other settings. However, the use of time-location sampling (24,25), a strategy known to help achieve representative samples of hard-to-reach populations helped to mitigate such potential bias. Third, because the data were collected through self-reporting, they may have been subject to recall or social-desirability biases. However, all interviews were conducted in confidential and safe environments by highly experienced interviewers with strong community rapport and involvement, facilitating accurate responses. Fourth, due to the lack of access to HCV RNA, we were not able to evaluate the number of participants with confirmed chronic HCV infection or with SVR. However, as studies of the natural history of HCV infection show, up to 85% of individuals infected with HCV will progress to chronic infection (57,58). In addition, the extremely low HCV treatment uptake within this cohort makes SVR a rare event.

SUMMARY

Our results show that almost one-half of self-reported HCV-seronegative women within a large cohort of women sex workers had not undergone an HCV test in the previous year, and hardly any women with HCV had accessed treatment. With increasing efficacy and tolerability of HCV drugs, these findings highlight the importance of facilitating and sustaining access to HCV testing, care and early treatment to all the population to improve not only individual and population outcomes, but also to decrease economic costs. In addition, given the high burden of HCV and large gaps in the HCV continuum of care among sex workers in Vancouver, future studies evaluating barriers and facilitators to HCV testing and care among this population are urgently needed to help inform public health efforts against HCV. Removal of criminal sanctions against sex work and drug use, as well as comprehensive interventions targeting these key populations, including mobile and culturally competent services, as well as harm reduction and addiction management will be critical to the success of any HCV treatment as a prevention strategy (52,56).

ACKNOWLEDGEMENTS: The authors thank all those who contributed their time and expertise to this project, particularly participants, AESHA community advisory board members and partner agencies. We wish to acknowledge Chrissy Taylor, Jennifer Morris, Tina Ok, Rachel Nicoletto, Julia Homer, Emily Leake, Rachel Croy, Emily Groundwater, Meenakshi Mannoe, Silvia Machat, Jasmine McEachern, Brittany Udall, Chris Rzepa, Jungfei Zhang, Xin (Eleanor) Li, Krista Butler, Peter Vann, Sarah Allan and Jill Chettiar for their research and administrative support. This research was supported by operating grants from the US National Institutes of Health (R01DA028648), the Canadian Institutes of Health Research (HHP-98835), the Canadian Institutes of Health Research/Public Health Agency of Canada (HEB-330155), and MacAIDS. KS is partially supported by a Canada Research Chair in Global Sexual Health and HIV/AIDS and Michael Smith Foundation for Health Research. JSGM is supported with grants paid to his institution by the British Columbia Ministry of Health and by the US National Institutes of Health (R01DA036307). MES is a Canadian Institutes of Health Research Bridge Fellow.

DISCLOSURES: JSGM has received limited unrestricted funding, paid to his institution, from AbbVie, Bristol-Myers Squibb, Gilead Sciences, Janssen, Merck and ViiV Healthcare.

REFERENCES

1. Webster DP, Klenerman P, Dusheiko GM. Hepatitis C. Lancet 2015;385:1124-35.
2. Kwong JC, Ratnasingham S, Campitelli MA, et al. The impact of infection on population health: Results of the Ontario burden of infectious diseases study. PloS One2012;7:e44103.
3. El Saadany S, Coyle D, Giulivi A, Afzal M. Economic burden of hepatitis C in Canada and the potential impact of prevention. Results from a disease model. Eur J Health Econ 2005;6:159-65.
4. Hsu PC, Buxton JA, Tu AW, Hill WD, Yu A, Krajden M. Publicly funded pegylated interferon-alpha treatment in British Columbia: Disparities in treatment patterns for people with hepatitis C. Can J Gastroenterol 2008;22:359-64.
5. Myers RP, Krajden M, Bilodeau M, et al. Burden of disease and cost of chronic hepatitis C infection in Canada. Can J Gastroenterol Hepatol 2014;28:243-50.
6. Federico CA, Hsu PC, Krajden M, et al. Patient time costs and out-of-pocket costs in hepatitis C. Liver Int 2012;32:815-25.
7. Backx M, Lewszuk A, White JR, et al. The cost of treatment failure: Resource use and costs incurred by hepatitis C virus genotype 1-infected patients who do or do not achieve sustained virological response to therapy. J Viral Hepat 2014;21:208-15.
8. McEwan P, Ward T, Bennett H, et al. Estimating the clinical and economic benefit associated with incremental improvements in sustained virologic response in chronic hepatitis C. PloS One 2015;10:e0117334.
9. Shah HA, Heathcote J, Feld JJ. A Canadian screening program for hepatitis C: Is now the time? CMAJ 2013;185:1325-8.
10. Remis R. Modelling the incidence and prevalence of hepatitis C infection and its sequelae in Canada. Ottawa: Public Health Agency of Canada, 2007.
11. Uhanova J, Tate RB, Tataryn DJ, Minuk GY. A population-based study of the epidemiology of hepatitis C in a North American population. J Hepatol 2012;57:736-42.
12. Xeroulis G, Inaba K, Stewart TC, et al. Human immunodeficiency virus, hepatitis B, and hepatitis C seroprevalence in a Canadian trauma population. J Trauma 2005;59:105-8.
13. Verna EC, Brown Jr RS. Hepatitis C virus and liver transplantation. Clin Liver Dis 2006;10:919-40.
14. Hill AM, Saleem J, Heath KA, Simmons B. Effects of sustained virological response (SVR) on the risk of liver transplant, hepatocellular carcinoma, death and re-infection: Meta-analysis of 129 studies in 23,309 patients with Hepatitis C infection. American Association for the Study of Liver Diseases (AASLD) Liver Meeting. Boston, November 7 to 11.
15. Grebely J, Matthews GV, Lloyd AR, Dore GJ. Elimination of hepatitis C virus infection among people who inject drugs through treatment as prevention: Feasibility and future requirements. Clin Infect Dis 2013;57:1014-20.
16. Martin NK, Vickerman P, Grebely J, et al. Hepatitis C virus treatment for prevention among people who inject drugs: Modeling treatment scale-up in the age of direct-acting antivirals. Hepatology 2013;58:1598-609.
17. Montaner JS, Hogg R, Wood E, et al. The case for expanding access to highly active antiretroviral therapy to curb the growth of the HIV epidemic. Lancet 2006;368:531-6.
18. Yehia BR, Schranz AJ, Umscheid CA, Lo Re V III. The treatment cascade for chronic hepatitis C virus infection in the United States: A systematic review and meta-analysis. PloS One 2014;9:e101554.

19. Rein DB, Smith BD, Wittenborn JS, et al. The cost-effectiveness of birth-cohort screening for hepatitis C antibody in U.S. primary care settings. Ann Intern Med 2012;156:263-70.
20. Shannon K, Kerr T, Marshall B, et al. Survival sex work involvement as a primary risk factor for hepatitis C virus acquisition in drug-using youths in a Canadian setting. Arch Pediatr Adolesc Med 2010;164:61-5.
21. Scorgie F, Nakato D, Harper E, et al. 'We are despised in the hospitals': Sex workers' experiences of accessing health care in four African countries. Cult Health Sex 2013;15:450-65.
22. Shannon K, Rusch M, Shoveller J, et al. Mapping violence and policing as an environmental-structural barrier to health service and syringe availability among substance-using women in street-level sex work. Int J Drug Policy 2008;19:140-7.
23. Shannon K, Bright V, Allinott S, et al. Community-based HIV prevention research among substance-using women in survival sex work: The Maka Project Partnership. Harm Reduct J 2007;4:20.
24. Clark JL, Konda KA, Silva-Santisteban A, et al. Sampling methodologies for epidemiologic surveillance of men who have sex with men and transgender women in Latin America: An empiric comparison of convenience sampling, time space sampling, and respondent driven sampling. AIDS Behav 2014;18:2338-48.
25. Stueve A, O'Donnell LN, Duran R, San Doval A, Blome J. Time-space sampling in minority communities: Results with young Latino men who have sex with men. Am J Public Health 2001;91:922-6.
26. Deering KN, Montaner JS, Chettiar J, et al. Successes and gaps in uptake of regular, voluntary HIV testing for hidden street- and off-street sex workers in Vancouver, Canada. AIDS Care 2014:1-8.
27. Hagan H, Campbell J, Thiede H, et al. Self-reported hepatitis C virus antibody status and risk behavior in young injectors. Public Health Rep 2006;121:710-9.
28. Kwiatkowski CF, Fortuin Corsi K, Booth RE. The association between knowledge of hepatitis C virus status and risk behaviors in injection drug users. Addiction 2002;97:1289-94.
29. Batona G, Gagnon MP, Simonyan DA, Guedou FA, Alary M. Understanding the intention to undergo regular HIV testing among female sex workers in Benin: A key issue for entry into HIV care. J Acquir Immune Defic Syndr 2015;68 Suppl 2:S206-12.
30. Xu J, Brown K, Ding G, et al. Factors associated with HIV testing history and HIV-test result follow-up among female sex workers in two cities in Yunnan, China. Sex Transm Dis 2011;38:89-95.
31. Wang Y, Li B, Pan J, Sengupta S, et al. Factors associated with utilization of a free HIV VCT clinic by female sex workers in Jinan City, Northern China. AIDS Behav 2011;15:702-10.
32. Surratt HL, O'Grady CL, Kurtz SP, Buttram ME, Levi-Minzi MA. HIV testing and engagement in care among highly vulnerable female sex workers: Implications for treatment as prevention models. J Health Care Poor Underserved 2014;25:1360-78.
33. Goldenberg SM, Liu V, Nguyen P, Chettiar J, Shannon K. International migration from non-endemic settings as a protective factor for HIV/STI risk among female sex workers in Vancouver, Canada. J Immigr Minor Health 2015;17:21-8.
34. O'Shaughnessy MV, Hogg RS, Strathdee SA, Montaner JS. Deadly public policy: What the future could hold for the HIV epidemic among injection drug users in Vancouver. Curr HIV/AIDS Rep 2012;9:394-400.
35. Heath K, Samji H, Nosyk B, et al. Cohort profile: Seek and treat for the optimal prevention of HIV/AIDS in British Columbia (STOP HIV/AIDS BC). Int J Epidemiol 2014;43:1073-81.
36. Barocas JA, Brennan MB, Hull SJ, Stokes S, Fangman JJ, Westergaard RP. Barriers and facilitators of hepatitis C screening among people who inject drugs: A multi-city, mixed-methods study. Harm Reduct J 2014;11:1.
37. Strauss SM, Munoz-Plaza C, Tiburcio NJ, et al. Barriers and facilitators to undergoing hepatitis C virus (HCV) testing through drug treatment programs. J Drug Issues 2008;38:1161-85.
38. Goldenberg SM, Chettiar J, Nguyen P, Dobrer S, Montaner J, Shannon K. Complexities of short-term mobility for sex work and migration among sex workers: Violence and sexual risks, barriers to care, and enhanced social and economic opportunities. J Urban Health 2014;91:736-51.
39. Guirgis M, Nusair F, Bu YM, Yan K, Zekry AT. Barriers faced by migrants in accessing healthcare for viral hepatitis infection. Intern Med J 2012;42:491-6.
40. Richter M, Chersich MF, Vearey J, Sartorius B, Temmerman M, Luchters S. Migration status, work conditions and health utilization of female sex workers in three South African cities. J Immigr Minor Health 2014;16:7-17.
41. Beattie TS, Mohan HL, Bhattacharjee P, et al. Community mobilization and empowerment of female sex workers in Karnataka State, South India: Associations with HIV and sexually transmitted infection risk. Am J Public Health 2014;104:1516-25.
42. Kerrigan DL, Fonner VA, Stromdahl S, Kennedy CE. Community empowerment among female sex workers is an effective HIV prevention intervention: A systematic review of the peer-reviewed evidence from low- and middle-income countries. AIDS Behav 2013;17:1926-40.
43. Parimi P, Mishra RM, Tucker S, Saggurti N. Mobilising community collectivisation among female sex workers to promote STI service utilisation from the government healthcare system in Andhra Pradesh, India. J Epidemiol Community Health 2012;66(Suppl 2):ii62-8.
44. Duff P, Tyndall M, Buxton J, Zhang R, Kerr T, Shannon K. Sex-for-crack exchanges: Associations with risky sexual and drug use niches in an urban Canadian city. Harm Reduct J 2013;10:29.
45. Shannon K, Rusch M, Morgan R, Oleson M, Kerr T, Tyndall MW. HIV and HCV prevalence and gender-specific risk profiles of crack cocaine smokers and dual users of injection drugs. Subst Use Misuse 2008;43:521-34.
46. DeBeck K, Kerr T, Li K, et al. Smoking of crack cocaine as a risk factor for HIV infection among people who use injection drugs. CMAJ 2009;181:585-9.
47. Nurutdinova D, Abdallah AB, Bradford S, O'Leary CC, Cottler LB. Risk factors associated with hepatitis C among female substance users enrolled in community-based HIV prevention studies. BMC Res Notes. 2011;4:126.
48. Grebely J, Raffa JD, Lai C, et al. Low uptake of treatment for hepatitis C virus infection in a large community-based study of inner city residents. J Viral Hepat 2009;16:352-8.
49. Bruggmann P. Accessing hepatitis C patients who are difficult to reach: It is time to overcome barriers. J Viral Hepat 2012;19:829-35.
50. Grebely J, Genoway KA, Raffa JD, et al. Barriers associated with the treatment of hepatitis C virus infection among illicit drug users. Drug Alcohol Depend 2008;93:141-7.
51. Myles A, Mugford GJ, Zhao J, Krahn M, Wang PP. Physicians' attitudes and practice toward treating injection drug users with hepatitis C: Results from a national specialist survey in Canada. Can J Gastroenterol 2011;25:135-9.
52. Harris M, Rhodes T. Hepatitis C treatment access and uptake for people who inject drugs: A review mapping the role of social factors. Harm Reduct J 2013;10:7.
53. Kerr T, Hayashi K, Ti L, Kaplan K, Suwannawong P, Wood E. The impact of compulsory drug detention exposure on the avoidance of healthcare among injection drug users in Thailand. Int J Drug Policy 2014;25:171-4.
54. Socias ME, Marshall BD, Aristegui I, et al. Factors associated with healthcare avoidance among transgender women in Argentina. Int J Equity Health 2014;13:81.
55. Kerr T, Small W, Wood E. The public health and social impacts of drug market enforcement: A review of the evidence. Int J Drug Policy 2005;16:210-20.
56. Bruggmann P. Treatment as prevention: The breaking of taboos is required in the fight against hepatitis C among people who inject drugs. Hepatology 2013;58:1523-5.
57. Micallef JM, Kaldor JM, Dore GJ. Spontaneous viral clearance following acute hepatitis C infection: A systematic review of longitudinal studies. J Viral Hepat 2006;13:34-41.
58. Thomson EC, Fleming VM, Main J, et al. Predicting spontaneous clearance of acute hepatitis C virus in a large cohort of HIV-1-infected men. Gut 2011;60:837-45.

4

Characterization of inflammatory bowel disease in elderly patients: A review of epidemiology, current practices and outcomes of current management strategies

Peter Stepaniuk BScHons[1], Charles N Bernstein MD[1,2],
Laura E Targownik MD MSHS[1,2], Harminder Singh MD MPH[1,2,3]

P Stepaniuk, CN Bernstein, LE Targownik, H Singh. Characterization of inflammatory bowel disease in elderly patients: A review of epidemiology, current practices and outcomes of current management strategies. Can J Gastroenterol Hepatol 2015;29(6):327-333.

The authors review and summarize the current literature regarding the epidemiology, clinical presentation and management of inflammatory bowel disease (IBD) in elderly patients.
Among elderly patients, the incidence of ulcerative colitis (UC) is higher than that of Crohn disease (CD). Elderly patients with a new diagnosis of UC are more likely to be male and have left-sided colitis. Elderly patients with a new diagnosis of CD are more likely to be female and have colonic disease. Conversely, increasing age at diagnosis has been associated with a lower likelihood of having any of a family history of IBD, perianal disease in CD and extraintestinal manifestations. Although response to drug therapies appears to be similar in elderly patients and younger individuals, the elderly are more likely to receive 5-aminosalicylic acid agents, and less likely to receive immunomodulators and biologics. Corticosteroid use in the elderly is comparable with use in younger individuals. The rates of surgical intervention appear to be lower for elderly CD patients but not elderly UC patients. Elderly individuals with UC are more likely to need urgent colectomy, which is associated with an increased mortality rate. Elective surgery is associated with similar outcomes among the elderly and young patients with IBD. Therefore, the use of immunomodulators and biologics, and earlier consideration of elective surgery for medically refractory disease in elderly patients with IBD, should be emphasized and further evaluated to prevent complications of chronic corticosteroid(s) use and to prevent emergency surgery.

Key Words: *Administrative data definition; Age-related differences*

La caractérisation des maladies inflammatoires de l'intestin chez les patients âgés : une analyse de l'épidémiologie, des pratiques courantes et des résultats des stratégies de prise en charge

Les auteurs ont analysé et résumé les publications récentes sur l'épidémiologie, la présentation clinique et la prise en charge des maladies inflammatoires de l'intestin (MII) chez les patients âgés.
Ainsi, chez les patients âgés, l'incidence de colite ulcéreuse (CU) est plus élevée que celle de la maladie de Crohn (MC). Un diagnostic de CU touche davantage les hommes et est davantage lié à une colite gauche, tandis qu'un diagnostic de MC touche davantage les femmes et est davantage lié à une maladie colique. En revanche, un âge plus avancé au diagnostic s'associe à une plus faible probabilité d'antécédents familiaux de MII, de MC périanale et de manifestations extra-intestinales. Même si la réponse à la pharmacothérapie semble similaire chez les personnes âgées et plus jeunes, les personnes âgées sont plus susceptibles de recevoir de l'acide 5-aminosalicylique et moins susceptibles de recevoir des immunomodulateurs et des produits biologiques. Leur utilisation de corticoïdes est toutefois comparable à celle des personnes plus jeunes. Le taux d'interventions chirurgicales semble plus faible chez les patients âgés atteints d'une MC, mais non chez ceux atteints d'une CU, qui risquent davantage de subir une colectomie urgente, associée à un taux de mortalité plus élevé. La chirurgie non urgente des MII donne des résultats similaires chez les personnes âgées et les patients plus jeunes. Par conséquent, il faut insister sur la possibilité d'utiliser des immunomodulateurs et des produits biologiques et d'envisager plus rapidement une chirurgie non urgente pour soigner une maladie réfractaire aux médicaments chez les personnes âgées atteintes d'une MII et l'envisager davantage pour prévenir les complications des corticoïdes chroniques et les chirurgies d'urgence.

Inflammatory bowel disease (IBD) is a collection of idiopathic inflammatory disorders that is becoming more prevalent in many countries around the world. There are two major subtypes of IBD, namely, Crohn disease (CD) and ulcerative colitis (UC). IBD is believed to arise in genetically predisposed individuals who develop an aberrant immune response to a yet to be defined environmental exposure (1). Although IBD is most commonly diagnosed in teenagers and young adults, it can present at any age. Because IBD has a minimal impact on life span, the numbers of elderly (most typically described as individuals >65 years of age) patients living with IBD is increasing (2). Thus, it is becoming more important to improve our knowledge base regarding the epidemiology and management of elderly individuals with IBD.

The present review summarizes the current literature regarding the epidemiology, clinical presentation, and the medical and surgical management of both CD and UC in elderly patients. We focused on evidence derived from analyses of population-based data sources, which are less likely to be affected by potential selection bias. We also discuss specific potential issues with the health care administrative data definition of IBD in the elderly because such data are commonly used to describe IBD epidemiology and clinical outcomes among those with IBD.

Search strategy

PubMed was searched to August 2014 using the following key words: (Inflammatory bowel disease OR IBD OR Crohn's disease OR ulcerative colitis) AND (elderly OR age OR epidemiology OR phenotype OR medication OR surgery OR administrative). Reference lists from the articles selected were reviewed for any additional relevant articles. Abstracts were reviewed and those reporting outcomes regarding IBD in the elderly were retrieved.

[1]*Internal Medicine, University of Manitoba;* [2]*University of Manitoba IBD Clinical and Research Centre;* [3]*Community Health Sciences, University of Manitoba, Winnipeg, Manitoba*
Correspondence: Dr Harminder Singh, University of Manitoba, 805-715 McDermot Avenue, Winnipeg, Manitoba R3E 3P4. e-mail Harminder.singh@med.umanitoba.ca

TABLE 1
Incidence rates of ulcerative colitis (UC) and Crohn disease (CD) in elderly patients

	Incidence rate in elderly* patients per 100,000	
Author (reference), study period; location	**UC**	**CD**
Lakatos et al (2), 1977-1981; Western Hungary	1.1	Virtually 0
Shivananda et al (4), 1991-1993; Europe (men)	10.8	2.9
Shivananda et al (4), 1991-1993; Europe (women)	5.3	2.0
Bernstein et al (9), 1989-1994; Manitoba	16.5	10.7
Piront et al (5), 1993-1996; Liège, Belgium	4.5	3.5
Loftus et al (7), 1990-2000; Minnesota, USA (men)	10.9	12.3
Loftus et al (7), 1990-2000; Minnesota, USA (women)	3.9	5.9
Gower-Rousseau et al (6), 1998-2006; Northern France	3.1	2.6
Bitton et al (11), 2001-2008; Quebec (men)	15.9	13.5
Bitton et al (11), 2001-2008; Quebec (women)	13.3	18.9
Lakatos et al (2), 2002-2008; Western Hungary	10.8	3.0

**Defined as >64 years of age in Shivananda et al (4); >60 years of age in Lakatos et al (2), Gower-Rousseau et al (6), Bernstein et al (9), Loftus et al (7) and Piront et al (5); and 70 to 79 years of age in Bitton et al (11)*

Epidemiology

Although the incidence of IBD is greatest in the second and third decades of life, up to 15% of cases are diagnosed after 65 years of age (3). There have been several studies that have examined the epidemiology of IBD in the elderly population, which have focused on elderly persons with a new diagnosis of IBD. A brief description of some of the key studies from Europe, the United States and Canada, along with their mode of data acquisition, is included below. Incidence rates of UC and CD from the various studies are listed in Table 1.

Shivananda et al (4) reported on an inception cohort of IBD diagnosed between October 1991 and September 1993 across 20 countries in Europe comparing the incidence of both UC and CD according to age, sex and geographical location. Physicians were asked to identify all new patients seen during the study period with symptoms consistent with IBD. Patient self-help groups were also targeted with exact methods of case finding varying at each site. IBD was diagnosed on the basis of endoscopic and/or radiographic evidence supported by mucosal biopsy wherever possible. The study found that for CD, men had incidence rates of 6.0, 3.2 and 2.9 per 100,000 in the 15 to 44, 45 to 64 and >64 age groups, respectively. Women had very similar incidence rates of 7.7, 3.0 and 2.0 per 100,000 in the different age groups, respectively. The incidence rates of UC in men were 11.2, 12.1 and 10.8 per 100,000, and in women were 10.7, 6.4 and 5.3 per 100,000 respectively, per age groups noted above. Thus, the incidence of UC in men was stable across age groups and there was a decline with age among women. However, in this era (1991 to 1993), in Europe, UC had a greater incidence at all ages compared with CD, with the greatest differences among the elderly population.

In Belgium, Piront et al (5) assessed the epidemiology of IBD in the province of Liège between June 1, 1993 and May 31, 1996. Data were collected from a questionnaire completed by gastroenterologists in the study area. IBD diagnosis was based on the criteria of Gower-Rosseau et al (6), which includes the opinion of two expert gastroenterologists based on clinical, radiographic, endoscopic and/or histological findings. Again, a higher annual incidence rate of UC (4.5 per 100,000) was reported among the elderly (≥60 years of age in this study) than that of CD in the same age category (3.5 per 100,000).

A population study using the *Registre des Maladies Inflammatoires Chroniques du Tube Digestif* (EPIMAD) registry from northern France with patient data from 1988 to 2008 was reported by Gower-Rousseau et al (6). The patient population consisted of all individuals ≥60 years of age diagnosed with IBD during the study period based on the opinion of two expert gastroenterologists on the basis of clinical, radiographic, endoscopic and/or histological findings. This registry was established by asking gastroenterologists to report the number of IBD patients in their care. The incidence rate of CD in the elderly (>60 years of age) patients was reported as 2.6 per 100,000 while the incidence rate of UC in the elderly was 3.1 per 100,000. Although the incidence rate of CD was comparable with other published studies, the incidence rate of UC in elderly patients appeared to be much lower than other studies (Table 1). Of the 11,724 patients diagnosed with IBD during the study period, 1058 (9%) were >60 years of age. These 1058 patients comprised 5% of all incident CD and 11% of all incident UC cases.

In Western Hungary, Lakatos et al (2) reported a population-based study examining incidence rates of IBD between 1977 and 2008. There were 914 patients with UC and 506 with CD diagnosed between January 1, 1977 and December 31, 2008. IBD patient data were collected every year from the general hospitals, gastroenterology outpatient units and family practitioners. The study report did not describe the criteria used for IBD diagnosis. The incidence rate of UC in elderly patients (defined as >60 years of age in this study) was reported to have increased from 1.09 in 1977 to 1981 to 10.8 per 100,000 from 2002 to 2007. The incidence rate of CD in the elderly also increased with CD being virtually absent in the elderly population until the 1990s, with the incidence rising to 3.04 per 100,000 from 2002 to 2007.

Loftus et al (7) reported IBD incidence rates in Olmstead County, Minnesota, USA, between 1990 and 2000. For this study, the investigators examined a central diagnostic database that included diagnoses generated from all outpatient visits, emergency room visits, hospitalizations, nursing home visits, surgical procedures, autopsy examinations and death certificates. Cases of CD were identified when they met at least two of the following criteria on two occasions two months apart: clinical history consistent with CD; endoscopic; radiological; laparotomy or histological findings of CD. A diagnosis of UC required on two occasions, separated by at least six months, diffusely granular or friable colonic mucosa on endoscopy and continuous mucosal involvement on endoscopy or barium studies. The incidence rates were standardized for the 2000 United States census population. For UC, the study reported overall incidence rates of 3.9 and 10.9 per 100,000 in the >60 years age group for women and men, respectively. For CD, the study reported overall incidence rates of 5.9 and 12.3 per 100,000 in the >60 years of age group for women and men, respectively. These incidence rates appear to be slightly higher than some of the other European studies reported during the same time period.

In a population-based study using administrative health care data from the central Canadian province of Manitoba, the overall annual incidence rates for CD and UC were 14.6 and 14.3 per 100,000, respectively, between 1989 and 1994 (8). These high rates were later corroborated by a multiprovince Canadian study (9). The diagnosis of IBD was based on administrative data definition, which was validated against chart review diagnosis of IBD, based on the Lennard-Jones (10) criteria, which use a combination of clinical features, radiographic, endoscopic and pathological criteria. In the elderly population (described as >60 years of age in this study), incidence rates were 10.7 and 16.5 per 100,000 for CD and UC, respectively. Both of these rates were higher than other studies, possibly owing to the fact that this was one of few studies that used administrative health care data to determine incidence rates. However, the trend of UC being more common in the elderly than CD again was evident (8).

Another Canadian study from Quebec reported some of the highest incidence rates of both CD and UC. Bitton et al (11) conducted a population-based cohort study using administrative health care data. In this study, physician claims and hospitalizations in Quebec using an IBD diagnostic code from January 1996 to December 2009 were examined. Patients who received at least one hospitalization or four physician billing claims with a CD or UC diagnostic code within a two-year period were classified as having IBD; the definition was not specifically validated in Quebec. The study found overall incidence

TABLE 2
Clinical phenotype of Crohn disease in elderly* patients

	Gower-Rousseau et al (6)					Lakatos et al (2)				Quezada et al (12)				
	Age, years					Age, years				Age, years				
Phenotype	<17	17–39	40–59	≥60	P	<19	19–60	>60	P	<17	17–40	41–59	≥60	P
Male sex	53	42	49	38	$<10^{-4}$	62	47	52		46	41	36	27	0.38
Family history of IBD	16	14	10	7	$<10^{-4}$	17.6	12.4	9.5		33	30	13	14	0.02
Perianal disease/anal fistula	5	3	4	7	0.0006	32.4	25.1	9.5	0.051	45	32	13	22	<0.01
Extraintestinal manifestations	20	12	13	7	$<10^{-4}$					27	28	27	18	0.82
Terminal ileum	14	17	17	10	$<10^{-4}$	31	33	38		20	38	37	20	<0.01
Colonic	20	25	40	65		24	37	62	†	20	18	38	55	
Ileocolonic	66	58	43	25		45	30	0		59	40	25	20	
Upper gastrointestinal						0	0.7	0		1	3	0	5	
Nonstricturing, nonpenetrating	72	66	69	78	0.004	62	57	38		26	34	56	64	<0.01
Stricturing	24	30	23	17		15	19	43	‡	31	31	33	14	
Penetrating	4	4	8	5		23	24	19		43	35	11	23	

*Data presented as % unless otherwise indicated. *Defined as >60 years of age in Lakatos et al (2), Gower-Rousseau et al (6) and Quezada et al (12); †Elderly versus pediatric: P=0.001; OR 5.06 (95% CI 1.81–14.1) and elderly versus adults: P=0.02; OR 2.79 (95% CI 1.13–6.88); ‡Elderly versus pediatric: P=0.005; OR 4.30 (95% CI 1.47–12.6) and elderly versus adults: P=0.01; OR 3.10 (95% CI 1.26–7.62). IBD Inflammatory bowel disease*

rates of 17.4 and 10.1 per 100,000 for CD and UC, respectively, from 2001 to 2008. Incidence and prevalence rates were further standardized using the 2001 Quebec census. For CD, the study found female incidence rates of 17.5, 18.9 and 15.5 and male incidence rates of 13.7, 13.5 and 12.8 in the 60 to 69, 70 to 79 and ≥80 years of age groups, respectively. For UC, the study found female incidence rates of 11.4, 13.3 and 12.8 and male incidence rates of 15.0, 15.9 and 14.8 in the 60 to 69, 70 to 79 and ≥80 years of age groups, respectively. These rates are comparable with the above-mentioned Canadian study using administrative data.

Overall, the incidence rate of IBD in the elderly varies across Europe, the United States and Canada, with higher rates reported in Canada. The rates of UC appear to be higher than rates of CD in the elderly, and the incidence rates of both UC and CD appear to be rising in the elderly in some studies, with the UC incidence rising at a faster rate than CD incidence. However, of the above studies that examined the incidence of IBD in the elderly, only the North American studies used administrative health care data to identify IBD patients. The European studies relied on physicians reporting on the number of IBD cases in their respective catchment areas among other techniques and, hence, make these data not directly comparable with the North American studies. More importantly, this raises the question of whether the relatively high incidence numbers reported from Canada are truly higher than that reported from other geographical areas or if this simply reflects the method of data collection. Physician reporting can be incomplete and could be contributing to lower rates in other studies. Conversely, case misclassification in administrative records may conceivably lead to an overestimation of the incidence of IBD. Also important is that studies have used somewhat different criteria for identification and diagnosis of IBD. Most studies use a combination of clinical, radiographic, endoscopic and pathological criteria, although the specific criteria varied for each study, which could also be contributing to differences in study findings.

CLINICAL PRESENTATION

CD

An overview of the clinical phenotype of CD at diagnosis in the elderly can be found in Table 2. In northern France, Gower-Rousseau et al (6) examined sociodemographic and the clinical characteristics of those diagnosed with IBD after 60 years of age versus those diagnosed at younger ages. Elderly CD patients were more likely to be female and to have anal fistulas while they were less likely to have a family history of IBD or extraintestinal manifestations. A slightly higher proportion of elderly patients were found to have nonstricturing and nonpenetrating CD behaviour compared with younger age groups who were more likely to present with penetrating CD. Elderly patients were also more likely to have isolated colonic disease and less likely to have ileocolonic disease when compared with the younger age groups (6).

In the study from Western Hungary (2), the elderly newly diagnosed with CD were less likely to report a family history of IBD and have perianal disease, and more likely to have their disease restricted to their terminal ileum or their colon (ie, less likely to have ileocolonic disease). Compared with their younger counterparts, elderly CD patients were also more likely to present with stricturing disease compared with the pediatric and adult populations. This is in contrast with the findings of Gower-Rousseau et al (6), who found a higher proportion of elderly with perianal disease/anal fistulas and with nonstricturing, nonpenetrating behaviour.

Quezada et al (12) examined the association between age at diagnosis and CD phenotype in a cross-sectional study involving adults with CD evaluated at the University of Maryland (Baltimore, USA) between July 2004 and April 2010. In this study, the elderly (defined as those ≥60 years of age at diagnosis) were less likely to have a family history of IBD and less likely to have perianal disease, similar to other studies. However, in this study, only 27% of the elderly CD population was male, which perhaps reflects that this cohort was a unique population. The elderly were found to be more likely to have colonic disease and less likely to have ileocolonic disease. Once again, the elderly were more likely to have nonstricturing and nonpenetrating disease behaviour compared with the younger age groups, similar to the French study (6).

In summary, most studies report that elderly patients diagnosed with CD are more likely to be female and have colonic disease compared with their younger counterparts. Furthermore, increasing age at diagnosis has been associated with a lower likelihood of family history of IBD, less perianal disease and extraintestinal manifestations.

UC

An overview of the clinical phenotype of UC at diagnosis in elderly patients is presented in Table 3. In the French study (6), elderly UC patients were more likely to be male and less likely to have a family history of IBD or extraintestinal manifestations. Elderly patients were more likely to have left-sided colitis whereas younger patients were more likely to have extensive colitis. These findings were also found in the study from Western Hungary by Lakatos et al (2).

In a study from Allegheny County, Pennsylvania, USA, UC patients were identified using *International Classification of Diseases, Ninth Revision* (ICD-9) codes from the medical records at 20 hospitals in the county, which includes the metropolitan area of Pittsburgh (13).

TABLE 3
Clinical phenotype of ulcerative colitis in elderly* patients

	Gower-Rousseau et al (6)					Lakatos et al (2)				Juneja et al (13)
	Age, years					Age, years				
Phenotype	<17	17–39	40–59	≥60	P	<19	19–60	>60	P	>64 years of age
Male sex	44	50	66	62	$<10^{-4}$	47	53	52		
Family history of inflammatory bowel disease	13	7	5	3	$<10^{-4}$	9.3	8.5	0.9	†	
Extraintestinal manifestations	5	3	3	3	<0.05					
Proctitis	31	50	48	29	$<10^{-4}$	23	26	21		28.4
Left-sided	27	26	30	45		40	50	61	‡	25.7
Extensive colitis	41	24	22	26		37	24	18	§	45.9

**Defined as >64 years of age in Juneja et al (13), and as >60 years of age in Lakatos et al (2) and Gower-Rousseau et al (6); †Elderly versus pediatric: P=0.007; OR 10.7 (95% CI 1.30–90.9) and elderly versus adults: P=0.006; OR 9.9 (95% CI 1.33–71.4); ‡Elderly versus pediatric: P=0.005; OR 2.38 (95% CI 1.30–4.36) and elderly versus adults: P=0.025 OR 1.61 (95% CI 1.06–2.44); §Pediatric versus elderly: P=0.003; OR 2.73 (95% CI 1.38–5.41) and pediatric versus adult: P=0.01; OR 1.89 (95% CI 1.15–3.11)*

TABLE 4
Summary of clinical features of inflammatory bowel disease (IBD) in elderly patients

Feature	Crohn disease	Ulcerative colitis
Sex	More commonly female	More commonly male
Family history of IBD	Less common	Less common
Perianal disease/anal fistulas	Less common	Not applicable
Extraintestinal manifestations	Less common	Less common
Localization of disease	Colonic	Left-sided colitis

Patient data were extracted for patients >65 years of age evaluated in the outpatient departments of these hospitals between January 1, 1991 and December 31, 2010. In contrast to other studies, Juneja et al (13) examined the phenotype of elderly patients with pre-existing IBD; 28.4% of patients had proctitis, 11.4% with proctosigmoiditis, 14.3% with left-sided/distal ulcerative colitis and 45.9% pancolitis. This relatively high proportion of patients with extensive colitis is in contrast to the findings of Gower-Rousseau et al (6) and Lakatos et al (2), who reported rates of extensive colitis to be 26% and 18%, respectively. The relatively low proportion of patients with left-sided colitis in this study may appear to contradict the findings of Gower-Rousseau et al (6) and Lakatos et al (2), who reported rates of left-sided colitis to be 45% and 61%, respectively. These findings may reflect the mode of data acquisition in these studies and, more importantly, evaluation of phenotype at disease diagnosis in the other studies versus that of prevalent IBD cases in the Pennsylvania study (13). A summary of the clinical features of IBD in elderly patients is presented in Table 4.

MEDICAL TREATMENT

The recommended treatment and their response rates for IBD in elderly patients are generally similar to that in younger individuals; however, the speed of response may be slower, which can be an important problem because of the tendency of rapid deterioration among the elderly without treatment (14). Elderly patients pose a unique challenge when determining therapy choice because patients >60 years of age have been not been included in most clinical trials (15). In addition, some authors have suggested that immunosuppressive therapy should be avoided in the elderly IBD population due to an increased baseline risk of infection and neoplastic complications (16). These issues reflect gaps in knowledge, which hinder the adoption of evidence-based treatment pathways in the elderly population with IBD. An overview of IBD medication usage in the elderly in clinical practice is listed in Table 5.

5-aminosalicylates

Drugs in the 5-aminosalicylate (ASA) class are commonly used for both treatment of mild-to-moderate active symptoms as well as maintenance therapy due to the safety of these drugs and their relatively benign side effect profile (17). Therefore, it may be expected that the use of 5-ASA would be favoured in the elderly. However, the prevalence of 5-ASA use among the elderly varies in different jurisdictions. In the Pennsylvania study (13), patterns of medication usage included use of a drug for more than six continuous months. They found that 5-ASA agents were the most common maintenance therapy in elderly IBD patients, with 44% of patients taking medications from this class. Although this study investigated medication usage in the entire elderly IBD population (both CD and UC), it should be noted that there were 54% more UC patients (n=243) in this study compared with CD patients (n=150) and, therefore, the true utilization of 5-ASA agents in the elderly CD population cannot be discerned from this study. In a population-based French study (18), there was a combined total of 841 elderly IBD patients with recorded exposure to medications. Patients were considered to be exposed if the medication (at least one prescription) was started at diagnosis or during follow-up. The elderly IBD population had relatively high rates of 5-ASA use, with 75% of patients exposed to this drug with similar rates in the CD and UC populations. Similar findings have also been reported from studies at tertiary care centres. For example, Bautista et al (19) performed a retrospective cohort study of patients ≥18 years of age who underwent abdominal surgery for IBD between 2000 and 2010 at a single tertiary academic centre in Milwaukee, Wisconsin, USA. Patient characteristics, including exposure to medications (5-ASA, corticosteroids, immunomodulators, biologics) within two months before surgery, were included in the analysis. The patients were stratified according to age, with 42 patients in the elderly (>64 years of age) age group. The patients in the elderly cohort all had pre-existing IBD with varying disease duration. In the elderly IBD cohort, 5-ASA use was higher compared with younger cohorts. Overall, only 23% of patients were exposed to this drug, significantly less than other reported studies. However, in this study, the elderly population consisted of only 42 patients (34 with CD and 8 with UC). These findings suggest that the use of 5-ASA agents is less common in the elderly CD population compared with the elderly UC population. The variation in the definition of 'exposure' in individual studies likely contributes to the difference in the prevalence of 5-ASA agents used in the different studies as does the inclusion of pre-existing IBD or elderly onset IBD. In addition, each study grouped CD and UC patients with varying proportions, making direct comparisons of medication usage in the elderly IBD population difficult. It is known that 5-ASA agents have a very limited role in CD, whereas they are first line for mild/moderate disease in UC in the general IBD population. Overall, 5-ASA agents continue to be one of the key drugs of choice in the management of IBD in the elderly, especially those with elderly onset IBD, and its use varies across North America and Europe.

Corticosteroids

Corticosteroid use in the treatment of IBD flares is common practice because these powerful drugs can be highly effective at managing acute symptoms. However, CD patients >50 years of age treated with corticosteroids may be at a significantly increased risk for the development of hypertension, hypokalemia and mental status changes (20). In the

TABLE 5
Medication usage in elderly inflammatory bowel disease patients

	Lakatos et al (2) (CD)				Lakatos et al (2) (UC)				Bautista et al (19)					Juneja et al (13)	Gower-Rousseau et al (6)
	Age, years				Age, years				Age, years					Age, years	Age, years
Medication	**<19**	**19–60**	**>60**	**P**	**<19**	**19–60**	**>60**	**P**	**18–35**	**36–49**	**50–64**	**>64**	**P**	**>64**	**>60**
Mesalamine									13	6	10	23	0.06	44	75
Prednisone	71.6	68.6	57.1		57.3	39.8	17	†	34	37	31	23	0.28	31.6	31
Immunomodulators	68.9	42.6	28.6	*	9.3	7.8	2.8		62	69	55	45	0.06	6.9	16
Biologics	9.5	9.5	0						42	37	38	19	0.03	2.6	3.6

*Data presented as % unless otherwise indicated. *Elderly versus pediatric: P=0.001; OR 0.18 (95% CI 0.06–0.52); †Pediatric versus elderly: P=0.001; OR 6.58 (95% CI 3.32–12.9), pediatric versus adult: P=0.003; OR 2.03 (95% CI 1.25–3.28), adult versus elderly: P=0.001; OR 3.24 (95% CI 1.91–5.49). CD Crohn disease; UC Ulcerative colitis*

study from Western Hungary (2), patients ≥55 years of age with either CD or UC were less likely to be exposed to corticosteroids than their younger counterparts. Only 17% of elderly UC patients received corticosteroids while 57.1% of elderly CD patients received them, although the study did not specify what constituted a medication trial. However, it should be noted that there were only 21 elderly CD patients in this study. In the study from Pensylvania (13), chronic corticosteroid use was identified in 31.6% of elderly IBD patients, which was similar to the findings in the French study (6), with 31% of elderly IBD patients exposed. Bautista et al (19) did not detect any significant difference in corticosteroid use in the elderly IBD population compared with younger cohorts, with 23% of elderly IBD patients exposed to corticosteroids. Overall, it appears that approximately one-quarter to one-third of elderly IBD patients are exposed to corticosteroids during their management, with only slightly less use compared with younger IBD patients. This appears to be independent of assessment among newly diagnosed elderly IBD patients or that among those with long-standing disease.

Biologics and immunomodulators

Biologics (adalimumab and infliximab) and immunomodulators (6-mercaptopurine, azathioprine or methotrexate) are generally reserved for individuals with severe IBD refractory to traditional therapies (21). Lakatos et al (2) and Bautista et al (19) both demonstrated decreased use of these drugs in the elderly IBD population compared with younger patients. Immunomodulator use varies, with 2.8% to 45% of elderly IBD patients exposed to these drugs whereas biologic use remains quite low, with exposure ranging from 0% to 19%, with many studies reporting use in <5% of elderly IBD patients (2,6,13,19). Again, this appears to be independent of assessment among elderly onset IBD or that among elderly patients with long-standing IBD.

A retrospective chart review cohort study that examined elderly patients (>60 years of age) treated with infliximab suggested that the response rate to biological agents is similar in the elderly to that in the younger patients (22). Small sample size studies, which suggest that absolute risk of complications and side effects of medications are higher in the elderly, have likely driven the lower use of immunomodulators and biologicals in elderly patients (23). Greater attention to the risks of infections and thromboembolism may mitigate these risks. In addition, alternative treatments, such as corticosteroids and urgent surgery, are also associated with higher absolute risk for complications in the elderly (20,24). Furthermore, thiopurines have the propensity to interact with a number of other drugs and, thus, with polypharmacy being a common practice in elderly patients, an updated medication list should be obtained from every patient on a thiopurine (25).

SURGICAL MANAGEMENT

The principles for surgical therapy in elderly patients with IBD do not differ from their younger counterparts and is largely reserved for patients refractory to medical management. However, the rates of surgical intervention appear to be different for elderly CD patients but not elderly UC patients.

The study from Allegheny County, Pennsylvania (13) also examined patterns of bowel surgery in individuals with IBD >65 years of age. Individuals with CD diagnosed at <65 years of age were found to be significantly more likely to undergo surgery (63.6%) than those diagnosed at ≥65 years of age (20.9%). This was attributed to elderly patients being diagnosed with CD earlier in life having had greater small bowel involvement compared with new-onset CD in the elderly population. Rates of surgical intervention in the elderly and younger UC populations were similar (39.5% versus 50.9% in the younger cohort [P not significant]).

In the study from Northern France (18) (841 IBD patients >60 years of age at diagnosis from 1988 to 2006, including 367 CD and 472 UC), the cumulative probability of first intestinal resection in elderly CD patients was 18% (95% CI 14% to 23%) at one year, 27% (95% CI 23% to 32%) at five years and 32% (95% CI 27% to 38%) at 10 years from the time of diagnosis. In elderly UC, the cumulative probability of colectomy was 4% (95% CI 2% to 6%) at one year, 8% (95% CI 5% to 11%) at five years and 8% (95% CI 6% to 12%) at 10 years from time of diagnosis. Elderly patients with ileal or ilealcolonic CD and those with stricturing or penetrating CD had an increased risk for intestinal resection while the use of corticosteroids was associated with a decreased likelihood of surgery. Age, sex, extraintestinal manifestations, perianal lesions and the use of immunomodulators did not influence this risk. Of elderly UC patients, those with extensive colitis and those exposed to corticosteroids were more likely to require surgery. Age, sex, extraintestinal manifestations and use of immunomodulators did not influence this risk in UC patients.

Multiple studies that have assessed complication rates in ileoanal pouch surgery for UC have found that the incidence of anastomotic leaks, pouch-related septic complications and ileal anal pouch failure rates do not differ between younger and older patients undergoing surgery (26-28). For example, a retrospective cohort study of patients ≥18 years of age who underwent abdominal surgery for IBD between 2000 and 2010 at a single tertiary academic centre in Milwaukee, Wisconsin (USA), 42 elderly IBD patients (≥65) were matched at least 1:1 to patients in each of three control groups (18 to 35, 36 to 49, 50 to 64 years of age) according to sex, disease type/location and type of surgery (19). They found that the postoperative complication rate was not significantly different between the elderly and younger cohorts (38% in 18 to 35; 39% in 36 to 49; 40% in 50 to 64; and 48% in >64). The only statistically significant risk factors for increased risk for complication were high Charlson comorbidity index, low preoperative hemoglobin level, total parenteral nutrition use and failed medical therapy (as an indication for surgery).

Targownik et al (29) examined colectomy in UC patients. The study's main purpose was to determine the incidence of colectomy in UC and identify risk factors associated with early and late colectomy. Patients were identified using a population-based data set that included all persons with UC in Manitoba from 1984 through 2010. Patients <25 years of age had higher rates of late colectomy (>90 days from diagnosis) compared with patients >65 years of age (HR 3.30 [95% CI 2.08 to 5.24]: P<0.0001). However, rates of early colectomy

(≤90 days from diagnosis date) were higher in patients >65 years of age (3.1%) compared with 1.6% of those <65 years of age (HR 2.00 [95% CI 1.33 to 3.55]; P=0.0079). This was postulated to be due to elderly UC patients presenting with more severe symptoms and being more prone to decompensation requiring urgent colectomy. Conversely, elective and semi-elective surgery for elderly UC patients may be less likely due to patient preference, or due to physician preference because of medical comorbidities, which would increase their risk for postoperative morbidity and mortality.

Not only are rates of early colectomy higher in elderly UC patients, but the mortality rate post-emergency colectomy also appears to be higher, as suggested by Ikeuchi et al (24): 26.7% of elderly patients (≥60 years of age) who underwent emergency colectomy died within 30 postoperative days whereas the mortality rate was only 2.6% in patients <60 years of age. Pneumonia and intra-abdominal sepsis were the most common causes of death post-emergency surgery. Mortality rates following elective surgery were comparable between the two groups, with a mortality rate of 0.78% in patients <60 years of age and 0.88% in elderly patients.

Comorbidities play a significant role in postoperative outcomes of IBD patients, as reported by Kaplan et al (30). This study found that as the number of comorbidities in surgical patients undergoing an IBD-related operation increased, so too did postoperative mortality. Congestive heart failure, liver disease, thromboembolic disease and renal disease were associated with a significant increase in mortality rate. Rates of postoperative death were highest for emergent surgery in the elderly population (65 to 80 years of age) with ≥2 comorbidities (20.6%). Elderly patients undergoing emergent surgery with <2 comorbidities had an 11.0% mortality rate. Mortality rates were lower in elderly patients undergoing elective IBD surgery with ≥2 comorbidities (7.7%) and with <2 comorbidities (2.8%).

Thus, elderly CD patients undergo surgery less often compared with their younger counterparts, whereas elderly UC patients have surgery at rates comparable with younger patients. However, the elderly are more likely to need urgent colectomy, which is associated with an increased mortality rate, especially if the patient has underlying comorbidities. Complication rates following elective surgery are comparable in elderly and younger patients. Surgery should be performed as soon as necessary because morbidity results from prolonged ineffective medical treatment. However surgery, which can be associated with increased mortality in elderly patients (31), may also be avoided by earlier and more aggressive use of immune therapies in the elderly.

ADMINISTRATIVE HEALTH CARE DATA DEFINITION OF IBD IN ELDERLY PATIENTS

As reviewed above, referral centre studies and population-based studies using administrative data can arrive at different conclusions. One of the challenges with analyzing administrative data is the possibility of miscoding of data entry such that some individuals counted as having IBD may not truly have IBD. Because the elderly are more likely to develop ischemic colitis and may be more likely to have complicated infectious colitis there is possibly a greater risk for misclassification of IBD among the elderly, than younger populations. ICD codes – the primary diagnostic coding system used in hospital settings and for physician outpatient visits – remain of vital importance in the generation of epidemiological data. While most administrative definitions of IBD have been validated across the population of IBD, these definitions may not necessarily perform as accurately for identifying elderly individuals with IBD. Other than the fact that the elderly may be more likely to be hospitalized or seek medical care for other colitides, the elderly may have a higher prevalence of other lower gastrointestinal symptoms including rectal bleeding, fecal urgency or diarrhea, which may be misattributed to IBD. This raises the concern that accepted algorithms for identification of IBD may be less specific in the elderly, leading to a reduced positive predictive value (PPV).

Currently, only one published study (Benchimol et al [32]) has examined the accuracy of ICD diagnostic codes used in administrative health care data (hospitalization discharge abstracts and physician claims data) among elderly IBD patients. This study was undertaken in Ontario, and included hospital discharge abstract data collected from hospitals and reported to the Canadian Institute for Health Information and also billing claims for all physician services in Ontario. The most accurate algorithm in Ontario to identify individuals with IBD included five physician claims and/or hospitalizations related to IBD within four years (sensitivity 76.8%; specificity 96.2%; PPV 81.4%; negative predictive value 95.0%). The addition of one pharmacy claim for an IBD-related medication improved algorithm accuracy in those >65 years of age, improving the PPV of the algorithm without sacrificing sensitivity. The sensitivity remained constant (59.3%) but the PPV increased from 64.0% without a pharmacy claim to 71.1% with a pharmacy claim and 76.3% to a pharmacy claim excluding 5-ASA for elderly with IBD. However, including a pharmacy claim as part of the definition will exclude those with milder disease who do not require any medications. Importantly, this study suggests that the accuracy and PPV of administrative health care data definition of IBD is lower in the elderly than in younger individuals. This study suggested that even with the most accurate algorithm, one-quarter of elderly patients identified to have IBD in the administrative data may not have IBD. Therefore, studies using this Canadian administrative data definition to estimate incidence rates may be overestimating incidence of IBD in the elderly. The Benchimol et al (32) study results also suggest that studies on elderly IBD using administrative health care data with single occurrence of ICD codes in the databases, should be viewed with caution because of the lower accuracy (PPV of 64%) of single ICD codes among the elderly. Our own recent study (33) suggests that while less accurate in UC, a single ICD-10 IBD code is satisfactory to identify elderly CD and UC hospitalized patients and, hence, the lower accuracy is predominantly when a single occurrence is encountered in the physician billing/outpatients datasets. More research is necessary to fully validate the administrative codes used to define CD and UC among the elderly in other jurisdictions.

CONCLUSIONS

IBD is an increasingly prevalent chronic immune disease affecting the elderly. The rates of UC appear to be higher than rates of CD in elderly patients, and the incidence rates of both UC and CD appear to be increasing, albeit modestly. Elderly newly diagnosed CD patients are more likely to be female and to have colonic disease and less likely to have a family history of IBD or perianal disease. Conversely, elderly patients newly diagnosed with UC are more likely to be male and have left-sided colitis and are also less likely to have a positive family history of IBD. The elderly are less likely to receive biological agents or immunomodulators or surgery. This may be secondary to concerns for risks for infection and cancer; or may reflect that the elderly do have more comorbidities and polypharmacy making the use of more medications or undergoing surgery riskier. However, because the risk for complications with corticosteroid use and that of death with urgent or emergent surgery is much higher in elderly than younger patients, optimizing treatment algorithms in the elderly with IBD is warranted. Considering the increased mortality with emergent surgery in the elderly, this is one scenario that needs to be avoided. Hence, an increase in the use of immunomodulators and biological agents and earlier elective surgery for medically refractory elderly patients with IBD should be encouraged. More research is necessary to validate the health care administrative data codes used to define CD and UC among the elderly and assess comparative incidence among the elderly in different geographical areas using standardized IBD case definition.

DISCLOSURES: Dr Charles Bernstein is supported, in part, by the Bingham Chair in Gastroenterology. He has consulted or served on advisory boards to AbbVie Canada. Shire Canada, Takeda Canada, Forrest Canada, Cubist Pharmaceuticals, Pfizer, Theradiag and Mylan Pharmaceuticals. He has received educational grants from AbbVie Canada, Shire Canada, Takeda Canada and Janssen Canada. He has been on speakers' bureau for AbbVie Canada and Shire Canada. Laura Targownik: Speaker's Panel for Janssen Canada, Takeda Canada, Shire Canada, Pfizer Canada. Grant Support from Pfizer Canada and AbbVie Canada. Advisory Boards for Takeda Canada, AbbVie Canada, Janssen Canada. Harminder Singh: Consultant: Medial Cancer Screening Ltd. Advisory Board: Pendopharm.

REFERENCES

1. Shanahan F. The microbiota in inflammatory bowel disease: Friend, bystander, and sometime-villain. Nutr Rev 2012;70(Suppl 1):S31-7.
2. Lakatos P, David G, Pandur T. IBD in the elderly population: Results from a population-based study in Western Hungary, 1977-2008. J Crohn's Colitis 2011;5:5-13.
3. Benchimol EI, Mack DR, Nguyen GC, et al. Incidence, outcomes, and health services burden of very early onset inflammatory bowel disease. Gastroenterology 2014;147:803-13,e7.
4. Shivananda S, Lennard-Jones J, Logan R, et al. Incidence of inflammatory bowel disease across Europe: Is there a difference between north and south? Results of the European Collaborative Study on Inflammatory Bowel Disease (EC-IBD) Gut 1996;39:690-7.
5. Piront P, Louis E, Latour P, Plomteux O, Belaiche J. Epidemiology of inflammatory bowel disease in the elderly in the province of Liége. Gastroentérologie Clin Biol 2008;25:157-61.
6. Gower-Rousseau C, Vasseur F, Fumery M, et al. Epidemiology of inflammatory bowel diseases: New insights from a French population-based registry (EPIMAD). Dig Liver Dis 2013;45:89-94.
7. Loftus CG, Loftus E V, Harmsen WS, et al. Update on the incidence and prevalence of Crohn's disease and ulcerative colitis in Olmsted County, Minnesota, 1940-2000. Inflamm Bowel Dis 2007;13:254-61.
8. Bernstein C. Epidemiology of Crohn's disease and ulcerative colitis in a central Canadian province: A population-based study. Am J Epidemiol 1999;149:916-24.
9. Bernstein CN, Wajda A, Svenson LW, et al. The epidemiology of inflammatory bowel disease in Canada: A population-based study. Am J Gastroenterol 2006;101:1559-68.
10. Lennard-Jones JE. Classification of inflammatory bowel disease. Scand J Gastroenterol 1989;24:2-6.
11. Bitton A, Vutcovici M, Patenaude V, Sewitch M, Suissa S, Brassard P. Epidemiology of inflammatory bowel disease in Québec: Recent trends. Inflamm Bowel Dis 2014;20:1770-6.
12. Quezada SM, Steinberger EK, Cross RK. Association of age at diagnosis and Crohn's disease phenotype. Age Ageing 2013;42:102-6.
13. Juneja M, Baidoo L, Schwartz MB, et al. Geriatric inflammatory bowel disease: Phenotypic presentation, treatment patterns, nutritional status, outcomes, and comorbidity. Dig Dis Sci 2012;57:2408-15.
14. Vavricka S, Rogler G. Treatment of severe ulcerative colitis: Differences in elderly patients? Dig Dis 2009;27:315-21.
15. Travis S. Is IBD different in the elderly? Inflamm Bowel Dis 2008;14:S12-S13.
16. Stallmach A, Hagel S, Gharbi A, et al. Medical and surgical therapy of inflammatory bowel disease in the elderly – prospects and complications. J Crohns Colitis 2011;5:177-88.
17. Schoepfer A, Bortolotti M, Pittet V, et al. The gap between scientific evidence and clinical practice: 5-aminosalicylates are frequently used for the treatment of Crohn's disease. Aliment Pharmacol Ther 2014;40:930-7.
18. Charpentier C, Salleron J, Savoye G, et al. Natural history of elderly-onset inflammatory bowel disease: A population-based cohort study. Gut 2014;63:423-32.
19. Bautista MC, Otterson MF, Zadvornova Y, et al. Surgical outcomes in the elderly with inflammatory bowel disease are similar to those in the younger population. Dig Dis Sci 2013;58:2955-62.
20. Akerkar GA, Peppercorn MA, Hamel MB, Parker RA. Corticosteroid-associated complications in elderly Crohn's disease patients. Am J Gastroenterol 1997;92:461-4.
21. Picco MF, Cangemi JR. Inflammatory bowel disease in the elderly. Gastroenterol Clin N Am 2009;38:447-62.
22. Moleski SM, Lindenmeyer CC, Kozuch PL. The use of infliximab in older inflammatory bowel disease patients. Gastroenterology 2011;142:S-361.
23. Ljung T. Infliximab in inflammatory bowel disease: Clinical outcome in a population based cohort from Stockholm county. Gut 2004;53:849-53.
24. Ikeuchi H, Uchino M, Matsuoka H, et al. Prognosis following emergency surgery for ulcerative colitis in elderly patients. Surg Today 2014;44:39-43.
25. Ha CY. Medical management of inflammatory bowel disease in the elderly: Balancing safety and efficacy. Clin Geriatr Med 2014;30:67-78.
26. Delaney C, Fazio VW, Remzi F, et al. Prospective, age-related analysis of surgical results, functional outcome, and quality of life after ileal-anal anastomosis. Ann Surg 2003;238:221-8.
27. Delaney C, Dadvand B, Remzi F, Church J, Fazio V. Functional outcome, quality of life, and complications after ileal pouch-anal anastomosis in selected septuagenarians. Dis Colon Rectum 2002;45:890-4.
28. Fazio VW, Tekkis PP, Remzi F, Lavery IC. Quantification of risk for pouch failure after ileal pouch anal anastomosis surgery. Ann Surg 2003;238:605-17.
29. Targownik LE, Singh H, Nugent Z, Bernstein CN. The epidemiology of colectomy in ulcerative colitis: Results from a population-based cohort. Am J Gastroenterol.2012;107:1228-35.
30. Kaplan G, Hubbard J. Risk of comorbidities on postoperative outcomes in patients with inflammatory bowel disease. Arch Surg 2011;146:959-64.
31. Bernstein CN, Nugent Z, Targownik LE, Singh H, Lix LM. Predictors and risks for death in a population-based study of persons with IBD in Manitoba. Gut 2014:1-9.
32. Benchimol EI, Guttmann A, Mack DR, et al. Validation of international algorithms to identify adults with inflammatory bowel disease in health administrative data from Ontario, Canada. J Clin Epidemiol 2014;67:887-96.
33. Stepaniuk P, Bernstein CN, Nugent Z, Singh H. Characterization of inflammatory bowel disease in elderly hospitalized patients in a large central Canadian Health Region. Can J Gastroenterol Hepatol 2015;29:274-8.

5

Post-transplant lymphoproliferative disorder in liver transplant recipients: Characteristics, management and outcome from a single-centre experience with >1000 liver transplantations

Khalid Mumtaz MD MSc[1], Nabiha Faisal MD[2], Max Marquez MD[2], Alicia Healey CNP[2], Leslie B Lilly MD MSc[2] Eberhard L Renner MD,FRCP[2]

K Mumtaz, N Faisal, M Marquez, A Healey, LB Lilly, EL Renner. Post-transplant lymphoproliferative disorder in liver transplant recipients: Characteristics, management and outcome from a single-centre experience with >1000 liver transplantations. Can J Gastroenterol Hepatol 2015;29(8):417-422.

BACKGROUND: The literature regarding post-transplant lymphoproliferative disorder (PTLD) in liver transplant recipients (LTRs) is limited.

OBJECTIVES: To study the incidence, predictors and outcomes of PTLD after liver transplantation in a single, large-volume centre.

METHODS: The charts of all LTRs (n=1372) in the authors' centre between January 2000 and June 2012 were retrospectively reviewed and those who developed PTLD were identified. Demographic, clinical and treatment data were prospectively collected. Responses to treatment, including complete response, no response, relapse and survival, were recorded.

RESULTS: The incidence of PTLD in LTRs was 32 in 1372 (2.3%). Overall, median survival was 37 months (range 0.5 to 195 months), with one-, three- and five-year survival rates of 81%, 74% and 60%, respectively. Epstein-Barr virus (EBV)-negative patients had a better mean (± SD) survival (95±79 months) than EBV-positive patients (41±42 months) (P=0.02). For stage I/II PTLD, one-, three- and five-year actuarial survival was 87%, 87% and 75%, compared with 50%, 30% and 0% for stage III/IV PTLD, respectively (P=0.001). In patients with complete response, median survival was 58 months (range 10 to 195 months); and one-, three- and five-year actuarial survival was 100%, 94% and 76%, respectively, after diagnosis of PTLD. Changing immunosuppression (IS) from calcineurin inhibitor to sirolimus at the time of diagnosis may have improved survival (seven of seven survivors) compared with only decreasing or stopping IS (14 of 25 survivors) (P=0.07).

CONCLUSIONS: This series from a single large-volume centre showed excellent short and long-term survival after PTLD in adult LTRs who were EBV negative, had early disease and showed complete response. Consistent with the known in vitro antiproliferative effect of sirolimus, switching IS from calcineurin inhibitor to sirolimus may improve survival.

Key Words: *Immunosuppression; Liver; Lymphoma; PTLD; Sirolimus; Transplant*

Les troubles lymphoprolifératifs après une transplantation du foie : les caractéristiques, la prise en charge et les résultats après plus de 1 000 transplantations dans un seul centre

HISTORIQUE : Les publications scientifiques sur les troubles lymphoprolifératifs après une transplantation (TLAP) chez des greffés du foie (GF) sont limitées.

OBJECTIFS : Étudier l'incidence, les prédicteurs et les résultats des TLAP chez les GF dans un seul centre à grand volume.

MÉTHODOLOGIE : Les auteurs ont effectué l'examen rétrospectif du dossier de tous les GF (n=1 372) de leur centre entre janvier 2000 et juin 2012 et extrait ceux qui ont présenté un TLAP. Ils ont fait une collecte prospective des données démographiques, cliniques et thérapeutiques. Ils ont saisi les réponses au traitement, y compris la réponse complète, l'absence de réponse, la récidive et la survie.

RÉSULTATS : Chez les GF, l'incidence de TLAP était de 32 cas sur 1 372 (2,3 %). Dans l'ensemble, la survie médiane était de 37 mois (plage de 0,5 à 195 mois) et le taux de survie au bout de un, trois et cinq ans de 81 %, 74 % et 60 %, respectivement. Les patients négatifs au virus d'Epstein-Barr (VEB) présentaient une meilleure survie moyenne (± ÉT) (95±79 mois) que ceux qui y étaient positifs (41±42 mois) (P=0,02). En cas de TLAP de stade I/II, la survie actuarielle au bout de un, trois et cinq ans s'élevait à 87 %, 87 % et 75 %, par rapport à 50 %, 30 % et 0 %, respectivement, pour la TLAP de stade III/IV (P=0,001). Chez les patients ayant une réponse complète, la survie moyenne était de 58 mois (plage de dix à 195 mois) et la survie actuarielle au bout de un, trois et cinq ans, de 100 %, 94 % et 76 %, respectivement, après le diagnostic de TLAP. Le passage du traitement immunosuppresseur (IS) par un inhibiteur de la calcineurine à un sirolimus au diagnostic a peut-être accru la survie (sept des sept survivants) par rapport à la diminution ou à l'abandon de l'IS (14 des 25 survivants) (P=0,07).

CONCLUSIONS : Cette série tirée d'un seul centre à grand volume a révélé une excellente survie à court et à long terme après un TLAP chez des GF adultes négatifs au VEB atteints peu après la transplantation et ayant une réponse complète. Conformément à l'effet antiprolifératif *in vitro* connu du sirolimus, le passage IS par un inhibiteur de la calcineurine au sirolimus améliore peut-être la survie.

Post-transplant lymphoproliferative disorder (PTLD) reflects uncontrolled B cell proliferation in the post-transplant setting, with pathological features ranging from polymorphic cellular expansion of lymphocytes of any size to monomorphic large-cell non-Hodgkin lymphomas (1). Compared with lymphoma in the general population, PTLD is characterized by increased extranodal involvement, a more aggressive clinical course and poorer response to conventional treatment (2,3).

Overall, the incidence of PTLD after solid organ transplantation ranges from 1% to 10%, with a mortality rate often exceeding 50% (4,5).The incidence varies with the type of allograft transplanted (6), the highest incidence (up to 30% lifetime) being observed in recipients of small bowel, heart and lung transplants (7,8). The prevalence of PTLD in liver transplant recipients (LTRs) ranges from 2% to 4% in adults (9,10), but is as high as 20% in pediatric series (10-13). PTLD

[1]*Wexner Medical Center, Gastroenterology, Hepatology and Nutrition Division, The Ohio State University, Columbus, Ohio, USA;* [2]*Liver Transplant Program/Multi-Organ Transplant Program, University Health Network, Toronto General Hospital, University of Toronto, Toronto, Ontario*
Correspondence: Dr Khalid Mumtaz, Wexner Medical Center, Gastroenterology, Hepatology and Nutrition Division, The Ohio State University, Columbus, Ohio 43210, USA. e-mail khalid.mumtaz@osumc.edu

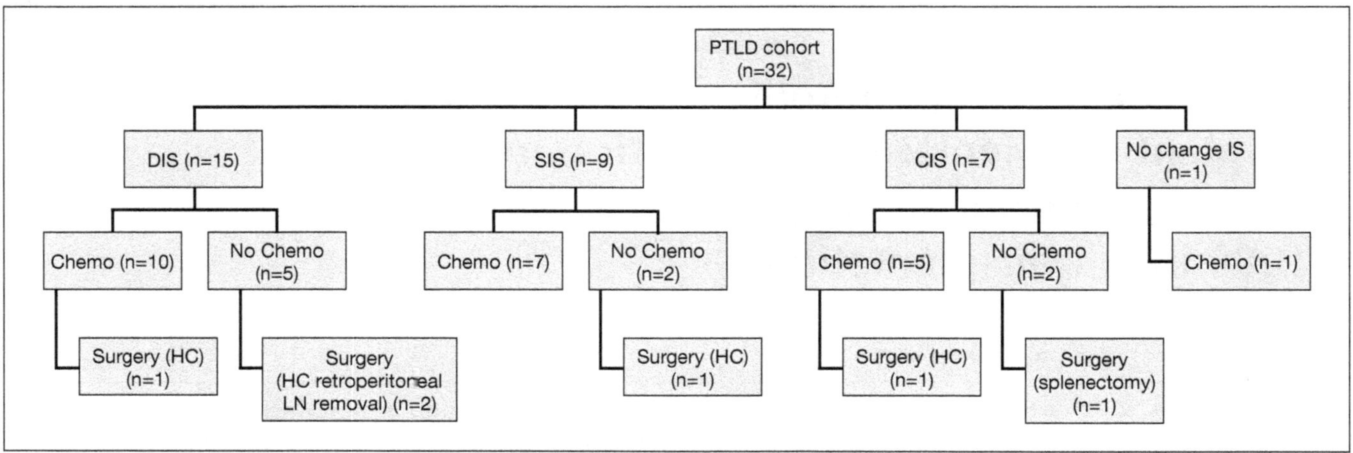

Figure 1) *Algorithm of various interventions used in the cohort of post-transplant lymphoproliferative disorder (PTLD) patients. Chemo Chemotherapy; CIS Changing immunosuppression to sirolimus; DIS Decrease in immunosuppression; HC Hemicolectomy; IS Immunosuppression; LN Lymph node; SIS Stopping immunosuppression*

typically manifests within the first year of liver tranplantation (LT), but may present as early as 20 days after LT and as late as several decades after the engraftment. However, the general concept is that the highest risk of PTLD development occurs in the early post-LT period, principally due to the higher doses of immunosuppression (IS) administered early. Early onset PTLD has been reported to have a more favourable outcome while late-onset disease is more likely to behave like an aggressive lymphoma (14).

Due to the limited understanding of the pathogenesis of PTLD, there is a lack of clear consensus in its management (15). Management options include reduction of IS, chemotherapy, such as combination of cyclophosphamide, hydroxy doxorubicin, vincristine (Oncovin, Eli Lilly, USA) prednisone (CHOP) and/or rituximab, as well as surgical resection in some cases (4-8,11-13).

Due to the small number of PTLD cases occurring at adult liver transplant centres, incidence, risk factors and outcomes of PTLD after adult LT remain debated. Therefore, we aimed to assess incidence, predictors and outcomes of PTLD after adult LT in our large-volume institution.

METHODS

The present study was approved by the authors' institutional ethics committee and conducted according to the Declaration of Helsinki.

The present study was a retrospective analysis of prospectively collected data from all adult patients who underwent LT at the Toronto General Hospital (Toronto, Ontario) between January 2000 and December 2012. Data were retrieved from the Organ Transplant Tracking Record (Hickman-Kenyon Systems, USA) (16), which is an internal web-based transplantation database linked to the electronic medical record for a solid organ transplant at the University Health Network, University of Toronto (Toronto, Ontario), which was instituted at Toronto General Hospital in 2000. To maintain group homogeneity, patients who underwent a liver re-transplant (re-LT) or a combined liver/non-liver solid organ transplant were excluded.

Diagnosis and follow-up of PTLD

Cytomegalovirus and Epstein-Barr virus (EBV) serology of the recipients before LT were recorded. Diagnosis and staging of PTLD was based on histological examination of biopsies or surgical specimens of tumours supplemented by computed tomography (CT) scans and/or gastrointestinal endoscopy, wherever indicated. WHO classification was used for defining PTLD, and lesions were classified as polymorphic or monomorphic disease (6). Results of immunohistochemistry for light-chain restriction, flow cytometry and in situ hybridization for EBV messenger RNA (EBER-1) were collected. Patients were staged according to the Ann Arbor staging system (17).

Serial CT scans of the brain, chest, abdomen and pelvis were used to assess response to therapy in all patients with PTLD. Bone marrow biopsies and gastrointestinal endoscopy were performed for diagnosis, staging and response assessment during follow-up when indicated.

IS

The IS protocol used in the authors' program has been reported elsewhere (18). Briefly, all patients received intravenous solumedrol (starting with 200 mg and tapered over five days), followed by oral prednisone 20 mg once daily, which was tapered and discontinued over a three-month period.

Thymoglobulin or basiliximab was used as induction per protocol in all living donor LT recipients, as well as in recipients with neurological and/or renal impairment to enable delaying calcineurin inhibitor (CNI) use. Maintenance IS (MIS) was CNI (cyclosporine A or tacrolimus) with or without an antiproliferative (mycophenolate mofetil, mycophenolate sodium or azathioprine) and was at the discretion of the responsible transplant hepatologist. Pre- or post-PTLD episodes of acute cellular rejection (ACR) were bioptically documented and graded according to the Banff system (19). Treatment typically consisted of boluses of intravenous solumedrol 500 mg daily for three (and up to five days) or of increasing MIS including oral prednisone, if the rejection was mild.

PTLD therapy

Therapy for PTLD included stopping antiproliferatives, if applicable, along with either: reduction of CNI dose; stopping all IS including CNI; or switching from CNI to sirolimus. The next step included one of following: CHOP with or without rituximab; rituximab alone; ABVD protocol (see below); radiation, and/or surgery. The CHOP regimen was given every 21 days for up to six cycles with or without concurrent rituximab (2,3). Rituximab was administered as a slow intravenous infusion weekly for four doses (2,3). The ABVD protocol consists of adriamycin, bleomycin, vinblastine and dacarbazine given monthly for four cycles (2). Surgery was reserved for anatomically limited disease with significant symptoms such as segmental small bowel PTLD presenting as small bowel obstruction. Radiotherapy was used for PTLD nonresponsive to chemotherapy and central nervous system involvement.

Definitions

Treatment response was assessed according to the revised International Workshop criteria (20) as follows: complete response (CR) was defined as complete disappearance of tumour (imaging, endoscopy) and absence of new lesions for at least four weeks after completing therapy; stable disease was defined as a decrease <50% or an increase <25% in tumour

TABLE 1

Demographics and descriptive results of post-transplant lymphoproliferative disorder (PTLD) patients (n=32)

Variable	
Age at liver transplantation, years, median (range)	44 (20–65)
Age at PTLD diagnosis, years, median (range)	53 (25–75)
Sex, male/female, n/n	18/14
Etiology of liver disease	
Autoimmune liver disease(s)	15 (47)
Hepatitis C virus disease	10 (31)
Alpha 1 antitrypsin deficiency	3 (9)
Others	4 (13)
DDLT/LDLT	27 (84)/5 (16)
Epstein-Barr virus recipient positive	26 (81)
Cytomegalvirus recipient positive	24 (75)
Thymoglobulin induction	5 (16)
Basiliximab induction	5 (16)
Cyclosporine	18 (56)
Tacrolimus	14 (44)
Mycophenolate	9 (28)
Azathioprine	9 (28)

Data presented as n (%) unless otherwise indicated. DDLT Deceased donor liver transplantation, LDLT Living donor liver transplantation

TABLE 2

Presentation of early and late-onset post-transplant lymphoproliferative disorder (PTLD) in adult liver transplant (LT) recipients

	PTLD onset		
Variable	Early (n=9)	Late (n=23)	P
Age at LT, years, mean ± SD	45±13	42±11	0.51
Age at PTLD diagnosis, mean ± SD	52±13	53±11	0.82
Sex, male/female, n/n	5/4	13/10	0.63
Time to PTLD, months, mean ± SD	20±42	93±56	–
Autoimmune liver disease	2 (22)	13 (56)	0.08
Cytomegalovirus seropositive	8 (89)	16 (70)	0.25
Epstein-Barr virus seropositive	6 (75)	20 (87)	0.18
EBER seropositive	8 (89)	9 (39)	0.02
Tacrolimus/cyclosporine, n/n	5/4	9/14	0.40
Diffuse large B cell lymphoma	7 (78)	14 (61)	0.35
CD20 tissue positive	8 (89)	17 (72)	0.45
CD45 tissue positive	4 (44)	11 (48)	0.94
Multiorgan involvement	7 (78)	15 (65)	0.49
Disseminated PTLD	2 (22)	3 (13)	0.52
Hodgkin disease	0 (0)	4 (17)	0.18
Non-Hodgkin lymphoma type	7 (78)	17 (74)	0.82
Monomorphic/polymorphic, n/n	5/3	12/6	0.83
Stage I & II/stage III & IV, n/n	6/3	18/5	0.50
Chemotherapy used	5 (56)	18 (78)	0.20
Complete response	7 (78)	18 (78)	0.97
Relapse of PTLD	1 (11)	4 (17)	0.90

Data presented as n (%) unless otherwise indicated. EBER Epstein-Barr virus messenger RNA

size, with no evidence of new lesions; partial response (PR) was defined as >50% size reduction of detectable lesions, without evidence of new lesions; and progressive disease was defined as enlargement of lesions by >25%, the occurrence of new lesions or worsening of clinical symptoms. Because the number of cases was limited, PR and progressive disease were analyzed together as no response. Relapse was defined as recurrence of PTLD at any time after having achieved a CR. Survival was reported from the time (in months) of diagnosis of PTLD to longest follow-up. The time between LT and PTLD diagnosis was defined as the period (in months) between engraftment and histopathological diagnosis of PTLD. Patients who presented with PTLD within the first 12 months post-LT were categorized as the 'early onset PTLD' group, and liver graft recipients who re-presented the disease beyond this time period after LT were 'late-onset PTLD' patients. Death was attributed to PTLD if either the patient died within six months postdiagnosis, or the patient died due to complications classified as a complication of the PTLD treatment (21).

Statistical analysis

Statistical analysis was performed using SPSS version 17.0 (IBM Corporation, USA). Descriptive analysis was performed for demographic and clinical features and results were presented as median with ranges for quantitative variables, and numbers (percentages) for qualitative variables. χ^2 and Fisher's exact tests were used for categorical variables, as appropriate, while the independent sample *t* test was used for numerical variables. The Kaplan-Meier method and log-rank test were used to estimate and compare cumulative survival (time from PTLD diagnosis to death from any cause). All P values were two-sided and considered to be statistically significant at P<0.05.

RESULTS

A total of 1372 patients received LT at the authors' institution during the study period; of these, 42 (3%) developed PTLD and 32 (2.3%) were eligible for the study. Of the 10 PTLD patients excluded, four underwent re-LT and six had combined liver/non-liver solid organ transplant.

Patient characteristics

The median age at the time of LT was 44 years (range 20 to 65 years). The most common etiology for LT was autoimmune liver disease (primary biliary cirrhosis [n=4]; autoimmune hepatitis [n=4]; primary sclerosing cholangitis [n=7]) in 15 (47%), followed by hepatitis C virus (HCV) in 10 (31%) PTLD patients. The details of EBV/cytomegalovirus serology, induction and MIS are presented in Table 1.

Before the diagnosis of PTLD, 14 (44%) patients developed ACR; three had experienced >1 episode of rejection. One patient had severe rejection, four had moderate and nine had mild ACR.

Median age at time of diagnosis of PTLD was 53 years (range 25 to 75 years). Median time between LT and diagnosis of PTLD was 60 months (range three to 240 months). The distribution of features of early and late-onset PTLD is elaborated in Table 2. PTLD was diagnosed with the help of radiological imaging in 22 (70%) patients and gastrointestinal endoscopy in 10 (30%). Abdominal viscera were involved in 42% (stomach 12%, small intestine 15%, large intestine and liver 6% each, and spleen 3%), followed by mesenteric and/or retroperitoneal lymph nodes 40%, cervical lymph nodes 9% and lungs 3%.

Diffuse large B cell lymphoma was the most common phenotype (n=21 [65%]), followed by Hodgkin disease (n=4 [13%]) and Burkitt type, diffuse large T cell type, and T and B cell mixed monoclonal lymphoma (n=2 each [6%]), and one monoclonal B cell type lymphoma.

All were tested for EBV in tissue samples and 17 (53%) were positive as determined by EBER in situ hybridization. Eight (89%) patients were EBER-positive in early onset PTLD compared with nine (39%) in late-onset PTLD (P=0.02).

Treatment modalities and responses

The most common early treatment strategy was decrease in IS (n=15), followed by stopping IS (n=9), changing to sirolimus (n=7) and no change in IS (n=1). Figure 1 summarizes the various treatment modalities in all 32 PTLD patients according to the treatment response. Among later interventions, chemotherapy (n=23 [72%]) was the most common followed by decrease in IS (n=5 [15%]) and surgery (n=4 [12%]). The details, in additon to outcomes of these interventions, are illustrated in Figure 2 and summarized in Table 3.

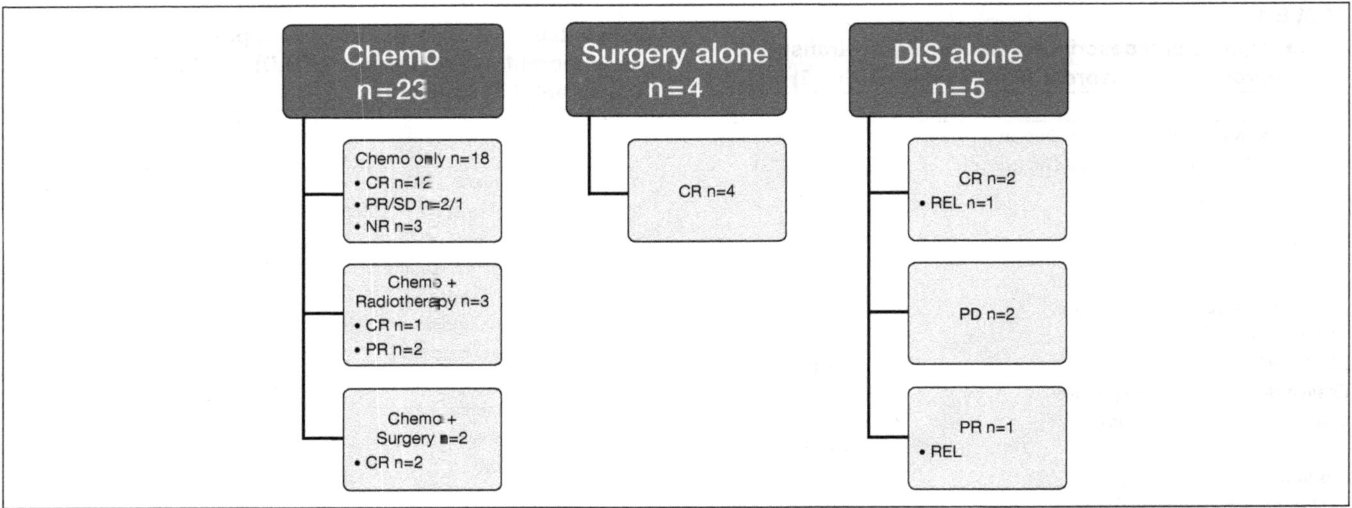

Figure 2) *Various interventions offered to post-tranplant lymphoproliferative disorder (PTLD) patients and their responses. Chemo Chemotherapy; CR Complete response; DIS Decrease in immunosuppression; NR No response; PD Progressive disease; PR Partial response; REL Relapse; SD Stable disease*

TABLE 3
Results of chemotherapy in patients with post-transplant lymphoproliferative disorder

Chemotherapy	CR	Relapse	NR	Survival, months, median (range)	Still alive
R-CHOP (n=8)	6	0	2	12 (1–39)	5
CHOP* (n=7)	6	1	1	75 (12–149)	5
Rituximab (n=7)	5	1	2	60 (4–142)	5
ABVD (n=3)	2	1	1	24 (3–48)	1

**Two patients were treated with cyclophosphamide, hydroxydoxorubicin, vincristine (Oncovin, Eli Lilly, USA) and prednisone (CHOP) after failure of rituximab monotherapy; ABVD Adriamycin, bleomycin, vinblastine and dacarbazine; CR Complete response; NR No response; R-CHOP Rituximab + CHOP*

Predictors of survival

Overall, median survival in all patients with PTLD was 37 months (range 0.5 to 195), and the one-, three- and five-year actuarial survival rates after diagnosis of PTLD were 81%, 74% and 50%, respectively (Figure 3). The 25 patients who achieved CR survived median of 58 months (10 to 195) months, and one-, three- and five-year actuarial survival after diagnosis of PTLD was 100%, 94% and 76%, respectively. All seven patients who did not respond to therapy died within one year of diagnosis of PTLD, with a median survival of only three months (range 0.5 to 12) (Figure 4). A better mean (± SD) survival was observed after diagnosis of PTLD in patients who were EBV negative (95±79 months) compared with those who were EBV positive (41±42 months; P=0.02). Stage I/II PTLD patients had a better overall survival (19 of 24 [79%]) compared with stage III/IV (two of eight [25%]) (P=0.05). Moreover, one-, three- and five-year cumulative survival among stage I/II was 87%, 87% and 75%, compared with 50%, 30% and 0%, respectively, in stage III/IV patients (P=0.001) (Figure 5).

Moreover, switching from CNI to sirolimus-based IS tended to lead to an improved survival, seven of seven patients surviving compared with decreasing and stopping IS (14 of 25 [56%]) (P=0.07) (Figure 6). However, this did not reach statistical significance, possibly due to the limited sample size. Finally, there were no differences in survival based on CD20 serology, early versus late PTLD, monomorphic versus polymorphic, and type of CNI use as MIS.

Cause of death

Eleven patients died in the present series. In nine patients, the cause of death was classified as PTLD related (among these, three experienced a relapse of PTLD before death and six had no response). One patient died from metastatic melanoma 39 months after achieving CR; another patient died a month after achieving CR with one cycle of rituximab + CHOP from graft failure and sepsis. Median survival from diagnosis of PTLD among the 11 deceased patients was 11 months (range 0.5 to 58 months) and mean survival 17±19 months, respectively.

DISCUSSION

We report a large series of PTLD after adult LT from a single centre. The observed incidence of PTLD in our program (2.3%) compares well with that previously reported by others in adult LTRs (2% to 3%) (14,22,23). The most intriguing observation of our retrospective analysis was that switching IS from a CNI to a sirolimus-based IS regimen may have improved survival after diagnosis of PTLD.

PTLD is a well-recognized complication of solid organ transplantation, with an incidence ranging in adults from 1% to 10% (4,5,23) and is associated with a significant mortality rate.

The highest rates of lymphomas (58% to 60%) were previously reported by others in graft recipients within the first 12 months post-LT, so-called 'early onset PTLD' (19). In contrast, late PTLD (occurring >12 months after LT) was observed in our series, with 72% of cases more frequently than early onset PTLD (28%). It is suggested by various authors that early and late PTLD consists of two different clinical and pathological entities (9,11,12). Early onset PTLD is characterized by being EBER negative (24) and CD20-positive, and more commonly involving the engrafted organ than late-onset PTLD. However, all of our cases were predominantly EBER positive (89%), and CD20 was equally distributed in the two groups. Also noteworthy, and in contrast to previous reports, none of our PTLD patients had involvement of the engrafted organ.

The most common underlying etiology leading to the need for LT in our PTLD patients was autoimmune liver disease (47%), followed by HCV (31%): previous reports suggest both of them as a possible risk factors for the development of PTLD (25-28). While HCV was the underlying cause of liver disease in approximately 30% of all LTs performed at our institution during the study period, underlying autoimmune liver diseases are clearly over-represented in the PTLD cohort. This may be attributable to the often higher level of IS required by these patients. Accordingly, Zimmermann et al (29) reported autoimmune liver disease and pre-LT prolonged steroid intake as independent risk factors predicting late PTLD development.

Not unexpectedly, we observed that early compared with late-stage PTLD (stage I/II versus II/IV) has a favourable cumulative survival (P=0.001). Nevertheless, a stage effect was absent in a previous retrospective review involving 27 pediatric patients (30). However, this report included heart and small intestine transplant patients, apart from a few with LT.

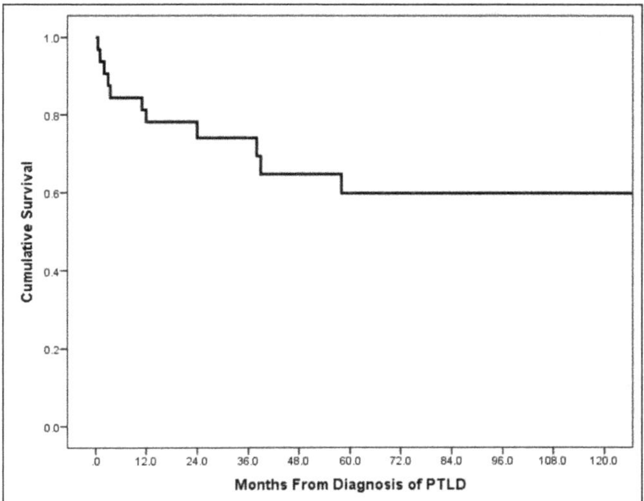

Figure 3) *Kaplan-Meier curve showing cumulative survival in 32 patients at one, three and five years after the diagnosis of post-transplant lymphoproliferative disorder (PTLD)*

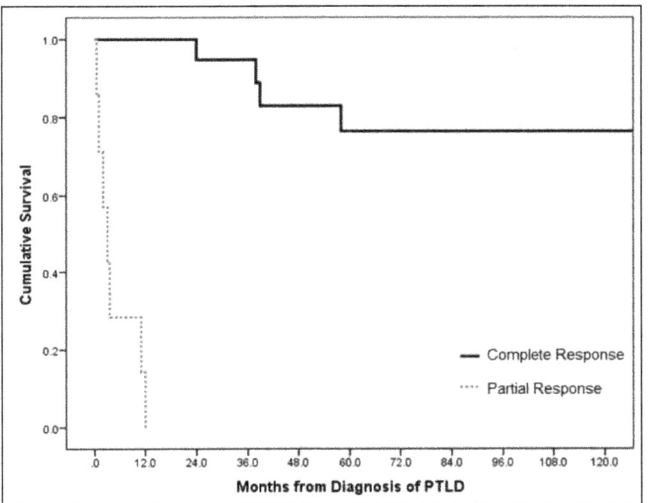

Figure 4) *Kaplan-Meier curve comparing survival among patients with complete response (n=25) and partial response (n=7) with post-transplant lymphoproliferative disorder (PTLD)*

There is no consensus as to the best approach to treatment of PTLD. This appears, at least in part, attributable to clinical variability of the disease, its frequent extranodal involvement and the wide spectrum of histological features. Reduction or cessation of IS is the most common initial approach, which was first reported by Starzl et al (31) in 1984, and has been practiced for several decades since (32,33). The hypothesis behind this strategy is that the recovery of the host's immune system will allow the development of cytotoxic T lymphocytes (CTL) against EBV with subsequent control of EBV-driven B cell proliferation. Another approach may be the use of mammalian target of rapamycin (mTOR) inhibitors such as rapamycin (sirolimus), which has a potent antiproliferative effect on in-vitro PTLD-derived cell lines and has been shown to inhibit the growth of solid tumours in a mouse model of PTLD without significantly compromising graft rejection (34,35). A multicentre clinical experience from European transplant centres supports the approach of using mTOR inhibitors in the management of PTLD following renal transplantation (36). While not reaching statistical significance, presumably due to the limited sample size, our observation of a trend toward improved survival with switching PTLD patients from CNI to sirolimus-based IS is entirely consistent with this renal transplant experience, and may warrant formal exploration in a controlled trial.

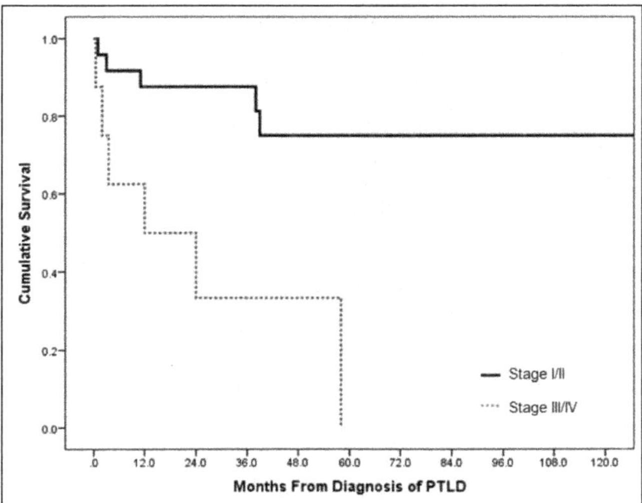

Figure 5) *Kaplan-Meier curve comparing the one-, three- and five-year survival in stage I/II post-transplant lymphoproliferative disorder (PTLD) compared with with stage III/IV PTLD*

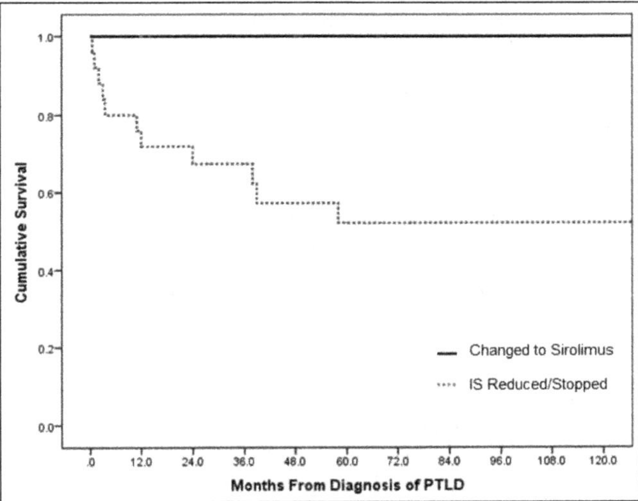

Figure 6) *Kaplan-Meier curve showing comparing changing immunosuppression (IS) to sirolimus approach with decrease in IS/stopping IS. PTLD Post-transplant lymphoproliferative disorder*

The present study clearly shows that short-term and long-term patient survival rates after diagnosis of PTLD are not as dismal as reported in other series (37,38). At a median follow-up of 60 months (range seven to 142 months), 19 of the 20 patients with CR were alive and without evidence of disease. The median overall survival after PTLD diagnosis was 37 months, with one-, three- and five-year cumulative survival rates of 81%, 74% and 60%, respectively. These survival rates are better than those reported in other series (25) and similar to those reported by Jain et al (15) from Pittsburgh, Pennsylvania (USA). Early onset PTLD in adults was associated with worse patient survival in one series (39), while late-onset PTLD was a poor prognostic factor in another series (11). However, we did not find any difference in terms of CR, mortality and survival at one, three and five years among early or late-onset PTLD.

A major limitation of our study was related to the rarity of the condition (ie, the limited sample size) and its retrospective nature. Moreover, we were unable to perform a logistic regression analysis due to small sample size. On the other hand, the heterogeneity of PTLD treatments makes it difficult to analyze a specific outcome of patients according to their treatment.

CONCLUSION

The incidence of PTLD in adult LT recipients was 2.3%. Early onset PTLD was characterized by a high percentage of EBER positivity; however, the other variables were comparable in early and late onset PTLD. Younger patients transplanted for various autoimmune liver diseases, such as primary biliary cirrhosis, primary sclerosing cholangitis and autoimmune hepatitis, appear to be at higher risk for PTLD. Patients with early stage PTLD, treatment with CHOP alone or in combination with rituximab and those achieving a CR have a favourable outcome. Most intriguingly, switching from CNI to sirolimus-based IS may improve survival and appears to warrant testing in a controlled trial.

DISCLOSURES: None of the authors have any commercial associations or other arrangements (eg, financial compensation received, patient-licensing arrangements, potential to profit, consultancy, stock ownership, etc) that may pose a conflict of interest in connection with this article.

REFERENCES

1. Nalesnik MA, Jaffe R, Starzl TE, et al. The pathology of post-transplant lympho-proliferative disorders occurring in the setting of cyclosporine A-prednisone immunosuppression. Am J Pathol 1988;133:173-92.
2. Elstrom RL, Andreadis C, Aqui NA, et al. Treatment of PTLD with rituximab or chemotherapy. Am J Transplant 2006;6:569-76.
3. Trappe R, Riess H, Babel N, et al. Chemotherapy for refractory and relapsed posttransplant lymphoproliferative disorders (PTLD) after treatment with a single-agent rituximab. Transplantation 2007;83:912-8.
4. Dror Y, Greenberg M, Taylor G, et al. Lympho-proliferative disorders after organ transplantation in children. Transplantation 1999;67:990-8.
5. Libertiny G, Watson CJ, Gray DW, Welsh KI, Morris PJ. Rising incidence of post-transplant lymphoproliferative disease in kidney transplant recipients. Br J Surg 2001;88:1330-4.
6. Campo E, Swerdlow SH, Harris NL, Pileri S, Stein H, Jaffe ES. The 2008 WHO classification of lymphoid neoplasms and beyond: Evolving concepts and practical applications. Blood 2011;117:5019-32.
7. Yousem SA, Randhawa R, Locker J. Post-transplant lympho-proliferative disorders in heart-lung transplant recipients: Primary presentation in the allograft. Hum Pathol 1989;20:361-9.
8. Douglas RS, Goldstein SM, Katowitz JA, et al. Orbital presentation of post-transplantation lympho-proliferative disorder: A small case series. Ophthalmology 2002;109:2351-5.
9. Domingo-Domenech E, De San Jose S, et al. Post-transplant lymphomas: A 20 year epidemiologic, clinical and pathologic study in a single center. Haematologica 2001;86:715-21.
10. Jain A, Nalesnik M, Reyes J et al. Post-transplant lymphoproliferative disorders in liver transplantation: A 20-year experience. Ann Surg 2002;236:429-36.
11. Khedmat H, Taheri S. Late onset post transplantation lymphoproliferative disorders: analysis of international data from 5 studies. Ann Transplant 2009;14:80-5.
12. Khedmat H, Taheri S. Early onset post transplantation lymphoproliferative disorders: Analysis of international data from 5 studies. Ann Transplant 2009;14:74-7.
13. Guthery SL, Heubi JE, Bucuvalas JC, et al. Determination of risk factors for EBV-associated post-transplant lympho-proliferative disorders in pediatric liver transplant recipients using objective case ascertainment. Transplantation 2003;75:987-93.
14. Dotti G, Fiocchi R, Motta T, et al. Lymphomas occurring late after solid organ transplantation: Influence of treatment on the clinical outcome. Transplantation 2002:74:1095-102.
15. Jain A, Mazariegos G, Kashyap R, et al. Pediatric liver transplantation. A single-center experience spanning 20 years. Transplantation 2002;73:941-7.
16. OTTR Chronic Care Solutions. <www.ottr.com/> (Accessed March 23, 2015).
17. Lister TA, Crowther D, Sutcliffe SB, et al. Report of a committee convened to discuss the evaluation and staging of patients with Hodgkin's disease: Cotswolds meeting. J Clin Oncol 1989;7:1630-6.
18. Selzner N, Girgrah N, Lilly L, et al. The difference in fibrosis progression of recurrent hepatitis C after live versus deceased donor liver transplantation is attributable to the difference in donor age. Liver Transpl 2008;14:1778-86.
19. Demetris A, Batts K, Dhillon A, Wight D, Williams J, Yamabe H. Banff schema for grading liver allograft rejection: An international consensus document. Hepatology 1997;25:658-63.
20. Cheson BD, Pfistner B, Juweid ME, et al. The International Harmonization Project on Lymphoma: Revised response criteria for malignant lymphoma. J Clin Oncol 2007;25:579-586.
21. Izadi M, Taheri S. Features, predictors and prognosis of lympho-proliferative disorders post-liver transplantation regarding disease presentation time: Report from the PTLD International survey. Ann Transplant 2011;16:39-47.
22. Leblond V, Choquet S. Lymphoproliferative disorders after liver transplantation. J Hepatol 2004;40:728-35.
23. Opelz G, Henderson R. Incidence of non-Hodgkin lymphoma in kidney and heart transplant recipients. Lancet 1993;342:1514-6.
24. Nelson BP, Nalesnik MA, Bahler DW, et al. Epstein-Barr virus-negative post-transplant lymphoproliferative disorders: A distinct entity? Am J Surg Pathol 2000;24:375-85.
25. Burra P, Buda A, Livi U, et al. Occurrence of post-transplant lymphoproliferative disorders among over thousand adult recipients: Any role for hepatitis C infection. Eur J Gastr Hepatol 2006;18:1065-70.
26. Duvoux C, Pageaux GP, Vanlemmens C, et al. Risk factors for lymphoproliferative disorders after liver transplantation in adults: An analysis of 480 patients. Transplantation 2002;74:1103-9.
27. Shpilberg O, Wilson J, Whiteside TL, Herberman RB. Pretransplant immunological profile and risk factor analysis of posttransplant lymphoproliferative disease development: The results of a nested matched case-control study. The University of Pittsburgh PTLD Study Group. Leuk Lymphoma 1999;36:109-21.
28. Walker RC, Marshall WF, Strickler JG, et al. Pretransplantation assessment of the risk of lymphoproliferative disorder. Clin Infect Dis 1995;20:1346-53.
29. Zimmermann T, Hoppe-Lotichius M, Tripkovic V, et al. Liver transplanted patients with preoperative autoimmune hepatitis and immunological disorders are at increased risk for post-transplant lymphoproliferative disease (PTLD). Eur J Intern Med 2010;21:208-15.
30. Donnelly LF, Frush DP, Marshall KW, et al. Lymphoproliferative disorders: CT findings in immunocompromised children. Am J Roentgenol 1998;171:725-31.
31. Starzl TE, Nalesnik MA, Porter KA, et al. Reversibility of lymphomas and lymphoproliferative lesions developing under cyclosporine steroid therapy. Lancet 1984;1:583-7.
32. Allen U, Preiksaitis J. Epstein-Barr virus and posttransplant lymphoproliferative disorder in solid organ transplant recipients. Am J Transplant 2009;9(Suppl 4):S87-S96.
33. Green M. Management of Epstein-Barr virus-induced posttransplant lymphoproliferative disease in recipients of solid organ transplantation. Am J Transplant 2001;1:103-8.
34. Nepomuceno RR, Balatoni CE, Natkunam Y et al. Rapamycin inhibits the interleukin 10 signal transduction pathway and the growth of Epstein-Barr virus B-cell lymphomas. Cancer Res 2003;63:4472-80.
35. Majewski M, Korecka M, Kossev P, et al. The immunosuppressive macrolide RAD inhibits growth of human Epstein-Barr virustransformed B lymphocytes in vitro and in vivo: A potential approach to prevention and treatment of posttransplant lymphoproliferative disorders. Proc Natl Acad Sci U S A 2000;97:4285-90.
36. Julio P. Post-transplant lymphoproliferative disorder – the potential of proliferation signal inhibitors. Nephrol Dial Transplant 2007;22(Suppl 1):i27-i35.
37. Ben-Ari Z, Amlot P, Lachmanan SR, et al. Posttransplant lymphoproliferative disorder in liver recipients: Characteristics, management, and outcome. Liver Transplant Surg 1999;5:184-91.
38. Glez-Chamorro A, Jimenez C, Moreno-Glez E, et al. Management and outcome of liver recipients with post-transplant lymphoproliferative disease. Hepato-Gastroenterology 2000;47:211-9.
39. Trofe J, Buell JF, Beebe TM et al. Analysis of factors that influence survival with post-transplant lymphoproliferative disorder in renal transplant recipients: The Israel Penn International Transplant Tumor Registry experience. Am J Transplant 2005;5(4 Pt 1):775-80.

6

Tax-deductible provisions for gluten-free diet in Canada compared with systems for gluten-free diet coverage available in various countries

Maria Ines Pinto-Sanchez MD MSc[1], Elena F Verdu MD PhD[1], Maria C Gordillo CPA CGA[2], Julio C Bai MD[3], Stephen Birch BA MSc DPhil[4], Paul Moayyedi MD BSc MB ChB PhD MPH[1], Premysl Bercik MD[1]

MI Pinto-Sanchez, EF Verdu, MC Gordillo, et al. Tax-deductible provisions for gluten-free diet in Canada compared with systems for gluten-free diet coverage available in various countries. Can J Gastroenterol Hepatol 2015;29(2):104-110.

Celiac disease affects 1% of the North American population, with an estimated 350,000 Canadians diagnosed with this condition. The disease is triggered by the ingestion of gluten, and a lifelong, strict gluten-free diet (GFD) is the only currently available treatment. Compliance with a strict GFD is essential not only for intestinal mucosal recovery and alleviation of symptoms, but also for the prevention of complications such as anemia, osteoporotic fractures and small bowel lymphoma. However, a GFD is difficult to follow, socially inconvenient and expensive. Different approaches, such as tax reduction, cash transfer, food provision, prescription and subsidy, have been used to reduce the additional costs of the GFD to patients with celiac disease. The current review showed that the systems in place exhibit particular advantages and disadvantages in relation to promoting uptake and compliance with GFD. The tax offset system used in Canada for GFD coverage takes the form of a reimbursement of a cost previously incurred. Hence, the program does not help celiac patients meet the incremental cost of the GFD – it simply provides some future refund of that cost. An ideal balanced approach would involve subsidizing gluten-free products through controlled vouchers or direct food provision to those who most need it, independently of 'ability or willingness to pay'. Moreover, if the cost of such a program is inhibitive, the value of the benefits could be made taxable to ensure that any patient contribution, in terms of additional taxation, is directly related to ability to pay. The limited coverage of GFD in Canada is concerning. There is an unmet need for GFD among celiac patients in Canada. More efforts are required by the Canadian medical community and the Canadian Celiac Association to act as agents in identifying ways of improving resource allocation in celiac disease.

Key Words: *Celiac disease; Gluten-free diet; Subsidy; Tax deduction*

Les déductions d'impôt applicables au régime sans gluten au Canada comparées aux couvertures offertes dans divers pays

La maladie cœliaque touche 1 % de la population nord-américaine, et on estime que 350 000 Canadiens sont diagnostiqués. Le seul traitement connu de cette maladie qui est déclenchée par l'ingestion de gluten est un régime sans gluten (RSG) rigoureux respecté tout au long de la vie. En effet, il est essentiel de respecter rigoureusement le RSG, non seulement pour le rétablissement de la muqueuse intestinale et le soulagement des symptômes, mais également pour la prévention de complications comme l'anémie, les fractures ostéoporotiques et le lymphome du grêle. Cependant, le RSG est difficile à respecter, peu pratique en société et coûteux. Diverses démarches, telles qu'une réduction d'impôt, un transfert d'argent, un approvisionnement, une prescription et une subvention, ont été privilégiées pour réduire les coûts supplémentaires attribuables au RSG chez les patients atteints d'une maladie cœliaque. La présente analyse révèle que les systèmes en place comportent certains avantages et désavantages lorsqu'il est question de promouvoir l'adoption et le respect du RSG. Le système de déduction fiscale utilisé au Canada pour couvrir le RSG consiste à rembourser des coûts déjà engagés. Ainsi, le programme n'aide pas les patients à payer les coûts supplémentaires du RSG, mais s'associe simplement à un futur remboursement de ces coûts. La démarche équilibrée et idéale consisterait à subventionner les produits sans gluten au moyen de coupons contrôlés ou d'approvisionnement alimentaire direct aux personnes qui en ont le plus besoin, quelle que soit « leur capacité ou leur volonté de payer ». De plus, si le coût d'un tel programme est prohibitif, la valeur des avantages pourrait être taxable afin que l'apport des patients soit directement lié à la capacité de payer, sous forme de taxe supplémentaire. La couverture limitée du RSG au Canada est inquiétante. Les besoins liés au RSG ne sont pas respectés chez les patients atteints de la maladie cœliaque au Canada. Le milieu canadien de la médecine et l'Association canadienne de la maladie cœliaque devront consentir plus d'efforts pour déterminer des moyens d'améliorer l'attribution des ressources dans le secteur de la maladie cœliaque.

Celiac disease (CD) is a chronic, small intestinal, immune-mediated disorder affecting 1% of the North American population (1), with an increasing incidence in the past 40 years (2). Studies outside of North America have revealed that CD is common, occurring in Caucasian as well as non-Caucasian populations. In Canada, serological screening has estimated that approximately 350,000 Canadians have this condition (3,4).

The disease is precipitated by the ingestion of gluten in individuals with genetic predisposition conferred by the presence of human leukocyte antigen-DQ2 or -DQ8. The only currently available treatment is the lifelong avoidance of gluten (5,6). The principal sources of dietary gluten are wheat, barley and rye. The term 'gluten-free' indicates a diet that contains gluten at such a low level as to be considered harmless for celiac patients (7). There are two types of gluten-free foods available for those requiring a gluten-free diet (GFD): naturally gluten-free foods, which are foods that naturally do not contain gluten (eg, fruits, vegetables, meat, eggs, rice and potatoes); and gluten-free substitutes, which are food products that have been manufactured as alternatives to gluten-containing counterparts, such as bread and pasta. Quality and palatability of gluten-free substitutes vary significantly among brands and products.

Compliance with a GFD will result in resolution of symptoms and repair of the intestinal damage over time in most individuals with CD (7). However, failure to adhere to the GFD carries risk of adverse

[1]*Department of Medicine, Farncombe Family Digestive Research Institute, McMaster University;* [2]*Kaizen Accounting, Hamilton, Ontario;* [3]*Hospital de Gastroenterología Dr C B Udaondo, Buenos Aires, Argentina;* [4]*Centre for Health Economics and Policy Analysis, Department of Clinical Epidemiology & Biostatistics, McMaster University, Hamilton, Ontario*

Correspondence: Dr Premysl Bercik, 1280 Main Street West, McMaster University Medical Centre, 3N9, Hamilton, Ontario L8S 4K1. e-mail bercikp@mcmaster.ca

TABLE 1
Example tax deduction system for a single gluten-free (GF) item and calculations for tax deduction

Example of a single item

1. Item: bread.
2. Number of items purchased: 52
3. Average cost of non-GF product: $3.49
4. Average cost of GF product: $6.99
5. Incremental cost (line 4 minus line 3): $6.99 - $3.49 = $3.50
6. Amount to claim (line 5 multiplied by line 2): $3.50 x 52 = $182.00

Example of the calculations for tax deduction

Item (Company, Country)	Items purchased, n	Non-GF food cost (comparable brand), $	Non-GF food amount, g	GF food cost, $	GF food amount, g	Cost difference, $
Polenta crisps (Dendy & Co, Canada)	1	2.50	400	4.99	120	4.24
Chicken nuggets (Ian's Natural Foods, USA)	10	6.00	870	7.99	224	64.45
Cereal (Nature's Path, USA)	22	3.49	500	4.79	325	55.47
GF Pantry cookies (Glutino, USA)	1	3.79		5.99	340	2.20
Pancake mix (Kinnikinnick Foods, Canada)	4	3.00	905	5.49	454	15.94
Bread (Udi's, USA)	18	2.50	600	5.99	340	82.32
Croutons (PaneRiso Foods, Canada)	1	2.89		4.99	175	2.10
Mini brownies (President's Choice, Canada)	1	2.69		5.99	280	3.30
Chocolate muffins (President's Choice, Canada)	2	3.99		5.99	372	4.00
Blueberry muffins (President's Choice, Canada)	1	3.99		5.99	372	2.00
Betty Crocker GF cake mix (General Mills, Canada)	3	3.99		5.99	425	6.00
Macaroni & cheese (Annie's, USA)	5	1.50		4.29	170	13.95
Candy bar (Glutino, USA)	2	2.79		4.49	125	3.40
Hotdog buns (Udi's, USA)	1	2.99	672	6.79	365	5.17
Spaghetti (Catelli, USA)	4	1.67	900	2.99	340	9.43
Genius bread (Glutino, USA)	3	2.50		5.49	395	24.60
Granola bar (Bakery on Main)	5	2.50		4.70	170	11.00
Hamburger buns (Udi's, USA)	3	2.99	672	5.79	295	13.43
Donuts (Glutino, USA)	1	2.49		4.44	340	1.95
Pretzels (Glutino, USA)	3	3.50		8.99	400	16.47
Pizza (Sarafina, Canada)	1	4.99	900	9.99	355	8.00
Rice crisps (President's Choice, Canada)	17	1.50		3.29	100	30.43
Banana loaf (President's Choice, Canada)	4	3.49		5.99	454	10.00
English muffins (Glutino, USA)	3	2.99		5.99	480	9.00
GF Pantry pizza mix (Glutino, USA)	5	3.29		5.99	624	13.50
Total						412.35

health consequences in celiac patients. These include increased risk for incidence of bone fractures (8-11), malignancies (small bowel adenocarcinoma, esophageal cancer, B cell and T cell non-Hodgkin lymphomas) and increased mortality (12-15). Furthermore, women with CD have an increased risk for infertility, spontaneous abortions, preterm deliveries and delivery of low-birth-weight infants (7). Evidence suggests the risk of complications is reduced in those who strictly adhere to a GFD (12-22).

Although products safe for consumption by celiac patients are labelled and certified as 'gluten-free', following a GFD can be cumbersome. Strict avoidance of gluten is challenging because of many hidden sources of gluten in processed food products, cosmetics and drugs. Moreover, celiac patients find that a GFD is difficult to follow, socially inconvenient (7,23) and more expensive than a regular diet. Gluten-free products can cost more than twice as much as 'regular' wheat-based products (24,25), with added extra costs such as shipping for some gluten-free specialty items. This increases the challenge for the celiac patient to comply with a GFD. To improve compliance among celiac patients, many jurisdictions provide programs to help meet the extra costs attached to living on a GFD. However, there is wide variation in the content and scope of the different approaches adopted. The objective of the present review was to assess how these approaches compare in terms of the expected level and distribution of impact on patient uptake and compliance with a GFD.

TAX REDUCTION FOR COSTS OF GLUTEN-FREE PRODUCTS

In Canada, celiac patients certified by a medical practitioner may claim the incremental cost associated with the purchase of gluten-free products compared with the cost of corresponding non-gluten-free products, as a medical expense to be deducted from their personal income before calculation of income tax (26). Table 1 shows an example of the calculation of a deduction. No allowance is made for the costs associated with finding (transport) and receiving (shipping) gluten-free products.

In the United States, if a taxpayer can establish the medical purpose of the diet, such as through a physician's diagnosis, the incremental cost of the diet is a medical care expense under section 213 (27-29). Rulings by the Internal Revenue Service suggest that if the total medical expenses for the year exceed 7.5% of the individual's adjusted gross income, then the incremental cost of the GFD, including the total cost of items that would not be purchased at all if the individual did not have CD, as well as other expenses related to transportation or shipping costs, are deductible (27,28).

TABLE 2
Estimated cost of gluten-free (GF) products in Argentina currency (AR$) and Canadian currency (CAD$) compared with regular wheat-containing products, and marginal cost for celiac patients

		Estimated cost				Marginal cost:
Product	Amount	GF, AR$*	GF, CAD$†	Non-GF, AR$*	Non-GF, CAD$†	GF – non-GF, CAD$
Oil	1 L	17.00	3.40	5.98	1.19	2.21
Corn starch	1.5 kg	49.29	3.60	28.89	5.77	2.17
Peas	2 cans	22.00	4.40	10.26	2.05	11.74
Cacao powder	800 g	70.00	14.00	23.95	4.79	9.21
Quince jam	1 kg	30.00	6.00	19.90	3.98	2.02
Rice flour	1 kg	30.00	6.00	30.00	6.00	0.00
Rice noodles	1 kg	60.00	12.00	60.00	12.00	0.00
Baking powder	400 g	35.00	7.00	6.64	1.32	5.67
Rice	1 kg	10.00	2.00	7.77	1.55	0.45
Bread mix	500 g	20.00	4.00	20.00	4.00	0.00
Total		322.00	64.40	213.19	42.65	21.75

**Least expensive price obtained from COTO, Argentina <http://www.cotodigital.com.ar/Home.asp>; †Currency exchange: 1 CAD$ = 5 AR$ as for official exchange in 2010*

If the total medical expenses are <7.5% of the adjusted gross income, then Flexible Spending Account (30) provisions can be used to pay for the incremental cost of the GFD (including related shipping expenses), providing a similar 'tax break' for these costs. Finally, for low-income families, the food stamp program (31) may help them buy the food they need to remain healthy.

The uses of tax offsets, or tax expenditures as they are referred to in public finance, have several features that may limit the impact of the approach on uptake and compliance with a GFD among celiac patients. First, the tax offset takes the form of a reimbursement (tax refund or reduction in taxes to be paid) of a cost previously incurred. Hence, the program does not help celiac patients meet the incremental cost of the GFD, it simply provides some future refund of that cost. The household resources, from which the incremental cost of the GFD must be paid, remain the same. In the case where the GFD is not affordable, the prospect of a future tax refund to cover part of this cost is of no help. Second, the value of the tax offset is directly related to the celiac patient's marginal rate of tax. If the patient does not have sufficient income to pay income tax, there is no tax refund to be claimed. In contrast, someone paying tax at a marginal rate of 50% will receive a tax refund equal to 50% of the incremental cost of the GFD. Therefore, although we may expect low-income patients with CD to have greater difficulty covering a GFD, the level of support for meeting these costs is greatest among higher-income celiac patients, for whom affordability of the GFD is less likely to be a problem. As a result, the impact of tax deductions on GFD uptake is unlikely to be substantial, even if celiac patients participate in the program claiming the deductions against tax that they qualify for. However, the inconvenience of retaining receipts for gluten-free products along with receipts for gluten-containing foods to determine the incremental cost of the GFD, together with problems arising from differences in packaged quantities and serving sizes between gluten-free and gluten-based brands, means that uptake of the tax deduction may be less than universal among celiac patients.

From a government perspective, the approach is relatively low cost, because it is delivered and administered through an existing bureaucracy (the taxation system) and, as indicated above, is unlikely to have a 100% uptake among eligible individuals. It also represents, at best, only partial reimbursement of the incremental costs of GFD and, in the case of non-income tax payers, no reimbursement whatsoever. However, there will likely be additional costs due to audit and enforcement to avoid claims based on inflated incremental costs associated with comparing high-cost varieties of gluten-free products with low-cost varieties of gluten-based products. Difficulties on the calculation of real incremental cost of gluten-free products are, in part, due to differences in package amounts and serving sizes of different brands.

The advantage of this system is that it constitutes the least costly approach compared with other options. The government does not need to deal with specific intermediaries because the procedures are performed during the regular tax claiming process.

MONEY PROVISION OR CASH TRANSFER

Some countries provide money to celiac patients. For example, in Argentina, a law passed in 2009 (32) provides celiac patients with the equivalent to $50 per month to offset the incremental costs of a GFD. Celiac patients are not required to provide proof of purchase of gluten-free food. Italy provides a monthly allowance to offset the incremental cost of gluten-free foods for those with CD. The allowance varies by region in accordance with local regulations and ranges from $184 to $200 per month. Gluten-free foods eligible to be purchased with the stipend are found on a list maintained by the Ministry of Health (33), which has established standards for a product being designated gluten free. In addition, CD patients receive extra time off work to offset the time taken to find and prepare gluten-free food. Furthermore, schools, hospitals and other public facilities must provide gluten-free food options.

This approach is more generous than the tax offset program because it provides upfront support of the incremental costs of the GFD equally to all celiac patients and, hence, it is more costly to the government. However, the GFD allowance is not 'earmarked' for spending only on gluten-free products. Celiac patients could, therefore, choose to simply reduce any existing spending on gluten-free products, effectively using the allowance in part or entirely on non-gluten-free products. As a policy aimed at increasing uptake and compliance with a GFD, any such substitution of allowance for existing GFD spending would represent an unintended 'leakage' of funds to other products. Moreover, if allowances are to be universal among all celiac patients, it implies that those allowances allocated to existing compliers have no impact on compliance. Instead, they simply transfer funds from taxpayers to compliant celiac patients. If compliance is correlated positively with patient income, then these allowances could be 'clawed back' by making them taxable. Those patients who already have the capacity to purchase GFD products would receive less than the full incremental cost of the GFD after payment of tax (34,35). Also, in some societies, administering a program of cash transfers involves risks of loss while transporting and distributing cash (34). Concern is being raised in countries such as Italy, where this system may impose a significant burden in terms of cost to the health care system, as opposed to the tax and benefits system.

DIRECT FOOD PROVISION

Some provinces in Argentina (36,37) and Spain (38) provide celiac patients with a monthly supply of specific gluten-free food. An example of a monthly food supply for CD patients in Argentina is shown in Table 2. The amount of food provided is based on the average consumption of gluten-free products per person (36). The program is means tested to concentrate resources on celiac patients who are least able to meet the incremental cost of GFD However, direct distribution and provision of food is costly, as is the administration and enforcement of means testing. The impact on uptake and compliance with a GFD is expected to exceed that of the other programs because it restricts patients from using program benefits for non-GFD products. In principle, patients could exchange gluten-free foods for other products, but there would appear to be a limited market for such exchange if all celiac patients receive the food allowance. In the presence of means testing, however, opportunities may exist for those patients who do not qualify for the food allowance because of their income level to exchange money or other products with those receiving the food allowance at prices below the market price of GFD.

Furthermore, testing for CD leads to program resources being targeted on those patients most at risk of not complying with the diet for affordability reasons, further enhancing the likely impact of the program on total uptake and compliance.

Price subsidy

In many countries, including New Zealand, the Republic of Ireland and the United Kingdom, gluten-free food is prescribed by physicians and, hence, the cost of the food is covered under the same arrangements as prescribed drugs. This usually involves some form of partial or total subsidy or fixed prepayment provision.

In New Zealand, specific gluten-free products are partially subsidized by the government under the Pharmac Part-Subsidy program for pharmaceuticals (39). Once the physician has submitted a form with the CD diagnosis confirmed by a duodenal biopsy, the patient can purchase gluten-free products at one-half of the retail price in a regular store, the other one-half being funded by the program. Examples of products and pricing available under the Pharmac part-subsidy in New Zealand are shown in Table 3 (39). In the Republic of Ireland, gluten-free foods fall under a Drug Payment Scheme, which sets a maximum monthly out-of-pocket expenditure that an individual or family would pay for prescription drugs, other medicine and certain appliances (as of December 31, 2012, that was €144 [approximately $194]). Typically, the general practitioner (GP) would prescribe specific gluten-free foods from a general medical services list that includes items such as bread, rolls, bread mixes and flour, pasta, breakfast cereals, pizza bases and baking powder. Eligible patients collect the prescribed gluten-free foods at the pharmacy. The United Kingdom also offers gluten-free foods under prescription using an approved list determined by the Advisory Committee on Borderline Substances (available in Coeliac UK's online Food and Drink Directory <www.coeliac.org.uk/gluten-free-diet-and-lifestyle/food-shopping/food-and-drink-directory/>).

The quantity of gluten-free food prescribed is largely based on a recommended number of units per month. For example, 400 g of gluten-free baguettes and 250 g gluten-free pasta would each count as one unit. The recommended number of allowable units varies according to age, sex and other factors, such as pregnancy or breastfeeding. The celiac society in the United Kingdom has established a strong position on items that can be prescribed under the plan, restricting the list to essential healthy products while excluding food considered not essentially 'healthy' such as gluten-free cake mixes and desserts. Of course, celiac patients can still purchase these items out of pocket at the store. Each item of prescribed gluten-free product carries a fixed prescription fee of £7.65. Elderly patients, children and people on welfare benefits are exempt from prescription charges. In addition, because a full list of prescribed gluten-free items would become expensive (eg, bread, pizza base, pasta and crackers would carry a charge for four items or approximately £30), fixed-price prepayment certificates can be purchased for periods of three months (£29.10) or one year (£104).

This approach has the potential to make the GFD more affordable to celiac patients provided they qualify for fee exemptions or purchase a prepayment certificate. However, the price of gluten-free products purchased at retail stores may be less than the per-item prescription charge (which is set annually in relation to National Health Service [NHS] plans as opposed to the cost of prescribed products). But this price difference is likely to be less than the subsidization arising from use of prepayment certificates and, hence, these arrangements will reduce the cost of the GFD. Whether this increases uptake and compliance with a GFD will also depend on the additional costs on individuals of having to attend the doctor's office for a prescription and the pharmacy for dispensation of the prescribed items. Patients may prefer to go directly to the grocery store if they can afford the nonsubsidized retail price, but people with low income have no other option. Collecting food from the pharmacy may involve social stigma for the patient while some doctors complain that by prescribing food items they risk resembling a 'grocery store' (40). Finally, the system involves coordination of several different providers from several intermediaries (GPs, pharmacists, third-party payers), which increases the administration costs of the program to the government.

TABLE 3
Example of gluten-free products coverage through a subsidy system in New Zealand

Items available on special foods schedule part subsidized	Size	Cost, including GST, $	Approximate retail, $
Orgran* corn & vegetable spirals	250 g	2.07	4.90
Orgran rice & millet spirals	250 g	2.30	5.20
Orgran buckwheat spirals	250 g	2.30	5.20
Orgran vegetable & rice spirals	250 g	2.07	4.90
Orgran rice & maize spirals	250 g	2.07	4.90
Orgran corn and vegetable shells	250 g	2.07	4.90
Orgran rice & corn spaghetti/noodles	250 g	2.07	5.20
Orgran rice & corn lasagne sheets	375 g	2.07	7.00
Orgran rice & corn penne	200 g	3.79	4.90
Orgran rice & corn macaroni	250 g	2.07	4.90
Orgran Italian-style spaghetti, long	220 g	2.07	5.20
Bakels† health bread mix	1 kg	4.65	13.30
Bakels low gluten bread mix	1 kg	4.83	12.20
Horleys† flour	2 kg	16.10	30.40
Horleys bread mix	1 kg	9.71	18.50
Simple† baking mix	1 kg	5.10	10.90

**Australia; †New Zealand. GST Goods and services tax*

In Switzerland, as well as under the NHS in Northern Ireland, Scotland and Wales, gluten-free products are provided through physician prescription without any patient charge. The principle is the same as in New Zealand and under the NHS in England; however, the level of subsidy is 100%. Recently, some changes have been introduced in Scotland with a Gluten-Free Food Additional Pharmaceutical Service being introduced on a limited period trial basis (33). The monthly prescription will no longer be provided by the GP, but from the pharmacist following referral from a GP or dietician stating the diagnosis and number of units of gluten-free food required per month. Pharmacists are responsible for assisting patients in selecting the most appropriate gluten-free foods, and agreeing with patients on the frequency of collection (41).

Under the scheme, the choice of gluten-free foods should be limited to a local formulary produced by NHS boards: "to optimise value for money and clinical effectiveness" (40). The scheme will be reviewed for its impact on "the clinical benefit for the patients concerned and cost effectiveness for NHS Scotland" (40,41) in comparison with the previous arrangements. Gluten-free food remains free to patients with a physician's prescription but access to the products is improved by providing the patient with direct access to the supplier (pharmacist) following an initial physician diagnosis. Funding for the new program is based on two elements. A fixed tariff of £125 per month is allocated to pharmacists for each eligible celiac patient, with an additional tariff based on the percentage corresponding to chronic medication service (41).

TABLE 4
Differences among available systems for gluten-free product coverage

	Tax deduction	Cash transfer	Food provision	Prescription	Subsidy	No coverage
Premium coverage	Partial	Total	Total	Partial	Partial/total	None
Example of countries with this system	Canada, United States, Netherlands, Portugal	Italy, Argentina, Uruguay, Finland, Greece, France, Norway, Belgium, Slovenia (child)	Some provinces in Argentina and Spain	New Zealand, Republic of Ireland, United Kingdom	Northern Ireland, Scotland, Czech Republic, Wales	Mexico, Spain (some provinces), Germany, Slovenia (adults)
Approximate amount (USD$/month)	Variable, depending on income	50.00 to 327.00	60.00	Maximum of 194.00	Maximum of 203.63	0.00
Advantages	Low cost for government. No new intermediaries	Convenient for patients. Less costly for government. Requires fewer intermediaries	Direct provision of treatment. Greater compliance expected	Less costly treatment with more purchasing power for other goods	Full coverage of treatment. Provides more purchasing power to patients	No cost for government
Disadvantages	Inconvenient for people with low income. Need to save receipts and calculations. Non-refundable tax credit subject to income declared; imperfect coverage, depending on product brand. Possible abuse by tax payers	If not controlled, could lead to other expenditures and consequently low treatment compliance. Risk of cash misuse	Expensive for government. Possible loss of quality in products provided (cheapest option)	Inconvenient for patients in getting the prescription from GP. Inconvenient for GP who feels like a 'grocery store'. Several intermediaries with increased marginal cost for government	Inconvenient for patients in getting the prescription from GP. Inconvenient for GP who feels like a 'grocery store'. Several intermediaries with increased marginal cost for government	No coverage of treatment for patients, who need to face the extra costs related to gluten-free diet

GP General practitioner

No coverage for gluten-free products

In many jurisdictions, including Mexico and most provinces in Spain, no financial support is provided for celiac patients' purchases of GFD. The celiac community in Mexico is currently actively campaigning for financial support from government to contribute to the extra costs of GFD (42).

A summary of main characteristics, advantages and disadvantages of each program is shown in Table 4.

HOW SOCIETY AND GOVERNMENT BENEFIT FROM EACH SYSTEM

Given that resources for health care are limited, governments must consider the impact of the different approaches on uptake and compliance with a GFD and, hence, patient outcomes, as well as any additional consequences for the health care system arising from changes in uptake and compliance.

Doing nothing is always an option for decision-makers and it may be attractive for those responsible for expenditure control. However, the option is likely to result in low levels of uptake and compliance and lead to additional demands on the health care system as celiac patients present with health problems. Compared with 'doing nothing', income support programs are aimed at providing more resources to patients for them to allocate however they like. They make people better off and, in economic terms, this generates an 'income effect' resulting in the individual being able to purchase more of any good or service than he previously purchased. Gluten-free products are not particularly favoured under this policy, because the price of gluten-free products remains the same relative to all other products. Hence, from an economics perspective, this policy does not make gluten-free products any more attractive than without the income supplement and the individual is able to spend the additional income in accordance with his relative preferences among all goods and services.

In contrast, price subsidies make individuals feel better off, because the amount of the subsidized gluten-free product purchased by the individual now costs less and so 'savings' accrue from this price reduction to spend on either more gluten-free foods or on other items. As with the income supplement, this is an 'income effect'. However, the subsidy has now made gluten-free products more attractive to patients because they are now less costly than others items in relative terms. The amount of other items you now must forgo to afford any particular gluten-free product has fallen. Thus, there is a 'substitution effect', which is an additional incentive to spend the 'savings' from the price reduction on gluten-free foods.

Investing resources in a price subsidy policy would be expected to have a greater impact on uptake and compliance than using the same resources for an income enhancement program. Similarly, price subsidies may be more effective than income enhancement in improving the distribution of uptake and compliance among the celiac disease population. Under income supplements, patients from poorer households may face greater immediate priorities for household spending such as accommodation, clothing and food for other family members whereas among better-off groups, the income supplements may contribute more to discretionary expenditures and, hence, may be more likely to be allocated to GFD. However, under price subsidies, the impact on uptake would be expected to produce more equitable distributions if individuals from poorer households are more sensitive and responsive to price changes in particular commodities because the savings made represent a much larger proportion of their household budget.

These expected outcomes are conditional on having few if any barriers to access to the programs. Complex or inconvenient methods of taking advantage of subsidies may mitigate the impact of those subsidies on affordability and, hence, adherence to a GFD. Similarly, if the methods used for direct provision stigmatize patients, the policy may be unacceptable to some potential beneficiaries and reduce the impact on uptake.

DISCUSSION AND CONCLUSION

The purpose of the present analysis was to review the different systems available to support GFD coverage as a treatment for CD patients. The review showed that the systems in place exhibit particular advantages and disadvantages in relation to promoting uptake and compliance with GFD. The analysis described in the present article demonstrated that the tax offsets for GFD creates inequity because they help least those in most need. If we define needs in terms of marginal capacity to benefit from health care, the health maximization model has the principle that those in equal needs should have equal access to service (43).

There is strong evidence that poor compliance with the GFD in CD patients places patients at risk for severe complications (7). For this reason, support for this treatment for those who are most in need (ie, those least able to meet the incremental cost of the GFD) is important. Increased demand of GFD in celiac patients is usually accompanied by better treatment compliance (7,44-46). Increasing coverage among those in need is likely to be associated with greater compliance and, hence, a decrease in complications due to the disease, which mitigates the cost of the program. A balanced approach would consist of subsidizing gluten-free products through controlled vouchers or direct food provision to those who most need it, independent of ability or willingness to pay. Moreover, if the cost of such a program is prohibitive, the value of the benefits could be made taxable to ensure that any patient contribution (in terms of additional taxation) is directly related to ability to pay.

We may question whether the GFD should be covered at all, given that a natural diet of fruits and vegetables is gluten free. However, individuals with low socioeconomic status consume a diet that heavily relies on wheat-containing products and that is low in fruit and vegetables (47,48). Programs could, therefore, be developed with the objective to generate changes in dietary habits in the long term, alongside immediate policies to promote uptake and compliance.

Economic evaluations of alternative approaches are needed to provide an empirical base for policy formulation and decision making, with the ultimate objective of achieving maximum uptake and compliance from whatever resources are committed to support celiac patients. International comparisons provide insights about alternative approaches and different contexts with countries learning from systems implemented elsewhere. This helps individual jurisdictions to assess their own policy options (49).

Finally, the limited coverage of GFD in Canada is concerning. This could reflect a lack of interest and understanding of the disease burden by governments. More efforts are required by the Canadian medical community and the Canadian Celiac Society to act as agents in identifying ways of improving resource allocation in celiac disease. There is definitely an unmet need for GFD among celiac patients in Canada.

DISCLOSURES: The authors have no financial disclosures or conflicts of interest to declare.

FUNDING: MIPS received a CIHR/CAG Post-Doctoral Fellowship for the period 2014-2016. PB received the Hamilton Health Sciences Research Early Career Award.

REFERENCES

1. Pulido O, Zarkadas M, Dubois S, et al. Clinical features and symptoms recovery on a gluten-free diet in Canadian adults with celiac disease. Can J Gastroenterol 2013;27:449-53.
2. Ludvigsson JF, Rubio-Tapia A, van Dyke CT, et al. Increasing incidence of celiac disease in a North American population. Am J Gastroenterol 2013;108:818-24.
3. Fasano A, Berti I, Gerarduzi T, et al. Prevalence of celiac disease in at-risk and not-at-risk groups in the US: A large multicenter study. Arch Intern Med 2003;163:286-92.
4. Statistics Canada. Canada' total population estimates 2013. <www.statcan.gc.ca/daily-quotidien/130926/dq130926a-eng.htm> (Accessed March 2014).
5. Zarkadas M, Case S. Celiac diasease and the gluten free diet: An overview. Topics Clin Nutr 2005;20:127-38.
6. Rashtak S, Murray JA. Celiac disease: New approaches to therapy. Alim Pharmacol Ther 2012;35:768-81.
7. Rubio-Tapia A, Hill I, Kelly CP, et al. ACG clinical guidelines: Diagnosis and management of celiac disease. Am J Gastroenterol 2013;108:656-76.
8. Akobeng AK, Thomas AG. Systematic review: Tolerable amount of gluten for people with coeliac disease. Aliment Pharmacol Ther 2008;27:1044-52.
9. Lucendo AJ, García-Manzanares A. Bone mineral density in adult coeliac disease: An updated review. Rev Esp Enferm Dig 2013;105:154-62.
10. Pinto-Sánchez MI, Mohaidle A, Baistrocchi A, et al. Risk of fracture in celiac disease: Gender, dietary compliance, or both? World J Gastroenterol 2011;17:3035-42.
11. Olmos M, Antelo M, Vazquez H, et al. Systematic review and meta-analysis of observational studies on the prevalence of fractures in coeliac disease. Dig Liver Dis 2008;40:46-53.
12. Ludvigsson JF. Mortality and malignancy in celiac disease. Gastrointest Endosc Clin N Am 2012;22:705-22.
13. West J, Logan RF, Smith CJ, et al. Malignancy and mortality in people with coeliac disease: Population based cohort study. BMJ 2004;329:716-9.
14. Askling J, Linet M, Gridley G, et al. Cancer incidence in a population based cohort of individuals hospitalized with celiac disease or dermatitis herpetiformis. Gastroenterology 2002;123:1428-35.
15. Corrao G, Corazza GR, Bagnardi V, et al. Mortality in patients with coeliac disease and their relatives: A cohort study. Lancet 2001;358:356-61.
16. Sategna-Guidetti C, Grosso SB, Grosso S, et al. The effects of 1-year gluten withdrawal on bone mass, bone metabolism and nutritional status in newly-diagnosed adult coeliac disease patients. Aliment Pharmacol Ther 2000;14:35-43.
17. Valdimarsson T, Lofman O, Toss G, et al. Reversal of osteopenia with diet in adult coeliac disease. Gut 1996;38:322-7.
18. Choi JM, Lebwohl B, Wang J, et al. Increased prevalence of celiac disease in patients with unexplained infertility in the United States. J Reprod Med 2011;56:199-203.
19. Ciacci C, Cirillo M, Auriemma G, et al. Celiac disease and pregnancy outcome. Am J Gastroenterol 1996;91:718-22.
20. Norgard B, Fonager K, Sorensen HT, et al. Birth outcomes of women with celiac disease: A nationwide historical cohort study. Am J Gastroenterol 1999;94:2435-40.
21. Ludvigsson JF, Montgomery SM, Ekbom A. Celiac disease and risk of adverse fetal outcome: A population-based cohort study. Gastroenterology 2005;129:454-63.
22. Khashan AS, Henriksen TB, Mortensen PB, et al. The impact of maternal celiac disease on birth weight and preterm birth: A Danish population based cohort study. Hum Reprod 2010;25:528-34.
23. Verrill L, Zhang Y, Kane R. Food label usage and reported difficulty with following a gluten-free diet among individuals in the USA with coeliac disease and those with noncoeliac gluten sensitivity. J Hum Nutr Diet 2013;26:479-87.
24. Lee A, Ng DL, Zivin J, Green P. Economic burden of a gluten-free diet. J Hum Nutr Diet 2007;20:423-30.
25. Tennyson CA, Simpson S, Lebwohl B, et al. Interest in medical therapy for celiac disease. Therap Adv Gastroenterol 2013;6:358-64.
26. Canada Revenue Agency Website: <www.cra-arc.gc.ca/tx/ndvdls/tpcs/ncm-tx/rtrn/cmpltng/ddctns/lns300-350/330/clc-eng.html> (Accessed March 2014).
27. US Internal Revenue Service. Medical and Dental Expenses <www.irs.gov/taxtopics/tc502.html> (Accessed November 19, 2014).
28. Medical and Dental Expenses (Including the Health coverage tax credit) Publication 502. Department of the Treasury Internal Revenue Service. <www.irs.gov/pub/irs-pdf/p502.pdf> (Accessed November 19, 2014).
29. Gluten free federal tax deduction. Celiac Sprue Association. <www.csaceliacs.info/tax_deductions.jsp> (Accessed March 2014).
30. Flexible spending accounts. Office of Personnel management U.S. <www.opm.gov/healthcare-insurance/flexible-spending-accounts/> (Accessed March 2014).
31. Supplemental Nutrition Assistance Program (SNAP). US Department of Agriculture, Food and Nutrition service. <www.fns.usda.gov/snap/supplemental-nutrition-assistance-program-snap> (Accessed March 2014).
32. La ley celiaca Argentina. <www.ley-celiaca.com.ar/> (Accessed March 2014).

33. Ministerio della salute –Senza glutine. Regolamento 41/2009 <www.salute.gov.it/portale/temi/p2_6.jsp?lingua=italiano&id=979&area=Alimenti%20particolari%20e%20integratori&menu=dietetici> (Accessed March 2014).
34. Cash transfer in programming emergencies. <www.unscn.org/layout/modules/resources/files/Cash_transfer_programming_in_emergencies.pdf> (Accessed March 2014).
35. Barham T, Maluccio JA. Eradicating diseases: The effect of conditional cash transfers on vaccination coverage in rural Nicaragua. J Health Econ 2009;28:611-21
36. Programa "Alimentación libre de Gluten" <www.larevistadelceliaco.com.ar/agenda/corrientes-salta-y-jujuy-entregan-alimentos-para-celiacos> (Accessed March 2014).
37. Determinación de gluten en alimentos y la implementación de un sistema de gestión de calidad para la acreditación de la técnica. <www.cequimap.com.ar/foro/Gliadina.pdf> (Accessed March 2014).
38. Asociacion de celiacos de Extremadura. Proyecto ayudas economicas. <www.celiacos-ex.com/Cargar/BD/PROY_AYUDA_ECONOM> (Accessed March 2014).
39. Pharmaceutical subsidy card. Ministry of Health New Zealand. <www.health.govt.nz/your-health/conditions-and-treatments/treatments-and-surgery/medications/pharmaceutical-subsidy-card> (Accessed March 2014).
40. Professional matters. Pharmaceut J 2013;291:582.
41. Details released for pharmacy gluten free-food service in Scotland. <www.pjonline.com/news/details_released_for_pharmacy_gluten_free_food_service_in_scotland> (Accessed March 2014).
42. Manifiesto celiaco Mexico. <http://manifiestoceliacomex.wordpress.com/manifiesto-celiaco-mexico/> (Accessed March 2014).
43. Morris S, Devlin N, Parkin D, et al. Economic Analysis in Health Care, 2nd edn. Hoboken: Wiley, 2007.
44. Tack GJ, Verbeek WH, Schreurs MW, et al. The spectrum of celiac disease: Epidemiology, clinical aspects and treatment. Nat Rev Gastroenterol Hepatol 2010;7:204-13.
45. Niewinski MM. Advances in celiac disease and gluten-free diet. J Am Diet Assoc 2008;108:661-72.
46. See J, Murray JA. Gluten-free diet: The medical and nutrition management of celiac disease. Nutr Clin Pract 2006;21:1-15.
47. De Irala-Estevez J, Groth M, Johansson L, et al. A systematic review of socio-economic differences in food habits in Europe: Consumption of fruit and vegetables. Eur J Clin Nutr 2000;54:706-14.
48. Clark AM, Duncan AS, Trevoy JE. Healthy diet in Canadians of low socioeconomic status with coronary heart disease: Not just a matter of knowledge and choice. Heart Lung 2011;40:156-63.
49. Parkin D. Comparing health service efficiency across countries. Oxford Rev Econ Policy 1989;5:75-88.

7

Comparative analysis of selected scales to assess prognosis in acute pancreatitis

Dorota Koziel PhD[2], Stanislaw Gluszek Prof[1,2], Jaroslaw Matykiewicz Md PhD[1,2], Piotr Lewitowicz MD PhD[2], Zuzanna Drozdzak MA[3]

D Koziel, S Gluszek, J Matykiewicz, P Lewitowicz, Z Drozdzak. Comparative analysis of selected scales to assess prognosis in acute pancreatitis. Can J Gastroenterol Hepatol 2015;29(6):299-303.

OBJECTIVE: To evaluate the utility of selected scales to prognosticate the severity and risk for death among patients with acute pancreatitis (AP) according to the revised Atlanta classification published in 2012.
METHODS: Prospective data regarding patients hospitalized due to AP were analyzed. The final analysis included a total of 1014 patients. The bedside index for severity in acute pancreatitis (BISAP), Panc 3 scores and Ranson scales were calculated using data from the first 24 h of admission.
RESULTS: Mild AP was diagnosed in 822 (81.1%) cases, moderate in 122 (12%) and severe in 70 (6.9%); 38 (3.7%) patients died. The main causes of AP were cholelithiasis (34%) and alcohol abuse (26.7%). Recurrence of AP was observed in 244 (24.1%) patients. In prognosticating the severity of AP, the most useful scale proved to be the Acute Physiology and Chronic Health Evaluation (APACHE) II (area under the curve [AUC] 0.724 [95% CI 0.655 to 0.793]), followed by BISAP (AUC 0.693 [95% CI 0.622 to 0.763]). In prognosticating a moderate versus mild course of AP, the CT severity index proved to be the most decisive (AUC 0.819 [95% CI 0.767 to 0.871]). Regarding prognosis for death, APACHE II had the highest predictive value (AUC 0.726 [95% CI 0.621 to 0.83]); however, a similar sensitivity was observed using the BISAP scale (AUC 0.707 [95% CI 0.618 to 0.797]).
CONCLUSIONS: Scoring systems used in prognosticating the course of the disease vary with regard to sensitivity and specificity. The CT severity index scoring system showed the highest precision in prognosticating moderately severe AP (as per the revised Atlanta criteria, 2012); however, in prognosticating a severe course of disease and mortality, APACHE II proved to have the greatest predictive value.

Key Words: *Acute pancreatitis; Scales; Severity of illness index*

L'analyse comparative d'échelles sélectionnées pour évaluer le pronostic de pancréatite aiguë

OBJECTIF : Évaluer l'utilité d'échelles sélectionnées pour établir le pronostic de gravité et de risque de décès chez les patients ayant une pancréatite aiguë (PA) selon la classification révisée de l'Atlanta publiée en 2012.
MÉTHODOLOGIE : Les chercheurs ont analysé les données prospectives relatives aux patients hospitalisés en raison d'une PA. L'analyse finale incluait 1 014 patients. Ils ont calculé l'indice clinique de gravité de pancréatite aiguë (ICGPA), les scores Panc 3 et les échelles de Ranson au moyen des données tirées des 24 premières heures suivant l'hospitalisation.
RÉSULTATS : Une PA bénigne a été diagnostiquée dans 822 cas (81,1 %), une PA modérée dans 122 cas (12 %) et une PA grave dans 70 cas (6,9 %), tandis que 38 patients (3,7 %) sont décédés. La cholélithiase (34 %) et la consommation excessive d'alcool (26,7 %) étaient les principales causes de PA. Les chercheurs ont observé une récurrence de PA chez 244 patients (24,1 %). Pour établir le pronostic de gravité de la PA, l'évaluation de l'état de santé chronique et physiologique aiguë (APACHE) II (aire sous la courbe [ASC] 0,724 [95 % IC 0,655 à 0,793]) était l'échelle la plus utile, suivie de l'ICGPA (ASC 0,693 [95 % IC 0,622 à 0,763]). Pour établir le pronostic d'une PA à évolution modérée ou bénigne, l'indice de gravité tomodensitométrique s'est révélé le plus décisif (ASC 0,819 [95 % IC 0,767 à 0,871]). À l'égard du pronostic de décès, le score APACHE II avait la valeur prédictive la plus élevée (ASC 0,726 [95 % IC 0,621 à 0,83]). Cependant, les chercheurs ont observé que l'échelle ICGPA (ASC 0,707 [95 % IC 0,618 à 0,797]) avait une sensibilité similaire.
CONCLUSIONS : Les systèmes de pointage pour établir le pronostic de l'évolution pathologique varient en matière de sensibilité et de spécificité. Le système d'indice de gravité tomodensitométrique établissait le pronostic de PA modérément bénin de la manière la plus précise (conformément aux critères révisés de l'Atlanta, 2012). Cependant, pour établir le pronostic de grave évolution de la maladie et de mortalité, l'échelle APACHE II avait la meilleure valeur prédictive.

Acute pancreatitis (AP) is the sudden inflammation of the pancreas and, to a varied degree, involves local tissues or distant organs. The morbidity rate in the adult Polish population has not been precisely specified; however, data from the National Health Insurance Fund and the National Department of Hygiene report it to be 70 to 76.2 per 100,000 (1,2). A prospective study (3) conducted in the Kielce region of Poland suggested that AP occurs in 99.9 per 100,000 inhabitants, with first-time events occurring in 79 per 100,000 inhabitants. Diagnosis of pancreatitis is based on clinical characteristics, primarily observation of the presence of two of three symptoms: abdominal pain typical of AP; at least threefold increase in serum lipase (or amylase) activity; and characteristic changes seen on computed tomography (CT) with contrast medium. More rarely, diagnosis is confirmed using magnetic resonance imaging or percutaneous ultrasound-guided pancreatic biopsy. Confirmation using CT is not necessary when the diagnosis has been based on the presence of abdominal pain and increase in enzyme activity (4). The accuracy of CT in prognosticating a severe course of AP is comparable with the clinical scoring systems; therefore, it is not recommended for evaluation of disease severity (5).

The majority of patients experience a mild form of the disease; however, in approximately 20% of patients, a severe form develops, in which the length of hospitalization is considerably longer and mortality is extremely high. Necrotic pancreatitis develops in 5% to 10% of patients (4). According to the revised Atlanta classification published in 2012 (4), there are three degrees of severity: mild, moderate and severe. Distinguishing the three AP categories enables the classification of patients into homogenous severity groups, in which the duration of hospitalization in the severe and moderate forms of the disease is similar; however, mortality due to severe AP is considerably higher (6).

Early identification of patients with a moderate or severe course of disease is very important because it enables optimization of treatment, leading to improved outcomes. Many prognostic scales and individual

[1]*Clinical General, Oncological and Endocrinological Surgery, Regional Hospital;* [2]*Faculty of Health Sciences, Jan Kochanowski University, Kielce;* [3]*Centre for Evaluation and Analysis of Public Policies, Jagiellonian University, Cracow, Poland*
Correspondence and reprints: Dr Dorota Koziel, Al IX Wieków Kielc 19, 25-317 Kielce, Poland. E-mail dorota.koziel@wp.pl

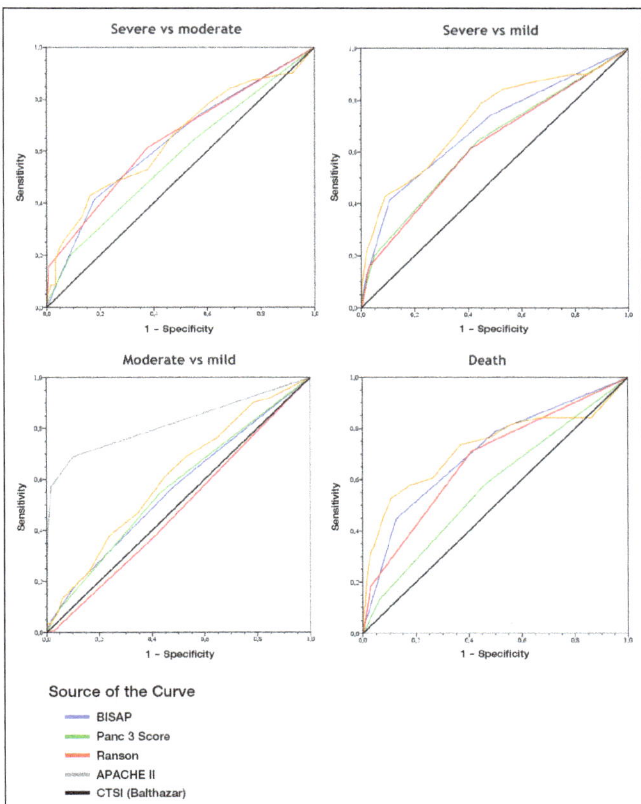

Figure 1) *Comparison of the selected scales area under the curve for predicting: severe (1) against moderate (0) form of acute pancreatitis (AP); severe (1) against mild (0) form of AP, moderate (1) against (mild) form of AP, and death (1). APACHE 11 Acute Physiology and Chronic Health Evaluation; BISAP Bedside index for severity in acute pancreatitis; CTSI Computed tomography severity index*

indicators are known, but their sensitivity and specificity, as well as their prognostic value, remain the subject of debate. One of the first multifactorial prognostic tests is the scale based on the Ranson criteria. The presence of >2 risk factors prognosticates a severe form of AP. The Acute Physiology and Chronic Health Evaluation (APACHE) II scoring system is characterized by high sensitivity and specificity. The final assessment reflects the sum of scores from 13 evaluated parameters, each of which may be scored on a scale of 0 to 4. An APACHE II score ≥8 prognosticates a severe course of disease. The application of the proposed scales is often limited by their complexity; furthermore, some hospitals may have only limited access to certain scales. The Panc 3 score is a simple scoring system that evaluates hematocrit, body mass index (BMI) and pleural effusion. It was proposed that hematocrit >44%, BMI >30 kg/m^2 and the presence of pleural effusion prognosticate a severe course of AP (7). Another simple, highly specific diagnostic test prognosticating a mild course of AP involves the evaluation of peritonitis, hematocrit value and creatinine level. According to Oskarsson et al (8) and Lankisch et al (9), a lack of peritonitis symptoms, hematocrit value <43% in men and <39.6% in women, and creatinine level ≤2 mg/dL (≤176.8 μmol/L) indicates a mild disease course. Another simple diagnostic test is the bedside index for severity in acute pancreatitis (BISAP). This index is calculated within the first 24 h after admission to hospital, and is based on five variables: blood urea nitrogen (azotemia) >25 mg/dL (>8.9 mmol/L); consciousness disorders (Glasgow coma scale <15); systemic inflammatory response syndrome (the presence of ≥2 of the following parameters: body temperature <36°C or >39°C; respiratory rate >20 breaths/min or partial pressure of carbon dioxide <32 mmHg; heart rate >90 beats/min; and white blood cell count <4.0×10^9/L or >12.0×10^9/L); age >60 years; and the presence of pleural effusion (10).

TABLE 1
Mean scores of selected scales according to severity of acute pancreatitis

	Acute pancreatitis		
Scale	**Severe**	**Moderately severe**	**Mild**
BISAP	1.27±0.992*	0.80±0.859*	0.60±0.712*
Panc 3	0.8571±0.767	0.6393±0.644	0.4818±0.589*
Ranson	0.7714±0.705*	0.3852±0.505	0.4404±0.557
APACHE II	7.9429±5.090*	5.4918±3.533*	4.2603±3.305*

*Data presented as mean ± SD. *Statistically significant at α=0.05 (one-way ANOVA using F-test for normally distributed variables or Welch for heteroskedasticity, followed by appropriate post hoc tests: Tamhane T2, Dunnett T3 and Games-Howell, which yielded consistent results). APACHE Acute Physiology and Chronic Health Evaluation; BISAP Bedside index for severity in acute pancreatitis*

Given the lack of the consensus regarding the predictive value of the above-mentioned scales, the present study aimed to evaluate the utility of the selected scales in prognosticating the severity and risk for death among patients with AP according to the revised Atlanta classification published in 2012 (4).

METHODS

Data from 1014 patients hospitalized due to AP from 16 surgical wards in the Kielce region of Poland between December 2010 and December 2011 were analyzed with regard to admission details, disease course and treatment outcome. The final sample should be regarded as complete because all cases of AP are presumably subject to inpatient care. The criterion for inclusion in the study was the diagnosis of AP according to the revised Atlanta classification. The Atlanta classification is based on a joint interpretation of the symptoms obtained in medical history, physical examination and targeted laboratory tests. The panel of diagnostic tests performed during treatment depended on the clinical state of the patient. Demographic characteristics, results of laboratory and imaging tests, the course and outcome of treatment, and duration of hospitalization, were reported in previously prepared documentation forms. The BISAP, Panc 3 score and Ranson scales were calculated based on data from the first 24 h of hospitalization in the ward, irrespective of the time of occurrence of the first symptoms of AP. The APACHE II score was calculated only for patients who required additional laboratory tests, which was usually related to a deteriorating state. Cut-off points for prognostication of mild, moderate and severe AP were BISAP >2, Panc 3 ≥2, Ranson ≥2 and APACHE ≥8. The CT severity index (CTSI) was performed and subsequently analyzed only in patients who required CT because of the following clinical indications (n=208): organ failure, fever or chills, long-lasting abdominal pain and recurrent increase in serum amylase activity. Results from CT performed between day 4 and day 7 of the disease were taken into consideration. In some patients, CT was not possible due to early death during the first hours of hospitalization.

Statistical analysis

In the first step of data analysis, the demographic and clinical characteristics of the patients in the sample were described. Subsequently, scores on BISAP, Ranson, Panc 3, APACHE II and CTSI scales were calculated for each patient (whenever possible). The mean score on each scale was calculated for each severity group of AP. ANOVA was used to verify whether the difference in means among severity groups were statistically significant. The assumption of normality of distribution was not met in these cases; however ANOVA is known to yield good results even when the normality assumption does not hold. The predictive validity of the scales was subsequently estimated and compared using the areas under the ROC curve (AUC) for four terminal nodes: severe AP (versus moderately severe and versus mild), moderately severe pancreatitis (versus mild) and death; sensitivity and specificity were calculated (Figure 1). In addition, ORs of developing a

TABLE 2
Comparative utility of selected scales in prognosticating severe and moderate courses of acute pancreatitis and death

Scale	Course of acute pancreatitis: Severe (versus moderate)	Severe (versus mild)	Moderately severe (versus mild)	Death
Bedside index for severity of acute pacreatitis	0.637 (0.554–0.720)*	0.693 (0.622–0.763)*	0.559 (0.503–0.615)*	0.707 (0.618–0.797)*
Panc 3	0.573 (0.488–0.659)	0.633 (0.561–0.705)*	0.564 (0.509–0.620)*	0.573 (0.477–0.669)
Ranson	0.646 (0.563–0.729)*	0.625 (0.553–0.698)*	0.480 (0.426–0.534)	0.678 (0.587–0.769)*
Computed tomography severity index	–	–	0.819 (0.767–0.871)*	–
Acute Physiology and Chronic Health Evaluation II	0.645 (0.561–0.729)*	0.724 (0.655–0.793)*	0.603 (0.550–0.655)*	0.726 (0.621–0.83)*

*Data presented as area under ROC curve (95% CI). *Statistically significant at α=0.05*

TABLE 3
Prediction of a severe and moderately severe course of acute pancreatitis (AP) and deaths according to selected parameters

Parameter	Severe course of AP (versus moderately severe or mild) OR (95% CI)	P	Moderately severe course of AP (versus mild) OR (95% CI)	P	Death OR (95% CI)	P
Age >65 years	1.850 (1.133–3.019)*	0.017	0.719 (0.466–1.108)	0.144	3.020 (1.564–5.833)*	0.001
Etiology						
Biliary	0.945 (0.564–1.585)	0.896	0.716 (0.470–1.091)	0.126	1.009 (0.509–1.997)	1.000
Alcoholic	0.801 (0.450–1.424)	0.048	1.501 (1.001–2.250)	0.063	0.609 (0.269–1.400)	0.268
Leukocytes >13×10^9/L	2.546 (1.550–4.181)*	0.001	2.149 (1.464–3.153)*	0.001	2.522 (1.299–4.896)*	0.006
C-reactive protein >150 mg/L	3.098 (1.879–5.108)*	0.001	5.006 (3.346–7.489)*	0.001	2.858 (1.473–5.543)*	0.003
BISAP >2	6.444 (2.558–16.238)*	0.001	3.151 (1.076–9.229)*	0.045	4.097 (1.163–14.437)*	0.051
APACHE II ≥8	6.782 (4.036–11.397)*	0.001	1.982 (1.160–3.386)*	0.015	9.316 (4.775–18.178)*	0.001
Panc 3 score ≥2	4.377 (2.285–8.386)*	0.001	1.937 (0.966–3.887)	0.082	2.313 (0.871–6.140)	0.090
Ranson ≥2	6.854 (3.217–14.602)*	0.001	0.275 (0.037–2.050)	0.236	7.374 (2.999–18.129)*	0.001

**Statistically significant at α=0.05, Fisher's exact test. APACHE Acute Physiology and Chronic Health Evaluation; BISAP Bedside index for severity in acute pancreatitis*

moderate form of AP (versus mild), severe form of AP (versus moderate or mild) as well as death were calculated according to patient characteristics (age, etiology of disease, C-reactive protein [CRP] levels and scores obtained from selected diagnostic scales). All the analyses were performed using SPSS version 21 (IBM Corporation, USA).

RESULTS

The present study enrolled 1014 patients hospitalized due to AP. Men constituted 62.7% of the study group; patients' mean age was 54 years and three months. Mild AP was diagnosed in 822 (81.1%) patients, moderate in 122 (12%) and severe in 70 (6.9%); 38 (3.7%) patients died. Necrotic pancreatitis confirmed by CT scan was diagnosed in 31 patients. The median hospitalization time in the examined group of patients was seven days. The main cause AP was cholelithiasis, diagnosed in 34% patients (52.6% of the total number of women and 23% of the total number of men). Alcohol was the cause of AP in 26.7% of cases (4.5% of all women and 39.9% of all men). AP in was diagnosed in 10 patients with pancreatic cancer (1% of the entire group), while AP as a complication of endoscopic retrograde cholangiopancreatography developed in 0.7% of all patients. In the remaining 37.6% of patients, the cause of AP was unknown. Recurrence of AP, as determined through patient interview, was observed in 244 patients (24.1%).

Table 1 presents the mean score results of the prognostic scales according to the severity of AP. The BISAP and APACHE II scale scores were statistically significantly related to the severity of AP (ANOVA and Welch ANOVA). Panc 3 scores differentiated mild from moderate and severe AP (α=0.05), but did not differentiate between moderately severe and severe AP). The Ranson scale score was only significant for severe AP.

Figure 1 presents a comparison of ROC curves generated for the prognosis of major clinical outcomes using the selected scales. Table 2 summarizes the predictive validity of the selected scales. Because the CTSI was calculated only for selected patients, the results for these scales were only presented when the number of observations was sufficient to make such estimations (namely, in distinguishing between moderate and mild AP).

In prognosticating a severe versus a mild course of AP (70 versus 822 cases), the most useful scale proved to be APACHE II (AUC 0.724 [95% CI 0.655 to 0.793]), followed by BISAP (AUC 0.693 [95% CI 0.622 to 0.763]).

In prognosticating a moderate versus mild course of AP (122 versus 822 cases), CTSI proved to be the most decisive (AUC 0.819 [95% CI 0.767 to 0.871]). Regarding the prognosis of death, the APACHE II scoring system had the highest predictive value (AUC 0.726 [95% CI 0.621 to 0.83]); however, a similar sensitivity was observed using the BISAP scale (AUC 0.707 [95% CI 0.618 to 0.797]). Unfortunately, distinguishing between a severe and moderately severe course of AP is very difficult, even when predictive scales are used; in this scenario, the predictive value of all the scales was relatively low and none of them crossed the threshold of AUC 0.65.

While detailed evaluation of the predictive validity, sensitivity and specificity of various predictive scales is crucial for preparing clinical guidelines in the future, many practicing physicians rely on practical and easily accessible information that is useful in their everyday clinical practice. For this purpose, two biomarkers (leukocytes and CRP), as well as patient age and etiology of AP, were evaluated in terms of their potential to predict serious clinical outcomes (Table 3).

All analyzed scales were able to significantly prognosticate severe AP (Table 3). Cut-off values for each scale to prognosticate the occurrence of severe AP were: BISAP >2 (OR 6.444); APACHE II ≥8 (OR 6.782); PANC 3 ≥2 (OR 4.377); and Ranson ≥2 (OR 6.854). Patients who score such high results on these scales are from four to almost seven times more likely to develop a severe course of AP, compared with patients who score lower than the cut-off points. Another meaningful risk factor is CRP level >150 mg/L at any time during the treatment, which is associated with a threefold increase in the risk of developing a severe AP, as well as leukocytes >13×10^9/L , which corresponds to a

2.5 times higher risk of severe AP. Age >65 years increases the risk of severe AP by 85%.

In evaluating the risk of developing moderate AP (compared with mild), composite scales can be helpful. Selected cut-off points of APACHE II ≥8 and BISAP >2 are, in fact, related to a greater risk of developing a moderate course of AP (from nearly twofold to threefold increase, respectively); however, these complex scales are unlikely to be used on patients who appear to be stable and undergoing the treatment without apparent complications. In such a case, CRP >150 mg/L may be very helpful because it is associated with a fivefold increase of risk. Another indication that mild AP is worsening to moderate might be leukocytosis (OR 2.149; P<0.006).

In the prediction of death, APACHE II ≥8 can be particularly helpful (OR 9.316), followed by Ranson ≥2 (OR 7.374); the BISAP scale is also useful (OR 3.151; P<0.045). Age is also acknowledged to be an independent risk factor of death during the course of AP. In the present analysis, age >65 years was significantly related with the risk for death (OR 3.02; P<0.001).

Patients with CRP levels >150 mg/L were approximately 2.5 more times likely to die than those whose levels of CRP never exceeded the threshold. Among single markers of death, leukocytosis may also be helpful. Analysis of collected clinical data indicated that leukocytes $>13\times10^9$/L carried a significant prognosis for death (OR 2.522).

The etiology of AP did not prove to be predictive of any of the clinical outcomes under investigation.

The results summarized in Table 3 suggest that composite scales provide a better estimate of whether a patient is at risk for a more severe course of AP and death. Single biomarkers, however, such as CRP and leukocytes, can also be helpful, especially in clinical settings where diagnostic capabilities are limited or time is crucial. For these reasons, a clinical application of these biomarkers should not be neglected, at least in some cases.

DISCUSSION

Reports in the literature have described the progressive increase in hospitalizations due to AP in recent years. Currently, there are no epidemiological data regarding the incidence of AP in Poland. Annual observation of AP cases enables highly reliable verification of the diagnosis and classification of AP severity according to the modified Atlanta classification. Results of our studies indicate that the AP incidence rate in the Kielce region is significantly higher (99.9 per 100,000) than reported in other countries. The high percentage of unexplained causes of AP reported in the present study is likely related to insufficient medical history taking concerning alcohol abuse, and inadequate diagnosis of microcholecystitis in some patients. A high incidence rate was also recorded for first episodes of AP (79.7 per 100,000) (3).

Recently, a revision of the Atlanta classification was published (4). The revision was prompted by advances in knowledge concerning the pathophysiology of the disease and improvement in imaging diagnostics. Due to the lack of uniform criteria for defining the severe form of AP used in the literature to date, it is difficult to fully evaluate the actual sensitivity of the markers applied in prognosticating the course of the disease. In the present study, we attempted to evaluate the utility of selected scales for prognosticating severe and moderate courses of AP (according to the revised Atlanta classification) and risk for death. To date, many reports have been published evaluating the effectiveness of the prognostic scales applied to the assessment of the risk of severe AP and death. However, there are few studies evaluating the utility of the scales in prognosticating moderate AP. The difficulty in using the scales is associated mainly with the complexity of the instruments applied. More than 80% of cases of AP have a mild course; therefore, simple indexes that enable rapid selection of patients for whom there is no need to perform costly and complicated procedures are needed. One such scale is the harmless acute pancreatitis score, which uses hematocrit and creatinine levels, and the presence or absence of peritoneal symptoms (9). In the present study, APACHE II scores ≥8 were successful in prognosticating severe AP and death; however, the CTSI index proved to be better at prognosticating a moderate course (AUC 0.819 [95% CI 0.767 to 0.871]). According to the current definition, moderate AP is characterized by transient organ failure and local and/or systemic complications. Until recently, these patients have most frequently been classified as having severe pancreatitis (11). However, recent insight into the pathophysiology AP, as well as the development of modern imaging techniques, has enabled standardized reporting on research and facilitated communication among clinicians (4). Furthermore, updated classification now recognizes two phases of AP (early and late) and has redefined the mild, moderate and severe severity criteria (4). In the first phase, the intensity of AP is related to multiple organ dysfunction, secondary to the inflammatory response, and not necessarily pancreatic necrosis. The scope of morphological changes in the pancreas and the peripancreatic tissues is not always proportional to the severity of the clinical course with or without organ dysfunction (12). The presence of necrosis indicates the risk of infection, but does not necessarily identify patients with a severe course of the disease (13). The present study confirms that CTSI is a good index for prognosticating local complications, but does not always prognosticate a severe course of the disease, and death.

The study confirmed previous observations that the use of APACHE II scale, even at the lowest cut-off point (>6), may be a good instrument for prognosticating organ failure and death (14). It is noteworthy that the use of APACHE II scale requires the evaluation of many variables, which considerably limits its usability. In addition, similar to other studies, its low sensitivity in the detection of local complications was confirmed (15).

The present study confirmed that, compared with APACHE II, the BISAP scale is easier to use, with similar sensitivity and specificity in prognosticating severe and moderate AP, and risk of death. Comprehensive population studies of early predictors of mortality due to AP did not show any statistically significant differences in predictive validity between BISAP and APACHE II (16) and, in the present study, AUCs for BISAP and APACHE II in prognosticating the risk of death were similar (0.707 versus 0.726). Recent Chinese studies (17,18) indicated the BISAP scale is valid instrument in prognosticating severe AP. Ease of data collection facilitates use of the BISAP in any hospital, for all patients. Increased mortality seen in patients who score high on the the APACHE II scale (scores eight times higher than the total population) may be attributed to selection of severely ill patients, in whom the use of APACHE II is more prevalent (16). This observation was also made in the present study. BISAP predicts severity in relation to immune response to damage or older age, while the Ranson scoring system seems to be more precise in predicting persistent organ dysfunction within 48 h (19). For the Ranson criteria, ≤2 positive variables predicted approximately 4% mortality, three or four positive variables predicted 10%, five or six variables predicted 36%, and seven or eight variables predicted >50% mortality (20). The present study confirmed a good predictive value of the Ranson scale in prognosticating severe AP and mortality; however, it was insufficient for predicting moderate AP (AUC 0.625; AUC 0.678 versus AUC 0.480). Previously research (21) has shown that prognosis of AP, based on Ranson and APACHE II scales, was burdened with a significant error. Most patients who developed severe AP were initially diagnosed with mild AP (21).

In the present study, sensitivity and specificity were evaluated on a simple scale (the Panc 3) that uses routine clinical parameters such as hematocrit, BMI, and the presence or absence of pleural effusion on radiological examination. The Panc 3 score is a sensitive instrument for prognosticating severe and moderate AP; however, no significant relationship was confirmed with prognosticating mortality. Considerably better results were obtained by the authors of the method, who confirmed that Panc 3 score prognosticates a severe course in 98% (7). A biomarker that may prognositicate severity of AP is CRP, which reaches maximum concentration in blood after 72 h.

However, there is no agreement concerning its cut-off value. Patients with CRP values <110 mg/L are at low risk for development of necrosis (22). Determination of the concentration of CRP in combination with multivariate prediction systems increases the effectiveness of diagnosing the disease. In the present study, CRP >150 mg/dL was correlated with mild pancreatitis. The white blood cell count is a simple test performed in patients with AP; in the present study, leukocytes $>13\times10^9$/L was statistically significantly predictive of severe and moderate disease, and death. Several clinical studies confirmed the role of cytokines and leukocytes in predicting the development and course of AP (23,24).

CONCLUSIONS

Our results, for the most part, confirm earlier observations concerning the effectiveness of the scales applied in prognosticating the severity of AP. A newly implemented revision in the classification of AP (4) now includes a 'moderate' course of the disease, in addition to mild and severe, thus necessitating the need for scales that distinguish among the three categories. It is suggested that there is a demand for studies of prognostic factors that take into acount the new classification (25).

It should be stated that none of the scales examined in the present study is a sufficiently valid instrument in prognosticating the course of AP. A higher sensitivity and specificity is usually related to scales that are used to evaluate more clinically severe patients, and require difficult-to-collect variables (eg, APACHE II); hence the necessity to seek new, simple and more sensitive prognostic indexes.

SUMMARY

Scoring systems used in the prognosis of the course of the disease vary with regard to sensitivity and specificity. The CTSI scoring system showed the highest precision in prognosticating moderately severe AP (Atlanta 2012); however, in prognosticating a severe course of the disease and mortality, APACHE II proved to possess the greatest predictive value.

AUTHOR CONTRIBUTIONS: D Koziel and S Gluszek designed the study; J Matykiewicz and P Lewitowicz performed the research; S Głuszek and D Kozieł analyzed the data; S Głuszek and D Kozieł wrote the manuscript; Z Drozdzak performed the statistical analysis.

REFERENCES

1. <https://prog.nfz.gov.pl/APP-JGP/Start.aspx> (Accessed September 9, 2012).
2. <http://www.statystyka.medstat.waw.pl/wyniki/TabelaEurostat2010szac.htm> (Accessed September 9, 2012).
3. Gluszek S, Koziel D. Prevalence and progression of acute pancreatitis in the Świętokrzyskie Voivodeship population. Pol Przegl Chir 2012;84:618-25.
4. Banks PA, Bollen TL, Dervenis C, et al. Acute Pancreatitis Classification Working Group. Classification of acute pancreatitis – 2012: Revision of the Atlanta classification and definitions by international consensus. Gut 2013;62:102-11.
5. Bollen TL, VK Singh, Maurer R, et al. A comparative evaluation of radiologic and clinical scoring systems in the early prediction of severity in acute pancreatitis. Am J Gastroenterol 2012;107:612-9.
6. Talukdar R, Clemens M, Vege SS. Moderately severe acute pancreatitis: Prospective validation of this new subgroup of acute pancreatitis. Pancreas 2012;41:306-9.
7. Brown A, James-Stevenson T, Dyson T, et al. The Panc 3 score: A rapid and accurate test for predicting severity on presentation in acute pancreatitis. J Clin Gastroenterol 2007;41:855-8.
8. Oskarsson V, Mehrabi M, Orsini N, et al. Validation of the harmless acute pancreatitis score in predicting nonsevere course of acute pancreatitis. Pancreatology 2011;11:464-8.
9. Lankisch PG, Weber-Dany B, Hebel K, et al. The harmless acute pancreatitis score: A clinical algorithm for rapid initial stratification of nonsevere disease. Clin Gastroenterol Hepatol 2009;7:702-5.
10. Gompertz M, Fernández L, Lara I, et al. Bedside index for severity in acute pancreatitis (BISAP) score as predictor of clinical outcome in acute pancreatitis: Retrospective review of 128 patients. Rev Med Chil 2012;140:977-83.
11. Vege SS, Gardner TB, Chari ST, et al. Low mortality and high morbidity in severe acute pancreatitis without organ failure: A case for revising the Atlanta classification to include "moderately severe acute pancreatitis". Am J Gastroenterol 2009;104:710-5.
12. Thoeni RF. The revised Atlanta classification of acute pancreatitis: Its importance for the radiologist and its effect on treatment. Radiology 2012;262:751-64.
13. Uomo G. Do we really need a new category of severity for patients with acute pancreatitis? JOP 2009;10:583-4.
14. Johnson CD, Abu-Hilal M. Persistent organ failure during the first week as a marker of fatal outcome in acute pancreatitis. Gut 2004;53:1340-4.
15. Gürleyik G, Emir S, Kiliçoglu G. Computed tomography severity index, APACHE II score, and serum CRP concentration for predicting the severity of acute pancreatitis. JOP 2005;6:562-7.
16. Wu BU, Johannes RS, Sun X, et al. The early prediction of mortality in acute pancreatitis: A large population-based study. Gut 2008;57:1698-703.
17. Chen L, Lu G, Zhou Q, et al. Evaluation of the BISAP score in predicting severity and prognoses of acute pancreatitis in Chinese patients. Int Surg 2013;98:6-12.
18. Wang AJ, Xu S, Hong JB, et al. The comparison of different clinical scoring systems for predicting prognosis in acute pancreatitis based on the revised Atlanta classification. Zhonghua Nei Ke Za Zhi 2013;52:668-71.
19. Papachristou GI, Muddana V, Yadav D, et al. Comparison of BISAP, Ranson's, APACHE-II, and CTSI scores in predicting organ failure, complications, and mortality in acute pancreatitis. Am J Gastroenterol 2010;105:435-41.
20. Spitzer AL, Barcia AM, Schell MT, et al. Applying Ockham's razor to pancreatitis prognostication: A four-variable predictive model. Ann Surg 2006;243:380-8.
21. Głuszek S, Matykiewicz D, Dudek A. Complications and prognosis in severe acute pancreatitis. Pol Przegl Chir 2006;78:139-50.
22. Bezmarević M, Kostić Z, Jovanović M et al. Procalcitonin and BISAP score versus C-reactive protein and APACHE II score in early assessment of severity and outcome of acute pancreatitis. Vojnosanit Pregl 2012;69:425-31.
23. Granger J, Remick D. Acute pancreatitis: Models, markers, and mediators. Shock 2005;(24 Suppl 1):45-51.
24. Moreno-Osset E, López A, de la Cueva L, et al. 99mTc-hexamethylpropylene amineoxime leukocyte scintigraphy in acute pancreatitis: An alternative to contrast-enhanced computed tomography? Am J Gastroenterol 2005;100:153-61.
25. Kapoor K, Banks PA. Early prognostic evaluation of acute pancreatitis: An on-going challenge. JOP 2013;14:109-11.

Biomarkers as potential treatment targets in inflammatory bowel disease: A systematic review

Travis B Murdoch MD MSc[1*] Sarah O'Donnell MD[2*], Mark S Silverberg MD PhD[2] Remo Panaccione MD[1]

TB Murdoch, S O'Donnell, MS Silverberg, R Panaccione. Biomarkers as potential treatment targets in inflammatory bowel disease: A systematic review. Can J Gastroenterol Hepatol 2015;29(4):203-208.

There is increasing interest in the concept of 'treat-to-target' in inflammatory bowel disease as a mechanism to standardize management and prevent complications. While clinical, radiographic and endoscopic treatment end points will figure prominently in this promising management paradigm, the role that noninvasive biomarkers will play is currently undefined. The goal of the present systematic review was to investigate the potential value of biomarkers as treatment targets in inflammatory bowel disease, with particular focus on those best studied: serum C-reactive protein (CRP) and fecal calprotectin. In Crohn disease, elevated CRP levels at baseline predict response to anti-tumour necrosis factor agents, and normalization is usually associated with clinical and endoscopic remission. CRP and hemoglobin levels can be used to help predict clinical relapse in the context of withdrawal of therapy. Ultimately, the authors conclude that currently available biomarkers should not be used as treatment targets in inflammatory bowel disease because they have inadequate operational characteristics to make them safe surrogates for clinical, endoscopic and radiographic evaluation. However, CRP and fecal calprotectin are important adjunctive measures that help alert the clinician to pursue further investigation.

Key Words: *Anti-tumour necrosis factor therapy; Biomarker; C-reactive protein; Crohn disease; Fecal calprotectin; Ulcerative colitis*

Les biomarqueurs comme cibles thérapeutiques potentielles en cas de maladies inflammatoire de l'intestin : une analyse systématique

On constate un intérêt croissant pour le concept de « traitement ciblé » (*treat-to-target*) en cas de maladies inflammatoires de l'intestin, utilisé comme mécanisme pour standardiser la prise en charge et prévenir les complications. Les indicateurs de traitement clinique, radiographique et endoscopique occupent une place de choix dans ce paradigme de prise en charge prometteur, mais le rôle des biomarqueurs non invasifs n'est toujours pas défini. La présente analyse systématique visait à examiner la valeur potentielle des biomarqueurs comme cibles thérapeutiques en cas de maladies inflammatoires de l'intestin, notamment celles qui sont les mieux étudiées, soit la protéine C réactive (PCR) sérique et la calpropectine fécale. Dans le cas de la maladie de Crohn, des taux de PCR élevés en début d'étude sont indicateurs d'une réponse aux agents des inhibiteurs du facteur de nécrose tumorale, et leur normalisation s'associe généralement à une rémission clinique et endoscopique. La PCR et les taux d'hémoglobine peuvent contribuer à prédire les récidives cliniques lors du sevrage thérapeutique. En définitive, les auteurs concluent que les biomarqueurs actuellement offerts ne devraient pas être utilisés comme cibles thérapeutiques en cas de maladies inflammatoires de l'intestin parce qu'en raison de leurs caractéristiques opérationnelles, ils ne peuvent pas se substituer en toute sécurité à l'évaluation clinique, endoscopique et radiographique. Cependant, la PCR et la calpropectine fécale sont des mesures complémentaires importantes qui indiquent au clinicien d'approfondir les examens.

The inflammatory bowel diseases (IBD), encompassing Crohn disease (CD) and ulcerative colitis (UC), remain a challenging group of conditions to treat. Despite many therapeutic modalities, there is uncertainty in terms of proper treatment end points. Endoscopic evaluation, combined with diagnostic imaging modalities, remain important methods to quantify the severity of disease and, by corollary, to determine response and remission following initiation of medical treatment. There is increasing recognition and uptake of the 'treat-to-target' paradigm – a concept borrowed from the rheumatology literature (1) – in guiding the management of IBD.

Biomarkers are defined as measureable substances derived from a biofluid or tissue specimens. This broad definition can include common laboratory work, such as hemoglobin concentration, or even molecular systems biology techniques, such as gene-expression profiling (eg, on peripheral blood cells). They are becoming especially important as less-invasive, cost-effective and resource-saving modalities to determine therapeutic response in IBD are developed. The aim of the present systematic review was to investigate the potential value of biomarkers as treatment targets in IBD, with discussion of their role in monitoring response to therapy and predicting relapse. We focus on assays that are widely available and used in current practice, especially serum C-reactive protein (CRP) and fecal calprotectin (fcalpro), with briefer mention of other assays (see Table 1 for summary of evidence).

METHODS

The present systematic review was performed under the aegis of the Selecting Therapeutic Targets in Inflammatory Bowel Disease (STRIDE) program, sponsored by the International Organization for the Study of Inflammatory Bowel Diseases. The systematic search strategy aimed to answer the question regarding the role of biomarkers as potential treatment targets in IBD. The initial search revealed 228 references that were manually screened by two of the study authors (SOD and TM). The search strategy used is presented in Appendix 1, and 50 relevant publications were found. Reference lists of studies and relevant review articles were reviewed. The search was limited to studies published before January 1, 2014.

SERUM BIOMARKERS

Summary

In CD and UC, CRP should not be used alone as a treatment target because it has inadequate operational characteristics to act as a surrogate for endoscopic, radiographic or clinical end points. CRP is a noninvasive adjunctive measure that can be used in both CD and UC to guide the need for further endoscopic or radiographic evaluation. Failure of CRP normalization following therapy initiation should prompt further endoscopic and/or radiographic evaluation, irrespective of symptoms. Other biomarkers, such as erythrocyte sedimentation rate (ESR), have inadequate evidence to support their use.

Authors who contributed equally to this work

[1]*Inflammatory Bowel Disease Clinic, Division of Gastroenterology, University of Calgary, Calgary, Alberta;* [2]*Zane Cohen Centre for Digestive Diseases, Mount Sinai Hospital, Toronto, Ontario*

Correspondence: Dr Remo Panaccione, Inflammatory Bowel Disease Clinic, Division of Gastroenterology, University of Calgary, 6th Floor, Teaching, Research and Wellness Building, 3280 Hospital Drive Northwest, Calgary, Alberta T2N 4Z6.e-mail rpanacci@ucalgary.ca

TABLE 1
Summary and evidence level of inflammatory bowel disease biomarkers

Blood biomarkers	Evidence level
Crohn disease	
There is a correlation between endoscopic but not histological disease activity and CRP levels in Crohn disease (6)	3b
A decrease in CRP is seen in anti-TNF therapy concomitant with increased response/remission to therapy in children and adults (10-14)	1b
Maintenance therapy with an anti-TNF agent is associated with continued normalization of CRP (12)	1b
Baseline elevation in CRP predicts anti-TNF responsiveness (8,9,17)	2b
Among responders to anti-TNF induction, normalization of CRP at week 14 is associated with maintenance of response (17)	3b
Elevation in CRP >5 mg/L, hemoglobin level ≤145 g/L, and leukocyte count >6x10^9/L predict relapse in patients on combination therapy in whom anti-TNF therapy is discontinued (25)	2b
Elevation of CRP >20 mg/L and hemoglobin level <120 g/L are predictors of relapse in whom azathioprine is withdrawn (22,24)	2b
Ulcerative colitis	
CRP is associated with active disease on endoscopy (7,37)	2b
Elevated CRP, low albumin and low hemoglobin levels are predictive of colectomy in acute severe ulcerative colitis (31-36)	2b
Few randomized controlled trials in ulcerative colitis have biomarkers as end points. Small studies have found that clinical response to therapy correlates with a reduction in CRP level (26,28,29)	1b
Fecal biomarkers	
Crohn disease	
Fcalpro is positively associated with endoscopic activity and Crohn disease activity index (6,43)	3b
Change in level correlates with change in endoscopic activity (47,48)	3b
Elevated fcalpro is associated with a higher one-year risk of relapse in Crohn disease (51,64)	2b
Ulcerative colitis	
Fcalpro correlates with endoscopic disease and healing (46,52,53)	2b
In trials of vedolizumab, changes in fcalpro correlate with Mayo score (22)	1b
Normalization of fcalpro predicts clinical response to therapy and sustained remission (51,53,54)	2a
In patients on maintenance infliximab, fcalpro was better than CRP at predicting relapse.	1b

CRP C-reactive protein; Fcalpro Fecal calprotectin; TNF Tumour necrosis factor

In CD, elevated CRP levels at baseline predict response to anti-tumour necrosis factor (TNF) agents, and normalization of CRP is modestly associated with clinical and endoscopic remission. CRP and hemoglobin levels can be used to help predict clinical relapse in the context of withdrawal of therapy.

In UC, the utility of CRP as an adjunctive measure is greatest in severe disease, in which persistent elevation, in addition to low albumin levels, is associated with the need for colectomy. CRP normalization is modestly correlated with clinical remission and mucosal healing.

Discussion

CRP: CRP is a 224-residue protein synthesized by the liver and named for its ability to precipitate the C-polysaccharide of *Streptococcus pneumoniae* (2). It was the first acute-phase reactant to be described, and is known to be elevated in most acute-phase responses, including infections, inflammatory diseases, trauma and malignancy. It is believed to be more accurate than other acute-phase reactants, such as ESR, which is susceptible to diurnal and nutritional variation (2), and lacks data supporting its use. Given these characteristics, as well as its popularization as a cardiovascular screening tool, CRP has become the acute-phase biomarker of choice for a broad range of diseases. CRP is of limited utility in some individuals because common polymorphisms are known to affect CRP levels and prevent an inflammation-induced rise in CRP (3,4). As such, documenting an elevated CRP level in the context of a flare is an important baseline to instruct the future utility of this assay.

CRP in CD: In CD, there is evidence that extraenteric inflammation according to computed tomography enterography correlates with CRP (5), and CRP is more often elevated in individuals with colonic involvement. There is modest correlation between endoscopic disease activity and CRP levels in CD (6,7).

The majority of randomized clinical trials (RCTs) in the past decade have included serial CRP measurements. Recent trials of biologic and nonbiologic therapy have increasingly used baseline elevated CRP level as an inclusion criterion to insure patients entered into these trials have an objective marker of inflammation. This practice largely stems from results of the phase II induction studies of certolizumab published in the mid-2000s, in which patients with elevated CRP levels were shown to have more robust response rates (8,9). A phase II study by Schreiber et al (9) demonstrated a high placebo response rate in subgroups of patients with low CRP and, in post hoc analysis, a significant effect of therapy only in the subgroup of patients with baseline CRP >10 mg/L. A further certolizumab induction trial by Sandborn et al (8) (Pegylated Antibody Fragment Evaluation in Crohn's Disease: Safety and Efficacy 1 [PRECISE 1]) stratified randomization of patients according to baseline CRP (≥ or <10 mg/L), and found a significant response among patients with an elevated CRP level. Both studies demonstrated a durable decrease in CRP among patients treated with 400 mg dosing of certolizumab over the study period.

Although these were the some of the first studies to demonstrate the role of CRP as an inclusion criterion, early trials of anti-TNF agents included serial measurement of CRP (10-12). In the original infliximab induction trial published by Targan et al (10), there was a significant decrease in CRP level among patients treated with infliximab (monoclonal cA2) at four weeks (eg, 16 mg/L in those treated with 5 mg/kg infliximab versus 2 mg/kg in placebo). Normal serum CRP levels were maintained among patients who received maintenance therapy with infliximab (12). The pediatric infliximab literature similarly suggests a decrease in CRP in association with infliximab therapy in CD (13). The Clinical Assessment of Adalimumab Safety and Efficacy Studied as Induction Therapy in Crohn's Disease (CLASSIC-1) trial, which explored induction with adalimumab in CD, reproduced the infliximab findings by showing a durable reduction in CRP over four weeks in patients treated with anti-TNF therapy (14).

Clinical trials exploring combination therapy have similarly included CRP as an outcome. The study by van Assche et al (15), which explored the discontinuation of immunosuppressants in CD patients on infliximab at six months, was particularly revealing as to the potential of this biomarker. The study authors demonstrated a significant increase in CRP in patients discontinuing immunosuppressants, although there was no difference in their primary outcome of patients requiring modification in infliximab dosing. Post hoc analysis demonstrated a correlation between CRP and serum trough levels of infliximab, and low trough levels were associated with higher median CRP levels. Analysis of data from the Study of Biologic and Immunomodulator Naive Patients in Crohn's Disease (SONIC) trial also demonstrated the best therapeutic results among patients with an elevated CRP level (16), as reported in monotherapy trials of anti-TNF agents.

Collectively, these trials raise several important questions. First, a majority of anti-TNF trials have included CRP measurement, and show significant decreases in CRP in the trial arm with higher rates of response and remission. However, little post hoc analysis has been performed to explore correlations between clinical and endoscopic end points of remission and CRP. Some of the best evidence comes from analysis of A Crohn's Disease Clinical Trial Evaluating Infliximab in a New Long-term Treatment regimen (ACCENT I) data by Reinisch et al (17). The central hypothesis was that CRP levels at baseline and 14 weeks are predictive of durable response/ remission, in this case between 14 weeks to one year after induction with infliximab. As shown in other trials, their results suggested an association between high baseline CRP and maintenance of remission in CD treated with infliximab. They also demonstrated that in responders, CRP <5 mg/L at week 14 was associated with maintenance of response (56.6% of patients versus 37% of patients with a CRP ≥5 mg/L) and, furthermore, that a CRP level drop from >5 mg/L to <5 mg/L (ie, normalization) was associated with significant probability of remaining in remission (55% versus 36%). Nonetheless, the ROC curve of baseline CRP as a predictor of response after 14 weeks suggests no optimal cut-off point and, thus, it is difficult to make firm clinical suggestions about the role of CRP in effecting a therapeutic change at 14 weeks.

A second question that arises about CRP from CD trials of anti-TNF agents is whether CRP is a general measure of systemic inflammation, versus a specific predictor of a disease phenotype that is responsive to anti-TNF agents. Recent investigation of novel biologic agents that target different inflammatory pathways, including the interleukin 23/Th17 axis and leukocyte adhesion, sheds further light on this question. A study assessing the use of ustekinumab for induction and maintenance therapy in refractory Crohn's disease explored use of ustekinumab, an anti-interleukin 12/23 p40 agent, in induction and maintenance of CD resistant to anti-TNF treatment (18). There was a significant reduction in mean CRP levels in patients receiving 6 mg/kg of ustekinumab induction versus placebo; reductions were sustained only in those receiving ustekinumab maintenance therapy. One reason for choosing anti-TNF-experienced patients in this trial was the demonstration of greater effect on CRP reduction in this subgroup (19). Further consistent with the hypothesis that CRP is a therapy-independent biomarker of response to therapy is the demonstration of changes in CRP in response to anti-integrin therapy. Studies suggest a significant decrease in CRP level in patients treated with natalizumab, among those with an elevated baseline CRP (20,21). This finding was not reproduced in the recent trial of vedolizumab (22).

In addition to biologic trials, thiopurine trials have confirmed CRP as a key biomarker predictive of relapse. Two studies exploring thiopurine withdrawal (23,24) demonstrated an elevated CRP level to be an independent predictor of relapse after withdrawal of azathioprine, in addition to a low hemoglobin level. Among patients treated with combination therapy, in whom infliximab is discontinued, CRP is also a predictor of relapse (25). However, the CRP cut-off used in these trials varied, between 5 mg/mL and 20 mg/L.

CRP in UC: Few RCTs assessing medical therapy in active UC have used CRP as an end point. An early multicentre prospective RCT designed to assess the efficacy, safety and tolerability of infliximab used CRP as a secondary end point (26). Eleven UC patients with severe, steroid-refractory disease were treated with a single dose of infliximab at 5 mg/kg, 10 mg/kg or 20 mg/kg, and the authors reported a decrease in CRP associated with clinical response. While >50% of patients enrolled to the Active Ulcerative Colitis (ACT) 1 and ACT 2 trials had an elevated CRP at enrollment, changes were not reported as a marker of response to treatment (27).

Other, smaller studies have included CRP measurement. A head-to-head trial of intravenous heparin and steroid therapy for moderate and severe UC in hospitalized patients included a secondary analysis focusing on changes in serological inflammatory indexes (28). In the first 10 days of treatment, a significant decrease in the CRP levels was observed in the steroid group, with no change in the heparin group. None of the heparin-treated patients showed an improvement in disease activity. An open-label randomized study involving 30 pediatric patients, comparing oral beclomethasone for eight weeks followed by oral mesalazine with oral mesalazine alone (29), demonstrated decreased ESR in both treatments arms by week 12; CRP level only dropped significantly in the beclomethasone group. There was concomitant reduction in clinical activity in the beclomethasone group.

The role of CRP in UC patients receiving anti-integrin therapy remains unclear. In patients receiving a single infusion of natalizumab, there was reduction in median CRP level two weeks postinfusion (30). However, the more recent induction and maintenance trials of vedolizumab did not include CRP level as an outcome (22).

In severe UC, baseline CRP elevation is a predictor of failure of medical therapy (31-33). A study involving 72 patients with steroid-refractory UC patients treated with oral or intravenous cyclosporin demonstrated a significant association of colectomy at three months with baseline CRP (11.8 mg/L in the colectomy group versus 5.5 mg/L) (34). However, a multivariate analysis found only the Ho index to be an independent predictor of colectomy, or treatment failure. Another retrospective chart review involving 135 patients with steroid-refractory UC treated with cyclosporine (35) found a CRP >45 mg/L to be predictive for colectomy. The Oxford score, designed following a prospective study involving hospitalized patients with acute severe UC, also includes a CRP >45 mg/L in patients with four to eight bowel movements per day as a predictor of colectomy (36). Elevated CRP level has been associated specifically with severe clinical activity, anemia and hypoalbuminemia, in addition to active disease on endoscopy (7,37).

Clinical disease activity scores have generally been used as primary end points in UC RCTs. Some of these clinical measures, such as the Truelove & Witts' criteria, Seo and Ho indexes, integrate biomarkers such as albumin, hemoglobin, ESR and CRP into their scoring. However, associations between biomarkers and response to therapy are generally not reported in RCTs.

Other peripheral blood biomarkers: Several other common laboratory tests have potential applicability in monitoring IBD, although they are not as well studied as CRP. Simple values derived from the complete blood count, including hemoglobin level and leukocyte counts, have demonstrated some utility in this context (23,25). In the aforementioned study by Louis et al (25), a hemoglobin level ≤145 g/L and a leukocyte count of $>6\times10^9$/L were both associated with decreased time-to-relapse in CD patients in whom infliximab therapy was stopped (HR 6.0 and 2.4, respectively). Interestingly, the HR was greater for hemoglobin than any other measured factor. Similarly, a study of azathioprine withdrawal found an association between a hemoglobin level <120 g/L and risk of relapse (23). Hypoalbuminemia and anemia are both associated with failure of medical therapy in severe ulcerative colitis (31). Nonetheless, there remains concern as to the sensitivity and specificity of these tests as biomarkers in IBD, given their variability in many disease states.

FECAL BIOMARKERS

Summary

In CD and UC, fcalpro should not be used alone as a treatment target because it has inadequate operational characteristics to act as a surrogate for endoscopic, radiographic or clinical end points. Fcalpro is a noninvasive adjunctive measure that can be used in both CD and UC to guide need for further endoscopic or radiographic evaluation.

In CD, fcalpro may especially useful in Crohn's colitis as a biomarker of response to therapy and to predict relapse. In UC, fcalpro likely has greater utility than CRP as an adjunctive measure to guide need for further investigation. Fcalpro has been shown to correlate with mucosal healing and response to induction therapy, and is predictive of loss of response to maintenance therapy in UC.

Discussion

Fcalpro: While blood-based biomarkers offer insight into systemic inflammation occurring as part of active IBD, fecal assays have the potential advantage of better reflecting inflammation at the mucosal level. Several different assays have been developed, most prominently those using calprotectin and lactoferrin as biomarkers. Both of these tests leverage the fact that actively inflamed bowel contains a large number of neutrophils and, furthermore, that the mucosal defects that occur during active inflammation result in spillover of neutrophils into stool. Calprotectin has become a more widely used assay, both because of better operational characteristics and the increasing availability of a rapid test.

Calprotectin is a calcium- and zinc-binding neutrophilic cystolic protein. It is an appropriate marker for mucosal inflammation, given that it is evenly distributed and stable in stool for up to one week (38,39). This protein can be measured using commercially available ELISA or more recently developed quantitative rapid tests (40), although there are data suggesting that quantitation is less accurate with a rapid test (PreventID Caldetect, Preventis GmbH, Germany) at values >15 μg/g (40). A major interest lies in the use of fcalpro to distinguish IBD from non-IBD (41), with studies reporting a sensitivity between 63% and 100%, and specificity of 48% to 100%, depending on trial design and fcalpro cut-off value (41). Several studies have demonstrated correlation between endoscopic/histological disease activity and fcalpro (42-45). This was first studied by Røseth et al (44) in patients with UC, who demonstrated increased correlation between endoscopic and histological grading of inflammation and fcalpro levels.

Fcalpro in CD: Recent work has suggested a correlation between fcalpro concentration and endoscopic disease activity in CD, and weaker associations with the CD activity index (6,46). Correlation between elevated fcalpro and active disease is strongest in Crohn's colitis (Spearman coefficient 0.80 versus 0.45 for all CD) (6). This finding has been reproduced in the pediatric population (43). There is evidence that supports use of fcalpro for monitoring response to anti-TNF therapy. Sipponen et al (47) demonstrated a drop in mean fcalpro levels after therapy with anti-TNF agents (from 1173 μg/g to 130 μg/g) (47); there was moderate correlation between change in fcalpro level and change in endoscopic activity using the CD Endoscopic Index of Severity (Spearman's rank-order correlation = 0.561), but not with ileal or colonic histological scores. Another study failed to demonstrate a significant change in fcalpro levels in patients who responded to medical therapy (48); however, both of these studies were limited by sample size (n=15 and n=11 CD patients, respectively). Although similar to CRP, fcalpro is now used as an objective measure of inflammation for inclusion in clinical trials of novel agents (49,50), few biologics trials report data regarding fcalpro in response to therapy.

There are data supporting the use of fcalpro to predict relapse in CD. In a prospective multicentre study, Gisbert et al (51) demonstrated that CD patients who relapsed within one year had significantly higher fcalpro levels at baseline. In this clinical context, fcalpro had a low sensitivity (28%) but was highly specific (93%). Collectively, it is difficult to suggest that fcalpro could replace endoscopic and clinical evaluation as a treatment target.

Fcalpro in UC: Fcalpro has been shown to correlate well with mucosal healing in ulcerative colitis (46,52,53). D'Haens et al (46) further reported that a fcalpro level >250 μg/g predicted active mucosal disease (sensitivity of 71% and specificity of 100%). Several studies have demonstrated that normalization of fcalpro predicts clinical response to medical therapies and sustained remission (51,53,54). One study (52) assessed the use of fcalpro as a marker of sustained remission in UC patients receiving maintenance infliximab. Of 87 patients included in the study, 30 (34.4%) were considered to be in sustained deep remission (defined as a partial Mayo score of 3 at all points and a Mayo endoscopic subscore 0 at week 52) and 13 (14.9%) to have relapsed. Fcalpro levels remained low (median 40 μg/g) throughout the study in those with sustained deep remission. Two consecutive fcalpro measurements of 300 μg/g at monthly intervals were identified as the best predictor of flare (61.5% sensitive and 100% specific). In that study, fcalpro level at time of relapse was significantly better than serum CRP concentration at predicting relapse. In a pediatric hospitalized severe UC population, clinical scores were most predictive of steroid nonresponse; CRP and fcalpro levels could predict steroid response, although neither were found to be predictive of infliximab response (33).

Until recently, most therapeutics trials investigating UC have not included fecal biomarkers as secondary end points. Two recent studies assessing vedolizumab in UC included fcalpro measurement (22,55). In the phase 3 trial in UC by Feagan et al (22), significant decreases in fcalpro were seen in the group receiving vedolizumab for induction and maintenance. An earlier vedolizumab dose-ranging study found decreases in the fcalpro corresponding with decrease in Mayo score; however, the sample size was inadequate to allow statistical analysis (55).

Lactoferrin: Lactoferrin is an iron-binding glycoprotein found in many body fluids, and is a major component of secondary granules of neutrophils (56). Similar to fcalpro, lactoferrin is very stable in feces over extended periods of time. However, some studies have suggested that fcalpro has better operational characteristics to distinguish IBD from non-IBD, and this, in part, has led to fcalpro becoming a more universally used and available biomarker (57,58).

A prospective study of UC patients in remission on mesalamine found fcalpro to be more sensitive and specific than fecal lactoferrin for predicting UC relapse (54). A cut-off value of 170 μg/g for fcalpro had a sensitivity of 76% and a specificity of 76% to predict relapse, while a cut-off value of 140 μg/g for lactoferrin had a sensitivity of 67% and a specificity of 68%. Others have shown similar results (51). There are data supporting the utility of lactoferrin for monitoring response to therapy (59); however, a major challenge remains significant overlap in fecal lactoferrin concentrations in patients with active and quiescent IBD (56).

FUTURE BIOMARKERS

With the advent of high-throughput systems and biology approaches, such as genomics, metagenomics and transcriptomics, it is likely that future biomarkers will result from integration of multiple metrics derived using these techniques. Gene expression profiling has found some utility in IBD; transcriptional profiles from biopsies of pediatric patients with severe UC predicts responsiveness to steroids (60). Polymorphisms in *FOXO3* have been associated with more severe CD, despite not being associated with disease development (61).

In other immune-mediated diseases, systems biology approaches have directly impacted clinical care. For instance, gene expression profiling of peripheral blood in heart transplant patients has great potential in monitoring for rejection (62). Proteomic profiling of blood plasmacytoid dendritic cells in systemic sclerosis patients demonstrated an association between CXCL4 levels and severity as well as pulmonary involvement (63). In the case of IBD, future studies should expand on the potential use of bioinformatics platforms to develop biomarkers for use as surrogates of endoscopic outcomes.

DISCLOSURES: The authors have no financial disclosures or conflicts of interest to declare.

CONCLUSION

As was suggested by consensus of the STRIDE program, currently available serum and fecal biomarkers do not possess adequate operational characteristics to make them stand-alone treatment targets in IBD; however, they should be considered important adjunctive measures to clinical, endoscopic and radiographic assessment. The present review demonstrated the continued value that biomarkers provide in alerting the clinician to a need for further investigation. As the strategy of 'treat-to-target' becomes more widespread in IBD management, it becomes crucial to define noninvasive measures that reliably reflect clinical and endoscopic disease activity. The development of more accurate biomarkers in IBD remains an important goal.

APPENDIX 1

The following search strategy was used on PubMed, with filters for only human studies, and only English results.
("calprotectin"[Title/Abstract] OR "lactoferrin"[Title/Abstract] OR "c-reactive protein"[Title/Abstract] OR "CRP"[Title/Abstract] OR "erythrocyte sedimentation rate"[Title/Abstract] OR "ESR"[Title/Abstract]OR "hemoglobin"[Title/Abstract] OR "bone mineral density"[Title/Abstract] OR "ferritin"[Title/Abstract] OR "erythrocyte sedimentation rate"[Title/Abstract] OR "ESR"[Title/Abstract]) AND ("ulcerative colitis"[Title/Abstract] OR "crohn disease"[Title/Abstract] OR "crohn's disease"[Title/Abstract] OR "crohns disease"[Title/Abstract] OR "inflammatory bowel disease"[Title/Abstract] OR "IBD"[Title/Abstract]) AND ("response"[Title/Abstract] OR "remission"[Title/Abstract] OR "Flare"[Title/Abstract]) AND (Clinical Trial[ptyp] OR Review[ptyp])

REFERENCES

1. Solomon DH, Bitton A, Katz JN, Radner H, Brown EM, Fraenkel L. Review: Treat to target in rheumatoid arthritis: Fact, fiction, or hypothesis? Arthritis Rheumatol 2014;66:775-82.
2. Pepys MB, Hirschfield GM. C-reactive protein: A critical update. J Clin Invest 2003;111:1805-12.
3. Szalai AJ, McCrory MA, Cooper GS, Wu J, Kimberly RP. Association between baseline levels of C-reactive protein (CRP) and a dinucleotide repeat polymorphism in the intron of the CRP gene. Genes Immun 2002;3:14-9.
4. Carlson CS, Aldred SF, Lee PK, et al. Polymorphisms within the C-reactive protein (CRP) promoter region are associated with plasma CRP levels. Am J Hum Genet 2005;77:64-77.
5. Colombel JF, Solem CA, Sandborn WJ, et al. Quantitative measurement and visual assessment of ileal Crohn's disease activity by computed tomography enterography: Correlation with endoscopic severity and C reactive protein. Gut 2006;55:1561-7.
6. Jones J, Loftus EV, Panaccione R, Chen LS. Relationships between disease activity and serum and fecal biomarkers in patients with Crohn's Disease. Clin Gastroenterol Hepatol 2008;6:1218-24.
7. Solem CA, Loftus EV, Tremaine WJ, Harmsen WS, Zinsmeister AR, Sandborn WJ. Correlation of C-reactive protein with clinical, endoscopic, histologic, and radiographic activity in inflammatory bowel disease. Inflamm Bowel Dis 2005;11:707-12.
8. Sandborn WJ, Feagan BG, Stoinov S, et al. Certolizumab pegol for the treatment of Crohn's disease. N Engl J Med 2007;357:228-38.
9. Schreiber S, Rutgeerts P, Fedorak RN, et al. A randomized, placebo-controlled trial of certolizumab pegol (CDP870) for treatment of Crohn's disease. Gastroenterology 2005;129:807-18.
10. Targan SR, Hanauer SB, van Deventer SJ, et al. A short-term study of chimeric monoclonal antibody cA2 to tumor necrosis factor alpha for Crohn's disease. Crohn's Disease cA2 Study Group. N Engl J Med 1997;337:1029-35.
11. Stack WA, Mann SD, Roy AJ, et al. Randomised controlled trial of CDP571 antibody to tumour necrosis factor-alpha in Crohn's disease. Lancet 1997;349:521-4.
12. Rutgeerts P, D'Haens G, Targan S, et al. Efficacy and safety of retreatment with anti-tumor necrosis factor antibody (infliximab) to maintain remission in Crohn's disease. Gastroenterology 1999;117:761-9.
13. Lamireau T, Cézard J-P, Dabadie A, et al. Efficacy and tolerance of infliximab in children and adolescents with Crohn's disease. Inflamm Bowel Dis 2004;10:745-50.
14. Hanauer SB, Sandborn WJ, Rutgeerts P, et al. Human anti-tumor necrosis factor monoclonal antibody (adalimumab) in Crohn's disease: The CLASSIC-I trial. Gastroenterology 2006;130:323-3.
15. van Assche G, Magdelaine-Beuzelin C, D'Haens G, et al. Withdrawal of immunosuppression in Crohn's disease treated with scheduled infliximab maintenance: A randomized trial. Gastroenterology 2008;1341861-8.
16. Colombel JF, Sandborn WJ, Reinisch W, et al. Infliximab, azathioprine, or combination therapy for Crohn's disease. N Engl J Med 2010;362:1383–95.
17. Reinisch W, Wang Y, Oddens BJ, Link R. C-reactive protein, an indicator for maintained response or remission to infliximab in patients with Crohn's disease: A post-hoc analysis from ACCENT I. Aliment Pharmacol Ther 2012;35:568-76.
18. Sandborn WJ, Feagan BG, Fedorak RN, et al. A randomized trial of ustekinumab, a human interleukin-12/23 monoclonal antibody, in patients with moderate-to-severe Crohn's disease. Gastroenterology 2008;135:1130-41.
19. Toedter GP, Blank M, Lang Y, Chen D. Relationship of C-reactive protein with clinical response after therapy with ustekinumab in Crohn's disease. Am J Gastroenterol 2009;104:2768-73.
20. Ghosh S, Goldin E, Gordon FH, et al. Natalizumab for active Crohn's disease. N Engl J Med 2003;348:24-32.
21. MacDonald JK, McDonald JW. Natalizumab for induction of remission in Crohn's disease. Cochrane Database Syst Rev 2007;(1):CD006097.
22. Feagan BG, Rutgeerts P, Sands BE, et al. Vedolizumab as induction and maintenance therapy for ulcerative colitis. N Engl J Med 2013;369:699-710.
23. Treton X, Bouhnik Y, Mary J-Y, et al. Azathioprine withdrawal in patients with Crohn's disease maintained on prolonged remission: A high risk of relapse. Clin Gastroenterol Hepatol 2009;7:80-5.
24. Lémann M, Mary J-Y, Colombel J-F, et al. A randomized, double-blind, controlled withdrawal trial in Crohn's disease patients in long-term remission on azathioprine. Gastroenterology 2005;128:1812-8.
25. Louis E, Mary J-Y, Vernier-Massouille G, et al. Maintenance of remission among patients with Crohn's disease on antimetabolite therapy after infliximab therapy is stopped. Gastroenterology 2012;142:63-70.e5.
26. Sands BE, Tremaine WJ, Sandborn WJ, et al. Infliximab in the treatment of severe, steroid-refractory ulcerative colitis: A pilot study. Inflamm Bowel Dis 2001;7:83-8.
27. Rutgeerts P, Sandborn WJ, Feagan BG, et al. Infliximab for induction and maintenance therapy for ulcerative colitis. N Engl J Med 2005;353:2462-76.
28. Panes J, Esteve M, Cabre E, et al. Comparison of heparin and steroids in the treatment of moderate and severe ulcerative colitis. Gastroenterology 2000;119:903-8.
29. Romano C, Famiani A, Comito D, Rossi P, Raffa V, Fries W. Oral beclomethasone dipropionate in pediatric active ulcerative colitis: A comparison trial with mesalazine. J Pediatr Gastroenterol Nutr 2010;50:385-9.
30. Gordon FH, Hamilton MI, Donoghue S, et al. A pilot study of treatment of active ulcerative colitis with natalizumab, a humanized monoclonal antibody to alpha-4 integrin. Aliment Pharmacol Ther 2002;16:699-705.
31. Kumar S, Ghoshal UC, Aggarwal R, Saraswat VA, Choudhuri G. Severe ulcerative colitis: Prospective study of parameters determining outcome. J Gastroenterol Hepatol 2004;19:1247-52.
32. Henriksen M, Jahnsen J, Lygren I, et al. C-reactive protein: A predictive factor and marker of inflammation in inflammatory bowel disease. Results from a prospective population-based study. Gut 2008;57:1518-23.
33. Turner D, Mack D, Leleiko N, et al. Severe pediatric ulcerative colitis: A prospective multicenter study of outcomes and predictors of response. Gastroenterology 2010;138:2282-91.

34. Aceituno M, Garcia-Planella E, Heredia C, et al. Steroid-refractory ulcerative colitis: Predictive factors of response to cyclosporine and validation in an independent cohort. Inflamm Bowel Dis 2008;14:347-52.
35. Cacheux W, Seksik P, Lémann M, et al. Predictive factors of response to cyclosporine in steroid-refractory ulcerative colitis. Am J Gastroenterol 2008;103:637-42.
36. Travis SP, Farrant JM, Ricketts C, et al. Predicting outcome in severe ulcerative colitis. Gut 1996;38:905-10.
37. Ricanek P, Brackmann S, Perminow G, et al. Evaluation of disease activity in IBD at the time of diagnosis by the use of clinical, biochemical, and fecal markers. Scand J Gastroenterol 2011;46:1081-91.
38. Røseth AG, Fagerhol MK, Aadland E, Schjønsby H. Assessment of the neutrophil dominating protein calprotectin in feces. A methodologic study. Scand J Gastroenterol 1992 27:793-8.
39. Abraham BP, Kane S. Fecal markers: Calprotectin and lactoferrin. Gastroenterol Clin North Am 2012;41:483-95.
40. Vestergaard TA, Nielsen SL, Dahlerup JF, Hornung N. Fecal calprotectin: Assessment of a rapid test. Scand J Clin Lab Invest 2008;68:343-7.
41. Burri E, Beglinger C. Faecal calprotectin – a useful tool in the management of inflammatory bowel disease. Swiss Med Wkly 2012;142:w13557.
42. Konikoff MR, Denson LA. Role of fecal calprotectin as a biomarker of intestinal inflammation in inflammatory bowel disease. Inflamm Bowel Dis 2006;12:524-34.
43. Bunn SK, Bisset WM, Main MJ, Gray ES, Olson S, Golden BE. Fecal calprotectin: Validation as a noninvasive measure of bowel inflammation in childhood inflammatory bowel disease. J Pediatr Gastroenterol Nutr 2001;33:14-22.
44. Røseth AG, Aadland E, Jahnsen J, Raknerud N. Assessment of disease activity in ulcerative colitis by faecal calprotectin, a novel granulocyte marker protein. Digestion 1997;58:176-80.
45. Limburg PJ, Ahlquist DA, Sandborn WJ. Fecal calprotectin levels predict colorectal inflammation among patients with chronic diarrhea referred for colonoscopy. Am J Gastroenterol 2000;95:2831-7.
46. D'Haens G, Ferrante M, Vermeire S, et al. Fecal calprotectin is a surrogate marker for endoscopic lesions in inflammatory bowel disease. Inflamm Bowel Dis 2012;18:2218-24.
47. Sipponen T, Savilahti E, Kärkkäinen P, et al. Fecal calprotectin, lactoferrin, and endoscopic disease activity in monitoring anti-TNF-alpha therapy for Crohn's disease. Inflamm Bowel Dis 2008;14:1392-8.
48. Wagner M, Peterson CGB, Ridefelt P, Sangfelt P, Carlson M. Fecal markers of inflammation used as surrogate markers for treatment outcome in relapsing inflammatory bowel disease. World J Gastroenterol 2008;14:5584-9.
49. Sandborn WJ, Feagan BG, Rutgeerts P, et al. Vedolizumab as induction and maintenance therapy for Crohn's disease. N Engl J Med 2013;369:711-21.
50. Hueber W, Sands BE, Lewitzky S, et al. Secukinumab, a human anti-IL-17A monoclonal antibody, for moderate to severe Crohn's disease: Unexpected results of a randomised, double-blind placebo-controlled trial. Gut 2012;61:1693-700.
51. Gisbert JP, Bermejo F, Perez-Calle J-L, et al. Fecal calprotectin and lactoferrin for the prediction of inflammatory bowel disease relapse. Inflamm Bowel Dis 2009;15:1190-8.
52. De Vos M, Louis EJ, Jahnsen J, et al. Consecutive fecal calprotectin measurements to predict relapse in patients with ulcerative colitis receiving infliximab maintenance therapy. Inflamm Bowel Dis 2013;19:2111-7.
53. Molander P, af Björkesten C-G, Mustonen H, et al. Fecal calprotectin concentration predicts outcome in inflammatory bowel disease after induction therapy with TNFα blocking agents. Inflamm Bowel Dis 2012;18:2011-7.
54. Yamamoto T, Shiraki M, Bamba T, Umegae S, Matsumoto K. Fecal calprotectin and lactoferrin as predictors of relapse in patients with quiescent ulcerative colitis during maintenance therapy. Int J Colorectal Dis 2014;29:485-91.
55. Parikh A, Leach T, Wyant T, et al. Vedolizumab for the treatment of active ulcerative colitis: A randomized controlled phase 2 dose-ranging study. Inflamm Bowel Dis 2012;18:1470-9.
56. Gisbert JP, McNicholl AG, Gomollon F. Questions and answers on the role of fecal lactoferrin as a biological marker in inflammatory bowel disease. Inflamm Bowel Dis 2009;15:1746-54.
57. Silberer H, Küppers B, Mickisch O, et al. Fecal leukocyte proteins in inflammatory bowel disease and irritable bowel syndrome. Clin lab 2005; 51:117-26.
58. Schröder O, Naumann M, Shastri Y Prospective evaluation of faecal neutrophil-derived proteins in identifying intestinal inflammation: Combination of parameters does not improve diagnostic accuracy of calprotectin. Aliment Pharmacol Ther 2007;26:1035-42.
59. Turner D, Leach ST, Mack D, et al. Faecal calprotectin, lactoferrin, M2-pyruvate kinase and S100A12 in severe ulcerative colitis: A prospective multicentre comparison of predicting outcomes and monitoring response. Gut 2010;59:1207-12.
60. Kabakchiev B, Turner D, Hyams J, et al. Gene expression changes associated with resistance to intravenous corticosteroid therapy in children with severe ulcerative colitis. PLoS ONE 2010;5:e13085.
61. Lee JC, Espéli M, Anderson CA, et al. Human SNP links differential outcomes in inflammatory and infectious disease to a *FOXO3*-regulated pathway. Cell 2013;155:57-69.
62. Pham MX, Teuteberg JJ, Kfoury AG, et al. Gene-expression profiling for rejection surveillance after cardiac transplantation. N Engl J Med 2010;362:1890-900.
63. van Roon JAG, Tesselaar K, Radstake TRDJ. Proteome-wide analysis and CXCL4 in systemic sclerosis. N Engl J Med 2014;370:1563-4.
64. D'Inca R, Dal Pont E, Di Leo V, et al. Can calprotectin predict relapse risk in inflammatory bowel disease? Am J Gastroenterol 2008;103:2007-14.

Venous thromboprophylaxis in gastrointestinal bleeding

Neel Malhotra MD, Nilesh Chande MD FRCPC

N Malhotra, N Chande. Venous thromboprophylaxis in gastrointestinal bleeding. Can J Gastroenterol Hepatol 2015;29(3):145-148.

OBJECTIVE: To study the use of venous thromboembolism (VTE) prophylaxis and the incidence of thrombotic events in patients with acute gastrointestinal (GI) bleeding.

METHODS: Individuals admitted with a primary diagnosis of a GI bleed along with any endoscopically confirmed source (over a two-year period) were included. Patient comorbidity and data regarding anticoagulation or antiplatelet agent use before hospitalization were collected, in addition to type of VTE prophylaxis and duration of treatment. The primary end point was the development of VTE (deep vein thrombosis or pulmonary embolism) within one year of presentation.

RESULTS: Data from 504 patients admitted with GI bleeding were eligible for review. The total number of VTE events was 20 (4%) while the mortality rate during hospitalization was 4.6%; 397 patients were not given VTE prophylaxis during their hospitalization. Of the patients who were given VTE prophylaxis, 68 received prophylactic heparin or heparin derivatives during their admission. One hundred sixty-five patients had at least one other significant risk factor for VTE including recent or subsequent surgery, past thrombotic event or malignancy. The incidence of thrombosis in those with significant risk factors for VTE was significantly higher than those without (8.5% versus 1.8%; P=0.0009). Overall, there was no significant difference in thrombotic events between individuals receiving pharmacological prophylaxis (1.2%) and those who did not (2.8%) (P=0.4).

CONCLUSION: Overall, VTE prophylaxis did not significantly affect thrombotic events in patients admitted for an active GI bleed.

Key Words: *Gastrointestinal bleeding; Venous thromboembolism prophylaxis*

La thromboprophylaxie veineuse en cas d'hémorragie gastro-intestinale

OBJECTIF : Étudier l'utilisation de la prophylaxie de la thromboembolie veineuse (TEV) et l'évaluation du risque d'incidents thrombotiques chez les patients ayant une hémorragie gastro-intestinale (GI) aiguë.

MÉTHODOLOGIE : Les personnes hospitalisées en raison d'un diagnostic primaire d'hémorragie GI dont la source est confirmée par endoscopie (sur une période de deux ans) ont été incluses dans l'étude. Les chercheurs ont colligé la comorbidité des patients et les données relatives à l'anticoagulation ou à l'utilisation d'antiplaquettaires avant l'hospitalisation, ainsi que le type de prophylaxie de la TEV et la durée du traitement. Le critère d'évaluation primaire était l'apparition d'une TEV (thrombose veineuse profonde ou embolie pulmonaire) dans l'année suivant le rendez-vous.

RÉSULTATS : Les chercheurs ont examiné les données de 504 patients hospitalisés en raison d'une hémorragie GI. Au total, il y a eu 20 incidents de TEV (4 %), et le taux de décès pendant l'hospitalisation s'élevait à 4,6 %. De plus, 397 patients n'ont pas reçu de prophylaxie de la TEV pendant leur hospitalisation. Chez les patients sous prophylaxie, 68 ont reçu de l'héparine prophylactique ou des dérivés de l'héparine pendant leur hospitalisation. Cent soixante-cinq patients présentaient au moins un autre facteur de risque important de TEV, y compris une chirurgie récente ou ultérieure, des incidents thrombotiques passés ou un cancer. L'incidence de thrombose chez les personnes ayant d'importants facteurs de risque de TEV était considérablement plus élevée que chez celles qui n'en présentaient pas (8,5 % par rapport à 1,8 %; P=0,0009). Dans l'ensemble, le nombre d'incidents thrombotiques ne différait pas beaucoup entre les personnes sous prophylaxie pharmacologique (1,2 %) et celles qui ne l'étaient pas (2,8 %) (P=0,4).

CONCLUSION : Dans l'ensemble, la prophylaxie de la TEV n'avait pas d'effet important sur les incidents thrombotiques chez les patients hospitalisés en raison d'une hémorragie GI.

Admissions for upper gastrointestinal (GI) bleeding in Canada account for 30 to 50 per 100,000 population per year (1). Patients presenting to hospital with GI bleeding often have an unpredictable clinical course, from being hemodynamically stable to requiring admission in the intensive care unit (ICU), emergency surgery or death. In Canada, the overall mortality rate for those found to have an upper GI bleed has been reoported to be 3.5% (1).

As a consequence of being hospitalized, all acutely ill patients, including those with GI bleed, are at increased risk for developing venous thromboembolism (VTE). The rate of these events in published studies varies between 1.0% and 1.9% (2). This is not a self-limiting problem because, once the diagnosis is made, future outcomes remain a concern. Pulmonary embolism (PE) alone is an independent risk factor for reduced survival within three months of onset, with a future risk for recurrence of VTE of up to 30% within 10 years (3). Up to 50% of VTEs are associated with right ventricular dysfunction, with the possibility of developing pulmonary hypertension (4). Patients hospitalized with GI bleeds fall within Virchow's triad due to long periods of immobility and venostasis.

As a result, there have been many studies showing the combined efficacy and safety of thromboprophylaxis in patients when clinically indicated (5). Most randomized controlled trials have studied low molecular weight heparin (6) and new oral anticoagulants in reducing the risk of VTE in patients with acute medical illnesses (7). Naturally, the development of deep vein thrombosis (DVT) or PE leads to an increased burden on the health care system. In the United States alone, the economic impact of recurrent VTE was reflected in average total annual health care cost for primary diagnoses of $10,804 per patient for DVT and $16,644 per patient for PE (8).

In 2011, the American College of Physicians strongly recommended pharmacological prophylaxis for VTE in hospitalized medical patients unless the assessed risk for bleeding outweighed the likely benefits (9). As the medical community trends toward adhering to guidelines, withholding VTE prophylaxis in higher-risk individuals is largely unsupported by the literature (10). However, caution still needs to be exercised in those with evidence of a GI bleed. To better examine the complications of pharmacological prophylaxis in this population, a recent study involving patients hospitalized with lower GI bleeding was completed (9). While it did not show a statistically significant difference in mortality, it did show an increased number transfusions (mean ± [SD] 3.47±2.36 versus 2.56±2.37; P=0.02) and length of ICU stay (6.07±3.87 days versus 4.41±3.19 days; P=0.02) with prophylactic subcutaneous heparin following admission. With some inherent risk and cost to the health care system, this raises the

Department of Medicine and Division of Gastroenterology, Western University/London Health Sciences Centre, London, Ontario
Correspondence: Dr Neel Malhotra, Department of Medicine, Western University, 800 Commissioners Road East, London, Ontario N6A 5W9. e-mail neel.malhotra@londonhospitals.ca

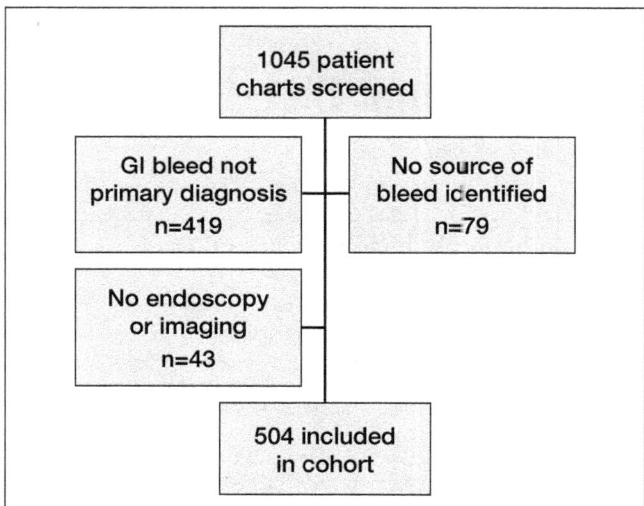

Figure 1) *Chart review process. GI Gastrointestinal*

TABLE 1
Demographic characteristics of the study group (n=504)

General characteristics	
Age, years, mean ± SD	70.5±15.3
Hospitalization, days, mean ± SD	5.5±7.3
Upper gastrointestinal bleeding	154 (30.6)
Lower gastrointestinal bleeding	350 (69.4)
Method of diagnosis	
Gastroscopy	350 (69.4)
Colonoscopy	146 (30)
Red blood cell scan	5 (1)
Computed tomography scan	3 (0.6)
Deep vein thrombosis prophylaxis	
Yes	107 (21.2)
No	397 (78.8)
Death	23 (4.6)
Deep vein thrombosis prophylaxis	7 (1.4)
No deep vein thrombosis prophylaxis	16 (3.2)

Data presented as n (%) unless otherwise indicated

question of whether this patient population is actually experiencing the perceived benefit of pharmacological DVT prophylaxis. Our aim was to conduct a retrospective cohort study to determine the overall use of VTE prophylaxis and the subsequent incidence of DVT/PE in patients admitted with a GI bleed.

METHODS

London Health Sciences Centre is a tertiary care centre located in London, Ontario. It operates two inpatient teaching hospitals and has a catchment of approximately two million in Southwestern Ontario. A retrospective chart review was performed on all inpatients admitted with a diagnosis of GI bleed between January 2007 and December 2008.

All patients >18 years of age were included, along with endoscopic evidence or radiological confirmation (computed tomography angiography or technetium-labelled red blood cell scan) of an active upper or lower GI bleed. Patients who developed evidence of bleeding while admitted to hospital for another reason were excluded. Only the initial event was considered in those with recurrent hospitalizations.

A standardized form was used to abstract data from electronic patient records. Demographic data, anticoagulation status, endoscopic findings and radiological evidence of VTE (DVT or PE) up to one year after discharge were recorded. This process was approved by Western University Research Ethics Board for Health Science Research Involving Human Subjects.

TABLE 2
Anticoagulation before admission

Antiplatelet	
Acetylsalicylic acid	186 (57.4)
Clopidogrel	12 (3.7)
Combination	44 (13.6)
Vitamin K antagonist	
Warfarin	36 (11.1)
Direct thrombin inhibitor	
Dabigatran	3 (0.9)
Low molecular weight heparin	
Dalteparin	3 (0.9)
Enoxaparin	2 (0.6)
Combination	
Antiplatelet agent + vitamin K antagonist	37 (11.4)
Low molecular weight heparin + vitamin K antagonist	1 (0.3)

Data presented as n (%)

Demographic data comprised patient age in additon to information regarding comorbidities that may have contributed to a thrombotic event including known history of thrombotic events, recent surgery (within three months), and diagnosis of malignancy or cirrhosis. Documenting anticoagulation status included antiplatelet agents, vitamin K antagonists, direct thrombin inhibitors, as well as heparin and heparin derivatives at the time of admission while hospitalized and after discharge. Any of these agents were considered to be protective against the development of thrombosis even if they were discontinued for a portion of the hospital stay.

The primary outcome was the incidence of VTE up to one year after discharge based those who received anticoagulant therapy for prophylaxis against VTE during hospitalization. The potential role of hypercoagulable states (malignancy, cirrhosis, recent surgery and past thrombosis) as risk factors for thromboembolic disease was used as a secondary outcome.

Statistical analysis

Standard descriptive analysis was performed to determine the source of GI bleeding along with the modality of endoscopic intervention most used for diagnosis. The frequency of VTE development, in additon to anticoagulation and hypercoagulable states, was also documented. Fisher's exact test with two-tailed P value using SAS/Stat Software version 9.1.3 (SAS Institute Inc, USA) was performed to analyze the primary outcome of VTE in those who received pharmacological prophylaxis comapred with those who did not. A subgroup analysis was performed to investigate the outcome of pharmacological prophylaxis in those with significant risk factors likely predisposing to thromboembolic complications.

RESULTS

After initial review, 1045 potential charts were identified. Of these, 419 had a primary diagnosis other than GI bleed leading to admission (Figure 1). Further screening found that 43 patients had no endoscopic procedure or imaging to confirm their diagnosis. Seventy-nine investigations were completed, but no source of bleeding was identified. This left 504 patient charts eligible for the present study. The demographic characteristics of the study group are summarized in Table 1. A higher number of patients were found to have a lower GI bleed (69.4%) compared with an upper GI bleed source (30.6%). In addition, pharmacological prophylaxis against VTE was only given in a minority (21.2%) of cases.

Table 2 summarizes the distribution of anticoagulant and antiplatelet agents before admission. The vast majority of patients were on some form of antiplatelet therapy (64.7%), with acetylsalicylic acid the most common agent (57.4%). For patients who were started on some form of VTE prophylaxis, the most common agents were heparin or low molecular weight heparin, as shown in Table 3. It should also be

TABLE 3
Pharmacological anticoagulation during hospital stay

Antiplatelet	
Acetylsalicylic acid	18 (16.8)
Clopidogrel	5 (4.7)
Combination	6 (5.6)
Vitamin K antagonist	
Warfarin	3 (2.8)
Direct thrombin inhibitor	
Dabigatran	0 (0)
Heparin or low molecular weight heparin	
Dalteparin	48 (44.9)
Enoxaparin	1 (0.9)
Heparin	19 (18.7)
Combination	
Antiplatelet agent + vitamin K antagonist	0 (0)
Antiplatelet agent + low molecular weight heparin	4 (3.7)
Antiplatelet agent + heparin	3 (2.8)

Data presented n (%)

noted that some patients were continued on their antiplatelet or anticoagulant medication while being treated.

Table 4 lists additional risk factors that played a role in the development of thromboembolic complications. Of these, history of malignancy was the most common (n=83) followed by evidence of cirrhosis (n=45). Less frequent risk factors included history of past thrombotic events (n=13) and recent surgery (within the past three months) (n=7).

Overall, the incidence of VTE in the present study was 4%. Using the Fisher's exact test, the overall incidence of DVT/PE was significantly higher in those with higher-risk conditions (P=0.0009). As shown in Table 5, the combined high/low risk group had a reduced incidence of thromboembolic complications (1.2%) with pharmacological prophylaxis, although not statistically significant (P=0.4). This difference was also observed in the subgroup of individuals with comorbidities, with the incidence of VTE being 1.8%.

DISCUSSION

It has been well established that acutely ill medical patients benefit from pharmacological prophylaxis for VTE (11). The difficulty in patients with evidence of GI bleed is the concern for recurrent bleeding and hemodynamic stability. Adding any antiplatelet or anticoagulant in these situations often increases concern among health care professionals, which was reflected in the present study by the low number of individuals who were given VTE prophylaxis (n=107). Interestingly, current evidence shows that there is no increase in mortality among patients with lower GI bleeding treated with VTE prophylaxis (10). However, there was a longer ICU stay and increased number of transfusions only within the first 24 h after admission (10). In a study investigating the clinical outcomes of patients with a recent diagnosis of VTE, 40% experienced a GI bleed 30 days previously (12). Three-month mortality was as high as 23%, which is a clinically important measure, indicating the importance of VTE prophylaxis.

Importantly, our study showed that thromboembolic complications do exist in patients admitted with a GI bleed. VTE is significantly more common among individuals with other risk factors predisposing them to hypercoagulable states (P=0.0009). Although not a statistically significant difference, we were also able to demonstrate that VTE prophylaxis decreased the incidence of thromboembolic events compared with those who did not receive it (1.2% versus 2.8%, respectively; P=0.4).

Other observational studies have reported that a large percentage of individuals with VTE were >65 years of age and had additional risk factors, including malignancy, for VTE (12). Patients in our study had a mean age of 70.5 years, with a mean length of hospitalization of 5.5 days. The most frequent comorbidities believed to confer additional risk for thromboembolism included malignancy (n=83), cirrhosis (n=45) and past thrombotic events (n=13).

TABLE 4
Comorbidities of patients who developed thrombotic conditions (n=165)

Comorbidity	
Malignancy	83
Past thromboembolic events*	13
Recent surgery†	7
Cirrhosis	45
Combination	
Malignancy – cirrhosis	11
Malignancy – past thromboembolic events	5
Malignancy – surgery	1

*Data presented as n. *Includes pulmonary embolism, deep vein thrombosis, aortic clot, apical thrombus and portal vein thrombosis; †Within three months*

TABLE 5
Incidence of thromboembolic disease (deep vein thrombosis [DVT]/pulmonary embolism [PE])

	Incidence of DVT/PE, %	P	OR (95%CI)
All patients (n=504)*	4.0		
No comorbidities	1.8		
Comorbidities	8.5	0.0009	5.1 (1.9–13.6)
Comorbidities (n=165)			
DVT prophylaxis	1.8		
No prophylaxis	6.7	0.72	0.7 (0.2–3.0)
No comorbidities (n=339)			
DVT prophylaxis	0.8		
No prophylaxis	0.8	0.11	0.2 (0.05–1.3)
Comorbidities + no comorbidities			
DVT prophylaxis	1.2		
No DVT prophylaxis	2.8	0.4	0.6 (0.2–1.6)

**Includes both with and without DVT prophylaxis*

Low-dose acetylsalicylic acid has been associated with an increased risk for upper GI bleeding (13). Our data indicate that antiplatelet therapy, specifically acetylsalicylic acid, was used most commonly (57.4%) by patients before presenting with upper or lower GI bleeding. Baseline demographic data showed that lower GI bleeds (69.4%) were more common than upper GI bleeds (30.6%). This supports the higher risk associated with acetysalicylic acid therapy in lower GI bleeds documented in a recent prospective study (14).

The lack of a significantly lower incidence of VTE observed in our cohort was likely multifactorial. First, there were low event rates observed in both groups. The study period could have been extended to determine the longer-term incidence of DVT/PE. Second, our teaching centre has a large catchment area with a significant referral base. Confirmation of thromboembolic events required documented radiological evidence and, for patients transferred back to their local hospital, some would be lost to follow-up, which may have led to an underestimation of our event rates.

Overall, it would be our recommendation to implement VTE prophylaxis on a case-by-case basis for patients admitted with an active GI bleed, particularly if they have other factors predisposing them to thrombosis. Further studies investigating complications of low-dose anticoagulant therapy are likely needed to adequately quantify the risk associated with therapy and to reassure future practice.

DISCLOSURES: The authors have no financial disclosures or conflicts of interest to declare.

REFERENCES

1. Targownik LE, Nabalamba A. Trends in management and outcomes of acute nonvariceal upper gastrointestinal bleeding: 1993-2003. Clin Gastroenterol Hepatol 2006;4:1459-66.
2. Guijarro R, San Roman C, Arcelus JI, et al. Bleeding and venous thromboembolism arising in acutely ill hospitalized medical patients. Findings from the Spanish National Discharge Database. Eur J Intern Med 2014;25:137-41.
3. Heit JA. Venous thromboembolism: Disease burden, outcomes and risk factors. J Thromb Haemost 2005;3:1611-7.
4. Kearon C. Natural history of venous thromboembolism. Circulation 2003;107:I22-I30.
5. Dentali F, Douketis JD, Gianni M, Lim W, Crowther MA. Meta-analysis: Anticoagulant prophylaxis to prevent symptomatic venous thromboembolism in hospitalized medical patients. Ann Intern Med 2007;146:278-88.
6. Samama MM, Cohen AT, Darmon JY, et al. A comparison of enoxaparin with placebo for the prevention of venous thromboembolism in acutely ill medical patients. Prophylaxis in Medical Patients with Enoxaparin Study Group. N Engl J Med 1999;341:793-800.
7. Roussin A. Effective management of acute deep vein thrombosis: Direct oral anticoagulants. Int Angiol June 13, 2014 (Epub ahead of print).
8. Spyropoulos AC, Lin J. Direct medical costs of venous thromboembolism and subsequent hospital readmission rates: An administrative claims analysis from 30 managed care organizations. J Manag Care Pharm 2007;13:475-86.
9. Qaseem A, Chou R, Humphrey LL, Starkey M, Shekelle P. Venous thromboembolism prophylaxis in hospitalized patients: A clinical practice guideline from the American College of Physicians. Ann Intern Med 2011;155:625-32.
10. Deutsch GB, Kandel AR, Knobel D, et al. Bleeding risk secondary to deep vein thrombosis prophylaxis in patients with lower gastrointestinal bleeding. J Intensive Care Med 2012;27:379-83.
11. Kahn SR, Lim W, Dunn AS, et al. Prevention of VTE in nonsurgical patients. In: Antithrombotic Therapy and Prevention of Thrombosis, 9th edn: American College of Chest Physicians Evidence-Based Clinical Practice Guidelines. Chest 2012;141:e195S-e226S
12. Nieto JA, Camara T, Gonzalez-Higueras E, et al. Clinical outcome of patients with major bleeding after venous thromboembolism. Findings from the RIETE Registry. Thromb Haemost 2008;100:789-96
13. Valkhoff VE, Sturkenboom MC, Hill C, Veldhuyzen van Zanten S, Kuipers EJ. Low-dose acetylsalicylic acid use and the risk of upper gastrointestinal bleeding: A meta-analysis of randomized clinical trials and observational studies. Can J Gastroenterol 2013;27:159-67.
14. Huang ES, Strate LL, Ho WW, Lee SS, Chan AT. A prospective study of aspirin use and the risk of gastrointestinal bleeding in men. PLoS One 2010;5:e15721.

10

Toward enteral nutrition in the treatment of pediatric Crohn disease in Canada: A workshop to identify barriers and enablers

Johan Van Limbergen MD FRCPCH PhD[1,2,15], Jennifer Haskett BSNH, PDt[1], Anne M Griffiths MD[3], Jeff Critch MD MSc[4], Hien Huynh MBBS FRCPC[5], Najma Ahmed MD[6], Jennifer C deBruyn MD[7], Robert Issenman MD FRCP[8], Wael El-Matary MD[9], Thomas D Walters MD[3], Cheryl Kluthe RN BScN[5], Marie-Eve Roy RD[10], Elizabeth Sheppard RN BN[4], Wallace V Crandall MD[11], Stan Cohen MD[12], Frank M Ruemmele MD PhD[13], Arie Levine MD[14], Anthony R Otley MD MSc FRCPC[1,15]

J Van Limbergen, J Haskett, AM Griffiths, et al. Toward enteral nutrition in the treatment of pediatric Crohn disease in Canada: A workshop to identify barriers and enablers. Can J Gastroenterol Hepatol 2015;29(7):351-356.

The treatment armamentarium in pediatric Crohn disease (CD) is very similar to adult-onset CD with the notable exception of the use of exclusive enteral nutrition (EEN [the administration of a liquid formula diet while excluding normal diet]), which is used more frequently by pediatric gastroenterologists to induce remission. In pediatric CD, EEN is now recommended by the pediatric committee of the European Crohn's and Colitis Organisation and the European Society for Paediatric Gastroenterology Hepatology and Nutrition as a first-choice agent to induce remission, with remission rates in pediatric studies consistently >75%. To chart and address enablers and barriers of use of EEN in Canada, a workshop was held in September 2014 in Toronto (Ontario), inviting pediatric gastroenterologists, nurses and dietitians from most Canadian pediatric IBD centres as well as international faculty from the United States and Europe with particular research and clinical expertise in the dietary management of pediatric CD. Workshop participants ranked the exclusivity of enteral nutrition; the health care resources; and cost implications as the top three barriers to its use. Conversely, key enablers mentioned included: standardization and sharing of protocols for use of enteral nutrition; ensuring sufficient dietetic resources; and reducing the cost of EEN to the family (including advocacy for reimbursement by provincial ministries of health and private insurance companies). Herein, the authors report on the discussions during this workshop and list strategies to enhance the use of EEN as a treatment option in the treatment of pediatric CD in Canada.

Key Words: *Crohn disease; Exclusive enteral nutrition; Nutritional therapy; Pediatrics*

Vers l'alimentation entérique pour le traitement de la maladie de Crohn en pédiatrie au Canada : un atelier pour déterminer les obstacles et les catalyseurs

L'arsenal thérapeutique de la maladie de Crohn (MC) pédiatrique est très similaire à celui de la MC qui se déclare chez les adultes, à l'exception importante de l'alimentation entérale exclusive (AEE [l'administration d'une préparation liquide qui exclut un régime normal]), plus utilisée par les gastroentérologues pédiatres pour induire une rémission. En cas de MC pédiatrique, l'AEE est désormais recommandée comme agent de première intention pour induire une rémission par le comité pédiatrique de l'*European Crohn's and Colitis Organisation* et par la Société européenne de gastroentérologie, d'hépatologie et de nutrition pédiatrique. Selon des études en pédiatrie, le taux de rémission est constamment supérieur à 75 %. Afin de décrire et d'examiner les catalyseurs et les obstacles à l'utilisation de l'AEE au Canada, un atelier a eu lieu en septembre 2014 à Toronto, en Ontario, auquel était invités les gastroentérologues, les infirmières et les diététistes spécialisés en pédiatrie de la plupart des centres de MII pédiatriques du Canada, ainsi que des conférenciers internationaux des États-Unis et de l'Europe ayant des compétences de recherche et de clinique dans la prise en charge diététique de la MC pédiatrique. Les participants à l'atelier ont classé l'exclusivité de l'alimentation entérale, les ressources en santé et les coûts comme les trois principaux obstacles à son utilisation. En revanche, les principaux catalyseurs étaient la standardisation et le partage des protocoles sur l'utilisation de l'alimentation entérale, l'assurance de ressources alimentaires suffisantes et la réduction des coûts de l'AEE pour les familles (y compris la promotion de son remboursement par les ministères de la santé provinciaux et les sociétés d'assurance privées). Dans le présent article, les auteurs rendent compte des échanges pendant l'atelier et dressent une liste des stratégies visant à améliorer l'utilisation de l'AEE comme option thérapeutique de la MC pédiatrique au Canada.

Crohn disease (CD) presents during childhood or adolescence in up to 25% of patients and is typically more extensive than adult-onset disease (1,2). Incidence rates in Canada are among the highest reported worldwide (3). Whereas teenagers with CD often progress toward penetrating complications in a rate similar to adult-onset CD, very early onset CD (<6 years of age at diagnosis, excluding monogenic forms of very early onset inflammatory bowel disease [IBD]) is associated with fewer hospitalizations and surgery (2,4). The treatment armamentarium

[1]*Division of Gastroenterology & Nutrition, Department of Pediatrics, IWK Health Centre;* [2]*Department of Microbiology and Immunology, Dalhousie University, Halifax, Nova Scotia;* [3]*Division of Pediatric Gastoenterology, Hepatology and Nutrition, Hospital for Sick Children, Toronto, Ontario;* [4]*Division of Pediatric Gastroenterology, Janeway Child Health Centre, St John's, Newfoundland & Labrador;* [5]*Division of Pediatric Gastroenterology, Stollery Children's Hospital, University of Alberta, Edmonton, Alberta;* [6]*Division of Pediatric Gastroenterology and Nutrition, Department of Pediatrics, Montreal Children's Hospital, McGill University, Montreal, Quebec;* [7]*Division of Gastroenterology and Nutrition, Department of Paediatrics, University of Calgary, Calgary, Alberta;* [8]*Division of Pediatric Gastroenterology and Nutrition, McMaster Children's Hospital, Hamilton, Ontario;* [9]*Section of Pediatric Gastroenterology, Department of Pediatrics, College of Medicine, Faculty of Health Sciences, University of Manitoba, Winnipeg, Manitoba;* [10]*Department of Pediatric Gastroenterology, Hepatology and Nutrition, Centre Hospitalier Universitaire Sainte-Justine, Montréal, Québec;* [11]*Division of Pediatric Gastroenterology, Hepatology and Nutrition, Nationwide Children's Hospital and The Ohio State University, Columbus, Ohio;* [12]*Children's Center for Digestive Health Care, Children's Healthcare of Atlanta, Atlanta, Georgia, USA;* [13]*Université Sorbonne Paris Cité, Université Paris Descartes, INSERM U989, Service de Gastroentérologie Pédiatrique, Hôpital Necker Enfants Malades, Paris, France;* [14]*Pediatric Gastroenterology and Nutrition Unit, Wolfson Medical Center, Sackler School of Medicine, Tel-Aviv University, Tel-Aviv, Israel;* [15]*Division of Gastroenterology, Faculty of Medicine, Dalhousie University, Halifax, Nova Scotia*

Correspondence: Dr Anthony R Otley, Department of Pediatrics, Dalhousie University, Division of Pediatric Gastroenterology & Nutrition, IWK Health Centre, 5850/5980 University Avenue, PO Box 9700, Halifax, Nova Scotia B3K 6R8. e-mail arotley@dal.ca

in pediatric CD is very similar to adult-onset CD with the exception of the frequent use of exclusive enteral nutrition (EEN [the administration of a liquid formula diet while excluding normal diet]), which is used more often by pediatric gastroenterologists. In pediatric CD, EEN is now recommended by the pediatric committee of the European Crohn's and Colitis Organisation (ECCO), and the European Society for Paediatric Gastroenterology Hepatology and Nutrition (ESPGHAN) as a first-choice agent to induce remission, with remission rates in pediatric studies consistently >75%, and is also commonly used in Japan and Australasia (5). The effectiveness of EEN was first described in adult patients with severe disease (6-9). In view of safety concerns of combination immunosuppressive therapy and because of increasing patient interest in dietary therapy, EEN is now also regaining momentum as a treatment option in adult CD (10-16).

Long-term dietary patterns and specific food items have been shown to influence risk of CD development in all age groups (17-19). Nutritional approaches once CD is established have included total parenteral nutrition, specific dietary exclusions, partial enteral nutition (EN) and avoidance of all dietary intake using EEN (19-22). Avoidance of all dietary intake using a complete nutritional alternative, such as EEN, has been shown to be superior to partial EN when the additional oral dietary intake is not controlled (23). Recent case series of successful use of specific exclusion diets (alone or in combination with EN) as well as requests from many patients to further develop evidence-based dietary therapy, require confirmation by controlled dietary intervention studies (24,25).

Achieving mucosal healing (or endoscopic improvement) has gained support as therapeutic target in CD and may dramatically reduce the risk for long-term disease complications including surgery (26). In small controlled studies, EEN has been shown to be superior to steroids in achieving mucosal healing, while being notably free from important adverse events (20,27-29). EEN has demonstrated efficacy at first induction of clinical remission as well as for subsequent flares, even in patients who have lost response to anti-tumour necrosis factor (TNF) therapy (20,25,29,30). Recently, the GROWTH CD study (a prospective, outcome-based study in newly diagnosed untreated pediatric patients with CD, conducted in Europe and Israel, established by the Porto Group of ESPGHAN, NIH NCT00711945) showed that in mild-to-moderate CD, EEN was superior to steroids both when considering remission according to the Pediatric Crohn's Disease Activity Index (PCDAI) (OR 5.8 [95% CI 1.8 to 18.3]) or combined normal PCDAI and C-reactive protein (OR 3.4 [95% CI 1.3 to 9]) (31).

The use of EN for maintenance of remission varies across IBD centres in terms of strategies used (eg, overnight versus supplemental day time, percentage of total daily calories, as monotherapy or in combination with immunomodulators/anti-TNF) (10,12,15,32-37). Research is ongoing with regard to the optimal maintenance strategy (eg, Cyclic Exclusive Enteral Nutrition as Maintenance Therapy for Pediatric Crohn's Disease [CD-HOPE; NCT02201693] by GETAID pédiatrique) but uncertainty with regard to its use during maintenance of remission could deter from using EEN as induction therapy. In adults and children with CD alike, quality of life using EEN improves even before mucosal healing is obtained (38-40).

The phenotype of CD that is suitable to be treated with EEN, particularly whether isolated colonic disease will respond, has long been a subject of debate (22,28,29,41-44). Clearly, any liquid diet may cause difficulties in terms of stool consistency and frequency, which can be troublesome in the presence of left-sided colonic disease (and, therefore, increase the PCDAI subscore). Afzal et al (42) reported that the achieved remission rate for isolated colonic disease (50% [seven of 14]) was less than in ileal (92% [11 of 12]) or ileocolonic disease (82% [32 of 39]) ($P=0.02$). However, subsequent case series did not report this different remission rate (22,41). In the study by Buchanan et al (41), isolated colonic disease (without upper gastrointestinal disease, as categorized according to the Vienna classification) achieved a remission rate of 79%, although not assessed prospectively by customary disease activity measures. Given these discrepancies, it would appear reasonable to consider use of EEN for all patients with luminal CD. Perianal CD or the presence of severe colonic disease are listed in the ECCO/ESPGHAN guidelines as factors that warrant consideration of other induction therapy (5).

In this era of balancing treatment-associated risks with maximum effectiveness, yet with appropriate attention for patient-reported outcomes, it is timely to address why EEN is not used more widely across North America (9,45). In fact, there are significant variations in the patterns of use and the acceptance of EEN by physicians between Canada and the United States: Canadian physicians show a greater use of EEN (46). Despite growing concerns about the safety profile of corticosteroids, these geographical variations have not changed over the past 15 years (45,46). The use of EEN appears to be influenced by the extent to which physicians are exposed to its use both in their training and in their current practice setting (46,47). To chart and address enablers and barriers of use of EEN in Canada, a workshop was held in September 2014 in Toronto (Ontario), inviting pediatric gastroenterologists, nurses and dietitians from most Canadian pediatric inflammatory bowel disease centres as well as international faculty from the United States and Europe with particular research and clinical expertise in the dietary management of pediatric CD. In the present report, we discuss the findings of this workshop dedicated to enhancing the use of EEN as a treatment option in the treatment of pediatric CD in Canada.

Preworkshop findings

Before the workshop, individual telephone interviews were conducted by a consumer insights professional with 11 patient families from various clinics across Canada to begin to explore the patient and family experience of choosing a treatment option. Of the 11 patients, seven had received some form of EN as part of their initial treatment. These data were grouped according to theme (Table 1) and formed part of the workshop discussions.

Workshop

Twenty pediatric stakeholders attended the one-day workshop, including three nurses, two dietitians and 15 pediatric gastroenterologists. Participants completed a premeeting assignment identifying experience in their pediatric practice with barriers and enablers to using EEN related to the following influencers: health system (internal and external), patient/family, EN, physician/care team-related or other. During the workshop, participants worked in groups and further refined and categorized the submitted responses, and discussed potential solutions to barriers and ways to enhance enablers (Table 2). These results were further ranked according to priority, highlighting similar barriers and enablers to the use of EEN as described in the literature.

Significant barriers to the use of EEN can be related to:

1. EEN mechanism of action: requirement of exclusivity
2. Patient/family adherence: oral versus nasogastric (NG) tube
3. Health care team/health care system/insurance

Consideration of priority of barriers and enablers to the use of EEN

After an iterative process of discussion, health care participants of the workshop consistently ranked: the exclusivity of EN; the health care resources; and cost implications as the top three barriers to its use (Table 2). Conversely, key enablers repeatedly mentioned included: standardization and sharing of protocols for use of EN (including approach in clinic to EEN, which is heavily influenced by the exposure of the health care team to the benefits of EEN during their training) as well as mode of delivery; ensuring sufficient dietetic resources; and reducing the cost of EEN to the family (including advocacy for reimbursement by provincial ministries of health and private insurance companies). As will be discussed below, emphasizing the need to completely avoid the prediagnosis oral diet by means of EEN, as a requirement to achieve successful induction of remission, can help address some of these barriers and enablers.

TABLE 1
Thematic summary of patient and family interviews

Factors/themes (with examples)	Considerations and impact on practice after discussion in workshop
Messaging from health care team	
• "Pharmacist said incidence of most side effects from steroids were 10% or lower". • Family opted for the steroid because they did not feel the efficacy of the EEN was explained	• Need for multidisciplinary education and conviction; ensure accurate and consistent messaging • Written information to ensure accurate recall by families
Parental assumptions and expectations	
• "at 14, no way would she do that" • "12 is a difficult, in-between age. Maybe if he was younger or older he would (been convinced to) have tried the [formula]."	• Importance of connecting parents with experienced parents • Involve social work or health psychology
Social concerns	
• Integration into school, activities, not eating • "EEN would be socially isolating" • "patient became emotional about not eating (worried about missing the food he liked; being different from his friends)"	• Importance of connecting patients to youth with EEN experience, use available resources (videos, camp/ social experience)
Guilt	
• Parents felt that he had already been through so much that they did not want to upset him further • "At 10 or 11, it was hard to imagine that he could only drink when his friends were eating"	• Focus on benefits of EEN, not only challenges • Importance of connecting parents with experienced parents • Involve social work or health psychology
Child as the decision maker	
• "Parents have to respect the wishes of their children (even very young children). The option of a steroid was the only one our son wanted to look at, so we had to go with his wishes." (Pt was 10 years old when EEN was offered) • "You can't make your teen do what they don't want to do"	• Be sure child is present and actively engaged in discussions regarding treatment • The child is a key player in the decision making, but they are not the only player – parental involvement is also important; a difficult decision to make alone • Engage supports – such as peers – connect with patient who has been on EEN
Adaptation	
• "It seems so traumatic at first, but you have to look ahead. There are so many possibilities for a good outcome." • "it is hard, but it will get a lot better" • "nervous but relieved [at decision to place NG tube]". The tube was in for 10½ weeks... Stayed in; changed 3 times. Very successful. She gained weight."	• Have families share their experiences and strategies

EEN Exclusive enteral nutrition; NG Nasogastric

The mechanism of action of EEN remains conjectural, but is suggested to involve modulation of gut microbial community (microbiota) composition, which is considered a critical factor in CD pathogenesis (40,48). Although the microbiota is rapidly altered to a limited degree by dietary changes, the overall community structure (and presumed metabolic function) has been shown to be resilient to short-term dietary interventions (eg, 10 days in the CAFE study by Wu et al [49]) and is linked with long-term dietary patterns (49-52).

TABLE 2
Summarized barriers and enablers from group exercise

Factor	Barriers	Enablers
Health system internal (hospital/ health authority)	• Insufficient clinic resources: allied health staff, knowledge, space*	• Adequate numbers of trained team members (nurses, dietitians, social work/psychology/child health) and dedicated space for teaching*
Health system external (provincial/ regional)	• Funding for supplies, formula	• Coverage for EEN supplies and formula* • Supportive home services
Patient/family	• Fear of NG tube and/or loss of food • Difficulty sustaining diet • Limited support to family/socialization	• Involving parents/family in feeding choice • Support of diet, acknowledging it may be difficult • Supportive dietitian throughout process
Enteral nutrition	• Exclusivity of enteral nutrition with no/ limited oral intake* • Cost of enteral nutrition* • Taste • NG tube	• Evidence-based/reduced need for steroids • Few side effects • Oral option possible; recipes
Physician/care team-related	• Lack of institutional experience or critical mass to "keep it going"* • Lack of standardization of enteral nutrition approach*	• Consistent and systematic approach to EEN (protocols, tools, talking points, defined roles for team members)* • Conviction of physician and team to support EEN • Quality review process • Resource sharing

**Barriers and enablers identified as highest priority. EEN Exclusive enteral nutirition; NG Nasogastric*

Although the clinical and biochemical parameters of disease remission are often achieved by four weeks of induction treatment, the relapse rates of up to 60% by the end of the first year may suggest that the 'state' of the microbiome was not sufficiently shifted (34,53,54). The duration of EEN varies substantially across published reports (9). This variation in practice can be linked with preferential use of oral versus NG tube-delivered EEN; although strict compliance with oral EEN (and avoidance of dietary intake) is achieved by many patients, for others it may be easier to sustain EEN via NG tube, with optimization of benefits beyond achieving normalization of inflammatory markers and clinical remission. The question of whether establishment of a new stable microbiome, away from the inflammation-associated 'state', can become a therapeutic goal, will need to be studied in prospective microbiome-focused trials. Gerasimidis et al (48) showed that EEN paradoxically reduces some of the presumed 'protective' features of the gut microbiota, such as community diversity and even particular bacterial species (eg, *Firmicutes*), making its mode of action difficult to infer (25,48). Recent reports describing specific exclusion diets have rekindled the debate of whether the strong therapeutic effect of EEN is mediated largely by the avoidance of putative dietary triggers (24,25). In other words, rather than actively inducing 'protective' microbiota changes, EEN may mediate remission by excluding dietary products that elicit pathological changes in gut microbiota composition/function and promote dysfunctional host-microbe interactions in the gut mucosa. Earlier trials of partial EN (50% of total daily calories) were not successful in inducing clinical remission; however, more recent evidence indicates that this is perhaps because no dietary changes were made in the remaining 50% of non-EN calories (23,55). Further support for the crucial role of exclusion of normal diet comes from the comparison of different types of dietary intervention (28). Although trials were not powered to the level of confirming noninferiority, comparison of remission rates for different compositions of formula used (eg, elemental versus polymeric, more/less long-chain triglycerides, etc) or additional additives (eg, glutamine), have not shown one type of formula to be superior (28). There is now an ongoing multicentre, randomized dietary intervention trial studying EEN versus EN plus a specific CD exclusion diet (NCT01728870 – clinicaltrials.gov) (25). Until the efficacy of these novel dietary interventions is shown, considering the benefits of EEN in terms of exclusion of other dietary intake can help overcome several barriers to its use.

To improve patient/family adherence, EEN could, thus, be presented as an effective treatment option by exclusion of the normal dietary intake, in addition to being an alternative to corticosteroid therapy. A consistent approach to discussing EEN in clinic significantly improves acceptance and compliance, as evident from the experience in many European centres (34,41,53). Furthermore, achieving steroid-free maintenance of remission has become an accepted treatment goal in pediatric CD due to its multitude of beneficial effects on growth, nutritional status, bone health and avoiding infectious adverse events, particularly in combination with other immunosuppressive agents (56-58). Clinicians, patients and their families embark on steroid-based induction therapy with a view to early discontinuation but, without anti-TNF therapy, 30% to 40% are steroid dependent after one year of follow-up in referral centre reports from the United States and Canada (56,57). Krupoves et al (57) showed there were no temporal differences in the rates of corticosteroid dependency in pediatric CD before 2000 (43.9%) and in subsequent years (39.4%; P=0.411), and no differences between the two pediatric IBD centres contributing to this study (Montreal [St-Justine Hospital]) and Ottawa), although anti-TNF therapy was used in only 5% of this cohort.

Clearly, choosing an alternative to steroids as induction therapy then becomes a logical approach to avoid steroid dependence. For induction, EEN may be administered orally or via NG tube. Oral feedings are more common in Europe, Australia and some United States centres, whereas many Canadian centres use NG tube (however, for example, at *Centre hospitalier universitaire Sainte-Justine* [Montreal, Quebec] the majority of patients successfully adhere to oral EEN) (9). In a retrospective comparative study on mode of delivery of EEN (based on physician preference), there was no difference in clinical outcome: both were equally effective (22,59). Up to 50% of patients may require an NG tube to complete a course of EN >6 weeks (41). It is important to discuss treatment options in case of difficulty with adherence or failure of EEN, as well as planning for the strategy to maintain remission. The guidelines of the North American Society for Pediatric Gastroenterology, Hepatology and Nutrition (NASPGHAN) IBD Committee refer to an induction period of at least eight weeks with EEN (9). Given that both the time to initial benefit and then to clinical remission can vary, the NASPGHAN report also suggests a period of three to four weeks of EEN before a decision is made about effectiveness (9).

Because the majority of treatment for pediatric CD occurs outside the hospital, the success of EEN therapy is highly dependent on adequate health care resources needed for a home EN program, notably the composition and commitment of a multidisciplinary team. Explaining how lifestyle choices (including diet) impact on general health and development of disease has become accepted practice. Dietary changes impacting on the family and social life are already part of living with IBD for many patients (19,60). Increasingly, patients and their families request, expect and accept dietary management as part of the treatment of IBD (60,61). Attitudes among health care staff to promote the use of EEN, the attending physician's exposure to dietary therapy during her/his training and the centre's experience are all key determinants of success of EEN (9,46,56,57). Nursing and dietitian support to deliver this evidence-based dietary management is, therefore, ideal for a successful EEN program. Access to psychological support can also be important for many patients and families to increase coping skills with the disease in general and EEN treatment in particular (9). Planning and discussing strategies to integrate EEN into the family, school and social life, significantly increase adherence to EEN.

The practical considerations of any successful home EN program include the determination of caloric and other nutrient requirements, determining the best method of administration (oral versus NG, also depending on health care system/insurance stipulation as discussed below), scheduled support during the induction period (eg, by means of a care pathway identifying the role of each team member), and addressing expectations around the time to clinical benefits and total duration of therapy (9).

EEN and its administration (formula, NG tube, and supplies and feeding pump) can be an expensive intervention. Differences in health care systems in each province of Canada affect at-home reimbursement coverage for the formula and supplies, with different programs often required for each. There are well-established programs in some provinces (eg, Alberta, Ontario and Quebec); however, restrictions, such as administration via NG tube (eg, Alberta and Ontario) or the need for home care nursing support can be a condition/barrier to funding. In turn, individual hospitals may operate their own programs in provinces in which a provincial home EN program is lacking. Because funding for health care is provincially managed, there is significant variation across Canada, which affects the uptake of EEN as a therapeutic option.

There has been some limited success in obtaining coverage with private insurance in Canada, but the process is often cumbersome and lengthy. For instance, clinics are required to write individual letters for patients to insurance companies to request financial support. This often requires two to three letters to receive what may still be limited coverage. Clearly, a positive response may justify the effort needed to seek support for home EN, but this often adds considerably to the workload of the multidisciplinary team delivering EEN. In cases where coverage is disputed, it has often been helpful to emphasize the role of the formula as the therapeutic intervention, and the therapeutic requirement for a six- to 12-week period to exclude all other dietary intake to achieve clinical remission.

CONCLUSIONS

EEN is an extremely safe but underused treatment for induction of remission in pediatric CD in North America. Guidelines from both the NASPGHAN IBD Committee as well as the recent ECCO/ESPGHAN guidelines recommend use of EEN as first-line induction therapy in pediatric CD. During this thematic workshop focused on improving the framework for successful implementation of EEN therapy in pediatric CD in Canada, the panel ranked the need for EEN, the health care resources needed for a home EN program and cost implications as the top three barriers to its use. Identifying and understanding the barriers enables us to work on targeted strategies to overcome them, and help clinics implement and improve their success using EEN. Overcoming the barriers is the next step in the process.

Until we improve our understanding of the environmental and dietary triggers of CD, the effectiveness of EN will continue to rely on exclusion of the 'prediagnosis' diet. A standardized yet individualized approach (ie, by considering the caloric and other nutrient requirements of each patient) will optimize the use of limited dietetic resources, ideally with additional support for home nutrition programs. Polymeric formulas (which tend to be less expensive and more palatable) may be better suited if the oral route is chosen, with the option of dietetic guidance to flavour the formula used to avoid taste fatigue. Reducing the cost of EEN to the family will require ongoing advocacy for reimbursement by provincial ministries of health and private insurance companies. Further research to enhance our understanding of the mechanisms of action and the optimal application of EEN (or partial EN with additional dietary modifications) is necessary. Until such time, EEN should be recommended and supported as a highly effective and safe treatment modality in CD.

ACKNOWLEDGEMENTS: JVL was supported by a NASPGHAN/CCFA Young Investigator Development Award (2013-2015), a Future Leaders in IBD research award 2014 and the Dalhousie Medical Research Foundation. Financial support was provided by Nestlé Health Science.

REFERENCES

1. Griffiths AM. Specificities of inflammatory bowel disease in childhood. Best Pract Res Clin Gastroenterol 2004;18:509-23.
2. Van Limbergen J, Russell RK, Drummond HE, et al. Definition of phenotypic characteristics of childhood-onset inflammatory bowel disease. Gastroenterology 2008;135:1114-22.
3. Benchimol EI, Fortinsky KJ, Gozdyra P, Van den Heuvel M, Van Limbergen J, Griffiths AM. Epidemiology of pediatric inflammatory bowel disease: A systematic review of international trends. Inflammatory Bowel Dis 2011;17:423-39.
4. Benchimol EI, Mack DR, Nguyen GC, et al. Incidence, outcomes, and health services burden of very early onset inflammatory bowel disease. Gastroenterology 2014;147:803-13.e7.
5. Ruemmele FM, Veres G, Kolho KL, et al. Consensus guidelines of ECCO/ESPGHAN on the medical management of pediatric Crohn's disease. J Crohns Colitis. 2014;8:1179-207.
6. O'Morain C, Segal AW, Levi AJ. Elemental diet as primary treatment of acute Crohn's disease: A controlled trial. BMJ 1984;288:1859-62.
7. O'Morain C, Segal AW, Levi AJ. Elemental diets in treatment of acute Crohn's disease. BMJ 1980;281:1173-5.
8. Gassull MA, Abad A, Cabre E, Gonzalez-Huix F, Gine JJ, Dolz C. Enteral nutrition in inflammatory bowel disease. Gut 1986;27(Suppl 1):76-80.
9. Critch J, Day AS, Otley A, King-Moore C, Teitelbaum JE, Shashidhar H. Use of enteral nutrition for the control of intestinal inflammation in pediatric Crohn disease. J Pediatr Gastroenterol Nutr 2012;54:298-305.
10. Kamata N, Oshitani N, Watanabe K, et al. Efficacy of concomitant elemental diet therapy in scheduled infliximab therapy in patients with Crohn's disease to prevent loss of response. Dig Dis Sci 2014. [Epub ahead of print]
11. Lee J, Allen R, Ashley S, et al. British Dietetic Association evidence-based guidelines for the dietary management of Crohn's disease in adults. J Human Nutr Dietetics 2014;27:207-18.
12. Li G, Ren J, Wang G, et al. Preoperative exclusive enteral nutrition reduces the postoperative septic complications of fistulizing Crohn's disease. Eur J Clin Nutr 2014;68:441-6.
13. Wall CL, Day AS, Gearry RB. Use of exclusive enteral nutrition in adults with Crohn's disease: A review. World J Gastroenterol 2013;19:7652-60.
14. Lomer MC, Gourgey R, Whelan K. Current practice in relation to nutritional assessment and dietary management of enteral nutrition in adults with Crohn's disease. J Human Nutr Dietetics 2014;27(Suppl 2):28-35.
15. Yamamoto T, Shiraki M, Nakahigashi M, Umegae S, Matsumoto K. Enteral nutrition to suppress postoperative Crohn's disease recurrence: A five-year prospective cohort study. Int J Colorectal Dis 2013;28:335-40.
16. Watanabe O, Ando T, Ishiguro K, et al. Enteral nutrition decreases hospitalization rate in patients with Crohn's disease. J Gastroenterol Hepatol 2010;25(Suppl 1):S134-7.
17. D'Souza S, Levy E, Mack D, et al. Dietary patterns and risk for Crohn's disease in children. Inflamm Bowel Dis 2008;14:367-73.
18. Hou JK, Abraham B, El-Serag H. Dietary intake and risk of developing inflammatory bowel disease: A systematic review of the literature. Am J Gastroenterol 2011;106:563-73.
19. Green TJ, Issenman RM, Jacobson K. Patients' diets and preferences in a pediatric population with inflammatory bowel disease. Can J Gastroenterol 1998;12:544-9.
20. Otley AR, Russell RK, Day AS. Nutritional therapy for the treatment of pediatric Crohn's disease. Expert Rev Clin Immunol 2010;6:667-76.
21. Levine A, Wine E. Effects of enteral nutrition on Crohn's disease: Clues to the impact of diet on disease pathogenesis. Inflamm Bowel Dis 2013;19:1322-9.
22. Rubio A, Pigneur B, Garnier-Lengliné H, et al. The efficacy of exclusive nutritional therapy in paediatric Crohn's disease, comparing fractionated oral vs. continuous enteral feeding. Aliment Pharmacol Ther 2011;33:1332-9.
23. Johnson T, Macdonald S, Hill SM, Thomas A, Murphy MS. Treatment of active Crohn's disease in children using partial enteral nutrition with liquid formula: A randomised controlled trial. Gut 2006;55:356-61.
24. Cohen SA, Gold BD, Oliva S, et al. Clinical and mucosal improvement with specific carbohydrate diet in pediatric Crohn disease. J Pediatr Gastroenterol Nutr 2014;59:516-21.
25. Sigall-Boneh R, Pfeffer-Gik T, Segal I, Zangen T, Boaz M, Levine A. Partial enteral nutrition with a Crohn's disease exclusion diet is effective for induction of remission in children and young adults with Crohn's disease. Inflamm Bowel Dis 2014;20:1353-60.
26. Ruemmele FM, Hyams JS, Otley A, et al. Outcome measures for clinical trials in paediatric IBD: An evidence-based, expert-driven practical statement paper of the paediatric ECCO committee. Gut 2015;64:438-46.
27. Borrelli O, Cordischi L, Cirulli M, et al. Polymeric diet alone versus corticosteroids in the treatment of active pediatric Crohn's disease: A randomized controlled open-label trial. Clin Gastroenterol Hepatol 2006;4:744-53.
28. Zachos M, Tondeur M, Griffiths AM. Enteral nutritional therapy for induction of remission in Crohn's disease. Cochrane Database Syst Rev 2007:CD000542.
29. Wilschanski M, Sherman P, Pencharz P, Davis L, Corey M, Griffiths A. Supplementary enteral nutrition maintains remission in paediatric Crohn's disease. Gut 1996;38:543-8.
30. Frivolt K, Schwerd T, Werkstetter KJ, et al. Repeated exclusive enteral nutrition in the treatment of paediatric Crohn's disease: Predictors of efficacy and outcome. Aliment Pharmacol Ther 2014;39:1398-407.
31. Levine A, Turner D, Pfeffer Gik T, et al. Comparison of outcomes parameters for induction of remission in new onset pediatric Crohn's disease: Evaluation of the porto IBD group "growth relapse and outcomes with therapy" (GROWTH CD) study. Inflamm Bowel Dis 2014;20:278-85.
32. Yamamoto T, Shiraki M. Long-term efficacy of enteral nutrition for patients with Crohn's disease. Dig Dis Sci 2012;57:3310-1; author reply 1.
33. Konno M, Takahashi M, Toita N, Fujiwara SI, Nojima M. Long-term therapeutic effect of maintenance enteral nutrition for Crohn's disease. Pediatr Int 2015;57:276-80.
34. Duncan H, Buchanan E, Cardigan T, et al. A retrospective study showing maintenance treatment options for paediatric CD in the first year following diagnosis after induction of remission with EEN: Supplemental enteral nutrition is better than nothing! BMC Gastroenterol 2014;14:50.
35. Navas-Lopez VM, Blasco-Alonso J, Maseri SL, et al. Exclusive enteral nutrition continues to be first line therapy for pediatric Crohn's disease in the era of biologics. An Pediatr (Barc) 2014: S1695-4033(14)00137-4.
36. Gupta K, Noble A, Kachelries KE, et al. A novel enteral nutrition protocol for the treatment of pediatric Crohn's disease. Inflamm Bowel Dis 2013;19:1374-8.
37. Alves A, Panis Y, Bouhnik Y, Pocard M, Vicaut E, Valleur P. Risk factors for intra-abdominal septic complications after a first ileocecal resection for Crohn's disease: A multivariate analysis in 161 consecutive patients. Dis Colon Rectum 2007;50:331-6.
38. Guo Z, Wu R, Zhu W, et al. Effect of exclusive enteral nutrition on health-related quality of life for adults with active Crohn's disease. Nutr Clinical Pract 2013;28:499-505.
39. Afzal NA, Van Der Zaag-Loonen HJ, Arnaud-Battandier F, et al. Improvement in quality of life of children with acute Crohn's disease does not parallel mucosal healing after treatment with exclusive enteral nutrition. Aliment Pharmacol Ther 2004;20:167-72.
40. Gerasimidis K, Nikolaou CK, Edwards CA, McGrogan P. Serial fecal calprotectin changes in children with Crohn's disease on treatment with exclusive enteral nutrition: Associations with disease activity, treatment response, and prediction of a clinical relapse. J Clin Gastroenterol 2011;45:234-9.
41. Buchanan E, Gaunt WW, Cardigan T, Garrick V, McGrogan P, Russell RK. The use of exclusive enteral nutrition for induction of remission in children with Crohn's disease demonstrates that disease phenotype does not influence clinical remission. Aliment Pharmacol Ther 2009;30:501-7.

42. Afzal NA, Davies S, Paintin M, et al. Colonic Crohn's disease in children does not respond well to treatment with enteral nutrition if the ileum is not involved. Dig Dis Sci 2005;50:1471-5.
43. Grogan JL, Casson DH, Terry A, Burdge GC, El-Matary W, Dalzell AM. Enteral feeding therapy for newly diagnosed pediatric Crohn's disease: A double-blind randomized controlled trial with two years follow-up. Inflamm Bowel Dis 2012;18:246-53.
44. Knight C, El-Matary W, Spray C, Sandhu BK. Long-term outcome of nutritional therapy in paediatric Crohn's disease. Clin Nutr (Edinburgh, Scotland) 2005;24:775-9.
45. Levine A, Milo T, Buller H, Markowitz J. Consensus and controversy in the management of pediatric Crohn disease: An international survey. J Pediatr Gastroenterol Nutr 2003;36:464-9.
46. Stewart M, Day AS, Otley A. Physician attitudes and practices of enteral nutrition as primary treatment of paediatric Crohn disease in North America. J Pediatr Gastroenterol Nutr 2011;52:38-42.
47. Day AS, Stephenson T, Stewart M, Otley AR. Exclusive enteral nutrition for children with Crohn's disease: Use in Australia and attitudes of Australian paediatric gastroenterologists. J Paediatr Child Health 2009;45:337-41.
48. Gerasimidis K, Bertz M, Hanske L, et al. Decline in presumptively protective gut bacterial species and metabolites are paradoxically associated with disease improvement in pediatric Crohn's disease during enteral nutrition. Inflamm Bowel Dis 2014;20:861-71.
49. Wu GD, Chen J, Hoffmann C, et al. Linking long-term dietary patterns with gut microbial enterotypes. Science 2011;334:105-8.
50. De Filippo C, Cavalieri D, Di Paola M, et al. Impact of diet in shaping gut microbiota revealed by a comparative study in children from Europe and rural Africa. Proc Natl Acad Sci U S A 2010;107:14691-6.
51. Ridaura VK, Faith JJ, Rey FE, et al. Gut microbiota from twins discordant for obesity modulate metabolism in mice. Science 2013;341:1241214.
52. Faith JJ, Guruge JL, Charbonneau M, et al. The long-term stability of the human gut microbiota. Science 2013;341:1237439.
53. Cameron FL, Gerasimidis K, Papangelou A, et al. Clinical progress in the two years following a course of exclusive enteral nutrition in 109 paediatric patients with Crohn's disease. Aliment Pharmacol Ther 2013;37:622-9.
54. de Bie C, Kindermann A, Escher J. Use of exclusive enteral nutrition in paediatric Crohn's disease in The Netherlands. J Crohns Colitis 2013;7:263-70.
55. Griffiths AM, Ohlsson A, Sherman PM, Sutherland LR. Meta-analysis of enteral nutrition as a primary treatment of active Crohn's disease. Gastroenterology 1995;108:1056-67.
56. Tung J, Loftus EV Jr, Freese DK, et al. A population-based study of the frequency of corticosteroid resistance and dependence in pediatric patients with Crohn's disease and ulcerative colitis. Inflamm Bowel Dis 2006;12:1093-100.
57. Krupoves A, Mack DR, Seidman EG, Deslandres C, Bucionis V, Amre DK. Immediate and long-term outcomes of corticosteroid therapy in pediatric Crohn's disease patients. Inflamm Bowel Dis 2011;17:954-62.
58. Lichtenstein GR, Feagan BG, Cohen RD, et al. Serious infection and mortality in patients with Crohn's disease: More than 5 years of follow-up in the TREAT registry. Am J Gastroenterol 2012;107:1409-22.
59. Terry A, Grogan JL, Casson DH, Dalzell AM, El-Matary W. Tube feeding therapy in paediatric Crohn's disease. Aliment Pharmacol Ther 2011;34:260-1; author reply 1.
60. Zallot C, Quilliot D, Chevaux JB, et al. Dietary beliefs and behavior among inflammatory bowel disease patients. Inflamm Bowel Dis 2013;19:66-72.
61. Richman E, Rhodes JM. Review article: Evidence-based dietary advice for patients with inflammatory bowel disease. Aliment Pharmacol Ther 2013;38:1156-71.

Inflammatory bowel disease patients are frequently nonadherent to scheduled induction and maintenance infliximab therapy: A Canadian cohort study

Christopher Ma MD*, Chad J Evaschesen MD FRCPC*, Grenvil Gracias MD, Vivian W Huang MD FRCPC, Darryl K Fedorak BSc, Karen I Kroeker MD FRCPC, Levinus A Dieleman MD PhD, Brendan P Halloran MD FRCPC, Richard N Fedorak MD FRCPC

C Ma, CJ Evaschesen, G Gracias, et al. Inflammatory bowel disease patients are frequently nonadherent to scheduled induction and maintenance infliximab therapy: A Canadian cohort study. Can J Gastroenterol Hepatol 2015;29(6):309-314.

BACKGROUND: Adherence to maintenance medication regimens in inflammatory bowel disease patients has traditionally been poor. Although infliximab has demonstrated efficacy in inducing and maintaining disease remission, adherence to regularly scheduled infliximab infusions is required to maintain therapeutic trough drug levels and prevent the development of anti-infliximab antibodies.

OBJECTIVES: To characterize patient adherence to regularly scheduled induction and maintenance infliximab infusions.

METHODS: A retrospective cohort study was conducted evaluating adult outpatients with Crohn disease or ulcerative colitis on an induction or maintenance regimen of regularly scheduled infliximab from 2008 to 2010 at the University of Alberta (Edmonton, Alberta). Nonadherence was defined by a discrepancy of >72 h between the scheduled date of infusion and the actual date of administration. Patients were defined as nonadherent if they received <80% of their infliximab infusions per schedule.

RESULTS: A total of 215 patients (173 Crohn disease, 42 ulcerative colitis) met the inclusion criteria. Patients received a median of 12.0 infliximab infusions (interquartile range 7.0 to 13.0) during the study period; 412 induction and 1837 maintenance infliximab infusions were administered. Of 140 patients, 109 (77.9%) were adherent to their infliximab induction regimen, while 68 of 215 (31.6%) were adherent to their infliximab maintenance regimen. One hundred ninety-eight of 215 (92.1%) patients received at least one delayed maintenance infliximab infusion and 20 (10.1%) received maintenance infusions, on average, >1 week late.

CONCLUSIONS: While three-quarters of patients are adherent to infliximab induction therapy, fewer than one-third remained adherent to their scheduled maintenance infliximab regimen.

Key Words: *Adherence; Crohn disease; Inflammatory bowel disease; Infliximab; Ulcerative colitis*

Souvent, les patients atteints d'une maladie inflammatoire de l'intestin ne respectent pas le traitement d'induction et d'entretien prévu à l'infliximab : une étude de cohorte canadienne

HISTORIQUE : En général, les patients atteints d'une maladie inflammatoire de l'intestin respectent peu leur traitement médicamenteux. Même si l'efficacité de l'infliximab à induire et maintenir une rémission est démontrée, il faut adhérer aux perfusions régulières d'infliximab afin de maintenir les taux thérapeutiques minimaux et de prévenir l'apparition d'anticorps anti-infliximab.

OBJECTIF : Caractériser l'adhérence des patients à des perfusions régulières d'induction et d'entretien à l'infliximab.

MÉTHODOLOGIE : Entre 2008 et 2010, des chercheurs de l'université de l'Alberta, à Edmonton, ont mené une étude de cohorte rétrospective pour évaluer des patients ambulatoires adultes atteints de la maladie de Crohn ou de la colite ulcéreuse suivant un traitement régulier d'induction ou d'entretien à l'infliximab. La non-adhérence au traitement était définie par un écart de plus de 72 heures entre la date de perfusion prévue et la date d'administration. Les patients étaient définis comme n'adhérant pas au traitement s'ils recevaient moins de 80 % de leurs perfusions d'infliximab par série.

RÉSULTATS : Au total, 215 patients (173 atteints de la maladie de Crohn, 42 de la colite ulcéreuse) respectaient les critères d'inclusion. Ils ont reçu une médiane de 12,0 perfusions d'infliximab (plage interquartile de 7,0 à 13,0) pendant la période de l'étude, soit 412 inductions et 1 837 traitements d'entretien. Des 140 patients, 109 (77,9 %) adhéraient au traitement d'induction à l'infliximab, et 68 sur 215 (31,6 %), au traitement d'entretien. Cependant, 198 des 215 patients (92,1 %) ont tardé avant de recevoir au moins une perfusion d'entretien à l'infliximab, dont 20 (10,1 %) avaient en moyenne une semaine de retard.

CONCLUSIONS : Les trois quarts des patients adhèrent au traitement d'induction à l'infliximab, mais moins du tiers continue d'adhérer au traitement d'entretien prévu.

Crohn disease (CD) and ulcerative colitis (UC) are chronic relapsing and remitting inflammatory bowel diseases (IBD). Similar to many chronic illnesses, lifelong therapy is typically required to maintain patients in remission. However, adherence to medical therapy in this cohort has traditionally been quite poor; a systematic review involving 4322 IBD patients found widely varying rates of nonadherence to maintenance medication regimens, including some studies reporting nonadherence rates as high as 72% (1). Multiple variables contribute to poor adherence, including illness-, treatment- and patient-related factors (2). Nonadherence in this population has been associated with poor outcomes and disease relapse (3); thus, compliance with therapy represents a critical component of the management plan.

Over the past decade, IBD patients with moderate-to-severe disease are increasingly being treated with biologic agents targeting tumour necrosis factor (TNF)-alpha, including infliximab. While infliximab has demonstrated efficacy in randomized controlled trials for the induction and maintenance of remission in both CD (4,5) and

**Authors who contributed equally to the work*
Division of Gastroenterology, University of Alberta, Edmonton, Alberta
Correspondence: Dr Richard N Fedorak, University of Alberta, 2-14A Zeidler Building, Edmonton, Alberta T6G 2X8. e-mail richard.fedorak@ualberta.ca

UC (6), its 'real-life' clinical effectiveness may be hampered by poor adherence. Specifically, secondary loss of response during maintenance therapy is a common phenomenon encountered in more than one-half of IBD patients receiving anti-TNF agents (7) with development of antidrug antibodies playing a central role. In patients treated with infliximab, missing or delaying regularly scheduled infusions may contribute to drug immunogenicity (8), decreased serum trough drug levels (9) and, ultimately, adverse clinical outcomes. In fact, frequently missing or delaying infliximab infusions can mimic an episodic administration regimen, which compared with fixed scheduled infusion protocols in subgroup analysis of the A Crohn's Disease Clinical Trial Evaluating Infliximab in a New Long-Term Treatment Regimen (ACCENT I) trial resulted in increased likelihood of antidrug antibody development (28% versus 9%), reduced likelihood of mucosal healing (18% versus 44%), increased hospitalizations (38% versus 23%), and reduced clinical response and remission rates (10).

Several authors have evaluated adherence to infliximab therapy in patients with IBD. Recently, Lopez et al (11) reviewed 13 studies evaluating adherence in this cohort and demonstrated varying rates of adherence ranging from 36% to 96%. Nevertheless, there are several limitations to the results from these studies. Most findings are extrapolated from pharmacy refill data or administrative databases (12), medical insurance claims (13) or self-reported questionnaires (14), which are inherently subject to under-reporting, recall bias and coding inaccuracy. Furthermore, they may not reflect true medication adherence because these types of data do not capture administration of delayed or early infusions. Defining nonadherence is also particularly challenging and highly heterogeneous: previous authors have used the number of missed or delayed infusions, missing refill prescriptions, medication possession ratio <80% or inadequate total infusions over a defined time period; however, these are all surrogate measures for assessing medication adherence.

Thus, in the present study, we assessed the 'true' adherence to regularly scheduled infliximab induction and maintenance therapy using manually searched medical chart review of scheduled and administered infliximab infusions. We define the proportion of patients who were nonadherent to infliximab therapy, characterize the incidence of early and delayed infusions and describe the variance around scheduled infusions. In a secondary analysis, we examined risk factors predicting nonadherence to long-term maintenance infliximab therapy.

METHODS

Study design and setting

The present retrospective cohort study was performed using data collected from CD and UC outpatients receiving maintenance infliximab therapy from August 2008 to August 2010, at the University of Alberta (Edmonton, Alberta). Patients were identified from the Division of Gastroenterology inflammatory bowel disease electronic database.

Patient population

Patients were eligible for inclusion if they met the following criteria: adult (>17 years of age); diagnosis of CD or UC; receiving either initial induction regimen of infliximab 5 mg/kg at weeks 0, 2 and 6; or advanced onto a regularly scheduled maintenance outpatient infliximab regimen. Disease phenotype, infliximab dose escalation, concurrent medical therapy and duration of infliximab before the study were not contraindications to inclusion.

Outcomes and definitions

The primary objectives of the present study were to: determine the proportion of IBD patients who are adherent to a regularly scheduled infliximab infusion protocol; and to characterize the incidence of early and delayed infliximab administration. The secondary objective was to assess risk factors predicting infliximab nonadherence.

For the present study, nonadherence to a single infliximab infusion was defined as a discrepancy of >72 h between the scheduled date of infliximab infusion and the actual date of administered infusion. There is no previous widely accepted definition for adherence to an individual infliximab infusion – a 72 h cut-off was decided on because this creates a six-day window for the patient to receive their infusion per protocol (three days before and three days after their scheduled infusion date). It was agreed that this was a sufficient time frame to accommodate statutory holidays, infusion clinic availability and minor personal reasons that may briefly delay or expedite the infusion schedule. A missed infliximab infusion was defined during maintenance therapy if the delay between administered infusions was >12 weeks.

Overall patient nonadherence was defined if <80% of infliximab infusions were received adherently. Thus, patients requiring a temporary delay in infliximab infusions due to infection, perianal disease, holidays, etc, were not deemed to be nonadherent to therapy. This cut-off was based on previous studies involving IBD patients in which definitions of nonadherence included medication possession ratio <80% and receiving <7 infliximab infusions in first year of therapy (seven infusions represents approximately 80% of the potential infusions for the first year of therapy, including three induction and six maintenance infusions) (15,16).

The scheduled date of infliximab infusion was determined by the patient's optimal infliximab infusion regimen (ie, the next scheduled infusion would be eight weeks from the previous administered infusion if the patients is receiving maintenance therapy every eight weeks). For induction therapy, the scheduled infusion dates were at weeks 0, 2 and 6. For maintenance therapy, the scheduled infusion dates were every eight weeks from the previous infusion. For patients requiring dose escalation during maintenance therapy and receiving infusions more frequently than every eight weeks, the scheduled date of infusion was adjusted accordingly.

Study follow-up time was defined from the date of first administered infusion to the date of last infusion received during the study inclusion period; total duration of infliximab therapy was determined from the date of first infliximab induction infusion.

Data collection

Data were extracted from official infliximab infusion records by two of the authors (CJE and GG) using a standardized case report form. These records included baseline patient demographic data, lifetime infliximab infusions received, infusion frequency (induction versus maintenance, maintenance infusion every eight weeks versus escalated/de-escalated schedule), actual date of administered infusion and previous medication exposure. Data were reviewed by two of the authors (CM and RNF).

Baseline patient data collected included sex, age, IBD diagnosis (CD versus UC), previous medical therapy (mesalamine or 5-aminosalicylate products, azathioprine or methotrexate) and total lifetime infliximab infusions received before inclusion. Subsequently, the scheduled infusion date was identified as described above and actual date of administered infliximab using the official infusion record. Clinical outcomes, including disease flares, hospital admissions and IBD-related surgeries, were not available in the infusion records (these outcomes were outside the scope of the present study).

Statistical methods

For continuous variables, mean, SD, median and interquartile range (IQR) were calculated. The variance in days between date of actual administered and date of scheduled infusion was calculated for each infusion.

A post hoc sensitivity analysis was performed. Because there is no consensus definition of nonadherence to an individual infliximab infusion, adherence to maintenance infliximab therapy was determined if the definition of adherence was relaxed to allow a discrepancy between scheduled and administered infusion of >7 days (from three days). A sensitivity analysis was also performed whereby patients receiving <6 maintenance infusions were excluded; this enhanced the power of the analysis to better account for temporary causes of nonadherence.

Univariate and multivariate logistic regression analyses were performed to examine the association between patient variables and

TABLE 1
Baseline patient demographics of 215 inflammatory bowel disease patients managed with induction or maintenance infliximab at the University of Alberta Inflammatory Bowel Disease Consultation and Research Clinic (Edmonton, Alberta) between 2008 and 2010

Characteristic	
Inflammatory bowel disease diagnosis	
Crohn disease	173 (80.5)
Ulcerative colitis	42 (19.5)
Infliximab dosing	
Induction dosing	140 (65.1)
Maintenance dosing	215 (100.0)
Infusion escalation during maintenance therapy	144 (67.0)
Sex	
Male	120 (55.8)
Female	95 (44.2)
Age, years, mean ± SD	
At study inclusion	41.4±13.7
At infliximab induction	40.8±13.7
Previous medications	
Mesalamine or 5-ASA products	209 (97.2)
Immunomodulators (azathioprine or methotrexate)	197 (91.6)
Infliximab Infusions, median (IQR)	
Lifetime infusions	12.0 (7.0–19.5)
Infusions during the study period	12.0 (7.0–13.0)
Follow-up, weeks, median (IQR)	
Study follow-up duration	80.1 (38.7–100.9)
Total time on infliximab	80.1 (38.7–138.7)

Data presented as n (%) unless otherwise indicated. ASA Aminosalicylate; IQR Interquartile range

nonadherence, with results expressed as OR with 95% CI. Age (defined as >40 and ≤40 years), sex, IBD diagnosis (CD versus UC), previous medication exposure, requirement for infusion escalation and lifetime infliximab infusions (defined as ≤7 infusions, eight to 12 infusions and ≥13 infusions) were a priori included in the regression model. Fistulizing disease was not included in the logistic regression model because it would not apply to patients with UC.

Statistical analysis was performed using SPSS version 22.0 (IBM Corporation, USA).

RESULTS

Baseline patient characteristics

Patient demographics are summarized in Table 1. A total of 215 patients (173 CD, 42 UC) met the inclusion criteria. One hundred forty (65.1%) patients received infliximab induction dosing during the study period and all patients received at least one maintenance infusion of infliximab. Patients received a median of 12.0 infliximab infusions (IQR 7.0 to 19.5) lifetime and were followed for a median duration of 80.1 weeks (IQR 38.7 to 100.9 weeks) during the study.

Forty-one (23.7%) CD patients had fistulizing disease; 144 patients (67.0%) required dose escalation of infliximab to more frequent infusions than every eight weeks (escalations to maintenance infusion every four to seven weeks, depending on the clinical circumstances and clinician judgement). Almost all patients had previous exposure to mesalamine, azathioprine or methotrexate.

Adherence with infliximab induction therapy

Adherence to infliximab induction infusions (5 mg/kg at weeks 0, 2 and 6) is summarized in Table 2. One hundred forty patients received a mean (± SD) of 2.9±0.3 infliximab induction infusions. A total of 412 infliximab induction infusions were administered.

TABLE 2
Patient adherence* to infliximab (IFX) induction therapy and characteristics of early and delayed IFX induction infusions in 140 inflammatory bowel disease patients

Variable	
IFX induction, n	
Patients receiving IFX induction therapy	140
Total administered IFX induction infusions	412
Adherence to induction IFX	
Patients adherent to IFX induction, n (%)	109 (77.9)
Adherent IFX induction infusions per patient, mean ± SD	2.7±0.6
Delayed induction infusions	
Patients receiving at least one delayed IFX infusion, n (%)	65 (46.4)
Late infusions per patient, mean ± SD	0.5±0.6
Late days per infusion, mean ± SD	1.5±1.8
Cumulative late days per patient, mean ± SD	2.1±4.1
Early induction infusions	
Patients receiving at least one early IFX infusion, n (%)	67 (47.9)
Early infusions per patient, mean ± SD	0.5±0.6
Early days per infusion, mean ± SD	1.2±1.6
Cumulative early days per patient, mean ± SD	1.1±2.8

**Defined as receiving >80% of induction infusions within 72 h of the regularly scheduled date of infusion at weeks 0, 2 and 6*

TABLE 3
Patient adherence* to infliximab (IFX) maintenance therapy and characteristics of early and delayed IFX maintenance infusions in 215 inflammatory bowel disease patients

IFX maintenance	
Patients receiving IFX maintenance therapy, n	215
IFX maintenance infusions, median (IQR)	9.0 (4.0–13.0)
Total administered IFX maintenance infusions, n	1837
Adherence to maintenance IFX	
Patients adherent to IFX maintenance, n (%)	68 (31.6)
Adherent IFX maintenance infusions per patient, mean ± SD	6.0±3.8
Total missed IFX infusions†, n (%)	32 (1.7)
Delayed maintenance infusions	
Patients receiving at least one delayed IFX infusion, n (%)	198 (92.1)
Late infusions per patient, mean ± SD	3.5±2.5
Late days per infusion, days, mean ± SD	3.3±4.6
Patients delaying infusions, on average, >7 days, n (%)	20 (10.1)
Early maintenance infusions	
Patients receiving at least one early IFX infusion, n (%)	171 (79.5)
Early infusions per patient, mean ± SD	2.1±1.8
Early days per infusion, days, mean ± SD	1.2±1.6

**Patient adherence defined as receiving >80% of maintenance infusions within 72 h of regularly scheduled date of infusion; †A maintenance infliximab infusion was considered to be missed if the delay between administered infusions was >12 weeks*

Of 140 patients, 109 (77.9%) were adherent to their infliximab induction infusions and received, on average, 2.7±0.6 induction infusions with adherence. Sixty-five (46.4%) patients received at least one delayed induction infusion, but mean delay time was short (1.5±1.8 days). A similar number of patients received at least one early infliximab induction infusion (n=67 [47.9%]).

Adherence to infliximab maintenance therapy

Adherence to infliximab maintenance therapy is summarized in Table 3. Two hundred fifteen patients received a total of 1837 infliximab maintenance infusions. Over the two-year study period, patients received a median of 9.0 maintenance infusions (IQR 4.0 to 13.0).

Sixty-eight of 215 (31.6%) patients were adherent to >80% of their infliximab maintenance infusions. On average, patients were adherent to 6.0±3.8 maintenance infliximab infusions over the follow-up period and, on average, accumulated 25.8±35.5 late days. Only 32 maintenance infliximab infusions were completely missed; however, nearly all (92.1% [198 of 215]) patients received at least one delayed maintenance infusion. Although the mean delay per infusion was 3.3±4.6 days, 10% (20 of 198) of patients received delayed infusions, on average, >1 week late.

Mean variance from the scheduled infliximab maintenance infusion was 4.0±4.6 days. As such, a patient would be expected to accrue a dropped infliximab maintenance infusion every 13.9 infusions (112.0 weeks) based on the standard maintenance regimen of infliximab infusion every eight weeks. Among the 20 patients experiencing a mean delay in administration of infliximab >7 days, a dropped infliximab maintenance infusion would be accrued every 33.3 weeks.

Sensitivity analysis

Exclusion of patients receiving <6 maintenance infusions in sensitivity analysis did not change overall adherence to therapy: only 46 of 145 (31.7%) patients in this cohort were adherent to their maintenance regimen. When the definition of nonadherence to an individual infliximab infusion was relaxed to allow a discrepancy between scheduled and administered infusion date of >7 days, 101 of 215 (46.5%) patients were adherent to >80% of their infliximab maintenance infusions.

Predictors of nonadherence during maintenance therapy

Univariate and multivariate ORs for risk factors predicting nonadherence to infliximab maintenance therapy are summarized in Appendix 1. Only male sex was associated with slightly increased risk for nonadherence in both univariate (OR 1.73 [95% CI 1.00 to 2.99]; P=0.049) and multivariate (OR 1.77 [95% CI 1.01 to 3.11]; P=0.046) regression analysis. No other factors reached statistical significance.

DISCUSSION

Infliximab has a central role in the management of patients with moderate-to-severe IBD and, although it has demonstrated efficacy in randomized clinical trials, real-life clinical effectiveness may be limited by poor adherence. Previous studies evaluating anti-TNF therapy compliance in this cohort have primarily used administrative data and surrogate markers of adherence such as medication possession ratio (15). While these measures capture substantial lapses in adherence, they lack the resolution to identify nonadherence on an infusion-to-infusion basis. Here, we present a large Canadian retrospective evaluation of 215 IBD patients treated with infliximab, and assess through detailed chart review, the true adherence to both induction and maintenance infliximab infusion regimens. We found that while 77.9% of patients were adherent to induction infusions, fewer than one-third of patients had sustained infusion adherence and 10% will experience delayed infusions, on average, >1 week from the ideal scheduled dosing protocol during maintenance therapy.

Adherence to infliximab was first examined by Kane et al (12) in 2006; they described 96% adherence to therapy after retrospectively reviewing 1185 infusions using administrative and pharmacy refill data. This incredibly high adherence rate relates to the use of 'no show' visits as the definition for nonadherence. Similarly, we found a very low incidence of completely missed infusions (1.7%). Other authors, using administrative database definitions for nonadherence including medication possession ratio <80% or <7 infusions during the first year of treatment, have also found varying rates of adherence, ranging from 57.1% to 79.8% (13,14,16).

In contrast to previous studies in which nonadherence was defined only by completely missed infusions, we believe that frequently delayed infliximab infusions are a reflection of poor patient compliance with the treatment regimen. Thus, we used a generous definition of adherence to both individual infusions (within six days of scheduled infusion) as well as overall adherence (80% of infusions received per schedule) to describe patient adherence to infliximab therapy with greater resolution. Other authors have used similarly constructed composite definitions of nonadherence (17). Our definition for nonadherence accounts for common barriers to compliance including patient inconvenience or statutory holidays (patients can schedule their infusion for the next available business day if the scheduled infusion date falls on a weekend or holiday and still be considered adherent to therapy) as well as unpredictable or excusable reasons for delaying infliximab infusions such as personal emergencies, infections or drug adverse effects.

Overall, adherence to anti-TNF therapy, given its necessity for scheduled intravenous administrations, may be slightly better compared with adherence to oral maintenance therapies including 5-aminosalicylate products and immunomodulators. A systematic review of oral therapy compliance in IBD patients by Jackson et al (1) found that most studies report approximately 30% to 45% nonadherence rates. This disparity may reflect differences in disease activity (highly active disease requiring biologic therapy may prompt better adherence), route and schedule of administration (intravenous versus oral, every eight week maintenance versus daily therapy), and more frequent clinician follow-up and monitoring. However, it should be highlighted that even among patients on infliximab therapy, there exists a significant rate of nonadherence. Exploring reasons for nonadherence was outside the scope of the present study but has been evaluated elsewhere (18).

Few studies have directly evaluated delays to individual scheduled infliximab infusions and are predominantly reported in abstract form. In a small retrospective cohort of 82 IBD patients on infliximab, Duncan et al (19) found that infusions were postponed in 14 of 82 (17%) patients over 12-month follow-up. Similarly, another abstract by Angelucci et al (20) found that 19.1% of infusions were delayed, median delayed administrations per patient was three infusions, and the median delay was 9.6 days (range one to 35 days). Both authors cite 'technical' reasons, including patient forgetfulness, inconvenience and intentional nonadministration as the reason for delayed infusion in >90% of cases. In our cohort, we report even higher rates of delayed infusions, delayed infusions per patient and total accumulated late days, most likely related to our increased sample size and duration of follow-up. Specific indication for delayed infliximab administration is not part of our infliximab infusion record. However, medically appropriate indications for delaying infusions, such as adverse events or infections, are accounted for by our definition of nonadherence. These specific indications for delaying infusions would presumably only affect a minority of the patient's total infusions and, thus, they should still be able to receive >80% of their infusions with adherence. In fact, Angelucci et al (20) report that only 3.3% of nonadherent infusions were due to adverse events in their cohort.

Delays and nonadherence with infliximab therapy may have significant implications on clinical outcomes in patients with IBD. In particular, we hypothesize that nonadherence has a substantial impact on long-term maintenance of clinical response and secondary loss of response. In both prospective (10) and retrospective (21) cohorts, it has been shown that patients managed with regularly scheduled infliximab compared with episodically administered therapy have improved outcomes for mucosal healing, disease-related hospitalizations and surgeries, and maintained clinical response. While all patients in our cohort were on scheduled therapy, prolonged and recurrent delays to infusion can mimic a more episodic pattern of administration. Unfortunately, clinical outcomes, including hospitalizations, surgeries and disease flares, are not tracked in our infliximab infusion records. Thus, we are unable to correlate infliximab adherence with disease outcomes, although this should be evaluated in future studies.

The mechanism behind primary and secondary loss of response to infliximab is influenced by subtherapeutic serum trough drug levels predisposing to the formation of neutralizing anti-infliximab antibodies (22). Recently, in a post hoc analysis of the ACCENT trial,

Cornillie et al (23) found that CD patients with week-14 infliximab trough level ≥3.5 μg/mL were 3.5 times more likely to achieve durable sustained response through week 54 of therapy (OR 3.5 [95% CI 1.1 to 11.4). Similar evidence exists in UC, in which detectable serum infliximab drug levels are significantly associated with clinical remission and decreased risk for colectomy (24). We hypothesize that delayed and nonadherent infliximab infusions increase the time during which patients have subtherapeutic infliximab trough levels and are predisposed to forming antidrug antibodies. Therapeutic drug monitoring with infliximab trough levels and antibody prevalence was not clinically available during the time of the study in our jurisdiction. Future studies should evaluate the relationship between delays in administered infliximab, therapeutic drug monitoring and, ultimately, clinical outcomes.

Nonadherence to infliximab maintenance therapy also has significant implications on health care utilization (25). Using the Integrated Health Care Information Service claims database, Kane et al (16) found significantly increased health care utilization costs among nonadherent CD patients on infliximab, who were 2.5 times more likely to require CD-related hospitalization, had 90% greater CD-related medical costs and 115% greater hospitalization costs compared with infliximab-adherent patients. Wan et al (26) estimated treatment costs using propensity-weighted regression models and reported similar results. Although nonadherent patients had expectedly reduced infliximab drug costs ($14,889 versus $28,289; P<0.001), they had substantially increased costs for hospitalizations ($17,634 versus $2,458; P<0.001), outpatient visits ($10,909 versus $7,357; P<0.001) and emergency room visits ($458 versus $236; P<0.001).

Early infusions did occur in our cohort, although this was less common and patients accumulated fewer early days compared with delayed infusions (6.9 early versus 25.8 late days). Although earlier infusion is likely preferable to delayed infusion to ensure ongoing maintenance of clinical response, excessively early administration may result in elevated infliximab trough levels, which have been shown to be associated with adverse events including dermatological reactions (27).

Definitive predictors of infliximab adherence have not been consistently reported across multiple studies (11,28). Kane et al (12) previously found that female patients were more likely to be nonadherent to infliximab but interestingly, the opposite trend was observed in our cohort, and nonadherent behaviours have been well described in male patients by other authors (29,30). Whether this relates to differences in patient population, financial drug coverage, temporal trends in infliximab use or sampling error is unclear. Kane et al (12) also reported that patients were more likely to be nonadherent for an infusion >18 weeks after induction; although not statistically significant in our regression analysis, the higher rate of adherence observed during induction therapy in our cohort compared with maintenance therapy suggests attrition of patient adherence behaviours over time. Additionally, we found that requirement for infusion escalation trended toward increased nonadherence; we hypothesize that common 'technical' reasons for delayed infusions such as forgetfulness and inconvenience are magnified in patients requiring more frequent infliximab infusions.

There were several limitations to the present study. First, although detailed review of infusion records enabled us to analyze infliximab drug adherence with more resolution than previous studies, we were limited in collecting other clinical parameters, including disease phenotype (particularly for CD), severity of disease and disease activity, clinical response and remission status, concurrent medical therapy and noninvasive biomarkers of inflammation (eg, C-reactive protein). This limited the power of our regression analysis for identifying predictors of nonadherence. A second limitation was that reasons for nonadherence were not available based on the infusion record. Although previous authors have defined patient-related 'technical' factors as being the predominant reason for nonadherence, we are unable to confirm this here. Medical reasons for delaying infliximab infusions such as opportunistic infections could not be captured in our data set, but we have attempted to minimize this potential misclassification bias using a composite definition of nonadherence as described above. While the criterion for defining adherence to an individual infliximab infusion may appear strict (within 72 h of schedule), this actually creates a six-day window for the patient to receive their infusion per protocol. Even when the definition for adherence was extended to seven days in the sensitivity analysis, fewer than one-half of patients were adherent.

Third, data are available to 2010 but may be less generalizable because experience with infliximab has evolved. Additionally, there may be referral bias because the present study was conducted at a tertiary care centre. In less specialized community gastroenterology practices, adherence to therapy may be even worse, particularly if infusion sites are not centralized, potentially amplifying the effects of preventable reasons associated with poor compliance (eg, inconvenience).

Finally, although we recognize that therapeutic drug monitoring will become a critical component of IBD management moving forward, data regarding serum infliximab trough levels and antibodies to infliximab were not available in our jurisdiction at the time of the study. Furthermore, clinical outcomes relevant to patients with IBD, including hospitalization, surgery and disease flare, are not recorded in our infliximab infusion database. These are major outcomes of interest associated with anti-TNF therapy adherence but were outside the scope of the present study and should be evaluated in future studies.

CONCLUSION

Although three-quarters of IBD patients were adherent to infliximab induction therapy, only one-third sustained adherence during maintenance regimens. Frequently, patients on infliximab experience delays to infusion, and future studies should evaluate whether these delays result in altered drug immunogenicity, increased anti-infliximab antibody formation or adverse patient outcomes.

DISCLOSURES: RNF has served as a consultant and advisory board member for AbbVie Canada Inc and Janssen Canada Inc, and received clinical/basic research grants from AbbVie and Janssen. KIK, LAD and BPH have served as consultants and speakers for AbbVie and Janssen. The remaining authors have no financial disclosures or conflicts of interest to declare.

FUNDING: This study was funded, in part, by the Alberta Innovates Health Solutions, Alberta Inflammatory Bowel Disease Consortium, and the Centre of Excellence for Gastrointestinal Inflammation and Immunity Research (CEGIIR).

AUTHOR CONTRIBUTIONS: RNF is acting as the article guarantor and contributed to study design, data collection, data analysis and manuscript editing. CM contributed to data analysis, manuscript drafting, and editing. CJE contributed to study design, data collection and data analysis. GG contributed to data collection and data analysis. VH, DKF, KIK, LAD and BPH contributed to manuscript editing. All authors have approved the final version of the manuscript, including the authorship list.

APPENDIX 1

Univariate and multivariate logistic regression analyses of risk factors predicting nonadherence* to infliximab maintenance therapy in 68 adherent and 147 nonadherent inflammatory bowel disease (IBD) patients

Risk factor predicting non-adherence to maintenance infliximab therapy	OR (95% CI)	
	Univariate	Multivariate
IBD diagnosis		
Crohn disease	1.00	1.00
Ulcerative colitis	0.74 (0.37–1.46)	0.62 (0.29–1.29)
Male sex	1.73 (1.00–2.99)	1.77 (1.01–3.11)
Age >40 years	1.43 (0.77–2.66)	1.37 (0.71–2.64)
Previous mesalamine/5-ASA exposure	0.87 (0.17–4.39)	0.64 (0.11–3.54)
Previous azathioprine or methotrexate	0.67 (0.26–1.78)	0.64 (0.22–1.84)
Requirement for infliximab infusion escalation	1.54 (0.86–2.74)	1.40 (0.67–2.94)
Lifetime infliximab infusions		
≤7	1.00	1.00
8–12	1.52 (0.71–3.26)	1.44 (0.65–3.15)
≥13	0.81 (0.43–1.53)	0.93 (0.43–2.04)

**Nonadherence defined as receiving <80% of infliximab maintenance infusions within 72 h of scheduled date. ASA Aminosalicylate*

REFERENCES

1. Jackson CA, Clatworthy J, Robinson A, Horne R. Factors associated with non-adherence to oral medication for inflammatory bowel disease: A systematic review. Am J Gastroenterol 2010;105:525-39.
2. Lopez-Sanroman A, Bermejo F. Review article: How to control and improve adherence to therapy in inflammatory bowel disease. Aliment Pharmacol Ther 2006;(24 Suppl 3):45-9.
3. Kane S, Huo D, Aikens J, Hanauer S. Medication nonadherence and the outcomes of patients with quiescent ulcerative colitis. Am J Med 2003;114:39-43.
4. Hanauer SB, Feagan BG, Lichtenstein GR, et al. Maintenance infliximab for Crohn's disease: The ACCENT I randomised trial. Lancet 2002;359:1541-9.
5. Sands BE, Anderson FH, Bernstein CN, et al. Infliximab maintenance therapy for fistulizing Crohn's disease. N Engl J Med 2004;350:876-85.
6. Rutgeerts P, Sandborn WJ, Feagan BG, et al. Infliximab for induction and maintenance therapy for ulcerative colitis. N Engl J Med 2005;353:2462-76.
7. Ma C, Huang V, Fedorak DK, et al. Crohn's disease outpatients treated with adalimumab have an earlier secondary loss of response and requirement for dose escalation compared to infliximab: A real life cohort study. J Crohn's Colitis 2014;8:1454-63.
8. Hanauer SB, Wagner CL, Bala M, et al. Incidence and importance of antibody responses to infliximab after maintenance or episodic treatment in Crohn's disease. Clin Gastroenterol Hepatol 2004;2:542-53.
9. Maser EA, Villela R, Silverberg MS, Greenberg GR. Association of trough serum infliximab to clinical outcome after scheduled maintenance treatment for Crohn's disease. Clin Gastroenterol Hepatol 2006;4:1248-54.
10. Rutgeerts P, Feagan BG, Lichtenstein GR, et al. Comparison of scheduled and episodic treatment strategies of infliximab in Crohn's disease. Gastroenterology 2004;126:402-13.
11. Lopez A, Billioud V, Peyrin-Biroulet C, Peyrin-Biroulet L. Adherence to anti-TNF therapy in inflammatory bowel diseases: A systematic review. Inflamm Bowel Dis 2013;19:1528-33.
12. Kane S, Dixon L. Adherence rates with infliximab therapy in Crohn's disease. Aliment Pharmacol Ther 2006;24:1099-103.
13. Carter CT, Leher H, Smith P, Smith DB, Waters HC. Impact of persistence with infliximab on hospitalizations in ulcerative colitis. Am J Manag Care 2011;17:385-92.
14. Kitney L, Turner JM, Spady D, et al. Predictors of medication adherence in pediatric inflammatory bowel disease patients at the Stollery Children's Hospital. Can J Gastroenterol 2009;23:811-5.
15. Carter CT, Waters HC, Smith DB. Effect of a continuous measure of adherence with infliximab maintenance treatment on inpatient outcomes in Crohn's disease. Patient Prefer Adherence 2012;6:417-26.
16. Kane SV, Chao J, Mulani PM. Adherence to infliximab maintenance therapy and health care utilization and costs by Crohn's disease patients. Adv Ther 2009;26:936-46.
17. Arranz EM, Arranz MDM, Herranz MJ, et al. Compliance to the scheduled treatment with infliximab. Experience in a tertiary care hospital. J Crohn's Colitis 2012;6:S156.
18. Horne R, Parham R, Driscoll R, Robinson A. Patients' attitudes to medicines and adherence to maintenance treatment in inflammatory bowel disease. Inflamm Bowel Dis 2009;15:837-44.
19. Duncan J, Sastrillo M, Baker J, et al. An audit of adherence to anti-TNF therapy in patients with inflammatory bowel disease. Gut 2011;60:A141-A2.
20. Angelucci E, Galletti B, De Nitto D, et al. Adherence to anti-tumor necrosis factor alpha therapy in inflammatory bowel diseases: An Italian monocentric experience. Gut 2011;60:A410.
21. Stein DJ, Ananthakrishnan AN, Issa M, et al. Impact of prior irregular infliximab dosing on performance of long-term infliximab maintenance therapy in Crohn's disease. Inflamm Bowel Dis 2010;16:1173-9.
22. Ben-Horin S, Chowers Y. Loss of response to anti-TNF treatments in Crohn's disease. Aliment Pharmacol Ther 2011;33:987-95.
23. Cornillie F, Hanauer SB, Diamond RH, et al. Postinduction serum infliximab trough level and decrease of C-reactive protein level are associated with durable sustained response to infliximab: A retrospective analysis of the ACCENT I trial. Gut 2014;63:1721-7.
24. Seow CH, Newman A, Irwin SP, Steinhart AH, Silverberg MS, Greenberg GR. Trough serum infliximab: A predictive factor of clinical outcome for infliximab treatment in acute ulcerative colitis. Gut 2010;59:49-54.
25. Carter CT, Waters HC, Smith DB. Impact of infliximab adherence on Crohn's disease-related healthcare utilization and inpatient costs. AdvTher 2011;28:671-83.
26. Wan GJ, Kozma CM, Slaton TL, Olson WH, Feagan BG. Inflammatory bowel disease: healthcare costs for patients who are adherent or non-adherent with infliximab therapy. J Med Economics 2014;17:384-93.
27. Huang V, Dhami N, Fedorak DK, et al. Infliximab trough levels are correlated with infliximab-associated adverse events. Gastroenterology 2014;146:S-1.
28. Fidder HH, Singendonk MM, van der Have M, Oldenburg B, van Oijen MG. Low rates of adherence for tumor necrosis factor-alpha inhibitors in Crohn's disease and rheumatoid arthritis: Results of a systematic review. World J Gastroenterol 2013;19:4344-50.
29. Sewitch MJ, Abrahamowicz M, Barkun A, et al. Patient nonadherence to medication in inflammatory bowel disease. Am J Gastroenterol 2003;98:1535-44.
30. Kaplan GG, Panaccione R, Hubbard JN, et al. Inflammatory bowel disease patients who leave hospital against medical advice: Predictors and temporal trends. Inflamm Bowel Dis 2009;15:845-51.

12

Fecal calprotectin use in inflammatory bowel disease and beyond: A mini-review

Bashaar Alibrahim MD[1,2], Mohammed I Aljasser MD FRCPC DABD[3], Baljinder Salh MB FRCPC[4]

B Alibrahim, MI Aljasser, B Salh. Fecal calprotectin use in inflammatory bowel disease and beyond: A mini-review. Can J Gastroenterol Hepatol 2015;29(3):157-163.

Given the number of inflammatory disorders affecting the gastrointestinal tract directly and indirectly, coupled with the considerable overlap with functional disorders, it is evident that more useful noninvasive diagnostic tests are required to aid with diagnosis. If these tests can also have some utility for individual patient follow-up in terms of disease activity and response to treatment, as well as providing forewarning of disease relapse, it would be extremely useful information for the clinician. One recently described test that may fulfill several of these attributes is based on leakage of a mononuclear cell cytoplasmic protein, calprotectin, along the intestinal tract, which can then be quantified in feces. This has been used to distinguish patients exhibiting symptoms of irritable bowel syndrome from patients with inflammatory bowel disease, with a measure of success greater than with currently used techniques. The present article summarizes the experience with this test used in inflammatory bowel disease, as well as a variety of gastrointestinal disorders.

Key Words: *Calprotectin; Fecal; Gastrointestinal disorders; Inflammatory bowel disease*

L'utilisation de la calprotectine fécale en cas de maladie inflammatoire de l'intestin et d'autres affections : une mini-analyse

Étant donné le nombre de troubles inflammatoires de l'intestin qui touchent directement et indirectement le tube digestif, de même que leur chevauchement considérable avec les troubles fonctionnels, il est évident qu'il faut plus de tests non effractifs utiles pour contribuer au diagnostic. Les cliniciens trouveraient également très utile que ces tests aient également une certaine utilité lors du suivi des patients, pour déterminer l'activité pathologique et la réponse au traitement, et pour de prévoir les récidives. Un test décrit récemment, qui respecterait plusieurs de ces caractéristiques, repose sur la fuite de la calpoprectine, une protéine cytoplasmique des cellules mononucléées située le long du tube digestif, qui peut ensuite être quantifié dans les selles. Elle a été utilisée pour distinguer les patients ayant des symptômes de syndrome du côlon irritable de ceux atteints d'une maladie inflammatoire de l'intestin, et la mesure de son succès est plus élevée que celle des techniques actuelles. Le présent article résume l'expérience de ce test chez les personnes atteintes d'une maladie inflammatoire de l'intestin, de même que de divers autres troubles gastro-intestinaux.

Gastrointestinal (GI) symptoms, such as abdominal pain and diarrhea, are common presenting symptoms in the general population. Determining the underlying cause of these symptoms is often challenging. The development of noninvasive diagnostic tools is useful to differentiate organic from functional bowel diseases and may reduce the need for unnecessary invasive procedures such as colonoscopy. One of these diagnostic tools is the measurement of fecal calprotectin (CPN), a major cytoplasmic protein in neutrophils (1).

Fecal CPN has gained popularity because of its high sensitivity and specificity in the diagnosis of inflammatory bowel disease (IBD) (2). In addition, it is a relatively simple and inexpensive test. However, it should be noted that many conditions have been shown to be associated with fecal CPN elevation, and that intestinal inflammation is a common feature in those conditions. The degree of elevation varies according to the cause, which may interfere with the diagnostic value of the test. Therefore, a broad knowledge of these conditions is essential for understanding the clinical utility of fecal CPN. The present review is aimed at highlighting the different causes associated with fecal CPN elevation.

HISTORY

The first description of CPN in the literature was in 1980 by Fagerhol et al (3) when it was named 'L1 protein'. Thereafter, it has been mentioned in the literature under different names such as MRP-8/14, calgranulin and cystic fibrosis antigen. In 1992, Roseth et al (4) developed the first method for isolating and quantifying CPN in stool using ELISA and rabbit anti-CPN. Many years later, an improved, commercially available and validated ELISA was developed, which measures CPN concentration in mg/kg rather than mg/L as in the original assay (5). Over the past two decades, fecal CPN was described to be a useful marker of several GI diseases. These include gastric cancer, colorectal adenoma or cancer, Crohn disease (CD) and ulcerative colitis (UC) (6).

FECAL CPN

CPN is a 36 kDa calcium-binding heterocomplex protein consisting of two heavy chains and one light chain. It belongs to the S-100 protein family and is derived predominantly from neutrophils and monocytes. CPN and its subunits appear to have mainly regulatory functions in inflammatory processes, as well as antimicrobial and antiproliferative activities (1). Being resistant to enzymatic degradation, it can be easily measured in stools with a commercially available ELISA immunoassay. Due to its high sensitivity and specificity, relative simplicity, quick turnaround time and long stability at room temperature (up to seven days), it has been increasingly used in the diagnostic process for IBD (4). The test can be performed on 50 mg to 100 mg of random stool samples that can be sent to the laboratory by mail because the protein is remarkably stable in stool. It costs approximately USD$100 per assay.

Fecal CPN concentrations in healthy individuals have been established in several studies. In the original study, the median stool CPN concentration in healthy adult was 2 mg/L, and the suggested cut-off for a positive test was 10 mg/L. In the newer assay, the suggested upper

[1]*Division of Internal Medicine, Department of Medicine, University of British Columbia, Vancouver, British Columbia;* [2]*Department of Medicine, King Faisal Special Hospital and Research Centre;* [3]*Department of Medicine, King Saud Bin Abdulaziz University for Health Sciences, King Abdulaziz Medical City, Riyadh, Saudi Arabia;* [4]*Division of Gastroenterology, Department of Medicine, University of British Columbia, Vancouver, British Columbia*

Correspondence: Dr Bashaar Alibrahim, Division of Internal Medicine, University of British Columbia, 5th Floor, 2775 Laurel Street, Vancouver, British Columbia V5Z 1M9. e-mail dr_bashare@hotmail.com

TABLE 1
Conditions associated with fecal calprotectin elevation

Inflammatory
Inflammatory bowel disease
Pouchitis
Graft rejection following intestinal transplant
Collagenous colitis
Graft versus host disease
Ankylosing spondylitis
Systemic sclerosis
Diverticular disease
Celiac disease
Pancreatitis
Infectious
HIV
Infectious diarrhea
Neoplastic
Colon cancer
Intestinal polyposis
Pancreatic cancer
Iatrogenic
Nonsteroidal anti-inflammatory drugs
Proton pump inhibitors
Radiotherapy
Miscellaneous
Liver cirrhosis
Food allergy
Gastroesophageal reflux disease
Cystic fibrosis
Young age

limit of normal has been increased by a factor of five, to 50 µg/g (7). However, the test appears to have better diagnostic precision for IBD at a cut-off of 100 µg/g than at 50 µg/g (2).

Levels of fecal CPN has been found to correlate well with radiolabelled white cell scanning, as well as histopathological and endoscopic features used in assessing IBD activity (8,9) (Box 1).

CONDITIONS ASSOCIATED WITH ELEVATION OF FECAL CPN

Many conditions have been linked to an elevation of fecal CPN (Table 1). However, the current clinical use of fecal CPN is focused on differentiating IBD from irritable bowel syndrome (IBS), monitoring disease activity for relapse and detecting disease recurrence postoperatively. The evidence for fecal CPN elevation in most of the other conditions listed is limited to case series and small observational studies.

IBD

Distinguishing between IBD and IBS: Fecal CPN has been shown to be of diagnostic value in IBD. The major systematic reviews (10,11) and meta-analyses (2,12) are shown in Table 2. In a meta-analysis of 30 studies including 5983 patients who underwent fecal CPN testing, Von Roon et al (2) found that the mean fecal CPN levels in patients with IBD were higher by 219.2 µg/g compared with control subjects (P<0.001). The pooled sensitivity and specificity of fecal CPN for distinguishing IBD from non-IBD patients from nine studies of adult and pediatric populations at a 50 µg/g cut-off was calculated to be 89% and 81%, respectively.

Disease activity and treatment response in IBD: Notably, fecal CPN has been shown to be a reliable marker for assessment of disease activity in IBD patients (13-17). In a study of patients with CD who underwent 140 ileocolonoscopies (13), fecal CPN correlated well with the simple endoscopic score for CD (Spearman rank correlation coefficient r=0.75). This correlation with endoscopic assessment was superior to that of C-reactive protein, blood leukocyte levels and CD activity index. Additionally, fecal CPN discriminated among the various subgroups of endoscopic activity index (inactive, mild, moderate and high activity). The sensitivity and specificity of fecal CPN for detecting endoscopic active disease (simple endoscopic score for CD ≥4) was 89% and 72%, respectively, at a cut-off value of 70 µg/g.

BOX 1

Fecal calprotectin is a useful test for organic causes of gastrointestinal symptoms, especially IBD. Its high sensitivity and simplicity make it a very useful screening tool in the primary care setting.

BOX 2

Proper use of fecal calprotectin has the potential to substantially reduce the numbers of unnecessary colonoscopies and the overall cost of care of patients with gastrointestinal symptoms.

Similarly, a study of 134 patients with UC showed a significant correlation with the Rachmilewitz endoscopic activity index (r=0.834) (18), which was superior to the Rachmilewitz clinical activity index, C-reactive protein and blood leukocyte for the detection of disease activity. Also, fecal CPN was able to discriminate among various groups of endoscopic activity indices. The sensitivity and specificity of fecal CPN for detection of endoscopic active disease (Rachmilewitz endoscopic activity index ≥4) was 93% and 71%, respectively, at a cut-off of 50 µg/g.

Fecal CPN has been investigated in assessing IBD patients' clinical response to treatment. A study of 38 patients (27 UC, 11 CD) presenting with disease relapse had fecal CPN measured at presentation, and at week 4 and week 8 after treatment (19). Treatment of relapse was individualized according to standard recommendations for management of IBD. UC patients showed a significant correlation of fecal CPN with the clinical score at week 4 after treatment (r=0.0424, P<0.01), while CD patients showed a significant correlation with the clinical score (Harvey-Bradshaw index) at week 8 after treatment (r=0.7804, P<0.01).

Predicting relapse in IBD: Several studies have indicated the value of CPN in predicting relapse in IBD, and have been the subject of a meta-analysis (20) (Table 3). This included a total of 672 patients of whom 318 had UC and 354 had CD. For CPN to predict relapse there was a pooled sensitivity and specificity of 78% and 73%, respectively. The data were comparable for both UC and colonic/ileocolonic CD.

Underscoring the value of regular monitoring of CPN, De Vos et al (21) determined levels every four weeks on patients with UC receiving infliximab, and found that the best predictor of a flare were two consecutive CPN levels >300 mg/kg. Those patients considered to be in deep remission (partial Mayo score <3 at all times and endoscopic Mayo score 0 at week 52) had consistently low CPN levels (<40 mg/kg).

Cost: With respect to cost, Waugh et al (10) conducted a comprehensive review regarding the economic value of fecal CPN in distinguishing IBD from IBS. They found that by reducing the number of unnecessary colonoscopies/gastroenterologist referrals, the use of fecal CPN test could lead to cost savings, especially in the secondary care setting. Furthermore, a recent cost-effectiveness study found that fecal CPN screening for a patient suspected to have IBD saved USD$417 per patient screened (22) (Box 2).

Subclinical disease: Interestingly, fecal CPN has been used to demonstrate the presence of subclinical intestinal inflammation in the first-degree relatives of CD patients (23). This particular study included 49 patients, 16 spouses and 151 first-degree relatives of CD patients. There was a statistically significant higher fecal CPN in the patients and first-degree relatives compared with the spouses and the control group (P<0.0001).

TABLE 2
Studies evaluating the performance of fecal calprotectin (CPN) in distinguishing inflammatory bowel disease (IBD) from irritable bowel syndrome (IBS)

Author/year	Study design	Subjects/studies, n/n	Sensitivity, %	Specificity, %	Cut-off value	Comments
Waugh et al, 2013 (10)	Systematic review	730/7	93	94	50 μg/g	Included many studies of different GI conditions. The seven studies included in this table specifically compared IBD with IBS
Jellema et al, 2011 (11)	Systematic review (primary care)	863/9	64–100	70–100	Multiple cut-offs ranging from 10 mg/L to 170 μg/g	When compared with other blood and fecal tests (CRP, ESR, IgG, ANCA, lactoferrin), CPN performed the best
Van Rheenen et al, 2010 (12)	Meta-analysis	670/6 (adults)	93	97	50 μg/g (Most studies)	CPN theoretically has the potential of reducing the number adults requiring colonoscopy by 67%.
Von Roon et al, 2007 (2)	Meta-analysis	5983/30	95	91	50 μg/g	IBD patients were compared with a heterogeneous group of patients with other GI conditions and showed a statistically significant higher level of CPN

ANCA Antineutrophil cytoplasmic antibody; CRP C-reactive protein; ESR Erythrocyte sedimentation rate; GI Gastrointestinal; IgG Immunoglobulin G

TABLE 3
Summary of the major studies correlating fecal calprotectin (CPN) with inflammatory bowel disease activity/relapse in adults

Author/year	Patients, n	Sensitivity, %	Specificity, %	Comments
Mooiweer et al, 2014 (17)	164 (83 CD, 74 UC, 7 unclassified)	86	72	The study used a cut-off of 140 mg/kg, and used the Mayo endoscopic activity score to assess mucosal inflammation
Faubion et al, 2013 (16)	264 (157 CD, 107 UC)	78	54	The sensitivity and specificity are for CPN at cut-off of 100 μg/g and SES-CD ≥4. The sensitivity and specificity in UC was not mentioned. However, there was a strong association with Mayo endoscopic score
Lobatón et al, 2013 (15)	123 UC	73.5	89.7	These values are for fecal CPN <250 μg/g to predict Mayo endoscopic score ≤1
Schoepfer et al, 2013 (14)	228 UC	91	90	The study used a cut-off of >57 μg/g to predict endoscopic active disease using modified Baron Score ≥2 points
Mao et al, 2012 (20)*	672 (meta-analysis of 6 studies; 318 UC, 354 CD)	78	73	This study had multiple CPN cut-off ranges, from 50 μg/g tp 340 μg/g. Identification of relapse was based on clinical activity indices or endoscopic findings.
Schoepfer et al, 2010 (13)	122 CD	89	72	These results are for CPN ≥70 μg/g correlated with SES-CD. The study also shows superior diagnostic accuracy compared with C-reactive protein, blood leukocyte and CD activity index

**This meta-analysis includes several earlier studies. CD Crohn disease, SES-CD Simple endoscopic score for CD; UC Ulcerative colitis*

Pouchitis

Patients undergoing restorative proctocolectomy for UC have a 40% lifetime risk of developing pouchitis (24), which is an inflammatory condition of the pouch with neutrophilic infiltration of the mucosa (25). The role of fecal CPN as a biomarker of this condition was assessed by a study of 54 patients who had undergone restorative proctocolectomy for UC and familial adenomatous polyposis (26). Six of 46 UC patients had pouchitis and pre-pouch ileitis, 13 had pouchitis alone and 27 were uninflamed. Of the eight familial adenomatous polyposis patients, one had pouchitis and pre-pouch ileitis, and seven had pouchitis alone. Fecal CPN >50 μg/g correlated significantly with high endoscopic inflammatory scores (scale of 0 to 6, depending on the number of acute macroscopic inflammatory features seen) (P<0.001). In addition, fecal CPN >50 μg/g correlated significantly with higher histological inflammation score (modification of the Moskowitz 12-point scoring system) (P<0.001).

Having established that fecal CPN levels are elevated in IBD, there are still a number of disorders that present with diarrhea that may impact on the interpretation of results. These are discussed below (Box 3).

Collagenous/microscopic colitis

Collagenous colitis is a disorder characterized by watery diarrhea, microscopic mucosal inflammation and deposition of collagen below the surface epithelium of colonic mucosa (27). There are generally no endoscopic abnormalities and colonic biopsy is needed to establish the diagnosis.

> **BOX 3**
> Inflammatory bowel disease is the only well-studied indication for fecal calprotectin test. However, because this test is a marker of gastrointestinal inflammation, it may have other potential uses.

The value of fecal CPN as a biomarker of collagenous colitis was evaluated by a study that involved 21 patients with active collagenous colitis, 12 patients with collagenous colitis in remission and 13 healthy controls (28). The median fecal CPN level in the active disease group was 80 μg/g compared with 26 μg/g in the disease remission group (P=0.025), and 6.25 μg/g in the control group (active versus control P=0.002).

In a recent study of 78 patients with chronic nonbloody diarrhea, 15 were found to have microscopic colitis (29). However, there was no correlation between this and CPN levels, indicating the importance of colonic biopsy for making this specific diagnosis.

Graft-versus-host disease

Diarrhea is common after hematopoietic stem cell transplant. The differential diagnosis includes infectious causes, chemo- or radiotherapy-associated toxicity, and GI graft-versus-host disease (GI-GVHD). Endoscopy remains the gold standard for the diagnosis of GI-GVHD.

However, the invasive nature of this procedure and the fact that this procedure may yield nonspecific findings, has led to a search for noninvasive markers. Fecal CPN may be a good marker due to the inflammatory nature of GVHD. This hypothesis was prospectively tested by Rodriguez-Otero et al (30), who recruited a total of 72 GVHD patients, of whom 51 had GI-GVHD. There were no statistically significant differences in fecal CPN levels between the GI-GVHD and non-GI-GVHD groups. The sensitivity of fecal CPN was only 30%. However, in this study, 21 of the 51 GI-GVHD patients had stage 1 GVHD, which has less inflammation than more advanced stages of GI-GVHD. Notably, higher concentrations of fecal CPN were strongly associated with steroid-resistant GVHD (P=0.00001).

A recent study of 59 patients with GVHD showed that mean fecal CPN in the patient with GI-GVHD was 500 mg/kg compared with 95 mg/kg in non GI-GVHD patients (P=0003) (31). Sensitivity and specificity were 100% and 81.8%, respectively at a cut-off of 160 mg/kg. However, the authors did not specify the stage of GVHD. These results were consistent with a prospective study of 23 post-stem cell transplant patients presenting with diarrhea, 11 of whom had colonoscopy-proven GI-GVHD (32). The fecal CPN level in patients with GI-GVHD was significantly higher than in patients with non-GI-GVHD diarrhea (P<0.001). The calculated sensitivity and specificity of fecal CPN were 83.3% and 90.9%, respectively, at a cut-off of 250 mg/kg. In this study, most of the patients presented with more advanced stages of GI-GVHD.

In an intriguing study hinting at the role of mucosal immunity in the development of acute GVHD (33), 24 pediatric allogeneic hematopoietic cell transplant recipients had their fecal CPNs measured on days 0, +5, +10 and +15. Surprisingly, a value of <424 mg/kg on day +10 was associated with a 77.8% incidence of acute GVHD.

In conclusion, fecal CPN is a good marker for advanced stages of GI-GVHD as well as steroid-resistant GI-GVHD, and reduced levels early on may predict the onset of acute GVHD.

Ankylosing spondylitis

Ankylosing spondylitis (AS) and IBD are known to share some clinicopathological features. These include chronic inflammation of unknown etiology, genetic overlap, and similar microscopic and macroscopic findings in gut inflammation (34-37). In addition, 5% to 10% of patients with AS have concurrent IBD (38).

A pilot prospective study of 39 patients with AS and 42 healthy controls showed a significant difference in the median fecal CPN in the AS group compared with the control group (P<0.001) (39). Another study showed that 68% of 210 AS patients had positive fecal CPN without associated GI symptoms (40). However, patients taking nonsteroidal anti-inflammatory drugs (NSAIDs), which can cause fecal CPN elevation, were not excluded. In fact, fecal CPN levels were found to be higher in the latter group of patients (40).

Systemic sclerosis

GI involvement in systemic sclerosis is known to have an inflammatory component (41). Therefore, a study of 81 systemic sclerosis patients was conducted to evaluate the role of fecal CPN as a biomarker of GI involvement of this disease (42). The study showed that 62 of 81 (76%) patients had a positive fecal CPN test using a cut-off of 50 µg/g. The mean CPN level was 174 µg/g.

Furthermore, mean fecal CPN was significantly higher in patients with pathological cineradiography results compared with patients with normal cineradiography (P=0.013).

Diverticular disease

Given the therapeutically beneficial effect of 5-aminosalicylic acid and the histologically inflammatory nature of diverticular disease, the potential use of fecal CPN as a marker for this disease was investigated (43). Tursi et al (44) used a semiquantitative test to assess fecal CPN elevation in 48 patients with newly diagnosed diverticular disease, 16 healthy controls and 16 patients with IBS. There was a significant fecal CPN elevation in patients with symptomatic uncomplicated diverticular disease and acute diverticulitis compared with the other groups. Fecal CPN levels normalized after treatment with mesalazine and rifaximin, in both symptomatic uncomplicated diverticular disease and acute diverticulitis.

Celiac disease

Celiac disease is a chronic immune-mediated disorder that primarily affects the GI tract. Histologically, celiac disease is characterized by villous atrophy and crypt cell hyperplasia, with both lamina propria and epithelial infiltration by lymphocytes, macrophages and plasma cells. One adult and two pediatric studies have correlated fecal CPN level with the inflammation in celiac disease.

In a study of 28 untreated adult celiac patients and 30 healthy volunteers (45), the mean fecal CPN levels were 45 µg/g and 36 µg/g, respectively. Although this difference was not statistically significant, it revealed a trend toward higher fecal CPN in celiac patients. Capone et al (46) assessed the correlation between CPN levels above a cut-off of 75 µg/g with symptoms, histology (Marsh grade) and tissue transglutaminase in 50 patients (46). They came to the conclusion that there was no signal for subclinical inflammation in celiac disease.

Fecal CPN was found to be significantly higher in 31 children with untreated celiac disease compared with 33 treated celiac disease patients and 34 normal controls (47). The mean fecal CPN was 117.2 µg/g in celiac patients compared with 3.7 µg/g for celiac children on a gluten-free diet and 9.6 µg/g in the control group (P=0.001). The second study included 29 children with newly diagnosed celiac disease and 10 healthy children (48). The mean fecal CPN was 13.4 mg/L in celiac patients compared with 4.3 mg/L in the controls (P<0.004).

Although there is no convincing evidence to support the use of fecal CPN as a diagnostic marker for celiac disease in the adult population, it may be considered as one of the possible causes of fecal CPN elevation in pediatric cases.

Pancreatitis

Pancreatic enzymes play a major role in digestion and absorption, and alterations of the normal secretion of these enzymes has been shown to have significant effect on intestinal ecology including bacterial overgrowth, change of bile acid absorption and modification of intestinal permeability (49-51). These changes may be expected to be associated with intestinal inflammation.

A study of 90 patients with different pancreatic diseases aimed to determine the association between pancreatic diseases and intestinal inflammation (52). This study included 20 patients with chronic pancreatitis and 11 patients with a history of recent acute pancreatitis. It was found that 55% and 40% of those patients, respectively, had high fecal CPN compared with only 4.3% of the normal control group. In this study, a semiquantitative method was used with a cut-off of only 15 µg/g, indicating that there is unlikely to be any confusion between IBD and pancreatic disorders.

Infectious diarrhea

Stool culture is considered to be the gold standard for the diagnosis of infectious diarrhea. However, the time required to obtain results (typically 48 h) and the relatively high cost of this test has led to a search for alternative diagnostic methods. Understandably, many stool biomarkers have been tested, of which fecal CPN showed the most promising results. A large multicentre study of 2383 patients with acute diarrhea showed a sensitivity and specificity of 83% and 87%, respectively, for fecal CPN in the diagnosis of acute bacterial diarrhea (53). The mean fecal CPN level for patients with culture-proven bacterial diarrhea was 142.84 mg/L compared with 16.8 mg/L in culture-negative patients. However, this study did not test patients for viral or parasitic infections.

Three viral infections were tested in a pediatric study that assessed fecal CPN elevation in patients with bacterial and viral diarrhea (54). These infections were rotavirus (52 patients), adenovirus (eight patients) and norovirus (31 patients). The mean fecal CPN levels were 89 µg/g, 95 µg/g and 93 µg/g, respectively, compared with 43 μg/g in the control group. The mean fecal CPN level for bacterial (*Salmonella*, *Campylobacter*) diarrhea patients was 754 μg/g, indicating a clear separation from viral diarrhea patients.

HIV

Immune activation is an important mechanism of CD4+ lymphocyte depletion in HIV infection, and is hypothesized to be due to gut flora translocation (55). A possible cause of gut flora translocation is damage of the mucosal barrier by intestinal inflammation and alteration of intestinal flora. To verify this, a study of 58 HIV-positive asymptomatic antiretroviral-naive individuals measured fecal CPN and the alteration of gut microbiota (56). The study showed that approximately one-half of the subjects (27 of 58) had a positive fecal CPN test at a cut-off 50 μg/g. Moreover, 34% (18 of 53) of the patients had levels >100 μg/g. It will be of interest to see how fecal CPN is affected by diarrhea from various causes in this patient population.

NEOPLASTIC CONDITIONS

Colorectal cancer and adenomatous polyps

One of the earliest areas of interest regarding fecal CPN was colorectal malignancy. A pilot study in 1993 by Roseth et al (57) found that fecal CPN was increased in stools in 50 of 53 patients with colorectal cancer (CRC) and in 32 of 40 patients with colorectal polyps.

However, data from the Norwegian Colorectal Cancer Prevention trial including 2321 patients (58) suggested a poorer sensitivity and specificity of fecal CPN compared with fecal occult blood test. Similarly, in a meta-analysis, Von Roon et al (2) found that the mean fecal CPN was higher by 132 μg/g in patients with CRC compared with those without CRC (P=0.18). They concluded that fecal CPN cannot be recommended as a screening test for CRC in the general population because of low-pooled sensitivity and specificity (36% and 71%, respectively) (Box 4).

Although fecal CPN is not recommended as a screening tool for CRC due to the availability of better screening tests, colorectal neoplasms should be considered as one of the causes of fecal CPN elevation.

Pancreatic cancer

The same mechanism of pancreatic enzyme insufficiency inducing intestinal inflammation may apply to pancreatic cancer. In the aforementioned study on pancreatitis (52), eight of 15 (53.3%) patients with pancreatic cancer had a positive fecal CPN level using a semi-quantitative method; 15 μg/g was used as the cut-off, making it of very limited value in this condition.

IATROGENIC

NSAIDs

NSAIDs have been shown to induce enteropathy at different levels of the GI tract. Some studies that assessed the inflammatory process associated with NSAIDs used fecal CPN to demonstrate this effect. In a study involving 312 patients taking different types of NSAIDs, 44% had a positive fecal CPN at a cut-off of 8.9 mg/L (59). The median CPN level was 7.3 mg/L compared with 2 mg/L in the control group (P<0.001). There were no significant differences between the different NSAIDs in the amplitude of CPN elevation. Another study demonstrated that the effect of NSAIDs on fecal CPN might be seen as early as a few days after the initiation of these medications (60).

Proton pump inhibitors

Although the exact mechanism is not clear, an incidental finding from a study involving 230 subjects to assess the normal range of fecal CPN in a late middle-aged population suggested that the use of proton pump inhibitors (PPIs) was associated with a higher level of fecal CPN (61). The mean fecal CPN level for those on PPIs was 78.16 μg/g compared with 30.9 μg/g for those who were not on PPIs (P<0.001); this finding appeared to be independent of the presence of reported dyspepsia.

> **BOX 4**
>
> Successful use of the fecal calprotectin test may rely not only on knowledge of potential causes of fecal CPN elevation, but also on disease and treatment-related changes within each individual patient.

Radiotherapy

Therapeutic radiotherapy to the pelvis induces acute GI symptoms in >90% of patients (62). Currently, there is no noninvasive method to measure radiotherapy-induced damage in the GI tract.

A study involving 59 patients who underwent pelvic radiotherapy measured fecal CPN at baseline and five weeks following radiotherapy (63). The mean fecal CPN at base line was 37.7 μg/g versus 62.9 μg/g at the fifth week post radiotherapy (P=0.01). However, this is unlikely to be any value in the management of these patients.

MISCELLANEOUS

Liver cirrhosis

In liver cirrhosis, numerous alterations in intestinal flora, mucosal barrier and immunological defense mechanisms have been described (64). The gut flora and bacterial translocation also play a role in hepatic encephalopathy (65). The role of fecal CPN as a marker of complications of liver cirrhosis was evaluated in a study of 61 patients with liver cirrhosis and 42 controls (66). The mean fecal CPN in the former group was 65.7 μg/g compared with a mean of 17.5 μg/g in the control group (P<0.001). Furthermore, this study showed that the degree of fecal CPN elevation is directly proportional to the severity of liver disease as assessed by the Child-Pugh score and the Model for End-stage Liver Disease scores, as well as hepatic encephalopathy and spontaneous bacterial peritonitis.

Food allergy

Food allergy is defined as an immune-mediated adverse reaction following the ingestion of an allergen; therefore, fecal CPN might be used as a marker of mucosal inflammation. A pediatric study of 281 children referred with GI symptoms other than diarrhea or abdominal pain included 49 patients with allergic colitis (67). This study showed that these patients had significantly higher fecal CPN compared with controls and patients with functional bowel disorders (P<0.001). Additionally, a reducing pattern toward normal fecal CPN level occurred after at least four weeks of exclusion diet (P<0.001).

Another study (68), conducted to determine the diagnostic accuracy of fecal CPN in the diagnosis of organic causes of diarrhea, suggested that food allergies have an association with high fecal CPN level. This study included 20 children diagnosed with chronic diarrhea attributed to cow's milk allergy or multiple food allergies, 17 (85%) of whom had elevated CPN levels.

Gastroesophageal reflux disease

Fecal CPN may also be a marker of mucosal inflammation of the upper GI tract. The aforementioned pediatric study of 281 children presented with different GI disorders (67) included 17 patients diagnosed with gastroesophageal reflux disease. The median fecal CPN value was 138 μg/g in the gastroesophageal reflux disease group compared with a median of 28 μg/g in the healthy control group (P<0.001).

Cystic fibrosis

There is some evidence of intestinal inflammation in cystic fibrosis (CF) supported by studies using gut lavage and endoscopic biopsy techniques (69,70). A study of 30 children with CF, 30 healthy controls and 15 IBD patients attempted to establish the incidence of intestinal inflammation in CF patients using fecal CPN and rectal nitric oxide production (71). All the CF patients were on pancreatic enzyme replacement before the study. The mean fecal CPN in CF patients was 219 μg/g versus 46 μg/g in the control group (P<0.01). The mean fecal CPN in the CF group was significantly lower than the mean value of the IBD group (309 μg/g) (P<0.001).

Graft rejection following intestinal transplant

Graft surveillance after intestinal transplant is difficult because of the need of serial invasive procedures and biopsies for the diagnosis of graft rejection. The ability to diagnose rejection noninvasively and early will be a major advance in the care of intestinal transplant patients. To find a better method of graft surveillance, a study collected serial stool samples for CPN level from 72 postintestinal transplant patients (64 children, 10 adults, two children underwent transplant twice) (72). The patients were divided into groups based on the biopsy results (normal, rejection, viral enteritis, bacterial enteritis, non-specific enteritis, indeterminate rejection and total mucosal loss). Fecal CPN level was significantly higher in the rejection, indeterminate rejection and total mucosal loss groups ($P<0.05$). However, the three stool samples taken at days 4, 7 and 14 before the index biopsies showing rejection showed no statistically significant different compared with the day of the biopsy. The study concluded that the exact role of fecal CPN in monitoring of graft rejection was unclear, and more frequent prospective sampling could perhaps demonstrate an advantage as an earlier indication of rejection.

Young age (<5 years)

A tendency toward a higher fecal CPN levels in infants and young children has been demonstrated. One study analyzed fecal CPN level of 115 presumably healthy children visiting local community centres for a routine examination (73). The children were divided into six age groups. The study showed abnormal mean fecal CPN levels for infants and young children up to five years of age (considering 50 µg/g as the upper limit of normal). The groups younger than one year of age had the highest mean fecal CPN levels (Box 5).

BOX 5

The noninvasive, inexpensive nature of the fecal CPN test, together with its excellent performance as a marker of inflammation, makes it a useful tool for intestinal inflammatory diseases research.

CONCLUSION

Although there are many conditions that may contribute to the elevation of fecal CPN level, the best available evidence currently only supports the use of fecal CPN as a screening and monitoring tool for IBD. We believe there may be an under-appreciated potential of other uses for fecal CPN as a diagnostic and follow-up tool for other GI conditions, where the individual patient acts as his/her own control. The simplicity and cost effectiveness of this test might help in decreasing the use of more invasive procedures and better planning for further investigation. However, more studies are clearly still needed to validate the utilization of fecal CPN use in different medical and surgical conditions.

DISCLOSURES: The authors have no financial disclosures or conflicts of interest to declare.

REFERENCES

1. Johne B, Fagerhol MK, Lyberg T, et al. Functional and clinical aspects of the myelomonocyte protein calprotectin. Mol Pathol 1997;50:113-23.
2. von Roon AC, Karamountzos L, Purkayastha S, et al. Diagnostic precision of fecal calprotectin for inflammatory bowel disease and colorectal malignancy. Am J Gastroenterol 2007;102:803-13.
3. Fagerhol MK, Dale I, Andersson T. A radioimmunoassay for a granulocyte protein as a marker in studies on the turnover of such cells. Bull Eur Physiopathol Respir 1980;16(Suppl):273-82.
4. Roseth AG, Fagerhol MK, Aadland E, et al. Assessment of the neutrophil dominating protein calprotectin in feces. A methodologic study. Scand J Gastroenterol 1992;27:793-8.
5. Ton H, Brandsnes, Dale S, et al. Improved assay for fecal calprotectin. Clin Chim Acta 2000;292:41-54.
6. Aadland E, Fagerhol MK. Faecal calprotectin: A marker of inflammation throughout the intestinal tract. Eur J Gastroenterol Hepatol 2002;14:823-5.
7. Konikoff MR, Denson LA. Role of fecal calprotectin as a biomarker of intestinal inflammation in inflammatory bowel disease. Inflamm Bowel Dis 2006;12:524-34.
8. Roseth AG, Aadland E, Jahnsen J, et al. Assessment of disease activity in ulcerative colitis by faecal calprotectin, a novel granulocyte marker protein. Digestion 1997;58:176-80.
9. Roseth AG, Schmidt PN, Fagerhol MK. Correlation between faecal excretion of indium-111-labelled granulocytes and calprotectin, a granulocyte marker protein, in patients with inflammatory bowel disease. Scand J Gastroenterol 1999;34:50-4
10. Waugh N, Cummins E, Royle P, et al. Faecal calprotectin testing for differentiating amongst inflammatory and non-inflammatory bowel diseases: Systematic review and economic evaluation. Health Technol Assess 2013;17:xv-xix, 1-211.
11. Jellema P, van Tulder MW, van der Horst HE, et al. Inflammatory bowel disease: A systematic review on the value of diagnostic testing in primary care. Colorectal Dis 2011;13:239-54.
12. van Rheenen PF, Van de Vijver E, Fidler V. Faecal calprotectin for screening of patients with suspected inflammatory bowel disease: Diagnostic meta-analysis. BMJ 2010;341:c3369.
13. Schoepfer AM, Beglinger C, Straumann A, et al. Fecal calprotectin correlates more closely with the Simple Endoscopic Score for Crohn's disease (SES-CD) than CRP, blood leukocytes, and the CDAI. Am J Gastroenterol 2010;105:162-9.
14. Schoepfer AM, Beglinger C, Straumann A, et al. Fecal calprotectin more accurately reflects endoscopic activity of ulcerative colitis than the Lichtiger Index, C-reactive protein, platelets, hemoglobin, and blood leukocytes. Inflamm Bowel Dis 2013;19:332-41.
15. Lobaton T, Rodriguez-Moranta F, Lopez A, et al. A new rapid quantitative test for fecal calprotectin predicts endoscopic activity in ulcerative colitis. Inflamm Bowel Dis 2013;19:1034-42.
16. Faubion WA Jr, Fletcher JG, O'Byrne S, et al. EMerging BiomARKers in Inflammatory Bowel Disease (EMBARK) study identifies fecal calprotectin, serum MMP9, and serum IL-22 as a novel combination of biomarkers for Crohn's disease activity: Role of cross-sectional imaging. Am J Gastroenterol 2013;108:1891-900.
17. Mooiweer E, Fidder HH, Siersema PD, et al. Fecal hemoglobin and calprotectin are equally effective in identifying patients with inflammatory bowel disease with active endoscopic inflammation. Inflamm Bowel Dis 2014;20:307-14.
18. Schoepfer AM, Beglinger C, Straumann A, et al. Ulcerative colitis: Correlation of the Rachmilewitz endoscopic activity index with fecal calprotectin, clinical activity, C-reactive protein, and blood leukocytes. Inflamm Bowel Dis 2009;15:1851-8.
19. Wagner M, Peterson CG, Ridefelt P, et al. Fecal markers of inflammation used as surrogate markers for treatment outcome in relapsing inflammatory bowel disease. World J Gastroenterol 2008;14:5584-9.
20. Mao R, Xiao YL, Gao X, et al. Fecal calprotectin in predicting relapse of inflammatory bowel diseases: A meta-analysis of prospective studies. Inflamm Bowel Dis 2012;18:1894-9.
21. De Vos M, Louis EJ, Jahnsen J, et al. Consecutive fecal calprotectin measurements to predict relapse in patients with ulcerative colitis receiving infliximab maintenance therapy. Inflamm Bowel Dis 2013;19:2111-7.
22. Yang Z, Clark N, Park KT. Effectiveness and cost-effectiveness of measuring fecal calprotectin in diagnosis of inflammatory bowel disease in adults and children. Clin Gastroenterol Hepatol 2014;12:253-62, e2.
23. Thjodleifsson B, Sigthorsson G, Cariglia N, et al. Subclinical intestinal inflammation: An inherited abnormality in Crohn's disease relatives? Gastroenterology 2003;124:1728-37.
24. Moskowitz RL, Shepherd NA, Nicholls RJ. An assessment of inflammation in the reservoir after restorative proctocolectomy with ileoanal ileal reservoir. Int J Colorectal Dis 1986;1:167-74.
25. Shepherd NA, Jass JR, Duval I, et al. Restorative proctocolectomy with ileal reservoir: Pathological and histochemical study of mucosal biopsy specimens. J Clin Pathol 1987;40:601-7.
26. Johnson MW, Maestranzi S, Duffy AM, et al. Faecal calprotectin: A noninvasive diagnostic tool and marker of severity in pouchitis. Eur J Gastroenterol Hepatol 2008;20:174-9.
27. Offner FA, Jao RV, Lewin KJ, et al. Collagenous colitis: A study of the distribution of morphological abnormalities and their histological detection. Hum Pathol 1999;30:451-7.

28. Wildt S, Nordgaard-Lassen I, Bendtsen F, et al. Metabolic and inflammatory faecal markers in collagenous colitis. Eur J Gastroenterol Hepatol 2007;19:567-74.
29. Larsson JK, Sjoberg K, Vigren L, et al. Chronic non-bloody diarrhoea: A prospective study in Malmo, Sweden, with focus on microscopic colitis. BMC Res Notes 2014;7:236.
30. Rodriguez-Otero P, Porcher R, Peffault de Latour R, et al. Fecal calprotectin and alpha-1 antitrypsin predict severity and response to corticosteroids in gastrointestinal graft-versus-host disease. Blood 2012;119:5909-17.
31. Chiusolo P, Metafuni E, Giammarco S, et al. Role of fecal calprotectin as biomarker of gastrointestinal GVHD after allogeneic stem cell transplantation. Blood 2012;120:4443-4.
32. Bastos Oreiro M, Castilla-Llorente C, de la Guia AL, et al. Fecal calprotectin in allogeneic stem cell transplantation for the diagnosis of acute intestinal graft versus host disease. Bone Marrow Transplant 2012;47:1241-2.
33. August KJ, Chiang KY, Qayed M, et al. Relative defects in mucosal immunity predict acute graft-versus-host disease. Biol Blood Marrow Transplant 2014;20:1056-9.
34. De Keyser F, Elewaut D, De Vos M, et al. Bowel inflammation and the spondyloarthropathies. Rheum Dis Clin North Am 1998;24:785-813, ix-x.
35. Leirisalo-Repo M, Turunen U, Stenman S, et al. High frequency of silent inflammatory bowel disease in spondylarthropathy. Arthritis Rheum 1994;37:23-31.
36. Taurog JD. The role of HLA-B27 in spondyloarthritis. J Rheumatol 2010;37:2606-16.
37. Salvarani C, Fries W. Clinical features and epidemiology of spondyloarthritides associated with inflammatory bowel disease. World J Gastroenterol 2009;15:2449-55.
38. Rosenbaum J, Chandran V. Management of comorbidities in ankylosing spondylitis. Am J Med Sci 2012;343:364-6.
39. Matzkies FG, Targan SR, Berel D, et al. Markers of intestinal inflammation in patients with ankylosing spondylitis: A pilot study. Arthritis Res Ther 2012;14:R261.
40. Klingberg E, Carlsten H, Hilme E, et al. Calprotectin in ankylosing spondylitis – frequently elevated in feces, but normal in serum. Scand J Gastroenterol 2012;47:435-44.
41. Manetti M, Neumann E, Muller A, et al. Endothelial/lymphocyte activation leads to prominent CD4+ T cell infiltration in the gastric mucosa of patients with systemic sclerosis. Arthritis Rheum 2008;58:2866-73.
42. Andreasson K, Scheja A, Saxne T, et al. Faecal calprotectin: A biomarker of gastrointestinal disease in systemic sclerosis. J Intern Med 2011;270:50-7.
43. Tursi A, Brandimarte G, Elisei W, et al. Assessment and grading of mucosal inflammation in colonic diverticular disease. J Clin Gastroenterol 2008;42:699-703.
44. Tursi A, Brandimarte G, Elisei W, et al. Faecal calprotectin in colonic diverticular disease: A case-control study. Int J Colorectal Dis 2009;24:49-55.
45. Montalto M, Santoro L, Curigliano V, et al. Faecal calprotectin concentrations in untreated coeliac patients. Scand J Gastroenterol 2007;42:957-61.
46. Capone P, Rispo A, Imperatore N, et al. Fecal calprotectin in coeliac disease. World J Gastroenterol 2014;20:611-2.
47. Balamtekin N, Baysoy G, Uslu N, et al. Fecal calprotectin concentration is increased in children with celiac disease: Relation with histopathological findings. Turk J Gastroenterol 2012;23:503-8.
48. Ertekin V, Selimoglu MA, Turgut A, et al. Fecal calprotectin concentration in celiac disease. J Clin Gastroenterol 2010;44:544-6.
49. Trespi E, Ferrieri A. Intestinal bacterial overgrowth during chronic pancreatitis. Curr Med Res Opin 1999;15:47-52.
50. Casellas F, Guarner L, Vaquero E, et al. Hydrogen breath test with glucose in exocrine pancreatic insufficiency. Pancreas 1998;16:481-6.
51. Madsen JL, Graff J, Philipsen EK, et al. Bile acid malabsorption or disturbed intestinal permeability in patients treated with enzyme substitution for exocrine pancreatic insufficiency is not caused by bacterial overgrowth. Pancreas 2003;26:130-3.
52. Pezzilli R, Barassi A, Morselli-Labate AM, et al. Fecal calprotectin and elastase 1 determinations in patients with pancreatic diseases: A possible link between pancreatic insufficiency and intestinal inflammation. J Gastroenterol 2007;42:754-60.
53. Shastri YM, Bergis D, Povse N, et al. Prospective multicenter study evaluating fecal calprotectin in adult acute bacterial diarrhea. Am J Med 2008;121:1099-106.
54. Chen CC, Huang JL, Chang CJ, et al. Fecal calprotectin as a correlative marker in clinical severity of infectious diarrhea and usefulness in evaluating bacterial or viral pathogens in children. J Pediatr Gastroenterol Nutr 2012;55:541-7.
55. Brenchley JM, Price DA, Schacker TW, et al. Microbial translocation is a cause of systemic immune activation in chronic HIV infection. Nat Med 2006;12:1365-71.
56. Gori A, Tincati C, Rizzardini G, et al. Early impairment of gut function and gut flora supporting a role for alteration of gastrointestinal mucosa in human immunodeficiency virus pathogenesis. J Clin Microbiol 2008;46:757-8.
57. Roseth AG, Kristinsson J, Fagerhol MK, et al. Faecal calprotectin: A novel test for the diagnosis of colorectal cancer? Scand J Gastroenterol 1993;28:1073-6.
58. Hoff G, Grotmol T, Thiis-Evensen E, et al. Testing for faecal calprotectin (PhiCal) in the Norwegian Colorectal Cancer Prevention trial on flexible sigmoidoscopy screening: Comparison with an immunochemical test for occult blood (FlexSure OBT). Gut 2004;53:1329-33.
59. Tibble JA, Sigthorsson G, Foster R, et al. High prevalence of NSAID enteropathy as shown by a simple faecal test. Gut 1999;45:362-6.
60. Meling TR, Aabakken L, Roseth A, et al. Faecal calprotectin shedding after short-term treatment with non-steroidal anti-inflammatory drugs. Scand J Gastroenterol 1996;31:339-44.
61. Poullis A, Foster R, Mendall MA, et al. Proton pump inhibitors are associated with elevation of faecal calprotectin and may affect specificity. Eur J Gastroenterol Hepatol 2003;15:573-4.
62. Khalid U, McGough C, Hackett C, et al. A modified inflammatory bowel disease questionnaire and the Vaizey Incontinence questionnaire are more sensitive measures of acute gastrointestinal toxicity during pelvic radiotherapy than RTOG grading. Int J Radiat Oncol Biol Phys 2006;64:1432-41.
63. Wedlake L, McGough C, Hackett C, et al. Can biological markers act as non-invasive, sensitive indicators of radiation-induced effects in the gastrointestinal mucosa? Aliment Pharmacol Ther 2008;27:980-7.
64. Garcia-Tsao G, Wiest R. Gut microflora in the pathogenesis of the complications of cirrhosis. Best Pract Res Clin Gastroenterol 2004;18:353-72.
65. Gupta A, Dhiman RK, Kumari S, et al. Role of small intestinal bacterial overgrowth and delayed gastrointestinal transit time in cirrhotic patients with minimal hepatic encephalopathy. J Hepatol 2010;53:849-55.
66. Gundling F, Schmidtler F, Hapfelmeier A, et al. Fecal calprotectin is a useful screening parameter for hepatic encephalopathy and spontaneous bacterial peritonitis in cirrhosis. Liver Int 2011;31:1406-15.
67. Berni Canani R, Rapacciuolo L, Romano MT, et al. Diagnostic value of faecal calprotectin in paediatric gastroenterology clinical practice. Dig Liver Dis 2004;36:467-70.
68. Carroccio A, Iacono G, Cottone M, et al. Diagnostic accuracy of fecal calprotectin assay in distinguishing organic causes of chronic diarrhea from irritable bowel syndrome: A prospective study in adults and children. Clin Chem 2003;49:861-7.
69. Smyth RL, Croft NM, O'Hea U, et al. Intestinal inflammation in cystic fibrosis. Arch Dis Child 2000;82:394-9.
70. Raia V, Maiuri L, de Ritis G, et al. Evidence of chronic inflammation in morphologically normal small intestine of cystic fibrosis patients. Pediatr Res 2000;47:344-50.
71. Bruzzese E, Raia V, Gaudiello G, et al. Intestinal inflammation is a frequent feature of cystic fibrosis and is reduced by probiotic administration. Aliment Pharmacol Ther 2004;20:813-9.
72. Mercer DF, Vargas L, Sun Y, et al. Stool calprotectin monitoring after small intestine transplantation. Transplantation 2011;91:1166-71.
73. Rugtveit J, Fagerhol MK. Age-dependent variations in fecal calprotectin concentrations in children. J Pediatr Gastroenterol Nutr 2002;34:323-4.

Retrospective study investigating the prevalence and clinical significance of hepatitis B virus precore and basal core promoter variants

Meaghan O'Brien MD FRCPC[1], Adara Casselman BSc[2], Gerry Smart BSc MLT[2], Ainsley Gretchen BSc[2], Kelly Kaita MD FRCPC[3], Kamran Kadkhoda PhD FCCM D(ABMM)[2,4]

M O'Brien, A Casselman, G Smart, A Gretchen, K Kaita, K Kadkhoda. Retrospective study investigating the prevalence and clinical significance of hepatitis B virus precore and basal core promoter variants. Can J Gastroenterol Hepatol 2015;29(8):e1-e6.

BACKGROUND: Hepatitis B virus (HBV) precore (PC) and basal core promoter (BCP) variants are well known; however, their prevalence in North America is unclear, especially among hepatitis B e antigen-negative patients.

OBJECTIVE: To investigate the prevalence of PC/BCP mutations and their clinical significance.

METHODS: One hundred twenty-eight patients positive for both hepatitis B surface antigen and hepatitis B e antibody were selected, and PC/BCP mutations were identified using a line probe assay. The subjects' charts were reviewed for race/ethnicity, HBV genotype, HBV viral load, sex, liver enzyme levels, imaging and biopsy results up to 10 years before the study.

RESULTS: The prevalence of PC and BCP variants were 47.6% and 62.5%, respectively. Older age was associated with aspartate aminotransferase-to-platelet index ratio (APRI) ≥0.7 (P=0.011) and abnormal imaging/biopsy results (P=0.0008). Although the presence of BCP variant(s) was associated with APRI ≥0.7 (P=0.029), it was not associated with abnormal imaging/biopsy results. The combination of age ≥50 years and the presence of BCP variant(s) was associated with abnormal imaging/biopsy results, suggestive of either cirrhosis or hepatocellular carcinoma (not observed with PC mutation). Neither sex or genotype, or median HBV viral load showed significant influence on any of these outcomes.

CONCLUSIONS: The present study suggests that the prevalence of PC and BCP mutations are higher than what has been previously reported. One potential explanation would be increased immigration in the past decade. Considering the potential public health and clinical implications of these variants, long-term multicentre and prospective studies could further unravel the uncertainty around these variants.

Key Words: *Canada; Clinical significance; Precore mutations*

Une étude rétrospective sur la prévalence et la signification clinique des variantes du gène précore et du promoteur du core basal du virus de l'hépatite B

HISTORIQUE : Les variantes du gène précore (PC) et du promoteur du core basal (PCB) du virus de l'hépatite B (VHB) sont bien connues. On n'en connaît toutefois pas la prévalence en Amérique du Nord, particulièrement chez les patients non porteurs de l'antigène e de l'hépatite B.

OBJECTIF : Examiner la prévalence et la signification clinique des mutations PC et PCB.

MÉTHODOLOGIE : Les chercheurs ont sélectionné 128 patients porteurs de l'antigène de surface de l'hépatite B et de l'anticorps e de l'hépatite B et ont repéré les mutations PC et PCB au moyen d'une sonde linéaire. Ils ont vérifié la race ou l'ethnie des patients, le génotype du VHB, la charge virale du VHB, leur sexe, leurs taux d'enzyme hépatique, les résultats de l'imagerie et des biopsies jusqu'à dix ans avant l'étude.

RÉSULTATS : La prévalence de variantes PC et PCB s'élevait à 47,6 % et à 62,5 %, respectivement. Un âge avancé était lié à un coefficient d'indexation entre l'aspartate aminotransférase et les plaquettes (APIR) d'au moins 0,7 (P=0,011) et à des résultats d'imagerie ou de biopsie anormaux (P=0,0008). Même si la présence de variante(s) des PCB s'associait à un APIR d'au moins 0,7 (P=0,029), il n'était pas relié à des résultats d'imagerie ou de biopsie anormaux. La combinaison d'un âge d'au moins 50 ans et de la présence de variante(s) PCB s'associait à des résultats d'imagerie et de biopsie anormaux, indicateurs d'une cirrhose ou d'un carcinome hépatocellulaire (non observé dans la mutation PC). Ni le sexe ni le génotype ni la charge virale médiane du VHB n'avaient d'influence significative sur ces résultats.

CONCLUSIONS : D'après la présente étude, la prévalence des mutations PC et PCB est plus élevée que ce qui a été déclaré par le passé. L'augmentation du taux d'immigration depuis dix ans pourrait contribuer à expliquer ce phénomène. Étant donné les conséquences potentielles de ces variantes pour la santé publique et les résultats cliniques, des études multicentriques et prospectives à long terme pourraient décortiquer les incertitudes entourant ces variantes.

Hepatitis B virus (HBV) contains a core gene that is divided into two parts – the precore (PC) and core regions. Mutations can occur in either region, culminating in a PC mutant or a basal core promoter (BCP) mutant, respectively. Mutations in the PC region (particularly at position G/A 1896) are primarily nonsense or frameshift mutations, or mutated initiation codons, whereas mutations in the BCP region (particularly dual mutations at positions 1762 and 1764, the so-called 'TA change') are primarily missense substitutions (1,2). These mutations lead to loss of expression of the hepatitis B e antigen (HBeAg) or a significant reduction in its production and secretion. The presence of HBeAg normally elicits an anti-HBeAg immune response. This immune response causes, for the most part, a desired reduction in HBV viral load, but selects variants that have little or no HBeAg expression such as those produced by PC or BCP mutations (1). To detect these mutations, a range of methodologies, including direct sequencing, restriction fragment length polymorphism (3) and line probe assay

[1]*Department of Internal Medicine, Upper River Valley Hospital, Waterville, New Brunswick;* [2]*Serology Section, Cadham Provincial Public Health Laboratory;* [3]*Section of Hepatology, Department of Internal Medicine, College of Medicine, Faculty of Health Sciences, University of Manitoba;* [4]*Department of Medical Microbiology & Infectious Diseases; and Department of Immunology, College of Medicine, Faculty of Health Sciences, University of Manitoba, Winnipeg, Manitoba*

Correspondence: Dr Kamran Kadkhoda, Serology Section, Cadham Provincial Public Health Laboratory, 750 William Avenue, Winnipeg, Manitoba R3C 3Y1. e-mail kamran.kadkhoda@gov.mb.ca

(LiPA) (4,5), have been developed, the latter being more sensitive in detecting variants than direct sequencing (5). Studies have shown a significant geographical variation in the prevalence of PC and BCP mutants, with some of the lowest prevalence rates being found in North America (6). Prevalence rates for PC mutants range from 0% to 100%, with most approximately 60%. Prevalence rates for BCP mutations range from 50% to 77%; however, these studies were either meta-analyses or performed outside of North America (6-8). The most recent comprehensive study investigating PC/BCP variants in North America was conducted in the United States by the HBV Epidemiology Study Group and was published in 2003 (9). Since then, no thorough studies have been performed, particularly in anti-HBe-positive patients. This is despite the fact that significant immigration to North America from endemic countries has occurred since then and may have culminated in the current findings. To date, only one study has assessed the prevalence of PC mutations in Canada (10), and that study only examined 30 Canadian Inuit patients. To date, no studies have investigated the prevalence of BCP mutations in Canada; however, there is evidence to suggest that the presence of mutants affects the natural history of HBV infection. Studies have shown that PC and BCP mutations may be associated with poorer clinical outcomes including a higher risk for cirrhosis and hepatocellular carcinoma (HCC), although results have been conflicting (11-15).

The primary aim of the present study, therefore, was to determine the prevalence of BCP and PC mutations in Manitoba in individuals chronically infected with HBV and who are anti-HBe positive. The study also aimed to determine whether a correlation exists between the presence of a mutant and HBV disease severity, including progression to cirrhosis and development of HCC.

METHODS

Study design

In the present study, 128 residual plasma specimens that were submitted to Cadham Provincial Laboratory (CPL, Winnipeg, Manitoba) from September 2010 through May 2014 for HBV viral load testing as well as HBV genotyping were chosen. Using the residual specimens, the prevalence of PC and BCP mutations were determined, and this information was linked to the subjects' other clinical and laboratory parameters. Most of subjects had a relatively long history of testing; however, past specimens were not available to be reanalyzed. CPL is the provincial public health laboratory in Manitoba that performs HBV viral load testing for the province, while referring out HBV genotyping requests to the National Microbiology Laboratory in Winnipeg (Manitoba).

Inclusion criteria were: age ≥18 years; hepatitis B surface antigen positive; HBeAg negative and anti-HBeAg positive; any HBV viral load test performed and, preferably, HBV genotype known (however, 30 samples without known HBV genotypes were genotyped in the authors' laboratory for the present study). Subjects coinfected with hepatitis C virus (HCV), hepatitis D virus and HIV were excluded. None of, sex, race/ethnicity and HBV genotype was used as inclusion or exclusion criteria. A retrospective chart review was performed to collect up to 10 years (where applicable) of additional clinical and laboratory information including race/ethnicity, sex, liver enzyme data (alanine aminotransferase [ALT], aspartate aminotransferase [AST]), platelet counts, imaging and liver biopsy results. Median HBV viral load, ALT and AST levels, and platelet counts were collected for each subject. Median AST-to-platelet index ratio (APRI) was calculated for each subject (where available) using the following formula:

$$\text{Median AST/ULN} \times 100/\text{median platelet count}$$

in which the upper limit of normal (ULN) for AST was defined as 32 U/L. Abnormal ALT level was based on the reference ranges provided by the testing laboratory: <25 U/L for females and <30 U/L for males. Only 17% of patients had any treatment history. Patients with fatty liver disease, autoimmune hepatitis and other non-HBV-related causes were excluded on chart review. Abnormal liver biopsy results were defined as fibrosis stage ≥2 (METAVIR F≥2) or HCC according to pathology report. Abnormal liver imaging was defined as a computed tomography scan, magnetic resonance imaging scan or an ultrasound radiology report of irregular hepatic surface/contour, or regenerative hepatic nodules with or without portal hypertension measured using Doppler. The study was approved by the University of Manitoba Research Ethics Board #H2014:125.

HBV viral load

Quantitative HBV DNA levels (or HBV viral loads) were previously determined at the CPL using COBAS AmpliPrep/COBAS TaqMan HBV Test (Roche Diagnostics GmbH, Germany) according to the manufacturers' instructions. This product requires 650 µL of plasma specimen, has a limit of detection of 20 IU/mL, a dynamic range of 20×10^8 IU/mL to 1.7×10^8 IU/mL, and covers all eight HBV genotypes (A through H). For consistency, only EDTA-plasma is accepted for this test.

HBV genotyping

HBV genotyping was referred to the National Microbiology Laboratory in Winnipeg, where genotyping is performed by either direct sequencing or INNO-LiPA HBV Genotyping kit (Innogenetics, Fujirebio, Japan). Thirty specimens had met inclusion criteria but did not have known HBV genotypes. These genotypes were determined using the INNO-LiPA HBV Genotyping kit (Innogenetics, Fujirebio). Briefly, HBV DNA was extracted from specimens (200 µL) using the easyMAG system (bioMérieux, France). Amplification was performed using Platinum Taq DNA Polymerase (Life Technologies, USA). The volume of water was adjusted because the $MgCl_2$ reagent is separate (final concentration 1.5 mM). The dNTPs were combined in equal volumes to yield a dNTP mixture of 25 mM each. The master mixes and cycling parameters were followed as per product insert for outer and nested amplification. The thermocycler used was an Eppendorf AG Mastercycler ep (Eppendorf, USA). The LiPA was performed on a Med/TEC AutoBlot 3000H (Fujirebio, Japan) following the automation protocol for this instrument.

HBV PC/BCP mutations

HBV PC (G1896A) and BCP (A1762T, G1764A) mutation testing was performed using the INNO-LiPA HBV PreCore kit (Innogenetics, Fujirebio). Briefly, HBV DNA was extracted from samples (200 µL) using easyMAG system (bioMérieux); the final elution volume was 25 µL. Amplification used Platinum Taq DNA Polymerase (Life Technologies) and 100 mM dNTP set. The volume of water was adjusted because the $MgCl_2$ reagent is separate (1.5 µL final concentration 1.5 mM). The dNTPs were combined in equal volumes to yield a dNTP mixture of 25 mM each. The master mixes and cycling parameters were followed as per product insert for outer and nested amplification. The thermocycler used was an Eppendorf AG Mastercycler ep (Eppendorf). LiPA was performed on a Med/TEC AutoBlot 3000H (Fujirebio) following the automation protocol for this instrument.

Statistical analyses

Data were analyzed using Analyse-it (Analyse-it Software, Ltd, United Kingdom) for Excel version 2.0 (Microsoft Corporation, USA). Statistical analyses were performed using the Fisher's exact test (including OR calculation and 95% CIs) and Mann-Whitney U test for categorical variables. Results were considered to be statistically significant at $P<0.05$.

RESULTS

One hundred twenty-eight specimens chosen from subjects who met the inclusion criteria were studied. Subjects belonged to three major ethnicities: Asian (71%), non-Hispanic white (10%) and African descent (19%). Of an initial 132 specimens, four were excluded because two did not have amplified DNA for genotyping and PC/BCP mutation testing and two were positive for HCV or hepatitis D virus. The age range was 18 to 80 years, with a median age of 40 years. Fifty-five percent of subjects were male and 45% were female. Of eight known HBV genotypes (A through H) only five were present: A

TABLE 1

Demographics, laboratory abnormalities and any abnormal liver findings according to imaging or biopsy versus precore (PC)/basal core promoter (BCP) mutations

	Mutation			
Variable	PC	BCP	P	Statistical test
Patient population	61/128 (47.6)	80/128 (62.5)		Not applicable
Age, years	41.27±13.37	41.25±13.42	0.099	Mann-Whitney
Male sex	37/61(60.6)	8/80 (60)	1.00	Fisher's exact
Abnormal median ALT level	30/61 (49.1)	46/79 (58.2)	0.66	Fisher's exact
Abnormal median AST level	16/61 (26.2)	26/79 (32.9)	0.65	Fisher's exact
ALT level, >1.5×upper limit of normal	29/61(47.5)	44/79 (55.6)	0.69	Fisher's exact
Platelet count, ×10^9/L	223±54	218±59	0.64	Mann-Whitney
APRI ≥0.7	8/61 (13.1)	17/78 (21.7)	0.37	Fisher's exact
Abnormal liver imaging and/or biopsy	22/59 (37.2)	32/75 (42.6)	0.80	Fisher's exact
Hepatitis B virus viral load, median, log$_{10}$ IU/mL	3.49	3.49		Not applicable

Data presented as n/n (%) or mean ± SD unless otherwise indicated. Abnormal liver biopsy results were defined as fibrosis stage ≥2 (METAVIR F≥2) or hepatocellular carcinoma according to the pathology report. Abnormal liver imaging was defined as a computed tomography scan, magnetic resonance imaging or an ultrasound radiology report of irregular hepatic surface/contour, or regenerative hepatic nodules with/without portal hypertension measured by Doppler. ALT Alanine aminotransferase; APRI Aspartate aminotransferase (AST)-to-platelet ratio index

(36.7%), B (17.1%), C (23.4%), D (14%) and E (8.5%). The distribution of HBV genotypes among different race/ethnicities were as follows: Asians: 35.1% A, 24.1% B, 31.8% C, 8.7% D and 0% E; non-Hispanic whites: 38.4% A, 0% B, 7.6% C, 53.8% D and 0% E; and African descent: 41.6% A, 0% B, 0% C, 12.5% D and 45.8% E. The median HBV viral load (log$_{10}$ IU/mL) in genotypes A to E were: 3.48, 3.49, 3.46, 3.48 and 3.48, respectively.

PC/BCP mutations and demographics

The prevalence of PC and BCP mutations were higher in the present study compared with those previously reported (15). The overall prevalence of PC and BCP mutations was 46.7% and 62.5%, respectively. The distribution of PC mutation (G1896A) in subjects with HBV genotypes A through E was 6.3%, 95.4%, 40%, 77.7% and 100%, respectively. The distribution of BCP (A1762T, G1764A) dual mutation in subjects with HBV genotypes A through E was: 61.7%, 36.3%, 80%, 66.6% and 63.6%, respectively. In Asian, African descent and non-Hispanic white racial/ethnic groups, the prevalence of PC mutation was: 46.1%, 54.1% and 46.1%, respectively. The prevalence of BCP dual mutation in the above racial/ethnic groups was: 62.6%, 62.5% and 61.5%, respectively. The mean (± SD) age of subjects with PC and BCP mutations was 41.27±13.37 and 41.25±13.42 years, respectively (P=0.99). The prevalence of PC and BCP mutations in males was 60.6% and 60%, respectively (P=1.00) (Table 1). It should be noted, however, that the estimated prevalence of HBV-related cirrhosis in Manitoba is up to 5%, while the incidence of HBV-related HCC is up to 1%.

PC/BCP mutations, HBV viral load and laboratory abnormalities

Of subjects with PC or BCP mutations, 49.1% and 58.2% had an abnormal median ALT value, respectively (P=0.66), whereas those with an abnormal median AST value were 26 2% and 32.9%, respectively (P=0.65). Median ALT values of ≥1.5×ULN were present in 47.5% and 55.6% of those with PC and BCP mutations, respectively (P=0.69). The mean platelet count in those with PC or BCP mutations was 233±54×10^9/L and 218±59×10^9/L, respectively (P=0.64). An APRI ≥0.7 was present in 13.1% and 21.7% of individuals with PC and BCP mutations, respectively (P=0.37) (OR 1.66 [95% CI 0.67 to 4.10]). Individuals with PC or BCP mutations had equal median HBV viral loads of 3.49 log$_{10}$ IU/mL (Table 1). Neither PC nor BCP mutations were significantly associated with median HBV viral load (data not shown).

PC/BCP mutations and abnormal liver imaging and/or biopsy

Abnormal liver imaging and/or biopsy results were found in 37.2% and 42.6% with PC and BCP mutations, respectively (P=0.80) (Table 1).

TABLE 2

Factors associated with an aspartate aminotransferase-to-platelet ratio index ≥0.7

Factor	Patients, n/n	OR (95% CI)	Statistical test	P
Sex		2.27 (0.77–6.69)	Fisher's exact	0.20
Male	14/69			
Female	5/56			
Hepatitis B virus genotype				
A	5/44	1.52 (0.51–4.5)	Fisher's exact	0.62
B	0/22	N/A	Fisher's exact	0.06
C	8/30	0.43 (0.16–1.17)	Fisher's exact	0.17
D	4/18	0.63 (0.18–2.11)	Fisher's exact	0.64
E	2/11	0.82 (0.16–4.02)	Fisher's exact	1.00
Age*	19/125	N/A	Mann-Whitney	0.011†
PC mutation	8/61	0.76 (0.28–2.02)	Fisher's exact	0.76
BCP mutation	17/78	5.12 (1.13–23.16)	Fisher's exact	0.029†
Age ≥50 years and BCP mutation	11/28	6.67 (0.79–56.42)	Fisher's exact	0.096

**Median difference 10.0 (95% CI 3.0 to 18.0), older age was significantly associated with an aspartate aminotransferase-to-platelet ratio ≥0.7; †Statistically significant. BCP Basal core promoter; N/A Not applicable; PC Precore*

Association of APRI ≥0.7 with age, sex and HBV genotype

An APRI ≥0.7 was chosen because those with APRI ≥0.7 had significantly more abnormal liver imaging and/or biopsy findings compared with those with APRI <0.7 (OR 2.4 [95% CI 1 to 5.7]; P=0.03). Additionally, in HCV-infected individuals, this cut-off has been deemed reasonable (16). Although 20.2% of males had an APRI ≥0.7, only 8.9% of females had an APRI ≥0.7; however, no statistical significance was found (OR 2.27 [95% CI 0.77 to 6.69]; P=0.20). HBV genotypes were compared for the presence of APRI ≥0.7 and no statistical significance was noted (Table 2). Age was an independent factor affecting APRI, with older age being associated with APRI ≥0.7 (median difference 10 years [95% CI 3.0 to 18.0]; P=0.011) (Table 2).

Association of APRI ≥0.7 with PC/BCP mutations and combination of age ≥50 years and BCP mutations

Individuals with a G1896A point mutation did not have a significantly higher proportion of APRI ≥0.7, but in fact showed a relative, but non-significant protective effect (OR 0.76 [95% CI 0.28 to 2.02]; P=0.76). However, those with a BCP dual mutation had a significantly higher

TABLE 3
Factors associated with abnormal liver imaging suggestive of cirrhosis/hepatocellular carcinoma and/or abnormal finding on liver biopsy

Factor	Patients, n/n	OR (95% CI)	Statistical test	P
Sex		2.23 (1.02-4.86)	Fisher's exact	0.059
Male	30/66			
Female	11/54			
Hepatitis B virus genotype				
A	13/43	1.20 (0.56–2.56)	Fisher's exact	0.77
B	6/21	1.23 (0.46–3.31)	Fisher's exact	0.87
C	15/29	0.55 (0.25–1.18)	Fisher's exact	0.18
D	6/17	0.96 (0.35–2.63)	Fisher's exact	1.00
E	1/10	3.63 (0.45–29.31)	Fisher's exact	0.35
Age*	41/120	N/A	Mann-Whitney	0.0008†
PC mutation	22/59	1.19 (0.58–2.43)	Fisher's exact	0.75
BCP mutation	32/75	2.13 (0.93–4.87)	Fisher's exact	0.098
Age ≥50 years AND BCP mutation	17/23	N/A‡	Fisher's exact	0.041†

**Median difference 9.0 (95% CI 4.0–14.0), older age was significantly associated with abnormal liver imaging suggestive of cirrhosis/hepatocellular carcinoma and/or abnormal finding on liver biopsy; †Statistically significant; ‡No patients in the wild type group. BCP Basal core promoter; NA Not applicable; PC Precore*

proportion of APRI ≥0.7 (OR 5.12 [95% CI 1.13 to 23.16]; P=0.029), suggesting that a BCP dual mutation is an independent and relatively strong factor in elevating APRI. Finally, a combination of BCP dual mutation and age ≥50 years showed a borderline nonstatistical significance with APRI ≥0.7 (OR 6.67 [95% CI 0.79 to 56.42];P=0.096) (Table 2).

Association of abnormal liver imaging and/or biopsy with age, sex and HBV genotype

Abnormal liver imaging and/or biopsy results were found in 45.4% of men, but only 20.3% of women had such results (OR 2.23 [95% CI 1.02 to 4.86]); however, the difference was nearly significant (P=0.059). Similar to what was observed with APRI ≥0.7, HBV genotypes did not affect liver imaging and/or biopsy results (Table 2). However, older age was associated with abnormal liver imaging and/or biopsy results, with a median difference of nine years (95% CI 4.0 to 14.0) (P=0.0008).

Association of abnormal liver imaging and/or biopsy with PC/BCP mutations and combination of age ≥50 years and BCP mutations

Those with a G1896A point mutation did not have a higher proportion of abnormal liver imaging and/or biopsy results compared with those without the mutation (P=0.75). Those with BCP dual mutation also did not have a significantly higher proportion of abnormal liver imaging and/or biopsy results; however, the nonsignificance was relatively marginal (P=0.098), with OR 2.13 (95% CI 0.93 to 4.87). Finally, the combination of age ≥50 years and BCP dual mutation showed statistical significance with abnormal liver imaging and/or biopsy results compared with those <50 years of age and no mutations at postions 1762 and 1764 (P=0.041) (Table 3). Lower age cut-offs applied did not yield statistical significance (data not shown).

DISCUSSION

The natural history of hepatitis B is associated with HBeAg loss. These patients will generally enter an inactive phase with a percentage of them becoming active with the development of either the PC or BCP mutation (17). Individuals in the inactive phase tend to have HBV viral loads <4.3 $\log_{10}$ IU/mL, normal liver enzyme levels, and no to minimal fibrosis on biopsy. However, those in the immune active phase may also have low-level HBV viral loads, especially when on nucleo(t)side analogues and may not have elevated liver enzyme levels. For this reason, we opted to use median viral load and liver enzyme levels to make a better assessment of disease activity over time, if any, in the context of PC/BCP mutations.

The most recent study to determine the prevalence of PC/BCP mutations in North America was performed more than a decade ago. Therefore, the present study serves as more recent and region-specific view on the prevalence of these mutations and sheds further light on their clinical significance. As mentioned in the Results section, the prevalence of both PC and dual BCP mutations was higher than previously reported (9). That study included HBeAg-positive patients but, as indicated earlier, the majority of chronically HBV-infected patients seroconvert to anti-HBe-positive status. This is especially true among immigrants from endemic regions who harbour HBV genotypes with a different distribution of PC/BCP mutations. Our study involved three main racial/ethnic groups: Asian, non-Hispanic white and African descent. The race/ethnicity distribution in our study was different than that of the United States study because no Hispanic whites participated in our study and the African descent subjects consisted mainly of recent immigrants from Africa, with 45% of them harbouring genotype E. It is also noteworthy to mention that the majority (71%) of our subjects were of Asian ancestry. There was an even age and sex distribution across the study population. The relatively high prevalence of PC and BCP mutations are of concern from both a clinical and public health perspective.

It appears that PC and BCP mutations are strategies to regain viral fitness in the form of replication capacity in lamivudine-resistant and tenofovir-resistant strains of HBV based on in vitro studies; however, no clinical cases have been reported with tenofovir resistance to date (18,19). From a clinical perspective, BCP mutations have been associated with nonresponse to interferon therapy (19). PC and BCP mutations have also been associated with fulminant hepatitis in HBV-noninfected individuals (21-26), making these mutations of concern from a clinical and public health perspective. Finally, these variants have also been associated with HCC (27-30).

In our study, median ALT and AST were not affected by PC or BCP mutations. Even when abnormal median ALT values 1.5×ULN were used, no significance was observed, suggesting that the fluctuation of liver enzyme levels over time could be due to more than the presence of a mutation. This was also the case for median platelet values as a loose measure of liver impairment. Even by combining AST and platelet count (ie, APRI and applying three different cut-offs [0.5, 0.6 and 0.7]), neither mutation independently affected APRI (data not shown for the first two cut-off values). This could, however, be due to the fact that LiPA is a sensitive method and detects variants even if they constitute a small subpopulation of the quasiviral pool combined with the fact that mutation detection was performed on samples collected between 2010 and 2014. However, the chart review was performed for a period of up to 10 years before the mutations analysis, when they potentially did not carry a mutation. By the same token, abnormal liver imaging and/or biopsy results were not independently affected by the mutations, perhaps because those changes typically occur later in the disease course and, for the most part, are preceded by liver enzyme and platelet count abnormalities; hence, APRI elevation (although the latter association may be weak). Although not statistically significant, but as shown in Table 1, liver enzyme abnormalities, lower platelet counts, abnormal APRI, and abnormal liver imaging and/or biopsy results were consistently more frequent in those with a BCP mutation than a PC mutation. The lack of significance may be due to the relatively cross-sectional nature of the present study and, perhaps, would have resulted in significant results had the study monitored subjects for a longer period because age was significantly associated with APRI ≥0.7 and abnormal liver imaging and/or biopsy results (Table 2). This could be explained by a greater chance of accumulating mutations as age increases.

Interestingly, despite previous reports (15), median HBV viral load (as a measure of viremia over time in the present study) was not significantly associated with either mutation, suggesting that median HBV viral load cannot be used as a stand-alone parameter to predict adverse outcomes. The contrasting result may be somewhat due to the fact that in the Risk Evaluation of Viral Load Elevation and Associated Liver Disease/Cancer-Hepatitis B Virus (REVEAL) (15) study, the design was different because only those with HBV viral loads >3.23 $\log_{10}$ IU/mL were tested for PC/BCP mutation, while we did not use such an inclusion criterion. Additionally, their study was limited to a certain race/ethnicity and only two HBV genotypes (B and C). However, consistently detectable levels of HBV viral load may increase the theoretical risk for further development of mutations because the higher the viral propagation (which is error prone), the higher the probability of developing a mutation, thereby exerting its adverse outcomes indirectly. This, at least in part, was reflected in a higher (although not significant) proportion of APRI ≥0.7, and significantly higher proportion of imaging and/or biopsy abnormalities in those with both older age (≥50 years) and BCP mutation (Table 2).

Male sex is known to be associated with HCC (17). In our study, individuals with abnormal liver imaging and/or biopsy results suggestive of advanced fibrosis, cirrhosis or HCC were grouped together. This may explain why statistical significance was achieved with either of the two sexes; although male sex showed both higher percentage with APRI ≥0.7 and/or liver imaging/biopsy abnormalities. This may partly be due to a sex bias in referrals and in further investigations (including HBV viral load, genotype, ALT/AST assessments and imaging) as recommended per guidelines for HCC screening or could be due to immunological differences between the two sexes leading to T-helper (Th)1/Th2 balance shifting toward Th1 in men. This could result in a greater chance of immune activation and liver tissue damage, which could lead to repair/regeneration and fibrosis as a consequence; however, this remains controversial (31-33).

Another interesting finding is the absence of an effect of HBV genotypes on clinical outcomes including genotype C which has been shown to have worse outcomes (16). There are eight genotypes of HBV (A through H [34-36]). Initially the genotypes were somewhat geographically distinct, with genotype A focused in northern Europe, North America, South America, Australia and Sub-Saharan Africa; genotype B focused in West Africa, southeast Asia, China and Japan; genotype C focused in Southeast and East Asia, Pacific Islands, Australian Aboriginals and Alaska; genotype D focused in the Mediterranean, Middle East, Eastern Europe, Indian subcontinent, Central Asia, Mongolia, South Africa, Arctic region, Somalia, Papua-New Guinea, Australian Aboriginals and Indonesia; genotype E focused in West Africa; genotype F focused in Alaska, South America and Central America; genotype G focused in the Europe and North America; and genotype H focused in Central America (36,37). Two additional genotypes, I and J, have recently been proposed (38,39). However, with more frequent and widespread migration occurring, the geographical distribution of the genotypes is likely shifting. The largest study to date investigating genotype prevalence in North America was by Chu et al in 2003 (9). This study examined 694 patients with chronic HBV infection and found that while all genotypes from A to G were seen, A and C were the most common, with a slight variation in prevalence based on ethnicity. Although none of the HBV genotypes found in our study were directly associated with abnormal laboratory, imaging or biopsy results, because the distribution of PC and BCP mutations varies with HBV genotypes, HBV genotype could be an indirect measure of outcomes, especially if used in combination with PC or BCP mutations as previously shown (40).

Our study shows that the prevalence of PC and BCP mutations in the largest subpopulation of chronically HBV-infected patients (anti-HBe patients) is relatively high; this could be attributed to immigration over the past decade. The high prevalence has unique and challenging ramifications from both clinical and public health perspective, requiring more attention and systematic monitoring. The present study was retrospective in nature, relatively small in size and limited in geographical span. Therefore, multicentre prospective observational studies could shed further light on the highly intricate relationship of HBV variants with their human host and may potentially benefit individual patient management.

KEY POINTS

- The prevalence of PC and BCP mutations in North America needs more recent studies in all jurisdictions.
- PC and BCP mutations are more common than previously known and tend to be, at least indirectly, associated with poor clinical outcomes.

DISCLOSURES: The authors have no financial disclosures or conflicts of interest to declare.

REFERENCES

1. Parekh S, Zoulim F, Ahn SH, et al. Genome replication, virion secretion, and e antigen expression of naturally occurring hepatitis B virus core promoter mutants. J Virol 2003;77:6601-12.
2. Okamoto H, Yotsumoto S, Akahane Y, et al. Hepatitis B viruses with precore region defects prevail in persistently infected hosts along with seroconversion to the antibody against e antigen. J Virol 1990;64:1298-303.
3. Lindh M, Furuta Y, Ljunggren KK, et al. Detection of hepatitis B virus precore TAG mutant by an amplification-created restriction site method. J Infect Dis 1995;171:194-7.
4. Grandjacques C, Pradat P, Stuyver L, et al. Rapid detection of genotypes and mutations in the pre-core promoter and the pre-core region of hepatitis B virus genome: Correlation with viral persistence and disease severity. J Hepatol 2000;33:430-9.
5. Hussain M, Chu CJ, Sablon E et al. Rapid and sensitive assays for determination of hepatitis B virus (HBV) genotypes and detection of HBV precore and core promoter variants. J Clin Microbiol 2003;41:3699-705.
6. Funk ML, Rosenberg DM, Lok AS. World-wide epidemiology of HBeAg-negative chronic hepatitis B and associated precore and core promoter variants. J Viral Hepat 2002;9:52-61.
7. Qin Y, Zhang J, Mao R, et al. Prevalence of basal core promoter and precore mutations in Chinese chronic hepatitis B patients and correlation with serum HBeAg titers. J Med Virol 2009;81:807-14.
8. Lin CL, Liao LY, Liu CJ, et al. Hepatitis B genotypes and precore/basal core promoter mutants in HBeAg-negative chronic hepatitis B. J Gastroenterol 2002;37:283-7.
9. Chu CJ, Keeffee EB, Han SH, et al. Prevalence of HBV precore/core promoter variants in the United States. Hepatology 2003;38:619-28.
10. Minuk GY, Orr PS, Brown R, et al. Pre-core mutant infections in the Canadian Inuit. J Hepatol 2000;33:781-4.
11. Juniastuti, Utsumi T, Aksono EB, et al. Predominance of precore mutations and clinical significance of basal core promoter mutations in chronic hepatitis B virus infection in Indonesia. Biomed Rep 2013;1:522-8.
12. Tseng TC, Liu CJ, Yang HC, et al. Higher proportion of viral basal core promoter mutant increases the risk of liver cirrhosis in hepatitis B carriers. Gut 2015;64:292-302.
13. Chu C-M, Lin C-C, Chen Y-C, et al. Basal core promoter mutation is associated with progression to cirrhosis rather than hepatocellular carcinoma in chronic hepatitis B virus infection. Br J Cancer 2012;107:2010-5.
14. Lin CL, Kao JH. Hepatitis B viral factors and clinical outcomes of chronic hepatitis B. J Biomed Sci 2008;15:137-45.
15. Chen CJ, Yang HI. Natural history of chronic hepatitis B REVEALed. J Gastroenterol Hepatol 2011;26:628-38.
16. McMahon BJ. The natural history of chronic hepatitis B virus infection. Hepatology 2009;49:S45-55.
17. Lin ZH, Xin YN, Dong QJ, et al Performance of the aspartate aminotransferase-to-platelet ratio index for the staging of hepatitis C-related fibrosis: An updated meta-analysis. Hepatology 2011;53:726-36.

18. Tacke F, Gehrke C, Luedde T, et al. Basal core promoter and precore mutations in the hepatitis B virus genome enhance replication efficacy of lamivudine-resistant mutants. J Virol 2004;78:8524-35.
19. Amini-Bavil-Olyaee S, Herbers U, Sheldon J, et al. The rtA194T polymerase mutation impacts viral replication and susceptibility to tenofovir in hepatitis B e antigen-positive and hepatitis B e antigen-negative hepatitis B virus strains. Hepatology 2009;49:1158-65.
20. Erhardt A, Reineke U, Blondin D, et al. Mutations of the core promoter and response to interferon treatment in chronic replicative hepatitis B. Hepatology 2000;31:716-25.
21. Hasegawa K, Huang J, Rogers SA, et al. Enhanced replication of a hepatitis B virus mutant associated with an epidemic of fulminant hepatitis. J Virol 1994;68:1651-9.
22. Sato S, Suzuki K, Akahane Y, et al. Hepatitis B virus strains with mutations in the core promoter in patients with fulminant hepatitis. Ann Intern Med 1995;122:241-8.
23. Kosaka Y, Takase K, Kojima M, et al. Fulminant hepatitis B: Induction by hepatitis B virus mutants defective in the precore region and incapable of encoding e antigen. Gastroenterology 1991;100:1087-94.
24. Omata M, Ehata T, Yokosuka O, et al. Mutations in the precore region of hepatitis B virus DNA in patients with fulminant and severe hepatitis. N Engl J Med 1991;324:1699-1704.
25. Liang TJ, Hasegawa K, Rimon N, et al. A hepatitis B virus mutant associated with an epidemic of fulminant hepatitis. N Engl J Med 1991;324:1705-9.
26. Karayiannis P, Alexopoulou A, Hadziyannis S, et al. Fulminant hepatitis associated with hepatitis B virus e antigen-negative infection: Importance of host factors. Hepatology 1995;22:1628-34.
27. Yuen MF, Tanaka Y, Mizokami M, et al. Role of hepatitis B virus genotypes Ba and C, core promoter and precore mutations on hepatocellular carcinoma: A case control study. Carcinogenesis 2004;25:1593-8.
28. Kao JH, Chen PJ, Lai MY, et al. Basal core promoter mutations of hepatitis B virus increase the risk of hepatocellular carcinoma in hepatitis B carriers. Gastroenterology 2003;124:327-34.
29. Chou YC, Yu MW, Wu CF, et al. Temporal relationship between hepatitis B virus enhancer II/basal core promoter sequence variation and risk of hepatocellular carcinoma. Gut 2008;57:91-7.
30. Tong MJ, Blatt LM, Kao JH, et al. Basal core promoter T1762/A1764 and precore A1896 gene mutations in hepatitis B surface antigen-positive hepatocellular carcinoma: A comparison with chronic carriers. Liver Int 2007;27:1356-63.
31. Miyaura H, Iwata M. Direct and indirect inhibition of Th1 development by progesterone and glucocorticoids. J Immunol 2002;168:1087-94.
32. Fairweather D, Frisancho-Kiss S, Rose NR. Sex differences in autoimmune disease from a pathological perspective. Am J Pathol 2008;173:600-9.
33. Girón-González JA, Moral FJ, Elvira J, et al. Consistent production of a higher TH1:TH2 cytokine ratio by stimulated T cells in men compared with women. Eur J Endocrinol 2000;143:31-6.
34. Okamoto H, Tsuda F, Sakugawa H, et al. Typing hepatitis B virus by homology in nucleotide sequence: Comparison of surface antigen subtypes. J Gen Virol 1988;69:2575-83.
35. Norder H, Hammas B, Löfdahl S, et al. Comparison of the amino acid sequences of nine different serotypes of hepatitis B surface antigen and genomic classification of the corresponding hepatitis B virus strains. J Gen Virol 1992;73:1201-8.
36. Arauz-Ruiz P, Norder H, Robertson BH, et al. Genotype H: A new Amerindian genotype of Hepatitis B virus revealed in Central America. J Gen Virol 2002;83:2059-73.
37. Cooksley WGE. Do we need to determine viral genotype in treating chronic hepatitis B? J Viral Hepat 2010;17:601-10.
38. Tran TT, Trinh TN, Abe K. New complex recombinant genotype of hepatitis B virus identified in Vietnam. J Virol 2008;82:5657-63.
39. Tatematsu K, Tanaka Y, Kurbanov F, et al. A genetic variant of hepatitis B virus divergent from known human and ape genotypes isolated from a Japanese patient and provisionally assigned to new genotype J. J Virol 2009;83:10538-47.
40. Yang HI, Yeh SH, Chen PJ, et al. Associations between hepatitis B virus genotype and mutants and the risk of hepatocellular carcinoma. J Natl Cancer Inst 2008;100:1134-43.

Prevalence of celiac disease and celiac autoimmunity in the Toba native Amerindian community of Argentina

Horacio Vázquez MD[1], María de la Paz Temprano RD[1], Emilia Sugai MS[1], Stella M Scacchi MS[1], Cecilia Souza MD[1], Daniel Cisterna MS[1], Edgardo Smecuol MD[1], María Laura Moreno MD[1], Gabriela Longarini MD[1], Roberto Mazure MD[1], María A Bartellini MS[1], Elena F Verdú MD[2], Andrea González RD[1], Eduardo Mauriño MD[1], Julio C Bai MD[1]

H Vázquez, M de la Paz Temprano, E Sugai, et al. Prevalence of celiac disease and celiac autoimmunity in the Toba native Amerindian community of Argentina. Can J Gastroenterol Hepatol 2015;29(8):431-434.

BACKGROUND: Celiac disease (CD) is mostly recognized among subjects with a Caucasian ethnic ancestry. No studies have explored conditions predisposing Amerindians to CD.

OBJECTIVE: To prospectively assess environmental, genetic and serological conditions associated with CD among members of the Toba native population attending a multidisciplinary sanitary mission.

METHODS: An expert nutritionist determined daily gluten intake using an established questionnaire. Gene typing for the human leukocyte antigen (HLA) class II alleles was performed on DNA extracted from peripheral blood (HLA DQ2/DQ8 haplotype). Serum antibodies were immunoglobulin (Ig) A tissue transglutaminase (tTG) and the composite deamidated gliadin peptides/tTG Screen test. Positive cases were tested for IgA endomysial antibodies.

RESULTS: A total of 144 subjects (55% female) were screened. The estimated mean gluten consumption was 43 g/day (range 3 g/day to 185 g/day). Genetic typing showed that 73 of 144 (50.7%) subjects had alleles associated with CD; 69 (94.5%) of these subjects had alleles for HLA DQ8 and four had DQ2 (5.5%). Four and six subjects had antibody concentrations above the cut-off established by the authors' laboratory (>3 times the upper limit of normal) for IgA tTG and deamidated gliadin peptides/tTG screen, respectively. Four of these had concomitant positivity for both assays and endomysial antibodies were positive in three subjects who also presented a predisposing haplotype.

CONCLUSION: The present study was the first to detect CD in Amerindians. The native Toba ethnic population has very high daily gluten consumption and a predisposing genetic background. We detected subjects with persistent CD autoimmunity and, at least, three of them fulfilled serological criteria for CD diagnosis.

Key Words: *Amerindians; Celiac autoimmunity; Celiac disease; Genetic predisposition; Gluten consumption; Tissue transglutaminase (tTG)*

La prévalence de maladie cœliaque et d'auto-immunité cœliaque dans la communauté amérindienne toba d'Argentine

HISTORIQUE : Ce sont surtout des Blancs qui sont atteints de la maladie cœliaque (MC). Aucune étude n'a porté sur les troubles prédisposant les Amérindiens à la MC.

OBJECTIF : Effectuer une évaluation prospective des troubles environnementaux, génétiques et sérologiques associés à la MC chez les membres du peuple amérindien toba qui fréquentaitent une mission sanitaire multidisciplinaire.

MÉTHODOLOGIE : Un expert en nutrition a déterminé la consommation quotidienne de gluten au moyen d'un questionnaire déterminé. Les chercheurs ont procédé au génotypage des allèles de classe II de l'antigène d'histocompatibilité humain (HLA) de l'ADN extrait du sang périphérique (haplotype HLA DQ2 ou DQ8). Les anticorps sériques étaient la transglutaminase tissulaire (tTG) de l'immunoglobuline (Ig) A et le test de dépistage composite des peptides de gliadine déamidés/tTG. Les cas positifs ont été testés pour dépister les anticorps anti-endomysium de l'IgA.

RÉSULTATS : Au total, 144 sujets (55 % de femmes) ont participé au dépistage. Leur consommation estimative moyenne de gluten était de 43 g/jour (plage de 3 g/jour à 185 g/jour). Le génotypage a révélé que 73 des 144 sujets (50,7 %) avaient des allèles associés à la MC, 69 d'entre eux (94,5 %), des allèles HLA DQ8 et quatre, des allèles DQ2 (5,5 %). Quatre et six sujets présentaient des concentrations d'anticorps supérieures aux seuils établis au laboratoire des auteurs (plus de trois fois le seuil supérieur normal) pour dépister les tTG IgA et les peptides de gliadine déamidés/tTG, respectivement. Quatre d'entre eux étaient également positifs aux deux tests, et les anticorps anti-endomysium étaient positifs chez trois sujets qui présentaient également un haplotype prédisposant.

CONCLUSION : La présente étude a été la première à déceler des cas de MC chez les Amérindiens. Le peuple amérindien toba consomme énormément de gluten au quotidien et présente des antécédents génétiques prédisposants. Les auteurs ont décelé des sujets présentant une auto-immunité persistante à la MC et au moins trois d'entre eux satisfaisaient les critères sérologiques du diagnostic de MC.

Celiac disease (CD) is a common autoimmune enteropathy, induced by dietary gluten in genetically predisposed individuals (1). The disease has been traditionally recognized among Caucasians; the estimated prevalence in the Western world is approximately 1% (2). CD is increasingly recognized in the Asian-Pacific region (3). In Latin America, there is a variable proportion of the population with European ancestry, with native communities having a diverse degree of mix with European colonizers (4).

The native Toba ethnic population comprises >60,000 individuals living with a cluster distribution in a forest named 'The Impenetrable' in Northeastern Argentina. This community experiences highly precarious social, economic, sanitary and educational conditions (5). In recent years, they have undergone a drastic change in dietary habits, with wheat replacing their ancestral food sources (6). This has mainly occurred as a consequence of governmental food aid programs aimed at improving nutritional conditions in the community. CD can only occur in individuals with certain class II human leukocyte antigen (HLA) molecules – namely, HLA DQ2 and/or DQ8. In this context, scarce information exists about the prevalence of HLA DQ2 and DQ8, and of CD in native South Americans (7).

[1]*Small Bowel Section, Department of Medicine, Hospital de Gastroenterología C Bonorino Udaondo. Buenos Aires, Argentina;* [2]*Farncombe Family Digestive Health Research Institute, McMaster University, Hamilton, Ontario*

Correspondence: Dr Elena Verdú, Farncombe Family Digestive Health Research Institute, McMaster University, 1200 Main Street West, HSC 3N5C, Hamilton, Ontario L8N 3Z5.e-mail verdue@mcmaster.ca

TABLE 1
Demographics, body mass index (BMI) and gluten consumption of the overall population and according to age (children versus adults)

Characteristic	Overall population	Children	Adults
Subjects, n	144	40	104
Age, years, median (range)	30 (3–79)	8 (3–17)	36 (19–79)
Sex, female/male, n/n	80/64	20/20	60/44
Body weight, kg, mean ± SEM	63.5±1.9	38.1±3.1	73.3±1.5
BMI, kg/m^2, mean ± SEM	26.1±0.5	28.5±0.5	19.8±0.8
Underweight, n (%)	9 (6)	5 (13)	4 (4)
Normal weight, n (%)	44 (31)	16 (37)	28 (27)
Overweight, n (%)*	91 (63)	19 (50)	72 (69)
Gluten consumption, g/day, mean ± SEM	43.1 (2.7)	27.5 (4.9)	48.9 (3.1)

**Overweight for children was considered to be weight above the 97% percentile*

Accordingly, we explored CD risk factors in members of the Toba community attending a multidisciplinary sanitary mission. Our specific aims were to estimate the consumption of gluten; determine the associated celiac genetic risk (ie, HLA DQ2/DQ8); and screen the population for CD.

METHODS

Definitions and overall study design

The present prospective population-based study has two major end points: diagnosis of CD autoimmunity and serological diagnosis of CD. CD autoimmunity was defined as two positive serological tests (≥3 times the upper limit of normal) in samples taken four months apart. Serological CD diagnosis was established in patients with positive ELISA serology and positive immunoglobulin (Ig)A endomysial antibodies (EmA). The study was approved by the Ethics and Scientific Boards of the C. Bonorino Udaondo Gastroenterology Hospital of Buenos Aires, Argentina. Written or expressed consent was obtained from enrolled subjects in the presence of the community leaders and/or their parents.

Study participants

From March 2013 through March 2014, four multidisciplinary missions by the same team were performed in the Toba population living in 'The Impenetrable', in the province of Chaco, Argentina. All subjects attending the health mission and fulfilling inclusion criteria were invited to participate in the study by a physician in the presence of the community leader. Individuals <3 years of age or having a previous diagnosis of CD were excluded, as well as those refusing blood sampling for cultural reasons. A full clinical history, physical examination and blood samples were obtained.

Nutritional analysis and gluten consumption

A nutritionist-expert in CD interviewed subjects at enrollment. Gluten consumption was estimated using a food frequency intake questionnaire (11) and compared with data obtained from a 48 h food recall recorded by the same nutritionist. The food frequency intake estimates the frequency of consumed foods during a specific period of time and identifies all foods containing wheat, rye and barley; the gluten content was calculated following previously described methods, taking into account only known gluten-containing products (11-14).

CD-specific serology

Serum samples obtained were kept frozen (–20°C) until analysis at Hospital de Gastroenterología C. Bonorino Udaondo. Serology tests included IgA antibodies to tissue transglutaminase (tTG) using ELISA (Quanta Lite h-tTG IgA, Inova Diagnostic Inc, USA), and the dual detection of IgA and IgG isotypes of antibodies to deamidated gliadin peptide (DGP) and tTG in a single assay (DGP/tTG Screen; Quanta Lite h-tTG/DGP Screen, Inova Diagnostic Inc). The cut-off value specified by the manufacturer for the ELISA assays were 20 U/mL but the cut-off used in the present study was based on that estimated by the authors in an earlier study (>3 times the upper limit of normal [60 U/mL for both tests]) (15). Positive ELISA tests were analyzed for IgA endomysial antibodies (IgA EmA) using immunofluorescence on primate esophagus substrate (dilution 1:5). Positive samples were retested after four months during a new visit to the mission to rule out transient positivity.

CD-related HLA genotyping

CD specific gene typing for the detection of HLA class II alleles was performed on DNA from a drop of whole blood obtained from the fingertip and collected on a paper filter card (Whatman 903 FTA, Whatman, USA). Samples were stored at room temperature and analyzed at the BioDiagene laboratory (Italy). A DQ-CD Typing Plus kit (BioDiagene, Italy) was used for the identification of HLA class II alleles: DQA1*0201,*03,*05, DQB1*02,*0301/04 and *0302.

Statistical analysis

Population data are reported as percentage, median and range or mean ± SEM, as appropriate. The Mann-Whitney test for independent samples was used for comparison of patients with and without CD autoimmunity.

RESULTS

Study population

A total of 144 subjects (55% female) were enrolled. Patients' characteristics are summarized in Table 1. Briefly, mean body mass index was above the range of normal weight and most subjects had normal weight or were overweight. Mean biochemical parameters were within normal range (data not shown). No IgA-deficient cases were detected. Genetic typing showed that 73 of 144 (50.7%) subjects had alleles associated with CD; 69 of these subjects (94.5%) had alleles for HLA DQ8 and 4 had DQ2 (5.5%).

Gluten consumption

The estimated mean gluten consumption for all enrolled subjects was almost double of that assessed for populations living in urban areas of Argentina (unpublished data) and in other parts of the world (16) (Table 1).

Identification of subjects with CD and persistent celiac autoimmunity

Two adults (both female, 23 and 38 years of age) and one child (male, 12 years of age) fulfilled serological criteria for CD (Table 2). One additional child (female) was excluded from enrollment due to previous CD diagnosis. From the newly CD diagnosed patients, one adult female had normal body mass index, the other was overweight and the child was obese. The patients were asymptomatic and biochemical parameters were normal (Table 3). High gluten consumption was detected in two of the three (114 g/day and 29 g/day for the adults, and 91 g/day for the child). The child was HLA DQ2+, and both adults were positive for HLA DQ8 (Table 2).

Three additional subjects (two HLA-DQ8+) had serum concentrations of CD antibodies above normal. In three, DGP/tTG was positive while both DGP/tTG and IgA tTG were positive in two. Repeat serology four months apart was positive in all three and, thus, they were considered to have persistent CD autoimmunity. Daily gluten intake was normal in two of these subjects, and high in one (Table 2).

DISCUSSION

We prospectively explored the consumption of gluten by the apparently healthy native Toba community, the frequency of HLA CD-predisposing genotypes, and the prevalence of CD autoimmunity and serological evidence of the disorder. The present study revealed a very high consumption of gluten, almost twofold higher than that of individuals of European ancestry living in urban areas of Argentina

TABLE 2
Individual estimation of gluten consumption, genetic human leukocyte antigen (HLA) typing and serological status in patients considered to have gluten autoimmunity* or celiac disease†

Subject: age (years), sex	Gluten consumption, g/day	HLA typing	IgA tTG, U/mL	DGP/tTG Screen test, U/mL	IgA EmA
1: 33, male	136	DR4-DQ8/ DQ7	12/9	80/101	–ve/–ve
2: 72, male	26	DQ7	37/32	80/101	–ve/–ve
3: 38, male	22	DQ8	120/63	109/77	–ve/–ve
4: 12, male	91	DR7-DQ2/ DQ7	>200/>200	>200/>200	+ve/ND
5: 38, female	114	DR4-DQ8/ DQ7	131/ND	100/ND	+ve/ND
6: 23, female	29	DQ8	>200/ND	60/ND	+ve/ND

**Patients 1, 2 and 3; †Patients 4, 5 and 6. Serum concentration of ELISA tests (immunoglobulin A [IgA], tissue transglutaminase [tTG] and deamidated gliadin peptides [DGP]/tTG Screen was re-tested in patients four months after the first positive result. Data above are presented as test 1/test 2 results. EmA Endomysial antibody; ND Not determined*

and in other parts of the world (16). This alimentary change was the product of a political decision implemented almost 30 years ago to improve alarming nutritional deficits in this population. The initial aim of the alimentary support appears to have achieved relative success because anthropometric evaluations revealed few underweight subjects and normal biochemical parameters. An unintentional consequence of such important alimentary intervention was that 50% and 70% of children and adults, respectively, were overweight or obese at the time of the study. The quality of food support, consisting mainly of wheat, may have played a role in this outcome.

Similar to native populations in Brazil and Chile (17), our study showed that only 3% of subjects had the HLA DQ2 alleles (prevalent in European populations), while 48% had the DQ8. This contrasts with the increased HLA-DQ2 allele frequency detected in Asia with high CD prevalence, as demonstrated in several recent studies (17-20).

A significant observation in the present study was the detection of CD autoimmunity in six subjects. Three of these subjects had concomitant positivity in all tests and, therefore, qualified for CD diagnosis. The patients were asymptomatic, did not exhibit weight loss and gluten consumption was very high in two of the cases. The results raise concerns about CD risk in the Toba population and its association with the quality of food aid programs targeted at them.

The present small study was based on serological prevalence due, in part, to some cultural resistance to the implementation of the screening program and to the difficulty of implementing endoscopy assessment in this setting. However, according to our experience and those of others, a positive IgA EmA test is almost absolutely predictive of CD enteropathy.

CONCLUSION

The Toba population has environmental and genetic risk factors for the development of CD. The present study detected a very high consumption of gluten as the result of government food aid programs largely based on wheat products. The genetic background was dominated by alleles codifying for DQ8 antigen. Three patients in the present study fulfilled criteria for CD and three others for persistent CD. One additional CD case had previously been identified in this community. These results raise important questions regarding the food quality in aid programs that should include a variety of non-gluten-containing grains and should prompt CD screening in these native populations.

TABLE 3
Demographic and some clinical characteristics of the three patients with serological diagnosis of celiac disease

Characteristic	Patient 1	Patient 2	Patient 3
Age, years	12	38	23
Sex	Male	Female	Female
Clinical categorization*	Asymptomatic	Asymptomatic	Asymptomatic
Body mass index, kg/m^2	25.7	26.6	23.8
Gluten consumption, g/day	91	114	29

**Clinical categorization was performed according to the Oslo nomenclature*

AUTHORSHIP STATEMENT: All authors approved the final version of the article, including the authorship list. Horacio Vázquez: research design, collection of patients, analysis of results. María de la Paz Temprano: acquisition of nutritional and alimentary data. Emilia Sugai: serological testing. Stella M Scacchi: acquisition of sera and processing, clinical laboratory testing. Cecilia Souza: acquisition of data in the pediatric population. Daniel Cisterna: genetic samples. Edgardo Smecuol: research design. María Laura Moreno: analysis of data. Gabriela Longarini: analysis of data. Roberto Mazure: research design. María A Bartellini: genetic samples. Elena F Verdú: drafted the manuscript and revised it critically. Andrea González: analysis of nutritional and alimentary information. Eduardo Mauriño: study design, analysis of results and critical review of the draft. Julio C Bai: study design, analysis of results, drafted the final manuscript, guarantor of the article.

ACKNOWLEDGEMENTS: The authors thank Dr Sonia Kupfer (University of Chicago, Chicago, Illinois) for scientific discussions and critical review regarding the genetic determinations and interpretations of the current study. They also thank the *Fundación Pequeños Gestos Grandes Logros*, all members of the multidisciplinary sanitary missions, Inova Diagnostic Inc (USA) for generously providing kits for serology, and BioDiagene Laboratory (Italy) for generously performing CD genetic typing analyses free of charge.

FUNDING SUPPORT: The present study was partially funded by the Consejo de Investigación, MSAL, Buenos Aires City Government and the *Asociación para el Estudio de Enfermedades del Intestino* (AEDEI). Source for salary support of researchers is provided by the Buenos Aires City Government as part of their research duties. EFV holds a Canada Research Chair and is funded by CIHR MOP 123282.

REFERENCES

1. Ludvigsson JF, Leffler DA, Bai JC, et al. The Oslo definitions for coeliac disease and related terms. Gut 2013;62:43-52.
2. Bai JC, Fried M, Corazza GR, et al. World Gastroenterology Organization global guidelines on celiac disease. Clin Gastroenterol 2013;47:121-6.
3. Makharia GK, Mulder CJ, Goh KL, et al. Issues associated with the emergence of coeliac disease in the Asia-Pacific region: A working party report of the World Gastroenterology Organization and the Asian Pacific Association of Gastroenterology. World Gastroenterology Organization-Asia Pacific Association of Gastroenterology Working Party on Celiac Disease. J Gastroenterol Hepatol 2014;29:666-77.
4. Pereyra MA, Ortiz-Agostinho CL, Nishitokukado I, et al. Prevalence of celiac disease in an urban area of Brazil with predominantly European ancestry. World J Gastroenterol 2006;12:6546-50.
5. Martinez G. Pluralismo médico y etnomedicina entre los Tobas (Qom) del río Bermejito (Chaco, Argentina). Desafíos y aportes para una gestión intercultural de la salud en el Impenetrable Chaqueño. Revista del Museo de Antropología 2011;4:195-210.
6. Bernio JC. Conquest by chainsaw <www.dandc.eu/en/article/logging-subtropical-dry-forest-deprives-indigenous-people-argentina-their-livelihood> (Accessed September 24, 2013).
7. Ramos da Rosa Utiyama S, Coelho Ribas JL, Mitsunori Nisihara R, et al. Celiac disease in native Indians from Brazil: A clinical and epidemiological survey. N Am J Med Sci 2010;2:138-42.

8. Peschken CA, Esdaile JM. Systemic lupus erythematosus in North American Indians: A population based study. J Rheumatol 2000;27:1884-91.
9. Houghton KM, Page J, Cabral DA, et al. Systemic lupus erythematosus in the pediatric North American Native population of British Columbia. J Rheumatol 2006;33:161-3.
10. Ferucci ED, Templin DW, Lanier AP. Rheumatoid arthritis in American Indians and Alaska Natives: A review of the literature. Semin Arthritis Rheum 2005;34:662-7.
11. Overbeek FM, Uil-DietermanI G, Moll W, et al. The daily gluten intake in relatives of patients with coeliac disease compared with that of the general Dutch population. Eur J Gastroenterol Hepatol 1997;9:1097-9.
12. Levingstone MB, Prentice AM, Strain JJ. Accuracy of weighed dietary records in studies of diet and health. Br Med J 1900;300:708-12.
13. Hopman EG, Pruijn R, Tabben EH, et al. Food questionnaire for the assessment of gluten intake by children 1 to 4 years old. J Pediatr Gastroenterol Nutr 2012;54:791-6.
14. Ministerio de Salud. SARA: Sistema de Análisis y Registro de Alimentos. Consulta de Composición Química de los Alimentos. <http://datos.dinami.gov.ar/produccion/sara/> (Accessed August 12, 2013).
15. Sugai E, Moreno ML, Hwang HJ, et al. Celiac disease serology in patients with different pretest probabilities: Is biopsy avoidable? World J Gastroenterol 2010;16:3144-52.
16. Lionetti E, Catassi C. Co-localization of gluten consumption and HLA-DQ2 and -DQ8 genotypes, a clue to the history of celiac disease. Dig Liv Dis 2014;46:1057-63.
17. Parada A, Araya M, Pérez-Bravo F, et al. Amerindian mtDNA haplogroups and celiac disease risk HLA haplotypes in mixed-blood Latin American patients. J Pediatr Gastroenterol Nutr 2011;53:429-34.
18. Catassi C, Doloretta Maccis M, Ratsch IM, et al. The distribution of DQ genes in the Saharawi population provides only a partial explanation for the high CD prevalence. Tissue Antigens 2001;58:402-6.
19. Shanmugalakshmi S, Balakrishnan K, Manoharan K, et al. HLA-DRB1*, -DQB1* in Piramalai Kallars and Yadhavas, two Dravidian-speaking castes of Tamil Nadu, South India. Tissue Antigens 2003;61:451-64.
20. Yuan J, Gao J, Li X, et al. The tip of the "celiac iceberg" in China: A systematic review and meta-analysis. PLoS One 2013;8:e81151.

Examining the clinical use of hemochromatosis genetic testing

Matthew B Lanktree MD PhD[1], Bruce B Lanktree MD, Guillaume Paré MD MSc[2], John S Waye PhD[2], Bekim Sadikovic PhD[2], Mark A Crowther MD[1]

MB Lanktree, BB Lanktree, G Paré, JS Waye, B Sadikovic, MA Crowther. Examining the clinical use of hemochromatosis genetic testing. Can J Gastroenterol Hepatol 2015;29(1):41-45.

BACKGROUND: Hereditary hemochromatosis leads to an increased lifetime risk for end-organ damage due to excess iron deposition. Guidelines recommend that genetic testing be performed in patients with clinical suspicion of iron overload accompanied by elevated serum ferritin and transferrin saturation levels.

OBJECTIVE: To evaluate guideline adherence and the clinical and economic impact of *HFE* genetic testing.

METHODS: The electronic charts of patients submitted for *HFE* testing in 2012 were reviewed for genetic testing results, biochemical markers of iron overload and clinical history of phlebotomy.

RESULTS: A total of 664 samples were sent for testing, with clinical, biochemical and phlebotomy data available for 160 patients. A positive C282Y homozygote or C282Y/H63D compound heterozygote test result was observed in 18% of patients. Patients with an at-risk *HFE* genotype had significantly higher iron saturation, serum iron and hemoglobin (P<0.001), without higher ferritin or liver enzyme levels. Fifty percent of patients referred for testing did not have biochemical evidence of iron overload (transferrin saturation >45% and ferritin level >300 µg/L). Patients were four times more likely to undergo phlebotomy if they were gene test positive (RR 4.29 [95% CI 2.35 to 7.83]; P<0.00001).

DISCUSSION: One-half of patients referred for testing did not exhibit biochemical evidence of iron overload. Many patients with biochemical evidence of iron overload, but with negative genetic test results, did not undergo phlebotomy. A requisition to determine clinical indication for testing may reduce the use of the *HFE* genetic test. Finally, improvement of current genetic test characteristics would improve rationale for the test.

CONCLUSION: A significant proportion of hemochromatosis genetic testing does not adhere to current guidelines and would not alter patient management.

Key Words: *Genetic testing; Hemochromatosis; Iron; Phlebotomy*

Examen de l'utilisation clinique du test de dépistage génétique de l'hématochromatose

HISTORIQUE : L'hématochromatose héréditaire s'associe à un plus grand risque à vie de dommage aux organes cibles causé par un dépôt excessif de fer. Les lignes directrices recommandent d'effectuer un test de dépistage génétique en cas de suspicion clinique de surcharge en fer accompagnée d'un taux élevé de ferritine sérique et d'une saturation de la transferrine.

OBJECTIF : Évaluer le respect des lignes directrices et les répercussions cliniques et économiques du test *HFE*.

MÉTHODOLOGIE : Les chercheurs ont examiné le dossier électronique des patients qui devaient subir un test *HFE* en 2012 afin de connaître leurs résultats génétiques, leurs marqueurs biochimiques de surcharge en fer et leurs antécédents cliniques de phlébotomie.

RÉSULTATS : Au total, 664 échantillons ont été envoyés au dépistage, et les chercheurs ont obtenu les données cliniques, les données biochimiques et la phlébotomie de 160 patients. Chez 18 % d'entre eux, le test a décelé une homozygotie C282Y ou une hétérozygotie composite C282Y et H63D. Les patients présentant un génotype *HFE* à risque avaient une saturation en fer, en fer sérique et en hémoglobine considérablement plus élevée (P<0,001), sans augmentation de leur taux de ferritine ou d'enzymes hépatiques. Cinquante pour cent des patients aiguillés pour le test ne présentaient pas de manifestations biochimiques de surcharge en fer (saturation en transferrine >45 % et taux de ferritine >300 µg/L). Les patients étaient quatre fois plus susceptibles de subir une phlébotomie si le test génétique était positif (RR 4,29 [95 % IC 2,35 à 7,83]; P<0,00001).

EXPOSÉ : La moitié des patients aiguillés pour subir le test ne présentaient pas de manifestations biochimiques de surcharge en fer. De nombreux patients qui présentaient des manifestations biochimiques de surcharge en fer, mais dont les résultats du test génétique étaient négatifs, n'ont pas subi de phlébotomie. Une réquisition pour établir l'indication clinique d'effectuer le test *HFE* peut en réduire l'utilisation. Enfin, en améliorant les caractéristiques du test génétique actuel, on en justifierait mieux l'utilisation.

CONCLUSION : Une proportion marquée de tests génétiques de l'hémochromatose ne respecte pas les lignes directrices et ne changerait pas la prise en charge des patients.

Hereditary hemochromatosis is a common genetic condition in which elevated total body iron stores leads to iron deposition in the liver, heart, pancreas, skin, joints, pituitary gland and testes, resulting in liver cirrhosis, heart failure, arthritis, diabetes, skin bronzing, endocrine abnormalities and cancer (1). In 1996, two missense mutations in the hemochromatosis gene (*HFE*) were identified to be a cause of hereditary hemochromatosis, leading to an increase in our understanding of iron metabolism, the pathophysiology of hemochromatosis and the genetic test that is still performed today (2).

The genetic test for hemochromatosis most commonly involves genotyping of a cysteine-to-tyrosine substitution at amino acid position 282 (C282Y) and a histidine-to-aspartic acid substitution at amino acid position 63 (H63D) in the *HFE* gene. A positive genetic test is defined as two copies (homozygosity) of the C282Y allele, or one C282Y allele with an H63D allele on the other chromosome, termed 'compound heterozygosity'. Fewer than 0.5% of individuals of European ancestry are homozygous for C282Y, with lower prevalence in other ethnicities (3).

The penetrance of the C282Y variant for iron accumulation leading to end-organ disease in one's lifetime, which also represents the positive predictive value of the genetic test, has been evaluated in prospective, retrospective and cross-sectional studies (4). Differences in study populations and disease definitions have yielded a wide range of estimates from 0% to 75% (4). Cross-sectional studies including large numbers of patients may underestimate lifetime risk because they may evaluate patients before they have developed the outcome and may rely on imperfect measures of disease diagnosis (5). Alternatively, prospective studies that follow small numbers of patients may identify disease that may not be clinically relevant (6). Moreover, penetrance

[1]*Department of Medicine;* [2]*Department of Pathology and Molecular Medicine, McMaster University, Hamilton, Ontario*
Correspondence: Dr Matthew B Lanktree, Department of Medicine, McMaster University, 1200 Main Street West, Room 3W10A-C, Hamilton, Ontario L8N 3Z5. E-mail matthew.lanktree@medportal.ca

is significantly different between men and premenopausal women, given the protection yielded by menses. Nevertheless, overall estimates from a meta-analysis of current clinical guidelines suggests a lifetime penetrance and, hence, positive predictive value of 13% (4).

The sensitivity of the current hemochromatosis genetic test is represented by the prevalence of individuals who test positive if they have developed end-organ iron overload damage. Estimates are again affected by the population observed and the gold standard used for diagnosis, but have ranged from 75% to 90% (4,7-9). Given that hemochromatosis exists in nonwhite populations, and the low prevalence of C282Y in nonwhite ethnicities, the sensitivity of the test is reduced in these populations (10). Non-C282Y *HFE* polymorphisms (such as H63D) have smaller effects on iron accumulation risk and may improve the sensitivity of a genetic test, but consequently negatively affect specificity because the penetrance of H63D is quite low (11). In general, patients with a negative genetic test may still develop iron overload, while 80% to 90% of individuals with a positive genetic test may never develop iron overload (4).

Over the past decade, considerable debate has occurred with regard to the appropriate clinical context in which to order genetic testing for hemochromatosis (11-13). Current guidelines published by the European Association for the Study of the Liver (4) and American Association for the Study of Liver Diseases (7) recommend against genetic testing for hemochromatosis in the absence of clinical suspicion for hemochromatosis, with the exception of siblings of confirmed *HFE* C282Y homozygotes with clinical hemochromatosis (ie, genetic screening of unaffected individuals is inappropriate). Hence, genetic testing of patients should be limited to individuals with elevated transferrin saturation (>45% in women and >50% in men) and serum ferritin level (>300 µg/L). Strategies investigating alternative causes of elevated ferritin levels and transferrin saturation followed by serial genetic investigations has also been proposed (13,14).

Phlebotomy can be both diagnostic and therapeutic because it removes excess iron, and the response to phlebotomy is an indicator of the quantity of iron present (4). Individuals with pathological iron overload will require removal of many grams of iron while those with other causes of elevated ferritin levels and transferrin saturation will generally experience a rapid fall in iron levels (4). Phlebotomy should be initiated in patients with suspected iron excess and continued until ferritin level has dropped to <50 µg/L, although no randomized control trials have been performed to support its use (4). Phlebotomy is safe in patients with advanced liver fibrosis or cirrhosis, and may improve arthropathies, diabetes, hypogonadism and hypothyroidism (4).

The cost of genetic testing for hemochromatosis has not changed substantially in the past decade despite immense improvements in genotyping technology. The cost of DNA extraction and single-nucleotide genotyping in a high-throughput research environment has plummeted, but because overhead costs, data storage and data integrity issues have remained constant, cost estimates from $150 to $500 per test have persisted (15).

To better understand the use of hemochromatosis genetic testing in current practice, we sought to identify the clinical and biochemical characteristics of patients submitted to a tertiary care laboratory in Hamilton, Ontario. Finally, we examined factors that may have impacted the clinician's decision to provide phlebotomy.

METHODS

Ethics approval to retrospectively review patient records was granted by the Hamilton Integrated Research Ethics Board (approval number 13-323-C). Data were collected from the electronic patient record (EPR) of Hamilton Health Sciences and included all patients referred for genetic testing for hemochromatosis in the 2012 calendar year. Hamilton Health Sciences is an academic tertiary care medical facility serving a catchment of 2.3 million Canadians. In total, the EPR of 620 patients were manually accessed to obtain genetic test results, biochemical markers of iron overload, biochemical markers of end-organ damage resulting from iron accumulation and clinical history of phlebotomy. Specifically, data included genotype at *HFE* C282Y and H63D, maximum ferritin, transferrin saturation, aspartate transaminase, alanine transaminase, gamma-glutamyl transpeptidase, alkaline phosphatase, glycated hemoglobin, fasting glucose, hemoglobin concentration and minimum mean corpuscular volume. Phlebotomy status was primarily ascertained by examining the 'reason for visit' field within the appointment scheduling system of the EPR, as well as manual review of the free-text consultation and progress notes. The type of physician who had made the referral was also available. Unfortunately, patients who had not required hospitalization for any reason, nor received evaluation by a hematologist or gastroenterologist, but were referred for genetic testing by a community primary care physician would not have clinical data available in the Hamilton Health Sciences EPR.

Data are expressed as counts and percentages for categorical variables, and mean ± SD or median (interquartile range) for continuous data depending on their distribution. Comparisons between groups were performed using χ^2 and ANOVA tests.

RESULTS

In total, 664 patients were submitted for hemochromatosis genetic testing, including nine repeated tests; 177 (26%) of patients were female. The observed minor allele frequency was 18% for C282Y and 20% for H63D, yielding 49 (8.1%) C282Y homozygotes and 49 (8.1%) C282Y/H63D compound heterozygotes, corresponding to an overall positive test rate of 16.2% (Table 1). Of 655 patients, 449 (68.5%) did not carry a C282Y allele. There was no significant difference in the rate of positive tests ordered by family physicians verses those ordered by specialists.

Of those referred for testing, clinical and biochemical data were available for 160 patients. Transferrin saturation and ferritin level data were available for 96 patients, and evidence of iron overload (measured transferrin saturation >45% and ferritin >300 µg/L) was present in 48 (50%). Comparing biochemical characteristics between those with positive hemochromatosis genetic testing results (either C282Y/C282Y or C282Y/H63D) had significantly higher transferrin saturation, hemoglobin and serum iron concentrations ($P<0.001$) (Table 2). There was no increase in either ferritin or liver enzyme levels in the group with a positive genetic test.

Phlebotomy was performed in 33 patients, 20 of whom were hemochromatosis gene test positive and 13 who were not. For a given ferritin level or transferrin saturation, patients were significantly more likely to undergo phlebotomy if they were gene test positive ($P<0.01$). The likelihood of undergoing phlebotomy for a given ferritin level and transferrin saturation is presented in decision-tree format in Figure 1. Examining the test characteristics of the hemochromatosis genetic test for subsequent need for phlebotomy in this clinically ascertained sample, in which genotype likely affected the decision to provide phlebotomy, the sensitivity for subsequent need for phlebotomy was 60% (95% CI 43% to 75%), with a specificity of 82% (95% CI 75% to 88%) and an RR of 4.29 (95% CI 2.35 to 7.83 [Fisher's exact $P<0.00001$]). The positive predictive value was 47% (95% CI 35% to 59%) and the positive likelihood ratio was 3.47 (95% CI 2.06 to 5.41).

DISCUSSION

The results of a retrospective analysis of 664 hemochromatosis genetic tests performed at a tertiary care academic medical centre in the 2012 calendar year indicates that a substantial proportion of hemochromatosis genetic testing was performed out of accordance with currently published guidelines. Additionally, hemochromatosis genetic testing appears to be affecting the decision to provide phlebotomy without solid evidence to support the practice. We believe that more stringent access to genetic testing for hemochromatosis may reduce the amount of unnecessary testing without adversely affecting patient care. Finally, improvements in the test characteristics of the hemochromatosis genetic test, and prospective randomized studies designed to appropriately direct its use, would lead to both cost savings and improvement in patient care.

TABLE 1
Genetic testing results

HFE genotype	Family physicians	Specialists*	Total
282Y/282Y	20 (7.1)	29 (9.0)	49 (8.1)
282Y/63D	22 (7.8)	27 (8.4)	49 (8.1)
Negative	241 (85.2)	265 (82.6)	506 (83.8)

*Data presented n (%). *Specialists include (in order of prevalence): gastroenterologists, hematologists and cardiologists. HFE Hemochromatosis gene*

Who should undergo genetic testing for hemochromatosis?
Current guidelines suggest confirmatory genetic testing in patients with clinical suspicion and biochemical evidence of iron accumulation (4,7). A diagnostic and therapeutic trial of phlebotomy may be indicated in these patients regardless of genetic testing results. Nevertheless, ferritin levels and transferrin saturations were available for only 96 patients sent for genetic testing and only one-half of these fulfilled biochemical criteria for possible iron overload. It is possible that more of the patients had ferritin and transferrin testing performed, which was unavailable for the present retrospective analysis; however, a large proportion of patients who are undergoing the genetic test do not have biochemical evidence of iron overload.

It is possible that a proportion of patients that received genetic testing without iron accumulation were first-degree relatives of homozygous C282Y hemochromatosis patients, and negative genetic testing results may reduce the need for serial transferrin and ferritin measurements in the future. This remains a poor explanation because 68.5% of those tested did not carry a C282Y allele, which is unlikely if they are a first-degree relative of homozygous C282Y hemochromatosis patients. The cost for ferritin and transferrin measurements is approximately $25 each (16). Additionally, first-degree relatives may still be at elevated risk if a non-*HFE* C282Y mechanism is the reason for iron accumulation.

So-called 'shotgun' investigations, in which multiple investigations are simultaneously ordered to cover a range of potential pathogenic mechanisms in patients with elevated liver enzyme levels, porphyria cutanea tarda, hepatocellular carcinoma, type 1 diabetes or well-defined chondrocalcinosis, may represent a portion of the hemochromatosis genetic testing. However, given the slow rate of disease progression of hemochromatosis, a staggered approach in which genetic testing follows measurement of ferritin levels and transferrin saturation appears to be more appropriate.

In general, our results suggest that a large proportion of current hemochromatosis genetic testing is unlikely to provide diagnostic certainty, change patient management or comply with current guidelines.

TABLE 2
Clinical characteristics according to *HFE* test result

	HFE test result	
Characteristic	Positive	Negative
n	42	117
Age, years	54.3±13.6	56.3±13.7
Female sex, n (%)	13 (28)	35 (30)
Maximum ferritin, µg/L	864±607	901±862
Maximum transferrin saturation, %	62±23*	46±20*
Maximum serum iron, umol/L	121±6.5*	21.5±7.7*
Maximum hemoglobin, g/L	155±9.4*	144±20*
Minimum mean corpuscular volume, fL	93.8±5.8	91±11
Maximum aspartate aminotransferase, U/L	68±20	1001±36
Maximum alanine aminotransferase, U/L	110±30	156±42
Maximum gamma-glutamyl transpeptidase, U/L	55±17	208±75
Maximum alkaline phosphatase, U/L	66±12	151±41
Viral hepatitis, n (%)	2 (4.7)	11 (9.4)
Received phlebotomy, n (%)	27 (64)*	12 (10)*

*Data presented as mean ± SD unless otherwise indicated. *P<0.001. HFE Hemochromatosis gene*

Efforts to improve education regarding the indications for hemochromatosis genetic testing may reduce the quantity of unnecessary testing. Finally, a laboratory requisition form for hemochromatosis genetic testing, including the indication for the test and measured ferritin levels and transferrin saturation, would enable more accurate evaluation of hemochromatosis genetic testing in the future and would prevent unnecessary testing.

Should the current *HFE* genetic test affect the decision to phlebotomize?
Clinicians need to incorporate information from history, physical investigations and patient preferences to evaluate the pros and cons of phlebotomy in a particular patient. Factors to be considered before offering phlebotomy are included in Box 1. The decision-tree in Figure 1 attests that the decision is not simply based on ferritin level, transferrin saturation and genotype. However, genotype does appear to affect the decision to provide phlebotomy in these data. To date, there exists a dearth of evidence that individuals with *HFE* C282Y-associated hemochromatosis respond differently to phlebotomy than those with non-*HFE* C282Y iron accumulation. Thus, while a positive test may increase the likelihood of pathogenic iron accumulation, a trial of phlebotomy is the gold standard for diagnosis.

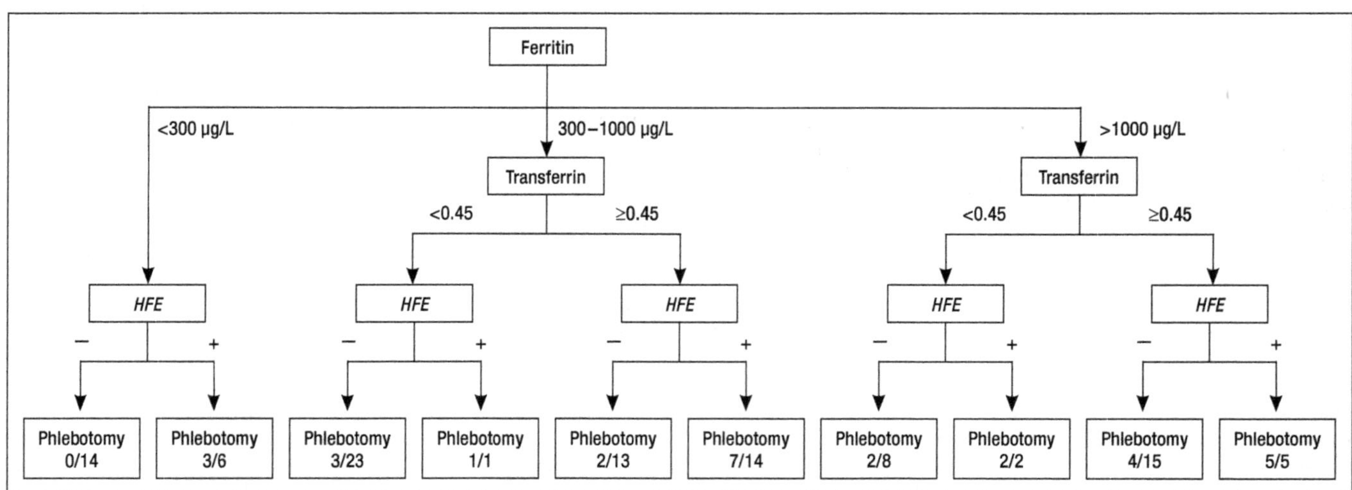

Figure 1) *Decision-tree analysis. HFE Hemochromatosis gene*

BOX 1
Factors affecting to the decision to phlebotomize

- Serum iron parameters (transferrin saturation, ferritin)
- Hemoglobin
- Presence of alternative explanation
  - o Alcohol use/abuse
  - o Metabolic syndrome
  - o Nonalcoholic steatohepatitis
  - o Viral hepatitis
- Inflammatory states (infectious, autoimmune)
- Noninvasive liver imaging (ultrasound, Fibroscan [Ecosens, France], computed tomography)
- Evidence of iron overload on liver biopsy
- Hemochromatosis gene (*HFE*) genotype
- Comorbidities
- Patient's goals of care
- Ease of access to facilities
- Tolerability of procedure (orthostasis)

Can hemochromatosis genetic test performance be improved?

Hemochromatosis is a complex disease with substantial genetic heterogeneity. While *HFE* C282Y is found in 80% of patients with clinical hemochromatosis (4), other genetic causes of hemochromatosis have been described. These include rare mutations and deletions in *HFE* undetected by standard clinical tests, and mutations in four other genes (hepcidin [*HAMP*], hemojuvelin [*HFE2*, previously *HJV*], transferrin receptor 2 [*TFR2*] and ferroportin [*SLC40A1*]) (17,18). Mutations that affect ferritin concentration (a common marker used to identify iron overload) but do not lead to iron accumulation, have been observed in the ferritin light chain (*FTL*) gene (19). Adding to the complexity, common polymorphisms in 14 additional genes have been associated with variation in iron metabolism in the general population and in hemochromatosis patients (10,20-22). Environmental exposures, including diet, alcohol and medication use, can also significantly alter the risk of pathological iron accumulation. Given the complexity, it is not surprising that the genotyping of only the two single-nucleotide polymorphisms in *HFE* yields a test that is not particularly sensitive nor specific for iron accumulation leading to end-organ damage. Test performance in non-European ethnicities is even worse given the low background frequency of the C282Y allele (10).

Schranz et al (13), suggested a staggered approach to testing with targeted sequencing of *HFE*, *TFR2*, *HFE2*, (*HJV*), *HAMP* and *SLC40A1* (13). Given the drastically declining cost of DNA sequencing via sequence capture and next-generation sequencing technologies, the production of a test that obtains all DNA variations in iron metabolism genes at a low cost is needed. A similar strategy has already been used for conditions such as hypertrophic cardiomyopathy and dyslipidemia (23,24). Barriers to the interpretation of next-generation sequencing data exist as many variants of unknown significance. Nevertheless, commercial testing using this strategy is already available <www.invitae.com/en/physician/condition-detail/CND0033/>.

Limitations of the current study

The most significant limitation of the current study was its retrospective chart review design. Data were extracted from the records of a selected hospital-based population, biasing the study toward the presence of more severely affected patients. Patients may have been instructed to attend blood donation clinics, which would not have entered our records as having undergone a phlebotomy. While a clinical trial randomly assigning patients to phlebotomy may not be ethically feasible, a prospectively collected cohort with stringent data collection protocols may mitigate bias in the evaluation of next-generation genetic tests for hemochromatosis.

CONCLUSIONS

The present retrospective chart review suggests that a large percentage of hemochromatosis genetic testing was performed on patients without iron accumulation and would not alter their diagnosis, prognosis or management. Patients with evidence of iron overload were more likely to receive phlebotomy in the presence of a positive test when those with negative tests, especially non-Europeans, may still benefit.

DISCLOSURES: The authors have no financial disclosures or conflicts of interest to declare.

REFERENCES

1. Pietrangelo A. Hereditary hemochromatosis – a new look at an old disease. N Engl J Med 2004;350:2383-97.
2. Feder JN, Gnirke A, Thomas W, et al. A novel MHC class I-like gene is mutated in patients with hereditary haemochromatosis. Nat Genet 1996;13:399-408.
3. Steinberg KK, Cogswell ME, Chang JC, et al. Prevalence of C282Y and H63D mutations in the hemochromatosis (*HFE*) gene in the United States. JAMA 2001;285:2216-22.
4. EASL clinical practice guidelines for *HFE* hemochromatosis. J Hepatol 2010;53:3-22.
5. Beutler E, Felitti VJ, Koziol JA, Ho NJ, Gelbart T. Penetrance of 845G→A (C282Y) *HFE* hereditary haemochromatosis mutation in the USA. Lancet 2002;359:211-8.
6. Olynyk JK, Hagan SE, Cullen DJ, Beilby J, Whittall DE. Evolution of untreated hereditary hemochromatosis in the Busselton population: A 17-year study. Mayo Clin Proc 2004;79:309-13.
7. Bacon BR, Adams PC, Kowdley KV, Powell LW, Tavill AS. Diagnosis and management of hemochromatosis: 2011 practice guideline by the American Association for the Study of Liver Diseases. Hepatology 2011;54:328-43.
8. Pietrangelo A, Caleffi A, Corradini E. Non-*HFE* hepatic iron overload. Semin Liver Dis 2011;31:302-18.
9. Asberg A, Hveem K, Thorstensen K, et al. Screening for hemochromatosis: High prevalence and low morbidity in an unselected population of 65,238 persons. Scand J Gastroenterol 2001;36:1108-15.
10. McLaren CE, McLachlan S, Garner CP, et al. Associations between single nucleotide polymorphisms in iron-related genes and iron status in multiethnic populations. PLoS One 2012;7:e38339.
11. Adams PC. H63D genotying for hemochromatosis: Helper or hindrance? Can J Gastroenterol Hepatol 2014;28:179-80.
12. Waalen J, Felitti VJ, Gelbart T, Beutler E. Screening for hemochromatosis by measuring ferritin levels: A more effective approach. Blood 2008;111:3373-6.
13. Schranz M, Talasz H, Graziadei I, et al. Diagnosis of hepatic iron overload: A family study illustrating pitfalls in diagnosing hemochromatosis. Diagn Mol Pathol 2009;18:53-60.
14. Aguilar-Martinez P, Grandchamp B, Cunat S, et al. Iron overload in HFE C282Y heterozygotes at first genetic testing: A strategy for identifying rare *HFE* variants. Haematologica 2011;96:507-14.
15. Chandrasekharan S, Pitlick E, Heaney C, Cook-Deegan R. Impact of gene patents and licensing practices on access to genetic testing for hereditary hemochromatosis. Genet Med 2010;12(4 Suppl):S155-70.
16. VanWagner LB, Green RM. Elevated serum ferritin. JAMA 2014;312:743-4.
17. Barton JC, Lafreniere SA, Leiendecker-Foster C, et al. *HFE*, *SLC40A1*, *HAMP*, *HJV*, *TFR2*, and *FTL* mutations detected by denaturing high-performance liquid chromatography after iron phenotyping and *HFE* C282Y and H63D genotyping in 785 HEIRS Study participants. Am J Hematol 2009;84:710-4.
18. Del-Castillo-Rueda A, Moreno-Carralero MI, Cuadrado-Grande N, et al. Mutations in the *HFE*, *TFR2*, and *SLC40A1* genes in patients with hemochromatosis. Gene 2012;508:15-20.
19. Yin D, Kulhalli V, Walker AP. Raised serum ferritin concentration in hereditary hyperferritinemia cataract syndrome is not a marker for iron overload. Hepatology 2014;59:1204-6.

20. Pelucchi S, Mariani R, Calza S, et al. *CYBRD1* as a modifier gene that modulates iron phenotype in *HFE* p.C282Y homozygous patients. Haematologica 2012;97:1818-25.
21. van der Harst P, Zhang W, Mateo Leach I, et al. Seventy-five genetic loci influencing the human red blood cell. Nature 2012;492:369-75.
22. McLaren CE, Garner CP, Constantine CC, et al. Genome-wide association study identifies genetic loci associated with iron deficiency. PLoS One 2011;6:e17390.
23. Johansen CT, Dubé JB, Loyzer MN, et al. LipidSeq: A next-generation clinical resequencing panel for monogenic dyslipidemias. J Lipid Res 2014;55:765-72.
24. Millat G, Chanavat V, Rousson R. Evaluation of a new NGS method based on a custom AmpliSeq library and Ion Torrent PGM sequencing for the fast detection of genetic variations in cardiomyopathies. Clin Chim Acta 2014;433:266-71.

Rate, delay and predictors of hepatitis C treatment in British Columbia

Alan Hoi Lun Yau MD, Terry Lee PhD, Alnoor Ramji MD FRCPC, Hin Hin Ko MD FRCPC

AHL Yau, T Lee, A Ramji, HH Ko. Rate, delay and predictors of hepatitis C treatment in British Columbia. Can J Gastroenterol Hepatol 2015;29(6):315-320.

BACKGROUND: The current treatment rate for chronic hepatitis C virus (HCV) infection is suboptimal despite the availability of efficacious antiviral therapy.

OBJECTIVE: To determine the rate, delay and predictors of treatment in patients with chronic HCV infection.

METHODS: A retrospective chart review of chronic HCV patients who were being evaluated at a tertiary hepatology centre in Vancouver, British Columbia, was performed.

RESULTS: One hundred sixty-four patients with chronic HCV infection who were assessed for treatment between February 2008 and January 2013 were reviewed. Treatment was initiated in 25.6% (42 of 164). In multivariate analyses, male sex (OR 7.90 [95% CI 1.35 to 46.15]) and elevated alanine aminotransferase (ALT) level (>1.5 times the upper limit of normal) (OR 3.10 [95% CI 1.32 to 7.27]) were positive predictors of treatment, whereas active smoking (OR 0.09 [95% CI 0.02 to 0.53]) and Charlson comorbidity index (per point increase) (OR 0.47 [95% CI 0.27 to 0.83]) were negative predictors of treatment. The most common reasons for treatment deferral were no or minimal liver fibrosis in 57.7% (n=30), persistently normal ALT levels in 57.7% (n=30) and patient unreadiness in 28.8% (n=15). The most common reasons for treatment noninitiation were patient refusal in 59.1% (n=26), medical comorbidities in 36.4% (n=16), psychiatric comorbidities in 9.1% (n=4) and decompensated cirrhosis in 9.1% (n=4). There was a statistically significant difference in the median time delay from HCV diagnosis to general practitioner referral between the treated and untreated patients (66.3 versus 119.5 months, respectively [P=0.033]). The median wait time from general practitioner referral to hepatologist consult was similar between the treated and untreated patients (1.7 months versus 1.5 months, respectively [P=0.768]). Among the treated patients, the median time delay was 6.8 months from hepatologist consult to treatment initiation.

CONCLUSIONS: The current treatment rate for chronic HCV infection remains suboptimal. Medical and psychiatric comorbidities represent a major obstacle to HCV treatment. Minimal hepatic fibrosis may no longer be a major reason for treatment deferral as more efficacious and tolerable antiviral therapies become available in the future. Greater educational initiatives for primary care physicians would promote early referral of patients. More nursing support would alleviate the backlog of patients awaiting treatment.

Key Words: *Boceprevir; Direct-acting antiviral agent; Hepatitis C; Telaprevir; Treatment*

Les taux, les retards et les prédicteurs de traitement de l'hépatite C en Colombie-Britannique

HISTORIQUE : Le taux de traitement actuel de l'infection par le virus de l'hépatite C (VHC) est sous-optimal, malgré l'existence d'une antivirothérapie efficace.

OBJECTIF : Déterminer le taux, les retards et les prédicteurs de traitement chez les patients atteints d'une infection chronique par le VHC.

MÉTHODOLOGIE : Les chercheurs ont effectué une analyse rétrospective des dossiers des patients atteints d'un VHC chronique évalués dans un centre tertiaire d'hépatologie de Vancouver, en Colombie-Britannique.

RÉSULTATS : Les chercheurs ont analysé 164 patients atteints d'une infection chronique par le VHC qui avaient été évalués en vue d'un traitement entre février 2008 et janvier 2013. Le traitement a été amorcé chez 25,6 % d'entre eux (42 sur 164). Aux analyses multivariées, le sexe masculin (RC 7,90 [95 % IC 1,35 à 46,15]) et un taux d'alanine aminotransférase (ALT) élevé (plus de 1,5 fois la limite supérieure de la normale (RC 3,10 [95 % IC 1,32 à 7,27]) étaient des prédicteurs positifs du traitement, tandis qu'un tabagisme actif (RC 0,09 [95 % IC 0,02 à 0,53]) et un index de comorbidité de Charlson (par point d'augmentation; RC 0,47 [95 % IC 0,27 à 0,83]) étaient des prédicteurs négatifs du traitement. Les principales raisons de reporter le traitement étaient l'absence ou le peu de fibrose hépatique chez 57,7 % des patients (n=30), des taux d'ALT toujours normaux chez 57,7 % des patients (n=30) et la réticence chez 28,8 % des patients (n=15). Les principales raisons de ne pas amorcer le traitement étaient le refus du patient dans 59,1 % des cas (n=26), des comorbidités médicales dans 36,4 % des cas (n=16), des comorbidités psychiatriques chez 9,1 % des patients (n=4) et une cirrhose décompensée chez 9,1 % d'entre eux (n=4). Il y avait une différence statistiquement significative entre les patients traités et non traités pour ce qui est de l'attente médiane entre le diagnostic de VHC et l'aiguillage par un omnipraticien (66,3 mois au lieu de 119,5 mois, [P=0,033]). Les patients traités et non traités présentaient un temps d'attente médian similaire entre l'aiguillage de l'omnipraticien et la consultation chez l'hépatologue (1,7 mois au lieu de 1,5 mois, [P=0,768]). Chez les patients traités, l'attente médiane était de 6,8 mois entre la consultation chez l'hépatologue et le début du traitement.

CONCLUSIONS : Le taux de traitement actuel de l'infection chronique par le VHC demeure sous-optimal. Les comorbidités médicales et psychiatriques constituent un obstacle majeur au traitement du VHC. Une fibrose hépatique minimale ne sera peut-être plus une raison importante pour reporter le traitement lorsque des antiviraux plus efficaces et plus faciles à tolérer seront mis en marché. Plus de projets de formation des médecins de première ligne favoriseraient l'aiguillage précoce du patient. Un plus grand soutien infirmier réduirait le nombre de patients en attente d'un traitement.

Chronic hepatitis C virus (HCV) infection has been estimated to affect 2% to 3% of the population worldwide (170 million individuals) (1) and 0.8% of Canadians (275,000 individuals) (2). In Canada, HCV-related morbidity and mortality increased 15% to 18% annually between 1994 and 2004 (3). The increasing medical and economic burden of HCV on the health care system has prompted the development of more efficacious treatment regimens. Two direct-acting antiviral agents – telaprevir and boceprevir – were approved in Canada in May 2011, in combination with pegylated interferon and ribavirin for the treatment of HCV genotype 1, with a sustained virological response (SVR) rate of 60% to 75% in treatment-naive patients (4).

Before the advent of protease inhibitors, when pegylated interferon and ribavirin were standard of care for chronic HCV, low rates of treatment uptake had been reported, ranging from 8.2% to 31.8% among United Startes (US) Veterans (5-11), 9.1% to 50.4% among US non-veterans (12-16), 16.0% to 37.8% in Canada (17-19), 21.3%

Department of Medicine, Division of Gastroenterology, University of British Columbia, Vancouver, British Columbia
Correspondence: Dr Hin Hin Ko, St Paul's Hospital, University of British Columbia, Pacific Gastroenterology Associates, 770-1190 Hornby Street, Vancouver, British Columbia V6Z 2K5.e-mail hinnih@gmail.com

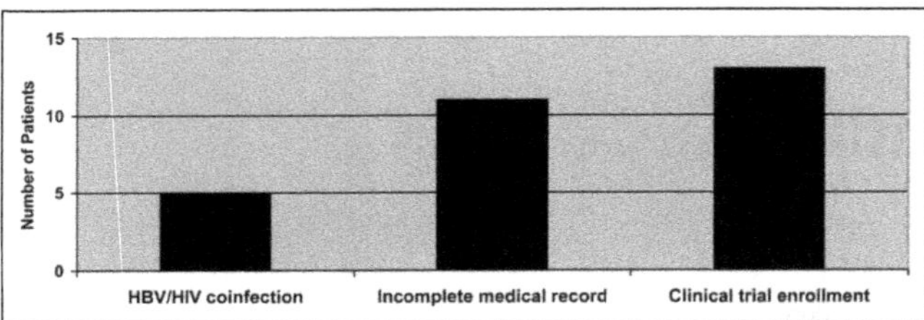

Figure 1) *Patients excluded from study participation. HBV Hepatitis B virus*

in Denmark (20), 16% in France (21), and <1% in Greece, Poland, Romania and Russia (21).

Despite the recent advances in chronic HCV treatment and the availability of efficacious antiviral agents, the current treatment rate is suboptimal. It is, therefore, important to better understand the barriers that result in low treatment rates. We performed a retrospective analysis of chronic HCV patients being referred to a tertiary hepatology centre affiliated with the University of British Columbia (Vancouver, British Columbia). Our aims were to determine the treatment rate; identify the predictors of treatment and the reasons for nontreatment; and determine the mean wait time from diagnosis to treatment initiation.

METHODS

Study population

A total of 164 patients with chronic HCV infection, who were evaluated for antiviral therapy between February 2008 and January 2013 by one of two hepatologists (HHK, AR) at a referral-based tertiary hepatology centre in downtown Vancouver, British Columbia, were reviewed. Most patients were from the greater Vancouver metropolitan area and were referred by their primary care providers. Patients who were eligible for treatment were offered pegylated interferon and ribavirin with or without telaprevir or boceprevir for 24 or 48 weeks, depending on the HCV genotype. All patients received comprehensive education and counselling about HCV by a specially trained nurse during their first office visit, and had regular follow-up appointments with both the nurses and physicians throughout treatment. Patients were excluded from the study if they were coinfected with hepatitis B virus (HBV) or HIV, enrolled in a clinical trial or if there was incomplete documentation in their medical record. The present study was approved by the University of British Columbia Institutional Review Board.

Data collection

A retrospective chart review was performed to collect information regarding patient demographics, medical history, details of HCV infection (route of transmission, genotype and previous treatment), psychiatric history (as reported by patients or referring physicians), social history and laboratory results (blood work, abdominal ultrasound, transient hepatic elastography [Fibroscan, Echosens, France] and liver biopsy). Patients with chronic HCV infection were identified from the electronic medical records using the *International Classification of Diseases, Ninth Revision* (ICD-9) ICD-9 diagnostic code 5731. Specific demographic information collected include age, sex, ethnicity and distance from home to hepatology centre. The distance from home to hepatology centre was defined as 'near' or 'far' based on a cut-off distance of 30 km. Medical comorbidity was quantified using the Charlson comorbidity index, in which 1 to 6 points were assigned for each of the 17 major medical conditions (22). Age was also taken into account in calculating the Charlson comorbidity index. Whether diagnostic tests (liver enzymes, HCV antibody, HCV RNA, HCV genotype and abdominal ultrasound) were ordered by the referring physicians within six to 12 months of referral were documented. Treatment initiation was the primary outcome of the study, and was defined as having received pegylated interferon and ribavirin with or without telaprevir or boceprevir for any duration of time. For patients in whom treatment was not started, the charts were reviewed to determine the reasons for treatment deferral or noninitiation. Finally, the time delay from HCV diagnosis to referral, from referral to hepatologist consult and from initial assessment by hepatologist to treatment initiation was determined.

Statistical analysis

Baseline characteristics of all patients were described using means and SDs for continuous data, and counts and percentages for categorical data. Demographic and clinical variables of treated and untreated patients were compared using t tests or Wilcoxon rank-sum tests for continuous data, and χ^2 or Fisher's exact tests for categorical data as appropriate. Factors that were statistically significant ($P<0.20$) from the univariate analyses and those found to be clinically important based on a literature review were selected as candidates to enter multivariate logistic regression models. Backward stepwise selection was performed such that variables with the largest P value were sequentially removed at each step from the model ending when all remaining variables had a two-sided $P<0.05$. Independent predictors of antiviral therapy initiation were identified with calculation of their respective OR and 95% CI. Data were analyzed using SAS version 9.3 (SAS Institute, USA).

RESULTS

Patient characteristics (Table 1)

A total of 164 patients with chronic HCV infection were assessed for treatment between February 2008 and January 2013. The mean age was 54.7 years, 69.5% (n=114) were men, 82.4% (n=131) were Caucasian and 66.5% (n=107) resided within 30 km of the tertiary hepatology centre. Twenty-nine patients were excluded due to HBV or HIV coinfection (n=5), incomplete medical record (n=11) or clinical trial enrollment (n=13) (Figure 1).

The most common risk factor for HCV infection was illicit drug use (43.9% [n=72]), followed by blood transfusion (21.3% [n=35]), tattooing (20.1% [n=33]) and sexual contact (4.9% [n=8]). Other risk factors were acupuncture, dental work, ear piercing, hemodialysis and needle stick injury (11.0% [n=18]). HBV and HIV coinfections were found in 0.6% (n=1) and 2.4% (n=4), respectively. Liver cirrhosis was present in 26.5% (n=40). The mean Charlson comorbidity index was 2.3.

Psychiatric conditions were present in 26.2% (n=43), with mood disorders 22.6% (n=37) being the most common, followed by anxiety disorders 7.9% (n=13) and psychiatric disorders 1.2% (n=2). Active use of tobacco, alcohol and illicit drugs were found in 40.3% (n=56), 35.7% (n=56) and 19.6% (n=19), respectively.

Liver enzyme levels were analyzed before treatment in 98.8% (n=162), with a mean alanine aminotransferase (ALT) level of 109.2 IU/L. HCV RNA was quantified before treatment in 84.1% (n=138), with a mean viral load of 5.7 $\log_{10}$ IU/mL. HCV genotype 1 was most commonly found in 70.6% (n=108), followed by genotype 3 in 15.7% (n=24) and genotype 2 in 10.5% (n=16).

Fibroscan was performed before treatment in 84.1% (n=138), whereas liver biopsy was performed before treatment in 27.4% (n=45). In patients with liver biopsy showing F0 (n=5), F1 (n=8), F2 (n=10), F3 (n=9) and F4 (n=3), the mean Fibroscan score was 6.1 kPA (range 3.6 to 9.4 kPA), 5.3 kPA (range 3.3 to 11.6 kPA), 9.5 kPA (range 5.3 to 13.4 kPA), 15.1 kPA (range 10.0 to 26.6 kPA [excluded an outlier of 75.0 kPA that was falsely elevated due to transaminitis], and 52.6 kPA (range 48.0 to 57.2 kPA), respectively. Higher Fibroscan scores appeared to predict the presence of portal hypertension.

Previous HCV treatment was documented in 27.4% (n=45), in which 35.6% (n=16) were relapsers, 11.1% (n=5) partial responders, 27.9% (n=12) were null responders and 22.2% (n=10) did not complete or tolerate the treatment.

Antiviral therapies initiated by hepatologists (Figure 2)

Treatment was initiated in 25.6% (42 of 164) of patients. Antiviral regimens consisted of pegylated interferon (n=42), ribavirin (n=42), boceprevir (n=10) and telaprevir (n=11). A total of 27 patients with

TABLE 1
Univariate logistic regression for treatment initiation

Characteristics	All (n=164)	Treated (n=42)	Deferred (n=52)	Not recommended (n=18)	P*	P†
Demographic						
Age‡, years, mean ± SD	54.7±9.6	54.6±8.6	52.4±9.7	57.6±7.5	0.420	0.125
Male sex‡	114 (69.5)	33 (78.6)	36 (69.2)	10 (55.6)	0.308	0.070
Ethnicity‡						
Caucasian	131 (82.4)	32 (80.0)	41 (82.0)	15 (88.2)	0.629	0.794
Asian	23 (14.5)	7 (17.5)	6 (12.0)	2 (11.8)		
Other§	5 (3.1)	1 (2.5)	3 (6.0)	0 (0.0)		
Distance to specialist						
Near (<30 km)	107 (66.5)	31 (75.6)	32 (61.5)	11 (64.7)	0.150	0.398
Far (>30 km)	54 (33.5)	10 (24.4)	20 (38.5)	6 (35.3)		
Medical						
Risk factors for HCV infection						
Illicit drug use	72 (43.9)	17 (40.5)	24 (46.2)	7 (38.9)	0.581	0.908
Blood transfusion (before 1992)	35 (21.3)	10 (23.8)	9 (17.3)	5 (27.8)	0.435	0.745
Sexual contact	8 (4.9)	1 (2.4)	3 (5.8)	1 (5.6)	0.626	0.514
Tattooing	33 (20.1)	14 (33.3)	9 (17.3)	3 (16.7)	0.072	0.189
Other¶	18 (11.0)	4 (9.5)	7 (13.5)	3 (16.7)	0.555	0.419
Cirrhosis‡	40 (26.5)	13 (33.3)	6 (11.8)	7 (43.8)	0.013	0.466
Charlson comorbidity index‡, mean ± SD	2.3±1.6	2.4±1.4	1.9±1.4	3.7±2.2	0.097	0.031
Psychiatric						
Any psychiatric disorders‡	43 (26.2)	9 (21.4)	12 (23.1)	7 (38.9)	0.849	0.161
Anxiety disorders	13 (7.9)	4 (9.5)	5 (9.6)	3 (16.7)	1.000	0.419
Mood disorders	37 (22.6)	8 (19.0)	8 (15.4)	6 (33.3)	0.638	0.231
Psychotic disorders	2 (1.2)	1 (2.4)	0 (0.0)	1 (5.6)	0.447	0.514
Social						
Smoking	56 (40.3)	11 (30.6)	17 (37.8)	11 (64.7)	0.497	0.019
Alcohol‡	56 (35.7)	14 (35.0)	16 (31.4)	6 (33.3)	0.715	0.902
Drugs‡	19 (19.6)	5 (20.8)	6 (20.0)	3 (27.3)	0.940	0.674
Disability	35 (21.3)	12 (28.6)	11 (21.2)	4 (22.2)	0.406	0.610
Diagnostic						
Liver enzyme measurement (before treatment)‡	162 (98.8)	42 (100.0)	51 (98.1)	17 (94.4)	–	–
Alanine aminotransferase, IU/L, mean ± SD	109.2±92.7	133.8±12.7	78.5±61.8	117.2±118.9	0.001	0.269
HCV RNA (before treatment), n (%)	138 (84.1)	42 (100.0)	42 (80.8)	13 (72.2)	–	–
Viral load, $\log_{10}$ IU/mL, mean ± SD	5.7±1.0	5.7±0.8	5.7±1.2	5.6±0.9	0.540	0.704
HCV genotype‡						
1	108 (70.6)	27 (65.9)	36 (72.0)	10 (76.9)	0.957	0.722
2	16 (10.5)	4 (9.8)	4 (8.0)	1 (7.7)		
3	24 (15.7)	8 (19.5)	8 (16.0)	1 (7.7)		
4	3 (2.0)	1 (2.4)	1 (2.0)	1 (7.7)		
5	0 (0.0)	0 (0.0)	0 (0.0)	0 (0.0)		
6	2 (1.3)	1 (2.4)	1 (2.0)	0 (0.0)		
Fibroscan** (before treatment)	138 (84.1)	31 (73.8)	49 (94.2)	15 (83.3)	–	–
Hepatic stiffness, kPA, mean ± SD	13.3±13.8	14.3±11.7	10.0±12.9	16.0±12.0	0.001	0.527
Liver biopsy stage (before treatment)	45 (27.4)	14 (33.3)	14 (26.9)	7 (38.9)	–	–
0	5 (12.5)	1 (8.3)	3 (23.1)	0 (0.0)	0.265	0.713
1	9 (22.5)	2 (16.7)	6 (46.2)	0 (0.0)		
2	10 (25.0)	4 (33.3)	2 (15.4)	2 (33.3)		
3	9 (22.5)	2 (16.7)	0 (0.0)	3 (50.0)		
4	7 (17.5)	3 (25.0)	2 (15.4)	1 (16.7)		
Therapeutic						
Previous treatment	45 (27.4)	15 (35.7)	9 (17.3)	4 (22.2)	–	–
Relapse	16 (35.6)	7 (46.7)	4 (44.4)	0 (0.0)	0.624	0.382
Partial response	5 (11.1)	2 (13.3)	0 (0.0)	0 (0.0)		
Null response	12 (27.9)	4 (26.7)	2 (22.2)	2 (66.7)		
Incomplete/intolerable	10 (22.2)	2 (13.3)	3 (33.3)	1 (33.3)		

*Data presented as n (%) unless otherwise indicated. Data not shown for untreated patients who refused treatment (n=26) or who were lost to follow up (n=26). *Comparison between treated and deferred; †Comparison between treated and not recommended; ‡Clinically important variable based on literature review; §Other ethnicity (Chinese, Indian, Native, Pakistanian, Vietnamese); ¶Other risk factors (acupuncture, dental work, ear piercing, hemodialysis, needle stick injury); **Echosens, France*

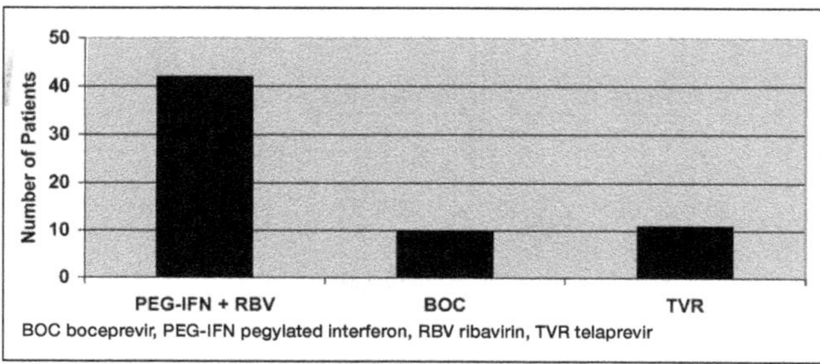

Figure 2) *Antiviral therapies initiated by hepatologists*

TABLE 2
Multivariate logistic regression for treatment initiation

Subgroup and predictor	OR (95% CI)*	P
Treated versus deferred		
ALT >1.5 times the upper limit of normal	3.10 (1.32–7.27)	0.009
Treated versus not recommended		
Male sex	7.90 (1.35–46.15)	0.022
Charlson comorbidity index (per point increase)	0.47 (0.27–0.83)	0.009
Smoking	0.09 (0.02–0.53)	0.008

**OR >1 indicates a higher probability of receiving treatment. ALT Alanine aminotransferase*

HCV genotype 1 were treated, six of whom did not receive triple therapy with a first-generation protease inhibitor. Among these six patients, five were assessed for treatment before the approval of boceprevir and telaprevir in May 2011, and one could not tolerate boceprevir or telaprevir due to anemia. Among untreated patients (122 of 164), 11.0% (18 of 164) were not suitable treatment candidates. The remaining 31.7% (52 of 164) patients had their treatment deferred, 15.9% (26 of 164) refused treatment and 15.9% (26 of 164) were lost to follow-up.

Univariate analyses for treatment initiation (Table 1)

In univariate analyses, patients who had their treatment deferred were more likely to have no liver cirrhosis (P=0.013), normal ALT levels (P=0.001), and a minimal or mild degree of liver fibrosis assessed by Fibroscan (P=0.001). In contrast, patients in whom treatment was not recommended were more likely to be active smokers (P=0.019) and had a higher Charlson comorbidity index (P=0.031).

Multivariate analyses for treatment initiation (Table 2)

In multivariate analyses, male sex (OR 7.90 [95% CI 1.35 to 46.15]) and elevated ALT level (>1.5 times the upper limit of normal) (OR 3.10 [95% CI 1.32 to 7.27]) were positive predictors of treatment initiation, whereas active smoking (OR 0.09 [95% CI 0.02 to 0.53]) and Charlson comorbidity index (per point increase) (OR 0.47 [95% CI 0.27 to 0.83]) were negative predictors of treatment initiation. An ALT threshold of 1.5 times the upper limit of normal was chosen because it was the minimum level required to obtain provincial coverage for HCV treatment.

Reasons for treatment deferral (Table 3)

Reasons for treatment deferral included no or minimal liver fibrosis in 57.7% (n=30), persistently normal ALT level in 57.7% (n=30), patient unreadiness in 28.8% (n=15), pending HCV investigations in 7.7% (n=4), breastfeeding in 3.8% (n=2), previous null responder in 1.9% (n=1) and hepatocellular carcinoma in 1.9% (n=2).

Reasons for treatment noninitiation (Table 4)

Reasons for treatment noninitiation included patient refusal in 59.1% (n=26), medical comorbidities in 36.4% (n=16), psychiatric comorbidities in 9.1% (n=4), decompensated cirrhosis in 9.1% (n=4), active substance use in 4.5% (n=2), lack of social supports in 4.5% (n=2), excessive active alcohol use in 4.5% (n=2) and previous nonadherence or noncompliance in 2.3% (n=1).

TABLE 3
Reasons for treatment deferral

	n (%)
No or minimal liver fibrosis	30 (57.7)
Persistently normal alanine aminotransferase level	30 (57.7)
Patient unreadiness	15 (28.8)
Pending hepatitis C investigations	4 (7.7)
Breastfeeding	2 (3.8)
Previous null responder	1 (1.9)
Hepatocellular carcinoma	1 (1.9)

TABLE 4
Reasons for treatment noninitiation*

	n (%)
Patient refusal	26 (59.1)
Medical comorbidities	16 (36.4)
Psychiatric comorbidities	4 (9.1)
Decompensated cirrhosis	4 (9.1)
Active substance use	2 (4.5)
Lack of social supports	2 (4.5)
Excessive active alcohol use	2 (4.5)
Previous nonadherence/noncompliance	1 (2.3)

**No reasons provided (n=26)*

Time delay from diagnosis to treatment initiation (Table 5)

There was a statistically significant difference in the median time lag from HCV diagnosis to general practitioner (GP) referral between the treated and untreated patients (66.3 versus 119.5 months, respectively [P=0.033]). The median wait time from GP referral to hepatologist consult was similar between the treated and untreated patients (1.7 months versus 1.5 months, respectively [P=0.768]). Among the treated patients, the median time delay was 66.1 months from diagnosis to treatment initiation, 7.9 months from GP referral to treatment initiation, and 6.8 months from hepatologist consult to treatment initiation.

Among the treated patients, the time delay from HCV diagnosis to GP referral and the wait time from hepatologist consult to HCV treatment both appear to shorten progressively over the years from February 2008 to January 2013. Unfortunately, when patients were stratified according to year of assessment, the number in each group was too small for statistical comparison (data not shown). Among the untreated patients, no meaningful trend in wait times was observed over the years.

Diagnostic tests ordered by primary care physicians (Figure 3)

Primary care physicians had ordered ALT tests within six months of referral in 75.6% (n=124), HCV antibody in 39.0% (n=64), HCV RNA in 51.8% (n=85), HCV genotype in 57.9% (n=95) and abdominal ultrasound within one year in 52.4% (n=86).

DISCUSSION

In the present study, the initiation rate of HCV treatment was 25.6%, which is comparable with previously reported rates of 16.0% to 37.8% in Canada (17-19). However, several potential confounders need to be considered. It should be noted that the actual treatment rate at our centre is higher than that reported in our study that excluded patients who enrolled in clinical trials. The treatment rate observed in our study could be higher than that of smaller centres due to the focus on viral hepatitis at our tertiary referral centre. Furthermore, our centre is staffed with two full-time nurses who are indispensable in facilitating treatment uptake with the currently approved triple therapy. In contrast, the treatment rate could be lower at our tertiary referral centre, which encounters a high proportion of difficult-to-treat patients with advanced age, multiple medical and psychiatric comorbidities, decompensated cirrhosis and hepatocellular carcinoma.

TABLE 5
Time interval from hepatitis C virus (HCV) diagnosis to treatment initiation

Interval, months	n	Treated, median (IQR)	n	Untreated, median (IQR)	P
HCV diagnosis to general practitioner referral	36	66.3 (2.8–123.1)	79	119.5 (20.5–203.5)	0.033
HCV diagnosis to HCV treatment	37	66.1 (17.1–123.5)		–	–
General practitioner referral to hepatologist consult	38	1.7 (0.6–2.7)	117	1.5 (0.8–2.5)	0.768
General practitioner referral to HCV treatment	38	7.9 (5.1–15.4)		–	–
Hepatologist consult to HCV treatment	42	6.8 (3.5–14.2)		–	–

IQR Interquartile range

Following the approval of boceprevir and telaprevir in May 2011, the treatment rate remained suboptimal (31.8% [21 of 66]) among HCV genotype 1 patients in our study. Similarly, Chen EY et al (23) reported a disappointingly low rate of treatment initiation of 18.7% in chronic HCV patients who were offered triple therapy. Although boceprevir and telaprevir have significantly improved SVR, they must be used in combination with pegylated interferon and ribavirin, and are associated with additional side effects (24-26). Furthermore, there is increasing recognition and anticipation among patients and physicians of the improved efficacy and tolerability of the newer antiviral regimens that will become available in the near future (23). These newer antiviral agents are desperately needed because HCV treatment will have minimal effects in reducing liver-related morbidity and mortality if the treatment rate remains low.

Elevated ALT level (>1.5 times the upper limit of normal) was found to be a positive predictor of treatment initiation in our study. This finding is expected because an ALT level >1.5 times the upper limit of normal is required for obtaining provincial drug coverage in British Columbia, in accordance with consensus guidelines that recommend antiviral therapy in chronic HCV patients with persistently high levels of ALT (18,20). In contrast with previous studies, age (18,20), ethnicity (11), presence of liver cirrhosis (6,18), viral load (11), genotype (20) and patient motivation (18) did not have a positive impact on treatment decision.

Active smoking was identified as a negative predictor of treatment initiation in the present study. It may represent a surrogate marker for other potential barriers to treatment such as lack of social support or low socioeconomic status. Charlson comorbidity index (per point increase) was also found to be a negative predictor of treatment initiation. This finding is not surprising because patients with advanced age and comorbid illnesses are less likely to tolerate the side effects of antiviral therapy. In contrast with previous studies, age (6,8,17,27), sex (12), ethnicity (6,8,13), advanced liver disease (8,27), psychiatric comorbidities (6,11,17,27), active alcohol use (5,6,12,17) and active substance use (6,11,17,19,20,27) did not have a negative impact on treatment decision.

The most common reasons for treatment deferral in our study were no or minimal liver fibrosis, persistently normal ALT level and patient unreadiness. Patients with normal ALT level and minimal liver fibrosis do not meet public reimbursement criteria and, thus, would have no choice but to defer therapy. The most common reasons for treatment noninitiation were patient refusal, medical comorbidities, psychiatric comorbidities and decompensated cirrhosis. As interferon-free antiviral regimens with improved safety and tolerability become available, the concern for side effects in patients with medical and psychiatric comorbidities will diminish and the treatment rate will increase accordingly. Our findings were in accordance with previous studies, which also reported other reasons for nontreatment including patient nonadherence to follow-up (5,10,12,14,17), patient concern about efficacy and side effects of medications (18,19,28), active alcohol and substance use (5,9,10,12,14,15,17-19,29), and lack of social supports (10,12,28).

The long delay from HCV diagnosis to GP referral may be due to several reasons. First, there may be gaps in primary care physicians' knowledge of HCV management (30). Second, significant barriers to specialist referral were reported by 50% of primary care physicians, including long wait times, long travel distances and substance dependence (30).

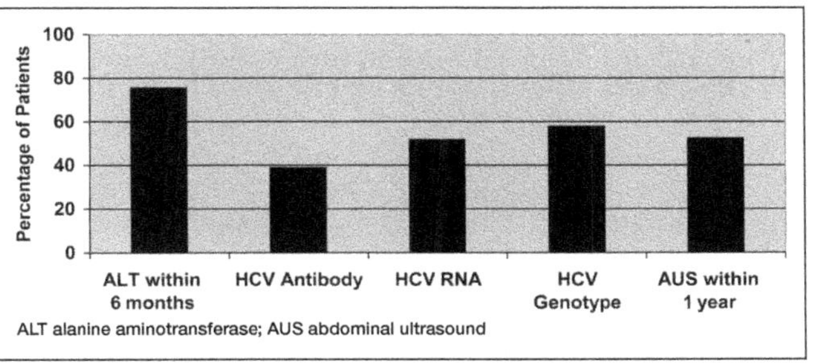

Figure 3) *Diagnostic tests ordered by primary care physicians. HCV Hepatitis C virus*

However, our study found that the wait time from GP referral to hepatologist consult was <2 months and that 66.5% of patients resided within 30 km of the tertiary hepatology centre. The short wait time from GP referral to hepatologist consult is perhaps a unique feature of our centre because both hepatologists make efforts to prioritize and assess patients with chronic HCV as soon as possible.

Furthermore, HCV treatment in patients with substance abuse has been shown to be feasible with ongoing support and monitoring through rehabilitation programs (10,17). Finally, some patients with HCV may not be interested in evaluation for treatment for some time after diagnosis. This would also explain the longer time delay from diagnosis to referral in untreated patients.

The prolonged wait time from hepatologist consult to HCV treatment may be due to several reasons. First, a significant proportion of our patients did not have all of the relevant investigations (eg, HCV viral load and genotype) ordered by the referring physicians before seeing the specialist. Second, many patients have medical and psychiatric comorbidities that require assessment and clearance by other consultants before initiation of antiviral therapy. Third, inadequate nursing support may be a contributing factor to the delay in treatment initiation following assessment by the hepatologist; however, the data points in our study were insufficient to detect any statistical significance. Our centre is staffed with only two full-time and nongovernment-funded nurses, each of whom is able to follow a limited number of patients at a certain time. This undoubtedly limits treatment uptake with pegylated interferon, ribavirin and the first-generation protease inhibitors, which are associated with complex dosing schedules and frequent adverse events.

Limitations of our study include its single-centre retrospective design and referral bias to a tertiary centre with a special interest in viral hepatitis. It is likely that the burden of mental health and socioeconomic obstacles varies among different centres. In addition, it should be noted that a significant number of patients were excluded due to incomplete data in the chart and enrollment in clinical trials. In fact, we estimated that only 11% of all patients being evaluated at our centre had complete retrospective data for study inclusion. Furthermore, case identification from our electronic medical records using the ICD-9 diagnostic code 5731 for viral hepatitis may have inadvertently missed some chronic HCV patients who were being evaluated for gastrointestinal concerns other than HCV. Nonetheless, we believe that the patient sample included in our study remains an accurate representation of the patient population seen at our tertiary referral centre. Finally, it is worth mentioning that the mean hepatic stiffness (based on Fibroscan) did not fully correlate with the fibrotic stage (based on liver biopsy), presumably due to such factors as the small number of patients with available test results, falsely elevated hepatic stiffness in the presence of transaminitis and sampling error of liver biopsy.

The landscape of HCV treatment is rapidly changing. As newer direct-acting antiviral agents with improved efficacy and tolerability are being approved, additional studies will be needed to re-examine the rate and predictors of antiviral therapy initiation. Finally, with the advent of all-oral interferon-free antiviral regimens with potentially fewer side effects, it is anticipated that more physicians will treat HCV and refer only the more complicated cases to hepatologists. Meanwhile, the role of nurses will gradually evolve into one of health promotion and disease prevention.

CONCLUSION

The current treatment rate for chronic HCV infection remains sub-optimal. Medical and psychiatric comorbidities represent a major obstacle to HCV treatment. The long delay from HCV diagnosis to GP referral should prompt greater educational efforts to enhance the knowledge of primary care physicians. More nursing support from the provincial health authorities would help alleviate the backlog of patients awaiting treatment. With the availability of all-oral, interferon-free antiviral regimens in the near future, it is hoped that patients who do not tolerate interferon-based therapy would benefit from these future treatment options and that primary care physicians would take on a greater role in providing HCV treatment to patients.

DISCLOSURES: The authors have no financial disclosures or conflicts of interest to declare.

REFERENCES

1. World Health Organization, Hepatitis C. World Health Organization Fact Sheet No. 164. <www.who.int/mediacentre/factsheets/fs164/en/> (Accessed July 26, 2013).
2. Remis R. Modeling the incidence and prevalence of hepatitis C infection and its sequelae in Canada, 2007 final report. Public Health Agency of Canada. 2007. <www.phac-aspc.gc.ca/sti-its-surv-epi/model/pdf/model07-eng.pdf> (Accessed July 26, 2013).
3. Myers RP, Liu MF, Shaheen AAM. The burden of hepatitis C virus infection is growing: A Canadian population-based study of hospitalizations from 1994 to 2004. Can J Gastroenterol 2008;22:381-7.
4. Ghany MG, Nelson DR, Strader DB, et al. An update on treatment of genotype 1 chronic hepatitis C virus infection: 2011 practice guideline by the American Association for the Study of Liver Diseases. Hepatology 2011;54:1433-44.
5. Butt AA, Wagener M, Shakil AO, et al. Reasons for non-treatment of hepatitis C in veterans in care. J Viral Hepat 2005;12:81-5.
6. Butt AA, Justice AC, Skanderson M, et al. Rate and predictors of treatment prescription for hepatitis C. Gut 2007;56:385-9.
7. Butt AA, McGinnis KA, Skanderson M, et al. Hepatitis C treatment completion rates in routine clinical care. Liver Int 2010;30:240-50.
8. Kanwal F, Hoang T, Spiegel BMR, et al. Predictors of treatment in patients with chronic hepatitis c infection – role of patient versus nonpatient factors. Hepatology 2007;46:1741-9.
9. Cawthorne CH, Rudat KR, Burton MS, et al. Limited success of HCV antiviral therapy in United States veterans. Am J Gastroenterol 2002;97:149-55.
10. Rowan PJ, Tabasi S, Abdul-latif M, et al. Psychosocial factors are the most common contraindications for antiviral therapy at initial evaluation in veterans with chronic hepatitis C. J Clin Gastroenterol 2004;38:530-4.
11. Backus LI, Boothroyd DB, Phillips BR, et al. Pretreatment assessment and predictors of hepatitis C virus treatment in US veterans coinfected with HIV and hepatitis C virus. J Viral Hepat 2006;13:799-810.
12. Morrill JA, Shrestha M, Grant RW. Barriers to the treatment of hepatitis C. Patient, provider, and system factors. J Gen Intern Med 2005;20:754-8.
13. Butt AA, Tsevat J, Leonard AC, et al. Effect of race and HIV co-infection upon treatment prescription for hepatitis C virus. Int J Infect Dis 2009;13:449-55.
14. Falck-Ytter Y, Kale H, Mullen KD, et al. Surprisingly small effect of antiviral treatment in patients with hepatitis C. Ann Intern Med 2002;136:288-92.
15. Rocca LG, Yawn BP, Wollan P, et al. Management of patients with hepatitis C in a community population: Diagnosis, discussions, and decisions to treat. Ann Fam Med 2004;2:116-24.
16. Cheung R, Mannalithara A, Singh G. Utilization and antiviral therapy in patients with chronic hepatitis C: Analysis of ambulatory care visits in the US. Dig Dis Sci 2010;55:1744-51.
17. Moirand R, Bilodeau M, Brissette S, et al. Determinants of antiviral treatment initiation in a hepatitis C-infected population benefiting from universal health care coverage. Can J Gastroenterol 2007;21:355-61.
18. Witkos M, Yi QL, Heathcote J, et al. Predictors of antiviral therapy in a post-transfusion cohort of hepatitis C patients. Can J Gastroenterol 2006;20:107-11.
19. Grebely J, Genoway KA, Raffa JD, et al. Barriers associated with the treatment of hepatitis C virus infection among illicit drug users. Drug Alcohol Depend 2008;93:141-7.
20. Hansen N, Obel N, Christensen PB, et al. Predictors of antiviral treatment initiation in hepatitis C virus-infected patients: A Danish cohort study. J Viral Hepat 2009;16:659-65.
21. Lettmeier B, Muhlberger N, Schwarzer R, et al. Market uptake of new antiviral drugs for the treatment of hepatitis C. J Hepatol 2008;49:528-36.
22. Librero J, Peiro S, Ordinana R. Chronic comorbidity and outcomes of hospital care: Length of stay, mortality, and readmission at 30 and 365 days. J Clin Epidemiol 1999;52:171-9.
23. Chen EY, Sclair SN, Czul F, et al. A small percentage of patients with hepatitis C receive triple therapy with boceprevir or telaprevir. Clin Gastroenterol Hepatol 2013;11:1014-20.
24. Kwo PY, Lawitz EJ, McCone J, et al. Efficacy of boceprevir, an NS3 protease inhibitor, in combination with peginterferon alfa-2b and ribavirin in treatment-naïve patients with genotype 1 hepatitis C infection (SPRINT-1): An open-label, randomized, multicenter phase 2 trial. Lancet 2010;376:705-16.
25. Poordad F, McCone J, Bacon BR, et al. Boceprevir for untreated chronic HCV genotype 1 infection. N Engl J Med 2011;364:1195-206.
26. Jacobson IM, McHutchison JG, Dusheiko G, et al. Telaprevir for previously untreated chronic hepatitis C virus infection. N Engl J Med 2011;364:2405-16.
27. Bini EJ, Brau N, Currie S, et al. Prospective multicenter study of eligibility for antiviral therapy among 4,084 U.S. veterans with chronic hepatitis C virus infection. Am J Gastroenterol 2005;100:1772-9.
28. Khokhar OS, Lewis JH. Reasons why patients infected with chronic hepatitis C virus choose to defer treatment: Do they alter their decision with time? Dig Dis Sci 2007;52:1168-76.
29. Shad J, Person J, Brann O, et al. How often are referred chronic hepatitis C patients candidates for anti-viral therapy? Hepatology 2000;32:283A.
30. Clark EC, Yawn BP, Galliher JM, et al. Hepatitis C identification and management by family physicians. Fam Med 2005;37:644-9.

Patient preference and willingness to pay for transient elastography versus liver biopsy: A perspective from British Columbia

Victoria Y Kan BSc MHA, Vladimir Marquez Azalgara MD MSc, Jo-Ann E Ford MScN,

WC Peter Kwan MD, Siegfried R Erb MD, Eric M Yoshida MD MHSc

VY Kan, V Marquez Azalgara, JE Ford, WCP Kwan, SR Erb, EM Yoshida. Patient preference and willingness to pay for transient elastography versus liver biopsy: A perspective from British Columbia. Can J Gastroenterol Hepatol 2015;29(2):72-76.

BACKGROUND: The cost of liver biopsy (LB) is publicly funded in British Columbia, while the cost of transient elastography (FibroScan [FS], Echosens, France) is not. Consequently, there is regional variation regarding FS access and monitoring of liver disease progression.

OBJECTIVE: To evaluate patient preference for FS versus LB and to assess the willingness to self-pay for FS.

METHODS: Questionnaires were distributed in clinic and via mail to LB-experienced and LB-naive patients who underwent FS at Vancouver General Hospital, Vancouver, British Columbia.

RESULTS: The overall response rate was 76%. Of the 422 respondents, 205 were LB-experienced. The mean age was 53.5 years, 50.2% were male, 54.7% were Caucasian, 38.2% had hepatitis C and 26.3% had an annual household income >$75,000. Overall, 95.4% of patients preferred FS to LB. FS was associated with greater comfort than LB, with the majority reporting no discomfort during FS (84.1% versus 7.8% for LB), no discomfort after (96.2% versus 14.6% LB) and no feelings of anxiety after FS explanation (78.2% versus 12.7% LB). FS was also associated with greater speed, with the majority reporting short test duration (97.2% versus 48.3% LB) and short wait for the test result (95.5% versus 30.2% LB). Most (75.3%) respondents were willing to self-pay for FS, with 26.3% willing to pay $25 to $49. Patients with unknown liver disease preferred LB (OR [FS preference] 0.20 [95% CI 0.07 to 0.53]).

CONCLUSIONS: FS was the preferred method of assessing liver fibrosis among patients, with the majority willing to self-pay. To ensure consistency in access, provincial funding for FS is needed. However, LB remains the procedure of choice for individuals with an unknown diagnosis.

Key Words: *Fibroscan; Liver biopsy; Patient; Preference; Transient elastography*

La préférence des patients et leur disposition à payer une élastographie transitoire plutôt qu'une biopsie du foie : le point de vue de la Colombie-Britannique

HISTORIQUE : Le régime public paie le coût de la biopsie du foie (BF) en Colombie-Britannique, mais pas celui de l'élastographie transitoire (FibroScan [FS], Echosens, France). Par conséquent, on observe des variations régionales en matière d'accès au FS et de surveillance de l'évolution des hépatopathies.

OBJECTIF : Évaluer la préférence des patients envers le FS ou la BF et leur disposition à payer eux-mêmes le FS.

MÉTHODOLOGIE : Des patients qui avaient déjà subi ou non une BF et qui se soumettaient à un FS au *Vancouver General Hospital*, en Colombie-Britannique, ont reçu un questionnaire en clinique et par la poste.

RÉSULTATS : Le taux de réponse globale s'est élevé à 76 %. De ces 422 répondants, 205 avaient déjà subi une BF. Ils avaient un âge moyen de 53,5 ans, 50,2 % étaient de sexe masculin, 54,7 % étaient blancs, 38,2 % avaient une hépatite C et 26,3 % avaient un revenu familial annuel de plus de 75 000 $. Dans l'ensemble, 95,4 % des patients préféraient le FS à la BF. Le FS causait moins d'inconfort que la BF, la majorité déclarant n'avoir ressenti aucun inconfort pendant le FS (84,1 % par rapport à 7,8 % pour la BF), aucun inconfort après le FS (96,2 % par rapport à 14,6 % pour la BF) et aucun sentiment d'anxiété après l'explication sur le FS (78,2 % par rapport à 12,7 % pour la BF). Le FS était également plus rapide à effectuer, la majorité le trouvant court (97,2 % par rapport à 48,3 % BF) et trouvant l'attente courte avant d'en obtenir les résultats (95,5 % par rapport à 30,2 % pour la BF). La plupart des répondants (75,3 %) étaient disposés à payer le FS, 26,3 % étant prêts à débourser de 25 $ à 49 $. Les patients dont l'hépatopathie n'était pas connue préféraient la BF (RC [préférence pour le FS] de 0,20 [95 % IC 0,07 à 0,53]).

CONCLUSIONS : Le FS était la méthode favorisée pour évaluer la fibrose hépatique chez les patients, dont la majorité était disposée à la payer. Pour garantir une uniformité d'accès, il faut assurer le financement provincial du FS. Cependant, la BF demeure l'intervention de choix pour les personnes dont on ne connaît pas le diagnostic.

Determining the appropriate treatment and management of chronic liver disease often relies on the degree of liver fibrosis. Although liver biopsy (LB) is the gold standard for liver fibrosis assessment, it is considered to be an invasive, resource-intensive and painful procedure that carries the risk of mild to severe complications (1-3). Furthermore, it is associated with several technical limitations including interobserver variability among pathologists and sampling variation (4-6).

Due to the drawbacks of LB, there has been significant focus on developing safer, noninvasive and more efficient means of evaluating liver fibrosis. Transient elastography (FibroScan [FS], Echosens, France) is one such noninvasive test that was approved for use in Canada in 2009. Since its approval, FS has been believed to be the preferred alternative to LB due to its noninvasive nature, painless and rapid procedure, and ease of use. It has also helped overcome some of the technical limitations of LB by sampling a greater volume of tissue (7) and yielding reproducible, operator-independent results (8). Furthermore, compared with LB and other tests of fibrosis, FS has demonstrated great diagnostic accuracy (9-11). One limitation of FS is its inability to accurately assess liver stiffness in obese patients (12-14).

Despite the limitations of FS, this modern and low-risk method of assessing liver fibrosis has been widely accepted among physicians across Canada (15,16). In fact, a survey conducted by Sebastiani et al (16) suggests that most Canadian physicians believe noninvasive methods, particularly FS, provide accurate staging of liver fibrosis. Furthermore, FS has been shown to have significantly reduced the need for LB among specialists, including gastroenterologists, hepatologists and infectious disease specialists, across Canada (15,16).

Division of Gastroenterology, Vancouver General Hospital, Vancouver, British Columbia

Correspondence: Dr Eric M Yoshida, Vancouver General Hospital, Division of Gastroenterology, Diamond Health Care Centre, 5153-2775 Laurel Street, Vancouver, British Columbia V5Z 1M9.e-mail eric.yoshida@vch.ca

LB and FS are both approved methods of fibrosis assessment in Canada; however, the cost of LB is covered under all provincial health care plans and the cost of FS is not. Quebec is the only province currently reimbursing for FS, with work in progress in British Columbia, Alberta, Ontario and Nova Scotia (16). As a result of this lack of public funding for FS, there is regional variation regarding access to FS and monitoring of liver disease progression. Specifically, FS clinics tend to be located at university hospitals and private practices rather than community-based hospitals, making physical accessibility an issue (16). Further adding to this issue of physical accessibility is the finding that the majority (59.6%) of Canadian specialists do not have a FS in their clinics, and 61.3% of these physicians do not have convenient access to FS whatsoever (16). Moreover, the amount patients are charged for FS can range from nil to hundreds of dollars depending on the site, thus posing financial barriers for some individuals.

To date, studies have shown FS to have high diagnostic accuracy compared with LB. Studies have also demonstrated overwhelming physician acceptance of FS as a novel, noninvasive method of assessing liver fibrosis. However, there are no data regarding the suitability of FS in terms of patient preference. Knowledge of their preference is of the utmost importance because patients are key stakeholders in the current efforts being made toward provincial reimbursement for FS. The present study, therefore, aimed to evaluate preference for FS versus LB among patients in British Columbia.

METHODS

The present study involved a survey of patients who underwent FS at Vancouver General Hospital (Vancouver, British Columbia) between March 1, 2011 and April 4, 2014. The questionnaire was distributed to patients via mail as well as in clinic over a period of three months (January 3 to April 4, 2014). Patients who underwent FS before the distribution of questionnaires in clinic (between March 1, 2011 and January 2, 2014) and had documentation of having previously undergone a LB in their clinic chart were mailed a questionnaire. Patients who underwent FS between January 3 and April 4, 2014, regardless of previous LB experience, were administered the questionnaire in clinic. The present study was approved by the Behavioural Research Ethics Board at the University of British Columbia, Vancouver, British Columbia.

The structured questionnaire assessed the following variables regarding patient experience with FS and LB (if applicable): patient discomfort during and after the test; anxiety experienced after having the test explained to them but before the test itself; perceived duration of the test; perceived duration of wait for test results; and time taken off work for testing. The questionnaire also assessed test preference, willingness to self-pay for the FS, and amount respondents were willing to pay. Brief descriptions of both tests were included in the questionnaires to mimic information that would normally be provided to patients, and to ensure patients (particularly those contacted through mail) were recalling the correct tests. For individuals who had never undergone LB, a more thorough description of the procedure and the standard instructions usually provided to a patient before the procedure were also included in the questionnaire. Finally, the questionnaire obtained demographic information from respondents including age, sex, annual household income, ancestry and liver disease history. Response options consisted of Likert-type scale responses, categorical response options, ordinal response options as well as fill-in-the-blank options. The questionnaire was available in both English and simplified Chinese.

Data analysis consisted of descriptive statistics including means and percentages of demographic data for all participants. Frequencies of each survey response were also tabulated as percentages. Analysis using the χ^2 test or Fisher's exact test, where applicable, evaluated any difference in responses between LB-experienced and LB-naive respondents. The t test was used to determine whether the hypothesis that >90% of patients would prefer FS to LB could be supported. Logistic regression analysis was used to evaluate any relationships between test preference and demographic characteristics, as well as willingness to pay and demographic characteristics. A two-tailed alpha level of significance of 0.05 was used for all statistical analyses.

TABLE 1
Respondent demographics

Characteristic	All respondents, n	Liver biopsy		P
		Naive	Experienced	
Age, years, mean ± SD	53.5±12.1	52.1±13.0	55.1±11.0	0.012
Sex				
Male	212 (50.2)	122 (56.2)	90 (43.9)	0.006
Female	206 (48.8)	91 (41.9)	115 (56.1)	
No response	4 (1.0)	4 (1.8)	0 (0)	
Liver disease				
Hepatitis C	161 (38.2)	61 (28.1)	100 (48.8)	<0.0001
Hepatitis B	116 (27.5)	79 (36.4)	37 (18.1)	
Don't know	48 (11.4)	32 (14.8)	16 (7.8)	
AIH	37 (8.8)	7 (3.2)	30 (14.6)	
NASH	27 (6.4)	21 (9.7)	6 (2.9)	
PBC	18 (4.3)	7 (3.2)	11 (5.4)	
Not disclosed	7 (1.7)	5 (2.3)	2 (1)	
No response	4 (1.0)	3 (1.4)	1 (0.5)	
Hemochromatosis	2 (0.5)	1 (0.5)	1 (0.5)	
Wilson disease	1 (0.2)	0 (0)	1 (0.5)	
ALD	1 (0.2)	1 (0.5)	0	
Ancestry				
Caucasian	228 (54.7)	93 (43.9)	135 (65.9)	0.001
East Asian	131 (31.4)	82 (38.7)	49 (23.9)	
South Asian	25 (6.0)	16 (7.6)	9 (4.4)	
Other Asian	13 (3.1)	9 (4.3)	4 (2.0)	
American Indian	10 (2.4)	5 (2.4)	5 (2.4)	
Not disclosed	5 (1.2)	3 (1.4)	2 (1.0)	
Black	3 (0.7)	3 (1.4)	0 (0)	
Hawaiian/Pacific Islander	2 (0.5)	1 (0.5)	1 (0.5)	
Annual household income, $				
0–24,999	82 (19.6)	45 (21.1)	37 (18.1)	0.40
25,000–49,999	89 (21.3)	50 (23.5)	39 (19.0)	
50,000–74,999	67 (16.0)	34 (16.0)	33 (16.1)	
≥75,000	110 (26.3)	55 (25.8)	55 (26.8)	
Not disclosed	70 (16.8)	29 (13.6)	41 (20.0)	

Data presented as n (%) unless otherwise indicated. AIH Autoimmune hepatitis; ALD Alcoholic liver disease; NASH Nonalcoholic steatohepatitis; PBC Primary biliary cirrhosis

RESULTS

The response rate in clinic was 99% (n=315) and via mail was 45% (n=107), yielding an overall response rate of 76% (422 of 558). Of the 422 respondents, 205 were LB-experienced. Table 1 summarizes the demographic characteristics of respondents. The mean age of all respondents was 53.5 years and one-half (50.2%) were male. The most common liver disease reported among respondents was hepatitis C (38.2%), followed by hepatitis B (27.5%) and "Don't know" (11.4%). The majority of respondents were Caucasian (54.7%) and East Asian (31.4%). Annual household income was fairly evenly distributed among all respondents, with the greatest proportion of respondents (26.3%) reporting an annual household income ≥$75,000. In addition, a substantial proportion of respondents (16.8%) preferred not to disclose their household income. There was a difference in sex and ancestry between the LB-naive and LB-experienced groups (P=0.006 and P=0.001, respectively). There was also a difference in age (P=0.012) and liver disease (P<0.0001) between the two groups. There was no difference in annual household income between the two groups (P=0.40).

TABLE 2
Patient experience with transient elastography* versus liver biopsy

	FibroScan (n=422)	Liver biopsy (n=205)
Level of discomfort experienced during the test		
None	355 (84.1)	16 (7.8)
Mild	61 (14.5)	55 (26.8)
Moderate	6 (1.4)	69 (33.7)
Significant	0 (0)	49 (23.9)
Severe	0 (0)	15 (7.3)
Blank response	0 (0)	1 (0.5)
Level of discomfort experienced after the test		
None	406 (96.2)	30 (14.6)
Mild	13 (3.1)	60 (29.3)
Moderate	2 (0.5)	63 (30.7)
Significant	1 (0.2)	38 (18.5)
Severe	0 (0)	13 (6.3)
Blank response	0 (0)	1 (0.5)
Level of anxiety experienced after test explanation before the test		
None	330 (78.2)	26 (12.7)
Mild	73 (17.3)	40 (19.5)
Moderate	17 (4.0)	78 (38.1)
Significant	1 (0.2)	52 (25.4)
Severe	0 (0)	8 (3.9)
Blank response	1 (0.2)	1 (0.5)
The test was too long		
No	410 (97.2)	99 (48.3)
Somewhat	11 (2.6)	65 (31.7)
Yes	0 (0)	39 (19.0)
Blank response	1 (0.2)	2 (1.0)
There was a long wait to receive the test results		
No	403 (95.5)	62 (30.2)
Somewhat	13 (3.1)	86 (42.0)
Yes	1 (0.2)	54 (26.3)
Blank response	5 (1.2)	3 (1.5)
Time taken off work to undergo the test, h		
<2	128 (30.3)	9 (4.4)
2–4	53 (12.6)	15 (7.3)
4–6	11 (2.6)	18 (8.8)
>6	19 (4.5)	61 (29.8)
Not applicable	209 (49.5)	99 (48.3)
Blank response	2 (0.5)	3 (1.4)

*Data presented as n (%). *FibroScan, Echosens, France*

Regarding patient experience, the majority (84.1%) of all respondents experienced no discomfort during the performance of FS. In comparison, few of the LB-experienced respondents (7.8%) felt no discomfort during the LB, while the majority (33.7%) experienced moderate discomfort. The majority (96.2%) of all respondents also reported experiencing no discomfort after FS, compared with only 14.6% of LB-experienced respondents feeling no discomfort after LB. Specifically, discomfort after LB was reported to be mild (29.3%) to moderate (30.7%) by the majority of LB-experienced respondents. The level of anxiety experienced after the FS explanation, but before the test, was reported to be 'none' by 78.2% of all respondents, compared with only 12.7% of LB-experienced respondents after LB explanation. The perceived duration of FS was not considered to be too long by 97.2% of respondents, and the perceived duration of LB was not considered to be too long by 48.3% of LB-experienced respondents. The majority (95.5%) of all respondents reported that the wait time was not too long for the FS result, while 30.2% of LB-experienced respondents reported that the wait time was not too long for the LB result. In terms of the amount of time taken off work for the tests, 30.3% reported taking <2 h off for FS while 4.5% reported taking >6 h off for FS. Only 4.4% of respondents reported taking <2 h off work for LB while 29.8% reported taking >6 h off. Table 2 presents a summary of the responses. There was no significant difference in the responses provided to questions relating to FS experience among LB-naive respondents when compared with LB-experienced respondents (Table 3).

TABLE 3
Survey responses regarding transient elastography* experience for liver biopsy-naive versus liver biopsy-experienced respondents

Survey	Liver biopsy			
responses	Naive	Experienced	Total	P
Level of discomfort experienced during FibroScan*				
None	190 (87.6)	165 (80.5)	355 (84.1)	0.106
Mild	24 (11.1)	37 (18.1)	61 (14.5)	
Moderate	3 (1.4)	3 (1.5)	6 (1.4)	
Significant	0 (0)	0 (0)	0 (0)	
Severe	0 (0)	0 (0)	0 (0)	
Level of discomfort experienced after FibroScan				
None	212 (97.7)	194 (94.6)	406 (96.2)	0.226
Mild	4 (1.8)	9 (4.4)	13 (3.1)	
Moderate	1 (0.5)	1 (0.5)	2 (0.5)	
Significant	0 (0)	1 (0.5)	1 (0.2)	
Severe	0 (0)	0 (0)	0 (0)	
Level of anxiety experienced after FibroScan explanation before the test				
None	172 (79.6)	158 (77.1)	330 (78.4)	0.778
Mild	35 (16.2)	38 (18.5)	73 (17.3)	
Moderate	9 (4.2)	8 (3.9)	17 (4.0)	
Significant	0 (0)	1 (0.5)	1 (0.2)	
Severe	0 (0)	0 (0)	0 (0)	
The FibroScan was too long				
No	213 (98.2)	197 (96.6)	410 (97.4)	0.369
Somewhat	4 (1.8)	7 (3.4)	11 (2.6)	
Yes	0 (0)	0 (0)	0 (0)	
There was a long wait to receive the FibroScan results				
No	208 (97.2)	195 (96.1)	403 (96.6)	0.677
Somewhat	6 (2.8)	7 (3.5)	13. (3.1)	
Yes	0 (0)	1 (0.5)	1 (0.2)	

*Data presented as n (%) unless otherwise indicated. *FibroScan, Echosens, France*

In terms of patient preference, the majority of respondents preferred FS to LB if both were available at no charge (95.4% versus 4.6%, respectively) (Figure 1). The results were similar among LB-experienced patients, with 95.1% preferring FS and 4.9% preferring LB. Of all respondents, 75.3% were willing to self-pay for FS even if LB was reimbursed by the province (Figure 2). The results were similar among LB-experienced patients (76.1%). Among all individuals who were willing to pay, the amount they were willing to pay varied without any particular amount being significantly preferred (Figure 3). However, the majority (26.3%) of respondents were willing to pay \$25 to \$49 for FS.

Logistic regression analysis was used to evaluate relationships between test preference (FS or LB) and demographic characteristics, including age, sex, previous biopsy experience, household income, liver disease and ancestry, and also to evaluate relationships between willingness to pay (yes or no) and the same demographic characteristic. There were no significant associations between test preference or willingness to pay and the demographic variables tested. However, the only variable significantly associated with lower odds of FS preference was unknown etiology of liver disease (OR 0.20 [95% CI 0.07 to 0.53]; P=0.001).

DISCUSSION

The present study was the first to evaluate preference for FS versus LB among patients. It demonstrates a greater preference for FS regardless of previous LB experience. In particular, patients tend to feel more

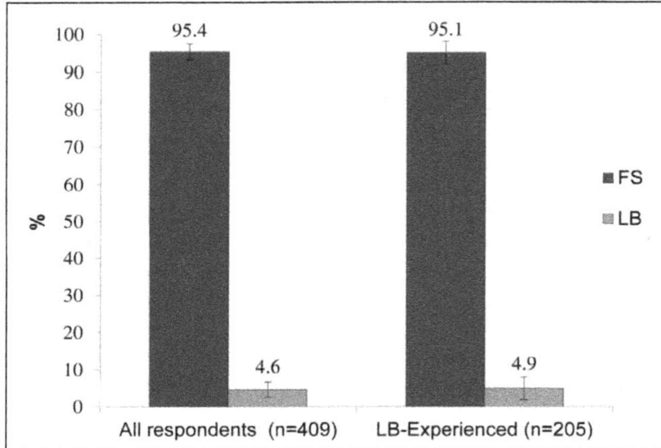

Figure 1) *Patient preference according to all respondents and liver biopsy (LB)-experienced respondents. FS FibroScan (Echosens, France)*

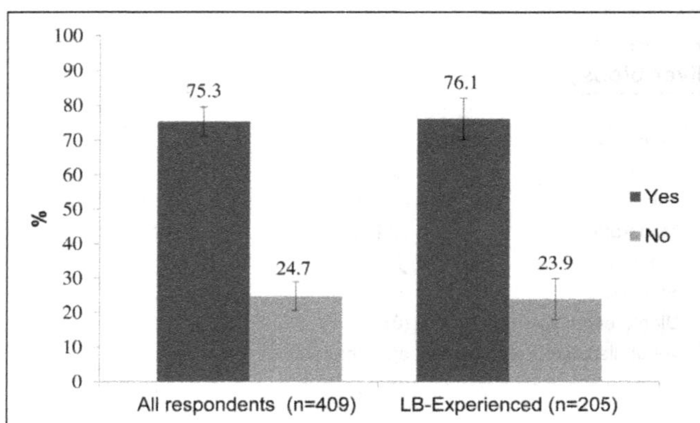

Figure 2) *Willingness to pay for transient elastography (FibroScan, Echosens, France) according to all respondents and liver biopsy (LB)-experienced respondents*

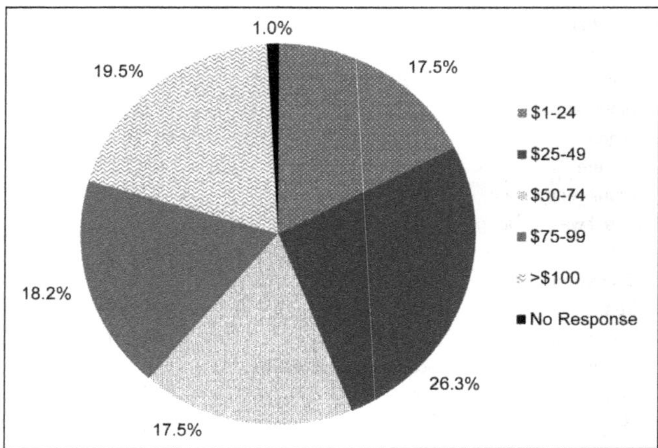

Figure 3) *Amount willing to pay for transient elastography (FibroScan, Echosens, France) among all respondents*

comfortable with FS than with LB, reporting little to no discomfort during and after the test, and less anxiety before the test. They also tended to associate less time with FS than with LB, reporting shorter test duration, shorter wait time for results and a reduced amount of time off work.

Our survey further reveals that an overwhelming majority (95.4%) of patients prefer FS to LB if both tests are available at no cost to them. This result did not change when only LB-experienced respondents were considered (95.1%), further establishing the clear preference for FS to LB. Seventy-five percent of patients reported being willing to self-pay to undergo FS if only LB was publicly funded. Again, previous LB experience did not affect the response rate. The reason for unwillingness to pay for FS was not explored in the present study. It was speculated that some of these individuals may be unwilling to pay based on principle due to the fact that provision of health care services in Canada is predominantly publicly funded. The results, nevertheless, suggest that the majority of patients prefer FS and would be willing to pay to undergo the test.

Interestingly, of the respondents who were willing to pay for FS, the majority (26.3%) report they would pay between $25 and $49. The response rates for other amount categories were fairly close, however, with the second highest percentage of respondents (19.5%) willing to pay >$100. These results do not suggest that a significant proportion of patients would pay exorbitant amounts for FS, nor do they suggest that they would pay negligible amounts. They do, however, suggest that the amount an individual is willing to pay is highly variable and may depend on a host of factors not explored in the present study.

Surprisingly, patients who reported having an unknown diagnosis of liver disease preferred LB (OR for FS preference 0.2 [95% CI 0.07 to 0.53]). This suggests that LB could still provide valuable information for this subset of patients, and is also consistent with the fact that LB is able to diagnose both liver disease and liver fibrosis, while FS can only diagnose liver fibrosis. Comparably, a survey of Canadian physicians discovered that autoimmune/cholestatic liver disease was the only liver disease category that was not reported to be associated with a reduction in LB practice or a tendency for higher use of noninvasive methods for assessing liver fibrosis (16). These findings may have significant implications for the use of LB in future clinical practice. For instance, LB may no longer be necessary for individuals with liver disease that can be diagnosed through serological tests or other less-invasive means. Instead, it may have greater utilization among individuals with liver disease of unknown etiology due to its ability to provide differential diagnoses. Furthermore, for individuals who require differential diagnosis but are hesitant to undergo LB, fibrosis assessment via FS may help guide the physician and patient's decision as to whether (or when) definitive diagnosis via LB is necessary.

Our study had several limitations. While the overall response rate (76%) was satisfactory, the response rate among participants being reached through mail was relatively low (45%). Although other studies of similar nature also obtained similar response rates (16), the low response via mail limited the data we could have obtained from the unique perspective of individuals who had the experience of both FS and LB. Additionally, the survey was conducted from one site in Vancouver, limiting the representation of other patient populations in British Columbia. Another disadvantage of conducting the survey at a single site is that the processes at the site may be distinct from other clinics. For instance, it is standard at the site for a patient to undergo FS and review their result with the physician on the same day. Thus, we acknowledge that not all sites operate in such a fashion. Recall bias is another limitation of the study because the survey relied on patient's self-assessed experience of both tests. This is particularly true for self-assessed reports on LB experience because LB could have been performed quite a few years before FS.

CONCLUSION

The present study evaluated preference for FS versus LB among patients. Because patients are key stakeholders in the current efforts being made toward implementing a reimbursement policy for FS in British Coumbia, knowledge of their preference is invaluable. Our findings demonstrate that the majority of patients preferred FS to LB, associating FS with increased comfort and speed compared with LB. This preference was further demonstrated by the willingness of the majority of patients to self-pay for FS if only LB were publicly funded. Despite this preference for FS, lack of convenient access to

FS among Canadian physicians, as demonstrated in previous studies, and varying costs of obtaining FS technology continue to be limiting factors for FS access among patients in British Columbia. Thus, these findings emphasize the need for provincial funding for FS to minimize regional variation regarding access to and availability of FS, as well as to ensure consistent monitoring of liver disease progression and management of chronic liver disease throughout the province. Furthermore, this could minimize the need for LB among a significant proportion of the population who do not wish to undergo, or are ineligible for, an invasive procedure. It is noted, however, that LB remained the procedure of choice for individuals with an unknown diagnosis.

DISCLOSURES AND ACKNOWLEDGEMENTS: None of the authors has a competing interest with regard to this work. The authors sincerely thank Mr and Mrs Arran and Ratana Stephens and Nature's Path Organic Foods for their generous support that facilitated the acquisition of our Fibroscan unit and ongoing liver disease research.

REFERENCES

1. Cadranel JF, Rufat P, Degos F. Practices of liver biopsy in France: Results of a prospective nationwide survey. For the Group of Epidemiology of the French Association for the Study of the Liver (AFEF). Hepatology 2000;32:477-81.
2. Firpi RJ, Soldevila-Pico C, Abdelmalek MF, Morelli G, Judah J, Nelson DR. Short recovery time after percutaneous liver biopsy: Should we change our current practices? Clin Gastroenterol Hepatol 2005;3:926-9.
3. Myers RP, Fong A, Shaheen AA. Utilization rates, complications and costs of percutaneous liver biopsy: A population-based study including 4275 biopsies. Liver Int 2008;28:705-12.
4. Robert M, Sofair AN, Thomas A, et al. A comparison of hepatopathologists' and community pathologists' review of liver biopsy specimens from patients with hepatitis C. Clin Gastroenterol Hepatol 2009;7:335-8.
5. Hølund B, Poulsen H, Schlichting P. Reproducibility of liver biopsy diagnosis in relation to the size of the specimen. Scand J Gastroenterol 1980;15:329-35.
6. Regev A, Berho M, Jeffers LJ, et al. Sampling error and intraobserver variation in liver biopsy in patients with chronic HCV infection. Am J Gastroenterol 2002;97:2614-8.
7. de Lédinghen V, Vergniol J. Transient elastography (FibroScan). Gastroenterol Clin Biol 2008;32:58-67.
8. Sandrin L, Fourquet B, Hasquenoph J, et al. Transient elastography: A new noninvasive method for assessment of hepatic fibrosis. Ultrasound Med Biol 2003;29:1705-13.
9. Shaheen A. FibroTest and FibroScan for the prediction of hepatitis C-related fibrosis: A systematic review of diagnostic test accuracy. Am J Gastroenterol 2007;102:2589-600.
10. Castéra L, Vergniol J, Foucher J, et al. Prospective comparison of transient elastography, Fibrotest, APRI, and liver biopsy for the assessment of fibrosis in chronic hepatitis C. Gastroenterology 2005;128:343-50.
11. Friedrich-Rust M, Ong M, Martens S, et al. Performance of transient elastography for the staging of liver fibrosis: A meta-analysis. Gastroenterology 2008;134:960-74.
12. Foucher J, Castéra L, Bernard P, et al. Prevalence and factors associated with failure of liver stiffness measurement using FibroScan in a prospective study of 2114 examinations. Eur J Gastroenterol Hepatol 2006;18:411-2.
13. Wong GL, Wong VW, Chim AM, et al. Factors associated with unreliable liver stiffness measurement and its failure with transient elastography in the Chinese population. J Gastroenterol Hepatol 2011;26:300-5.
14. Castéra L, Foucher J, Bernard P, et al. Pitfalls of liver stiffness measurement: A 5-year prospective study of 13,369 examinations. Hepatology 2010;51:828-35.
15. Aljawad M, Yoshida EM, Uhanova J, Marotta P, Chandok N. Percutaneous liver biopsy practice patterns among Canadian hepatologists. Can J Gastroenterol 2013;27:e31-4.
16. Sebastiani G, Ghali P, Wong P, Klein MB, Deschenes M, Myers RP. Physicians' practices for diagnosing liver fibrosis in chronic liver diseases: A nationwide, Canadian survey. Can J Gastroenterol Hepatol 2014;28:23-30.

Prevalence and predictors of sleep disturbance among liver diseases in long-term transplant survivors

Mamatha Bhat MD MSc[1*], Jonathan M Wyse MD MSc[2*], Erica Moodie PhD[3], Peter Ghali MD MSc[1], Nir Hilzenrat MD[2], Philip Wong MD MSc[1], Marc Deschênes MD[1]

M Bhat, JM Wyse, E Moodie, et al. Prevalence and predictors of sleep disturbance among liver diseases in long-term transplant survivors. Can J Gastroenterol Hepatol 2015;29(8):440-444.

BACKGROUND: Patients with cirrhosis are known to experience sleep disturbance, which negatively impacts health-related quality of life.

OBJECTIVE: To assess the prevalence and predictors of sleep disturbance before and after liver transplantation (LT).

METHODS: Both pre- and post-LT patients were administered the Basic Nordic Sleep Questionnaire. The primary outcome was overall sleep satisfaction; the secondary outcomes were sleep latency and sleep duration.

RESULTS: Eighty-three patients participated pre-LT and 273 post-LT. Overall, participants having completed both pre- and post-LT questionnaires reported satisfactory sleep 61% of the time before LT and 65% of the time after LT. However, on review of all questionnaires, patients with alcoholic liver disease (ETOH) experienced dramatically less sleep disturbance (OR 0.13 [95% CI 0.03 to 0.60]) post-LT, whereas those with hepatitis C remained without improvement (OR 0.90 [95% CI [0.38 to 2.15]). On logistic regression, patients with ETOH had statistically less sleep satisfaction pre-LT (OR 5.8 [95% CI 1.0 to 40.5]) and significantly better sleep satisfaction post-LT (OR 0.50 [95% CI 0.20 to 1.00]) compared with those with hepatitis C. In addition, both ETOH and other conditions had significantly better sleep latency than hepatitis C patients.

CONCLUSIONS: Sleep parameters for patients who undergo LT for hepatitis C do not improve following LT as much as they do in patients transplanted for ETOH. Following LT, patients transplanted for ETOH are significantly more satisfied with their sleep than those transplanted for hepatitis C. Physicians should address and manage sleep quality after LT, so as to ultimately improve quality of life.

Key Words: *Sleep; Liver transplant; Hepatitis C; BNSQ (Basic Nordic Sleep Questionnaire)*

La prévalence et les prédicteurs des troubles du sommeil chez les survivants d'une transplantation hépatique de longue date

HISTORIQUE : Les patients atteints de cirrhose se caractérisent par des troubles du sommeil qui nuisent à leur qualité de vie liée à la santé.

OBJECTIF : Évaluer la prévalence et les prédicteurs de troubles du sommeil avant et après une transplantation hépatique (TH).

MÉTHODOLOGIE : Des patients ont participé au questionnaire de base sur le sommeil dans les pays nordiques avant ou après une TH. La satisfaction générale vis-à-vis du sommeil était le résultat principal, tandis que le délai d'endormissement et la durée du sommeil étaient les résultats secondaires.

RÉSULTATS : Quatre-vingt-trois patients ont participé au questionnaire avant la TH et 273, après la TH. Dans l'ensemble, les participants qui avaient rempli les deux questionnaires ont déclaré avoir un sommeil satisfaisant 61 % du temps avant la TH et 65 % du temps après la TH. Cependant, à l'analyse de tous les questionnaires, les patients atteints d'une stéatose hépatique non alcoolique (ETOH) éprouvaient beaucoup moins de troubles du sommeil (RC 0,13 [95 % IC 0,03 à 0,60]) après la TH, tandis que le sommeil de ceux qui étaient atteints d'hépatite C ne s'améliorait pas (RC 0,90 [95 % IC [0,38 à 2,15]). À la régression logistique, les patients atteints d'ETOH étaient statistiquement moins satisfaits vis-à-vis de leur sommeil avant la TH (RC 5,8 [95 % IC 1,0 à 40,5]) et beaucoup plus satisfaits vis-à-vis de leur sommeil après la TH (RC 0,50 [95 % IC 0,20 à 1,00]) que ceux qui étaient atteints d'hépatite C. De plus, tant l'ETOH que les autres problèmes de santé s'associaient à une réduction considérable du délai d'endormissement par rapport aux patients atteints d'hépatite C.

CONCLUSIONS : Les paramètres du sommeil chez les patients qui subissent une TH en raison de l'hépatite C ne s'améliorent pas autant après la TH que chez les patients greffés en raison de l'ETOH. Après la TH, les patients greffés en raison de l'ETOH sont considérablement plus satisfaits de leur sommeil que ceux greffés en raison de l'hépatite C. Les médecins devraient évaluer et prendre en charge la qualité du sommeil après la TH afin de finir par améliorer la qualité de vie.

As long-term survival after liver transplantation (LT) continues to improve, there has been greater focus on issues related to health-related quality of life (HRQoL) (1-4). Sleep disturbance in patients with cirrhosis has a clear negative impact on HRQoL (5,6) and even survival (7). However, it is uncertain how sleep parameters are affected following LT.

Prevalence of sleep disturbance in cirrhotic patients without hepatic encephalopathy has been reported to be as high as 47.7% (8). The mechanisms of sleep disturbance in cirrhotic patients are not well elucidated, but appear temporally related to the liver disease itself (8). The absence of circadian oscillation in plasma melatonin results in a lack of synchronicity between internal and social rhythms (9). A case report involving a patient with hepatitis C cirrhosis attested to this lack of melatonin oscillation pretransplant, with full restoration of circadian rhythm following LT (10). One would assume that restoration of circadian rhythms would enhance sleep quality after LT. Three studies measuring HRQoL in the post-LT period, although not directly focusing on sleep disturbance itself, found inconsistent results. One reported sleep as the most common problem in post-LT patients (45%) (11), a second found sleep quality improved back to the level of the general population (12) and a third (13) suggested less sleep disturbance, although impairments that remained in excess of 20% relative to the individual's premorbid status.

We performed a retrospective observational study to determine the prevalence and risk factors for sleep disturbance before and after LT in long-term survivors (>6 months). We hypothesized that patients with hepatitis C virus (HCV) infection, which is universally recurrent following LT, would sleep less well than those whose liver disease did not recur. The primary outcome of our study was overall self-reported sleep satisfaction; the secondary outcomes were sleep latency (time required to fall asleep) and total sleep duration.

**Authors who contributed equally*

[1]*Division of Gastroenterology, McGill University Health Centre;* [2] *Division of Gastroenterology, Sir Mortimer B Davis-Jewish General Hospital;* [3] *Department of Epidemiology, Biostatistics, & Occupational Health, McGill University, Montreal, Quebec*

Correspondence: Dr Mamatha Bhat, Division of Gastroenterology, Royal Victoria Hospital, 687 Pine Avenue West, Montreal, Quebec H3A 1A1.e-mail mamatha.bhat@mail.mcgill.ca

METHODS

All patients seen at the outpatient LT clinic of the McGill University Health Centre (Montreal, Quebec), between July 1998 and December 2012 were invited to participate in the present study. Patients were administered the Basic Nordic Sleep Questionnaire (BNSQ) (14) at each outpatient visit pre-LT and every follow-up visit post-LT (Appendix 1 [go to www.pulsus.com]). Not all patients completed questionnaires at each visit because they had the right to refuse at any time. For pre-LT questionnaires, the survey completed closest to transplant was chosen. If a patient was already transplanted when first approached for the study, only post-LT questionnaires existed. The questionnaire furthest away from LT was chosen as being the most representative of the patient's post-LT sleep health.

The BNSQ is a quantitative measure of subjective sleep complaints, and has been validated in several studies (14-20). Three items from this questionnaire have been shown to most closely correlate with objective sleep polysomnography (21,22) and were, therefore, included in the analysis:

- Overall sleep satisfaction (binary: yes, no)
- Sleep latency (binary: time to fall asleep ≥30 min; yes, no)
- Sleep duration (binary: total sleep duration ≥6 h; yes, no)

Patients were classified into three categories based on their liver disease: HCV, alcoholic (ETOH) and other. Demographic and clinical characteristics, such as age, sex, body mass index (BMI) and Model for End-stage Liver Disease (MELD) scores, were obtained. MELD scores were calculated based on laboratory data immediately pre-LT. The number of postoperative days spent in hospital was also included as a surrogate for perioperative complications at LT. At the time of each questionnaire, information regarding all medication use, including sleep aids, was recorded. Sleep aids included all benzodiazepines, such as lorazepam and diazepam, as well as quetiapine. The generic as well as pharmaceutical names were used to interrogate the transplant database as to the use of these sleep aids.

The present study was approved by the Institutional Review Board at the McGill University Health Centre, Montreal, Quebec.

Statistical analysis

Demographic and clinical characteristics such as sex, age, and bilirubin levels were summarized using descriptive statistics (medians and ranges for continuous covariates, proportions for binary covariates). Subject information was summarized by selecting the visit closest to the time of LT in the pre- and postoperative period, so that at most one visit per patient was included in each of the pre-LT and post-LT summary of characteristics.

The analyses were performed using a linear or a logistic regression model (multivariable analysis). To make use of all the information, the missing data were imputed using the MICE package in R and the analyses are based on 25 imputations. Because the analysis was exploratory in nature, all tests of association were performed at the 0.05 level of significance.

RESULTS

Patient characteristics

A total of 83 patients participated before LT (32 HCV, nine ETOH, 42 other). The 'other' category included patients with cirrhosis secondary to hepatitis B (n=4), nonalcoholic steatohepatitis or cryptogenic cirrhosis (n=18), primary biliary cirrhosis (n=5), primary sclerosing cholangitis (n=9), autoimmune hepatitis (n=2), polycystic liver disease (n=2) and acute liver failure due to thrombosed hepatic artery (n=2). Their demographic characteristics are listed in Table 1. There was a significantly greater percentage of men among patients with ETOH (89.9%) than among patients with HCV (75%) or other liver diseases (52%). HCV patients used more sleep aids than those with ETOH or other chronic liver diseases (12.5% versus 0% and 4.8%, respectively). No patients completed a questionnaire before repeat transplant.

A total of 273 patients participated following LT (73 HCV, 53 ETOH, 147 other). The 'other' category included patients with cirrhosis secondary to hepatitis B (n=35), nonalcoholic steatohepatitis or cryptogenic cirrhosis (n=48), hemochromatosis (n=6), primary biliary cirrhosis (n=21), primary sclerosing cholangitis (n=24), Wilson disease (n=3), autoimmune hepatitis (n=3), polycystic liver disease (n=1), acute liver failure due to thrombosed hepatic artery (n=3), Caroli disease (n=1), drug-induced liver failure (n=1) and alpha-1-antitrypsin deficiency (n=1). Again, individuals with ETOH tended to be male (86.8%) compared with HCV (76.7%) or other (67.3%). All groups of patients used sleep aids to a similar extent. There was no difference in median MELD score at the time of transplant. All groups of patients stayed in hospital for a similar median number of postoperative days.

TABLE 1
Demographic characteristics of pre- and post-transplant patients

	Liver transplant indication		
Characteristic	**Hepatitis C**	**Alcoholic**	**Other**
Pretransplant			
n	32	9	42
Age, years	55.25 (28.9–66.7)	58.20 (40.9–64.0)	58.55 (23.3–71.6)
Male sex, %	75	89.9	52.4
Using sleep aid, %	12.5	0	4.76
Days before transplant	122 (1–2659)	71 (14–947)	138 (3–2660)
Post-transplant			
n	73	53	147
Age, years	55.1 (37.7–74.0)	57.2 (42.3–70.8)	57.8(18.0–73.1)
Male sex, %	76.7	86.8	67.3
Second transplant, %	9.5	1.8	12.9
Using sleep aid, %	15	16.9	13.6
MELD score	20 (10–34)	20 (8–38)	20 (6–62)
Days in hospital stay post-transplant	20 (6–81)	20 (8–141)	21 (7–376)
Total days in hospital stay	26 (6–140)	30 (9–234)	30 (7–376)
Days since transplant	868 (31–5892)	1510 (56–6263)	1427 (19–6738)

Data presented as median (range) unless otherwise indicated. Sleep outcomes (%) according to patient type, before and after surgery: information from the visit nearest to time of transplant before surgery and nearest to (but no less than) six months after surgery (at most one measure per person in each of the pre- and postoperative periods). MELD Model for End-stage Liver Disease

Sleep satisfaction pre-LT and post-LT

When examining the 57 patients with both pre-LT and post-LT questionnaires completed, 61% pre-LT patients reported satisfactory sleep experience in the past month versus 65% in the post-LT period (OR 1.17 [95% CI 0.53 to 2.56]). Participants reported taking ≥30 min to fall asleep 36% of the time before and 25% after LT, respectively (OR 0.60 [95% CI 0.35 to 1.01]). Patients reported staying asleep for ≤6 h 45% of the time before and 34% after LT, respectively (OR 0.65 [95% CI 0.39 to 1.08]).

All questionnaires completed pre-LT and post-LT across types of liver disease were then evaluated, using at most one preoperative and one postoperative measure per person. The ETOH group improved dramatically from most sleep disturbance pre-LT (66.7%) to least sleep disturbance post-LT (20.7%, OR 0.13 [95% CI 0.03 to 0.60]). HCV patients experienced no improvement in sleep disturbance, with 40% reporting sleep disturbance pre-LT and 38.4% post-LT (OR 0.90 [95% CI 0.38 to 2.15]). Similarly, patients with other types of liver disease had no differences in reported sleep disturbance pre-LT at 33.1% versus post-LT (27.2%, OR 0.74 [95% CI 0.35 to 1.57]).

Those with HCV had sleep latency of 47.8 min pre-LT and 37.1 min post-LT. Sleep duration was 6.9 h and 6.95 h pre-LT and post-LT, respectively. Patients with ETOH had the shortest sleep latency

TABLE 2
Predictors of poor pretransplant sleep according to logistic regression analysis

Parameter	Sleep*		
	Latency	Duration	Satisfaction
Age	0.97 (0.92–1.02)	1.01 (0.97–1.06)	0.98 (0.93–1.02)
Female sex	3.53 (1.16–11.51)†	0.51 (0.19–1.37)	3.83 (1.27–12.65)†
Sleep aid	2.07 (0.29–18.32)	1.29 (0.21–10.27)	2.13 (0.30–15.54)
Primary Dx of ETOH (relative to HCV)	0.19 (0.009–1.47)	0.72 (0.14–3.78)	5.82 (1.04–40.58)†
Primary Dx of other liver diseases (relative to HCV)	0.50 (0.16–1.54)	0.93 (0.34–2.53)	0.64 (0.20–1.95)
Body mass index	0.95 (0.84–1.07)	0.97 (0.88–1.07)	0.99 (0.87–1.11)
Model for End-stage Liver Disease score	0.97 (0.90–1.04)	0.99 (0.93–1.05)	0.91 (0.83–0.97)†

*Data presented as OR (95% CI). *Latency, duration and satisfaction as binary variables (note: an OR <1 suggests shorter sleep latency, longer sleep duration and more sleep satisfaction); †Associations with P<0.05. Dx Diagnosis; ETOH Alcoholic liver disease; HCV Hepatitis C virus*

TABLE 3
Predictors of poor post-transplant sleep according to logistic regression analysis

Parameter	Sleep*		
	Latency	Duration	Satisfaction
Age	0.99 (0.97–1.03)	0.97 (0.94–0.99)†	1.00 (1.00–1.01)
Female sex	2.10 (1.11–3.96)†	1.54 (0.83–2.93)	1.40 (0.80–2.70)
Sleeping aid	0.92 (0.37–2.08)	0.55 (0.26–1.16)	2.30 (1.10–4.90)†
Primary Dx ETOH, relative to HCV	0.40 (0.16–0.95)†	1.37 (0.63–3.02)	0.50 (0.20–1.00)†
Primary Dx of other liver diseases, relative to HCV	0.47 (0.25–0.90)†	1.10 (0.59–2.01)	0.60 (0.30–1.10)
Body mass index	1.01 (0.963–1.06)	1.02 (0.97–1.07)	1.00 (1.00–1.10)
Model for End-stage Liver Disease score	0.99 (0.955–1.04)	0.97 (0.94–1.01)	1.00 (0.90–1.00)
Hospital discharge	0.99 (0.97–1.00)	1.00 (0.99–1.02)	1.00 (0.98–1.00)

*Data presented as OR (95% CI) for the association between each of three sleep outcomes – latency, duration and satisfaction – and diagnostic group before and at least 180 days after liver transplantation. Questionnaires were administered at least six months after transplant. When >1 was available, the questionnaire furthest away from liver transplantation was selected. An OR <1 suggests shorter sleep latency, longer sleep duration and more sleep satisfaction. *Latency, duration and satisfactory as binary; †Associations with P<0.05. Dx Diagnosis; ETOH Alcoholic liver disease; HCV Hepatitis C virus*

pre-LT (21.1 min) and 26.3 min post-LT, while mean sleep duration was 6.8 h pre-LT and 7.4 h post-LT.

Predictors of sleep parameters (Tables 2 and 3)

Logistic regression modelling was performed using, at most, one preoperative and one postoperative measure per person.

Pre-LT: Pre-LT patients with ETOH were significantly less likely to report satisfactory sleep than those with HCV (OR 5.8 [95% CI 1.0 to 40.5]). Patients with HCV trended toward poor sleep satisfaction, as well as worse sleep latency and duration compared with both ETOH and other. Pre-LT patients with stage 3 encephalopathy took 119 min (95% CI 29.3 min to 210.3 min) longer than those with stage 1 encephalopathy to fall asleep.

Post-LT: Post-LT patients who underwent LT for ETOH slept much better than those transplanted for HCV, with statistically significant less sleep disturbance (OR 0.50 [95% CI 0.20 to 1.00]).

Both ETOH and other patients had significantly less sleep latency than HCV patients (OR 0.4 [95% CI 0.2 to 0.9] and OR 0.5 [95% CI 0.2 to 0.9], respectively) (22).

DISCUSSION

In the present retrospective study, we used a validated questionnaire (21-23) to evaluate sleep quality before and after LT. Sleep disturbance significantly affects quality of life, and is known to be prevalent in up to 48% of patients with advanced liver disease (7). This is postulated to be due to disrupted circadian rhythms in cirrhotic patients, caused by constantly elevated melatonin levels (24-26). Increased melatonin levels are hypothesized to alter function of the suprachiasmatic nucleus (27-29). Additional physiological processes that affect sleep in cirrhotic patients include endocrine abnormalities leading to melatonin derangement (29), decreased activation of the retinohypothalamic system (30), central nervous system effects of gut-derived toxins (31), and less sensory input due to decreased activity level and light exposure (29). These phenomena translate into abnormal sleep actigraphy with delayed bedtime, late wake-up time and evening chronotypology (8). There is also animal evidence of deranged sleep patterns in cirrhosis. Male rats with pharmacologically induced cirrhosis had progressively decreased total wake time, along with increased total slow wave and rapid eye movement sleep (32).

In our study, we found that overall sleep parameters did not improve following LT, which was likely driven by the HCV group (the largest) based on subsequent analysis. Specifically, when reviewing all questionnaires, patients with ETOH experienced dramatically less sleep disturbance following LT compared with pre-LT (OR 0.13 [95% CI 0.03 to 0.60]). In contrast, HCV patients remained without improvement following LT (OR 0.90 [95% CI 0.38 to 2.15]). On logistic regression, patients transplanted for ETOH had statistically less sleep satisfaction pre-LT (OR 5.8 [95% CI 1.0 to 40.5]) and significantly better sleep satisfaction post-LT (OR 0.50 [95% CI 0.20 to 1.00]) compared with those transplanted for HCV. The HCV group had worse sleep latency, duration and satisfaction after LT compared with the ETOH group. This was consistent with our a priori hypothesis that universal HCV recurrence in long-term LT survivors would result in decreased sleep quality as shown in other areas of HRQoL (33-37). This may be due to liver dysfunction caused by the virus, or may be a direct viral cerebral effect (38,39) and can be considered to be an extrahepatic manifestation of HCV. In contrast, the improvement and overall better sleep parameters in ETOH patients may be attributed to the resolution of disease in most patients as opposed to the universal recurrence of HCV after LT.

Stage 3 encephalopathy was also a predictor of decreased sleep quality before LT in our study. Patients with stage 3 encephalopathy took 119 min longer to fall asleep than patients with stage 1 encephalopathy, which was likely a reflection of worse circadian physiology. LT then corrected circadian rhythms, with patients with stage 3 encephalopathy before LT having a quality of sleep similar to that of patients with stage 1 encephalopathy. Not surprisingly, use of a sleep aid post-LT was independently associated with better sleep satisfaction.

Various quality of life questionnaires incorporate sleep as a parameter. The Liver Disease Quality of Life instrument has been developed and validated in patients with advanced, chronic liver disease (40). It incorporates not only aspects related to sleep, but also multiple other items such as liver disease-related symptoms, their effects on activities of daily living, concentration, memory and sexual functioning. The sleep-related questions pertained principally to the quality of sleep and fatigue as a consequence. A better HRQoL has been shown to predict better survival in patients with cirrhosis (7). A survey of pediatric LT recipients found that poor sleep quality after LT had a significant impact on HRQoL (41). In a smaller cross-sectional study of pre-LT and post-LT patients, 73% of pre-LT and 77% of post-LT patients complained of poor sleep quality (42). The different findings in our study may be due to our use of a standardized sleep questionnaire, with different definitions as to what constitutes impaired sleep latency and duration.

The BNSQ is comprised of 21 different questions, with 27 total items covering sleep complaints. These include difficulty in initiating and maintaining sleep, general sleep habits, subjective sleep quality, use of sleep-inducing medication, excessive daytime sleepiness, napping and snoring. The questionnaire takes 5 min to 10 min to complete and is self-reported. Participants assign scores for most of these items, ranging from 1 (mild or infrequently present) to 5 (severe or frequently present). Other questions relating to sleep duration and sleep latency require specific durations or times to be assigned. The BNSQ has been administered to various patient populations, including those with chronic conditions (23), and has been confirmed to be reproducible and reliable. The three items from this questionnaire (overall sleep satisfaction, sleep latency, and sleep duration) that we chose to evaluate have been shown to most closely correlate with objective sleep polysomnography (21,22).

Our observational study had certain limitations. The questionnaires were not administered systematically, and questionnaire responses were analyzed in a retrospective manner.

Additionally, there may have been a selection bias in terms of the type of patient who agreed to complete a questionnaire before LT; however, it is likely nondifferential across types of liver disease. Use of self-reported questionnaires for sleep quality assessment may lead to recall bias that can misclassify patients. Physical symptoms may dominate a patient's quality of life before LT, with sleep disturbance being less important. However, because LT recipients improve physically, patients may report sleep disturbance more when physical issues are less consuming, possibly underestimating the benefit of orthotopic LT on sleep quality. Sleep quality is notoriously difficult to measure, and future work should likely combine multiple modalities of sleep assessment. These may include wrist actigraphy (8,43) (measures arm movement to infer sleep and wakefulness) and/or polysomnography (which measures cerebral and muscle activity, and eye movement). The present study did not have the statistical power to examine more subgroups of patients and could not explore other liver diseases. Finally, no control group was examined. However, a large population-based Canadian study examining mental health, including sleep disturbance, using a self-reported questionnaire found a prevalence of 15.6% to 15.9% for individuals 45 to 74 years of age and 12.6% for the those 35 to 44 years of age (44). Our prevalence post-LT of inadequate sleep quality was considerably higher and highlights the need for further work in this field.

CONCLUSION

Patients who undergo LT for HCV continue to experience higher rates of sleep disturbance after LT, possibly due to the universal recurrence of HCV. The advent of highly effective HCV therapies may provide an opportunity to both prevent development of recurrent liver disease with disturbed circadian pattern and prevent the damaging effects of HCV directly on the brain's circadian rhythms, thereby improving sleep quality in this patient subpopulation. Patients transplanted for ETOH, a disease that recurs much less often, have significantly improved sleep parameters after LT. Physicians should remain alert to this domain of HRQoL in long-term LT survivors, and diligently question patients about sleep quality. Sleep disturbance after LT should be addressed and managed, so as to ultimately improve HRQoL for these patients.

ACKNOWLEDGEMENT: The authors thank Myriam Fernandez for her management of the transplant database.

DISCLOSURES: The authors have no financial disclosures or conflicts of interest to declare.

FUNDING: CIHR fellowship to M Bhat.

REFERENCES

1. OPTN/SRTR 2013 Annual Data Report: Liver. Am J Transpl 2015;15(Suppl 2):1-28.
2. Santos RJ, Miyazaki MC, Domingos NA, Valerio NI, Silva RF, Silva RC. Patients undergoing liver transplantation: Psychosocial characteristics, depressive symptoms, and quality of life. Transplant Proc 2008;40:802-4.
3. van Ginneken BT, van den Berg-Emons RJ, Kazemier G, Metselaar HJ, Tilanus HW, Stam HJ. Physical fitness, fatigue, and quality of life after liver transplantation. Eur J Appl Physiol 2007;100:345-53.
4. Russell RT, Feurer ID, Wisawatapnimit P, Salomon RM, Pinson CW. The effects of physical quality of life, time, and gender on change in symptoms of anxiety and depression after liver transplantation. J Gastrointest Surg 2008;12:138-44.
5. Roth T, Ancoli-Israel S. Daytime consequences and correlates of insomnia in the United States: Results of the 1991 National Sleep Foundation Survey. II. Sleep 1999;(22 Suppl 2):S354-8.
6. Krasnoff JB, Vintro AQ, Ascher NL, Bass NM, Dodd MJ, Painter PL. Objective measures of health-related quality of life over 24 months post-liver transplantation. Clin Transplant 2005;19:1-9.
7. Kanwal F, Gralnek IM, Hays RD, et al. Health-related quality of life predicts mortality in patients with advanced chronic liver disease. Clin Gastroenterol Hepatol 2009;7:793-9.
8. Cordoba J, Cabrera J, Lataif L, Penev P, Zee P, Blei AT. High prevalence of sleep disturbance in cirrhosis. Hepatology 1998;27:339-45.
9. Martino ME, Romero-Vives M, Fernandez-Lorente J, De Vicente E, Barcena R, Gaztelu JM. Sleep electroencephalogram alterations disclose initial stage of encephalopathy. Methods Find Exp Clin Pharmacol 2002;(24 Suppl D):119-22.
10. Cordoba J, Steindl P, Blei AT. Melatonin arrhythmia is corrected after liver transplantation. Am J Gastroenterol 2009;104:1862-3.
11. Lowe D, O'Grady JG, McEwen J, Williams R. Quality of life following liver transplantation: A preliminary report. J R Coll Physicians Lond 1990;24:43-6.
12. Bonsel GJ, Essink-Bot ML, Klompmaker IJ, Slooff MJ. Assessment of the quality of life before and following liver transplantation. First results Transplantation 1992;53:796-800.
13. Tarter RE, Van Thiel DH, Hegedus AM, Schade RR, Gavaler JS, Starzl TE. Neuropsychiatric status after liver transplantation. J Lab Clin Med 1984;103:776-82.
14. Partinen M, Gislason T. Basic Nordic Sleep Questionnaire (BNSQ): A quantitated measure of subjective sleep complaints. J Sleep Res Jun 1995;4(S1):150-5.
15. Lindblom N, Kivinen S, Heiskala H, Laakso ML, Kaski M. Sleep disturbances in aspartylglucosaminuria (AGU): A questionnaire study. J Inherit Metab Dis 2006;29:637-46.
16. Eriksen W, Bjorvatn B, Bruusgaard D, Knardahl S. Work factors as predictors of poor sleep in nurses' aides. Int Arch Occup Environ Health 2008;81:301-10.
17. Strang-Karlsson S, Raikkonen K, Kajantie E, et al. Sleep quality in young adults with very low birth weight – the Helsinki study of very low birth weight adults. J Pediatr Psychol 2008;33:387-95.
18. Devine EB, Hakim Z, Green J. A systematic review of patient-reported outcome instruments measuring sleep dysfunction in adults. Pharmacoeconomics 2005;23:889-912.
19. Vaari T, Engblom J, Helenius H, Erkkola R, Polo-Kantola P. Survey of sleep problems in 3421 women aged 41-55 years. Menopause Int 2008;14:78-82.
20. Sarti CD, Chiantera A, Graziottin A, et al. Hormone therapy and sleep quality in women around menopause. Menopause 2005;12:545-51.

21. Kawada T, Suzuki S, Aoki S, Ogawa M. Relationship between subjective sleep rating and objective sleep parameters: A case study. Environ Res 1993;60:136-44.
22. Armitage R, Trivedi M, Hoffmann R, Rush AJ. Relationship between objective and subjective sleep measures in depressed patients and healthy controls. Depress Anxiety 1997;5:97-102.
23. Janson C, De Backer W, Gislason T, et al. Increased prevalence of sleep disturbances and daytime sleepiness in subjects with bronchial asthma: A population study of young adults in three European countries. Eur Respir J 1996;9:2132-8.
24. Steindl PE, Finn B, Bendok B, Rothke S, Zee PC, Blei AT. Disruption of the diurnal rhythm of plasma melatonin in cirrhosis. Ann Intern Med 1995;123:274-7.
25. Montagnese S, Middleton B, Mani AR, Skene DJ, Morgan MY. Sleep and circadian abnormalities in patients with cirrhosis: Features of delayed sleep phase syndrome? Metab Brain Dis 2009;24:427-39.
26. Cordoba J, Dupuis J, Gottstein J, Blei AT. Stenosis of a portacaval anastomosis affects circadian locomotor activity in the rat: A multivariable analysis. Am J Physiol 1997;273(6 Pt 1):G1218-25.
27. Steindl PE, Gottstein J, Blei AT. Disruption of circadian locomotor activity in rats after portacaval anastomosis is not gender dependent. Hepatology 1995;22:1763-8.
28. Zee PC, Mehta R, Turek FW, Blei AT. Portacaval anastomosis disrupts circadian locomotor activity and pineal melatonin rhythms in rats. Brain Res 1991;560:17-22.
29. Montagnese S, Middleton B, Mani AR, Skene DJ, Morgan MY. On the origin and the consequences of circadian abnormalities in patients with cirrhosis. Am J Gastroenterol 2010;105:1773-81.
30. Reichenbach A, Fuchs U, Kasper M, el-Hifnawi E, Eckstein AK. Hepatic retinopathy: Morphological features of retinal glial (Muller) cells accompanying hepatic failure. Acta Neuropathol 1995;90:273-81.
31. Steindl PE, Coy DL, Finn B, Zee PC, Blei AT. A low-protein diet ameliorates disrupted diurnal locomotor activity in rats after portacaval anastomosis. Am J Physiol 1996;271(4 Pt 1):G555-560.
32. Jimenez-Anguiano A, Diaz-Medina V, Farfan-Labonne BE, et al. Modification of sleep architecture in an animal model of experimental cirrhosis. World J Gastroenterol 2009;15:5176-80.
33. Feurer ID, Wright JK, Payne JL, et al. Effects of hepatitis C virus infection and its recurrence after liver transplantation on functional performance and health-related quality of life. J Gastrointest Surg 2002;6:108-15.
34. Paterson DL, Gayowski T, Wannstedt CF, et al. Quality of life in long-term survivors after liver transplantation: Impact of recurrent viral hepatitis C virus hepatitis. Clin Transplant 2000;14:48-54.
35. Gayowski T, Wagener MM, Marino IR, Singh N. Quality of life and functional status of liver transplant recipients with recurrent viral hepatitis C. Transplant Proc 1999;31:1386-7.
36. Bona MD, Rupolo G, Ponton P, et al. The effect of recurrence of HCV infection of life after liver transplantation. Transpl Int 1998;(11 Suppl 1):S475-479.
37. De Bona M, Ponton P, Ermani M, et al. The impact of liver disease and medical complications on quality of life and psychological distress before and after liver transplantation. J Hepatol 2000;33:609-15.
38. Forton DM, Allsop JM, Main J, Foster GR, Thomas HC, Taylor-Robinson SD. Evidence for a cerebral effect of the hepatitis C virus. Lancet 2001;358:38-9.
39. Forton DM, Thomas HC, Murphy CA, et al. Hepatitis C and cognitive impairment in a cohort of patients with mild liver disease. Hepatology 2002;35:433-9.
40. Gralnek IM, Hays RD, Kilbourne A, et al. Development and evaluation of the Liver Disease Quality of Life instrument in persons with advanced, chronic liver disease – the LDQOL 1.0. Am J Gastroenterol 2000;95:3552-65.
41. Fredericks EM, Dore-Stites D, Calderon SY, et al. Relationship between sleep problems and health-related quality of life among pediatric liver transplant recipients. Liver Transpl 2012;18:707-15.
42. Rodrigue JR, Nelson DR, Reed AI, Hanto DW, Curry M. Fatigue and sleep quality before and after liver transplantation. Prog Transplant 2010;20:221-33.
43. NIH State-of-the-Science Conference Statement on manifestations and management of chronic insomnia in adults. NIH Consens State Sci Statements. June 13 to 15, 2005;22:1-30.
44. Tjepkema M. Insomnia. Health Rep 2005;17:9-25.

19

A study investigating the association of dermatological and infusion reactions to infliximab and infliximab trough levels

Vivian Wai-Mei Huang MSc MD FRCPC[1], Neil Dhami BSc[1], Darryl Fedorak BSc[1], Connie Prosser PhD[2], Carol Shalapay MLT[2], Karen Ivy Kroeker MSc MD FRCPC[1], Brendan Phillip Halloran MD FRCPC[1], Levinus Albert Dieleman MD PhD[1], Richard Neil Fedorak MD FRCPC[1]

VWM Huang, N Dhami, D Fedorak, et al. A study investigating the association of dermatological and infusion reactions to infliximab and infliximab trough levels. Can J Gastroenterol Hepatol 2015;29(1):35-40.

BACKGROUND: Although infliximab is an effective therapy for inflammatory bowel disease (IBD), it is associated with dermatological events and infusion reactions. It is not known whether a relationship between these adverse events (AEs) and infliximab trough levels (ITLs) exists.

OBJECTIVES: To report the prevalence of infliximab-associated AEs in IBD patients receiving stable maintenance infliximab therapy, and to correlate ITLs with dermatological and infusion reactions to infliximab.

METHODS: Adult IBD patients receiving stable maintenance infliximab therapy were recruited from the University of Alberta Infusion Clinic (Edmonton, Alberta). ITLs were measured in blood samples collected before infusion, and the patients' records were reviewed for dermatological and infusion reactions to infliximab.

RESULTS: One-quarter (18 of 71 [25.4%]) of patients experienced dermatological or infusion reactions to infliximab: nine (12.7%) dermatological events and nine (12.7%) infusion reactions. The median ITL was similar among patients with and without these AEs (7.2 µg/mL [interquartile range (IQR) 2.0 µg/mL to 13.3 µg/mL] versus 6.6 µg/mL [IQR 3.2 µg/mL to 12.7 µg/mL]; P=0.648). The median ITL of patients who experienced infusion reactions (2.0 µg/mL [IQR 0.1 µg/mL to 5.7 µg/mL]) was lower than that of patients who experienced no such AEs (6.6 µg/mL [IQR 3.2 µg/mL to 12.7 µg/mL]; P=0.008]) and lower than that of patients who experienced dermatological AEs (13.3 µg/mL [IQR 8.8 µg/mL to 17.4 µg/mL]; P<0.001).

CONCLUSION: One-quarter of IBD outpatients receiving stable maintenance infliximab therapy experienced dermatological and infusion reactions. Low ITLs were correlated with infusion reactions, and normal or high ITLs with dermatological events.

Key Words: *Adverse events; Infliximab; Infusion reaction; Trough level*

Les nadirs d'infliximab sont corrélés avec les événements indésirables liés à ce médicament chez les patients ambulatoires atteints d'une maladie inflammatoire de l'intestin sous traitement d'entretien à l'infliximab

HISTORIQUE : L'infliximab est un traitement efficace des maladies inflammatoires de l'intestin (MII), mais il s'associe à des événements dermatologiques et à des réactions aux perfusions. On ne sait pas s'il y a un lien entre ces événements indésirables (ÉI) et les nadirs d'infliximab.

OBJECTIFS : Rendre compte de la prévalence d'ÉI liés à l'infliximab chez les patients atteints d'une MII qui reçoivent un traitement d'entretien stable à l'infliximab et les corréler avec les réactions aux perfusions d'infliximab.

MÉTHODOLOGIE : Les chercheurs ont recruté des adultes atteints d'une MII sous traitement d'entretien stable de l'infliximab à la clinique de perfusion de l'université de l'Alberta, à Edmonton. Ils ont mesuré les nadirs dans des échantillons de sang prélevés avant les perfusions et examiné le dossier des patients pour déterminer s'ils avaient eu des réactions dermatologiques à l'infliximab et des réactions aux perfusions de ce médicament.

RÉSULTATS : Le quart des patients (18 sur 71 [25,4 %]) a présenté des réactions dermatologiques à l'infliximab ou aux perfusions de ce médicament, soit neuf événements dermatologiques (12,7 %) et neuf réactions aux perfusions (12,7 %). Le nadir médian était similaire chez les patients ayant subi ou non ces ÉI (7,2 µg/mL [plage interquartile (PIQ) 2,0 µg/mL à 13,3 µg/mL] par rapport à 6,6 µg/mL [PIQ 3,2 µg/mL à 12,7 µg/mL]; P=0,648). Le nadir médian des patients qui avaient des réactions aux perfusions (2,0 µg/mL [PIQ 0,1 µg/mL à 5,7 µg/mL]) était plus bas que celui des patients qui n'en avaient pas subi (6,6 µg/mL [PIQ 3,2 µg/mL à 12,7 µg/mL]; P=0,008]) et que celui des patients qui avaient subi des ÉI de nature dermatologique (13,3 µg/mL [PIQ 8,8 µg/mL à 17,4 µg/mL]; P<0,001).

CONCLUSION : Le quart des patients ambulatoires atteints d'une MII sous traitement d'entretien stable à l'infliximab a subi des réactions dermatologiques et aux perfusions. Des nadirs bas étaient liés aux réactions aux perfusions, et des nadirs normaux ou élevés, aux réactions dermatologiques

Anti-tumour necrosis factor (TNF) agents have transformed the management of inflammatory bowel diseases (IBDs) and other autoimmune diseases. Despite its efficacy as a therapeutic agent, there have been increasing reports of infliximab-associated adverse events, some of which appear to be paradoxical immune-mediated inflammatory disorders (1,2). Theses adverse events include dermatological conditions, such as psoriasis, eczema, erythema multiforme, lupus-like syndrome and nonmelanoma skin cancer (3-7), and infusion reactions, both acute and delayed (8,9). The management of these adverse events include conservative therapy, use of corticosteroids and adjustment of infliximab therapy. These are all attempts to overcome hypothesized mechanisms of action, and are often unsuccessful. In severe cases of adverse reactions to infliximab, patients often must discontinue the medication (9-14).

Pathophysiological mechanisms that have been postulated to contribute to infliximab-induced adverse events differ depending on the adverse event; however, these reactions are believed to be mediated through the immune system. Pathogenesis may include the development of antibodies to infliximab (ATI), either immunoglobulin (Ig) E resulting in a hypersensitivity type reaction (15,16) or IgG resulting in the formation of antibody/antigen complexes (1,8,17-20), or induction of autoantibodies such as antinuclear, antidouble-stranded DNA and antihistone (1,10,13,19). Dysregulation of the immune system and cytokine responses may also be involved (2,4-6,14,21-24). Given the importance of the pharmacodynamics of infliximab, the level of infliximab in the blood may be central to the generation of infliximab-induced side effects. However, to date, there have been no published

[1]*Division of Gastroenterology, Department of Medicine;* [2]*Division of Medical Biochemistry, Department of Laboratory Medicine and Pathology, University of Alberta, Edmonton, Alberta*

Correspondence: Dr Vivian Wai-Mei Huang, 2-14A Zeidler Building, University of Alberta, Edmonton, Alberta T6G 2X8. e-mail vwhuang@ualberta.ca

studies investigating the correlation between infliximab trough levels (ITLs) and infliximab-associated adverse events among IBD patients receiving stable maintenance infliximab therapy.

The aims of the present study were to report the prevalence of infliximab-associated dermatological and infusion reaction events among IBD outpatients receiving stable maintenance infliximab therapy, and to correlate ITLs with infliximab-associated adverse events.

METHODS

Study design and setting

The present cross-sectional study was conducted at the University of Alberta IBD Infliximab Infusion Clinic (Edmonton, Alberta) in 2013.

Participants

Consecutive IBD outpatients receiving infliximab maintenance therapy were systematically and prospectively identified at the infliximab infusion clinic and invited to participate in the present study. Patients were included in the study if they met the following criteria: a known endoscopic and histological and/or radiological diagnosis of Crohn disease (CD) or ulcerative colitis (UC); primary responders to the initial induction regimen of infliximab 5 mg/kg at weeks 0, 2 and 6 (defined as a decrease in their Harvey-Bradshaw index after induction therapy by 3 points, or a decrease in their partial Mayo score to 0 or 1); receiving stable maintenance infliximab infusions (defined as infusions subsequent to the third induction dose or subsequent to the first infusion after any dose or interval change); and having complete records of infliximab infusions since initial induction. There were no specific exclusion criteria. All consented patients had blood for the determination of ITLs drawn immediately before their infliximab infusion, and underwent a detailed history and chart review to confirm infliximab-associated adverse events.

Outcomes

The primary outcome of the present study was the prevalence of infliximab-associated dermatological and infusion reaction adverse events among IBD outpatients receiving stable maintenance infliximab therapy. The secondary outcome was the correlation between ITLs and infliximab-associated adverse events.

Data sources and definitions

ITL: ITLs were determined on serum samples collected within 30 min before the beginning of the infusion. Serum was separated and frozen within 4 h of collection. An ELISA method (Immunodiagnostik, Germany) was used to quantify levels of free infliximab. Results are reported down to 0.4 μg/mL, with an interassay precision of 8% at 1.8 μg/mL, 9% at 8.6 μg/mL and 20% at 12.9 μg/mL.

Demographics and infliximab-associated adverse events: At the authors' centre, each infliximab infusion is characterized with the following: clinical disease activity scores (Harvey-Bradshaw index for CD, partial Mayo for UC); health care provider documentation of infliximab-associated adverse events; and patient completion of a questionnaire of infliximab-associated adverse events.

Data were extracted from several sources by two of the authors (VH and ND) using a standardized case report form: electronic medical records; infusion questionnaires completed by the patient at the time of each infusion; nurse-generated infusion reports produced at the time of each infusion; and physician-generated clinic letters at the time of each infusion or subsequent clinic visits.

Infliximab-associated adverse events were classified as dermatological or infusion reactions. Dermatological adverse events were documented by a gastroenterologist and, when possible, confirmed by a dermatologist. Infusion reactions were documented by the infusion nurse and confirmed by a gastroenterologist. Acute infusion reactions were defined as any adverse reaction occurring during or within 24 h of an infusion, while delayed infusion reactions were defined as any adverse reaction occurring between 24 h and 14 days after an infusion. The data were reviewed by four of the authors (VH, ND, KK and RF).

Demographic patient data included sex, age, type of disease, premedication with solumedrol or diphenhydramine, and concomitant immunosuppression (azathioprine, 6-mercaptopurine or methotrexate). Infliximab treatment characteristics were obtained from the above listed sources and included: infliximab dose in mg/kg; infliximab dosing interval in weeks; cumulative infliximab dose in mg; and number of infliximab infusions.

Time to infliximab-associated adverse event was calculated by determining the number of weeks from the first stable maintenance infliximab infusion until the first report of the infliximab-associated adverse event. Censoring occurred if the patient reported no infliximab-associated adverse events; time to censoring was calculated by determining the number of weeks from the first stable maintenance infliximab infusion until the ITL measurement.

Study size

A total of 100 consecutive IBD patients receiving stable maintenance infliximab therapy were recruited and consented to participate. However, the final evaluable sample size (n=71) was determined by subsequent chart review, which excluded 29 patients due to incomplete infusion records.

Statistical methods

Continuous variables were presented as median (interquartile range [IQR]) due to the small sample size and nonparametric distributions. Nonparametric Mann-Whitney and Kruskall-Wallis tests were used to determine significant differences between median values. For categorical variables, proportions were calculated and comparison between subgroups was performed using Fisher's exact test; $P<0.05$ was considered to be statistically significant.

Ethics

The present study was approved by the Health Research Ethics Board of the University of Alberta. Patients consented to provide a blood sample and allow review of their medical records.

RESULTS

Patient characteristics

Table 1 summarizes the demographic and infliximab treatment characteristics of the 71 evaluable patients. The number of men (39 of 71 [45.1%]) and women (32 of 71 [54.9%]) were similar, and there were more patients with CD (46 of 71 [64.8%]) than patients with UC (25 of 71 [35.2%]). The median duration of time on stable maintenance infliximab therapy was 71.9 weeks (IQR 30.0 to 126.7 weeks), with a median time since last infusion of 7.0 weeks (IQR 4.1 to 8.0 weeks).

Infliximab-associated adverse events

More than one-quarter (18 of 71 [25.4%]) of study participants experienced either a confirmed infliximab-associated dermatological event (nine of 71 [12.7%]) or an infusion reaction (nine of 71 [12.7%]) while receiving maintenance infliximab therapy (Figure 1). Of the nine patients who experienced dermatological events, two (22.0%) had psoriasis and seven (77.8%) had persistent nonpsoriatic skin eruptions that had developed while on infliximab and were attributed to infliximab. Of the nine patients with infusion reactions, five (55.6%) experienced acute infusion reactions and four (44.4%) experienced delayed infusion reactions.

Comparing patients who experienced adverse events with those who did not, the median age was similar (Table 2). There were more women (12 of 18 [66.7%]) than men (six of 18 [33.3%]) in the group with adverse events ($P=0.046$). The proportions of CD and UC were similar between groups, with 62.3% (33 of 53) of patients without adverse events and 72.2% (13 of 18) of patients with adverse events having a diagnosis of CD, and 37.7% (20 of 53) of patients without adverse events and 27.8% (five of 18) of patients with adverse events having a diagnosis of UC ($P=0.445$).

There were slightly more patients on concomitant therapy among those without adverse events (40 of 53 [75.5%]) than among those

TABLE 1
Demographics of inflammatory bowel disease outpatients receiving stable maintenance infliximab therapy at the Infliximab Infusion Clinic, University of Alberta, Edmonton, Alberta (n=71)

Characteristic	
Age, years, median (IQR)	38.0 (26.0–47.0)
Sex	
Female	39 (54.9)
Male	32 (45.1)
Disease	
Crohn disease	46 (64.8)
Ulcerative colitis	25 (35.2)
Premedication	
No	28 (39.4)
Yes	43 (60.6)
Concomitant medication	
No	21 (29.6)
Yes	50 (70.4)
Infliximab history	
Duration on stable dose, weeks, median (IQR)	71.9 (30.0–126.7)
Dosing interval	
<6 weeks	29 (40.8)
≥6 weeks	42 (59.2)
Dosing interval, weeks, median (IQR)	7.0 (4.1–8.0)
Dose per infusion, mg/kg, median (IQR)	5.3 (4.8–5.7)

Data presented as n (%) unless otherwise indicated. IQR Interquartile range

with adverse events (10 of 18 [55.6%]) (P=0.110). There were slightly more patients receiving premedication among those who experienced adverse events (12 of 18 [66.7%]) than among those without adverse events (31 of 53 [58.5%]) (P=0.540).

ITLs and infliximab-associated adverse events

The median infliximab dose at the time of trough collection was 5.3 µg/mL (IQR 4.8 to 5.7 µg/mL), with a median time since last infusion of 7.0 weeks (IQR 4.1 to 8.0 weeks). The median time to infliximab-associated adverse event was 38.2 weeks (IQR 18.0 to 94.7 weeks) and the median time to censorship was 73.1 weeks (IQR 34.0 to 143.1 weeks). As shown in Table 2, the proportion of patients who experienced dermatological or infusion reactions was similar among those receiving infliximab <6 weeks interval (31.0%) and those receiving infliximab at ≥6 weeks interval (21.4%) (P=0.360).

As shown in Figure 2A, patients who experienced any infliximab-associated adverse event during stable maintenance infliximab therapy had a median ITL of 7.2 µg/mL (IQR 2.0 to 13.3 µg/mL), which was similar to the median ITL of patients who experienced no adverse events of 6.6 µg/mL (IQR 3.2 to 12.7 µg/mL) (P=0.648).

Dermatological adverse events: As shown in Figure 2B, patients who experienced infliximab-associated dermatological adverse events had a median ITL of 13.3 µg/mL (IQR 8.8 to 17.4 µg/mL) while the patients who experienced no adverse events had a median ITL of 6.6 µg/mL (IQR 3.2 to 12.7 µg/mL) (P=0.058). The ITL of the two infliximab-induced psoriasis cases were 9.9 µg/mL and 13.3 µg/mL.

Infusion reaction adverse events: In contrast, Figure 2C shows that patients who experienced infliximab-associated infusion reactions had a significantly lower median ITL of 2.0 µg/mL (IQR 0.1 to 5.7 µg/mL) compared with the median ITL in those who experienced no adverse events of 6.6 µg/mL (IQR 3.2 to 12.7 µg/mL) (P=0.008). Separating the infusion reactions into acute and delayed, the median ITL of those who experienced acute infusion reactions was lower (0.60 µg/mL [IQR 0.20 to 3.1 µg/mL]) compared with those who experienced delayed infusion reactions (3.5 µg/mL [IQR 1.0 to 6.9 µg/mL]), but the difference did not reach statistical significance (P=0.611). However, the

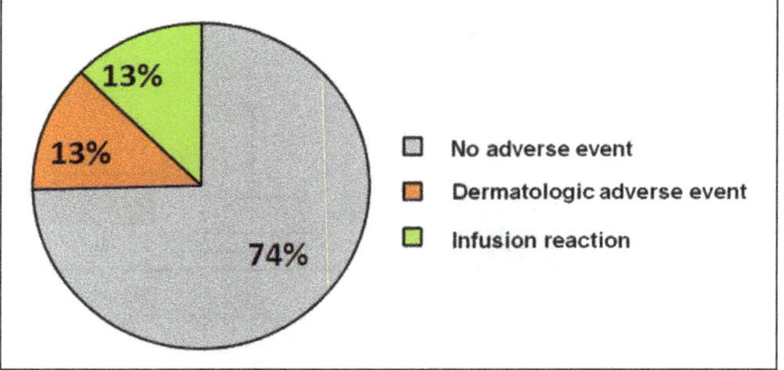

Figure 1) *More than one-quarter of inflammatory bowel disease patients receiving stable maintenance infliximab therapy developed infliximab-associated adverse events*

TABLE 2
Demographics of inflammatory bowel disease outpatients receiving stable maintenance infliximab therapy according to development of dermatological or infusion reaction to infliximab at the Infliximab Infusion Clinic, University of Alberta, Edmonton, Alberta (n=71)

	Adverse event*		
Characteristic	No	Yes	P
Total number	53 (74.6)	18 (25.4)	–
Age, years, median (IQR)	36.0 (24.0–47.0)	37.0 (29.0–42.0)	0.832
Sex			
Female	27 (50.9)	12 (66.7)	0.247
Male	26 (49.1)	6 (33.3)	
Disease			
Crohn disease	33 (62.3)	13 (72.2)	0.445
Ulcerative colitis	20 (37.7)	5 (27.8)	
Premedication			
No	22 (41.5)	6 (33.3)	0.540
Yes	31 (58.5)	12 (66.7)	
Concomitant medication			
No	13 (24.5)	8 (44.4)	0.110
Yes	40 (75.5)	10 (55.6)	
Infliximab history			
Duration on stable dose (weeks)	73.1 (34.0–143.1)	38.2 (18.0–94.7)	0.455
Dosing interval			
<6 weeks (n=29)	20 (69.0)	9 (31.0)	0.360
>6 weeks (n=42)	33 (78.6)	9 (21.4)	
Dosing interval, weeks, median (IQR)	7.0 (4.1–8.0)	5.6 (4.0–7.6)	0.839
Dose per infusion, mg/kg, median (IQR)	5.3 (4.8–5.7)	5.1 (4.9–5.5)	0.195

*Data presented as n (%) unless otherwise indicated. *Dermatological or infusion reaction to infliximab. IQR Interquartile range*

ITL in those who experienced acute infusion reactions (0.60 µg/mL [IQR 0.20 to 3.1 µg/mL]) was significantly lower than the median ITL of those who experienced no adverse reactions (6.6 µg/mL [IQR 3.2 to 12.7 µg/mL]) (P=0.0167). The median ITL in those who had delayed infusion reactions (3.5 µg/mL [IQR 1.0 to 6.9 µg/mL]) was slightly lower than the median ITL of those who had no adverse reactions (6.6 µg/mL [IQR 3.2 to 12.7 µg/mL]), but the difference did not reach statistical significance (P=0.163).

Finally, as shown in Figure 2D, patients who experienced any infusion reactions had a significantly lower ITL, with a median of 2.0 µg/mL (IQR 0.1 to 5.7 µg/mL), compared with patients who experienced dermatological events, with a median of 13.3 µg/mL (IQR 8.8 to 17.4 µg/mL) (P<0.001).

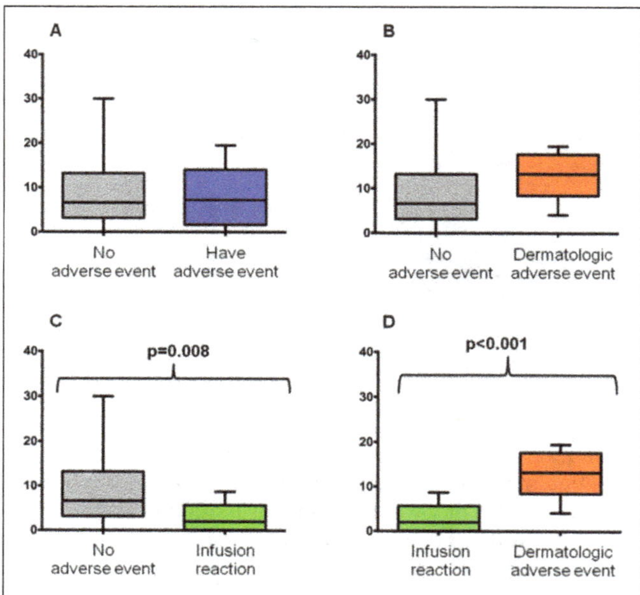

Figure 2) *Infliximab trough level was correlated with type of infliximab-associated adverse event. No adverse event (n=53), experience adverse event (n=18) (***A***); dermatological adverse event (n=9) (***B***); infusion reaction (n=9) (***C***); infusion reactions versus dermatological adverse events (***D***). Median, 25th percentile to 75th percentile, minimum and maximum values for infliximab trough levels (µg/mL) are plotted*

DISCUSSION

Infliximab is one of the cornerstone therapies for refractory CD and UC. In the present study, we demonstrated that more than one-quarter of IBD outpatients receiving stable maintenance infliximab therapy experience infliximab-associated dermatological or infusion reaction adverse events. In addition, we have shown a correlation between ITLs and the type of infliximab-associated adverse event: patients with dermatological adverse events had significantly higher ITLs, while patients with infusion reactions had significantly lower ITLs compared with patients who did not experience adverse events.

Dermatological adverse events

Paradoxical skin eruptions of eczema and psoriasis are common in patients receiving anti-TNF therapy, with up to 39% of reported cases of anti-TNF-induced psoriasis occurring in IBD patients (5,25-27). Among studies specifically investigating IBD patients treated with anti-TNF therapies, the main focus has been on the induction or exacerbation of psoriasiform lesions, with reported incidence rates between 5% and 22% (1,4). The literature varies due to different inclusion criteria and definition of the skin lesions of interest.

In our study, we included both inflammatory skin lesions and any de novo skin eruptions that occurred during stable maintenance infliximab therapy, and estimated a prevalence of 12.7% for dermatological adverse events and 2.8% for anti-TNF-induced psoriasis. In a retrospective study conducted by *Groupe d'Etude Thérapeutique des Affections Inflammatoires du tube Digestif* (GETAID) centres in France, Rahier et al (12) estimated an incidence of 5% for inflammatory skin lesions, with 2% for psoriasiform lesions and 3% for eczematiform lesions. In a systematic analysis conducted in Madrid (Spain), Guerra et al (28) diagnosed 21 of 1294 IBD patients with anti-TNF-α-induced psoriasis, 14 (67%) of whom were being treated with infliximab; they reported a cumulative incidence of 1.62% (95% CI 1.06% to 2.47%). Of the 21 patients, 15 were female and 17 had CD. Eighteen patients developed new-onset psoriasis and, in virtually every case, the onset of psoriasis occurred during maintenance anti-TNF therapy. A systematic literature review of 69 published cases of IBD patients with infliximab-induced psoriasis, performed by Denadai et al (23), and a case series of 30 IBD patients with anti-TNF-related psoriasiform lesions conducted by Cullen et al (29) yielded similar findings with female sex and CD predominance. Similar to these studies, our study also showed a female sex and CD predominance among patients who experienced dermatological adverse events. These inflammatory skin lesions can be treated topically; however, up to 43% of patients require withdrawal of anti-TNF therapies due to uncontrolled skin lesions (12,24,29).

Although the underlying mechanism of infliximab-induced psoriasis is unknown, there is general consensus in the literature regarding a possible role for alterations in cytokine levels such as interferon (IFN)-α (1,2,24). It is proposed that anti-TNF-α agents, such as infliximab, decrease TNF-α inhibition of IFN-α production (23). Tillack et al (24) histologically examined psoriasiform skin lesions in anti-TNF treated IBD patients, and characterized them as having infiltrates with INF-α-expressing cells. Consequently, the abnormally high levels of IFN-α resulting from decreased TNF-α inhibition can initiate psoriatic lesions. Although we did not measure cytokine levels in our study, the ITLs of patients who expereinced dermatological adverse events were higher than the ITLs of patients who did not.

Regarding dermatological adverse events, a limitation of the present study was the inclusion of skin eruptions documented by patient self-completed questionnaires; however, these data were only included if reported at least twice by patients, and confirmed by the gastroenterologist and, when possible, by a dermatologist, to have developed during infliximab therapy and attributed to infliximab therapy.

Based on our study findings and the pathophysiological mechanisms proposed in the literature, a management strategy for patients presenting with difficult-to-control infliximab-associated dermatological adverse events is to first measure ITLs. If the patient is clinically in remission and, if the ITLs are high, the infliximab dose should be lowered accordingly. If this dose adjustment and conservative measures are ineffective, then switching to another therapy may be required.

Infusion reactions

Infusion reactions to infliximab have been reported to occur in 5% to 10% of all infusions among IBD patients (30,31). When Steenholdt et al (32) examined 25 (8%) of 315 IBD patients who experienced acute severe infusion reaction to infliximab, they found that IgG ATI levels were highly positive in 19 of 20 patients (95%) after the reactions, but that IgE ATI was negative in all patients with reactions (32). The importance of ATI is emphasized in the systematic review and meta-analysis by O'Meara et al (33), which estimated risk ratios for any acute infusion reaction (RR 2.4 [95% CI 1.5 to 3.8]; P<0.001) and severe infusion reactions (RR 5.8 [95% CI 1.7 to 19]; P=0.004) that are higher in ATI-positive compared with ATI-negative patients.

Delayed infusion reactions are believed to be type III immune complex-mediated reactions with formation of antigen-antibody complexes that deposit in the tissues (8). Interestingly, ATI may not be as important in the mechanism of action for delayed infusion reactions because O'Meara et al (33) showed that the RR of delayed hypersensitivity reactions is not significantly different between ATI-positive and ATI-negative patients (RR 2.8 [95% CI 0.2 to 33]; P=0.4) (33).

In our study, patients who experienced delayed infusion reactions had slightly higher ITLs than patients who experienced acute infusion reactions, and slightly lower ITLs than patients who had no adverse events. Interestingly, patients who experienced acute infusion reactions had statistically significantly lower ITLs than patients who had no adverse events.

Regarding infusion reactions, a limitation of our study was the lack of antinuclear, anti-double-stranded DNA, antihistone, IgE levels, and of ATI levels in patients who developed infusion reactions. These measurements may have been beneficial to identify patients with drug-induced lupus, and also to confirm whether reactions occurred due to IgE-mediated versus ATI mechanisms.

A limitation of the present study was the small number of dermatological and infusion reactions to infliximab. However, the prevalence of these adverse events is similar to that reported in previous studies, as mentioned above.

Another limitation was that at the time of the present study, we only had access to ELISA-based ITLs, but not to ATI levels. However, given the finding that the patients who experienced infusion reactions had lower ITLs, it would be important to determine whether they have ATIs. Both the ITL and the ATI level would be important to know in patients who develop adverse events such as infusion reactions to infliximab. Therefore, for future research studies, and also for clinical practice, it would be important to measure ATIs in patients who experience infusion reactions or who have low ITLs.

A strength of the present study was that it prospectively analyzed ITLs in all patients receiving stable maintenance infliximab therapy who consented to participate during the study period, regardless of whether they experienced dermatological or infusion reactions. This is in comparison with study designs that analyze ITLs in patients who have levels drawn for specific reasons (eg, loss of response or adverse events) because the latter study design would introduce selection bias into the analysis.

Based on the results of the present study, we propose a strategy to manage infusion reactions by measuring ITLs in patients who develop infusion reactions; if the ITL is low, ATIs should be measured with consideration to switch drugs if the patient has a high level of ATI.

CONCLUSION

Our study has shown that ITLs are correlated with infliximab-associated adverse events. Measurement of the ITL in patients who develop dermatological and infusion reactions to infliximab, with subsequent adjustment of infliximab dose may be a strategy to manage these infliximab-associated adverse events.

ACKNOWLEDGEMENTS: VH was supported by the Alberta Innovates Health Solutions (AIHS) Clinician Researcher Fellowship Award. DF was supported by the Cecile Mactaggart Summer Student Research Fund. This study was supported by the Alberta IBD Consortium and by the Centre of Excellence for Gastrointestinal Inflammation and Immunity Research (CEGIIR). RNF is acting as the article guarantor and contributed to study design, data analysis and manuscript editing. VH contributed to study design, data collection, data analysis and manuscript drafting and editing. ND contributed to study design, data collection and manuscript editing. DF, CP and CS contributed to data collection and manuscript editing. KIK, LAD and BPH contributed to data analysis and manuscript editing. All authors have approved the final version of the manuscript.

DISCLOSURES: RNF, KIK and LAD have served as speakers and consultants for AbbVie Canada Inc and Janssen Canada Inc. The remaining authors have no financial disclosures or conflicts of interest to declare. This work was presented at Digestive Diseases Week 2014 (May 3, 2014, AGA Forum – Adverse Events Related to IBD Therapy).

FUNDING SUPPORT: Alberta Innovates Health Solutions Clinician Researcher Fellowship. Centre of Excellence for Gastrointestinal Inflammation and Immunity Research, Alberta IBD Consortium.

REFERENCES

1. Fiorino G, Danese S, Pariente B, Allez M. Paradoxical immune-mediated inflamamtion in inflammatory bowel disease patients receiving anti-TNF-a agents. Autoimmun Rev 2014;13:15-9.
2. Niess JH, Danese S. Anti-TNF and skin inflammation in IBD: A new paradox in gastroenterology? Gut 2014;63:533-5.
3. Moran GW, Lim AW, Bailey JL et al. Dermatological complications of immunosuppressive and anti-TNF therapy in inflammatory bowel disease. Aliment Pharmacol Ther 2013;38:1002-24.
4. Mocci G, Marzo M, Papa A, Armuzzi A, Guidi L. Dermatological adverse reactions during anti-TNF treatments: Focus on inflammatory bowel disease. J Crohns Colitis 2013:1-11.
5. Collamer AN, Guerrero KT, Henning JS, Battafarano DF. Psoriatic skin lesions induced by tumor necrosis factor antagonist therapy: A literature review and potential mechanisms of action. Arthritis Rheum 2008;59:996-1001.
6. Collamer AN, Battafarano DF. Psoriatic skin lesions induced by tumor necrosis factor antagonist therapy: Clinical features and possible immunopathogenesis. Semin Arthritis Rheum 2010;40:233-40.
7. Hernández MV, Sanmartí R, Cañete JD, et al. Cutaneous adverse events during treatment of chronic inflammatory rheumatic conditions with tumour necrosis factor antagonists. Study based on the Spanish Registry of Adverse Events of Biological Therapies in Rheumatic Diseases. Arthritis Care Res 2013;65:2024-31.
8. Mayer L, Young Y. Infusion reactions and their management. Gastroenterol Clin North Am 2006;35:857-66.
9. Moss AC, Fernandez-Becker N, Jo Kim K, Cury D, Cheifetz AS. The impact of infliximab infusion reactions on long-term outcomes in patients with Crohn's disease. Aliment Pharmacol Ther 2008;28:221-7.
10. Yanai H, Shuster D, Calabrese E, Mlynarsky L, Tumuluri S, Cohen RD. The incidence and predictors of lupus-like reaction in patients with IBD treated with anti-TNF therapies. Inflamm Bowel Dis 2013:19:2778-86.
11. Hawryluk EB, Linskey KR, Duncan LM, Nazarian RM. Broad range of adverse cutaneous eruptions in patients on TNF-alpha antagonists. J Cutan Pathol 2012;39:481-92.
12. Rahier JF, Buche S, Biroulet LP, et al. Severe skin lesions cause patients with inflammatory bowel disease to discontinue anti-tumor necrosis factor therapy. Clin Gastroenterol Hepatol 2010;8:1048-55.
13. Verma HD, Scherl EJ, Jacob VE, Bosworth BP. Anti-nuclear antibody positivity and the use of certolizumab in inflammatory bowel disease patients who have had arthralgias or lupus-like reactions from infliximab or adalimumab. J Dig Dis 2011;12:379-83.
14. Denadai R, Teixeira FV, Saad-Hossne R. The onset of psoriasis during the treatment of inflammatory bowel diseases with infliximab: Should biological therapy be suspended? Arq Gastroenterol 2012;49:172-6.
15. Brennan PJ, Rodriguez Bouza T, Hsu FI, Sloane DE, Castells MC. Hypersensitivity reactions to mAbs: 105 desensitizations in 23 patients, from evaluation to treatment. J Allergy Clin Immunol 2009;124:1259-66.
16. Hong DI, Bankova L, Cahill KN, Kyin T, Castells MC. Allergy to monoclonal antibodies: Cutting-edge desensitization methods for cutting-edge therapies. Expert Rev Clin Immunol 2012;8:43-54.
17. Casteele NV, Gils A, Singh S, et al. Antibody response to infliximab and its impact on pharmacokinetics can be transient. Am J Gastroenterol 2013;108:962-71.
18. Matucci A, Pratesi S, Petroni G, et al. Allergological in vitro and in vivo evaluation of patients with hypersensitivity reactions to infliximab. Clin Exp Allergy 2013;43:659-64.
19. Vultaggio A, Maggi E, Matucci A. Immediate adverse reactions to biologicals: From pathogenic mechanisms to prophylactic management. Curr Opin Allergy Clin Immunol 2011;11:262-8.
20. Vultaggio A, Matucci A, Nencini F, Pratesi S, Maggi E. Skin testing and infliximab-specific antibodies detection as a combined strategy for preventing infusion reaction. Intern Emerg Med 2011;7:77-9.
21. Roux CH, Brocq O, Leccia N, et al. New-onset psoriatic palmoplantaris pustulosis following infliximab therapy: A class effect? J Rheumatol 2007;34:434-7.
22. de Gannes GC, Ghoreishi M, Pope J, et al. Psoriasis and pustular dermatitis triggered by TNF-{alpha} inhibitors in patients with rheumatologic conditions. Arch Dermatol 2007;143:223-31.
23. Denadai R, Teixeira FV, Steinwurz F, Romiti R, Saad-Hossne R. Induction or exacerbation of psoriatic lesions during anti-TNF-α therapy for inflammatory bowel disease: A systematic literature review based on 222 cases. J Crohns Colitis 2013;7:517-24.
24. Tillack C, Ehmann LM, Friedrich M, et al. Anti-TNF antibody-induced psoriasiform skin lesions in patients with inflammatory bowel disease are characterised by interferon-expressing Th1 cells and IL-17A/IL-22-expressing Th17 cells and respond to anti-IL-12/IL-23 antibody treatment. Gut 2014;63:567-77.
25. Kip KE, Swoger JM, Grandinetti LM, Barrie AM III, Greer JB, Regueiro MD. Tumor necrosis factor-{alpha} antagonist-associated psoriasis in inflammatory diseases. Inflamm Bowel Dis 2013;19:1164-72.
26. Shmidt E, Wetter DA, Ferguson SB, Pittelkow MR. Psoriasis and palmoplantar pustulosis associated with tumor necrosis factor. J Am Acad Dermatol 2012;67:e179-e185.

27. Ko JM, Gottlieb AB, Kerbleski JF. Induction and exacerbation of psoriasis with TNF-blockade therapy: A review and analysis of 127 cases. J Dermatol Treat 2009;20:100-8.
28. Guerra I, Algaba A, Pérez-Calle JL, et al. Induction of psoriasis with anti-TNF agents in patients with inflammatory bowel disease: A report of 21 cases. J Crohns Colitis 2012;6:518-23.
29. Cullen G, Kroshinsky D, Cheifetz AS, Korzenik JR. Psoriasis associated with anti-tumour necrosis factor therapy in inflammatory bowel disease: A new series and a review of 120 cases from the literature. Aliment Pharmacol Ther 2011;34:1318-27.
30. Hanauer SB, Feagan BG, Lichtenstein GR et al. Maintenance infliximab for Crohn's disease: The ACCENT I randomised trial. Lancet 2002;359:1541-9.
31. Cheifetz A, Smedley M, Martin S, et al. The incidence and management of infusion reactions to infliximab: A large center experience. Am J Gastroenterol 2003;98:1315-24.
32. Steenholdt C, Svenson M, Bendtzen K, Thomsen OO, Brynskov J, Ainsworth MA. Severe infusion reactions to infliximab: Aetiology, immunogenicity and risk factors in patients with inflammatory bowel disease. Aliment Pharmacol Ther 2011;34:51-8.
33. O'Meara S, Nanda KS, Moss AC. Antibodies to infliximab and risk of infusion reactions in patients with inflammatory bowel disease. Inflamm Bowel Dis 2014;20:1-6.

Polyethylene glycol versus sodium picosulfalte bowel preparation in the setting of a colorectal cancer screening program

Omar Kherad MD MPH[1*], Sophie Restellini MD[2*], Myriam Martel Bsc[3], Alan N Barkun MD MSc[3,4]

O Kherad, S Restellini, M Martel, A Barkun. Polyethylene glycol versus sodium picosulfate bowel preparation in the setting of a colorectal cancer screening program. Can J Gastroenterol Hepatol 2015;29(7):384-390.

BACKGROUND: Adequate bowel preparation for colonoscopy is an important predictor of colonoscopy quality.

OBJECTIVE: To determine the difference in terms of effectiveness between different existing colon cleansing products in the setting of a colorectal cancer screening program.

METHODS: The records of consecutive patients who underwent colonoscopy at the Montreal General Hospital (Montreal, Quebec) between April 2013 and April 2014 were retrospectively extracted from a dedicated electronic digestive endoscopic institutional database.

RESULTS: Overall, 2867 charts of patients undergoing colonoscopy were assessed, of which 1130 colonoscopies were performed in a screening setting; patients had adequate bowel preparation in 90%. Quality of preparation was documented in only 61%. Bowel preparation was worse in patients receiving sodium picosulfate (PICO) alone compared with polyethylene glycol, in a screening setting (OR 0.3 [95% CI 0.2 to 0.6]). Regardless of the preparation type, the odds of achieving adequate quality cleansing was 6.6 for patients receiving a split-dose regimen (OR 6.6 [95% CI 2.1 to 21.1]). In multivariable analyses, clinical variables associated with inadequate bowel preparation in combined population were use of PICO, a nonsplit regimen and inpatient status. The polyp detection rate was very high (45.6%) and was correlated with withdrawal time.

CONCLUSION: Preparation quality needs to be more consistently included in the colonoscopy report. Split-dose regimens increased the quality of colon cleansing across all types of preparations and should be the preferred method of administration. Polyethylene glycol alone provided better bowel cleansing efficacy than PICO in a screening setting but PICO remains an alternative in association with an adjuvant.

Key Words: *Bowel preparation; Colonoscopy; Quality; Screening setting*

Une préparation intestinale de polyéthylèneglycol comparée à celle de picosulfate de sodium dans un programme de dépistage du cancer colorectal

HISTORIQUE : Une bonne préparation intestinale en vue de la coloscopie est un important prédicteur de la qualité de l'intervention.

OBJECTIF : Déterminer la différence d'efficacité entre divers produits de nettoyage du côlon dans le cadre d'un programme de dépistage du cancer colorectal.

MÉTHODOLOGIE : Les chercheurs ont examiné les patients consécutifs qui ont subi une coloscopie à l'Hôpital général de Montréal, au Québec, entre avril 2013 et avril 2014. Ils les ont extraits d'une base de données rétrospective de l'établissement consacrée aux endoscopies digestives.

RÉSULTATS : Les chercheurs ont évalué le dossier de 2 867 patients qui avaient subi une coloscopie, dont 1 130 dans le cadre d'un dépistage. Au total, 90 % des patients avaient eu une bonne préparation coloscopique. La qualité de la préparation n'était consignée que chez 61 % des patients. La préparation intestinale était moins bonne chez les patients qui avaient reçu du picosulfate de sodium (PICO) seul au lieu du polyéthylèneglycol dans le cadre du dépistage (rapport de cotes [RC] 0,3 [95 % IC 0,2 à 0,6]). Quel que soit le type de préparation, la possibilité d'obtenir un nettoyage de bonne qualité était 6,6 fois plus élevée chez les patients qui recevaient une dose divisée (RC 6,6 [95 % IC 2,1 à 21,1]). Dans les analyses multivariables, les variables cliniques associées à une préparation intestinale inadéquate dans une population combinée étaient l'utilisation de PICO, l'administration en une seule dose et l'hospitalisation. Le taux de dépistage de polypes, très élevé (45,6 %), était corrélé avec le délai d'attente.

CONCLUSION : Il faut consigner plus systématiquement la qualité des préparations dans le rapport de coloscopie. La division des doses accroît la qualité du nettoyage du côlon dans tous les types de préparation et devrait être favorisée. L'efficacité du nettoyage intestinal par le polyéthylèneglycol seul était préférable au PICO dans un cadre de dépistage, mais le PICO demeure une solution s'il s'associe à un adjuvant.

Colorectal cancer (CRC) is the third most common cancer in the world, with nearly 1.4 million new cases diagnosed in 2012, and one of the leading causes of death from cancer in Canada and worldwide (1-3). CRC screening has the potential for being the most cost-effective cancer screening in terms of both prevention and early detection (4).

There exist no convincing data establishing the superiority of any one colon cleansing product over another, with conflicting summary data in the literature as summarized in a recent meta-analysis (5). It appears that some of these products are generally well tolerated according, in part, to patient characteristics. Following black boxing, and in some jurisdictions withdrawal of sodium phosphate due to rare but significant renal toxicity (6), two of the main products that share the market in Canada include polyethylene glycol (PEG) and a preparation of sodium picosulfate plus magnesium oxide (PICO).

PEG preparations have long been proven safe and efficacious (7-9); however, the main issue with PEG is its relatively low patient adherence, mainly due to its taste and the large volume patients must consume (3 L to 4 L). It has been estimated that at least 5% to 15% of patients do not complete the preparation due to the large volume and/or poor palatability (10). As an alternative, low-volume

*Authors who contributed equally

[1]Internal Medicine Department, La Tour Hospital and University of Geneva; [2]Department of specilaties of Internal Medicine, Divison of Gastroenterology, Geneva University Hospital, Geneva, Switzerland; [3]Division of Gastroenterology; [4]Department of Epidemiology and Biostatistics and Occupational Health, McGill University Health Centre, McGill University, Montreal, Quebec

Correspondence: Dr Alan N Barkun, Division of Gastroenterology, The McGill University Health Centre, Montreal General Hospital site, 1650 Cedar Avenue, Room D16.125, Montreal, Quebec H3G 1A4. e-mail alan.barkun@muhc.mcgill.ca

hyperosmolar preparations have recently emerged. PICO has been approved in Canada since 2004 and in the United States since 2012. The dual-action bowel cleansing preparation containing sodium picosulfate and magnesium citrate acts as both stimulant and osmotic laxative to clean the colon. Compared with sodium phosphate, PICO revealed comparable cleansing efficacy but with better tolerance (11), while its comparative efficacy with PEG remains unclear with conflicting results (8,12-14).

Regardless of the type of preparation, splitting the volume over two days appears to be more effective and improves tolerance (15). The superiority of split-dosing may relate to the shortened interval of time between the last intake of preparation product and the colonoscopy (16). Aligned with this theory, the European Society of Gastrointestinal Endoscopy and the US Multi-Society Task Force (US MSTF) on CRC recommend a split-dose regimen for morning colonoscopy and a same-day regimen for afternoon colonoscopy (17,18).

To the best of our knowledge, the comparative effectiveness of these different preparations in a specific population screening program setting is not known. Of note, the need for repeat procedures because of a poor preparation carries a significant cost burden, especially in the context of a population-based CRC screening program with limited resource availability and may affect the subsequent uptake of colonoscopy in healthy individuals (19). It was, therefore, necessary to identify the optimal bowel preparation in terms of effectiveness in a 'real life' setting, to make recommendations for the optimal future deployment of the Quebec Ministry of Health and Social Services population-based CRC provincial screening program (PQDCCR). The Montreal General Hospital, site of the McGill University Health Centre (MUHC-MGH, Montreal, Quebec) has been chosen as a pilot site to ensure the performance of colonoscopies follows established standards of care before full program deployment; therefore, we reviewed colon preparation performances of patients undergoing a screening colonoscopy on this site.

The primary objective of the present study was to assess bowel cleanliness after different existing colonoscopy preparations in a cancer screening program setting. We also aimed to identify independent predictors of inadequate bowel preparation and polyp detection rate (PDR) by assessing different indicators, in addition to colonic preparations, such as age, sex, American Society of Anesthesiologists (ASA) score, patient status (inpatient versus outpatient) and indication for colonoscopy.

METHODS

A cross-sectional study using an anonymized endoscopic electronic database, including consecutive patients referred for colonoscopy, was conducted. The MUHC-MGH is a tertiary care institution with a five-room endoscopy unit staffed by 12 medical and surgical endoscopists, performing >6500 colonoscopies yearly. All patients receive an information sheet on colon cancer screening, colonoscopy and specific bowel preparation instructions beforehand, while consent is obtained by the endoscopist on the day of the procedure. Endoscopists use a commercial computerized endoscopic report generator (Endoworks, Olympus Inc, USA) modified to fulfill PQDCCR-recommended quality indicator collection (20). For the purpose of the present study, the corresponding endoscopic patient database over a one-year period spanning April 2013 to April 2014 was accessed. The records of consecutive patients were retrospectively extracted and analyzed when the type of products for the bowel preparation was documented. Also tracked were several validated preprocedural, intraprocedural and postprocedural quality indicators. The study protocol was approved by an independent ethics committee.

Definition

Quality indicators extracted were separated as follows: preprocedural indicators (appropriate indication, colonic preparation, clinical characteristics [sex, age, ASA score, inpatients]); intraprocedural indicators (bowel preparation quality, rates of cecal intubation, PDR, colonoscopy withdrawal time); and postprocedure indicators (completeness of colonoscopy and complication rates including perforation and bleeding).

The primary outcome was preparation adequacy. At colonoscopy, gastroenterologists at the MGH assess colonic cleaning using the following modified validated scale recommended by the PQDCCR, which is a modification of the Aronchick scale (21): excellent – no fecal matter or nearly none in the colon, small to moderate amounts of liquid easily suctioned; good – no fecal matter or nearly none in the colon, large amounts of liquid easily suctioned; fair – some semisolid stool that could be suctioned or washed away with difficulty; and poor – semisolid stool that could not be suctioned or washed away, complete colonoscopy impossible

According to the established criteria, bowel preparations were classified as 'excellent' or 'good' for the procedure to be considered valid and meet the criteria of an adequate preparation (the definition was embedded in the reporting system, providing point-of-care referencing). A pre-planned dichotomization was performed by grouping excellent or good bowel preparation versus fair, poor (ie, excellent or good = adequate versus fair or poor = inadequate preparation), consistent with US MSTF operational recommendations (18).

'Screening indication' was defined as a colonoscopy performed for subjects without any symptoms who are at average or high risk for colonic cancer. Colonoscopy performed after immunochemical fecal occult blood test was also considered to be a 'screening indication'.

Different regimens of colon cleansing product groups were compared in terms of efficacy and reported as follows: PEG versus others; PEG versus PICO; split versus non split; PEG split versus PEG non split; all preparations + adjuvant versus all preparations; PEG + adjuvant versus PEG; and PICO + adjuvant versus PICO.

Adjuvants reported were magnesium (Mg) citrate or bisacodyl. Of note, PICO used at MGH already contains Mg citrate. Therefore, PICO + Mg citrate means addition of Mg citrate to a pre-existing PICO + Mg citrate solution.

Statistical analyses

Descriptive variables are presented as means ± SDs for continuous variables and proportions with 95% CIs for categorical variables. Inferential testing was performed using χ^2 tests, Fischer's tests or *t* tests, where appropriate. The effect of different colon cleansing product groups on preparation adequacy was first examined by performing an unadjusted analysis, using χ^2 testing. Results are reported with ORs and 95% CIs. The product group/preparation adequacy association was then stratified according to each potential available confounder in turn (age, sex, indication of colonoscopy, ASA score, inpatient), allowing the assessment of potential effect modifiers and confounding at this stage. If no significant heterogeneity was detected among strata, a pooled Mantel-Haenszel OR for preparation adequacy comparing cleansing product group was produced and compared with the crude OR to assess the extent of confounding in a table. Multiple logistic regressions of outcomes on combinations of potential risk factors were performed with variables selected through forward fitting. Likelihood ratio tests were performed to assess the goodness of fit of different models.

Separate multivariable regression analyses were performed to adjust for all potential confounders and to identify potential independents predictors for an inadequate bowel preparation and PDR. For both analyses, stratified analysis according to the indication was performed. The software used for all statistical computations was SPSS version 20.0 (IBM Corporation, USA).

Sample size

Assuming a satisfactory rate of 90%, the estimated number of 3000 participants was significantly higher than the 213 patients needed per colonoscopic preparation group to reach 80% power to detect a between-group difference in efficacy of 10% at a two-sided significance level of 0.05.

TABLE 1
Patient characteristics and procedure description

	Missing, n	Overall (n=2867)	Screening (n=1130)	Non-screening (n=1737)	P*
Patient characteristics					
Age, years, mean ± SD	17	60.4±13.5	59.5±9.9	61.0±15.4	<0.01
18–25		39 (1.4)	0 (0)	39 (2.3)	<0.01
25–50		498 (17.5)	156 (13.9)	342 (19.8)	
50–75		1927 (67.6)	900 (80.1)	1027 (59.5)	
>75		386 (13.5)	68 (6.0)	318 (18.4)	
Male sex	11	1432 (50.1)	579 (51.4)	853 (49.3)	0.3
Inpatients		108 (3.8)	0 (0)	108 (6.2)	<0.01
ASA score mean ± SD	468	1.42±0.52	1.34±0.47	1.47±0.54	<0.01
ASA 1		1414±58.9	602±65.8	812±54.7	<0.01
ASA 2		953±39.7	312±34.1	641±43.2	
ASA 3		31±1.3	1±0.1	30±2.0	
ASA 4		1±0.1	0±0	1±0.1	
Procedure description					
Type of bowel preparation	0				
PEG		1720 (60)	633 (56.0)	1087 (62.6)	<0.01
PICO		1124 (39.2)	496 (43.9)	628 (36.2)	
Enema		20 (0.7)	0 (0)	20 (1.2)	
Adjuvants alone		3 (0.1)	1 (0.1)	2 (0.1)	
Use of adjuvants		368 (12.8)	158 (14.0)	210 (12.1)	0.1
PEG + Mg citrate		121 (4.2)	63 (5.6)	58 (3.3)	0.1
PICO + Mg citrate		238 (8.3)	91 (8.1)	147 (8.5)	
PEG + bisacodyl		4 (0.1)	2 (0.2)	2 (0.1)	
PICO + bisacodyl		5 (0.2)	2 (0.2)	3 (0.2)	
Split regimen	579	119 (4.2)	55 (6.2)	64 (4.6)	0.1
Quality of preparation	1100				
Documented		1767(61.6)	748 (66.2)	1019 (58.6)	
Adequate		1484 (84.0)	672 (89.8))	812 (79.7)	<0.01
Withdrawal time, min, mean ± SD	1478	9.4±4.3	9.76±4.4	9.20±4.2	0.02
Polyp detection rate	0	1255 (43.8)	515 (45.6)	740 (42.6)	0.1
Cecal intubation	57	2699 (96.0)	1071 (96.1)	1628 (96.0)	0.8
Complication	1	13 (0.5)	4 (0.4)	9 (0.5)	0.5

*Data presented as n (%) unless otherwise indicated. *Comparisons made between the screening and non-screening setting: χ^2 test, Fischer test and* t *test analyses. ASA American Society of Anesthesiology; Mg Magnesium; PEG Polyethylene glycol; PICO Sodium picosulfate*

RESULTS

Patients' characteristics and procedure description

A total of 6304 colonoscopies were performed at the MUHC-MGH from April 1, 2013 to April 30, 2014. Colonoscopies for which the type of bowel preparation was not identified (n=3437) were excluded, leaving 2867 colonoscopies for 2730 patients (50.1% men, mean age 60.4 years, 96.2% outpatients). Of note, colonoscopy indications and patient characteristics (age and sex) did not differ in the excluded group.

Most (67.6%) patients were between 50 and 75 years of age, without significant acute illness as expressed by the mean ASA score <2 (Table 1). Colonoscopies were performed in a screening setting in 1130 patients (39.4%) versus 1737 (60.6%) for other indications. The indications for a colonoscopy performed outside the screening setting are described in Figure 1. In an analysis restricted to endoscopy for screening, patients characteristics differed significantly compared with the nonscreening group because patients were younger (mean [± SD] age 59.5±9.9 versus 61±15.4 years; P<0.01) and in better health (mean ASA score 1.34±0.47 versus 1.47±0.54; P<0.01) (Table 1).

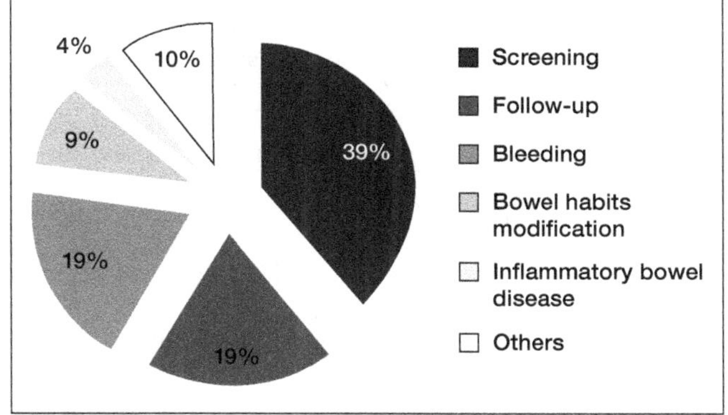

Figure 1) *Indications for colonoscopy*

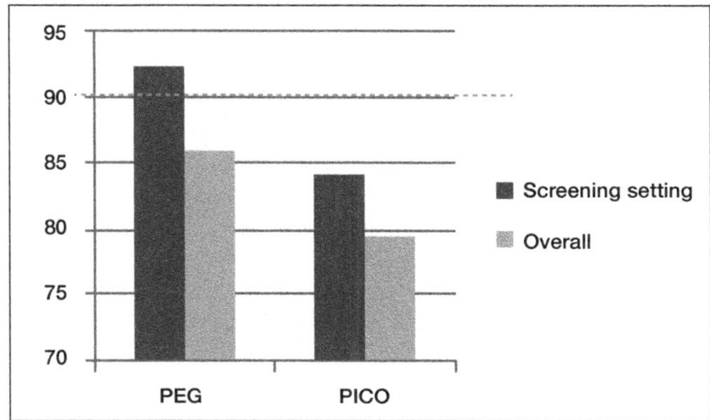

Figure 2) *Percentage of adequate preparation with the two most commonly used preparations in screening setting and overall. PEG Polyethylene glycol; PICO Sodium picosulfate*

The two most commonly used products both overall and in the screening setting were PEG- and PICO-based preparations (60% and 39.2%, respectively, overall, and 56.1% and 43.9%, respectively, for a screening indication). Adjuvant and split-dose regimens were used more rarely (12.8% and 4.2%, respectively). Colonoscopies were completed and the cecum intubated in 96%, with a complication rate of 0.5%.

Quality of preparation

Overall, quality of preparation was adequate in 84%, reaching 89.8% in the screening setting. Of note, the quality of preparation was documented in only 61.6%. The crude effectiveness of PEG and PICO overall and in the screening setting are depicted in Figure 2.

Head-to-head comparisons of different products and regimens are shown in Table 2 in crude analysis and after adjustment for the previously identified confounders in turn (ASA score, sex, age, indication), with no effect modifiers detected at this stage. Bowel preparation was worse in patients receiving PICO compared with PEG, even after adjustment for ASA score, sex, age and indication of colonoscopy (OR 0.3 [95% CI 0.2 to 0.4]) (Table 3). A similar finding was noted when restricting the analysis to screening indications (OR 0.3 [95% CI 0.2 to 0.6]). Overall, PEG in a split-dosing regimen yielded significantly greater proportions of patients with adequate preparations (OR 7.0 [95% CI 1.7 to 29.1]). Split-dose regimens, regardless of the product used and the indication of colonoscopy, achieved better quality preparation in the overall population (OR 6.6 [95% CI 2.1 to 21.1]), with a trend noted in colonoscopies for screening indications (OR 5.8 [95% CI 0.8 to 43.6]) (P=0.08) (Table 3). There were no clinically relevant differences in bowel preparation between regimens with use

TABLE 2
Head-to-head comparison of different bowel preparations in terms of quality of preparation, adjusted for potential confounders in turn

Bowel preparation	Overall OR* (95% CI)	Screening setting OR* (95% CI)
PEG versus others	1.2 (0.9–1.5)	1.8 (1.1–2.8)
Adjusted for ASA score	1.4 (1.1–1.9)	1.9 (1.1–3.4)
Adjusted for sex	1.2 (0.0–1.5)	1.8 (1.1–3.0)
Adjusted for age	1.2 (0.9–1.5)	1.8 (1.1–2.9)
Adjusted for indication	1.2 (0.9–1.6)	–
PICO versus PEG	0.6 (0.5–0.8)	0.5 (0.3–0.7)
Adjusted for ASA score	0.3 (0.2–0.4)	0.2 (0.2–0.4)
Adjusted for sex	0.6 (0.5–0.8)	0.5 (0.4–0.7)
Adjusted for age	0.6 (0.4–0.8)	0.5 (0.4–0.7)
Adjusted for indication	0.5 (0.4–0.7)	–
Split versus nonsplit	6.6 (2.0–21.0)	6.0 (0.8–44.4)
Adjusted for ASA score	6.7 (2.1–21.6)	5.8 (0.8–44.7
Adjusted for sex	6.4 (2.0–20.4)	5.7 (0.8–41.8)
Adjusted for age	6.6 (2.0–21.0)	6.1 (0.8–45.4)
Adjusted for indication	6.7 (2.1–21.5)	–
PEG split versus PEG	7.9 (1.9–32.6)	†
Adjusted for ASA score	7.0 (1.7–28.9)	
Adjusted for sex	7.8 (1.9–32.1)	
Adjusted for age	7.9 (1.9–32.4)	
Adjusted for indication	8.0 (1.9–33.6)	–
PICO + adjuvants versus PICO	2.1 (1.3–3.5)	2.4 (1.0–5.8)
Adjusted for ASA score	0.9 (0.3–2.8)	2.2 (0.2–20.6)
Adjusted for sex	2.1 (1.3–3.5)	2.4 (1.0–5.7)
Adjusted for age	2.1 (1.3–3.6)	2.5 (1.0–59)
Adjusted for indication	2.2 (1.3–3.8)	–

*Statistical analysis: χ^2 analysis. *Crude and Mantel Haenszel OR reported after testing homogeneity of strata; †No subject received the preparation in one group. ASA American Society of Anesthesiology; PEG Polyethylene glycol; PICO Sodium picosulfate; Split Split-dose regimen*

of adjuvants, except for PICO, which yielded significantly greater proportions of patients with adequate preparartions in both the combined population and screening settings (OR 2.1 [95% CI 1.3 to 3.5], and OR 2.4 [95% CI 1.0 to 5.8], respectively).

Independent predictors of inadequate bowel preparation in the combined population are described in Table 4. In addition to PICO product use and non-split-dose regimen, other predictors of inadequate preparation in multivariable analysis were inpatient and indication of colonoscopy (Table 4). Bowel preparation was lowest in patients referred for bleeding and bowel habits modification.

In the screening setting, the sole independent predictor of inadequate bowel preparation was the use of PICO (Table 5). In multivariable analysis, use of PICO remained independently and significantly predictive, yielding significantly greater proportions of patients with inadequate preparations compared with PEG (OR 7.4 [95% CI 4.3 to 12.6]).

PDR

PDR was high (43.8%), and even higher in the screening setting (45.6%) with age >75 years, male sex, ASA score and indication of colonoscopy as significant preprocedural indicator predictors in univariable analysis (Tables 6 and 7).

PDR was not associated with quality of preparation; however, PDR was associated with withdrawal time. The odds of detecting a polyp was three times higher when withdrawal time was >6 min in both the combined population and screening setting (OR 2.8 [95% CI 1.9 to 4.2, and OR 3.3 [95% CI 1.7 to 6.6], respectively) even in multivariable analysis after adjustment for quality of preparation, age, ASA score and indication for colonoscopy (Tables 6 and 7).

TABLE 3
Multivariable analysis of different bowel preparations in terms of quality of preparation in screening setting and overall

	Overall OR* (95% CI)	P	Screening OR† (95% CI)	P
PICO vs PEG	0.3 (0.2–0.4)	<0.01	0.3 (0.2–0.6)	<0.01
Split regimen vs non split	6.6 (2.1–21.1)	<0.01	5.8 (0.8–43.6)	0.08

**Adjusted for American Society of Anesthesiology (ASA), sex, age, indication; †Adjusted for ASA, sex, age. PEG Polyethylene glycol; PICO Sodium picosulfate; vs Versus*

There were no differences in complete colonoscopy rates (intubation of cecum) as a result of bowel cleanliness (95.4% versus 97.1%; P=0.2), in the adequate and inadequate cleanliness group, respectively.

Mean withdrawal time was, however, significantly shorter in the adequate versus inadequate cleanliness group (difference 1.2±0.5 min; absolute times: 8.4±4.1 min versus 9.6±6.3 min (P=0.03).

DISCUSSION

Our study reflects actual practice at a large academic institution participating in population screening for CRC. The major findings are that quality of preparation meets the standardized quality threshold of 90% in screening examinations. PEG is the most commonly used product and appears to be superior to PICO in terms of cleansing effectiveness in a screening setting, even after adjustment for potential confounding. Additionally, regardless of the type of product used, split dosing achieved more satisfactory cleansing results but it was not yet the preferred method of colon cleansing at MUHC-MGH, with a very low rate of utilization in 2013. Use of adjuvants only improved bowel cleansing when using PICO; however, there were low rates of low-volume PEG preparation use. PDR was high but was not associated with the quality of the preparation. PDR was highly correlated with withdrawal time. These results remain very pertinent, even in an era of split-dosing preference, and appropriate predominance because a large proportion of patients endoscoped before 10:30 refuse to get up early to take the second of a split-dose regimen at MGH-MUHC (personal communication, Dr AN Barkun, Montreal, Quebec).

For years, PEG products have been proven safe and efficacious (7) and may be the ideal current preparation, with low side effects. However, the large volume (4 L) and the salty taste impinge on compliance and effectiveness of cleansing, representing barriers to performing successful colonoscopy in healthy individuals, particularly in a screening setting (22). In an effort to improve patient compliance, a reduced-volume lavage regimen comprised of 2 L of PEG plus four tablets of bisacodyl has been introduced, with conflicting results on efficacy in comparison with high-volume PEG (5,23). A recent meta-analysis concluded that data remain lacking to confidently conclude as to the comparative role of low-volume PEG preparations (24).

As an alternative to PEG products, PICO became an interesting lower volume, better tasting alternative. High patient acceptability and preference have contributed to its wide adoption across the country over the past 10 years (12). However, inconsistent results exist as to its relative efficacy (13,25,26). Our study revealed that the likelihood of achieving adequate bowel cleansing was lowest with PICO compared with PEG, both in the overall colonoscopy population and in those procedures performed solely in the screening setting. This result is consistent with studies performed in the United Kingdom (14,27). More recent randomized controlled trials also compared PEG with PICO. When administered as a day-before preparation, bowel cleansing by PICO was noninferior compared with PEG, with more favourable tolerance (28-30). Of note, these results differ from our study because the PEG regimen utilized as comparator was systematically a low-volume regimen with an adjuvant administered the day before the colonoscopy. When compared with 4 L PEG

TABLE 4
Independent predictors of inadequate bowel preparation overall

	Multivariable analysis* OR (95% CI)	P
Bowel preparation		
PEG	1.00 (reference)	
PICO versus PEG	4.0 (2.7–6.1)	<0.01
Split regimen	0.1 (0.05–0.5)	<0.01
Age >75 years	0.9 (0.5–1.5)	0.6
Male sex	0.8 (0.6–1.1)	0.2
ASA score	–	
Inpatient	2.8 (1.4–5.6)	<0.01
Indication for colonoscopy		
Screening	1.00 (reference)	
Follow-up	1.4 (0.8–2.4)	0.2
Bleeding	2.3 (1.4–3.8)	<0.01
Bowel habits modification	2.2 (1.2–3.9)	<0.01
Inflammatory bowel disease	1.1 (0.4–3.0)	0.8
Others	3.0 (1.7–5.3)	<0.01

**Adjusted for all variables listed in the table. ASA American Society of Anesthesiology; PEG Polyethylene glycol; PICO Sodium picosulfate*

TABLE 5
Independent predictors of inadequate bowel preparation in a screening setting

	Multivariable analysis* OR (95% CI)	P
Bowel preparation		
Polyethylene glycol	1.00 (reference)	–
Sodium picosulfate	7.4 (4.3–12.6)	<0.01
Age >75 years	1.8 (0.7–4.2)	0.2
Male sex	1.3 (0.8–2.3)	0.3
ASA score	1.3 (0.7–2.1)	0.4

**Adjusted for all variables listed in the table. ASA American Society of Anesthesiology*

TABLE 6
Independent predictors of polyp detection rate overall

	Univariable analysis OR (95% CI)	P	Multivariable analysis* OR (95% CI)	P
Quality of preparation	1.3 (0.9–1.6)	0.2	1.2 (0.8–1.8)	0.5
Age >75 years	1.6 (1.3–2.0)	<0.01	1.7 (0.9–3.0)	0.06
Male sex	1.7 (1.5–2.0)	<0.01	1.2 (0.8–1.8)	0.3
ASA score	1.2 (1.1–1.4)	<0.01	1.5 (1.1–2.2)	0.02
Inpatient	1.0 (0.7–1.5)	0.9	0.5 (0.2–1.5)	0.3
Indication of colonoscopy				
Screening	1.00 (reference)	–	1.00 (reference)	–
Follow-up	1.8 (1.4–2.2)	<0.01	1.7 (1.0–2.7)	0.04
Bleeding	0.8 (0.6–0.9)	0.01	0.6 (0.4–1.0)	0.05
Bowel habits	0.6 (0.4–0.8)	<0.01	0.5 (0.3–0.9)	0.04
modification	0.2 (0.1–0.4)	<0.01	0.3 (0.1–0.8)	0.02
IBD	0.7 (0.5–0.9)	0.02	0.7 (0.4–1.3)	0.3
Others				
Withdrawal time (>6 min)	2.9 (2.2–3.9)	<0.01	2.8 (1.9–4.2)	<0.01

**Adjusted for all variables listed in the table. ASA American Society of Anesthesiology; IBD Inflammatory bowel disease*

TABLE 7
Independent predictors of polyp detection rate in screening setting

	Univariable analysis OR (95% CI)	P	Multivariable analysis* OR (95% CI)	P
Quality of preparation	1.4 (0.8–2.2)	0.2	1.3 (0.5–2.9)	0.5
Age >75 years	2.0 (1.2–3.3)	<0.01	1.4 (0.3–6.7)	0.6
Male sex	1.8 (1.4–2.3)	<0.01	1.1 (0.5–2.1)	0.8
ASA score	0.9 (0.6–1.1)	0.2	1.6 (0.8–3.1)	0.1
Withdrawal time (>6 min)	3.2 (1.9–5.4)	<0.01	3.3 (1.7–6.6)	<0.01

**Adjusted for all variables listed in the table. ASA American Society of Anesthesiology*

in a split-dose regimen, PICO appears to be less effective and may not offer clear advantages, at least in a non-split-dose regimen (31). Results of a recent meta-analysis show a trend toward benefit for PEG versus PICO without reaching statistical significance (5). Absence of information regarding the amount of PEG administered and patient's tolerance in our study precludes any firm conclusion as to a direct comparison between PEG low volume and PICO preparations. In our study, use of adjuvants only improved bowel cleansing when using PICO. A randomized control trial suggested that PICO associated with bisacodyl provides enhanced colon cleansing compared with PICO alone (11). PICO appears, therefore, acceptable especially if used with adjuvant such as bisacodyl.

European and American guidelines suggest splitting the dose to increase bowel cleanliness and tolerance (17,18). Interestingly, our observational data showed that the endoscopy department at MUHC-MGH had not yet adopted the widespread use of split-dose regimens. In fact, only 5% of bowel preparations were administrated in a split regimen between April 2013 and April 2014. However, even in our data, split-dose regimens (particularly with PEG) achieved significantly greater proportions of patients with adequate preparations, emphasizing the increased efficacy attributable to a decreased runway time and the greater cleansing ability of PEG-based preparations (16). As a continuous total quality management process, this information has been transmitted to all endoscopists at MUHC-MGH to alter practice accordingly, and adherence to the unit recommendations of split dosing, at least for patients colonoscoped after 10:30, is being enforced more strictly. The diffusion of the European and American guidelines on bowel cleansing should further enhance this practice (17,18).

The PDR was high (43.8%), and even higher in the screening setting (45.6%), meeting international targets because PDR is a close and reliable proxy of adenoma detection rate (32,33). Intriguingly, the quality of preparation was not associated with the PDR in contrast to results from a previous multicentre European study (34). Reasons for this are unclear but may reflect, in part, varying operator expertise and differing patient characteristics. Of note, cleansing quality in the European study was based on three categories and authors used low-quality cleansing as the reference category. Because endoscopists at the MUHC-MGH do not always adopt the recommended PQDCCR reporting terminology to express the quality of preparation, we dichotomized cleansing quality by grouping excellent or good bowel into adequate preparation versus fair and poor classifications as inadequate – paralleling the actual repeat colonscopy interval recommendation as put forth by the US MSTF (18). Therefore, we cannot determine the impact of poor preparation versus others, which may also explain, in part, the discrepancy in our results. Additionally, the detection of histopathologically confirmed cancer was not significantly compromised by poorly prepared examinations in the European work (34). In another United States study, inadequate preparation quality hindered the detection of small polyps only, while there was no association between quality of the preparation and the detection of significant polyps (>9 mm) (35). Size and histopathological characteristics of polyps detected at MUHC-MGH were not accessible for the present study, precluding any conclusion.

The sole independent predictor of polyp detection in our study was withdrawal time. Polyp detection was greater for endoscopists with an average withdrawal time exceeding 6 min, which is consistent with reports from experts, who recommend that the withdrawal phase, exclusive of time for biopsy and polypectomy, should average at least 6 min to 10 min (32). Prevalence rates of polyps varied among screening centres and appeared to be higher for examiners spending more time performing the examination (32,36). Our study establishes a strong association between withdrawal time and PDR even after adjustment for quality preparation and should prompt endoscopists to follow this recommendation. However, endoscopists face enormous pressures to increase efficiency by decreasing colonoscopy procedure time. In addition, some question the definition of withdrawal time and its use as as quality indicator (37).

The PQDCCR has developed recommendations to monitor the technical performance of colonoscopy and for continuous quality improvement in colonoscopy (38). A recent audit assessing the quality of reports at MGH-MUHC suggested documentation of these indicators in everyday practice remained incomplete in a small series (39). In our study, quality indicators of the procedure were adequately documented in comparison with this recent audit, mainly the preprocedure indicators. However, the quality of the bowel preparation was not recorded in 40% of the reports in our analysis. The reason for this may be selection bias in the initial report, coupled with the recent adoption of a scoring system that is more difficult to use. This group without bowel cleansing data was slightly different, mainly in terms of colonoscopy indication (bleeding), hospital setting (inpatient), use of preparation type (PICO was more utilized) and split-dose regimen. We can only make the assumption that endoscopists are aiming at least to document quality indicators, and do not have time to split the dose in certain circumstance such as inpatient bleeding. The fact that PICO was used more often in situations may skew definitive conclusion on its efficacy.

Study limitations and strengths

The present study provides very important data for the Quebec Ministry of Health in the future deployment of the PDRCC because it provides a snapshot of contemporary bowel cleansing products in an unselected population and in a real-life setting. Strengths include its large sample size, real-life setting and use of a validated electronic data-capturing software (20). The separate assessment among patients for screening and nonscreening indications, among in- and outpatient settings, and the careful analyses of bowel preparation information and quality indicators using a priori defined multivariable analyses are all strengths of the present analysis.

The generalizability of these results (to other countries and secondary or primary hospitals) may be a limitation of the study, particularly for non-Caucasian and for-profit practice environments. Besides, inherent in observational research, such as the present study, is the potential for confounding. Despite adjustment for a range of known and potential confounding variables, we cannot exclude the possibility that our findings can partly be accounted for by residual confounding. In particular, more specific data regarding mode of administration (low versus high volume), differences in timing of colonoscopy, dietary and hydration instructions, and patient support may all contribute to variability in outcomes.

In addition, the present study, similar to all audits, was dependent on the accuracy with which the colonoscopists recorded information into the completed electronic report, with a risk of information bias that was at least partially overcome by the local presence of a research assistant who identified the endoscopic reports and validated the accuracy of information.

However, more than one-half of our patients did not have the bowel preparation type documented and, thus, were initially excluded. In addition, a large group of patients were missing quality-of-bowel-cleansing data, reflecting the difficulty to disseminate quality indicators in everyday practice. Furthermore, electronic records unfortunately did not report information on patient tolerance of the preparation; this information is critical for screening programs because it may affect the acceptability of colonoscopy by healthy individuals and the uptake of colonoscopy in screening programs, also precluding firm conclusions regarding patient tolerance among different products. We hope that the results of our study and the future deployment of the PQDCCR may act as catalyst for quality improvement in colonoscopy by facilitating an automatic computerized data collection infrastructure. Even with a computerized endoscopic report generator, documentation of some key quality fields is still not optimal. Preformatted text with compulsory fields and drop-down menus embedded in the electronic reporting system should facilitate reporting.

CONCLUSION

To standardize and optimize bowel cleansing, we recommend using a split-dose regimen as much as possible for the full deployment of PQDCCR; the use of a split-dose regimen, irrespective of bowel preparations, yielded significantly greater proportions of patients with clean bowels. Endoscopists should then been informed and encouraged to use a split-dose regimen, and this practice monitored. PICO alone appeared to be less effective than PEG in a non-split-dose regimen, such as in our study. However, PICO would be an alternative, particularly if repeated examination will be required, with use of an adjuvant.

Head-to-head in randomized comparisons between PEG and PICO with different modes and timing of administration (high volume, low volume, split-dose regimen according timing of colonoscopy) are lacking. Further comparative studies are warranted to accurately determine the relative efficacy of these two colorectal cleansing agents before colonoscopy, particularly in a screening setting. The true role of adding an adjuvant to PICO compared with PEG also warrants further study.

DISCLOSURES: The authors have no financial disclosures or conflicts of interest to declare.

REFERENCES

1. International Agency for Research on Cancer. GLOBOCAN 2012: Estimated Cancer Incidence, Mortality and Prevalence Worldwide in 2012. <http://globocan.iarc.fr> (Accessed December 13, 2013).
2. Leddin DJ, Enns R, Hilsden R, et al. Canadian Association of Gastroenterology position statement on screening individuals at average risk for developing colorectal cancer: 2010. Can J Gastroenterol 2010;24:705-14.
3. Jemal A, Bray F, Center MM, Ferlay J, Ward E, Forman D. Global cancer statistics. CA Cancer J Clin 2011;61:69-90.
4. Edwards BK, Ward E, Kohler BA, et al. Annual report to the nation on the status of cancer, 1975-2006, featuring colorectal cancer trends and impact of interventions (risk factors, screening, and treatment) to reduce future rates. Cancer 2010;116:544-73.
5. Belsey J, Crosta C, Epstein O, et al. Meta-analysis: The relative efficacy of oral bowel preparations for colonoscopy 1985-2010. Aliment Pharmacol Ther 2012;35:222-37.
6. Hookey LC, Depew WT, Vanner S. The safety profile of oral sodium phosphate for colonic cleansing before colonoscopy in adults. Gastrointest Endosc 2002;56:895-902.
7. DiPalma JA, Brady CE III, Stewart DL, et al. Comparison of colon cleansing methods in preparation for colonoscopy. Gastroenterology 1984;86:856-60.
8. Saunders BP, Masaki T, Fukumoto M, Halligan S, Williams CB. The quest for a more acceptable bowel preparation: Comparison of a polyethylene glycol/electrolyte solution and a mannitol/picolax mixture for colonoscopy. Postgrad Med J 1995;71:476-9.
9. Hangartner PJ, Munch R, Meier J, Ammann R, Buhler H. Comparison of three colon cleansing methods: Evaluation of a randomized clinical trial with 300 ambulatory patients. Endoscopy 1989;21:272-5.

corticosteroid therapy in hospital (14). However, comprehensive studies comparing hospitalization costs for UC pre and post the infliximab era are lacking.

Accordingly, we conducted a population-based study to identify the primary drivers of in-hospital cost for UC and to assess whether these factors changed following the introduction of infliximab.

METHODS

The present study was a population-based costing analysis of medical and surgical hospitalization admissions for adults (18 years of age) with UC in the Calgary Health Zone (CHZ) from January 1, 2001 to December 31, 2009.

Data sources

The Discharge Abstract Database used by Alberta Health Services (AHS) captures all diagnostic and procedural codes that occur during a hospital admission. AHS' Data Integration, Measurement and Reporting department (DIMR) identified patients coded using the *International Classification of Diseases, Ninth Revision, Clinical Modification* (ICD-9-CM) and the *Tenth Revision* (ICD-10-CM). All patients who were admitted with UC (ICD-9-CA code 556.9; ICD-10-CM code K51) from January 1, 2001 to December 31, 2009 in the CHZ were captured; this method of identifying admissions has been previously validated (15). The CHZ contains the City of Calgary and >20 surrounding villages, towns, smaller cities and hamlets, and had a population of 1,408,606 in 2011. AHS is a single-payer, publically funded health care system that is responsible for all medical and surgical care within the CHZ.

The total cost for each individual admission was attained from Activity and Costing – AHS Finance Department and through the Physician Claims Database. This database includes claims submitted for payment by Alberta physicians for services provided to registrants of the Alberta Health Care Insurance Plan, a universal plan that covers >99% of Alberta residents (16). The total costs associated with each hospital admission account for both direct and indirect costs. Direct costs are all costs associated with direct patient care. These include – but are not limited to – nonphysician salaries, drugs, equipment depreciation cost, and the allocated costs for the nursing units and supporting care areas (ie, diagnostic imaging, cardiac labs, ambulatory services). Indirect costs are expenses that account for hospital overhead (eg, administration, support services, site utilities, human resources, information technology services, etc). Virtually all (99.7%) of the admissions had matched finance data and were included in the analysis (the remaining 0.3% were excluded).

Study population

The DIMR database identified 1062 UC patients who had either a colectomy or flare admission. If patients had >1 UC-related admission, a single admission was randomly selected for analysis; this method of analysis has previously been validated (15). After chart review of the 1062 admissions, a total of 318 patients were excluded from the study because the charts were unavailable (n=85), they had Crohn disease (n=54), they did not have IBD (n=42), they underwent a previous colectomy (n=72) or UC was not the primary reason for admission (n=65). The remaining 744 admissions were then stratified into one of three admission categories: responsive to inpatient medical therapy (n=309), emergent colectomy during admission (n=227) and elective colectomy during admission (n=208). These admissions were then submitted to the AHS Finance Department for retrieval of costing data. Two additional admissions were excluded due to a lack of matching cost data, resulting in a study population of 742 patients.

Data extraction

Each patient underwent a comprehensive chart review to capture the following data: age at admission; residence; admit and discharge dates; duration of flare; date of UC diagnosis; disease extent; sex; smoking; UC-validated comorbidities (17,18); UC medications including corticosteroids, 5-aminosalicylic acid, azathioprine, infliximab, adalimumab, budesonide and methotrexate; disease activity (stool frequency, blood in stool and hemoglobin level at admission); in-hospital complications both medical and surgical (assigned according to Clavien class [19]); and length of stay.

The data attained through the chart review were used to stratify patients into the three categories: responsive to medical management (flare); elective colectomy; or emergent colectomy. Patients were deemed responsive to inpatient medical therapy if they came to hospital with a UC-related flare and were discharged without surgery; during analysis, this was the referent group. Emergent colectomy was defined as a UC flare that required an unplanned colectomy. Elective colectomy was defined as admission for a scheduled colectomy.

Statistical analysis

Due to the skewed nature of the cost data, the annual median costs, with interquartile ranges for each of the admission categories (ie, responsive to inpatient medical management, elective colectomy, and emergent colectomy) were calculated. All costs were adjusted to 2013 Canadian dollars using the Consumer Price Index (20).

Linear regression was used to determine the effect that the different admission types had on the annual trend in hospital costs incurred for the admissions. Admission type was modelled as a categorical variable with 'responsive to inpatient medical management' as the referent level. The hospital costs were logarithmically transformed because of their skewed distribution and adjusted for the following covariates: year of admission as continuous variable; age stratified according to tercile (18 to 31, 32 to 47 and ≥48); sex; residence as defined as residing within or outside the CHZ; smoking as current, ex-smoker, never or unknown; comorbidities as 0 or ≥1; disease extent as pancolitis versus left-sided/undetermined; flare duration (<2 weeks, two to eight weeks, >8 weeks or unknown); disease severity as presence of blood in stool or stool frequency (>5 or ≤5 days), and hemoglobin level at admission (>100 g/L versus ≤100 g/L); in-hospital complications; UC medications at admission (5-aminosalicylic acid, prednisone, azathioprine,) and/or during hospital (infliximab); and length of stay. All of these clinically relevant covariates were a priori included in the adjusted analysis.

Using linear regression, the annual cost increase was calculated for each UC admission type with the regression model including the adjustment for confounders. The beta coefficients were exponentiated to give the final annual percentage increase and their corresponding 95% CIs. A Joinpoint analysis assessed for significant inflection points for both the mean and median cost of the aggregate data over the study period. The a priori analysis assessed for a significant cost inflection point after 2005 when infliximab was introduced (14). The statistical analysis was performed using SAS version 9.3 (SAS Institute, USA). The Joinpoint analysis was performed with Joinpoint Regression Program version 4.1.0 (21).

RESULTS

Patient characteristics stratified according to type of admission are presented in Table 1. The median costs for each of the three admission types were: UC flare $5,499 (interquartile range [IQR] $3,374 to $8,904), elective colectomy $14,316 (IQR $11,932 to $18,331) and emergent colectomy $23,698 (IQR $17,981 to $32,385). The inflation-adjusted cost of all admissions increased by 6.0% (95% CI 4.5% to 7.5%) per year (Table 2). The median costs stratified according to admission type per year are presented in Figure 1. Significant Joinpoints were not identified in the median or mean costs.

UC patients who underwent an elective colectomy had a 179.8% (95% CI 151.6% to 211.1%) increase in cost versus those who were medically responsive to in-patient medical management after adjusting for covariates. Those who underwent emergent colectomies were significantly higher with a 211.1% (95% CI 183.2% to 241.6%) increase in cost (Table 2).

Other independent predictors of hospitalization costs included: age (≥48 years versus 18 to 31 years: 14.0% [95% CI 4.2% to 24.6%]); in-hospital complication (Clavien II/III/IV/V: 31.7% [95% CI 20.6%

TABLE 1
Characteristics of the study population

Characteristic	Total cohort (n=742)	Flare (n=307)	Colectomy Elective (n=208)	Colectomy Emergent (n=227)	P*
Age, years, tercile					
18–31	33 (244)	40 (122)	28 (59)	28 (63)	<0.001†
32–47	34 (251)	36 (109)	31 (65)	34 (77)	
≥48	33 (247)	25 (76)	40 (84)	38 (87)	
Sex					
Male	58 (431)	52 (160)	66 (137)	59 (134)	0.008
Female	42 (311)	48 (147)	34 (71)	41 (93)	
Residence (Alberta)					
Calgary Health Zone	77 (568)	88 (271)	64 (133)	72 (164)	<0.001
Non-Calgary Health Zone	23 (174)	12 (36)	36 (75)	28 (63)	
Smoking status					
Current	9 (63)	11 (31)	8 (17)	7 (15)	0.041
Ex-smoker	30 (215)	29 (82)	25 (50)	37 (83)	
Never	61 (430)	60 (171)	67 (134)	56 (125)	
Missing data	(n=34)	(n=23)	(n=7)	(n=4)	
Comorbidities					
0	47 (346)	51 (156)	49 (102)	39 (88)	0.016
≥1	53 (396)	49 (151)	51 (106)	61 (139)	
Extent of disease					
Left-sided	31 (205)	45 (113)	23 (46)	21 (46)	<0.001
Pancolitis	69 (464)	55 (140)	77 (150)	79 (174)	
Missing (undetermined)	(n=73)	(n=54)	(n=12)	(n=7)	
Ulcerative colitis flare length, weeks					
<2	15 (104)	19 (57)	1 (1)	21 (46)	<0.001
2–8	44 (293)	49 (150)	21 (31)	50 (112)	
>8	41 (276)	32 (97)	78 (113)	29 (66)	
Missing (undetermined)	(n=69)	(n=3)	(n=63)	(n=3)	
Blood in stool and stool frequency >5/day					
No	25 (149)	22 (58)	42 (54)	17 (37)	<0.001
Yes	75 (455)	78 (205)	58 (75)	83 (175)	
Missing data‡	(n=138)	(n=44)	(n=79)	(n=15)	
In-hospital complication					
0 or Clavien I	79 (584)	93 (285)	78 (162)	60 (137)	<0.001
Clavien II/III/IV/V	21 (158)	7 (22)	22 (46)	40 (90)	
5-aminosalicylic acid (history)§					
No	33 (247)	42 (129)	29 (59)	26 (59)	<0.001
Yes	67 (492)	58 (178)	71 (146)	74 (168)	
Missing data	(n=3)		(n=3)		
Prednisone (history)§					
No	33 (246)	49 (149)	18 (36)	27 (61)	<0.001
Yes	67 (493)	51 (158)	82 (169)	73 (166)	
Missing data	(n=3)		(n=3)		
Azathioprine/6-mercaptopurine (history)§					
No	77 (572)	89 (273)	62 (127)	76 (172)	<0.001
Yes	23 (167)	11 (34)	38 (78)	24 (55)	
Missing data	(n=3)		(n=3)		
Infliximab (history)§					
No	95 (700)	98 (301)	92 (189)	93 (210)	0.002
Yes	5 (39)	2 (6)	8 (16)	7 (17)	
Missing data	(n=3)		(n=3)		
Infliximab (in-hospital)¶					
No	93 (683)	88 (271)	100 (204)	92 (208)	<0.001
Yes	7 (54)	12 (36)	0 (0)	8 (18)	
Missing data	(n=5)		(n=4)	(n=1)	
Hemoglobin, g/L					
≤100	16 (115)	17 (51)	9 (17)	21 (47)	0.002
>100	84 (601)	83 (250)	91 (177)	79 (174)	
Missing data	(n=26)	(n=6)	(n=14)	(n=6)	
Length of stay, days, median (interquartile range)	10 (7–19)	6 (4–10)	9 (8–13)	22 (16–31)	<0.001**

*Data presented as % (n) unless otherwise indicated. *Fisher's exact test, unless otherwise indicated; †Row mean scores test; ‡Defined as missing data regarding stool frequency, the presence of blood in stool or both; §Defined as medication being taken at the time of admission to hospital or in the past; no refers to no record of drug use in the medical chart; ¶Defined as medication prescribed during the hospital stay; no refers to no record of drug use in the medical chart; **Kruskal-Wallis test*

TABLE 2
Linear regression analysis of costs

Variable	Comparison	Percent cost change (95% CI)	P
Year	Per year	6.0 (4.5 to 7.5)	<0.0001
Admission type	Elective versus flare	179.8 (151.6 to 211.1)	<0.0001
Admission type	Emergency versus flare	211.1 (183.2 to 241.6)	<0.0001
Age, years	32 to 47 versus 18 to 31	−0.6 (−8.2 to 7.6)	0.879
Age, years	≥48 versus 18 to 31	14.0 (4.2 to 24.6)	0.004
Sex	Male versus female	0.7 (−5.7 to 7.6)	0.826
Residence	Non-CHZ versus CHZ	−7.2 (−14.2 to 0.5)	0.065
Smoking*	Current versus never	−11.6 (−21.4 to −0.5)	0.040
Smoking	Ex versus never	−5.1 (−12.2 to 2.4)	0.178
Comorbidity	Yes versus no	2.0 (−4.7 to 9.1)	0.565
Disease extent	Pancolits versus left-sided/undetermined	11.4 (3.8 to 19.6)	0.003
Flare length†	2 to 8 weeks versus <2 weeks	8.9 (−1.4 to 20.2)	0.094
Flare length	>8 weeks versus <2 weeks	8.9 (−2.2 to 21.2)	0.120
Blood in stool/stool frequency	>5/day versus ≤5/day or no blood	1.0 (−7.4 to 10.1)	0.821
Blood in stool/stool frequency	Undetermined versus ≤5/day or no blood	−10.3 (−19.4 to −0.2)	0.046
In-hospital complication	Clavien II/III/IV/V versus 0 or Clavien I	31.7 (20.6 to 43.8)	<0.0001
5-aminosalicylic acid (history)	Yes versus no	−2.3 (−9.2 to 5.1)	0.530
Prednisone (history)	Yes versus no	1.1 (−6.6 to 9.4)	0.795
Azathioprine (history)	Yes versus no	−1.0 (−9.1 to 7.9)	0.825
Infliximab (history)	Yes versus no	−9.9 (−22.6 to 4.9)	0.177
Infliximab (in-hospital)	Yes versus no	69.5 (49.2 to 92.5)	<0.0001
Hemoglobin	>100 versus ≤100	−5.7 (−13.8 to 3.0)	0.192
Length of stay	Per day	1.7 (1.5 to 1.9)	<0.0001

**Smoking included missing indicator level (not shown); †Flare length included missing indicator level (not shown). CHZ Calgary Health Zone (Alberta)*

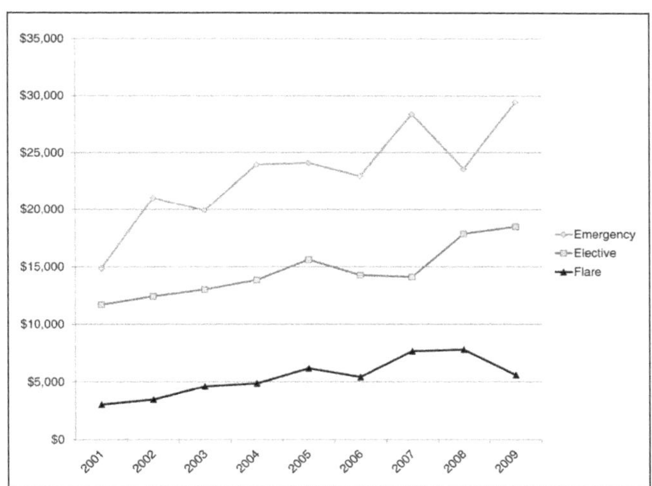

Figure 1) *Temporal trends from 2000 to 2009 of median costs stratified according to each admission type: medically responsive flare, elective colectomy and emergent colectomy*

to 43.8%]); disease extent (pancolitis versus left-sided: 11.4% [95% CI 3.8% to 19.6%]); smoking (current versus never: −11.6% [95% CI −21.4% to −0.5%]); infliximab prescribed in hospital (69.5% [95% CI 49.2% to 92.5%]); and length of stay (1.7% per day [95% CI 1.5% to 1.9%]).

DISCUSSION

It has previously been shown that during our study period, the rates of colectomy fell in the CHZ, particularly the rate of elective colectomies (9). Despite the reduction in colectomies, in-hospital costs for all UC patients increased by 6.0% per year. After adjusting for disease severity, surgical admissions were associated with the highest magnitude of increased hospital costs. Emergent operations are associated with a greater length of stay and interventions performed in hospital due to higher occurrence of in-hospital complications and longer recovery periods (5,7). Other drivers of cost were age, disease severity, in-hospital complications, length of stay, smoking status and in-hospital prescription of infliximab. By identifying some of the factors that contributed to the rising cost of UC care, we may consider alternative ways to mitigate these costs while providing appropriate patient care.

The present study evaluated the temporal trends of UC hospitalization costs in Canada since the introduction of infliximab in 2005 for the treatment of hospitalized UC patients refractory to intravenous corticosteroid therapy (14). Since 2005, infliximab has become a mainstay in-hospital rescue agent for UC patients. A previous study demonstrated a sharp uptake of in-hospital prescriptions of infliximab in the CHZ (9). In the current study, infliximab was an independent risk factor for increasing hospital costs after adjusting for patient factors, disease severity and surgery. However, an expected inflection point of rising hospital costs after 2005 was not observed, which suggests that multiple factors contributed to increased hospital costs and/or money spent on infliximab may have been offset by alternative cost saving.

Moreover, infliximab is a modifiable cost factor that could prove to be a significant cost savings to hospital budgets. In Alberta, in-hospital use of infliximab is paid for by the hospital budget, whereas outpatient infliximab use is covered by provincial drug plans or by private drug coverage. Thus, timelier introduction of infliximab in the outpatient setting would avoid this expense in hospital and, additionally, may reduce the need for hospitalizations or lead to an elective colectomy. Among UC patients with fulminant colitis that requires infliximab in hospital, an alternative method of funding anti-tumour necrosis factor therapy that does not impact patient care should be implemented. Policy changes could lead to shifting in-hospital infliximab costs to drug insurance plans, or existing biologic infusion centres could be expanded to include in-hospital care where additional efficiencies may be generated.

In Canada, from 2000 to 2009, annual health care inflation rate rose by an average of approximately 1.6% per year (20). Costs for in-hospital admissions for UC increased significantly faster than the inflation rate. This is due, in part, to the overall rise in health care costs that are outpacing the national inflation rate. Over time, hospitals have been admitting older and sicker patients with a greater number of chronic comorbidities, which has led to an increase in-hospital expenditures (22). In addition, medical technology and the labour force have been found to be significant cost drivers in the United States (23). In Alberta, the salary of unionized health care work force increased by 2% to 3% per year during our study period (24). Also, patients are staying longer in emergency departments due to bed shortages in hospital, and delays exceeding 12 h have been shown to increase cost by 11% (25). These factors may have played a part in the increasing expenditures.

A few limitations of our study should be considered (26). While we were able to explore both direct and indirect in-hospital costs, we were unable to differentiate whether the direct or indirect costs were the major cost drivers. In addition, we did not account for outpatient costs that patients incurred to the health care system. However, the purpose of the present study was to evaluate the evolution of in-hospital costs independent of the effect of outpatient management. By focusing on in-hospital costs, our study informs health care resource allocation planning for inpatient care, but is not generalizable to outpatient management. Because we undertook a retrospective chart review, some clinical factors, such as outpatient medication use, were missing in some patients. While the linkage to the AHS Finance Department that reported both direct and indirect hospital costs was 99.7% complete, indirect outpatient costs (eg, loss of work productivity) were not assessed in our study. Finally, administrative data were used to initially identify UC patients. While administrative data are subject to misclassification errors, we have previously validated this approach (15).

CONCLUSION

Studying temporal trends of in-hospital costs and identifying factors that drive these costs are important because UC is an expensive disease with a lifelong burden on patients and to the health care system. We studied a large population-based cohort of UC patients using rigorous study design including confirmation of clinical data through chart review and accurate linking to direct and indirect in-hospital costing data. We demonstrated that surgical admissions, infliximab and disease severity drives costs upward. Costs of hospitalizations for UC patients are steadily increasing and, thus, it is prudent to find ways to mitigate these costs without compromising patient care.

ACKNOWLEDGEMENTS: Dr Kaplan is supported through a New Investigator Award from the Canadian Institute of Health Research and a Population Health Investigator Award from Alberta-Innovates Health-Solutions. This study was based, in part, on data provided by Alberta Health. The interpretation and conclusions contained herein are those of the researchers and do not necessarily represent the views of the Government of Alberta. Neither the Government nor Alberta Health express any opinion in relation to this study. Stephanie Coward is supported by an Eyes High Doctoral Recruitment Scholarship from the University of Calgary.

DISCLOSURES: Dr Kaplan has served as a speaker for Janssen, Schering-Plough, Abbott, and UCB Pharma. He has participated in advisory board meetings for Jansen, Abbott, Merck, Schering-Plough, Shire, and UCB Pharma. Dr Kaplan has received research support from Merck, Abbott and Shire. Dr Panaccione has served as a speaker, a consultant and an advisory board member for Abbott Laboratories, Merck, Schering-Plough, Shire, Centocor, Elan Pharmaceuticals, and Procter and Gamble. He has served as a consultant and speaker for Astra Zeneca. He has served as a consultant and an advisory board member for Ferring and UCB. He has served as a consultant for Glaxo-Smith Kline and Bristol-Myers Squibb. He has served as a speaker for Byk Solvay, Axcan, Janssen and Prometheus. He has received research funding from Merck, Schering-Plough, Abbott Laboratories, Elan Pharmaceuticals, Procter and Gamble, Bristol-Myers Squibb and Millennium Pharmaceuticals. He has received educational support from Merck, Schering-Plough, Ferring, Axcan and Janssen. Dr Ghosh has served as a speaker for Merck, Schering-Plough, Centocor, Abbott, UCB Pharma, Pfizer, Ferring, and Procter and Gamble. He has participated in ad hoc advisory board meetings for Centocor, Abbott, Merck, Schering-Plough, Proctor and Gamble, Shire, UCB Pharma, Pfizer and Millennium. He has received research funding from Procter and Gamble, Merck and Schering-Plough. Dr Cynthia Seow has served as a speaker for Janssen and Warner Chilcott. She has participated in advisory board meetings for Janssen and Abbvie. Dr Seow has received research support from Merck. Dr Yvette Leung has received research support from Janssen. She has served as a speaker for Jansen. She has participated in advisory board meetings for Abbott, Jansen and Shire. The other authors do not have relevant conflict of interests to declare.

AUTHOR CONTRIBUTIONS: Stephanie Coward was involved with the study concept and design, acquisition of data, analysis and interpretation of data, and drafting of the manuscript. Steven Heitman, Neel Datta, Mark Swain, Remo Panaccione, Cynthia Seow, Yvette Leung, Subrata Ghosh were involved with interpretation of data and critical revision of the manuscript. James Hubbard was involved with data analysis and critical revision of the manuscript. Marie-Claude Proulx, Scott Zimmer and Rob Myers were involved with acquisition of data, and interpretation of data. Fiona Clement was involved with statistical analysis and critical revision of the manuscript. Gilaad G Kaplan was involved with the study concept and design, statistical analysis, acquisition of data, analysis and interpretation of data, drafting of the manuscript, funding and study supervision.

FUNDING SOURCES: This research is supported by The Alberta IBD Consortium, which is funded by an AHFMR Interdisciplinary Team Grant. AHFMR is now Alberta Innovates – Health Solutions. The funders had no role in study design, data collection and analysis, decision to publish or preparation of the manuscript.

REFERENCES

1. Loftus EV Jr. Clinical epidemiology of inflammatory bowel disease: Incidence, prevalence, and environmental influences. Gastroenterology 2004;126:1504-17.
2. Molodecky NA, Soon IS, Rabi DM, et al. Increasing incidence and prevalence of the inflammatory bowel diseases with time, based on systematic review. Gastroenterology 2012;142:46-54
3. Frolkis AD, Dykeman J, Negron ME, et al. Risk of surgery for inflammatory bowel diseases has decreased over time: A systematic review and meta-analysis of population-based studies. Gastroenterology 2013;145:996-1006.
4. Kaplan GG, Hubbard J, Panaccione R, et al. Risk of comorbidities on postoperative outcomes in patients with inflammatory bowel disease. Arch Surg 2011;146:959-64.
5. Kaplan GG, McCarthy EP, Ayanian JZ, et al. Impact of hospital volume on postoperative morbidity and mortality following a colectomy for ulcerative colitis. Gastroenterology 2008;134:680-7.
6. Soon IS, Wrobel I, deBruyn JC, et al. Postoperative complications following colectomy for ulcerative colitis in children. J Pediatr Gastroenterol Nutr 2012;54:763-8.
7. de Silva S, Ma C, Proulx MC, et al. Postoperative complications and mortality following colectomy for ulcerative colitis. Clin Gastroenterol Hepatol 2011;9:972-80.
8. de Zeeuw S, Ahmed Ali U, Donders RA, et al. Update of complications and functional outcome of the ileo-pouch anal anastomosis: Overview of evidence and meta-analysis of 96 observational studies. Int J Colorectal Dis 2012;27:843-53.
9. Kaplan GG, Seow CH, Ghosh S, et al. Decreasing colectomy rates for ulcerative colitis: A population-based time trend study. Am J Gastroenterol 2012;107:1879-87.
10. Kappelman MD, Rifas-Shiman SL, Porter CQ, et al. Direct health care costs of Crohn's disease and ulcerative colitis in US children and adults. Gastroenterology 2008;135:1907-13.
11. Bernstein CN, Longobardi T, Finlayson G, Blanchard JF. Direct medical cost of managing IBD patients: A Canadian population-based study. Inflamm Bowel Dis 2012;18:1498-508.
12. Rocchi A, Benchimol EI, Bernstein CN, et al. Inflammatory bowel disease: A Canadian burden of illness review. Can J Gastroenterol 2012;26:811-7.

13. Bernstein CN, Papineau N, Zajaczkowski J, et al. Direct hospital costs for patients with inflammatory bowel disease in a Canadian tertiary care university hospital. Am J Gastroenterol 2000;95:677-83.
14. Jarnerot G, Hertervig E, Friis-Liby I, et al. Infliximab as rescue therapy in severe to moderately severe ulcerative colitis: A randomized, placebo-controlled study. Gastroenterology 2005;128:1805-11.
15. Ma C, Crespin M, Proulx MC, et al. Postoperative complications following colectomy for ulcerative colitis: A validation study. BMC Gastroenterol 2012;12:39.
16. Data Disclosure Handbook. In: Alberta Health and Wellness; 2003:1-15.
17. Charlson ME, Pompei P, Ales KL, MacKenzie CR. A new method of classifying prognostic comorbidity in longitudinal studies: Development and validation. J Chron Dis 1987;40:373-83.
18. Deyo RA, Cherkin DC, Ciol MA. Adapting a clinical comorbidity index for use with ICD-9-CM administrative databases. J Clin Epidemiol 1992;45:613-9.
19. Dindo D, Demartines N, Clavien PA. Classification of surgical complications: A new proposal with evaluation in a cohort of 6336 patients and results of a survey. Ann Surg 2004;240:205-13.
20. Statistics Canada. Consumer Price Index. <www.statcan.gc.ca/tables-tableaux/sum-som/l01/cst01/econ09a-eng.htm> (Accessed August 3, 2014).
21. National Cancer Institute. Joinpoint Regression Program Version 4.1.0. <http://surveillance.cancer.gov/joinpoint/> (Accessed April 1, 2014).
22. Steiner CA, Friedman B. Hospital utilization, costs, and mortality for adults with multiple chronic conditions, Nationwide Inpatient Sample, 2009. Prev Chronic Dis 2013;10:E62.
23. Hay JW. Hospital cost drivers: an evaluation of 1998-2001 state-level data. Am J Managed Care 2003;9 Spec No 1:SP13-24.
24. Alberta Health Services. Collective Agreements and Bargaining Updates. <www.albertahealthservices.ca/8611.asp> (Accessed November 1, 2013).
25. Huang Q, Thind A, Dreyer JF, Zaric GS. The impact of delays to admission from the emergency department on inpatient outcomes. BMC Emerg Med 2010;10:16.
26. Molodecky NA, Panaccione R, Ghosh S, et al. Challenges associated with identifying the environmental determinants of the inflammatory bowel diseases. Inflamm Bowel Dis 2011;17:1792-9.

The effectiveness of ultrasound surveillance for hepatocellular carcinoma in a Canadian centre and determinants of its success

Korosh Khalili MD[1], Ravi Menezes PhD[1], Tae Kyoung Kim MD[1], Leyla Kochak Yazdi MD[1], Hyun-Jung Jang MD[1], Suraj Sharma MD[2], Jordan Feld MD[2], Morris Sherman MD[2]

K Khalili, R Menezes, TK Kim, et al. The effectiveness of ultrasound surveillance for hepatocellular carcinoma in a Canadian centre and determinants of its success. Can J Gastroenterol Hepatol 2015;29(5):267-273.

BACKGROUND: The effectiveness of surveillance for hepatocellular carcinoma (HCC) using ultrasound (US) in North America has been questioned due to the predominance of patients of Caucasian ethnicity and larger body habitus.

OBJECTIVE: To determine the effectiveness of US surveillance for HCC in a Canadian hepatobiliary centre and to identify independent variables associated with early detection of tumour(s).

METHODS: A retrospective review of patients with first HCC in a US surveillance population at the authors' hospital yielded 201 patients (over a 10.5-year period). Patients were categorized into three groups: regular surveillance (frequency of surveillance ≤12 months [n=109]); irregular surveillance (frequency of surveillance >12 months [n=38]); or first surveillance (tumour detected on first scan [n=54]). The Milan criteria for transplantation and Barcelona Clinic Liver Cancer (BCLC) staging system were used as outcome measures. Effective surveillance was defined as tumour detection within Milan criteria or curative BCLC stages 0 and A; its association with multiple patient- and disease-related variables was tested.

RESULTS: When using the Milan criteria as outcome, 84 of 109 (77%) regular surveillance patients, 23 of 38 (61%) irregular surveillance patients and 40 of 54 (74%) first surveillance patients had tumours meeting the transplantation criteria. The difference between regular and irregular surveillance was statistically significant (P=0.03). When using the BCLC staging system, 87 of 109 (80%) regular surveillance patients, 26 of 38 (68%) irregular surveillance patients and 41 of 54 (76%) first surveillance patients had their tumours detected in BCLC curative stages (0 and A; P=0.11). Regular surveillance was the only variable significantly associated with detection of tumour(s) within the Milan criteria (OR 2.76 [95% CI 1.10 to 6.88]). Tumours detected more recently were more likely to be <2 cm in size (BCLC stage 0; OR 2.38 [95% CI 1.07 to 5.31]).

CONCLUSION: A high rate of HCC surveillance success was achieved using US alone when performed regularly in a specialized hepatobiliary centre.

Key Words: *Canada; Effectiveness; Hepatocellular carcinoma; Surveillance; Ultrasound*

L'efficacité de la surveillance échographique du carcinome hépatocellulaire dans un centre canadien et les déterminants de son succès

HISTORIQUE : L'efficacité de la surveillance du carcinome hépatocellulaire (CHC) par échographie (ÉG) en Amérique du Nord a été remise en question en raison de la prédominance de patients de race blanche d'un type morphologique plus imposant.

OBJECTIF : Déterminer l'efficacité de la surveillance ÉG du CHC dans un centre hépatobiliaire canadien et les variables indépendantes associées à la détection précoce des tumeurs.

MÉTHODOLOGIE : Une analyse rétrospective des patients ayant un premier CHC dans une population sous surveillance ÉG à l'hôpital des auteurs a révélé 201 patients atteints (sur une période de 10,5 ans). Les patients ont été classés en trois groupes : surveillance régulière (fréquence de la surveillance ≤12 mois [n=109]), surveillance irrégulière (fréquence de la surveillance >12 mois [n=38]) ou première surveillance (tumeur décelée à la première ÉG [n=54]). Les chercheurs ont utilisé les mesures de résultats suivantes : critères de Milan pour la transplantation et la classification de la clinique pour le cancer du foie de Barcelone (acronyme BCLC en anglais). Ils ont défini une surveillance efficace comme la détection de tumeurs conformément aux critères de Milan ou à la classification BCLC aux stades curatifs 0 et A. Ils en ont vérifié l'association avec des variables pathologiques touchant de multiples patients.

RÉSULTATS : Lorsque les critères de Milan servent de mesure de résultat, 84 des 109 patients sous surveillance régulière (77 %), 23 des 38 patients sous surveillance irrégulière (61 %) et 40 des 54 patients sous première surveillance (74 %) avaient des tumeurs respectant les critères de transplantation. La différence entre une surveillance régulière et irrégulière était statistiquement significative (P=0,03). Lorsqu'on utilise les stades de classification BCLC, 87 des 109 patients sous surveillance régulière (80 %), 26 des 38 patients sous surveillance irrégulière (68 %) et 41 des 54 patients sous première surveillance (76 %) se sont fait déceler une tumeur aux stades curatifs de la classification BCLC (0 et A; P=0,11). La surveillance régulière était la seule variable associée de manière significative à la détection des tumeurs selon les critères de Milan (RC 2,76 [95 % IC 1,10 à 6,88]). Les tumeurs décelées plus récemment étaient plus susceptibles d'être de dimension inférieure à 2 cm (stade de BCLC 0; RC 2,38 [95 % IC 1,07 à 5,31]).

CONCLUSION : L'ÉG seule s'associait à un taux élevé de réussite de la surveillance du CHC lorsqu'elle était effectuée régulièrement dans un centre hépatobiliaire spécialisé.

Surveillance of populations at risk for hepatocellular carcinoma (HCC) using ultrasound (US) has been suggested to improve survival in a randomized controlled trial, is cost effective, and is recommended by American and international associations dedicated to the study of liver disease (1-5). It is, nevertheless, an imperfect test and its effectiveness has been questioned in the North American setting, in which patient size may limit sound transmission (6,7). The method of HCC surveillance in North America is variable, with serum alphafetoprotein (AFP), computed tomography (CT) and magnetic resonance imaging (MRI) also being used as primary surveillance tools (8,9).

Approximately 3000 patients undergo active HCC surveillance in our combined hospitals. We aimed to determine the proportion of patients whose tumour was detected at a potentially curable stage using US surveillance alone. Furthermore, we wanted to know whether there were any patient-, liver- or tumour-related factors that were associated with success of surveillance (ie, was the detection of the tumour in a curative stage). The aim of the present study was, therefore, twofold: to determine the effectiveness of US surveillance for HCC in a specialized Canadian liver disease centre; and to determine independent variables that were associated with success of surveillance.

[1]*Department of Medical Imaging;* [2]*Department of Gastroenterology, University of Toronto, Toronto, Ontario*

Correspondence: Dr Korosh Khalili, Department of Medical Imaging, University Health Network, Princess Margaret Hospital, 610 University Avenue, Room 3-964, Toronto, Ontario M5G 2M9. e-mail korosh.khalili@uhn.ca

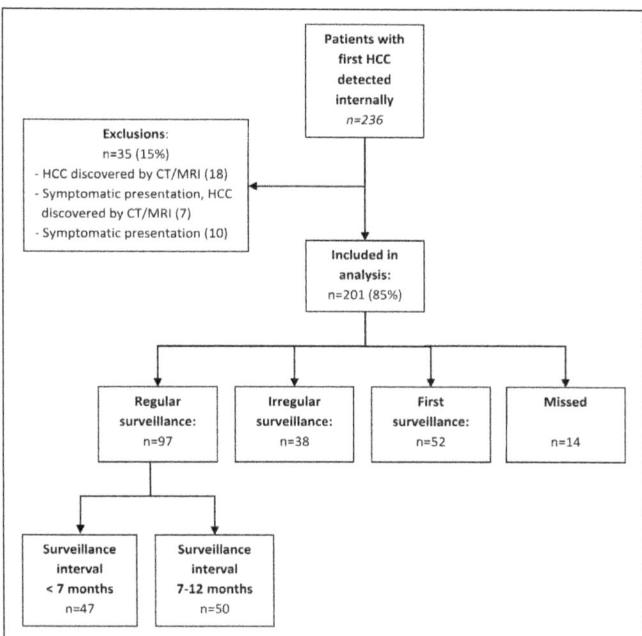

Figure 1) *Patient population. CT Computed tomography; HCC Hepatocellular carcinoma; MRI Magnetic resonance imaging*

METHODS

The present study was approved by the institution's research ethics board and informed consent was waived. Electronic patient charts, available at one of the surveillance centres (Toronto Western Hospital, Toronto, Ontario), were retrospectively reviewed to identify all 236 patients with a new diagnosis of HCC between January 2000 and June 2010, and whose tumour was discovered first at the authors' centre. Patients who presented symptomatically or whose tumour was found using other modes of imaging were excluded (Figure 1). The reasons for other modes of imaging were: long-term follow-up of a different, ultimately benign nodule detected by US surveillance (10); symptomatic presentation (7); imaging for unrelated disease (5); surveillance by CT for previous HCC >2 years from the original tumour (3).

A total of 201 patients were included in the analysis; their characteristics are summarized in Table 1. The diagnosis of HCC was confirmed for all patients using the following criteria: updated American Association for Study of Liver Disease (AASLD) HCC management guidelines' imaging diagnostic criteria of one positive contrast-enhanced imaging scan in at-risk patients; positive histopathology from core biopsy or explant specimens; or recurrence of tumour after treatment (3). All relevant imaging was directly and retrospectively reviewed by a fellowship-trained abdominal imager with expertise in hepatobiliary imaging (12 years of practice) to ensure compliance with the latest imaging criteria and to confirm staging. The reviewer was blinded to the surveillance intervals.

Determining effectiveness

The Milan criteria for treatment of HCC by transplantation was used as an outcome measure with successful US surveillance defined as fulfilling the criteria of one nodule <5 cm in size, or three nodules <3 cm in size, with no vascular invasion or distant metastases (10). The Barcelona Clinic Liver Cancer (BCLC) staging classification was also used as an additional outcome measure (11). Successful surveillance was defined as detection of tumour using US in potentially curative stages 0 (one nodule <2 cm) or A (one resectable nodule of any size or three nodules <3 cm) of disease. Failure was defined as presentation in the palliative stages B to D.

For each patient, reports and images of all US scans were reviewed to determine the date of the earliest positive surveillance US of a nodule. This scan was noted as the 'detection scan' and was used

TABLE 1
Patient demographics (n=201)

Age, years	
Mean ± SD	62.3±10.3
Median (range)	61 (31–86)
Male sex	159 (79)
Ethnicity	
African	7 (4)
Asian, East/Southeast	49 (24)
Asian, South	10 (5)
Caribbean	2 (1)
Caucasian	125 (62)
Latin American	3 (2)
Middle Eastern	5 (3)
Cause of liver disease	
Hepatitis C	97 (48)
Hepatitis B	42 (21)
More than one cause	30 (15)
Alcohol	19 (9)
Nonalcoholic fatty liver disease	8 (4)
Primary biliary cirrhosis	2 (1)
Alagille syndrome	1
Autoimmune hepatitis	1
Budd-Chiari syndrome	1
Child-Pugh score	
0 (not cirrhotic)	2 (1)
A	142 (71)
B	48 (24)
C	9 (5)
Surveillance intervals	
First scan detection	54 (23)
<7 months	59 (25)
7–12 months	50 (21)
13–24 months	23 (10)
>24 months	15 (6)
BCLC stage	
0	62 (31)
A	104 (52)
B	16 (8)
C	12 (6)
D	7 (3)

Data presented as n (%) unless otherwise indicated. BCLC Barcelona Clinic Liver Cancer

to determine tumour stage. If a nodule ≥1 cm had been detected by a surveillance US but followed by imaging and subsequently proven to be malignant, the tumour stage at the time of the original detection scan, rather than at time of proof of malignancy, was used. The surveillance US images were reviewed to ensure that the nodule(s) detected corresponded to that characterized as malignant on contrast-enhanced imaging; if it did not, the malignant nodule was designated either as 'missed' by surveillance (if within three months of the surveillance US) or detected by other means and excluded (if ≥3 months of the surveillance US). Nodules <1 cm were not examined for the purposes of the present study because these were conventionally followed (3).

Missed tumours

If a surveillance US was negative and a HCC was detected within three months by other means (imaging or AFP, including a repeat US scan), it was defined as 'missed' by surveillance. Three months was used as a cut-off because a fast-growing malignant tumour may be truly undetectable on one surveillance image but present on the next. 'Missed' nodules were omitted from statistical analysis of effectiveness of surveillance (above) and also of potential determinants of successful

TABLE 2
Comparison of surveillance interval using Milan criteria for liver transplantation

		Milan criteria		
Surveillance	Total, n	Within	Outside	Missed
Regular*	109	84 (77)	16 (15)	9 (8)
Irregular†	38	23 (61)	12 (32)	3 (8)
First	54	40 (74)	12 (22)	2 (4)

*Data presented as n (%) unless otherwise indicated. *Surveillance intervals ≤12 months; †Surveillance intervals >12 months*

TABLE 3
Comparison of surveillance interval and tumour stage determined using the Barcelona Clinic Liver Cancer (BCLC) staging system

		BCLC stage			
Surveillance	Total, n	0	A	B to D	Missed
Regular*	109	36 (33)	51 (47)	13 (12)	9 (8)
Irregular†	38	8 (21)	18 (47)	9 (24)	3 (8)
First	54	14 (26)	27 (50)	11 (20)	2 (4)

*Data presented as n (%) unless otherwise indicated. *Frequency of surveillance ≤12 months; †Frequency of surveillance >12 months*

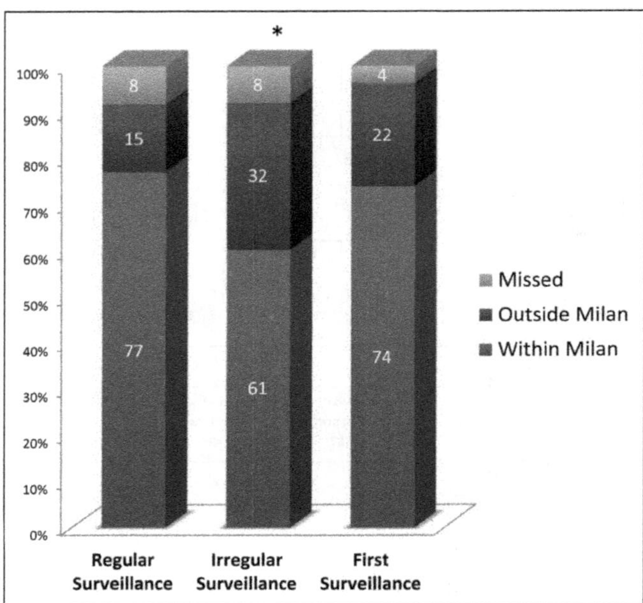

Figure 2) *Hepatocellular carcinoma ultrasound intervals according to Milan criteria for liver transplantation. *Indicates statistically significant difference in distribution between regular surveillance (frequency ≤12 months) and irregular (frequency >12 months) surveillance intervals (P=0.03)*

surveillance (below) because it would not be possible to determine what proportion of these would be detected by next surveillance US still in a curable stage.

The contribution of AFP

The charts of all patients were reviewed and level and dates of serum AFP measurements were recorded. Only measurements within 90 days of the final surveillance US were included (from 90 days before to 90 days after). If multiple measurements were available before the last surveillance scan, then the highest level was used. If multiple measurements were available after the detection of tumour, the closest measurement to the date of US was used. An AFP level >20 mg/mL was defined as positive.

Potential determinants of successful surveillance

For the present analysis, patients whose tumours were detected on the first surveillance US scan were omitted because theirs were representative of a screening (prevalence) and not of a surveillance (incidence) population. On the first surveillance scan, a tumour may be detected in the advanced stage just for the reason that there was no previous surveillance. On subsequent surveillance scans, a tumour may be detected in an advanced stage due to patient- or tumour-related variables that led to failure of surveillance.

The following variables were tested for an association with successful surveillance: age, sex, ethnicity, cause of liver disease, severity of liver disease, year of detection, location of residence and surveillance frequency. Due to the limited size of the study population, variables were broadly grouped where appropriate. Ethnicity was divided into two groups of Caucasian (European descent) and others (including East/Southeast Asian, African, Middle-Eastern, South-Asian, Caribbean and Latin American). Causes of liver disease included chronic hepatitis B virus infection, chronic hepatitis C virus infection, other causes and multiple causes (ie, patients with more than one cause of disease). This latter category (multiple causes) was created because these patients are more likely to be closely followed and screened. The Child-Pugh score was used for severity of liver disease, with noncirrhotic patients grouped with Child's A versus patients with Child's B and C scores. Year of detection was divided into two groups of 2000 to 2005 and 2006 to 2010, to determine the effect of implementation of the first AASLD HCC guidelines in 2005 in addition to installation of the latest generation of scanners at the authors' institution in 2006. Patients were divided based on their residence into metropolitan (greater Toronto area) versus beyond metropolitan groups to determine whether there was a difference in urban versus rural populations. Finally, surveillance frequency was grouped into ≤12 months (regular) and >12 months (irregular). The authors acknowledge that a six-month surveillance interval is the standard of care; however, because the present study covers a substantial period before publication of the AASLD guidelines in 2005, 12-month follow-up interval may still have been considered to be appropriate by some clinicians. Therefore, for the sake of consistent terminology, both within the present study and with other studies that cover a similar period of examination (9), frequency of ≤12 months has been defined as 'regular surveillance'.

Statistical methods

A two-tailed Fisher's exact test was used to compare effectiveness between regular and irregular surveillance groups. χ^2 analysis, *t* test (for age) and unadjusted logistic regression were used to identify potential determinants of successful surveillance. Patients' region of residence were derived using the Postal Code Conversion File + Version 5 (12). Postal Code Conversion File + Version 5 is a series of files created by Statistics Canada based on the most recent Canadian census data and assigns geographical identifiers based on postal codes. P<0.05 was considered to indicate a significant association. Analyses were performed using SAS version 9.2 (SAS Institute, USA) and SPSS version 17.0 (IBM Corporation, USA) for Windows (Microsoft Corporation, USA).

RESULTS

Effectiveness of US surveillance

The stage of tumour at detection by US using end points of Milan criteria and BCLC staging system are listed in Tables 2 and 3, and graphically depicted in Figures 2 and 3, respectively. Seventy-seven percent of tumours were discovered within the Milan criteria through regular US surveillance. Similarly, 80% of patients undergoing regular surveillance had their tumour detected by US in BCLC curative stages of 0 (33%) and A (47%).

When comparing the proportion of patients within and outside the Milan criteria for transplantation, there was a statistically significant difference between patients undergoing regular surveillance versus

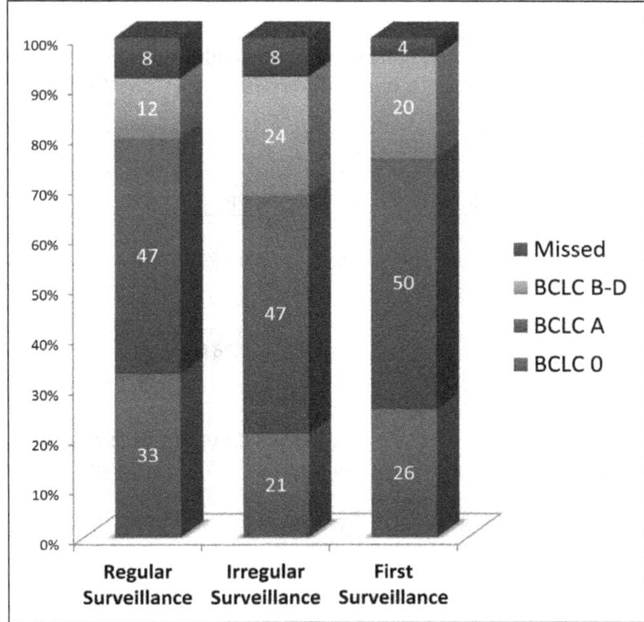

Figure 3) *Hepatocellular carcinoma ultrasound intervals according to the Barcelona Clinic Liver Cancer (BCLC) staging system. Regular surveillance: frequency ≤12 months; Irregular surveillance: frequency >12 months*

those undergoing irregular surveillance (P=0.03). The difference between regular and irregular surveillance did not reach significance when using the BCLC staging system as the end point (P=0.11).

Missed tumours

Of the 201 patients undergoing US surveillance who developed HCC, 14 (7%) had their tumour missed by surveillance US but detected through other means (six by CT, four by a second US and four by AFP) within three months of surveillance. The repeat US were all for unrelated reasons. Thirteen of the patients had a single HCC, and one had three HCCs. The median size of missed tumours was 2.4 cm (range 0.8 cm to 7.8 cm). The BCLC stage of the missed tumours were stage 0 (four of 14 [29%]), stage A (eight of 14 [57%]) and stage C (two of 14 [14%]).

Role of AFP

Of the 201 patients undergoing US surveillance, 129 (64%) had serum AFP levels measured within 180 days of US (from 90 days before to 90 days after). Of these 129 patients, 69 (53%) had serum AFP measured before or on the same day as the surveillance US. Only 26 of 69 (38%) of these measurements were performed after 2005 (AASLD guidelines published) and some may have been performed on the same day as the US scan due to a positive US.

The results of serology are shown in Table 4. The overall sensitivity of AFP detection of tumour within Milan criteria was 32% using a threshold of 20 ng/mL. AFP was able to detect tumour within three months of a negative surveillance US (ie, missed) in four patients, all of whom were undergoing regular surveillance. Because 70 of 129 patients with available serum AFP were undergoing regular US

TABLE 5
Univariate and multivariate analysis of potential determinants of success of hepatocellular carcinoma ultrasound surveillance using Milan criteria for liver transplantation

	Milan criteria, n (%)			OR (95% CI)	
	Outside (n=25)	Within (n=110)	P	Unadjusted	Adjusted
Sex					
Female	8 (32)	21 (19)	0.156	0.50 (0.19–1.32)	
Male	17 (68)	89 (81)		1.00	
Ethnicity					
Caucasian	16 (64)	66 (60)	0.712	1.00	
Other	9 (36)	44 (40)		1.19 (0.48–2.92)	
Surveillance interval					
≤12	13 (52)	84 (76)	0.014	2.98 (1.21–7.33)	2.76 (1.10–6.88)
>12	12 (48)	26 (24)		1.00	1.00
Cause(s) of hepatocellular carcinoma					
Hepatitis B	4 (16)	27 (25)	0.806	1.88 (0.44–7.94)	
Hepatitis C	12 (48)	51 (46)		1.18 (0.37–3.82)	
Combined	4 (16)	14 (13)		0.97 (0.22–4.31)	
Other/unknown	5 (20)	18 (16)		1.00	
Age, years					
Mean ± SD	65.8±9.2	62.1±10.3	0.099*	0.96 (0.92–1.01)	0.97 (0.93–1.01)
Median	66 (50–86)	61 (31–84)			
Detection year					
2000 to 2005	11 (44)	45 (41)	0.777	1.00	
2006 to 2010	14 (56)	65 (59)		1.14 (0.47–2.73)	
Child-Pugh class					
A	8 (32)	28 (25)	0.504	1.38 (0.54–3.54)	
B/C	17 (68)	82 (75)		1.00	
Residence area					
Beyond metropolitan	4 (16)	18 (16)	0.965	1.03 (0.32–3.35)	
Metropolitan area	21 (84)	92 (84)		1.00	

TABLE 4
Serum alpha-fetoprotein (AFP) levels in hepatocellular carcinomas within and outside Milan criteria for liver transplantation

	Milan criteria, n (%)	
AFP, ng/mL	Within	Outside
<20	67 (68)	8 (27)
20–199	23 (23)	10 (33)
200–400	6 (6)	4 (13)
>400	3 (3)	8 (27)

surveillance, the addition of AFP resulted in detection of HCC in four of 70 (6%) of patients undergoing regular US surveillance.

Potential determinants of success of surveillance by US

The results of univariate analysis using Milan transplantation criteria are summarized in Table 5. Regular surveillance was the only statistically significant variable associated with detection of tumour meeting the Milan criteria (P=0.01). Because there were only 25 patients outside the Milan criteria, only two variables with the lowest P values (surveillance interval and age) were used in the multivariate analysis. Regular surveillance remained significantly associated with detection meeting Milan criteria (OR 2.76 [95% CI 1.10 to 6.88]).

The results of univariate analysis using the BCLC staging system as the end point are summarized in Table 6. In none of the variables tested was there a statistically significant correlation with detection in the palliative stages of tumour (BCLC B to D). Table 7 shows the results when using BCLC stage 0 (ie, detection in very early stage of tumour [<2 cm] as an end point). Univariate analysis revealed that the year of detection (2000 to 2005 versus 2006 to 2010) was the only significant variable associated with discovery of tumour <2 cm (stage 0; P=0.02). In an adjusted multivariate analysis, tumours discovered after 2006 had a significantly higher probability of being in BCLC stage 0 (OR 2.38 [95% CI 1.07 to 5.31]).

Six-month versus 12-month surveillance strategy was not analyzed because the former may not have been considered to be the standard of care before publication of the AASLD guidelines in 2005. However, it was noted that of the 166 patients with a previous surveillance history, there was a small rise in the proportion of patients who underwent six-month surveillance after 2005 (38 of 87 [44%]) compared with 2005 and before (21 of 60 [35%]).

TABLE 6
Univariate analysis of potential determinants of success of hepatocellular cancer ultrasound surveillance in the detection of tumour in curable stages (Barcelona Clinic Liver Cancer [BCLC] stages 0 or A)

	BCLC stage			
	B/C/D (n=22)	0/A (n=113)	P	Unadjusted OR (95% CI)
Sex				
Female	7 (32)	22 (19)	0.255	0.52 (0.19–1.42)
Male	15 (68)	91 (81)		1.00
Ethnicity				
Caucasian	14 (64)	68 (60)	0.761	1.00
Other	8 (36)	45 (40)		1.16 (0.45–2.99)
Surveillance interval, months				
≤12	13 (59)	84 (74)	0.146	2.01 (0.78–5.18)
>12	9 (41)	29 (26)		1.00
Cause(s)				
Hepatitis B	2 (9)	29 (26)	0.402	3.05 (0.51–18.34)
Hepatitis C	12 (55)	51 (45)		0.90 (0.26–3.12)
Combined	4 (18)	14 (12)		0.74 (0.16–3.47)
Other/unknown	4 (18)	19 (17)		1.00
Age, years				
Mean ± SD	65.8±9.2	62.2±10.3	0.125	0.96 (0.92–1.01)
Median (range)	65 (50–86)	61 (31–84)		
Year of detection				
2000–2005	10 (45)	46 (41)	0.679	1.00
2006–2010	12 (55)	67 (59)		1.21 (0.48–3.04)
Child-Pugh class				
A	9 (41)	27 (24)	0.099	2.21 (0.85–5.72)
B/C	13 (59)	86 (76)		1.00
Residence area				
Beyond metropolitan	3 (14)	19 (17)	0.770	1.28 (0.34–4.76)
Metropolitan	19 (86)	94 (83)		1.00

Data presented as n (%) unless otherwise indicated

DISCUSSION

The present study showed that in a diverse Canadian population, regular US surveillance performed in a hepatobiliary centre resulted in detection of tumour in curative stages in at least 80% of patients. In addition, at least 77% of patients met the Milan criteria for liver transplantation. Furthermore, regular surveillance resulted in detection of tumour in the curable stages in a significantly higher proportion of patients than in irregular surveillance or screening (first surveillance) populations. In other words, regular surveillance for HCC resulted in stage migration. Tumours detected by other means within three months of negative US surveillance were designated as 'missed' in the present study because some of these would likely be detected within a curable range on a future surveillance scan; therefore, the true effectiveness of US in our centre is likely higher.

Our results show that US surveillance in North America is as effective as the best published rates. A recent meta-analysis of prospective studies assessing the effectiveness of US surveillance for HCC reported a combined sensitivity of 63% for detection of tumour within Milan criteria for transplantation (13). However, most studies used in this particular analysis were old, with only three of 12 study analyses performed in the past decade, and only one after 2006. Imaging technology is continually improving and important sonographic innovations, such as harmonic and compound imaging, have become part of standard-imaging US scanners over the past decade (14). A review of prospective and retrospective US surveillance studies with or without AFP published over the past two years show a sensitivity of 69% to 88% for detection of tumour within Milan criteria (15-19). Furthermore, detection of very early HCC (BCLC stage 0, <2 cm) is reported to be 8% to 43% (15-19). Our rates of 77% for detection of HCC within Milan criteria and 33% for detection of very early HCC are well within this reported range.

Regular surveillance was the only independent variable associated with detection of disease in early, curable stages. Regular surveillance increased the odds of detecting tumour within Milan criteria by a factor of 2.76 (95% CI 1.10 to 6.88). Just as importantly, there was no significant difference in sensitivity of US surveillance between Caucasian and non-Caucasian ethnicity. The present study was the first to directly demonstrate similar sensitivities within different ethnic groups; previously, this was inferred by comparing effectiveness rates published in Europe and Asia (1,3). Furthermore, a significantly higher proportion of tumours were detected in the very early BCLC stage of 0 (<2 cm) in the latter half of our study period; that is, from 2006 to 2010 than in 2000 to 2005 (P=0.02). This is an important finding because treatment of tumours <2 cm in size result in five-year survival rates of up to 90% (5). The reason for increased detection of smaller tumours in 2006 to 2010 may be because of improved US technology; all of our scanners were upgraded in 2006 to the latest generation of high-end equipment. It was also due to increased frequency of

TABLE 7
Univariate analysis of potential determinants of success of hepatocellular cancer ultrasound surveillance in the detection of tumour in very early stage (Barcelona Clinic Liver Cancer [BCLC] stage 0)

	BCLC stage			OR (95% CI)	
Determinant	A/B/C/D (n=91)	0 (n=44)	P	Unadjusted	Adjusted
Sex					
Female	17 (19)	12 (27)	0.255	1.63 (0.70–3.81)	
Male	74 (81)	32 (73)		1.00	
Ethnicity					
Caucasian	55 (60)	27 (61)	0.918	1.00	
Other	36 (40)	17 (39)		0.96 (0.46–2.01)	
Surveillance interval, months					
≤12	62 (68)	35 (80)	0.167	1.82 (0.77–4.28)	1.88 (0.77–4.61)
>12	29 (32)	9 (20)		1.00	1.00
Cause(s)					
Hepatitis B	21 (23)	10 (23)	0.142	3.18 (0.76–13.24)	2.70 (0.63–11.64)
Hepatitis C	38 (42)	25 (57)		4.39 (1.18–16.33)	4.11 (1.07–15.79)
Combined	12 (13)	6 (14)		3.33 (0.70–15.86)	2.93 (0.60–14.41)
Other/unknown	20 (22)	3 (7)		1.00	1.00
Age, years					
Mean	63.2±10.0	61.8±10.6	0.441	0.99 (0.95–1.02)	
Median	63 (31–86)	60.5 (37–82)			
Year of detection					
2000-2005	44 (48)	12 (27)	0.020	1.00	1.00
2006-2010	47 (52)	32 (73)		2.50 (1.14–5.45)	2.38 (1.07–5.31)
Child-Pugh class					
A	26 (29)	10 (23)	0.472	1.36 (0.59–3.15)	
B/C	65 (71)	34 (77)		1.00	
Residence area					
Beyond metropolitan	14 (15)	8 (18)	0.680	1.22 (0.47–3.18)	
Metropolitan	77 (85)	36		1.00	

surveillance because the proportion of patients undergoing six-month surveillance intervals rose from 38% to 44% after the publication of the first HCC management guidelines from the AASLD (20).

In the present study, we used tumour stage at time of detection by US rather than tumour stage at time of treatment. This was done to assess US alone as a surveillance test in a Canadian setting. In a companion study involving the same cohort to assess effectiveness of our surveillance program compared with patients referred to our own institution, we used tumour stage at time of treatment (21). Occasionally, the surveillance test is effective in detection but the work-up strategy fails to diagnose a HCC as malignant. Sometimes a false-positive surveillance test for a benign nodule leads to incidental detection of a HCC by the work-up imaging (such as CT or MRI). Finally, some HCCs are detected on CT/MRI for patients undergoing follow-up imaging for indeterminate nodules detected elsewhere by surveillance US. These scenarios explain why, in the present study, in which US surveillance effectiveness was assessed, tumour stage is BCLC 0/A 83% (Table 1), whereas in our companion study, in which effectiveness of the comprehensive surveillance program was measured, it was 75%.

The present study demonstrates the limited role of serum AFP levels in a surveillance setting. Using a threshold of 20 ng/mL, AFP was able to detect 73% of tumours beyond Milan criteria, but only 32% of tumours within Milan. More importantly, in only 6% of patients did AFP detect a tumour when US was negative. The present study demonstrates that when US is performed effectively, most tumours are detected before reaching a differentiation sufficiently advanced, or a volume sufficiently large, to produce even a low threshold of AFP. The present study supports the AASLD recommendation of elimination of AFP in the setting of regular US surveillance (3). We hope that this result further promotes the abandonment of AFP as a surveillance test, because even in our own practice, we have found multiple examples of noncompliance with the AASLD guidelines in this regard.

Studies that assess clinical performance of US in HCC surveillance demonstrate acceptable sensitivity (15-19). However, studies that assess sensitivity of US with explant correlation show an unacceptably low performance, leading some to advocate its replacement by CT or MRI (7,22). Why the discrepancy? First, the per-patient sensitivity – rather than per-lesion sensitivity – is relevant to US surveillance because all patients with positive US undergo CT or MRI, in which additional nodules can be detected. Second, the assumption with explant correlation studies is that every HCC is significant and, therefore, its detection affects patient survival. However, this is a faulty assumption; it is tumour biology – rather than its mere presence – that affects patient survival (23). Multiple studies have shown that markers of aggressive tumour behaviour, such as poor histological differentiation, serum AFP, vascular invasion and tumour size are all stronger predictors of poor survival outcomes (23-26). The fact that multiple proposed expansions of the Milan criteria for transplantation allows for inclusion of additional tumours is a reflection of the relative lower clinical significant of tumour number (27). Finally, the slow growth rate of early HCC allows for multiple chances for US to detect a nodule; repeated application of the test improves its overall sensitivity.

Strengths and limitations

The present study was the first to investigate HCC surveillance effectiveness using US in Canadian patients with HCC. A major strength was the direct correlation of tumour to the original surveillance scan. By anatomically reviewing the exact location of every HCC to the surveillance US, we ensured that every HCC was truly detected by the US or otherwise designated as missed. Furthermore, the stringent inclusion criteria for the diagnosis of HCC and staging through review of the imaging and clinical features ensured that every nodule included fulfilled current criteria for malignancy and staging systems. Moreover, we accounted for missed tumours that were found by other means. However, the retrospective nature of the present study subjects it to inherent bias. For example, HCCs detected by surveillance in our centre but characterized elsewhere could not be included because the imaging was not available for staging. Our study shows a stage migration through regular surveillance, but we cannot demonstrate a direct survival benefit. In addition, the sample size of 135 patients used for uni- and multivariate analyses limited our ability to demonstrate small but significant differences. Finally, our results are reflective of an academic US department in a hepatobiliary centre with the latest generation of scanners, uniformity of scanning standards, and with continual training for sonographers and direct physician supervision.

CONCLUSION

The present study showed that US surveillance, when regularly performed in expert centres, is an efficacious tool resulting in the detection of HCC in treatable stages in 80% of patients. It also shows no difference in performance of US surveillance in Caucasian versus other ethnic populations, thereby validating its use as the surveillance tool in North America. Finally, the exclusion of serum AFP from HCC surveillance is supported because it improves the detection rate by only 6%.

DISCLOSURES: The authors have no financial disclosures or conflicts of interest to declare.

REFERENCES

1. Omata M, Lesmana LA, Tateishi R, et al. Asian Pacific Association for the Study of the Liver consensus recommendations on hepatocellular carcinoma. Hepatol Int 2010;4:439-74.
2. Zhang BH, Yang BH, Tang ZY. Randomized controlled trial of screening for hepatocellular carcinoma. J Cancer Res Clin Oncol 2004;130:417-22.
3. Bruix J, Sherman M. Management of hepatocellular carcinoma: An update. Hepatology 2011;53:1020-2.
4. Sherman M, Bain V, Villeneuve JP, et al. The management of chronic viral hepatitis: A Canadian consensus conference 2004. Can J Gastroenterol 2004;18:715-28.
5. EASL-EORTC clinical practice guidelines: Management of hepatocellular carcinoma. J Hepatol 2012;56:908-43.
6. Singal AG, Conjeevaram HS, Volk ML, et al. Effectiveness of hepatocellular carcinoma surveillance in patients with cirrhosis. Cancer Epidemiol Biomarkers Prev 2012;21:793-9.
7. Federle MP. Use of radiologic techniques to screen for hepatocellular carcinoma. J Clin Gastroenterol 2002;35(5 Suppl 2):S92-S100.
8. Sarkar M, Stewart S, Yu A, Chen MS, Nguyen TT, Khalili M. Hepatocellular carcinoma screening practices and impact on survival among hepatitis B-infected Asian Americans. J Viral Hepat 2012;19:594-600.
9. Davila JA, Morgan RO, Richardson PA, Du XL, McGlynn KA, El-Serag HB. Use of surveillance for hepatocellular carcinoma among patients with cirrhosis in the United States. Hepatology 2010;52:132-41.
10. Mazzaferro V, Regalia E, Doci R, et al. Liver transplantation for the treatment of small hepatocellular carcinomas in patients with cirrhosis. N Engl J Med 1996;334:693-9.
11. Forner A, Hessheimer AJ, Isabel Real M, Bruix J. Treatment of hepatocellular carcinoma. Crit Rev Oncol Hematol 2006;60:89-98.
12. Statistics Canada. Geography Division. Postal code conversion file. Ottawa: Statistics Canada, 2011. <www.easl.eu/research/our-contributions/clinical-practice-guidelines/detail/easl-eortc-clinical-practice-guidelines-management-of-hepatocellular-carcinoma/report/1> (Accessed January 2, 2014).
13. Singal A, Volk ML, Waljee A, et al. Meta-analysis: Surveillance with ultrasound for early-stage hepatocellular carcinoma in patients with cirrhosis. Aliment Pharmacol Ther 2009;30:37-47.
14. Irshad A, Anis M, Ackerman SJ. Current role of ultrasound in chronic liver disease: Surveillance, diagnosis and management of hepatic neoplasms. Curr Probl Diagn Radiol 2012;41:43-51.
15. Kuo YH, Lu SN, Chen CL, et al. Hepatocellular carcinoma surveillance and appropriate treatment options improve survival for patients with liver cirrhosis. Eur J Cancer 2010;46:744-51.
16. Qian MY, Yuwei JR, Angus P, Schelleman T, Johnson L, Gow P. Efficacy and cost of a hepatocellular carcinoma screening program at an Australian teaching hospital. J Gastroenterol Hepatol 2010;25:951-6.
17. Trinchet JC, Chaffaut C, Bourcier V, et al. Ultrasonographic surveillance of hepatocellular carcinoma in cirrhosis: A randomized trial comparing 3- and 6-month periodicities. Hepatology 2011;54:1987-97.
18. Santi V, Trevisani F, Gramenzi A, et al. Semiannual surveillance is superior to annual surveillance for the detection of early hepatocellular carcinoma and patient survival. J Hepatol 2010;53:291-7.
19. Noda I, Kitamoto M, Nakahara H, et al. Regular surveillance by imaging for early detection and better prognosis of hepatocellular carcinoma in patients infected with hepatitis C virus. J Gastroenterol 2010;45:105-12.
20. Bruix J, Sherman M. Management of hepatocellular carcinoma. Hepatology 2005;42:1208-36.
21. Khalili K, Menezes R, Yazdi LK, et al. Hepatocellular carcioma in a large Canadian urban centre: Stage at treatment and its potential determinants. Can J Gastroenterol Hepatol 2014;28:150-4.
22. Taouli B, Krinsky GA. Diagnostic imaging of hepatocellular carcinoma in patients with cirrhosis before liver transplantation. Liver Transpl 2006;12(11 Suppl 2):S1-7.
23. DuBay D, Sandroussi C, Sandhu L, et al. Liver transplantation for advanced hepatocellular carcinoma using poor tumour differentiation on biopsy as an exclusion criterion. Ann Surg 2011;253:166-72.
24. Wai CT, Woon WA, Tan YM, Lee KH, Tan KC. Younger age and presence of macrovascular invasion were independent significant factors associated with poor disease-free survival in hepatocellular carcinoma patients undergoing living donor liver transplantation. Transplant Proc 2012;44:516-9.
25. Marelli L, Grasso A, Pleguezuelo M, et al. Tumour size and differentiation in predicting recurrence of hepatocellular carcinoma after liver transplantation: External validation of a new prognostic score. Ann Surg Oncol 2008;15:3503-11.
26. Grasso A, Stigliano R, Morisco F, et al. Liver transplantation and recurrent hepatocellular carcinoma: Predictive value of nodule size in a retrospective and explant study. Transplantation 2006;81:1532-41.
27. Yao FY. Liver transplantation for hepatocellular carcinoma: Beyond the Milan criteria. Am J Transplant 2008;8:1982-9.

Mild clinical behaviour of Crohn disease in elderly patients in a Latin American country: A case-control study

Jesús K Yamamoto-Furusho MD PhD MSc[1,2], Andrea Sarmiento-Aguilar MD[1]

JK Yamamoto-Furusho, A Sarmiento-Aguilar. Mild clinical behaviour of Crohn disease in elderly patients in a Latin American country: A case-control study. Can J Gastroenterol Hepatol 2015;29(8):435-439.

BACKGROUND: Crohn disease is characterized by fluctuating clinical behaviour, which is influenced by various factors. There are no data from Latin America that evaluate the clinical behaviour of Crohn disease in elderly patients.

OBJECTIVE: To evaluate the clinical course of elderly onset Crohn disease compared with younger onset in the Mexican population.

METHODS: The present analysis was a case-control study that included 132 patients with a histopathological diagnosis of Crohn disease between 1983 and 2013 in an inflammatory bowel disease clinic of a tertiary care centre. Statistical analysis was performed using SPSS version 17 (IBM Corporation, USA) and descriptive statistics, χ^2 and Fisher's exact test for categorical variables and Student's *t* test for numerical variables. Univariate and multivariate analysis were performed to identify associated risk factors and OR was calculated.

RESULTS: A total of 132 patients (73 men and 59 women) were divided into two groups according to age at diagnosis: 27 cases (>60 years of age) and 105 controls (≤60 years of age). Factors influencing the clinical course of Crohn disease in the elderly were: female sex (OR 2.55 [95% CI 1.06 to 6.10]; P=0.02); colonic location (OR 0.22 [95% CI 0.03 to 0.89]; P=0.02); mild clinical behaviour of disease (OR 10.08 [95% CI 3.74 to 27.17]; P=0.0001); response to medical treatment (OR 2.85 [95% CI 1.08 to 7.48]; P=0.02); frequent use of sulfasalazine (OR 4.46 [95% CI 1.22 to 16.28]; P=0.03); less use of azathioprine (OR 0.38 [95% CI 0.13 to 1.03]; P=0.04); and long-term remission (OR 4.96 [95% CI 1.70 to 14.48]; P=0.002).

CONCLUSION: Elderly patients with Crohn disease had a mild clinical course characterized by the lack of escalation to immunosuppressive and anti-tumour necrosis factor therapy, as well as long-term remission.

Key Words: *Clinical behaviour; Crohn disease; Elderly; Geriatric*

Le comportement clinique bénin de la maladie de Crohn chez les patients âgés d'un pays d'Amérique latine : une étude cas-témoins

HISTORIQUE : Le comportement clinique de la maladie de Crohn fluctue en raison de divers facteurs. Aucune donnée en provenance d'Amérique latine ne l'évalue chez les patients âgés.

OBJECTIF : Évaluer l'évolution clinique de la maladie de Crohn qui se manifeste chez les personnes âgées par rapport à celle qui se manifeste chez des personnes plus jeunes au sein de la population mexicaine.

MÉTHODOLOGIE : La présente analyse était une étude cas-témoins composée de 132 patients ayant obtenu un diagnostic histopathologique de maladie de Crohn entre 1983 et 2013 au sein de la clinique de maladies inflammatoires de l'intestin d'un centre de soins tertiaires. Les chercheurs ont procédé à l'analyse statistique au moyen de la version 17 du SPSS (IBM Corporation, États-Unis), ont établi les variables qualitatives au moyen des statistiques descriptives, du test du χ^2 et du test exact de Fisher et ont établi les variables numériques à l'aide du test *t* de Student. Ils ont effectué des analyses univariée et multivariée pour déterminer les facteurs de risque connexes et ont calculé le risque relatif.

RÉSULTATS : Au total, 132 patients (73 hommes et 59 femmes) ont été répartis entre deux groupes en fonction de leur âge au diagnostic, soit 27 cas (plus de 60 ans) et 105 sujets témoins (60 ans ou moins). Les facteurs qui influaient sur l'évolution clinique de la maladie de Crohn chez les personnes âgées étaient le sexe féminin (RR 2,55 [95 % IC 1,06 à 6,10]; P=0,02), le foyer dans le côlon (RR 0,22 [95 % IC 0,03 à 0,89]; P=0,02), le comportement clinique bénin de la maladie (RR 10,08 [95 % IC 3,74 à 27,17]; P=0,0001), la réponse au traitement médical (RR 2,85 [95 % IC 1,08 à 7,48]; P=0,02), l'utilisation fréquente de sulfasalazine (RR 4,46 [95 % IC 1,22 à 16,28]; P=0,03), la moins grande utilisation d'azathioprine (RR 0,38 [95 % IC 0,13 à 1,03]; P=0,04) et la rémission prolongée (RR 4,96 [95 % IC 1,70 à 14,48]; P=0,002).

CONCLUSION : L'évolution clinique de la maladie était bénigne chez les personnes âgées, ne s'aggravait pas jusqu'à nécessiter un traitement immunosuppresseur et la prise d'inhibiteur du facteur de nécrose tumorale alpha, et s'associait à une rémission prolongée.

Inflammatory bowel disease (IBD) includes Crohn disease (CD) and ulcerative colitis (UC). Little is known regarding its multifactorial and etiopathogenic mechanisms, of which three main factors – genetics, immunity and the environment – are involved (1). The course of CD is characterized by fluctuating clinical behaviour, which is influenced by various factors including hospitalization rates, treatment response, postsurgical recurrence, relapses, exacerbations and older age at diagnosis.

In 2014, the world's population was estimated to be 7.2 billion and, from this number, 8% appeared to be older adults (2); however, by 2050, the number of older persons worldwide is projected to more than double to two billion (3). Population aging is occurring in every country, although each country is at a different stage of this transition (4). For example, by 2050, Mexico's aged are projected to represent 27.7% of the population (5), while the same demographic in the United States is expected to represent 20.3% (6). This fast-growing elderly population represents part of an age group in which important physiological changes occur, including immune deficiency (peculiar to old age), increased frequency of comorbidities and polypharmacy. These characteristics place this age group at higher risk, which undoubtedly affects the clinical course of CD. Despite the fact that fewer than one-third of epidemiological studies have documented a bimodal distribution of IBD incidence with a second peak between 60 and 70 years of age (7), elderly patients with CD should not be treated with aggressive

[1]*Inflammatory Bowel Disease Clinic, Department of Gastroenterology. Instituto Nacional de Ciencias Médicas y Nutrición Salvador Zubirán, Mexico City, Mexico;* [2]*Pan American Crohns and Colitis Organisation (PANCCO)*

Correspondence: Dr Jesús K Yamamoto-Furusho, Head of the Inflammatory Bowel Disease Clinic, Department of Gastroenterology, Instituto Nacional de Ciencias Médicas y Nutrición Salvador Zubirán. President of PANCCO: Vasco de Quiroga 15, Colonia Sección XVI, Tlalpan, CP 14000, Mexico City, Mexico.e-mail kazuofurusho@hotmail.com

therapy because it has been proposed that the disease process may not represent the same disease process present in younger patients (8). Furthermore, 10% to 15% of patients with IBD will be diagnosed at >60 years of age (9-13) (65% during the sixth decade of life, 25% during the seventh decade and 10% during the eighth decade [12,13]). Women are more likely than men to be diagnosed with IBD after 60 years of age (14-18), and there is a higher incidence of UC than CD (14). These phenotypic differences among age groups reflect variations in mucosal immune response, composition of the microbiome, genetics and/or environmental risk factors that are not fully understood (18). Interestingly, very-early onset (zero to six years of age) and elderly onset CD are characterized by the predominance of pure colonic disease (L2, according to the Montreal classification) (18-26).

Several studies have proposed that elderly patients diagnosed with CD have a less aggressive clinical course compared with those diagnosed at a younger age. This suggests that gastrointestinal tract changes occur with aging and produce dietary shifts among older individuals, alterations in gastrointestinal motility and gastric pH due to mucosal atrophy, increased intestinal permeability and changes in the gut microbiota associated with aging that may influence host-inflammatory responses (27-30). For example, the risk for surgery decreases with older age at diagnosis, disease distribution and history of cigarette smoking (31). In CD, the prevalence of diarrhea, abdominal pain, extraintestinal manifestations, weight loss and fever decreases in older or elderly patients (18,32). A large population-based cohort study involving a French population that included 841 IBD patients concluded that the clinical course is mild in elderly onset IBD patients due to disease behaviour, and was reported to remain stable in 91% of patients with elderly onset CD after a median follow-up period of six years (33). Other studies have concluded that the clinical manifestations of the first flare of CD are similar in the >60 years of age and younger age groups (18), characterized by the predominance of inflammatory behaviour (18,21,34,35). To date, no data from genetic studies have been published regarding elderly onset IBD variants to further define the contribution of specific gene associations with elderly onset IBD; however, the role of genetic factors is believed to be greater in pediatric-onset than in late-onset IBD (18). It has also been found that older patients with IBD may have an increased susceptibility to gastrointestinal infection because the response to stress in the setting of acute inflammation may be altered or blunted (28,30,36).

There are some key features considered for elderly onset CD: inflammatory disease behaviour, colonic or ileocolonic disease location, uncommon family history of IBD, and reduced progression to penetrating or stricturing disease phenotypes (37). Prompted by the lack of data from Latin America evaluating the clinical behaviour of CD in the elderly, the aim of the present study was to evaluate the clinical course of elderly compared with younger-onset CD in the Mexican population.

METHODS

The present analysis was a retrospective case-control study that included 132 patients with histopathological diagnosis of CD between 1983 and 2013 in the IBD clinic of a tertiary care centre (National Institute of Medical Sciences and Nutrition "Salvador Zubirán", Mexico). Clinical records were reviewed and a database including the following variables was constructed: sex, age, place of birth, place of residence, weight, height and body mass index; family history of IBD and family or personal history of other immune-mediated diseases; pack-years of smoking (number of cigarettes smoked daily multiplied by the years of smoking divided by 20); history of appendectomy or tonsillectomy; intake of nonsteroidal anti-inflammatory drugs; thrombosis and its location (upper limbs, lower limbs, pulmonary thromboembolism, acute coronary syndrome or vascular brain disease); age at diagnosis; disease phenotype (inflammatory, stricturing or fistulizing); location (terminal ileum, ileocolonic, colonic, upper digestive tract); clinical course (initially active followed by long-term remission, intermittent activity (less or one relapse a year) or continuous activity (≥2 relapses per year); extraintestinal manifestations such as arthritis, arthralgia, ankylosing spondylitis, sacroiliitis, sclerosing cholangitis, pyoderma gangrenosum, erythema nodosum or uveitis; number of hospitalizations; treatment response and reasons for lack of response response and surgical treatment. Other variables included were: post-surgical recurrence; current CD treatment; and clinical activity or remission of the disease at the time of evaluation. The present research was approved by the local ethics committee.

Statistical analysis

Demographic, clinical, and laboratory characteristics are presented as mean ±SD, median and range. The Fisher's exact probability test was used to compare categorical variables when the number of expected subjects was <5 and by the χ^2 test otherwise. The unpaired *t* test was used to compare differences in the means of continuous variables. Nonparametric variables in independent samples were compared using the Mann-Whitney U test. ORs and 95% CIs were calculated using univariate and multivariate analyses adjusted for age, sex, extent of disease, Mayo Score, C-reactive protein level and current medical treatment. P≤0.05 was considered to be statistically significant and Bonferroni correction for P value was applied for multiple comparison calculated as α/n. All statistical analyses were performed using SPSS version 17.0 (IBM Corporation, USA).

RESULTS

A total of 132 patients were evaluated 73 (55.3%) men and 59 (44.7%) women divided in two groups: 27 (20.5%) who were diagnosed at >60 years of age (cases) and 105 (79.5%) who were diagnosed at ≤60 years of age (controls). Detailed demographic and clinical characteristics are summarized in Table 1 and 2, respectively.

CD patients >60 years of age

From the elderly patients, the case group consisted of 27 individuals (10 men and 17 women). The presence of steroid dependency was 3.8% and steroid resistance 0.8%; thiopurine-resistance 0.8% and intolerance to thiopurines 2.3%; and 72.7% undewent intestinal surgical resection. The reasons for surgery were bowel fibrostenotic stricturing (33.7%), bowel perforation (12.7%), intra-abdominal abscess (10.3%), penetrating disease (9%), dysplasia (5%), cancer (1.5%) and toxic megacolon (0.5%).

CD patients ≤60 years of age

The control group included 105 patients (63 men and 42 women). The presence of steroid dependency was 18.5% and steroid-resistance 7%; thiopurine resistance 15% and intolerance to thiopurines 2.5%. All patients underwent intestinal surgical resection for the following reasons: bowel fibrostenotic stricturing (35.2%), penetrating disease (28.57%), intra-abdominal abscess (20.5%), bowel perforation (10.5%), dysplasia (2.3%), cancer (1.5%) and toxic megacolon (1.43%).

Factors influencing the clinical course of CD in elderly patients

The univariate analysis found that the following factors influenced the clinical course of CD in elderly patients: female sex (OR 2.55 [CI 95% 1.06 to 6.10]; P=0.02); a less frequent colonic location (OR=0.22 CI 95% CI 0.03 to 0.89]; P=0.02); mild clinical behaviour of disease (OR 10.08 [95% CI 3.74 to 27.17]; P=0.0001) and a better response to the medical treatment (OR 2.85 [95% CI 95% 1.08 to 7.48]; P=0.02), frequent use of sulfasalazine (OR 4.46 95% CI 1.22 to 16.28]; P=0.03); low use of azathioprine (OR 0.38 [95% CI 95% 0.13 to 1.03]; P=0.04) and higher clinical remission rate (OR 4.96 [95% CI 1.70 to 14.48]; P=0.002) (Table 3).

Significant statistical trends were found in the following factors: elderly patients present pulmonary thromboembolism more frequently (OR 6.53 [95% CI 1.01 to 40.67]; P=0.05), as well as a predominant terminal-ileum location (OR 2.03 [95% CI 0.83 to 4.94]; P=0.09) and a more common inflammatory pattern (OR 0.48 [95% CI 0.18 to 1.24]; P=0.09).

TABLE 1
Demographic variables of patients with Crohn disease

Variable	Cases (n=27)	Controls (n=105)
Sex		
Male	10 (37)	63 (59.05)
Female	17 (63)	42 (40)
Family history of inflammatory bowel disease	0 (0)	1 (0.95)
Smoker	10 (37.03)	47 (44.76)
Autoimmune concomitant disease	3 (11.11)	13 (12.38)
Appendectomy	7 (25.92)	24 (22.85)
Tonsillectomy	3 (11.11)	18 (17.14)

Data presented as n (%)

In the multivariate analysis, the following factors influenced the clincal course of CD in elderly patients: a less frequent colonic location (P=0.021, OR=0.130), a mild clinical course of disease (P=0.0001, OR=0.151), less use of azathioprine (P=0.027, OR=0.255) and inflammatory pattern (P=0.01, OR=0.193).

DISCUSSION

The present study clearly showed a milder clinical course of CD in elderly patients, and was the first performed in the Latin American population. We found that women were predominantly affected, similar to a previous study (the EPIMAD registry [19]) and one study from the United States Military Health Care Population (38). Smoking was associated with increased risk for CD and worse outcomes over the disease course (18,39,40) and, interestingly, our study found a higher proportion of smoking patients in the young group than in the elderly group, which could have also influenced the different outcomes. In the present study, CD patients had a less frequent colonic location and had a more common inflammatory pattern, compared with the French registry, which found that pure colonic disease and inflammatory behaviour were the most frequent phenotypes (14). A study from Hungary (41) found pure colonic disease and a more common stenosing pattern. A retrospective study concluded that older and younger patients underwent surgery in similar percentages (83% versus 77%) (42), a finding similar to that reported in our population (66.6% versus 74.28%).

The elderly patients in the present study exhibited mild clinical disease behaviour, characterized by initially active disease followed by long-term clinical remission, better response to medical treatment and more use of sulfasalazine and less use of azathioprine. These findings were also reported in other studies. For example, the Hungarian study (41) found that their elderly population required less systemic steroids compared with their younger groups. A French study (43) concluded that immunosuppressants were more frequently required in the childhood-onset group than in the elderly onset patients. In this study, it was also found that older adults were less likely to require immunosuppressants or readmission for CD flares compared with younger patients (44). It has even been concluded that corticosteroid response is similar in older and younger patients hospitalized for IBD, but older corticosteroid-responsive patients are less likely to be treated with an anti-tumour necrosis factor agent than younger patients (45). Furthermore, care providers should be aware of polypharmacy and its potential for drug interactions because it has been shown that the prevalence of medication use is higher among patients with IBD than matched members of the general population, particularly the use of analgesic and psychiatric drugs (46). Regarding the extraintestinal manifestations of IBD, thromboembolic events represent a major cause of morbidity and mortality (47,48), and it is known that IBD is an independent risk factor for thromboembolic phenomena (49). One study also found a more frequent prevalence of venous thromboembolism (TE) in elderly patients with IBD (6.15%) compared with the control group (1.62%) (50). This correlates with the fact that our elderly patients tended to present with pulmonary TE more frequently. Other studies have concluded that the most frequent thromboembolic complications are deep vein thromboses and pulmonary emboli, and that this event can occur as a postsurgical complication or spontaneous event (48).

This less agressive clinical behaviour is gradually occupying an important place among the diverse factors that influence the decision-making process regarding the therapeutic approach to an elderly patient

TABLE 2
Clinical characteristics of Crohn disease (CD) patients

Variable	Cases (n=27)	Controls (n=105)
CD pattern		
Inflammatory	7 (25.92)	44 (41.9)
Fistulizing	10 (37.03)	30 (28.57)
Stricturing	10 (37.03)	37 (35.23)
CD location		
Terminal ileum	11 (40.74)	26 (24.76)
Ileocolonic	12 (44.44)	47 (44.76)
Colonic	2 (7.4)	27 (25.71)
Upper digestive tract	2 (7.4)	7 (6.6)
Clinical course		
Initially active followed by long-term remission	19 (70.37)	21 (20)
Intermittent activity	5 (18.51)	51 (48.57)
Continuous activity	2 (7.4)	27 (25.71)
Extraintestinal manifestations		
Arthritis	1 (3.7)	6 (5.71)
Arthralgias	3 (11.11)	23 (21.9)
Ankylosing spondylitis	4 (14.81)	8 (7.61)
Sclerosing cholangitis	0 (0)	1 (0.95)
Pyoderma gangrenosum	0 (0)	3 (2.85)
Erythema nodosum	0 (0)	2 (1.9)
Uveitis	1 (3.7)	0 (0)
Lack of response to medical treatment	10 (37.03)	70 (66.6)
Surgical treatment	18 (66.6)	78 (74.28)
Postsurgical recurrence	1 (3.7)	10 (9.52)
Current treatment		
Sulfasalazine	4 (14.81)	6 (5.71)
Mesalazine	4 (14.81)	23 (21.9)
Azathioprine	6 (22.22)	31 (29.52)
Prednisone	4 (14.81)	25 (23.8)
Methotrexate	0 (0)	1 (0.95)
Biologic therapy	0 (0)	2 (1.9)

Data presented as n (%)

TABLE 3
Factors influencing the course of Crohn disease in the elderly population

Variable	Univariate analysis	Multivariate analysis
Female sex	P=0.02 OR=2.55	P=0.54
Colonic location	P=0.02 OR=0.22	P=0.03 OR=0.13
Mild clinical course of disease	P=0.0001 OR=10.08	P=0.0001 OR=0.15
Response to medical treatment	P=0.02 OR=2.85	P=0.31
Use of sulfasalazine	P=0.03 OR=4.46	P=0.54
Use of azathioprine	P=0.04 OR=0.38	P=0.02 OR=0.25
Clinical remission	P=0.002 OR=4.96	P=0.22

diagnosed with CD. Nevertheless, it should not be considered without taking into account all other aspects that play a role in each specific case. For example, one current approach to drug therapy in the elderly is to 'start low; go slow' and then reassess their candidacy for more aggressive therapy (biologics, apheresis, surgery), and not treat or exclude patients on the basis of age alone (51). In conclusion, elderly patients with CD had a mild disease course characterized by long-term remission, less use of aggressive therapy, such as thiopurines and anti-tumour necrosis factor agents, and better response to medical treatment.

DISCLOSURES: The authors have no financial disclosures or conflicts of interest to declare.

CONCLUSIONS

Elderly patients had a less aggressive clinical course compared with younger patients, and commonly achieved remission without the need of escalating the treatment beyond aminosalycilates.

There are no data from Latin America that evaluate the clinical behaviour of CD in elderly patients. The present study showed that elderly patients with CD had a mild clinical course characterized by the lack of escalation to immunosuppressive and anti-tumour necrosis factor therapy, as well as long-term remission.

REFERENCES

1. Shanahan F. The microbiota and inflammatory bowel disease: Friend, bystander and sometime villain. Nutr Rev 2012;70:S31-S37.
2. 2014 World Population Data Sheet. Population Reference Bureau. <www.prb.org/pdf14/2014-world-population-data-sheet_eng.pdf> (Accesed October 11, 2014).
3. United Nations. Population ageing and sustainable development. Department of Economic and Social Affairs. <www.un.org/en/development/desa/population/publications/ pdf/popfacts/PopFacts_2014-4.pdf> (Accesed October 11, 2014).
4. World Health Organization. Good Health adds life to years. Global Brief for World Health Day 2012. <http://whqlibdoc.who.int/hq/2012/WHO_DCO_WHD_2012.2_eng.pdf?ua=1> (Accesed October 11, 2014)
5. Proyecciones de la población de México 2005-2050. Consejo Nacional de Población.<www.conapo.gob.mx/es/CONAPO/Documento_Metodologico_Proyecciones_2010_2050 > (Accesed October 13, 2014)
6. Wiener JM, Tilly J. Population ageing in the United States of America, International Epidemiological Association 2002. Int J Epidemiol 2002;31:776-781.
7. Molodecky NA, Soon IS, Rabi DM, et al. Increasing incidence and prevalence of the inflammatory bowel disease with time, based on systematic review. Gastroenterology 2012;142:46-54.
8. Katz S, Pardi DS. Inflammatory bowel disease of the elderly: Frequently asked questions (FAQs). Am J Gastroenterol 2011;11:1889-97.
9. Kelsen J, Baldassano RN. Inflammatory bowel disease: The difference between children and adults. Inflamm Bowel Dis 2008;14:S9-S11.
10. Travis S. Is IBD different in the elderly? Inflamm Bowel Dis 2008;14:S12-S13.
11. Bernstein CN, Wajda A, Svenson LW, et al. The epidemiology of inflammatory bowel disease in Canada: A population-based study. Am J Gastroenterol 2006;101:1559-68.
12. Loftus EV Jr, Silverstein MD, Sandborn WJ, et al. Crohn's disease in Olmsted County, Minnesota, 1940-1993: Incidence, prevalence, and survival. Gastroenterology 1998;114:1161-8.
13. Loftus EV Jr, Silverstein MD, Sandborn WJ, et al. Ulcerative colitis in Olmsted County, Minnesota, 1940-1993: Incidence, prevalence, and survival. Gut 2000;46:336-43.
14. Charpentier C, Salleron J, Savoye G, et al. Natural history of elderly-onset inflammatory bowel disease: A population-based cohort study. Gut 2013;63:423-4.
15. Mamula P, Telega GW, Markowitz JE, et al. Inflammatory bowel disease in children 5 years of age and younger. Am J Gastroenterol 2002;97:2005-10.
16. Shivananda S, Lennard-Jones J, Logan R, et al. Incidence of inflammatory bowel disease across Europe: Is there a difference between north and south? Results of the European Collaborative Study on Inflammatory Bowel Disease (EC-IBD). Gut 1996;39:690-7.
17. Gupta N, Bostrom AG, Krischner BS, et al. Gender differences in presentation and course of disease in pediatric patients with Crohn disease. Pediatrics 2007;120:1418-25.
18. Ruel J, Ruane D, Mehandru S, Gower-Rosseau C, Colombel JF. IBD across the age spectrum – is it the same disease? Nat Rev Gastroenterol Hepatol 2014;11:88-98.
19. Gower-Rousseau C, Vasseur F, Fumery M, et al. Epidemiology of inflammatory bowel diseases: New insights from a French population-based registry (EPIMAD). Dig Liver Dis 2013;45:89-94.
20. Guariso G, Gasparetto M, Visonà Dalla Pozza L, et al. Inflammatory bowel disease developing in paediatric and adult age. J Pediatr Gastroenterol Nutr 2010;51:698-707.
21. Vernier-Massouille G, Balde M, Salleron J, et al. Natural history of pediatric Crohn's disease: A population-based cohort study. Gastroenterology 2008;135:1106-13.
22. Heyman MB, Kirschner BS, Gold BD, et al. Children with early-onset inflammatory bowel disease (IBD): Analysis of a pediatric IBD consortium registry. J Pediatr 2005;146:35-40.
23. Paul T, Birnbaum A, Pal DK, et al. Distinct phenotype of early childhood inflammatory bowel disease. J Clin Gastroenterol 2006;40:583-6.
24. Langholz E, Munkholm P, Krasilnikoff PA, Binder V. Inflammatory bowel diseases with onset in childhood. Clinical features, morbidity, and mortality in a regional cohort. Scand J Gastroenterol 1997;32:139-47.
25. Freeman HJ. Comparison of longstanding pediatric-onset and adult-onset Crohn's disease. J Pediatr Gastroenterol Nutr 2004;39:183-6.
26. Müller KE, Lakatos PL, Arató A, et al. Incidence, Paris classification and follow-up in a nationwide, incident cohort of pediatric patients with inflammatory bowel disease. J Pediatr Gastroenterol Nutr 2013;57:576-82.
27. Britton E, McLaughlin JT. Ageing and the gut. Proc Nutr Soc 2013;72:173-7.
28. Schiffrin EJ, Morley JE, Donnet-Hughes A, Guigoz Y. The inflammatory status of the elderly: The intestinal contribution. Mutat Res 2010;690:50-6.
29. Biagi E, Candela M, Fairweather-Tait S, Franceschi C, Brigidi P. Aging of the human metaorganism: The microbial counterpart. Age 2012;34:247-67.
30. Ha CY, Katz S. Clinical implications of ageing for the management of IBD. Nat Rev Gastroenterol Hepatol 2013;11:128-38.
31. Tremaine WJ, Timmons LJ, Loftus EV Jr, et al. Age at onset of inflammatory bowel disease and the risk of surgery for non-neoplastic bowel disease. Aliment Pharmacol Ther 2007;25:1435-41.
32. Malaty HM, Fan X, Opekun AR, Thibodeaux C, Ferry GD. Rising incidence of inflammatory bowel disease among children: A 12-year study. J Pediatr Gastroenterol Nutr 2010;50:27-31.
33. Charpentier C, Salleron J, Savoye G, et al. Natural history of elderly-onset inflammatory bowel disease: A population-based cohort study. Gut 2013;63:423-32.
34. Kugathasan S, Judd RH, Hoffmann RG, et al. Epidemiologic and clinical characteristics of children with newly diagnosed inflammatory bowel disease in Wisconsin: A statewide population-based study. J Pediatr 2003;143:525-31.
35. Sauer CG, Kugathasan S. Pediatric inflammatory bowel disease: Highlighting pediatric differences in IBD. Med Clin North Am 2010;94:35-52.
36. Cambier J. Immunosenescence: A problem of lymphopoiesis, homeostasis, microenvironment, and signaling. Immunol Rev 2005;205:5-6.
37. Ha C, Katz S. Elderly-onset IBD: A milder disease? Nat Rev Gastroenterol Hepatol 2013;10:264-65.
38. Betteridge JD, Armbruster SP, Maydonovitch C, Verappan GR. Inflammatory bowel disease prevalence by age, gender, race, and geographic location in the U.S. military health care population. Inflamm Bowel Dis 2013;19:1421-7.
39. Clegg A, Young J, Iliffe S, Rikkert MO, Rockwood K. Frailty in elderly people. Lancet 2013;381:752-62.

40. Karczewski J, Poniedzialek B, Rzymski P, Rychlewska-Hanczewska A, Adamski Z, Wiktorowicz K. The effect of cigarette smoking on the clinical course of inflammatory bowel disease. Prz Gastroenterol 2014;9:153-9.
41. Lakatos PL, David G, Pandur T, et al. IBD in the elderly population: Results from a population-based study in Western Hungary, 1977-2008. J Crohns Colitis 2011;5:5-13.
42. Wagtmans MJ, Verspaget HW, Lamers CB, Van Hogezand RA. Crohn's disease in the elderly: A comparison with young adults. J Clin Gastroenterol 1998;27:129-33.
43. Pigneur B, Seksik P, Viola S, et al. Natural history of Crohn's disease: Comparison between childhood- and adult-onset disease. Inflamm Bowel Dis 2010;16:953-61.
44. Heresbach D, Alexandre JL, Bretagne JF, et al. Crohn's disease in the over-60 age group: A population based study. Eur J Gastroenterol Hepatol 2004;16:657-64.
45. Weber NK, Bruining DH, Loftus EV Jr, et al. Comparative outcomes of younger and older hospitalized patients with inflammatory bowel disease treated with corticosteroids. Inflamm Bowel Dis 2013;19:2644-51.
46. Buckley JP, Kappelman MD, Allen JK, Van Meter SA, Cook SF. The burden of comedication among patients with inflammatory bowel disease. Inflamm Bowel Dis 2013;19:2725-36.
47. Dolapcioglu C, Soylu A, Kendir T, et al. Coagulation parameters in inflammatory bowel disease. Int J Clin Exp Med 2014;7:1442-8.
48. Talbot RW, Heppell J, Dozois RR, Beart RW Jr. Vascular complications of inflammatory bowel disease. Mayo Clin Proc 1986;61:140-5.
49. Cosnes J, Beaugerie L, Carbonnel F, Gendre JP. Smoking cessation and the course of Crohn's disease: An intervention study. Gastroenterology 2001;120:1093-9.
50. Miehsler W, Reinisch W, Valic E, et al. Is inflammatory bowel disease an independent and disease specific risk factor for thromboembolism? Gut 2004;53:542-8.
51. Suárez FC, Vera Mendoza MI, Amo San Román L, et al. Risk of thromboembolic phenomena in patients with inflammatory bowel disease. Gastroenterol Hepatol 2014;35:634-9.

Resident trainees do not affect patient satisfaction in an outpatient gastroenterology clinic: A prospective study conducted in a Canadian gastroenterology clinic

Mayur Brahmania MD FRCPC, Madison Young MD, Chetty Muthiah MD FRCPC, Alexandra Ilnyckyj MD FRCPC, Donald Duerksen MD FRCPC, Dana C Moffatt MD FRCPC

M Brahmania, M Young, C Muthiah, A Ilnyckyj, D Duerksen, DC Moffatt. Resident trainees do not affect patient satisfaction in an outpatient gastroenterology clinic: A prospective study conducted in a Canadian gastroenterology clinic. Can J Gastroenterol Hepatol 2015;29(7):363-368.

BACKGROUND: There is little literature regarding how a gastroenterology trainee affects a patient's interpretation of care during outpatient clinic visits. Improving patient satisfaction is desirable and benefits may include enhanced patient compliance as well as providing trainees with areas for improvement.

OBJECTIVES: To evaluate patient satisfaction in an outpatient gastroenterology clinic when seen by a trainee and attending physician versus an attending physician alone. The secondary objective was to evaluate physician characteristics that play a role in creating a positive clinical experience.

METHODS: A randomized prospective survey study was conducted over an 11-month period (July 2012 to June 2013) at St Boniface Hospital (Winnipeg, Manitoba). Two gastroenterology fellows (postgraduate year 4 and 5) and nine internal medicine residents (postgraduate year 1 to 3) comprised the 'trainee' role, while three academic clinicians comprised the 'attending' role. Patients included individuals seen for an initial consultation and were >18 years of age.

RESULTS: A total of 211 patients comprised the final study group, with 118 in the attending group and 93 in the trainee group. In univariate analysis, patients more often had a very good experience when seen by an attending physician alone versus a trainee and attending physician (73% versus 56%; P=0.016); however, on multivariate analysis, there was no significant difference in patient satisfaction (OR 0.89; P=0.931). Physician factors found to be associated with high patient satisfaction on multivariate analysis included: addressing all patient concerns (OR 27.56; P=0.021); giving the patient a preliminary diagnosis (OR 78.02; P=0.006); and feeling the physician was thorough (OR 72.53; P=0.029).

CONCLUSIONS: The present study did not reveal a difference in patient satisfaction if a patient sees an attending physician alone or with a trainee. Moreover, to improve patient satisfaction in a gastroenterology clinic, physicians should address all patient concerns, provide a preliminary diagnosis and appear to be thorough in their assessment. Further work to increase patient awareness on the role of residents in teaching hospitals is warranted to further promote careers in gastroenterology.

Key Words: *Gastroenterology residents; Health care quality; Medical education; Patient care; Patient satisfaction*

Les résidents ne nuisent pas à la satisfaction des patients dans une clinique ambulatoire de gastroentérologie : une étude prospective au sein d'une clinique canadienne de gastroentérologie

HISTORIQUE : Peu de publications scientifiques portent sur l'effet qu'ont les stagiaires en gastroentérologie sur l'interprétation que les patients se font des soins lors de leur rendez-vous ambulatoire. Il est souhaitable d'améliorer la satisfaction des patients, ce qui peut favoriser une meilleure adhésion au traitement et la transmission de conseils aux stagiaires pour qu'ils s'améliorent.

OBJECTIFS : Évaluer la satisfaction des patients dans une clinique de gastroentérologie ambulatoire lorsqu'ils sont vus par un stagiaire et un médecin ou par un médecin seulement. L'objectif secondaire consistait à évaluer les caractéristiques du médecin qui contribuaient à une expérience clinique positive.

MÉTHODOLOGIE : Les chercheurs ont réalisé une étude prospective aléatoire d'une durée de 11 mois (juillet 2012 à juin 2013) à l'Hôpital Saint-Boniface de Winnipeg, au Manitoba. Deux aspirants spécialistes en gastroentérologie (en 4e et 5e année) et neuf résidents en médecine interne (de 1re à 3e année) représentaient les stagiaires, et trois cliniciens universitaires, les médecins. Les patients, qui avaient tous plus de 18 ans, étaient vus pour une première consultation.

RÉSULTATS : Au total, 211 patients ont formé les groupes d'étude définitifs, dont 118, le groupe du médecin, et 93, le groupe des stagiaires. À l'analyse univariée, les patients vivaient plus souvent une très bonne expérience auprès d'un médecin seul qu'auprès d'un stagiaire et un médecin (73 % par rapport à 56 %; P=0,016). Cependant, à l'analyse multivariée, on n'observait pas de différence significative en matière de satisfaction des patients (RR 0,89; P=0,931). Les facteurs liés aux médecins qui s'associaient à une forte satisfaction des patients à l'analyse multivariée incluaient la réponse à toutes les inquiétudes des patients (RR 27,56; P=0,021), la transmission d'un diagnostic préliminaire aux patients (RR 78,02; P=0,006) et l'impression que le médecin transmettait une évaluation approfondie (RR 72,53; P=0,029).

CONCLUSIONS : La présente étude n'a pas révélé de différence dans la satisfaction des patients lorsqu'ils voient un médecin ou un médecin et un stagiaire. Pour accroître la satisfaction des patients à la clinique de gastroentérologie, les médecins devraient répondre à toutes les inquiétudes des patients, fournir un diagnostic préliminaire et sembler effectuer une évaluation approfondie. Des mesures s'imposent pour mieux sensibiliser les patients au rôle des résidents dans les hôpitaux universitaires afin de promouvoir des carrières en gastroentérologie.

Patient satisfaction is a quality indicator of health care delivery used to improve clinical effectiveness. However, patient satisfaction is influenced by many factors including technical factors, interpersonal factors and meeting patient expectations. Improving patient satisfaction has desirable effects such as reducing complaints, enhancing compliance and preventing recurrent physician visits (1-4). The field of

Department of Medicine, Division of Gastroenterology, University of Manitoba, St Boniface Hospital, Winnipeg, Manitoba
Correspondence: Dr Mayur Brahmania, Department of Medicine, Division of Gastroenterology, University of Manitoba, 804F-175 McDermot Avenue, Winnipeg, Manitoba R3E 3P4. e-mail mbrahmania@gmail.com

gastroenterology (GI) incorporates both office- and procedure-based patient contact. To date, most studies have focused on enhancing patient satisfaction in the procedural component of the specialty focusing on endoscopy, purgatives and sedation techniques (5-7). Although these areas are continually evolving, GI outpatient clinics provide another area in which patient satisfaction initiatives can be undertaken. Recently, Larkin et al (8) showed that patient satisfaction in a GI clinic was highest when the physician was deemed to be a good listener and an effective treatment plan was explained. Interestingly, long wait times before the visit and clinic esthetics had no effect on overall satisfaction. Currently, there are no studies investigating the influence trainees have on patient satisfaction in an outpatient GI clinic. The education of GI trainees is key to developing future health care professionals who can continue research, education and care for the aging population. Thus, it would be beneficial for training programs to be aware of elements of care in which a trainee's education can improve and also possibly identify patient needs during consultation. Accordingly, the primary objective of our study was to evaluate patient satisfaction in an outpatient GI clinic when seen by a trainee and attending physician compared with an attending physician alone. Our secondary objective was to determine physician characteristics that play a role in patient satisfaction in a GI clinic.

METHODS

Patients and definitions

The present analysis was a prospective single-blinded study conducted at St Boniface Hospital, a tertiary care centre, located in Winnipeg, Manitoba, over an 11-month period (July 2012 to June 2013). The clinic is an outpatient facility for university-based resident teaching and clinical care. The study was conducted over a 120-day clinic period. Eleven trainees, including two GI fellows (postgraduate year [PGY] 4 and 5) and nine internal medicine residents (three PGY 1, three PGY 2 and three PGY 3) comprised the 'trainee' population. The GI fellows each had one half-day clinic per week, and internal medicine residents on GI rotation attended two half-day clinics per week. All academic clinicians at the hospital (n=3) participated in the present study comprising the 'attending' role. The Canadian health care system has no merit-based compensation programs and, thus, no specific information regarding a specific attending or trainee satisfaction score was gathered. The trainee or attending physician saw patients in chronological order based on appointment time. In other words, no discrimination regarding the type of case (ie, irritable bowel syndrome or celiac disease) or patient characteristic, etc, seen by a trainee was made. Trainees initially saw the patient individually and recommended their impressions and plans to the patient. Trainees then reviewed care plans with the attending physician in a separate private room, and both parties would then proceed to reaffirm the trainees plan or further clarify the history, physical examination or management plan. The attending physician would move on to see the next patient if the trainee was still in consultation to continue clinic flow.

Participation was entirely voluntary and, if agreeable, a survey was provided by the scheduling assistant after consultation. Patients were offered enrollment sequentially from July 2012 to June 2013. Attending physicians and trainees were blinded to patient participation, and results were kept confidential and anonymous. No deviation from normal care was undertaken if the patient was enrolled in the study. Attending physicians and trainees were not aware of the components of the survey aside from two of the authors (MB [trainee], DM [attending]). Incomplete surveys were counted as a refusal to participate. Patients included those seen for an initial consultation and were >18 years of age. Exclusion criteria included patients being seen for follow-up, and those with physical or intellectual disabilities that would make the survey difficult to complete. Demographic characteristics, such as age, sex, education level and location of residence, were gathered in addition to factors related to the clinic experience (time in waiting room, experience with booking clerk, time from referral to consultation, etc).

Survey components

In 1999, the Accreditation Council for Graduate Medical Education (ACGME) established the 'general competencies' for patient care. These included domains of patient care, medical knowledge, practice-based learning and improvement, interpersonal and communication skills, professionalism and systems-based practice. The ACGME competency mandate requires all residency programs to train and evaluate resident performance in these areas (9). Similarly, in 2005, the Royal College of Physicians and Surgeons of Canada developed a framework (CanMEDS) based on seven roles that all physicians must embody to be the 'ideal' physician. Areas included: Medical Expert, Communicator, Collaborator, Manager, Health Advocate, Scholar and Professional. These seven roles are also used to train and evaluate residents and medical students in all realms of medicine in Canada (10). The authors developed their own GI-specific survey (Appendix 1) incorporating the key features of the ACGME/CanMEDS roles, reflecting the ideal qualities in a physician as well as questions adapted from subsections of three standardized patient satisfaction surveys that have been previously validated: the Patient Satisfaction Questionnaire Short-Form (PSQ-18) developed by the RAND Corporation (The RAND Corporation is a nonprofit institution that helps improve policy and decision making through research and analysis) (11); the Consumer Assessment of Health Plans Survey originally commissioned by the Agency for Healthcare Research and Quality (12); and the Visit Specific Satisfaction Questionnaire-9 (VSQ-9) (13). Questions from these surveys formed the body of the study that was of primary interest. Answers were recorded using a five-point Likert scale or in yes/no format primarily to encourage participation by offering an easy and quick survey acknowledging many patients are returning to work, paying for metered parking or trying to meet other appointments. The study received Research Ethics Board approval from the University of Manitoba and St Boniface Hospital, and was conducted in accordance with all national and institutional guidelines. No external funding was obtained for the present study.

Statistical analysis

Descriptive statistics used frequencies and percentages for the survey questions. Survey responses of the two groups were compared using a Mann-Whitney U test for ordinal variables and a χ^2 test for nonordinal variables. To investigate predictors of patient satisfaction, multivariate logistic regression was used. Six potential predictors of patient satisfaction in the logistic regression model were examined: time spent with the patient, all concerns and questions addressed, receiving an adequate diagnosis and explanation, confidence in physician, perceived competence of physician and perceived thoroughness of physician. Finally, to investigate whether patient satisfaction differed between the groups after accounting for the effect of the six predictors, a final ordinal logistic regression model that included a term for the group effect as well as the potential predictors was examined. In addition to these variables of interest, all models adjusted for patient characteristics including age, sex and education. All P values were two-sided; statistical significance was defined as $P<0.05$. Data were analyzed using the R statistical software version 2.14.1 (R, USA).

RESULTS

Baseline demographics and patient characteristics

A total of 272 patients volunteered to complete the survey after initial consultation. Sixty-one patient surveys were excluded due to incomplete responses (ie, ≥1 question[s] not answered). Of the 61 incomplete surveys, 27 were from the attending group and 34 were from the attending and trainee group ($P>0.05$). A total of 211 patients comprised the final study group, with 118 in the attending group and 93 in the trainee group (Figure 1).

Table 1 highlights patient characteristics. There was a significant difference in age, with the attending group seeing a higher percentage of patients who were >65 years of age (19% versus 11%) and between the 46 and 55 years of age (30% versus 16%), while the trainee and

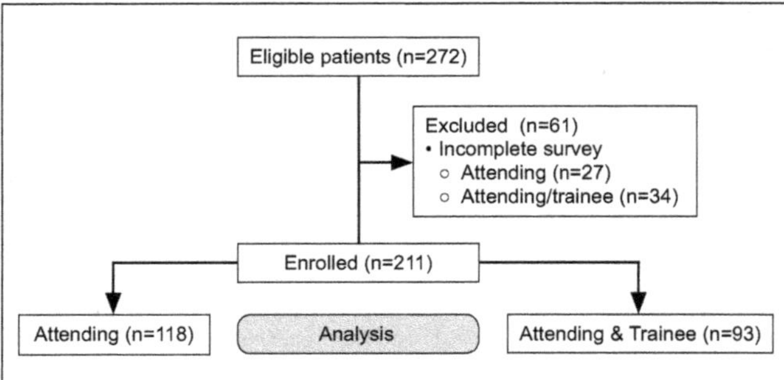

Figure 1) *Patient enrollment*

TABLE 1
Baseline demographics and patient characteristics

	Physician		
Characteristic	**Attending (n=118)**	**Trainee/ attending (n=93)**	**P***
Age, years			
18–25	16 (14)	12 (13)	0.019
26–35	12 (10)	21 (23)	
36–45	19 (16)	24 (26)	
46–55	35 (30)	15 (16)	
56–65	14 (12)	11 (12)	
>65	22 (19)	10 (11)	
Male sex, %	37.9	40.2	0.88
Residence (Winnipeg, Manitoba)	84 (71)	79 (85)	0.018
Education			
Elementary school	4 (3)	0 (0)	0.001
High school	45 (38)	23 (25)	
Some postsecondary	37 (31)	24 (26)	
University/college degree	27 (23)	40 (43)	
Postgraduate degree	5 (4)	6 (6)	
Referral by family physician	38 (32)	20 (22)	0.084
Time to consult, months			
<1	17 (14)	10 (11)	0.453
3	40 (34)	29 (31)	
6	38 (32)	35 (38)	
9	13 (11)	14 (15)	
>12	10 (8)	5 (5)	
Duration of symptoms, months			
<1	9 (8)	9 (10)	0.301
3	10 (8)	6 (6)	
6	16 (14)	11 (12)	
9	15 (13)	4 (4)	
>12	68 (58)	63 (68)	

*Data presented as n (%) unless otherwise indicated. *Mann-Whitney U-test for ordinal variables and χ^2 test for categorical variables*

attending group saw patients between 26 and 35 (23% versus 10%) and 36 and 45 years of age (26% versus 16%) (P=0.019). There was also a significant difference in place of residence because the trainee and attending groups saw patients from Winnipeg at a greater percentage (85% versus 71%) (P=0.018). Education level was also significantly different between the two groups, with the trainee and attending group seeing patients who had some university/college degree (43% versus 23%), while the attending group saw more patients that had some high school education (38% versus 25%) (P=0.001). There were no differences between the two groups in referral pattern according to general practitioner (P=0.084), time to consult (P=0.453) and duration of symptoms (P=0.301).

TABLE 2
Univariate analysis of effect of trainee on patient satisfaction

	Physician		
	Attending (n=118)	**Trainee/ attending (n=93)**	**P***
Overall patient satisfaction			
Not good whatsoever	0 (0)	1 (1)	0.016
Bad	2 (2)	1 (1)	
Neutral	7 (6)	7 (8)	
Good	23 (19)	32 (34)	
Very good	86 (73)	52 (56)	
Patient has an idea of the cause of symptom			
Not at all	46 (39)	25 (27)	0.051
Not much	18 (15)	12 (13)	
Not sure	24 (20)	24 (26)	
A bit	12 (10)	15 (16)	
Yes	18 (15)	17 (18)	
Time spent with patient, min			
≤5	3 (3)	1 (1)	<0.001
10	24 (20)	8 (9)	
15	41 (35)	20 (22)	
20	32 (27)	30 (32)	
≥25	18 (15)	34 (37)	
Greeted appropriately	117 (99)	91 (98)	0.427
Physician introduction	117 (99)	90 (97)	0.209
Physician role explained	112 (95)	81 (87)	0.044
Physician could have spent more time with you	6 (5)	3 (3)	0.507
Would return to clinic (yes)	115 (97)	91 (98)	0.853

*Data presented as n (%) unless otherwise indicated. *Mann-Whitney U-test for ordinal variables and χ^2 test for categorical variables*

Patients reported a higher level of 'very good' overall patient satisfaction when seen by an attending alone versus an attending and trainee (73% versus 56%) (P=0.016) (Table 2). Conversely, patients reported a higher level of 'good' overall patient satisfaction when seen by an attending and trainee versus attending physician alone (34% versus 19%). Patients had neutral, bad or not good-at-all experiences <10% of the time during the study period. Time spent with the patient was also significantly different between the two groups, with the trainee/attending group spending a greater proportion of time >25 min with a patient (37% versus 15%) (P≤0.001). In both groups, at least 98% of patients were greeted appropriately, 95% were satisfied with the time physicians spent with them, and at least 97% of patients would return for clinic follow-up.

In multivariate analysis, three variables were associated with overall patient satisfaction: addressing all patient concerns (OR 27.56; P=0.021) (Table 3), giving the patient a preliminary diagnosis (OR 78.02; P≤0.006) and feeling the physician was thorough with the evaluation (OR 72.53; P=0.029). After accounting for the six variables chosen on physician characteristics (time spent, confidence, competence, addressing concerns, giving a diagnosis and being thorough), patient satisfaction was no different when seen by an attending alone versus attending and trainee (OR 0.89; P=0.931) (Table 3).

DISCUSSION

The importance of patient satisfaction has been previously shown to improve compliance with treatment and improved health outcomes (1-4). Satisfaction studies involving the endoscopic aspect of GI have found technical factors do not have a role in patient satisfaction, stressing the importance of communication and developing a therapeutic alliance with the patient (14-16). In our prospective study, we found no difference in patient satisfaction when seen by trainees and attending versus attending alone because most patients had a good to

TABLE 3
Results of overall patient satisfaction: multivariate logistic regression*

	Model A†		Model B‡	
Predictor	**OR (95% CI)**	**P**	**OR (95% CI)**	**P**
Time spent with patient	0.46 (0.04–5.56)	0.536	0.49 (0.03–7.69)	0.611
Confidence in physician	9.96 (0.49–100)	0.139	8.70 (0.44–100)	0.156
Competence in physician	0.08 (0–7.69)	0.277	0.08 (0–7.69)	0.276
All concerns addressed	27.56 (1.67–458)	0.021	27.33 (1.64–454)	0.021
Given a preliminary diagnosis	78.02 (3.45–1762)	0.006	81.39 (3.07–2161)	0.009
Thoroughness of physician	72.53 (1.56–3384)	0.029	72.74 (1.55–3415)	0.029
Group (trainee/attending versus attending)	–	–	0.89 (0.06–13.74)	0.931

**Both models are adjusted for age, sex and education; †Model A examines the predictors of increased patient satisfaction with clinical experience; ‡Model B examines the group effect adjusting for all predictors in model A*

very good experience (92% attending versus 90% trainee and attending) with the consultation process. However, patients did rate their experience as 'very good' more often when seen by an attending physician alone; however, when the groups 'very good' and 'good' were grouped together in multivariate analysis, this difference between attending and trainee/attending was not apparent (P=0.931) (Table 3). The difference observed in univariate analysis may have been the result of the attending physicians seeing older, less educated patients and those from an area outside of our hospital's city, where physicians may be held in a higher social regard/authority. However, this was accounted for in our multivariate model and it did not remain statistically significant, suggesting an unmeasured variable accounting for this difference not captured by our quantitative survey-based analysis. As stated above, no special selection of patients occurred (ie, trainees seeing more interesting or less complex cases). If we are to believe that a difference exists between a 'very good' and 'good' consultation process, a set of criteria must be established to define this entity; however, this currently does not exist in any survey or quality benchmark and was not the purpose of our study. Nevertheless, we must acknowledge this difference and surmise the inexperience of trainees or personal manners may be playing a role, which has been previously been shown in a study by Yancy et al (17) and other studies (18-20) examining patient satisfaction in trainee ambulatory clinics not involving GI. Previous studies also have shown resident training level correlates with satisfaction, with junior residents scoring less than more senior residents. Reasons for this again include inexperience and the limited ability to deal with complex medical care seen in academic referral centres (21). In our study, we had PGY1 to PGY5 residents participating, and it is unclear whether junior residents had an effect because we could not account for this from our data due the survey design. If this were a significant issue, the difference between the trainee and attending group would have strengthened, which was not the case in our study.

Ways to improve trainee performance and enhance patient satisfaction have been studied using techniques such as receiving psychosocial training before patient care, enhancing supervision during important clinical decision-making time points and improving nonverbal communication styles (22-24). In our study, we found three factors significantly correlating with overall patient satisfaction (addressing patient concerns, providing a diagnosis and appearing to be thorough with the consultation). Therefore, most of the satisfaction patients derive from a physician visit relates to the history and physical examination rather than physician 'kindness', wait times or if the physician was punctual. These factors emphasize the value patients place on good communication skills. Interestingly, arriving at the 'correct' clinical diagnosis may not mean as much to patients as it does to the physician. However, helping patients understand their disease or reduce their anxiety may have its own therapeutic value. For example, providing a patient with a preliminary diagnosis (ie, "you most likely have irritable bowel syndrome" and addressing any concerns/questions a patient may have during consultation, ie, "Do I have cancer?" was strongly correlated with patient satisfaction (OR 27.56). This was independent of which provider group the patients had seen. This has important implications as training programs have to ensure trainees have sufficient knowledge to confidently deliver a diagnosis, relative to their training level and be able to acknowledge their limitations while aiming for high patient satisfaction.

We must stress that other important factors likely contribute to overall patient satisfaction and these may occur even before the patient sees a physician. In our study, we found significant correlations between the interactions with the booking clerk and overall satisfaction; however, in the multivariate model, these were not significant. This suggests that many factors contribute to creating a positive or negative experience in clinic and, of course, confounders outside of the scope of the present study undoubtedly impact patient satisfaction.

Our study had several limitations. First, two authors knew the components of the survey (one was an attending physician [DM] and the other a trainee [MB]), which may have led to a change in clinician behaviour (ie, Hawthorne effect). However, sustaining a changed behaviour over a prolonged period would have been difficult and, given the relatively small size of our institution, the study design and concept would have been difficult to complete if we excluded physicians involved in the study design. Second, the study may have limited generalizability because patients were from one discipline (ie, GI); therefore, the results may not be generalizable to other specialties. However, given that patients seen in this clinic were referred from primary care community physicians, we believe these results would likely be similar at other university-based residency programs with trainee clinics, irrespective of location/specialty. Third, although our survey has not been validated in other studies, we believe that no optimal questionnaire existed to answer what patient satisfaction meant to the patient; therefore, we developed our own questionnaire using components of the ACGME/CanMEDs and above-mentioned questionnaires. A future prospective study validating the study is ongoing. Finally, almost all patients were satisfied or very satisfied, raising the question as to whether selection bias occurred and unsatisfied patients refused to complete the questionnaire. However, if a patient was unsatisfied, the survey would have provided an opportunity to express these concerns or negative feelings in an anonymous fashion because surveys were completed in a private area and an area for free-text comments was also offered. Conversely, it is possible that a 'social-desirability bias' occurred and patients were attempting to 'please' the physicians and trainees by giving generally positive scores (25). We did attempt to limit these types of bias by assuring patient confidentiality and by giving surveys after the consult. We acknowledge the predictors of overall satisfaction we found (ie, addressing patient concerns, etc) could actually be considered an individual component of satisfaction, thus, they may be ultimately correlated with 'overall' satisfaction.

CONCLUSION

In the present study, we identified specific elements of the physician patient interaction appearing to improve patient satisfaction during a first GI consultation. Addressing all patient concerns and providing a preliminary diagnosis while appearing thorough in the consultation process all contribute to a positive patient experience. Further work in validating a GI-specific patient satisfaction survey is needed; however, trainees should be reminded the emphasis on patient care is solely not on the medical management of disease but also the social aspects the patient's life. Our study showed that GI trainees do not affect patient satisfaction in GI clinics and, thus, gastroenterologists should welcome trainees into routine clinical practice to enhance learning while also promoting interest in GI.

ACKNOWLEDGEMENTS: The authors thank Angela Vandendrissche for data collection and Matt Kowgier for statistical analysis.

FUNDING: This study was internally funded.

DISCLOSURES: The authors have no financial disclosures or conflicts of interest to declare.

AUTHOR CONTRIBUTIONS: Mayur Brahmania: study concept and design; acquisition of data; analysis and interpretation of data; drafting of the manuscript; critical revision of the manuscript for important intellectual content; statistical analysis; administrative, technical or material support; study supervision). Madison Young: acquisition of data; analysis and interpretation of data). Chetty Muthiah: study concept and design; drafting of the manuscript; critical revision of the manuscript for important intellectual content; statistical analysis. Alexandra Ilnyckyj: critical revision of the manuscript for important intellectual content. Donald Duerksen: critical revision of the manuscript for important intellectual content. Dana Moffatt: study concept and design; acquisition of data; analysis and interpretation of data; drafting of the manuscript; critical revision of the manuscript for important intellectual content; statistical analysis; administrative, technical or material support; study supervision.

APPENDIX 1

SECTION 1 – PATIENT DEMOGRAPHICS

Please mark the appropriate box for each question or fill in the blank where needed

1. Age
   ☐ 18-25 ☐ 26-35 ☐ 36-45
   ☐ 46-55 ☐ 56-65 ☐ >65
2. Gender
   ☐ Female ☐ Male
3. Highest education level
   ☐ Elementary School
   ☐ High School
   ☐ Some Post Secondary Education
   ☐ University/College Degree
   ☐ Post Graduate Degree
4. Marital status
   ☐ Single ☐ Married ☐ Common law
   ☐ Divorced ☐ Widowed
5. City of residence:
   ☐ Winnipeg ☐ Other

SECTION 2 – CLINIC EXPERIENCE

1. Did you get parking easily? *(please check one)*
   ☐ Yes ☐ No
2. How would you rate your experience with the booking clerk?
   1 2 3 4 5
   Not Pleasant ... Very Pleasant
3. For how long did you have your gastrointestinal symptoms/problem before you saw the doctor?
   1 2 3 4 5
   <1mo 3mo 6mo 9mo >1 year
4. Do you have an idea about the cause of your gastrointestinal symptoms/problem?
   1 2 3 4 5
   Not at all ... Yes
5. Who referred you to the GI specialist?
   ☐ Family Doctor ☐ Nurse practitioner ☐ Another specialist
6. How long did you wait for this appointment after the referral?
   1 2 3 4 5
   <1mo 3mo 6mo 9mo > 1year
7. How long was your wait in the clinic today prior to seeing the doctor?
   1 2 3 4 5
   <5min 15min 30min 45min >1hr
8. How long was your trip to the hospital clinic?
   1 2 3 4 5
   <5min 15min 30min 45min >1hr

SECTION 3 – DOCTOR EXPERIENCE

Which physician(s) did you see in the clinic? *(please check one)*

☐ Gastroenterologist (1 doctor) – If seen by one doctor, go to section 3A

☐ Gastroenterologist and Trainee (2 doctors) – If seen by two doctors, go to section 3B

SECTION 3A – DOCTOR ONLY EXPERIENCE

(Please fill in this section if you were seen by one doctor only)

1. How much time did the doctor spend with you?
   <5min 10min 15min 20min >25min
2. Were you and family/friends greeted appropriately?
   ☐ Yes ☐ No
3. Did the doctor introduce her/himself appropriately?
   ☐ Yes ☐ No
4. Did the doctor explain her/his role to you?
   ☐ Yes ☐ No
5. Were your questions/concerns addressed?
   1 2 3 4 5
   None were addressed ... All concerns addressed
6. Do you feel that the doctor could have spent more time with you?
   ☐ Yes ☐ No

IF Yes: How much more time would be reasonable ______________

7. Were you given a diagnosis or explanation of the possible causes of symptoms and was it explained well to you?
   1 2 3 4 5
   No diagnosis/Not explained ... Well explained
8. How satisfied were you with your overall experience with the doctor?
   1 2 3 4 5
   Not satisfied at all ... Very satisfied
9. How confident do you feel with the doctor's diagnosis/management plan?
   1 2 3 4 5
   Not confident at all ... Very confident
10. How competent do you feel the doctor was?
    1 2 3 4 5
    Not competent at all ... Very competent
11. How thorough do you feel the doctor was?
    1 2 3 4 5
    Not thorough at all ... Very thorough
12. Would you return to this clinic?
    ☐ Yes ☐ No
13. How was your experience overall?
    1 2 3 4 5
    Not good at all ... Very good

SECTION 3B – DOCTOR AND TRAINEE EXPERIENCE

(Please fill in this section if you were seen by two doctors)

1. How much time did the trainee and doctor spend with you?
   <5min 10min 15min 20min >25min
2. Were you and family/friends greeted appropriately?
   ☐ Yes ☐ No
3. Did the doctor and trainee introduce themselves appropriately?
   ☐ Yes ☐ No
4. Did the trainee and doctor explain their roles to you?
   ☐ Yes ☐ No

5. Were your questions/concerns addressed?
   1 2 3 4 5
   None were addressed ... All concerns addressed
6. Do you feel that the doctor and trainee could have spent more time with you?
   ☐ Yes ☐ No
   IF Yes: How much more time would be reasonable____________
7. Were you given a diagnosis or explanation of the possible causes of symptoms and was it explained well to you?
   1 2 3 4 5
   No diagnosis/Not explained ... Well explained
8. Did the trainee's diagnosis differ from the gastroenterologist?
   ☐ Yes ☐ No
9. How satisfied were you with your overall experience with the trainee and doctor?
   1 2 3 4 5
   Not satisfied at all ... Very satisfied
10. How confident do you feel with the doctor's diagnosis/management plan?
   1 2 3 4 5
   Not confident at all ... Very confident
11. How competent do you feel the trainee and doctor were?
   1 2 3 4 5
   Not competent at all ... Very competent
12. How thorough do you feel the trainee and doctor were?
   1 2 3 4 5
   Not thorough at all ... Very thorough
13. Would you return to this clinic?
   ☐ Yes ☐ No
14. Would you like to have a trainee see you again prior to the doctor again if give the choice?
   ☐ Yes ☐ No
15. How was your experience overall?
   1 2 3 4 5
   Not good at all ... Very good

SECTION 4

1. Any additional comments?

____
____
____

REFERENCES

1. Frances V, Korsch BM, Morris MJ. Gaps in doctor-patient communications. N Engl J Med 1969;280:535-40.
2. Davis MS. Variation in patients' compliance with doctors' orders: Medical practice and doctor-patient interaction. Psychiatry Med 1971;2:31-54.
3. Kasteler J, Kane RL, Olsen DM, et al. Issues underlying prevalence of "doctor-shopping" behavior. J Health Soc Behav 1976;17:329-39.
4. Vaccarino JM. Malpractice: The problem in perspective. JAMA 1977;238:861-3.
5. Johnson DA, Barkun AN, Cohen LB, et al. US Multi-Society Task Force on Colorectal Cancer. Optimizing adequacy of bowel cleansing for colonoscopy: Recommendations from the US multi-society task force on colorectal cancer. Gastroenterology 2014;147:903-24.
6. Ko HH, Zhang H, Telford JJ, Enns R. Factors influencing patient satisfaction when undergoing endoscopic procedures. Gastrointest Endosc 2009;69:883-91.
7. Leffler DA, Bukoye B, Sawhney M, et al. Development and validation of the PROcedural Sedation Assessment Survey (PROSAS) for assessment of procedural sedation quality. Gastrointest Endosc 2015;81:194-203.
8. Larkins AS, Windsor AV, Trebble TM. An evaluation of patient attitudes to the gastroenterology outpatient experience. Eur J Gastroenterol Hepatol 2013;25:44-55.
9. Hechtel L, Chang T, Tseng E, et al. ACGME Outcome Project. Common Program Requirements: general competencies. February 2007. <www.acgme.org/outcome/comp/compCPRL.asp> (Accessed May 4, 2015).
10. Frank JR, ed. The CanMEDS 2005 Physician Competency Framework. Better standards. Better physicians. Better care. Ottawa: The Royal College of Physicians and Surgeons of Canada, 2005.
11. RAND Health. Patient satisfaction questionnaire. Instructions for scoring the PSQ-18. April 2007. <www.rand.org/health/surveys_tools/psq/index.html> (Accessed May 4, 2015).
12. Consumer Assessment Surveys. Agency for Healthcare Research and Quality. September 2012. <https://cahps.ahrq.gov/surveys-guidance/cg/visit/index.html> (Accessed May 4, 2015).
13. Davies AR, Ware JE. GHAA's Consumer Satisfaction Survey and User's Manual, 2nd edn. 1991.
14. Jackson JL, Osgard E, Fincher RK. Resident participation in flexible sigmoidoscopy does not affect patient satisfaction. Am J Gastroenterol 2000;95:1563-6.
15. Qureshi M, Shafqat F, Ahmed S, Niazi TK, Khokhar N. Factors affecting patient satisfaction during endoscopic procedures. J Coll Physicians Surg Pak 2013;23:775-9.
16. McEntire J1, Sahota J, Hydes T, Trebble TM. An evaluation of patient attitudes to colonoscopy and the importance of endoscopist interaction and the endoscopy environment to satisfaction and value. Scand J Gastroenterol 2013;48:366-73.
17. Yancy WS Jr, Macpherson DS, Hanusa BH, et al. Patient satisfaction in resident and attending ambulatory care clinics. J Gen Intern Med 2001;16:755-62.
18. Huynh M, Lee AD, Miller LM, Davis S, Feldman SR, McMichael A. Patients' satisfaction with dermatology residents. South Med J 2012;105:520-3.
19. Monk SM, Nanagas MT, Fitch JL, Stolfi A, Pickoff AS. Comparison of resident and faculty patient satisfaction surveys in a pediatric ambulatory clinic. Teach Learn Med 2006;18:343-7.
20. Issa MA, Kim CH. Patient satisfaction with residents vs. attending following fluoroscopy-guided pain injections. Pain Med 2012;13:185-9.
21. Nakar S, Levi D, Rosenberg R, Vinker S. Patient attitudes to being treated by junior residents in the community. Patient Educ Couns 2010;78:111-6.
22. Smith RC, Lyles JS, Mettler JA, et al. A strategy for improving patient satisfaction by the intensive training of residents in psychosocial medicine: A controlled, randomized study. Acad Med 1995;70:729-32.
23. Farnan JM, Petty LA, Georgitis E, et al. A systematic review: The effect of clinical supervision on patient and residency education outcomes. Acad Med 2012;87:428-42.
24. Griffith CH, Wilson JF, Langer S, Haist SA. House staff nonverbal communication skills and standardized patient satisfaction. J Gen Intern Med 2003;18:170-4.
25. Dalton D, Ortegren, M. Gender differences in ethics research: The importance of controlling for the social desirability response bias. J Business Ethics 2011;103:73-93.

Endoscopy in Canada: Proceedings of the National Roundtable

Noah Switzer MD[1], Elijah Dixon MD BSc MScEpi FRCSC FACS[2], Jill Tinmouth MD PhD[3], Nori Bradley MD[4], Melina Vassiliou MD MEd FRCSC[5], Steve Schwaitzberg MD FACS[6], Anthony Gomes MD LMCC FRCSC[7], James Ellsmere MD MSc FRCSC[8], Chris de Gara MB MS FRCS FACS FFSTEd[1]

N Switzer, E Dixon, J Tinmouth, et al. Endoscopy in Canada: Proceedings of the National Roundtable. Can J Gastroenterol Hepatol 2015;29(5):259-265.

This 2014 roundtable discussion, hosted by the Canadian Association of General Surgeons, brought together general surgeons and gastroenterologists with expertise in endoscopy from across Canada to discuss the state of endoscopy in Canada. The focus of the roundtable was the evaluation of the competence of general surgeons at endoscopy, reviewing quality assurance parameters for high-quality endoscopy, measuring and assessing surgical resident preparedness for endoscopy practice, evaluating credentialing programs for the endosuite and predicting the future of endoscopic services in Canada. The roundtable noted several important observations. There exist inadequacies in both resident training and the assessment of competency in endoscopy. From these observations, several collaborative recommendations were then stated. These included the need for a formal and standardized system of both accreditation and training endoscopists.

Key Words: *Credentialing; Endoscopy; Quality assurance parameters; Resident training*

L'endoscopie au Canada : les délibérations de la table ronde nationale

Une table ronde, organisée en 2014 par l'Association canadienne des chirurgiens généraux, a réuni les chirurgiens généraux et les gastroentérologues ayant des compétences en endoscopie du Canada entier afin de discuter du statut de l'endoscopie au Canada. Cette table ronde visait à évaluer les compétences des chirurgiens généraux en endoscopie, l'examen des paramètres de contrôle de la qualité pour garantir des endoscopies de qualité, la mesure et l'évaluation de la préparation des résidents en chirurgie à effectuer des endoscopies, l'évaluation des programmes d'agrément à l'endosuite et l'avenir des services endoscopiques au Canada. La table ronde a fait ressortir plusieurs observations importantes. Il y a des inadéquations dans la formation des résidents et l'évaluation des compétences en endoscopie. Plusieurs recommandations consensuelles sont ressorties de ces observations, y compris un système officiel et standardisé d'agrément et de formation des endoscopistes.

This 2014 roundtable discussion, hosted by the Canadian Association of General Surgeons (CAGS) and chaired by Dr Elijah Dixon, president of CAGS, brought together general surgeons and gastroenterologists with expertise in endoscopy from across Canada to discuss the state of endoscopy in Canada. The major themes for the roundtable discussion were defining competence in endoscopy and identifying the current major issues facing endoscopists in Canada. The goals of the roundtable included the following:

- Review the current patterns of practice for endoscopy in Canada, including urban and rural areas.
- Review how general surgeons are currently being trained in endoscopy.
- Review what guidelines regarding endoscopy in Canada currently exist.
- Review the role of endoscopy in the 'community' general surgeon practice.
- Understand the role general surgeons play in the provision of endoscopic services in Canada now and in the future.
- Identify issues that may impair the ability of general surgeons to provide endoscopic services in Canada.
- Strike an endoscopy taskforce and identify either a chair or two co-chairs who will develop a plan of action for the future to help general surgeons provide high-quality endoscopic services throughout Canada in a collaborative fashion with their gastroenterologist colleagues.

ENDOSCOPY IN CANADA AND CURRENT ISSUES FOR THE GENERAL SURGEON

Dr C de Gara: Department of Surgery, University of Alberta, Edmonton, Alberta

General surgeons perform approximately 50% of the colonoscopies in Canada (Figure 1) (1). The remainder of the distribution of endoscopy services is provided, in large part, by gastroenterologists, but also by other specialists including general internists and family physicians.

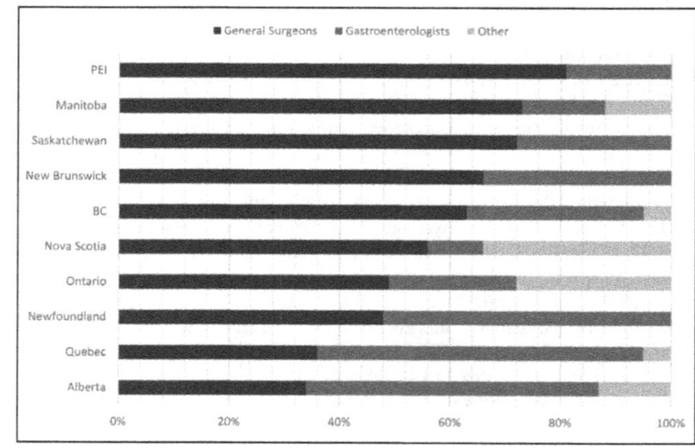

Figure 1) *Provincial division of labour for the delivery of colonoscopy. BC British Columbia; PEI Prince Edward Island. Adapted with permission from reference 1*

[1]*Department of Surgery, University of Alberta, Edmonton;* [2]*Department of Surgery, University of Calgary, Calgary, Alberta;* [3]*Division of Gastroenterology, University of Toronto, Toronto, Ontario;* [4]*Department of Surgery, University of British Columbia, Vancouver, British Columbia;* [5]*Department of Surgery, McGill University, Montreal, Quebec;* [6]*Department of Surgery, Harvard Medical School, Boston, Massachusetts, USA;* [7]*Division of Surgery, University of Alberta, Lethbridge, Alberta;* [8]*Department of Surgery, Dalhousie University, Halifax, Nova Scotia*

Correspondence: Dr Chris de Gara, Department of Surgery, University of Alberta, 2-590 ECHA, 11405-87 Avenue Northwest, Edmonton, Alberta T6G 2W6 e-mail cdegara@ualberta.ca

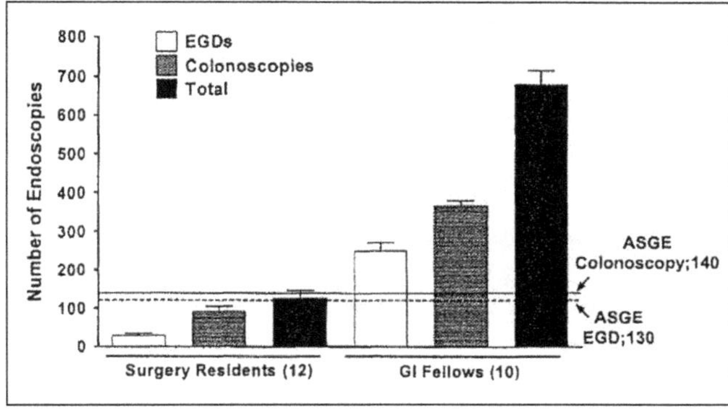

Figure 2) *Gastroenterology (GI) fellows versus surgical residents on achieving minimum numbers. ASGE American Society for Gastrointestinal Endoscopy; EGD Esophagogastroduodenoscopy. Reproduced with permission from reference 2*

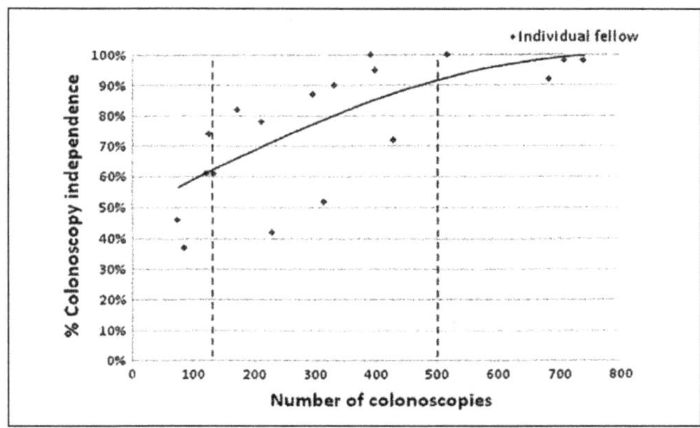

Figure 3) *Number of colonoscopies required for minimum standard of competency. Reproduced with permission from reference 3*

The suggested definition of competency in colonoscopy is defined as a cecal intubation rate (CIR) >90%; however, there are many other measures of competence including adenoma detection rate (ADR), sedation usage, withdrawal time, interval colorectal cancer detection rates and surveillance recommendations. Although it has been widely debated whether absolute procedural numbers alone equate to adequate training in endoscopy, measuring the competency of trainees from both general surgery and gastroenterology residency training programs ultimately becomes more of a numbers game.

While no Royal College of Physicians and Surgeons of Canada minimum figures exist for the number of endoscopic procedures performed, the guidelines established by the American Society for Gastrointestinal Endoscopy (ASGE) recommend a minimum of 130 endoscopies and 140 colonoscopies (2) (Figure 2). Using that mark as a threshold, there is evidence to suggest that surgical residents consistently fail to meet these most basic requirements, while gastroenterology fellows easily achieve significantly higher numbers than the stated threshold (2). In fact, even this minimum threshold has been shown to be far too low – Spier et al (3) showed that after 140 colonoscopies, no gastroenterology fellow achieved a CIR >90% (Figure 3). Rather, it required 500 colonoscopies before all of the fellows achieved this competency standard. Therefore, the question this poses is, are general surgeons sufficiently competent to provide endoscopy services?

How this difference in Canadian training programs translates into practice is debatable. A review of the Canadian literature reveals lack of consensus on the issue, with some studies reporting no difference in missed colorectal neoplasm rates among specialties (4), and others showing that nonsurgeons/nongastroenterologists performing endoscopy to be an independent risk factor for new or missed colorectal cancers (5,6). Still others report that colonoscopy performed by nongastroenterologists (including surgeons) is independently associated with subsequent incident colorectal cancer rates (7). Other studies have reported that surgeons have higher rates of interval cancers, recommend colonoscopy at inappropriately short intervals, and are more likely to require the assistance of and costs associated with anesthesiologists for colonoscopy compared with gastroenterologists (8-10).

There remain many questions moving forward in the discussion of competency training and endoscopy in Canada. How and who should be assessing the qualitative and quantitative measures of competence including CIRs and ADRs, sedation usage, withdrawal times, interval colorectal cancer detection rates and surveillance advice times? Should Canada implement a group similar to the Joint Advisory Group on Gastrointestinal Endoscopy (JAG) established in the United Kingdom (UK)? JAG sets standards for individual endoscopists in colonoscopy training, and in quality assurance of both the endoscopy units and of the training courses. Ultimately, opening discussion forums, such as this retreat, among experts who provide endoscopic services will lead to these questions being answered.

QUALITY OF COLONOSCOPY: EVIDENCE AND THE CANCER CARE ONTARIO PERSPECTIVE

Dr J Tinmouth: Division of Gastroenterology, University of Toronto, Toronto, Ontario

Cancer Care Ontario recently released a report entitled, 'Guidelines for Colonoscopy Quality Assurance in Ontario', with the objective of providing the basis for a quality-assurance program for all colonoscopy procedures in Ontario (11). This was an evidence-based review conducted by the Program in Evidence-Based Care, focusing on providing recommendations for three key aspects of colonoscopy: training and maintenance of competency for endoscopists; institutional quality-assurance parameters and performance indicators; and auditable outcomes for colonoscopy.

Training and maintenance of competency

The guidelines recommend minimum training periods for credentialing in endoscopy based on guidance for credentialing in endoscopy, as recommended by the Canadian Association of Gastroenterology. For gastroenterologists, a formal subspecialty training program (≥2 years) must be completed for credentialing, while for other physicians, notably surgical residents, at least six months of technical training and knowledge acquisition in colonoscopy is recommended. These training periods were selected as the minimum believed to be required to successfully achieve three vital parameters: ≥85% CIR; ≥300 colonoscopies; and cognitive proficiency in aspects of the procedure (knowledge of the indications, contraindications and surveillance recommendations). However, the training period and volume criteria alone cannot ensure competence, demonstrated proficiency in the cognitive aspects of colonoscopy and in the technical aspects of colonoscopy (eg, cecal intubation) is required.

To maintain competency, a minimum of 200 colonoscopies per year are then recommended, with a minimum CIR of 95% in patients with adequate bowel preparation and no obstructive lesions. The evidence supporting this recommendation stems from reports that low-volume endoscopists have higher complication rates and may have higher rates of missed colorectal cancers (6,8,12-14). Additional endoscopic training is recommended for physicians who have been away from practice for three years or who have performed <6 months of endoscopy in the past five years (15).

Institutional recommendations

The guideline lists quality assurance parameters for patient assessment before the procedure, infection control, periprocedural monitoring with conscious sedation, resuscitation capability and acceptable endoscope quality.

Performance indicators and auditable outcomes

A distinction is drawn in the report between quality indicators, where sufficient evidence exists to recommend targets, and auditable

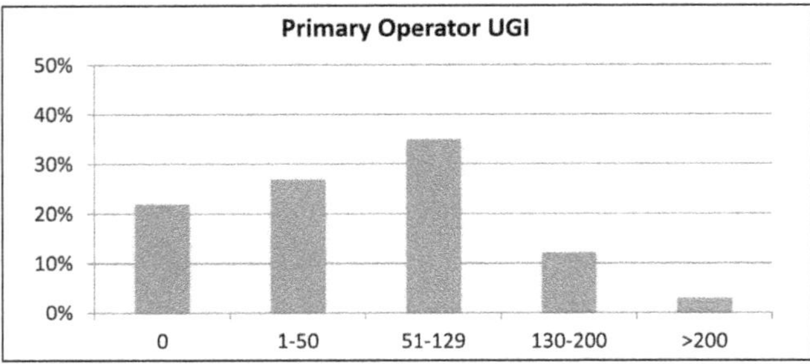

Figure 4) *Volume of upper gastrointestinal (UGI) procedures performed by Canadian surgical residents across all years of training. Adapted with permission from the Canadian Association of General Surgeons Residency Committee Survey*

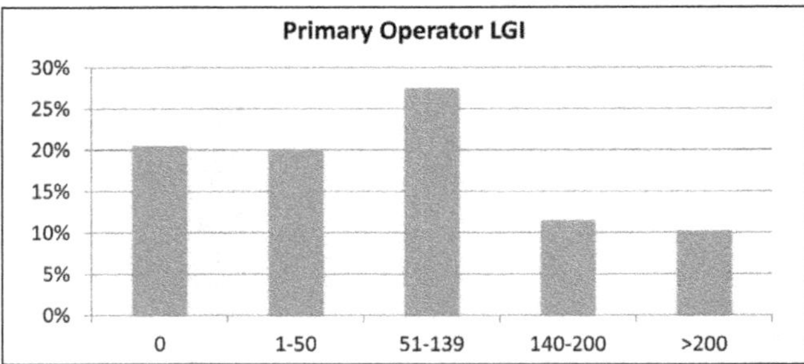

Figure 5) *Volume of lower gastrointestinal (LGI) procedures performed by Canadian surgical residents across all years of training. Adapted with permission from the Canadian Association of General Surgeons Residency Committee Survey*

outcomes, where there is insufficient evidence to recommend targets; however, monitoring should still be performed.

The three quality indicators included are CIR, bleeding rates post-polypectomy and perforation rates. The targeted values are: CIR ≥90% unadjusted (for bowel preparation, indication, strictures, previous colonic surgery or severe colitis); or ≥95% adjusted, postpolypectomy bleeding rate <1 per 100 colonoscopies with polypectomy and perforation rate <1 per 1000 overall.

The four recommended auditable outcomes included are ADR, polypectomy rate, bowel preparation, withdrawal time and postcolonoscopy cancer detection rates. Setting targets for these parameters in Ontario was not believed to be appropriate because there was insufficient evidence and/or because further auditing of the local experience was required to establish appropriate context-specific targets. For example, ADR and PR have been shown to be highly variable in the literature, making setting a specific target challenging (16-18). Withdrawal time (19) was not recommended as an auditable outcome because it was believed to inconsistently reflect high-quality endoscopy and because there were other more robust quality indicators/ auditable outcomes available.

ENDOSCOPY IN CANADA: THE GENERAL SURGERY RESIDENT PERSPECTIVE

Dr N Bradley: Department of Surgery, University of British Columbia, Vancouver, British Columbia

The CAGS Residents Committee, with the support of the CAGS Executive and Education Committees, assessed resident experience in endoscopy training in Canadian general surgery training programs. An online survey was distributed to all program directors at English-speaking residency programs. The goal was to characterize current endoscopy curricula, training, supervision, support, competence and perception of readiness for practice. One hundred fifty-six of 463 (34%) Canadian general surgery residents, from all English-speaking programs, participated in the study. There was approximately equal representation across residency years postgraduate year (PGY) 1 through PGY5/6.

The majority (93%) of respondents reported that their program encompassed a formal endoscopy rotation. The curriculum itself varied, consisting of either a formal gastroenterology or endoscopy block (24%) versus longitudinal training (33%), or a combination of the two (31%). The length of the formal endoscopy training in Canada varied among the programs: 12 weeks (37%), eight weeks (40%) and four weeks (17%). Most often, PGY2 (66%) was the clinical year in which the formal gastroenterology rotation occurred.

Across all years, 84% of applicants failed to meet the minimum requirement set forth by the ASGE for upper gastrointestinal (UGI) procedural numbers. Twenty-two percent of respondents have yet to complete a single UGI endoscopy as the primary operator, 27% completed one to 50, 35% completed between 51 and 129, 12% completed between 130 and 200, and only 3% completed >200 (Figure 4).

Only 22% of respondents met the minimum criteria set forth by the ASGE for lower gastrointestinal (LGI) procedural numbers. Almost one-quarter (21%) of respondents had yet to complete a single LGI endoscopy as the primary operator, 20% competed between one and 50, 28% completed 51 to 139, 12% completed 140 to 200, and 10% completed >200 (Figure 5).

Graduating residents (n=26) were also polled specifically regarding their comfort with endoscopy quality indicators and procedures. The overwhelming majority (97% to 100%) of graduating residents were comfortable with the cognitive components of endoscopy including indications, obtaining consent and sedation administration. Most (88% to 100%) graduates reported being prepared for basic UGI endoscopy tasks such as UGI endoscopy to the second part of the duodenum, biopsies and synoptic reporting. However, more advanced therapeutic procedures, such as variceal banding, esophageal stent insertion, esophageal dilation and endoscopic retrograde cholangiopancreatography were outside the 'comfort zone' of 58% to 92% of graduates. With respect to colonoscopy, the overwhelming majority of graduating residents were prepared to perform basic motor skills independently, including reaching the cecum 85% of the time (81%), completing a screening colonoscopy in ≤30 min (88%) and a withdrawal time of >6 min (96%). Similarly, 81% to 100% of residents stated being comfortable with detecting polyps or other pathology, obtaining biopsies, tattooing lesions, using flexible or rigid sigmoidoscopy, and synoptic reporting of LGI endoscopy.

However, graduating residents were not confident with their skills in emergent procedures such as foreign body extraction and management of acute gastrointestinal bleed; 4% to 50% of graduates were comfortable only with back-up support available. Finally, 39% of graduating residents believed that the endoscopy training they received as residents had not prepared them for their planned surgical practice.

This study was limited by its 34% participation rate, subjective responses and underestimation of procedure volumes throughout an entire general surgery residency. However, it highlights the variability of current endoscopy training within general surgery residency programs and raises the important point that graduating residents appear to feel underprepared for independent practice even though they appear to be comfortable with basic UGI and LGI skills. Increasing emphasis should be placed on competency-based training, credentialing requirements and quality indicators. A standardized approach to endoscopy training across Canadian general surgery residency programs is needed.

MEASURING COMPETENCE IN FLEXIBLE ENDOSCOPY

Dr M Vassiliou: Department of Surgery, McGill University, Montreal, Quebec

There are several work-based assessment and simulation tools that can be used to measure performance in endoscopy: The Global Assessment of Gastrointestinal Endoscopic Skills (GAGES), Direct Observation of Procedural Skills (DOPS), the Mayo Colonoscopy Skills Assessment Tool (MCSAT) and Assessment of Competency in Endoscopy (ACE).

GAGES, now a part of Fundamentals of Endoscopic Surgery (FES) Society of American Gastrointestinal and Endoscopic Surgeons (SAGES)

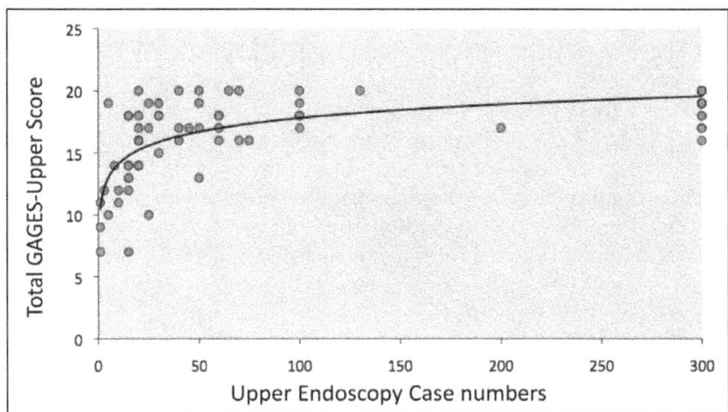

Figure 6) *Using the Global Assessment of Gastrointestinal Endoscopic Skills (GAGES) to access proficiency at upper gastrointestinal endoscopy. Reproduced with permission from reference 21*

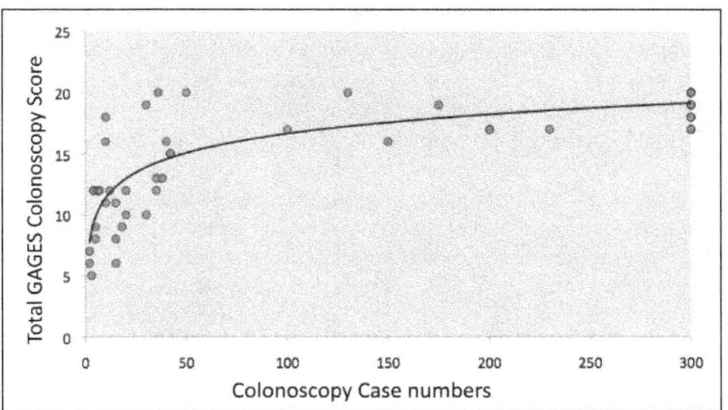

Figure 7) *Using the Global Assessment of Gastrointestinal Endoscopic Skills (GAGES) to access proficiency at lower gastrointestinal endoscopy. Reproduced with permission from reference 21*

task force, stems from a multicentre, multidisciplinary trial aimed to develop tools for evaluating basic flexible endoscopic skills and to demonstrate their reliability and validity (20). The fundamental skills required for flexible endoscopy were identified and then separated into two global assessments: GAGES Upper Endoscopy (GAGES-UE) and GAGES Colonoscopy (GAGES-C). GAGES-UE is divided into five categories: intubation of the esophagus; scope navigation; ability to maintain a clear endoscopic field; instrumentation; and quality of the examination. GAGES-C is also divided into five categories: scope navigation; use of strategies; ability to maintain a clear endoscopic field; instrumentation; and quality of examination (20).

There are two important outcomes that emerged from this trial. First, both GAGES tools have excellent inter-rater reliability between observers and attending surgeons, implying that attending surgeons and gastroenterologists can complete the assessments without the need for an observer. Second, GAGES has the ability to differentiate between novice and experienced endoscopists, and can incorporate the experience of the operator into the assessment. When using the GAGES score to determine competency, it appears that proficiency in basic endoscopy can be achieved in approximately 50 UGI endoscopies and between 75 and 100 colonoscopies (21) (Figures 6 and 7).

DOPS is the assessment tool used in the UK, required for trainees to apply for certification by the JAG (22). This tool moves away from a numbers-based experience toward a more structured approach based on competence. The DOPS assessment is broken down into four major categories: assessment; consent; communication, safety and sedation; endoscopic skills during insertion and withdrawal; and diagnostic and therapeutic ability (22). The significant drawbacks of the assessment tool were that it was difficult to evaluate the degree of case difficulty appropriately and, interestingly, DOPS was negatively correlated with number of procedures and, thus, more procedures undertaken equated to poorer scores on formal tests, which raises concern as to its validity as an assessment of competency tool.

MCSAT separates endoscopy into its core motor and cognitive skill elements in a bedside clinical competency assessment (23). It consists of 14 survey items, five of which assess core competencies including hands-on participation, colonoscope advancement, loop reduction, mucosal visualization and therapeutic manoeuvres. In addition, overall motor skills were graded, along with the depth of advancement. Cognitive skills were also evaluated through six survey parameters: indications, sedation, pain management, landmark recognition, tool selection and pathology identification (23). Scores in the MCSAT reached the literature-accepted minimum competency levels at approximately 275 colonoscopies per trainee, with improvement being evident up to 400 procedures (23). The limitation of the MCSAT are that only gastroenterologists, with unclear rater training at a single centre reviewed it, thereby precluding its widespread reproducibility and validation.

The ACE, endorsed by the ASGE, for the assessment of colonoscopy and esophagogastroduodenoscopy (EGD), built on the MCSAT model of performance-based rather than a numbers-based assessment (24). It includes specific survey items believed missed by the MCSAT model, such as ADR and breaks down broad tasks, such as safe colonoscope advancement, into specifics such as tip control, steering and lumen identification. The Colonoscopy Skills Assessment Tool is based on 11 survey items focusing on specific tasks, and two survey items being based on overall assessment. The EGD Skills Assessment Tool is based on seven survey items and two overall assessment items. The ACE is meant to be used in a continuous fashion to enable identification of certain skill deficiencies and plateaus and, therefore, is recommended to be performed in regular intervals, a suggested 10% of total procedures by the trainee (24).

Simulation

The value of simulators for assessment in training needs to be further researched. There exist limited data demonstrating predictive validity for flexible endoscopy simulators. Unfortunately, they lack the reality aspect, making it hard to reproduce a real-life experience for the trainee.

ACQUISITION AND MEASUREMENT OF COMPETENCY: AMERICAN COLLEGE OF SURGEONS/SAGES

Dr S Schwaitzberg: Department of Surgery, Cambridge Health Alliance and Harvard Medical School, Cambridge, Massachusetts, USA

SAGES FES™: FES defines a surgical endoscopist as a surgeon who has the knowledge and technical skill to use flexible endoscopy to provide care for patients with common gastrointestinal diseases. This ability includes all of the following: an understanding of the indications and contraindications for performing UGI and LGI endoscopy; the accurate recognition and management of normal and abnormal findings in the gastrointestinal tract; the recognition and management of complications from performing gastrointestinal endoscopy; the safe performance of UGI and LGI endoscopy including complete navigation of the esophagus, stomach, proximal duodenum and colon; mucosal inspection and recognition of lesions that may require surgery; tissue acquisition using biopsy or polypectomy; management of periprocedural bleeding; and placement of a percutaneous endoscopic gastrostomy (PEG).

The curriculum is structured into the five-year general surgery training program, with the recognition that flexible endoscopy is only one component of an entire surgical residency. FES provides a stepwise, milestone-based curriculum that includes both didactic and hands-on training. There are suggested cognitive and technical milestones for each level of general surgery residency; however, individual residents at any point may complete milestones in their residency.

The resources required for a training program to provide FES curriculum are minimal. Experts in flexible gastrointestinal endoscopy, either surgical endoscopists or gastroenterologists, are required as teachers. Training programs should have flexible endoscopy simulators (inanimate, animate or computer based) and should provide a concentrated clinical rotation in flexible gastrointestinal endoscopy.

At the PGY1/2 level, early experience is dedicated to didactic materials to provide a basic understanding of gastrointestinal diseases and the anatomy of the gastrointestinal tract as perceived by flexible endoscopic techniques. The technical milestones are simulation or clinical tutorial based – basic scope manipulation: one-handed wheel deflection; control of suction, irrigation and insufflation; and passage of instruments through the working channel. Attention is focused on basic principles of characteristics of endoscopes, equipment set-up, troubleshooting and care.

At the PGY 2/3 level, a direct role in patient preparation and management during flexible endoscopic procedures is sought. The cognitive milestones of patient preparation, moderate sedation and management of findings with follow-up on pathology are all key elements. A dedicated one-month flexible endoscopy rotation is typically completed during this time. Cognitive milestones include indications for and contraindications to UGI and LGI flexible endoscopy, and periprocedural patient management. Technical milestones include simulation exposure or clinical tutorial, dedicated endoscopy experience, intraoperative endoscopy and intensive care unit endoscopy.

At the senior experience for PGY 4/5 levels, the curriculum is divided, with the first part encompassing improvements in both cognitive and technical skills. The cases will be monitored by GAGES, with the goal to achieve a minimum score of 18. Cognitive milestones include image differentiation of normal/abnormal pathology, understanding intraoperative and postoperative gastrointestinal anatomy, understanding appropriate use of endoscopy, recognition and management of complications and, ultimately, outcome measures. Technical milestones include intraoperative endoscopy, intesive care unit endoscopy and continued endoscopic experience for technical retention.

The second part of the senior experience prepares the resident for postgraduate clinical practice in endoscopy with exposure to the tools/adjuncts for therapeutic endoscopic interventions. During this final experience, residents will have additional mandatory requirements beyond part 1 of the senior experience: document procedural numbers for Royal College proficiency requirements; obtain FES certification; and have endoscopy cases monitored by GAGES to continue to consistently achieve minimum score of 18.

ACQUISITION AND MEASUREMENT OF COMPETENCY: JAG AND OTHER OPTIONS

Dr A Gomes: Division of Surgery, University of Alberta, Lethbridge, Alberta

The JAG was established in 1994 under the auspices of the Academy of Medical Royal Colleges (25). Its aim was to establish UK standards for individual endoscopists and trainees, with the goal of providing the highest quality endoscopic services.

JAG performs quality assurance credentialing of individual endoscopy units for screening and fitness of training. For a unit to pass a JAG visit, it must provide evidence of level B or better for the following domains of the endoscopy Global Rating Scale: clinical quality; quality of the patient experience; and training and workforce. This includes an assessment of the environment, decontamination facilities and processes.

JAG accreditation is not mandatory but is necessary to work in the UK screening program and is beginning to become the minimum standard for endosuite privileges. JAG endoscopist certification is for endoscopists in training in all specialties, with specific certifications available in UGI endoscopy (both basic and therapeutic), sigmoidoscopy, colonoscopy (basic and full), gastrointestinal stenting, pediatric colonoscopy, endoscopic retrograde cholangiopancreatography, endoscopic ultrasound and small bowel enteroscopy.

The certification process

JAG developed an endoscopy training system known as 'JETS'. JETS is a registry portal and basic skills course booking resource for trainees that provides a framework for training and auditing progress. Trainees register for JETS as the first step in their certification process, which submits their key endoscopic data. They are provided performance outputs required to chart their competence progression.

The formative data regarding the trainee is then collected via DOPS, which is a preceptor-based evaluation tool of endoscopy skills (22). Once the trainee has completed ≥200 procedures and met the provisional JAG certification criteria, including CIR ≥90%, the trainee can receive provisional acceptance. Full eligibility for endoscopy is not granted until ≥300 procedures are completed and additional full colon certification criteria are met.

Of note, JAG-accredited endoscopy units are mandated to provide equal access to endoscopy training to all trainees regardless of specialty. This allows trainees to learn from experts in both specialties with equal opportunity to caseloads. A survey by Hammond et al (26) in 2012, reported that two of three surgical trainees from all UK training regions admit to lacking access to endosuite, with no scheduled training lists. The authors concluded that surgical trainees are failing to achieve national standards due to deficiencies in endoscopy training.

ISSUES/ROADBLOCKS AND GOING FORWARD

Dr J Ellsmere: Department of Surgery, Dalhousie University, Halifax, Nova Scotia

Ensuring the delivery of high-quality endoscopy to Canadians is of upmost importance. We need to ensure that residents who are going to provide gastroenterological surgical services are competent endoscopists at the end of their postgraduate training. This will likely require a combination of increased endoscopy exposure, simulation-based learning and adopting objective structured assessments of training. In addition, there is a need for an endoscopy auditing system in Canada, in which endoscopy services can be evaluated at an individual, institutional and provincial level.

Delivery of gastrointestinal endoscopy services

The delivery of gastrointestinal endoscopy services remains in flux. Some groups have suggested that the delivery of endoscopy services be the domain of dedicated endoscopists. However, across Canada, the need for physicians who provide more than one type of service is increasingly recognized to be important to the effective delivery of health care services. Individuals providing endoscopy services should be expected to provide acute care, such as acute surgical care in the case of surgical endoscopists, or acute medical care in the case of medical endoscopists.

Regionalization versus community-based endoscopy

Should the delivery of endoscopy services be regionalized similar to cancer care or is there an opportunity to shift routine endoscopy services from larger tertiary care centres to the community-based heath care centres? Given the increasing evidence that routine endoscopy services can be effectively delivered at lower cost at ambulatory care centres, it appears that a more distributed care delivery system is the better of the two strategies.

Rebranding of general surgery

General surgery as a specialty is in need of rebranding. There exists a training misconception that surgical residents are limited to two months of exposure to gastrointestinal care, contrasted with gastroenterology fellows who deveote two years to training. This is a fallacy because most general surgery residents devote the majority of their postgraduate training – usually between five and eight years – to managing gastrointestinal diseases. This misconception could be avoided if the specialty of general surgery was rebranded to general and gastrointestinal surgery.

Subspecializing resident training

With the trend of subspecializing in general surgery, it should not be expected that all general surgery residents be competent endoscopists.

For breast and endocrine surgeons, for example, colonoscopy is not a fundamental tool in their practice. With the extra assessment necessary to provide endoscopy training for a program, it may be a cost-effective measure to have the residents self-select themselves, similar to choosing a fellowship training program. If they want to provide endoscopy services in their practice, then they are responsible to ensure that, over their training, they take advantage of endoscopy opportunities and perform the necessary numbers to meet the current ASGE guidelines, and undergo the necessary assessments for credentialing.

The numbers game

Assessing competency should not be based solely on numbers alone but, rather, should incorporate simulated, bedside or outcomes-based assessments. However, numbers are still important. The existing ASGE numbers are reasonable goals for training competent, basic endoscopists. The proposal of increasing the minimum to 300 colonoscopies in training will prove more challenging and, in many situations, exceedingly difficult to achieve during residency.

Going forward, there a several important steps CAGS should take to ensure that endoscopy remains an integral part of Canadian surgeons scope of practice including: forming a CAGS Endoscopy Taskforce; working with our gastroenterology colleagues to improve endoscopic services; and rebranding general surgery to general and gastrointestinal surgery. It is also important that Canadian surgery training programs adopt a formal endoscopy training curriculum and that the surgical community work to adopt a universal strategy for auditing operators who provide these important services to Canadians.

WORKING GROUP OBSERVATIONS AND DISCUSSION

Roundtable discussion with participants

Observations: The roundtable discussion focused on a number of major observations concerning the state of endoscopy in Canada. The most obvious finding is that the effective delivery of endoscopic services in Canada is an ongoing concern: to general surgeons, gastroenterologists and public health. There appears to be a disparity in endoscopic services delivered nationwide, with the assessment of competency in endoscopy being limited.

Training in endoscopy continues to be a significant topic of discussion. There is a sense of inconsistency within the training of endoscopy. There is a perception both among trainers and surgical trainees that endoscopy training is lacking, as evidenced by the opinions obtained by the resident survey.

Recommendations

The endoscopy taskforce formulated several recommendations based on these observations.

First and foremost, quality endoscopy services must include the complete spectrum of quality care inclusive of a proper diagnosis, treatment and follow-up. Endoscopy is a component of the larger assessment and evaluation of disorders of the gastrointestinal tract.

There needs to be a system of accreditation of endoscopists and endoscopic units, which would include practice audit and competency assessments. Competency in endoscopy should start with the appropriate training of trainees and should extend into the arena of the practising endoscopist. Postgraduate training needs to be formalized and standardized; this includes structured curriculum to acquire both the technical and cognitive components of endoscopy. Furthermore, opportunities for advanced training and enhancing endoscopic skills for practitioners need to be in place.

CONCLUSION

The Proceedings of General Surgeons and the Provision of Quality Endoscopy Services in Canada, held in Toronto, Ontario, was a collaborative effort between general surgeons and gastroenterologists to ensure Canadians are receiving high-quality endoscopic services.

ACKNOWLEDGEMENTS: Canadian Association of General Surgeons (CAGS): Debrah Wirtzfeld, Manitoba; Michael Harrington, Saskatchewan; Paul Hardy, Alberta; Stephen Hiscock, British Columbia; Angus MacIver, Ontario; Andy McFadden, Ontario; Mark O'Driscoll, Ontario; Garth Warnock, British Columbia; Allan Okrainec, Ontario; Phil Smith, Nova Scotia; David Urbach, Ontario, Mark Walsh, Nova Scotia; Les Wasilewski, Nova Scotia; Brent Zabolotny, Manitoba; Chris Vinden, Ontario. Canadian Association of Gastroenterology: Jamie Gregor, Ontario; Lawrence Hookey, Ontario. Cancer Care Ontario: Linda Rabeneck, Robin McLeod. Other: Jolanta Karpinski, Royal College of Physicians and Surgeons of Canada.

REFERENCES

1. Reid SR, George GL, Warnock, et al. Canadian Association of General Surgeons statement on endoscopy. Can J Surg 2011;54:295.
2. Asfaha S, Alqahtani S, Hilsden RJ, MacLean AR, Beck PL. Assessment of endoscopic training of general surgery residents in a North American health region. Gastrointest Endosc 2008;68:1056-62.
3. Spier BJ, Benson M, Pfau PR, et al. Colonoscopy training in gastroenterology fellowships: Determining competence. Gastrointest Endosc 2010;71:319-24.
4. Singh H, Nugent Z, Demers AA, Bernstein CN. Rate and predictors of early/missed colorectal cancers after colonoscopy in Manitoba: A population-based study. Am J Gastroenterol 2010;105:2588-96.
5. Bressler B, Paszat LF, Chen Z, et al. Rates of new or missed colorectal cancers after colonoscopy and their risk factors: A population-based analysis. Gastroenterology 2007;132:96-102.
6. Baxter NN, Sutradhar R, Forbes SS, et al. Analysis of administrative data finds endoscopist quality measures associated with postcolonoscopy colorectal cancer. Gastroenterology 2011;140:65-72.
7. Rabeneck L, Paszat LF, Saskin R. Endoscopist specialty is associated with incident colorectal cancer after a negative colonoscopy. Clin Gastroenterol Hepatol 2010;8:275-9.
8. Rex DK, Rahmani EY, Haseman JH, et al. Relative sensitivity of colonoscopy and barium enema for detection of colorectal cancer in clinical practice. Gastroenterology 1997;112:17-23.
9. Mysliwiec PA, Brown ML, Klabunde CN, Ransohoff DF. Are physicians doing too much colonoscopy? A national survey of colorectal surveillance after polypectomy. Ann Intern Med 2004;14:264-71.
10. Khiani VS, Soulos P, Gancayco J, Gross CP. Anesthesiologist involvement in screening colonoscopy: Temporal trends and cost implications in the medicare population. Clin Gastroenterol Hepatol 2012;10:58-64 e1.
11. Tinmouth J, Kennedy E, Baron D, et al. Guideline for Colonoscopy Quality Assurance in Ontario. September 9, 2013 Cancer Care Ontario. Toronto.
12. Rabeneck L, Paszat LF, Hilsden RJ, et al. Bleeding and perforation after outpatient colonoscopy and their risk factors in usual clinical practice. Gastroenterology 2008;135:1899-906,1906, e1.
13. Wexner SD, Garbus JE, Singh JJ; SCSO Group. A prospective analysis of 13,580 colonoscopies. Reevaluation of credentialing guidelines. Surg Endosc 2001;15:251-61.
14. Shah HA, Paszat LF, Saskin R, Stukel TA, Rabeneck L. Factors associated with incomplete colonoscopy: A population-based study. Gastroenterology 2007;132:2297-303.
15. Segnan N, Patnick J, Karsa Lv; Commission E, eds. In: European guidelines for quality assurance in colorectal cancer screening and diagnosis, 2010, Luxembourg: Publications Office of the European Union.
16. Rex DK, Petrini JL, Baron TH, et al. Quality indicators for colonoscopy. Gastrointest Endosc 2006;63(4 Suppl):S16-28.
17. Baxter NN, Sutradhar R, Forbes S, et al. Can quality indicators for colonoscopy be derived from administrative data? Identifying predictors of new or missed cancers after colonoscopy. Gastrointest Endosc 2009;69:AB120. (Abst)
18. Sawhney MS, Cury MS, Neeman N, et al. Effect of institution-wide policy of colonoscopy withdrawal time > or = 7 minutes on polyp detection. Gastroenterology 2008;135:1892-8.

19. Lieberman D. A call to action – measuring the quality of colonoscopy. N Engl J Med 2006;355:2588-9.
20. Vassiliou MC, Kaneva PA, Poulose BK, et al. Global Assessment of Gastrointestinal Endoscopic Skills (GAGES): A valid measurement tool for technical skills in flexible endoscopy. Surg Endosc 2010;24:1834-41.
21. Vassiliou MC, Kaneva PA, Poulose BK, et al. How should we establish the clinical case numbers required to achieve proficiency in flexible endoscopy? Am J Surg 2010;199:121-5.
22. Barton JR, Corbett S, van der Vleuten CP; English Bowel Cancer Screening Programme; UK Joint Advisory Group for Gastrointestinal Endoscopy. The validity and reliability of a Direct Observation of Procedural Skills assessment tool: Assessing colonoscopic skills of senior endoscopists. Gastrointest Endosc 2012;75:591-7.
23. Sedlack RE. Training to competency in colonoscopy: Assessing and defining competency standards. Gastrointest Endosc 2011;74:355-66,e1-2.
24. ASGE Training Committee, Sedlack RE, Coyle WJ, et al. ASGE's assessment of competency in endoscopy evaluation tools for colonoscopy and EGD. Gastrointest Endosc 2014;79:1-7.
25. JAG accreditation System. 2013; <www.jagaccreditation.org/> (Accessed August 1, 2014).
26. Hammond JS, S WNF, Lund JN and Barton JR. Surgical endoscopy training: the Joint Advisory Group on gastrointestinal endoscopy national review. Frontline Gastroenterol. 2012;4:20-4.

Increased eligibility for treatment of chronic hepatitis C infection with shortened duration of therapy: Implications for access to care and elimination strategies in Canada

Sergio M Borgia MD MSc[1,2], Adenike Rowaiye MBA MSc[1]

SM Borgia, A Rowaiye. Increased eligibility for treatment of chronic hepatitis C infection with shortened duration of therapy: Implications for access to care and elimination strategies in Canada. Can J Gastroenterol Hepatol 2015;29(3):125-129.

BACKGROUND: All oral, highly effective direct-acting antiviral combinations, such as sofosbuvir-ledipasvir, have recently been licensed in Canada but cost as much as $67,000 for a 12-week course of therapy, representing a major economic barrier to predominately single-payer health care systems such as that found in Ontario. In hepatitis C virus (HCV) genotype 1 noncirrhotic patients with a baseline viral load of $<6\times10^6$ IU/mL, treatment with sofosbuvir-ledipasvir can be shortened to eight weeks without compromising ≥95% efficacy. The number of HCV-infected patients in Ontario eligible for shortened therapy, and the associated cost savings, are unknown. The authors propose that treating every patient with shortened therapy, regardless of baseline viral load, would lead to significant public cost savings and collateral efficiencies, enabling increased HCV treatment capacity and cure.

METHODS: The present study designed a three-part model to investigate the cost of cure per patient and cost savings per patient under three eligibility pathways: conservative, permissive and ideal. In the conservative model, every patient is treated for 12 weeks regardless of baseline viral load, whereas in the permissive model, patients with a baseline viral load $<6\times10^6$ IU/mL are treated for eight weeks. In the ideal model, every patient receives eight weeks of therapy regardless of baseline viral load. Relapsed patients are retreated for 12 weeks. Data obtained from the Ontario Public Health Laboratory were used to validate the model and generate the outcomes.

RESULTS: In Ontario, 75.34% of HCV genotype 1 patients had a baseline viral load of $<6\times10^6$ IU/mL and were eligible for shortened therapy. The cost of cure per patient in the ideal model was $47,328.44, representing a 29% reduction in the cost of curative therapy and 3.5 weeks of shortened treatment duration compared with the conservative model. The ideal model generated a cost savings per patient of $3,855.17 (8% reduction in treatment cost) and 0.7 weeks of shorter therapy compared with the permissive model, and was the shortest and most efficient while maintaining a cure rate ≥95%.

CONCLUSIONS: These results demonstrate that recommendations for a shortened treatment course of eight weeks using all-oral direct-acting antivirals in HCV genotype 1 noncirrhotic patients, regardless of baseline viral load, affords significant public cost savings and, on a population level, offers opportunities for expanded HCV treatment and cure.

Key Words: *Cost; Cost effectiveness; Economic analysis; Hepatitis C; Modelling; Shortened therapy; Sofosbuvir-ledipasvir; Treatment*

Une plus grande admissibilité à un traitement plus court de l'infection par l'hépatite C : les répercussions sur l'accès aux soins et les stratégies d'élimination au Canada

HISTORIQUE : Toutes les associations antivirales orales à action directe hautement efficaces, comme le sofosbuvir-lédipasvir, ont récemment été homologuées au Canada, mais peuvent coûter jusqu'à 67 000 $ pour un cycle de 12 semaines, ce qui constitue un énorme obstacle économique à des systèmes de santé qui sont surtout à payeur unique, comme celui de l'Ontario. Chez les patients non cirrhotiques atteints du virus de l'hépatite C (VHC) de génotype 1 ayant une charge virale de départ inférieure à 6×10^6 UI/mL, un traitement au sofosbuvir-lédipasvir peut être raccourci à huit semaines sans compromettre une efficacité de 95 % ou plus. On ne sait pas combien de patients de l'Ontario infectés par le VHC sont admissibles à ce type de traitement ni les économies qui s'y rapportent. Les auteurs stipulent qu'un traitement plus court par patient, indépendamment de la charge virale de départ, entraînerait des économies considérables pour la santé publique et une efficacité accessoire, ce qui accroît la capacité thérapeutique et la guérison du VHC.

MÉTHODOLOGIE : La présente étude en trois parties visait à examiner le coût pour guérir un patient et les économies par patient en vertu de trois voies d'admissibilité : prudent, permissif et idéal. Dans le modèle prudent, chaque patient est traité pendant 12 semaines, indépendamment de sa charge virale de départ, tandis que dans le modèle permissif, les patients ayant une charge virale de départ inférieure à 6×10^6 UI/mL sont traités pendant huit semaines. Dans le modèle idéal, chaque patient est traité huit semaines, indépendamment de sa charge virale de départ. Les patients en rechute sont traités de nouveau pendant 12 semaines. Les données obtenues auprès du Laboratoire de santé publique de l'Ontario ont permis de valider le modèle et de produire des résultats.

RÉSULTATS : En Ontario, 75,34 % des patients atteints du VHC de génotype 1 ayant une charge virale de base inférieure à 6×10^6 UI/mL étaient admissibles à un traitement plus court. Dans un modèle idéal, le coût pour guérir un patient s'élevait à 47 328,44 $, soit une réduction de 29 % du coût du traitement curatif et une réduction du traitement de 3,5 semaines par rapport au modèle prudent. Le modèle idéal produisait des économies par patient de 3 855,17 $ (réduction de 8 % du coût du traitement) et raccourcissait le traitement de 0,7 semaine par rapport au modèle permissif. C'était le plus court et le plus efficace, et il maintenait un taux de guérison de 95 % ou plus.

CONCLUSIONS : Ces résultats démontrent que les recommandations en vue d'un cycle de traitement plus court de huit semaines au moyen d'antiviraux à action directe par voie orale chez des patients non cirrhotiques atteints du VHC de génotype 1, indépendamment de leur charge virale de départ, permettent à la santé publique de réaliser d'importantes économies et, en population, d'accroître le taux de traitement et de guérison du VHC.

Chronic hepatitis C virus (HCV) infection represents an important worldwide public health problem and is a leading cause of cirrhosis, hepatocellular carcinoma and liver transplantation (1-3). Worldwide, it is estimated that 170 million people are infected, with three million to four million incident cases annually, corresponding to a mean seroprevalence of approximately 3% (4,5). In Canada, the prevalence of HCV infection is unknown; the Public Health Agency of Canada has reported that approximately 1% of the population was anti-HCV

[1]*William Osler Health System, Brampton Civic Hospital, Brampton;* [2]*Division of Infectious Diseases, Department of Medicine, McMaster University, Hamilton, Ontario*

Correspondence: Dr Sergio M Borgia, William Osler Health System, 2100 Bovaird Street East, Brampton, Ontario L6R 3J7. e-mail sergio.borgia@williamoslerhs.ca

positive in 2011, of which approximately 0.75% would be expected to be viremic (6).

The burden of HCV in Canada is increasing based on previous modelling studies; however, the precise magnitude of this increase has not been well characterized (7,8). Recent studies have attempted to refine the future burden of HCV disease in Canada and suggest that between now and 2035, cases of compensated and decompensated cirrhosis are expected to peak in 2031 at 36,210 and 3380 cases, respectively (9). Compared with 2013, cases of compensated cirrhosis, decompensated cirrhosis, hepatocellular carcinoma and liver-related deaths are expected to increase 89%, 80%, 205% and 160%, respectively, by 2035 (9). In Ontario, HCV infection is associated with the greatest years of life lost due to premature mortality and health-adjusted life years – more than any other infectious disease (10).

Previously, the treatment of chronic HCV required prolonged courses of pegylated interferon-alfa in combination with ribavirin for 24 to 48 weeks, and offered cure rates as measured by a sustained virological response (SVR) of approximately 40% to 50% (11). For HCV genotype 1 infections, the development of 'first-generation' direct-acting antiviral agents (DAAs) represented a major advancement in SVR rates – from 60% to 75% – depending on the combination of pre-existing viral and host prognostic factors (12-17). Very recently, the treatment of HCV infection has again witnessed enormous improvement, with SVR rates >95% in uncomplicated patients using all-oral DAA combinations such as sofosbuvir-ledipasvir (SOF-LDV), simeprevir plus sofosbuvir, and ombitasvir/paritaprevir/r plus dasabuvir ± ribavirin (18-23). The currently available all-oral HCV therapies achieve nearly all of the necessary conditions for a positive societal impact with regard to burden of HCV disease including shorter, safer, easier and more effective treatment, with no demonstrable clinical concerns of viral resistance or adverse side effects.

The improvement in curative therapy for the majority of HCV patients in Canada will have a transformative effect on the way these patients receive liver-related care, and has outpaced the ability of the Canadian health care system to effectively and quickly respond to the challenge of treating the increased number of patients required to make a meaningful impact on prevention, treatment delay and the future burden of HCV disease. Still, the number of individuals with HCV infection treated in Canada has remained relatively fixed at approximately 5000 patients per year (9).

Rationale

New therapies, such as SOF-LDV, are expensive ($67,000 for a 12-week course) for governments and prohibitive for private-paying patients. Hence, under the fiscal constraints of a predominantly single-payer health care system, such as that found in Canada, the most significant barrier to HCV treatment has become access to funding for curative therapy across a broad spectrum of patients. Although presently under funding review by provincial ministries of health, there are no current access options for these latest and most effective all-oral HCV treatments for the vast majority of treatment-eligible patients who lack private or third-party insurance.

The ability to shorten HCV therapy confers significant benefit to individual patients and to the overall impact on the capacity of a health care system challenged by suboptimal infrastructure and cost-containment concerns. Preliminary data suggest that six weeks of therapy may be the duration limit (24,25). Recent pivotal trials have demonstrated that daily therapy with SOF-LDV for 12 weeks achieves a very high cure rate (95%), with an exceedingly low 1% relapse rate as measured by SVR in noncirrhotic patients with HCV genotype 1 infection (20).

These same studies demonstrated that for the majority of patients with a baseline HCV RNA level <6.0×10^6 (<6M) IU/mL treatment may be shortened to eight weeks without compromising SVR efficacy and with only minimal increases in the relapse rate. The registration trial that led to these recommendations examined relapse rates across a wide range of baseline viral loads; relapse rates were not appreciably impacted up to and including a baseline viral load of 10M IU/mL. The 6M IU/mL threshold of offering shortened treatment confers a very conservative relapse safety margin (20,26).

Retreatment using SOF-LDV, or other all-oral DAA combinations, for 12 weeks in the few patients who relapse after eight weeks of therapy may be expected to achieve very high SVR rates given that patients who have previously failed a sofosbuvir regimen have been universally successfully retreated with a 12-week regimen of SOF-LDV, although this remains to be confirmed (27).

These findings suggest a positive impact on the number of patients in Ontario living with HCV infection who may be cured using a shorter duration of therapy. Presently, the number, or proportion, of patients in Ontario that would benefit from a reduction in treatment duration remains unknown.

Objectives

The present investigation sought to quantify the number of individuals in Ontario with HCV genotype 1 infection who may benefit from a shorter duration of therapy based on their pretreatment baseline viral load. Data were obtained from databases of the Public Health Ontario Laboratory (PHOL). These data were used to construct and test a mathematical model of cost efficiency based on the discounted cost of shorter therapy ('the model'). Given the very few number of people who fail to achieve an SVR after treatment duration of either eight or 12 weeks, a more liberal recommendation to treat all noncirrhotic patients, regardless of baseline HCV viral load, with the shorter eight-week regimen suggests a simpler and viable cost-effective strategy. The authors suggest that offering a 12-week retreatment course to relapsed patients is both fair and ethical. Thus, based on population estimates of the number of individuals infected with HCV who fall above and below the 6M IU/mL threshold, the present hypothesis recommending a short, eight-week treatment course in every noncirrhotic HCV genotype 1-infected patient, regardless of baseline viral load, would be expected to produce significant cost savings as predicted in the model. This recommendation has implications to increase treatment capacity and shorten the time horizon required to ensure a robust population impact on the advancement of liver health, prevent liver-related complications and, ultimately, contribute to the achievement of an elimination strategy for HCV infection in Ontario over the next 20 years and beyond.

METHODS

The present cost and cost savings exploratory analysis was based on specific input values from a previous study (26). Table 1 contains the input data used in the generation of the model. Some assumptions were made and used in the model structure, namely those contained in the information provided by an exploratory analysis of 500 patients in a community HCV database that revealed 80% of HCV genotype 1 patients had a baseline viral load of <6M IU/mL (Borgia, data on file) and the baseline viral load values contained in the Ontario provincial dataset that approximately 75% of patients have a baseline viral load of <6M IU/mL and approximately 25% of patients have a baseline viral load of ≥6M IU/mL. This 75:25 ratio formed the basis of the present model. Also, the model assumed that either the population in question consists of patients without cirrhosis or that the presence of patients with cirrhosis does not alter the proportions of baseline viral loads, and that baseline HCV viral load is not correlated with the presence or absence of underlying cirrhosis (28). The model construct was developed using a spreadsheet (Excel 2010 version 14, Microsoft Corporation, USA). Three models and their rationale, were considered as follows:

1. Conservative model: Every patient is treated with LDV-SOF for 12 weeks regardless of baseline viral load. Physicians may choose to treat every patient with 12 weeks of therapy based on uncertainty of patients' true baseline viral loads between 1M IU/mL and 9M IU/mL. Also, treating physicians may be hesitant to risk relapse with shortened therapy in cases in which they may foresee not having access to a retreatment course or particularly for noncirrhotic patients with stage 3 fibrosis. By applying the input values and

TABLE 1
Input values used to generate models

Parameter	Therapy, weeks 12	8
Cure rate, <6 million IU/mL at baseline	96	97
Cure rate, ≥6 million IU/mL at baseline	94	90
Relapse rate, <6 million IU/mL at baseline	2	2
Relapse rate, ≥6 million IU/mL at baseline	1	10
Cost for therapy, $	67,000	
One-week therapy, $	5583.33333	

Data presented as % unless otherwise indicated. Cure and relapse rates from product monograph tables (HARVONI [ledipasvir/sofosbuvir, Gilead Sciences Inc, Canada])

TABLE 2
Hepatitis C virus (HCV) genoptype 1 patients according to viral load and sex

HCV viral load	Sex: Female	Male	Not reported	Total
<6 million IU/mL	2746 (79.18)	5088 (73.30)	308 (77.39)	8142 (75.34)
≥6 million IU/mL	722 (20.82)	1853 (26.70)	90 (22.61)	2665 (24.66)
Total	3468 (100.00)	6941 (100.00)	398 (100.00)	10807 (100.00)

Data presented as n (%). Public Health Ontario Laboratory HCV type 1 viral load dataset April 26, 2010 to November 7, 2014

TABLE 3
Output values for models

Output	Value $	Weeks
Curative cost per patient, conservative	67,000.00	12
Curative cost per patient, permissive	51,183.61	9.2
Curative cost per patient, ideal	47,328.44	8.5
Cost savings per patient, conservative/permissive	15,816.39	2.8
Cost savings per patient, conservative/ideal	19,671.56	3.5
Cost savings per patient, permissive/ideal	3,855.17	0.7

assumptions, the curative cost per patient (CPP) for the conservative model can be calculated using the following formula:

$$CPP_{conservative} = 12\ (x + y)\ *5583.33333/(x + y)$$

in which x and y are the number of patients with viral load of <6M IU/mL and ≥6M IU/mL, respectively.

2. Permissive model: Patients are treated as per the LDV-SOF product monograph and Health Canada recommendations. Noncirrhotic HCV-infected patients with a baseline viral load ≥6M IU/mL receive LDV-SOF for 12 weeks whereas noncirrhotic patients with <6M IU/mL receive eight weeks of therapy. Only relapsed patients who received eight weeks of therapy are retreated. By applying the input values and assumptions, the curative CPP for the permissive model can be calculated using the following formula:

$$CPP_{permissive} = ([8.24x + 12y]*5583.33333)/(x + y)$$

3. Ideal model: All noncirrhotic patients are treated with LDV-SOF for eight weeks with all relapsed patients retreated for an additional 12 weeks of therapy. By applying the input values and assumptions, the curative CPP for the ideal model can be calculated using the following formula:

$$CPP_{ideal} = ([8.24x + 9.2y]*5583.33333)/(x + y)$$

It is expected that changes in the treatment duration would have an impact on the overall cost of curative therapy for HCV patients. To evaluate the impact of changing the treatment duration, the CPP for the three models were compared. The main outcome of interest was to maximize the cost savings, using the shortest mean duration, while maintaining a cure rate ≥95%. The cost savings estimate is used to determine the return on each of the models. The cost savings per patient (CSPP) between the conservative and the permissive model is estimated using the following formula:

$$CSPP_{conservative/permissive} = (3.76x)*5583.33333/(x + y)$$

The CSPP between the conservative and the ideal model is estimated using the following formula:

$$CSPP_{conservative/ideal} = ((3.76x + 2.8y)*5583.33333)/(x + y)$$

The CSPP between the permissive and the ideal model is estimated using the following formula:

$$CSPP_{permissive/ideal} = 2.8y*5583.33333/(x + y)$$

Supplementary data and description of the model are provided in Appendix 1.

RESULTS

To test the model, an analysis was performed using the Ontario patient-level data from the PHOL datasets for patients falling above and below the 6M IU/mL thresholds for shortened therapy. The source data provided by PHOL consists primarily of two data sources: an HCV viral load dataset and an HCV genotype dataset. The information stored in the PHOL discrete datasets includes duplicate data for patient identification characteristics, sex, and all available viral loads including non-genotype 1 and mixed genotype infections as measured by the COBAS TaqMan HCV Test, v2.0 assay (Roche Molecular Systems, USA). Clinical information is not collected and was not considered in this analysis. The two datasets were combined and linked to produce a total of 49,123 unique patients, of whom 10,807 were infected with HCV genotype 1 and were genotyped between April 26, 2010 and November 7, 2014. Further analysis revealed that 8142 genotype 1 patients (75.34%) had a viral load of <6M IU/mL and 2665 genotype 1 patients (24.66%) had a viral load of ≥6M IU/mL (29). Table 2 provides information on some characteristics of patients in the Ontario patient-level datasets.

For the conservative model, the CPP remained unchanged ($67,000). For the permissive model, the CPP was $51,183.61, representing a 24% reduction in overall cost of therapy for HCV patients compared with the conservative model. The permissive model reduced the therapy duration by 2.8 weeks compared with the conservative model. Under the ideal model, the CPP was $47,328.44, representing a 29% reduction in the overall cost of therapy for HCV patients and 3.5 weeks of shortened treatment duration compared with the conservative model. The ideal model generated a cost savings per patient of $3,855.17 compared with the permissive model and an 8% reduction in the cost of treatment over the permissive model. For every $1.00 spent on the ideal model, $1.08 will be spent on the permissive model and $1.42 will be spent on the conservative model. The ideal model is the most economically efficient of the three models and also offers the shortest mean treatment duration, while still maintaining a cure rate ≥95%. The magnitude of the effectiveness of the ideal model over the permissive model can be expressed by stating that one additional patient can be cured for approximately every 13 HCV patients treated. In a health care system with a fixed budget for HCV treatment, the ideal model would allow for more HCV patients to be treated. Table 3 summarizes the outputs from the three models using genotype 1 patients in Ontario.

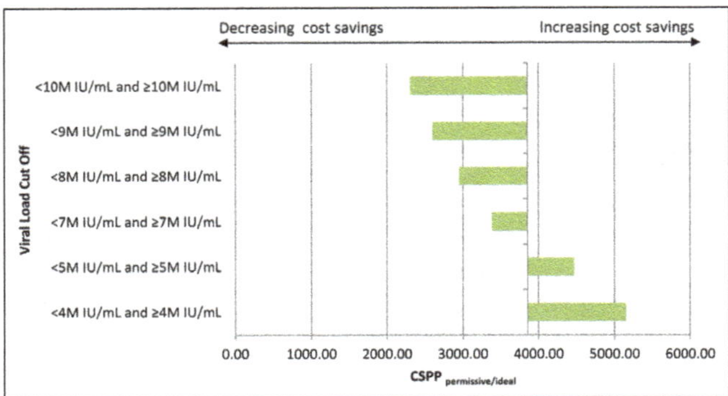

Figure 1) *Cost savings per patient (CSPP) permissive/ideal with different hepatitis C virus viral load cut-offs*

Sensitivity analysis

Based on data indicating that shortened therapy with SOF-LDV for eight weeks is equally effective as 12 weeks for baseline viral load thresholds up to 10M IU/mL, a sensitivity analysis was conducted to determine how changes in the baseline viral load threshold to offer shortened therapy would affect the $CSPP_{permissive/ideal}$ model. The analysis was performed using seven different viral load cut-off thresholds provided in the PHOL patient-level dataset. Baseline viral load cut-offs in 1M IU/mL increments ranging between 4M IU/mL to 10M IU/mL were considered.

Changes in the $CSPP_{permissive/ideal}$ varied from $5,154.21 for a viral load cut-off of 4M IU/mL to $2,307.32 for a viral load cut-off of 10 M IU/mL. Overall, there are cost savings irrespective of the viral load cut-off value. The $CSPP_{permissive/ideal}$ model is most sensitive to a viral load threshold of 10M; all viral load cut-offs between 10 M IU/mL and 6 M IU/mL had lesser effects on the $CSPP_{permissive/ideal}$ model. Viral load thresholds between 4M IU/mL and 6M IU/mL result in increased cost savings on $CSPP_{permissive/ideal}$, as shown in Figure 1.

DISCUSSION

The results of the present modelling analysis demonstrate a significant gain in efficiency in both cost and time to cure patients chronically infected with HCV in Ontario, and offers substantial collateral advantages. Compared with the permissive model, our ideal model saves $3,855.17 per patient and shortens therapy by 0.7 weeks per patient. Of the estimated 5000 patients treated yearly in Canada, not all of whom achieve an SVR on older therapy, these savings represent an additional 3500 treatment weeks that would accommodate 412 additional patients yearly who could be not just treated, but almost all cured. It should not escape notice that, in addition to medication cost savings, there would be significant additional savings found in hospital, laboratory, nursing and physician costs that contribute to overall cost burden.

Presently, aside from SOF-LDV, there are no other all-oral treatment options that allow for treatment duration <12 weeks in patients with HCV genotype 1 infection; thus, any prorated discounts of other all-oral genotype 1 regimens are not possible at this time.

On a broader population level, shortening therapy for all genotype 1 noncirrhotic HCV patients to eight weeks with SOF-LDV will enable more patients to be integrated into HCV treatment opportunities over time. The estimated 251,990 HCV viremic patients in Canada represent a viable reservoir in whom to target the efficiencies predicted by our model using a shorter eight-week course of therapy. Other provinces deliberating funding criteria for HCV treatment may benefit from similar modelling analyses.

There are some limitations associated with our model. First, the model hypothesis was based on a relatively small provincial community database of 500 HCV patients with an 80:20 ratio of genotype 1 patients with a baseline HCV viral load <6M IU/mL. Reassuringly, the number of unique genotype 1 patients with a viral load <6M IU/mL (75.34%) contained in the entire Ontario provincial database and used to test the model was very close to our community estimate. Nevertheless, the patient-level data contained in the PHOL datasets may not be fully representative of the Ontario HCV population due to missed linkage analysis, errors in patient coding information and unforeseen data entry errors. This model does not take into account the trivial incremental cost of the addition of ribavirin if necessary in patients requiring retreatment nor does it consider additional retreatment after a second relapse, however rare that necessity. Finally, the model is tied to the manufacturer's pre-negotiated list price of $67,000 for SOF-LDV (Harvoni, Gilead Sciences Inc, USA), which may not be the price that provincial governments ultimately pay for a course of curative therapy; however, the relative savings over total expenditures would be preserved and, in general, the model will afford savings regardless of the final price based on prorated discounts for shorter treatment durations. A recent analysis has shown that the cost of treatment per SVR, defined as the cost per regimen divided by the SVR rate of that regimen, is a valuable metric to frame discussions concerning access to treatment funding (30). In fact, the cost per SVR of 94% for a 12-week course of therapy for SOF-LDV is approximately $71,277 – significantly lower than the cost per SVR of previously funded treatments containing first-generation DAA agents in combination with interferon and ribavirin (30). However, we believe it useful to also examine retail drug costs because it aids in framing cost analyses such as ours. In an era in which the efficacy benchmark has been recalibrated and SVR rates are >95%, a drug's retail price and its 'cost per SVR' will be very similar. Our model builds on the 'cost per SVR' discussion and supports even greater efficiencies ($47,328.44 CPP_{ideal}) by taking into account the shorter treatment duration and the specific relapse and retreatment costs.

These results demonstrate that recommendations for a shortened treatment course of eight weeks using all-oral DAAs in genotype-1 noncirrhotic HCV-infected patients, regardless of baseline viral load, confers significant public cost savings and, on a population level, offers opportunities for expanded HCV treatment and cure.

ACKNOWLEDGEMENTS: The authors thank Alex Marchand-Austin and Vica Dang at Public Health Ontario (PHOL) and Homie Razavi from the Center for Disease Analysis (CDA), Louisville, Colorado, USA.

DISCLOSURES: This study was supported by an unrestricted grant from Gilead Sciences Canada, manufacturers of antiviral therapies for HCV. The sponsor had no input into the design, conduct or analysis of this study, drafting of the manuscript, nor the decision to submit the manuscript for publication. Sergio Borgia has received research support, consulting, and/or speaking fees from Gilead Sciences, Merck Canada, Roche Canada, Vertex, Janssen, and Abbvie. Nike Rowaiye has no declared disclosures. Homie Razavi is an employee of the Centre for Disease Analysis and Alex Marchand-Austin and Vica Dang are employees of PHOL.

APPENDIX 1: DERIVATION OF MODELS

Curative cost per patient (CPP) for conservative model

$CPP_{conservative} = 12(x + y) * 5583.33333/(x + y)$

Where x = Total patients with (<6 M IU/mL)

y = Total patients with (≥6 M IU/mL)

(x + y) = Total patients with viral load

12 (x + y) = 12 weeks MULTIPLIED BY Total patients with viral load

5583.33333/(x + y) = Cost per week DIVIDED BY Total patients with viral load

Curative cost per patient (CPP) for permissive model

$CPP_{permissive} = ([8.24x + 12y] * 5583.33333)/(x + y)$

Where x = Total patients with (<6M IU/mL)

y = Total patients with (≥6M IU/mL)

(x + y) = Total patients with viral load

8.24x = 8x + (0.02x * 12) = 8 weeks MULTIPLIED BY Total patients with (<6M IU/mL) PLUS relapse patients (2% of total patients with [<6M IU/mL] MULTIPLIED BY 12 weeks)

12y = 12 weeks MULTIPLIED BY Total patients with (≥6M IU/mL)

5583.33333/(x + y) = Cost per week DIVIDED BY Total patients with viral load

APPENDIX 1 – CONTINUED

Curative cost per patient (CPP) ideal model

CPP_{ideal} = ([8.24x + 9.2y]*5583.33333)/(x + y)

Where x = Total patients with (<6M IU/mL)

y = Total patients with (≥6M IU/mL)

(x + y) = Total patients with viral load

8.24x = 8x + (0.02x * 12) = 8 weeks MULTIPLIED BY total patients with (<6M IU/mL) PLUS relapse patients (2% of total patients with (<6M IU/mL) MULTIPLIED BY 12 weeks)

9.2y = 8y + (0.1y * 12) = 8 weeks MULTIPLIED BY Total patients with (≥6M IU/mL) PLUS relapse patients (10% of total patients with (≥6M IU/mL) MULTIPLIED BY 12 weeks)

5583.33333/(x + y) = Cost per week DIVIDED BY Total patients with viral load

Cost savings per patient (CSPP) = Cost per patient for conservative model MINUS cost per patient for permissive model

$CSPP_{conservative/permissive}$
= 12(x + y) * 5583.33333/(x + y) – ([8.24x + 12y] * 5583.33333)/(x + y))
= 12x + 12y * 5583.33333/(x + y) – 8.24x – 12y * 5583.33333/(x + y)
= (12x – 8.24x + 12y – 12y) * 5583.33333/(x + y)
= (3.76x)* 5583.33333/(x + y)

Cost savings per patient (CSPP) = Cost per patient for conservative model MINUS cost per patient ideal model

$CSPP_{conservative/ideal}$
= 12(x + y) * 5583.33333/(x + y) – ([8.24x + 9.2y] * 5583.33333)/(x + y))
= 12x + 12y * 5583.33333/(x + y) – 8.24x – 9.2y * 5583.33333/(x + y)
= (12x – 8.24x + 12y – 9.2y) * 5583.33333/(x + y)
= (3.76x + 2.8y) * 5583.33333)/(x + y)

Cost savings per patient (CSPP) = Cost per patient for permissive model MINUS Cost per patient ideal model

$CSPP_{permissive/ideal}$
= ([8.24x + 12y] * 5583.33333)/(x + y) – ([8.24x + 9.2y] * 5583.33333)/[x + y])
= (8.24x – 8.24x + 12y – 9.2y) * 5583.33333/(x + y)
= 2.8y * 5583.33333/(x + y)

REFERENCES

1. Alberti A, Chemello L, Benvegnu L. Natural history of hepatitis C. J Hepatol 1999;31:17-24.
2. Koike K, Tsutsumi T, Miyoshi H, et al. Molecular basis for the synergy between alcohol and hepatitis C virus in hepatocarcinogenesis. J Gastroenterol Hepatol 2008;23:S87-91.
3. Perz JF, Armstrong GL, Farrington LA, et al. The contributions of hepatitis B and hepatitis C virus infection to cirrhosis and primary liver cancer worldwide. J Hepatol 2006;45:529-38.
4. Dore GJ, Ward J, Thursz M. Hepatitis C disease burden and strategies to mange the burden. J Viral Hepat 2014;21(Suppl 1):1-4.
5. Mohd Hanafiah K, Groeger J, Flaxman AD, Wiersma ST. Global epidemiology of hepatitis C virus infection: New estimates of age-specific antibody to HCV seroprevalence. Hepatology 2013;57:1333-42.
6. Trubnikov M. Developing estimates of prevalent and undiagnosed HCV infection in Canada in 2011. <www.youtube.com/watch?v=234ru9AFA6U&list=PLy0zwf7_pKrXSh9bx6XxznPv65S44nN6q&noredirect=12014> (Accessed January 9, 2015).
7. Remis RS. Modelling the incidence and prevalence of hepatitis C infection and its sequelae in Canada, 2007: Final report. Community Acquired Infections Division, Centre for Communicable Diseases and Infection Control, Public Health Agency of Canada 2009. <http://epe.lac-bac.gc.ca/100/200/301/phac-aspc/modeling_hepatis_c_infection-e/HP40-39-2009E.pdf> (Accessed January 9, 2015)
8. Zou S, Tepper M, El Saadany S. Prediction of hepatitis C burden in Canada. Can J Gastroenterol 2000;14:575-80.
9. Myers RP, Krajden M, Bilodeau M, et al. Burden of disease and cost of chronic hepatitis C infection in Canada. Can J Gastroenterol Hepatol 2014;28:243-50.
10. Kwong JC, Crowcroft NS, Campitelli MA, et al. Ontario Burden of Infectious Disease Study (ONBOIDS): An OAHPP/ICES Report. Toronto: Ontario Agency for Health Protection and Promotion; Institute for Clinical Evaluative Sciences; 2010.
11. McHutchison JG, Lawitz EJ, Shiffman ML, et al. Peginterferon alfa-2b or alfa-2a with ribavirin for treatment of hepatitis C infection. N Engl J Med 2009;361:580-93.
12. Asselah T, Marcellin P. New direct-acting antivirals' combination for the treatment of chronic hepatitis C. Liver Int 2011;31(Suppl 1):68-77.
13. Jacobson IM, McHutchison JG, Dusheiko G, et al. Telaprevir for previously untreated chronic hepatitis C virus infection. N Engl J Med 2011;364:2405-16.
14. Poordad F, McCone J Jr, Bacon BR, et al. Boceprevir for untreated chronic HCV genotype 1 infection. N Engl J Med 2011;364:1195-206.
15. Bacon BR, Gordon SC, Lawitz E, et al. Boceprevir for previously treated chronic HCV genotype 1 infection. N Engl J Med 2011;364:1207-17.
16. Zeuzem S, Andreone P, Pol S, et al. Telaprevir for retreatment of HCV infection. N Engl J Med 2011;364:2417-28.
17. Sherman KE, Flamm SL, Afdhal NH, et al. Response-guided telaprevir combination treatment for hepatitis C virus infection. N Engl J Med 2011;365:1014-24.
18. Afdal N, Zeuzem S, Kwo P, et al. Ledipasvir and sofosbuvir for untreated HCV genotype 1 infection. N Engl J Med 2014;370:1889-98.
19. Afdal N, Reddy KR, Nelson DR et al. Ledipasvir and sofosbuvir for previously treated HCV genotype 1 infection. N Engl J Med 2014;370:1483-93.
20. Kowdley K, Gordon S, Reddy KR, et al. Ledipasvir and sofosbuvir for 8 or 12 weeks for chronic HCV without cirrhosis. N Engl J Med 2014;370:1879-88.
21. Feld JJ, Kowdley KV, Coakley E, et al. Treatment of HCV with ABT-450/r-ombitasvir and dasabuvir with ribavirin. N Engl J Med 2014;370:1594-603.
22. Zeuzem S, Jacobson IM, Baykal T, et al. Retreatment of HCV with ABT-450/r-ombitasvir and dasabuvir with ribavirin. N Engl J Med 2014;370:1604-14
23. Poordad F, Hezode C, Trinh R, et al. ABT-450/r-ombitasvir and dasabuvir with ribavirin for hepatitis C with cirrhosis. N Engl J Med 2014;370:1973-82.
24. Lawitz E, Poordad F, Gutierrez JA, et al. C-SWIFT: Grazoprevir (MK-5172) + elbasvir (MK-8742) + sofosbuvir in treatment-naive patients with hepatitis C virus genotype 1 infection, with and without cirrhosis, for durations of 4, 6, or 8 weeks. Abstract 33. The 65th Annual Meeting of the American Association for the Study of Liver Diseases (AASLD). Boston, November 7 to 11, 2014.
25. Kohli A, Osinusi A, Sims Z, et al. Virological response after 6 week triple-drug regimens for hepatitis C: A proof-of-concept phase 2A cohort study. Lancet 2015 January 12 (Epub ahead of print).
26. HARVONI™, Gilead Sciences Inc. Foster City, CA Dated October 14, 2012 [product monograph]. <http://www.gilead.ca/pdf/ca/harvoni_pm_english.pdf> (Accessed November 17, 2015).
27. Wyles DL, Pockros PJ, Yang JC, et al. Retreatment of patients who failed prior sofosbuvir-based regimens with all oral fixed-dose combination ledipasvir/sofosbuvir plus ribavirin for 12 weeks. Abstract 235. The 65th Annual Meeting of the American Association for the Study of Liver Diseases (AASLD). Boston, November 7 to 11, 2014.
28. Zhu Y, Chen S. Antiviral treatment of hepatitis C virus infection and factors affecting efficacy. World J Gastroenterol 2013;19:8963-73.
29. Public Health Ontario Laboratories (PHOL). Examination of HCV Viral Load in Ontario Population. PHOL HCV Type 1 Viral Load dataset April 26, 2010 to November 7, 2014 [Data on file].
30. Shafran S. The hepatitis C genotype paradox: Cost per treatment is increasing, but cost per cure is decreasing. Can J Gastroenterol Hepatol 2015;29:46-8.

Prospective evaluation of gastric neurostimulation for diabetic gastroparesis in Canada

Himanish Panda BHSc[1], Philip Mitchell MD[2], Michael Curley MD[1], Michelle Buresi MD[1], Lynn Wilsack MSc[1], Christopher N Andrews MD MSc[1]

H Panda, P Mitchell, M Curley, M Buresi, L Wilsack, CN Andrews. Prospective evaluation of gastric neurostimulation for diabetic gastroparesis in Canada. Can J Gastroenterol Hepatol 2015;29(4):198-202.

BACKGROUND: The efficacy of gastric neurostimulation therapy for diabetic gastroparesis (GP) in a 'real-life' Canadian setting has not been assessed.

AIMS: To assess changes in health-related quality of life (QoL), weekly vomiting frequency (WVF), total symptom score (TSS) and health care utilization 12 months before and after gastric neurostimulator implantation in a diabetic GP cohort.

METHODS: Medication-refractory diabetic GP patients (n=7, four female, mean age 42 years) were prospectively recruited from 2008 to 2012. QoL scores were self-administered and obtained at baseline, 24 and 48 weeks postimplantion. WVF and TSS were assessed similarly. Health care usage, measured as hospitalization frequency and medication cost, was obtained six and 12 months before and after implant. Changes from baseline to six and 12 months for all outcomes were compared.

RESULTS: The mean (± SD) QoL according to EuroQol was significantly better at 24 weeks after the baseline measurement (baseline 29±5, 24 weeks 52±7; P=0.03). The mean improvement in TSS was significantly better at one year postintervention (baseline score 35±5 versus 12 months 27±3; P=0.03). Changes in Short-Form 36 Health Survey and WVF were not significant. Days of GP-related hospitalization were highly variable but decreased from a median of 71 days (range 0 to 227 days) to 29 days (range two to 334 days) one year before and after surgery, respectively (P=0.735). Outpatient medication costs did not decrease to a significant extent.

CONCLUSION: Gastric neurostimulation for diabetic GP appeared to show some beneficial palliative effects overall in the present small open-label series, but the effect is highly variable among patients, and placebo effect cannot be ruled out.

Key Words: *Canada; Diabetic gastroparesis; Enterra therapy; Neurostimulation*

L'évaluation prospective de la neurostimulation gastrique pour la gastroparésie diabétique au Canada

HISTORIQUE : L'efficacité du traitement par neurostimulation gastrique pour la gastroparésie (GP) diabétique n'a pas été évaluée dans un milieu canadien « réel ».

OBJECTIFS : Évaluer les modifications à la qualité de vie liée à la santé (QdV), la fréquence de vomissements hebdomadaires (FVH), le score global de symptômes (SGS) et l'utilisation des services de santé 12 mois avant et après l'implantation d'un neurostimulateur gastrique dans une cohorte de GP diabétique.

MÉTHODOLOGIE : Les patients atteints de GP diabétique réfractaire aux médicaments (n=7, quatre femmes, âge moyen de 42 ans) ont fait l'objet d'une étude prospective réalisée de 2008 à 2012. Les scores de QdV on été obtenus au moyen de questionnaires auto-administrés colligés en début d'étude, 24 et 48 semaines après l'implantation. La FVH et le SGS ont été évalués de la même façon. L'utilisation des soins de santé, mesurée par la fréquence d'hospitalisations et le coût des médicaments, a été mesurée six et 12 mois avant et après l'implantation. Les chercheurs ont comparé les modifications entre le début de l'étude et six et 12 mois plus tard à l'égard de tous les résultats.

RÉSULTATS : D'après le questionnaire EuroQol, la QdV moyenne (± ÉT) était plus importante 24 semaines après la mesure de départ (29±5 en début d'étude, 52±7 au bout de 24 semaines; P=0,03). L'amélioration moyenne du SGS était beaucoup plus substantielle un an après l'intervention (score de départ 35±5 par rapport à 27±3 au bout de 12 mois; P=0,03). Les changements au questionnaire court en 36 questions et la FVH n'étaient pas significatifs. Les jours d'hospitalisation liés à la GP étaient hautement variables, mais ont reculé d'une médiane de 71 jours (plage de 0 à 227 jours) à 29 jours (plage de deux à 334 jours) un an avant et après l'opération, respectivement (P=0,735). Le coût des médicaments en milieu ambulatoire n'a pas diminué de manière significative.

CONCLUSION : La neurostimulation gastrique pour la GP diabétique semble avoir des effets palliatifs globaux bénéfiques dans cette petite cohorte ouverte, mais l'effet est hautement variable, et on ne peut pas écarter un effet placebo.

Gastroparesis (GP) is a disorder manifested by delayed gastric emptying in the absence of mechanical obstruction (1,2). Patients with severe GP experience frequent hospital admissions and have a poor quality of life (QoL) (1). There are many causes of GP, although the disorder is most often associated with longstanding diabetes and its neuropathic sequelae. Clinical presentations of GP range from mild to severe, and can vary in any combination of nausea, vomiting, epigastric pain and distress, abdominal fullness or bloating, early satiety, heartburn and loss of appetite (3).

Traditional management of GP has centred around supportive management (4). Reduction of symptoms is typically achieved with prokinetic and antiemetic medication to stimulate and coordinate gastroduodenal motility and treat nausea, respectively (4). Feeding jejunostomy to bypass the stomach and maintain nutrition is a palliative strategy; however, symptoms may persist.

Gastric electrical stimulation (GES) was developed for the treatment of patients with chronic, drug-refractory nausea and vomiting secondary to GP. The Enterra Therapy system (Medtronic, USA) is an implantable device that works with high-frequency stimulation and provides pulses of low-energy stimulation to the stomach antrum (5,6).

Enterra therapy is approved by Health Canada as a class 3 device for use in the treatment of chronic intractable (drug-refractory) nausea and vomiting (7). It has also been approved by the United States Food and Drug Administration under the Humanitarian Device Exemption for the same indication (8). In the United States (US), this exemption does not require the manufacturer to show clinical evidence of efficacy, but need only show that the device does not pose an unreasonable or significant risk for illness or injury. The Humanitarian Device Exemption is intended to benefit patients by treating a disease that affects <4000 individuals in the US per year (9).

[1]*Division of Gastroenterology;* [2]*Division of General Surgery, Cumming School of Medicine, University of Calgary, Calgary, Alberta*
Correspondence: Dr Christopher N Andrews, University of Calgary, 6D34, 3280 Hospital Drive Northwest, Calgary, Alberta T2N 4Z6. e-mail candrews@ucalgary.ca

A small pilot study of Enterra therapy was undertaken in Calgary (Alberta) from 2008 to 2012. The aim of the present study was to assess changes in health-related QoL before and after implantation of Enterra therapy in patients with severe drug-refractory GP secondary to diabetes mellitus (DM) and to compare health care utilization and symptoms for the year before and the year following implantation.

METHODS

The present analysis was a single-centre prospective pilot study of Enterra therapy for a small, highly selected population of diabetic patients with overwhelming symptoms due to medically refractory GP.

Subjects

Patients residing in southern Alberta between 18 and 65 years of age with insulin-treated DM and GP confirmed previously with gastric emptying scintigraphy (>10% retention at 4 h), who were demonstrably refractory to other GP treatment were invited to participate in the present study. Each participant had a clinical diagnosis of diabetic GP for at least one year, with a severity requiring at least two prokinetic and/or antiemetic medications or requiring jejunal feeding by any route; no alternative causes for symptoms found on gastroscopy within the past year; and was deemed medically fit to undergo surgery. Exclusion criteria included any clinical evidence (including physical examination or electrocardiography) that could pose surgical or other safety risks to the patient, previous gastrointestinal surgery, use of narcotics that could not be stopped before entering the study, and pregnant or breastfeeding women. Any patient on peritoneal dialysis or with an intra-abdominal transplant (or being considered for same) was also excluded. The present study was approved the Conjoint Health Research Ethics Review Board at the University of Calgary, Alberta. Each participant gave written informed consent before study procedures were performed. The study was funded through the local health board and received no funding from the neurostimulator manufacturer.

Implant procedure

Enterra therapy was implanted using laparoscopic lead placement into the gastric antral muscularis, with a small open skin incision to create the pocket for the pulse generator. The surgery required approximately 60 min of operating room time, using a standard laparoscopic tray. The procedure is technically straightforward and is described elsewhere (10). Patients generally required a one to two day inpatient stay after implantation.

Health care costs

Cumulative health care costs incurred in the 12 months before Enterra implantation were compared with those incurred from the day of the Enterra placement surgery and 12 months thereafter. These data were obtained from Alberta Health Services (AHS).

The number of inpatient hospitalization days was determined through use of the hospital separations and physicians claims databases at AHS using a diagnostic-related grouping methodology to ensure only GP-related stays were included. Outpatient drug costs were calculated by assigning a value for each drug (per tablet or single dose) based on AHS and Alberta Blue Cross data when available, and on prices listed on www.CanadaDrugs.com (June 2013) when Alberta Blue Cross data were unavailable, and then multiplying by the quantity of the drug prescribed. Only drugs related to GP, specifically prokinetics and antiemetics, were included in the present analysis.

Questionnaires

QoL questionnaires: At baseline, 12, 24 and 48 weeks postimplant, subjects answered three QoL questionnaires: Functional Living Index-Emesis (FLIE), a validated assessment of QoL specifically related to nausea and vomiting in the previous five days (11); the 36-item Short-Form Health Survey (SF-36) questionnaire (12,13), a validated measure of global QoL with both mental and physical domains; and the EuroQol 5D-3L (EQ-5D) with visual analogue scale (EQ-VAS), a preference-based QoL questionnaire that has been used in patients with gastrointestinal diseases (14,15). All questionnaires were self-administered by the patients.

Symptom questionnaires: At baseline, 12, 24 and 48 weeks postimplant, subjects completed a vomiting diary for the two weeks before the visit, a total symptom score (TSS) of GP symptom severity determined by the sum of the severity ratings of six symptom subscores (vomiting, nausea, early satiety, bloating, postprandial fullness and epigastric pain). Symptoms were rated on a five-point scale as 0 = absent, 1 = mild (not influencing usual activities), 2 = moderate (diverting from, but not requiring modifications of, usual activities), 3 = severe (influencing usual activities severely enough to require modifications) and 4 = extremely severe (requiring bed rest). This is a frequently used score in the GES literature (16-18).

Data and statistical analysis

The primary outcome measures of mean change in FLIE, SF-36 (mental and physical composite scores), EQ-5D (scored with USA TTO dataset) and EQ-VAS from baseline to 24 weeks and 48 weeks postimplant were compared using the Wilcoxon signed-rank test. The means of number of days hospitalized and outpatient medication costs for the year before and the year postintervention were compared using Wilcoxon signed-rank test and Student's *t* test, respectively.

For secondary outcome measurements of weekly vomiting frequency and TSS, differences in means were compared using the Wilcoxon signed-rank test; $P<0.05$ was considered to be statistically significant. In addition, records were analyzed for evidence of surgical complications and changes in use of prokinetic and antiemetic medications.

RESULTS

In total, seven medication-refractory diabetic GP patients (four female, mean age 42 years) were prospectively recruited in the Calgary area between 2008 and 2012. No surgical complications directly related to the implantation occurred. All implanted units were functional during the study. Four of seven patients have died at the time of the present report: three within two years of implantation and one died two years and two months postimplantation; however, none died during the study period. Causes of death varied, although all were related to diabetic sequelae.

QoL measures

The mean FLIE scores postimplant showed numerical improvement shortly after intervention, then trended back to baseline values by week 48 ($P=0.753$). Note that in this metric, larger values indicate a better QoL. Neither SF-36 physical nor mental scores showed significant improvement at any time postimplant with respect to baseline ($P=0.173$ at week 48 for the SF-36 physical scores; $P=0.172$ at week 48 for SF-36 mental scores).

EQ-5D scores showed improvement from baseline, from a mean of 0.387 (where 1 is considered perfect health) to 0.553 at 24 weeks and 0.550 at 48 weeks, but neither measure achieved statistical significance. EQ-VAS scores were generally low and well below 100, which represents the best possible health imaginable. The baseline score was 29, increasing significantly to 52 at week 24 ($P=0.034$) and 46 at week 48 ($P=0.058$). Data are shown in Figure 1.

Symptoms

Although weekly vomiting frequency decreased substantially from an average baseline score of 27.9 to 9.6, 4.4, and 6.8 at weeks 12, 24, and 48, respectively, the improvement was not statistically significant despite the apparent clinical benefit ($P=0.116$ at week 48). The mean TSS score showed a significant improvement 48 weeks postimplantation compared with baseline ($P=0.046$). The averaged TSS score at week 48 was 27.2 compared with baseline (35.0). Note that in this metric, smaller values indicate improvement. Data are shown in Figure 2.

Health care usage

Hospitalization length was highly variable among the seven patients, as shown in Figure 3. The total number of days hospitalized decreased

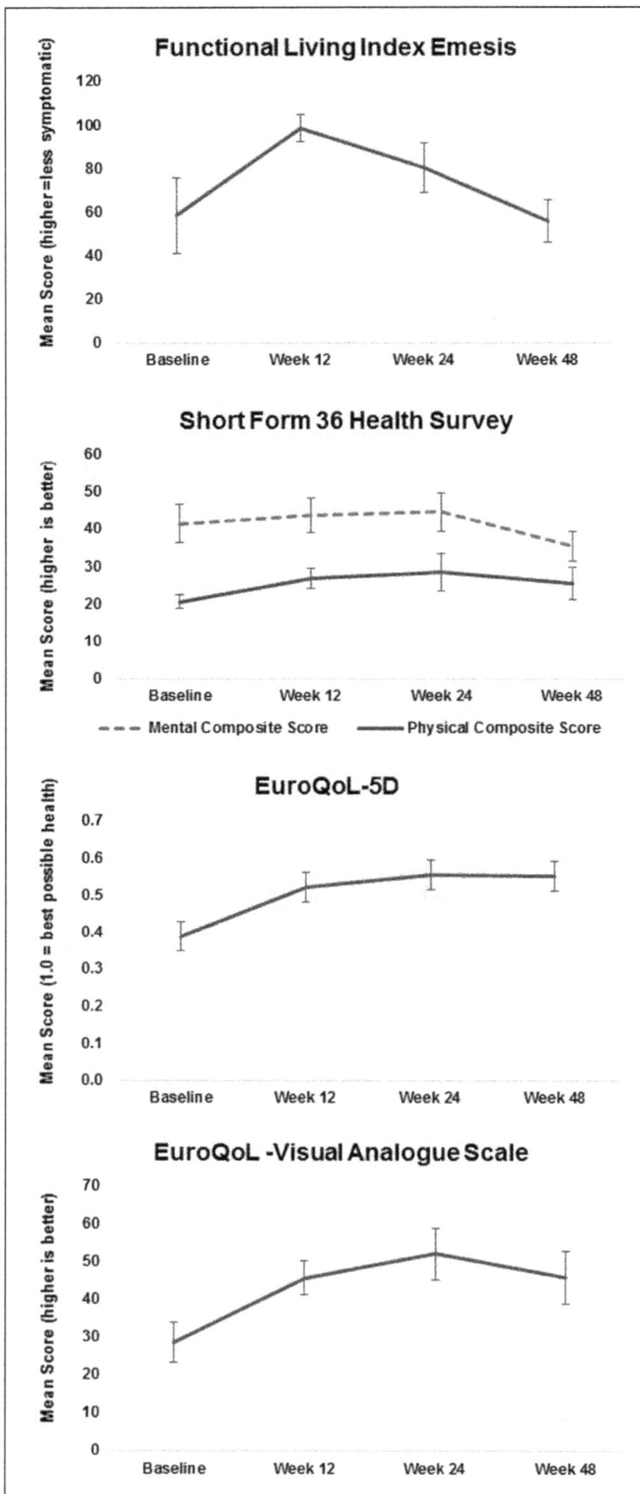

Figure 1) *Quality of life questionnaire scores*

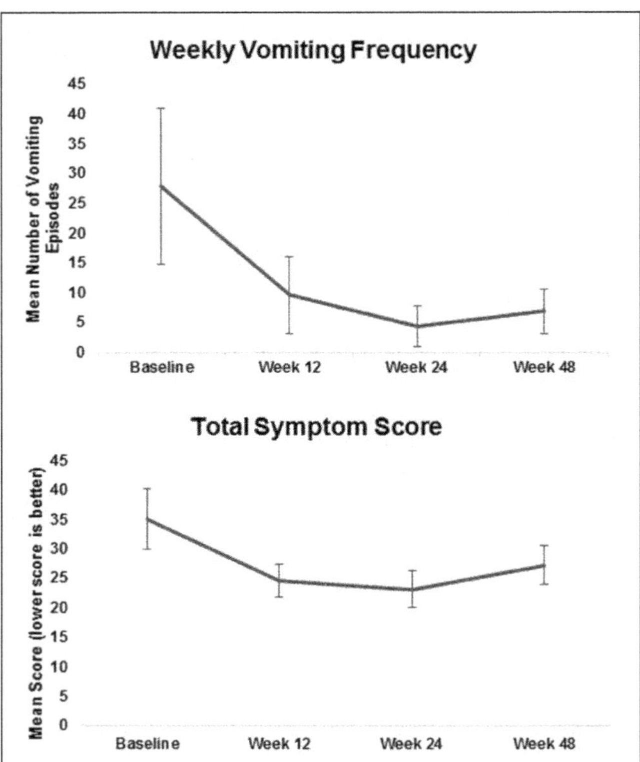

Figure 2) *Gastroparesis symptoms measures*

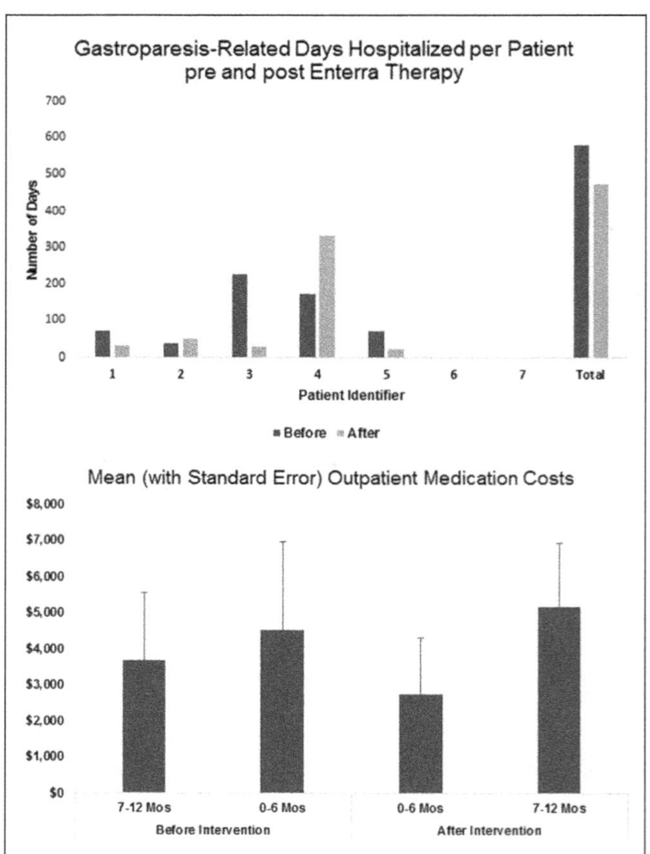

Figure 3) *Health care-related costs, before and after Enterra (Medtronic, USA) implantation. Mos Months*

after the intervention, going from 769 days (median 72 days) days in the year before surgery to 621 days (median 40 days) in the year after the surgery. Total number of days hospitalized due to GP symptoms as the primary reason also decreased after the intervention, going from 580 days (median 71 days) in the year before surgery to 473 days (median 29 days) in the year postintervention. Given the variability, these reductions in hospital usage were not significant (P=0.735).

Mean outpatient medication costs were $8,199 for the year before and $7,945 for the year postimplant. There was a 39% decrease in mean medication costs when comparing costs at six months pre- and postintervention ($4,517 and $2,764, respectively); however, this difference was not significant (P=0.10). All patients were using prokinetics and antiemetics at 12 months postimplantation.

DISCUSSION

The present study is the first to report the outcomes of Enterra therapy in the 'real-world' setting since approval in Canada. From a qualitative perspective, it is clear that some patients improved significantly, whereas others did not appear to benefit. There was a general trend toward improvements in symptoms and QoL overall, although most end points did not reach statistical significance.

The present study had two main limitations, the first of which was small sample size. There were several reasons for this including the relative rarity of the disease and exclusions for other comorbidities. Because individuals with end-stage diabetic complications, such as GP, almost invariably have other complications, such as nephropathy and peripheral neuropathy, many patients failed screening due to the need for peritoneal dialysis and/or the use of narcotics, respectively. Although our group has performed other Enterra implantations for out-of-province patients, they were not included in the present study because we would not have been able to capture the hospitalization and medication use data. The other limitation of the present study was that it was not blinded and, thus, had no control for the placebo effect. Previous controlled studies have shown similar levels of improvement in groups with and without the intervention, suggesting the presence of a placebo effect (1,19). However, given that all previous medical interventions failed to alleviate symptoms in this cohort, at least some of the positive results observed in the present study could be attributed to the implanted neurotransmitter.

The mechanism of action of GES is not completely understood (20). Contrary to common misperception, Enterra is not a 'pacemaker' analogous to those used in cardiac applications, although the hardware appears similar. Although entrainment ('pacing') of the intrinsic gastric electrical activity is possible, it requires more energy than an implantable battery can provide and, in fact, can cause significant patient discomfort (21). The low-energy, high-frequency stimulation used in the Enterra system does not appear to work by improving gastric emptying; it is, thus, not a true pacemaker and neurostimulator is a more appropriate term. It was previously believed that it may function via vagal afferent stimulation and subsequently increased central descending inhibition (22), similar to spinal cord stimulation for chronic pain (23); however, the evidence is conflicting because high-frequency GES also works in patients with vagotomy (24). Other animal studies have suggested it may work through modulation of gastric spinal afferent nerves (25).

The American Gastroenterological Association Technical Review on the diagnosis and treatment of GP quoted an unpublished study of patients with severe GP that estimated health care costs to be an average of USD$6,972 per patient per month (4). Most expenditures were attributed to requirements for hospitalization and temporary or long-term use of intravenous alimentation (parenteral nutrition). In addition, diagnostic tests for patients with presumed GP are associated with significant costs, especially for endoscopy and gastric emptying scintigraphy tests (4). The cost of implanting the Enterra therapy system in the US is approximately USD$30,000, and is covered by Medicare as well as some private health insurance companies (26). The total cost of the device hardware per patient is approximately CAD$12,000. Costs of diagnostic tests, surgical intervention (laparoscopy or laparotomy) to implant the device, hospitalization and follow-up are supplemental.

Although a substantial literature base assessing the efficacy of GES for GP has accumulated, all of the studies (with the exception of two) have been uncontrolled open-label case series similar to the present analysis. The two published randomized sham-controlled studies (in which the device was implanted and patients were blinded as to whether the device was turned on or off), were both negative in meeting their primary end point (19,27). Two meta-analyses have subsequently reviewed this literature, suggesting that the benefits of GES for GP are substantial for symptom reduction and gastric emptying, with the caveat of the limitations of uncontrolled studies (28,29). This is a major caveat, however, because the numbers of patients in the uncontrolled studies were significantly greater than those in the controlled studies and would disproportionately favour positive conclusions.

The Alberta Heritage Foundation for Medical Research completed a health technology assessment on GES in 2006, which concluded that there was insufficient evidence to determine the efficacy of the therapy, and that routine use of the technology was not recommended (26). Based on current data since that time, and including the results of the present study, it is unlikely that those conclusions would be altered. To our knowledge, GES is not regularly offered publicly or privately anywhere in Canada.

However, more recent guidance from the United Kingdom National Institute for Health and Care Excellence (NICE) concluded that current evidence on the efficacy and safety of GES for GP is adequate to support the use of this procedure (30). The Committee concluded that the evidence of efficacy was adequate only after prolonged debate about the design of the available randomized trials, and also noted the possibility of a placebo response in most of the evidence. Furthermore, the NICE Committee also acknowledged that GP can be a highly debilitating condition with very few treatment options, and took into account anecdotal patient commentaries describing substantial improvements in QoL with GES. Clinicians are advised to inform patients considering the procedure that some patients do not receive any benefit from it (30). This NICE recommendation deviates from their typical guidance, which is generally significantly more evidence focused. It suggests they have adopted a similar stance to the Humanitarian Device Exemption given to GES in the US, whereby efficacy is not required to be proven for devices used in small populations with significant suffering.

CONCLUSION

In the present small study involving an end-stage diabetic GP population, the palliative effect on symptoms was witnessed and sustained to one year; however, the magnitude of the placebo effect was not estimable due to the study design. There did not appear to be any signal suggesting any improvement in hospitalization rate or medication costs, but the study was almost certainly underpowered given the variability of these measures. Furthermore, not all patients experienced a beneficial effect. The low morbidity of the implantation procedure and the modest cost compared with hospitalization or chronic medical therapy for GP make it reasonable to consider as a last-resort option; however, there is no certainty, based on the present study or others, that either symptoms or costs will improve for any given patient. Thus, any further funding for GES should be directed toward specialized multidisciplinary clinics with a focus on research and improvement of the technology.

DISCLOSURES: The authors have no financial disclosures or conflicts of interest to declare.

REFERENCES

1. Abell TL, Bernstein RK, Cutts T, et al. Treatment of gastroparesis: A multidisciplinary clinical review. Neurogastroenterol Motil 2006;18:263-83.
2. Bielefeldt K. Gastroparesis: Concepts, controversies, and challenges. Scientifica (Cairo) 2012;2012:424802.
3. Bortolotti M. Gastric electrical stimulation for gastroparesis: A goal greatly pursued, but not yet attained. World J Gastroenterol 2011;17:273-82.

4. Parkman HP, Hasler WL, Fisher RS. American Gastroenterological Association medical position statement: Diagnosis and treatment of gastroparesis. Gastroenterology 2004;127:1589-91.
5. McCallum RW, Lin Z, Forster J, Roeser K, Hou Q, Sarosiek I. Gastric electrical stimulation improves outcomes of patients with gastroparesis for up to 10 years. Clin Gastroenterol Hepatol 2011;9:314-9,e1.
6. McKenna D, Beverstein G, Reichelderfer M, Gaumnitz E, Gould J. Gastric electrical stimulation is an effective and safe treatment for medically refractory gastroparesis. Surgery 2008;144:566-72.
7. Health Canada Medical Devices Active License Listing, License Number 60264. <http://webprod5.hc-sc.gc.ca/mdll-limh/start-debuter.do?lang=eng> (Accessed July 17, 2014).
8. U.S. Food and Drug Administration. Medical Devices | Device Approvals, Denials and Clearances, H990014, Enterra™ Therapy System, formerly named Gastric Electrical Stimulation (GES) System. <http://www.accessdata.fda.gov/scripts/cdrh/cfdocs/cftopic/pma/pma.cfm?num=H990014> (Accessed July 23, 2014).
9. U.S. Food and Drug Administration. Medical Devices | Humanitarian Device Exemption. <http://www.fda.gov/medicaldevices/deviceregulationandguidance/howtomarketyourdevice/premarketsubmissions/humanitariandeviceexemption/default.htm> (Accessed September 16, 2014).
10. Brody F, Nam A, Drenon E, Ali A, Soffer E. Laparoscopic insertion of gastric electrodes for electrical stimulation. J Laparoendosc Adv Surg Tech A 2007;17:1-6.
11. Martin AR, Pearson JD, Cai B, Elmer M, Horgan K, Lindley C. Assessing the impact of chemotherapy-induced nausea and vomiting on patients' daily lives: A modified version of the Functional Living Index-Emesis (FLIE) with 5-day recall. Support Care Cancer 2003;11:522-7.
12. Velanovich V. Quality of life and symptomatic response to gastric neurostimulation for gastroparesis. J Gastrointest Surg 2008;12:1656-62.
13. Ware JE Jr, Gandek B. Overview of the SF-36 Health Survey and the International Quality of Life Assessment (IQOLA) Project. J Clin Epidemiol 1998;51:903-12.
14. Konig HH, Ulshofer A, Gregor M, et al. Validation of the EuroQol questionnaire in patients with inflammatory bowel disease. Eur J Gastroenterol Hepatol 2002;14:1205-15.
15. Ratcliffe J, Longworth L, Young T, Bryan S, Burroughs A, Buxton M. Assessing health-related quality of life pre- and post-liver transplantation: A prospective multicenter study. Liver Transpl 2002;8:263-70.
16. Abell TL, Van Cutsem E, Abrahamsson H, et al. Gastric electrical stimulation in intractable symptomatic gastroparesis. Digestion 2002;66:204-12.
17. Lin Z, Forster J, Sarosiek I, McCallum RW. Treatment of diabetic gastroparesis by high-frequency gastric electrical stimulation. Diabetes Care 2004;27:1071-6.
18. van der Voort IR, Becker JC, Dietl KH, Konturek JW, Domschke W, Pohle T. Gastric electrical stimulation results in improved metabolic control in diabetic patients suffering from gastroparesis. Exp Clin Endocrinol Diabetes 2005;113:38-42.
19. McCallum RW, Snape W, Brody F, Wo J, Parkman HP, Nowak T. Gastric electrical stimulation with Enterra therapy improves symptoms from diabetic gastroparesis in a prospective study. Clin Gastroenterol Hepatol 2010;8:947-54.
20. Hasler WL. Methods of gastric electrical stimulation and pacing: A review of their benefits and mechanisms of action in gastroparesis and obesity. Neurogastroenterol Motil 2009;21:229-43.
21. Familoni BO, Abell TL, Voeller G, Salem A, Gaber O. Electrical stimulation at a frequency higher than basal rate in human stomach. Dig Dis Sci 1997;42:885-91.
22. Sogabe M, Okahisa T, Tsujigami K, et al. Ultrasonographic assessment of gastric motility in diabetic gastroparesis before and after attaining glycemic control. J Gastroenterol 2005;40:583-90.
23. Stevens JE, Jones KL, Rayner CK, Horowitz M. Pathophysiology and pharmacotherapy of gastroparesis: current and future perspectives. Expert Opin Pharmacother 2013;14:1171-86.
24. McCallum R, Lin Z, Wetzel P, Sarosiek I, Forster J. Clinical response to gastric electrical stimulation in patients with postsurgical gastroparesis. Clin Gastroenterol Hepatol 2005;3:49-54.
25. Ouelaa W, Ghouzali I, Langlois L, et al. Gastric electrical stimulation decreases gastric distension-induced central nociception response through direct action on primary afferents. PLoS One 2012;7:e47849.
26. Moga C, Harstall C. Health Technology Assessment Report #37: Gastric Electrical Stimulation (Enterra™ Therapy System) for the Treatment of Gastroparesis. Edmonton, Alberta: Alberta Heritage Foundation for Medical Research, 2006.
27. Abell T, McCallum R, Hocking M, et al. Gastric electrical stimulation for medically refractory gastroparesis. Gastroenterology 2003;125:421-8.
28. O'Grady G, Egbuji JU, Du P, Cheng LK, Pullan AJ, Windsor JA. High-frequency gastric electrical stimulation for the treatment of gastroparesis: A meta-analysis. World J Surg 2009;33:1693-701.
29. Chu H, Lin Z, Zhong L, McCallum RW, Hou X. Treatment of high-frequency gastric electrical stimulation for gastroparesis. J Gastroenterol Hepatol 2012;27:1017-26.
30. National Institute for Health and Care Excellence (NICE). Interventional Procedure Guidance 489. Gastroelectrical Stimulation for Gastroparesis. <www.nice.org.uk/guidance/ipg489/> (Accessed October 1, 2014).

Renal impairment in cirrhosis unrelated to hepatorenal syndrome

Gavin Low MBChB MRCS FRCR[1,3], Graeme JM Alexander MD FRCP[2], David J Lomas MA MB BChir FRCP FRCR[1]

G Low, GJM Alexander, DJ Lomas. Renal impairment in cirrhosis unrelated to hepatorenal syndrome. Can J Gastroenterol Hepatol 2015;29(5):253-257.

Renal impairment is common in liver disease and may occur as a consequence of the pathophysiological changes that underpin cirrhosis or secondary to a pre-existing unrelated insult. Nevertheless, the onset of renal impairment often portends a worsening prognosis. Hepatorenal syndrome remains one of the most recognized and reported causes of renal impairment in cirrhosis. However, other causes of renal impairment occur and can be classified into prerenal, intrinsic or postrenal, which are the subjects of the present review.

Key Words: *Acute kidney injury; Cirrhosis; Intrinsic renal impairment; Postrenal impairment; Prerenal impairment*

Une atteinte rénale causée par la cirrhose, mais non liée au syndrome hépatorénal

L'atteinte rénale est courante dans le contexte d'une maladie hépatique. Elle peut découler des changements physiopathologiques de la cirrhose ou d'un problème préexistant non lié. Elle sous-tend souvent un pronostic plus sombre. Le syndrome hépatorénal demeure l'une des causes d'atteinte rénale les plus connues et les plus signalées en cas de cirrhose. La présente analyse porte sur les autres causes possibles, qui sont d'étiologie pré-rénale, intrinsèque ou post-rénale.

Patients with cirrhosis are susceptible to renal impairment as a result of pre-existing circulatory and neurohormonal imbalances, and exposure to precipitating factors. Hampel et al (1) found that acute renal impairment occurred in 24.7% (23 of 96) of hospitalized patients with cirrhosis. In an intensive care unit setting, the incidence of acute renal impairment in cirrhotic patients was reported to be 15.1% (144 of 932) by Peron et al (2) and 39.2% (73 of 186) by du Cheyron et al (3). Terra et al (4) found that 27% (29 of 106) of cirrhotic patients with sepsis unrelated to spontaneous bacterial peritonitis (SBP) developed renal impairment compared with only 8% (eight of 100) of those without infection. Combining the findings of the four aforementioned studies, Garcia-Tsao et al (5) calculated that acute renal impairment occurred in approximately 19% of patients with cirrhosis. In contrast, chronic renal failure is present in 1% of individuals with cirrhosis (5). A variety of etiologies are implicated in the development of renal impairment (Figure 1). These have an impact on clinical presentation, treatment options and prognosis. In a prospective study investigating 562 consecutive patients with cirrhosis and renal impairment (6), three-month survival was 73% for intrinsic renal impairment, 46% for hypovolemia-induced prerenal impairment, 31% for infection-related renal impairment and 15% for hepatorenal syndrome (HRS). The onset of renal impairment in cirrhosis is an important prognostic indicator (7-12). A systematic review of 118 studies by D'Armico et al (10) found that measures of renal impairment (serum creatinine, blood urea nitrogen/azotemia) were strong predictors of mortality in decompensated cirrhosis. In a systematic review of 74 studies, Fede et al (11) found that cirrhotic patients with renal impairment had a >7-fold increased risk for death within one year compared with cirrhotic patients without renal impairment. The one-month and 12-month mortality for patients with renal impairment other than HRS were 56% and 36%, respectively. Renal impairment correlated with death on univariate analysis in 29 studies and on multivariate analysis in 13 studies (11). When serum creatinine was evaluated, it correlated with death on univariate analysis in 16 studies and on multivariate analysis in two studies (11). Serum creatinine level is a variable in calculating the Model for End-stage Liver Disease score – a recognized predictor of the three-month mortality risk and a method used for allocating liver transplants (13,14).

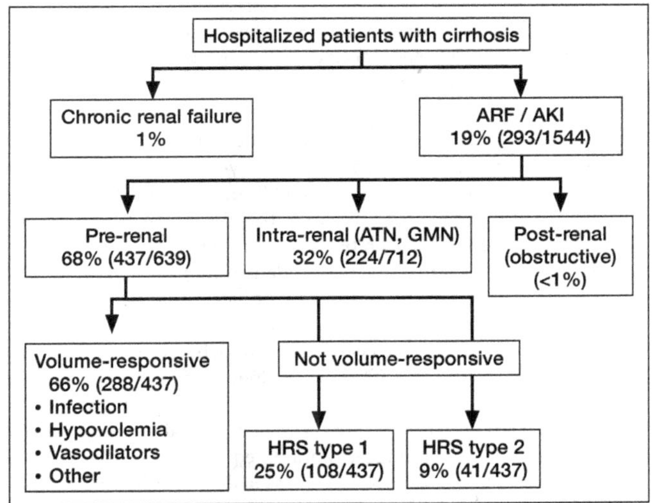

Figure 1) *Types of renal impairment in hospitalized patients with cirrhosis. AKI Acute kidney injury; ARF Acute renal failure; ATN Acute tubular necrosis; GMN Glomerulonephritis; HRS Hepatorenal syndrome. Adapted and reproduced with permission from reference 5*

Traditionally, renal failure is defined by a serum creatinine threshold of ≥1.5 mg/dL (≥133 μmol/L). However, serum creatinine is not a reliable marker of renal function in cirrhosis because a normal serum creatinine level does not exclude mild disease. Cirrhotic patients may show a lower baseline serum creatinine level than normal due to: reduced endogenous creatinine synthesis from liver dysfunction; decreased conversion of creatine to creatinine as a consequence of

[1]*Department of Radiology;* [2]*Division of Gastroenterology & Hepatology, Addenbrooke's Hospital, Cambridge University Hospitals NHS Foundation Trust, United Kingdom;* [3]*University of Alberta, Edmonton, Alberta*
Correspondence: Dr Gavin Low, University of Cambridge School of Clinical Medicine, Box 218, Cambridge Biomedical Campus, Cambridge, CB2 0QQ United Kingdom.e-mail timgy@yahoo.com

TABLE 1
Acute Kidney Injury Network classification of acute kidney injury (AKI)

AKI stage	Serum creatinine criteria
1	Increase in serum creatinine level ≥26.4 μmol/L (≥0.3 mg/dL) within 48 h or an increase ≥150% to 200% (1.5- to twofold) from baseline
2	Increase in serum creatinine level to >200% to 300% (>2–3-fold) from baseline
3	Increase in serum creatinine level to >300% (>3-fold) from baseline or serum creatinine ≥354 μmol/L (≥4.0 mg/dL) with an acute increase of ≥44 μmol/L (≥0.5 mg/dL) or initiation of renal replacement therapy

Reproduced and adapted from reference 20

TABLE 2
Revised diagnostic criteria for hepatorenal syndrome (International Ascites Club, 2007)

- Cirrhosis with ascites
- Serum creatinine >133 μmol/L (1.5 mg/dL)
- No improvement in serum creatinine (decrease to a level of ≤133 μmol/L) after ≥2 days with diuretic withdrawal and volume expansion with albumin. The recommended dose of albumin is 1 g/kg of body weight/day up to a maximum of 100 g/day
- Absence of shock
- No current or recent treatment with nephrotoxic drugs
- Absence of parenchymal kidney disease as indicated by proteinuria >500 mg/day, microscopic hematuria (>50 red blood cells per high-power field) and/or abnormal renal ultrasonography

Reproduced with permission from reference 24

reduced skeletal muscle mass from malnutrition; drug-related increased tubular secretion of creatinine; and laboratory-based underestimations of serum creatinine in the setting of hyperbilirubinemia (15). Cystatin C is an alternative method for estimating the glomerular filtration rate (GFR). It may be more accurate than serum creatinine in cirrhotic patients because cystatin C is independent of hepatic function; it is not influenced by muscle mass, sex, race or diet, and tubular secretion of cystatin C is not increased in cirrhosis (16). Several cystatin C-based GFR equations have been described such as the Larsson equation, Hoek equation and Chronic Kidney Disease Epidemiology Collaboration (CKD-EPI) cystatin C equation (17-19). In a study involving 72 patients with cirrhosis, Mindikoglu et al (15) compared a combined CKD-EPI creatinine-cystatin C equation with that of 24 h urinary creatinine clearance, Cockcroft-Gault and other creatinine-and/or cystatin C-based GFR equations (15). Using nonradiolabelled iothalamate clearance as the gold standard for measuring GFR, a significantly higher accuracy was found for the CKD-EPI creatinine-cystatin C equation compared with the other GFR-based equations. However, the study noted that the diagnostic performance of the CKD-EPI creatinine-cystatin C equation was worse in cirrhotic patients than in those without cirrhosis.

The definition of renal dysfunction was revised by the Acute Kidney Injury Network in 2005, and the term 'acute renal failure' was replaced by 'acute kidney injury' (AKI) (20). Table 1 describes the AKI classification and staging system. To frame AKI in the context of cirrhosis, a working group comprising members of the International Ascites Club and the Acute Dialysis Quality Initiative was formed in 2010. The working group defined AKI in cirrhosis as encompassing all causes of acute renal impairment in which the serum creatinine level increased by ≥50% (≥1.5-fold) from baseline or in which the serum creatinine level rose by ≥26.4 μmol/L (≥ 0.3 mg/dL) in <48 h (21). This revised definition has higher sensitivity (65.4% versus 53.8%) but reduced specificity (61.8% versus 85.3%) for predicting in-hospital mortality compared with traditional criteria (22). Chronic kidney disease (CKD) in cirrhosis was defined as a GFR <60 mL/min for >3 months, calculated using the Modification of Diet in Renal Diseases-6 formula (21). Finally, acute-on-CKD was defined as AKI occurring in the context of pre-existing CKD (21).

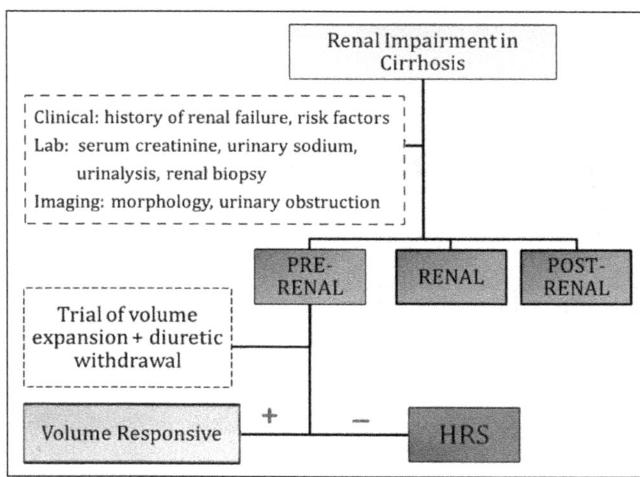

Figure 2) *The authors' simplified diagnostic algorithm for the clinical work-up of renal impairment in patients with cirrhosis. HRS Hepatorenal syndrome*

The most common causes of renal impairment in cirrhosis are hypovolemia-induced prerenal AKI, acute tubular necrosis (ATN) and HRS (5). HRS is the most lethal of these causes of renal impairment, with prognosis typically ranging from weeks to months (14,23). It is a functional renal impairment due to profound renal arterial vasoconstriction precipitated by unregulated circulatory and neurohormonal alterations that underpin cirrhosis. While the topic of HRS is outside the scope of the present review, the revised diagnostic criteria for HRS is included (Table 2) because it allows HRS to be recognized and differentiated from other causes of renal impairment and vice versa (24). Type 1 HRS is a rapid, progressive, renal impairment defined by a doubling of the serum creatinine to a level >2.5 mg/dL (>226 μmol/L) in <2 weeks (24). Type 2 HRS is moderate renal impairment (serum creatinine >1.5 mg/dL [>133 μmol/L] and up to 2.5 mg/dL [226 μmol/L]) with a steady progressive course that evolves over weeks to months (24). In the present article, the causes of renal impairment other than HRS, collectively termed 'non-HRS' renal impairment are discussed, which are divided into prerenal, intrinsic renal and postrenal causes. A simplified diagnostic algorithm for the clinical work-up of renal impairment in cirrhosis is illustrated in Figure 2.

PRERENAL IMPAIRMENT

Prerenal impairment is the most common type of renal dysfunction in cirrhosis, accounting for 68% of AKI cases (5), and may be due to either hypovolemia-induced prerenal impairment (66%) or HRS (34%) (5). Patients with cirrhosis are predisposed to develop prerenal impairment due to underlying circulatory disturbances such as arterial underfilling (that occurs as a result of splanchnic vasodilation). These pathophysiological changes are mild in hypovolemia-induced prerenal impairment, but severe in HRS, in which neurohormonal activation results in renal vasoconstriction and severely diminished renal perfusion. These two causes of prerenal impairment can be differentiated by the response to volume expansion. Hypovolemia-induced prerenal impairment responds to volume expansion with rapid normalization of renal function. Conversely, HRS is insensitive to volume expansion, with no improvement in renal function. Albumin is superior to saline for volume expansion, and the recommended dose of intravenous albumin is 1 g/kg of body weight/day up to a maximum of 100 g per day (24,25).

In hypovolemia-induced prerenal impairment, excessive fluid loss may be the result of various causes including vomiting, diarrhea, sodium and water restriction, gastrointestinal hemorrhage, large-volume paracentesis

(LVP) and excessive diuretic therapy. Diarrhea may be a side effect of lactulose therapy in patients with hepatic encephalopathy. Diuretic-induced prerenal impairment occurs in 15% to 25% of patients with cirrhosis and ascites (26). Nephrotoxic drugs, such as nonsteroidal anti-inflammatory drugs, angiotensin-converting enzyme inhibitors and angiotensin II inhibitors, may precipitate prerenal failure by interfering with vasoactive mechanisms that modulate renal perfusion. Infection is another significant cause of prerenal impairment and may precede septic shock. This precipitates renal impairment due to systemic vasodilation, arterial hypotension and diminished renal perfusion. Up to 10% of cases of SBP are associated with septic shock (27).

Measures should be instituted in cirrhotic patients to prevent intravascular volume depletion and, consequently, reduce the risk for prerenal impairment, including careful attention to fluid balance, judicious use of diuretics, avoidance of nephrotoxic drugs (especially nonsteroidal anti-inflammatory drugs and aminoglycoside antibiotics), avoidance of excessive lactulose therapy, administration of intravenous albumin after LVP and prompt treatment of infection. A meta-analysis of 17 randomized trials involving 1225 patients with cirrhosis and tense ascites (25) found that albumin reduced mortality and morbidity in patients undergoing LVP when compared with alternative treatments such as dextran, gelatin, hydroxyethyl starch and hypertonic saline. The incidence of postparacentesis complications, such as circulatory dysfunction (OR 0.39 [95% CI 0.27 to 0.55]), hyponatremia (OR 0.58 [95% CI 0.39 to 0.87]) and death (OR 0.64 [95% CI 0.41 to 0.98]) were lower in patients who were administered albumin compared with those who underwent alternative treatments. Given the evidence, the American Association for the Study of Liver Diseases recommends that albumin infusion (6 g/L to 8 g/L of ascitic fluid removed) be performed for LVP (≥5 L) but that albumin may not be necessary for a single paracentesis of <4 L to 5 L (28). The European Association for the Study of the Liver also recommends the use of albumin (8 g/L of ascitic fluid removed) after LVP (≥5 L) (29). In a randomized controlled trial involving 1126 patients with cirrhosis and SBP, Sort et al (30) found that the addition of albumin to cefotaxime (antibiotic) reduced the risk of renal impairment compared with treatment by cefotaxime alone. The incidence of renal impairment was 10% in the albumin plus cefotaxime group compared with 33% in the cefotaxime only group (P=0.01). Furthermore, the in-hospital and three-month mortality rates were lower in the albumin plus cefotaxime group compared with the cefotaxime only group (10% compared with 29% [P=0.01] and 22% compared with 41% [P=0.03]). A more recent study involving 38 episodes of SBP in 28 patients suggested that albumin should be given when the serum creatinine level is >88.4 μmol/L or bilirubin >68.4 μmol/L but was not necessary in patients who did not meet this criteria (31). The American Association for the Study of Liver Diseases recommends that patients with SBP that have a serum creatinine level >1 mg/dL (88.4 μmol/L), blood urea nitrogen >30 mg/dL (10.7 mmol/L) or total bilirubin >4 mg/dL (68.4 μmol/L) should receive 1.5 g of albumin per kg body weight within 6 h of detection and 1 g/kg on day 3 (28). In contrast, the European Association for the Study of the Liver recommends that all patients that develop SBP should be treated with broad-spectrum antibiotics and albumin (29).

INTRINSIC RENAL IMPAIRMENT

Intrinsic renal impairment may be caused by abnormalities in the renal tubules (eg, ATN), the glomeruli (eg, glomerulonephritis) or interstitium (eg, interstitial nephritis). It should be suspected in patients with proteinuria (>500 mg/dL [>5 g/L]) and/or hematuria (>50 red cells per high-powered field). A careful review of the medical history, including comorbidities and drug history, is often helpful in achieving the correct diagnosis. Renal biopsy is reserved for indeterminate cases in which obtaining histological diagnosis may alter patient management or prognosis. Ideally, the decision to biopsy should be taken following expert consultation with a renal physician. Indications for renal biopsy in cirrhosis patients as reported by Francoz et al (32) include: abnormal duration of acute renal impairment or failure of recovery with specific therapy; acute renal impairment with suspicion of superimposed CKD in candidates for liver transplantation; suspicion of uncommon intrinsic kidney disease (systemic disease, immune-allergic-induced drug toxicity, thrombotic microangiopathy) with potential curative intervention; and candidates for liver transplantation that have CKD with a GFR of either 15 mL/min to 30 mL/min (systematic biopsy) or 30 mL/min to 60 mL/min (biopsy if suspicious for parenchymal disease as indicated by proteinuria >500 mg/day, microhematuria and/or a recognized cause of CKD such as hepatitis C virus [HCV] or hepatitis B virus [HBV], diabetes or history of hypertension) (32).

ATN is the most common cause of intrinsic renal impairment in patients with cirrhosis. According to separate reports, ATN accounts for <32%, 41.7% and 44.4% of cases of renal impairment in cirrhosis (5,33,34). ATN may be either ischemic or toxic, while the clinical course is divided into initiation (hours to weeks), maintenance (one to two weeks) and recovery stages. Hypovolemia-induced prerenal impairment may progress into ischemic ATN in severe cases. Aminoglycoside antibiotics are the most common cause of toxic ATN, and others include nonsteroidal anti-inflammatory drugs, calcineurin inhibitor immunosuppression and radioiodine contrast agents. While radioiodine contrast agents are safe in patients with normal renal function, those with borderline renal function are at an increased risk for nephrotoxicity. In a prospective study involving 404 patients with cirrhosis, Prakash et al (34) found that precipitating factors for ATN included sepsis (most commonly SBP followed by urinary tract infection) in 61.4%, hypovolemia (most commonly gastrointestinal bleed followed by septic shock) in 36.4% and nephrotoxic drugs in 2.3%. In ATN, the tubular ability to reabsorb sodium and concentrate urine is impaired, leading to high urinary sodium (> 40 mEq/L), high fractional excretion of sodium (2%) and low urinary osmolality (<350 mOsm/kg) (5). Conversely, patients with HRS have low urinary sodium (<20 mEq/L), low fractional excretion of sodium (<1%) and elevated urinary osmolality (>500 mOsm/kg) (5). Nevertheless, differentiation can be challenging in patients with cirrhosis because diuretic therapy can lead to a urinary sodium >10 mEq/L while a fractional sodium excretion ≤1% is occasionally found in ATN (5). Granular casts may be found in both ATN and HRS, but epithelial casts favour ATN (5,35).

Chronic HBV and HCV cirrhosis are associated with a variety of glomerular diseases such as membranoproliferative glomerulonephritis (most common glomerular disease induced by HCV) with or without cryoglobulinemia, membranous nephropathy (most common glomerular disease induced by HBV) and focal glomerulosclerosis (36,37). Abnormalities of immunoglobuilin (Ig)A have been reported in patients with alcohol-related cirrhosis, including mesangial IgA deposits and development of secondary IgA nephropathy (38,39). Elevated circulating serum IgA is common in patients with alcohol-related liver disease, while mesangial IgA renal deposits are reported at autopsy in 36% of those with cirrhosis (40,41). Interstitial nephritis is most commonly due to drug-induced hypersensitivity reactions. Renal impairment may also be due to comorbidities, such as diabetes and atherosclerosis, which may lead to diabetic nephropathy and atherosclerotic-related glomerulosclerosis. Studies suggest that a significant proportion of cirrhotic patients have underlying atherosclerotic disease, as noted by the high incidence of coronary artery disease (20% to 25%) in this cohort (32,42,43). Letiexhe et al (44) reported that the incidence of diabetes may be as high as 25% in HCV cirrhosis. The clinical diagnosis of diabetic nephropathy is dependent on the detection of albuminuria and declining renal function. Albuminuria is diagnosed based on abnormal findings in two of three urine samples collected over a three- to six-month period. Depending on the urinary albumin excretion rate, it is staged as either microalbuminuria (30 mg/24 h to 299 mg/24 h) or macroalbuminuria (≥300 mg/24 h) (45,46). In a diabetic patient with cirrhosis, differentiating diabetic nephropathy from other causes of intrinsic renal impairment is often challenging. Patients may have multiple risk factors for renal impairment (eg, diabetes, hypertension, HCV, etc), more than one type of renal disease may exist concurrently, while nondiabetic renal disease may be found in diabetic patients (47-51). Findings that are regarded as suspicious for a nondiabetic cause of intrinsic renal impairment in a diabetic patient include: hematuria; absence of coexisting diabetic retinopathy; nephrotic range proteinuria (3.5 g/24 h) in patients with diabetes <5 years in duration; active urinary sediment; and rapid deterioration of GFR in patients with previously stable renal function (49-52). In such cases, a renal biopsy may be performed to obtain histological diagnosis (52).

POSTRENAL IMPAIRMENT

Postrenal impairment due to urinary tract obstruction is an uncommon cause of renal dysfunction, accounting for <1% of AKI cases in cirrhosis and 5% to 10% of AKI cases in the general population (5,53-55). The presence of hydronephrosis on imaging differentiates postrenal from prerenal or intrinsic renal causes of kidney dysfunction. Ultrasound is used most often to assess for hydronephrosis because it is widely available, portable, inexpensive and free from ionizing radiation. Computed tomography and magnetic resonance imaging may be performed in cases in which the ultrasound examination is incomplete or to re-evaluate equivocal sonographic findings. Urinary tract obstruction has many causes. In a retrospective study of 888 patients admitted to intensive care from 1998 to 2010, Hamdi et al (53) found that malignancy represented one-half of all causes of postrenal AKI. Prostate cancer was the most common malignancy (55%) while pelvic cancers as a group accounted for 90% of all malignancies. Other causes of postrenal impairment include benign prostatic hypertrophy (21%), urinary tract calculi (10%) and, less often, retroperitoneal fibrosis, urethral stricture, postradiation ureteritis, bladder dysfunction and papillary necrosis. Resolution of acute urinary tract obstruction may be associated with postobstructive diuresis (POD) as defined by a urine output >4 L per day. Close monitoring is required during this stage because excessive urinary losses can lead to hypovolemia and electrolyte disturbance. Hamdi et al (53) reported that 63% of intensive care patients with severe postrenal AKI had POD following release of urinary obstruction. The authors found that POD occurrence was predictive of renal recovery while the absence of POD was predictive of CKD. Predictors of POD occurrence before release of urinary obstruction were a higher serum creatinine concentration, a higher serum bicarbonate concentration and the presence of urinary retention ($P<0.05$). Predictors of CKD included lower blood hemoglobin and serum bicarbonate concentrations on admission, a longer time from admission to the release of urinary obstruction and the absence of POD ($P<0.05$).

CONCLUSION

Renal impairment in patients with cirrhosis unrelated to hepatorenal syndrome encompasses prerenal, intrinsic and postrenal causes. Regardless of etiology, the development of renal impairment in cirrhosis is a marker of poor patient outcome. Hypovolemia-induced prerenal impairment can be differentiated from HRS because only the former shows a response to volume replacement. Intrinsic renal impairment should be suspected in patients with hematuria and/or proteinuria. The presence of hydronephrosis indicates a postrenal cause of the renal impairment.

GRANT SUPPORT: NIHR CBRC funding (National Institute of Health Research UK, Cambridge Biomedical Research Centre).

DISCLOSURES: The authors have no financial disclosures or conflicts of interest to declare.

REFERENCES

1. Hampel H, Bynum GD, Zamora E, El-Serag HB. Risk factors for the development of renal dysfunction in hospitalized patients with cirrhosis. Am J Gastroenterol 2001;96:2206-10.
2. Peron JM, Bureau C, Gonzalez L, et al. Treatment of hepatorenal syndrome as defined by the international ascites club by albumin and furosemide infusion according to the central venous pressure: A prospective pilot study. Am J Gastroenterol 2005;100:2702-7.
3. du Cheyron D, Bouchet B, Parienti JJ, Ramakers M, Charbonneau P. The attributable mortality of acute renal failure in critically ill patients with liver cirrhosis. Intens Care Med 2005;31:1693-9.
4. Terra C, Guevara M, Torre A, et al. Renal failure in patients with cirrhosis and sepsis unrelated to spontaneous bacterial peritonitis: Value of MELD score. Gastroenterology 2005;129:1944-53.
5. Garcia-Tsao G, Parikh CR, Viola A. Acute kidney injury in cirrhosis. Hepatology (Baltimore, Md) 2008;48:2064-77.
6. Martin-Llahi M, Guevara M, Torre A, et al. Prognostic importance of the cause of renal failure in patients with cirrhosis. Gastroenterology 2011;140:488-96,e4.
7. de Carvalho JR, Villela-Nogueira CA, Luiz RR, et al. Acute Kidney Injury Network criteria as a predictor of hospital mortality in cirrhotic patients with ascites. J Clin Gastroenterol 2012;46:e21-6.
8. Lopes JA, Melo MJ, Costa AC, et al. Acute kidney injury and in-hospital mortality in critically ill patients with cirrhosis: A cohort study. Gut 2012;61:955-6.
9. Weismuller TJ, Prokein J, Becker T, et al. Prediction of survival after liver transplantation by pre-transplant parameters. Scand J Gastroenterol 2008;43:736-46.
10. D'Amico G, Garcia-Tsao G, Pagliaro L. Natural history and prognostic indicators of survival in cirrhosis: A systematic review of 118 studies. J Hepatol 2006;44:217-31.
11. Fede G, D'Amico G, Arvaniti V, et al. Renal failure and cirrhosis: A systematic review of mortality and prognosis. J Hepatol 2012;56:810-8.
12. Belcher JM, Garcia-Tsao G, Sanyal AJ, et al. Association of AKI with mortality and complications in hospitalized patients with cirrhosis. Hepatology (Baltimore, Md) 2013;57:753-62.
13. Kamath PS, Kim WR. The model for end-stage liver disease (MELD). Hepatology (Baltimore, Md) 2007;45:797-805.
14. Alessandria C, Ozdogan O, Guevara M, et al. MELD score and clinical type predict prognosis in hepatorenal syndrome: Relevance to liver transplantation. Hepatology (Baltimore, Md) 2005;41:1282-9.
15. Mindikoglu AL, Dowling TC, Weir MR, Seliger SL, Christenson RH, Magder LS. Performance of chronic kidney disease epidemiology collaboration creatinine-cystatin C equation for estimating kidney function in cirrhosis. Hepatology (Baltimore, Md) 2014;59:1532-42.
16. Mindikoglu AL, Weir MR. Current concepts in the diagnosis and classification of renal dysfunction in cirrhosis. Am J Nephrol 2013;38:345-54.
17. Larsson A, Malm J, Grubb A, Hansson LO. Calculation of glomerular filtration rate expressed in mL/min from plasma cystatin C values in mg/L. Scand J Clin Lab Invest 2004;64:25-30.
18. Hoek FJ, Kemperman FA, Krediet RT. A comparison between cystatin C, plasma creatinine and the Cockcroft and Gault formula for the estimation of glomerular filtration rate. Nephrol Dialysis Transpl 2003;18:2024-31.
19. Inker LA, Schmid CH, Tighiouart H, et al. Estimating glomerular filtration rate from serum creatinine and cystatin C. N Engl J Med 2012;367:20-9.
20. Mehta RL, Kellum JA, Shah SV, et al. Acute Kidney Injury Network: Report of an initiative to improve outcomes in acute kidney injury. Crit Care 2007;11:R31.
21. Wong F, Nadim MK, Kellum JA, et al. Working party proposal for a revised classification system of renal dysfunction in patients with cirrhosis. Gut 2011;60:702-9.
22. Noronha Ferreira C, Rodrigues T, Cortez-Pinto H, et al. The new definition of acute kidney injury in patients with cirrhosis: A critical look. Gut 2012;61:1513.
23. Schepke M, Appenrodt B, Heller J, Zielinski J, Sauerbruch T. Prognostic factors for patients with cirrhosis and kidney dysfunction in the era of MELD: Results of a prospective study. Liver Int 2006;26:834-9.
24. Salerno F, Gerbes A, Gines P, Wong F, Arroyo V. Diagnosis, prevention and treatment of hepatorenal syndrome in cirrhosis. Gut 2007;56:1310-8.

25. Bernardi M, Caraceni P, Navickis RJ, Wilkes MM. Albumin infusion in patients undergoing large-volume paracentesis: A meta-analysis of randomized trials. Hepatology (Baltimore, Md) 2012;55:1172-81.
26. Gentilini P, Laffi G, La Villa G, Romanelli RG, Blendis LM. Ascites and hepatorenal syndrome during cirrhosis: Two entities or the continuation of the same complication? J Hepatol 1999;31:1088-97.
27. Moreau R, Lebrec D. Review article: Hepatorenal syndrome – definitions and diagnosis. Aliment Pharmacol Ther 2004;(20 Suppl 3):24-8.
28. Runyon BA. Introduction to the revised American Association for the Study of Liver Diseases Practice Guideline management of adult patients with ascites due to cirrhosis 2012. Hepatology (Baltimore, Md) 2013;57:1651-3.
29. EASL clinical practice guidelines on the management of ascites, spontaneous bacterial peritonitis, and hepatorenal syndrome in cirrhosis. J Hepatol 2010;53:397-417.
30. Sort P, Navasa M, Arroyo V, et al. Effect of intravenous albumin on renal impairment and mortality in patients with cirrhosis and spontaneous bacterial peritonitis. N Engl J Med 1999;341:403-9.
31. Sigal SH, Stanca CM, Fernandez J, Arroyo V, Navasa M. Restricted use of albumin for spontaneous bacterial peritonitis. Gut 2007;56:597-9.
32. Francoz C, Glotz D, Moreau R, Durand F. The evaluation of renal function and disease in patients with cirrhosis. J Hepatol 2010;52:605-13.
33. Moreau R, Durand F, Poynard T, et al. Terlipressin in patients with cirrhosis and type 1 hepatorenal syndrome: A retrospective multicenter study. Gastroenterology 2002;122:923-30.
34. Prakash J, Mahapatra AK, Ghosh B, Arora P, Jain AK. Clinical spectrum of renal disorders in patients with cirrhosis of liver. Ren Fail 2011;33:40-6.
35. Gines P, Schrier RW. Renal failure in cirrhosis. N Engl Med 2009;361:1279-90.
36. Kamar N, Alric L, Izopet J, Rostaing L. Hepatitis C virus and kidney disease. Clin Res Hepatol Gastroenterol 2013:37:328-33.
37. Li P, Wei RB, Tang L, Wu J, Zhang XG, Chen XM. Clinical and pathological analysis of hepatitis B virus-related membranous nephropathy and idiopathic membranous nephropathy. Clin Nephrol 2012;78:456-64.
38. Novak J, Julian BA. Sugars and alcohol: IgA-associated renal diseases in alcoholic cirrhosis. Kidney Int 2011;80:1252-4.
39. Tissandie E, Morelle W, Berthelot L, et al. Both IgA nephropathy and alcoholic cirrhosis feature abnormally glycosylated IgA1 and soluble CD89-IgA and IgG-IgA complexes: Common mechanisms for distinct diseases. Kidney Int 2011;80:1352-63.
40. Sancho J, Egido J, Sanchez-Crespo M, Blasco R. Detection of monomeric and polymeric IgA containing immune complexes in serum and kidney from patients with alcoholic liver disease. Clin Exp Immunol 1982;47:327-35.
41. Sinniah R. Heterogeneous IgA glomerulonephropathy in liver cirrhosis. Histopathology 1984;8:947-62.
42. Kalaitzakis E, Rosengren A, Skommevik T, Bjornsson E. Coronary artery disease in patients with liver cirrhosis. Dig Dis Sci 2010;55:467-75.
43. Tiukinhoy-Laing SD, Rossi JS, Bayram M, et al. Cardiac hemodynamic and coronary angiographic characteristics of patients being evaluated for liver transplantation. Am J Cardiol 2006;98:178-81.
44. Letiexhe MR, Scheen AJ, Gerard PL, et al. Insulin secretion, clearance, and action on glucose metabolism in cirrhotic patients. J Clin Endocrinol Metabol 1993;77:1263-8.
45. Standards of medical care in diabetes – 2013. Diabet Care 2013;36(Suppl 1):S11-66.
46. Gross JL, de Azevedo MJ, Silveiro SP, Canani LH, Caramori ML, Zelmanovitz T. Diabetic nephropathy: Diagnosis, prevention, and treatment. Diabetes Care 2005;28:164-76.
47. Sumida K, Ubara Y, Hoshino J, et al. Hepatitis C virus-related kidney disease: Various histological patterns. Clin Nephrol 2010;74:446-56.
48. Ramesh Prasad GV, Bastacky S, Johnston JR. Diabetic nephropathy with concurrent hepatitis C virus infection related membranoproliferative glomerulonephritis. Clin Nephrol 1999;51:50-3.
49. Zhuo L, Ren W, Li W, Zou G, Lu J. Evaluation of renal biopsies in type 2 diabetic patients with kidney disease: A clinicopathological study of 216 cases. Int Urol Nephrol 2013;45:173-9.
50. Hironaka K, Makino H, Ikeda S, Haramoto T, Ota Z. Nondiabetic renal disease complicating diabetic nephropathy. J Diabetes Complications 1991;5:148-9.
51. Yaqub S, Kashif W, Hussain SA. Non-diabetic renal disease in patients with type-2 diabetes mellitus. Saudi J Kidney Dis Tranpl 2012;23:1000-7.
52. Gonzalez Suarez ML, Thomas DB, Barisoni L, Fornoni A. Diabetic nephropathy: Is it time yet for routine kidney biopsy? World J Diabetes 2013;4:245-55.
53. Hamdi A, Hajage D, Van Glabeke E, et al. Severe post-renal acute kidney injury, post-obstructive diuresis and renal recovery. BJU Int 2012;110(11 Pt C):E1027-34.
54. Khan IH, Catto GR, Edward N, Macleod AM. Acute renal failure: Factors influencing nephrology referral and outcome. QJM 1997;90:781-5.
55. Liano F, Pascual J. Epidemiology of acute renal failure: A prospective, multicenter, community-based study. Madrid Acute Renal Failure Study Group. Kidney Int 1996;50:811-8.

Postoperative resource utilization and survival among liver transplant recipients with Model for End-stage Liver Disease score ≥40: A retrospective cohort study

Filipe S Cardoso MD[1], Constantine J Karvellas MD SM[1,2], Norman M Kneteman MD MSc[3], Glenda Meeberg RN MsN[4], Pedro Fidalgo MD[1], Sean M Bagshaw MD MSc[1]

FS Cardoso, CJ Karvellas, NM Kneteman, G Meeberg, P Fidalgo, SM Bagshaw. Postoperative resource utilization and survival among liver transplant recipients with Model for End-stage Liver Disease score ≥40: A retrospective cohort study. Can J Gastroenterol Hepatol 2015;29(4):185-192.

BACKGROUND: Cirrhotic patients with Model for End-stage Liver Disease (MELD) score ≥40 have high risk for death without liver transplant (LT).
OBJECTIVE: To evaluate these patients' outcomes after LT.
METHODS: The present study analyzed a retrospective cohort of 519 cirrhotic adult patients who underwent LT at a single Canadian centre between 2002 and 2012. Primary exposure was severity of liver disease measured by MELD score at LT (≥40 versus <40). Primary outcome was duration of first intensive care unit (ICU) stay after LT. Secondary outcomes were duration of first hospital stay after LT, rate of ICU readmission, re-LT and survival rates.
RESULTS: On the day of LT, 5% (28 of 519) of patients had a MELD score ≥40. These patients had longer first ICU stays after LT (14 versus two days; P<0.001). MELD score ≥40 at LT was independently associated with first ICU stay after LT ≥10 days (OR 3.21). These patients had longer first hospital stays after LT (45 versus 18 days; P<0.001); however, there was no significant difference in the rate of ICU readmission (18% versus 22%; P=0.58) or re-LT rate (4% versus 4%; P=1.00). Cumulative survival at one month, three months, one year, three years and five years was 98%, 96%, 90%, 79% and 72%, respectively. There was no significant difference in cumulative survival stratified according to MELD score ≥40 versus <40 at LT (P=0.59).
CONCLUSIONS: Cirrhotic patients with MELD score ≥40 at LT utilize greater postoperative health resources; however, they derive similar long-term survival benefit from LT.

Key Words: *Cirrhosis; End-stage liver disease; Liver transplant; MELD*

L'utilisation des ressources après l'opération et la survie des greffés du foie ayant un score MELD de 40 ou plus : une étude rétrospective de cohorte

HISTORIQUE : Les patients cirrhotiques ayant un score MELD (acronyme anglais de modèle de maladie hépatique en phase terminale) de 40 ou plus risquent fort de mourir sans avoir reçu de transplantation hépatique (TH).
OBJECTIF : Évaluer les résultats de ces patients après une TH.
MÉTHODOLOGIE : La présente étude portait sur une cohorte de 519 patients cirrhotiques adultes qui ont subi une TH dans un seul centre canadien entre 2002 et 2012. L'exposition primaire était la gravité de la maladie hépatique mesurée par le score MELD à la TH (40 ou plus par rapport à moins de 40). Le résultat primaire était la durée du premier séjour en soins intensifs (USI) après la TH, le taux de réadmissions à l'USI, une nouvelle TH et les taux de survie.
RÉSULTATS : Le jour de la TH, 5 % des patients (28 sur 519) avaient un score MELD de 40 ou plus. Ces patients séjournaient plus longtemps à l'USI après la TH (14 jours par rapport à deux; P<0,001). Leur score MELD de 40 ou plus à la TH s'associait de manière indépendante à un premier séjour à l'USI de dix jours ou plus après la TH (RC 3,21). Le premier séjour hospitalier de ces patients était plus long après la TH (45 jours par rapport à 18; P<0,001), mais il n'y avait pas de différence significative dans le taux de réadmission à l'USI (18 % par rapport à 22 %; P=0,58) ou de nouvelle TH (4 % par rapport à 4 %; P=1,00). La survie cumulative au bout d'un mois, de trois mois, d'un an, de trois ans et de cinq ans s'élevait à 98 %, 96 %, 90 %, 79 % et 72 %, respectivement. Il n'y avait pas de différence significative dans la survie cumulative stratifiée selon le score MELD de 40 ou plus par rapport à celui de moins de 40 à la TH (P=0,59).
CONCLUSIONS : Les patients cirrhotiques ayant un score MELD de 40 ou plus à la TH utilisaient plus de ressources de santé après l'opération, mais tiraient des bienfaits similaires de la TH à long terme sur le plan de la survie.

The Model for End-stage Liver Disease (MELD), a marker of chronic liver disease severity based on the patient's serum bilirubin and creatinine levels, and international normalized ratio (1), has proven to be an accurate predictor of wait list for liver transplant (LT) three-month mortality (2).

Adoption of MELD to select and prioritize patients on the wait list for LT has contributed to a reduction in the number of patients waiting for LT, shorter wait times, an increase in the number of LT procedures and lower mortality (3).

Despite the continuous character of MELD, transplant authorities have decided to restrict its score to a maximum of 40. However, recent data have shown that patients with a MELD score ≥40 have higher wait list mortality than candidates with a lower MELD score (4,5).

While patients with end-stage liver disease and a high MELD score represent considerable resource expenditure for the health system, especially if transplanted, their outcomes have been reported to be acceptable (6-8).

We hypothesized that patients with a MELD score ≥40 at LT would consume greater health resources post-LT, but would achieve similar short- and long-term survival rates. Accordingly, our primary objective was to determine whether patients with a biochemical MELD score ≥40 at LT would have a longer duration of first intensive care unit (ICU) stay after LT. Our secondary objectives were to evaluate these patients' duration of first hospital stay after LT, rate of ICU readmission, re-LT rate, and one-month, three-month, one-year, three-year and five-year survival rates.

[1]*Division of Critical Care Medicine;* [2]*Division of Gastroenterology, Department of Medicine, Faculty of Medicine and Dentistry;* [3]*Division of Transplantation, Department of Surgery, University of Alberta;* [4]*Liver Transplant Program, Alberta Health Services, Edmonton, Alberta*
Correspondence: Dr Constantine J Karvellas, Division of Critical Care Medicine, Division of Hepatology, 1-40 Zeidler Ledcor Building, University of Alberta, 130 University Drive, Edmonton, Alberta T6G 2X8.e-mail dean.karvellas@ualberta.ca

METHODS

The present study adhered to the STROBE statement for observational studies (9). The local ethics committee approved the present study before commencement. The requirement for individual informed consent was waived.

Design, setting and participants

The present analysis was a single-centre, retrospective cohort study, which included all adult (≥18 years of age) cirrhotic patients who underwent LT at a Canadian transplant centre (University of Alberta Hospital [Edmonton, Alberta]) between January 1, 2002 and June 30, 2012. Patients were excluded if: they had concomitant transplant with another organ (eg, liver, kidney); their primary diagnosis was acute liver failure; or they were lost to follow-up.

Operational definitions

Cirrhosis was defined as bridging fibrosis on previous liver biopsy or a composite of clinical signs and findings provided by laboratory test results, endoscopy and radiological imaging (10).

Complications of cirrhosis included infection, variceal bleeding, hepatic encephalopathy, hepatorenal syndrome and hepatopulmonary syndrome. Although infection is not a specific complication of cirrhosis, it was considered as such due to its high prevalence and potential to alter disease course, being a common cause of acute decompensation and increased mortality (11). This definition included spontaneous bacterial peritonitis, bloodstream infection, urinary tract infection and pneumonia. Variceal bleeding was defined as any confirmed episode of acute bleeding originating from esophageal or gastric varices due to portal hypertension (12). Hepatic encephalopathy was defined as any acute confusional state in a patient with underlying liver disease, after excluding any potentially confounding metabolic, infectious or neurological disorders (13). Hepatorenal syndrome was defined as acute kidney injury in a patient with advanced liver disease in the absence of an identifiable cause based on the most recent criteria of the European Association for the Study of the Liver (14). Hepatopulmonary syndrome was defined as an oxygenation defect caused by pulmonary vascular dilation in the setting of portal hypertension, with the diagnosis being made by contrast-enhanced transthoracic echocardiography (15).

All MELD scores were calculated according to the United Network for Organ Sharing recommendations (16), without adjusting for serum sodium level or standardized exception points (17).

Variables

The primary exposure was the severity of end-stage liver disease measured by biochemical MELD score at LT (≥40 versus <40). The primary outcome was the duration of first ICU stay after LT. A prolonged first ICU stay following LT was defined as ≥10 days based on previous literature (8,18). The secondary outcomes were the duration of first hospital stay after LT, rate of ICU readmission, re-LT rate, and one-month, three-month, one-year, three-year and five-year patient cumulative survival rates.

Data collection

The LT program at the University of Alberta Hospital started in 1989 and, since 1995, has maintained a dedicated computerized database of all cases using the Organ Transplant Tracking Record (HKS Medical Information Systems, USA).

Eligible patients were initially identified using the Organ Transplant Tracking Record. Data regarding patients' age, sex, race, body mass index, etiology of liver disease and its complications, comorbidities, laboratory parameters, the need for pretransplant ICU admission, severity aggregate scores (Sequential Organ Failure Assessment, Child-Turcotte-Pugh and MELD), time between listing for and receipt of LT, Donor Risk Index parameters (19), operative requirements of red blood cells and platelets, and outcomes were extracted from that database and from patients' medical records.

Statistical analysis

Statistical analysis was performed using SPSS version 20.0 (IBM Corporation, USA). Categorical variables were presented as frequencies (percentages) and continuous variables as mean ± SD, if normally distributed, or median (interquartile range [IQR]) if non-normally distributed. In the event of missing values, data were not replaced or estimated.

Univariable analysis of outcomes was performed using χ^2 or Fisher's exact test (<5 events) for categorical variables, and Student's *t* test (parametric) or Mann-Whitney test (nonparametric) for continuous variables; P<0.05 was considered to be statistically significant for all comparisons.

Crude survival analysis was performed using the Kaplan-Meier estimator (with Breslow test) and adjusted survival analysis with Cox proportional-hazards regression. In this context, potentially confounding factors were selected based on previous literature (20) and clinical rationale.

Logistic regression was performed to study the effect of a MELD score ≥40 at LT on the probability of experiencing a prolonged first ICU stay following LT, after adjustment for other patients', donors' and perioperative covariates. Variables initially included in the model were selected based on the following three features: minimum frequency of 85% of the total number of cases under analysis (21); P<0.15 on univariable analysis; and clinical rationale. A backward stepwise selection of variables was performed to build the final models. Potentially collinear variables were excluded. Models' goodness of fit and discrimination were assessed using the χ^2 statistic (with correspondent degrees of freedom) and the area under ROC curve, respectively.

RESULTS

Patients', donors' and perioperative characteristics

Of the 603 LT procedures performed at the University of Alberta Hospital during the study period, 519 met the eligibility criteria. Median follow-up time for the entire cohort was 3.6 years (IQR 1.5 to 6.4 years).

Patients' pre-LT characteristics for the entire cohort and stratified according to MELD score categories at LT (≥40 versus <40) are shown in Table 1.

On the day of LT, median MELD score was 15 (IQR 11 to 23); at that same time, 5% (28 of 519) of patients had a MELD score ≥40, with a correspondent median score of 43 (IQR 41 to 46). The proportions of these patients for the periods 2002 to 2006 and 2007 to 2012 were similar (4% versus 7%; P=0.20). While median age (54 versus 53 years) and male sex proportion (68% versus 68%) were similar between the two MELD score categories (P>0.50 for both comparisons), patients with a MELD score ≥40 at LT had a marginally significant greater median body mass index (26 kg/m^2 versus 25 kg/m^2; P=0.047).

The most common indications for LT were hepatitis C (32% versus 23%) and hepatic malignancy (11% versus 22%), and their proportions were similar between the two MELD score categories (P>0.20 for both comparisons). Patients with a MELD score ≥40 at LT were more likely to have infection (67% versus 39%; P=0.005), hepatic encephalopathy (89% versus 59%; P=0.001) and hepatorenal syndrome (54% versus 14%; P<0.001) before LT. ICU admission before LT was required for 13% (69 of 519) of patients and occurred more commonly in patients with a MELD score ≥40 at transplant (82% versus 9%; P<0.001). Organ support was necessary for the majority of patients in the ICU: mechanical ventilation in 76% (34 of 45); vasopressors in 71% (32 of 45); and renal replacement therapy in 69% (31 of 45). Among patients in ICU, there were nonsignificant trends toward greater use of vasopressors (88% versus 62%) and renal replacement therapy (88% versus 59%) in those with a MELD score ≥40 at transplant (P=0.09 for both comparisons). On the day of LT, patients in ICU had a mean (± SD) Sequential Organ Failure Assessment score of 16±4, which was similar for the two MELD score categories (17 versus 16; P=0.14).

Donors' and perioperative characteristics for the entire cohort and stratified according to MELD score categories at LT (≥40 versus <40) are

TABLE 1
Analysis of patients' pre-liver transplant (LT) characteristics for the entire cohort and according to Model for End-stage Liver Disease (MELD) score (≥40 versus <40) categories at LT

Characteristic	Total	MELD score <40 (n=491)	MELD score ≥40 (n=28)	P
Demographics				
Age, years, median (IQR) (n=519)	54 (48–59)	54 (48–59)	53 (48–59)	0.57
Male sex, n (%) (n=519)	353 (68)	334/491 (68)	19/28 (68)	0.99
Race (Caucasian) , n (%) (n=515)	441 (86)	419/487 (86)	22/28 (79)	0.27
Body mass index, kg/m^2, median (IQR) (n=517)	25 (22–28)	25 (22–28)	26 (23–31)	0.047
Indication for LT, n (%) (n=519)				
Hepatitis C	124 (24)	115/491 (23)	9/28 (32)	0.29
Hepatitis B	13 (3)	10/491 (2)	3/28 (11)	0.028
Hepatic malignancy	109 (21)	106/491 (22)	3/28 (11)	0.23
Primary biliary cirrhosis or primary sclerosing cholangitis	96 (18)	90/491 (18)	6/28 (21)	0.68
Alcohol	70 (14)	68/491 (14)	2/28 (7)	0.41
Nonalcoholic steatohepatitis or cryptogenic	41 (8)	39/491 (8)	2/28 (7)	1.00
Autoimmune	20 (4)	20/491 (4)	0/28 (0)	0.62
Other	46 (9)	43/491 (9)	3/28 (11)	0.73
Complications of cirrhosis, n (%)				
Infection (n=465)	190 (41)	172/438 (39)	18/27 (67)	0.005
Variceal bleeding (n=517)	235 (45)	220/489 (45)	15/28 (54)	0.38
Hepatic encephalopathy (n=519)	312 (60)	287/491 (59)	25/28 (89)	0.001
Grades 3 to 4	90 (17)	73/491 (15)	17/28 (61)	<0.001
Hepatorenal syndrome (n=456)	75 (16)	61/430 (14)	14/26 (54)	<0.001
Hepatopulmonary syndrome (n=454)	12 (3)	12/428 (3)	0/26 (0)	1.00
Comorbidities, n (%)				
Coronary artery disease (n=519)	24 (5)	24/491 (5)	0/28 (0)	0.63
Chronic obstructive pulmonary disease (n=519)	22 (4)	22/491 (5)	0/28 (0)	0.62
Diabetes mellitus (n=519)	110 (21)	103/491 (21)	7/28 (25)	0.61
Chronic kidney diease (GFR <60 mL/min) (n=374)	78 (21)	70/351 (20)	8/23 (35)	0.09
Laboratory parameters (day of LT)				
Hemoglobin*, g/L, mean ± SD (n=188)	100±20	101±20	83±6	<0.001
International normalized ratio, median (IQR) (n=519)	1.4 (1.2–1.7)	1.3 (1.2–1.6)	2.8 (2.3–3.3)	<0.001
Albumin, g/L, median (IQR) (n=486)	34 (30–39)	34 (30–39)	38 (33–44)	0.001
Bilirubin, µmol/L, median (IQR) (n=519)	50 (28–133)	46 (27–104)	757 (515–907)	<0.001
Creatinine, µmol/L, median (IQR) (n=519)	86 (68–112)	84 (67–109)	117 (100–196)	<0.001
Sodium, mmol/L, median (IQR) (n=510)	136 (133–138)	136 (133–138)	138 (134–139)	0.07
pH*, mean ± SD (n=188)	7.39±0.06	7.39±0.07	7.40±0.05	0.64
Lactate, mmol/L, median (IQR) (n=188)	1.7 (1.2–2.7)	1.7 (1.2–2.6)	2.7 (1.4–4.2)	0.09
Intensive care unit stay before LT, n (%) (n=519)	69 (13)	46/491 (9)	23/28 (82)	<0.001
Mechanical ventilation† (n=45)	34 (76)	23/29 (79)	11/16 (69)	0.43
Vasopressors† (n=45)	32 (71)	18/29 (62)	14/16 (88)	0.09
Renal replacement therapy† (n=45)	31 (69)	17/29 (59)	14/16 (88)	0.09
Severity aggregate scores (day of LT)				
Sequential Organ Failure Assessment*†, mean ± SD (n=45)	16±4	16±4	17±3	0.14
Child-Turcotte-Pugh, median (IQR) (n=519)	10 (8–12)	9 (7–12)	13 (12–13)	< 0.001

**Normal distribution; †Data available for 45 of 69 patients. IQR Interquartile range; GFR Glomerular filtration rate*

presented in Appendix 1. Median time between listing for and receipt of LT was significantly lower for patients with a MELD score ≥40 at LT (1.3 versus 6.2 months; P<0.001). Due to their liver disease severity, these patients received an organ from a national source more often (64% versus 36%; P=0.003). The median Donor Risk Index score was similar for the two MELD score categories (1.5 versus 1.4; P=0.19).

Primary outcome

Overall, median length of first ICU stay following LT was two days (IQR one to six days) days. Patients with a MELD score ≥40 at LT had significantly greater median length of first ICU stay after LT (14 days [IQR five to 24 days) versus two days (IQR one to five days); P<0.001 (Figure 1).

Factors associated with a prolonged first ICU stay after LT

Univariable analysis for patients', donors' and perioperative characteristics stratified according to the duration of first ICU stay after LT is presented in Table 2.

The predicted probability of experiencing a prolonged ICU stay after LT was associated with the continuous MELD score at LT (Figure 2). On unadjusted analysis, a MELD score ≥40 at LT was significantly associated with a length of first ICU stay after LT ≥10 days (OR 6.73 [95% CI 3.07 to 14.7]; P<0.001) (Table 3). After adjusting for demographics (age and sex), pre-LT complications of cirrhosis (high-grade [3 to 4] hepatic encephalopathy, hepatorenal syndrome and hepatopulmonary syndrome), ICU stay before LT, standardized quality assessment of the liver received (Donor Risk Index) and the volume of red blood cells used during

TABLE 2
Analysis of patients', donors' and perioperative characteristics for the entire cohort and according to duration of the first intensive care unit stay after liver transplant (≥10 days versus <10 days)

	Intensive care unit stay		
Characteristic	**<10 days (n=427)**	**≥10 days (n=86)**	**P**
Demographics			
Age, years, median (IQR) (n=519)	54 (48–59)	54 (50–59)	0.42
Male sex, n/n (%) (n=519)	292/427 (68)	57/86 (66)	0.70
Race (Caucasian), n/n (%) (n=515)	369/424 (87)	68/85 (80)	0.09
Body mass index, kg/m^2, median (IQR) (n=517)	24 (22–28)	25 (23–30)	0.045
Indication for liver transplant, n/n (%) (n=519)			
Hepatitis C	107/427 (25)	17/86 (20)	0.30
Hepatitis B	12 /427 (3)	1/86 (1)	0.71
Hepatic malignancy	87/427 (20)	19/86 (22)	0.72
Primary biliary cirrhosis or primary sclerosing cholangitis	79/427 (19)	16/86 (19)	0.98
Alcohol	56/427 (13)	12/86 (14)	0.83
Complications of cirrhosis, n/n (%)			
Infection (n=465)	150/382 (39)	39/77 (51)	0.06
Variceal bleeding (n=517)	194/425 (46)	38/86 /44)	0.81
Hepatic encephalopathy (n=519)	246/427 (58)	63/86 (63)	0.007
Grades 3 to 4	56/427 (13)	34/86 (40)	<0.001
Hepatorenal syndrome (n=456)	54/372 (15)	21/78 (27)	0.008
Hepatopulmonary syndrome (n=454)	6/370 (2)	6/78 (17)	0.003
Comorbidities, n/n (%)			
Coronary artery disease (n=519)	20/427 (5)	4/86 (5)	1.00
Chronic obstructive pulmonary disease (n=519)	19/427 (4)	3/86 (4)	1.00
Diabetes mellitus (n=519)	93/427 (22)	17/86 (20)	0.68
Chrnic kidney disease (GFR <60 mL/min) (n=374)	60/300 (20)	16/69 (23)	0.56
Laboratory parameters (day of liver transplant)			
Hemoglobin*, g/L, mean ± SD (n=188)	101±20	94±19	0.037
International normalized ratio, median (IQR) (n=519)	1.3 (1.2–1.7)	1.6 (1.2–2.8)	<0.001
Albumin, g/L, median (IQR) (n=486)	34 (30–39)	36 (31–41)	0.08
Bilirubin, µmol/L, median (IQR) (n=519)	47 (27–105)	96 (36–456)	<0.001
Creatinine, µmol/L, median (IQR) (n=519)	86 (67–112)	88 (71–122)	0.62
Sodium, mmol/L, median (IQR) (n=510)	136 (133–138)	136 (133–139)	0.65
pH* (n=88), mean ± SD	7.39±0.06	7.39±0.07	0.70
Lactate, mmol/L, median (IQR) (n=188)	1.6 (1.2–2.5)	2.6 (1.4–4.1)	0.002
Intensive care unit stay before liver transplant, n/n (%) (n=519)	39/427 (9)	30/86 (35)	<0.001
Mechanical ventilation† (n=45)	14/22 (64)	20/23 (87)	0.09
Vasopressors† (n=45)	13/22 (59)	19/23 (83)	0.11
Renal replacement therapy† (n=45)	11/22 (50)	20/23 (87)	0.011
Severity aggregate scores (day of liver transplant)			
Sequential Organ Failure Assessment*†, mean ± SD (n=45)	14±3	18±3	0.001
Child-Turcotte-Pugh, median (IQR) (n=519)	9 (7–11)	12 (9–13)	<0.001
Model for End-stage Liver Disease score (n=519)	15 (10-22)	22 (11-35)	<0.001
Donor risk index (n=455)	1.4 (1.2–1.7)	1.5 (1.2–1.7)	0.78
Red blood cells in operating room, units, median (IQR) (n=519)	2 (0–5)	5 (0–12)	<0.001
Platelets in operating room, units, median (IQR) (n=519)	0 (0–4)	2 (0–7)	<0.001

**Normal distribution; †Data available for 45 of 69 patients. IQR Interquartile range; GFR Glomerular filtration rate*

the LT surgery, the OR for this association remained significant (OR 3.21 [95% CI 1.12 to 9.20]; P=0.030).

In the same multivariable analysis, other factors that showed a significant association with a prolonged first ICU stay after LT were high-grade hepatic encephalopathy (OR 2.29 [95% CI 1.14 to 4.60]; P=0.020), hepatopulmonary syndrome (OR 4.86 [95% CI 1.23 to 19.2]; P=0.024) and preoperative ICU admission (OR 2.42 [95% CI 1.02 to 5.71]; P=0.044) before LT. Additionally, transfusion of ≥5 units of red blood cells during the LT surgery showed a nonsignificant trend toward a higher likelihood of a prolonged first ICU stay after LT (OR 1.75 [95% CI 0.97 to 3.16]; P=0.06).

Secondary outcomes

Overall, the median length of first hospital stay following LT was 19 days (IQR 12 to 36 days). Patients with a MELD score ≥40 at LT experienced significantly greater median length of first hospital stay after LT (45 days [IQR 26 to 66 days]) versus 18 days [IQR 11 to 34 days]; P<0.001) (Figure 1).

For the entire follow-up period, the rate of ICU readmission was 22% (113 of 513). Of all ICU readmissions, 36% (41 of 113) represented at least a second readmission. Overall, median time to first ICU readmission was 16 days (IQR five to 346 days). The rate of ICU readmission was similar for the two MELD score categories (18% [five of 28] versus 22% [108 of 485]; P=0.58).

TABLE 3
Logistic regression analysis: variables associated with a length of first intensive care unit stay after liver transplant (LT) ≥10 days

	Unadjusted		Model 1		Model 2	
Variable	OR (95% CI)	P	OR (95% CI)	P	OR (95% CI)	P
Age, years	1.01 (0.99–1.04)	0.37	1.01 (0.99–1.04)	0.30	1.01 (0.98–1.04)	0.61
Male sex	0.91 (0.56–1.49)	0.70	0.90 (0.54–1.50)	0.69	0.85 (0.46–1.56)	0.59
Biochemical MELD score at LT (≥40)	6.73 (3.07–14.7)	<0.001	6.86 (3.12–15.1)	<0.001	3.21 (1.12–9.20)	0.030
Hepatic encephalopathy (grades 3 to 4)	4.33 (2.59–7.25)	<0.001			2.29 (1.14–4.60)	0.020
Hepatorenal syndrome	2.17 (1.22–3.87)	0.009			0.81 (0.37–1.74)	0.58
Hepatopulmonary syndrome	5.06 (1.59–16.1)	0.006			4.86 (1.23–19.2)	0.024
Intensive care unit admission before LT	5.33 (3.07–9.26)	<0.001			2.42 (1.02–5.71)	0.044
Donor risk index	1.08 (0.57–2.07)	0.81			0.76 (0.36–1.60)	0.46
Red blood cells in operating room (≥5 units)	3.08 (1.92–4.96)	<0.001			1.75 (0.97–3.16)	0.06

Model 1 goodness of fit (n=513): χ^2 22 (degrees of freedom ([Df], 3); P<0.001; Area under the ROC curve (AUROC) 0.61 (95% CI 0.54 to 0.67). Model 2 goodness of fit (n=399): χ^2 49 (Df, 9); P<0.001; AUROC 0.75 (95% CI 0.68 to 0.81). MELD Model for End-stage Liver Disease

For the entire period of follow-up, the re-LT rate was 4% (18 of 519). Causes of re-LT were hepatic artery thrombosis (56% [10 of 18]), primary nonfunction (17% [three of 18]), acute rejection (6% [one of 18]), chronic rejection (6% [one of 18]) and others (17% [three of 18]). The re-LT rate was similar for the two MELD score categories (4% [one of 28] versus 4% [17 of 491]; P=1.00).

During the study period, 139 deaths occurred. Causes of death were recurrence of underlying liver disease (32% [45 of 139]), sepsis (24% [33 of 139]), cardiovascular events (18% [25 of 139]), de novo malignancy (12% [16 of 139]), chronic rejection (4% [five of 139]) and others (11% [15 of 139]).

Kaplan-Meier cumulative survival at one month, three months, one year, three years and five years was 98%, 96%, 90%, 79% and 72%, respectively. Kaplan-Meier curves were similar for the two MELD score categories (P=0.59 [Figure 3]). After adjusting for confounding factors, including age, sex, etiology of liver disease (hepatitis C versus other) and ICU admission before LT, a MELD score ≥40 at LT was not significantly associated with survival (HR 0.76 [95% CI 0.33 to 1.73]; P=0.51). Advanced age was the only factor significantly associated with worse survival after LT (HR 1.03 per incremental year [95% CI 1.01 to 1.06]; P=0.002 (model not shown [n=519]: χ^2 25 [degrees of freedom 6]; P<0.001).

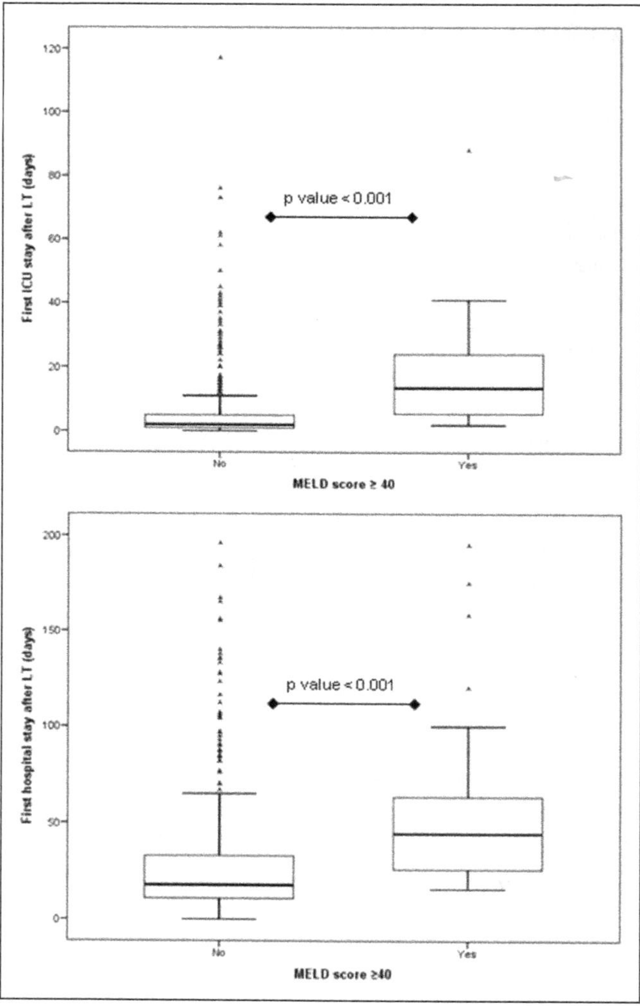

Figure 1) *Median length of first intensive care unit and hospital stays according to biochemical Model for End-stage Liver Disease (MELD) score categories (≥40 versus <40) at liver transplant (LT)*

DISCUSSION

The purpose of the present study was to characterize post-LT outcomes of patients with end-stage liver disease and a MELD score ≥40 at LT.

Key results

In a large single-centre cohort of adult cirrhotic patients who underwent LT, we found that patients with a biochemical MELD score ≥40 at LT had significantly greater length of first ICU and hospital stays following LT, but similar short- and long-term survival. We also found that a biochemical MELD score ≥40 at LT, high-grade hepatic encephalopathy, hepatopulmonary syndrome and ICU admission before LT, were all independently associated with a length of first ICU stay after LT ≥10 days.

Comparison with previous studies

Our findings that a MELD score ≥40 at LT was significantly associated with a prolonged first ICU stay after LT, but not with survival, are consistent with results from previous studies. Oberkofler et al (8) found that a MELD score >23 was significantly associated with a length of ICU stay after transplant >10 days, but not with survival. Conclusions of this study may not be readily comparable with ours because they used the MELD corrected for exceptions (hepatocellular carcinoma and hepatopulmonary syndrome), instead of the biochemical MELD, which resulted in greater median MELD scores (19 [IQR eight to 40]) and, possibly, overestimation of liver disease severity. Similarly to our study, Alexopoulos et al (7) reported, for patients with an unspecified MELD score ≥40 at LT, Kaplan-Meier cumulative survival at one year and three years of 89% and 77%, respectively. In contrast to our study, they included patients receiving both LT and kidney transplants, and excluded those who received a split graft; however, outcomes for these patients have been reported to be similar to those of whole liver-only recipients (22,23). Further evidence

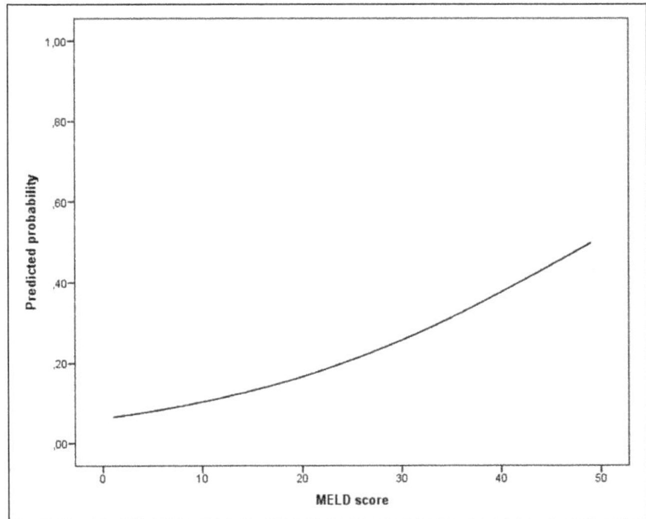

Figure 2) *Unadjusted predicted probabilities of a length of first intensive care unit stay after liver transplant ≥10 days according to continuous biochemical Model for End-stage Liver Disease (MELD) score at liver transplant*

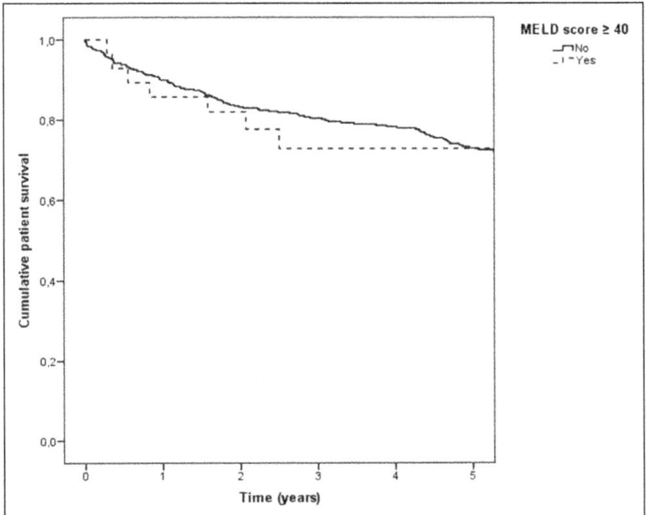

Figure 3) *Kaplan-Meier patient survival curve for biochemical Model for End-stage Liver Disease (MELD) score categories (≥40 versus <40) at liver transplant (P=0.59)*

regarding the absence of a significant association between MELD score and survival after LT was provided by Sharma et al (5), who found that cirrhotic patients with a biochemical MELD score ≥40 at listing had a post-LT survival similar to transplant recipients for fulminant hepatic failure, the highest status group of patients for liver allocation. Overall, these results consistently indicate that patients with a MELD score ≥40 at LT are at higher risk for a longer duration of the first ICU stay postoperatively; however, their short- and long-term survival is expected to be similar to other LT recipients.

The significant association of pre-LT high-grade hepatic encephalopathy with a prolonged first ICU stay after LT found in our cohort has not been reported in the literature. In fact, pre-LT hepatic encephalopathy has been associated with post-LT neurocognitive changes (24,25), but its role as a predictor of outcomes after LT has not been well characterized. In a study involving cirrhotic patients admitted to the ICU for high-grade hepatic encephalopathy, but without receiving LT, Fichet et al (26) reported a median length of stay of six days (IQR two to 10 days) and an ICU mortality (35% [25 of 71]) greater than what has been reported for other cirrhotic patients admitted to the ICU.

Although we found pre-LT hepatopulmonary syndrome to be significantly associated with a prolonged first ICU stay after LT, possibly in the context of these patients' high dependence on oxygen, outcomes of hepatopulmonary syndrome reported in the literature are controversial, with some studies suggesting acceptable outcomes (27,28) and others suggesting that these patients experience poor outcomes after LT (29,30).

The association between ICU admission before LT and the duration of hospital stay after transplant has been studied. Smith et al (31) found that ICU admission before LT was independently associated with a longer hospital stay after transplant; however, this study did not specifically evaluate the utilization of ICU resources postoperatively. Oberkofler et al (8) did not find any significant association between ICU admission before LT and a length of ICU stay after LT >10 days.

While we found a nonsignificant trend between a consumption of ≥5 units of red blood cells during transplant surgery and a prolonged first ICU stay after LT, Oberkofler et al (8) reported that patients who underwent LT and were transfused with >7 units of red blood cells during the procedure had a significantly higher likelihood of experiencing a length of ICU stay after LT of >10 days.

Overall, these results suggest that patients with one or more factors found to independently increase the risk for a prolonged first ICU stay after LT should be expected to consume a greater volume of hospital resources; therefore, specific institutional policies should be developed to improve the efficiency of care provided to them.

Study limitations

The results of the present study need to be considered in the context of the following limitations. First, it was a retrospective analysis of prospectively collected data from a single transplant centre and may be prone to selection bias. Second, individual transplant centres, especially from different countries, have diverse groups of patients and practices regarding enlistment for transplant, allocation of organs, and medical and surgical management strategies; therefore, results may, in part, reflect those specific realities. Despite these limitations, our study was one of the few to characterize the health resource utilization, and short- and long-term outcomes of cirrhotic patients with a MELD score ≥40 at LT. In future studies, an effort should be made to standardize the most valuable markers of post-LT morbidity and survival to further help decision-making on organ allocation and peri-LT care.

CONCLUSIONS

While cirrhotic patients with a MELD score ≥40 at LT experienced significantly greater length of first ICU and hospital stays after LT, they derived similar long-term survival benefit. Despite representing a higher burden of care and costs for the health system, these patients seemed to similarly benefit from LT.

ACKNOWLEDGEMENTS: Dr Bagshaw is supported by a Canada Research Chair in Critical Care Nephrology and Clinical Investigator Award from Alberta Innovates – Health Solutions. Drs Cardoso and Fidalgo are supported by an unrestricted educational grant donated by Gambro Inc.

DISCLOSURES: The authors have no financial disclosures or conflicts of interest to declare.

CONSENT: Given that this was a retrospective study, the Institutional Review Board at the University of Alberta waved the requirement for informed consent. While this was a retrospective study and no patients were in contact with study investigators, ethical standards were in accordance with the Helsinki Declaration of 1975, as revised in 2000 and 2008.

APPENDIX 1

Analysis of donors' and perioperative characteristics for the entire cohort and according to Model for End-stage Liver Disease (MELD) score (≥40 versus <40) categories at liver transplant (LT)

		MELD score		
	Total (n=519)	<40 (n=491)	≥40 (n=28)	P
Time between listing for and receipt of LT, months, median (IQR) (n=519)	5.6 (1.6–12.6)	6.2 (1.8–13.2)	1.3 (0.4–4.1)	<0.001
Donor risk index, median (IQR) (n=455)	1.4 (1.2–1.7)	1.4 (1.2–1.7)	1.5 (1.3–1.8)	0.19
Split graft, n (%) (n=519)	70 (13)	69/491 (14)	1/28 (4)	0.16
Live donor, n (%) (n=519)	64 (12)	63/491 (13)	1/28 (4)	0.23
Cold ischemia time*, h, mean ± SD (n=517)	5.5±2.9	5.4±2.9	7.1±2.8	0.003
Organ from national source, n/n (%) (n=519)	195 (38)	177/491 (36)	18/28 (64)	0.003
Red blood cells in operating room, units, median (IQR) (n=519)	2 (0–6)	2 (0–5)	8 (4–13)	<0.001
Platelets in in operating room, units, median (IQR) (n=519)	0 (0–5)	0 (0–4)	5 (2–10)	<0.001

**Normal distribution. IQR Interquartile range*

REFERENCES

1. Edwards EB, Harper AM. Application of a continuous disease severity score to the OPTN liver waiting list. Clin Transpl 2001:19-24.
2. Wiesner RH, Mcdiarmid SV, Kamath PS, et al. MELD and PELD: Application of survival models to liver allocation. Liver Transpl 2001;7:567-80.
3. Bernardi M, Gitto S, Biselli M. The MELD score in patients awaiting liver transplant: Strengths and weaknesses. J Hepatol 2011;54:1297-306.
4. Wiesner R, Edwards E, Freeman R, et al. Model for End-stage Liver Disease (MELD) and allocation of donor livers. Gastroenterology 2003;124:91-6.
5. Sharma P, Schaubel DE, Gong Q, Guidinger M, Merion RM. End-stage liver disease candidates at the highest Model for End-stage Liver Disease scores have higher wait-list mortality than status-1A candidates. Hepatology 2012;55:192-8.
6. Shawcross DL, Austin MJ, Abeles RD, et al. The impact of organ dysfunction in cirrhosis: Survival at a cost? J Hepatol 2012;56:1054-62.
7. Alexopoulos S, Matsuoka L, Cho Y, et al. Outcomes after liver transplantation in patients achieving a Model for End-stage Liver Disease score of 40 or higher. Transplantation 2013;95:507-12.
8. Oberkofler CE, Dutkowski P, Stocker R, et al. Model for End-stage Liver Disease (MELD) score greater than 23 predicts length of stay in the ICU but not mortality in liver transplant recipients. Crit Care 2010;14:R117.
9. Von Elm E, Altman DG, Egger M, Pocock SJ, Gotzsche PC, Vandenbroucke JP. Strengthening the Reporting of Observational Studies in Epidemiology (STROBE) statement: Guidelines for reporting observational studies. BMJ 2007;335:806-8.
10. Moreau R, Jalan R, Gines P, et al. Acute-on-chronic liver failure is a distinct syndrome that develops in patients with acute decompensation of cirrhosis. Gastroenterology 2013;144:1426-37, e1-9.
11. Arvaniti V, D'amico G, Fede G, et al. Infections in patients with cirrhosis increase mortality four-fold and should be used in determining prognosis. Gastroenterology 2010;139:1246-56;56:e1-5.
12. Garcia-Tsao G, Sanyal A J, Grace ND, Carey W. Prevention and management of gastroesophageal varices and variceal hemorrhage in cirrhosis. Hepatology 2007;46:922-38.
13. Blei AT, Cordoba J. Hepatic encephalopathy. Am J Gastroenterol 2001;96:1968-76.
14. European Association for the Study of the Liver. EASL clinical practice guidelines on the management of ascites, spontaneous bacterial peritonitis, and hepatorenal syndrome in cirrhosis. J Hepatol 2010;53:397-417.
15. Machicao VI, Fallon MB. Hepatopulmonary syndrome. Semin Respir Crit Care Med 2012;33:11-6.
16. UNOS. MELD/PELD Calculator Documentation. <www.unos.org/docs/MELD_PELD_Calculator_Documentation.pdf> (Accessed February 11, 2015).
17. Massie AB, Caffo B, Gentry SE, et al. MELD exceptions and rates of waiting list outcomes. Am J Transplant 2011;11:2362-71.
18. Karvellas C J, Lescot T, Goldberg P, et al. Liver transplantation in the critically ill: A multicenter Canadian retrospective cohort study. Crit Care 2013;17:R28.
19. Feng S, Goodrich NP, Bragg-Gresham JL, et al. Characteristics associated with liver graft failure: The concept of a donor risk index. Am J Transplant 2006;6:783-90.
20. Singal AK, Hmoud BS, Guturu P, Kuo YF. Outcome after liver transplantation for cirrhosis due to alcohol and hepatitis C: Comparison to alcoholic cirrhosis and hepatitis C cirrhosis. J Clin Gastroenterol 2013;47:727-33.
21. Groenwold RH, Donders AR, Roes KC, Harrell FE Jr, Moons KG. Dealing with missing outcome data in randomized trials and observational studies. Am J Epidemiol 2012;175:210-7.
22. Doyle MB, Maynard E, Lin Y, et al. Outcomes with split liver transplantation are equivalent to those with whole organ transplantation. J Am Coll Surg 2013;217:102-12.
23. Ruiz R, Kunitake H, Wilkinson AH, et al. Long-term analysis of combined liver and kidney transplantation at a single center. Arch Surg 2006;141:735-41.
24. Sotil EU, Gottstein J, Ayala E, Randolph C, Blei AT. Impact of preoperative overt hepatic encephalopathy on neurocognitive function after liver transplantation. Liver Transpl 2009;15:184-92.
25. Kanwal F, Chen D, Ting L, et al. A model to predict the development of mental status changes of unclear cause after liver transplantation. Liver Transpl 2003;9:1312-9.
26. Fichet J, Mercier E, Genee O, et al. Prognosis and 1-year mortality of intensive care unit patients with severe hepatic encephalopathy. J Crit Care 2009;24:364-70.
27. Deberaldini M, Arcanjo AB, Melo E, et al. Hepatopulmonary syndrome: Morbidity and survival after liver transplantation. Transplant Proc 2008;40:3512-6.
28. Gupta S, Castel H, Rao RV, et al. Improved survival after liver transplantation in patients with hepatopulmonary syndrome. Am J Transplant 2010;10:354-63.
29. Arguedas MR, Abrams GA, Krowka MJ, Fallon MB. Prospective evaluation of outcomes and predictors of mortality in patients with hepatopulmonary syndrome undergoing liver transplantation. Hepatology 2003;37:192-7.
30. Schiffer E, Majno P, Mentha G, et al. Hepatopulmonary syndrome increases the postoperative mortality rate following liver transplantation: A prospective study in 90 patients. Am J Transplant 2006;6:1430-7.
31. Smith JO, Shiffman ML, Behnke M, et al. Incidence of prolonged length of stay after orthotopic liver transplantation and its influence on outcomes. Liver Transpl 2009;15:273-9.

Neutrophil infiltration and the distribution of intestinal metaplasia is associated with metachronous gastric cancer following endoscopic submucosal dissection

Takafumi Sugimoto MD[1,2], Yutaka Yamaji MD[2], Kosuke Sakitani MD[2], Yoshihiro Isomura MD[2], Shuntaro Yoshida MD[2], Atsuo Yamada MD[2], Yoshihiro Hirata MD[2], Keiji Ogura MD[3], Makoto Okamoto MD[4], Kazuhiko Koike MD[2]

T Sugimoto, Y Yamaji, K Sakitani, et al. Neutrophil infiltration and the distribution of intestinal metaplasia is associated with metachronous gastric cancer following endoscopic submucosal dissection. Can J Gastroenterol Hepatol 2015;29(6):321-325.

BACKGROUND: Endoscopic submucosal dissection (ESD) of early gastric cancer is a minimally invasive procedure. However, the risk for metachronous cancers after successful cancer treatment remains high and the risk factors for metachronous cancers have not been elucidated.
OBJECTIVE: To evaluate the risk factors for metachronous gastric cancers after ESD with a long-term follow-up.
METHODS: A total of 155 consecutive patients (119 men, 36 women, mean age 68.9 years) were treated with ESD between September 2000 and September 2009. Biopsy specimens were obtained from the greater curvature of the antrum and middle corpus to evaluate gastric mucosal status, including *Helicobacter pylori*, intestinal metaplasia (IM) and neutrophil infiltration (NI) before ESD. Follow-up endoscopy after ESD was scheduled at two and six months, one year and annually thereafter. *H pylori* eradication was recommended when possible.
RESULTS: The median follow-up period was 4.2 years. Metachronous gastric cancers were found in 23 of 155 patients (3.5% per year). No local recurrences were observed. The cumulative incidence of metachronous gastric cancer was significantly high in IM and NI in the corpus (P=0.0093 and P=0.0025, respectively [log-rank test]). The ORs for IM and NI in the corpus were 2.65 and 3.06, respectively, according to the Cox proportional hazards model (P=0.024 and P=0.0091, respectively).
CONCLUSIONS: The presence of IM and NI in the corpus was closely related to the development of metachronous gastric cancer after ESD.

Key Words: *Endoscopic submucosal dissection;* Helicobacter pylori; *Intestinal metaplasia; Neutrophil infiltration; Stomach neoplasms*

La distribution de l'infiltration de neutrophiles et de la métaplasie intestinale associée au cancer gastrique métachrone après une dissection endoscopique des tissus sous-muqueux

HISTORIQUE : La dissection endoscopique des tissus sous-muqueux (DETS) du cancer gastrique précoce est une intervention à effraction minimale. Cependant, le risque de cancers métachrones demeure élevé après un traitement fructueux du cancer, et les facteurs de risque de cancers métachrones ne sont pas établis.
OBJECTIF : Évaluer les facteurs de risque des cancers gastriques métachrones après une DETS au moyen d'un suivi à long terme.
MÉTHODOLOGIE : Au total, 155 patients consécutifs (119 hommes et 36 femmes, d'un âge moyen de 68,9 ans) ont subi une DETS entre septembre 2000 et septembre 2009. Les échantillons de biopsie ont été prélevés dans la grande courbure de l'antre et le corps moyen pour évaluer l'état de la muqueuse gastrique, y compris l'*Helicobacter pylori*, la métaplasie intestinale (MI) et l'infiltration de neutrophiles (IN) avant la DETS. Une endoscopie de suivi après la DETS était planifiée au bout de deux mois, de six mois, d'un an, puis tous les ans par la suite. Dans la mesure du possible, l'éradication de l'*H pylori* était recommandée.
RÉSULTATS : La période de suivi médiane était de 4,2 ans. Les médecins ont découvert des cancers gastriques métachrones chez 23 des 155 patients (3,5 % par année). Ils n'ont observé aucune récurrence locale. L'incidence cumulative de cancers gastriques métachrones était très élevée en présence de MI et d'IN du corps de l'estomac (P=0,0093 et P=0,0025, respectivement [test de Mantel-Haenszel]). Les rapports de cote de MI et d'IN dans le corps de l'estomac correspondaient à 2,65 et à 3,06, respectivement, conformément au modèle des risques proportionnels de Cox (P=0,024 et P=0,0091, respectivement).
CONCLUSIONS : La présence de MI et d'IN dans le corps de l'estomac était étroitement liée à l'apparition de cancer gastrique métachrone après une DETS.

The rate of diagnosis of early gastric cancer has increased due to improved diagnostic procedures such as endoscopic examinations including chromoendoscopy and narrow-band imaging (1,2). Endoscopic resection (ER), which includes endoscopic mucosal resection (EMR) and endoscopic submucosal dissection (ESD), has been established as a less-invasive treatment for early gastric cancer without concomitant lymph node metastasis (3-5). ESD has become accepted for en bloc resection, which is beneficial for curative treatment to avoid local recurrence (5-10). Although ER contributes to preservation of the majority of the stomach, metachronous cancers developing at other sites in the stomach may occur more frequently after ER than those after surgical partial gastrectomy (11). Several studies have reported that the incidence of metachronous cancers after ER is 2.4% to 14% (12,13). The residual gastric mucosa after ER is believed to be a high-risk microenvironment.

Helicobacter pylori is one of the most important risk factors for gastric cancer (14-17). Inflammation caused by *H pylori* infection may play an initiating role in gastric carcinogenesis (18). Based on this hypothesis, intestinal-type gastric adenocarcinoma develops through multistep mucosal changes from superficial gastritis to atrophic

[1]*Department of Gastroenterology, Kyoundo Hospital;* [2]*Department of Gastroenterology, Graduate School of Medicine, the University of Tokyo;* [3]*Department of Gastroenterology, Tokyo Metropolitan Police Hospital;* [4]*Department of Gastroenterology, JR Tokyo General Hospital, Tokyo, Japan*
Correspondence and reprints: Dr Takafumi Sugimoto, Department of Gastroenterology, Graduate School of Medicine, University of Tokyo, 7-3-1 Hongo, Bunkyo-ku, Tokyo 113-8655, Japan. e-mail sugimot-tky@umin.ac.jp

TABLE 1
Baseline characteristics of 155 patients

	Total	Metachronous (n=23)	No recurrence (n=132)	P
Demographic characteristics				
Sex, male/female, n	119/36	20/3	99/33	0.19
Age, years, mean (range)	68.9 (50–83)	69.2 (57–80)	68.8 (50–83)	0.84
Follow-up period, years (range)	4.23 (1.02–10.8)	4.11 (1.05–7.35)	4.25 (1.02–10.8)	0.78
Outcome (alive/dead)	149/6	23/0	126/6	0.16
Helicobacter pylori status after endoscopic submucosal dissection, n (%)				0.17
Negative	25 (16.1)	1 (4.3)	24 (18.1)	
Eradication	100 (64.5)	17 (74.0)	83 (62.9)	
Persistent	30 (19.4)	5 (21.7)	25 (19.0)	
Histological characteristics, n (%)				
Intestinal metaplasia in the antrum	101 (65.2)	15 (65.2)	86 (65.2)	0.99
Intestinal metaplasia in the corpus	74 (47.7)	15 (65.2)	59 (44.7)	0.068
Neutophil infiltration in the antrum	30 (19.4)	8 (34.8)	22 (16.7)	0.056
Neutophil infiltration in the corpus	50 (32.3)	15 (65.2)	35 (26.5)	0.0004
Pepsinogen, ng/mL, mean (n=127)				
I	44.1	27.2	47.2	0.17
II	16.8	13.3	17.5	0.24
I/II	2.61	2.10	2.71	0.15
Tumour characteristics				
Location, upper/middle/lower, n	14/65/76	1/13/9	13/52/67	0.27
Size, mm, mean	15.7	20.3	14.9	0.0048
Macroscopic type, protruded/depressed, n	88/67	12/11	76/56	0.63
Pathology, adenoma/cancer, n	31/124	5/18	26/106	0.82

gastritis, intestinal metaplasia (IM) and dysplasia (18). Eradication of *H pylori* is a highly useful method of preventing primary and metachronous gastric cancers (19-21). We have reported previously that IM and neutrophil infiltration (NI) are high-risk microenvironments and risk factors for primary gastric cancers (22). However, the risk factors for metachronous cancers are unclear. Thus, the aim of the present study was to evaluate the risk factors for metachronous gastric cancers after ESD in a long-term follow-up.

METHODS

Patients

A total of 201 consecutive patients were treated using ESD at the University of Tokyo Hospital (Tokyo, Japan) between September 2000 and September 2009. All demographic data and endoscopic findings obtained before ESD were evaluated retrospectively from patient records. The location, macroscopic types and histological findings of the tumours were categorized according to the Japanese classification of gastric carcinoma (23). Expanded ESD criteria have been proposed by Gotoda et al (24). All patients were followed up for at least one year. Forty-six patients were excluded from the study: 19 did not meet the ESD specimen histology criteria and required additional surgery, five had remnant stomach cancers, nine were followed for up to one year, four did not have *H pylori* information and, in nine patients, ESD was performed before enrollment. The present study was approved by the Institutional Review Board of the University of Tokyo.

Disease assessment

Experienced endoscopists performed the gastrointestinal endoscopy procedures, and gastric cancer was diagnosed by histology. Biopsy specimens were obtained from the greater curvature of the antrum and middle corpus to evaluate gastric mucosal status including *H pylori*, IM and NI before ESD (22). Both a rapid urease test (Helicocheck, Otsuka Pharmaceuticals, Japan) and histopathological examination were performed. A positive result on at least one test was deemed to be evidence of *H pylori* infection. Blood samples were also collected before ESD to measure serum levels of pepsinogen I and II using ELISA kits (LS test Eiken Kagaku, Inc, Japan). ESD was performed as described previously (8,9). Follow-up endoscopy after ESD for detection of metachronous gastric cancers was scheduled at two and six months, one year and annually thereafter. Lesions detected within one year of initial ESD were regarded as synchronous multiple lesions because microcancers may have been missed at the time of ESD (12,25). Metachronous gastric cancers were defined as new gastric cancers in different areas from the initial lesion and occurring at least one year after the initial ESD. Eradication of *H pylori* was recommended when possible. Metachronous gastric cancer was defined as the development of a new carcinoma in areas other than the primary gastric cancer site at least one year after ESD. *H pylori*-positive patients were treated with triple therapy consisting of 200 mg clarithromycin, 750 mg amoxicillin and 30 mg lansoprazole twice daily for one week after ER. Patients in whom *H pylori* was not eradicated were treated with second-line therapy consisting of 250 mg metronidazole instead of clarithromycin (26). Eradication was confirmed by a negative ^{13}C-urea breath test (8,9).

Statistical analyses

Study subjects were categorized according to the presence of the metachronous gastric cancer. All analyses were performed using JMP version 9 (SAS Institute, USA). Student's *t* tests were used for intergroup comparison of mean age, follow-up period and tumour size. The other patient clinical characteristics and histopathological characteristics of the gastric mucosa and cancers were compared using the χ^2 test or Fisher's exact test, as appropriate. Relative risks for metachronous gastric cancer were calculated using the Cox proportional hazards model. Differences were considered to be statistically significant at the 5% probability level.

RESULTS

Baseline characteristics of the study subjects

Baseline clinical characteristics of the patients are summarized in Table 1. A total of 155 patients (119 men, 36 women, mean age 68.9 years) were followed for up to 10.8 years (mean 4.2 years). Six patients died during the study period, but none of the deaths were related to gastric cancer. Twenty-three (14%) patients developed metachronous gastric cancers (3.5% per year).

TABLE 2
Cumulative incidence of metachronous gastric cancer, univariate analysis*

	OR (95% CI)	P
Demographic characteristics		
Sex	1.78 (0.61–7.55)	0.32
Age	1.02 (0.96–1.08)	0.60
Helicobacter pylori status after ESD		0.64
Negative	1	
Eradication	2.56 (0.52–46.3)	0.30
Persistent	2.50 (0.40–48.3)	0.36
Histological characteristics		
Intestinal metaplasia in the antrum	0.99 (0.43–2.46)	0.97
Intestinal metaplasia in the corpus	3.00 (1.29–7.51)	0.010
Neutrophil infiltration in the antrum	1.89 (0.76–4.39)	0.16
Neutrophil infiltration in the corpus	3.47 (1.51–8.63)	0.0034
Pepsinogen (measured in 127 patients)		
I	0.98 (0.96–1.01)	0.11
II	0.96 (0.92–1.02)	0.18
I/II	0.83 (0.59–1.17)	0.28
Tumour characteristics		
Location		0.29
Upper	1	
Middle	2.26 (0.45–41.2)	0.38
Lower	1.21 (0.23–22.4)	0.85
Tumour size	1.05 (1.01–1.09)	0.021
Macroscopic type (protruded/depressed)	0.79 (0.35–1.83)	0.58
Pathology (cancer/adenoma)	1.18 (0.47–3.59)	0.73

**Cox proportional hazards model. ESD Endoscopic submucosal dissection*

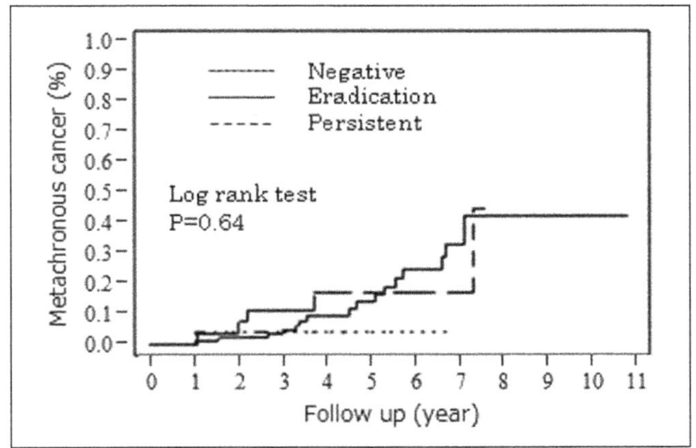

Figure 1) *The cumulative incidence of metachronous gastric cancer.* Helicobacter pylori *status after endoscopic submucosal dissection*

Of the 155 patients, 25 were *H pylori* negative, 27 had *H pylori* eradicated before ESD, 73 received eradication therapy and 30 experienced continuing *H pylori* infection. IM at the greater curvature of the antrum and middle corpus was observed in 101 (65.1%) and 74 (47.7%) patients, respectively. NI was observed in 30 (19.4%) and 50 (32.3%) patients, respectively.

Factors related to metachronous gastric cancers

The cumulative incidence of metachronous gastric cancer is summarized in Table 2, and Figures 1 and 2. The ORs for metachronous gastric cancers in IM and NI in the corpus were 3.00 (P=0.010) and 3.47 (P=0.0034). IM and NI in the corpus and tumour size were risk factors. Successful *H pylori* eradication did not contribute significantly to the reduction of risk for

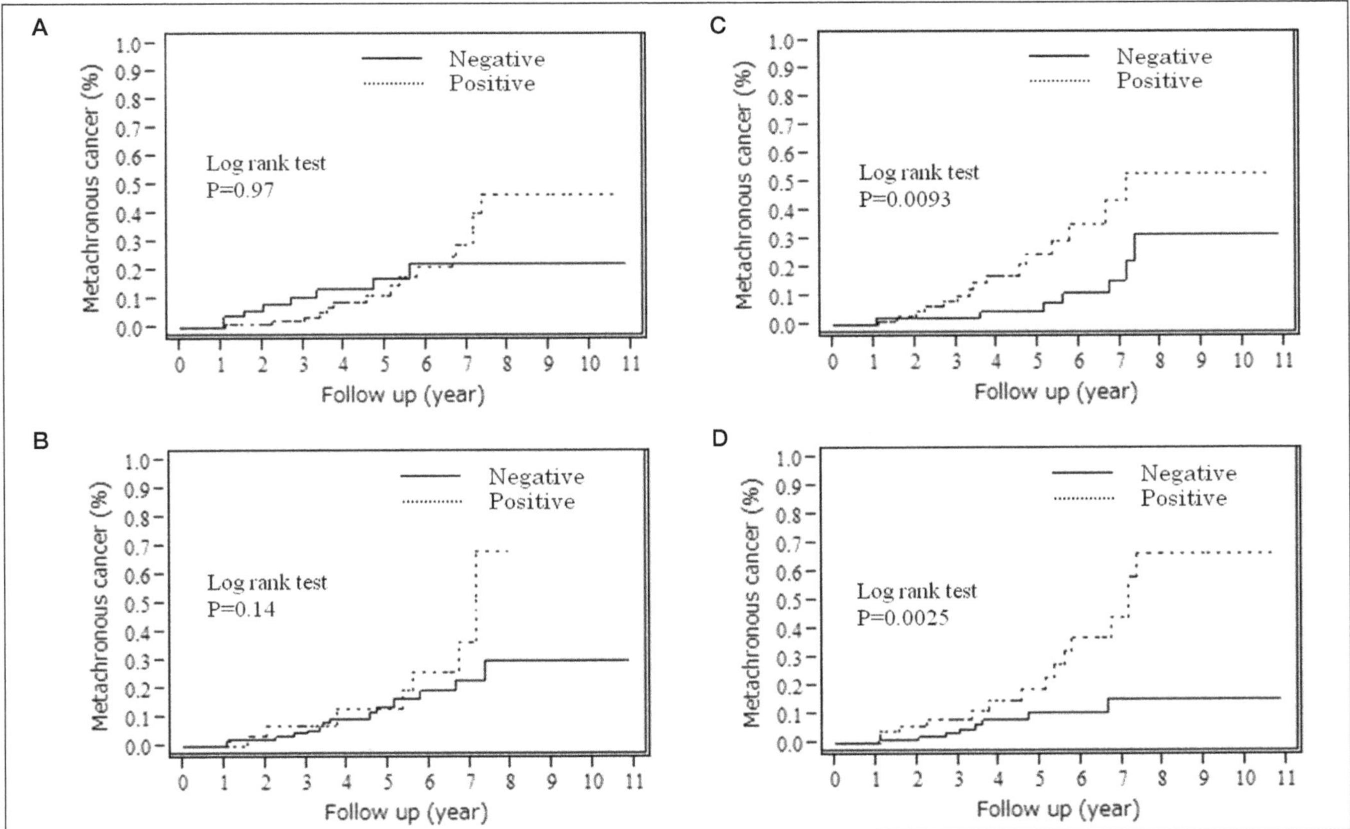

Figure 2) *Cumulative incidence of metachronous gastric cancer.* **A** *Intestinal metaplasia in the antrum.* **B** *Intestinal metaplasia in the corpus.* **C** *Neutrophil infiltration in the antrum.* **D** *Neutrophil infiltration in the corpus*

TABLE 3
Multivariate analysis* for metachronous gastric cancer

Variable	OR (95% CI)	P
Tumour size	1.03 (0.99–1.07)	0.20
Intestinal metaplasia in the corpus	2.65 (1.13–6.66)	0.024
Neutrophil infiltration in the corpus	3.06 (1.32–7.64)	0.0091

Cox proportional hazards model

metachronous gastric cancer in the present analysis (OR 2.56 [95% CI 0.52 to 46.3]; P=0.30).

Multivariate analysis for metachronous gastric cancer

The multivariate analysis of tumour size, IM and NI in the corpus showed that IM and NI in the corpus were independent risk factors for metachronous gastric cancer (Table 3). The ORs for metachronous gastric cancer were 2.65 (95% CI 1.13 to 6.66; P=0.024) in the IM-positive group in the corpus and 3.06 (95% CI 1.32 to 7.64; P=0.0091) in the NI-positive group in the corpus.

DISCUSSION

The results of the present single-centre study involving 155 consecutive patients showed that IM and NI in the corpus were related to metachronous gastric cancer. IM and NI were high-risk microenvironments and risk factors for primary and metachronous gastric cancers. Not only *H pylori* eradication but also annual follow-up endoscopy may be important for improving the prognosis of patients with metachronous gastric cancer.

Although many studies of residual gastric cancer have been conducted (27,28), few reports of metachronous gastric cancers after ESD for early gastric cancer are available. The rate of recurrence of early gastric cancer in the gastric stump has been estimated to be 1% to 3% (13,29). An increased recurrence rate would be expected in patients treated with ESD, in proportion to the larger area of gastric mucosa remaining in these patients. Other studies have reported that the annual incidence of metachronous gastric cancer after EMR is 2.5% to 4% (12,20). In the present study, the annual incidence was approximately 3.5%.

In their randomized controlled study, Fukase et al (20) reported that eradicating *H pylori* reduced the risk for developing metachronous gastric cancer in patients treated with EMR. In our previous study, we reported that IM, NI and the gastritis pattern are related to developing primary gastric cancer (22). In the present study, our data revealed that IM and NI were also related to metachronous gastric cancer; however, no significant difference in *H pylori* status was observed. It has been suggested that a larger sample size would demonstrate a significant reduction in the incidence of metachronous gastric cancer after eradicating *H pylori*. Some studies have reported a significant reduction in cancer development following *H pylori* eradication in subgroups with mild atrophy or antrum-dominant atrophy (30,31). The preventive effect of *H pylori* eradication against gastric cancer may depend on the degree of baseline gastric mucosal atrophy and IM. Many patients in our study had IM, which may represent a point of no return at which the development of gastric cancer can no longer be prevented by eradicating *H pylori*.

In a previous study, we reported that NI in the corpus was related to prevalent diffuse-type gastric cancer (22). In that study, the diffuse type of the metachronous cancers were found in only two cases. A prospective study with a larger number is needed to elucidate the difference between diffuse type and intestinal type cancers.

The annual incidence of metachronous gastric cancer in the present study was 3.5%. The cumulative prevalence of gastric remnant cancer after surgical partial gastrectomy for early gastric cancer is reported to be 2.4% at five years. Patients with early gastric cancer after ESD treatment have been identified as a high-risk group for metachronous gastric cancer. If *H pylori* is eradicated, the NI status of gastric mucosa improves (32). However, IM of the gastric mucosa cannot easily be improved (33-36). Endoscopists should recognize the characteristics of metachronous gastric cancer and special attention should be devoted to detection of early gastric cancer.

Six patients died during the study period, but none of these deaths were related to gastric cancer. All patients with metachronous gastric cancer were treated completely with ESD or surgery. Annual endoscopy follow-up may be useful for detection of metachronous gastric cancer during the early stage and prevention of gastric cancer death.

A limitation of our study may be a retrospective cohort study in consecutive patients. A prospective study is needed to determine whether baseline histological characteristics can be predictive factors for metachronous gastric cancer and be used to determine the intensity of surveillance endoscopy among this patient population.

CONCLUSION

The presence of IM and NI in the corpus was closely related to the development of metachronous gastric cancer after ESD.

DISCLOSURES: The authors have no financial disclosures or conflicts of interest to declare.

ACKNOWLEDGEMENTS: The authors thank the gastroenterologists at the University of Tokyo Hospital for their endoscopic procedure. They are grateful to all the pathologists for their histopathological examinations on our subjects.

REFERENCES

1. Curvers WL, van den Broek FJ, Reitsma JB, et al. Systematic review of narrow-band imaging for the detection and differentiation of abnormalities in the esophagus and stomach (with video). Gastrointest Endosc 2009;69:307-17.
2. Polkowski M. Endoscopic diagnosis and treatment of upper gastrointestinal tumors. Endoscopy 2008;40:862-7.
3. Kakushima N, Fujishiro M. Endoscopic submucosal dissection for gastrointestinal neoplasms. World J Gastroenterol 2008;21;14:2962-7.
4. Yamamoto H. Technology insight: Endoscopic submucosal dissection of gastrointestinal neoplasms. Nat Clin Pract Gastroenterol Hepatol 2007;4:511-20.
5. Gotoda T. Endoscopic resection of early gastric cancer. Gastric Cancer 2007;10:1-11.
6. Soetikno R, Kaltenbach T, Yeh R, et al. Endoscopic mucosal resection for early cancers of the upper gastrointestinal tract. J Clin Oncol 2005;23:4490-8.
7. Takenaka R, Kawahara Y, Okada H, et al. Risk factors associated with local recurrence of early gastric cancers after endoscopic submucosal dissection. Gastrointest Endosc 2008;68:887-94.
8. Isomoto H, Shikuwa S, Yamaguchi N, et al. Endoscopic submucosal dissection for early gastric cancer: A large-scale feasibility study. Gut 2009;58:331-6.
9. Sugimoto T, Okamoto M, Mitsuno Y, et al. Endoscopic submucosal dissection is an effective and safe therapy for early gastric neoplasms: A multicenter feasible study. J Clin Gastroenterol 2012;46:124-9.
10. Jang JS, Choi SR, Qureshi W, et al. Long-term outcomes of endoscopic submucosal dissection in gastric neoplastic lesions at a single institution in South Korea. Scand J Gastroenterol 2009;44:1315-22.
11. Choi KS, Jung HY, Choi KD, et al. EMR versus gastrectomy for intramucosal gastric cancer: comparison of long-term outcomes. Gastrointest Endosc 2011;73:942-8.
12. Nasu J, Doi T, Endo H, et al. Characteristics of metachronous multiple early gastric cancers after endoscopic mucosal resection. Endoscopy 2005;37:990-3.
13. Hosokawa O, Kaizaki Y, Watanabe K, et al. Endoscopic surveillance for gastric remnant cancer after early cancer surgery. Endoscopy 2002;34:469-73.
14. Nomura A, Stemmermann GN, Chyou PH, et al. *Helicobacter pylori* infection and gastric carcinoma among Japanese Americans in Hawaii. N Engl J Med 1991;325:1132-6.
15. Parsonnet J, Friedman GD, Vandersteen DP, et al. *Helicobacter pylori* infection and the risk of gastric carcinoma. N Engl J Med 1991;325:1127-31.

16. Eslick GD, Lim LL, Byles JE, et al. Association of *Helicobacter pylori* infection with gastric carcinoma: A meta-analysis. Am J Gastroenterol 1999;94:2373-9.
17. Uemura N, Okamoto S, Yamamoto S, et al. *Helicobacter pylori* infection and the development of gastric cancer. N Engl J Med 2001;345:784-9.
18. Correa P. Human gastric carcinogenesis: A multistep and multifactorial process – First American Cancer Society Award Lecture on Cancer Epidemiology and Prevention. Cancer Res 1992;52:6735-40.
19. Wong BC, Lam SK, Wong WM, et al. *Helicobacter pylori* eradication to prevent gastric cancer in a high-risk region of China: A randomized controlled trial. JAMA 2004;291:187-94.
20. Fukase K, Kato M, Kikuchi S, et al. Effect of eradication of *Helicobacter pylori* on incidence of metachronous gastric carcinoma after endoscopic resection of early gastric cancer: An open-label, randomised controlled trial. Lancet 2008;372:392-7.
21. Ogura K, Hirata Y, Yanai A, et al. The effect of *Helicobacter pylori* eradication on reducing the incidence of gastric cancer. J Clin Gastroenterol 2008;42:279-83.
22. Sakitani K, Hirata Y, Watabe H, et al. Gastric cancer risk according to the distribution of intestinal metaplasia and neutrophil infiltration. J Gastroenterol Hepatol 2011;26:1570-5.
23. Association JGC. Japanese classification of gastric carcinoma: Third English edition. Gastric Cancer 2011;14:101-12.
24. Gotoda T, Yanagisawa A, Sasako M, et al. Incidence of lymph node metastasis from early gastric cancer: Estimation with a large number of cases at two large centers. Gastric Cancer 2000;3:219-25.
25. Mitsudomi T, Watanabe A, Matsusaka T, et al. A clinicopathological study of synchronous multiple gastric cancer. Br J Surg 1989;76:237-40.
26. Maehata Y, Nakamura S, Fujisawa K, et al. Long-term effect of *Helicobacter pylori* eradication on the development of metachronous gastric cancer after endoscopic resection of early gastric cancer. Gastrointest Endosc 2012;75:39-46.
27. Kaneko K, Kondo H, Saito D, et al. Early gastric stump cancer following distal gastrectomy. Gut 1998;43:342-4.
28. Ohashi M, Katai H, Fukagawa T, et al. Cancer of the gastric stump following distal gastrectomy for cancer. Br J Surg 2007;94:92-5.
29. Takeda J, Toyonaga A, Koufuji K, et al. Early gastric cancer in the remnant stomach. Hepatogastroenterology 1998;45:1907-11.
30. Shiotani A, Uedo N, Iishi H, et al. Predictive factors for metachronous gastric cancer in high-risk patients after successful Helicobacter pylori eradication. Digestion 2008;78:113-9.
31. Yanaoka K, Oka M, Ohata H, et al. Eradication of *Helicobacter pylori* prevents cancer development in subjects with mild gastric atrophy identified by serum pepsinogen levels. Int J Cancer 2009;125:2697-703.
32. Tari A, Kitadai Y, Sumii M, et al. Basis of decreased risk of gastric cancer in severe atrophic gastritis with eradication of *Helicobacter pylori*. Dig Dis Sci 2007;52:232-9.
33. Ito M, Haruma K, Kamada T, et al. *Helicobacter pylori* eradication therapy improves atrophic gastritis and intestinal metaplasia: A 5-year prospective study of patients with atrophic gastritis. Aliment Pharmacol Ther 2002;16:1449-56.
34. Kodama M, Murakami K, Okimoto T, et al. *Helicobacter pylori* eradication improves gastric atrophy and intestinal metaplasia in long-term observation. Digestion 2012;85:126-30.
35. Forbes GM, Warren JR, Glaser ME, et al. Long-term follow-up of gastric histology after *Helicobacter pylori* eradication. J Gastroenterol Hepatol 1996;11:670-3.
36. Satoh K, Kimura K, Takimoto T, et al. A follow-up study of atrophic gastritis and intestinal metaplasia after eradication of *Helicobacter pylori*. Helicobacter 1998;3:236-40.

Hospital volume and other risk factors for in-hospital mortality among diverticulitis patients: A nationwide analysis

Michael J Diamant MD MSc[1], Stephanie Coward MSc[1,2], W Donald Buie MD[3], Anthony MacLean MD[3], Elijah Dixon MD MSc[3], Chad G Ball MD[3], Samuel Schaffer MD[4], Gilaad G Kaplan MD MPH[1,2]

MJ Diamant, S Coward, WD Buie, et al. Hospital volume and other risk factors for in-hospital mortality among diverticulitis patients: A nationwide analysis. Can J Gastroenterol Hepatol 2015;29(4):193-197.

BACKGROUND: Previous studies have found that a higher volume of colorectal surgery was associated with lower mortality rates. While diverticulitis is an increasingly common condition, the effect of hospital volume on outcomes among diverticulitis patients is unknown.
OBJECTIVE: To evaluate the relationship between hospital volume and other factors on in-hospital mortality among patients admitted for diverticulitis.
METHODS: Data from the Nationwide Inpatient Sample (years 1993 to 2008) were analyzed to identify 822,865 patients representing 4,108,726 admissions for diverticulitis. Hospitals were divided into quartiles based on the volume of diverticulitis cases admitted over the study period, adjusted for years contributed to the dataset. Mortality according to hospital volume was modelled using logistic regression adjusting for age, sex, race, comorbidities, health care insurance, admission type, calendar year, colectomy, disease severity and clustering. Risk estimates were expressed as adjusted ORs with 95% CIs.
RESULTS: Patients at high-volume hospitals were more likely to be admitted emergently, undergo surgical treatment and have more severe disease. In-hospital mortality was higher among the lowest quartile of hospital volume compared with the highest volume (OR 1.13 [95% CI 1.05 to 1.21]). In-hospital mortality was increased among patients admitted emergently (OR 2.58 [95% CI 2.40 to 2.78]) as well as those receiving surgical treatment (OR 3.60 [95% CI 3.42 to 3.78]).
CONCLUSIONS: Diverticulitis patients admitted to hospitals with a low volume of diverticulitis cases had an increased risk for death compared with those admitted to high-volume centres.

Key Words: *Colectomy; Diverticulitis; Hospital volume; Mortality*

Le volume hospitalier et d'autres facteurs de risque de mortalité hospitalière chez les patients atteints de diverticulite : une analyse nationale

HISTORIQUE : Des études antérieures ont établi qu'un plus fort volume de chirurgie colorectale s'associait à une diminution du taux de mortalité. La diverticulite est de plus en plus fréquente, mais on ne connaît pas l'effet du volume hospitalier sur l'issue des patients qui en sont atteints.
OBJECTIF : Évaluer le lien entre le volume hospitalier et d'autres facteurs sur la mortalité hospitalière des patients hospitalisés en raison d'une diverticulite.
MÉTHODOLOGIE : Les chercheurs ont analysé les données du *Nationwide Inpatient Sample* (1993 à 2008) pour en extraire 822 865 patients représentant 4 108 726 hospitalisations en raison d'une diverticulite. Ils ont réparti les hôpitaux en quartiles selon le volume de cas de diverticulite admis pendant la période de l'étude, rajusté compte tenu des années de participation au fichier de données. Ils ont établi le modèle de mortalité selon le volume hospitalier au moyen de la régression logistique rajustée en fonction de l'âge, du sexe, de la race, des comorbidités, de l'assurance-maladie, du type d'admission, de l'année civile, de la colectomie, de la gravité de la maladie et des regroupements. Ils ont exprimé le risque sous forme de rapport de cote (RC) rajusté ayant un indice de confiance de 95 %.
RÉSULTATS : Les patients des hôpitaux à fort volume étaient plus susceptibles d'être admis en urgence, de subir un traitement chirurgical et d'avoir une maladie plus grave. La mortalité hospitalière était plus élevée dans le plus petit quartile de volume hospitalier que dans celui ayant le volume le plus élevé (RC 1,13 [95 % IC 1,05 à 1,21]). La mortalité hospitalière était plus élevée chez les patients admis d'urgence (RC 2,58 [95 % IC 2,40 à 2,78]) et chez ceux ayant subi un traitement chirurgical (RC 3,60 [95 % IC 3,42 à 3,78]).
CONCLUSIONS : Les patients atteints de diverticulite admis dans des hôpitaux ayant un faible volume de cas de diverticulite présentaient un risque de décès plus élevé que ceux qui étaient admis dans des centres à fort volume.

In the United States, diverticular disease accounted for approximately 314,000 hospitalizations in 2006 (1), while in 2005, the age-adjusted hospitalization rate for diverticulitis was 75.5 per 100,000 persons (2). Diverticulitis develops in 10% to 25% of patients with diverticulosis (3), and is more common in elderly individuals (4). The severity of diverticulitis spans from mild (controlled by conservative medical management) to life-threatening (requiring surgical intervention) (5). Approximately 15% to 20% of patients with diverticulitis will develop complications such as abscess or fistula formation, obstruction, bleeding or perforation (3,6).

Consequently, management of diverticulitis depends on the severity of the disease, with established guidelines recommending non-operative treatment with dietary modification and administration of oral or intravenous antibiotics in cases of diverticulitis without the above complications (7). Percutaneous drainage, in conjunction with intravenous antibiotics, is the treatment of choice in patients with large diverticular abscesses (7,8). Between 10% and 45% of patients who develop diverticulitis will undergo surgical management (9,10), with approximately 25% of patients requiring surgery during their initial hospital admission for diverticulitis (10).

Previous studies have found lower mortality rates at high-volume hospitals during hospitalization or postoperatively for patients with inflammatory bowel disease (11-13). On the other hand, Karanicolas et al (14) examined the effects of both surgeon and hospital volumes on mortality from colorectal resection for various reasons, and found that while surgeons seeing a higher-volume caseload was associated with a

[1]*Departments of Medicine;* [2]*Community Health Sciences; and* [3]*Surgery, University of Calgary, Calgary, Alberta;* [4]*Global Medical Affairs, Shire Canada, Toronto, Ontario*
Correspondence: Dr Gilaad G Kaplan, Departments of Medicine and Community Health Sciences, University of Calgary, 6D17, 3280 Hospital Drive Northwest, Calgary, Alberta T2N 4N1 e-mail ggkaplan@ucalgary.ca

lower mortality rate, an association between hospital volume and mortality rates was not observed. Furthermore, Weber et al (15) found that diverticulosis and diverticulitis patients undergoing sigmoid colon resection were more likely to receive laparoscopic versus open sigmoid colectomies in high-volume hospitals, although a difference in mortality between the two types of surgery was not demonstrated (16). Thus, the effect of hospital volume on mortality related to surgical and medical management of diverticulitis remains unknown.

The objectives of the present study were to use a nationally representative database of hospital discharges in the United States to determine the effect of hospital volume on mortality for all inpatients with diverticulitis, as well as to evaluate other risk factors for mortality on patients hospitalized for diverticulitis.

METHODS

Study design and patient population

The Nationwide Inpatient Sample (NIS) is a collection of inpatient records from a cross-section of 1000 acute care hospitals across the United States (US) representing approximately eight million admissions per year, which provides a 20% stratified sample of US hospitals. The database is populated using stratified random sampling to ensure it accounts for approximately 90% of hospital admissions and is representative of the US population. The NIS contains deidentified demographic information, diagnostic and procedure codes based on the *International Classification of Diseases, Ninth Revision, Clinical Modification* (ICD-9-CM), outcomes, and hospital characteristics. The NIS is compiled and validated by the Agency for Healthcare and Research Quality (Rockville , Maryland, USA).

Data collection and outcomes

ICD-9-CM codes (562.01, 562.03, 562.11 or 562.13) were used to identify 822,865 patients admitted with diverticulitis in all diagnostic positions from 1993 to 2008. Patients who underwent surgery for their diverticulitis were identified with the following ICD-9-CM codes: 45.71, 45.72, 45.73, 45.74, 45.75, 45.76, 45.79, 45.8, 45.81, 45.82 and 45.83. Patients with colostomies (codes 46.1, 46.11, 46.13) and ileostomies (46.2, 46.20, 46.21, 46.22, 46.23, 46.24) were excluded from the surgery group because the vast majority (>99%) did not have concurrent coding of resection operations during admission. Additionally, all patients were stratified according to disease severity as defined by the Hinchey classification (17), which categorizes diverticulitis according to an assessment of peritoneal contamination. Patients with ICD-9-CM codes 567.22, 569.5 and 614.3 reflecting the presence of a pericolic, mesenteric or walled-off pelvic abscess were identified as Hinchey I or II. Patients with ICD-9-CM codes 567, 567.21, 567.9 and 614.5 reflecting the presence of either generalized purulent or fecal peritonitis were identified as Hinchey III or IV.

The primary comparison of interest was hospital volume, which was determined by dividing the hospitals into quartiles based on the total number of diverticulitis patients admitted over the total study period. The variable number of years each hospital contributed to the dataset was taken into account. The median number of diverticulitis admissions at a single hospital was 440 (interquartile range 216 to 793). The primary outcome of interest in the present study was in-hospital mortality.

Statistical analysis

Descriptive statistics with Wald confidence limits, with NIS sample weights taken into account, were used to summarize patient characteristics in each quartile of hospital volume. A complete case analysis was used for the comparison of all patient characteristics, covariates and outcomes.

All risk estimates are represented by adjusted ORs created using logistic regression with a marginal generalized estimating equation extension. The generalized estimating equations account for clustering of responses by hospital (ie, to account for the potential correlation of outcomes between patients from the same hospital). The sampling weights for each observation were included in all analyses. For all multivariable analyses, the following variables were included as covariates: age, categorized as <40, 40 to 64 and ≥65 years of age; sex; race and ethnicity classified as non-Hispanic white, black, Hispanic and other/unavailable (because multiple states purposely do not collect information on race); comorbidity index scores, which were based on the presence of 30 illnesses and modelled continuously (18); health insurance status, classified as Medicare, Medicaid, private insurance, self-pay or other; admission type, classified as elective, urgent (which requires immediate attention for care or treatment) or emergent (which denotes medical intervention required as a result of severe, potentially fatal or disabling conditions); disease severity determined by ICD-9 codes reflecting Hinchey classification, categorized as no Hinchey score, Hinchey I or II, and Hinchey III or IV; calendar year of admission, modelled as a continuous variable; and whether the patient underwent a colectomy as surgical treatment. A sensitivity analysis was performed in which only patients who had diverticulitis as the primary diagnosis on admission were included in the analysis.

All analyses were performed using SAS version 9.2 (SAS Institute, USA); associations with P<0.05 were considered to be statistically significant.

RESULTS

Between 1993 and 2008, a total of 822,865 patients who were admitted for diverticulitis were captured in the NIS sample. This represents an estimated 4,108,726 admissions for diverticulitis nationwide over the course of the study period. The characteristics of patients admitted for diverticulitis, stratified according to hospital volume, are summarized in Table 1. Data were missing only for patient sex (n=1051), insurance type (n=2081) and admission type (n=78,069). Of note, patients at high-volume hospitals were more likely to be admitted emergently and were more likely to undergo surgical treatment.

Patient mortality according to hospital volume

Table 2 lists the raw mortality rate, and unadjusted and adjusted OR of mortality according to hospital volume among diverticulitis patients. A total of 80,731 admissions had data missing from the adjusted model, leaving data representing 3,719,102 admissions nationwide. Patients admitted to the lowest-volume hospitals were 1.13 times more likely to die (95% CI 1.05 to 1.21) during hospital admission than patients admitted to higher-volume hospitals. A similar trend was observed in patients in the second lowest quartile of hospital volume (OR 1.12 [95% CI 1.04 to 1.20]).

For the sensitivity analysis, restricting the analysis to patients that had diverticulitis listed as their primary diagnosis on admission, a total of 2,638,832 admissions were included in the final adjusted model. The trend was similar to the comparison of hospital volume among all individuals with diverticulitis: patients admitted to the lowest and second lowest quartiles of hospital volume were 1.17 (95% CI 1.06 to 1.29) and 1.16 (95% CI 1.05 to 1.28) times more likely to die during their admission compared with high-volume hospitals, respectively.

Other predictors of mortality in diverticulitis

Table 3 lists other factors, aside from hospital volume, that are associated with mortality among diverticulitis patients. The odds of mortality were greater with increased age (≥60 versus <40 years of age: OR 13.53 [95% CI 10.28 to 17.82]), greater burden of comorbidities (OR 1.45 [95% CI 1.43 to 1.47]), having a type of health insurance other than a private provider (Medicare versus private: OR 1.55 [95% CI 1.42 to 1.69]), being admitted to hospital urgently or emergently versus electively (OR 1.90 [95% CI 1.76 to 2.06] and OR 2.58 [95% CI 2.40 to 2.78], respectively), undergoing surgical treatment (OR 3.60 [95% CI 3.42 to 3.78]), having an ICD-9 code corresponding to a more severe disease according to the Hinchey classification (Hinchey I and II versus 0, OR 1.18 [95% CI 1.11 to 1.27]; Hinchey III and IV versus 0, OR 3.07 [95% CI 2.82 to 3.34]), and being admitted to a hospital in the Northeastern US. The odds of mortality was decreased among female patients (OR 0.90 [95% CI 0.87 to 0.94]), individuals of

TABLE 1
Characteristics of patients admitted for diverticulitis between 1993 and 2008, according to hospital volume quartile

	All hospitals (n=4,108,726)	Hospital volume			
		1st quartile	2nd quartile	3rd quartile	4th quartile
Age*, years					
<40	7.0 (6.9–7.1)	6.8 (6.7–6.9)	7.2 (7.1–7.3)	7.1 (7.0–7.2)	6.9 (6.8–7.0)
40–59	38.6 (38.4–38.7)	36.5 (36.3–36.7)	39.0 (38.7–39.1)	39.4 (39.2–39.6)	39.4 (39.2–39.6)
≥60	54.4 (54.3–54.5)	56.7 (56.5–56.9)	53.9 (53.7–54.1)	53.4 (53.2–53.6)	53.7 (53.5–53.9)
Sex*					
Male	39.2 (39.1–39.3)	38.0 (37.8–38.2)	39.0 (38.8–39.2)	39.6 (39.4–39.8)	40.2 (39.9–40.4)
Female	60.8 (60.7–60.9)	62.0 (61.8–62.2)	61.0 (60.8–61.2)	60.4 (60.2–60.6)	59.8 (59.6–60.1)
Race/ethnicity*					
Non-Hispanic white	63.8 (63.7–64.0)	58.5 (58.2–58.7)	59.5 (59.3–59.7)	65.3 (65.1–65.5)	72.2 (72.0–72.4)
Black	5.4 (5.4–5.5)	5.0 (4.9–5.1)	5.7 (5.6–5.8)	5.0 (4.9–5.1)	6.0 (5.9–6.1)
Hispanic	5.4 (5.3–5.4)	5.2 (5.1–5.3)	5.4 (5.3–5.5)	4.2 (4.1–4.3)	6.7 (6.6–6.8)
Unavailable	25.4 (25.3–25.5)	31.4 (31.1–31.6)	29.4 (29.2–29.6)	25.5 (25.3–25.6)	15.1 (14.9–15.2)
Comorbidity score*, mean	1.78 (1.77–1.78)	1.77 (1.77–1.78)	1.80 (1.79–1.80)	1.77 (1.76–1.78)	1.77 (1.77–1.78)
Health insurance*					
Private	36.5 (36.4–36.6)	31.5 (31.3–31.7)	36.3 (36.1–36.6)	38.8 (38.6–39.0)	39.4 (39.2–39.63)
Medicare	53.7 (53.6–53.8)	56.5 (56.3–56.8)	53.3 (53.1–53.5)	52.3 (52.0–52.5)	52.7 (52.5–53.0)
Medicaid	4.1 (4.1–4.2)	5.5 (5.4–5.6)	4.3 (4.2–4.3)	3.7 (3.7–3.8)	3.0 (2.9–3.1)
Self-pay	3.3 (3.3–3.4)	3.9 (3.9–4.0)	3.4 (3.3–3.5)	3.2 (3.2–3.3)	2.7 (2.7–2.8)
Other	2.3 (2.3–2.4)	2.6 (2.5–2.7)	2.7 (2.6–2.7)	2.0 (1.9–2.0)	2.1 (2.1–2.2)
Admission type*					
Elective	20.5 (20.5–20.7)	21.5 (21.3–21.7)	20.8 (20.7–21.0)	20.1 (19.9–20.2)	19.9 (19.7–20.1)
Urgent	24.1 (24.0–24.2)	30.2 (30.0–30.4)	25.2 (25.0–25.4)	21.6 (21.4–21.8)	19.6 (19.5–19.9)
Emergent	55.3 (55.2–55.4)	48.3 (48.1–48.6)	54.0 (53.7–54.2)	58.4 (58.1–58.6)	60.4 (30.2–60.6)
Hinchey classification*					
0	90.3 (90.3–90.4)	91.8 (91.6-91.9)	90.5 (90.4–90.6)	89.9 (89.7-90.0)	89.2 (89.1–89.4)
I or II	8.1 (8.0–8.1)	6.7 (6.6–6.8)	7.8 (7.7–7.9)	8.5 (8.4–8.6)	9.2 (9.1–9.3)
III or IV	1.6 (1.6–1.6)	1.5 (1.5–1.6)	1.7 (1.6–1.7)	1.6 (1.6–1.7)	1.6 (1.5–1.6)
Surgical treatment*					
Partial or total colectomy	20.8 (20.7–20.9)	17.4 (17.3–17.6)	20.8 (20.7–21.0)	22.1 (21.9–22.3)	23.0 (22.8–23.2)

*Data presented as % (95% CI) unless otherwise indicated. *P<0.0001*

TABLE 2
Mortality rate, unadjusted and adjusted OR of mortality of diverticulitis patients according to hospital volume quartile

Hospital volume	Mortality rate, % (95% CI)	Unadjusted* OR (95% CI)	Adjusted† OR (95% CI)
1st quartile (lowest)	1.80 (1.74–1.86)	0.94 (0.90–0.99)	1.13 (1.05–1.21)
2nd quartile	1.90 (1.83–1.96)	1.00 (0.95–1.04)	1.12 (1.04–1.20)
3rd quartile	1.89 (1.83–1.95)	0.99 (0.95–1.04)	1.06 (0.98–1.14)
4th quartile (highest)	1.90 (1.83–1.95)	1.00	1.00

**Unadjusted analysis does not account for the effect of clustering; †Adjusted for age, sex, race, comorbidity index score, health insurance, admission type, Hinchey classification, calendar year of admission, colectomy and effect of clustering*

Hispanic background versus non-Hispanic white (OR 0.85 [95% CI 0.77 to 0.94]), and with admission to a hospital later in the study period (OR 0.96 [95% CI 0.95 to 0.96] per year).

DISCUSSION

The present population-based study using data representing >4 million hospital admissions for diverticulitis in the US highlights a few important associations. First, patients admitted to low-volume hospitals had a greater odds for mortality compared with patients admitted to hospitals that encountered a higher caseload of diverticulitis. Second, increasing age and comorbidity burden, being admitted emergently, increasing disease severity according to the Hinchey classification, undergoing surgery, receiving health insurance from a source other than a private provider, and being admitted to a hospital in the Northeastern US were all associated with an increased odds of mortality.

Although the present study was the first to examine the relationship between hospital volume and mortality in diverticulitis patients, many of the findings are consistent with previous research examining the effect of hospital volume in other patient populations. Previous studies have found an increased mortality risk among patients treated in lower-volume hospitals for inflammatory bowel disease (12), as well as increased mortality risk among patients undergoing surgery for inflammatory bowel disease in low-volume hospitals (11,13,19). Similar to findings in Etzioni et al (20), we also demonstrated an increased risk for mortality among patients who were either older, undergoing surgery, or admitted emergently or urgently, as well as a temporal trend of decreased mortality over the course of the study period. Our finding of a greater risk for mortality among diverticulitis patients with insurance coverage other than a private provider has also been previously demonstrated (21,22). Although Ho et al (22) also found no increased risk for mortality among black patients compared with non-Hispanic white patients, others studies have demonstrated an increased risk for mortality among African Americans (21,23,24). Furthermore, to our knowledge, the present study was the first to use ICD-9-CM diagnostic codes in an administrative database to stratify diverticulitis patients according to disease severity according to the Hinchey classification.

TABLE 3
Demographic and hospital-level predictors of in-hospital mortality in patients hospitalized for diverticulitis

Characteristic	Mortality rate, % (95% CI)	Unadjusted OR (95% CI)*	Adjusted OR (95% CI)†
Age, years			
<40	0.12 (0.09–0.15)	1.00	1.00
40–59	0.49 (0.46–0.51)	4.19 (3.25–5.40)	3.48 (2.66–4.55)
≥60	3.07 (3.02–3.13)	26.99 (21.02–36.64)	13.53 (10.28–17.82)
Sex			
Male	1.63 (1.59–1.68)	1.00	1.00
Female	2.03 (1.99–2.07)	1.25 (1.21–1.29)	0.90 (0.87–0.94)
Race/ethnicity			
Non-Hispanic white	2.03 (1.99–2.07)	1.00	1.00
Black	1.93 (1.80–2.06)	0.95 (0.88–1.02)	1.00 (0.91–1.09)
Hispanic	1.28 (1.18–1.39)	0.63 (0.58–0.68)	0.85 (0.77–0.94)
Other/unavailable	1.59 (1.53–1.64)	0.78 (0.75–0.81)	0.95 (0.91–1.01)
Comorbidity score‡	0.65 (0.62–0.68)	1.45 (1.43–1.47)	1.45 (1.43–1.47)
Health insurance			
Private	0.51 (0.49–0.54)	1.00	1.00
Medicare	2.97 (2.92–3.02)	5.94 (5.62–6.27)	1.55 (1.42–1.69)
Medicaid	1.11 (1.00–1.23)	2.19 (1.95–2.45)	1.90 (1.66–2.19)
Self-pay	0.58 (0.49–0.67)	1.14 (0.97–1.34)	1.44 (1.19–1.74)
Other	1.00 (0.85–1.14)	1.95 (1.67–2.29)	1.93 (1.62–2.29)
Admission type			
Elective	1.41 (1.37–1.46)	1.00	1.00
Urgent	1.76 (1.69–1.82)	1.25 (1.19–1.31)	1.90 (1.76–2.06)
Emergent	2.17 (2.13–2.22)	1.55 (1.49–1.61)	2.58 (2.40–2.78)
Surgery			
No surgical treatment	1.43 (1.40–1.46)	1.00	1.00
Partial or total colectomy	3.54 (3.45–3.63)	2.53 (2.45–2.61)	3.60 (3.42-3.78)
Hinchey clasification			
0	1.70 (1.67–1.73)	1.00	1.00
I or II	2.40 (2.28–2.52)	1.42 (1.35–1.50)	1.18 (1.11–1.26)
III or IV	8.63 (8.15–9.11)	5.46 (5.12–5.82)	3.07 (2.82–3.34)
Region			
Northeast	2.12 (2.05–2.19)	1.00	1.00
Midwest	1.74 (1.69–1.80)	0.82 (0.78–0.86)	0.82 (0.76–0.88)
South	1.74 (1.69–1.78)	0.82 (0.78–0.85)	0.79 (0.74–0.84)
West	2.05 (1.97–2.13)	0.97 (0.92–1.02)	0.83 (0.76–0.91)
Quarter of admission			
January to March	1.90 (1.85–1.95)	1.00	1.00
April to June	1.81 (1.75–1.87)	0.95 (0.91–1.00)	0.96 (0.92–1.02)
July to September	1.78 (1.72–1.84)	0.94 (0.90–0.98)	0.96 (0.91–1.01)
October to December	1.99 (1.92–2.05)	1.05 (1.01–1.10)	1.07 (1.01–1.13)
Calendar year	N/A	0.96 (0.96–0.97)	0.96 (0.95–0.96)

**Unadjusted analysis does not account for effect of clustering; †Adjusted for age, sex, race, comorbidity index score, health insurance, admission type, Hinchey classification, calendar year of admission, colectomy and effect of clustering; ‡Mortality rate, unadjusted OR and adjusted OR for each additional point on comorbidity score. N/A Not applicable*

There are a few different explanations that could account for the greater mortality risk in low-volume centres as opposed to high-volume hospitals. In high-volume hospitals, physicians may be more familiar with therapeutic regimens, more likely to recognize early and potentially serious complications, and may recognize the need for surgical intervention and operate earlier (11). High-volume hospitals may also be able to provide more comprehensive intensive care, allow for more multidisciplinary support, and employ more experienced allied health care professionals and nursing staff (12). Hospital volume may influence outcomes for patients undergoing surgery in several ways, including surgeon experience, patient selection for operative treatment and postoperative care (13).

It should be noted that our adjusted risk estimates demonstrated a high risk for mortality among patients in low-volume hospitals, despite the fact that the raw mortality rates were higher among patients in high-volume centres compared with low-volume hospitals. However, high-volume centres also had a greater proportion of patients admitted emergently, who underwent surgery and with greater disease severity, all of which are associated with an increased risk for mortality and were included as covariates in the adjusted model. Moreover, the association of a higher risk for mortality among patients in lower-volume hospitals was also demonstrated when restricting analyses solely to patients who had a primary diagnosis of diverticulitis.

There were several limitations to the present study. As is the case with all studies using administrative databases, the accuracy of the data and the potential for misclassification of exposure and outcome are potential issues (25), although the outcome of in-hospital mortality is relatively robust. Additionally, we attempted to control for disease severity by using the Hinchey classification and patients' presentation on admission (such as urgent or emergent), the latter of which is also prone to misclassification. Due to the limitations of ICD-9-CM coding, we could only assign patients into strata of Hinchey classification (despite there being five levels of the classification). This method of adjustment has not been previously validated and is, thus, a limitation of our attempt to adjust for severity of diverticulitis. Also, because the NIS extracts information from hospital discharge abstracts, it is exceedingly difficult to accumulate long-term data in terms of outcomes postdischarge. Furthermore, because NIS data are deidentified, multiple admissions by the same individual for diverticulitis could not be accounted for in the data analysis. Despite the limitations of NIS data, they do allow for a national determination of risk factors and aspects of health care delivery associated with better outcomes.

CONCLUSION

Using a nationwide database of hospital admissions, we established that diverticulitis patients admitted to hospitals that encounter a low volume of diverticulitis cases have an increased risk for death compared with those admitted to high-volume centres. The present study highlights the need for further research into determining which mechanisms, practices and models of care are directly responsible for better outcomes in the treatment of diverticulitis.

ACKNOWLEDGEMENTS: The authors acknowledge the intellectual contribution to the manuscript by Linnette Yen at Shire, Wayne, PA, USA. Dr Kaplan and Dr Dixon are supported by New Investigator awards from the Canadian Institute of Health Research and Population Health Investigator awards from Alberta-Innovates Health Solutions. Shire Development LLC was a source of funding for this study and reviewed the manuscript.

DISCLOSURES: Shire Development LLC, Wayne, PA, USA provided funding for the study. Gilaad Kaplan has served as a speaker and received honoraria from Merck, Schering-Plough, Janssen, Abbott and UCB Pharma. He has participated in advisory board meetings and received honoraria from Abbott, Merck, Schering-Plough, Janssen, Shire and UCB Pharma. Dr Kaplan has received research support from Merck, Abbott and Shire. Dr Kaplan and Dr Dixon are supported by New Investigator awards from the Canadian Institute of Health Research and Population Health Investigator awards from Alberta-Innovates Health Solutions. Samuel Schaffer is an employee of Shire. The remaining authors have no financial disclosures or conflicts of interest to declare.

CONTRIBUTIONS: Dr Gilaad Kaplan and Michael Diamant were involved in all aspects of the study, including conception, design, data collection, analysis and writing. They had full access to all of the data in the study and take responsibility for the integrity of the data and the accuracy of the data analysis. All other authors contributed substantially to data interpretation, study write-up and final approval of the version to be published.

REFERENCES

1. DeFrances CJ, Lucas CA, Buie VC, Golosinskiy A. 2006 National Hospital Discharge Survey. Natl Health Stat Report 2008;5:1-20.
2. Nguyen GC, Sam J, Anand N. Epidemiological trends and geographic variation in hospital admissions for diverticulitis in the United States. World J Gastroenterol 2011;17:1600-5.
3. Parks TG. Natural history of diverticular disease of the colon. Clin Gastroenterol 1975;4:53-69.
4. Kyle J, Adesola AO, Tinckler LF, de Beaux J. Incidence of diverticulitis. Scand J Gastroenterol 1967;2:77-80.
5. Lopez DE, Brown CV. Diverticulitis: The most common colon emergency for the acute care surgeon. Scand J Surg 2010;99:86-9.
6. Waugh JM, Walt AJ. Current trends in the surgical treatment of diverticulitis of the sigmoid colon. Surg Clin North Am 1962;42:1267-76.
7. Rafferty J, Shellito P, Hyman NH, et al. Practice parameters for sigmoid diverticulitis. Dis Colon Rectum 2006;49:939-44.
8. Telem DA, Buch KE, Nguyen SQ, et al. Current recommendations on diagnosis and management of right-sided diverticulitis. Gastroenterol Res Pract 2009;2009:359485.
9. Parks TG, Connell AM. The outcome in 455 patients admitted for treatment of diverticular disease of the colon. Br J Surg 1970;57:775-8.
10. Peppas G, Bliziotis IA, Oikonomaki D, Falagas ME. Outcomes after medical and surgical treatment of diverticulitis: A systematic review of the available evidence. J Gastroenterol Hepatol 2007;22:1360-8.
11. Ananthakrishnan AN, McGinley EL, Binion DG. Does it matter where you are hospitalized for inflammatory bowel disease? A nationwide analysis of hospital volume. Am J Gastroenterol 2008;103:2789-98.
12. Nguyen GC, Steinhart AH. Nationwide patterns of hospitalizations to centers with high volume of admissions for inflammatory bowel disease and their impact on mortality. Inflamm Bowel Dis 2008;14:1688-94.
13. Kaplan GG, McCarthy EP, Ayanian JZ, et al. Impact of hospital volume on postoperative morbidity and mortality following a colectomy for ulcerative colitis. Gastroenterology 2008;134:680-7.
14. Karanicolas PJ, Dubois L, Colquhoun PH, et al. The more the better?: The impact of surgeon and hospital volume on in-hospital mortality following colorectal resection. Ann Surg 2009;249:954-9.
15. Weber WP, Guller U, Jain NB, et al. Impact of surgeon and hospital caseload on the likelihood of performing laparoscopic vs open sigmoid resection for diverticular disease: A study based on 55,949 patients. Arch Surg 2007;142:253-9.
16. Siddiqui MR, Sajid MS, Khatri K, et al. Elective open versus laparoscopic sigmoid colectomy for diverticular disease: A meta-analysis with the Sigma trial. World J Surg 2010;34:2883-901.
17. Hinchey, EJ, Schaal, PG, Richards GK. Treatment of perforated diverticular disease of the colon. Adv Surg 1978;12:85-109.
18. Elixhauser A, Steiner C, Harris DR, Coffey RM. Comorbidity measures for use with administrative data. Med Care 1998;36:8-27.
19. Kaplan GG, Hubbard J, Panaccione R, et al. The risk of comorbidities on postoperative outcomes in inflammatory bowel disease patients. Arch Surg 2011;146:959-64.
20. Etzioni DA, Mack TM, Beart RW Jr, Kaiser AM. Diverticulitis in the United States: 1998-2005: Changing patterns of disease and treatment. Ann Surg 2009;249:210-7.
21. Lidor AO, Gearhart SL, Wu AW, Chang DC. Effect of race and insurance status on presentation, treatment, and mortality in patients undergoing surgery for diverticulitis. Arch Surg 2008;143:1160-5.
22. Ho VP, Nash GM, Feldman EN, et al. Insurance but not race is associated with diverticulitis mortality in a statewide database. Dis Colon Rectum 2011;54:559-65.
23. Alavi K, Cervera-Servin JA, Sturrock PR, et al. Racial differences in short-term surgical outcomes following surgery for diverticulitis. J Gastrointest Surg 2012;16:613-21.
24. Schneider EB, Haider A, Sheer AJ, et al. Differential association of race with treatment and outcomes in Medicare patients undergoing diverticulitis surgery. Arch Surg 2011;146:1272-6.
25. Molodecky NA, Panaccione R, Ghosh S, Barkema HW, Kaplan GG. Challenges associated with identifying the environmental determinants of the inflammatory bowel diseases. Inflamm Bowel Dis 2011;17:1792-9.

Nonmedicinal interventions in nonalcoholic fatty liver disease

Manuela G Neuman PhD[1,2], Radu M Nanau BSc[1], Lawrence B Cohen MD MSc[3]

MG Neuman, RM Nanau, LB Cohen. Nonmedicinal interventions in nonalcoholic fatty liver disease. Can J Gastroenterol Hepatol 2015;29(5):241-252.

Unhealthy diet and lack of physical exercise are responsible for fat accumulation in the liver, which may lead to liver disease. Histologically, the severity of the disease has two stages: nonalcoholic fatty liver disease (NAFLD) and nonalcoholic steatohepatitis (NASH). NAFLD is defined by the presence of steatosis with no evidence of cellular injury such as hepatocyte ballooning. NASH is a distinct entity from NAFLD, and is characterized by the presence of inflammation with hepatocytes damage, with or without fibrosis. While several therapeutic strategies have been proposed to improve this condition, the present review aims to discuss nonmedicinal interventions used to reduce liver involvement or to prevent the disease altogether. The authors investigated dietary patterns and vitamin deficiencies associated with NAFLD, and their role in enhancing disease severity. Additionally, they reviewed the role of exercise and the use of interventions, such as as intragastric balloon and bariatric surgery, for improving disease progression. The authors propose monitoring disease progression or repair by following changes in cytoadipokine levels.

Key Words: *Adipokine; Bariatric surgery; Cytokeratine 18; Intragastric balloon placement; Nonalcoholic fatty liver disease; Nonalcoholic steatohepatitis; Noninvasive biomarkers*

Les interventions non médicales en cas de stéatose hépatique non alcoolique

Un régime malsain et le manque d'exercice physique sont responsables de l'accumulation de graisse dans le foie et peuvent favoriser une maladie hépatique. Sur le plan histologique, la gravité de la maladie se divise en deux phases : la stéatose hépatique non alcoolique (SHNA) et la stéatohépatite non alcoolique (SNA). La SHNA se définit par la présence de stéatose, sans manifestation de lésions cellulaires comme l'hypertrophie hépatocytaire. La SNA est une entité distincte de la SHNA et se caractérise par la présence d'inflammation et de lésions hépatocytaires, accompagnées ou non de fibrose. Il existe plusieurs stratégies thérapeutiques pour soigner cette maladie, mais la présente analyse porte sur les interventions non médicales utilisées pour réduire l'atteinte hépatique ou prévenir la maladie. Les auteurs ont exploré les profils diététiques et les carences en vitamines associées à la SHNA ainsi que leur rôle dans l'aggravation de la maladie. Ils ont également analysé le rôle de l'exercice et le recours à des interventions comme la chirurgie intragastrique à ballonnet et la chirurgie bariatrique pour en limiter la progression. Les auteurs proposent de surveiller la progression ou la réparation de la maladie après des modifications au taux de cytoadipokines.

Nonalcoholic fatty liver disease (NAFLD) is a chronic liver condition characterized by insulin resistance, frequently associated with impaired glucose intolerance or type 2 diabetes mellitus, and hepatic fat accumulation ranging from simple steatosis to severe steatohepatitis (NASH) (1,2). NAFLD is believed to be the most common cause of cryptogenic cirrhosis (1). The diagnosis of NAFLD is based on evidence of hepatic fatty infiltration, shown either by imaging or histology in the absence of other causes for secondary hepatic fat accumulation such as significant alcohol consumption, use of medication or presence of hereditary disorders known to produce hepatic lipidosis, or viral hepatitis B or C (1). NASH is mediated by other risk factors such as comorbid obesity and type 2 diabetes mellitus, in the presence or absence of the metabolic syndrome (1).

Morphological hallmarks of NAFLD-nonalcoholic steatohepatitis (NASH) include the severity of steatosis, hepatocyte ballooning, lobular inflammation, portal granulocytic inflammation, Mallory-Denk hyaline bodies and satellitosis (2,3). The pathology subcommittee of the Clinical Research Network for NASH designed and validated a histological feature scoring system for the full spectrum of lesions of NAFLD. This group evaluated 14 histological features and, after analysis, proposed a NAFLD activity score. This activity score includes class 1, which is simple steatosis; class 2, steatosis with lobular inflammation; class 3, the additional presence of ballooned hepatocytes; and class 4, the presence of either Mallory-Denk hyaline bodies or fibrosis. These stages were correlated with increasing severity of disease and likelihood of progression to cirrhosis (2).

NASH is primarily a mitochondrial disease arising from the inability of the mitochondria to adapt to an oversupply of fat (3,4). In addition to the mitochondrial dysfunction and oxidative stress in NASH, the translocation of gut-derived endotoxin to the portal vein initiates liver injury (3). These processes stimulate inflammatory responses. Lipotoxicity plays a key role in the pathogenesis of NAFLD. Dysregulation of hepatic metabolism – a result of lipotoxicity signalling and inflammatory processes – illustrates the cytokine imbalance that leads to liver injury (3).

Approximately 30% to 40% of patients with NAFLD develop NASH. NASH is commonly associated with perisinusoidal and perivenular fibrosis that may progress to cirrhosis (5). Moreover, it is estimated that 10% to 30% of patients with NAFLD develop cirrhosis after 10 years (3). Additionally, NASH represents the second or third leading indication for liver transplant in North American and is projected to become the leading indication in the next 10 to 20 years (6). Based on data from United States adult liver transplantation databases, 2004 to 2013, Wong et al (6) concluded that the number of adults with NASH awaiting liver transplant has almost tripled.

Potential therapeutic interventions in NAFLD/NASH must involve a good understanding of the molecular mechanisms of the disease that may reduce hepatic steatosis and the development of necroinflammation/fibrosis by reversing defects at three levels including: reducing substrate supply for lipogenesis from excess dietary triglycerides or from excessive lipolysis and free fatty acid flux to the liver from insulin-resistant adipose tissue; activating key molecular steps that

[1]*In Vitro Drug Safety and Biotechnology;* [2]*Department of Pharmacology and Toxicology, Faculty of Medicine, University of Toronto;* [3]*Division of Gastroenterology, Sunnybrook Health Sciences Centre, Department of Medicine, Faculty of Medicine, University of Toronto, Toronto, Ontario*
Correspondence: Dr Manuela G Neuman, Department of Pharmacology and Toxicology and International Health, Faculty of Medicine, University of Toronto, In Vitro Drug Safety and Biotechnology, Banting Institute, 100 College Street, Lab 217, Toronto, Ontario M5G 0A3.
e-mail manuela.neuman@utoronto.ca

TABLE 1
Mean serum 25-hydroxyvitamin D levels reported in selected studies

Author (reference)	Study sample	Association with NAFLD
Targher et al (15)	60 patients with biopsy-proven NAFLD; 60 healthy controls	51.0±22 nmol/L in NAFLD versus 74.5±15 nmol/L in control (P<0.001)
Pirgon et al (16)	87 obese adolescents (45 NAFLD patients and 43 non-NAFLD obese controls); 30 lean controls	29.5±18.4 in obese NAFLD versus 41.0±17.9 in obese non-NAFLD versus 48.1±22.2 ng/mL in lean controls
Dasarathy et al (17)	148 NAFLD patients; 39 healthy controls	21.2±10.4 ng/mL in NAFLD versus 35.7±6.0 ng/mL in control (P=0.002)
Hao et al (18)	514 subjects with normal liver function tests and with body mass index ≥18.5 kg/m² to <25 kg/m² (76 NAFLD patients and 438 non-NAFLD controls)	13.46±4.65 ng/mL in NAFLD versus 15.65±5.89 ng/mL in non-NAFLD controls (P=0.002)
Küçükazman et al (19)	211 subjects undergoing NAFLD examination (154 NAFLD patients and 57 non-NAFLD controls)	12.3±8.9 ng/dL in NAFLD versus 20±13.6 ng/dL in non-NAFLD controls (P<0.001)

NAFLD Nonalcoholic fatty liver disease

stimulate fatty acid oxidation and/or inhibit hepatic lipogenesis (ie, AMP-activated protein kinase); or by ameliorating the inflammation cascade generated by mitochondrial dysfunction from fat overload (ie, activation of Kupffer cells, local production of cytokines, induction of apoptosis, etc) (7-12).

Dietary intervention, the current standard of care for NAFLD and NASH, primarily reduces substrate supply (fat and carbohydrate overload) with modest and variable secondary improvements on hepatic molecular steps and local inflammation. The present review discusses the roles of nutrition and exercise as well as intragastic balloon placement on the evolution and progression of NAFLD/NASH, with an emphasis on changes in serum levels of adipokines (primarily adiponectin and leptin) as a result of these interventions.

DIETARY PATTERNS IN NAFLD

Several dietary components have often been found to be lacking in NAFLD patients. For example, insufficient intake of vitamin C, vitamin K, folate and omega-3 fatty acids were risk factors for NAFLD, while their increased consumption was protective (13,14). In particular, low serum 25-hydroxyvitamin D (25[OH]D_3) levels were measured among NAFLD patients compared with controls in several studies. These are summarized in Table 1 (15-19).

Serum 25(OH)D_3 levels may independently predict NAFLD (18,20). The association between lower serum 25(OH)D_3 levels and increased NAFLD incidence was maintained after controlling for age, sex, body mass index (BMI), creatinine, calcium, homeostasis model assessment-insulin resistance and the presence of the metabolic syndrome (15). Decreased serum 25(OH)D_3 levels further predicted histological severity of hepatic steatosis, hepatocyte ballooning, necroinflammation and fibrosis (15,17,18).

Vitamin D status was assessed as insufficient in 50.6% and as deficient in 17.3% in a sample of 156 adolescents with NAFLD. Inadequate serum 25(OH)D_3 levels at 17 years of age was a risk factor for NAFLD after adjusting for sex, race, physical activity, BMI and insulin resistance (21). Vitamin D supplementation was associated with increased serum 25(OH)D_3 levels, and this was accompanied by decreases in serum malondialdehyde and high-sensitive C-reactive protein (CRP) levels in a NAFLD sample (22). Inadequate intake of antioxidant vitamin C was noted in NAFLD patients (mean [± SD] 84.3±43.1 mg/day) when compared with non-NAFLD control individuals (144.2±63.1 mg/day) (P=0.0001). A deficient level of vitamin E was also described (5.4±1.9 mg/day in NASH versus 8.7±2.9 mg/day in control; P=0.0001) (23).

Food components whose increased consumption is associated with NAFLD include fruits, nuts and grilled meat (14,23). In particular, a higher Western dietary pattern (high intakes of fast food, red meat, processed meats, full-fat dairy products, fried potatoes, refined cereals, cakes and biscuits, confectionery, soft drinks, sauces and dressings) at 14 years was associated with a greater risk for NAFLD, assessed by liver ultrasound, at 17 years (OR 1.59 [95% CI 1.17to 2.14]; P<0.005) in a large sample of adolescents followed since birth (NAFLD was present in 15.2%). This was strongly linked to BMI at 14 years of age, such that this association was predominantly observed among obese adolescents. In contrast, a healthy dietary pattern (high in whole grains, fruit, vegetables, legumes, fish, fibre, folic acid and most micronutrients, and low in energy derived from total fat, saturated fat and refined sugar) at 14 years of age was protective against NAFLD development by 17 years of age (OR 0.63 [95% CI 0.41 to 0.96]; P=0.033) (24). In contrast, a recent cross-sectional study found similar consumption of Western diet high in fat and sodium intake and low in intakes of suboptimal micronutrients between 74 patients with biopsy-proven NAFLD and 27 healthy controls (25).

Higher BMI and waist circumference predicted a higher incidence of NAFLD (42.6%) in a sample of 82 obese children. Total carbohydrate intake was higher in NAFLD patients (P<0.001), while the intake of saturated fats was proportional to the degree of steatosis (13). Fried food consumption was associated with hepatic steatosis in a cross-sectional study of 74 overweight adolescents (P=0.04). Total fat intake (P=0.03) and the percentage of daily energy intake derived from fat (P=0.02) were associated with hepatic steatosis in this sample. Daily consumption of fibre was associated with lower visceral obesity (P=0.03) but had no influence on hepatic steatosis (26).

Overnutrition increases adipose tissue and results in insulin resistance, which favours high rates of free nonesterified fatty acid flux to the liver. This adds to the liver content of triacylglycerol, while triacylglycerol metabolites lead to a lipotoxic environment (27).

PHYSICAL ACTIVITY

Significantly lower proportions of NAFLD patients met recommendations for physical activity compared with healthy controls (P=0.02). Levels of physical activity were similar between NASH patients and subjects with simple steatosis. However, levels of physical activity were even lower among individual who also suffer from diabetes mellitus (25,28). On the other hand, regular physical exercise decreased the risks of having elevated aspartate aminotransferase (AST) levels and of developing NAFLD in a large sample of apparently healthy, nondiabetic adults (29). Table 2 describes some of the results of various interventions using physical exercise in NAFLD patients (29-43).

A recent study suggests that short-term aerobic training (AT) can help reduce the risk for NAFLD progression by targeting hepatic lipid composition; this effect appears to be mediated by adiponectin (38). Physical activity was also shown to have beneficial effects on intrahepatic triglyceride content, but not on the hepatic lipoprotein kinetics (35). Both AT and resistance training led to decreases in intrahepatic lipid levels, independent of weight loss (30,38,44,45). The combination of AT and resistance training generally led to better results than either intervention alone (32). Physical exercise was also associated with lower intrahepatic lipid levels independent of weight loss in a meta-analysis (46).

A multidisciplinary program of dietary and exercise advice for developing individualized goals was found to have a positive effect on stabilizing BMI, and in improving total and low-density lipoprotein cholesterol as well as improving serum alanine aminotransferase (ALT) and AST levels for up to one-year follow-up in a prospectively

TABLE 2
Physical activity

Study; sample population	Physical activity regimen	Outcome
Bae et al (29); 72,359 healthy Korean adults without diabetes	Physical exercise (30 min/day, 3 times/week) for 3 months (n=12,967) No exercise (n=59,392)	Physical exercise: lower odds of having elevated AST level (OR 0.85 [95% CI 0.74–0.99]) and ALT (OR 0.74 [95% CI 0.67–0.81])
Hallsworth et al (30); 19 sedentary adults with clinically defined NAFLD	RT for 8 weeks (n=11) Continued normal treatment (n=8)	RT: significant reduction in liver lipids (P<0.05), improvements in lipid oxidation, glucose control and HOMA-IR
Bhat et al (31); 42 NAFLD patients	Regular AT (30 min/day for at least 5 days/week)	AT: decreased insulin resistance, BMI, waist circumference and ALT levels (P<0.01 for all), and improved NASH scores
de Piano et al (32); 58 obese adolescents (28 with NAFLD)	Interdisciplinary weight-loss therapy for 1 year AT alone (n=29) AT + RT (n=29)	AT: reduced body mass, BMI and fat mass in non-NAFLD patients AT: reduced body mass, BMI, fat mass and visceral fat in NAFLD patients AT + RT: improvement in body mass, BMI, fat mass, glycemia, total cholesterol and low-density lipoprotein-cholesterol in non-NAFLD patients AT + RT: improvement in body mass, BMI, fat mass, glycemia, total cholesterol and low-density lipoprotein-cholesterol, and subcutaneous fat in NAFLD patients AT + RT: higher magnitude of changes in total cholesterol, low-density lipoprotein-cholesterol, ALT and adiponectin compared to AT alone in NAFLD
Fealy et al (33); 13 obese NAFLD patients	Treadmill walking for 60 min/day on 7 consecutive days	AT: decreased CK-18 (558.4±106.8 vs 323.4±72.5 U/L; P<0.01) and ALT levels (30.2±5.1 vs 24.3±4.8 U/L; P<0.05) AT: increased whole body fat oxidation (49.3±6.1 mg/min vs 69.4±7.1 mg/min; P<0.05) and circulating sFasL levels (66.5±6.0 pg/mL vs 63.0±5.7 pg/mL; P<0.06)
Grønbæk et al (34); 117 obese NAFLD children	Moderate exercise (1 h/day) with restricted energy intake for 10 weeks	Training: weight loss and improvement in ultrasonographic liver steatosis, liver fat content and insulin resistance
Sullivan et al (35); 18 obese NAFLD patients	Exercise training (0–60 min for 5 days/week) for 16 weeks (n=12) No intervention (n=6)	Exercise training: decreased intrahepatic triglyceride content (10.3±4.6%; P<0.05), with no influence on body weight, percent body fat, and very low density lipoprotein triglyceride and apolipoprotein B-100 secretion rates
Al-Jiffri et al (36); 100 type 2 diabetes male patients with NAFLD	Physical training (3 times/week for 12 weeks) combined with dietary measures Dietary measures only	Physical training + dietary measures: decreases in ALP (P=0.0015), ALT (P=0.0013), AST (p=0.0027), γ-GTP (p=0.0056), HOMA-IR (P=0.0083) and BMI (P=0.0094) Dietary measures only: no changes
Bacchi et al (37); 31 sedentary adults with NAFLD and type 2 diabetes	Physical exercise for 4 months AT (n=14) RT (n=17)	Both AT and RT equally effectively reduced hepatic fat content (P<0.001 vs baseline), led to disappearance of hepatic steatosis (defined as hepatic fat content >5.56%) in almost a quarter of patients, increased insulin sensitivity during euglycemic clamp, and decreased total body fat mass, visceral adipose tissue, superficial subcutaneous abdominal adipose tissue and hemoglobin A1c
Haus et al (38); 17 obese NAFLD patients	Short-term AT program (60 min/day of treadmill walking at 85% of maximal heart rate) for 7 consecutive days	AT increased the liver polyunsaturated lipid index (P<0.05), insulin sensitivity (P<0.05), high molecular weight adiponectin levels (P<0.05) and maximal oxygen consumption (P<0.05) AT reduced reactive oxygen species production during oral glucose tolerance test
Khaoshbaten et al (39); 90 NAFLD patients	Medical treatment after AT (30 min/day, 3 times/week for 3 months) (n=45) Medical treatment alone (n=45)	Medical treatment + AT: decreased AST (41.5±2.7 at baseline vs 29±9.5 IU/L at 3 months; P=0.006) and ALT (61.1±3.6 at baseline vs 44.9±2.4 IU/L at 3 months; P=0.01)
Malin et al (40); 13 obese NAFLD patients	AT for 7 days (60 min/day at 85% maximum heart rate)	AT: reduced insulin resistance (P<0.05) and circulating fetuin-A levels (P<0.02)
Oh et al (41); 212 obese, middle-age men (19.8% had abnormal liver function and suspicious liver fibrosis)	Exercise training program without any dietary restriction for 12 weeks (n=108) Dietary restriction program (n=104)	Both regimens reduced body weight, waist circumference and visceral adipose tissue area, serum ALT and γ-GTP levels, and insulin resistance; dietary restriction program superior Exercise training increased adiponectin levels Exercise training reduced serum levels of inflammation and oxidative stress markers such as ferritin and thiobarbituric acid reactive substances in subjects with suspected liver fibrosis
Oh et al (42); 169 obese NAFLD patients	MVPA weight reduction for 12 weeks <150 min/week (n=40) 150–250 min/week (n=42) ≥250 min/week (n=87)	The degree of hepatic steatosis decreased more significantly in the ≥250 min/week group compared to the <250 min/week groups ≥250 min/wk MVPA associated with more pronounced decreases in abdominal visceral adipose tissue, levels of ferritin and lipid peroxidation, along with a significant increase in adiponectin levels than <250 min/week MVPA
Zelber-Sagi et al (43); 64 NAFLD patients without secondary liver disease	RT 3 times/week for 3 months (n=33) Home stretching (n=31)	RT reduced hepatorenal-ultrasound index (P=0.017), total, trunk and android fat, serum ferritin and total cholesterol levels RT increased lean body mass No difference in AST, ALT and γ-GTP levels

ALP Alkaline phosphatase; ALT Alanine aminotransferase; AST Aspartate aminotransferase; AT Aerobic training; BMI Body mass index; CK-18 Cytokeratin-18; γ-GTP Gamma-glutamyl transpeptidase; HOMA-IR Homeostasis model assessment-insulin resistance; MVPA Moderate to vigorous intensity physical activity; NAFLD Nonalcoholic fatty liver disease; NASH Nonalcoholic steatohepatitis; RT Resistance training; sFas Soluble Fas; vs versus

tracked pediatric NAFLD population (47). Another multidisciplinary intervention consisting of personalized diet, physical exercise and behaviour therapy for three months led to a mean weight loss of 8%, improvement in liver function tests and a decrease in the liver fat content in 12 NAFLD patients included in an open-label nonrandomized study (48). Weight loss >10% was an additional predictor of disease remission in a sample of NAFLD patients undergoing an intervention program designed to increase energy expenditure and reduce caloric intake using lifestyle behavioural changes (49).

A recent study assessed the degree of weight loss necessary to improve the markers of hepatic function and insulin resistance in 100 type 2 diabetes male patients with NAFLD. Physical training combined with dietary measures led to improvements in liver function tests and insulin resistance, while no such changes were observed in the control group receiving dietary measures alone (36). The addition of AT to medical treatment led to an improvement in serum ALT and AST levels, along with an improvement in liver echogenicity compared with medical treatment alone (39).

The severity of NASH is inversely associated with the ability to oxidize fat (50). Exercise and diet modification were shown to reduce the degree of steatosis in nonobese NAFLD patients (51). NAFLD severity was correlated with intramuscular adipose tissue content in a cross-sectional study ($P<0.01$), suggesting that skeletal muscle fat accumulation may influence the pathogenesis and severity of NASH (52). Changes in intramuscular adipose tissue content were correlated with changes in weight ($P<0.05$), BMI ($P<0.05$), subcutaneous fat area ($P<0.01$), triglycerides ($P<0.05$), fasting plasma glucose ($P<0.01$), insulin ($P<0.05$), homeostasis model assessment-insulin resistance ($P<0.05$), the quantitative insulin sensitivity check index ($P<0.01$), and histopathological assessments such as changes in steatosis and lobular inflammation ($P<0.01$ for both) (52). Improved hepatic steatosis with reduced insulin resistance, ALT levels and serum interleukin 6 (IL-6) levels were associated with voluntarily or electrically contracted quadriceps and hamstrings, independent of changes in muscle mass (53). A short-term AT program decreased serum markers of hepatic apoptosis in a small sample of obese NAFLD patients. Fealy et al (33) proposed that changes in the proapoptotic environment following short-term exercise are mediated by improved insulin sensitivity and increased oxidative capacity.

Fetuin-A is a liver protein believed to be associated with NAFLD and type 2 diabetes. Seven days of AT reduced insulin resistance ($P<0.05$) and circulating fetuin-A levels ($P<0.02$) in obese adults with clinically diagnosed NAFLD, with a correlation between these two parameters ($P<0.04$). These occurred independent of changes in body weight. This study suggests that improvements in glucose tolerance in patients with NAFLD after physical exercise may be mediated by lowering fetuin-A levels, particularly in skeletal muscles (40).

ADIPOKINES AS BIOMARKERS OF NASH AND OBESITY IN ADOLESCENTS

Noninvasive routine laboratory tests in NAFLD patients have been described in recent reviews (54,55) and a large study (56). In the present work, we focus on serum adipokines levels. Adipokines, such as adiponenctin, leptin and ghrelin, were measured in samples of obese children and adolescents, with or without NAFLD. These findings are presented in Table 3 (32,57-61).

The main adipokines studied in NAFLD are adiponectin, leptin and ghrelin. Differences in adiponectin levels between controls and NAFLD patients are generally small. Among NAFLD patients, adiponectin levels are generally decreased in NASH compared with simple steatosis, such that decreased adiponectin levels can be used to predict NAFLD progression (62). On the other hand, leptin levels are higher in NAFLD patients compared with controls (57). Blood leptin levels reflect total body fat, and correlate with fibrosis and insulin resistance (63).

Adiponectin, resistin and retinol-binding protein-4 could be used to differentiate between steatosis with elevated serum ALT levels and nonsteatotic obese patients. Adiponectin and resistin were significantly lower and retinol-binding protein-4 was significantly higher in obese children with advanced steatosis compared with obese children without liver steatosis. Leptin was not a good predictor of liver steatosis or hepatopathic obesity (57).

Adiponectin levels were negatively correlated with the NAFLD activity score in a recent meta-analysis (64). In this study, low adiponectin levels were associated with NAFLD progression to NASH, with nonsignificant differences in adiponectin levels between healthy controls and patients with simple steatosis. In contrast, leptin levels were elevated in NAFLD patients compared with nonsteatotic controls in a sample of obese children (57). Elevated leptin levels could further be used to predict significant fibrosis ($F\geq2$), and classify NAFLD patients according to the level of fibrosis (65).

The relationship between interdisciplinary therapy consisting of various forms of physical exercise, clinical, nutritional and psychological interventions, and changes in serum adipokines levels was assessed in samples of obese adolescents. Improvements in the levels of noninvasive biomarkers and clinical characteristics of disease progression were observed, particularly among patients with NAFLD receiving multidisciplinary intervention (32,61).

USE OF INTRAGASTRIC BALLOON AND CHANGES IN ADIPOKINE LEVELS

Combining the use of intragastric balloon with physical exercise led to more substantial weight loss, especially fat mass, in obese patients (66). Aside from weight loss, use of intragastric balloon (Bioenteric, Inamed Health, USA) led to improvement in the metabolic syndrome parameters such as diabetes mellitus, hypertension, dyslipidemia and fatty liver in compliant patients (67). After six months of intragastric balloon placement, 77.5% of subjects no longer met the diagnostic criteria for the metabolic syndrome in a sample of 40 previously overweight or obese patients with the metabolic syndrome. This procedure further reduced the percentage of truncal, android, gynoid and total fat ($P=0.0001$), and improved lung function parameters (68). In a small sample of obese patients treated with the BioEnterics intragastric balloon, the amount of weight loss experienced during the first month may be an important predictor of weight loss up to one year after balloon removal (18 months since baseline) (69). Table 4 describes the consequence of intragastric balloon placement in obese patients with NAFLD (70-73). Compared with baseline, anthropometric and biochemical measurements were improved at the time of balloon removal.

Studies show that weight loss is associated with histological improvements in obese NAFLD patients. The BioEnterics intragastric balloon has shown promising results in terms of weight loss in morbidly obese individuals and in individuals with treatment-resistant obesity. The balloon is filled with fluid to provide a sensation of satiety, thus reducing food ingestion during the time the balloon is used. Furthermore, it was conjectured that the short-term presence of the intragastric balloon would create behavioural routine whereby lower food amounts are consumed even after the balloon is removed, usually within six months of placement (70).

BioEnterics intragastric balloon placement was associated with improved eating patterns (74). Treatment with intragastric balloon for six months followed by 12 months of behavioural modification was associated with more pronounced weight loss and reversal of the metabolic syndrome compared with 12 months of behavioural modification alone in a small sample of obese adults (75). A reduction in hunger and an increase in sense of satiety was observed among nine obese children using a gastric balloon system (Obalon, Obalon Therapeutics, USA) for three months (73).

Leptin

Treatment with intragastric balloon placement, low-calorie diet (1500 kcal/day) and physical exercise was associated with decreased leptin levels, which correlated with weight loss (76). There were significant changes in leptin levels in both patient groups throughout the 12-month study period in a sample of 43 obese Caucasian patients

TABLE 3
Adipokine levels in children and adolescents

Study and sample population	Changes in adipokines	Associations and diagnostic performance
de Piano et al (32) 58 postpubertal obese adolescents randomly assigned to AT or AT + RT AT group: 15 patients without NAFLD and 14 with NAFLD AT + RT group: 15 patients without NAFLD and 14 with NAFLD	Change in adiponectin at 1 year: −0.13±1.73 in AT without NAFLD, −0.13±2.78 in AT with NAFLD, 2.57±2.50 in AT + RT without NAFLD (P<0.05 vs AT), 2.69±2.54 μg/mL in AT + RT with NAFLD (P<0.05 vs AT) Change in leptin at 1 year: −7.13±12.68 in AT without NAFLD, −12.17±17.22 in AT with NAFLD, −9.19±13.61 in AT + RT without NAFLD (P<0.05 vs AT), −11.95±13.08 μg/mL in AT + RT with NAFLD Change in adiponectin/leptin at 1 year: 0.72±1.56 in AT without NAFLD, 1.02±1.19 in AT with NAFLD, 0.68±1.30 in AT + RT without NAFLD, 0.40±1.85 in AT + RT with NAFLD	n/a
Boyraz et al (57) Group I: 63 obese children with liver steatosis Group II: 12 obese children with elevated serum ALT activity from group I Group III: 85 obese children without liver steatosis	Mean adiponectin: 2.7±0.7 in group I, 2.5±0.4 in group II and 4.7±1.1 μg/mL in group III (P<0.001 vs group I and P<0.001 vs group II) Mean resistin: 8.5±3.2 in group I, 8.5±3.2 in group II and 15.0±3.9 ng/mL in group III (P<0.001 vs group I and P<0.001 vs group II) Mean RBP4: 33.2±7.5 in group I, 35.4±3.3 in group II and 35.4±3.3 μg/mL in group III (P<0.001 vs group I and P<0.001 vs group II) Mean leptin: 27.4±11.9 in group I, 27.4±11.9 in group II and 27.4±11.9 μg/mL in group III (P=0.560 vs group I and P=0.681 vs group II)	Adiponectin: sensitivity of 84.21% and specificity of 63.64% for advanced liver steatosis at cut-off 2.56 μg/mL Resistin: sensitivity of 36.8% and specificity of 95.5% for advanced liver steatosis at cut-off 5.2 ng/mL RBP4: sensitivity of 84.20% and a specificity of 68.20% for advanced liver steatosis at cut-off 35 μg/mL Adiponectin: sensitivity of 100% and specificity of 83.53% to differentiate hepatopathic obese children at cut-off 3.2 μg/mL Resistin: sensitivity of 100% and specificity of 77.65% to differentiate hepatopathic obese children at cut-off 12.0 ng/mL RBP4: sensitivity of 100% and a specificity of 92.94% to differentiate hepatopathic obese children at cut-off 26 μg/mL
Koot et al (58) 119 severely obese children (47% prevalence of steatosis)	Mean adiponectin: 7.8±2.8 μg/mL at baseline Mean leptin: 33.9±6 32.4 at baseline	Adiponectin not associated with steatosis in univariate analysis (OR 0.96 [95% CI 0.84–1.09]; P=0.52) Leptin associated with steatosis in multivariate analysis (OR 1.04 [95% CI 1.01–1.09]; P=0.03)
Pacifico et al (59) 44 obese children with NAFLD 44 obese children without NAFLD	Leptin: mean 19.5 in NAFLD vs 20.8 μg/L in non-NAFLD Adiponectin: mean 9.0 in NAFLD vs 12.9 μg/L in non-NAFLD (P<0.05)	
Klein et al (60) 106 middle school students of varying body mass index	Mean adiponectin: 12.2±4.9 μg/mL Mean adiponectin: lower in males 10±4.2 vs females 12.8±4.1 μg/mL (P<0.05)	
Sanches et al (61) 79 obese adolescents (33 with NAFLD and 46 without NAFLD) Interdisciplinary (clinical, nutritional and psychological) therapy, exercise (60 min/day for 3 days/week) and physiotherapy for 1 year	Change in adiponectin at 1 year: 1.61±3.09 in non-NAFLD and 1.66±2.09 μg/mL in NAFLD Change in leptin at 1 year: mean −15.06 (P<0.05 vs baseline) in non-NAFLD and mean −16.17 μg/mL in NAFLD (P<0.05 vs baseline) Change in leptin/adiponectin ratio at 1 year: mean −3.36 (P<0.05 vs baseline) in non-NAFLD and mean −3.49 in NAFLD (P<0.05 vs baseline) Change in PAI-1 at 1 year: −3.58±5.81 (P<0.05 vs baseline) in non-NAFLD and −3.85±6.50 ng/mL in NAFLD (P<0.05 vs baseline)	

AT Aerobic training; ALT Alanine aminotransferase; n/a Not applicable; NAFLD Nonalcoholic fatty liver disease; PAI-1 Plasminogen activator inhibitor-1; RBP4 Retinol-binding protein-4; RT Resistance training; vs Versus

(divided into morbidly and nonmorbidly obese based on a BMI cut-off of 40 kg/m^2) (77). Throughout the study period, leptin levels remained below baseline values. Leptin levels decreased beginning at month 1 (mean 21.1 ng/mL versus mean 28.7 ng/mL at baseline), with slight oscillations thereafter (mean 17.4 ng/mL to 19.0 ng/mL) among morbidly obese individuals (BMI ≥40 kg/m^2). Leptin levels were lower at months 1 and 6 (mean 11.8 ng/mL and 10.5 ng/mL, respectively, versus mean 25.1 ng/mL at baseline) among obese individuals (BMI <40 kg/m^2), with a slight rise approaching month 12 (mean 17.5 ng/mL) (77). Leptin significantly decreased during the first month in another small sample of obese patients undergoing intragastric balloon placement (median 67.1 ng/dL at baseline, 53.7 at month 1, 39.3 at month 3 and 26.5 ng/dL at month 6; P=0.0353) (78).

Leptin levels decreased throughout the study in the balloon group (P=0.05 at month 6 and P=0.04 at month 10 versus baseline) and remained relatively stable in the control group among obese patients treated with the BioEnterics intragastric balloon, low-calorie diet (1500 kcal/day) and physical exercise, or low-calorie diet and physical exercise alone (76). Leptin levels decreased throughout the study period in a small sample of nonmorbidly obese patients with intragastric balloon (27.8±3.7 ng/mL at baseline versus 18.7±2.7 ng/mL at month 6; P=0.024) (79). Leptin was significantly decreased (30.4±17.2 μg/L versus 14.9±15.5 μg/L; P<0.001) after six months in 22 obese individuals with the balloon (80).

Ghrelin

Ghrelin is a peptide hormone that stimulates food intake. Abnormalities in meal-related peptides, such as ghrelin, are associated with binge eating disorders (81). In normal individuals, ghrelin levels rise before meals and fall following meals. However, in obese individuals, ghrelin

TABLE 4
Intragastric balloon placement

Study and sample population	Association with nonalcohoilic fatty liver disease
Lee et al (70) 18 obese NASH patients Bioenterics* intragastric balloon + exercise + diet for 6 months (n=8) Sham balloon (500 mL of saline solution) + exercise + diet (n=10)	Balloon: reductions in mean body mass index (1.52 kg/m^2 vs 0.8 kg/m^2; P=0.0008) and median NAFLD activity scores (2 vs 4; P=0.0.03), with a trend toward improved steatosis scores (P=0.075) compared with sham No differences with respect to lobular inflammation, hepatocellular ballooning or fibrosis scores No significant changes in AST or ALT levels
Tai et al (71) 28 obese patients Bioenterics* intragastric balloon placement for 6 monhts	Improvement in body mass index (mean 32.4±3.7 kg/m^2 vs 28.5±3.7 kg/m^2; P<0.01), waist circumference (mean 101.9±8.9 cm vs 90.6±9.3 cm; P<0.01), glucose (median 93.5 mg/dL vs 91.0 mg/dL; P<0.01), AST level (median 33.0 IU/L vs 23.0 IU/L; P<0.01), ALT level (median 49.0 IU/L vs 22.0 IU/L; P<0.01), triglyceride (median 149.0 mg/dL vs 88.5 mg/dL; P<0.01), cholesterol (median 200.0 vs 186 mg/dL; P=0.13), high-density lipoprotein-cholesterol (median 45.0 vs 52.0 mg/dL; P=0.01), low-density lipoprotein-cholesterol (median 119.5 vs 114.0 mg/dL; P=0.03), the metabolic syndrome (64.3% vs 32.1%; P=0.01) at 6 months compared with baseline
Folini et al (72) 40 obese subjects Intragastric balloon or gastric banding for 6 months (n=24) Diet for 6 months (n=16)	Intragastric balloon or gastric banding: lower ALT (P=0.02), AST (P=0.03), total cholesterol (P=0.007), low-density lipoprotein-cholesterol (P=0.03) and fat-free mass (P=0.01) at the end of the study compared with diet
Nobili et al (73) 9 obese children Obalon† balloon for 3 months	Significant improvements in ALT, total and low-density lipoprotein cholesterol, insulin, HOMA-IR and uric acid levels No significant changes in ultrasonographic examination

**Inamed Health, USA; †Obalon Therapeutics, USA. ALT Alanine aminotransferase; AST Aspartate aminotransferase; HOMA-IR Homeostasis model assessment-insulin resistance; NAFLD Nonalcoholic fatty liver disease; NASH Nonalcoholic steatohepatitis; vs Versus*

levels are lower than in healthy individuals, a phenomenon believed to be the result rather than the cause of overeating (81). Obesity was shown to actually lead to lower ghrelin levels. Ghrelin significantly increased (240.5±101.5 µg/L versus 335.8±149.2 µg/L; P<0.002) after six months in 22 obese individuals with intragastric balloon treatment (80). Among obese individuals (BMI <40 kg/m^2), ghrelin levels peaked around month 3 (mean 1346.2 pg/mL versus mean 958.3 pg/mL at baseline), and then decreased steadily to near baseline levels by month 12 (mean 922.6 pg/mL). Ghrelin levels remained relatively stable through the first six months of the study among morbidly obese individuals (BMI ≥40 kg/m^2), with a drop toward month 12 (mean 742.6 pg/mL versus mean 948.3 pg/mL at baseline) (77).

Blood ghrelin levels were measured every 15 min for 1 h after breakfast at the start of treatment, after 13 weeks and after 26 weeks in a sample of 40 obese patients randomly assigned to a 13-week period of sham or balloon, followed by 13 weeks of balloon. There were no differences in plasma ghrelin levels (either fasting or trough levels) between groups. Ghrelin levels did not change from baseline to end of treatment (mean fasting 725.9±152.2 pg/mL and trough 596.0±105.3 pg/mL at baseline versus mean fasting 773.4±113.8 pg/mL and trough 669.3±128.4 pg/mL at six months) despite weight loss (82).

Weight loss was more pronounced in the balloon group at six months (17.1±8.0 kg versus 3.2±6.4 kg) in a sample of 21 obese subjects who were treated with a BioEnterics intragastric balloon, low-calorie diet (1500 kcal) and physical exercise, and 15 controls treated with low-calorie diet and physical exercise alone. In the balloon group, ghrelin levels increased at one month (P=0.006 versus baseline) and slowly decreased thereafter until they approached baseline levels at three months after balloon removal. Ghrelin levels were relatively stable throughout the study in the control group. Based on these findings, balloon treatment is associated with transiently elevated ghrelin levels in obese patients (76).

Weight reduction was correlated with plasma ghrelin levels in 17 nonmorbidly obese patients. Plasma ghrelin levels decreased throughout the study (3.2±0.4 ng/mL at baseline versus 1.9±0.1 ng/mL at month 6; P=0.021). In this study, weight loss associated with intragastric balloon correlated with plasma ghrelin variations (79). Ghrelin was significantly increased (240.5±101.5 µg/L versus 335.3±149.2 µg/L; P<0.002) after six months in 22 obese individuals with intragastric balloon placement (80).

Weight loss occurred in both groups in a randomized, double-blinded, sham-controlled trial of four months' duration in morbidly obese patients treated with either intragastric balloon placement or sham operation. However, there were no significant differences between the balloon group and the control group in terms of degree of weight loss. These changes were not mediated by changes in plasma ghrelin levels, because ghrelin levels did not fluctuate between fasting and postprandial conditions, and between the two treatments (mean fasting ghrelin 934.4±199.2 pg/mL at baseline versus 947.1±195.1 pg/mL at day 30 in the balloon group and mean fasting ghrelin 970.1±125.2 pg/mL at baseline versus 962.0±93.9 pg/mL at day 30 in the control group) (83).

Adiponectin

Adiponectin levels remained unchanged throughout the study in a sample of obese patients treated with intragastric balloon, low-calorie diet (1500 kcal) and physical exercise (76). Adiponectin levels showed no significant difference in other small samples of obese patients undergoing intragastric balloon placement (78,80). A transient increase in adiponectin levels was observed in obese controls treated with low-calorie diet (1500 kcal) and physical exercise without intragastric balloon placement (P=0.045 at six months versus baseline) (76). Adiponectin levels did not change significantly in another study (78). In contrast, Mion et al (79) found increasing adiponectin levels (6.6±0.5 ng/mL at baseline versus 7.8±0.8 ng/mL at six months; P=0.037).

BARIATRIC SURGERY AND CHANGES IN CIRCULATING LIVER ENZYME LEVELS, INFLAMMATORY MARKERS AND ADIPOKINE LEVELS

An additional nonmedicinal intervention that can be used in NAFLD patients is bariatric surgery. Bariatric surgery is believed to ameliorate some of the abnormalities associated with NAFLD and, thus, lead to an improvement in NAFLD. This is believed to involve improvements in circulating liver enzyme levels, decreases in inflammatory marker levels, and changes in adipokines levels such as increases in adiponectin and decreases in leptin (84,85). Improvements in steatosis, lobular inflammation, chronic portal inflammation and steatohepatitis were noted in a sample of NAFLD patients in which liver biopsy were analyzed pre- and postbariatric surgery (86). Roux-en-Y gastric bypass surgery was shown to decrease the metabolic syndrome parameters, as well as the prevalence of comorbidities associated with obesity, including fatty liver and steatosis. This effect was observed in both sexes and among all individuals 18 to 65 years of age (87). The effects of bariatric surgery in obese populations, with or without NAFLD/NASH, are described in Table 5 (88-97).

Liver enzymes

The effects of bariatric surgery on circulating levels of liver enzymes were assessed in several studies (88-93). ALT, AST and gamma-glutamyl transpeptidase (γ-GTP) levels are generally lower at follow-up (six to 12 months after the surgical intervention) compared with baseline levels (ie, before surgery) in samples of obese patients (90,93,98). Lower serum AST and ALT levels were maintained at two and 10 years follow-up in a large sample of obese subjects who underwent bariatric surgery. ALT reduction was proportional to weight loss (99). In another study, bariatric surgery led to decreased γ-GTP levels at follow-up, but had no effects on ALT and AST levels (88). Elsewhere, near normalization of ALT was observed in a sample of morbidly obese patients undergoing bariatric surgery, while mean serum γ-GTP levels decreased to levels below those of controls. No significant differences were observed in AST levels between controls and NAFLD patients, and between levels before and after bariatric surgery within the NAFLD subsample (91,92).

Roux-en-Y gastric bypass was associated with improved NAFLD parameters, and these effects were maintained for at least five years. Decreased AST levels following Roux-en-Y gastric bypass were observed in association with a decrease in NAFLD fibrosis score (1.142±1.261 at baseline versus 0.066±1.027 at 12 months; P=0.0394) in a large sample of 1236 obese patients (98). Bariatric surgery produced significant histopathological improvements in terms of steatosis (P<0.001), ballooning degeneration (P<0.001), Mallory-Denk hyaline bodies (P=0.005), glycogen nuclei (P=0.001), lobular inflammation (P<0.001), portal inflammation (P=0.005) and fibrosis (P<0.001) (88).

Adipokines

In general, serum adiponectin levels are lower (2.2±1.7 μg/mL in patients versus 4.1±2.5 μg/mL in control; P<0.001) and serum leptin levels are elevated (28.1±18.4 ng/mL in patients versus 4.7±3.9 ng/mL in control; P<0.001) in obese patients compared with nonobese controls (100). Bariatric surgery increases adiponectin levels and decreases leptin levels compared with presurgery levels. For example, mean serum leptin level was elevated, while mean serum adiponectin level was lower in a small sample of obese patients undergoing laparoscopic sleeve gastrectomy compared with nonobese controls. Leptin levels decreased, while adiponectin levels increased within one month following this intervention. These changes were maintained six months after the procedure (100). Roux-en-Y gastric bypass led to decreasing leptin levels in a small sample of morbidly obese patients (91). Adiponectin levels increased and leptin levels decreased in a time-dependent manner through month 12 in a sample of obese patients undergoing laparoscopic adjustable gastric banding (93). In another study, normalization of both adiponectin and leptin levels to values similar to those observed in controls were observed in morbidly obese patients following bariatric surgery (92).

Decreased adiponectin levels at baseline were observed in other samples of obese patients compared with nonobese controls. Following bariatric surgery, adiponectin levels increased in a time-dependent manner (90,94,96,97,101). Interestingly, no further increases were observed beyond four weeks in a sample of 108 morbidly obese NAFLD patients undergoing bariatric surgery. However, adiponectin levels continued to increase through six months in NASH patients (102).

Bechmann et al (64) reported that obesity is negatively correlated with adiponectin levels, while adiponectin levels are inversely correlated with the NAFLD activity score. Low adiponectin levels (<13.5 mg/L in one study and <23 mg/L in another study) were a strong predictor of NAFLD severity elsewhere (95,103). Adiponectin levels were lower in NASH compared with simple steatosis. Adiponectin levels further decreased progressively with increasing steatosis severity and with more severe lobular inflammatory activity (95). Lower adiponectin levels were also found in patients with fibrosis compared with those without fibrosis (101).

Leptin levels increased in parallel with increasing steatosis severity and with the severity of lobular inflammatory activity. A trend for higher leptin levels in patients with more advanced fibrosis was also noted (95). Leptin levels did not correlate with fibrosis in another study (101).

Other adipokines measured in NAFLD patients include ghrelin, visfatin and resistin. Ghrelin levels were correlated with lobular inflammation but not with steatosis severity, and were an independent predictor of NASH (95,104). Ghrelin levels decreased significantly in a sample of obese patients who underwent laparoscopic sleeve gastrectomy and dieatry changes (127.5±96.9 ng/L at baseline, 62.6±60.9 ng/L at six months and 87.5±59.1 ng/L at 12 months) (105). Mean serum visfatin levels decreased at six weeks' follow-up in a small sample of severely obese patients undergoing gastric bypass (94). Serum resistin levels were elevated at six months, but decreased to below baseline levels by 12 months after laparoscopic adjustable gastric banding (93).

Inflammatory markers

Elevated mean blood high-sensitivity CRP and interleukin (IL)-6 levels were present at baseline among obese patients versus lean controls (CRP: 26.19±23.17 g/L versus 1.71±2.27 g/L, respectively; P<0.001, and IL-6: 3.84±1.67 pg/mL versus 0.92±0.47 pg/mL, respectively; P<0.001). IL-6 and high-sensitivity CRP levels decreased by three months after bariatric surgery (97). Significant decreases in CRP levels were observed after six months in a sample of obese patients undergoing laparoscopic adjustable gastric banding. These differences were even more pronounced at 12 months (93). Serum IL-6 levels decreased within six months of bariatric surgery in a small sample of severely obese patients. Similarly, hepatic IL-6 messenger RNA expression was decreased at six months (106). IL-18, soluble tumour necrosis factor (TNF) receptor 2 and CRP levels decreased at 12 months' follow-up compared with baseline in a small sample of morbidly obese subjects undergoing gastric bypass (96).

Elevated mean TNF-α levels at baseline were observed in a sample of morbidly obese patients compared with lean controls (2.10±1.86 pg/mL versus 0.86±0.67 pg/mL; P<0.001). TNF-α levels initially increased at three months, but stabilized to near-baseline levels by month 12 after bariatric surgery in one study (97). Significant decreases in TNF-α were observed after six months elsewhere (93). Serum TNF-α levels were undetectable in another study, while hepatic TNF-α messenger RNA expression did not change (106).

FRUCTOSE

Fructose is found in a wide variety of processed foods and beverages (107). A relationship is believed to exist between high fructose consumption and NAFLD development and progression. While providing fructose as a dietary supplement for seven days in children of type 2 diabetes patients or controls, Lê et al (108) observed that a diet high in fructose is associated with dyslipidemia, especially in individuals with a family history of type 2 diabetes. A high-fructose diet significantly increased intrahepatocellular lipids, intramyocellular lipids, very low-density lipoprotein-triacylglycerols and fasting hepatic glucose output in both groups. However, higher intrahepatocellular lipids and total triacylglycerols, and lower whole-body insulin sensitivity occurred in children with a family history of type 2 diabetes (108). High acute fructose consumption from soft drinks further led to elevated endotoxin levels, which are associated with the presence of hepatic steatosis. High chronic fructose consumption was associated with persistently elevated endotoxin levels, suggesting that the association between fructose consumption and liver steatosis is mediated, at least in part, by endotoxin (109).

TABLE 5
Circulating liver enzyme levels, adipokines and inflammatory markers in patients undergoing bariatric surgery

Study; sample population	Changes following bariatric surgery
Vargas et al (88); 26 morbidly obese patients (96.1% had NASH) undergoing Roux-en-Y gastric bypass with a modified Fobi-Capella technique (follow-up after 16.3±3 months)	Mean (± SD) serum AST: 0.35±0.09 μkat/L during presurgery vs 0.36±0.17 μkat/L after surgery (P=0.862) Mean serum ALT: 0.49±0.20 μkat/L during presurgery vs 0.37±0.31 μkat/L after surgery (P=0.143) Mean serum γ-GTP: 40.2±17.4 during presurgery vs 19.2±12.8 IU/L after surgery (P<0.001)
Cazzo et al (89); 63 obese subjects undergoing Roux-en-Y gastric bypass surgery	Mean (± SD) serum ALT: 30.7±17.1 mg/dL at baseline vs 20.7±7.2 mg/dL at month 12 (P<0.0001) Mean serum AST: 25.7±10.6 mg/dL at baseline vs 21.2±5.4 mg/dL at month 12 (P=0.0005)
Carazo et al (90); 60 morbidly obese patients undergoing bariatric surgery	Mean (± SD) plasma AST: 29.8±3.3 IU/L before surgery vs 20.4±1.0 IU/L at month 12 (P=0.0070) Mean plasma ALT: 31.0±3.1 IU/L before surgery vs 19.3±1.5 IU/L at month 12 (P=0.001) Mean plasma γ-GTP: 37.7±8.0 IU/L before surgery vs 17.5±2.1 IU/L at month 12 (P<0.001) Mean plasma adiponectin: 39.9±6.0 μg/mL before surgery vs 60.0±2.1 μg/mL at month 12 (P=0.002)
Tai et al (91); 21 morbidly obese patients undergoing Roux-en-Y gastric bypass	Median serum AST: 27.0 IU/L before surgery vs 27.0 IU/L at month 12 (P=0.66) Median serum ALT: 34.0 IU/L before surgery vs 24.0 IU/L at month 12 (P<0.01) Median serum γ-GTP: 28.0 IU/L before surgery vs 12.0 IU/L at month 12 (P<0.01) Median serum leptin: 29.0 μg/L before surgery vs 4.1 μg/L at month 12 (P<0.01)
Felipo et al (92); 47 mordibly obese patients undergoing bariatric surgery (evaluated before and 18±5 months after surgery) 45 controls	Mean (± SD) serum AST: 20±4.0 IU/L in controls, 23.5±15 IU/L in simple steatosis patients before surgery, 23±95 IU/L in simple steatosis patients after surgery, 24.5±16 IU/L in NASH patients before surgery and 20.5±6 IU/L in NASH patients after surgery Mean serum ALT: 17±6.0 IU/L in controls, 31±17 IU/L in simple steatosis patients before surgery (P<0.05 vs control) , 23±17 IU/L in simple steatosis patients after surgery, 37±16 IU/L in NASH patients before surgery (P<0.05 vs control) and 20±12 IU/L in NASH patients after surgery Mean serum γ-GTP: 40±10 IU/L in controls, 39±26 IU/L in simple steatosis patients before surgery, 19±15 IU/L in simple steatosis patients after surgery (P<0.05 vs before surgery), 45±19 IU/L in NASH patients before surgery and 23±11 IU/L in NASH patients after surgery (P<0.05 vs before surgery) Mean serum adiponectin: 11±4 i μg/mL n controls, 7±2 μg/mL in simple steatosis patients before surgery (P<0.05 vs control), 16±10 μg/mL in simple steatosis patients after surgery (P<0.001 vs before surgery), 6±2 μg/mL in NASH patients before surgery (P<0.05 vs control) and 12±7 μg/mL in NASH patients after surgery (P=0.008 vs before surgery) Mean serum leptin: 13±9 ng/mL in controls, 69±25 ng/mL in simple steatosis patients before surgery (P<0.001 vs control), 20±18 ng/mL in simple steatosis patients after surgery (P<0.001 vs before surgery), 54±22 ng/mL in NASH patients before surgery (P<0.001 vs control) and 24±11 ng/mL in NASH patients after surgery (P<0.001 vs before surgery)
Moschen et al (93); 30 severely obese patients undergoing laparoscopic adjustable gastric banding	Mean (± SD) serum AST: 30.1±2.6 IU/L at baseline, 26.8±2.4 IU/L at month 6 (P<0.05) and 23.9±1.2 IU/L at month 12 (P<0.05) Mean serum ALT: 35.1±5.6 IU/L at baseline, 25.4±3.5 IU/L at month 6 (P<0.05) and 20.8±1.6 IU/L at month 12 (P<0.01) Mean serum γ-GTP: 34.5±4.0 IU/L at baseline, 26.1±2.4 IU/L at month 6 (P<0.05) and 21.8±3.2 IU/L at month 12 (P<0.001) Mean serum adiponectin: 7.46±0.67 ng/mL at baseline, 8.65±0.84 ng/mL at month 6 (P<0.05) and 8.95±0.81 ng/mL at month 12 (P<0.01) Mean serum leptin: 27.4±1.38 μg/mL at baseline, 18.3±1.58 μg/mL at month 6 (P<0.01) and 15.15±1.50 μg/mL at month 12 (P<0.001) Mean serum resistin: 4.26±0.21 pg/mL at baseline, 5.18±0.26 pg/mL at month 6 (P<0.05) and 3.37±0.19 pg/mL at month 12 (P<0.01) Mean serum CRP: 0.86±0.08 at baseline, 0.63±0.06 at month 6 (P<0.01) and 0.42±0.05 mg/dL at month 12 (P<0.001) Mean serum TNF-α: 2.36±0.08 pg/mL at baseline, 1.77±0.06 pg/mL at month 6 (P<0.05) and 0.8±0.03 pg/mL at 12 (P<0.01)
Hosseinzadeh-Attar et al (94); 35 severely obese patients undergoing gastric bypass	Mean (± SD) serum adiponectin: 36.5±11 ng/mL at baseline vs 41.3±11 ng/mL at week 6 (P<0.01) Mean serum visfatin: 5±3.5 ng/mL at baseline vs 3.4±3.2 ng/mL at week 6 (P<0.01)
Machado et al (95); 82 morbidly obese individuals with biopsy-proven NAFLD (13.4% had NASH) undergoing bariatric surgery	Mean (± SD) serum adiponectin: 22.9±9.8 ng/mL overall, 16.2±6.6 ng/mL in NASH and 24.1±9.9 ng/mL in no NASH (P=0.014 vs NASH) Mean serum leptin: 20.1±6.9 ng/mL overall, 22.2±6.8 ng/mL in NASH and 19.1±6.9 ng/mL in no NASH (P=0.291 vs NASH) Mean serum ghrelin: 19.9±11.5 pg/mL overall, 17.9±7.0 pg/mL in NASH and 20.2±12.3 pg/mL in no NASH (P=0.544 vs NASH)
Vilarrasa et al (96); 65 morbidly obese subjects undergoing gastric bypass	Median serum adiponectin: 11.5 μg/mL at baseline vs 23.4 μg/mL at month 12 (P<0.01) Median serum IL-18: 229.8 pg/mL at baseline vs 168.9 pg/mL at month 12 (P<0.01) Median serum sTNFR1: 2.23 ng/mL at baseline vs 2.27 ng/mL at month 12 Median serum sTNFR2: 5.07 pg/mL at baseline vs 4.32 pg/mL at month 12 (P<0.01) Median serum CRP: 7.9 mg/L at baseline vs 0.91 mg/L at month 12 (P<0.01)
Illán-Gómez et al (97); 60 morbidly obese women undergoing gastric bypass 30 lean controls	Mean (± SD) blood adiponectin: 5.82±2.93 μg/mL at baseline vs 7.64±3.74 μg/mL at month 3 (P<0.001) Mean blood IL-6: 3.84±1.67 pg/mL at baseline vs 3.36±1.53 pg/mL at month 3 (P<0.01) Mean blood hsCRP: 26.19±23.17 g/L at baseline vs 13.15±17.95 g/L at month 3 (P<0.001)

ALT Alanine aminotransferase; AST Aspartate aminotransferase; CRP C-reactive protein; γ-GTP Gamma-glutamyl transpeptidase; hsCRP High-sensitivity CRP; IL Interleukin; NAFLD Nonalcoholic fatty liver disease; NASH Nonalcoholic steatohepatitis; sTNFR Soluble tumour necrosis factor receptor; vs Versus

The association between elevated fructose consumption and NAFLD risk appears to be influenced by the actual amount of fructose consumed. Higher fructose intake (highest intake quartile: 29.2 g/day to 88.0 g/day) was not associated with NAFLD in an older Finnish population compared with lower fructose intake (lowest intake quartile: 2.2 g/day to 15.2 g/day). A possible explanation for these findings is that the levels of fructose intake were similar to the average population levels (110).

Because lifestyle interventions such as physical activity and dietary modifications represent an important first line of treatment in NAFLD patients, limiting fructose intake is an important area of research (111). Decreased fructose intake led to decreased intrahepatic fat content in a small sample of NAFLD patients at six months. However, this was coupled with reduced intake of glucose and sucrose, as well as reduced overall carbohydrate intake and energy consumption, such that no definitive conclusion can be drawn with regards to decreasing fructose alone (112).

In a recent study, consuming isocaloric diets with high fructose or high glucose content did not cause significant changes in the hepatic concentration of triacylglycerols or the serum levels of liver enzymes in a sample of healthy overweight men, with no differences between treatments. However, when high fructose or high glucose was administered as part of a hypercaloric diet, significant increases in these parameters occurred. This study suggests that overnutrition is associated with NAFLD risk factors, and not the levels of specific macronutrients (113). These findings are corroborated by a recent systematic review and meta-analysis (114) that showed no clear evidence that markers of hepatotoxicity are associated with excessive fructose intake per se, but rather with excessive energy intake. A separate systematic review and meta-analysis in controlled feeding trials shows that deriving a higher percentage of energy from fructose is not associated with a higher NAFLD risk in healthy controls. High fructose consumption coupled with excess energy intake is associated with elevated intrahepatocellular lipids and ALT levels, again suggesting that it is overnutrition rather than overconsumption of fructose that predisposes healthy individuals to NAFLD (115). Because restricting caloric intake, as a whole, is one of the main interventions aimed at weight loss in NAFLD patients, it is unclear what benefit, if any, would lower fructose intake have in the absence of lower overall energy intake (116).

Obesity was a risk factor for NAFLD at 17 years of age in a large sample of adolescents followed since birth. A higher energy-adjusted fructose intake at 14 years of age was associated with a higher risk of NAFLD at 17 years of age among obese adolescents, and this association was maintained after adjusting for confounding variables. This study shows that fructose rather than total sugar consumption is a risk factor for developing NAFLD in obese adolescents (117). Decreasing fructose intake reduced markers of liver dysfunction in a small sample of NAFLD children and adolescents (118). Sullivan et al (119) show in a small sample of NAFLD, obese controls and lean controls that fructose absorption and metabolism was more effective between NAFLD patients and lean control, but similar in NAFLD and obese controls. Children with NAFLD were more sensitive to dyslipidemia occurring in response to dietary fructose intake than children without NAFLD (120). While fructose alone is not generally believed to cause NAFLD development, fructose metabolites may influence NAFLD progression, and fructose intake may favour NAFLD progression on a background of pre-existing risk factors such as obesity, the metabolic syndrome or diabetes (121,122). In view of the continued increase of the number of cases of NAFL-NASH , Chalasani et al (123) published the practice guidelines of the American Association for the Study of Liver Diseases, American College of Gastroenterology, and the American Gastroenterological Association.

CONCLUSIONS

The current practice for diagnosis of patients with NASH is to perform a liver biopsy and imaging. The present review defined novel criteria based on noninvasive biomarkers that optimized noninvasive diagnosis of liver disease in patients with NAFLD. Active inflammation of the liver (steatohepatitis) should be excluded first by blood tests that should assess adipokine and pro-inflammatory cytokine levels.

The clinical evaluation should also involve a personalized evaluation of laboratory data. Monitoring adipokine levels may inform the clinician on the changes in the severity of the liver disease in time as well as of the efficacy of the intervention. Aditionally, we suggest that these biomarkers may also be useful in improving the assessment of noninvasive fibrosis.

Dietary habits may promote steatohepatitis directly by modulating hepatic triglyceride accumulation and antioxidant activity, as well as indirectly by affecting insulin sensitivity and postprandial triglyceride metabolism. Our findings provide further rationale for more specific alimentary and physical activity interventions, particularly in nonobese, nondiabetic, normolipidemic NASH patients. In addition, the review highlights the importance of other procedures, such as gastric balloon and bariatric surgery, in improving ouotcomes. Finally, we recommend continued collaboration between clinicians and laboratory, which will benefit patients.

ACKNOWLEDGMENTS: In Vitro Drug Safety and Biotechnology and Mahaffy's innovative grant – Sunnybrook Health Science Centre, Toronto, Canada supported the work.

DISCLOSURES: All authors have read the *Journal's* policy on conflicts of interest and have none to declare. Also, all authors have read the *Journal's* policy on authorship and there are no conflicts of interest.

REFERENCES

1. Neuman MG, Hilzenrat N, Cohen L, Winkler RE, Nanau R. Multiple factors involved in nonalcoholic hepatitis pathogenesis. Int J Hepatol 2012;2012:429805.
2. Kleiner DE, Brunt EM, Van Natta M, et al. Design and validation of a histological scoring system for nonalcoholic fatty liver disease. Hepatology 2005;41:1313-21.
3. Neuman MG, French SW, French BA, et al. Alcoholic and non-alcoholic steatohepatitis. Exp Mol Pathol 2014;97:492-510.
4. Begriche K, Igoudjil A, Pessayre D, Fromenty B. Mitochondrial dysfunction in NASH: Causes, consequences and possible means to prevent it. Mitochondrion 2006;6:1-28.
5. Farrell GC, Larter CZ. Nonalcoholic fatty liver disease: From steatosis to cirrhosis. Hepatology 2006;43:S99-112.
6. Wong RJ, Aguilar M, Cheung R, et al. Nonalcoholic steatohepatitis is the second leading etiology of liver disease among adults awaiting liver transplantation in the United States. Gastroenterology 2014; pii: S0016-5085(14)01474-7. doi: 10.1053/j.gastro.2014.11.039.
7. Neuman MG. Apoptosis in diseases of the liver. Crit Rev Clin Lab Sci 2001;38:109-66.
8. Ben-Ari Z, Schmilovotz-Weiss H, Belinki A, et al. Circulating soluble cytochrome C in liver disease as a marker of apoptosis. J Intern Med 2003;254:168-75.
9. Browning JD, Horton JD. Molecular mediators of hepatic steatosis and liver injury. J Clin Invest 2004;114:147-52.
10. Reddy JK, Rao MS. Lipid metabolism and liver inflammation. II. Fatty liver disease and fatty acid oxidation. Am J Physiol Gastrointest Liver Physiol 2006;290:G852-8.
11. Weickert MO, Pfeiffer AF. Signalling mechanisms linking hepatic glucose and lipid metabolism. Diabetologia 2006;49:1732-41.
12. Miele L, Forgione A, La Torre G, et al. Serum levels of hyaluronic acid and tissue metalloproteinase inhibitor-1 combined with age predict the presence of nonalcoholic steatohepatitis in a pilot cohort of subjects with nonalcoholic fatty liver disease. Transl Res 2009;154:194-201.

13. Papandreou D, Karabouta Z, Pantoleon A, Rousso I. Investigation of anthropometric, biochemical and dietary parameters of obese children with and without non-alcoholic fatty liver disease. Appetite 2012;59:939-44.
14. Han JM, Jo AN, Lee SM, et al. Associations between intakes of individual nutrients or whole food groups and non-alcoholic fatty liver disease among Korean adults. J Gastroenterol Hepatol 2014;29:1265-72.
15. Targher G, Bertolini L, Scala L, et al. Associations between serum 25-hydroxyvitamin D_3 concentrations and liver histology in patients with non-alcoholic fatty liver disease. Nutr Metab Cardiovasc Dis 2007;17:517-24.
16. Pirgon O, Cekmez F, Bilgin H, Eren E, Dundar B. Low 25-hydroxyvitamin D level is associated with insulin sensitivity in obese adolescents with non-alcoholic fatty liver disease. Obes Res Clin Pract 2013;7:e275-83.
17. Dasarathy J, Periyalwar P, Allampati S, et al. Hypovitaminosis D is associated with increased whole body fat mass and greater severity of non-alcoholic fatty liver disease. Liver Int 2014;34:e118-27.
18. Hao YP, Ma XJ, Luo YQ, et al. Serum vitamin D is associated with non-alcoholic fatty liver disease in Chinese males with normal weight and liver enzymes. Acta Pharmacol Sin 2014;35:1150-6.
19. Küçükazman M, Ata N, Dal KA, et al. The association of vitamin D deficiency with non-alcoholic fatty liver disease. Clinics (Sao Paulo) 2014;69:542-6.
20. Seo JA, Eun CR, Cho H, et al. Low vitamin D status is associated with nonalcoholic fatty liver disease independent of visceral obesity in Korean adults. PLoS One 2013;8:e75197.
21. Black LJ, Jacoby P, She Ping-Delfos WC, et al. Low serum 25-hydroxyvitamin D concentrations associate with non-alcoholic fatty liver disease in adolescents independent of adiposity. J Gastroenterol Hepatol 2014;29:1215-22.
22. Sharifi N, Amani R, Hajiani E, Cheraghian B. Does vitamin D improve liver enzymes, oxidative stress, and inflammatory biomarkers in adults with non-alcoholic fatty liver disease? A randomized clinical trial. Endocrine 2014;47:70-80.
23. Musso G, Gambino R, De Michieli F, et al. Dietary habits and their relations to insulin resistance and postprandial lipemia in nonalcoholic steatohepatitis. Hepatology 2003;37:909-16.
24. Oddy WH, Herbison CE, Jacoby P, et al. The Western dietary pattern is prospectively associated with nonalcoholic fatty liver disease in adolescence. Am J Gastroenterol 2013;108:778-85.
25. Da Silva HE, Arendt BM, Noureldin SA, Therapondos G, Guindi M, Allard JP. A cross-sectional study assessing dietary intake and physical activity in Canadian patients with nonalcoholic fatty liver disease vs healthy controls. J Acad Nutr Diet 2014;114:1181-94.
26. Mollard RC, Sénéchal M, MacIntosh AC, et al. Dietary determinants of hepatic steatosis and visceral adiposity in overweight and obese youth at risk of type 2 diabetes. Am J Clin Nutr 2014;99:804-12.
27. Veena J, Muragundla A, Sidgiddi S, Subramaniam S. Non-alcoholic fatty liver disease: Need for a balanced nutritional source. Br J Nutr 2014;112:1858-72.
28. Gerber L, Otgonsuren M, Mishra A, et al. Non-alcoholic fatty liver disease (NAFLD) is associated with low level of physical activity: A population-based study. Aliment Pharmacol Ther 2012;36:772-81.
29. Bae JC, Suh S, Park SE, et al. Regular exercise is associated with a reduction in the risk of NAFLD and decreased liver enzymes in individuals with NAFLD independent of obesity in Korean adults. PLoS One 2012;7:e46819.
30. Hallsworth K, Fattakhova G, Hollingsworth KG, et al. Resistance exercise reduces liver fat and its mediators in non-alcoholic fatty liver disease independent of weight loss. Gut 2011;60:1278-83.
31. Bhat G, Baba CS, Pandey A, Kumari N, Choudhuri G. Life style modification improves insulin resistance and liver histology in patients with non-alcoholic fatty liver disease. World J Hepatol 2012;4:209-17.
32. de Piano A, de Mello MT, Sanches Pde L, et al. Long-term effects of aerobic plus resistance training on the adipokines and neuropeptides in nonalcoholic fatty liver disease obese adolescents. Eur J Gastroenterol Hepatol 2012;24:1313-24.
33. Fealy CE, Haus JM, Solomon TP, et al. Short-term exercise reduces markers of hepatocyte apoptosis in nonalcoholic fatty liver disease. J Appl Physiol (1985) 2012;113:1-6.
34. Grønbæk H, Lange A, Birkebæk NH, et al. Effect of a 10-week weight loss camp on fatty liver disease and insulin sensitivity in obese Danish children. J Pediatr Gastroenterol Nutr 2012;54:223-8.
35. Sullivan S, Kirk EP, Mittendorfer B, Patterson BW, Klein S. Randomized trial of exercise effect on intrahepatic triglyceride content and lipid kinetics in nonalcoholic fatty liver disease. Hepatology 2012;55:1738-45.
36. Al-Jiffri O, Al-Sharif FM, Abd El-Kader SM, Ashmawy EM. Weight reduction improves markers of hepatic function and insulin resistance in type-2 diabetic patients with non-alcoholic fatty liver. Afr Health Sci 2013;13:667-72.
37. Bacchi E, Negri C, Targher G, et al. Both resistance training and aerobic training reduce hepatic fat content in type 2 diabetic subjects withnonalcoholic fatty liver disease (the RAED2 Randomized Trial). Hepatology 2013;58:1287-95.
38. Haus JM, Solomon TP, Kelly KR, et al. Improved hepatic lipid composition following short-term exercise in nonalcoholic fatty liver disease. J Clin Endocrinol Metab 2013;98:E1181-8.
39. Khaoshbaten M, Gholami N, Sokhtehzari S, Monazami AH, Nejad MR. The effect of an aerobic exercise on serum level of liver enzymes and liver echogenicity in patients with non alcoholic fatty liver disease. Gastroenterol Hepatol Bed Bench 2013;6:S112-6.
40. Malin SK, Mulya A, Fealy CE, et al. Fetuin-A is linked to improved glucose tolerance after short-term exercise training in nonalcoholic fatty liver disease. J Appl Physiol (1985) 2013;115:988-94.
41. Oh S, Tanaka K, Warabi E, Shoda J. Exercise reduces inflammation and oxidative stress in obesity-related liver diseases. Med Sci Sports Exerc 2013;45:2214-22.
42. Oh S, Shida T, Yamagishi K, Tanaka K, So R, Tsujimoto T, Shoda J. Moderate to vigorous physical activity volume is an important factor for managingnon-alcoholic fatty liver disease: A retrospective study. Hepatology 2015;61:1205-15.
43. Zelber-Sagi S, Buch A, Yeshua H, et al. Effect of resistance training on non-alcoholic fatty-liver disease: A randomized-clinical trial. World J Gastroenterol 2014;20:4382-92.
44. Johnson NA, Sachinwalla T, Walton DW, et al. Aerobic exercise training reduces hepatic and visceral lipids in obese individuals without weight loss. Hepatology 2009;50:1105-12.
45. Lee S, Bacha F, Hannon T, Kuk JL, Boesch C, Arslanian S. Effects of aerobic versus resistance exercise without caloric restriction on abdominal fat, intrahepatic lipid, and insulin sensitivity in obese adolescent boys: A randomized, controlled trial. Diabetes 2012;61:2787-95.
46. Keating SE, Hackett DA, George J, Johnson NA. Exercise and non-alcoholic fatty liver disease: A systematic review and meta-analysis. J Hepatol 2012;57:157-66.
47. DeVore S, Kohli R, Lake K, Nicholas L, Dietrich K, Balistreri WF, Xanthakos SA. A multidisciplinary clinical program is effective in stabilizing BMI and reducing transaminase levels in pediatric patients with NAFLD. J Pediatr Gastroenterol Nutr 2013;57:119-23.
48. Scaglioni F, Marino M, Ciccia S, et al. Short-term multidisciplinary non-pharmacological intervention is effective in reducing liver fat content assessed non-invasively in patients with nonalcoholic fatty liver disease (NAFLD). Clin Res Hepatol Gastroenterol 2013;37:353-8.
49. Wong VW, Chan RS, Wong GL, et al. Community-based lifestyle modification programme for non-alcoholic fatty liver disease: A randomized controlled trial. J Hepatol 2013;59:536-42.
50. Croci I, Byrne NM, Choquette S, et al. Whole-body substrate metabolism is associated with disease severity in patients with non-alcoholic fatty liver disease. Gut 2013;62:1625-33.
51. Jin YJ, Kim KM, Hwang S, et al. Exercise and diet modification in non-obese non-alcoholic fatty liver disease: Analysis of biopsies of living liver donors. J Gastroenterol Hepatol 2012;27:1341-7.
52. Kitajima Y, Hyogo H, Sumida Y, et al. Severity of non-alcoholic steatohepatitis is associated with substitution of adipose tissue in skeletal muscle. J Gastroenterol Hepatol 2013;28:1507-14.
53. Kawaguchi T, Shiba N, Maeda T, et al. Hybrid training of voluntary and electrical muscle contractions reduces steatosis, insulin resistance, and IL-6 levels in patients with NAFLD: A pilot study. J Gastroenterol 2011;46:746-57.
54. Castéra L, Vilgrain V, Angulo P. Noninvasive evaluation of NAFLD. Nat Rev Gastroenterol Hepatol 2013;10:666-75.
55. Neuman MG, Cohen LB, Nanau RM. Biomarkers in non-alcoholic fatty liver disease. Can J Gastroenterol Hepatol 2014;28:607-14.
56. Kim D, Kim WR, Kim HJ, Therneau TM. Association between noninvasive fibrosis markers and mortality among adults with nonalcoholic fatty liver disease in the United States. Hepatology 2013;57:1357-65.

57. Boyraz M, Cekmez F, Karaoglu A, Cinaz P, Durak M, Bideci A. Serum adiponectin, leptin, resistin and RBP4 levels in obese and metabolic syndrome children with nonalcoholic fatty liver disease. Biomark Med 2013;7:737-45.
58. Koot BG, van der Baan-Slootweg OH, Bohte AE, et al. Accuracy of prediction scores and novel biomarkers for predicting nonalcoholic fatty liver disease in obese children. Obesity (Silver Spring) 2013;21:583-90.
59. Pacifico L, Bezzi M, Lombardo CV, et al. Adipokines and C-reactive protein in relation to bone mineralization in pediatric nonalcoholic fatty liver disease. World J Gastroenterol 2013;19:4007-14.
60. Klein M, Iazzettii L, Speiser P, et al. Alanine transferase: An independent indicator of adiposity related comorbidity risk in youth. J Diabetes September 30, 2014. doi: 10.1111/1753-0407.12221 (Epub ahead of print).
61. Sanches PL, de Piano A, Campos RM, et al. Association of nonalcoholic fatty liver disease with cardiovascular risk factors in obese adolescents: The role of interdisciplinary therapy. J Clin Lipidol 2014;8:265-72.
62. Polyzos SA, Toulis KA, Goulis DG, Zavos C, Kountouras J. Serum total adiponectin in nonalcoholic fatty liver disease: A systematic review and meta-analysis. Metabolism 2011;60:313-26.
63. Caro JF, Sinha MK, Kolaczynski JW, Zhang PL, Considine RV. Leptin: The tale of an obesity gene. Diabetes 1996;45:1455-62.
64. Bechmann LP, Kocabayoglu P, Sowa JP, et al. Free fatty acids repress small heterodimer partner (SHP) activation and adiponectin counteracts bile acid-induced liver injury in superobese patients with nonalcoholic steatohepatitis. Hepatology 2013;57:1394-406.
65. Fitzpatrick E, Mitry RR, Quaglia A, Hussain MJ, DeBruyne R, Dhawan A. Serum levels of CK18 M30 and leptin are useful predictors of steatohepatitis and fibrosis in paediatric NAFLD. J Pediatr Gastroenterol Nutr 2010;51:500-6.
66. Mazure RA, Salgado G, Villarreal P, Cobo B, Valencia A, Culebras JM. Intragastric balloon and multidisciplinary team. Nutr Hosp 2009;24:282-7.
67. Yasawy MI, Al-Quorain AA, Hussameddin AM, Yasawy ZM, Al-Sulaiman RM. Obesity and gastric balloon. J Family Community Med 2014;21:196-9.
68. Mafort TT, Madeira E, Madeira M, et al. Six-month intragastric balloon treatment for obesity improves lung function, body composition, and metabolic syndrome. Obes Surg 2014;24:232-40.
69. Dogan UB, Gumurdulu Y, Akin MS, Yalaki S. Five percent weight lost in the first month of intragastric balloon treatment may be a predictor for long-term weight maintenance. Obes Surg 2013;23:892-6.
70. Lee YM, Low HC, Lim LG, et al. Intragastric balloon significantly improves nonalcoholic fatty liver disease activity score in obese patients with nonalcoholic steatohepatitis: A pilot study. Gastrointest Endosc 2012;76:756-60.
71. Tai CM, Lin HY, Yen YC, et al. Effectiveness of intragastric balloon treatment for obese patients: One-year follow-up after balloon removal. Obes Surg 2013;23:2068-74.
72. Folini L, Veronelli A, Benetti A, et al. Liver steatosis (LS) evaluated through chemical-shift magnetic resonance imaging liver enzymes in morbid obesity; effect of weight loss obtained with intragastric balloon gastric banding. Acta Diabetol 2014;51:361-8.
73. Nobili V, Corte CD, Liccardo D, et al. Obalon intragastric balloon in the treatment of paediatric obesity: A pilot study. November 14, 2014. doi: 10.1111/ijpo.268 (Epub ahead of print).
74. Genco A, Maselli R, Frangella F, et al. Effect of consecutive intragastric balloon (BIB®) plus diet versus single BIB® plus diet on eating disorders not otherwise specified (EDNOS) in obese patients. Obes Surg 2013;23:2075-9.
75. Fuller NR, Pearson S, Lau NS, et al. An intragastric balloon in the treatment of obese individuals with metabolic syndrome: A randomized controlled study. Obesity (Silver Spring) 2013;21:1561-70.
76. Konopko-Zubrzycka M, Baniukiewicz A, Wróblewski E, et al. The effect of intragastric balloon on plasma ghrelin, leptin, and adiponectin levels in patients with morbidobesity. J Clin Endocrinol Metab 2009;94:1644-9.
77. Nikolic M, Boban M, Ljubicic N, et al. Morbidly obese are ghrelin and leptin hyporesponders with lesser intragastricballoon treatment efficiency: Ghrelin and leptin changes in relation to obesity treatment. Obes Surg 2011;21:1597-604.
78. Sekino Y, Imajo K, Sakai E, et al. Time-course of changes of visceral fat area, liver volume and liver fat area duringintragastric balloon therapy in Japanese super-obese patients. Intern Med 2011;50:2449-55.
79. Mion F, Napoléon B, Roman S, et al. Effects of intragastric balloon on gastric emptying and plasma ghrelin levels in non-morbid obese patients. Obes Surg 2005;15:510-6.
80. Bužga M, Evžen M, Pavel K, et al. Effects of the intragastric balloon MedSil on weight loss, fat tissue, lipid metabolism, and hormones involved in energy balance. Obes Surg 2014;24:909-15.
81. Geliebter A, Yahav EK, Gluck ME, Hashim SA. Gastric capacity, test meal intake, and appetitive hormones in binge eating disorder. Physiol Behav 2004;81:735-40.
82. Mathus-Vliegen EM, Eichenberger RI. Fasting and meal-suppressed ghrelin levels before and after intragastric balloons and balloon-induced weight loss. Obes Surg 2014;24:85-94.
83. Martinez-Brocca MA, Belda O, Parejo J, et al. Intragastric balloon-induced satiety is not mediated by modification in fasting or postprandial plasma ghrelin levels in morbid obesity. Obes Surg 2007;17:649-57.
84. Lassailly G, Caïazzo R, Pattou F, Mathurin P. Bariatric surgery for curing NASH in the morbidly obese? J Hepatol 2013;58:1249-51.
85. Sasaki A, Nitta H, Otsuka K, et al. Bariatric surgery and non-alcoholic fatty liver disease: Current and potential future treatments. Front Endocrinol (Lausanne) 2014;5:164.
86. Taitano AA, Markow M, Finan JE, Wheeler DE, Gonzalvo JP, Murr MM. Bariatric surgery improves histological features of nonalcoholic fatty liver disease and liver fibrosis. J Gastrointest Surg 2015;19:429-36.
87. Andrade-Silva SG, Caranti DA, Sallet JA, Leal LP, Leal AJ, Dâmaso AR. Age and gender may influence the results of Roux-en-Y gastric bypass? Metabolic syndrome parameters. Arq Gastroenterol 2014;51:171-9.
88. Vargas V, Allende H, Lecube A, et al. Surgically induced weight loss by gastric bypass improves non alcoholic fatty liver disease in morbid obese patients. World J Hepatol 2012;4:382-8.
89. Cazzo E, Jimenez LS, Pareja JC, Chaim EA. Effect of Roux-en-y gastric bypass on nonalcoholic fatty liver disease evaluated through NAFLD fibrosis score: A prospective study. Obes Surg November 8, 2014 (Epub ahead of print).
90. Carazo A, León J, Casado J, et al. Hepatic expression of adiponectin receptors increases with non-alcoholic fatty liver disease progression in morbid obesity in correlation with glutathione peroxidase 1. Obes Surg 2011;21:492-500.
91. Tai CM, Huang CK, Hwang JC, et al. Improvement of nonalcoholic fatty liver disease after bariatric surgery in morbidly obese Chinese patients. Obes Surg 2012;22:1016-21.
92. Felipo V, Urios A, García-Torres ML, et al. Alterations in adipocytokines and cGMP homeostasis in morbid obesity patients reverse after bariatric surgery. Obesity (Silver Spring) 2013;21:229-37.
93. Moschen AR, Molnar C, Wolf AM, et al. Effects of weight loss induced by bariatric surgery on hepatic adipocytokine expression. J Hepatol 2009;51:765-77.
94. Hosseinzadeh-Attar MJ, Golpaie A, Janani L, Derakhshanian H. Effect of weight reduction following bariatric surgery on serum visfatin and adiponectin levels in morbidly obese subjects. Obes Facts 2013;6:193-202.
95. Machado MV, Coutinho J, Carepa F, Costa A, Proença H, Cortez-Pinto H. How adiponectin, leptin, and ghrelin orchestrate together and correlate with the severity of nonalcoholic fatty liver disease. Eur J Gastroenterol Hepatol 2012;24:1166-72.
96. Vilarrasa N, Vendrell J, Sánchez-Santos R, et al. Effect of weight loss induced by gastric bypass on proinflammatory interleukin-18, soluble tumour necrosis factor-alpha receptors, C-reactive protein and adiponectin in morbidly obese patients. Clin Endocrinol (Oxf) 2007;67:679-86.
97. Illán-Gómez F, Gonzálvez-Ortega M, Orea-Soler I, et al. Obesity and inflammation: Change in adiponectin, C-reactive protein, tumour necrosis factor-alpha and interleukin-6 after bariatric surgery. Obes Surg 2012;22:950-5.
98. Caiazzo R, Lassailly G, Leteurtre E, et al. Roux-en-Y gastric bypass versus adjustable gastric banding to reduce nonalcoholic fatty liver disease: A 5-year controlled longitudinal study. Ann Surg 2014;260:893-8.
99. Burza MA, Romeo S, Kotronen A, et al. Long-term effect of bariatric surgery on liver enzymes in the Swedish Obese Subjects (SOS) study. PLoS One 2013;8:e60495.

100. Umemura A, Sasaki A, Nitta H, Otsuka K, Suto T, Wakabayashi G. Effects of changes in adipocyte hormones and visceral adipose tissue and the reduction of obesity-related comorbidities after laparoscopic sleeve gastrectomy in Japanese patients with severe obesity. Endocr J 2014;61:381-91.
101. Nazal L, Riquelme A, Solís N, et al. Hypoadiponectinemia and its association with liver fibrosis in morbidly obese patients. Obes Surg 2010;20:1400-7.
102. Kahraman A, Sowa JP, Schlattjan M, et al. Fetuin-A mRNA expression is elevated in NASH compared with NAFL patients. Clin Sci (Lond) 2013;125:391-400.
103. Pirvulescu I, Gheorghe L, Csiki I, et al. Noninvasive clinical model for the diagnosis of nonalcoholic steatohepatitis in overweight and morbidly obese patients undergoing bariatric surgery. Chirurgia (Bucur) 2012;107:772-9.
104. Estep M, Abawi M, Jarrar M, et al. Association of obestatin, ghrelin, and inflammatory cytokines in obese patients with non-alcoholic fatty liver disease. Obes Surg 2011;21:1750-7.
105. Bužga M, Zavadilová V, Holéczy P, et al. Dietary intake and ghrelin and leptin changes after sleeve gastrectomy. Wideochir Inne Tech Malo Inwazyjne 2014;9:554-61.
106. Moschen AR, Molnar C, Geiger S, et al. Anti-inflammatory effects of excessive weight loss: Potent suppression of adipose interleukin 6 and tumour necrosis factor alpha expression. Gut 2010;59:1259-64.
107. Bray GA, Popkin BM. Calorie-sweetened beverages and fructose: What have we learned 10 years later. Pediatr Obes 2013;8:242-8.
108. Lê KA, Ith M, Kreis R, et al. Fructose overconsumption causes dyslipidemia and ectopic lipid deposition in healthy subjects with and without a family history of type 2 diabetes. Am J Clin Nutr 2009;89:1760-5.
109. Jin R, Willment A, Patel SS, et al. Fructose induced endotoxemia in pediatric nonalcoholic fatty liver disease. Int J Hepatol 2014;2014:560620.
110. Kanerva N, Sandboge S, Kaartinen NE, Männistö S, Eriksson JG. Higher fructose intake is inversely associated with risk of nonalcoholic fatty liver disease in older Finnish adults. Am J Clin Nutr 2014;100:1133-8.
111. Yang M, Gong S, Ye SQ, et al. Non-alcoholic fatty liver disease in children: Focus on nutritional interventions. Nutrients 2014;6:4691-705.
112. Volynets V, Machann J, Küper MA, et al. A moderate weight reduction through dietary intervention decreases hepatic fat content in patients with non-alcoholic fatty liver disease (NAFLD): A pilot study. Eur J Nutr 2013;52:527-35.
113. Johnston RD, Stephenson MC, Crossland H, et al. No difference between high-fructose and high-glucose diets on liver triacylglycerol or biochemistry in healthy overweight men. Gastroenterology 2013;145:1016-25.
114. Chung M, Ma J, Patel K, Berger S, Lau J, Lichtenstein AH. Fructose, high-fructose corn syrup, sucrose, and nonalcoholic fatty liver disease or indexes of liver health: A systematic review and meta-analysis. Am J Clin Nutr 2014;100:833-49.
115. Chiu S, Sievenpiper JL, de Souza RJ, et al. Effect of fructose on markers of non-alcoholic fatty liver disease (NAFLD): A systematic review and meta-analysis of controlled feeding trials. Eur J Clin Nutr 2014;68:416-23.
116. Carvalhana S, Machado MV, Cortez-Pinto H. Improving dietary patterns in patients with nonalcoholic fatty liver disease. Curr Opin Clin Nutr Metab Care 2012;15:468-73.
117. O'Sullivan TA, Oddy WH, Bremner AP, et al. Lower fructose intake may help protect against development of nonalcoholic fatty liver in adolescents with obesity. J Pediatr Gastroenterol Nutr 2014;58:624-31.
118. Mager DR, Iñiguez IR, Gilmour S, Yap J. The Effect of a Low Fructose and Low Glycemic Index/Load (FRAGILE) dietary intervention on indices of liver function, cardiometabolic risk factors, and body composition in children and adolescents with nonalcoholic fatty liver disease (NAFLD). JPEN J Parenter Enteral Nutr 2015;39:73-84.
119. Sullivan JS, Le MT, Pan Z, et al. Oral fructose absorption in obese children with non-alcoholic fatty liver disease. Pediatr Obes June 24, 2014. doi: 10.1111/ijpo.238 (Epub ahead of print).
120. Jin R, Le NA, Liu S, et al. Children with NAFLD are more sensitive to the adverse metabolic effects of fructose beverages than children without NAFLD. J Clin Endocrinol Metab 2012;97:E1088-98.
121. Abdelmalek MF, Lazo M, Horska A, et al. Higher dietary fructose is associated with impaired hepatic adenosine triphosphate homeostasis in obese individuals with type 2 diabetes. Hepatology 2012;56:952-60.
122. Nomura K, Yamanouchi T. The role of fructose-enriched diets in mechanisms of nonalcoholic fatty liver disease. J Nutr Biochem 2012;23:203-8.
123. Chalasani N, Younossi Z, Lavine JE et al. The diagnosis and management of non-alcoholic fatty liver disease: Practice guideline by the American Association for the Study of Liver Diseases, American College of Gastroenterology, and the American Gastroenterological Association. Hepatology 2012;55:2005-23.

Transient elastography in Canada: Current state and future directions

Mohammed Aljawad MD[1,2], Sanjeev Sirpal MD JD[3,4], Eric M Yoshida MD[5], Natasha Chandok MD MPH[1,3]

M Aljawad, S Sirpal, EM Yoshida, N Chandok. Transient elastography in Canada: Current state and future directions. Can J Gastroenterol Hepatol 2015;29(7):373-376.

BACKGROUND: Transient elastography (TE) is a safe and effective technology to noninvasively assess hepatic fibrosis in patients with numerous liver conditions. TE is not readily available to all Canadians, and data regarding how this technology is incorporated into clinical practice are lacking.
OBJECTIVE: To describe TE practices in Canada, and to identify strategies to optimize access and usage.
METHODS: All Canadian centres with TE devices were invited to complete a survey after obtaining purchasing data from the national distributor of the device. Descriptive statistics were generated.
RESULTS: Forty-two devices were available in Canada as of January 2015. Seventy-one percent are used in academic settings, 74% are hospital based and 26% are in private clinics. The test is performed by trained nurses in 48% of centres, physicians in 19%, technicians in 9.5% and by any member of the health care team in 19%. Nineteen percent of centres provide satellite clinics to perform the test. While the majority of the centres perform the test at no additional cost to patients, 29% charge a variable fee.
CONCLUSION: In Canada, most TE devices are used in academic and/or hospital-based settings, thus limiting access to this technology to many patients. A sizeable minority of centres mandate patients pay variable out-of-pocket fees. Satellite clinics offered by some centres could increase access, but are not widespread. The lack of uniformity with TE practices in Canada suggests that a national policy is needed.

Key Words: *Access; Chronic liver disease; Equity; Liver fibrosis; Transient elastography*

L'élastographie transitoire au Canada : la situation actuelle et les futures orientations

HISTORIQUE : L'élastographie transitoire (ÉT) est une technologie sécuritaire et efficace pour évaluer la fibrose hépatique de manière non effractive chez des patients atteints de divers problèmes hépatiques. L'ÉT n'est pas offerte à tous les Canadiens, et on ne possède pas de données sur l'intégration de cette technologie à la pratique clinique.
OBJECTIF : Décrire les pratiques d'ÉT au Canada et établir des stratégies pour en optimiser l'accès et l'usage.
MÉTHODOLOGIE : Tous les centres canadiens possédant des dispositifs d'ÉT ont été invités à participer à un sondage après avoir obtenu les données d'achat du distributeur national. Des statistiques descriptives en ont été tirées.
RÉSULTATS : Quarante-deux dispositifs étaient en usage au Canada en janvier 2015. De ce nombre, 71 % sont utilisés en milieu universitaire, 74 % en milieu hospitalier et 26 % dans des cliniques privées. Le test est effectué par des infirmières formées dans 48 % des centres, par des médecins dans 19 % des centres, par des techniciens dans 9,5 % des centres et par un membre de l'équipe soignante dans 19 % des centres. De plus, 19 % des centres étaient dotés de cliniques satellites pour effectuer le test. La majorité des centres effectuent le test sans frais supplémentaires pour le patient, mais 29 % demandent des sommes variables.
CONCLUSION : Au Canada, la plupart des dispositifs d'ÉT sont utilisés en milieu universitaire ou hospitalier, ce qui en limite l'accès pour de nombreux patients. Une minorité importante de centres exige des frais variables aux patients. Les cliniques satellites de certains centres pourraient en accroître l'accès, mais ne sont pas répandues. L'absence d'uniformité à l'égard des pratiques d'ÉT au Canada donne à penser qu'une politique nationale s'impose.

Fibrosis is an important outcome in liver diseases because it is the main contributing factor leading to cirrhosis and subsequent complications from portal hypertension. In addition to its prognostic significance, staging liver fibrosis has an important role in guiding therapy in several liver diseases. Historically, liver biopsy was the only definitive method to quantify fibrosis. Although still considered to be the gold standard, liver biopsy is an imperfect, invasive test that is associated with several inherent limitations. Aside from the potential risks of the procedure, interpretation of liver biopsy is subject to sampling error and inter/intraobserver variability among pathologists (1-3). The need for noninvasive ways to assess fibrosis has led to the development of several blood- and imaging-based tests. Although appealing to health care providers and patients, there are advantages and disadvantages to the use of such tests and, ultimately, the use of many are hindered by high cost and poor access (Table 1).

Transient elastography (TE), approved for use by Health Canada in 2009, is an effective imaging-based noninvasive technique to stage liver fibrosis in various liver conditions. The diagnostic accuracy of TE has been established in several studies. For instance, a meta-analysis of 50 studies by Friedrich-Rust et al (4) showed mean areas under the ROC curve for the diagnosis of significant fibrosis, severe fibrosis and cirrhosis of 0.84, 0.89 and 0.94, respectively ($P<0.05$) (4). In general, an area under the ROC curve of 1.0 is a perfect test whereas 0.5 indicates a poor test that is slightly better than random chance in predictive ability and 0.8 is considered to be a good test.

Aside from staging fibrosis, recent studies suggest prognostic significance of mean elastography measurements using TE. For example, Wang et al (5) demonstrated that a liver stiffness measurement ≥17 kPa is an independent predictor of progression of portal hypertension in patients with well compensated cirrhosis. In addition, TE can be used to monitor liver fibrosis serially over time and, when the test is performed by the patient's health care provider, results can be instantly reported to patients and immediately impact clinical management.

TE has gained wide acceptance among Canadian gastroenterologists and hepatologists. In our previous study, TE resulted in a perceived 50% reduction in the need for percutaneous liver biopsy to stage liver disease (6). Similar to our observation, Sebastiani et al (7) reported concordant results in a survey-based study of hepatologists, gastroenterologists and infectious disease specialists in Canada. Although multiple professional societies have incorporated TE in the

[1]*Department of Medicine, Western University, London, Ontario;* [2]*Multiorgan Transplant Unit, King Fahad Specialist Hospital, Dammam, Saudi Arabia;* [3]*Division of Gastroenterology, William Osler Health Centre, Brampton;* [4]*Faculty of Medicine, University of Toronto, Toronto, Ontario;* [5]*Division of Gastroenterology, University of British Columbia, Vancouver, British Columbia*
Correspondence: Dr Natasha Chandok, Division of Gastroenterology, University of Western Ontario, 339 Windermere Road, London, Ontario N6A 5A5. e-mail dr.n.chandok@gmail.com

TABLE 1
Summary of noninvasive methods to assess hepatic fibrosis

Test	Description	Advantage(s)	Disadvantage(s)
Imaging			
Transient elastography (FibroScan, EchoSens, France)	Low-frequency and low-amplitude vibrations pass through a limited area of the liver. The wave speed measurement by pulse-echo ultrasound is converted to an estimate of elastography	1. Easy to learn and perform 2. Immediate results 3. Moderate to high accuracy	1. Limited availability 2. Reimbursment for physicians ± usually lacking 3. Technical difficulty in obese patients 4. False higher results in some circumstances
Magnetic resonance elastography	Low-frequency vibrations pass through the entire liver and the speed is measured by magnetic resonance imaging spin echo sequence to convert to an estimate of elastography	1. High accuracy 2. Examines the entire liver 3. Can be used in obese patients	1. Limited availability 2. High cost 3. Time consuming 4. Contraindicated in claustrophobic subjects, or iron overloaded liver
Serum biomarkers			
AST/ALT ratio (indirect marker of fibrosis)		1. Readily available and easy to calculate 2. Inexpensive	1. Nonspecific 2. Poor accuracy
Acoustic radiation force impulse imaging (indirect marker of fibrosis)	(AST elevation*/platelet count) x 100	1. Readily available and easy to calculate 2. Inexpensive 3. Moderate accuracy	1. Nonspecific 2. Results affected by various comorbidities
FibroTest (LabCorp, USA) (indirect marker of fibrosis)	Components of this mathematical calculated test include patient sex, age, gamma globulin, alpha-2-macroglobulin, haptoglobin, apolipoprotein A1 and total bilirubin levels	1. Moderate to high accuracy to distinguish insignificant and advanced fibrosis	1. Limited availability 2. Cost to laboratory or patient 3. False positive results in hyperbilirubinemia from any cause 4. Delay in test results
Hepascore (Quest Diagnostics, USA) (indirect marker of fibrosis)	Components of this score include patient sex, age, bilirubin, gamma-glutamyl transpeptidase, hyaluronic acid and alpha-2 macroglobulin levels	1. Moderate to high accuracy to distinguish insignificant and advanced fibrosis	1. Limited availability 2. High cost 3. False-positive results in hyperbilirubinemia from any cause 4. Delay in test results
FibroSpect II (*Société des Produits Nestlé SA*, Switzerland) (direct marker of fibrosis)	Components of this test include serum hyaluronic acid, tissue inhibitor of metalloproteinase-1 and alpha-2 macroglobulin levels	1. Moderate to high accuracy to distinguish insignificant and advanced fibrosis	1. Limited availability 2. Cost 3. Delay in test results

**Aspartate aminotransferase (AST) level divided by the upper limit of normal for the laboratory. ALT Alanine aminotransferase*

management of viral hepatitis and other chronic liver diseases, the test remains underutilized nationally due to lack of availability and access (7,8). The goal of the present study was to describe current TE practices in Canada, and to identify strategies to optimize access and use of this technology.

METHODS

TE device locations and clinic contact details were obtained from the national distributor of FibroScan (Echosens, France) in Canada (KNS Canada Inc) as well as publically available medical directories. All centres were contacted by telephone or fax to acquire information pertinent to TE practice, presence and number of satellite clinics covered by the same device and patient fees (if any). Descriptive statistics were generated.

RESULTS

All 42 TE clinics were contacted between September 2014 and January 2015; of these, a response was obtained from 29 (69%) centres (Table 2). Wait times for TE only ranged from one to 52 weeks, although eight centres reported internally variable wait periods corresponding to standard duration to see the hepatologist or gastroenterologist. Provincial distribution of the devices is shown in Table 3. Thirty-six of the nation's 42 devices are in the provinces of Ontario (n=13 [31%]), Quebec (n=9 [21%]), British Columbia (n=8 [19%]) and Alberta (n=6 [14%]). Seventy-four percent (n=31) of devices are in hospital settings, including 71% (n=30) in University centres and the remaining 26% (n=11) in privatized clinics. Six centres (21% of respondent clinics) charged patients a fee for the test, ranging from \$80 to \$125, whereas 79% performed the test at no additional cost to patients. Sixty percent of centres reported results immediately to patients after completion of TE and 40% of surveyed centres provide results to referring physicians only. Of respondent centres, TE was in use for a mean of 3.5 days per week (range one to five days (Table 2).

DISCUSSION

TE is a novel technology that can play an important role in the noninvasive assessment of liver fibrosis, which is essential to the management of patients with chronic liver diseases. The present study revealed that there are only 42 devices confined to Canada's largest cities – a miniscule number considering the national burden of liver disease and generally lengthy wait times reported as the standard waitlist period with the hepatologist/gastroenterologist who would conduct the test as part of a routine consultation, to approximately one year for TE test alone in Nova Scotia. These data, in conjunction with previous published studies, demonstrates abysmal access to TE and the resultant previously reported underutilization of the test is, thus, not surprising.

Barriers, in fact, exist at multiple levels that impact the use of TE in Canada. It is possible to categorize these barriers at the system, practitioner and patient level to facilitate further analysis. At the system level, the limited availability of the technology requires improved integration among external community-based treatment settings.

TABLE 2
Summary of survey results

Transient elastography variable	
Device setting	
Academic/community	30 (71)/12 (29)
Hospital/private clinic	31 (74)/11 (26)
Response rate (n=42 centres)	29 (69)
Satellite clinics offered	
Yes	6 (20.7)
No	23 (79.3)
Primary operator	
Nurse	17 (59)
Physician	5 (17)
Technician	2 (7)
Any of the above	5 (17)
Wait time, weeks, mean (range)	5.7 (1–52)
Usage, days/week, mean (range)	3.5 (1–5)
Fee for patient	
Yes	6 (20.7)
No	23 (79.3)
Results given to: %	
Patient ± referring physician	60
Referring physician only	40

Data presented as n (%) unless otherwise indicated

Successful use of this technology, therefore, necessitates a revision of community-based models of care that integrate assertive engagement and shared utilization of limited resources. Additionally, current reimbursement policies serve as a systemic barrier to widespread use of this technology. In France, *la Haute Authorité de Santé* has approved reimbursement for TE and FibroTest (LabCorp, USA) since 2007, while in Canada, TE is currently reimbursed only in Quebec (7). The implications of no physician reimbursement for TE across Canada's remaining nine provinces renders acquisition and maintenance of a TE device financially unfeasible for many practitioners, especially considering that initial costs of current models are approximately $100,000.

At the practitioner level, there are significant hindrances to the use of this technology. First, wait times and the lack of local devices render widespread, routine use problematic. Furthermore, a nationwide Canadian survey revealed that individual physician characteristics impact their willingness to use TE (7). For instance, Sebastiani et al (7) showed that older physicians used more noninvasive methods for liver fibrosis than younger respondents ($P=0.02$). Interestingly, hepatologists and infectious diseases specialists used more noninvasive methods for liver fibrosis (particularly TE) than gastroenterologists, although this difference was not statistically significant. The results of the current study showed remarkable variability in practice patterns and access to TE among the various provinces and in urban versus rural settings. In addition to the limited number of devices available in Canada, the devices are mainly located in tertiary liver centres and selected private clinics in densely populated urban areas. While satellite clinics may help to address the lack of availability of TE in lower-density populations and reduced wait times for patients within and outside of urban areas, only six centres (14% of respondent centres) in Canada are performing TE in secondary locations. Clearly, there may be logistic barriers to operating satellite clinics not addressed in the present study, including the fear of damage to the equipment in transport, lack of available trained personnel and lack of external resources to support such endeavours.

At the patient level, the primary barriers to the routine use of TE in Canada are limited access and variable costs associated with access. Inasmuch as all these factors play significant roles in serving as barriers to TE use, the limited availability of the technology is perhaps the most significant. As of January 2015, approximately 70% Canada's 42 devices are located in three provinces, while more than one-half

TABLE 3
Distribution of FibroScan (Echosens, France) devices according to province and city in Canada (n=42)

Province	FibroScan devices, n
Ontario (n=13 [31%])	
Toronto	6
Ottawa	2
Hamilton	2
London	1
Brampton	1
Sudbury	1
Quebec (n=9 [21.4%])	
Montreal	6
Quebec city	2
Saint-Hyacinthe	1
British Columbia (n=8 [19%])	
Vancouver	5
Victoria	1
Prince George	1
Kelowna	1
Alberta (n=6 [14.3%])	
Edmonton	3
Calgary	3
Saskatchewan (n=3 [7%])	
Saskatoon	2
Regina	1
Nova Scotia (n=1 [2.4%])	
Halifax	1
New Brunswick (n=1 [2.4%])	
Moncton	1
Manitoba (n=1 [2.4%])	
Winnipeg	1

TABLE 4
Per capita availability of FibroScan (Echosens, France) according to Canadian province

Province	FibroScan devices, n	Population, $\times 10^6$	Per capita availability
Ontario	13	13.6	0.96
Quebec	9	8.2	1.10
British Columbia	8	4.6	1.74
Alberta	6	4.1	1.46
Saskatchewan	3	1.1	2.73
Manitoba	1	1.3	0.77
Nova Scotia	1	0.9	1.11
New Brunswick	1	0.8	1.25
Newfoundland	0	0.5	0

are in the five largest cities. Although a population-based argument can be made for even more TE devices in these cities, the current distribution leaves a considerable geographical area with no access to TE. A comparison of the per capita availability of FibroScan according to province is presented in Table 4.

Few (11 of 42 [26%]) of the devices are located in private clinics while the majority are situated within hospitals. To improve access to TE, 21% (six of 29) the respondent centres offer satellite clinics for the test where one machine is transported to offsite locations. Of the respondent centres, 7% (two of 29) are private clinics dedicated solely to perform TE for referring physicians while 55% (16 of 29) perform TE as part of a standard patient assessment. Dedicated TE clinics may conceivably help minimize wait times and improve test access, although this remains to be proven at the patient level because patients will presumably wait additional time to obtain follow-up with their referring physicians. Direct and indirect costs to patients for TE is another likely impediment. Our results indicate that nearly 21% of centres charge patients an out-of-pocket fee for test completion (range $80 to $125). While approximately 75% of patient respondents in a survey published by Kan et al (9) expressed a willingness to pay for TE, indirect costs, such as transportation, parking, time away from work, etc, likely contribute significantly to overall patient-associated costs. Given Canada's single-tier heath care system with central principles of equity and accessibility, a broader question regarding lack of universal coverage for a medically necessary test and fairness in variability of direct costs requires further consideration, especially given that many patients with chronic liver diseases have low socioeconomic backgrounds.

From a financial prospective, a cost analysis study from the United States published in 2000 (10) showed that performing percutaneous liver biopsy with no complication costs US$1,033 per patient, which is highly comparable with the cost in Canada. In the case of a complication of liver biopsy requiring hospitalization, the median direct cost can reach $4,579 according to a 2008 Canadian population-based study (11). Because liver fibrosis is a dynamic process, significant proportions of patients require surveillance assessment of fibrosis stage in periodic intervals, which adds to total cost, and renders liver biopsy a poor diagnostic option from the patient and provider perspective. On the other hand, the cost of purchasing a current model of the TE device with a variable yearly maintenance fee is approximately $100,000. This amount would suggest TE would be revenue neutral once it is used appropriately in lieu of percutaneous liver biopsy in just 100 patients. It follows that in a British cost-effectiveness analysis, TE was more cost effective than percutaneous liver biopsy in an economic model involving 1000 patients (12).

Aside from the cost benefits and clear clinical utility of TE, Kan et al (9) demonstrated that patients strongly favour TE for the assessment of liver fibrosis. In that study, questionnaires were distributed to patients who had and had not previously undergone percutaneous liver biopsies to evaluate patient preference for TE versus liver biopsy, and to assess the willingness to self-pay for TE. Overall, an astounding 95.4% of patients preferred TE to percutaneous liver biopsy, citing benefits of improved comfort and short duration to receive results. These results, coupled with our findings, further support the need for systemic policy change at the provincial level to facilitate patient access to TE.

In Canada, liver disease is associated with significant personal and health care costs, and current mortality rates are largely underestimated, in part due to variance in reporting and inaccurate *International Classification of Diseases* coding (13). The staging of liver fibrosis is the single most important factor impacting the prognosis of patients with chronic liver diseases and it has a major role in management decisions. Inasmuch as there is a clear need for early diagnosis and staging of liver fibrosis, in 20% of patients with chronic liver diseases, the diagnosis of cirrhosis is made on presentation of the first episode of hepatic decompensation (14). Hence, the effective implementation of noninvasive liver fibrosis staging, such as via TE technology, will enable preclinical management and will facilitate effective long-term planning for these patients.

Another issue that will impact the widespread availability of TE in Canada may be competition with radiologists for newer technologies that will assess elastography using currently available magnetic resonce imaging or ultrasound systems. However, the advent of new radiological technologies, such as acoustic radiation force impulse imaging, creates an opportunity for interdisciplinary collaboration in the noninvasive assessment of liver disease. Our study highlights that there are significant barriers to widespread use of this technology within Canada, and hopefully serves as the nidus for spearheading effective policy changes at the provincial levels to render it more accessible.

DISCLOSURES: The authors have no financial disclosures or conflicts of interest to declare.

REFERENCES

1. Regev A, Berho M, Jeffers LJ, et al. Sampling error and intraobserver variation in liver biopsy in patients with chronic HCV infection. Am J Gastroenterol 2002;97:2614-8.
2. Olsson R, Hagerstrand I, Broome U, et al. Sampling variability of percutaneous liver biopsy in primary sclerosing cholangitis. J Clin Pathol 1995;48:933-5.
3. Robert M, Sofair AN, Thomas A, et al. A comparison of hepatopathologists' and community pathologists' review of liver biopsy specimens from patients with hepatitis C. Clin Gastroenterol Hepatol 2009;7:335-8.
4. Friedrich-Rust M, Ong MF, Martens S, et al. Performance of transient elastography for the staging of liver fibrosis: A meta-analysis. Gastroenterology 2008;134:960-74.
5. Wang JH, Chuah SK, Lu SN, et al. Baseline and serial liver stiffness measurement in prediction of portal hypertension progression for patients with compensated cirrhosis. Liver Int 2014;34:1340-8.
6. Aljawad M, Yoshida EM, Uhanova J, et al. Percutaneous liver biopsy practice patterns among Canadian hepatologists. Can J Gastroenterol 2013;27:e31-4.
7. Sebastiani G, Ghali P, Wong P, et al. Physicians' practices for diagnosing liver fibrosis in chronic liver diseases: A nationwide, Canadian survey. Can J Gastroenterol Hepatol 2014;28:23-30.
8. Chandok N. Out with the old and in with the new: Hepatic fibrosis assessment in Canada. Can J Gastroenterol Hepatol 2014;28:20.
9. Kan VY, Marquez Azalgara V, Ford JA, et al. Patient preference and willingness to pay for transient elastography versus liver biopsy: A perspective from British Columbia. Can J Gastroenterol Hepatol 2015;29:72-6.
10. Wong JB, Koff RS. Watchful waiting with periodic liver biopsy versus immediate empirical therapy for histologically mild chronic hepatitis C. A cost-effectiveness analysis. Ann Intern Med 2000;133:665-75.
11. Myers RP, Fong A, Shaheen AA. Utilization rates, complications and costs of percutaneous liver biopsy: A population-based study including 4275 biopsies. Liver Int 2008;28:705-12.
12. Stamuli E, Kruger J, Hutton J. Cost-effectiveness of ultrasound elastography in the assessment of liver fibrosis. Economic report for the NHS Centre for Evidence-based Purchasing. 2009:CEP08053.
13. Desai S, Peltekian KM. Canadian mortality rates for liver disease: Taking a closer look at ICD coding. Can J Public Health 2004;95:198-200.
14. Garcia-Tsao G, Friedman S, Iredale J, et al. Now there are many (stages) where before there was one: In search of a pathophysiological classification of cirrhosis. Hepatology 2010;5:1445-9.

Does the level of reproductive knowledge specific to inflammatory bowel disease predict childlessness among women with inflammatory bowel disease?

Vivian W Huang MSc MD FRCPC[1], Hsiu-Ju Chang MSc[1], Karen I Kroeker MSc MD FRCPC[1], Karen J Goodman PhD[1], Kathleen M Hegadoren RN PhD[2], Levinus A Dieleman MD PhD[1], Richard N Fedorak MD FRCPC[1]

VW Huang, H Chang, KI Kroeker, et al. Does the level of reproductive knowledge specific to inflammatory bowel disease predict childlessness among women with inflammatory bowel disease? Can J Gastroenterol Hepatol 2015;29(2):95-103.

BACKGROUND: Women with inflammatory bowel disease (IBD) may choose to remain childless due to a lack of IBD-specific reproductive knowledge.

OBJECTIVES: To examine the effects of IBD-specific reproductive knowledge and discussion of family planning with a physician on childlessness among women with IBD.

METHODS: Female IBD patients 18 to 45 years of age completed the Crohn's and Colitis Pregnancy Knowledge questionnaire (CCPKnow), and answered questions regarding reproductive history, plans to have children and discussion of family planning with a physician. CCPKnow scores were grouped according to poor (0 to 7), adequate (8 to 10), good (11 to 13) and very good (14 to 17).

RESULTS: Of 434 eligible women, 248 (57.1%) completed the questionnaires. Of these 248 women, 51.6% were childless and, among these, 12.9% were voluntarily childless and 12.1% were trying to become pregnant. Childless women had a lower median CCPKnow score than women with children (6.0 versus 8.0; P=0.001). After adjusting for current age and marital status, each one point increase in the CCPKnow score corresponded to 8% lower odds of childlessness (OR 0.92 [95% CI 0.86 to 0.99]), 9% lower odds of voluntary childlessness (OR 0.91 [95% CI 0.79 to 1.0]) and 20% higher odds of trying to become pregnant (OR 1.2 [95% CI 1.0 to 1.4]). Discussion of family planning with a gastroenterologist corresponded to 72% lower odds of a poor CCPKnow score (OR 0.28 [95% CI 0.15 to 0.53]) and of voluntary childlessness (OR 0.28 [95% CI 0.057 to 1.3]).

CONCLUSION: In the present study, higher IBD-specific reproductive knowledge lowered the odds of childlessness among women with IBD. Discussion of family planning with a physician was associated with higher CCPKnow scores and lower odds of voluntary childlessness.

Key Words: *Childlessness; Family planning; Inflammatory bowel disease; Patient knowledge; Physician knowledge*

Le niveau de connaissances sur la procréation en cas de maladie inflammatoire de l'intestin est-il prédictif de l'infécondité chez les femmes atteintes de cette maladie?

HISTORIQUE : Les femmes atteintes d'une maladie inflammatoire de l'intestin (MII) peuvent choisir de ne pas avoir d'enfants en raison d'un manque de connaissances sur la procréation en cas de MII.

OBJECTIFS : Examiner les effets des connaissances sur la procréation liée aux MII et des discussions sur la planification familiale avec un médecin au sujet de l'infécondité chez les femmes atteintes d'une MII.

MÉTHODOLOGIE : Des patientes atteintes d'une MII de 18 à 45 ans ont rempli le questionnaire CCPKnow sur les connaissances de la grossesse en cas de maladie de Crohn et de colite ulcéreuse et ont répondu à des questions sur les antécédents reproductifs, les projets d'avoir des enfants et les discussions sur la planification familiale avec un médecin. Les scores du CCPKnow étaient classés comme mauvais (0 à 7), adéquats (8 à 10), bons (11 à 13) et très bons (14 à 17).

RÉSULTATS : Sur 434 femmes admissibles, 248 (57,1 %) ont rempli le questionnaire. De ce nombre, 51,6 % n'avaient pas d'enfants, dont 12,9 % avaient fait ce choix volontairement et 12,1 % tentaient d'en avoir. Les femmes sans enfants présentaient un score CCPKnow médian plus faible que celles qui en avaient (6,0 par rapport à 8,0; P=0,001). Après rajustement compte tenu de l'âge et de la situation conjugale, chaque augmentation d'un point au score CCPKnow correspondait à une diminution de 8 % de la probabilité de ne pas avoir d'enfant (RC 0,92 [95 % IC 0,86 à 0,99]), à une diminution de 9 % de la décision volontaire ne pas en avoir (RC 0,91 [95 % IC 0,79 à 1,0]) et à une augmentation de 20 % de la probabilité d'essayer d'avoir un enfant (RC 1,2 [95 % IC 1,0 à 1,4]). Des discussions sur la planification familiale avec un gastroentérologue entraînaient une diminution de 72 % de la probabilité d'obtenir un mauvais score CCPKnow (RC 0,28 [95 % IC 0,15 à 0,53]) et de décider volontairement ne pas avoir d'enfant (RC 0,28 [95 % IC 0,057 à 1,3]).

CONCLUSION : Dans la présente étude, de meilleures connaissances sur la procréation liée aux MII réduisait la possibilité de ne pas avoir d'enfant chez les femmes atteintes d'une MII. Les discussions sur la planification familiale avec un médecin s'associaient à des scores CCPKnow plus élevés et à une plus faible probabilité de décider volontairement de ne pas avoir d'enfants.

Inflammatory bowel disease (IBD) affects patients in their reproductive adult years. Management of IBD can, therefore, be challenging when integrating the effects of IBD and its treatment on fertility, pregnancy, breastfeeding, and maternal and neonatal outcomes. Women with IBD have been reported to have fewer children, with reported ranges of 14% to 36% of women with IBD choosing not to have children, a decision termed 'voluntary childlessness' (1-4), compared with reported ranges of 2.5% to 28% in the general population (3). Previous reports have documented that the reasons women with IBD choose voluntary childlessness include concerns related to medication-associated teratogenicity (2,3), the impact of pregnancy on their IBD (1), the risk of passing IBD to their child (1,2) and not being able to care for their child (1). However, many of these concerns are unsubstantiated because women with IBD, in general, have similar fertility to the general population, and are able to carry successful pregnancies and care for their infants, especially if their IBD is in remission (5-11).

Several studies have reported deficits in IBD-specific reproductive knowledge among women with IBD and investigators have postulated that voluntary childlessness is due to lack of knowledge (2,4,12,13). However, the association between IBD-specific reproductive knowledge and childlessness among women with IBD is not well characterized. Current age and marital status may influence a woman's decision regarding having children; these variables have not been studied previously with respect to IBD-specific reproductive knowledge and

[1]*Division of Gastroenterology, Faculty of Medicine;* [2]*Faculty of Nursing, University of Alberta, Edmonton, Alberta*
Correspondence: Dr Vivian W Huang, 130 University Campus, Zeidler Ledcor Centre, Edmonton, Alberta T6G 2X8. e-mail vwhuang@ualberta.ca

childlessness, including voluntary childlessness. IBD-specific reproductive knowledge may also influence whether a childless woman tries to become pregnant, a behaviour that reveals whether childlessness is voluntary. The association between knowledge and trying to become pregnant has not previously been examined among women with IBD. In addition, there has also been little exploration of the association of physician-patient discussion of family planning with knowledge and childlessness among patients.

The primary objective of the present study was to examine the effect of IBD-specific reproductive knowledge level (adjusted for age and marital status) on childlessness, including but not limited to voluntary childlessness, among women with IBD. The secondary objectives were to examine the effects of family planning discussion with a physician on IBD-specific reproductive knowledge among women with IBD and voluntary childlessness among childless women with IBD.

METHODS

Study design

A cross-sectional survey of women with IBD was designed to assess their knowledge of IBD-specific reproductive issues and to assess their plans with regard to having children.

Setting and participants

The study was conducted at the Inflammatory Bowel Disease Consultation and Research Clinic at the University of Alberta Hospital (Edmonton, Alberta). An electronic health records database was used to systematically identify all female patients with IBD (Crohn disease, ulcerative colitis, indeterminate colitis) 18 to 45 years of age who attended the clinic between January 2010 and June 2013. The years for inclusion criteria for clinic attendance were selected to reflect the most recent interaction between patient and physician. There were no predetermined exclusion criteria in terms of IBD type, medications or reproductive history.

A package of printed material consisting of an information and consent form, the study questionnaires and a stamped, postage-paid return envelope was either mailed to the home address on file or handed directly to patients during a clinic visit. Patients who had not returned the self-administered questionnaire and survey within three months were sent either a reminder by mail or, if an e-mail address was available, by e-mail with a link to an Internet-based version of the questionnaire. A second reminder via telephone occurred approximately three months after the first reminder. The number of eligible women identified from the electronic database and the subsequent response rate determined the sample size for the present study.

Data sources and variable definitions

Crohn's and Colitis Pregnancy Knowledge score: The tool used to measure IBD-specific reproductive knowledge was the Crohn's and Colitis Pregnancy Knowledge survey (CCPKnow) (14). CCPKnow scores, calculated by summing the correct responses to each question, range from 0 to 17, and respondents were categorized into poor (0 to 7), adequate (8 to 10), good (11 to 13) and very good (14 to 17) CCPKnow knowledge levels.

Reproductive history: *history of pregnancy and childlessness*: Women were divided into categories according to history of pregnancy: ever pregnant or never been pregnant; and according to history of having biological children: childless (no live births, adopted children only); or have children. Women who had adopted children only were included in the childless group, and women who had both biological and adopted children were included in the have children group because many of the knowledge topic target issues relating to the physiology of not only carrying a biological pregnancy, but also of delivery, postpartum and breastfeeding.

Family planning decisions: *voluntary childlessness and trying to become pregnant*: Women were asked about their current plans to have children with response options: do not want to have children; would like adopted children only; trying to become pregnant; have children, would like more; have children, completed family; do not have children, would like to later on; and unsure. Childless women were subcategorized as having voluntary childlessness if they were childless at the time of the survey and if they had selected 'do not want to have children' or 'would like adopted children only'. Childless women were also categorized as 'trying to become pregnant' if they were childless at the time of study completion and if they selected the response 'trying to become pregnant'.

Other study variables

The classification of subjects as having discussed family planning with a physician was based on responses to the question "who have you discussed family planning with?", with response options being: Gastroenterologist; General medicine specialist; Obstetrician; Family physician; IBD nurse; Pharmacist; Family and friends; Other; or Have not discussed with anyone. The questionnaire also asked patients questions regarding demographics, factors that could affect knowledge, fertility and pregnancy, IBD history, and reproductive history and current family planning status.

Exposure (independent) variables and outcome (dependent) variables: To estimate the effects of IBD-specific reproductive knowledge on childlessness, the CCPKnow score was analyzed as the exposure variable, with childlessness as the outcome variable. To estimate the effects of discussion of family planning on IBD-specific reproductive knowledge and voluntary childlessness, discussion of family planning was analyzed as the exposure variable, with CCPKnow score and voluntary childlessness as the outcome variables.

Methods used to reduce bias

To minimize selection bias, all eligible patients in the database were systematically invited to participate and multiple attempts were made to reach each patient. Using detailed questions to accurately classify reproductive history, family planning status and intentions with respect to having children minimized measurement bias. In addition, a standard measure of IBD-specific reproductive knowledge was obtained through the CCPKnow survey.

Statistical methods

The present analysis aimed to estimate the effects of IBD-specific reproductive knowledge on childlessness in two ways: by comparing the median CCPKnow scores of childless women to women with children; and by estimating ORs for the dose-response association between the CCPKnow score in one point increments and childlessness. The analysis aimed additionally to estimate the effects of having discussed family planning on IBD-specific reproductive knowledge and voluntary childlessness.

For CCPKnow scores and other continuous variables, medians and interquartile ranges (IQRs) were calculated, and medians were compared across subgroups using nonparametric Mann-Whitney and Kruskall-Wallis tests. For categorical variables, frequency distributions of categories were tabulated and differences in distributions were compared across subgroups. The statistical hypothesis that the distributions did not differ was tested using the χ^2 test. P-values for the null hypothesis of no difference are reported for the comparison of medians and frequency distributions.

When analyzed as an outcome variable, IBD-specific reproductive knowledge was dichotomized as poor versus adequate/good/very good based on the CCPKnow score. To estimate the effect of independent variables on either childlessness or poor CCPKnow scores, multivariable logistic regression was used to obtain ORs and 95% CIs adjusted for selected covariates. Covariates were selected based on whether they could affect both the independent variable of interest and the outcome of interest, without being affected by the independent variable. The resulting multivariable models were repeated with stratification according to age and marital status because the prevalence of childlessness was lower among older women and among married women. SPSS version 21.0 (IBM Corporation, USA) for Windows 2012 (Microsoft Coporation, USA) was used for all data analysis.

TABLE 1
Prevalence of childlessness according to selected study variables among 248 female inflammatory bowel disease (IBD) patients*

Category	Total patients (n=248) n (% of total)	Prevalence of childlessness n/n (% of category)	P (χ^2)
Current age, years			
18 to 24	49 (19.8)	47/49 (95.9)	<0.001
25 to 29	50 (20.2)	34/50 (68.0)	
30 to 34	56 (22.6)	24/56 (42.9)	
35 to 39	47 (19.0)	12/47 (25.5)	
40 to 45	46 (18.5)	11/46 (23.9)	
Marital status			
Partnered	173 (69.8)	66/173 (38.2)	<0.001
Divorced	8 (3.2)	2/8 (25.0)	
Single	67 (27.0)	60/67 (89.6)	
Ethnicity (n=230)†			
Caucasian	216 (93.9)	118/216 (54.6)	0.39
Non-Caucasian	14 (6.1)	6/14 (42.9)	
Education			
Grade 1 to grade 12	56 (22.6)	29/56 (51.8)	0.26
College/university	157 (63.3)	76/157 (48.4)	
Graduate degree	11 (4.4)	7/11 (63.6)	
Professional school degree	5 (2.0)	2/5 (40.0)	
Other	19 (7.7)	14/19 (73.7)	
Employment (n=229)†			
Unemployed	28 (12.2)	13/28 (46.4)	0.02
Part-time	46 (20.1)	24/46 (52.2)	
Full-time	117 (51.1)	73/117 (62.4)	
Other	38 (16.6)	13/38 (34.2)	
Family history of IBD			
Yes	107 (43.1)	52/107 (48.6)	0.41
No	141 (56.9)	76/141 (53.9)	
History of pregnancy			
Yes	136 (54.8)	17/136 (12.5)	<0.001
No	112 (45.2)	111/112 (99.7)	
Timing of pregnancy in relation to IBD diagnosis (n=135)†			
Before only	36 (26.7)	4/36 (11.1)	0.50
After only	78 (57.8)	11/78 (14.1)	
Before and after	21 (15.6)	1/21 (4.8)	
Type of IBD			
Crohn disease	150 (60.5)	76/150 (50.7)	0.49
Ulcerative colitis	88 (35.5)	45/88 (51.1)	
Indeterminate	10 (4.0)	7/10 (70.0)	
Age at IBD diagnosis, years			
<18	65 (26.2)	49/65 (75.4)	<0.001
≥18	183 (73.8)	79/183 (43.2)	
Duration of IBD, years			
0 to 5	57 (23.0)	33/57 (57.9)	0.28
≥5	191 (77.0)	95/191 (49.7)	
History of medications for IBD			
5-aminosalicylates/sulfasalazine			
Yes	218 (87.9)	113/218 (51.8)	0.85
No	30 (12.1)	15/30 (50.0)	
Steroids			
Yes	216 (87.1)	114/216 (52.8)	0.34
No	32 (12.9)	14/32 (43.8)	
Methotrexate			
Yes	42 (16.9)	19/42 (45.2)	0.36
No	206 (83.1)	109/206 (52.9)	
Azathioprine/6-mercaptopurine			
Yes	180 (72.6)	97/180 (53.9)	0.24
No	68 (27.4)	31/68 (45.6)	
Biologics (infliximab, adalimumab)			
Yes	117 (47.2)	62/117 (53.0)	0.68
No	131 (52.8)	66/131 (50.4)	
History of surgery for IBD			
Ostomy			
Yes	24 (9.7)	18/24 (75.0)	0.016
No	224 (90.3)	110/224 (49.1)	
Colectomy			
Yes	17 (6.9)	13/17 (76.5)	0.034
No	231 (93.1)	115/231 (49.8)	
History of perianal disease			
Yes	54 (24.0)	29/54 (53.7)	0.99
No	92 (76.0)	92/171 (53.8)	
Have discussed family planning with a physician			
Yes	154 (62.1)	72/154 (46.8)	0.050
No	94 (37.9)	56/94 (59.6)	
Have discussed family planning with a gastroenterologist			
Did not discuss	94 (37.9)	56/94 (59.6)	0.016
Gastroenterologist	105 (42.3)	43/105 (41.0)	
Nongastroenterologist	49 (19.8)	29/49 (59.2)	
CCPKnow level			
Poor (0 to 7)	131 (52.8)	78/131 (59.4)	0.070
Adequate (8 to 10)	49 (19.8)	21/49 (42.9)	
Good (11 to 13)	48 (19.4)	21/48 (43.8)	
Very good (14 to 17)	20 (8.1)	8/20 (40.0)	
CCPKnow level - dichotomized			
Poor (0 to 7)	131 (52.8)	78/131 (59.4)	0.008
Adequate (≥8)	117 (47.2)	50/117 (42.7)	

**From the IBD Consultation and Research Clinic, University of Alberta Hospital (Edmonton, Alberta), 2010 to 2013; †Percentages calculated using the total number of patients with responses as the denominator. CCPKnow Crohn's and Colitis Pregnancy Knowledge survey*

Ethical considerations

The present study and the study materials were approved by the Health Research Ethics Board of the University of Alberta. Patients were informed that participation in the study was voluntary. Paper study materials included wet-ink signed consent forms while electronic study materials included an electronic consent form. Patients received a $10 coffee gift card on return of the completed study materials.

RESULTS

Participants

Of the 434 eligible women with IBD in the clinic database, 252 of 434 (58.1%) completed the questionnaire. Four incomplete questionnaires were excluded, yielding 248 of 434 (57.1% response rate) evaluable subjects, with their demographic data summarized in Table 1. The median age was 32.0 years (IQR 26.0 to 38.0 years), with 69.8% (173 of 248) of the respondents indicating they had a partner (married/in a long-term relationship). Crohn disease was present in 60.5% (150 of

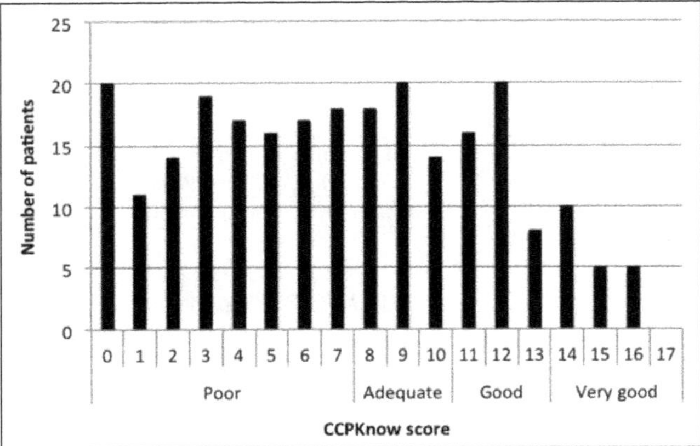

Figure 1) *Distribution of Crohn's and Colitis Pregnancy Knowledge survey (CCPKnow) scores measuring inflammatory bowel disease (IBD)-specific reproductive knowledge among 248 female IBD patients from the IBD Consultation and Research Clinic, University of Alberta Hospital (Edmonton, Aberta), 2010 to 2013 (n=248). Poor n=131/248 (52.8%); Adequate n=49/248 (19.8%); Good n=48/248 (19.4%); Very good n=20/248 (8.1%)*

248), ulcerative colitis in 35.0% (88 of 248) and indeterminate colitis in 4.0% (10 of 248) of the respondents. The median age at diagnosis was 21.0 years (IQR 17.0 to 27.0 years) and the median duration of disease was 9.0 years (IQR 5.0 to 15.0 years).

Reproductive history: pregnancy history and childlessness, voluntary childlessness and trying to become pregnant

Slightly more than one-half (128 of 248 [51.6%]) of the women were childless. Seven women indicated they had adopted children, of whom three had both adopted and biological children, and four had adopted children only. Only one of these four women had ever been pregnant (she underwent a therapeutic abortion). The prevalence of childlessness among each category examined is shown in Table 1. The prevalence of childlessness was higher among patients who were diagnosed with IBD before 18 years of age (P<0.001), had a history of ostomy (P<0.016) or colectomy (P=0.034), and among those working full-time (P=0.02), those with poor CCPKnow scores (P=0.008) and those without partners (P=0.001).

Among the 128 childless women, 14 (10.9%) indicated they had ever sought a fertility specialist or had undergone fertility treatment. There was little difference between women who had children and childless women in the proportion who had sought a fertility specialist or fertility treatment (14 of 120 [11.7%] versus 14 of 128 [10.9%]; P=0.86).

Seventeen of the 128 childless women indicated they had ever been pregnant, with a total of 22 pregnancies. Five (29.4%) of these 17 women had been pregnant only before they were diagnosed with IBD, 11 (64.7%) only after they were diagnosed with IBD, and one (5.9%) both before and after being diagnosed with IBD; 13 of these 17 (76.5%) women experienced miscarriage/stillbirth/spontaneous abortions and nine (52.9%) underwent therapeutic abortions.

Of the 128 childless women, 116 answered the question about their current plans to have children. Of these 116 childless women, 15 (12.9%) were categorized as having voluntary childlessness and 14 (12.1%) were categorized as trying to become pregnant.

The 15 voluntarily childless women had an older median age than the remainder of the childless women (36.0 years [IQR 30.0 to 45.0 years versus 26.0 years [IQR 22.0 to 32.0 years]; P=0.001). They were diagnosed at an older median age (22.0 years [IQR 20.0 to 26.0 years] versus 18.0 years [IQR 15.0 to 22.0 years]; P=0.007), and had a longer median duration of disease (13.0 years [IQR 6.0 to 20.0 years] versus 8.0 years [IQR 5.0 to 12.0 years]; P=0.032). There were no clear differences in disease type or disease characteristics (perianal disease, ostomy, colectomy) compared with other childless women. Three of these 15 (20.0%) women had ever been pregnant; all three underwent therapeutic abortions.

The 14 childless women who were trying to become pregnant had an older median age than the remainder of the childless women (34.0 years [IQR 31.0 to 37.0 years] versus 26.0 years [IQR 21.0 to 32.0 years]; P<0.001). Their median age at diagnosis was similar to that of the remainder of the childless women (19.0 years [IQR 17.0 to 28.0 years] versus 19.0 years [IQR 15.0 to 22.0 years]; P=0.19). There were no clear differences in disease type or disease characteristics (perianal disease, ostomy, colectomy) compared with other childless women. Four (28.6%) women had ever been pregnant, all of whom experienced miscarriages.

CCPKnow scores

Figure 1 illustrates the distribution of CCPKnow scores. The median CCPKnow score was 7.0 (IQR 3.0 to 11.0), with more than one-half (131 of 248 [52.8%]) of the women having poor CCPKnow scores. Table 2 shows the median CCPKnow scores and the proportion with poor CCPKnow scores among women in each category. The median CCPKnow score was 4.0 points higher among women who had ever been pregnant (P<0.001), 2.0 points higher among women who had children (P<0.001) and 4.0 points higher among women who had discussed family planning with a physician (P<0.001).

The association of IBD-specific reproductive knowledge level with childlessness, voluntary childlessness and trying to become pregnant

As shown in Table 1, the prevalence of childlessness was 16.8% higher among women with poor CCPKnow scores than among women with adequate or higher CCPKnow scores (78 of 131 [59.5%] versus 50 of 117 [42.7%], respectively; P=0.008). As shown in Table 2, childless women had a median CCPKnow score that was 2.0 points lower than the median CCPKnow score of women with children (6.0 [IQR 3.0 to 9.0] versus 8.0 [IQR 5.0 to 11.0], respectively; P=0.001). As shown in Table 3, multivariable logistic regression estimated an OR adjusted for current age and marital status of 0.92 (95% CI 0.86 to 0.99) for the effect of IBD-specific reproductive knowledge on childlessness; thus, every one point increase in the CCPKnow score corresponded to an 8% lower odds of childlessness.

Given that there were large differences in the prevalence of childlessness across groups defined according to age and marital status, the logistic regression model was repeated within strata of age and marital status. Table 3 shows that the observed inverse association between the CCPKnow score and childlessness was strongest among women ≤34 years of age who were partnered; in contrast, the CCPKnow score was not clearly associated with childlessness in single women ≤34 years of age or in older women with partners (there were too few single women >34 years of age to examine this relationship in this group).

Compared with the nonvoluntarily childless women, the 15 voluntarily childless women had a similar median CCPKnow score (4.0 [IQR 2.0 to 10.0] versus 6.0 [IQR 3.0 to 9.0]; P=0.985) and a similar prevalence of poor CCPKnow scores (10 of 15 [66.7%] versus 60 of 101 [59.4%]; P=0.592) (Table 2). Multivariable logistic regression estimated an OR adjusted for current age in years and marital status of 0.91 (95% CI 0.79 to 1.0) for the effect of CCPKnow score on voluntary childlessness among childless women; thus, every one point increase in the CCPKnow score corresponded to 9% lower odds of voluntary childlessness among childless women (Table 4).

Compared with the childless women who were not trying to become pregnant, the 14 childless women who were trying to become pregnant had a higher median CCPKnow score (10.0 [IQR 9.0 to 12.0] versus 5.0 [IQR 2.0 to 9.0]; P=0.001) and a lower prevalence of poor CCPKnow scores (three of 14 [21.4%] versus 67 of 102 [65.7%]; P=0.002) (Table 2). As shown in Table 4, multivariable logistic regression estimated an OR adjusted for current age in years and marital status of 1.2 (95% CI 1.0 to 1.4) for the effect of CCPKnow on trying to become pregnant; thus, every one point increase in CCPKnow score corresponded to 20% higher odds of trying to become pregnant among childless women.

TABLE 2

Description of inflammatory bowel disease (IBD)-specific reproductive knowledge levels in 248 female IBD patients*

Category	CCPKnow score, median (IQR)	P†	Poor CCPKnow score n/n (%)	P‡
Total (n=248)	7.0 (3.0–11.0)	–	131/248 (52.8)	–
Age, years				
18 to 24	4.0 (2.0–8.0)	0.001	34/49 (69.4)	0.04
25 to 29	8.0 (4.0–12.0)		23/50 (46.0)	
30 to 34	9.0 (6.0–12.0)		23/56 (41.1)	
35 to 39	6.0 (3.0–9.0)		27/47 (57.4)	
40 to 45	7.0 (4.0–11.0)		24/46 (52.2)	
Marital status				
Partnered	8.0 (4.0–11.0)	<0.001	75/173 (43.4)	<0.001
Divorced	8.0 (5.0–10.0)		4/8 (50.0)	
Single	5.0 (2.0–7.0)		52/67 (77.6)	
Ethnicity				
Caucasian	7.0 (3.0–11.0)	0.498	113/216 (52.3)	0.384
Non-Caucasian	5.0 (4.0–11.0)		9/14 (64.3)	
Education status				
Grade 1 to grade 12	4.0 (2.0–8.0)	0.004	39/56 (69.6)	0.007
College/university	8.0 (4.0–11.0)		79/157 (50.3)	
Graduate degree	9.0 (8.0–11.0)		2/11 (18.2)	
Professional degree	9.0 (8.0–14.0)		1/5 (20.0)	
Other	7.0 (5.0–10.0)		10/19 (52.6)	
Employment status				
Unemployed	6.0 (3.0–9.0)	0.38	18/28 (64.3)	0.47
Part-time	8.0 (3.0–12.0)		21/46 (45.7)	
Full-time	7.0 (4.0–11.0)		61/117 (52.1)	
Other	7.0 (4.0–10.0)		21/38 (55.3)	
Family history of IBD				
Yes	8.0 (4.0–11.0)	0.39	54/107 (50.5)	0.52
No	7.0 (3.0–11.0)		77/141 (54.6)	
Type of IBD				
Crohn's disease	7.0 (4.0–11.0)	0.37	77/150 (51.3)	0.51
Ulcerative colitis	7.0 (4.0–11.0)		47/88 (53.4)	
Indeterminate	3.0 (1.0–10.0)		7/10 (70.0)	
Age at IBD diagnosis, years				
<18	6.0 (3.0–9.0)	0.027	40/65 (61.5)	0.10
≥18	7.0 (4.0–11.0)		91/183 (49.7)	
Duration of IBD, years				
<5	6.0 (3.0–10.0)	0.218	34/57 (59.6)	0.24
≥5	7.0 (4.0–11.0)		97/191 (50.8)	
History of pregnancy				
No	6.0 (3.0–9.0)	<0.001	70/112 (62.5)	0.006
Yes	8.0 (5.0–11.0)		61/136 (44.9)	
Timing of pregnancy in relation to IBD diagnosis (n=135)				
Before only	5.0 (2.0–8.0)	0.001	23/36 (63.9)	0.011
After only	10.0 (7.0–12.0)		27/78 (34.6)	
Before and after	7.0 (4.0–10.0)		11/21 (52.4)	
Have children				
No	6.0 (3.0–9.0)	<0.001	78/128 (60.9)	0.008
Yes	8.0 (5.0–11.0)		53/120 (44.2)	
Voluntary childlessness				
No	6.0 (3.0–9.0)	0.985	60/101 (59.4)	0.592
Yes	4.0 (2.0–10.0)		10/15 (66.7)	
Childless and trying to become pregnant				
No	5.0 (2.0–9.0)	0.002	67/102 (65.7)	0.002
Yes	10.0 (9.0–12.0)		3/14 (21.4)	
Discussed family planning with a physician				
Yes	8.0 (5.0–11.0)	<0.001	66/154 (42.9)	<0.001
No	4.0 (2.0–8.0)		65/94 (69.1)	
Have discussed family planning with a gastroenterologist				
Did not discuss	4.0 (2.0–8.0)	<0.001	65/94 (69.1)	<0.001
Gastroenterologist	9.0 (7.0–12.0)		34/105 (32.4)	
Non-gastroenterologist	6.0 (3.0–9.0)		32/49 (65.3)	

Continued in next column

**Median Crohn's and Colitis Pregnancy Knowledge survey (CCPKnow) score and the proportion with a poor CCPKnow score according to selected study variables, IBD Consultation and Research Clinic, University of Alberta Hospital (Edmonton, Alberta), 2010 to 2013. †Mann-Whitney test; ‡χ^2 test. IQR Interquartile range*

Secondary objective: The effect of having discussed family planning with a physician on IBD-specific reproductive knowledge

Among the respondents, 62.1% (154 of 248) reported having discussed family planning with a physician (Table 1). As shown in Table 2, women who had discussed family planning with a physician had a median CCPKnow score that was 4.0 units higher than the median CCPKnow score of those who had not (8.0 [IQR 5.0 to 11.0] versus 4.0 [IQR 2.0 to 8.0], respectively; P<0.001). In addition, as shown in Table 2, women who discussed family planning with their gastroenterologist had a median CCPKnow score that was 5.0 points higher than the median CCPKnow score of those who had not discussed family planning with any physician, and 3.0 points higher than that of those who had discussed family planning with a non-gastroenterologist only (P<0.001). Conversely, the proportion with poor CCPKnow scores was 36.7% higher among women who had not discussed family planning with any physician, and 32.9% higher among women who had discussed family planning with a non-gastroenterologist only, compared with the proportion with poor CCPKnow scores among women who had discussed family planning with their gastroenterologist (65 of 94 [69.1%] versus 32 of 49 [65.3%] versus 34 of 105 [32.4%]; P<0.001).

As shown in Table 5, multivariable logistic regression estimated an OR adjusted for current age and marital status of 0.44 (95% CI 0.25 to 0.77) for the effect of having discussed family planning with a physician on having a poor CCPKnow score; thus, discussion of family planning with a physician corresponded to 56% lower odds of having a poor CCPKnow score. Table 5 also shows that discussion of family planning with a gastroenterologist corresponded to 72% lower odds of having a poor CCPKnow score, while discussion with a non-gastroenterologist did not show any effect.

Given that there were large differences in the prevalence of childlessness across groups defined according age and marital status, the logistic regression model was repeated within strata of age and marital status. Table 6 shows that the observed inverse association between discussion of family planning and having a poor CCPKnow score was strongest among women >34 years of age. In contrast, discussion of family planning was not clearly associated with the CCPKnow score in women ≤34 years of age (there were too few single women >34 years of age to examine this relationship in this group).

TABLE 3
Multivariable logistic regression model for estimating the effect of inflammatory bowel disease (IBD)-specific reproductive knowledge* on childlessness

	OR	95% CI
All respondents		
CCPKnow score, per unit increase	0.92	0.86–0.99
Marital status		
Single (n=62)	1.0	–
Partnered† (n=181)	0.21	0.08–0.53
Current age, per year increase	0.86	0.82–0.91
Women ≤34 years of age		
CCPKnow score, per unit increase	0.88	0.79–0.97
Marital status		
Single (n=62)	1.0	–
Partnered (n=91)	0.17	0.051–0.56
Current age, per year increase	0.80	0.71–0.90
Women >34 years of age		
CCPKnow score, per unit increase	0.99	0.88–1.1
Marital status		
Single (n=5)	1.0	–
Partnered (n=88)	0.52	0.078–3.5
Current age, per year increase	0.96	0.83–1.1
Women ≤34 years of age AND partnered		
CCPKnow score, per unit increase	0.87	0.78–0.98
Current age, per year increase	0.81	0.70–0.91
Women >34 years of age AND partnered		
CCPKnow score, per unit increase	0.98	0.86–1.1
Current age, per year increase	0.99	0.83–1.1
Women ≤34 years of age AND single‡		
CCPKnow score, per unit increase	0.92	0.70–1.2
Current age, per year increase	0.78	0.60–1.0

*Analysis stratified according to age and marital status among 248 female IBD patients from the IBD Consultation and Research Clinic, University of Alberta Hospital (Edmonton, Alberta), 2010 to 2013. *According to Crohn's and Colitis Pregnancy Knowledge survey [CCPKnow] score; †Partnered includes married/long-term relationship and divorced; ‡Unable to complete for women >34 years of age who are single because n=5*

Secondary objective: The effect of discussion of family planning with a physician on childlessness, voluntary childlessness and trying to become pregnant

There was a difference in the prevalence of childlessness across groups based on discussion of family planning (Table 2) because 59.6% (56 of 94) of those who had not discussed family planning with a physician were childless, compared with 41.0% (43 of 105) of these who had discussed family planning with their gastroenterologist, and 59.2% (29 of 49) of those who had discussed family planning with a nongastroenterologist (P=0.016).

Among the 116 childless women who responded to the question on their decisions regarding having children, the prevalence of voluntary childlessness was similar among those who had discussed family planning with a physician and those who had not discussed family planning with a physician (eight of 67 [11.9%] versus seven of 49 [14.3%]; P=0.71). As shown in Table 4, multivariable logistic regression estimated an OR adjusted for current age and marital status of 0.28 (95% CI 0.057 to 1.3) for the effect of having discussed family planning with a gastroenterologist on voluntary childlessness among childless women. Thus, discussion of family planning with a gastroenterologist corresponded with 72% lower odds of voluntary childlessness among childless women. However, due to small numbers, the wide CI leaves uncertainty regarding the direction of this effect. Multivariable logistic regression was not performed for the effect of discussion of family planning with a physician on trying to become pregnant because all of the 48 childless women who had not discussed family planning with a physician indicated they were not trying to become pregnant (48 of 48 [100.0%]).

TABLE 4
Multivariable logistic regression model for estimating the effect of infalmmatory bowel disease (IBD)-specific reproductive knowledge* and having discussed family planning with a physician on voluntary childlessness and on trying to become pregnant

Variable	OR	95%CI
Effect of CCPKnow on voluntary childlessness		
CCPKnow score	0.91	0.79–1.0
Current age, years	1.2	1.1–1.3
Marital status		
Single	1.0	–
Partnered†	1.2	0.27–5.2
Effect of discussion of family planning on voluntary childlessness‡		
Have discussed family planning with a physician		
No	1.0	–
Gastroenterologist	0.28	0.057 – 1.3
Nongastroenterologist	0.26	0.043 – 1.5
Current age, years	1.2	1.1 – 1.3
Marital status		
Single	1.0	–
Partnered	1.6	0.32–7.6
Effect of CCPKnow on trying to become pregnant		
CCPKnow score	1.2	1.0–1.4
Current age, years	1.1	1.0–1.2
Marital status		
Single	1.0	–
Partnered	7.1	0.81–62

*Data from 116 childless female IBD patients from the IBD Consultation and Research Clinic, University of Alberta (Edmonton, Alberta) Hospital, 2010 to 2013. *According to Crohn's and Colitis Pregnancy Knowledge survey (CCPKnow) score; †Includes married/long-term relationship and divorced; ‡Unable to complete for the effect of discussion of family planning with a physician on trying to become pregnant because all 48 women who had not discussed family planning were not trying to become pregnant*

DISCUSSION

The present analysis is the largest reported study of women with IBD assessing IBD-specific reproductive knowledge and investigating relationships among IBD-specific knowledge, discussion of family planning and childlessness. Previous studies reported results from smaller numbers of respondents, or from a combination of male and female patients with IBD (3). The response rate in the present study was relatively high (57.1%), as was the prevalence of childlessness (51.6%) in the study population. More than one-half (52.8%) of the participants had poor IBD-specific reproductive knowledge scores according to the CCPKnow survey; this is higher than the percentage of patients with poor knowledge (44.8%) reported in the study by Selinger et al (14), which first used the CCPKnow survey.

The present study estimated a clear inverse dose-response gradient for the relationship of IBD-specific reproductive knowledge, as measured by the CCPKnow survey, and childlessness; that is, each one point increase in the CCPKnow score corresponded to lower odds of childlessness. Knowledge regarding reproductive issues in IBD, however, was not the only factor associated with childlessness in the present study population. Other factors associated with childlessness included current age, marital status, age at diagnosis and duration of disease.

TABLE 5
Multivariable logistic regression model for estimating the effect of having discussed family planning with a physician on having poor inflammatory bowel disease (IBD)-specific reproductive knowledge*

	OR	95% CI
Discussed family planning with any physician		
Have discussed family planning with a physician		
No (n=94)	1.0	–
Yes (n=154)	0.44	0.25–0.77
Marital status		
Single (n=67)	1.0	–
Partnered† (n=181)	0.22	0.10–0.48
Current age, per year increase	1.0	0.98–1.1
Discussed family planning with a gastroenterologist		
Have discussed family planning with a physician		
No (n=94)	1.0	–
Gastroenterologist (n=105)	0.28	0.15–0.53
Nongastroenterologist (n=49)	1.0	0.48–2.2
Marital status		
Single (n=67)	1.0	–
Partnered† (n=181)	0.24	0.11–0.52
Current age, per year increase	1.0	0.98–1.1

*Data from 248 female IBD patients from the IBD Consultation and Research Clinic, University of Alberta Hospital (Edmonton, Alberta), 2010 to 2013. *According to Crohn's and Colitis Pregnancy Knowledge survey score; †Includes married/long-term relationship and divorced*

TABLE 6
Multivariable logistic regression model for estimating the effect of having discussed family planning with a physician on having poor inflammatory bowel disease (IBD)-specific reproductive knowledge*

	OR	95% CI
Among childless women		
Have discussed family planning with a physician		
No (n=56)	1.0	–
Gastroenterologist (n=43)	0.31	0.12–0.78
Nongastroenterologist (n=29)	1.3	0.46–3.9
Marital status		
Single (n=60)	1.0	–
Partnered† (n=68)	0.34	0.14–0.84
Current age (per year increase)	1.0	0.93–1.1
Among women who have children		
Have discussed family planning with a physician		
No (n=38)	1.0	–
Gastroenterologist (n=62)	0.32	0.13–0.81
Nongastroenterologist (n=20)	0.82	0.26–2.5
Marital status		
Single (n=7)	1.0	–
Partnered† (n=113)	0.10	0.011–0.99
Current age (per year increase)	1.1	1.0–1.2
Among women ≤34 years of age		
Have discussed family planning with a physician		
No (n=59)	1.0	–
Gastroenterologist (n=68)	0.35	0.15–0.84
Nongastroenterologist (n=28)	1.4	0.48–4.0
Marital status		
Single (n=62)	1.0	–
Partnered† (n=93)	0.20	0.088–0.46
Current age (per year increase)	1.0	0.91–1.1

Continued in next column

TABLE 6 – CONTINUED
Multivariable logistic regression model for estimating the effect of having discussed family planning with a physician on having poor inflammatory bowel disease (IBD)-specific reproductive knowledge*

	OR	95% CI
Among women >34 years of age†		
Have discussed family planning with a physician		
No (n=35)	1.0	–
Gastroenterologist(n=37)	0.19	0.063–0.57
Nongastroenterologist (n=21)	0.60	0.18–2.0
Marital status		
Single (n=5)	1.0	–
Partnered† (n=88)	0.74	0.069–7.9
Current age (per year increase)	0.91	0.79–1.0
Among women ≤34 years of age AND partnered		
Have discussed family planning with a physician		
No (n=23)	1.0	–
Gastroenterologist (n=54)	0. 40	0.13–1.2
Nongastroenterologist (n=16)	1.6	0.4.3–6.0
Current age (per year increase)	1.0	0.90–1.1
Among women >34 years of age AND partnered		
Have discussed family planning with a physician		
No (n=30)	1.0	–
Gastroenterologist (n=37)	0.19	0.065–0.57
Nongastroenterologist (n=21)	0.61	0.18–2.0
Current age (per year increase)	0.91	0.80–1.0
Among women ≤34 years of age AND single‡		
Have discussed family planning with a physician		
No (n=36)	1.0	–
Gastroenterologist (n=14)	0.30	0.067–1.3
Nongastroenterologist (n=12)	1.01	0.18–5.9
Current age (per year increase)	0.98	0.85–1.1

*Analysis stratified according to childlessness, age and marital status among 248 female IBD patients from the IBD Consultation and Research Clinic, University of Alberta Hospital (Edmonton, Alberta), 2010 to 2013. *According to Crohn's and Colitis Pregnancy Knowledge survey score; †Includes married/long-term relationship and divorced; ‡Unable to complete for women >34 years of age who are single because n=5*

While age and marital status are inevitably strong determinants of childlessness, the goal of the present analysis was to estimate the extent to which IBD-specific reproductive knowledge is associated with childlessness among women with IBD, independent of other determinants of childlessness. Even after adjusting for the confounding effects of current age and marital status, both of which encompass a wide range of factors that affect childlessness, our results still showed an inverse effect gradient of increasing levels of IBD-specific reproductive knowledge associated with decreasing odds of childlessness. This inverse relationship was strongest among women ≤34 years of age who were partnered, and was not clearly associated with childlessness in single women ≤34 years of age; thus, reproductive knowledge was strongly associated with childlessness once a stable partnered relationship was established.

Although it may be theorized that women who have been pregnant and who have children may have higher IBD-specific reproductive knowledge gained through the experience of having children, this study estimated higher IBD-specific reproductive knowledge scores among women who had discussed family planning with a physician compared with women who had not done so, irrespective of whether they had children. In addition, the present study estimated that the effect of having discussed family planning with a physician is dependent on the physician type. Those who had discussed family planning with their gastroenterologist had much higher CCPKnow scores than those who had spoken with a nongastroenterologist only or who had not spoken with any physician. The odds of having poor CCPKnow scores were 72% lower among women who had discussed family planning with their gastroenterologist, compared with those who had not discussed family planning with any physician. The effect of discussion of family planning with a gastroenterologist on IBD-specific reproductive knowledge remained strong even after stratification according to childlessness, age and marital status. This finding is important becuase it suggests that higher IBD-specific reproductive knowledge does not necessarily require the personal experience of having children, and that discussion with a physician, especially a gastroenterologist may influence IBD-specific reproductive knowledge.

Previous studies have estimated that approximately 20% to 30% of women with IBD discuss family planning with a physician (13,21), although Marri et al (1) reported that 68% of their study population had discussed reproductive issues with a physician. In this study population, 62.1% of women indicated that they previously had discussed family planning with a physician, with 42.3% indicating they had discussed family planning with their gastroenterologist.

While the cross-sectional design of the present study does not rule out the possibility that women who derive their knowledge from other sources seek family planning discussions with physicians, the results suggest the potential for improving women's IBD-specific reproductive knowledge through increasing the engagement of physicians in discussing family planning with their IBD patients. Further studies differentiating physician IBD-specific reproductive knowledge levels are required because nongastroenterologist physicians may have lower IBD-specific reproductive knowledge.

The present study also estimated the effect of IBD-specific reproductive knowledge on voluntary childlessness and trying to become pregnant among childless women with IBD. When childless women who did not want to have children were classified as having 'voluntary childlessness', the 'voluntary childlessness' rate (15 of 116 [12.9%]) was slightly lower than reported in previous studies (1,2,4,12). Higher IBD-specific reproductive knowledge did not correspond to lower odds of voluntary childlessness, but did correspond to higher odds of trying to become pregnant among childless women with IBD. Due to the cross-sectional nature of the study, however, it is unclear if childless women who were trying to become pregnant had higher CCPKnow scores due to actively seeking information because they were trying to become pregnant, or if they decided to try to become pregnant due to higher knowledge levels.

Although we had a relatively high response rate and a large number of respondents, there are several limitations to the present study. The patient population was sampled from a single tertiary care centre and, therefore, may represent a patient population that may be more knowledgeable than the general female IBD population. However, this university-based centre services a large referral base that encompasses Western Canada, and the patient population is thus, quite diverse in terms of demographic factors and IBD disease history. We attempted to obtain a representative sample of women with IBD by systematically recruiting all female IBD patients in the target age group who had been seen in the IBD clinic within the past three years regardless of their disease characteristics or reproductive history. Women who responded may have been more knowledgeable or interested in the study than those who did not respond and, thus, may not represent all women with IBD. Also, despite having a relatively large sample size compared with previous studies, results estimated for subgroups of childless women (voluntary childless; and those who were trying to become pregnant) had reduced precision. In addition, although we captured highest level of education as a marker of socioeconomic status, we did not capture financial socioeconomic status, which may have an effect on knowledge, concerns and interaction with health care providers. Nevertheless, our results show how complex this topic is, and reveal questions that warrant further research.

Another limitation of the present cross-sectional survey was its inability to establish the time sequence of events. The present study cannot determine what a woman's IBD-specific reproductive knowledge was when she made the decision to remain voluntarily childless or to have children. It is possible that women who made the decision to remain childless earlier in their lives have lower IBD-specific reproductive knowledge because they believed they did not need to seek this knowledge. Our data pattern, which shows that the women in the youngest age group (18 to 24 years) and those who were single at time of the study had the lowest median CCPKnow scores, supports this possibility. In addition, the finding that childless women who were trying to become pregnant had higher CCPKnow scores also suggests that they may have sought out the knowledge because they were trying to become pregnant.

When assessing knowledge, personal experience may play a role, and women with a history of pregnancy had higher CCPKnow scores and a lower prevalence of poor CCPKnow levels. Nevertheless, the finding that discussion of family planning with a physician, especially a gastroenterologist, corresponded to higher CCPKnow scores and lower prevalence of poor CCPKnow level knowledge suggests that regardless of personal experience, discussion with a physician is associated with better IBD-specific reproductive knowledge.

It should be noted that the study questionnaire did not capture whether the patient or a physician initiated discussions of family planning, or exactly what the discussion encompassed. Physician-patient discussion of family planning could vary from a brief discussion about contraception or fertility, to lengthy discussions about reproductive issues in IBD, and are likely to vary substantially across health care providers. These would be important pieces of information to capture in future studies. However, even with this limitation, our study does indicate that a discussion of family planning, reported as such by the patient, is associated with higher knowledge scores, and identifies discussion of family planning as a potentially modifiable factor that may influence patient knowledge and childlessness. Additionally, future studies should examine whether men with IBD also have similar concerns and, thus, contribute to decisions regarding voluntary childlessness in couples in which one or both partners has IBD.

Overall, the present study found that poor IBD-specific reproductive knowledge is strongly associated with increased childlessness among women with IBD and describes an association of increasing IBD-specific reproductive knowledge with decreasing odds of childlessness. A potentially modifiable factor that was associated with better knowledge and lower odds of childlessness among women with IBD was discussion of family planning with a physician, specifically with a gastroenterologist. Future research investigating this topic should focus on elucidating the complex factors involved in voluntary childlessness among women with IBD, and on identifying effective approaches for physicians who care for women with IBD to discuss family planning issues specific to their concerns.

ACKNOWLEDGEMENTS: Vivian Huang was supported to perform this research by the Alberta Innovates Health Solutions (AIHS) Clinician Research Fellowship Award. Additional support for the research was provided by the Alberta Inflammatory Bowel Disease Consortium, the Centre of Excellence for Gastrointestinal Inflammation and Immunity Research (CEGIIR), and the Cecile Mactaggart Summer Studentship. RNF acts as the article guarantor and contributed to the study design, data collection, data analysis, and manuscript drafting and editing. VH contributed to study design, data collection, data analysis, and manuscript drafting and editing. HJC contributed to data analysis and manuscript editing. KIK contributed to study design, data collection and manuscript editing. KG contributed to study design, data analysis and manuscript editing. KH and LAD contributed to study design and manuscript editing. All authors approved the final version of the manuscript.

DISCLOSURES: The authors have no financial disclosures or conflicts of interest to declare.

REFERENCES

1. Marri SR, Ahn C, Buchman AL. Voluntary childlessness is increased in women with inflammatory bowel disease. Inflamm Bowel Dis 2007;13:591-9.
2. Mountifield R, Bampton P, Prosser R, et al. Fear and fertility in inflammatory bowel disease: A mismatch of perception and reality affects family planning decisions. Inflamm Bowel Dis 2009;15:720-5.
3. Tavernier N, Fumery M, Peyrin-Biroulet L, et al. Systematic review: Fertility in non-surgically treated inflammatory bowel disease. Aliment Pharmacol Ther 2013:38:847-53.
4. Mountifield RE, Prosser R, Bampton P, et al. Pregnancy and IBD treatment: This challenging interplay from a patients' perspective. J Crohns Colitis 2010;4:175-82.
5. Willoughby CP, Truelove C. Ulcerative colitis and pregnancy. Gut 1980;21:469-74.
6. Mayberry JF, Weterman IT. European survey of fertility and pregnancy in women with Crohn's disease: A case control study by European collaborative group. Gut 1986;27:821-5.
7. Khosla R, Willoughby CP, Jewell DP. Crohn's disease and pregnancy. Gut 1984;25:52-6.
8. Lindhagen T, Bohe M, Ekelund G, et al. Fertility and outcome of pregnancy in patients operated on for Crohn's disease. Int J Colorectal Dis 1986;1:25-7.
9. Dr Domabl FT, Burton IL, Goligher JC. Crohn's disease and pregnancy. BMJ 1972;3:550-3.
10. Hudson M, Flett G, Sinclair TS, et al. Fertility and pregnancy in inflammatory bowel disease. Int J Gynecol Obstet 1997;58:229-37.
11. Manosa M, Navarro-Llavat M, Marin L, et al. Fecundity, pregnancy outcomes, and breastfeeding in patients with inflammatory bowel disease: A large cohort survey. Scan J Gastroenterol 2013;48:427-32.
12. Selinger CP, Eaden J, Selby W, et al. Inflammatory bowel disease and pregnancy: Lack of knowledge is associated with negative views. J Crohns Colitis 2013;7:e206-213.
13. Toomey D, Waldron B. Family planning and inflammatory bowel disease: The patient and the practitioner. Fam Pract 2013;30:64-8.
14. Selinger CP, Eaden J, Selby W, et al. Patients' knowledge of pregnancy-related issues in inflammatory bowel disease and validation of a novel assessment tool ('CCPKnow'). Aliment Pharmacol Ther 2012;36:57-63.
15. Gawron LM, Hammond C, Keefer L. Documentation of reproductive health counseling and contraception in women with inflammatory bowel diseases. Patient Educ Counsel 2014;94:134-7.
16. Julsgaard M, Norgaard M, Hvas CL, et al. Self-reported adherence to medical treatment prior to and during pregnancy among women with ulcerative colitis. Inflamm Bowel Dis 2011;17:1573-80.

Partially covered self-expandable metal stents versus polyethylene stents for malignant biliary obstruction: A cost-effectiveness analysis

Alan N Barkun MD MSc[1,2], Viviane Adam MSc[1], Myriam Martel BSc[1], Khalid AlNaamani MD BSc[3], Peter L Moses MD[4]

AN Barkun, V Adam, M Martel, K AlNaamani, PL Moses. Partially covered self-expandable metal stents versus polyethylene stents for malignant biliary obstruction: A cost-effectiveness analysis. Can J Gastroenterol Hepatol 2015;29(7):377-383.

BACKGROUND/OBJECTIVE: Partially covered self-expandable metal stents (SEMS) and polyethylene stents (PES) are both commonly used in the palliation of malignant biliary obstruction. Although SEMS are significantly more expensive, they are more efficacious than PES. Accordingly, a cost-effectiveness analysis was performed.

METHODS: A cost-effectiveness analysis compared the approach of initial placement of PES versus SEMS for the study population. Patients with malignant biliary obstruction underwent an endoscopic retrograde cholangiopancreatography to insert the initial stent. If the insertion failed, a percutaneous transhepatic cholangiogram was performed. If stent occlusion occurred, a PES was inserted at repeat endoscopic retrograde cholangiopancreatography, either in an outpatient setting or after admission to hospital if cholangitis was present. A third-party payer perspective was adopted. Effectiveness was expressed as the likelihood of no occlusion over the one-year adopted time horizon. Probabilities were based on a contemporary randomized clinical trial, and costs were issued from national references. Deterministic and probabilistic sensitivity analyses were performed.

RESULTS: A PES-first strategy was both more expensive and less efficacious than an SEMS-first approach. The mean per-patient costs were US$6,701 for initial SEMS and US$20,671 for initial PES, which were associated with effectiveness probabilities of 65.6% and 13.9%, respectively. Sensitivity analyses confirmed the robustness of these results.

CONCLUSION: At the time of initial endoscopic drainage for patients with malignant biliary obstruction undergoing palliative stenting, an initial SEMS insertion approach was both more effective and less costly than a PES-first strategy.

Key Words: *Biliary obstruction; Cost-effectiveness; Gastrointestinal cancer; Stent*

Les endoprothèses de métal auto-expansibles partiellement couvertes par rapport aux endoprothèses de polyéthylène en cas d'obstruction biliaire maligne : une analyse coût-efficacité

HISTORIQUE ET OBJECTIF : Les endoprothèses de métal auto-expansibles (EMAE) partiellement couvertes et les endoprothèses de polyéthylène (EPÉ) sont toutes deux souvent utilisées pour pallier une obstruction biliaire maligne. Même si les EMAE sont beaucoup plus coûteuses, elles sont plus efficaces que les EPÉ. C'est pourquoi les chercheurs ont effectué une analyse coût-efficacité.

MÉTHODOLOGIE : Dans une analyse coût-efficacité, les chercheurs ont comparé l'EPÉ initiale à l'EMAE initiale auprès de la population à l'étude. Les patients ayant une obstruction biliaire maligne ont subi une cholangiopancréatographie rétrograde endoscopique pour insérer la première endoprothèse. Si l'insertion échouait, ils subissaient une cholangiographie transhépatique percutanée. En cas d'occlusion de l'endoprothèse, une EPÉ était insérée à la reprise de la cholangiopancréatographie rétrograde endoscopique, soit en milieu ambulatoire, soit après l'hospitalisation pour cholangite. Les chercheurs ont adopté la perspective du tiers payeur. L'efficacité était exprimée comme la probabilité d'absence d'occlusion dans l'horizon prévu d'un an. Les probabilités reposaient sur un essai clinique aléatoire contemporain tandis que les coûts étaient extrapolés des références nationales. Les chercheurs ont effectué des analyses de sensibilité déterministes et probabilistes.

RÉSULTATS : La stratégie d'installation d'une EPÉ initiale était à la fois plus coûteuse et moins efficace que celle de l'installation d'une EMAE initiale. Les coûts moyens par patient étaient de 6 701 $US pour l'EMAE initiale et de 20 671 $US pour l'EPÉ initiale, qui s'associaient à des probabilités d'efficacité de 65,6 % et de 13,9 %, respectivement. Les analyses de sensibilité ont confirmé la validité des résultats.

CONCLUSION : Au drainage endoscopique initial chez les patients ayant une obstruction biliaire maligne qui se faisaient installer une endoprothèse palliative, l'insertion initiale d'une EMAE était à la fois plus efficace et moins coûteuse que celle d'une EPÉ.

The management of patients with malignant biliary obstruction (MBO) focuses significantly on the relief of patient symptoms achieved with endoscopic drainage using different stent technologies (1,2). Polyethylene stents (PES) remain widely used, even though the newer self-expandable metal stent (SEMS) types have demonstrated improved efficacy (2,3), both covered and uncovered (4-7).

Although there are many reasons for their continued popularity, the choice of PES in the palliation of distal MBO is, no doubt, in part explained by its ease of insertion and favourable upfront cost compared with SEMS (8), with little consideration for downstream expenses attributable to differences in efficacy. Several previous analyses have suggested that SEMS were both clinically superior and more cost effective than PES, even though they do not prolong survival. Specifically, advantages of SEMS have been reported in selected health care settings and in patients with longer anticipated life expectancy (5,9-14). However, even these conclusions have been questioned by some (5,8). Since the publication of most of these reports, the advent of partially covered SEMS (15), coupled to the evolution of biliary endoscopic techniques (16,17), now justifies a more contemporary cost-effectiveness analysis that is based on a contemporary randomized clinical trial (RCT) (6).

[1]*Division of Gastroenterology;* [2]*Division of Epidemiology, Biostatistics and Occupational Health, McGill University Health Centre, McGill University, Montreal, Quebec;* [3]*Division of Gastroenterology, Hepatology and Liver Transplantation, The Armed Forces Hospital, Muscat, Oman;* [4]*Division of Gastroenterology and Hepatology, University of Vermont, Burlington, Vermont, USA*

Correspondence: Dr Alan N Barkun, Division of Gastroenterology, Room D7-346, 1650 Cedar Avenue, Montreal General Hospital site, McGill University Health Centre, Montreal, Quebec H3G 1A4. e-mail alan.barkun@muhc.mcgill.ca

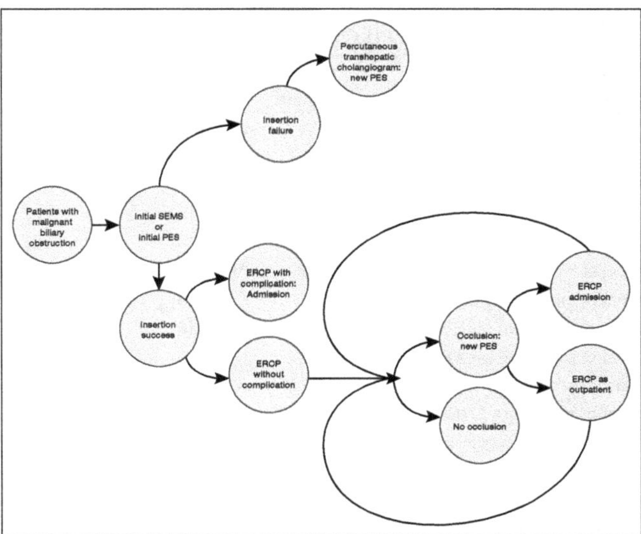

Figure 1) *Influence diagram of the adopted model structure. ERCP Endoscopic retrograde cholangiopancreatography; PES Polyethylene stent; SEMS Partially covered self-expandable metal stent*

METHODS

Overview

An incremental cost-effectiveness decision tree to compare the use of initial PES versus initial SEMS using TreeAge Pro Suite Healthcare module 2015 (18) for palliative patients deemed inoperable with MBO during a 12-month follow-up period was constructed. The outcome was the cost per stent occlusion averted. A third-party-payer perspective was adopted.

Model design

Although the RCT included one-, three-, six-, nine- and 12-month visits following randomization, these were not included in the base-case analysis because they would represent closer follow-up than in a real-life setting. At the beginning of the model, in each branch, after a first consultation an endoscopic retrograde cholangiopancreatography (ERCP) is performed to insert initially either a PES (initial PES) or an SEMS (initial SEMS). If the insertion fails, a percutaneous transhepatic cholangiogram (PTC) is performed, and it is the end of the model for this path. Figure 1 illustrates the structure of the model across a 12-month time horizon following randomization. The model allows for a possible post-ERCP adverse event state as well (pancreatitis, cholangitis, bleeding). In follow-up, the only adverse event that was modelled for was the most common one of interval stent occlusion. The treatment of stent occlusion correspondingly required a stent change to a PES, either on an outpatient basis or with an admission to hospital because of associated cholangitis. Stent occlusion could occur between each planned outpatient visit at any time over the 12-month time frame, termed time horizon. After the 12 months, the patient arrives at a terminal node and a cost per no occlusion is computed by the model for each branch. No patient death node was specifically included in the structure of the model.

Probabilities

The probabilities of failure of stent insertion (initial SEMS or initial PES) were extracted from the literature (2). All other probabilities in the model were based on a contemporary RCT of 85 patients comparing PES to SEMS (6). The inclusion criteria applied in this RCT are listed in Table 1. All of the probabilities used in the model are listed in Table 2.

Costs and lengths of stay

The national United States (US) market price of the two types of stents used for PTC or ERCP were provided by Boston Scientific, USA (19).

TABLE 1
Inclusion and exclusion criteria

Inclusion criteria
Age ≥18 years
Extrahepatic biliary obstruction by any malignant process, extending no more proximal than 1 cm below the common hepatic ductal bifurcation, as measured at the time of enrollment endoscopic retrograde cholangiopancreatography
A Karnofsky Performance Scale Index >60%
Patient physically capable and mentally willing to comply with the protocol requirements including stent insertion as well as the required three-monthly clinical follow-up visits
Signed voluntary informed consent
Exclusion criteria
Jaundice secondary to primary intrahepatic biliary obstruction (eg, patient had underlying chronic liver disease)
Active cholangitis
International normalized ratio >3.0
Multiple strictures requiring therapy
Biliary anatomy inappropriate for the available sizes of the biliary stents used in the study
Previous attempt at curative surgical resection of the malignant obstructing tumour
Patient has participated in a protocol involving an investigational drug or device 90 days before proposed entry into the present study
Allergy to any component of the individual stents, or their materials/delivery systems
Active alcohol or drug abuse
Unstable social circumstances that may preclude follow-up
Inability to understand and execute informed consent

All physicians fees were derived from the national database of the American Medical Association (20). Cost of hospitalization for cholangitis was based on specific admissions included in the Nationwide Inpatient Sample (NIS) 2008 (21). This national hospitalization database comprises eight million hospitalizations occurring in >1000 different hospitals located in 42 states of the US. Hospitalizations of patients who died during the hospital stay or who were <18 years of age were excluded from the analysis. Only hospitalizations that were recorded with Medicare, Medicaid or a private insurance as the primary payer were included. All the charges associated with the *International Classification of Diseases* 9-CM code '576.1: cholangitis' were selected. Costs were computed based on the average charges combined with a cost-to-charge ratio, which was, for the most part, specific to the hospital where the hospitalization took place; if not, a recommended (22) group average cost-to-charge ratio was applied. To obtain valid national cost estimates, discharge weights were used in the computations. The short time horizon of 12 months covering the entire study adopted obviates the need for discounting. All dollar values were expressed in 2014 US dollars utilizing the consumer price index for the medical care services published by the US Department of Labor (23). Indirect costs were not considered in the present analysis. All cost and length of stay estimates are presented in Table 3.

Cost-effectiveness analysis

The effectiveness is expressed as the probability of maintained patency occurring over the full duration of the model (one year with successful initial stent insertion). The costs are the sum of the cost items listed for all the 12 months. Results of the deterministic analysis are reported as cost, effectiveness, cost-effectiveness ratio and incremental cost-effectiveness ratio (ICER).

Sensitivity and threshold analyses

Deterministic and probabilistic sensitivity analyses were performed. All bounds used for the sensitivity analyses of the probabilities were

TABLE 2
Probability estimates of the model

Description of probability	Point estimate	Bound Low	Bound High
Initial polyethylene stent			
Failure of insertion at baseline	0.05	0.02	0.09
Cholangitis before 1 month and before first occl	0.195	0.102	0.34
Cholangitis before 1 month and after first occl	0.286	0.082	0.641
Cholangitis between 1 and 3 months	0.25	0.0456	0.699
Cholangitis between >3 and 6 months	0.286	0.082	0.641
Cholangitis between >6 and 9 months	0	0	0.5
Cholangitis between >9 and 12 months	1	0.342	1
Occl before 1 month	0.171	0.085	0.313
Occl between 1 and 3 months	0.138	0.055	0.306
Occl between >3 and 6 months	0.411	0.216	0.64
Occl between >6 and 9 months	0	0	0.435
Occl between >9 and 12 months	0.667	0.208	0.939
Initial partially covered self-expandable metal stent			
Failure of insertion at baseline	0.0597	0.03	0.1
Cholangitis before 1 month and before first occl	0.049	0.014	0.161
Cholangitis before 1 month and after first occl	0.5	0.095	0.906
Cholangitis between 1 and 3 months	0.5	0.095	0.906
Cholangitis between >3 and 6 months	0	0	0.5
Cholangitis between >6 and 9 months	0	0	0.5
Cholangitis between >9 and 12 months	0	0	0.5
Occl before 1 month	0.049	0.014	0.161
Occl between 1 and 3 months	0.074	0.021	0.234
Occl between >3 and 6 months	0.111	0.031	0.328
Occl between >6 and 9 months	0	0	0.269
Occl between >9 and 12 months	0.111	0.03	0.564

Source of the probabilities: The two probabilities of failure of the insertion were issued from the literature (2). All the other probabilities were based on a randomized clinical trial whose patients were issued from the six participating North American university centres mentioned in the method section (6). The bounds for the sensitivity analysis were based on the 95% CIs. occl Occlusion

TABLE 3
Cost and length of stay (LOS) estimates of the model

Description of cost/ LOS variable	Point estimate	Source	Bound Low	Bound High
Price of initial PES	115	BS (18)	60	230
Price of initial SEMS	2,413	BS (18)	1000	5000
Cost of outpatient follow-up	83	AMA (19)	40	170
Cost of inpatient follow-up	107	AMA (19)	50	220
Cost of outpatient first consultation	139	AMA (19)	70	280
Cost of inpatient first consultation	122	AMA (19)	60	250
Cost of emergency consultation	66	AMA (19)	30	140
Cost of physician fees PTC	467	AMA (19)	230	940
Cost of physician fees ERCP	483	AMA (19)	240	970
Cost of procedure PTC	2,471	BS (18)	1200	5000
Cost of procedure ERCP	1,170	BS (18)	500	2400
Cost of admission for cholangitis	15,100	NIS (20)	5000	30,000
LOS of cholangitis	5.5	NIS (20)	1	12

All costs are in 2014 United States dollars. The bounds for the sensitivity analysis were based on the SD for the LOS and cost of an admission for cholangitis. For all the other cost variables, the low (high) bounds were based on the point estimates divided (multiplied) by 2. AMA American Medical Association; BS Boston Scientific (USA); ERCP Endoscopic retrograde cholangiopancreatography; NIS National Inpatient Sample; PES Polyethylene stent; PTC Percutaneous transhepatic cholangiogram; SEMS Partially covered self-expandable metal stent

TABLE 4
Cost-effectiveness analysis report

Stent	Cost	IC	Eff	IE	CER	ICER
Initial SEMS	6,701	0	0.6561	0	10,213	0
Initial PES	20,671	13,971	0.1385	−0.5176	149,243	Dominated*

*All costs are expressed in 2014 United States dollars. *Incremental cost-effectiveness ratio (ICER) = − US$26,993. CER Cost-effectiveness ratio; Eff Effectiveness (probability of no occlusion during the consecutive 12 months); IC Incremental cost; IE Incremental effectiveness; PES Polyethylene stent; SEMS Partially covered self-expandable metal stent*

based on the 95% CIs. The ranges for the length of stay and cost of an admission for cholangitis were based on the standard deviations of the NIS database (21). All other cost ranges were equal to the point estimates plus 100% or less 50%.

One-way sensitivity analyses were performed on all variables used in the model to investigate the robustness of the results and to determine which factors influenced most these results.

A Tornado diagram was chosen to represent how the variation of the value of an input variable of the model across a pre-set range influences the results of the model. The widest bars of the Tornado diagram identify the variables that most changed the ICER (beyond 20%). As part of the sensitivity analysis, the influence of the time horizon chosen on the model results was specifically assessed. Threshold analyses were also performed to observe the cut-off points for which the conclusion of the model could change. Probabilistic sensitivity analysis was performed running 10,000 Monte Carlo simulations based on the uncertainty distributions of all the variables and a chosen maximal willingness to pay (WTP) value of US$50,000. A WTP represents the pre-fixed maximum dollar value that is deemed to be acceptable spending for a given treatment. Results of the probabilistic analysis are reported as an incremental cost-effectiveness scatter plot. It illustrates the probability that a stent is cost effective compared with the other one, given a WTP.

RESULTS

Base-case analysis

The report of the results (Table 4) shows that using initial SEMS over a period of 12 months for patients with MBO is a strategy significantly less expensive but also more efficacious; indeed, the initial SEMS approach dominates the initial PES strategy. On average, it would save US$13,971 per patient to opt for the initial SEMS strategy while at the same time the patient would have a 50% decrease in probability of biliary stent occlusion during the 12 consecutive month follow-up after the index ERCP. Therefore, the average cost per no occlusion is significantly more attractive with initial SEMS (US$10,213) than for initial PES (US$149,243).

Sensitivity analyses

One-way deterministic sensitivity analyses: Any variable used in the model does not change the final choice of strategy: initial PES is always dominated, even if the point estimates vary inside their respective plausible ranges. The Tornado diagram (Figure 2) shows that within the plausible ranges of all the variables of the model, the cost of hospitalization for cholangitis most influences the ICER value. It is only when the variables are made to assume values outside their predetermined ranges that there is a change in the dominance of the SEMS approach. Indeed, threshold analysis shows that the cost of initial SEMS would have to increase above US$16,240 for initial PES to be no longer dominated. The other threshold values for other variables are even less clinically plausible.

Regardless of the selected time horizon, the initial PES approach is still dominated by the initial SEMS approach (Figure 3). Adding the costs of three-monthly follow-up visits from the first to the 12th month (as in the RCT), do not alter the conclusion with a corresponding ICER of US$26,700 and an average cost strategy of US$7,060 (initial SEMS) versus US$20,880 (initial PES).

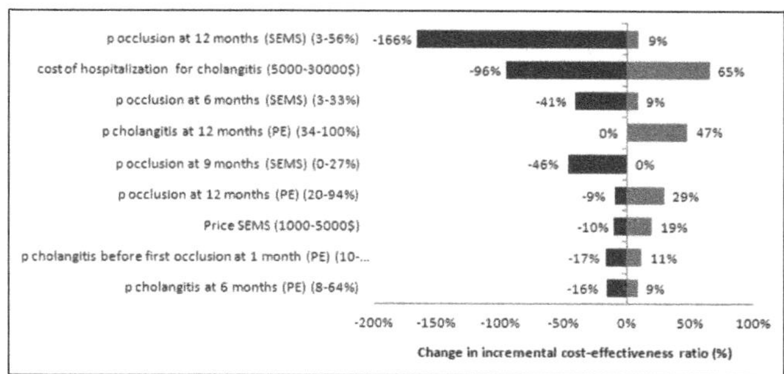

Figure 2) *Tornado diagram for variations of ≥20% and from the base-case incremental cost-effectiveness ratio (ICER) estimate. p Probability of; PES Polyethylene stent; SEMS Partially covered self expandable metal stent*

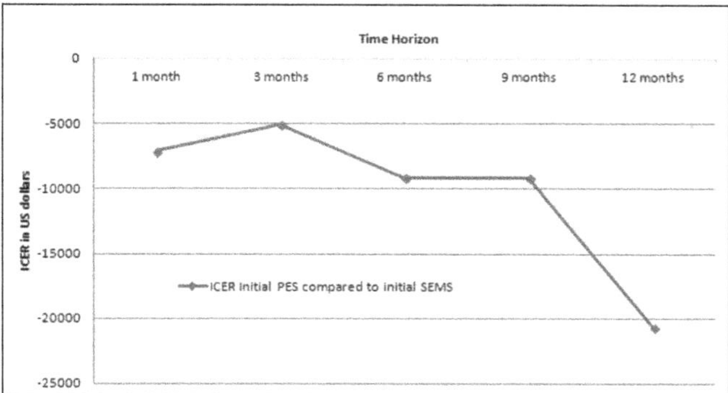

Figure 3) *Incremental cost-effectiveness ratio (ICER) as a function of the adopted time horizon for the cost-effectiveness model. PES Polyethylene stent; SEMS Partially covered self-expandable metal stent*

Probabilistic sensitivity analyses: The Monte-Carlo analysis was based on 10,000 simulations that each vary simultaneously all the variables of the model according across their adopted ranges. It represents the uncertainty about the estimates of costs and measure of effectiveness. The incremental cost-effectiveness scatter plot is presented in Figure 4: each point represents the incremental cost and effectiveness pair from the simulation results for initial SEMS relative to initial PES. The ellipse circumscribes 95% of the cloud of the results of cost and effectiveness points when simulations are performed according to the Monte Carlo analysis. All points below the WTP threshold are the simulation iterations where the initial SEMS is preferred given a WTP of US$50,000: they represent 90% of the 10,000 simulation iterations. It expresses the probabilities of preference for initial SEMS or PES, across a range of possible WTP threshold values. For every WTP threshold, the initial SEMS approach is the preferred management compared with initial PES.

DISCUSSION

The present analysis demonstrates the economic dominance of a SEMS-first approach with robust sensitivity analysis. Such a broad, overriding conclusion has not previously been adopted by other cost-effectiveness analyses and, thus, warrants a discussion of our methodology and a review of the literature.

Considering the short clinically pertinent survival time and the outcome of 'cost per no occlusion', we adopted a cost-effectiveness decision tree approach whose structure was dictated by the RCT on which it is based (6). In accordance with the concept of a theoretical decision model, we simplified some possible events: initial insertion failures are immediately followed by a PTC and a final node. The outcomes of these few patients until the 12th month was not taken into account because it would not have any impact on our comparison; this group of patients is similar between both approaches. We considered

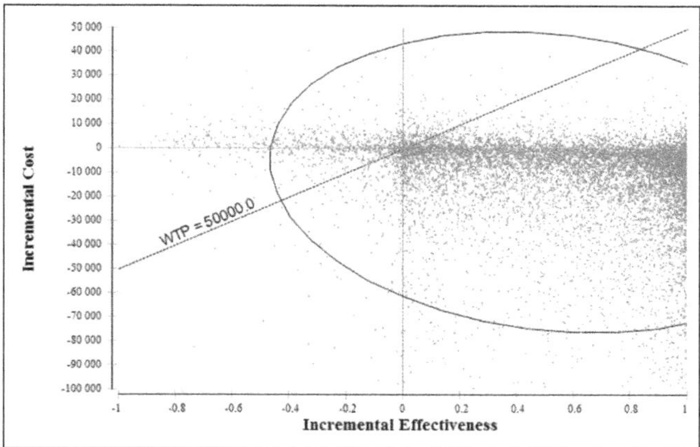

Figure 4) *Incremental cost-effectiveness scatter plot of initial partially covered self-expandable metal stent (SEMS) versus initial polyethylene stent (PES). All costs expressed in 2014 United States dollars. The willingness-to-pay (WTP) threshold (US$50,000) is the line that intersects the origin and the initial PES strategy is represented by the origin (0;0)*

the probability of all adverse events, but specifically modelled for the length of stay and costs attributable to bouts of cholangitis because it is by far the most common in this context (24); other authors have done similarly in past published economic analyses (9,14). We chose not to include death in the measure of effectiveness because this outcome introduces a lot of uncertainty (in addition to the estimation of related costs). Finally, the literature has reported no difference in mortality when comparing the use of each of the two types of stents (5,9,10,12-14,25,26).

Because the clinical trial data were available up to 12 months, it was, therefore, logical to fix the time horizon at one year, as has been performed in most cost-effectiveness analyses (9,13,27). This chosen time horizon is also consistent with the median survival duration observed in recent cohorts of patients with MBO treated with metallic stents (28).

All efficacy assumptions were based on real outcomes information observed in a contemporary RCT comparing both stent technologies (6), an important decision that is supported by the evolution of biliary endoscopic techniques and contemporary supportive care in this patient group. Although such a choice of data source decreases the external validity of the point estimates, it allows for inclusion of more uniform and clinical, currently valid assumptions; in fact, five of the seven studies included in an authoritative meta-analysis by Moss et al (2) were ≥15 years of age (2). Moreover, the RCT on which our analysis is based likely carries sufficient generalizability because patients were included from six participating North American university centres. The robustness of the cost variables of the model is based on the national representativeness of the recent data we used from the NIS 2008 (21) and that we updated for 2011, and are both, here too, more contemporary and more generalizable to a Western setting than many previous reports. In fact, as in efficacy information, many published cost analyses assessing malignant biliary stenting are significantly older, bringing into question the clinical relevance of those previous results. As examples, Prat et al (12) and Davids et al (25) based their analysis on costs markedly inferior to today's purchase costs (for example, US$20 for PES); the SEMS, at that time, also referring to an uncovered stent technology. In addition, most published cost models have not assessed relevant estimates for a North American practice with US data, focusing rather on local national costs and currencies (7,11,14,15,27,29). Other limitations are the inclusion in many analyses of sole stent costs without consideration other expenses (5,13,29,30). The summary review by Moss et al (31) in 2007 computed cost-effectiveness results in which three of the four studies were published before the turn of the century; moreover, their ICER calculation is very approximate because it is based on the assumption that the stent price is the only difference between the two interventions. Because the patients included in the RCT (6) we used as reference were followed every three months during the study, we also added the cost of the follow-up visits from the first to the 12th month in another version of our base-case model. The comparison of the two approaches remained, however, in the same proportion.

The ICER was expressed as cost per occlusion (ie, ERCP) averted as was performed by Moss et al (31) and others (25,29,32), rather than quality-adjusted life-year (QALY) as was performed by Arguedas et al (9), who recognized the limited reliability of this measure with the available information for palliative patients with MBO. Furthermore, the ICERs expressed as cost per QALY are sometimes more difficult to interpret over short time horizons than if the ICER is based on outcomes reflecting more concrete end points. In such clinical contexts, the unit of efficacy can be chosen specific to the medical condition (28,33-37), as has been the case for MBO (11,12,14,25,27,29,31).

To confirm robustness of the results, we performed both deterministic and probabilistic sensitivity analyses consistent with existing methodological recommendations (38-41). The ranges of the probability variables were guided by the data from our RCT (6), while we used wide intervals for the cost estimates (Table 3), identifying only highly improbable, clinically irrelevant thresholds values. Because patient survival and time horizon are critical individual factors, we also assessed the impact of the duration of the time horizon, and tested the relevance of the adopted 12 months. Regardless of the time horizon chosen between one and 12 months, the initial SEMS approach remained dominant. The probabilistic sensitivity analysis leads to similar conclusions. Representative gastrointestinal disease literature have fixed values for WTP in the clinical context of gastrointestinal bleeding (33-37), gastroesophageal reflux disease (42) or adverse events in critically ill patients (43). However, we did not find reference to any set WTP that could apply to patients suffering from MBO. In fact, the choice of a threshold for a WTP often remains subjective (37,44-46). An arbitrary WTP was therefore estimated at US$50,000 based on the magnitude of the average cost of the treatment of occlusion, as suggested by Enns et al (37).

Our conclusions differ from most past economic analyses in that the increased cost effectiveness of a SEMS-first strategy (if not dominance) is found to apply regardless of patient survival or cost settings. A review of published cost analyses comparing plastic to metal stents in patients stented for distal MBO explain the reasons for such a disparity: other cost studies are derived from direct RCT results, comparing plastic with uncovered metal stents (11,12,25,27). Three RCTs (11,12,27) tabulated total costs, which were higher in the metal stent group, without calculating ICERs and without performing any sensitivity analysis. Moss et al (31) performed an extrapolated ICERs for these three studies (11,12,27), based on the assumption that, in each individual trial, the only difference in unit costs between both interventions was the stent price. They concluded that uncovered SEMS had significantly higher patency rates than PES as early as four months after insertion. A later RCT by Kaassis et al (27) concluded that placement of a PES in a patient with hepatic metastases and a metal stent in patients without hepatic metastases was cost effective solely based on extrapolations of median survival data. Three more recent RCTs (5,7,47) yielded direct cost analyses; however, once again, no modelling or sensitivity analyses were performed that recommended that the more effective uncovered SEMS be used in unresectable patients with malignant common bile duct strictures, who survived a median of 4.5 months, while less costly PES were preferable in the one-third of patients who have distant metastases.

Two true cost-effectivenesss analyses with appropriate modelling and sensitivity analyses were performed, both >10 years ago. In the first, Yeoh et al (13) concluded in 1999 that the choice of stent should be guided by the relative local costs of ERCP and metal stents, and by the prognosis of the patient, whereby metal stents were preferable for patients surviving >4 months. The authors, however, did not carry out a true cost-effectiveness analysis, as highlighted by others (9). Later in 2002, Arguedas et al (9) completed a very thorough and methodologically sound cost-effective analysis, with costs drawn from the University of Alabama (Birmingham, Alabama, USA) in 1999 and even considered the costs of home care, while using QALYs as the unit of effectiveness as discussed earlier. The authors concluded initial endoscopic placement of an uncovered metal stent was a cost-saving strategy compared to initial PES placement, particularly in patients expected to survive >6 months.

Principal differences between these previous cost-effective analyses and our own that explain the differences in findings include a higher incidence of cholangitis associated with stent obstruction in the plastic group of our recent RCT, and observed real-life probabilities at the different time points of follow-up. Moreover, as already suggested above, contemporary costs, including equipment and admissions to hospital are much greater (for example US$8,333 in the Arguedas et al [9] analysis versus our US$13,768, or the per diem of €480 in the Soderlund et al study (7) versus our US$2,500).

A more recent study by Hamada et al (15) compared a one-step versus an unusual two-step SEMS (covered and uncovered) insertion strategy in which a nasobiliary catheter is first inserted, while using local institutional charges. They concluded on the superiority of a SEMS-first approach but did not perform sensitivity analyses.

Because the majority of biliary stents inserted worldwide remain PES, an effort will need to be made to consider using the SEMS technology at the initial endoscopic procedure. In fact, decision makers need to consider the downstream cost effectiveness of an approach as opposed to the upfront cost of a piece of equipment.

Limitations of our analysis include the realization that many of the probabilities we used are based on small groups of patients whose numbers decrease gradually as patients flow through the 12-month trial duration, including complication rates (although sensitivity analyses include broad published ranges of these). The probability values in the last months of the model are therefore the most uncertain. It is also possible that very discreet subgroups of patients may benefit from an initial PES insertion; however, this appears to be unlikely based on the robustness and generalizability imparted by the uniform findings on multiple sensitivity analyses in our study. Also, the RCT data are principally based on 10 mm partially covered SEMS and 10 Fr plastic stents. Moreover, the SEMS technology adopted for the present cost analysis was that of a partially covered stent; however, the relative efficacy of covered versus uncovered SEMS remains controversial (1,48). It is important to note that partially covered SEMS are more expensive than uncovered SEMS and these latter stents are commonly used in clinical practice. Therefore, had the model included uncovered SEMS they may have been the preferred strategy and, at a minimum, partially covered SEMS would most likely not have been dominant. Similarly, because of the source data for this analysis, other relevant comparators were not assessed and this is a limitation. The costs adopted are US-based and reflect not only US currency but also a US fee structure; this realization needs to be kept in mind when the results are generalized to a Canadian setting. Moreover, our data, based on palliative patients, do not allow us to confidently conclude on the choice of stent for temporary biliary drainage while awaiting a possible curative surgery as has been increasingly practiced, with some surgeons preferring initial plastic stent insertion. Two recent cost-minimization studies examining in part this issue suggest the cost effectiveness of covered SEMS over PES and/or DoubleLayer PES (10,30), although a policy of routine preoperative drainage has not been shown to improve and may even worsen subsequent outcomes (48-50).

DISCLOSURES: Dr Alan Barkun has served as consultant for Olympus and Cook; speaker for AstraZeneca, and has received research funding from Cook and Boston Scientific. Viviane Adam, Myriam Martel and Dr Peter Moses have no conflict of interest to declare. The sponsor participated in no aspects of study design, data collection, analysis or the interpretation of the results. Nor did they participate in any way in the writing of the report, nor in the decision to submit the manuscript for publication.

CONCLUSION

Using a cost-effectiveness analysis modelled principally based on a recent RCT, a PES-first strategy was both more expensive and less efficacious than that of a SEMS-first approach in palliating patients with distal MBO. The average cost was US$6,701 for the SEMS-first and US$20,674 for PES-first, and associated with effectiveness probabilities of experiencing no occlusion over the 12 months of 65.6% and 13.9%, respectively. Extensive sensitivity analyses confirmed the robustness of these conclusions. The present analysis, based on contemporary estimates of both effectiveness and costs, expands on previous, now less-pertinent cost-effectiveness analyses, suggesting that in 2013, at the time of initial endoscopic drainage for patients with MBO undergoing palliative stenting, insertion of a partially covered SEMS is thus both more effective and less costly than that of a PES, regardless of anticipated survival or cost settings.

REFERENCES

1. Almadi MA, Barkun AN, Martel M. No benefit of covered vs uncovered self-expandable metal stents in patients with malignant distal biliary obstruction: a meta-analysis. Clin Gastroenterol Hepatol 2013;11:27-37,e21.
2. Moss AC, Morris E, Mac Mathuna P. Palliative biliary stents for obstructing pancreatic carcinoma. Cochrane Database Syst Rev 2006(2):CD004200.
3. Hong WD, Chen XW, Wu WZ, Zhu QH, Chen XR. Metal versus plastic stents for malignant biliary obstruction: An update meta-analysis. Clin Res Hepatol Gastroenterol 2013;37:496-500.
4. Isayama H, Yasuda I, Ryozawa S, et al. Results of a Japanese multicenter, randomized trial of endoscopic stenting for non-resectable pancreatic head cancer (JM-test): Covered Wallstent versus DoubleLayer stent. Dig Endosc 2011;23:310-5.
5. Katsinelos P, Paikos D, Kountouras J, et al. Tannenbaum and metal stents in the palliative treatment of malignant distal bile duct obstruction: A comparative study of patency and cost effectiveness. Surg Endosc 2006;20:1587-93.
6. Moses PL, Alnaamani KM, Barkun AN, et al. Randomized trial in malignant biliary obstruction: Plastic vs partially covered metal stents. World J Gastroenterol 2013;19:8638-46.
7. Soderlund C, Linder S. Covered metal versus plastic stents for malignant common bile duct stenosis: A prospective, randomized, controlled trial. Gastrointest Endosc 2006;63:986-95.
8. Levy MJ, Baron TH, Gostout CJ, Petersen BT, Farnell MB. Palliation of malignant extrahepatic biliary obstruction with plastic versus expandable metal stents: An evidence-based approach. Clin Gastroenterol Hepatol 2004;2:273-85.
9. Arguedas MR, Heudebert GH, Stinnett AA, Wilcox CM. Biliary stents in malignant obstructive jaundice due to pancreatic carcinoma: A cost-effectiveness analysis. Am J Gastroenterol 2002;97:898-904.
10. Kahaleh M, Brock A, Conaway MR, et al. Covered self-expandable metal stents in pancreatic malignancy regardless of resectability: A new concept validated by a decision analysis. Endoscopy. 2007;39:319-24.
11. Knyrim K, Wagner HJ, Pausch J, Vakil N. A prospective, randomized, controlled trial of metal stents for malignant obstruction of the common bile duct. Endoscopy 1993;25:207-12.
12. Prat F, Chapat O, Ducot B, et al. A randomized trial of endoscopic drainage methods for inoperable malignant strictures of the common bile duct. Gastrointest Endosc 1998;47:1-7.
13. Yeoh KG, Zimmerman MJ, Cunningham JT, Cotton PB. Comparative costs of metal versus plastic biliary stent strategies for malignant obstructive jaundice by decision analysis. Gastrointest Endosc 1999;49(4 Pt 1):466-71.
14. Yoon WJ, Ryu JK, Yang KY, et al. A comparison of metal and plastic stents for the relief of jaundice in unresectable malignant biliary obstruction in Korea: An emphasis on cost-effectiveness in a country with a low ERCP cost. Gastrointest Endosc 2009;70:284-9.
15. Hamada T, Nakai Y, Isayama H, et al. One- and two-step self-expandable metal stent placement for distal malignant biliary obstruction: A propensity analysis. J Gastroenterol 2012;47:1248-56.
16. Isayama H, Nakai Y, Kawakubo K, et al. Endoscopic retrograde cholangiopancreatography for distal malignant biliary stricture. Gastrointest Endosc Clin N Am 2012;22:479-90.
17. Ridtitid W, Rerknimitr R. Management of an occluded biliary metallic stent. World J Gastrointest Endosc 2012;4:157-61.
18. TreeAge Software Inc. Healtcare Module. Williamstown: USA 2015. <www.treeage.com/> (Accessed March 2015).
19. Boston Scientific. <www.bostonscientific.com/home.bsci> (Accessed March 2015).
20. American Medical Association. <www.ama-assn.org/> (Accessed March 2015).
21. Nationwide Inpatient Sample (NIS) 2008. Healthcare Cost and Utilization Project. <www.hcup-us.ahrq.gov/databases.jsp> (Accessed March 2015).
22. HCUP. Introduction to the HCUP Nationwide Inpatient Sample 2008: Agency of Healthcare Research and Quality Project HCaU;2010.
23. U.S. Bureau of Labor Statistics. <www.bls.gov/> (Accessed March 2015).
24. Chan G, Barkun J, Barkun AN, et al. The role of ciprofloxacin in prolonging polyethylene biliary stent patency: A multicenter, double-blinded effectiveness study. J Gastrointest Surg 2005;9:481-8.
25. Davids PH, Groen AK, Rauws EA, Tytgat GN, Huibregtse K. Randomised trial of self-expanding metal stents versus polyethylene stents for distal malignant biliary obstruction. Lancet 1992;340:1488-92.
26. Mukai T, Yasuda I, Nakashima M, et al. Metallic stents are more efficacious than plastic stents in unresectable malignant hilar biliary strictures: a randomized controlled trial. J Hepatobiliary Pancreat Sci 2013;20:214-22.
27. Kaassis M, Boyer J, Dumas R, et al. Plastic or metal stents for malignant stricture of the common bile duct? Results of a randomized prospective study. Gastrointest Endosc 2003;57:178-82.
28. Pola S, Muralimohan R, Cohen B, Fehmi SM, Savides TJ. Long-term risk of cholangitis in patients with metal stents for malignant biliary obstruction. Dig Dis Sci 2012;57:2693-6.
29. Schmassmann A, von Gunten E, Knuchel J, et al. Wallstents versus plastic stents in malignant biliary obstruction: Effects of stent patency of the first and second stent on patient compliance and survival. Am J Gastroenterol 1996;91:654-9.
30. Chen VK, Arguedas MR, Baron TH. Expandable metal biliary stents before pancreaticoduodenectomy for pancreatic cancer: A Monte-Carlo decision analysis. Clin Gastroenterol Hepatol 2005;3:1229-37.
31. Moss AC, Morris E, Leyden J, MacMathuna P. Do the benefits of metal stents justify the costs? A systematic review and meta-analysis of trials comparing endoscopic stents for malignant biliary obstruction. Eur J Gastroenterol Hepatol 2007;19:1119-24.
32. Tham TC, Carr-Locke DL, Vandervoort J, et al. Management of occluded biliary Wallstents. Gut 1998;42:703-7.
33. Al-Sabah S, Barkun AN, Herba K, et al. Cost-effectiveness of proton-pump inhibition before endoscopy in upper gastrointestinal bleeding. Clin Gastroenterol Hepatol 2008;6:418-25.
34. Barkun AN, Adam V, Sung JJ, et al. Cost effectiveness of high-dose intravenous esomeprazole for peptic ulcer bleeding. Pharmacoeconomics 2010;28:217-30.
35. Barkun AN, Herba K, Adam V, et al. The cost-effectiveness of high-dose oral proton pump inhibition after endoscopy in the acute treatment of peptic ulcer bleeding. Aliment Pharmacol Ther 2004;20:195-202.
36. Barkun AN, Herba K, Adam V, et al. High-dose intravenous proton pump inhibition following endoscopic therapy in the acute management of patients with bleeding peptic ulcers in the USA and Canada: A cost-effectiveness analysis. Aliment Pharmacol Ther 2004;19:591-600.
37. Enns RA, Gagnon YM, Rioux KP, Levy AR. Cost-effectiveness in Canada of intravenous proton pump inhibitors for all patients presenting with acute upper gastrointestinal bleeding. Aliment Pharmacol Ther 2003;17:225-33.
38. Guidelines for the Economic Evaluation of Health Technologies: Canada: Ottawa: Canadian Agency for Drugs and Technologies in Health, 3rd edn, 2006.
39. Guide to the Methods of Technology Appraisal: National Institute for Health and Clinical Excellence: June 2008.
40. A format for submission of clinical and economic evidence of pharmaceuticals in support of formulary consideration: Foundation

for Managed Care Pharmacy – Academy of Managed Care Pharmacy, 2009.
41. Drummond MF, Jefferson TO. Guidelines for authors and peer reviewers of economic submissions to the BMJ. The BMJ Economic Evaluation Working Party. BMJ 1996;313:275-83.
42. Briggs AH, Goeree R, Blackhouse G, O'Brien BJ. Probabilistic analysis of cost-effectiveness models: Choosing between treatment strategies for gastroesophageal reflux disease. Med Decis Making 2002;22:290-308.
43. Barkun AN, Adam V, Martel M, Bardou M. Cost-effectiveness analysis: Stress ulcer bleeding prophylaxis with proton pump inhibitors, H_2 receptor antagonists. Value Health 2013;16:14-22.
44. Ubel PA, Hirth RA, Chernew ME, Fendrick AM. What is the price of life and why doesn't it increase at the rate of inflation? Arch Intern Med 2003;163:1637-41.
45. Ariely D. Predictably Irrational: The Hidden Forces that Shape our Decisions. 1st Harper Perennial edn. New York: Harper Perennial; 2010.
46. Weintraub WS, Cohen DJ. The limits of cost-effectiveness analysis. Circ Cardiovasc Qual Outcomes 2009;2:55-8.
47. Isayama H, Komatsu Y, Tsujino T, et al. A prospective randomised study of "covered" versus "uncovered" diamond stents for the management of distal malignant biliary obstruction. Gut 2004;53:729-34.
48. Saleh MM, Norregaard P, Jorgensen HL, Andersen PK, Matzen P. Preoperative endoscopic stent placement before pancreaticoduodenectomy: A meta-analysis of the effect on morbidity and mortality. Gastrointest Endosc 2002;56:529-34.
49. Sewnath ME, Karsten TM, Prins MH, et al. A meta-analysis on the efficacy of preoperative biliary drainage for tumors causing obstructive jaundice. Ann Surg 2002;236:17-27.
50. van der Gaag NA, Rauws EA, van Eijck CH, et al. Preoperative biliary drainage for cancer of the head of the pancreas. N Engl J Med 2010;362:129-37.

An update on the management of chronic hepatitis C: 2015 Consensus guidelines from the Canadian Association for the Study of the Liver

Robert P Myers MD MSc[1]*, Hemant Shah MD MScCH HPTE[2]*, Kelly W Burak MD MSc[1], Curtis Cooper MD[3], Jordan J Feld MD MPH[2]*

RP Myers, H Shah, KW Burak, C Cooper, JJ Feld. An update on the management of chronic hepatitis C: 2015 Consensus guidelines from the Canadian Association for the Study of the Liver. Can J Gastroenterol Hepatol 2015;29(1):19-34.

Chronic hepatitis C remains a significant medical and economic burden in Canada, affecting nearly 1% of the population. Since the last Canadian consensus conference on the management of chronic hepatitis C, major advances have occurred that warrant a review of recommended management approaches for these patients. Specifically, direct-acting antiviral agents with dramatically improved rates of virological clearance compared with standard therapy have been developed and interferon-free, all-oral antiviral regimens have been approved. In light of this new evidence, an update to the 2012 Canadian Association for the Study of the Liver consensus guidelines on the management of hepatitis C was produced. The present document reviews the epidemiology of hepatitis C in Canada, preferred diagnostic testing approaches and recommendations for the treatment of chronically infected patients with the newly approved antiviral agents, including those who have previously failed peginterferon and ribavirin-based therapy. In addition, recommendations are made regarding approaches to reducing the burden of hepatitis C in Canada.

Key Words: *Dasabuvir; Direct-acting antivirals; Guideline; Hepatitis C; Interferon; Ledipasvir; Ombitasvir; Paritaprevir; Peginterferon; Simeprevir; Sofosbuvir; Ribavirin; Therapy; Treatment*

Mise à jour sur la prise en charge de l'hépatite C chronique : les lignes directrices consensuelles 2015 de l'Association canadienne pour l'étude du foie

L'hépatite C chronique demeure un fardeau médical et économique important au Canada, car il touche près de 1 % de la population. Depuis la dernière conférence consensuelle canadienne sur la prise en charge de l'hépatite C chronique, on a réalisé des progrès marqués qui justifient une analyse des démarches de prise en charge recommandées. Notamment, on a mis au point des antiviraux à action directe au taux de clairance virologique bien supérieur à celui du traitement standard et on a homologué des antiviraux sans interféron par voie orale. À la lumière de ces nouvelles données probantes, l'Association canadienne pour l'étude du foie a mis à jour les lignes directrices consensuelles 2012 sur la prise en charge de l'hépatite C. Le présent document traite de l'épidémiologie de l'hépatite C au Canada, des démarches et des recommandations favorisées pour traiter les patients atteints d'une infection chronique à l'aide des nouveaux antiviraux homologués, y compris les patients qui n'avaient pas répondu à un traitement à l'interféron pégylé et à la ribavirine. Il contient également des recommandations sur les approches pour réduire le fardeau de l'hépatite C au Canada.

The present guidelines were written to assist physicians and other health care professionals in the management of patients with chronic hepatitis C virus (HCV) infection. They were drafted by Canadian HCV experts at the request of the Executive Committee of the Canadian Association of the Study of the Liver (CASL). The document was made available for review by CASL members and a revised draft based on this feedback was submitted to the Executive Committee of CASL for approval. The information contained within the present guidelines represents a synthesis of evidence from the published literature and scientific abstract presentations available at the time of writing with supplementation by the expert opinions of the authors. Any recommendations should be considered preferred approaches to care rather than strict standards. In some cases, off-label use of regimens are recommended based on the authors' opinions. To more fully characterize the quality of evidence supporting these recommendations, we have assigned a class (reflecting benefit versus risk) and level (assessing strength of certainty) of evidence as adapted from the American College of Cardiology and the American Heart Association Practice Guidelines (1,2), and as used in similar practice guidelines of CASL (3) and the American Association for the Study of Liver Diseases (4) (Table 1). No funding was provided to the authors for this work.

Since the last update of the CASL management guidelines for chronic hepatitis C (CHC) in 2012 (3), major advances have occurred including: the approval of novel direct-acting antiviral agents (DAAs) used with pegylated interferon (PEG-IFN) that have improved efficacy and tolerability compared with first-generation DAAs and/or standard PEG-IFN-based therapy (5-7); and the approval of all-oral, IFN-free, DAA combination therapies with markedly improved efficacy and tolerability and activity beyond just HCV genotype 1 (5,8-15). The current document was developed as an update to previous guidelines with a focus on the management of HCV-infected patients rather than an exhaustive review of CHC or HCV screening. Future guidelines will include 'special populations' with CHC, including people who use injection drugs (PWIDs), incarcerated individuals, patients with decompensated cirrhosis, those pre- or post-transplantation, and patients with HIV/HCV coinfection (for whom relevant guidelines have recently been published by the Canadian Institute of Health Research HIV Trials Network) (16). Due to the rapidity of advances in this field, recommendations in the present document will be updated regularly as new information emerges and novel agents are approved.

*Authors who contributed equally to this work.

[1]Liver Unit, Division of Gastroenterology and Hepatology, University of Calgary, Calgary, Alberta; [2]Toronto Western Hospital Liver Centre, University of Toronto, Toronto; [3]Division of Infectious Diseases, University of Ottawa, Ottawa, Ontario

Correspondence: Dr Robert P Myers, Liver Unit, University of Calgary, 6D22, Teaching, Research and Wellness Building, 3280 Hospital Drive Northwest, Calgary, Alberta T2N 4Z6. e-mail rpmyers@ucalgary.ca

TABLE 1
Grading system for recommendations

Classification	Description
Class of evidence	
Class 1	Conditions for which there is evidence and/or general agreement that a given diagnostic evaluation, procedure or treatment is beneficial, useful and effective
Class 2	Conditions for which there is conflicting evidence and/or a divergence of opinion about the usefulness/efficacy of a diagnostic evaluation, procedure or treatment
Class 2a	Weight of evidence/opinion is in favour of usefulness/efficacy
Class 2b	Usefulness/efficacy is less well established by evidence/opinion
Class 3	Conditions for which there is evidence and/or general agreement that a diagnostic evaluation, procedure/treatment is not useful/effective and in some cases may be harmful
Grade of evidence	
Level A	Data derived from multiple randomized clinical trials or meta-analyses
Level B	Data derived from a single randomized trial, or nonrandomized studies
Level C	Only consensus opinions of experts, case studies or standard-of-care

EPIDEMIOLOGY OF HEPATITIS C IN CANADA

CHC remains a significant medical and economic burden in Canada (17-19). In the Canadian Health Measures Survey (20), Statistics Canada and the Public Health Agency of Canada reported an estimated anti-HCV prevalence of 0.5% (95% CI 0.3% to 0.9%) or approximately 138,600 (95% CI 55,800 to 221,300) anti-HCV-positive individuals in Canada. However, these figures are likely underestimates because the Canadian Health Measures Survey excluded several high-risk populations including incarcerated individuals, Aboriginals and PWIDs (20). In fact, a recent modelling study suggests that approximately 252,000 Canadians (uncertainty interval 178,000 to 315,000) were chronically infected in 2013 (18). The peak prevalence was estimated to have occurred in 2003, with approximately 260,000 infected individuals. It has been estimated that approximately 60% of HCV cases in Canada are among current or former PWIDs, 20% are among infected immigrants and 11% have received contaminated blood products (21). Of the nearly 8000 incident cases in Canada in 2007, approximately 80% likely occurred via sharing of injecting equipment, and most of the remainder among immigrants from endemic countries (21). There is wide variation in estimates of the number of HCV-infected individuals who remain undiagnosed. Modelling data from the Public Health Agency of Canada estimated that 79% of individuals were diagnosed in 2003 (21); however, the CMHS found that only 30% of anti-HCV-positive individuals were aware of their infection (20).

Genotype 1 infection is the most prevalent genotype in Canada, representing 65% of infected individuals (56% genotype 1a, 33% genotype 1b, and 10% with an unspecified subtype or mixed infection) (22). The genotype 1 subtype is of relevance for some of the new antiviral regimens due to differing efficacy between genotypes 1a and 1b. Genotypes 2 and 3 account for approximately 14% and 20% of infections in Canada, respectively, whereas genotypes 4, 5 and 6 are very rare (<1% of all infections) (22).

Although the overall prevalence of CHC is declining, complications of CHC are increasing due to aging of the infected population and progression of liver fibrosis (17-19). Modelling data suggest that by 2035, cases of decompensated cirrhosis, hepatocellular carcinoma (HCC) and liver-related mortality will increase by 80%, 205% and 160%, respectively, compared with 2013 levels (Figure 1) (18). Similarly, annual direct costs associated with CHC (excluding the cost of antiviral therapies) are expected to rise from an estimated $161 million in 2013 to >$258 million at the peak in 2032 (18). Given the alarming estimates of future disease burden, more accurate information regarding the incidence and prevalence of CHC and its sequelae is required to inform health care planning and the allocation of resources. The identification of undiagnosed cases and the dissemination of effective antiviral therapies should be prioritized to reduce complications of this disease (23).

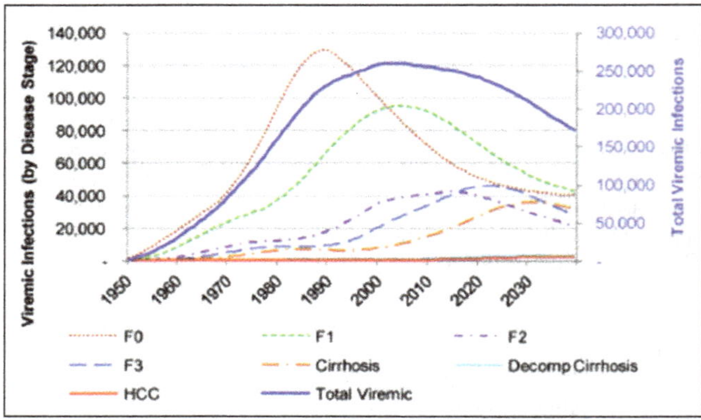

Figure 1) *Modelled incidence of hepatitis C-related sequelae in Canada, 1950 to 2035. Estimates are not mutually exclusive. Reproduced with permission from Myers et al (18). Decomp Decompensated; HCC Hepatocellular carcinoma*

Recommendations:

1. A large population-based seroprevalence survey should be conducted to accurately define the prevalence of hepatitis C in Canada. The design of the study should include populations with an increased risk of hepatitis C, particularly PWIDs, incarcerated individuals and immigrants from endemic countries (Class 2a, Level C).
2. To reduce the future burden of HCV-related morbidity and mortality in Canada, strategies for case identification, harm reduction and disease management – including but not limited to antiviral therapy – should be developed and implemented (Class 2a, Level C).

ANTIVIRAL THERAPY

The primary objective of anti-HCV therapy is complete eradication of the virus, termed a sustained virological response (SVR). SVR has traditionally been defined as undetectable serum HCV RNA at least 24 weeks following the end of treatment (SVR24) (24); however, recent data suggest that earlier assessment at 12 weeks after treatment (SVR12) is sufficient to define this outcome (25). Once achieved, an SVR is considered to be a long-term cure of the virus because late relapses are rare (26,27). SVR is associated with long-term health benefits including improved quality of life (28,29), extrahepatic manifestations of HCV (eg, cryoglobulinemic vasculitis) (30), liver histology (31,32), HCC incidence (33), liver-related morbidity and mortality (34-36), and all-cause mortality (33).

The landscape of antiviral treatment for hepatitis C is changing rapidly. Until recently, the standard therapy was the combination of PEG-IFN and ribavirin (RBV), usually administered for 48 weeks in patients with genotype 1, 4, 5 and 6, and 24 weeks in those with genotypes 2 and 3 (3). Dual therapy achieves SVR rates of 40% to 50% in patients with genotype 1 and approximately 80% in those with genotypes 2, 3, 5 and 6. Results for HCV genotype 4 are intermediate (3). In 2011, the first DAAs, boceprevir (BOC) and telaprevir (TVR), were approved for treatment of HCV genotype 1 in combination with PEG-IFN and RBV. These nonstructural (NS) 3/4A protease inhibitors (PIs) substantially increase rates of SVR in both treatment-naive and previously treated patients compared with dual therapy (37-41). However, they are associated with significant toxicity, complex regimens involving response-guided therapy (RGT), drug-drug interactions (DDIs), and low response rates in patients with cirrhosis and previous treatment failures. In addition, BOC and TVR required coadministration with PEG-IFN and RBV for 24 to 48 weeks, markedly increased the cost of therapy, and are associated with the emergence of resistance-associated variants (RAVs) in the majority of patients who fail treatment (3). The subsequent approval of DAAs with improved efficacy and tolerability, shorter treatment durations, and the option of PEG-IFN- and RBV-free therapy, represents a major advance in the field.

The treatment of CHC is complex and resource intensive. Contributing factors include the high prevalence of psychiatric comorbidities in HCV-infected individuals (eg, depression and addictions), multiple modes of drug administration, side effects, and the requirement for careful on-treatment monitoring of symptoms and laboratory tests. The most successful model to deliver comprehensive CHC care is via a multidisciplinary approach including experienced physicians, nurses and allied health professionals (eg, psychologists, psychiatrists, addiction specialists and social workers). Currently in Canada, a relatively small number of physicians treat CHC, leading in some cases to prolonged wait times for patients to be adequately evaluated and treated. These deficiencies in access to care are greater in rural and remote communities, despite a high HCV prevalence in many regions with limited health care capacity. Moreover, public funding for treatment nurses – who have represented a vital component of the management team – is not universally available. To achieve a meaningful reduction in the future burden of CHC, it will be vital to expand treatment capacity via additional training and funding of experienced personnel and enhanced access to publically funded antiviral therapies (42). With the advent of all-oral antiviral regimens that have few contraindications, minimal toxicity and short treatment courses, the number of patients that can be treated should increase dramatically. However, team-based management will still be necessary to achieve this goal.

Recommendation:

3. Increased resources are necessary to improve hepatitis C treatment capacity in Canada, including the training of expert treaters and public funding for treatment nurses (Class 2a, Level C).

INDICATIONS AND CONTRAINDICATIONS TO ANTIVIRAL TREATMENT

All patients with CHC should be considered candidates for antiviral treatment. The decision of if and when to initiate therapy should be based on the balance between the perceived benefits and risks of treatment and the wishes of the individual patient. Factors to consider include the probability of SVR and the likelihood of progression to advanced liver disease without viral eradication, the presence of extrahepatic manifestations of CHC, the patient's anticipated tolerability of treatment and the life expectancy of the patient. The prospect of new therapies with expected benefits over currently available treatments should also be considered. In light of these issues, prompt initiation of treatment should be considered in certain patient subgroups, especially those with advanced liver fibrosis (F3 or F4 according to the METAVIR classification [bridging fibrosis or cirrhosis]) (43). These patients are at the highest risk of HCV-related complications including liver failure and HCC. Treatment of patients with mild to moderate fibrosis (F1 or F2) should also be considered because progression to more advanced stages is associated with a reduced likelihood of SVR. Moreover, viral eradication in patients at risk of infecting others (eg, PWIDs who continue to share injecting equipment) may reduce the incidence of new infections (44). The curative nature of HCV therapy means that those who achieve SVR before developing cirrhosis do not require long-term follow-up. There are additional benefits to SVR beyond liver disease prevention, including improved quality of life (28,29) and a reduction in all-cause mortality (33). Patients with extrahepatic manifestations of CHC including cryoglobulinemic vasculitis, porphyria cutanea tarda and glomerulonephritis should be considered for treatment regardless of their underlying liver disease severity because these conditions typically respond to viral eradication (30).

TABLE 2
Contraindications for treatment with peginterferon and ribavirin

Absolute contraindications	Pregnancy
Strong, but not absolute, contraindications	Alcohol abuse
	Hepatic decompensation
	Coronary artery disease
	Solid organ transplantation (except liver)
Relative contraindications	Major depression
	Major psychosis
	Autoimmune disease
	Injection drug use
	Renal failure (including dialysis)
Contraindications that are no longer contraindications	Normal alanine aminotransferase
	Stable methadone maintenance
	Neutropenia, anemia or thrombocytopenia
	Controlled seizure disorder
	Older than 65 years of age
	Alcohol use

Table reproduced with permission from (3)

There are very few absolute contraindications to treatment with PEG-IFN- and RBV- based therapy. As postmarketing experience with these therapies has grown, many conditions previously regarded as absolute contraindications are now considered relative, and some may be present only temporarily (Table 2) (3). In most cases, treatment of these patients with PEG-IFN and RBV requires considerable expertise and, therefore, patients with relative contraindications should be treated in expert centres. Contraindications to the recently approved, all-oral regimens are distinctly uncommon.

Recommendations:

4. All patients with chronic HCV infection should be considered candidates for antiviral therapy (Class 1, Level A).
5. Antiviral treatment should be strongly considered in patients with evidence of liver fibrosis (Class 1, Level A).
6. Patients with extrahepatic manifestations of HCV should be considered for antiviral therapy (Class 1, Level A).

PRETREATMENT ASSESSMENT

Routine assessment

The routine assessment of HCV-infected patients should include risk factors for viral acquisition (eg, injection drug use, receipt of potentially contaminated blood products or tissues, and origin from a high-prevalence region), signs and symptoms of advanced liver disease (eg, jaundice, ascites, encephalopathy, portal hypertension-related hemorrhage) or extrahepatic manifestations of CHC, presence of cofactors that may accelerate disease progression (eg, alcohol use, obesity,

TABLE 3
Routine testing of patients with chronic hepatitis C virus (HCV)*

Category of Testing	Tests	Comments
Confirmation and characterization of chronic infection	HCV RNA	Confirms chronicity and baseline for treatment responses
	HCV genotype and subtype	Directs choice of treatment regimen
Assessment of liver disease	Complete blood count	Thrombocytopenia may indicate cirrhosis and portal hypertension. Platelets needed for APRI calculation
	Alanine aminotransferase	Normal value does not preclude significant fibrosis
	Asparatate aminotransferase	Asparatate aminotransferase needed for calculation of APRI
	Gamma-glutamyl transferase	
	Alkaline phosphatase	
	Bilirubin	Elevated bilirubin or INR, or hypoalbuminemia may indicate significant liver dysfunction
	INR (or prothrombin time)	
	Albumin	
	Creatinine	Renal dysfunction increases ribavirin-related hemolytic anemia and may impact drug pharmacodynamics
	Abdominal ultrasound	May suggest cirrhosis, in which case, serves as a baseline for hepatocellular carcinoma surveillance
Viral coinfections	Immunoglobulin G anti-HAV	If negative, vaccinate against hepatitis A
	HBsAg	Exclude hepatitis B coinfection
	Hepatitis B surface antibody	If negative (and HBsAg-negative), vaccinate against hepatitis B
	anti-HIV	Exclude HIV coinfection
Exclude other causes of liver disease†	Alpha-1-antitrypsin	Alpha-1-antitrypsin deficiency
	Ceruloplasmin	Wilson disease.
	Ferritin, serum iron, total iron-binding capacity	Iron overload
	Antinuclear antibody	Autoimmune hepatitis
	Smooth muscle antibody	
	Antimitochrondrial antibody	Primary biliary cirrhosis
	Immunoglobulin G	Often elevated in autoimmune hepatitis and cirrhosis of any cause
	Immunoglobulin A	Often elevated in fatty liver and alcoholic liver disease
	Immunoglobulin M	Often elevated in primary biliary cirrhosis
Contraindications to treatment	Serum or urine β-human chorionic gonadotropin	Exclude pregnancy in women of reproductive age
	Electrocardiogram	If >50 years of age or history of cardiac disease
	Thyroid-stimulating hormone	Exclude thyroid disease, which may be exacerbated by interferon
	Fundoscopy	Exclude retinopathy in patients >50 years of age or with hypertension or diabetes mellitus if interferon is to be prescribed

**Confirmed anti-HCV antibody positive; †Suggested tests only. Tailor testing to individual case. Anti-HAV Antibodies to hepatitis A virus; APRI Aspartate aminotransferase/platelet ratio index; HBsAg Hepatitis B surface antigen; INR International normalized ratio*

coinfections) and potential contraindications to IFN-based therapy (Table 2), which would favour the use of an IFN-free regimen. Necessary laboratory testing includes virological tests to confirm and characterize HCV infection, liver biochemistry, abdominal ultrasound, an assessment of fibrosis stage and tests to rule out coinfections, direct appropriate vaccination and identify contraindications to treatment. In patients with abnormal liver biochemistry, serological tests to exclude coexisting liver diseases should be considered (Table 3).

Virological testing

Approximately one-quarter of patients infected with HCV will clear the virus spontaneously (45). Therefore, chronic HCV infection must be confirmed in all anti-HCV-positive individuals using a sensitive HCV RNA test. HCV RNA detection and quantification using real-time polymerase chain reaction assays is standard due to their sensitivity, specificity, accuracy and broad dynamic range. Results should be expressed in IU/mL and normalized to the WHO international standard. Quantitative assays with a lower limit of detection of approximately 10 IU/mL to 15 IU/mL are recommended. HCV RNA test results should be available within a timely fashion (within seven days) to facilitate management decisions. The rapid identification of failing treatment will reduce patient exposure to costly therapies and potential toxicity, and likely limit the development of RAVs.

The HCV genotype should be assessed in all patients because it has important implications for the decision to initiate treatment and the choice of regimen. With PEG-IFN and RBV therapy, knowledge of only the main genotype (1 to 6) was necessary. However, knowledge of the subtype is now critical, particularly for genotype 1, because of the differing genetic barriers to resistance of HCV subtypes 1a and 1b for many classes of DAAs (46,47). For some DAAs, additional testing (eg, for the Q80K polymorphism [see below]) and/or alternative treatment based on subtype (eg, the use of RBV) may be required.

Recommendations:

7. HCV RNA, genotype, and subtype testing (ie, 1a versus 1b) are essential to the management of patients with chronic hepatitis C (Class 1, Level A).
8. HCV RNA testing should be performed using a sensitive quantitative assay (lower limit of detection of ≤10 IU/mL to 15 IU/mL) with a broad dynamic range. Standardized results should be expressed in IU/mL and be available within a maximum of seven days to facilitate management decisions (Class 1, Level A).

Assessment of liver disease severity

Assessment of the severity of hepatic fibrosis is vital for determining the prognosis of HCV-infected patients and the necessity of antiviral treatment. Identification of patients with cirrhosis is particularly important due to their increased risk of hepatic complications, reduced likelihood of treatment response, and their requirement for surveillance for HCC and esophageal varices. Although the diagnosis of cirrhosis is obvious in some cases based on routine tests (eg, a nodular shrunken liver, splenomegaly or portal hypertensive collaterals on ultrasound), traditionally, liver biopsy has been the reference method for staging fibrosis, determining the severity of other histological lesions (eg, necroinflammation, steatosis) and ruling out coexistent liver diseases (eg, iron overload). Various validated scoring systems have demonstrated sufficient reproducibility and interobserver variability to justify clinical use (eg, METAVIR, Scheuer, Ishak, and Knodell's Hepatic Activity Index) (48). However, liver biopsy has several limitations, including invasiveness and the potential for serious complications including hemorrhage (approximately one in 1000) and death (approximately one in 10,000) (49,50), sampling error and variability in pathological interpretation, high cost, limited availability in many centres, and the difficulty of repeating biopsies to monitor temporal changes in fibrosis. In light of these limitations, numerous noninvasive alternatives to biopsy have been developed (51) including serum markers (eg, the aspartate aminotransferase/platelet ratio index [52]), FibroTest (FibroSure, LabCorp, USA) (53), transient elastography (TE; FibroScan, Echosens, France) (54-57) and other imaging-based tools (58,59).

Although not universally available, a wealth of literature has confirmed that these noninvasive tools can be used instead of liver biopsy to stage HCV-related fibrosis at acceptable levels of accuracy and reproducibility. In a recent survey of Canadian specialists who manage patients with chronic liver disease (60), TE was the primary mode of fibrosis assessment in HCV-infected individuals in 53% of respondents, followed by liver biopsy in 37%. Nearly one-half of respondents estimated that these noninvasive alternatives have reduced their use of liver biopsy by over 50%. In general, these tests are highly accurate for diagnosing cirrhosis and have acceptable, but lower, performance for moderate to severe fibrosis (F2 or greater). The identification of mild fibrosis (F1) and the differentiation between individual stages is poor; however, these limitations also apply to liver biopsy. Emerging data have also demonstrated a correlation between these tests and HCV-related clinical outcomes (61-63), their cost-effectiveness compared with biopsy (64) and responsiveness to viral eradication (65,66). Future studies are necessary to determine the minimal clinically important changes in these markers to facilitate serial monitoring of fibrosis.

Recommendations:

9. Liver fibrosis assessment is vital to the management of patients with CHC (Class 1, Level A).
10. Acceptable methods of fibrosis assessment include liver biopsy, TE (FibroScan) and serum biomarker panels (eg, FibroTest), either alone or in combination. All jurisdictions should provide access to at least one accurate, noninvasive method to assess fibrosis (Class 1, Level A).
11. Alternatively, cirrhosis can be confidently diagnosed in some patients with clear clinical or radiographic evidence (Class 2a, Level C).

Utility of interleukin 28B testing

Genome-wide association studies have identified single nucleotide polymorphisms (SNPs) near the interleukin 28B (IL28B) gene on chromosome 19 that are strongly associated with both spontaneous and treatment-induced HCV clearance (67-70). Patients with the favourable CC genotype at rs12979860 have a more than twofold likelihood of spontaneous HCV clearance compared with heterozygotes (CT) and homozygotes (TT) (67). The CC genotype is also associated with an approximately twofold increase in SVR to PEG-IFN and RBV therapy compared with the unfavourable SNPs in patients with HCV genotype 1 (68,70). The relevance in genotypes 2 and 3 and in treatment-experienced patients is less clear. There is marked ethnic variation in the prevalence of the IL28B genotypes.

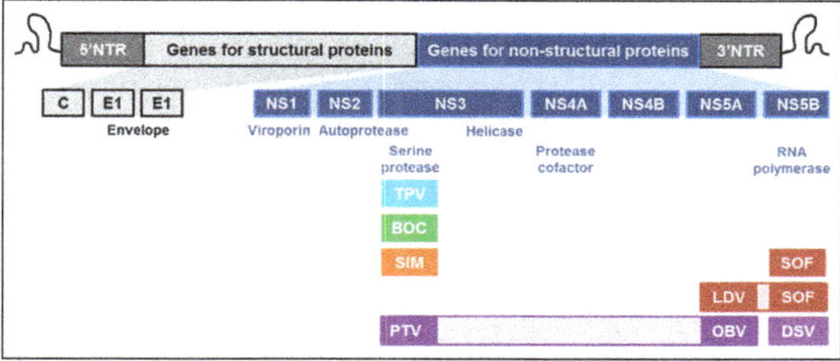

Figure 2) *Hepatitis C virus genome and the polyprotein targets of newly approved, direct-acting antiviral agents. Note: Sofosbuvir (SOF) is a nucleotide nonstructural protein (NS)5B polymerase inhibitor and dasabuvir (DSV) is a non-nucleoside polymerase inhibitor. BOC Boceprevir; LDV Ledipasvir; OBV Ombitasvir; PTV Paritaprevir; SIM Simeprevir; TPV Telaprevir*

The CC genotype is highly prevalent in Asians, but relatively uncommon in Africans, while Caucasians and Hispanics have an intermediate prevalence (68). Similar associations have been reported for the rs8099917 SNP (favourable allele = T and unfavourable allele = G) (71), and for the recently described IFN-lambda 4 (IFNL4) SNP ss46915590 (favourable allele = T and unfavourable allele = ΔG) (72). For simplicity, further discussion will refer to the rs12979860 SNP.

The impact of the IL28B genotype on treatment success is lower when treatment includes DAAs. Patients with the CC genotype have a very high rate of SVR when treated with DAAs plus PEG-IFN and RBV, reaching 98% with sofosbuvir (SOF)-based triple therapy for HCV genotype 1 (5). DAAs lead to a greater relative increase in SVR in non-CC patients. While the IL28B genotype is of limited importance with respect to SVR rates with IFN-free regimens (8,15), whether patients with the favourable IL28B genotype will be able to shorten therapy or use fewer DAAs is unclear.

Recommendations:

12. The IL28B genotype may provide valuable information regarding the likelihood of SVR depending on the HCV genotype and therapy under consideration (Class 2b, Level A).
13. A nonfavourable IL28B genotype does not preclude antiviral therapy (Class 1, Level A).

DAAs

Multiple steps in the HCV life cycle have proven attractive targets for novel pharmacological therapies (Figure 2). Particularly promising agents target the NS3/4A serine protease, the NS5B RNA-dependent RNA polymerase and the NS5A protein (73). The first DAAs approved by Health Canada for the treatment of HCV genotype 1 were the NS3/4A PIs, BOC and TVR. A second-generation PI, simeprevir (SIM), was approved in 2013 for use in combination with PEG-IFN and RBV for genotype 1. In 2013, the first HCV nucleotide polymerase inhibitor, SOF, was approved for use in combination with PEG-IFN and RBV for genotypes 1 and 4 and with RBV alone for genotypes 2 and 3. In 2014, the single-tablet regimen of SOF combined with the NS5A inhibitor ledipasvir (LDV) was approved for patients with HCV genotype 1, including those previously treated with BOC and TVR. In addition, the combination of the ritonavir-boosted PI paritaprevir (PTV_R), the NS5A inhibitor ombitasvir (OBV), and the non-nucleoside polymerase inhibitor dasabuvir (DSV) with or without RBV was approved for patients with HCV genotype 1. Given the markedly improved efficacy and tolerability of these regimens, all patients would benefit from IFN-free therapy. Therefore, these newly approved agents are recommended as first-line therapy for all indications throughout these guidelines. However, access to IFN-free regimens is not universal across Canada. Whether to initiate therapy with an IFN-containing regimen or wait for the availability of all-oral regimens is an individualized decision that must consider the patient's wishes, the urgency of therapy, the severity of liver disease, the anticipated tolerability of PEG-IFN, the likelihood of SVR and the expected timeline for access to IFN-free regimens.

TREATMENT-NAIVE PATIENTS WITH HCV GENOTYPE 1 (TABLE 4)

PEG-IFN-free regimens

SOF/LDV: The nucleotide polymerase inhibitor SOF (400 mg) has been combined with the NS5A inhibitor LDV (90 mg) in a single tablet regimen (SOF/LDV) administered once daily. This combination was evaluated in treatment-naive patients in the open-label ION-1 (8) and ION-3 (10) phase 3 trials with a primary end point of SVR12. In the ION-1 study, which included patients with compensated cirrhosis (16%), participants were randomly assigned to 12 or 24 weeks of SOF/LDV with or without weight-based RBV (8). Among patients who received SOF/LDV for 12 weeks, SVR12 rates were 97% (211 of 217) and 99% (211 of 214) in those who received and did not receive RBV, respectively. In the 24-week treatment arms, SVR12 rates were 99% (215 of 217) in RBV-treated patients compared with 98% (212 of 217) in those who received SOF/LDV alone. There were no statistically significant differences between treatment arms or pretreatment characteristics that were predictive of response. Among the 136 cirrhotic patients, SVR12 rates ranged from 94% to 100%, with no differences between 12 and 24 weeks or with or without RBV. The IL28B genotype was not predictive of response; SVR12 rates ranged from 97% to 99% among patients with the unfavourable non-CC genotype. Only one patient experienced virological breakthrough on therapy and two patients relapsed. All three of these patients had NS5A resistance, but no SOF resistance was detected by deep sequencing. Although the majority of patients complained of at least one adverse event, 93% were mild to moderate in severity with the most common being fatigue, headache, insomnia and nausea. Adverse events were more common in patients randomized to receive RBV. No patient receiving SOF/LDV alone had a hemoglobin concentration <100 g/L.

In the ION-3 study (10), treatment-naive, noncirrhotic patients with HCV genotype 1 were randomly assigned to eight weeks of SOF/LDV with or without weight-based RBV, or SOF/LDV alone for 12 weeks. Among the 215 patients randomly assigned to SOF/LDV for eight weeks, 202 (94%) achieved SVR12, compared with 201 of 216 (93%) who received SOF/LDV/RBV for eight weeks, and 206 of 216 (95%) who received SOF/LDV for 12 weeks. The relapse rates were 5% (SOF/LDV) and 4% (SOF/LDV/RBV) in the eight-week treatment arms and 1% in the 12-week treatment arm. Although the 12-week regimen had a lower relapse rate, treating all patients for an additional four weeks would lead to overtreatment of the majority of individuals. Therefore, a post hoc analysis of baseline viral load was conducted to identify patients in whom an eight-week regimen would suffice (74). In this analysis, patients with an HCV RNA level <6 million IU/mL had a 2% relapse rate in both the eight-week (two of 123) and 12-week (two of 131) SOF/LDV treatment arms, and SVR12 rates of 97% (119 of 123) and 96% (126 of 131), respectively. However, in patients with a baseline viral load ≥6 million IU/mL, those treated for only eight weeks with SOF/LDV had a 10% (nine of 92) relapse rate versus only 1% (one of 85) if treated for 12 weeks. Corresponding SVR12 rates were 90% (83 of 92) and 94% (80 of 85), respectively. Based on these findings, Health Canada and the United States Food and Drug Administration (FDA) have recommended an eight-week regimen of SOF/LDV in treatment-naive, noncirrhotic patients with baseline HCV RNA <6 million IU/mL and 12 weeks in patients with a higher viral load (74).

In addition to baseline viral load, the impact of baseline RAVs on treatment response was examined (10). Although 15 of 23 relapsers (65%) to SOF/LDV had NS5A-resistant variants detected at the time of relapse (present at baseline in nine patients), SOF resistance was not identified. Among 116 patients (18%) with NS5A resistance at baseline, 90% achieved SVR12, suggesting a minimal impact of baseline NS5A RAVs on treatment response with SOF/LDV.

Recommendations:

14. In noncirrhotic, treatment-naive patients with HCV genotype 1, SOF/LDV should be given for eight weeks (Class 1, Level B).
15. In noncirrhotic, treatment-naive patients with genotype 1 and baseline HCV RNA ≥6 million IU/mL, extension of SOF/LDV therapy to 12 weeks can be considered (Class 1, Level C).
16. In cirrhotic, treatment-naive patients with genotype 1, SOF/LDV should be given for 12 weeks (Class 1, Level B).

PTV_R/OBV/DSV ± RBV: The PI PTV is given with low-dose ritonavir (PTV_R) to permit once-daily dosing. PTV_R (150 mg/100 mg) and the NS5A inhibitor OBV (25 mg) are coformulated in a single tablet taken as two tablets once daily. This tablet is combined with the non-nucleoside polymerase inhibitor DSV (250 mg) taken as one tablet twice daily. Placebo or the combination of the three DAAs plus ritonavir (referred to as the '3D' regimen) and weight-based RBV was given for 12 weeks to treatment-naive, noncirrhotic patients with HCV genotype 1 in the phase 3 SAPPHIRE-I trial (15). Patients randomly assigned to placebo subsequently received active treatment. Of 473 patients who started active therapy, 455 (96%) achieved SVR12, clearly superior to a historical control of TVR-based triple therapy in a similar patient population (estimated SVR12 of 78%). SVR12 did

TABLE 4
Treatment-naive patients with hepatitis C virus (HCV) genotype 1

Population	Recommended	Alternative (IFN-free)	Alternative (IFN-containing)	Not recommended
Genotype 1a, noncirrhotic	SOF/LDV × 8–12 weeks*	SOF/SIM × 12 weeks	SOF/PEG/RBV × 12 weeks	PEG/RBV
	PTV_R/OBV/DSV/RBV × 12 weeks		SIM/PEG/RBV × 24 weeks (if Q80K–)	PEG/RBV/BOC or TVR
				SIM/PEG/RBV × 24 weeks (if Q80K+)
Genotype 1b, noncirrhotic	SOF/LDV × 8–12 weeks*	SOF/SIM × 12 weeks	SOF/PEG/RBV × 12 weeks	PEG/RBV
	PTV_R/OBV/DSV × 12 weeks		SIM/PEG/RBV × 24 weeks	PEG/RBV/BOC or TVR
Genotype 1a, cirrhotic	SOF/LDV × 12 weeks	SOF/SIM × 12 weeks	SOF/PEG/RBV × 12 weeks	PEG/RBV
	PTV_R/OBV/DSV/RBV × 12 weeks		SIM/PEG/RBV × 24–48 weeks (if Q80K–)	PEG/RBV/BOC or TVR
				SIM/PEG/RBV × 24 weeks (if Q80K+)
Genotype 1b, cirrhotic	SOF/LDV × 12 weeks	SOF/SIM × 12 weeks	SOF/PEG/RBV × 12 weeks	PEG/RBV
	PTV_R/OBV/DSV/RBV × 12 weeks		SIM/PEG/RBV × 24 weeks	PEG/RBV/BOC or TVR

**In noncirrhotic, treatment-naive patients with HCV genotype 1a or 1b, treat with sofosbuvir/ledipasvir (SOF 400 mg/LDV 90 mg) once daily (one tablet) for eight weeks. Consider 12 weeks of treatment if baseline HCV RNA ≥6 million IU/mL. + Positive; – Negative; BOC Boceprevir; DSV Dasabuvir (250 mg) one tablet twice daily; IFN Interferon; PEG Peginterferon alfa-2a (180 μg subcutaneously/week) or peginterferon alfa-2b (1.5 μg/kg/week); PTV_R/OBV Paritaprevir (150 mg)/ritonavir (100 mg)/ombitasvir (25 mg) two tablets once daily; Q80K Simeprevir-associated resistance variant at position 80; RBV Ribavirin (weight-based dosing: 1000 mg daily if <75 kg; 1200 mg daily if ≥75 kg); SIM Simeprevir (150 mg once daily); SOF Sofosbuvir (400 mg once daily); TVR Telaprevir*

not differ between patients with HCV genotype 1a (95% [307 of 322]) or 1b (98% [148 of 151]). The only baseline factor associated with response was body mass index (BMI). Obese patients (BMI ≥30 kg/m^2) had an SVR12 rate of 91.5% compared with 97% in patients with a lower BMI. There was no difference in response according to IL28B genotype, fibrosis stage, baseline HCV RNA level, ethnicity or RBV dose modification.

Safety of the regimen was evaluated by comparing with patients randomly assigned to placebo (15). Adverse events were more common in those on active treatment (88%); however, 73% of placebo-treated patients also experienced at least one adverse event. Severe adverse events (2.1%), in particular those requiring drug discontinuation (<1%), were rare. The most common side effects were fatigue and headache, but were no more frequent with active treatment than placebo. Nausea, pruritus, insomnia, diarrhea and asthenia were reported more frequently in patients on active treatment. Total bilirubin elevations were seen in 2.8% of patients on this regimen, likely due to RBV-related hemolysis and inhibition of the bilirubin transporters OATP1B1 and OATP1B3 by PTV_R. No episodes of hepatotoxicity were reported. Grade 2 anemia (hemoglobin 80 g/L to 100 g/L) was reported in 5.8% of patients treated with this regimen including RBV. In 5.5% of patients, the RBV dose was modified due to adverse events, but no impact on the rate of SVR12 was observed in these patients.

The TURQUOISE-II phase 3 trial evaluated the PTV_R/OBV/DSV plus RBV regimen (12 versus 24 weeks) in treatment-naive and treatment-experienced patients with compensated cirrhosis (13). Among treatment-naive patients, the rates of SVR12 were similar between the 12- and 24-week arms (94% [81 of 86] versus 95% [70 of 74]) and by genotype subtype (genotype 1a: 12 versus 24 weeks: 92% [59 of 64] versus 93% [52 of 56]; and genotype 1b: 100% in both the 12-week [22 of 22] and 24-week [18 of 18] groups). SVR12 rates among previously treated patients from TURQUOISE-II are discussed below.

To evaluate the importance of RBV administration with PTV_R/OBV/DSV, the PEARL-III and PEARL-IV phase 3 trials were conducted in treatment-naive, noncirrhotic patients with genotypes 1b and 1a, respectively (75). In PEARL-III, patients with HCV genotype 1b were randomly assigned to receive PTV_R/OBV/DSV alone (n=209) or with RBV (n=210) for 12 weeks. Only three of 419 patients in the trial failed treatment; the SVR12 rate was 99% in both groups. In the PEARL-IV trial, of 205 patients with HCV genotype 1a randomly assigned to receive PTV_R/OBV/DSV alone for 12 weeks, 185 (90%) achieved SVR12; this rate was significantly lower than that observed in patients treated with PTV_R/OBV/DSV plus RBV (97% [97 of 100]), emphasizing the importance of RBV coadministration when this regimen is prescribed to patients with HCV genotype 1a (75).

Recommendations:

17. In treatment-naive patients with HCV genotype 1a infection, with or without cirrhosis, and for those with genotype 1b infection and cirrhosis, coformulated PTV_R/OBV/DSV should be given with weight-based RBV for 12 weeks (Class 1, Level A).
18. In noncirrhotic, treatment-naive patients with genotype 1b infection, coformulated PTV_R/OBV/DSV should be given without RBV for 12 weeks (Class 1, Level A).

SOF and SIM: SOF (400 mg daily) was combined with the second-generation PI SIM (150 mg daily) with or without RBV for 12 or 24 weeks in the phase 2 COSMOS study (76). The study was divided into two cohorts: cohort 1 included 80 null responders with mild fibrosis (F0 to F2) and cohort 2 included 87 treatment-naive and null responders with advanced fibrosis (F3 and F4). HCV RNA was suppressed on treatment in all patients, but six patients relapsed. The overall SVR12 rate was 92% (154 of 167), with similar results in cohorts 1 and 2 (90% [72 of 80] versus 94% [82 of 87], respectively). The SVR12 rates did not differ between 12 and 24 weeks of treatment, with or without RBV, or in treatment-naive versus treatment-experienced patients (95% [38 of 40] versus 91% [116 of 127]). The presence of a polymorphism at position 80 with a substitution of a K (lysine) for Q (glutamine), referred to as the 'Q80K' polymorphism, which is associated with reduced activity of SIM and found almost exclusively in patients with HCV genotype 1a (see below) (77,78), did not impact the rate of SVR12 (76). Although four of the six relapsers had genotype 1a infection and the Q80K polymorphism at baseline, 88% (51 of 58) of patients with this polymorphism still achieved SVR12. In this small trial, the regimen was well tolerated; headache, fatigue and nausea were the most commonly reported side effects. Only four patients (2%) discontinued treatment due to adverse events. Although the results from this trial are encouraging, given its small sample size and the availability of other effective and less expensive all-oral antiviral regimens, this regimen should be considered as a second-line option until further data emerge.

Recommendation:

19. In treatment-naive patients with HCV genotype 1a or 1b infection, with or without cirrhosis, SOF (400 mg daily) and SIM (150 mg daily) should be given for 12 weeks without RBV (Class 1, Level B).

PEG-IFN-containing regimens

Given the efficacy and markedly improved tolerability of SOF or SIM combined with PEG-IFN and RBV compared with TVR- or BOC-based regimens, the latter first-generation PIs should no longer be used except in rare circumstances where treatment is urgent and access to newer agents is not available. The use of BOC and TVR is reviewed in the 2012 version of the present guidelines (3).

SOF, PEG-IFN and RBV: SOF (400 mg daily) was combined with PEG-IFN and RBV for 12 weeks in patients with HCV genotypes 1, 4, 5 and 6 in the uncontrolled, open-label, phase 3 NEUTRINO trial (5). Among patients with HCV genotype 1, the SVR12 rate was 89% (261 of 292). Although a higher proportion of patients with genotype 1a achieved SVR12 than those with genotype 1b (92% [206 of 225] versus 82% [54 of 66]), this difference was not statistically significant. In multivariate analysis, the presence of cirrhosis and a non-CC IL28B genotype were the only predictors of virological failure. The SVR12 rate was 92% (252 of 273) in noncirrhotic patients versus 80% (43 of 54) in patients with compensated cirrhosis. The SVR12 rate was 98% (93 of 95) in patients with the IL28B CC genotype, compared with 87% (202 of 232) in those with a non-CC genotype. Although the side effect profile appeared similar to that of PEG-IFN and RBV dual therapy, the uncontrolled nature of the study precluded a clear evaluation of safety. However, only 2% of patients discontinued treatment due to an adverse event. Among the 28 patients who relapsed (9% of the cohort), resistance to SOF was not detected by deep sequencing (5).

Recommendation:

20. In patients with HCV genotype 1a or 1b, with or without cirrhosis, SOF (400 mg daily) should be given with PEG-IFN plus weight-based RBV for 12 weeks (Class 1, Level B).

SIM, PEG-IFN and RBV: In the QUEST-1 and QUEST-2 phase 3 trials (6,7), conducted in North America and Europe, respectively, the second-generation PI SIM (150 mg once daily) was combined with PEG-IFN and weight-based RBV for 12 weeks followed by an additional 12 or 36 weeks of PEG-IFN plus RBV and compared with PEG-IFN plus RBV for 48 weeks in patients with HCV genotype 1. Patients randomly assigned to triple therapy who had HCV RNA <25 IU/mL at week 4 and undetectable HCV RNA at week 12 continued PEG-IFN plus RBV for 12 additional weeks and then stopped all treatment. Patients who did not meet these early response criteria continued PEG-IFN and RBV for an additional 36 weeks (ie, 48 weeks total). In pooled data from these trials, the SVR12 rate in the SIM/PEG-IFN/RBV groups was 80% (419 of 521), significantly higher than in patients receiving PEG-IFN and RBV alone (50% [132 of 264]) (6,7). In total, 88% (459 of 521) of patients in the SIM/PEG-IFN/RBV groups qualified for shortened therapy and 88% (405 of 459) of these patients achieved SVR12. In the two trials, of the 12% (62 of 521) of patients who did not qualify for shortened therapy, the SVR12 rate was 32% despite up to 36 weeks of additional PEG-IFN and RBV. SVR12 rates differed according to baseline fibrosis level, decreasing from 84% (317 of 378) in patients with F0 to F2 fibrosis to 60% (29 of 48) in those with cirrhosis. The IL28B genotype was also important, with SVR12 rates of 95% (144 of 152) in CC patients treated with triple therapy compared with 80% (63 of 79) with PEG-IFN and RBV alone, and 75% (275 of 369) in patients with a non-CC genotype who received triple therapy compared with 37% (69 of 185) in the control arm.

The most important predictor of response was the presence of the Q80K polymorphism at baseline (described above). In pooled data from these trials (6,7), the SVR12 rate with SIM-based triple therapy was 58% (49 of 84) in patients with genotype 1a and Q80K; no different than that seen in the PEG-IFN and RBV control arm (52% [23 of 44]). In contrast, among patients with genotype 1a infection without Q80K, the SVR12 rate was 84% (138 of 165), similar to that seen in patients with genotype 1b infection (85% [228 of 267]) and significantly higher than found in the control arms (43% [36 of 83] in genotype 1a without Q80K and 53% [70 of 133] in genotype 1b). In these trials, the Q80K polymorphism was present at baseline in 34% of patients with genotype 1a infection and available sequencing data, but in only one of 400 patients with genotype 1b (6,7). Rates of Q80K positivity among patients with HCV genotype 1a in Canada have been reported to be as high as 47% (79).

SIM was well tolerated in these trials (6,7). In pooled data across the SIM study program (80), the main adverse events seen more frequently in SIM-treated patients were rash (mostly mild) seen in 23% of patients (versus 17% of controls) and photosensitivity in 3.3% (versus 0.5% of controls). Total bilirubin elevation, which is due to inhibition of biliary transporters and RBV-related hemolytic anemia, was observed in 7.9% of patients (versus 2.8% of controls). Notably, the incidence of anemia was similar among patients treated with SIM-based triple therapy versus PEG-IFN and RBV alone.

Recommendations:

21. In patients with HCV genotype 1b infection and patients with genotype 1a infection without the Q80K polymorphism, SIM (150 mg daily) should be given with PEG-IFN plus weight-based RBV for 12 weeks followed by an additional 12 weeks of PEG-IFN plus RBV (Class 1, Level A).
22. Patients with genotype 1a infection must be tested for the Q80K polymorphism before starting therapy with SIM, PEG-IFN and RBV. Patients with the Q80K polymorphism should be treated with an alternative regimen (Class 1, Level A).
23. RGT should not be used with SIM, PEG-IFN and RBV. Patients who have HCV RNA ≥25 IU/mL at week 4 or detectable HCV RNA at week 12 should stop all therapy given the low probability of SVR and the need for prolonged exposure to PEG-IFN and RBV (Class 2b, Level C).

TREATMENT-EXPERIENCED PATIENTS WITH HCV GENOTYPE 1 (TABLE 5)

PEG-IFN-free regimens

Patients who have failed IFN-based therapy should be categorized as relapsers (undetectable HCV RNA during treatment with reappearance of HCV RNA within six months of stopping therapy), partial responders (decline of at least 2 $\log_{10}$ IU/mL in HCV RNA without ever achieving undetectable HCV RNA during therapy), or null responders (<2 $\log_{10}$ IU/mL decline in HCV RNA during therapy; or breakthrough [increase by >1 $\log_{10}$ IU/mL in HCV RNA above nadir despite ongoing antiviral therapy]) (3). Patients with an unknown previous response should be managed as null responders. As in treatment-naive patients, all previously treated patients with HCV genotype 1 would benefit from all-oral DAA regimens rather than those containing IFN because these patients, with the exception of relapsers, have demonstrated poor IFN responsiveness.

SOF/LDV: The single tablet regimen of SOF/LDV was evaluated in treatment-experienced patients in the ION-2 phase 3 trial (9). The study included relapsers (56%) and nonresponders (44%), including patients who had failed PEG-IFN/RBV dual therapy (48%) or in combination with a PI (52%). Patients were randomly assigned to receive 12 or 24 weeks of treatment with or without weight-based RBV. The SVR12 rate was 94% (102 of 109) in patients who received 12 weeks of SOF/LDV and 96% (107 of 111) in those who also received RBV. The SVR12 rate in patients who received 24 weeks of SOF/LDV therapy was 99% (218 of 220 overall) whether the patients also received RBV. Virological relapse occurred in 4% to 6% of patients treated for 12 weeks, but in none treated for 24 weeks. The SVR12 rate in patients with compensated cirrhosis (20% of each treatment arm) treated for 12 weeks with SOF/LDV alone was 86% (19 of 22) versus 82% (18 of 22) in those who also received RBV. In cirrhotic patients treated for 24 weeks (with or without RBV), the SVR12 rate was 100% (44 of 44). No baseline or on-treatment predictors of relapse were identified in patients with cirrhosis. There were no differences in SVR12 rates according to receipt of RBV, previous antiviral regimen (PEG-IFN/RBV versus PEG-IFN/RBV plus a PI), or previous treatment response (relapse versus nonresponse). Among the 62 patients (14%) with detectable NS5A resistance at baseline, 55 (89%) achieved an SVR12. All 11 patients who relapsed had detectable NS5A resistance at the time of relapse, but SOF-associated resistance was not detected. Among patients previously treated with a PI-containing regimen, 71% had NS3/4A resistance at baseline and 98% of these patients achieved an SVR12 (9). Tolerability of SOF/LDV was similar to that observed in the ION-1 and ION-3 studies (see above) (8,10); more adverse events were reported in patients treated with RBV.

Based on the higher rates of response observed in the ION-2 trial among previous treatment failure patients with compensated cirrhosis treated for 24 versus 12 weeks, Health Canada and the FDA have recommended a 24-week regimen of SOF/LDV in this patient subgroup. However, a subsequent and significantly larger randomized trial from France (the SIRIUS trial) (81) suggested that a 12-week regimen of SOF/LDV plus weight-based RBV is as effective as a 24-week SOF/LDV regimen in patients with cirrhosis who had failed both PEG-IFN/RBV and triple therapy including a PI. Specifically, 74 of 77 patients (96%) randomly assigned to SOF/LDV/RBV for 12 weeks had an SVR12 (4% relapse rate) compared with 75 of 77 patients (97%) randomly assigned to SOF/LDV alone for 24 weeks (3% relapse rate). Furthermore, in a pooled analysis of data from the SIRIUS trial and six other phase 2 and 3 studies that included 352 treatment-experienced patients with cirrhosis (82), 12 weeks of SOF/LDV/RBV resulted in a similar SVR12 rate to 24 weeks of SOF/LDV alone (96% versus 98%).

Recommendations:

24. In noncirrhotic patients with HCV genotype 1 who have failed previous therapy with PEG-IFN and RBV, with or without a PI, SOF/LDV without RBV should be given for 12 weeks (Class 1, Level B).
25. In cirrhotic patients with genotype 1 who have failed previous therapy with PEG-IFN and RBV, with or without a PI, SOF/LDV and weight-based RBV should be given for 12 weeks (Class 1, Level A).

PTV_R/OBV/DSV and RBV: The combination of PTV_R/OBV/DSV with weight-based RBV was evaluated in treatment-experienced patients without cirrhosis in the SAPPHIRE-II phase 3 trial (14). Among 297 patients randomly assigned to PTV_R/OBV/DSV plus RBV regimen for 12 weeks, 286 (96%) achieved SVR12. No pre- or on-treatment predictors of response were identified. The SVR12 rate was

TABLE 5
Treatment-experienced patients with hepatitis C virus (HCV) genotype 1

Population	Recommended	Alternative (IFN-free)	Alternative (IFN-containing)	Not recommended
Genotype 1a, noncirrhotic	SOF/LDV × 12 weeks	SOF/SIM × 12 weeks†	SOF/PEG/RBV × 12 weeks	PEG/RBV
	PTV_R/OBV/DSV/RBV × 12 weeks		SIM/PEG/RBV × 24–48 weeks (if Q80K–)†‡	PEG/RBV/BOC or TVR
				SIM/PEG/RBV (if Q80K+)
Genotype 1b, noncirrhotic	SOF/LDV × 12 weeks	SOF/SIM × 12 weeks†	SOF/PEG/RBV × 12 weeks	PEG/RBV
	PTV_R/OBV/DSV × 12 weeks		SIM/PEG/RBV × 24–48 weeks†,‡	PEG/RBV/BOC or TVR
Genotype 1a, cirrhotic	SOF/LDV/RBV × 12 weeks	SOF/LDV × 24 weeks	SOF/PEG/RBV × 12 weeks	PEG/RBV
	PTV_R/OBV/DSV/RBV × 12–24 weeks*	SOF/SIM × 12 weeks†	SIM/PEG/RBV × 24–48 weeks (if Q80K–)†‡	PEG/RBV/BOC or TVR
				SIM/PEG/RBV if Q80K+)
Genotype 1b, cirrhotic	SOF/LDV/RBV × 12 weeks	SOF/LDV × 24 weeks	SOF/PEG/RBV × 12 weeks	PEG/RBV
	PTV_R/OBV/DSV/RBV × 12 weeks	SOF/SIM × 12 weeks†	SIM/PEG/RBV × 24–48 weeks†‡	PEG/RBV/BOC or TVR

**Patients with HCV genotype 1a, cirrhosis and previous null response should receive 24 weeks of PTV_R/OBV/DSV/RBV (paritaprevir//ritonavir/ombitasvir/dasabuvir/ribavirin) if treated with this regimen. Relapsers and partial responders with genotype 1 and cirrhosis can be treated for 12 weeks with PTV_R/OBV/DSV/RBV; †Simeprevir (SIM)-containing regimens should not be given to patients who have failed previous therapy with a protease inhibitor; ‡Previous null responders with genotype 1a or 1b should not be treated with SIM/ Peginterferon alfa-2a or peginterferon alfa-2b (PEG)/RBV regardless of the presence or absence of cirrhosis. Previous relapsers should be treated for 24 weeks total (12 weeks of SIM/PEG/RBV followed by 12 weeks of PEG/RBV) if HCV RNA <25 IU/mL at week 4 and undetectable at week 12. Otherwise, all treatment should be discontinued. Partial responders should be treated for 48 weeks total (12 weeks of SIM/PEG/RBV followed by 36 weeks of PEG/RBV) if HCV RNA <25 IU/mL at week 4 and undetectable at weeks 12 and 24; otherwise, all treatment should be discontinued. + Positive; – Negative; BOC Boceprevir; DSV: 250 mg one tablet twice daily; IFN Interferon; PEG: Peginterferon alfa-2a (180 μg subcutaneously/week) or peginterferon alfa-2b (1.5 μg/kg/week); PTV_R/OBV: 150 mg/100 mg/25 mg, two tablets once daily; Q80K SIM-associated resistance variant at position 80; RBV weight-based dosing: 1000 mg daily if <75 kg; 1200 mg daily if ≥75 kg; SIM: 150 mg once daily; SOF Sofosbuvir (400 mg once daily); SOF/LDV SOF 400 mg/ledipasvir 90 mg once daily (one tablet); TVR Telaprevir*

similar between patients with genotype 1a (96% [166 of 173]) and 1b (97% [119 of 123]), and did not differ between relapsers (95% [82 of 86]), partial responders (100% [65 of 65]) and null responders (95% [139 of 146]). RAVs to one or more of the three DAAs in the regimen were detected in five of the seven patients with post-treatment relapse (14).

In the TURQUOISE-II trial (13), PTV_R/OBV/DSV plus RBV regimen for 12 or 24 weeks was evaluated in 380 patients with compensated cirrhosis, of whom 58% had previously failed PEG-IFN and RBV therapy. Among patients with genotype 1b, the SVR12 rate was 99% (67 of 68) with 12 weeks of therapy and 100% (51 of 51) with 24 weeks. Response rates did not differ according to treatment duration or previous treatment history. In patients with genotype 1a infection, SVR12 rates were 89% (124 of 140) with 12 weeks and 94% (114 of 121) with 24 weeks of therapy; this difference was not statistically significant. There was no difference between the 12- and 24-week study arms among treatment-naive, cirrhotic patients with genotype 1a (12 versus 24 weeks: 92% [59 of 64] versus 93% [52 of 56]), previous relapsers (93% [14 of 15] versus 100% [13 of 13]) or partial responders (100% [11 of 11] versus 100% [10 of 10]). However, among null responders with genotype 1a, the 24-week arm was superior to 12 weeks of treatment (93% [39 of 42] versus 80% [40 of 50]) (13).

To determine the importance of RBV in noncirrhotic, treatment-experienced patients with HCV genotype 1b infection, the PEARL-II study randomly assigned patients to receive PTV_R/OBV/DSV with or without RBV for 12 weeks (83). All 91 patients (100%) who received PTV_R/OBV/DSV alone achieved SVR12 compared with 97% (85 of 88) randomly assigned to also receive RBV.

There is expected to be overlap between RAVs due to PI-based therapies. Because the PTV_R/OBV/DSV regimen contains a PI and other regimens with documented activity in these patients are available (ie, SOF/LDV) (9), this regimen should not be used in patients who have failed another PI (eg, TVR, BOC or SIM).

Recommendations:

26. In noncirrhotic, treatment-experienced patients with HCV genotype 1a infection, coformulated PTV_R/OBV/DSV should be given with weight-based RBV for 12 weeks (Class 1, Level A).
27. In noncirrhotic, treatment-experienced patients with genotype 1b infection, coformulated PTV_R/OBV/DSV should be given without RBV for 12 weeks (Class 1, Level A).
28. In cirrhotic patients with genotype 1a infection and a history of previous null response to PEG-IFN and RBV, coformulated PTV_R/OBV/DSV should be given with RBV for 24 weeks (Class 1, Level B).
29. In patients who have failed therapy with another PI, coformulated PTV_R/OBV/DSV should not be given due to the potential for cross-resistance with PTV (Class 2b, Level C).

SOF and SIM: As previously described, SOF (400 mg daily) was combined with the PI SIM (150 mg daily) with or without RBV for 12 or 24 weeks in the phase 2 COSMOS study (76). Cohort 1 included 80 null responders with mild fibrosis (F0 to F2) and cohort 2 included 47 null responders (plus 40 treatment-naive patients) with advanced fibrosis (F3 and F4). Overall, 116 of 127 null responders (91%) achieved an SVR12, not significantly different from that observed among treatment-naive subjects (95% [38 of 40]). SVR12 rates among null responders were similar regardless of fibrosis severity (F0 to F2: 90% [72 of 80] versus F3: 96% [23 of 24] versus F4: 91% [21 of 23]), treatment duration or receipt of RBV. Given the expected cross-resistance between other PIs and SIM, patients who previously failed treatment with these agents were excluded from the study (76). Because of the small sample size of this trial and the availability of other effective and less expensive IFN-free regimens, this combination should be considered as a second line option until further data emerge.

Recommendations:

30. In patients with HCV genotype 1a or 1b infection, with or without cirrhosis, who have failed previous therapy with PEG-IFN and RBV, SOF (400 mg daily) and SIM (150 mg daily) should be given without RBV for 12 weeks (Class 1, Level B).
31. The combination of SOF and SIM should not be used in patients who have failed therapy with another PI (Class 2b, Level C).

PEG-IFN-containing regimens

Given the efficacy and markedly improved safety and tolerability of SOF and SIM combined with PEG-IFN and RBV compared to TVR or BOC-based regimens, these first generation PIs should no longer be used except in rare circumstances (see above).

SOF, PEG-IFN and RBV: Experience with the use of SOF (400 mg) in combination with PEG-IFN and RBV in patients who have failed IFN-based therapy is limited. Nevertheless, Health Canada and the United States FDA have approved this regimen for treatment-experienced patients. Based on a modelling approach, the FDA projected an SVR12 rate of 78% in PEG-IFN and RBV treatment failures if retreated with SOF plus PEG-IFN/RBV for 12 weeks. In the NEUTRINO phase 3 trial of treatment-naive patients (5), 52 patients with HCV genotype 1 had characteristics typical of the treatment-experienced population (ie, advanced fibrosis [F3 and F4], a non-CC IL28B genotype and high baseline viral load [≥800,000 IU/mL]). Thirty-seven of these patients (71%) achieved SVR12 with 12 weeks of SOF/PEG-IFN/RBV (74). Although this regimen is also untested in patients who have failed therapy with a PI, the absence of cross-resistance between the PIs and SOF suggests that these patients should respond similarly to those who failed treatment with PEG-IFN/RBV alone.

Recommendation:

32. In patients with HCV genotype 1a or 1b infection, with or without cirrhosis, who have failed previous therapy with PEG-IFN and RBV with or without a PI, SOF (400 mg daily) should be given with PEG-IFN plus weight-based RBV for 12 weeks (Class 2b, Level C).

SIM, PEG-IFN and RBV: SIM (150 mg daily) has been evaluated in combination with PEG-IFN and weight-based RBV for 12 weeks followed by an additional 12 to 36 weeks of PEG-IFN and RBV in patients with HCV genotype 1 who failed IFN-based therapy in two trials. The phase 3 PROMISE study (84) included relapsers, whereas the phase 2b ASPIRE trial (85) also included partial and null responders. In the PROMISE trial (84), an RGT approach identical to that used in treatment-naive patients was evaluated (see above). Treatment with triple therapy was more effective than PEG-IFN and RBV dual therapy (SVR12: 79% [206 of 260] versus 36% [48 of 133]) in these relapsers. The majority of SIM-treated patients (93% [241 of 260]) were eligible to shorten treatment from 48 to 24 weeks and 83% of these patients (200 of 241) achieved SVR12. In patients with undetectable HCV RNA at week 4 (77% of the cohort), the SVR12 rate was 87% (173/200), compared with 60% in those with HCV RNA <25 IU/mL but detectable at week 4. Among patients who did not qualify for shortened therapy, the SVR12 rate was 40% (six of 15) despite 48 weeks of treatment. Of the 39 patients with compensated cirrhosis randomly assigned to triple therapy, 29 (74%) achieved SVR12 compared with five of 19 (26%) in the control arm. As reported in treatment-naive subjects from the QUEST-1 and QUEST-2 trials (6,7), patients with HCV genotype 1a and the Q80K mutation at baseline did not benefit from SIM treatment (SVR12 rates in the simeprevir and placebo arms: 47% [14 of 30] versus 30% [six of 20], respectively). RAVs similar to those selected by TVR and BOC emerged in most patients (90%) who did not achieve SVR12 in the SIM arm (84).

In the phase 2b ASPIRE trial (85), 462 patients who failed PEG-IFN/RBV therapy (40% relapsers, 35% partial responders and 25% null responders) were randomly assigned to receive SIM (100 mg or 150 mg or placebo) for 12, 24 or 48 weeks in combination with PEG-IFN and RBV for 48 weeks. In patients treated with SIM 150 mg daily for 12 weeks, SVR24 rates were 77% (20 of 26) in relapsers, 65% (15 of 23) in partial responders and 44% (eight of 18) in null responders; all superior to rates in the control arm (37%, 19% and 9%, respectively). Among null responders with cirrhosis (across all SIM 150 mg arms), 31% (four of 13) achieved SVR24 with SIM compared with neither of two patients treated with PEG-IFN and RBV.

Recommendations:

33. In patients with HCV genotype 1b or genotype 1a infection without the Q80K polymorphism who relapsed to PEG-IFN and RBV, SIM (150 mg daily) should be given with PEG-IFN and weight-based RBV for 12 weeks followed by PEG-IFN plus RBV for an additional 12 weeks. All therapy should be discontinued in patients who have HCV RNA >25 IU/mL at week 4 or detectable HCV RNA at week 12 (Class 1, Level A).
34. In patients with previous partial or null response, alternative regimens should be considered given the low probability of SVR and the need for prolonged exposure to PEG-IFN and RBV with this regimen (Class 2b, Level B).

PATIENTS WITH HCV GENOTYPE 2 (TABLE 6)

SOF and RBV

In the phase 3 FISSION trial (5), SOF (400 mg daily) was administered in combination with weight-based RBV for 12 weeks to treatment-naive patients with HCV genotype 2. Patients randomly assigned to the control arm received a 24-week course of PEG-IFN and RBV (800 mg daily). Patients with cirrhosis accounted for approximately 20% of the study population. The SVR12 rates in the SOF/RBV and PEG-IFN/RBV arms were 97% (68 of 70) and 78% (52 of 67), respectively. The incidence of adverse events was consistently lower among patients who received SOF/RBV, particularly the influenza-like symptoms and depression characteristic of IFN-based therapy. In the phase 3 VALENCE trial (12), 32 treatment-naive patients with HCV genotype 2 received a 12-week regimen of SOF and weight-based RBV. As observed in the FISSION study (5), all but one of these patients (97%) achieved an SVR12. The response rate did not differ between cirrhotic (100% [two of two]) and noncirrhotic patients (97% [29 of 30]). In the phase 3 POSITRON trial (11), 143 IFN-ineligible patients with HCV genotype 2 were randomly assigned to receive SOF and weight-based RBV for 12 weeks or placebo. The majority of patients in this trial had contraindications to or refused IFN therapy; only 7% had previously

TABLE 6
Patients with hepatitis C virus genotype 2

Population	Recommended	Alternative (IFN-free)	Alternative (IFN-containing)	Not recommended
Treatment-naive	SOF/RBV × 12 weeks	None	SOF/PEG/RBV × 12 weeks	PEG/RBV/PI
			PEG/RBV × 24 weeks*	SOF/LDV
				PTV_R/OBV/DSV ± RBV
				SOF/SIM
Treatment-experienced, noncirrhotic	SOF/RBV × 12 weeks	None	SOF/PEG/RBV × 12 weeks	PEG/RBV
Treatment-experienced, cirrhotic	SOF/PEG/RBV × 12 weeks	SOF/RBV × 16 weeks*	None	PEG/RBV/PI
				SOF/LDV
				PTV_R/OBV/DSV ± RBV
				SOF/SIM

**Clinically inferior regimen. DSV Dasabuvir (250 mg) one tablet twice daily; IFN Interferon; PEG Peginterferon alfa-2a (180 µg subcutaneously/week) or peginterferon alfa-2b (1.5 µg/kg/week); PI Protease inhibitor (eg, boceprevir, telaprevir or simeprevir); PTV_R/OBV Paritaprevir (150 mg)/ritonavir (100 mg)/ombitasvir (25 mg) two tablets once daily; RBV Ribavirin (weight-based dosing [1000 mg daily if <75 kg; 1200 mg daily if ≥75 kg] if combined with sofosbuvir (SOF); 800 mg daily if used in dual therapy with PEG); SIM Simeprevir (150 mg daily); SOF: 400 mg daily; SOF/LDV SOF 400 mg/ledipasvir 90 mg once daily (one tablet)*

failed IFN-based treatment. Among 109 patients with genotype 2 treated with SOF/RBV for 12 weeks, 101 patients (93%) achieved an SVR12, similar to results observed in the FISSION and VALENCE trials (5,12). SVR12 rates did not differ between patients with and without cirrhosis (94% [16 of 17] versus 92% [85 of 92]).

SOF (400 mg daily) and weight-based RBV has also been studied in treatment-experienced patients with HCV genotype 2 in the VALENCE (12) and FUSION (11) phase 3 trials. In VALENCE (12), 37 of 41 (90%) treatment-experienced patients had an SVR12 following a 12-week course of SOF/RBV. In the FUSION trial (11), 68 patients who had previously failed an IFN-containing regimen (approximately 75% due to relapse), were randomly assigned to receive SOF/RBV for either 12 or 16 weeks. Overall, an SVR12 was observed in 86% (31 of 36) of patients treated for 12 weeks versus 94% (30 of 32) treated for 16 weeks, although this difference was not statistically significant. In patients without cirrhosis, high rates of SVR12 were observed regardless of treatment duration (12 versus 16 weeks: 96% [25 of 26] versus 100% [23 of 23]). However, lower rates of response were observed among patients with cirrhosis (12 versus 16 weeks: 60% [six of 10] versus 78% [seven of nine]). Although this difference was not statistically significant, the poor response rate among patients treated for 12 weeks suggests that a 16-week regimen is preferred if SOF/RBV is prescribed to this patient subgroup, particularly in IFN-ineligible subjects. Tolerability of SOF/RBV was similar to that observed in the FISSION and POSITRON studies (5,11).

SOF, PEG-IFN and RBV

SOF, PEG-IFN, and weight-based RBV for 12 weeks has been studied in patients with HCV genotype 2 who failed previous therapy in an open-label phase 2 study (86). Among 23 patients with HCV genotype 2 (14 with cirrhosis), an SVR12 rate of 96% (22 of 23) was observed. High rates of response were observed among cirrhotic (93% [13 of 14]) and noncirrhotic patients (100% [nine of nine]). Among the entire study population (n=47), which also included 24 patients with genotype 3, three patients discontinued RBV due to anemia and one patient discontinued all therapy due to pain. Serious adverse events occurred in four patients (9%); the majority were considered due to PEG-IFN or RBV and none due to SOF.

Recommendations:

35. In treatment-naive patients with HCV genotype 2, SOF (400 mg daily) should be given with weight-based RBV for 12 weeks (Class 1, Level A).
36. In noncirrhotic, treatment-experienced patients with genotype 2, SOF (400 mg daily) should be given with weight-based RBV for 12 weeks (Class 1, Level A).
37. In IFN-eligible, treatment-experienced patients with genotype 2 and cirrhosis, SOF (400 mg daily) should be given with PEG-IFN and weight-based RBV for 12 weeks. In IFN-ineligible patients, SOF (400 mg daily) should be given with weight-based RBV for 16 weeks (Class 1, Level B).

PATIENTS WITH HCV GENOTYPE 3 (TABLE 7)

SOF and RBV

In the phase 3 FISSION trial (5), SOF (400 mg daily) in combination with weight-based RBV for 12 weeks or PEG-IFN/RBV (800 mg daily) for 24 weeks were administered to 359 treatment-naive patients with HCV genotype 3. Overall, an SVR12 was observed in 56% (102 of 183) of patients randomly assigned to receive SOF/RBV compared with 63% (110 of 176) in those treated with PEG-IFN/RBV. This difference was not statistically significant. In light of the suboptimal responses observed with a 12-week SOF/RBV regimen in this trial, the VALENCE trial examined a 24-week course in patients with HCV genotype 3 (12). Among treatment-naive patients, 94% (99 of 105) achieved an SVR12; responses did not differ between cirrhotic (92% [12 of 13]) and noncirrhotic patients (95% [87 of 92]).

SOF/RBV combination therapy has also been studied in treatment-experienced patients with HCV genotype 3. In the FUSION phase 3 trial (11), 127 patients who had failed previous treatment were randomly assigned to 12 or 16 weeks of SOF and weight-based RBV. Overall, SVR12 rates were 30% (19 of 64) and 62% (39 of 63) in the 12- and 16-week groups, respectively. The presence of cirrhosis was a strong negative predictor of response in patients treated for 12 weeks; only 19% (five of 26) of cirrhotic patients and 37% (14 of 38) of noncirrhotic patients had an SVR12 with this regimen. In the 16-week treatment arm, SVR12 rates were 61% (14 of 23) among patients with cirrhosis and 63% (25 of 40) in those without cirrhosis. In this trial, the primary mode of treatment failure was relapse, which was observed among 66% (42 of 64) of patients treated for 12 weeks and 38% (24 of 63) of those treated for 16 weeks. Therefore, the VALENCE trial examined a longer course (24 weeks) of SOF/RBV therapy in 145 treatment-experienced patients with HCV genotype 3 (12). Among 98 noncirrhotic patients in this trial, an SVR12 was observed in 85 (87%). However, only 62% (29 of 47) of patients with cirrhosis had an SVR12. Based on these data, alternative treatment options are necessary in cirrhotic, treatment-experienced patients with HCV genotype 3.

SOF, PEG-IFN and RBV

SOF, PEG-IFN and weight-based RBV administered for 12 weeks was studied in patients with HCV genotype 3 who failed previous therapy in a small, open-label phase 2 study (86). Among 24 patients, 12 of whom had cirrhosis, an SVR12 rate of 83% (20 of 24) was observed. There was no difference in response between cirrhotic and noncirrhotic patients (83% [10 of 12] in both groups).

SOF/LDV plus RBV

The single tablet regimen of SOF/LDV has been studied in patients with HCV genotype 3 in the open-label, phase 2, ELECTRON-2 trial conducted in two centres in New Zealand (87). In this study, 51 treatment-naive patients (16% with cirrhosis) were randomly assigned to 12 weeks of SOF/LDV with or without weight-based RBV. Fifty treatment-experienced patients (44% with cirrhosis) all received SOF/LDV plus RBV. Among treatment-naive patients, SVR12 rates were 64% (16 of 25) in the SOF/LDV group and 100% (26 of 26) in those who received SOF/LDV plus RBV. In treatment-experienced patients

TABLE 7
Patients with hepatitis C virus genotype 3

Population	Recommended	Alternative (IFN-free)	Alternative (IFN-containing)	Not recommended
Treatment-naive, noncirrhotic	SOF/RBV × 24 weeks	SOF/LDV/RBV × 12 weeks	SOF/PEG/RBV × 12 weeks PEG/RBV × 24 weeks*	PEG/RBV/PI PTV_R/OBV/DSV ± RBV SOF/SIM
Treatment-naive, cirrhotic	SOF/RBV × 24 weeks	SOF/LDV/RBV × 12 weeks	SOF/PEG/RBV × 12 weeks	PEG/RBV
Treatment-experienced, noncirrhotic	SOF/RBV × 24 weeks	SOF/LDV/RBV × 12 weeks	SOF/PEG/RBV × 12 weeks	PEG/RBV/PI
Treatment-experienced, cirrhotic	SOF/PEG/RBV × 12 weeks	SOF/RBV × 24 weeks* SOF/LDV/RBV × 12 weeks	None	PTV_R/OBV/DSV ± RBV SOF/SIM

**Approved, but clinically inferior regimen. DSV Dasabuvir (250 mg) one tablet twice daily; IFN Interferon; PEG Peginterferon alfa-2a (180 µg subcutaneously/week) or peginterferon alfa-2b (1.5 µg/kg/week); PI Protease inhibitor (eg, boceprevir, telaprevir or simeprevir); PTV_R/OBV Paritaprevir (150 mg)/ritonavir (100 mg)/ombitasvir (25 mg) two tablets once daily; RBV Ribavirin (weight-based dosing [1000 mg daily if <75 kg; 1200 mg daily if ≥75 kg] if combined with sofosbuvir (SOF); 800 mg daily if used in dual therapy with PEG); SIM Simeprevir (150 mg daily); SOF: 400 mg daily; SOF/LDV SOF (400 mg)/ledipasvir (90 mg) once daily (one tablet)*

treated with SOF/LDV/RBV for 12 weeks, noncirrhotic patients had higher SVR12 rates than those with cirrhosis (89% [25 of 28] versus 73% [16 of 22]). LDV has limited activity against genotype 3 HCV in vitro (88); therefore, although SOF/LDV is a potential therapeutic option in these patients, additional data in diverse populations are necessary before it can be recommended as first-line therapy over other SOF-containing regimens.

Recommendations:

38. In treatment-naive patients and noncirrhotic treatment-experienced patients with HCV genotype 3, SOF (400 mg daily) should be given with weight-based RBV for 24 weeks (Class 1, Level B).
39. In cirrhotic, treatment-experienced patients with HCV genotype 3, SOF (400 mg daily) should be given with PEG-IFN and weight-based RBV for 12 weeks (Class 1, Level B).

PATIENTS WITH HCV GENOTYPES 4, 5 AND 6 (TABLE 8)

There are limited data to guide treatment decision-making for patients with HCV genotypes 4, 5 or 6 due to the small numbers of patients enrolled in phase 3 clinical trials. In Canada, these genotypes are present in <1% of cases (22). Although the first-generation PIs, BOC and TVR, do not have clinically significant activity against genotypes 4, 5 or 6, SOF (5) and SIM (89) have activity against all of these genotypes. However, due to a paucity of published data, Health Canada and the United States FDA have approved only SOF for the treatment of HCV genotype 4.

PEG-IFN-free regimens

PTV_R/OBV ± RBV: The fixed-dose combination of the ritonavir-boosted, NS3/4A PI PTV_R and the NS5A inhibitor OBV was studied in patients with HCV genotype 4 in the PEARL-I study (90). Treatment-naive patients were randomly assigned to receive PTV_R/OBV with or without weight-based RBV for 12 weeks; all treatment-experienced patients received RBV. Nearly all patients (93%) in this study had mild fibrosis (F0 to F2) and none had cirrhosis. Among subjects who received PTV_R/OBV plus RBV, all treatment-naive (42 of 42) and treatment-experienced patients (41 of 41) achieved an SVR12. However, the SVR12 rate was lower (91% [40 of 44]) among treatment-naive patients randomly assigned to the RBV-free regimen, suggesting that RBV is necessary with this drug combination. The safety profile of PTV_R/OBV plus RBV was similar to that observed in patients with HCV genotype 1 who were also treated with DSV (14,15).

SOF/LDV: The single tablet regimen of SOF/LDV was evaluated in patients with HCV genotype 4 in a single-center, open-label phase 2a trial (National Institutes of Allergy and Infectious Diseases SYNERGY) (91). Twenty-one patients (38% treatment-experienced; 40% with cirrhosis) received SOF/LDV for 12 weeks. Among 20 patients who completed the post-treatment follow-up period, 19 (95%) achieved SVR12. No patient discontinued treatment due to an adverse event. In a similar, open-label study conducted among 25 patients with HCV genotype 6 (92% treatment-naive; 8% with cirrhosis; 80% IL28B CC genotype) from two centres (ELECTRON-2) (87), a 12-week regimen of SOF/LDV resulted in an SVR12 rate of 96% (24 of 25). Although in vitro data suggest that SOF/LDV should be efficacious in patients with HCV genotype 5 (88), it cannot currently be recommended in this patient subgroup until clinical trial data are available.

SOF and RBV: The all-oral combination of SOF (400 mg daily) and weight-based RBV for 12 or 24 weeks was studied in a randomized trial conducted among 103 Egyptian patients with HCV genotype 4 (52% treatment-experienced; 17% with compensated cirrhosis) (92). Among treatment-naive subjects, the SVR12 rates in the 12- and 24-week treatment arms were similar (84% [21 of 25] versus 92% [22 of 24]). Whereas noncirrhotic patients had similar responses in the 12- and 24-week treatment arms (86% [19 of 22] versus 90% [19 of 21]), patients with cirrhosis appeared to benefit from prolonged therapy (SVR12 in 12- versus 24-week arms: 67% [two of three] versus 100% [three of three]); however, the sample size was limited. Among treatment-experienced patients (41% nonresponders), a 24-week regimen was superior overall (SVR12 rates in 12- versus 24-week arms: 70% [19 of 27] versus 89% [24 of 27]) and in noncirrhotic patients (73% [16 of 22] versus 95% [20 of 21]). In patients with cirrhosis, SVR12 rates in the 12- and 24-week treatment groups were 60% (three of five) and 67% (four of six), respectively (92). These results were supported by a small trial of Egyptian persons living in the United States treated with SOF and weight-based RBV for 12 or 24 weeks (93). In treatment-naive patients, the SVR12 rate was 79% (11 of 14) in patients treated for 12 weeks and 100% (14 of 14) in those treated for 24 weeks. In treatment-experienced patients, corresponding SVR12 rates were 59% (10 of 17) and 87% (13 of 15).

PEG-IFN-containing regimens

SOF, PEG-IFN and RBV: In the phase 2 ATOMIC study (94), SOF (400 mg once daily) was administered for 24 weeks in combination with PEG-IFN/RBV to a small number of patients with HCV genotypes 4 and 6. SVR12 rates of 82% (nine of 11) in patients with genotype 4 and 100% (five of five) in genotype 6 were observed, supporting the antiviral activity of this regimen. In the phase 3 NEUTRINO study (5), a small subset of patients with HCV genotypes 4 (n=28), 5 (n=1) and 6 (n=5) received this regimen for a shorter 12-week treatment period, and SVR12 rates of 96% (27 of 28) in patients with genotype 4 and 100% (six of six) for genotypes 5 and 6 were reported. The one patient with genotype 4 who failed to achieve an SVR12 had cirrhosis and relapsed after cessation of therapy. The tolerability was similar to that observed historically among patients treated with PEG-IFN and RBV.

SIM, PEG-IFN and RBV: The RESTORE study was a phase 3, single-arm, open-label trial that evaluated SIM with PEG-IFN/RBV in 35 treatment-naive and 72 treatment-experienced patients with HCV genotype 4 (95). All patients received 12 weeks of triple therapy followed by 12 or 36 weeks of PEG-IFN and RBV dual therapy. Treatment-naive and relapser patients were eligible for RGT (an additional 12 weeks of PEG-IFN and RBV dual therapy if HCV RNA <25 IU/mL at week 4 and undetectable at week 12; otherwise, an additional 36 weeks) while partial and null responders received 36 weeks of dual therapy (48 weeks total). Overall, 65% (70 of 107) of patients achieved SVR12 (83% [29 of 35] of treatment-naive patients, 86% [19 of 22] of relapsers, 60% [six of 10] of partial responders and 40% [16 of 40] of null responders). The majority of patients (89% of treatment-naive and 91% of relapsers) met criteria for shortened therapy and SVR12 rates of 94% and 95% were observed in these groups, respectively. Safety was similar to that observed in other phase 3 trials of SIM/PEG-IFN/RBV therapy (6,7).

Recommendations:

40. Patients with HCV genotype 4 should be treated with coformulated PTV_R/OBV plus weight-based RBV or SOF/LDV alone for 12 weeks (Class 1, Level B).
41. Patients with HCV genotype 5 should be treated with SOF (400 mg daily) and PEG-IFN plus weight-based RBV for 12 weeks (Class 1, Level B).
42. Patients with HCV genotype 6 should be treated with SOF/LDV for 12 weeks (Class 1, Level B).

ANTIVIRAL RESISTANCE

Emergence of RAVs must be considered with all DAA-based therapies. Due to the high replication rate of HCV and the low fidelity of its RNA-dependent RNA polymerase, new variants emerge continuously (96-98). HCV circulates as a large of population of related viruses known as quasispecies. Variants with mutations that lead to DAA resistance emerge by chance and are present at low frequencies

TABLE 8
Patients with hepatitis C virus (HCV) genotypes 4, 5 and 6

Population	Recommended	Alternative (IFN-free)	Alternative (IFN-containing)	Not recommended
Genotype 4	PTV_R/OBV/RBV × 12 weeks	SOF/RBV × 24 weeks	SOF/PEG/RBV × 12 weeks	PEG/RBV
	SOF/LDV × 12 weeks		SIM/PEG/RBV × 24-48 weeks*	PEG/RBV/BOC or TVR
Genotype 5	SOF/PEG/RBV × 12 weeks	None	None	PTV_R/OBV/DSV ± RBV
Genotype 6	SOF/LDV × 12 weeks	None	SOF/PEG/RBV × 12 weeks	

**Treatment-naive and previous relapser patients with HCV genotype 4 should be treated for 24 weeks total (12 weeks of simeprevir/ peginterferon alfa-2a or peginterferon alfa-2b/ribavirin [SIM/PEG/RBV] followed by 12 weeks of PEG/RBV) if HCV RNA <25 IU/mL at week 4 and undetectable at week 12. Otherwise, all treatment should be discontinued. Partial and null responders with HCV genotype 4 should be treated for 48 weeks total (12 weeks of SIM/PEG/RBV followed by 36 weeks of PEG/RBV) if HCV RNA <25 IU/mL at week 4 and undetectable at weeks 12 and 24; otherwise, all treatment should be discontinued. BOC Boceprevir; DSV Dasabuvir (250 mg) one tablet twice daily; IFN Interferon; PEG Peginterferon alfa-2a (180 µg subcutaneously/week) or peginterferon alfa-2b (1.5 µg/kg/week); PTV_R/ OBV Paritaprevir (150 mg)/ritonavir (100 mg)/ombitasvir (25 mg) two tablets once daily; RBV: weight-based dosing (1000 mg daily if <75 kg; 1200 mg daily if ≥75 kg); SIM: 150 mg once daily; SOF Sofosbuvir (400 mg once daily); SOF/LDV SOF(400 mg)/ledipasvir (90 mg) once daily (one tablet); TVR telaprevir*

before DAA exposure. With DAA exposure, these resistant variants have a selective advantage over wild-type virus and will emerge as the dominant strains in the quasispecies. The probability that resistance will emerge with particular DAAs depends on their genetic barrier to resistance. This barrier usually reflects the number of nucleotide substitutions that must occur for high-level resistance to emerge. For example, the common PI mutation, R155K, requires two substitutions in a genotype 1b virus, but a single substitution in a genotype 1a virus and, consequently, this variant is much more common in patients with genotype 1a (99). In addition to the genetic barrier, the fitness of the RAV is important. A RAV that replicates very poorly is unlikely to emerge on therapy and will be quickly suppressed by wild-type virus once selective drug pressure is removed (97,98). For example, the S282T variant that confers resistance to SOF has extremely low replicative fitness and, as a result, has been identified only rarely in patients during SOF therapy and quickly disappears on treatment cessation (100). In contrast, many variants resistant to NS5A inhibitors are very fit and compete well with wild-type virus (88,101). As a result, NS5A-resistant variants are found in 10% to 15% of genotype 1 patients before drug exposure and persist long after therapy is discontinued in patients who fail an NS5A inhibitor-containing regimen (8,9).

Strategies to overcome resistance include avoiding DAA monotherapy and DAA dose reductions, maximizing adherence, combining DAAs with nonoverlapping resistance profiles, choosing DAAs with high barriers to resistance, and combining DAAs with PEG-IFN and RBV (96). NS5A inhibitors (eg, LDV, OBV), non-nucleoside polymerase inhibitors (eg, DSV) and NS3/4A PIs (eg, TVR, BOC, SIM) have low barriers to resistance (88). However, when potent agents of multiple classes are combined, on-treatment virological failure is extremely rare (eg, one patient of 473 treated with PTV_R/OBV/DSV plus RBV in the SAPPHIRE-I trial) and post-treatment relapse is very uncommon (eg, seven of 463 patients in this trial) (15). However, resistance to two or all three classes of drugs has been identified in almost all patients with virological failure on this combination. LDV-resistant variants are also uncommon, but present at the time of relapse in most patients who fail SOF/LDV combination therapy (8,9).

There are no data to support pretreatment resistance testing. In patients who have failed a DAA-containing regimen, it is reasonable to assume that resistance to that DAA is present at the time of retreatment. Therefore, a regimen containing DAAs without overlapping resistance should be selected in this situation. For example, in patients who have failed TVR or BOC, SOF/LDV combination therapy is very effective. In the ION-2 trial (9), 159 of 163 patients (98%) with persistent PI resistance at treatment initiation achieved an SVR12 with this regimen. Although RAVs may return to pretreatment levels after prolonged duration off therapy, there are no data on treating patients with PI resistance with a PI-containing regimen (eg, PTV_R/OBV/DSV). Therefore, this approach should not be adopted, particularly given the presence of other proven alternatives (ie, SOF/LDV).

NS5A resistance is of slightly more concern because NS5A inhibitors are a component of most all-oral regimens (88). In patients with baseline NS5A resistance, 90% achieved SVR12 with SOF/LDV in the ION-1 (8) and ION-3 (10) trials. Although this SVR rate was slightly lower than in patients without baseline resistance, the differences were not statistically significant and, therefore, pretreatment identification of resistance would not change management. Detailed baseline sequencing was not performed on all patients treated with the PTV_R/OBV/DSV regimen in the phase 3 trials (13-15,75,83); however, the rates of virological failure were low, suggesting that baseline NS5A resistance is unlikely to be a major issue. Whether retreatment of patients with emergent NS5A resistance with an NS5A-inhibitor-containing regimen will be effective remains to be determined.

Recommendations:

43. DAAs should not be used as monotherapy (Class 1, Level B).
44. Dosage reductions of DAAs should not be used to manage treatment-related side effects (Class 2a, Level C).
45. Adherence with DAAs should be maximized to reduce the likelihood of resistance (Class 2a, Level C).
46. Patients who failed therapy with a PI in the past should be treated with a regimen that does not contain a PI (Class 1, Level B).
47. With the exception of testing for Q80K in patients being considered for treatment with SIM, PEG-IFN and RBV, there is no role for baseline resistance testing with current DAA regimens (Class 1, Level A).

DDIs

Before the initiation of any DAA, potential DDIs must be considered, including those attributable to prescription and over-the-counter pharmaceuticals and herbal preparations. Identification of potential interactions requires knowledge of the metabolism of these agents. All currently available HCV PIs (TVR, BOC, SIM, PTV) are inhibitors and substrates of Cytochrome P450 3A4 (CYP3A4). Ritonavir, which is used to increase exposure and allow for once-daily dosing of PTV, is also an inhibitor and substrate of CYP3A4. Therefore, PIs are contraindicated with medications that are potent inducers of CYP3A4/5, which would reduce plasma concentrations and the therapeutic effect of the PI, and for those highly dependent on CYP3A4/5 for clearance, in which elevated plasma concentrations are associated with serious and/or life-threatening events (ie, a narrow therapeutic index). Other drug-metabolizing pathways are involved in individual PI handling that may affect DDIs. NS5A inhibitors and nucleotide polymerase inhibitors have fewer known DDIs than PIs; however, before starting therapy, all concomitant medications should be reviewed. Reference to an online updated database of DDIs is recommended before starting therapy (eg, http://www.hep-druginteractions.org).

Recommendation:

48. All prescription, over-the-counter and herbal medications should be reviewed for possible interactions with DAAs before starting therapy (Class 1, Level C).

FUTURE THERAPEUTIC OPTIONS

Numerous additional antiviral agents are under investigation in various stages of clinical development, from phase 1 though premarketing approval. Promising DAAs include NS3/4A PIs (eg, asunaprevir, grazoprevir, sovaprevir, vedroprevir), NS5A inhibitors (eg, daclatasvir, GS-5816, elbasvir, ACH-3102 and samatasvir), and non-nucleoside (eg, beclabuvir and GS-9669) and nucleotide NS5B polymerase inhibitors (eg, MK-3682 and ACH-3422). As new data regarding these agents emerge, including their receipt of regulatory approval, these HCV management guidelines will be updated.

ACKNOWLEDGEMENTS: The authors would like to acknowledge members of the CASL Executive and CASL membership, particularly Drs Eberhard Renner, Rick Schreiber, Gisela Macphail, and Jerome Simon for helpful comments regarding this manuscript.

DISCLOSURES: Production of these guidelines was not supported by any third party. The authors report the following potential conflicts of interest: Dr Kelly Burak: Consulting fees from Astellas, Gilead Sciences, Janssen and Novartis. Speaking fees from Astellas. Research support from Bayer, Bristol-Myers Squibb and Oncozyme. Dr Curtis Cooper: Consulting fees from AbbVie, Boehringer-Ingelheim, Bristol-Myers Squibb, Gilead Sciences, Janssen, Merck, Roche Canada and Vertex. Speaking fees from AbbVie, Boehringer-Ingelheim, Gilead Sciences, Janssen and Roche Canada. Research support from AbbVie, Gilead Sciences and Vertex. Dr Jordan Feld: Consulting fees from AbbVie, Boehringer-Ingelheim, Bristol-Myers Squibb, Gilead Sciences, Janssen, Merck and Theravance. Research support from AbbVie, Boehringer-Ingelheim, Gilead Sciences, Janssen and Merck. Dr Robert Myers: Consulting fees from AbbVie, Boehringer-Ingelheim, Bristol-Myers Squibb, Gilead Sciences, GlaxoSmithKline, Janssen, Merck, Roche Canada and Vertex. Speaking fees from AbbVie, Boehringer-Ingelheim, Gilead Sciences, Janssen, Roche Canada and Vertex. Research support from Gilead Sciences and Vertex. Dr Hemant Shah: Consulting fees from AbbVie, Bristol-Myers Squibb, Gilead Sciences, Janssen and Merck. Speaking fees from AbbVie, Gilead Sciences, Janssen, Merck, Roche Canada and Vertex. Grant support from Boehringer-Ingelheim and Janssen.

REFERENCES

1. Shiffman RN, Shekelle P, Overhage JM, et al. Standardized reporting of clinical practice guidelines: a proposal from the Conference on Guideline Standardization. Ann Intern Med 2003;139:493-8.
2. American Heart Association. <http://my.americanheart.org/idc/groups/ahamah-public/@wcm/@sop/documents/downloadable/ucm_319826.pdf> (Accessed February 1, 2012).
3. Myers RP, Ramji A, Bilodeau M, Wong S, Feld JJ. An update on the management of hepatitis C: Consensus guidelines from the Canadian Association for the Study of the Liver. Can J Gastroenterol 2012;26:359-75.
4. Ghany MG, Nelson DR, Strader DB, Thomas DL, Seeff LB. An update on treatment of genotype 1 chronic hepatitis C virus infection: 2011 practice guideline by the American Association for the Study of Liver Diseases. Hepatology 2011;54:1433-44.
5. Lawitz E, Mangia A, Wyles D, et al. Sofosbuvir for previously untreated chronic hepatitis C infection. N Engl J Med 2013;368:1878-87.
6. Manns M, Marcellin P, Poordad F, et al. Simeprevir with pegylated interferon alfa 2a or 2b plus ribavirin in treatment-naive patients with chronic hepatitis C virus genotype 1 infection (QUEST-2): A randomised, double-blind, placebo-controlled phase 3 trial. Lancet 2014;384:414-26.
7. Jacobson IM, Dore GJ, Foster GR, et al. Simeprevir with pegylated interferon alfa 2a plus ribavirin in treatment-naive patients with chronic hepatitis C virus genotype 1 infection (QUEST-1): A phase 3, randomised, double-blind, placebo-controlled trial. Lancet 2014;384:403-13.
8. Afdhal N, Zeuzem S, Kwo P, et al. Ledipasvir and sofosbuvir for untreated HCV genotype 1 infection. N Engl J Med 2014;370:1889-98.
9. Afdhal N, Reddy KR, Nelson DR, et al. Ledipasvir and sofosbuvir for previously treated HCV genotype 1 infection. N Engl J Med 2014;370:1483-93.
10. Kowdley KV, Gordon SC, Reddy KR, et al. Ledipasvir and sofosbuvir for 8 or 12 weeks for chronic HCV without cirrhosis. N Engl J Med 2014;370:1879-88.
11. Jacobson IM, Gordon SC, Kowdley KV, et al. Sofosbuvir for hepatitis C genotype 2 or 3 in patients without treatment options. N Engl J Med 2013;368:1867-77.
12. Zeuzem S, Dusheiko GM, Salupere R, et al. Sofosbuvir and ribavirin in HCV genotypes 2 and 3. N Engl J Med 2014;370:1993-2001.
13. Poordad F, Hezode C, Trinh R, et al. ABT-450/r-ombitasvir and dasabuvir with ribavirin for hepatitis C with cirrhosis. N Engl J Med 2014;370:1973-82.
14. Zeuzem S, Jacobson IM, Baykal T, et al. Retreatment of HCV with ABT-450/r-ombitasvir and dasabuvir with ribavirin. N Engl J Med 2014;370:1604-14.
15. Feld JJ, Kowdley KV, Coakley E, et al. Treatment of HCV with ABT-450/r-ombitasvir and dasabuvir with ribavirin. N Engl J Med 2014;370:1594-603.
16. Hull M, Shafran S, Tseng A, et al. CIHR Canadian HIV Trials Network Co-Infection and Concurrent Diseases Core: Updated Canadian adult guidelines for the treatment of hepatitis C infection in HIV-hepatitis C coinfected patients. Can J Infect Dis Med Microbiol 2014;25:311-20.
17. Myers RP, Liu M, Shaheen AA. The burden of hepatitis C virus infection is growing: A Canadian population-based study of hospitalizations from 1994 to 2004. Can J Gastroenterol 2008;22:381-7.
18. Myers RP, Krajden M, Bilodeau M, et al. Burden of disease and cost of chronic hepatitis C infection in Canada. Can J Gastroenterol Hepatol 2014;28:243-50.
19. Schanzer DL, Paquette D, Lix LM. Historical trends and projected hospital admissions for chronic hepatitis C infection in Canada: A birth cohort analysis. CMAJ 2014;Open 2:E139-144.
20. Rotermann M, Langlois K, Andonov A, Trubnikov M. Seroprevalence of hepatitis B and C virus infections: Results from the 2007 to 2009 and 2009 to 2011 Canadian Health Measures Survey. Health Rep 2013;24:3-13.
21. Remis RS. Modelling the incidence and prevalence of hepatitis C infection and its sequelae in Canada, 2007. Final Report. <http://www.phac-aspc.gc.ca/sti-its-surv-epi/model/pdf/model07-eng.pdf>. (Accessed February 1, 2012).
22. Antonishyn NA, Ast VM, McDonald RR, et al. Rapid genotyping of hepatitis C virus by primer-specific extension analysis. J Clin Microbiol 2005;43:5158-63.
23. Shah HA, Heathcote J, Feld JJ. A Canadian screening program for hepatitis C: Is now the time? CMAJ 2013;185:1325-8.
24. Pearlman BL, Traub N. Sustained virologic response to antiviral therapy for chronic hepatitis C virus infection: A cure and so much more. Clin Infect Dis 2011;52:889-900.
25. Yoshida EM, Sulkowski MS, Gane EJ, et al. Concordance of sustained virological response 4, 12, and 24 weeks post-treatment with sofosbuvir-containing regimens for hepatitis C virus. Hepatology 2014 [Epub ahead of print].
26. Maylin S, Martinot-Peignoux M, Moucari R, et al. Eradication of hepatitis C virus in patients successfully treated for chronic hepatitis C. Gastroenterology 2008;135:821-9.
27. Swain MG, Lai MY, Shiffman ML, et al. A sustained virologic response is durable in patients with chronic hepatitis C treated with peginterferon alfa-2a and ribavirin. Gastroenterology 2010;139:1593-601.
28. Myers RP, Cooper C, Sherman M, et al. Outcomes of chronic hepatitis C therapy in patients treated in community versus academic centres in Canada: Final results of APPROACH (a prospective study of peginterferon alfa-2a and ribavirin at academic and community centres in Canada). Can J Gastroenterol 2011;25:503-10.
29. Neary MP, Cort S, Bayliss MS, Ware JE Jr. Sustained virologic response is associated with improved health-related quality of life in relapsed chronic hepatitis C patients. Semin Liver Dis 1999;(19 Suppl 1):77-85.
30. Cacoub P, Ratziu V, Myers RP, et al. Impact of treatment on extra hepatic manifestations in patients with chronic hepatitis C. J Hepatol 2002;36:812-8.
31. Poynard T, McHutchison J, Manns M, et al. Impact of pegylated interferon alfa-2b and ribavirin on liver fibrosis in patients with chronic hepatitis C. Gastroenterology 2002;122:1303-13.
32. Poynard T, McHutchison J, Davis GL, et al. Impact of interferon alfa-2b and ribavirin on progression of liver fibrosis in patients with chronic hepatitis C. Hepatology 2000;32:1131-7.

33. van der Meer AJ, Veldt BJ, Feld JJ, et al. Association between sustained virological response and all-cause mortality among patients with chronic hepatitis C and advanced hepatic fibrosis. JAMA 2012;308:2584-93.
34. Veldt BJ, Heathcote EJ, Wedemeyer H, et al. Sustained virologic response and clinical outcomes in patients with chronic hepatitis C and advanced fibrosis. Ann Intern Med 2007;147:677-84.
35. Backus LI, Boothroyd DB, Phillips BR, et al. A sustained virologic response reduces risk of all-cause mortality in patients with hepatitis C. Clin Gastroenterol Hepatol 2011;9:509-516, e501.
36. Singal AG, Volk ML, Jensen D, Di Bisceglie AM, Schoenfeld PS. A sustained viral response is associated with reduced liver-related morbidity and mortality in patients with hepatitis C virus. Clin Gastroenterol Hepatol 2010;8:280-8, e281.
37. Jacobson IM, McHutchison JG, Dusheiko G, et al. Telaprevir for previously untreated chronic hepatitis C virus infection. N Engl J Med 2011;364:2405-16.
38. Sherman KE, Flamm SL, Afdhal NH, et al. Response-guided telaprevir combination treatment for hepatitis C virus infection. N Engl J Med 2011;365:1014-24.
39. Zeuzem S, Andreone P, Pol S, et al. Telaprevir for retreatment of HCV infection. N Engl J Med 2011;364:2417-28.
40. Poordad F, McCone J Jr, Bacon BR, et al. Boceprevir for untreated chronic HCV genotype 1 infection. N Engl J Med 2011;364:1195-206.
41. Bacon BR, Gordon SC, Lawitz E, et al. Boceprevir for previously treated chronic HCV genotype 1 infection. N Engl J Med 2011;364:1207-17.
42. MacParland SA, Bilodeau M, Grebely J, et al. The 3rd Canadian Symposium on Hepatitis C Virus: Expanding care in the interferon-free era. Can J Gastroenterol Hepatol 2014;28:481-7.
43. Intraobserver and interobserver variations in liver biopsy interpretation in patients with chronic hepatitis C. The French METAVIR Cooperative Study Group. Hepatology 1994;20:15-20.
44. Hellard M, Rolls DA, Sacks-Davis R, et al. The impact of injecting networks on hepatitis C transmission and treatment in people who inject drugs. Hepatology 2014;60:1861-70.
45. Micallef JM, Kaldor JM, Dore GJ. Spontaneous viral clearance following acute hepatitis C infection: A systematic review of longitudinal studies. J Viral Hepat 2006;13:34-41.
46. Sarrazin C, Kieffer TL, Bartels D, et al. Dynamic hepatitis C virus genotypic and phenotypic changes in patients treated with the protease inhibitor telaprevir. Gastroenterology 2007;132:1767-77.
47. Sherman KE, Sulkowski MS, Zoulim F, et al. Follow-up of SVR durability and viral resistance in patients with chronic hepatitis C treated with telaprevir-based regimens: Interim analysis of the EXTEND study (Abstract 248). Hepatology 2011;54(Suppl 4):485A.
48. Bedossa P. Liver biopsy. Gastroenterol Clin Biol 2008;32:4-7.
49. Myers RP, Fong A, Shaheen AA. Utilization rates, complications and costs of percutaneous liver biopsy: A population-based study including 4275 biopsies. Liver Int 2008;28:705-12.
50. Rockey DC, Caldwell SH, Goodman ZD, Nelson RC, Smith AD. Liver biopsy. Hepatology 2009;49:1017-44.
51. Guha IN, Myers RP, Patel K, Talwalkar JA. Biomarkers of liver fibrosis: What lies beneath the receiver operating characteristic curve? Hepatology 2011;54:1454-62.
52. Shaheen AA, Myers RP. Diagnostic accuracy of the aspartate aminotransferase-to-platelet ratio index for the prediction of hepatitis C-related fibrosis: A systematic review. Hepatology 2007;46:912-21.
53. Poynard T, Imbert-Bismut F, Munteanu M, et al. Overview of the diagnostic value of biochemical markers of liver fibrosis (FibroTest, HCV FibroSure) and necrosis (ActiTest) in patients with chronic hepatitis C. Comp Hepatol 2004;3:8.
54. Friedrich-Rust M, Ong MF, Martens S, et al. Performance of transient elastography for the staging of liver fibrosis: A meta-analysis. Gastroenterology 2008;134:960-974.
55. Myers RP, Elkashab M, Ma M, Crotty P, Pomier-Layrargues G. Transient elastography for the noninvasive assessment of liver fibrosis: A multicentre Canadian study. Can J Gastroenterol 24:661-70.
56. Myers RP, Pomier-Layrargues G, Kirsch R, et al. Feasibility and diagnostic performance of the FibroScan XL probe for liver stiffness measurement in overweight and obese patients. Hepatology 2012;55:199-208.
57. Castera L. Transient elastography and other noninvasive tests to assess hepatic fibrosis in patients with viral hepatitis. J Viral Hepat 2009;16:300-14.
58. Venkatesh SK, Ehman RL. Magnetic resonance elastography of liver. Magn Reson Imaging Clin N Am 2014;22:433-46.
59. Ferraioli G, Tinelli C, Dal Bello B, et al. Accuracy of real-time shear wave elastography for assessing liver fibrosis in chronic hepatitis C: A pilot study. Hepatology 2012;56:2125-33.
60. Sebastiani G, Ghali P, Wong P, et al. Physicians' practices for diagnosing liver fibrosis in chronic liver diseases: A nationwide Canadian survey. Can J Gastroenterol Hepatol 2014;28:23-30.
61. Vergniol J, Foucher J, Terrebonne E, et al. Noninvasive tests for fibrosis and liver stiffness predict 5-year outcomes of patients with chronic hepatitis C. Gastroenterology 2011;140:1970-9, e1971-3.
62. Vergniol J, Boursier J, Coutzac C, et al. Evolution of noninvasive tests of liver fibrosis is associated with prognosis in patients with chronic hepatitis C. Hepatology 2014;60:65-76.
63. Pang JX, Zimmer S, Niu S, et al. Liver stiffness by transient elastography predicts liver-related complications and mortality in patients with chronic liver disease. PLoS One 2014;9:e95776.
64. Steadman R, Myers RP, Leggett L, et al. A health technology assessment of transient elastography in adult liver disease. Can J Gastroenterol 2013;27:149-58.
65. Poynard T, Ngo Y, Munteanu M, et al. Biomarkers of liver injury for hepatitis clinical trials: A meta-analysis of longitudinal studies. Antivir Ther 2010;15:617-31.
66. Vergniol J, Foucher J, Castera L, et al. Changes of non-invasive markers and FibroScan values during HCV treatment. J Viral Hepat 2009;16:132-40.
67. Thomas DL, Thio CL, Martin MP, et al. Genetic variation in IL28B and spontaneous clearance of hepatitis C virus. Nature 2009;461:798-801.
68. Ge D, Fellay J, Thompson AJ, et al. Genetic variation in IL28B predicts hepatitis C treatment-induced viral clearance. Nature 2009;461:399-401.
69. Kurbanov F, Abdel-Hamid M, Latanich R, et al. Genetic polymorphism in IL28B is associated with spontaneous clearance of hepatitis C virus genotype 4 infection in an Egyptian cohort. J Infect Dis 2011;204:1391-4.
70. Tanaka Y, Nishida N, Sugiyama M, et al. Genome-wide association of IL28B with response to pegylated interferon-alpha and ribavirin therapy for chronic hepatitis C. Nat Genet 2009;41:1105-9.
71. Suppiah V, Moldovan M, Ahlenstiel G, et al. IL28B is associated with response to chronic hepatitis C interferon-alpha and ribavirin therapy. Nat Genet 2009;41:1100-4.
72. Prokunina-Olsson L, Muchmore B, Tang W, et al. A variant upstream of IFNL3 (IL28B) creating a new interferon gene IFNL4 is associated with impaired clearance of hepatitis C virus. Nat Genet 2013;45:164-71.
73. Kronenberger B, Zeuzem S. New developments in HCV therapy. J Viral Hepat 2012;19(Suppl 1):48-51.
74. Gilead Sciences Inc. Product Monograph: Harvoni (Ledipasvir/Sofosbuvir) Tablets 90 mg/400 mg Antiviral Agent. <www.gilead.ca/pdf/ca/harvoni_pm_english.pdf>. October 14, 2014. (Accessed November 6, 2014).
75. Ferenci P, Bernstein D, Lalezari J, et al. ABT-450/r-ombitasvir and dasabuvir with or without ribavirin for HCV. N Engl J Med 2014;370:1983-92.
76. Lawitz E, Sulkowski MS, Ghalib R, et al. Simeprevir plus sofosbuvir, with or without ribavirin, to treat chronic infection with hepatitis C virus genotype 1 in non-responders to pegylated interferon and ribavirin and treatment-naive patients: The COSMOS randomised study. Lancet 2014;384:1756-65.
77. Bae A, Sun SC, Qi X, et al. Susceptibility of treatment-naive hepatitis C virus (HCV) clinical isolates to HCV protease inhibitors. Antimicrob Agents Chemother 2010;54:5288-97.
78. Paolucci S, Fiorina L, Piralla A, et al. Naturally occurring mutations to HCV protease inhibitors in treatment-naive patients. Virol J 2012;9:245.
79. Andonov A, Kadkhoda K, Osiowy C, Kaita K. Pretreatment resistance to hepatitis C virus protease inhibitors boceprevir/telaprevir in hepatitis C virus subgenotype 1a-infected patients from Manitoba. Can J Gastroenterol 2013;27:414-6.
80. Antiviral Drugs Advisory Committee Meeting Briefing Document. Simeprevir (TMC435) Treatment of Patients with Chronic Hepatitis C. NDA 205123. October 24, 2013. <www.fda.gov/downloads/AdvisoryCommittees/CommitteesMeetingMaterials/Drugs/AntiviralDrugsAdvisoryCommittee/UCM371624.pdf>. (Accessed November 17, 2014).

81. Bourliere M, Bronowicki JP, De Ledinghen V, et al. Ledipasvir/sofosbuvir fixed dose combination is safe and efficacious in cirrhotic patients who have previously failed protease-inhibitor based triple therapy (Abstract LB-6). Hepatology 2014;60:13-4.
82. Bourliere M, Sulkowski M, Omata M, et al. An integrated safety and efficacy analysis of >500 patients with compensated cirrhosis treated with ledipasvir/sofosbuvir with or without ribavirin (Abstract 82). Hepatology 2014;60:239A.
83. Andreone P, Colombo MG, Enejosa JV, et al. ABT-450, ritonavir, ombitasvir, and dasabuvir achieves 97% and 100% sustained virologic response with or without ribavirin in treatment-experienced patients with HCV genotype 1b infection. Gastroenterology 2014;147:359-65, e351.
84. Forns X, Lawitz E, Zeuzem S, et al. Simeprevir with peginterferon and ribavirin leads to high rates of SVR in patients with HCV genotype 1 who relapsed after previous therapy: A phase 3 trial. Gastroenterology 2014;146:1669-79, e1663.
85. Zeuzem S, Berg T, Gane E, et al. Simeprevir increases rate of sustained virologic response among treatment-experienced patients with HCV genotype-1 infection: a phase IIb trial. Gastroenterology 2014;146:430-41, e436.
86. Lawitz E, Poordad F, Brainard DM, et al. Sofosbuvir with peginterferon-ribavirin for 12 weeks in previously treated patients with hepatitis C genotype 2 or 3 and cirrhosis. Hepatology 2014 [Epub ahead of print].
87. Gane EJ, Hyland RH, An D, et al. High efficacy of LDV/SOF regimens for 12 weeks for patients with HCV genotype 3 or 6 infection (Abstract LB-11). Hepatology 2014;60:22.
88. Gao M. Antiviral activity and resistance of HCV NS5A replication complex inhibitors. Curr Opin Virol 2013;3:514-20.
89. Moreno C, Berg T, Tanwandee T, et al. Antiviral activity of TMC435 monotherapy in patients infected with HCV genotypes 2-6: TMC435-C202, a phase IIa, open-label study. J Hepatol 2012;56:1247-53.
90. Pol S, Reddy KR, Baykal T, et al. Interferon-free regimens of ombitasvir and ABT-450/r with or without ribavirin in patients with HCV genotype 4 infection: PEARL-I study results (Abstract 1928). Hepatology 2014;60:1129A.
91. Kapoor R, Kohli A, Sidharthan S, et al. All oral treatment for genotype 4 chronic hepatitis C infection with sofosbuvir and ledipasvir: Interim results from the NIAID SYNERGY Trial (Abstract 240). Hepatology 2014;60:321A.
92. Esmat GE, Shiha G, Omar RF, et al. Sofosbuvir plus ribavirin in the treatment of Egyptian patients with chronic genotype 4 HCV infection (Abstract 959). Hepatology 2014;60:662A-663A.
93. Ruane PJ, Ain D, Stryker R, et al. Sofosbuvir plus ribavirin for the treatment of chronic genotype 4 hepatitis C virus infection in patients of Egyptian ancestry. J Hepatol [Epub ahead of print].
94. Kowdley KV, Lawitz E, Crespo I, et al. Sofosbuvir with pegylated interferon alfa-2a and ribavirin for treatment-naive patients with hepatitis C genotype-1 infection (ATOMIC): An open-label, randomised, multicentre phase 2 trial. Lancet 2013;381:2100-7.
95. Moreno C, Hezode C, Marcellin P, et al. Once-daily simeprevir (TMC435) with peginterferon/ribavirin in treatment-naive or treatment-experienced chronic HCV genotype 4-infected patients: Final results of a phase III trial (Abstract). J Hepatol 2014;60:S535.
96. Feld JJ. Interferon-free strategies with a nucleoside/nucleotide analogue. Semin Liver Dis 2014;34:37-46.
97. Halfon P, Sarrazin C. Future treatment of chronic hepatitis C with direct acting antivirals: is resistance important? Liver Int 2012;32(Suppl 1):79-87.
98. Vermehren J, Sarrazin C. The role of resistance in HCV treatment. Best Pract Res Clin Gastroenterol 2012;26:487-503.
99. Susser S, Welsch C, Wang Y, et al. Characterization of resistance to the protease inhibitor boceprevir in hepatitis C virus-infected patients. Hepatology 2009;50:1709-18.
100. Svarovskaia ES, Dvory-Sobol H, Parkin N, et al. Infrequent development of resistance in genotype 1-6 hepatitis C virus-infected subjects treated with sofosbuvir in phase 2 and 3 clinical trials. Clin Infect Dis 2014;59:1666-74.
101. Kohler JJ, Nettles JH, Amblard F, et al. Approaches to hepatitis C treatment and cure using NS5A inhibitors. Infect Drug Resist 2014;7:41-56.

Randomized controlled trial comparing outcomes of video capsule endoscopy with push enteroscopy in obscure gastrointestinal bleeding

Dev S Segarajasingam MBBS FRACP[1], Stephen C Hanley MD PhD[2], Alan N Barkun MDCM FRCPC MSc[3,4], Kevin A Waschke MDCM FRCPC[3], Pascal Burtin MD[5], Josée Parent MDCM FRCPC[3], Serge Mayrand MDCM FRCPC[3], Carlo A Fallone MDCM FRCPC[3], Gilles Jobin MD[6], Ernest G Seidman MDCM FRCPC[3], Myriam Martel BSc[3]

DS Segarajasingam, SC Hanley, AN Barkun, et al. Randomized controlled trial comparing outcomes of video capsule endoscopy with push enteroscopy in obscure gastrointestinal bleeding. Can J Gastroenterol Hepatol 2015;29(2):85-90.

BACKGROUND: Optimal management of obscure gastrointestinal bleeding (OGIB) remains unclear.

OBJECTIVE: To evaluate diagnostic yields and downstream clinical outcomes comparing video capsule endoscopy (VCE) with push enteroscopy (PE).

METHODS: Patients with OGIB and negative esophagogastroduodenoscopies and colonoscopies were randomly assigned to VCE or PE and followed for 12 months. End points included diagnostic yield, acute or chronic bleeding, health resource utilization and crossovers.

RESULTS: Data from 79 patients were analyzed (VCE n=40; PE n=39; 82.3% overt OGIB). VCE had greater diagnostic yield (72.5% versus 48.7%; P<0.05), especially in the distal small bowel (58% versus 13%; P<0.01). More VCE-identified lesions were rated possible or certain causes of bleeding (79.3% versus 35.0%; P<0.05). During follow-up, there were no differences in the rates of ongoing bleeding (acute [40.0% versus 38.5%; P not significant], chronic [32.5% versus 45.6%; P not significant]), nor in health resource utilization. Fewer VCE-first patients crossed over due to ongoing bleeding (22.5% versus 48.7%; P<0.05).

CONCLUSIONS: A VCE-first approach had a significant diagnostic advantage over PE-first in patients with OGIB, especially with regard to detecting small bowel lesions, affecting clinical certainty and subsequent further small bowel investigations, with no subsequent differences in bleeding or resource utilization outcomes in follow-up. These findings question the clinical relevance of many of the discovered endoscopic lesions or the ability to treat most of these effectively over time. Improved prognostication of both patient characteristics and endoscopic lesion appearance with regard to bleeding behaviour, coupled with the impact of therapeutic deep enteroscopy, is now required using adapted, high-quality study methodologies.

Key Words: *Anemia; Enteroscopy; Obscure GI bleeding; Obscure GI hemorrhage; Small bowel bleeding; Videocapsule*

Un essai aléatoire et contrôlé comparant les résultats de l'endoscopie par vidéocapsule à l'entéroscopie poussée en cas de saignement gastro-intestinal occulte

HISTORIQUE : On ne connaît pas encore la prise en charge optimale du saignement gastro-intestinal occulte (SGIO).

OBJECTIF : Évaluer le rendement diagnostique et les résultats cliniques en aval de l'endoscopie par vidéocapsule (EVC) par rapport à l'entéroscopie poussée (EP).

MÉTHODOLOGIE : Les patients ayant un SGIO et dont les œsophagogastroduodénoscopies et les coloscopies étaient négatives ont été répartis au hasard entre une EVC ou une EP et suivis pendant 12 mois. Les critères d'évaluation incluaient le rendement diagnostique, le saignement aigu ou chronique, l'utilisation des ressources de santé et les permutations.

RÉSULTATS : Les chercheurs ont analysé les données de 79 patients (EVC n=40; EP n=39; 82,3 % de SGIO manifeste). L'EVC s'associait au meilleur rendement diagnostique (72,5 % par rapport à 48,7 %; P<0,05), notamment dans l'intestin grêle distal (58 % par rapport à 13 %; P<0,01). Plus de lésions déterminées par l'EVC étaient classées comme les causes possibles ou établies du saignement (79,3 % par rapport à 35,0 %; P<0,05). Pendant le suivi, il n'y avait pas de différences dans le taux de saignement continu (aigu [40,0 % par rapport à 38,5 %; P non significatif], chronique [32,5 % par rapport à 45,6 %; P non significatif]) ni dans l'utilisation des ressources de santé. Moins de patients ayant d'abord subi une EVC ont été permutés en raison d'un saignement continu (22,5 % par rapport à 48,7 %; P<0,05).

CONCLUSIONS : La décision de commencer par l'EVC comportait un avantage diagnostique important par rapport à l'EP chez les patients ayant un SGIO, notamment pour ce qui est de la détection de lésions de l'intestin grêle, et influait sur la certitude clinique et les examens subséquents de l'intestin grêle, sans différences subséquentes sur les résultats des saignements ou l'utilisation des ressources lors du suivi. Ces observations remettent en question la pertinence clinique de déceler de nombreuses lésions endoscopiques ou d'en traiter la plupart avec efficacité au fil du temps. Il faut désormais poser un meilleur pronostic des caractéristiques des patients et de l'apparence des lésions endoscopiques en regard du comportement du saignement, et tenir compte des effets de l'entéroscopie thérapeutique profonde à l'aide de méthodologies adaptées d'études de haute qualité.

Although esophagogastroduodenoscopy and colonoscopy provide a diagnosis in up to 95% of cases of gastrointestinal bleeding (1,2), the remaining 5% experience obscure gastrointestinal bleeding (OGIB). Although relatively rare, OGIB nevertheless prompts significant utilization of health care resources. Reports document an average of up to seven diagnostic tests per patient (3), five hospitalizations and 46 units of blood transfused before definitive diagnosis (4). Further complicating the matter is the debate regarding the optimal imaging modality in the diagnostic workup (5).

Video capsule endoscopy (VCE) has been touted as a useful tool for the evaluation of OGIB, and is recommended by several gastroenterology societies – including the American Gastroenterological Association, American Society for Gastrointestinal Endoscopy and European Society of Gastrointestinal Endoscopy – for patients with OGIB after initial negative esophagogastroduodenoscopy and colonoscopy (which can first be repeated in selected cases) (5,6). These recommendations are based on summary evidence that demonstrated VCE to be equivalent or superior to other imaging modalities with respect to diagnostic

[1]*Department of Gastroenterology and Hepatology, Sir Charles Gairdner Hospital and School of Population Health, The University of Western Australia, Crawley, Australia;* [2]*Division of General Surgery;* [3]*Division of Gastroenterology;* [4]*Department of Epidemiology, Biostatistics and Occupational Health, McGill University Health Centre, McGill University, Montreal, Quebec;* [5]*Endoscopy Unit, Gustave Roussy Institute of Oncology, Villejuif, Greater Paris, France;* [6]*Division of Gastroenterology, Montreal University, Montreal, Quebec*

Correspondence: Dr Alan N Barkun, Division of Gastroenterology, McGill University Health Centre, Montreal General Hospital site, 1650 Cedar Avenue, room D16.125, Montréal, Quebec H3G 1A4 e-mail alan.barkun@muhc.mcgill.ca

yield in patients with OGIB (7), including individuals presenting with iron deficiency anemia (an occult type of OGIB) (8). A subsequent randomized clinical trial comparing VCE and push enteroscopy (PE) as first-line imaging tests after a negative gastroscopy and colonoscopy demonstrated a greater diagnostic yield for VCE, but similar clinical outcomes at 12 months. The initial VCE strategy was associated with a reduced need for alternative exploration (9). A more recent randomized clinical trial confirmed the diagnostic superiority of VCE over small bowel follow-through, but also showed equivalent clinical outcomes (10). The International Conference on Capsule Endoscopy (11) has recommended further clinical trials to assess the impact of VCE on clinical outcomes.

Accordingly, we performed a randomized controlled trial (RCT) comparing VCE and PE with regard to diagnostic yield and to a series of clinically meaningful bleeding outcomes.

METHODS

Trial design

Patients were randomly assigned to VCE or PE, which was the most widely available small bowel endoscopic approach at the time the trial was designed and funding secured.

It was hypothesized that VCE would have a higher diagnostic yield than PE, leading to less acute and/or chronic bleeding than PE over time. Furthermore, it was hypothesized that these improved outcomes would also manifest as resolution of anemia, with less transfusion, hospitalization and surgical intervention rates. Other objectives were to determine the safety profile and crossover rates of the respective imaging modalities.

Study population

Between March 2006 and December 2009, all consecutive adult (≥18 years of age) patients referred to the McGill University Health Centre (Montreal, Quebec) with OGIB were considered for randomization.

OGIB was defined as either overt health care professional-witnessed melena (occult blood-positive black tarry stools) or hematochezia (maroon- or red-coloured stools) within the past six months, associated with an acute drop in hemoglobin level ≥15 g/L (on two separate blood tests) or occult bleeding (persistent or recurrent iron deficiency anemia [hemoglobin <100 g/L for six months with a mean corpuscular volume <80 μm^3, hypoferritinemia and iron saturation <15% without evidence of associated hematological or inflammatory disease), requiring transfusion or iron supplementation, in the context of repeated fecal occult blood-positive tests (5). At the time of randomization, patients had already undergone negative esophagogastroduodenoscopy, colonoscopy and small bowel follow-through studies.

Exclusion criteria were significant cardiopulmonary disease preventing endoscopic examination, presence of an implanted electromedical device (a contraindication for VCE at the time of trial planning), pregnancy, previous gastrectomy or vagotomy, suspected obstruction or fistula, gastrointestinal motility disorders, intestinal diverticula, extensive Crohn enteritis, contraindication to erythromycin or polyethylene glycol ingestion, or recent VCE or PE examinations. To increase the generalizability of the findings, the use of acetylsalicylic acid, anti-inflammatory, anticoagulant or antiplatelet medications were not among the exclusion criteria.

Procedures

Procedures (VCE or PE) were undertaken within seven days of randomization. All patients received a 2 L polyethylene glycol oral preparation solution 12 h before their study. Patients undergoing small bowel VCE (Given Imaging, Israel) received oral simethicone 80 mg and erythromycin 500 mg before the study. Capsule recording devices were retrieved 8 h after capsule ingestion. Images were read by a trained technician and all examinations reviewed in their entirety by an experienced physician (EGS). Abdominal radiography was performed within days after the study if the capsule did not reach the cecum and evacuation of the capsule was in doubt. Patients underwent PE under conscious sedation with a standard push videoenteroscope (Olympus Corporation, Japan) passed as far as possible into the small bowel. Photographic documentation of all lesions was performed. The investigator recorded the distance reached by the enteroscope and the presence of any lesions, as well as their location and relation to bleeding (none/possible/certain) using a standardized report sheet specifically designed for the present trial. Therapy was performed consistent with standard of care.

Outcomes

The primary outcome of the present study was lesion detection rate. Also assessed were lesions deemed to be responsible for OGIB so that the potential implications of the interpretation of detected lesions on clinical end points could be addressed. Principal secondary outcomes included bleeding outcomes at follow-up and study completion. Ongoing (defined as persistent or recurrent bleeding) included acute bleeding (defined as melena, hematochezia or a drop in hemoglobin >15 g/L over 24 h) and chronic bleeding (defined as a hemoglobin level <100 g/L or a drop in hemoglobin >20 g/L within one visit, or a drop in hemoglobin >10 g/L over two consecutive visits). Improvement was defined as an increase in hemoglobin over two successive visits in the absence of acute or chronic bleeding. A stable condition was defined as the absence of improvement, acute or chronic bleeding.

Crossovers could only occur after patients had undergone the procedure to which they had been randomly assigned and was dictated by the clinical course. Immediate crossover (before ongoing rebleeding occurred) was permitted only in patients who underwent VCE initially, and had a lesion identified as needing endoscopic therapy.

Other secondary outcomes included diagnostic yield relative to lesion location, health resource utilization including additional imaging procedures, transfusions and iron supplementation, hospitalizations and surgical interventions, and adverse events related to study procedures.

Demographic data were collected at baseline. Follow-up visits occurred every month for six months and again at 12 months. Patients were questioned about OGIB recurrence, transfusion requirements, iron supplementation, hospitalizations and surgical interventions, recent blood tests, crossover and adverse events. Patient satisfaction measures were recorded at baseline and the one-, six- and 12-month visits. Questionnaires for missed visits were completed either by mail or by telephone. Unscheduled visits were also recorded.

Randomization

Patients were randomly assigned to undergo VCE or PE. Randomization occurred in a 1:1 proportion, stratified according to overt versus occult bleeding as defined above, using serially numbered sealed envelopes and blocks of eight. Due to the nature of the procedures, blinding of patients and treating physicians was not possible. Treating physicians were made aware of the procedure results. The study was approved by the McGill University Institutional Review Board and the study was registered at clinicalgov.com (NCT01424254). All patients provided informed written consent and all authors had access to the study data and had reviewed and approved the final manuscript.

Statistical analysis and sample size

Based on representative detection rates available at the time of planning of the trial (12), 40 patients in each group were required to demonstrate a significant difference in lesion detection rates with a statistical power of 90% and an alpha of 0.05. Descriptive statistics were performed for demographic and clinical characteristics of patients at baseline. Comparisons of outcomes were performed using inferential testing for categorical (χ^2 test: diagnostic yields, imaging procedures, iron supplementation, hospitalizations, surgical interventions) and continuous variables (t test: transfusion requirements, hospital length of stay). All statistical analyses were performed using SAS version 9.2 (SAS Institute Inc, USA) and data are presented as mean ± SD or percentage and 95% CI. Because of the expected high rate of crossovers in both directions (even if this was only allowed after initial testing), the primary analysis

TABLE 1
Baseline characteristics

Characteristic	VCE (n=40)	PE (n=39)
Age, years	68.4±15.2	68.6±14.0
Female sex	40.0 (24.1–55.9)	33.3 (17.9–48.8)
Duration of diagnosis, months	24.8±32.7	25.8±49.6
Overt bleeding	82.5 (69.5–94.8)	82.1 (69.5–94.7)
Previous investigations, %		
Esophagogastroduodenoscopy	1.3±0.7	1.5±1.0
Colonoscopy	1.5±0.8	1.8±1.7
Small bowel follow-through	100	100
Angiography	46.2 (14.8–77.5)	28.6 (1.5–55.6)
Nuclear scans	85.0 (67.9–100.0)	69.6 (49.2–89.9)
Hemoglobin, g/L	103±18	102±15
Iron, µmol/L	12±16	8±7
Ferritin, µg/L	69±84	59±101
Transfusions	5.8±6.3	8.6±6.4
Iron supplementation	47.5 (31.3–36.7)	38.5 (22.5–54.4)
Antiplatelet	5.0 (0.0–12.1)	5.1 (0.0–12.4)
Anticoagulant	7.5 (0.0–16.0)	15.4 (3.5–27.2)
Anti-inflammatory	2.5 (0.0–7.6)	7.7 (0.0–16.4)
Acetylsalicylic acid	22.5 (9.0–36.0)	25.6 (11.3–40.0)
Proton pump inhibitor	57.5 (41.5–73.5)	61.5 (45.6–77.5)
Calcium channel blocker	27.5 (13.0–42.0)	33.3 (17.9–48.8)
Comorbidities	2.7±1.4	2.4±1.8

*Data presented as mean ± SD or % (95% CI). *All patients underwent esophagogastroduodenoscopy, colonoscopy and small bowel follow through before enrollment in the present trial. N/A Not applicable; PE Push enteroscopy; VCE Video capsule endoscopy*

TABLE 2
Diagnostic yields from video capsule endoscopy (VCE) and push enteroscopy (PE)

	VCE (n=40)	PE (n=39)	P
Lesion location	72.5	48.7	0.03
Stomach	17.5	33.3	0.11
Duodenum	12.5	12.8	0.97
Proximal jejunum	50.0	10.3	<0.01
Distal jejunum	40.0	2.6	<0.01
Proximal ileum	27.5	N/A	
Distal ileum	32.5	N/A	
Colon	12.5	N/A	
Lesions per patient, mean ± SD	1.9±1.8	0.6±0.7	<0.01
Lesion as possible/certain cause	79.3	35.0	<0.01
Lesion type*			
Vascular anomaly	65.5	36.8	0.05
Ulcer	24.1	15.8	0.49
Erosion	6.9	21.1	0.15
Mucosal change	17.2	15.8	0.90
Polyp/tumour	17.2	5.3	0.22
Stricture	0.0	5.3	0.22
Diverticulum	0.0	5.3	0.22
Celiac disease†	3.5	0.0	0.41
Lymphangiectasia	6.9	0.0	0.24
Endoscopic therapy	N/A	4.3	

*Data presented as % unless otherwise indicated. *The sum of lesion types may exceed 100% because patients may have been found to have more than one type of lesion; †Not confirmed on subsequent duodenal biopsies or tissue transglutaminase testing. N/A Not applicable*

was conducted on an intention-to-treat basis; also performed was an exploratory secondary analysis on a per-protocol basis. A threshold of 0.05 was adopted for statistical significance.

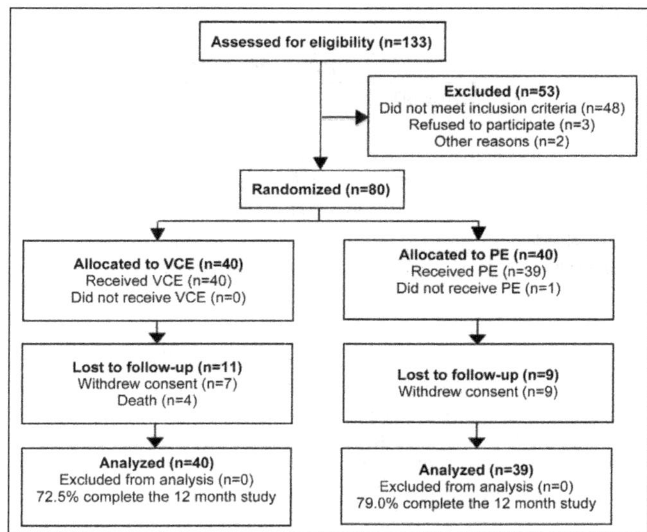

Figure 1) *Study flowchart. PE Push enteroscopy; VCE Video capsule endoscopy*

RESULTS

Study population

Enrollment occurred over the first 45 months, starting March 2006, with the final patient completing follow-up in January 2011. Over the study period, 133 patients were assessed and 80 were enrolled, with 40 allocated to each group (Figure 1). One patient randomly assigned to PE did not undergo PE, underwent no other procedures and was withdrawn from the study. Most (82.3%) presented with overt OGIB. Baseline characteristics are presented in Table 1; demographics were comparable between the two groups. Of note, patients had been initially diagnosed with OGIB, on average, >2 years before enrollment in the study, and most had undergone extensive endoscopic and imaging studies to ascertain the origin of the bleeding before enrollment (although all required recent evidence of bleeding for enrollment within the past six months, as defined above).

Primary outcome

Lesion detection rates for initial diagnostic studies (ie, before any potential crossover) are shown in Table 2. While ≥1 lesions were detected in most patients, VCE detected lesions (regardless of relation to bleeding) in significantly more patients than PE (72.5% versus 48.7%; P=0.03) and also in significantly more locations per patient (1.9±1.8 versus 0.6±0.7, respectively; P<0.01). Furthermore, while the diagnostic yield was comparable between the two modalities with respect to the stomach and duodenum, more patients had lesions detected in the jejunum by VCE than by PE (57.5% versus 12.8%; P<0.01). VCE also detected lesions in the ileum and colon (regions not reached by PE) for which no comparisons could be made.

Overall, the lesions identified by VCE were more likely to be deemed causes of bleeding than those identified by PE (possible/certain: 79.3% versus 35.0%; P=0.01; certain: 17.2% versus 0.0%; P=0.05). The types of lesions identified between the two imaging modalities were comparable.

Secondary outcomes

Patients were followed for a mean (± SD) of 9.6±4.0 months (Table 3). Over the course of the study, the rates of patients reporting either ongoing acute or chronic bleeding at at least one follow-up visit did not differ between the VCE and PE groups (acute: 40.0% versus 38.5%; P not significant; chronic: 32.5% versus 45.6%, respectively; P>0.05). Patients experienced a mean of 0.9±1.6 documented bleeding episodes over the course of the study and, again, this outcome did not differ between VCE and PE patients (0.8±1.3 versus 1.1±1.9; P>0.05).

TABLE 3
Bleeding rates

Variable	VCE (n=40)	PE (n=39)	P
Follow-up, months, mean ± SD	9.8±4.3	9.5±3.6	0.23
Any acute bleeding*	40.0	38.5	0.89
Any chronic bleeding†	32.5	45.6	0.31
Bleeding episodes, mean ± SD	0.8±1.3	1.1±1.9	0.96
Study completion	77.5	79.5	0.73
Acute bleeding*	15.0	10.3	0.53
Chronic bleeding†	2.5	10.3	0.16
Stable	65.0	74.4	0.37
Improved	7.5	2.6	0.32
Death	10.0	0.0	0.05

*Data presented as % unless otherwise indicated. *Defined as melena, hematochezia or a drop in hemoglobin >15 g/L over 24 h; †Defined as a hemoglobin level <100 g/L or a drop in hemoglobin >20 g/L within one visit, or a drop in hemoglobin >10 g/L over two consecutive visits. PE Push enteroscopy; VCE Video capsule endoscopy*

TABLE 4
Health resource utilization

Variable	VCE (n=40)	PE (n=39)	P
Follow-up, months, mean ± SD	9.8±4.3	9.5±3.6	0.23
Hospitalization	40.0	53.9	0.22
Hospitalization, days, mean ± SD	9.4±5.8	11.7±20.0	0.22
Blood transfusion	42.5	57.9	0.17
Blood transfusion, units, mean ± SD	2.8±4.3	3.0±5.6	0.65
Iron supplementation	82.5	84.2	0.84
Surgical intervention	5.0	5.1	0.98
Cross-over	22.5	48.7	0.01

Data presented as % unless otherwise indicated. PE Push enteroscopy; VCE Video capsule endoscopy

Most (78.2%) patients completed the one-year follow-up period. At study completion, the majority of patients were found to have a 'stable condition' as predefined with respect to their gastrointestinal bleeding (65.0% versus 74.4% for VCE versus PE; P>0.05). No bleeding outcome categorizations at study completion differed between the two study groups (Table 3).

Over the duration of follow-up, nearly one-half (46.8%) of patients were hospitalized due to gastrointestinal bleeding, with a length of stay of 10.7±15.5 days. One-half of patients also required blood transfusions over that time, with a mean of almost three units of blood transfused (2.9±5.0 units) per patient. Most (83.3%) patients also required iron supplementation (Table 4). Again, there were no significant differences between the study groups.

Endoscopic treatment was performed in 7.7% of patients (three of 39) at the initial procedure, which was, of course, only applicable to patients having been randomly assigned to PE.

Patient satisfaction was assessed after undergoing the study procedure (Table 5). While patient satisfaction was uniformly very high, there was a significant difference in satisfaction with office wait times between patients undergoing VCE and PE (91.4% versus 67.7%; P<0.05).

More than one-third (35.4%) of patients crossed over due to ongoing bleeding. Crossovers occurred significantly more often toward VCE: that is, among those who were originally randomly assigned to undergo PE (22.5% versus 48.7%; P=0.01). Significantly more lesions were detected by VCE in patients initially randomly assigned to PE than vice versa (10.0 versus 35.9%; P<0.01). After considering crossovers, overall lesion detection rates did not differ between VCE-first and PE-first groups (75.0% versus 71.8%; P not significant). Similarly, there was no difference in terms of identifying causes of bleeding (possible/certain: 79.3% versus 57.0%; P not significant; certain: 17.2% versus 10.7%; P not significant).

TABLE 5
Postprocedural patient satisfaction*

Variable	VCE (n=40)	PE (n=39)	P
Wait for appointment	94.3	85.9	0.20
Office wait time	91.4	67.7	<0.05
Physician professionalism	97.1	100.0	0.32
Physician technical skills	100.0	100.0	N/A
Staff professionalism	100.0	97.6	0.31
Explanations	100.0	100.0	N/A
Overall	100.0	100.0	N/A
Comfort during procedure	94.1	85.3	0.23
Comfort after procedure	94.1	85.3	0.23
Willingness to repeat procedure	100.0	100.0	N/A

*Data presented as % unless otherwise indicated. *Excellent/very good versus good, fair or poor. N/A Not applicable; PE Push enteroscopy; VCE Video capsule endoscopy*

In follow-up, after allowing for crossovers, endoscopic treatment rates (with PE) were noted in 5.0% (two of 40) of those initially randomly assigned to VCE and 7.7% (three of 39) among the PE group (ie, there were no additional treatments after the initial PE procedure in this latter group).

There were a total of five patients who, at the initial examination, were found to have lesions deemed to be a certain cause of OGIB. All were initially randomly assigned to VCE and all crossed over for potential endoscopic therapy. Two patients exhibited bleeding angiodysplasia lesions identified on VCE, one in the distal ileum and one at multiple sites. Neither underwent an endoscopic treatment on PE because in one case the endoscopist believed no treatment was necessary and in the other only a diverticulum could be identified at PE. A third patient had an ulcerated bleeding mass seen in the distal jejunum on VCE, but the lesion could not be reached on PE and underwent surgery. A fourth patient had an ulcerated, bleeding, stenosing lesion identified at the level of the mid-small bowel on VCE. On PE, a partially obstructing bleeding lesion was identified and biopsied. This lesion was found to be an adenocarcinoma; the patient underwent en bloc resection of portions of jejunum and transverse colon. The final patient was found to have active bleeding in the jejunum on VCE, without identification of a specific lesion. On crossover, the patient underwent electrocoagulation of the culprit lesions and remained bleed free thereafter.

Four patients (two in each group) required surgical intervention related to their gastrointestinal bleeding (5.0% versus 5.1%; P not significant), with no further rebleeding. One patient initially randomly assigned to VCE underwent en bloc resection of a T4 adenocarcinoma identified on VCE and biopsied on subsequent PE, as described above. Another patient (randomly assigned to PE) underwent right hemicolectomy for a cecal adenocarcinoma that was originally missed on prestudy colonoscopy. A third patient (randomly assigned to PE) underwent a nondiagnostic intraoperative enteroscopy but did not rebleed. The final of these four patients (randomly assigned to PE) underwent small bowel resection for an undetermined benign lesion at another hospital and completed the study.

While mortality was significantly higher in the VCE group (10.0% versus 0.0%; P=0.04), none of these deaths could be directly attributed to VCE or to gastrointestinal bleeding (one due to hepatic failure, one due to bladder cancer, two due to renal cancer).

DISCUSSION

VCE has become the recommended imaging modality of choice in OGIB patients after a negative gastroscopy and colonoscopy (5), certainly in patients with inactive bleeding (6), with a favourable cost-effectiveness profile (13,14).

A meta-analysis by Triester et al (7) showed that VCE was superior to PE, small bowel follow-through, computed tomography enteroclysis and magnetic resonance imaging with respect to detection of a bleeding source in OGIB. A head-to-head comparison even confirmed VCE to be as good as intraoperative enteroscopy with respect to diagnostic yield (15). Comparative data regarding more contemporary deep enteroscopic methods (single-, double-balloon and spiral enteroscopy) were not widely available at the time.

In a meta-analysis of patients with iron deficiency anemia, Koulaouzidis et al (8) suggested that small bowel VCE provides a diagnotic yield reaching 66.6% (95% CI 61.0% to 72.3%) (8); in particular, more vascular, inflammatory, and mass or tumour lesions are detected. Recent RCTs assessing the diagnostic yields of VCE have focused on different types of small bowel endoscopic capsules with no one technology exhibiting superior results (16-18).

Limitations of these comparisons are the choice of diagnostic yield as outcome, rather than true, more downstream patient end points. An initial high-quality, true clinical outcomes trial by de Leusse et al (9) confirmed the diagnostic advantage of VCE over PE (50% versus 25%; P=0.02), with a corresponding significant reduction in need for further imaging in patients investigated with VCE as first-line modality. However, overall diagnostic yield (after counting all crossovers), clinical outcomes and therapeutic impact did not differ between groups. A subsequent trial by Laine et al (10) found that VCE had a greater diagnostic yield than small bowel follow-through (30% versus 7% [difference 23%]; 95% CI 11% to 36%), although again, clinical outcomes including rebleeding, transfusion and hospitalization rates did not differ.

We report the first RCT trial to compare true clinical outcomes in patients with OGIB investigated initially by VCE or PE, with crossover dictated by clinical course, allowing for differentiation of the true clinical impact of each management approach. In fact, unlike the study by de Leusse et al (9), we compared clinical outcomes following a single diagnostic study (VCE versus PE) as opposed to an immediate crossover in the event of a negative initial study (VCE ± PE versus PE ± VCE). Moreover, patients were only allowed to crossover after the primary end point was reached and, in fact, only after ongoing bleeding was noted except if the index VCE identified a lesion requiring therapeutic PE (9). This is an important difference from the de Leusse et al (9) trial, given its study design (cross-over if negative study), analysis, and that study's high crossover rates in both directions (50% and 76%). These differences make it more difficult to draw conclusions about clinical outcomes that may not be attributed to a single investigation. The present report also differs from that of Laine et al (10) in that VCE is compared with PE as opposed to small bowel follow-through – an investigation that has recently fallen out of favour given its limited diagnostic yield (7). We believe our trial is, thus, more representative of a true clinical scenario and yields more clinically meaningful results, with its own limitations, of course, as discussed below. Our study population reflected a 'real-life' patient group with characteristics similar to other reports in that the patients had undergone numerous previous diagnostic studies, having been under investigation for an average of two years.

We found VCE to have a significantly better diagnostic yield than PE (72.5% versus 48.7%; P<0.05), consistent with previous reports (38% to 93% for VCE) (19). Interestingly, a significant number of patients (36.7%) were found to have lesions in the stomach, duodenum and colon – regions that are visualized by standard endoscopy, as has been previously noted (5,6,13), leading to the recommendation that esophagogastroduodenoscopy and colonoscopy should be repeated whenever possible as part of the diagnostic workup, ideally immediately after an acute bleed (20). The cost effectiveness of this approach has more recently been questionned by some (21). Most (40.5%) lesions were found in the jejunum and ileum, and VCE was significantly better than PE at visualizing lesions in these more distal regions not amenable to routine endoscopy (58% versus 13%; P<0.01). A review of the recent literature suggests that small bowel disease is the cause of OGIB in 40% to 70% of cases (22-26).

Interestingly, lesions identified by VCE were more likely to be scored as possible or certain sources of OGIB (79.3% versus 35.0%; P<0.01), as has been previously reported, especially in older patients (24,25,27). Although VCE has a diagnostic advantage, PE can be used as both a diagnostic and therapeutic tool. However, despite the fact that one-half of patients who underwent PE had lesions identified as a possible source of OGIB, only 7.7% underwent any intervention. Moreover, despite the therapeutic advantage of PE, more patients crossed over to VCE than to PE. In fact, only 22.5% of patients who crossed over after initial VCE study (and 5.0% of all patients initially randomly assigned to VCE) underwent any endoscopic treatment. These findings merit consideration and likely reflect the limited accessibility of distal small bowel lesions by PE, and the nature of the identified lesions that may not be judged to be a cause of OGIB, at least in a population with a mix of occult and overt bleeding similar to that observed in our and most trials. Endoscopic intervention is associated with a 27% to 85% success rate with respect to prevention of further OGIB (27). However the contribution of any therapeutic intervention is difficult to assess when the majority of untreated individuals also do not experience any further OGIB (9,10). In fact, it is the hope that future research can lead to improved selection of patients, such as those with overt bleeding (5,6,28), and that endoscopic recognition of lesions is required to optimize the downstream impact of any VCE findings.

While PE was the diagnostic test of choice at the time of the study, new endoscopic techniques, namely deep enteroscopy (single balloon, double balloon, spiral enteroscopy) are evolving and may represent signficant improvements over PE (6). It has been reported that PE can only visualize approximately one-third of the small bowel, providing a diagnosis in less than one-half of cases of OGIB (29-31). A recent meta-analysis by Teshima et al (32) showed similar diagnostic yields for VCE and double balloon enteroscopy (62% versus 56%, respectively); with improved double-balloon enteroscopy yield when performed in patients with a positive VCE (OR 1.79 [95% CI 1.09 to 2.96]). It may be that the optimal role of VCE is to select patients who will then undergo further therapeutic intervention using contemporary, more effective deep enteroscopic therapeutic techniques.

While it could be argued that our RCT was underpowered to detect differences in clinical outcomes, our results match those found in the literature (9,10). Although representative, the patients' duration of symptoms (despite recent recurrence or persistence of bleeding within six months of randomization) may have influenced the results. Indeed, recent data suggest a greater yield for investigation initiated soon after an episode of bleeding (33). It has been noted that the diagnostic yield of VCE is higher in patients with overt rather than occult OGIB (33), even more so during or soon after an active bleed (15,34). Importantly, Leung et al (28) recently reported an RCT of 60 patients with overt OGIB with mean follow-up of 48.5 months. The diagnostic yield of performing an immediate VCE was significantly higher than that of an urgent angiography (53.3% versus 20.0%; P=0.016), but the cumulative rates of rebleeding in the angiography and VCE group were 33.3% and 16.7%, respectively (P=0.10); again raising questions about the clinical pertienence of many of the findings.

Other criticisms could include the generalizability of the patient population. In fact, patients had experienced symptoms, on average, for two years. Furthermore, there was a low recent ongoing intake of nonsteroidal anti-inflammatory drugs, probably consistent with the chronicity of symptoms and discontinuation of such possible offending medications. The low intervention rate noted also emphasizes the chronicity of symptoms in most of these patients.

We believe the present RCT more clearly characterizes the diagnostic impact of endoscopic findings on subsequent clinical management than previously. The findings suggest a higher diagnostic yield for VCE compared with PE, but with no significant alterations in bleeding outcomes or health care utilization over a 12-month follow-up period. Additional high-quality RCTs, perhaps focusing on patients with overt bleeding while better characterizing the appearance of endoscopic lesions that bear true prognostic significance, in association with the use of contemporary deep enteroscopic techniques, are needed.

AUTHOR CONTRIBUTIONS: Conception and design of the study: Alan N Barkun, Pascal Burtin and Myriam Martel. Generation, collection, assembly, analysis and/or interpretation of data: Alan N Barkun, Dev S Segarajasingam, Stephen C Hanley, Kevin A Waschke, Pascal Burtin, Josée Parent, Serge Mayrand, Carlo A Fallone, Gilles Jobin, Ernest G Seidman and Myriam Martel. Drafting or revision of the manuscript: Alan N Barkun, Dev S Segarajasingam, Stephen C Hanley, Kevin A Waschke, Pascal Burtin, Josée Parent, Serge Mayrand, Carlo A Fallone, Gilles Jobin, Ernest G Seidman and Myriam Martel. Approval of the final version of the manuscript: Alan N Barkun, Dev S Segarajasingam, Stephen C Hanley, Kevin A Waschke, Pascal Burtin, Josée Parent, Serge Mayrand, Carlo A Fallone, Gilles Jobin, Ernest G Seidman and Myriam Martel.

FUNDING: Support from the American Society for Gastrointestinal Endoscopy, American College of Gastroenterology, and Canadian Institutes for Health Research. Support for DS Segarajasingam from the University of Western Australia Overseas Medical Fellowships (Eva KA Nelson). Dr EG Seidman is supported by a Canada Research Chair.

DISCLOSURES: Dr AN Barkun is a consultant for Olympus. Dr EG Seidman has received research study support, is on the medical advisory board and speakers' bureau for Given Imaging Inc. Dev S Segarajasingam, Stephen C Hanley, Kevin A Waschke, Pascal Burtin, Josée Parent, Serge Mayrand, Carlo A Fallone, Gilles Jobin and Myriam Martel have no conflicts of interests to declare related to this article.

REFERENCES

1. Lewis BS. Small intestinal bleeding. Gastroenterol Clin North Am 2000;29:67-95, vi.
2. Zuckerman GR, Prakash C, Askin MP, Lewis BS. AGA technical review on the evaluation and management of occult and obscure gastrointestinal bleeding. Gastroenterology 2000;118:201-21.
3. Foutch PG, Sawyer R, Sanowski RA. Push-enteroscopy for diagnosis of patients with gastrointestinal bleeding of obscure origin. Gastrointest Endosc 1990;36:337-41.
4. Flickinger EG, Stanforth AC, Sinar DR, MacDonald KG, Lannin DR, Gibson JH. Intraoperative video panendoscopy for diagnosing sites of chronic intestinal bleeding. Am J Surg 1989;157:137-44.
5. Raju GS, Gerson L, Das A, Lewis B. American Gastroenterological Association (AGA) Institute medical position statement on obscure gastrointestinal bleeding. Gastroenterology 2007;133:1694-6.
6. Fisher L, Lee Krinsky M, Anderson MA, et al. The role of endoscopy in the management of obscure GI bleeding. Gastrointest Endosc 2010;72:471-9.
7. Triester SL, Leighton JA, Leontiadis GI, et al. A meta-analysis of the yield of capsule endoscopy compared to other diagnostic modalities in patients with obscure gastrointestinal bleeding. Am J Gastroenterol 2005;100:2407-18.
8. Koulaouzidis A, Rondonotti E, Giannakou A, Plevris JN. Diagnostic yield of small-bowel capsule endoscopy in patients with iron-deficiency anemia: A systematic review. Gastrointest Endosc 2012;76:983-92.
9. de Leusse A, Vahedi K, Edery J, et al. Capsule endoscopy or push enteroscopy for first-line exploration of obscure gastrointestinal bleeding? Gastroenterology 2007;132:855-62.
10. Laine L, Sahota A, Shah A. Does capsule endoscopy improve outcomes in obscure gastrointestinal bleeding? Randomized trial versus dedicated small bowel radiography. Gastroenterology 2010;138:1673-1680 e1; quiz e11-2.
11. Pennazio M, Eisen G, Goldfarb N. ICCE consensus for obscure gastrointestinal bleeding. Endoscopy 2005;37:1046-50.
12. Ell C, Remke S, May A, Helou L, Henrich R, Mayer G. The first prospective controlled trial comparing wireless capsule endoscopy with push enteroscopy in chronic gastrointestinal bleeding. Endoscopy 2002;34:685-9.
13. Ladas SD, Triantafyllou K, Spada C, et al. European Society of Gastrointestinal Endoscopy (ESGE): Recommendations (2009) on clinical use of video capsule endoscopy to investigate small-bowel, esophageal and colonic diseases. Endoscopy 2010;42:220-7.
14. Marmo R, Rotondano G, Rondonotti E, et al. Capsule enteroscopy vs. other diagnostic procedures in diagnosing obscure gastrointestinal bleeding: A cost-effectiveness study. Eur J Gastroenterol Hepatol 2007;19:535-42.
15. Hartmann D, Schmidt H, Bolz G, et al. A prospective two-center study comparing wireless capsule endoscopy with intraoperative enteroscopy in patients with obscure GI bleeding. Gastrointest Endosc 2005;61:826-32.
16. Dolak W, Kulnigg-Dabsch S, Evstatiev R, et al. A randomized head-to-head study of small-bowel imaging comparing MiroCam and EndoCapsule. Endoscopy 2012;44:1012-8.
17. Choi EH, Mergener K, Semrad C, et al. A multicenter, prospective, randomized comparison of a novel signal transmission capsule endoscope to an existing capsule endoscope. Gastrointest Endosc 2013;78:325-332.
18. Pioche M, Vanbervliet G, Jacob P, et al. Prospective randomized comparison between axial- and lateral-viewing capsule endoscopy systems in patients with obscure digestive bleeding. Endoscopy 2014;46:479-84.
19. Tang SJ, Haber GB. Capsule endoscopy in obscure gastrointestinal bleeding. Gastrointest Endosc Clin N Am 2004;14:87-100.
20. Cooper GS, Kou TD, Wong RC. Use and impact of early endoscopy in elderly patients with peptic ulcer hemorrhage: A population-based analysis. Gastrointest Endosc 2009;70:229-35.
21. Rondonotti E, Marmo R, Petracchini M, de Franchis R, Pennazio M. The American Society for Gastrointestinal Endoscopy (ASGE) diagnostic algorithm for obscure gastrointestinal bleeding: Eight burning questions from everyday clinical practice. Dig Liver Dis 2013;45:179-85.
22. Mylonaki M, Fritscher-Ravens A, Swain P. Wireless capsule endoscopy: A comparison with push enteroscopy in patients with gastroscopy and colonoscopy negative gastrointestinal bleeding. Gut 2003;52:1122-6.
23. Ginsberg GG, Barkun AN, Bosco JJ, et al. Wireless capsule endoscopy: August 2002. Gastrointest Endosc 2002;56:621-4.
24. Tacheci I, Deviere J, Kopacova M, Douda T, Bures J, Van Gossum A. The importance of upper gastrointestinal lesions detected with capsule endoscopy in patients with obscure digestive bleeding. Acta Gastroenterol Belg 2011;74:395-9.
25. Jensen DM. Current diagnosis and treatment of severe obscure GI hemorrhage. Gastrointest Endosc 2003;58:256-66.
26. Levi Z, Gal E, Vilkin A, Chonen Y, Belfer RG, Fraser G, Niv Y. Fecal immunochemical test and small bowel lesions detected on capsule endoscopy: Results of a prospective study in patients with obscure occult gastrointestinal bleeding. Eur J Gastroenterol Hepatol 2011;23:1024-8.
27. Raju GS, Gerson L, Das A, Lewis B. American Gastroenterological Association (AGA) Institute technical review on obscure gastrointestinal bleeding. Gastroenterology 2007;133:1697-717.
28. Leung WK, Ho SS, Suen BY, et al. Capsule endoscopy or angiography in patients with acute overt obscure gastrointestinal bleeding: A prospective randomized study with long-term follow-up. Am J Gastroenterol 2012;107:1370-6.
29. Landi B, Tkoub M, Gaudric M, et al. Diagnostic yield of push-type enteroscopy in relation to indication. Gut 1998;42:421-5.
30. Pennazio M, Arrigoni A, Risio M, Spandre M, Rossini FP. Clinical evaluation of push-type enteroscopy. Endoscopy 1995;27:164-70.
31. Bouhnik Y, Bitoun A, Coffin B, Moussaoui R, Oudghiri A, Rambaud JC. Two way push videoenteroscopy in investigation of small bowel disease. Gut 1998;43:280-4.
32. Teshima CW, Kuipers EJ, van Zanten SV, Mensink PB. Double balloon enteroscopy and capsule endoscopy for obscure gastrointestinal bleeding: An updated meta-analysis. J Gastroenterol Hepatol May 2011;26:796-801.
33. Yamada A, Watabe H, Kobayashi Y, Yamaji Y, Yoshida H, Koike K. Timing of capsule endoscopy influences the diagnosis and outcome in obscure-overt gastrointestinal bleeding. Hepatogastroenterology 2012;59:676-9.
34. Pennazio M, Santucci R, Rondonotti E, et al. Outcome of patients with obscure gastrointestinal bleeding after capsule endoscopy: Report of 100 consecutive cases. Gastroenterology 2004;126:643-53.

Yield of screening colonoscopy in renal transplant candidates

Turki AlAmeel MBBS FRCPC FACP[1], Bahaa Bseiso MD[1], Meteb M AlBugami MBBS FRCPC FACP[2], Sami AlMomen MRCP(UK)[1], Lee Stuart Roth MD FRCPC DABIM FACP MSc[3]

T AlAmeel, B Bseiso, MM AlBugami, S AlMomen, LS Roth. Yield of screening colonoscopy in renal transplant candidates. Can J Gastroenterol Hepatol 2015;29(8):423-426.

BACKGROUND: Cardiovascular disease is the most common cause of death among patients with end-stage renal disease undergoing maintenance dialysis. Renal transplantation offers a survival advantage to patients with end-stage renal disease; it is also associated with a three- to fivefold increase in the risk of developing a neoplasm.
OBJECTIVE: To determine the yield of screening colonoscopy among patients with chronic kidney disease who were considered for renal transplantation.
METHODS: Patients were included if they were ≥50 years of age, had chronic kidney disease and were being considered for renal transplantation. They underwent a screening colonoscopy that was performed as part of their pretransplant workup. Data from December 2008 to May 2014 were collected retrospectively for all eligible patients.
RESULTS: During the study period, 433 patients were considered for renal transplantation. Of these, 170 underwent colonoscopies as part of their pretransplant workup. One was excluded because of previous history of colon cancer. Of the 169 procedures performed, ≥1 polyp(s) was diagnosed in 24%. The most common pathological diagnoses were hyperplastic polyp or normal colonic tissue. Fifteen (37%) patients had tubular adenomas and one patient had a sessile serrated adenoma. Advanced adenomas, defined as villous, tubulovillous or high-grade dysplasia, were found in four patients. Adenocarcinoma was diagnosed in one patient.
CONCLUSION: In a population of asymptomatic potential kidney transplant recipients ≥50 years of age, the prevalence of colorectal adenomatous polyps was 24%. Colonoscopy appeared to be useful as a screening tool in potential transplant recipients.

Key Words: *Colonoscopy; Colorectal cancer; Renal transplant; Screening*

Le rendement de la coloscopie de dépistage chez les candidats à la transplantation rénale

HISTORIQUE : La maladie cardiovasculaire est la principale cause de décès chez les patients atteints d'une néphropathie au stade terminal sous dialyse d'entretien. La transplantation rénale rallonge la survie des patients atteints d'une néphropathie rénale. Toutefois, elle triple le risque de néoplasme, et peut même aller jusqu'à le quintupler.
OBJECTIF : Déterminer le rendement de la coloscopie de dépistage chez les patients atteints d'une néphropathie chronique chez qui une transplantation rénale était envisagée.
MÉTHODOLOGIE : Les patients étaient inclus dans l'étude s'ils avaient au moins 50 ans, s'ils avaient une néphropathie chronique et s'ils étaient évalués en vue d'une transplantation rénale. Ils ont subi une coloscopie de dépistage dans le cadre de leur bilan avant la transplantation. Pour tous les patients admissibles, les chercheurs ont recueilli rétrospectivement les données colligées entre décembre 2008 et mai 2014.
RÉSULTATS : Pendant la période de l'étude, une transplantation rénale a été envisagée chez 433 patients. De ce nombre, 170 ont subi une coloscopie dans le cadre de leur bilan avant la transplantation. Un a été exclu en raison d'antécédents de cancer du côlon. Dans 24 % des 169 interventions effectuées, au moins un polype a été diagnostiqué. Les diagnostics pathologiques les plus courants étaient des polypes hyperplasiques ou des tissus coliques normaux. Quinze patients (37 %) avaient des adénomes tubulaires et un patient, un adénome festonné sessile. On a découvert des adénomes avancés, définis comme des dysplasies villeuses, tubulovilleuses ou de haut grade, chez quatre patients. Un adénocarcinome a été diagnostiqué chez un patient.
CONCLUSION : Dans une population de greffés du rein potentiels et asymptomatiques de 50 ans ou plus, la prévalence de polypes adénomateux colorectaux s'élevait à 24 %. La coloscopie semblait utile comme outil de dépistage chez des greffés potentiels.

Renal transplantation offers a survival advantage to patients with end-stage renal disease (ESRD) compared with maintenance dialysis (1). It is also associated with a three- to fivefold increase in the risk for developing a neoplasm (2). The majority of these malignancies are nonmelanoma skin cancers and lymphoproliferative disorders. The risk of developing colorectal cancer also increases after solid organ transplantation (3). Compared with the general population, the Swedish cancer registries found the incidence of colorectal cancer to be approximately twice as high in transplant recipients (4). In addition, there was a propensity for right-sided cancers, with a standardized incidence ratio of 3.3 versus 1.8 for left-sided colon cancers.

In the general population, screening for colorectal cancer with flexible sigmoidoscopy or colonoscopy has been shown to reduce overall mortality from colorectal cancer. Only colonoscopy was associated with reduced mortality from proximal colorectal cancer (5). In average-risk individuals, most guidelines recommend starting screening at 50 years of age and propose colonoscopy as an option for screening (6-8).

Among patients with ESRD on maintenance dialysis, cardiovascular disease is the most common cause of death. Cancer deaths, however, remain proportionately lower. Malignancy accounts for up to 10% of all deaths in the United States, with a rate of 2.6 deaths per 1000 patient-years (9).

Cancer screening in the general population is effective when it yields survival benefits without incurring high costs or significant complications. In dialysis patients, early detection may not improve mortality; therefore, cancer screening should be individualized based on the patient's overall status and life expectancy (10).

No studies investigating the benefits or harms of screening kidney transplant candidates for colorectal cancer have been published. Accordingly, we determined the yield of screening colonoscopy among

[1] *Division of Gastroenterology;* [2]*Multi-Organ Transplant Center, King Fahad Specialist Hospital-Dammam, Dammam, Saudi Arabia;* [3]*William Osler Health System, Toronto, Ontario*
Correspondence: Dr Lee Stuart Roth, 117-2291 Kipling Avenue, Toronto, Ontario M9W 4L6.e-mail lroth@tsh.to

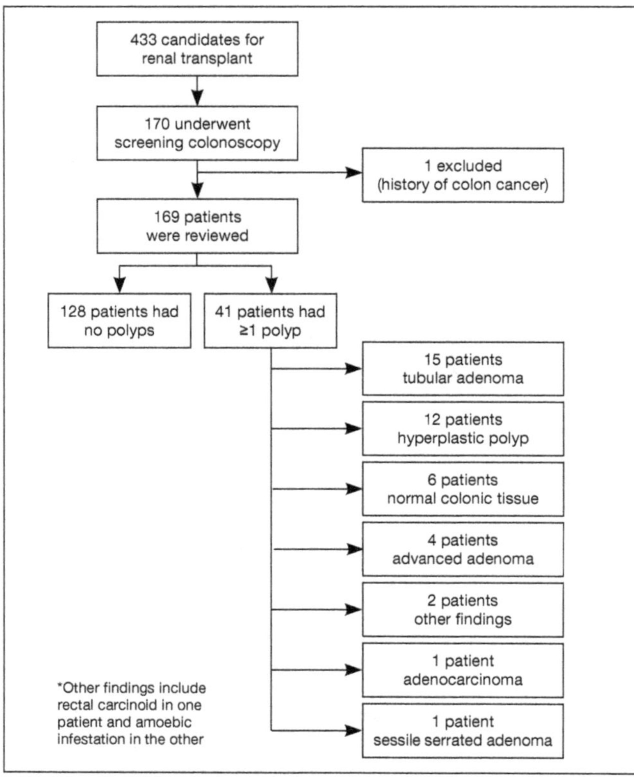

Figure 1) *Enrollment and outcomes*

patients with chronic kidney disease who were considered for renal transplantation.

METHODS

Patients were included if they were ≥50 years of age, had chronic kidney disease and were being considered for renal transplantation. Candidates underwent clinical history and physical examination, transthoracic echocardiogram and psychosocial evaluation. They underwent a screening colonoscopy that was performed as part of their pretransplant workup. Patients were excluded if they had a clinical indication for diagnostic colonoscopy such as rectal bleeding, family history of familial adenomatous polyposis or history of colon cancer. A review of patients' medical records, including procedure notes and pathology reports, was performed and data from all eligible patients were retrospectively analyzed. All procedures were performed by nine experienced gastroenterologists who performed >1000 colonoscopies before their involvement with the study patients. Patients underwent moderate sedation with propofol or other sedative agents given by an anesthesiologist. Colonoscopies were performed in the endoscopy unit of a tertiary care centre. The study was approved by Institutional Review Board of King Fahad Specialist Hospital, located in Dammam, Saudi Arabia.

RESULTS

During the study period (September 2008 to May 2014), 433 patients with chronic kidney disease were considered for renal transplantation. The majority of these patients were on peritoneal or hemodialysis by the time they were referred for transplant. In most cases, the etiology of chronic kidney disease could not be identified due to the advanced stage of illness. Of these, 263 were deemed not suitable for transplant after preliminary workup. Reasons for exclusion from transplant included advanced cardiac or pulmonary disease, severe psychiatric illness and poor social support, which can lead to poor post-transplant follow-up. The remaining 170 patients underwent colonoscopy as part of their pretransplant workup. All procedures were completed to the cecum. One patient was excluded due to a history of colon cancer (Figure 1). The characteristics of the study population are summarized in Table 1.

TABLE 1
Characteristics of the study population

Age, years, mean (range)	57.9 (50–74)
Sex, male/female, %	61/49
Polyps, n/n total (%)	41/169 (24)
Advanced adenomas, n/n total (%)	4/169 (2)
Polyps >1 cm, n/n total (%)	8/169 (5)

Of the 169 procedures performed, 128 revealed no evidence of polyps. Forty-one (24%) patients had ≥1 polyp(s) diagnosed at the time of colonoscopy. The majority (n=25 [60.9%]) of patients had only one polyp; eight (19.5%) had two; four (9.75%) had three; and ≥4 polyps were found in four (9.75%) patients (Figure 2). In eight patients, the polyps were estimated to be >10 mm in size. The remainder of subjects had subcentimetre size polyps.

The most common pathological diagnosis was tubular adenoma without high-grade dysplasia in 15 (37%) patients. Twelve patients had hyperplastic polyps. In six patients, pathological examination of the removed polyps revealed normal colonic tissue. One patient had a sessile serrated adenoma. Advanced adenoma, defined as villous, tubulovillous or high-grade dysplasia, was found in four patients (Figure 3).

Adenocarcinoma was diagnosed incidentally in one patient. Other findings on pathological examination of resected polyps included rectal carcinoid in one patient and amoebic infestation in another. No complications were reported in the study population.

DISCUSSION

We found that routine colonoscopy in patients with ESRD considered for renal transplantation with no other clinical indication for colonoscopy resulted in identification of polyps in 24% of cases. This is similar to the 22% polyp detection rate found in the general asymptomatic population in Saudi Arabia (11). In our study, one of 169 (0.59%) patients was found to have adenocarcinoma in the colon.

In patients with ESRD, renal transplantation offers the best treatment. However, the long-term use of immunosuppressive therapy has been associated with increased risk for solid organ tumours (12). Identifying major comorbidities before transplantation may improve overall survival. Early detection of adenomatous polyps may reduce morbidity in post-transplant candidates.

Colorectal cancer is the second leading cause of death due to malignancy. It has an incidence that is two to three times higher in transplant recipients than the general population (13,14). Transplant recipients who develop colorectal cancer are often younger at diagnosis and experience poorer outcomes compared with the general population (15,16).

Colorectal cancer can present relatively early after transplant. In one study investigating 5600 solid organ transplant recipients with mean follow-up of 9.3 years (16), 40 patients developed colorectal cancer. The median time from transplant to cancer diagnosis was 6.6 years and 30% were diagnosed <5 years post-transplant. Screening in renal transplant candidates achieves two important goals: scarce resources will be well utilized; and that transplantation of recipients with active cancer will be avoided to prevent harm (17).

Screening methods for colon cancer can be broadly divided into noninvasive tests and structural examinations. Noninvasive tests include stool testing for fecal occult blood or fecal immunochemical testing, and structural examinations, which can include partial or full examinations of the colon, colonoscopy, flexible sigmoidoscopy and computed tomography colonography (18).

In the general population, the incidence of colorectal cancer was lower among men and women who underwent screening colonoscopy compared with those who had no history of endoscopy. Screening colonoscopy was associated with reduced mortality from both proximal

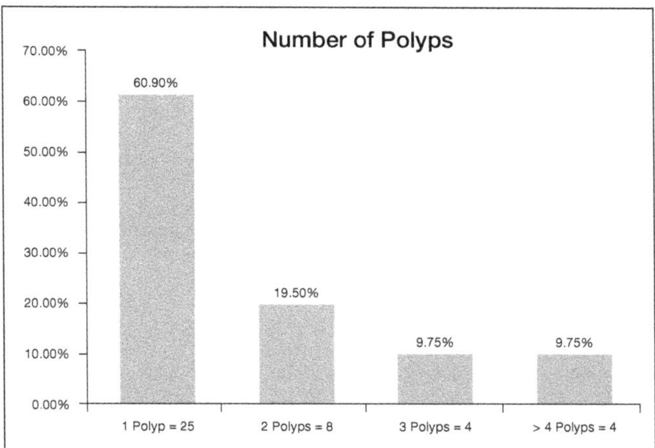

Figure 2) *Percentage of patients with ≥1 polyp(s)*

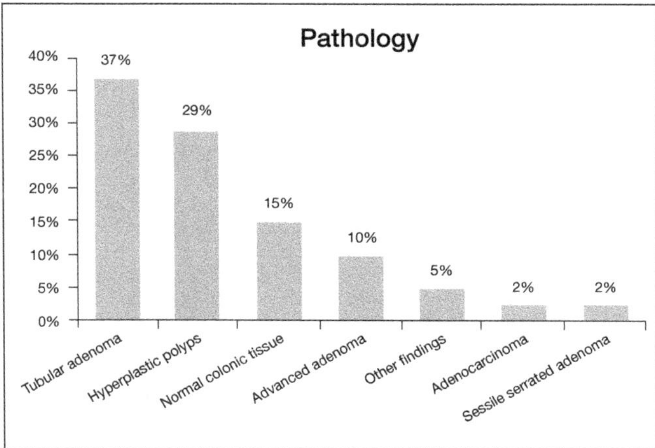

Figure 3) *Percentage of different pathological findings*

and distal colorectal cancer (5). Colonoscopy, however, is not without risks. There is a one in 1000 risk for major complications such as bleeding and perforation. Risk factors for a complication include older age, male sex, polypectomy and having the procedure performed by a low-volume endoscopist (19). Such complications can lead to significant morbidity, and even mortality, in patients with ESRD.

On the other hand, noninvasive tests have their limitations. In an Australian study investigating the prevalence of colon cancer and advanced adenoma in renal transplant recipients (12), fecal hemoglobin testing had poor sensitivity (31.0%) for advanced neoplasia. Nonetheless, the performance of the test was similar to what was found in the general population.

To our knowledge, the present study was the first to investigate the yield of screening colonoscopy in renal transplant candidates. The rate of significant findings, including adenomatous polyps, in our patient population was 13.6%. Other studies examined the yield of colon cancer screening in renal transplant recipients. The prevalence of advanced adenoma or colorectal neoplasia was 13% (12).

Our study had limitations. All procedures were performed by experienced gastroenterologists, which may have led to a lower complication rate. In a population-based study, nongastroenterologist endoscopists had a higher perforation rate compared with gastroenterologists (20). Lower complication results in our study may also have resulted from having anesthesiologists managing sedation in these patients. Our patients were chosen from a referral centre, which raises the possibility of selection bias. Due to the advanced stage they presented with, we could not identify the underlying cause of their kidney disease.

Renal transplant is an expensive treatment. Adding screening colonoscopy with an anesthesiologist providing sedation to the cost makes it even more expensive. Our study did not address the cost effectiveness of this approach.

On the basis of our results and current recommendations for the general population, we suggest that patients with chronic kidney disease, ≥50 years of age who are at average risk for colon cancer and being considered for renal transplant should undergo a screening test. The choice of screening test, however, should be individualized based on the patient's preference, and the risks and benefits profile of that particular patient.

STUDY HIGHLIGHTS

Current knowledge:

- Renal transplantation offers survival advantage to patients with ESRD.
- Screening for colorectal cancer in the general population reduces overall mortality from colorectal cancer.
- Colorectal cancer is the second most common malignancy in transplant recipients.
- The yield of screening colonoscopy among patients with chronic kidney disease who were considered for renal transplantation is unknown.

What is new:

- One-quarter of renal transplant candidates had polyps found on screening colonoscopy.
- The rate is similar to what has been reported in the general population.
- The choice of colon cancer screening test in renal transplant candidates should be individualized based on the patient's preference, and the risks and benefits profile.

AUTHOR CONTRIBUTIONS: Turki AlAmeel designed the study, reviewed the literature and drafted the manuscript. Bahaa Bseiso contributed to the design of the study, reviewed the literature, collected the data and contributed to the drafting of the manuscript. Meteb AlBugami contributed to the design of the study, collected the data and reviewed the manuscript. Sami AlMomen contributed to the design of the study and reviewed the manuscript. Lee Roth contributed to the design of the study, performed the statistical analysis and contributed to the drafting of the manuscript.

DISCLOSURES: The authors have no financial disclosures or conflicts of interest to declare.

REFERENCES

1. Schnuelle P, Lorenz D, Trede M, Van Der Woude FJ. Impact of renal cadaveric transplantation on survival in end-stage renal failure: Evidence for reduced mortality risk compared with hemodialysis during long-term follow-up. J Am Soc Nephrol 1998;9:2135-41.
2. Birkeland SA, Løkkegaard H, Storm HH. Cancer risk in patients on dialysis and after renal transplantation. Lancet 2000;355:1886-7.
3. AlBugami M, Kiberd B. Malignancies: Pre and post transplantation strategies. Transplant Rev (Orlando) 2014;28:76-8.
4. Adami J, Gäbel H, Lindelöf B, et al. Cancer risk following organ transplantation: A nationwide cohort study in Sweden. Br J Cancer 2003;89:1221-7.
5. Nishihara R, Wu K, Lochhead P, et al. Long-term colorectal-cancer incidence and mortality after lower endoscopy. N Engl J Med 2013;369:1095-105.
6. Levin B, Lieberman DA, McFarland B, et al; American Cancer Society Colorectal Cancer Advisory Group; US Multi-Society Task Force; American College of Radiology Colon Cancer Committee. Screening and surveillance for the early detection of colorectal cancer and adenomatous polyps, 2008: A joint guideline from the

American Cancer Society, the US Multi-Society Task Force on Colorectal Cancer, and the American College of Radiology. CA Cancer J Clin 2008;58:130-60.
7. U.S. Preventive Services Task Force. Screening for colorectal cancer: U.S. Preventive Services Task Force recommendation statement. Ann Intern Med 2008;149:627-37.
8. Qaseem A, Denberg TD, Hopkins RH Jr, et al; Clinical Guidelines Committee of the American College of Physicians. Screening for colorectal cancer: A guidance statement from the American College of Physicians. Ann Intern Med 2012;156:378-86. (Erratum in: Ann Intern Med 2012;157:152).
9. Collins AJ, Foley RN, Chavers B, et al. United States Renal Data System 2011 Annual Data Report: Atlas of chronic kidney disease & end-stage renal disease in the United States. Am J Kidney Dis 2012;59(Suppl 1):A7,e1-420.
10. Holley JL. Screening, diagnosis, and treatment of cancer in long-term dialysis patients. Clin J Am Soc Nephrol 2007;2:604-10.
11. Almadi MA, Alharbi O, Azzam N, Wadera J, Sadaf N, Aljebreen AM. Prevalence and characteristics of colonic polyps and adenomas in 2654 colonoscopies in Saudi Arabia. Saudi J Gastroenterol 2014;20:154-61.
12. Collins MG, Teo E, Cole SR, et al. Screening for colorectal cancer and advanced colorectal neoplasia in kidney transplant recipients: Cross sectional prevalence and diagnostic accuracy study of faecal immunochemical testing for haemoglobin and colonoscopy. BMJ 2012;345:e4657.
13. Vajdic CM, McDonald SP, McCredie MR, et al. Cancer incidence before and after kidney transplantation. JAMA 2006;296:2823-31.
14. Kasiske BL, Snyder JJ, Gilbertson DT, Wang C. Cancer after kidney transplantation in the United States. Am J Transplant 2004;4:905-13.
15. Miao Y, Everly JJ, Gross TG, et al. De novo cancers arising in organ transplant recipients are associated with adverse outcomes compared with the general population. Transplantation 2009;87:1347-59.
16. Johnson EE, Leverson GE, Pirsch JD, Heise CP. A 30-year analysis of colorectal adenocarcinoma in transplant recipients and proposal for altered screening. J Gastrointest Surg 2007;11:272-9.
17. Kiberd BA, Keough-Ryan T, Clase CM. Screening for prostate, breast and colorectal cancer in renal transplant recipients. Am J Transplant 2003;3:619-25.
18. Levin B, Lieberman DA, McFarland B, et al; American Cancer Society Colorectal Cancer Advisory Group; US Multi-Society Task Force; American College of Radiology Colon Cancer Committee. Screening and surveillance for the early detection of colorectal cancer and adenomatous polyps, 2008: A joint guideline from the American Cancer Society, the US Multi-Society Task Force on Colorectal Cancer, and the American College of Radiology. Gastroenterology 2008;134:1570-95.
19. Rabeneck L, Paszat LF, Hilsden RJ, et al. Bleeding and perforation after outpatient colonoscopy and their risk factors in usual clinical practice. Gastroenterology 2008;135:1899-906,1906.e1.
20. Bielawska B, Day AG, Lieberman DA, Hookey LC. Risk factors for early colonoscopic perforation include non-gastroenterologist endoscopists: A multivariable analysis. Clin Gastroenterol Hepatol 2014;12:85-92.

Adenoma detection rates decline with increasing procedural hours in an endoscopist's workload

Majid A Almadi MBBS, FRCPC, MSc[1,2], Maida Sewitch PhD[3], Alan N Barkun MD CM FRCPC MSc[1,3], Myriam Martel BSc[1], Lawrence Joseph PhD[3]

MA Almadi, M Sewitch, AN Barkun, M Martel, L Joseph. Adenoma detection rates decline with increasing procedural hours in an endoscopist's workload. Can J Gastroenterol Hepatol 2015;29(6):304-308.

BACKGROUND: Operator fatigue may negatively influence adenoma detection (AD) during screening colonoscopy.

OBJECTIVE: To better characterize factors affecting AD, including the number of hours worked, and the number and type of procedures performed before an index screening colonoscopy.

METHODS: A retrospective cohort study was conducted involving individuals undergoing a screening colonoscopy at a major tertiary care hospital in Montreal, Quebec. Individuals were identified using an endoscopic reporting database; AD was identified by an electronic chart review. A hierarchical logistic regression analysis was performed to determine the association between patient- and endoscopist-related variables and AD.

RESULTS: A total of 430 consecutive colonoscopies performed by 10 gastroenterologists and two surgeons were included. Patient mean (± SD) age was 63.4±10.9 years, 56.3% were males, 27.7% had undergone a previous colonoscopy and the cecal intubation rate was 95.7%. The overall AD rate was 25.7%. Age was associated with AD (OR 1.06 [95% CI 1.03 to 1.08]), while female sex (OR 0.44 [95% CI 0.25 to 0.75]), an indication for average-risk screening (OR 0.47 [95% CI 0.27 to 0.80]) and an increase in the number of hours during which endoscopies were performed before the index colonoscopy (OR 0.87 [95% CI 0.76 to 0.99]) were associated with lower AD rates. On exploratory univariable analysis, a threshold of 3 h of endoscopy time performed before the index colonoscopy was associated with decreased AD.

CONCLUSION: The number of hours devoted to endoscopies before the index colonoscopy was inversely associated with AD rate, with decreased performance possibly as early as within 3 h. This metric should be confirmed in future studies and considered when optimizing scheduling practices.

Key Words: *Adenoma detection; Endoscopy workload; Screening colonoscopy*

Le taux de détection des adénomes inversement proportionnel aux heures d'interventions par l'endoscopiste

HISTORIQUE : La fatigue de l'opérateur peut nuire à la détection des adénomes (DA) pendant les coloscopies de dépistage.

OBJECTIF : Mieux caractériser les facteurs influant sur la DA, y compris le nombre d'heures de travail ainsi que le nombre et le type d'interventions effectuées avant une coloscopie de dépistage de référence.

MÉTHODOLOGIE : Les chercheurs ont réalisé une étude rétrospective de cohorte auprès de personnes qui subissaient une coloscopie de dépistage dans un grand hôpital de soins tertiaires de Montréal, au Québec. Ils ont trouvé ces personnes dans une base de données des endoscopies. Ils ont examiné les dossiers électroniques pour en extraire les DA, puis ont effectué une analyse de régression logistique hiérarchique afin de déterminer l'association entre les variables liées aux patients et aux endoscopiques et les DA.

RÉSULTATS : Au total, 430 coloscopies consécutives effectuées par dix gastroentérologues et deux chirurgiens ont été incluses dans l'étude. Les patients avaient un âge moyen (± ÉT) de 63,4±10,9 ans, 56,3 % étaient de sexe masculin, 27,7 % avaient déjà subi une coloscopie et 95,7 % avaient subi une intubation cæcale. Le taux global de DA s'élevait à 25,7 %. L'âge s'associait aux DA (RR 1,06 [95 % IC 1,03 à 1,08]), tandis que le sexe féminin (RR 0,44 [95 % IC 0,25 à 0,75]), une indication de dépistage constituant un risque moyen (RR 0,47 [95 % IC 0,27 à 0,80]) et une augmentation du nombre d'heures d'interventions endoscopiques avant la coloscopie de référence (RR 0,87 [95 % IC 0,76 à 0,99]) s'associaient à une baisse du taux de DA. À l'analyse univariable exploratoire, un seuil de trois heures d'interventions endoscopiques avant la coloscopie de référence s'associait à une diminution des DA.

CONCLUSION : Le nombre d'heures consacrées aux endoscopies avant la coloscopie de référence était inversement proportionnel au taux de DA, le rendement pouvant diminuer moins de trois heures après le début des interventions endoscopiques. Il faudrait confirmer cette mesure lors de futures études et en tenir compte pour optimiser les horaires de travail.

Colorectal cancer (CRC) ranks third in cancer-related mortality for men and women, with an estimated 96,830 new cases and >50,310 deaths in the United States in 2014 (1). In Canada, CRC is estimated to rank second in incidence in men and third in women (2). Recent trends demonstrate a declining incidence and mortality from CRC (1,3,4), and screening is believed to play a major role in these declines owing to the broad practice of opportunistic screening colonoscopy (1,3,4). Data from the Nurses' Health Study (5) and the Health Professionals Follow-up cohort (6) demonstrate that both colonoscopy and sigmoidoscopy are associated with a reduced incidence of distal CRC, with screening colonoscopy also being associated with a reduction in the incidence of proximal CRC.

CRC screening programs require allocation of significant resources (7), and improving the performance of colonoscopy as a screening tool has been the focus of many regulatory bodies and gastroenterology societies. There have been concerns about the rates of detection and removal of polyps in colonoscopies performed in Quebec, where they were found to be below sex-specific benchmarks (8). Furthermore, it was noted that the completion rates of colonoscopies as well as the reporting of key quality metrics were suboptimal, at best, in Manitoba (9). Colonoscopy-related quality measures have been identified to establish standards in the performance of colonoscopies; among these is the adenoma detection (AD) rate (ADR). Identifying factors that prognosticate AD are being targeted with the aim of optimizing the

[1]*Division of Gastroenterology, The McGill University Health Centre, Montreal General Hospital, McGill University, Montreal, Quebec;* [2]*Division of Gastroenterology, King Khalid University Hospital, King Saud University, Riyadh, Saudi Arabia;* [3]*Division of Clinical Epidemiology, The McGill University Health Centre, Montreal General Hospital, McGill University, Montreal, Quebec*

Correspondence: Dr Alan N Barkun, Montreal General Hospital, 1650 Cedar Avenue, #D7-148, Montreal, Quebec H3G 1A4. e-mail alan.barkun@muhc.mcgill.ca

TABLE 1
Characteristics of the 430 patients included in the study

Characteristic	
Male sex	56.3
Age, years, mean ± SD	63.4±10.9
Previous colonoscopy	27.7
Previous polyp removal	26.8
Average-risk screening indication	45.5
Family history of colorectal cancer	29.2
Hereditary nonpolyposis colorectal cancer	1.4
Familial adenomatous polyposis	1.9

Data presented as % unless otherwise indicated

effectiveness of colonoscopy. Some studies suggest that operator fatigue, either as a function of the time of day (10), queue position (11,12) or performing emergency procedures the night before the colonoscopy (13), may affect quality (14) and the detection of polyps (15) and adenomas (16), while others have not found this association (17,18). These metrics, as well as others, have been stressed in the clinical practice guideline by Tinmouth et al (19) for colonoscopy quality assurance in Ontario.

Accordingly, we sought to identify factors that affect AD during screening colonoscopy. More specifically, we examined the relationship between AD and the numbers of endoscopy hours and procedures performed before the index colonoscopy. We hypothesized that the numbers of hours worked and procedures performed, as well as the type of procedure mix, may influence AD.

METHODS

Patient and operator populations

The study cohort was identified using an endoscopic reporting database of individuals seen at a major tertiary care hospital in Montreal, Quebec. Both surgeons and gastroenterologists staff the endoscopy service. On average, 12,000 colonoscopies and gastroscopies are performed annually, of which since 2008, 65% to 75% are colonoscopies. The study population included consecutive individuals who underwent a CRC screening colonoscopy (average and high risk, initial and follow-up) as indicated in the endoscopy report between June 1 and August 25, 2009. Individuals who underwent flexible sigmoidoscopy or for whom the colonoscopy indication was documented on the endoscopy report as other than CRC screening were excluded.

Data collection

Three trained research assistants abstracted data from electronic colonoscopic reports (Endoworks, Olympus Corporation, USA) of procedures performed during the study period, as well as from the corresponding electronic hospital pathology reports. All data were entered into standardized electronic case report forms.

Endoscopy sessions extended from morning (08:00) until late afternoon (16:00) and were staffed by the same endoscopists with no set ratio of colonoscopies to other endoscopic procedures. Endoscopies were performed by 12 attending staff (10 gastroenterologists, two colorectal surgeons) and the endoscopy lists did not include acute or hospitalized patients.

Collected information included demographic (age, sex) and clinical data (family history of CRC, previous colonoscopy, previous polypectomy and estimated CRC risk according to the endoscopist clinical assessment), and colonoscopic findings (quality of the bowel preparation, cecal intubation, photodocumentation of the cecum, the number of polyps detected and polyp location). For incomplete colonoscopies, the reason and level of the colon reached were recorded. The scoping activity of the endoscopist on the index day up to the point of the index colonoscopy was analyzed including the time devoted to performing any procedure (gastroscopy, colonoscopy or endoscopic ultrasound) and their numbers. Colonoscopies that were performed between 08:00 and 12:00 were considered to be morning, and those performed between 12:01 and 16:00 were considered to be afternoon procedures.

TABLE 2
Characteristics and findings of screening colonoscopies

Variable	Percentage (95% CI) or mean ± SD
Complete colonoscopy*	96.3 (94.0–97.9)
Cecal intubation	95.7 (93.4–97.7)
Photodocumentation of the cecum	72.1 (67.8–76.4)
Bowel preparation quality	
Good	86.3 (83.0–89.5)
Fair	9.1 (6.3–11.8)
Poor	3.7 (1.9–5.5)
Not documented	0.9 (0.02–1.8)
Polyp detected on index colonoscopy	40.9 (36.2–45.6)
Number of polyps	0.8±1.7
Adenoma detected on current colonoscopy	25.7 (21.4–29.9)
Cancer	0.07 (0.0–2.0)
Advanced adenoma (from total adenomas)	28.9 (21.5–36.2)
Colonoscopies performed in the morning	70.9 (66.6–75.2)
Colonoscopies performed in the afternoon	28.8 (24.5–33.1)
Timing of colonoscopy not documented	0.2 (0.0–0.7)

**Discrepancy between the cecal intubation rate and the colonoscopy completion rate may be related to incomplete documentation*

AD was defined as a colonoscopy in which at least one adenoma was identified and determined based on a review of electronically filed pathology reports. The Institutional Review Board at the McGill University Health Centre (Montreal, Quebec) approved the study.

Statistical analysis

Sample size calculation was based on an a priori baseline adenoma prevalence in the centre's population of 30%. Using the rule of 10 outcome events per predictor variable (20), and the authors' desire to include up to 12 variables in the multivariable model, it was estimated that 400 screening colonoscopies would be needed to provide sufficient accuracy.

Data analysis included descriptive statistics computed for continuous variables including means and SDs, and minimum and maximum values. Percentages were used for categorical variables. A hierarchical logistic regression model was used to examine the association between independent variables and AD. A two-level data structure was used for patient- and endoscopist-level clustering. Patient age and sex, and estimated CRC risk were adjusted according to the endoscopist's clinical assessment, previous colonoscopy, previous polypectomy, family history of CRC and bowel preparation quality.

An estimate of ADR variability within endoscopist speciality (gastroenterologist or surgeons) were computed from random intercepts for endoscopists and the specialty (surgeon or gastroenterologist) at the endoscopist level.

An exploratory univariable analysis was also performed to determine the optimal time threshold in hours for AD, using ROC curve analysis; $P \leq 0.05$ was considered to be statistically significant. All analyses were performed using SAS version 9.2 (SAS Institute, USA).

RESULTS

The authors identified 430 consecutive eligible patients who underwent a colonoscopy for a screening indication (June 1 to August 25, 2009). The characteristics of the 430 patients are summarized in Table 1. The mean (± SD) age was 63.4±10.9 years and 56.3% were male. The bowel preparation quality was rated to be good in 86.3% of procedures and cecal intubation was achieved in 95.7% (Table 2). In total, the ADR was 25.7% (95% CI 21.4% to 29.9%) in the study population. The mean time from the beginning of the endoscopy session to the time

TABLE 3
Results of between-group comparisons for detection of at least one adenoma on screening colonoscopy

Variable	Adenoma Yes (n=106)*	Adenoma No (n=307)*
Male sex	67.0 (57.9–76.1)	51.5 (45.8–57.1)
Age, years, mean ± SD	67.0±11.3	62.2±10.5
History of colonoscopy	26.9 (16.9–37.0)	28.7 (22.3–35.1)
Average-risk screening indication	34.0 (24.4–43.6)	49.1 (43.3–55.0)
Family history of colorectal cancer	29.6 (17.1–42.2)	29.8 (22.9–36.8)
Incomplete colonoscopy	2.2 (0.0–6.6)	1.9 (0.0–4.7)
Previous polyp removed	43.5 (28.6–58.4)	21.8 (15.2–28.4)
Cecum intubated	97.2 (94.0–1.00)	95.1 (92.6–97.5)
Good	92.5 (87.3–97.6)	85.3 (81.4–89.3)
Fair	4.7 (0.6–8.8)	9.5 (6.2–12.7)
Poor	1.9 (0.0–4.5)	4.2 (2.0–6.5)
Photodocumentation of the cecum	82.1 (74.7–89.5)	68.7 (63.5–73.4)
Polyp number, mean ± SD	2.2±2.0	0.3±1.3
Hours to colonoscopy, mean ± SD	2.3±1.8	2.9±2.2
Number of colonoscopies before the index colonoscopy	3.1 (2.5–3.7)	4.0 (3.6–4.4)
Number of endoscopic procedures before the index colonoscopy	4.6 (3.9–5.3)	5.5 (5.0–6.0)
Endoscopy in the morning	77.4 (69.3–85.5)	68.7 (63.5–73.4)
Male endoscopist	85.9 (79.1–92.6)	91.9 (88.8–94.9)
Surgical endoscopist specialty	30.2 (21.3–39.1)	38.1 (32.7–43.6)

*Data presented as % (95% CI) unless otherwise indicated. *Missing information for 17 patients*

of the index colonoscopy was 163±125 min, the number of endoscopic procedures (gastroscopies, colonoscopies or endoscopic ultrasounds) before the index colonoscopy was 5.3±4.0 endoscopies and 3.7±3.5 colonoscopies. The majority (70.9%) of colonoscopies were performed in the morning.

Adenomas were detected more often in men (67.0% versus 51.5%), and patients who were older (67.0 versus 62.2 years) had a history of polyp removal (43.5% versus 21.8%) and a higher number of polyps detected at the index colonoscopy (2.2 versus 0.3). The no adenoma group had a higher proportion of average-risk individuals (49.1%) compared with the group with adenomas (34.0%) (Table 3).

In univariable analysis, the significant predictors (at P=0.10 level) included sex, age, average-risk indication, previous polyp removal, good bowel preparation, photodocumentation of the cecum, number of polyps detected, hours to colonoscopy, morning endoscopy and male endoscopist. On multivariable analysis, the only factor positively associated with AD was age (OR 1.06 [95% CI 1.32 to 1.08]) per year. AD was lower in women (OR 0.44 [95% CI 0.25 to 0.75]) and in patients with average-risk screening (OR 0.47 [95% CI 0.27 to 0.80]). AD also decreased with an increase in the number of scoping hours elapsed until the index colonoscopy (OR 0.87 [95% CI 0.76 to 0.99]) per hour.

A statistically significant difference was noted in the AD when comparing colonoscopies performed ≤3 h versus >3 h after the start of the endoscopy session (P=0.03) (Figure 1). No optimal cut-off could be identified using ROC curve analysis (area under the curve = 0.58).

DISCUSSION

Multiple quality indicators have been proposed to optimize a common standard of practice with the aim of maximizing the detection of adenomas during screening colonoscopy and decreasing the incidence of CRC. Indeed, the ADR may be the most important quality indicator because it has been associated with interval CRC (21), and both CRC incidence and mortality (22).

The ADR in our study was comparable with that reported by other large groups of endoscopists (23-25). A variant definition of the ADR

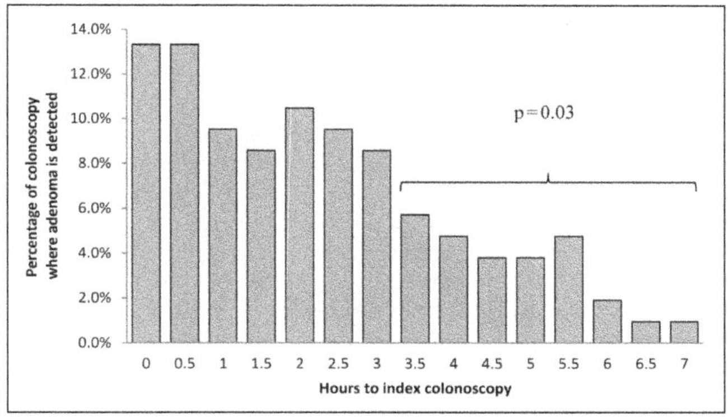

Figure 1) *Percentage of colonoscopies in which adenomas were detected from the beginning of the endoscopy session (h)*

is the proportion of adenomas detected per patient (26). Our definition of the ADR does not account for the presence of >1 adenoma per patient, which may be a shortcoming; nonetheless, we opted to use the traditional definition because of its broad adoption in the literature, which permits benchmarking with other studies (27,28). There are limitations to the use of the ADR as a means for evaluating the performance of endoscopists because it may not be dependable (29) and it has been suggested that a large number of procedures (ie, 500 colonoscopies) may be required for it to be reliable (30).

Among patient characteristics, we observed that age predicted increased risk for adenoma(s) (OR 1.06 [95% CI 1.32 to 1.08]); in contrast, AD was lower in women OR 0.44 (95% CI 0.25 to 0.75). Both findings are consistent with previous studies (31-33). In univariable analysis, we also noted an association between AD and a history of polypectomy (43.5% versus 21.8%).

There are numerous studies that have demonstrated that increased withdrawal time is associated with an increased ADR (23,34-36). We did not have withdrawal times for most endoscopists because time recording had not yet been implemented in a standardized way or widely disseminated as a quality measure during the study period in our institution. In a recent retrospective study, in which time recording was implemented, there was a statistically nonsignificant increase in polyps detected, and these were mostly small nonadenomatous polyps with no cancer potential (36,37). Withdrawal times likely reflect a characteristic of the endoscopist and the degree of care and scrutiny that he/she takes in examining the colon. With overall improvement in colonoscopy quality and the adoption of other quality markers, withdrawal time may or may not remain an independent predictor and, thus, we do not believe that its absence invalidates the results of the present analysis, especially considering the high ADR in our study.

Importantly, we found that fewer adenomas were detected as the time to the index colonoscopy increased in multivariable analysis (OR 0.87 [95% CI 0.76 to 0.99]), confirming our hypothesis. We showed a difference in AD when comparing colonoscopies performed ≤3 h versus >3 h following the start of the endoscopy session using an exploratory univariable analysis; however, ROC curve analysis failed to reveal an optimal cut-off, perhaps because of inadequate statistical power. Possible explanations for the existence of such a threshold include operator fatigue and pressure for keeping the procedure scheduling on time. This finding was noted independently of bowel preparation and other aforementioned patient-related factors by using hierarchical logistic regression adjustment. Our results mirror some previous studies showing that the polyp detection rate decreased with time (16), although no threshold values have previously been proposed. Harewood et al (14) demonstrated that colonoscopy insertion times increased and cecal intubation rates decreased with successive procedures, supporting operator fatigue, although this study did not demonstrate a significant decline in lesion detection, which may be due to differences in staffing of morning and afternoon sessions (14), in which no endoscopist worked >4 h and colonoscopies comprised only 33% of a given endoscopy list. Our study differs in that endoscopy sessions extended from morning until late afternoon staffed by the same endoscopist, with no set ratio in colonoscopies to other endoscopic procedures, possibly further favouring the detection of a threshold phenomenon. Lee et al (12) found that each elapsed hour in the day was associated with a 4.6% reduction in polyp detection as well as 5.4% reduction in polyp detection with each increase in the colonoscopy queue position (12); however, the authors attributed an inability to demonstrate a statistically significant decrease to the small number of procedures occupying a queue position ≥8 (12). A decrease in polyp detection was not observed in studies in which endoscopy sessions were limited to 3 h (10) or half-day schedules (17,18), further raising the notion of operator fatigue as noted by Spiegel (38). In a study by Kaneshiro et al (15), the effect of decreased AD as the day progressed persisted even after using visual reminders.

Recently, a study from the English Bowel Cancer Screening Programme (11) speculated endoscopist fatigue as a factor for decreased ADR because it was found that the earlier the colonoscopy in the procedural list, the higher the ADR. Operator fatigue has also been shown to affect AD during screening colonoscopy when performed by endoscopists who performed emergent on-call procedures the night before performing screening colonoscopies when compared with those who were not on call (30% versus 39%, respectively; P=0.043) (13).

Interestingly, operator fatigue has been noted in the surgical literature in which a randomized trial of frequent breaks for surgeons during their operations resulted in a reduction in the surgeon's serum cortisol levels and fewer intraoperative events when compared with the usual practice of no breaks (39). In the cardiology literature, adverse outcomes were noted in patients undergoing primary percutaneous coronary interventions during off hours (40), and were again observed in those undergoing nonurgent percutaneous coronary interventions later in the day when compared with those performed earlier (41). Fatigue has also been a focus of the aviation industry, in which there are established regulations to countermeasure its occurrence (42).

Other endoscopist characteristics have been examined in the literature. Neither sex nor endoscopist specialty were shown to affect AD in the present study (Table 3), a finding similar to that reported by Bannert et al (43), which included 52,506 screening colonoscopies performed by 196 endoscopists over a period of approximately 3.5 years. A recent population-based study from Alberta (44) found that primary care physicians performing colonoscopies achieved some of the quality benchmarks, including cecal intubation rates and ADRs, but included a mixture of patients who were symptomatic and others who were undergoing colonoscopies.

Adler et al (31) demonstrated an association between the number of continuing medical education meetings attended by the physician and ADR. This may be a reflection of the importance of endoscopist personality traits, and how this characteristic may affect individual performance parameters. Additional operator factors are also likely to play a role in ADR, such as the gazing pattern of the endoscopic image by the operator, as recently proposed (45).

Strengths of the present study include the elimination of the possibility of a Hawthorne effect due its the retrospective nature. We included all patients meeting eligibility criteria during the study period to reduce the potential for selection bias. Possible additional limitations include the lack of information regarding the date of the last colonoscopy and the CRC risk based on the endoscopists' assessment rather than an objective measure. Although both of these variables may induce a misclassification bias, the finding of a lower proportion of average-risk individuals with adenomas (34.0% versus 49.1%) is consistent with previous reports (46). Furthermore, in addition to a minor amount of missing data (reported in Table 3), additional information regarding some potential confounders were unavailable such as the presence of the metabolic syndrome (47-49), smoking (50-52), body mass index (53) and socioeconomic status (54), preventing more extensive patient-level adjustment for AD predictive factors. In addition, the present study was conducted at a single institution, which may limit generalizability, even though multiple endoscopy disciplines were represented.

CONCLUSION

Patient characteristics and increased time from start of the endoscopy session until the index colonoscopy were associated with AD, with increased time to index colonoscopy negatively impacting AD. Operator fatigue may, thus, play a role, and has implications for optimal endoscopy session scheduling. Confirmatory studies are required to better characterize the effect of prolonged endoscopy sessions on the detection of adenomas during screening colonoscopy and identify an operational threshold.

ACKNOWLEDGEMENTS: The authors extend their sincere appreciation to the Deanship of Scientific Research at King Saud University for its funding of this research through the Research Group Project number RGP-VPP-279.

DISCLOSURES: Majid Almadi, Maida Sewitch, Myriam Martel and Lawrence Joseph have no financial disclosures or conflicts of interest to declare. Dr Barkun is a consultant for Boston Scientific Inc, Olympus Canada Inc and Cook Inc. He has also received 'at arms-length' grant funding from both Boston Scientific Inc and Cook Inc.

REFERENCES

1. Siegel R, Ma J, Zou Z, Jemal A. Cancer statistics, 2014. CA Cancer J Clin 2014;64:9-29.
2. Canadian Cancer Society's Advisory Committee on Cancer Statistics. Canadian Cancer Statistics 2014. Toronto: OCCS.
3. Jemal A, Siegel R, Xu J, Ward E. Cancer statistics, 2010. CA Cancer J Clin 2010;60:277-300.
4. Edwards BK, Ward E, Kohler BA, et al. Annual report to the nation on the status of cancer, 1975-2006, featuring colorectal cancer trends and impact of interventions (risk factors, screening, and treatment) to reduce future rates. Cancer 2010;116:544-73.
5. Siegel R, Desantis C, Jemal A. Colorectal cancer statistics, 2014. CA Cancer J Clin 2014;64:104-17.
6. Nishihara R, Wu K, Lochhead P, et al. Long-term colorectal-cancer incidence and mortality after lower endoscopy. N Engl J Med 2013;369:1095-105.
7. Major D, Bryant H, Delaney M, et al. Colorectal cancer screening in Canada: Results from the first round of screening for five provincial programs. Current Oncol 2013;20:252-7.
8. Sewitch MJ, Jiang M, Fon Sing M, Barkun A, Joseph L. Screening polypectomy rates below quality benchmarks: A prospective study. World J Gastroenterol 2014;20:16300-5.
9. Singh H, Kaita L, Taylor G, Nugent Z, Bernstein C. Practice and documentation of performance of colonoscopy in a central Canadian health region. Can J Gastroenterol Hepatol 2014;28:185-90.
10. Munson GW, Harewood GC, Francis DL. Time of day variation in polyp detection rate for colonoscopies performed on a 3-hour shift schedule. Gastrointest Endosc 2011;73:467-75.
11. Lee TJ, Rees CJ, Blanks RG, et al. Colonoscopic factors associated with adenoma detection in a national colorectal cancer screening program. Endoscopy 2014;46:203-11.
12. Lee A, Iskander JM, Gupta N, et al. Queue position in the endoscopic schedule impacts effectiveness of colonoscopy. Am J Gastroenterol 2011;106:1457-65.
13. Benson M, Grimes I, Gopal D, et al. Influence of previous night call and sleep deprivation on screening colonoscopy quality. Am J Gastroenterol 2014;109:1133-7.
14. Harewood GC, Chrysostomou K, Himy N, Leong WL. Impact of operator fatigue on endoscopy performance: Implications for procedure scheduling. Dig Dis Sci 2009;54:1656-61.
15. Kaneshiro M, Ho A, Chan M, Cohen H, Spiegel BM. Colonoscopy yields fewer polyps as the day progresses despite using social influence theory to reverse the trend. Gastrointest Endosc 2010;72:1233-40.

16. Chan MY, Cohen H, Spiegel BM. Fewer polyps detected by colonoscopy as the day progresses at a Veteran's Administration teaching hospital. Clin Gastroenterol Hepatol 2009;7:1217-23.
17. Freedman JS, Harari DY, Bamji ND, et al. The detection of premalignant colon polyps during colonoscopy is stable throughout the workday. Gastrointest Endosc 2011;73:1197-206.
18. Gurudu SR, Ratuapli SK, Leighton JA, Heigh RI, Crowell MD. Adenoma detection rate is not influenced by the timing of colonoscopy when performed in half-day blocks. Am J Gastroenterol 2011;106:1466-71.
19. Tinmouth J, Kennedy EB, Baron D, et al. Colonoscopy quality assurance in Ontario: Systematic review and clinical practice guideline. Can J Gastroenterol Hepatol 2014;28:251-74.
20. Vittinghoff E, McCulloch CE. Relaxing the rule of ten events per variable in logistic and Cox regression. Am J Epidemiol 2007;165:710-8.
21. Kaminski M, Regula J, Kraszewska E, et al. Quality indicators for colonoscopy and the risk of interval cancer. N Engl J Med 2010:1795-803.
22. Corley DA, Jensen CD, Marks AR, et al. Adenoma detection rate and risk of colorectal cancer and death. N Engl J Med 2014;370:1298-306.
23. Barclay RL, Vicari JJ, Doughty AS, Johanson JF, Greenlaw RL. Colonoscopic withdrawal times and adenoma detection during screening colonoscopy. N Engl J Med 2006;355:2533-41.
24. Shaukat A, Oancea C, Bond JH, Church TR, Allen JI. Variation in detection of adenomas and polyps by colonoscopy and change over time with a performance improvement program. Clin Gastroenterol Hepatol 2009;7:1335-40.
25. Dellon ES, Lippmann QK, Sandler RS, Shaheen NJ. Gastrointestinal endoscopy nurse experience and polyp detection during screening colonoscopy. Clin Gastroenterol Hepatol 2008;6:1342-7.
26. Church J. Adenoma detection rate and the quality of colonoscopy: The sword has two edges. Dis Colon Rectum 2008;51:520-3.
27. Sanaka MR, Deepinder F, Thota PN, Lopez R, Burke CA. Adenomas are detected more often in morning than in afternoon colonoscopy. Am J Gastroenterol 2009;104:1659-64.
28. Peters SL, Hasan AG, Jacobson NB, Austin GL. Level of fellowship training increases adenoma detection rates. Clin Gastroenterol Hepatol 2010;8:439-42.
29. Saini SD, Schoenfeld P, Vijan S. Can the adenoma detection rate reliably identify low-performing endoscopists? Results of a modeling study. Dig Dis Sci 2013;58:1856-62.
30. Do A, Weinberg J, Kakkar A, Jacobson BC. Reliability of adenoma detection rate is based on procedural volume. Gastrointest Endosc 2013;77:376-80.
31. Adler A, Wegscheider K, Lieberman D, et al. Factors determining the quality of screening colonoscopy: A prospective study on adenoma detection rates, from 12,134 examinations (Berlin colonoscopy project 3, BECOP-3). Gut 2013;62:236-41.
32. Qumseya BJ, Coe S, Wallace MB. The effect of polyp location and patient gender on the presence of dysplasia in colonic polyps. Clin Transl Gastroenterol 2012;3:e20.
33. Corley DA, Jensen CD, Marks AR, et al. Variation of adenoma prevalence by age, sex, race, and colon location in a large population: Implications for screening and quality programs. Clin Gastroenterol Hepatol 2013;11:172-80.
34. Gellad ZF, Weiss DG, Ahnen DJ, Lieberman DA, Jackson GL, Provenzale D. Colonoscopy withdrawal time and risk of neoplasia at 5 years: Results from VA Cooperative Studies Program 380. Am J Gastroenterol 2010;105:1746-52.
35. Simmons DT, Harewood GC, Baron TH, et al. Impact of endoscopist withdrawal speed on polyp yield: Implications for optimal colonoscopy withdrawal time. Aliment Pharmacol Ther 2006;24:965-71.
36. Lee TJ, Blanks RG, Rees CJ, et al. Longer mean colonoscopy withdrawal time is associated with increased adenoma detection: Evidence from the Bowel Cancer Screening Programme in England. Endoscopy 2013;45:20-6.
37. Taber A RJ. Effect of simply recording colonoscopy withdrawal time on polyp and adenoma detection rates. Gastrointest Endosc 2010;71:782-6.
38. Spiegel BM. Does time of day affect polyp detection rates from colonoscopy? Gastrointest Endosc 2011;73:476-9.
39. Engelmann C, Schneider M, Kirschbaum C, et al. Effects of intraoperative breaks on mental and somatic operator fatigue: A randomized clinical trial. Surg Endosc 2011;25:1245-50.
40. Glaser R, Naidu SS, Selzer F, et al. Factors associated with poorer prognosis for patients undergoing primary percutaneous coronary intervention during off-hours: Biology or systems failure? JACC Cardiovascular interventions 2008;1:681-8.
41. Cantor WJ, Strauss BH, Graham MM, et al. Time of day and outcomes of nonurgent percutaneous coronary intervention performed during working hours. Am Heart J 2010;159:1133-8.
42. Caldwell JA, Mallis MM, Caldwell JL, et al. Fatigue countermeasures in aviation. Aviation, space, and environmental medicine 2009;80:29-59.
43. Bannert C, Reinhart K, Dunkler D, et al. Sedation in screening colonoscopy: Impact on quality indicators and complications. Am J Gastroenterol 2012;107:1837-48.
44. Kolber MR, Wong CK, Fedorak RN, Rowe BH. Prospective study of the quality of colonoscopies performed by primary care physicians: The Alberta Primary Care Endoscopy (APC-Endo) Study. PLoS One 2013;8:e67017.
45. Almansa C, Shahid MW, Heckman MG, Preissler S, Wallace MB. Association between visual gaze patterns and adenoma detection rate during colonoscopy: A preliminary investigation. Am J Gastroenterol 2011;106:1070-4.
46. Barret M, Boustiere C, Canard JM, et al. Factors associated with adenoma detection rate and diagnosis of polyps and colorectal cancer during colonoscopy in France: Results of a prospective, nationwide survey. PLoS One 2013;8:e68947.
47. Sato Y, Nozaki R, Yamada K, Takano M, Haruma K. Relation between obesity and adenomatous polyps of the large bowel. Dig Endosc 2009;21:154-7.
48. Siddiqui A, Pena Sahdala HN, Nazario HE, et al. Obesity is associated with an increased prevalence of advanced adenomatous colon polyps in a male veteran population. Dig Dis Sci 2009;54:1560-4.
49. Kaaks R, Toniolo P, Akhmedkhanov A, et al. Serum C-peptide, insulin-like growth factor (IGF)-I, IGF-binding proteins, and colorectal cancer risk in women. J Natl Cancer Inst 2000;92:1592-600.
50. Botteri E, Iodice S, Raimondi S, Maisonneuve P, Lowenfels AB. Cigarette smoking and adenomatous polyps: A meta-analysis. Gastroenterology 2008;134:388-95.
51. Giovannucci E, Colditz GA, Stampfer MJ, et al. A prospective study of cigarette smoking and risk of colorectal adenoma and colorectal cancer in U.S. women. J Natl Cancer Inst 1994;86:192-9.
52. Giovannucci E, Rimm EB, Stampfer MJ, et al. A prospective study of cigarette smoking and risk of colorectal adenoma and colorectal cancer in U.S. men. J Natl Cancer Inst 1994;86:183-91.
53. Hassan C, Pickhardt PJ, Marmo R, Choi JR. Impact of lifestyle factors on colorectal polyp detection in the screening setting. Dis Colon Rectum 2010;53:1328-33.
54. Singh SM, Paszat LF, Li C, He J, Vinden C, Rabeneck L. Association of socioeconomic status and receipt of colorectal cancer investigations: A population-based retrospective cohort study. CMAJ 2004;171:461-5.

Dysphagia among adult patients who underwent surgery for esophageal atresia at birth

Valérie Huynh-Trudeau MD, Stéphanie Maynard MD, Tatjana Terzic MD, Geneviève Soucy MD, Mickael Bouin MD PhD

V Huynh-Trudeau, S Maynard, T Terzic, G Soucy, M Bouin. Dysphagia among adults patients who underwent surgery for esophageal atresia at birth. Can J Gastroenterol Hepatol 2015;29(5):91-94.

BACKGROUND: Clinical experiences of adults who underwent surgery for esophageal atresia at birth is limited. There is some evidence that suggests considerable long-term morbidity, partly because of dysphagia, which has been reported in up to 85% of adult patients who undergo surgery for esophageal atresia. The authors hypothesized that dysphagia in this population is caused by dysmotility and/or anatomical anomalies.

OBJECTIVE: To determine the motor and anatomical causes of dysphagia.

METHODS: A total of 41 adults, followed at the Esophageal Atresia Clinic at *Hôpital Saint-Luc* (Montreal, Quebec), were approached to particpate in the present prospective study. Evaluation was completed using upper endoscopy, manometry and barium swallow for the participants who consented. The medical charts of respondents were systematically reviewed from the neonatal period to 18 years of age to assess medical and surgical history.

RESULTS: All 41 patients followed at the clinic consented and were included in the study. Dysphagia was present in 73% of patients. Esophagogastroduodenoscopy was performed in 32 patients: hiatal hernia was present in 62% (n=20); esophageal diverticulum in 13% (n=4); macroscopic Barrett esophagus in 31% (n=10); and esophagitis in 19% (n=6). Histological esophagitis was present in 20% and intestinal metaplasia in 10%. There were no cases of dysplagia or adenocarcinoma. Esophageal manometry was performed on 56% of the patients (n=23). Manometry revealed hypomotility in 100% of patients and included an insufficient number of peristaltic waves in 96%, non-propagating peristalsis in 78% and low-wave amplitude in 95%. Complete aperistalsis was present in 78%. The lower esophageal sphincter was abnormal in 12 (52%) patients, with incomplete relaxation the most common anomaly. Of the 41 patients, 29 (71%) consented to a barium swallow, which was abnormal in 13 (45%). The anomalies found were short esophageal dilation in 28%, delay in esophageal emptying in 14%, diverticula in 14% and stenosis in 7% of patients. There was more than one anomaly in 14% of patients.

CONCLUSION: Dysphagia was a highly prevalent symptom in adults who underwent surgery for esophageal atresia. The present study is the first to demonstrate that motor and anatomical abnormalities may be implicated in causes of dysphagia in this population. Furthermore, these anomalies may be demonstrated with simple investigations such as endoscopy, manometry and barium swallow.

Key Words: *Barium swallow; Barrett esophagus; Dysphagia; Endoscopy; Esophageal atresia*

La dysphagie chez les patients adultes ayant été opérés à la naissance en raison d'une atrésie œsophagienne

HISTORIQUE : On a peu d'expérience clinique auprès des adultes qui ont été opérés en raison d'une atrésie œsophagienne à la naissance. Selon certaines données, la morbidité à long terme est considérable, en partie à cause de la dysphagie qui s'observe chez jusqu'à 85 % des patients adultes opérés en raison d'une atrésie œsophagienne. Les auteurs postulent qu'au sein de cette population, la dysphagie est causée par la dysmotilité ou les anomalies anatomiques.

OBJECTIF : Déterminer les causes motrices et anatomiques de la dysphagie.

MÉTHODOLOGIE : Au total, 41 adultes, suivis à la clinique d'atrésie œsophagienne de l'Hôpital Saint-Luc de Montréal, au Québec, ont été invités à participer à la présente étude prospective. Les chercheurs ont effectué l'évaluation par endoscopie œsogastroduodénale, manométrie et déglutition barytée chez les participants qui y ont consenti. Ils ont examiné systématiquement le dossier médical des répondants entre la période néonatale et l'âge de 18 ans afin d'évaluer leurs antécédents médicaux et chirurgicaux.

RÉSULTATS : Les 41 patients suivis à la clinique ont consenti et participé à l'étude. De ce nombre, 73 % avaient une dysphagie et 32 patients ont subi une œsophagogastroduodénoscopie : hernie hiatale chez 62 % (n=20), diverticule œsophagienne chez 13 % (n=4), œsophage de Barrett macroscopique chez 31 % (n=10) et œsophagite chez 19 % (n=6). De plus, 20 % avaient une œsophagite histologique et 10 % une métaplasie intestinale, mais on n'observait aucun cas de dysplagie ou d'adénocarcinome. Au total, 56 % des patients (n=23) ont subi une manométrie œsophagienne, qui a révélé une hypomotilité chez 100 % d'entre eux. Cette manométrie indiquait un nombre insuffisant d'ondes péristaltiques chez 96 % d'entre eux, une absence de propagation du péristaltisme chez 78 % d'entre eux et une faible amplitude des ondes chez 95 % d'entre eux. On constatait un apéristaltisme complet chez 78 % des patients. Le sphincter œsophagien inférieur était anormal chez 12 patients (52 %), la relaxation incomplète du sphincter étant l'anomalie la plus courante. Sur les 41 patients, 29 (71 %) ont consenti à une déglutition barytée, qui était anormale chez 13 patients (45 %). Les anomalies observées étaient une faible dilatation œsophagienne chez 28 %, un retard de la vidange œsophagienne chez 14 %, un diverticule chez 14 % et une sténose chez 7 % des patients. Enfin, 14 % des patients présentaient plus d'une anomalie.

CONCLUSION : La dysphagie était un symptôme hautement prévalent chez les adultes qui avaient été opérés en raison d'une atrésie œsophagienne. La présente étude est la première à démontrer que les anomalies motrices et anatomiques peuvent contribuer à la dysphagie au sein de cette population. De plus, ces anomalies peuvent être démontrées par de simples examens comme l'endoscopie, la manométrie et la déglutition barytée.

Esophageal atresia is a congenital anomaly defined by an interruption of the continuity of the esophagus with or without a communication with the trachea. This condition occurs in one in 2500 to 3000 live births (1) and is the most common congenital anomaly of the esophagus. There are five types of esophageal atresia; the most common is type C, with distal tracheoesophageal fistula occurring in 85% of cases. The majority of children with atresia undergo surgery at birth. Even if the continuity of the esophagus is anatomically replaced, the children are prone to several gastrointestinal problems including dysphagia and gastroesophageal reflux (GER) (2). Dysphagia is observed

Hôpital Saint-Luc, Montreal, Quebec

Correspondence: Dr Valérie Huynh-Trudeau, Hôpital Saint-Luc, 1058 Rue Saint-Denis, Montreal, Quebec H2X 3J4. e-mail vhtrudeau@gmail.com

in more than one-half of children who undero surgery for esophageal atresia at birth, and may cause malnutrition and failure to thrive (3-5).

Clinical experiences of adults who undergo surgery for esophageal atresia at birth are limited. There is some evidence that suggests considerable long-term morbidity, partly because of dysphagia, which has been reported in up to 85% of adult patients who undergo surgery for esophageal atresia (6,7). Studies have shown that different anomalies may be implicated including dysmotility, stenosis, esophagitis and cancer (5). In fact, esophageal atresia is accompanied by intrinsic and extrinsic anomalies of neurological maturation, possibly leading to dysmotility. In addition, surgical repair may lead to stenosis and associated diverticula, which may contribute to dysphagia (5). However, there are no studies to guide the necessary investigations.

Hôpital Saint-Luc is a tertiary centre located in Montreal, Quebec, with a specialized clinic for adults who undergo surgery for esophageal atresia. We have previously demonstrated that dysphagia is a common symptom in this population (8,9); however, the etiology remains unclear. Our hypothesis was that dysphagia in this population is caused by dysmotility and/or anatomical anomalies. The objective of the current study was to determine the motor and anatomical causes of dysphagia.

METHODS

Forty-one adult patients followed at the Esophageal Atresia Clinic at the *Hôpital Saint-Luc* were approached to participate in the present prospective study. After providing signed consent, all participants underwent an interview and clinical assessment. Evaluation was completed using upper endoscopy, manometry and barium swallow for individuals who consented. Medical charts of respondents were systematically reviewed from the neonatal period to 18 years of age to assess medical and surgical history.

Interview and questionnaires

All participants were interviewed by an investigator and answered a standard questionnaire regarding gastrointestinal symptoms. Information regarding swallowing difficulties with different food textures, food impaction, dietary limitations, adaptive dietary habits and GER symptoms was requested. Dysphagia was defined as swallowing difficulties with one or multiple food textures and/or food impaction ≥1 per month. Criteria for GER were the presence of pyrosis and/or regurgitation ≥1 per month. Adaptive dietary habits included drinking water while eating, avoiding certain foods and usually eating slower than other people.

Upper endoscopy and histology

After an overnight fast, esophagogastroduodenoscopy was performed under conscious sedation using an endoscope (EG2990i 2.8, Pentax Canada). Particular attention was devoted to the distal esophagus to locate the esophagogastric junction and the presence of mucosal injury. All abnormal endoscopic findings were reported. A fibrotic rim was defined as a narrowing of the esophagus, while a stricture was defined the inability to pass the endoscope through the narrowing (5). Esophagitis was graded according to the Los Angeles Classification (10). Barrett esophagus (11) was suspected when the squamocolumnar junction was ≥5 mm proximal to the esophagogastric junction. Biopsies were obtained for histological examination when an endoscopic anomaly was noted.

All biopsies were fixed overnight in Tissuefix (Qiagen, Canada) and subsequently embedded in paraffin. Six slices (3 µm each) were obtained for each biopsy and stained with hematoxylin phloxine saffron. The slides were reviewed by two pathologists blinded to the patients' clinical data. All cases were evaluated for the presence of esophagitis, metaplasia and dysplasia.

Esophageal manometry

Stationary low-compliance perfusion manometry was performed using a round, four-lumen catheter. Lower esophageal sphincter (LES) pressure measurements were performed using the four distal openings of the catheter at a recording speed of 2.5 mm/s. The tip of the catheter was positioned in the stomach and then slowly withdrawn in 1.0 cm increments. LES pressure was recorded at midexpiration and end-expiration. Values were calculated as the mean of the three pressure channel readings. Contractions of the esophagus were recorded with the four pressure channels positioned 5 cm, 10 cm, 15 cm and 20 cm above the LES; 10 swallows of 5 mL each were then given at 30 s intervals. The measurement of each peristaltic parameter represented the mean of the 10 sequential swallows, and both amplitude and duration were individually determined for the different recording site above the LES. The data were transferred to a specifically designed software program (Gastrosoft, Synectics Medical Inc, USA) for analysis.

Barium swallow

Barium swallow was performed according to a routine protocol. Participants stood for the examination. The fluoroscope (Philips, USA) was positioned and moved to capture the different phases of deglutition. Bolus transit patterns were assessed and anomalies were recorded in terms of intraesophageal stasis, dilation, diverticulum, stenosis and mass. All boluses were confirmed to have been cleared from the esophagus before the next bolus was administered.

Statistical analysis

Data were collected in a database and analyzed using SPSS (IBM Corporation, USA). Differences between groups were analyzed using the χ^2 test; P=0.05 was considered to be statistically significant.

Ethics

The Ethics Committee of *Hôpital Saint-Luc* approved the present study and written informed consent was obtained from each patient. The study met the requirements of the Declaration of Helsinki.

RESULTS

Patients

All 41 patients followed at the clinic consented to participate and were included in the study; patient characteristics are summarized in Table 1. Esophageal atresia with distal tracheoesophageal fistula (type C, gross classification) was the most prevalent (85%) anomaly. At least one associated malformation was present in 26 (63%) patients. Eighty percent had a primary anastomosis and the others had either a gastric pull-up or a colonic interposition. The most frequent complications during childhood were failure to thrive (49%) and recurrent anastomotic stricture (29%).

All patients completed the questionnaire on gastrointestinal symptoms. Dysphagia was present in 30 (73%) patients (Table 1). Among these patients, dysphagia with solid food was present in 41%, with dry food in 32%, soft food in 10%, thick liquids in 10% and liquids in 10%. Blockages were present in six (15%) patients, but the episodes were self-resolved. Adaptive dietary habits were reported by 78% of the patients, the most common (61%) of which was to drink plenty between bites. Eating slower than other people was reported by 49% and avoidance of certain types of food by 15%. Twenty-two percent did not need to take precautions when eating. GER symptoms were recorded in 12 (29%) patients, including 17% under medical treatment and 12% without treatment. Pyrosis was present in 20% of patients and regurgitation in 24%.

Endoscopic and histological findings (Table 2)

Esophagogastroduodenoscopy was performed in 32 (78%) patients. An esophageal anastomotic scar was recognizable in most (n=26 [81%]) cases of primary anastomosis; no strictures were found. Hiatal hernia was present in 20 (62%) patients, esophageal diverticulum in four (13%), macroscopic Barrett esophagus in 10 (31%) and esophagitis in six (19%), all of whom had grade A esophagitis. One-half of the patients with esophagitis and 80% of patients with macroscopic Barrett esophagus had no GER symptoms.

Biopsies were obtained in presence of macroscopic Barrett esophagus (n=10) and/or esophagitis (n=6). A total of 12 (38%) patients had

TABLE 1
Characteristics of the study participants with surgically repaired esophageal atresia and the prevalence of selected gastrointestinal symptoms (n=41)

Age, years, mean (range)	25 (18–44)
Male sex	23 (56)
Body mass index, kg/m², mean (range)	22 (15–38)
Dysphagia	30 (73)
Swallowing difficulties*	25 (61)
Blockage*	6 (15)
Adaptive dietary habits	32 (78)
Gastroesophageal reflux symptoms	12 (29)
With medical treatment	7 (17)
Without medical treatment	5 (12)

*Data presented as n (%) unless otherwise indicated. *One patient had both swallowing difficulties and a blockage*

TABLE 2
Endoscopic findings in adult patients with surgically repaired esophageal atresia (n=32)

Endoscopic finding	
Esophagus	
Esophagitis	6 (19)
Stricture	0 (0)
Esophageal diverticulum	4 (13)
Endoscopically suspected Barrett esophagus	10 (31)
Stomach	
Hiatal hernia	20 (62)
Aberrent pancreas	8 (25)

Data presented as n (%)

biopsies taken. No biopsies were taken in patients with a normal-appearing esophagus. Biopsies of the distal esophagus were examined to assess the presence of esophagitis, metaplasia, dysplasia or adenocarcinoma. Histological evidence of esophagitis was present in two patients who underwent biopsies without endoscopic esophagitis. Intestinal metaplasia was found in four of the 10 patients with endoscopically suspected Barrett esophagus. There was no dysplasia or adenocarcinoma in any biopsy specimen (Table 3).

Esophageal manometry (Table 4)

Esophageal manometry was performed in 23 (56%) patients and, of these, dysphagia was present in 13 (57%). Manometry revealed hypomotility in 100% of the patients and included an insufficient number of peristaltic waves in 96%, nonpropagating peristalsis in 78% and low wave amplitude in 95%. Complete aperistalsis was present in 78%. The LES was abnormal in 12 (52%) patients; the most common anomaly was incomplete relaxation and the upper esophageal sphincter was always normal.

Barium swallow (Table 5)

Of the 41 patients, 29 (71%) consented to a barium swallow, which was abnormal in 13 (45%). The anomalies found were: short esophageal dilation in eight (28% [four upstream and three downstream of the anastomosis]); delay in esophageal emptying in four (14% [two associated with aperistalsis in the inferior esophagus, one with an ileocolic transposition and one with a dilation]); diverticula in four (14% [one at the anastomosis level, three in the inferior esophagus]); and stenosis in two (7% [one in the proximal esophagus associated with a dilatation and one at the esophagocolic junction of a ileocolic transposition]). There was more than one anomaly in four (14%) patients. Dysphagia occurred in 48% of the 29 patients. Anatomical anomalies according to barium swallow were more prevalent in patients with dysphagia than in patients without dysphagia; however, this difference was not statistically significant (64% versus 27%; P=0.07).

TABLE 3
Histological findings in esophageal biopsies from adult patients with surgically repaired esophageal atresia (n=12)

Histological finding	
Esophagitis	8 (67)
Gastric metaplasia	6 (50)
Intestinal metaplasia	4 (33)
Dysplasia	0 (0)
Adenocarcinoma	0 (0)

Data presented as n (%)

TABLE 4
Manometric findings in adult patients with surgically repaired esophageal atresia (n=23)

Manometric finding	
Superior esophageal sphincter	
Normal	23 (100)
Abnormal coordination and relaxation	0 (0)
Esophageal body	
Normal	0 (0)
Hypomotility	23 (100)
Aperistaltism	18 (78)
High amplitude waves	3 (14)
Abnormal duration	0 (0)
Lower esophageal sphincter	
Normal	11 (48)
Hypotonia*	4 (17)
Hypertonia	1 (4)
Abnormal relaxation*	8 (40)

*Data presented as n (%). *One patient had hypertonia and abnormal relaxation*

TABLE 5
Barium swallow findings in adult patients with surgically repaired esophageal atresia (n=29)

Barium swallow finding	
Normal	16 (55)
Esophageal dilation	8 (28)
Delay in esophageal emptying	4 (14)
Esophageal diverticulum	4 (14)
Esophageal stenosis	2 (7)

Data presented as n (%)

DISCUSSION

In many studies, dysphagia is a common symptom in adults who undergo surgery for esophageal atresia (5,7). In our study, dysphagia was present in 73% of the cohort. To our knowledge, the present study was the first to investigate the causes of dysphagia in this population. We conducted a prospective study among adults followed in the esophageal atresia clinic using a standard questionnaire, gastroesophageal duodenoscopy, manometry and barium swallow.

The participation rate for endoscopy was 78%, which is high considering that some patients had already undergone an endoscopy during their pediatric follow-up and that this test was not prompted by a new symptom. In our study, the prevalence of histological esophagitis was 20%, which is slightly below the reported prevalence in adults who have undergone surgery for esophageal atresia (between 25% and 90%) (5). This could be explained by the fact that most of our population had undergone surgical antireflux therapy (37%) or were on antireflux medication (34%) compared with other studies (10% in the study by Sistonen et al [5]). There were no cases of cancer in our study, probably because of the young mean age of our population.

Manometry was performed in 56% of patients. All of the manometry findings in our study were abnormal. Hypomotility was demonstrated in 100% of patients and 78% had aperistalsis. The function of the LES was abnormal in 52% and primarily related to incomplete relaxation (40%). The high prevalence of hypomotility caused by abnormalities intrinsic to the disease and surgical dissection during repair may, in part, explain dysphagia in adults who undergo surgery for esophageal atresia. Patients without dysphagia, despite dysmotility, are probably asymptomatic through adaptation (8).

Barium swallow was performed in 71% of participants and abnormalities were demonstrated in 45%. Abnormalities included dilations (28%), delays in esophageal emptying (14%), diverticula (14%) and stenoses (7%). Anatomical abnormalities were more common in patients with dysphagia than in those without dysphagia, although not a statistically significant difference (64% versus 27%; P=0.07). These anatomical abnormalities may play a role in the occurrence of dysphagia in this population (12) and the absence of a statistically significant difference was probably due to a lack of power.

CONCLUSION

Dysphagia is a highly prevalent symptom in adults who undergo surgery for esophageal atresia. The present study was the first to demonstrate that motor and anatomical abnormalities could be implicated as causes of dysphagia in this population. Furthermore, these anomalies may be demonstrated with simple investigations such as endoscopy, manometry and barium swallow.

DISCLOSURES: The authors have no financial disclosures or conflicts of interest to declare.

REFERENCES

1. Harmon CM, Coran AG. Congenital anomalies of the esophagus. In: Pediatric Surgery, 5th edn, St-Louis: Mosby, 1998;941-67.
2. Gottrand F, Sfeir R, Coopman S, Deschildre A, Michaud L. Atrésie de l'oesophage : devenir des enfants opérés. Archives de pédiatrie 2008;15:1837-42.
3. Koivusalo A, Pakarinen MP, Rintala RJ. The cumulative incidence of significant gastroesophageal reflux in patients with esophageal atresia with a distal fistula – a systematic clinical, pHmetric, and endoscopic follow-up study. J Pediatr Surg 2007;42:370-4.
4. Krug E, Bergmeijer JH, Dees J, de Krijger R, Mooi WJ, Hazebroek FW. Gastroesophageal reflux and Barrett's esophagus in adults born with esophageal atresia. Am J Gastroenterol 1999;94:2825-8.
5. Sistonen SJ, Koivusalo A, Lindahl H, Pukkala E, Rintala RJ, Pakarinen MP. Esophageal morbidity and function in adults with repaired esophageal atresia with tracheoesophageal fistula: A population-based long-term follow-up study. Ann Surg 2010;251:1167-73.
6. Tovar JA, Diez Pardo JA, Murcia J, et al. Ambulatory 24-hour manometric and pH metric evidence of permanent impairment of clearance capacity in patients with esophageal atresia. J Pediatr Surg 1995;30:1224-31.
7. Tomaselli V, Volpi ML, Dell'Agnola CA, et al. Long-term evaluation of esophageal function in patients treated at birth for esophageal atresia. Pediatr Surg Int 2003;19:40-3.
8. Maynard S, Bouin M. L'atrésie de l'œsophage : une nouvelle maladie de l'adulte? Le clinicien 2011;51-4.
9. Maynard S, Bouin M. Follow-up of adult patients with repaired esophageal atresia: How, when, and for how long? Dis Esophagus 2013;26:422-4.
10. Lundell LR, Dent J, Bennett JR, et al. Endoscopic assessment of oesophagitis: Clinical and functional correlates and further validation of the Los Angeles classification. Gut 1999;45:172-80.
11. Fitzgerald RC, di Pietro M, Ragunath K et al. Bristish Society of Gastroenterology guidelines on the diagnosis and management of Barrett's oesophagus. Gut 2014;63:7-42.
12. Cho YK, Choi MG, Oh SN, et al. Comparison of bolus transit patterns identified by esophageal impedance to barium esophagram in patients with dysphagia. Dis Esophagus 2012;25:17-25.

Adult lactose digestion status and effects on disease

Andrew Szilagyi MD FRCPC

A Szilagyi. Adult lactose digestion status and effects on disease. Can J Gastroenterol Hepatol 2015;29(3):149-156.

BACKGROUND: Adult assimilation of lactose divides humans into dominant lactase-persistent and recessive nonpersistent phenotypes.
OBJECTIVES: To review three medical parameters of lactose digestion, namely: the changing concept of lactose intolerance; the possible impact on diseases of microbial adaptation in lactase-nonpersistent populations; and the possibility that the evolution of lactase has influenced some disease pattern distributions.
METHODS: A PubMed, Google Scholar and manual review of articles were used to provide a narrative review of the topic.
RESULTS: The concept of lactose intolerance is changing and merging with food intolerances. Microbial adaptation to regular lactose consumption in lactase-nonpersistent individuals is supported by limited evidence. There is evidence suggestive of a relationship among geographical distributions of latitude, sunhine exposure and lactase proportional distributions worldwide.
DISCUSSION: The definition of lactose intolerance has shifted away from association with lactose maldigestion. Lactose sensitivity is described equally in lactose digesters and maldigesters. The important medical consequence of withholding dairy foods could have a detrimental impact on several diseases; in addition, microbial adaptation in lactase-nonpersistent populations may alter risk for some diseases. There is suggestive evidence that the emergence of lactase persistence, together with human migrations before and after the emergence of lactase persistence, have impacted modern-day diseases.
CONCLUSIONS: Lactose maldigestion and lactose intolerance are not synonymous. Withholding dairy foods is a poor method to treat lactose intolerance. Further epidemiological work could shed light on the possible effects of microbial adaptation in lactose maldigesters. The evolutionary impact of lactase may be still ongoing.

Key Words: *Adaptation; Evolution; Intolerance; Lactose*

La digestion du lactose chez les adultes et ses effets sur les maladies

HISTORIQUE : L'assimilation du lactose par les adultes divise les humains en phénotypes lactase-persistant dominant et non persistant récessif.
OBJECTIFS : Examiner trois paramètres médicaux de la digestion du lactose, soit l'évolution du concept d'intolérance au lactose, les effets possibles de l'adaptation microbienne des populations lactase-non persistantes sur les maladies et la possibilité que l'évolution de la lactase ait eu une incidence sur des modèles de distribution pathologique.
MÉTHODOLOGIE : Les chercheurs ont effectué une analyse d'articles dans PubMed et Google Scholar et une analyse manuelle pour obtenir un examen narratif du sujet.
RÉSULTATS : Le concept d'intolérance au lactose évolue et fusionne avec les intolérances alimentaires. Peu de données probantes appuient le concept d'adaptation microbienne à la consommation régulière de lactose chez les personnes lactase-non persistantes. Des données probantes laissent supposer un lien entre la répartition géographique de latitude et d'exposition au soleil et la distribution proportionnelle de la lactase dans le monde.
EXPOSÉ : La définition d'intolérance au lactose s'est éloignée de la maldigestion du lactose. La sensibilité au lactose est décrite tout autant chez les personnes qui digèrent bien le lactose que chez celles qui le digèrent mal. L'évitement des produits laitiers pourrait contribuer à plusieurs maladies. De plus, l'adaptation microbienne des populations lactase-non persistantes pourrait modifier le risque lié à certaines maladies. Des données laissent supposer que l'émergence de persistance à la lactase, conjointement aux migrations humaines qui l'ont précédée et l'ont suivie, a eu un effet sur les maladies modernes.
CONCLUSIONS : Les termes maldigestion du lactose et intolérance au lactose ne sont pas synonymes. Il n'est pas conseillé d'éviter les produits laitiers pour traiter l'intolérance au lactose. D'autres recherches épidémiologiques pourraient jeter la lumière sur les effets possibles de l'adaptation microbienne des personnes qui digèrent mal le lactose. La lactase agit peut-être encore sur l'évolution.

Generally, humans are the only mammals who consume the milk of other animals. Lactose in mother's milk is digested by infants of all human races and ethnic groups with the exception of a rare congenital absence of the intestinal enzyme lactase (1,2). However, in adults, lactase (or lactase-phlorizin hydrolase) diminishes to approximately 10% of original levels in two-thirds to three-quarters of the human population (3). The loss of intestinal brush border lactase occurs in an inconsistent fashion (4), and begins at variable ages depending on race and ethnicity (3). Approximately 5% of cow's milk consists of lactose.

The lactase gene (*LCT*) is found on chromosome 2q21 (5). Its transcription is controlled by a gene (MCM6) in cis position in exon 13, approximately 14 kb upstream from LCT. The ability in adults to digest lactose is a dominant trait known as lactase persistence (LP). Those who cannot digest lactose (recessive trait) are described as lactase nonpersistent (LNP) (3). The first gene identified to control transcription of lactase was described by Ennatah et al (6), who found the variant C/T-13910 approximately 14 kb away from *LCT*. In this polymorphism, substitution of thiamine for cytosine enables continued lactose digestion in adulthood. Another polymorphism G/A-22018A, was also described but is not as strongly associated with lactose digestion, at least in north Europeans (6). G/A-22018A is, however, the dominant control for lactase persistance in northern Chinese populations (7). Subsequently, several other polymorphisms were described in Africa (8,9) and the Middle East (10).

The emergence of the ability to digest lactose is a relatively recent event in human history, occurring 7500 to 10,000 years ago (9). There are those who believe that lactase remains under evolutionary pressure (11). The emergence of LP divides the entire human population into those who can and those who cannot digest lactose in adulthood. Diseases affecting the brush border (eg, celiac disease, *Giardia*, bacterial overgrowth, viral gastroenteritis, radiation and others) can lead to secondary lactose maldigestion. Whether natural or as a result of small bowel disease, loss of intestinal lactase leads to a failure to split the disaccharide lactose into its monosaccharide components of glucose and galactose. As such, lactose reaches the colon and is then metabolized by intestinal bacteria.

There are two historical questions about the impact of lactose digestion/maldigestion on humans. First, is the teleological question 'why' some individuals develop LP status. The second is 'how' does the

Jewish General Hospital, Department of Medicine, Division of Gastroenterology, McGill University School of Medicine, Montreal, Quebec
Correspondence: Dr Andrew Szilagyi, Jewish General Hospital, Department of Medicine, Division of Gastroenterology, McGill University School of Medicine, 3755 Cote St Catherine Road, E110, Montreal, Quebec H3T 1E2 e-mail aszilagy@jgh.mcgill.ca

divide affect humans? The present narrative provides an overview of different levels at which dairy foods and milk consumption intersects with medical relevance.

WHY IS THE WORLD'S POPULATION DIVIDED INTO TWO PHENOTYPES OF LACTOSE DIGESTION?

Before the emergence of lactose digestion, all populations were believed to be LNP (12). More than 100,000 years ago, ancestors migrated from Africa into Europe and eastward into Russia, Siberia, India, China and the South Pacific. From Siberia during the last ice age, an existing land mass of Beringia allowed people to immigrate to the Americas (13,14). Approximately 7500 to 10,000 years ago, lactose digestion emerged. There are several hypotheses as to why this occurred; however, two major ones dominate. In Europe, the calcium assimilation hypothesis is based on the observation that there is a steep north-south gradient in LP/LNP distributions with LNP increasing toward the south. Flatz and Rotthauwe (15) suggested that due to lack of sunshine and skin synthesis of vitamin D the ability to digest lactose allowed greater amounts of calcium assimilation from dairy foods, especially raw milk. This genetic advantage could reduce the prevalence of rickets. Around the same time frame, lactose digestion also emerged in Africa and the Middle East. However, the postulated environmental pressure was the practice of pastoralism and herding (16-18), which may also have allowed greater fluid ingestion in arid places such as desserts (18). The pastoral/herding model is referred to as the gene culture coevolution hypothesis. An analytical computer model of these two hypotheses provided evidence favouring the gene culture coevolution hypothesis over the calcium assimilation hypothesis. However, in Europe, the calcium assimilation model could not be excluded (19). The gene culture coevolution hypothesis in Europe suggests that LP dominance began in central Europe and populations then migrated both north and eastward. In more modern times, north and western European populations migrated into the new worlds of the Americas and Australia. Therefore, the modern day distributions of LP and LNP comprised predivide and postdivide human migrations with some intermingling of the two phenotypes.

How does the dichotomy of lactose digestion impact humans?

If we accept that the evolution of lactase persistence divides humans into two phenotypes with respect to dairy food and milk consumption, it is intuitive that these would have consequences. There are three broad areas in which the genetic divide could intersect with health issues. These include lactose intolerance, the effect of dairy food consumption on various diseases and the possible impact of the genetic evolution of LP status, together with pre- and posthuman migrations on the propensity for some diseases.

Lactose intolerance

The most clinically evident and researched effect is the concept of lactose intolerance. In the second half of the 20th century, when lactase was first described, it was hoped that a biochemical explanation for symptoms of irritable bowel syndrome (IBS) would be detected. In LNP individuals, ingestion of large quantities of lactose in a single session often induces cramps, abdominal bloating, flatus and, at times, diarrhea and even vomiting (20). The diagnosis of lactose maldigestion initially based on direct duodenal biopsies was too invasive. More practical indirect tests based on failure of rise of blood glucose (lactose tolerance test; measure of absorbed glucose) or a specific increase of measured breath hydrogen (measure of bacterial metabolism of maldigested lactose) after oral lactose loading are applied to clinical evaluation of lactose maldigestion. It is also used to evaluate lactose intolerance through various symptom scores at the time of testing (21-23). These indirect tests have been validated against both intestinal biopsies (24,25) as well as against (at least) the north European C/T-13910 polymorphism for lactase production (26).

Through extensive population-based studies, several findings are worth noting. Patients with IBS generally do not experience a higher frequency of lactose maldigestion than the background population (27). The frequency of lactose intolerance, although most likely worse in lactose maldigestion individuals, is seen almost as frequently in LP populations (27,28). These observations have led to a more recent concept of lactose sensitivity. This term refers to additional symptoms beyond the above outlined gastrointestinal symptoms (eg, headache, depression). Lactose sensitivity is increased but also becomes independent of race and ethnicity in patients with inflammatory bowel disease (IBD) (29,30). Moreover, some of these persons meet criteria for IBS and also may have other carbohydrate sensitivities. This novel concept has been incorporated in the hypothesis and utilization of a multicarbohydrate-restrictive diet (Fermentable Oligo- Di- Mono-saccharides and Polyols [FODMAP]) (31).

In LNP persons, the putative mechanisms of symptoms relate to osmotic accentuation of gastrointestinal motility and bacterial metabolism of maldigested lactose. A similar mechanism to LNP may affect LP persons with food sensitivities (32,33). The long-term consequences of the FODMAP diet have yet to be elucidated, and short-term use may be more suitable and easier to follow.

Nevertheless, the notion of the symptomatic consequences of lactose intolerance is a well-established obsession in the public view. This is reflected in the >1 million Internet searches regarding lactose intolerance. However, it has been shown in multiple studies that minute quantities of lactose in medication is imperceptible (34-36), small doses of lactose consumed (eg, 1 g, 9 g, 10 g) are tolerated by most LNP persons (37), and double-blinded trials of lactose in one to two cups of milk per day separated in time are symptomatically indistinguishable by LNP persons (38,39).

In individuals believing themselves to be lactose intolerant, minor psychological traits (eg, anxiety, depression and somatization) are prevalent. In addition, a nocebo effect for lactose (an expectation of adverse outcome with previous knowledge of the ingested substance) has been described (32,36).

In fact, at an National Institutes of Health conference held in 2010, the main conclusions about the health effect of lactose intolerance was that the consequent avoidance of dairy foods, which could lead to several medical problems. Among these were most emphasized was the possibility of poor skeletal outcome with osteoporosis and fractures due to reduced calcium intake. There were also possibilities that dairy foods have a protective role for some features of the metabolic syndrome and several gastrointestinal cancers (notably colorectal cancer, but others as well) (40).

Treatments for lactose-intolerant individual include altered forms of dairy food ingestion (eg, yogurt, low-lactose cheeses). The use of lactose-reduced milk or extraneous lactase enzyme ingestion could also be helpful. A more recent method, not yet widely applied, is the use of intestinal microbial adaptation in lactose-intolerant persons who are also maldigesters (41,42).

EFFECT OF MILK AND DAIRY FOODS ON DISEASE

The second effect of interest is the consumption of dairy foods and milk on different diseases. The LP/LNP divide has two different inherent characteristics that could influence disease risk. First, LNP populations consume lower quantities of milk (Figure 1) (43). This may be due to development of symptoms of lactose intolerance when larger quantities of lactose are consumed intermittently or may be cultural because the world's LNP populations live in Asia, Africa or Pacific locales. Second, when LNP persons do consume milk and lactose-containing dairy foods regularly over prolonged periods, bacterial adaptation occurs (41). This also has two consequences. Adapted LNP persons could consume more dairy foods without significant symptoms, and a modified colonic microflora over long periods may influence the development of various diseases. Adaptation to carbohydrates in LNP persons occurs via colonic bacterial modification. A formal description of this phenomenon was provided by Hertzler and Savaiano (41).

The definition includes altered bacterial response to lactose challenge with reduced hydrogen production after a period of regular lactose consumption and a variable reduction in aspects of lactose

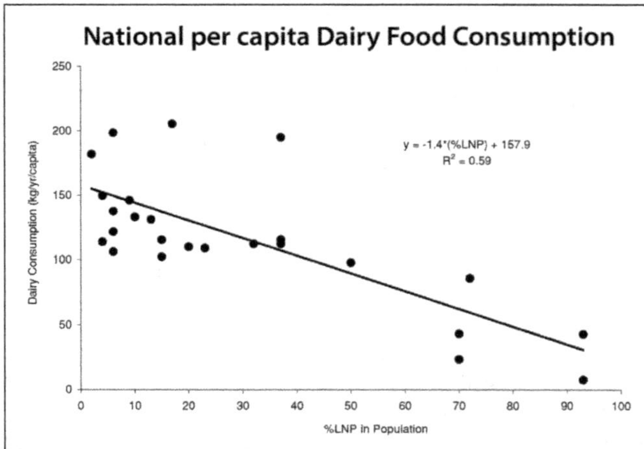

Figure 1) *Dairy food consumption for is plotted against the percentage of the population in various countries that are lactase nonpersistent (LNP). Twenty-five countries are represented and national total dairy food consumption was assessed up to March 2007. Reproduced with permission from Shrier et al (43)*

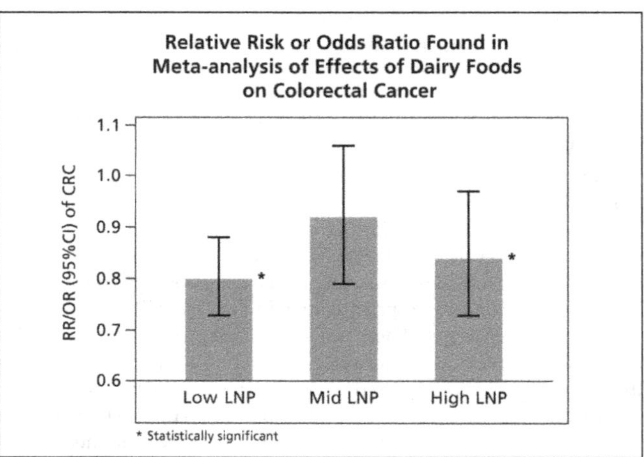

Figure 2) *Summary of relative risks of the effects of dairy food consumption on colorectal cancer (CRC) are shown with 95% CIs presented as bar graphs from three geographical areas of the world divided into groups of countries with high (≥80%), low (≤20%)and mid (average 50% [21% to 79%]) lactase-nonpersistent (LNP) proportions of the population. Modest reduction of CRC rates are found in both high and low LNP countries. In studies using data from intermediate LNP countries, the effects of dairy foods on CRC are diminished and are not statistically significant. The graph is based on meta-analyses of relevant studies from 1975 to 2005. Modified from Szilagyi et al (56) with permission*

intolerance symptoms. Other prebiotics derived from lactose are studied in an attempt to reduce symptoms by promoting colonic adaptation of lactose intolerance in LNP persons (42).

An equally important aspect supporting the existence of microfloral adaptation is the emergence of 'probiotic'-type bacteria in the colon, shown first in in vitro studies (44-46). There is also limited in vivo evidence for this (47), but the prebiotic effect of lactose is largely limited to LNP and not LP subjects (48,49). The amount of lactose consumed at each meal plays a role in the spillover of lactose into the colon (50). It is not clear whether other mammals can also adapt in this manner because in adult humans at least, lactase is a nonadaptable enzyme to increased lactose ingestion (51). After cessation of lactose intake, de-adaptation is quite rapid after approximately two to three days (41). The altered microflora with increased lactic acid-producing types could lead to short-chain fatty acid production (52). The lactic acid and acetate, in turn, could promote other short-chain fatty acids, such as butyrate, through activation of butyrogenic bacteria (53). The dominant emerging species of bifidobacteria and lactobacilli, as well as other probiotics, have been studied for anticarcinogenic and immune-modifying effects (54).

These two additional characteristics, together with inherent active elements in milk and dairy foods (55), could independently or in concert modify dairy food impact on various diseases. The interactions are complex and there is a discrepancy between ecological and patient level-determined risk for disease. For example, on an ecological level, prostate and colorectal cancer, and ulcerative colitis rates are higher in populations consuming dairy foods, while those for stomach cancer are rates are lower (43).

In addition, ovarian, breast and lung cancers and Crohn disease (CD) appear to be under-represented in a significantly high proportion of LNP populations (43).

However, in patient-level studies, variable relationships emerge between dairy food and milk consumption and risk for these same diseases. The best example is the modest protective effect of dairy foods against colorectal cancer, contrary to ecologically defined expectations (56). In fact, dairy foods may protect against most gastrointestinal cancers (57). Although there are many active molecules in milk that could be anticarcinogenic (58), the putative protection against colorectal cancer is through calcium (59). At least in Western societies, the addition of vitamin D is also believed to be protective (60). When studies from different geographical locales in the world are grouped into relative high, low or mixed proportions of LP/LNP distributions, however, dairy foods and milk provide statistically significant reduced relative rates only in high and low LP/LNP (Figure 2) (56). The high dairy food-related intake of calcium is not achieved in LNP populations as proposed in high LP populations. Also, at the time of the analyses, vitamin D was not added to milk in most high-proportion LNP populations. This study supports a hypothetical role for colonic adaptation to lactose as a possible additional active anticarcinogenic agent through manipulation of the microbiome.

In the case of other various gastrointestinal cancers, however, mechanisms of apparent protection may be different and the role of prebiotics is even less clear. Furthermore, adapted LNP individuals could consume higher quantities of dairy foods and face exposure to other active elements.

Other cancers also have controversial relationships with dairy food consumption. Of these, prostate cancer has been more aligned with a detrimental effect, possibly through the injurious impact of calcium (61-63). It is of interest that, despite an increased intake of dairy foods in individuals with dominant TT genotype for lactase compared with CT or the CC recessive genotype, there was no relationship to prostate cancer (64). However, when phenotype of lactose intolerance was examined, there were fewer LNP individuals in a group with prostate cancer (65).

The deleterious effect of dairy foods on testicular cancer has been reported (66,67); however, a risk-reductive effect on bladder cancer, which may be more evident in Asians, has also been noted (68). Probiotics have shown some protection against bladder cancer (69,70), suggesting that a lactose prebiotic effect may have been responsible for the observed differences between Europeans and Asians.

Originally, Cramer (71) postulated that because rates of ovarian cancer were low in populations with high proportion of LNP individuals, galactose in LP individuals could be toxic to the ovaries. However, the notion that dairy food consumption increases the risk of ovarian cancer has been highly controversial and, currently, the role of galactose has not been shown (72-77).

Breast cancer, which is more common in high dairy-consuming nations, has a variable relationship to milk and dairy food consumption. Initial data suggested increased risk with high fat and presence of estrogen compounds in milk cows (78). More recent studies were unable to confirm risk and suggested even protection by low-fat dairy foods (79,80). The relationship between milk consumption and lung cancer has not been established (81).

TABLE 1
Hypothesized effects of dairy food consumption on various diseases

Promotes	Protects	Independent
Prostate cancer* (61-65)	Colorectal cancer* (56,57)	Breast cancer* (78-80)
Ovarian cancer* (71-77)	Bladder cancer (68)	Lung cancer* (81)
Testicular cancer (66,67)	Stomach cancer† (57)	Crohn disease* (88-90)
Ulcerative colitis* (86,87)	Osteoporosis (83,84)	Atherosclerosis (98-100)
Cataracts (96)	Obesity (93,94)	
	Hypertension (97)	
	Type II diabetes (95)	

**More common in lactase-persistent societies; †More common in lactase-nonpersistent societies (43). The other diseases are believed to be more common in 'Western' societies. The list is based on patient level studies and most are based on meta-analyses or large single reports. Given the variability of studies and meta-analytical outcomes, these relationships are estimates and do change. For example, there are studies that suggest that low-fat dairy foods may protect against breast cancer (80). Similarly, unpasteurized milk may protect against Crohn disease (90)*

Among other non-neoplastic diseases, IBD (CD and idiopathic ulcerative colitis [UC]) is more common in high dairy food- and milk-consuming nations, although only UC reaches a statistically significant level. High-proportion LNP nations tended to have lower IBD rates (43). On the patient level, the results from different studies do not specifically implicate dairy products (82), although low calcium and low vitamin D are risks for osteopenia (83,84), and associated with increased risk for colorectal cancer (85). There are fewer cohort studies examining the impact of dairy food consumption on IBD. Existing reports are conflicting. A study from Japan epidemiologically links a number of 'Western' foods, including dairy products, with the emergence of IBD in that country (86). An earlier study from France examined 'harmful' foods (including dairy products) that were excluded from diet of patients with IBD (87). The increased prevalence of lactose intolerance and sensitivity in CD was noted above (30). However, other studies did not find dairy foods to be associated with development or aggravation of IBD (88,89), and at least one study found pasteurized milk to be protective for CD (90). The north European lactase genetic polymorphism correlation with IBD has been evaluated in very few studies. A German study failed to find any increased rates of IBD in CC or GG genotypes evaluated in patients and controls (91). However, the TT dominant genotype was found to be related to CD risk in Caucasians from New Zealand (92).

Several other non-neoplastic diseases, such as obesity (93,94), the metabolic syndrome (95) and cataracts (96), and cardiovascular disorders such as hypertension (97) and atherosclerosis (98-100), have been evaluated for a relationship with milk and dairy food consumption. Results generally have also been controversial. Table 1 outlines the relationships of diseases with dairy food consumption.

Although the effects of consumption of dairy foods on various diseases have been described, there is no clear outcome for most. Gastrointestinal cancers, especially colon cancer, appear to derive benefit from dairy products, but prostate and ovarian cancer appear to have increased risk. The failure to show conclusive outcomes are due to several factors, including variations in data-collection methods, variation in populations, and focus variability in studies (eg, different foods versus dairy products and their derivatives), all of which are likely to play a role in accounting for different outcomes. The specific elements in dairy products and milk, which potentially affect each disease, may also vary. The question from the perspective of the present review is whether ignoring differences between LP and LNP individuals may play a role in outcome of studies?

Potential impact of the role of dosage or a threshold effect of putative pathogenic elements in dairy foods may depend on LP or LNP status. In the genetic relationship between lactase and prostate cancer (64), there was a quantitative difference in dairy food intake among the three haplotypes, but a discrepancy in prostate cancer prevalence. In these cases, CC genotypes consume the least and TT the most dairy products; however, total intakes of these products were not incorporated in the analysis. In this regard, studies of diseases (not putatively affected by microbial flora) may incorporate adapted LNP patients able to consume more dairy foods.

In diseases in which microbial flora are relevant to pathogenesis, adapted LNP persons may be protected compared with LP and unadapted LNP persons. These aspects have not been adequately evaluated in relation to disease.

The possible impact of LP evolution on disease development

The third and most intriguing impact of the lactase dichotomy on diseases may be related to the emergence of LP status 7500 to 10,000 years ago. Timing of population migration (ie, before and after the emergence of LP) reflect modern day lactase distributions of populations.

An interesting observation was reported in 1980 by Garland and Garland (101), who suggested that latitude – specifically, increased vitamin D skin synthesis through greater exposure to sunlight – reduced mortality of colon cancer. Subsequently, at least 23 cancers and other diseases were noted to have reduced risk toward the equator (102-104). In addition, reports accumulated on the antineoplastic and immunomodulating effects of vitamin D (105,106), and beneficial effects of vitamin D were described in other cancers and autoimmune diseases.

In parallel with the sunshine hypothesis, a few groups also linked reduced disease risk with increasing proportions of LNP status of the populations (13,71,107,108). These reports, however, were largely overshadowed by the importance attributed to modifying effects of vitamin D. Later, seven 'Western' lifestyle diseases were confirmed to negatively relate with increasing population LNP frequency (43). The direction of disease risk reduction mimicked that observed with diminishing latitude and increasing sunshine exposure. The relationship of these two variables raised the question as to why this should be so. Did sunshine exposure play a role in evolution of LP status? As outlined above, in Europe, such a relationship could not be entirely excluded (17); however, a parallel evolution of LP in a spotty fashion in Africa supports pastoralism and animal husbandry as the dominant environmental pressure for the event (109). Nevertheless LP/LNP north-south gradients do exist at a much reduced level in Africa (17), and also in India (110) and China (7). However, this process, at least in Asia, is attributed mostly to population migrations after the emergence of LP, rather than to de novo evolution (111,112). Similarly, LP migrations to the new worlds of the Americas and Australia took place long after LP emergence in Europe. In this paradigm, latitude and sunshine are relatively fixed variables and LP/LNP distributions change overtime as further migrations of LNP populations occur in the current and last centuries.

Diseases also change over time due to altered environment and altered host susceptibility. It is interesting that many modern diseases (increases in the past 50 years) appear to originate and be prevalent in 'Western' societies. These include cancers, autoimmune diseases and those affiliated with the metabolic syndrome. These 'Western' lifestyle hypothesized diseases are also increasing in areas with previously low rates which are becoming more industrialized. The changing lifestyles support 'Western' lifestyle-associated environment changes in pathogenesis of these diseases.

Nevertheless, the apparent interactions of latitude, sunshine and LP/LNP distributions suggest mechanisms that may act together or independently to describe particular geographical patterns of disease. Sunshine putatively acts largely through vitamin D, which impacts many diseases through immune and neoplastic interventions. The role of LP/LNP is likely different and more complex.

CD and UC offer a model to study relationships between the two worldwide disease-modifying risk variables. There are a number of features of IBD that qualifies it as a worthy model, reflecting possible interactions.

IBD became an important disease in the past 75 years and rates are increasing in Westernized societies. Also, IBD has been linked both with sunshine (113), vitamin D (114) and lactase distributions (43,108). In general, UC precedes CD by approximately 15 to 20 years (115).

IBD is currently attributed to an interaction between genetic susceptibility, altered intestinal microflora and disturbance in control of the immune response (116). There are currently 163 genes attributed to susceptibility or protection against either of the main forms of IBD, with approximately two-thirds interacting with both diseases (117). Furthermore genetic polymorphisms associated with IBD have also been linked to some 24 other diseases, many of which are autoimmune mediated but others are not (eg, colorectal cancer) (118).

Finally, in the past two to four years, substantial new information emerged both on world distributions of LP/LNP (119) and the incidence rates of IBD (120).

An epidemiological re-evaluation of the relationships of both forms of IBD with latitude, sunshine and LP/LNP distributions revealed that modest to moderate correlations with all three variables can be found over a 30-year period (median year 2000) (121). This study suggested that LNP and high sunshine exposure appear to reduce risk of disease worldwide. When analysis is divided separately into European and non-European theatres, it becomes clear that global effects are largely driven by correlations in Europe. However, in non-European locations, the correlations with latitude and sunshine are somewhat weaker and LNP status has a greater risk-reducing effect. Because both national IBD rates and LP/LNP distributions were largely estimated, the actual magnitude of effect sizes are less reliable. However, it is argued that independent regional data on rates of the two changing variables (IBD and lactase) were published without bias of expected correlations by multiple authors. Second, even if average national values were obtained for large countries in North America and Asia, there is no dispute that LP frequency dominates in the former and LNP in the latter. This observation gives lateral polarity (a global west to east gradient from North America to east) to any possible effect of LP/LNP status. It is thus argued that any relationship that significantly correlates disease rate with LP/LNP status echoes a potential influence of the event that occurred millennia ago.

The relationship of IBD to both sunshine and LP/LNP distributions suggests there may indeed be interactions between these two variables. Even if sunshine has a role in evolution of LP status, the mechanism of effect of lactase distributions on diseases would be expected to differ from that of sunshine and vitamin D and is likely to be complex. One of the most likely influences of lactase evolution is the co-evolution of other genes that could contribute to other diseases. There are a number of these hypothesized.

The human leukocyte antigen system for immunity has been correlated with dominant lactase frequency in Europe (17). It is postulated that the nuclear oligomerization domain (NOD2) system evolved together with lactase dominance, also in Europe. This system is involved in preventing bacteria associated with unpasteurized milk drinking from invading the host. These bacteria included *Mycobacterium bovis*, *Listeria*, *Escherichia coli* and others (122,123).

The cystic fibrosis sodium transport receptor may also have evolved in Europe (124-126). It is postulated that the heterozygous mutation of this polymorphism may have protected against diarrhea caused by agents incurred by milk drinking.

Other genetic polymorphisms have also been hypothesized to have coevolved with lactase (127,128). The generalizability of the IBD model to show an evolutionary role of lactase in shaping of modern-day 'Western' diseases is unproven. However, the described similarities in risk reduction for several other diseases by both sunshine exposure and increasing LNP proportion of the population does suggest that these observations are not merely coincidental.

Mechanisms other than direct genetic alterations involve the vastly different environment in low-latitude countries where infectious diseases still impact dominant disease profiles. Other differences may be economic and relate to gross national products as well as distribution of wealth in the country. This latter mechanism would impact on access to health care by the local populations, perhaps resulting in under-reporting or under-recognition of various diseases.

SUMMARY AND CONCLUSIONS

In the present review, three interconnected effects of the dichotomous phenotypic division of humans into lactose digesters and maldigesters were emphasized.

To date, the major focus of research on human health has been the effect of lactose intolerance. In this context, there are two paradigm shifts. First, the definition of lactose intolerance has been broadened. It is no longer acceptable to interchange the terms lactose intolerance and lactose maldigestion. Malabsorption may be secondary to intestinal diseases but the large majority of the world is genetically divided in adulthood into LP and LNP populations. While lactose intolerance still occurs in LNP, care must be taken to determine whether symptoms occur during testing only, whether symptoms are self-reported without testing or whether symptoms are independent of lactose maldigestion status. In this latter category, lactose intolerance may be part of the broader concept of food sensitivities that can affect LP and LNP approximately equally. Second, with the shift in definitions, the greater medical issue is the failure to consume dairy products, which could lead to medical problems as outlined at a National Institutes of Health conference on lactose intolerance (40).

The second medical issue, which generally has not been emphasized, is the possible impact of the different ways that LP and LNP handle consumption of dairy products. These interactions are likely complex because of the multiple active elements in milk and different influence on possible disease pathogenesis. In general, diseases that would be affected by larger intakes of dairy foods (and in which the putative pathogenic factor is not lactose) could also impact adapted LNP persons who are now able to consume larger quantities of dairy products (eg, diseases such as prostate and testicular cancer). However, diseases in which a favourably modified intestinal flora may act against the disease, dairy food and/or milk-consuming LNP persons may be somewhat protected (eg, diseases such as colorectal cancer and bladder cancer).

It would be of interest to evaluate dairy foods from this perspective because consumption of lactose by LNP in the modern world may represent one of the largest natural use of a prebiotic that can be quantified. In the wake of important population differences in the microbiome and pattern differences among several human diseases (129-131), the impact of dairy food consumption on microbial flora between LP and LNP may be of further interest.

Third, the relationship among latitude, sunshine and LP/LNP distributions explains the observation that low latitude, high sunshine exposure and increasing LNP population proportions mimic similar apparent disease rate reductions. However, the relationship of these two different variables makes it highly unlikely that the evolution of LP status effect is coincidental. The relationship with lactase is likely to be quite complex.

The possibility that the evolution of dominant lactase persistence appears to provide a background for the geographic spread of 'Western' lifestyle diseases is intriguing. The ability to digest lactose generally is believed to have arisen as a need to improve nutrition in populations that have adopted a pastoral life style, or to protect against bone disease (in Europe). The co-evolution of other genes to accommodate the new source of nutrition likely accompanied this great divide. In modern times, the changing industrialized lifestyles, along with changes in diet, likely impact on existent genes through alterations of the colonic microenvironment leading to development of neoplastic and immune-mediated diseases. Some of these questions are outlined in Table 2.

TABLE 2
Remaining questions concerning lactose, lactase and disease interactions

Because bacterial metabolism is postulated to be associated with both bacterial and immunological response to maldigested lactose in intolerant lactose maldigesters, what is the mechanism of lactose sensitivity in lactose digesters?
Because colonic bacterial adaptation is a real phenomenon in lactose maldigesters consuming regular dairy foods, does such a potential prebiotic effect influence host health over the long term?
Is there a real impact of lactase status on outcome of studies examining effects of dairy foods on different diseases, given the outlined difference of handling dairy food consumption by lactase-persistent compared with lactase-nonpersistent populations?
Is the inflammatory bowel disease model representative of a possible genetic evolutionary mechanism that responds to current dietetic environmental pressures predisposing to some diseases?
Given modern-day migrations of lactase-nonpersistent populations, dominance of the lactase persistent genotype and changing progressive 'Western' type diet, should we expect further evolutionary changes accompanying these environmental pressures?

ACKNOWLEDGEMENT: This review was funded by a general academic grant from the Department of Medicine of the Jewish General Hospital, Montreal, Quebec.

REFERENCES

1. Saviahti A, Launiala K, Kuitunen P. Congenital lactase deficiency. A clinical study on 16 patients. Arch Dis Child 1983;58:246-52.
2. Uchida N, Sakamoto O, Irie M, et al. Two novel mutations in the lactase gene in a Japanese infant with congenital lactase deficiency. Tohoku J Exp Med 2012;227:69-72.
3. Swallow DM. Genetics of lactase persistence and lactose intolerance. Ann Hum Genet 2003;37:197-219.
4. Maiuri L, Raia V, Potter J, et al. Mosaic pattern of lactase expression by villous enterocytes in human adult-type hypolactasia. Gastroenterology 1991;100:359-69.
5. KruseTA, Bolund L, Grzeschik KH, et al. The human lactase-phlorizin gene is located on chromosome 2. FEBS Lett 1988;240:123-6.
6. Enattah NS, Sahi T, Savilahti E, Terwilliger JD, Peltonen L, Järvelä I. Identification of a variant associated with adult-type hypolactasia. Nat Genet 2002;30:233-7.
7. Xu L, Sun H , Zhang X, et al. The G-22018 A allele matches the lactase persistence phenotype in northern Chinese populations. Scand J Gastroenterol 2010;45:168-74.
8. Ingram CJE, Elamin MF, Mulcare CA, et al. A novel polymorphism associated with lactose tolerance in Africa: Multiple causes for lactase persistence? Hum Genet 2007;120:779-88.
9. Tishkoff SA, Reed FA, Ranciaro A, et al. Convergent adaptation of human lactase persistence in Africa and Europe. Nat Genet 2007;39:31-40.
10. Imtiaz F, Savilahti E, Sarnesto A, et al.The T/G 13915 variant upstream of the lactase gene (LCT) is the founder allele of lactase persistence in an urban Saudi population. J Med Genet 2007;44:4e89.
11. Bersaglieri T, Sabeti PC, Patterson N, et al. Genetic signatures of strong recent positive selection at the lactase gene. Am J Hum Genet 2004;74:1111-20.
12. Burger J, Kirchner M, Bramanti B, Haak W, Thomas MG. Absence of the lactase-persistence-associated allele in early Neolithic Europeans. Proc Natl Acad Sci USA. 2007;104: 3736-41.
13. Forster P, Matsumura S. Did early humans go north or south. Science 2005;308:965-6.
14. Malyarchuk B, Derenko M, Denisova G, et al. Ancient links between Siberians and native Americans revealed by subtyping the Y chromosome haplogroup Q1A. J Hum Genet 2011;56:583-8.
15. Flatz G, Rotthauwe HW. Lactose nutrition and natural selection. Lancet 1973;2:76-7.
16. Simoons FJ. Primary adult lactose intolerance and the milking habit: a problem in biologic and cultural interrelations. II: A culture historical hypothesis. Am J Dig Dis 1970;15:695-710.
17. Simoons FJ. The geographic hypothesis and lactose malabsorption: A weighing of the evidence. Am J Dig Dis 1978;23:963-80.
18. Holden C, Mace R. Phylogenetic analysis of the evolution of lactose digestion in adults. Hum Biol 1997;69:605-28.
19. Gerbault P, Moret C, Currat M, Sanchez-Mazas A. Impact of selection and demography on the diffusion of lactase persistence. PLoS One, 2009;4:e6369doi:10.1371/journal.pone.0006369. 20. Misselwitz B, Pohl D, Fruhauf H, Fried M, Vavricka S, Fox M. Lactose malabsorpyion and intolerance: Pathogenesis, diagnosis and treatment. United Eur Gastroenterol J 2013;1:151-9.
21. Newcomer AD, McGill DB, Thomas PJ, Hofmann AF. Prospective comparison of indirect methods for detecting lactase deficiency. N Engl J Med 1975;293:1232-6.
22. Arola H. Diagnosis of hypolactasia and lactose malabsorption. Scand J Gastroeneterol 1994;202:26-35.
23. Waud JP, Matthews SB, Campbell AK. Measurement of breath hydrogen and methane, together with lactase genotype, defines the current best practice for investigation of lactose sensitivity. Ann Clin Biochem 2008;45(Pt 1):50-8.
24. Furnari M, Bonfanti D, Parodi A, et al. A comparison between lactose breath test and quick test on duodenal biopsies for diagnosing lactase deficiency in patients with self- reported lactose intolerance. J Clin Gastroenterol 2013;47:148-52.
25. Enattah NS, Kuokkanen M, Forsblom C, et al. Correlation of intestinal disaccharidase activities with the C/T-13910 variant and age. World J Gastroenterol 2007;13:3508-12.
26. Marton A, Xue X, Szilagyi A. Meta-analysis: The diagnostic accuracy of lactose breath hydrogen or lactose tolerance tests for predicting the north European lactase polymorphism C/T – 13910. Aliment Pharmacol Ther 2012;35:429-40.
27. Farup PG, Monsbakken KW, Vandvik PO. Lactose malabsorption in a population with irritable bowel syndrome: Prevalence and symptoms. A case-control study. Scand J Gastroenterol 2004;7:645-9.
28. Szilagyi A, Malolepszy P, Yesovitch S, et al. Inverse dose effect of pretest dietary lactose intake on breath hydrogen results and symptoms in lactase non persistent subjects. Dig Dis Sci 2005;50:2178-82.
29. Mishkin S. Dairy sensitivity, lactose malabsorption, and elimination diets in inflammatory bowel disease. Am J Clin Nutr 1997;65:564-7.
30. Eadala P, Mathews SB, Waud JP, Gren JT, Campbell AK. Association of lactose sensitivity with inflammatory bowel disease – demonstrated by analysis of genetic polymorphism, breath gases and symptoms. Aliment Pharmacol Ther 2011;34:735-46.
31. Gibson PR, Shepherd SJ. Personal view: Food for thought – Western lifestyle and susceptibility to Crohn's disease. The FODMAP hypothesis. Aliment Pharmacol Ther 2005;21:1399-409.
32. Yang J, Fox M, Cong Y, et al. Lactose intolerance in irritable bowel syndrome patients with diarrhea: The roles of anxiety, activation of the innate mucosal system and visceral sensitivity. Aliment Pharmacol Ther 2014;39:302-11.
33. Gibson PR Shepherd SJ. Evidence-based dietary management of functional gastrointestinal symptoms: The FODMAP approach. J Gastroenterol Hepatol 2010;25:252-8.
34. Montalto M, Gallo L, Santoro L, et al. Low-dose lactose in drugs neither increases breath hydrogen excretion nor causes gastrointestinal symptoms. Aliment Pharmacol Ther 2008;28:1003-12.
35. Casellas F, Aparici A, Casaus M, Rodríguez P, Malagelada JR.Subjective perception of lactose intolerance does not always indicate lactose malabsorption. Clin Gastroenterol Hepatol 2010;8:581-6.
36. Tomba C, Baldassarri A, Coletta M, Cesana BM, Basilisco G. Is the subjective perception of lactose intolerance influenced by the psychological profile? Aliment Pharmacol Ther 2012;36:660-9.
37. Hertzler SR, Huynh BC, Savaiano DA. How much lactose is low lactose. J Am Diet Assoc 1996;96:243-6.
38. Suarez FL, Savaiano DA, Levitt MD. A comparison after the consumption of milk or lactose-hydrolyzed milk by people with self-reported severe lactose intolerance. N Engl J Med 1995;333:1-4.
39. Suarez FL, Savaiano D, Arbisi P, Levitt MD. Tolerance to the daily ingestion of two cups of milk by individuals claiming lactose intolerance. Am J Clin Nutr 1997;65:1502-6.
40. Suchy FJ, Brannon PM, Carpenter TO, et al. National institute of health consensus development conference: Lactose intolerance in health. Ann Intern Med 2010;152:792-6.

41. Hertzler SR, Savaiano DA. Colonic adaptation to daily lactose feeding in lactose maldigesters reduces lactose intolerance. Am J Clin Nutr 1996;64:232-6.
42. Savaiano DA, Ritter A, Klaenhammer TR, et al. Improving lactose digestion and symptoms of lactose intolerance with a novel galacto-oligosaccharide (RP-G28): A randomized, double-blind clinical trial. Nutr J 2013;12:160-9.
43. Shrier I, Szilagyi A, Correa JA. Impact of lactose containing foods and the genetics of lactase on diseases: An analytical review of population data. Nutr Cancer 2008;60:292-300.
44. Jiang, T; Savaiano, DA. In vitro lactose fermentation by human colonic bacteria is modified by *Lactobacillus acidophilus* supplementation. J Nutr 1997;1237:1489-95.
45. Jiang T, Savaiano DA. Modification of colonic fermentation by bifidobacteria and pH in vitro. Impact on lactose metabolism, short-chain fatty acid, and lactate production. Dig Dis Sci 1997;42:2370-7.
46. Makivuokko HA, Saarinen MT, Ouwehand AC, Rautonen NE. Effects of lactose on colon microbial community structure and function in a four-stage semi-continuous culture system. Biosci Biotechno Biochem 2006;70:2056-63.
47. Ito M, Kimura M. Influence of lactose on faecal microflora in lactose maldigesters. Microb Ecol Health Dis 1993;6:73-6.
48. Szilagyi A. Review article: Lactose – a potential prebiotic. Aliment Pharmacol Ther 2002;16:1591-602.
49. Szilagyi A, Shrier I, Heilpern D, et al. Differential impact of lactose/lactase phenotype on colonic microflora. Can J Gastroenterol 2010;24:373-9.
50. Oku T, Nakamura S, Ichinose M. Maximum permissive dose of lactose and lactitol for transitory diarrhea, and utilizable capacity for lactose in Japanese female adults. J Nutr Sci Vitaminol 2005;51:51-7.
51. Gilat T, Russo S, Gelman-Malachi E, Aldor TA. Lactase in man: A nonadaptable enzyme. Gastroenterology 1972;62:1125-7.
52. He T, Venema K, Priebe MG, Welling GW, Brummer R.-J, Vonk RJ. The role of colonic metabolism in lactose intolerance. Eur J Clinical Invest 2008;38:541-7.
53. Duncan SH, Louis P, Flint HJ. Lactate-utilizing bacteria, isolated from human feces, that produce butyrate as a major fermentation product. Appl Environ Microbiol 2004;70:5810-7.
54. Commane D, Hughes R, Short C, Rowland I. The potential mechanisms involved in the anti-carcinogenic action of probiotics. Mut Res 2005;591:276-89.
55. Severin S, Wenshui X. Milk biologically active components as nutraceuticals. Rev Crit Rev Food Sci Nutr 2005;45:645-56.
56. Szilagyi A, Nathwani U, Vinokuroff C, Correa JA, Shier I. Evaluation of relationships among national colorectal cancer mortality rates, genetic lactase non persistence status and per capita yearly milk and milk product consumption. Nut Cancer 2006;55:151-6.
57. Park Y, Leitzmann MF, Subar, et al. Dairy food, calcium, and risk of cancer in the NIH-AARP diet and health study. Arch Intern Med 2009;169:391-401.
58. Norat T, Riboli E. Dairy products and colorectal cancer: A review of possible mechanisms and epidemiological evidence. Eur J Clin Nutr 2003;57:1-17.
59. Grau MV, Baron JA, Sandler RS, et al. Prolonged effect of calcium supplementation on risk of colorectal adenomas in a randomized trial. J Natl Cancer Inst 2007;99:129-36.
60. Grau MV, Baron JA, Sandler RS, et al. Vitamin D, calcium supplementation, and colorectal adenomas: Results of a randomized trial. J Natl Cancer Inst 2003;95:1765-71.
61. Torfadottir JE, Steingrimsdottir L , Mucci L, et al. Milk intake in early life and risk of advanced prostate cancer. Am J Epidemiol 2012;175:144-53.
62. Pettersson A, Kasperzyk JL, Kenfield SA, et al. Milk and dairy consumption among men with prostate cancer and risk of metastases and prostate cancer death. Cancer Epidemiol Biomarkers Prev 2012;21:428-36.
63. Song Y, Chavarro JE, Cao Y, et al. Whole milk intake is associated with prostate cancer-specific mortality among U.S. male physicians. J Nutr 2013;143:189-96.
64. Travis RC, Appleby PN, Siddiq A, et al. Genetic variation in lactase gene, dairy product intake and risk for prostate cancer in the European prospective investigation into cancer and nutrition. Int J Cancer 2013;132:1901-10.
65. Agarwal MM, Rana SV, Mandal AK, et al. Lactose intolerance in prostate cancer patients: Incidence and associated factors. Scand J Gastroeneterol 2008;43:270-6.
66. Davies TW, Palmer, CR, Lipscombe JM. Adolescent milk, dairy product and fruit consumption and testicular cancer. Br J Cancer 1998;74:657-60.
67. Stang A, Ahrens W, Baumgrdt-Elms C. Adolescent milk fat and galactose consumption and testicular cell cancer. Cancer Epidemiol Biomark Prev 2006;15:2189-95.
68. Mao QQ, Dai Y, Lin YW, Qin J, Xie LP, Zheng XY. Milk consumption and bladder cancer risk: Meta-analysis of published epidemiological studies. Nut Cancer 2011;63:1263-71.
69. Hayatsu H, Hayatsu T. Suppressing effect of *Lactobacillus casei* administration on the urinary mutagenicity arising from ingestion of fried ground beef in the human. Cancer Lett 1993;73:173-9.
70. Aso Y, Akaza H, Kotake T, Tsukamoto T, Imai K, Naito S. Preventive effect of a *Lactobacillus casei* preparation on the recurrence of superficial bladder cancer in a double-blind trial. The BLP study group. Eur Urol 1995;27:104-9.
71. Cramer DW. Lactase persistence and milk consumption as determinants of ovarian cancer risk. Am J Epidemiol 1989;130:904 -10.
72. Qin LQ, Xu JY, Wang PY, et al. Milk/dairy products consumption, galactose metabolism and ovarian cancer: Meta-analysis of epidemiological studies. Eur J Cancer Prev 2005;14:13-9.
73. Larsson SC, Orsini N, Wolk A. Milk, milk products and lactose intake and ovarian cancer risk: A meta-analysis of epidemiological studies. Int J Cancer 2006;118:431-41.
74. Genkinger JM, Hunter, DJ, Spiegelman D, et al. Dairy products and ovarian cancer: A pooled analysis of 12 cohort studies. Cancer Epidemiol Biomarkers Prev 2006;15:364-72.
75. Faber MT, Jensen A, Søgaard M, et al. Use of dairy products, lactose, and calcium and risk of ovarian cancer – results from a Danish case-control study. Acta Oncol 2012;51:454-64.
76. Mommers M, Schouten LJ, Goldbohm RA, van den Brandt PA. Dairy consumption and ovarian cancer risk in the Netherlands Cohort Study on Diet and Cancer. Br J Cancer 2006;94:165-70.
77. Schulz M, Nöthlings U, Allen N, et al. No association of consumption of animal foods with risk of ovarian cancer. Cancer Epidemiol Biomark Prev 2007;16:852-5.
78. Ganmaa D, Sato A. The possible role of female sex hormones in milk from pregnant cows in the development of breast, ovarian and corpus uteri cancers. Med Hypotheses 2005;65:1028-37.
79. Krog V, Berrino F, Sieri S, et al. Meat, eggs, dairy products, and risk of breast cancer in the European Prospective Investigation into Cancer and Nutrition (EPIC) cohort. Am J Clin Nutr 2009;90:602-12.
80. Dong JY, Zhang L, He K, Qin LQ. Dairy consumption and risk of breast cancer: A meta-analysis of prospective cohort studies. Breast Cancer Res Treat 2011;127:23-31.
81. Sankaranarayanan R, Varghese C, Duffy SW, Padmakumary G, Day NE, Nair MK. A case-control study of diet and lung cancer in Kerala, south India. Int J Cancer 1994;58:644-9.
82. Richman E, Rhodes JM. Review article: Evidence-based dietary advice for patients with inflammatory bowel disease. Aliment Pharmacol Ther 2013;38:1156-71.
83. Vernia P, Loizos P, Di Giuseppantonio I, Amore B, Chiappini A, Cannizzaro S. Dietary calcium intake in patients with inflammatory bowel disease. J Crohn's Colitis 2014;8:312-7.
84. Ghishan FK, Kiela PR. Advances in the understanding of mineral and bone metabolism in inflammatory bowel diseases. Am J Physiol Gastrointest Liver Physiol 2011;300:G191-201.
85. Ananthakrishnan AN, Cheng SC, Cai T, et al. Association between reduced plasma 25-hydroxy vitamin D and increased risk of cancer in patients with inflammatoy bowel diseases. Clin Gastoenterol Hepatol 2014;12:821-7.
86. Asakura H, Suzuki K, Kitahora T, Morizane T. Is there a link between food and intestinal microbes and the occurrence of Crohn's disease an ulcerative colitis. J Gastroenterol Hepatol 2008;23:1794-801.
87. Joachim G. The relationship between habits of food consumption and reported reactions to food in people with inflammatory bowel disease – testing the limits. Nutr Health 1999;13:69-83.
88. Gilat T, Hacohen D, Lilos P Langman MJ. Childhood factors in ulcerative colitis and Crohn's disease. An international cooperative study. Scand J Gastroenterol 1987;22:1009-24.

89. Jantchou P, Morois S, Clavel-Chapelon F, Boutron-Ruault MC, Carbonnel F. Animal protein intake and risk of inflammatory bowel disease: The E3N prospective study. Am J Gastroenterol 2010;105:2195-201.
90. Abubakar I, Myhill DJ, Hart AR, et al. A case-control study of drinking water and dairy products in Crohn's disease – further investigation of the possible role of *Mycobacterium avium* paratuberculosis. Am J Epidemiol 2007;165:776-83.
91. Buning C, Ockenaga J, Kruger S, et al. The C/C (−13910) and G/G (−22018) genotypes for adult-type hypolactasia are not associated with inflammatory bowel disease. Scand J Gastroenterol 2003;38:538-42.
92. Nolan DJ, Han DY, Lam WJ, Morgan AR, Fraser AG, Tapsell LC. Genetic adult lactase persistence is associated with risk of Crohn's disease in a New Zealand population. BMC Res Notes 2010;3:339-43.
93. Kettunen J, Silander K, Saarela O, et al. European lactase persistence genotype shows evidence of association with increase in body mass index. Hum Mol Genet 2010;19:1129-36.
94. Chen M, Pan A, Malik VS, Hu FB. Effects of dairy intake on body weight and fat: meta-analysis of randomized controlled trials. Am J Clin Nutr 2012;96:735-47.
95. Louie JC, Flood VM, Rangan AM, et al. Higher regular fat dairy consumption is associated with lower incidence of metabolic syndrome but not type 2 diabetes. Nutr Metab Cardiovasc Dis 2013;23:816-21.
96. Birlouez-Aragon I, Ravelontseheno L, Villate-Cathelineau B, Cathelineau G, Abitbol G. Disturbed galactose metabolism in elderly and diabetic humans is associated with cataract formation. J Nutr 1993;123:1370-6.
97. Nicklas TA, Haiyan Q, Hughes SO, et al. Self-percieved lactose intolerance results in lower intakes of calcium and dairy foods and is associated with hypertension and diabetes in adults. Am J Clin Nutr 2011;94:191-8.
98. Nettleton JA, Steffen LM, Loehr LR, Rosamond WD, Folsom AR. Incident heart failure is associated with lower whole-grain intake and greater high-fat dairy and egg intake in the Atherosclerosis Risk in Communities (ARIC) study. J Am Diet Assoc 2008;108:1881-7.
99. Huth PJ, Park KM. Influence of dairy product and milk fat consumption on cardiovascular disease risk: A review of the evidence. Adv Nutr 2012;3:266-85.
100. Louie JC, Flood VM, Burlitsky G, Rangan AM, Gill TP, Mitchell P. Dairy consumption and the risk of 15-year cardiovascular disease mortality in a cohort of older Australians. Nutrients 2013;5:441-54.
101. Garland CF, Garland FC. Do sunlight and vitamin D reduce the likelihood of colon cancer? Int J Epidemiol 1980;9:227-31.
102. Grant WB. A meta-analysis of second cancers after a diagnosis of non-melanoma skin cancer: Additional evidence that solar ultraviolet –B irradiance reduces the risk of internal cancers. J Steroid Biochem Mol Biol 2007;103:668-74.
103. Mohr SB, Garland CF, Gorham ED, Grant WB, Garland FC. Ultraviolet B and incidence rates of leukemia worldwide. Am J Prev Med 2011:41:68-74.
104. Simpson S Jr, Blizzard L, Otahal P, Van der Mei I, Taylor B. Latitude is significantly associated with the prevalence of multiple sclerosis: A meta-analysis. J Neurol Neurosurg Psychiatry 2011;82:1132-41.
105. Zitterman A. Vitamin D in preventive medicine: Are we ignoring the evidence? Br J Nutr 2003;89:552-72.
106. Cantorna MT, Mahon BD. Mounting evidence for vitamin D as an environmental factor affecting autoimmune disease prevalence. Exp Biol Med (Maywood) 2004;229:1136-42.
107. International Agency for Research Group. Dietary fiber, transit – time, faecal bacteria, steroids, and colon cancer in two Scandinavian populations. Lancet 1977;2:207-11.
108. Nanji AA, Denardi FG. Primary adult lactose intolerance protects against development of inflammatory bowel disease. Med Hypotheses 1986;19:1-6.
109. Itan Y, Powell A, Beaumont MA, Burger J, Thomas MG. The origins of lactase persistence in Europe. PLoS Comput Biol 2009;5:e1000491. doi:10.1371/journal.pcbi.1000491.
110. Babu J, Kumar S, Babu P, et al. Frequency of lactose malabsorption among healthy southern and northern Indian populations by genetic analysis and lactose hydrogen breath and tolerance tests. Am J Clin Nutr 2010;91:140-6.
111. Heyer E, Brazier L , Segurel L, et al. Lactase persistence in central Asia: Phenotype, genotype, and evolution. Hum Biol 2011;83:379-92.
112. Gallego Romero I, Basu Mallick C, Liebert A, et al. Herders of Indian and European cattle share their predominant allele for lactase persistence. Mol Biol Evol 2012;29:249-60.
113. Peyrin-Biroulet I, Oussalah A, Bigard M-A. Crohn's disease: The hot hypothesis. Med Hypotheses 2009;73:94-6.
114. Garg M, Lubel JS, Sparrow MP, Holt SG, Gibson PR. Review article: Vitamin D and inflammatory bowel disease – established concepts and future directions. Aliment Pharmacol Ther 2012;36:324-44.
115. Ng Sc, Bernstein CN, Vatn MH, et al. Geographical variability and environmental risk factors in inflammatory bowel disease. Gut 2013;62:630-49.
116. Abraham C, Cho JH. Inflammatory bowel disease. N Engl J Med 2009;361:2066-78.
117. Ananthakrishnan AN, Hailiang H, Nguyen DD, Sauk J, Yajnik V, Xavier RJ. Differential effect of genetic burden on disease phenotypes in Crohn's disease and ulcerative colitis: Analysis of a North American cohort. Am J Gastroenterol 2014;109:395-400.
118. Lees CW, Barrett JC, Parkes M, Satsangi J. New IBD genetics: Common pathways with other diseases. Gut 2011;60:1739-53.
119. Itan Y, Jones BL, Ingram CJE, Swallow DM, Thomas MG. A worldwide correlation of lactase persistence phenotype and genotypes. BMC Evolutionary Biol 2010;10:36 <www.biomedcentral.com/1471-2148/10/36> (Accessed March 6, 2015).
120. Molodecky NA, Soon IS, Rabi D, et al. Increasing incidence and prevalence of the inflammatory bowel diseases with time, based on a systematic review. Gastroenterology 2012;142:46-54.
121. Szilagyi A, Leighton H, Burstein B, Xue X. Latitude, sunshine, and human lactase phenotype distributions may contribute to geographic patterns of modern disease: The inflammatory bowel disease model. Clin Epidemiol2014;6:183-98.
122. Gasche C, Nemeth M, Grundtner P, Willheim-Polli C, Ferenci P, Schwarzenbacher R. Evolution of Crohn's disease-associated *Nod2* mutations. Immunogenetics 2008;60:115-20.
123. Naser SA , Arce M, Khaja, et al. Role of ATG16L, NOD2 and IL23R in Crohn's disease pathogenesis. World J Gastroenterol 2012;18:412-24.
124. Alfonso-Sanchez MA, Perez-Miranda AM, Garcia-Obregon S, Pena JA. An evolutionary approach to the high frequency of the delta F508 CFTR mutation in European populations. Med Hypotheses 2010;74:989-92.
125. Modiano G, Ciminellui BM, Pignatti PF. Cystic fibrosis and lactase persistence: A possible correlation. Eur J Hum Genet 2007;15:255-9.
126. Madry E, Fidler E, Sobczynska-Tomaszewska A, et al. Mild *CFTR* mutations and genetic predisposition to lactase persistence in cystic fibrosis. Eur J Hum Genet 2011;19:748-52.
127. Truedsson M, Carlson J, Simren M, Ohlsson B. Polymorphism in the oxytocin promoter region in patients with lactase non-persistence is not related to symptoms. BMC Gastroenterol 2009;9:90-107.
128. Sacerdote C, Guarrera S, Smith GD, et al. Lactase persistence and bitter taste response: Instrumental variables and Mendelian randomization in epidemiologic studies of dietary factors and cancer risk. Am J Epidemiol 2007;166:576-81.
129. Wu GD, Chen J, Hoffmann C, et al. Linking long-term dietary patterns with gut microbial enterotypes. Science 2011;334:105-8.
130. Arumugam M, Raes J, Pelletier E, et al. Enterotypes of the human gut microbiome. Nature 2011;473:174-80.
131. Hecht G. AGA's approach to the microbiome. 2014;146:1119-21.

Clinicians' guide to the use of fecal calprotectin to identify and monitor disease activity in inflammatory bowel disease

Brian Bressler MD[1], Remo Panaccione MD[2], Richard N Fedorak MD[3], Ernest G Seidman MD[4]

B Bressler, R Panaccione, RN Fedorak, EG Seidman. Clinicians' guide to the use of fecal calprotectin to identify and monitor disease activity in inflammatory bowel disease. Can J Gastroenterol Hepatol 2015;29(7):369-372.

BACKGROUND: Objective monitoring of the severity of inflammation in patients with inflammatory bowel disease (IBD) is an essential part of disease management. However, repeat endoscopy to define extent and severity of inflammation is not practical. Fecal calprotectin (FC) is a biomarker that can be used as a surrogate test to distinguish inflammatory from noninflammatory gastrointestinal disease.

METHODS: A targeted search of the literature regarding FC, focusing primarily on the past three years, was conducted to develop practical clinical guidance on the current utility of FC in the routine management of IBD patients.

RESULTS: It is recommended that samples for FC testing be obtained from the first bowel excretion of the day. FC testing should be used as standard of care to accurately confirm inflammation and 'real-time' disease activity when a clinician suspects an IBD flare. Although FC is a reliable marker of inflammation, its role in routine monitoring in improving long-term outcomes has not yet been fully assessed. Based on available evidence, the authors suggest the following cut-off values and management strategies: when FC levels are <50 μg/g to 100 μg/g, quiescent disease is likely and therapy should be continued; when FC levels are >100 μg/g to 250 μg/g, inflammation is possible and further testing (eg, colonoscopy) is required to confirm inflammation; and when FC levels are >250 μg/g, active inflammation is likely and strategies to control inflammation should be initiated (eg, optimizing current therapies or switching to an alternative therapy).

DISCUSSION: FC is a useful biomarker to accurately assess the degree of inflammation and should be incorporated into the management of patients with IBD.

Key Words: *Disease activity; Fecal calprotectin; Inflammatory bowel disease; Monitoring*

Le guide du clinicien sur l'utilisation de la calprotectine fécale afin de déterminer et de surveiller l'activité des maladies inflammatoires de l'intestin

HISTORIQUE : Il est essentiel de surveiller objectivement la gravité des maladies inflammatoires de l'intestin (MII) dans le cadre de leur prise en charge. Cependant, il n'est pas pratique de faire une nouvelle endoscopie pour définir l'étendue et la gravité de l'inflammation. La calprotectine fécale (CF) est un biomarqueur qui peut remplacer l'endoscopie pour distinguer les maladies gastro-intestinales inflammatoires de celles qui ne le sont pas.

MÉTHODOLOGIE : Les chercheurs ont réalisé une analyse bibliographique sur la CF axée sur les trois années précédentes afin de rédiger des orientations cliniques sur l'utilité de la CF dans la prise en charge habituelle des patients atteints d'une MII.

RÉSULTATS : Pour effectuer les tests de CF, il est recommandé de prélever la première excrétion intestinale de la journée. Lorsqu'un clinicien craint une récidive de MII, les tests de CF doivent servir de norme de soins pour confirmer avec précision l'inflammation et l'activité de la maladie « en temps réel ». Même si la CF est un marqueur fiable d'inflammation, son rôle dans la surveillance systématique en vue d'améliorer les résultats cliniques à long terme n'a pas encore été pleinement évalué. D'après les données disponibles, les auteurs proposent les valeurs seuils et les stratégies de prise en charge suivantes : lorsque les taux de CF sont sous les 50 μg/g à 100 μg/g, la maladie est probablement en rémission et le traitement doit être maintenu. Lorsque les taux de CF se situent entre plus de 100 μg/g et 250 μg/g, l'inflammation est possible et d'autres tests (p. ex., coloscopie) s'imposent pour confirmer l'inflammation. Enfin, lorsque les taux de CF sont supérieurs à 250 μg/g, l'inflammation active est probable, et il faut adopter des stratégies pour contrôler l'inflammation (p. ex., optimisation des traitements en place ou transition vers un autre traitement).

EXPOSÉ : La CF est un biomarqueur utile pour évaluer le degré d'inflammation avec précision. Elle devrait faire partie de la prise en charge des patients atteints d'une MII.

Objective monitoring of the extent and severity of inflammation in patients with inflammatory bowel disease (IBD) is an essential part of effective disease management. Specifically, it is used to determine whether new symptoms reflect a change in inflammatory activity and, thus, require an optimization of therapy. While direct endoscopic examination of mucosal inflammation is the gold standard, frequent endoscopy is not always practical or feasible. Therefore, the use of a surrogate biomarker, such as fecal calprotectin (FC), can provide significant advantages in the management of these patients.

In IBD, inflammation with infiltration of the intestinal mucosa by neutrophils and macrophages, leads to secretion of cytosolic proteins, which can be detected in stool (1). Calprotectin is a calcium and zinc-binding protein that constitutes 60% of neutrophil cytosolic protein (2). It is stable in feces for up to one week at room temperature. Thus, FC may provide an accurate assessment of the inflammatory burden in the gut.

An ideal biomarker would effectively and accurately distinguish inflammatory from noninflammatory disease, correlate with endoscopic inflammation, demonstrate response to therapy and predict relapse (1,3-7). In addition, the biomarker test would be readily available, easy to use and affordable. FC meets many of these criteria, is available throughout Canada and has the potential to significantly enhance IBD care. However, although changes in FC demonstrate good sensitivity and specificity in IBD patients, clinicians should be aware that elevated FC levels may be found in several non-IBD conditions (Table 1) (1,3).

[1]*Department of Medicine, Division of Gastroenterology, St Paul's Hospital, Vancouver, British Columbia;* [2]*Department of Medicine, University of Calgary, Calgary;* [3]*Division of Gastroenterology, University of Alberta, Edmonton, Alberta;* [4]*Division of Gastroenterology, McGill University Health Centre, Montreal, Quebec*

Correspondence: Dr Brian Bressler, University of British Columbia, 1190 Hornby Street, Suite 770, Vancouver, British Columbia V6Z 2K5. e-mail brian_bressler@hotmail.com

TABLE 1
Factors and conditions associated with elevated fecal calprotectin levels (1,3)

Infectious	Inflammatory conditions
• Bacterial dysentery	• Inflammatory bowel disease
• *Giardia lamblia*	• Autoimmune enteropathy
• *Helicobacter pylori* gastritis	• Cirrhosis
• Infectious diarrhea	• Cystic fibrosis
• Viral gastroenteritis	• Diverticulitis
Neoplasms	• Eosinophilic colitis/enteritis
• Colonic and gastric polyps	• Gastroesophageal reflux disease
• Colorectal cancer	• Juvenile polyp
• Gastric carcinoma	• Microscopic colitis
• Intestinal lymphoma	• Peptic ulcer
	• Untreated celiac disease
Drugs	**Other**
• NSAIDs	• Age <5 years
• PPIs	• Untreated food allergy

NSAIDs Nonsteroidal anti-inflammatory drugs; PPIs Proton pump inhibitors

Blood-based biomarkers, such as C-reactive protein (CRP) are often more acceptable, but they have limited sensitivity, with at least 50% of patients with active ulcerative colitis (UC) having normal CRP levels (8). CRP also has limited specificity, particularly in patients with infections, or rheumatoid or other autoimmune disorders (1). While CRP may be useful in some IBD patients, elevated FC levels demonstrate a significantly superior correlation with active disease compared with CRP measurements (8-13).

The present user's guide will outline the pragmatic elements a clinician should be aware of relating to the use of FC in practice.

When should samples be collected for FC testing?

FC levels decrease with increasing time between bowel movements; therefore, it is recommended that samples be obtained from the first bowel movement of the day (14). Because of the day-to-day variability in FC levels (14), retrieving samples at similar times over two consecutive days is preferable.

What test should be used?

The standard test to assess FC levels in IBD patients is the ELISA. This method has been used for >20 years, but is time consuming, labour intensive and requires specialized laboratory equipment (7,15). The Bühlmann Quantum Blue Reader (Alpco Immunoassays, USA) device offers an alternative method to the standard ELISA technique. It demonstrates an excellent correlation with standard ELISA assays and provides accurate results within 30 min (7). The ELISA method is cost-efficient for high throughput when running multiple samples simultaneously.

FC testing is available in all provinces in Canada, with the majority of tests being performed using ELISA. However, provincial funding is currently available only in Alberta and Quebec, while in other provinces, the cost must be covered by the patient or a third party.

Should FC levels be measured in asymptomatic IBD patients?

Although FC is a reliable marker of inflammation, the role of routine monitoring of FC levels in clinically asymptomatic patients has not been fully assessed. It is not yet known whether a strategy of routine monitoring coupled with early optimization of therapy on detection of relapse would improve health outcomes (16,17). Nevertheless, this may be a clinically useful approach and warrants further investigation.

When should FC levels be measured, as the standard of care, to monitor IBD patients?

In the context of a suspected flare, assessing FC levels enables accurate and expedient confirmation of inflammation and 'real-time' disease activity. Several recent trials and meta-analyses have established that FC is an accurate surrogate marker of active endoscopic disease in IBD patients, with high sensitivity (70% to 100%) and specificity (44% to 100%) (8,13,18-21). In general, FC levels are highest among patients with active disease, less so in patients with quiescent disease and lowest in non-IBD control subjects (21).

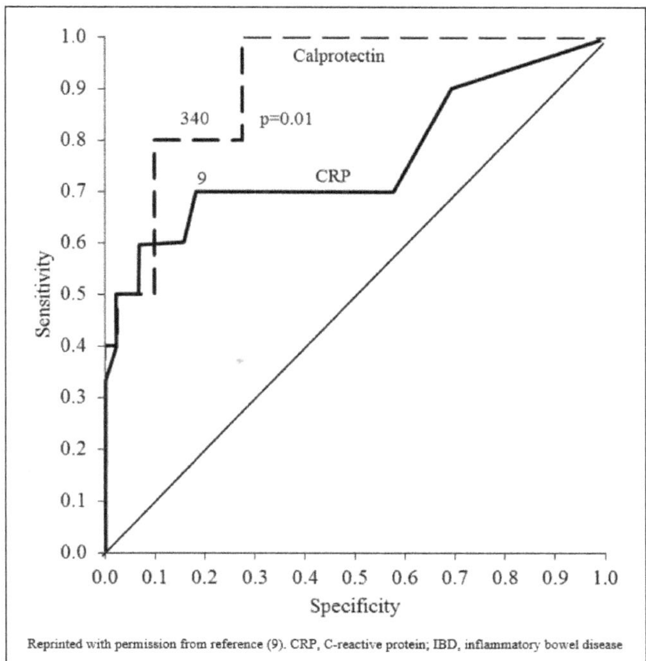

Figure 1) *ROC curve for fecal calprotectin in predicting relapse in inflammatory bowel disease patients*

What are the potential future applications of FC in IBD care?

Routine use of FC levels to monitor risk for flare in quiescent IBD, response to treatment or postoperative recurrence is not currently recommended, although these indications do warrant further investigation.

Studies suggest that in patients with quiescent disease, serial FC measurements showing increasing levels may be useful to actively monitor the risk for flare (8,17,22,23). In one study involving 53 consecutive IBD patients who were followed for 12 months (9), elevated FC levels (>340 µg/g) indicated an 18-fold higher risk for relapse (Figure 1). Conversely, reductions in FC levels have been associated with better response to treatment in IBD patients (24-27).

As a biomarker of inflammation, FC has been shown to be responsive to treatment intensification. A randomized trial involving patients with quiescent UC found that FC levels were lowered by increasing the dose of mesalamine (28). Moreover, relapse occurred sooner in patients with FC levels >200 µg/g compared with those with lower FC levels (P=0.01). This suggests that FC has a role in the monitoring of UC patients, one that enables the clinician to respond to elevated levels to prevent a clinical flare.

Using FC levels to monitor patients for postoperative recurrence could decrease the need for follow-up colonoscopy at six to 12 months and help optimize IBD outcomes. The predictive value of FC for postoperative recurrence has been established in patients who have undergone ileocolonic resection for Crohn disease (29-32). However, FC levels in the postoperative patient may not be as consistent as they are in other clinical situations (32).

How are test results interpreted?

Suggested potential cut-off values and courses of action are provided in Figure 2; however, individual patient variability exists. Therefore, in a particular patient, rising FC levels over time are the best indicator of

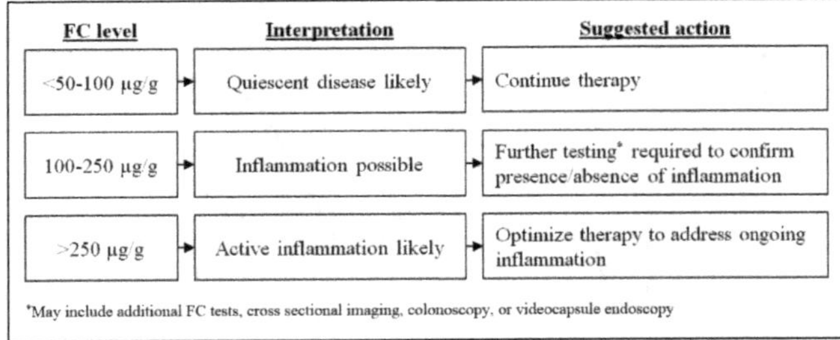

Figure 2) *Interpreting fecal calprotectin (FC) test results*

increased disease activity for that individual. Another approach to improve the reliability of FC testing would be to obtain a baseline FC level during a period of known active inflammation. An elevated FC level in this context will help clinicians determine whether this is a reliable biomarker for monitoring inflammation in an individual patient.

Some controversy remains as to optimal cut-off values. Results vary depending on the specific test used (33). The sensitivity and specificity of FC as a marker for active disease was found to differ at various cut-off values. A meta-analysis of 13 studies (n=1471) compared cut-off FC levels of 50 µg/g, 100 µg/g and 250 µg/g, and found that with higher levels the sensitivity decreased, while the specificity increased (19). In patients with a pretest probability of active disease of 66% (endoscopy-positive cases), an elevated FC level increased the probability to 90%, whereas a negative FC test decreased the probability to 26% (19). Cut-off values and management strategies based on the currently available evidence are shown in Figure 2.

ACKNOWLEDGEMENTS: The authors thank AbbVie Canada for their generous support of the development of the manuscript, as well as Pauline Lavigne and Steven Portelance for editorial assistance.

DISCLOSURES: Dr Bressler has received research grants and/or served as a consultant and/or speaker for: AbbVie, Alba Therapeutics, Alpco, Amgen, BMS, Celltrion, Ferring, Genentech, GSK, Hospira, Janssen, Pendopharm, Shire, Takeda and Warner Chilcott. Dr Panaccione has received research grants and/or served as a consultant and/or speaker for: Abbott, AbbVie Amgen, Aptalis, AstraZeneca, Baxter, Eisai, Ferring, Janssen, Merck, Schering-Plough, Shire, Centocor, Elan, Glaxo-Smith Kline, UCB, Pfizer, Bristol-Myers Squibb, Warner Chilcott, Takeda, Prometheus, Proctor and Gamble, and Millenium. Dr Seidman has received research funding or served as a consultant and/or speaker for: AbbVie, Alpco, Aptavis, Covidien, Janssen, Prometheus, Shire, Takeda and Vertex. Dr Fedorak has received research grants and/or served as a consultant and/or speaker for: AbbVie/Abbott, Alba, Bristol-Myers Squibb, Celltrion, Centoco, Ferring Pharma, GSK, Genentec, Janssen, Merck, Milllennium, Novartis, Pfizer, Proctor & Gamble, Roche, Shire, VSL#3, and is an owner/shareholder with Metabolomic Technologies Inc.

REFERENCES

1. Kopylov U, Rosenfeld G, Bressler B, Seidman E. Clinical utility of fecal biomarkers for the diagnosis and management of inflammatory bowel disease. Inflamm Bowel Dis 2014;20:742-56.
2. Roseth AG, Fagerhol MK, Aadland E, Schjonsby H. Assessment of the neutrophil dominating protein calprotectin in feces. A methodologic study. Scand J Gastroenterol 1992;27:793-8.
3. van Rheenen PF, Van de Vijver E, Fidler V. Faecal calprotectin for screening of patients with suspected inflammatory bowel disease: Diagnostic meta-analysis. BMJ 2010;341:c3369.
4. von Roon AC, Karamountzos L, Purkayastha S, et al. Diagnostic precision of fecal calprotectin for inflammatory bowel disease and colorectal malignancy. Am J Gastroenterol 2007;102:803-13.
5. Costa F, Mumolo MG, Bellini M, et al. Role of faecal calprotectin as non-invasive marker of intestinal inflammation. Dig Liver Dis 2003;35:642-7.
6. Tibble JA, Sigthorsson G, Foster R, Forgacs I, Bjarnason I. Use of surrogate markers of inflammation and Rome criteria to distinguish organic from nonorganic intestinal disease. Gastroenterology 2002;123:450-60.
7. Sydora MJ, Sydora BC, Fedorak RN. Validation of a point-of-care desk top device to quantitate fecal calprotectin and distinguish inflammatory bowel disease from irritable bowel syndrome. J Crohns Colitis 2012;6:207-14.
8. Lewis JD. The utility of biomarkers in the diagnosis and therapy of inflammatory bowel disease. Gastroenterology 2011;140:1817-26 e2.
9. Kallel L, Ayadi I, Matri S, et al. Fecal calprotectin is a predictive marker of relapse in Crohn's disease involving the colon: A prospective study. Eur J Gastroenterol Hepatol 2010;22:340-5.
10. Schoepfer AM, Beglinger C, Straumann A, et al. Fecal calprotectin more accurately reflects endoscopic activity of ulcerative colitis than the Lichtiger Index, C-reactive protein, platelets, hemoglobin, and blood leukocytes. Inflamm Bowel Dis 2013;19:332-41.
11. Schoepfer AM, Beglinger C, Straumann A, et al. Fecal calprotectin correlates more closely with the Simple Endoscopic Score for Crohn's disease (SES-CD) than CRP, blood leukocytes, and the CDAI. Am J Gastroenterol 2010;105:162-9.
12. Schoepfer AM, Beglinger C, Straumann A, et al. Ulcerative colitis: Correlation of the Rachmilewitz endoscopic activity index with fecal calprotectin, clinical activity, C-reactive protein, and blood leukocytes. Inflamm Bowel Dis 2009;15:1851-8.
13. Sipponen T, Savilahti E, Kolho KL, et al. Crohn's disease activity assessed by fecal calprotectin and lactoferrin: Correlation with Crohn's disease activity index and endoscopic findings. Inflamm Bowel Dis 2008;14:40-6.
14. Lasson A, Stotzer PO, Ohman L, et al. The intra-individual variability of faecal calprotectin: A prospective study in patients with active ulcerative colitis. J Crohns Colitis 2015;9:26-32.
15. Elkjaer M, Burisch J, Voxen Hansen V, et al. A new rapid home test for faecal calprotectin in ulcerative colitis. Aliment Pharmacol Ther 2010;31:323-30.
16. van Rheenen P. Do not read single calprotectin measurements in isolation when monitoring your patients with inflammatory bowel disease. Inflamm Bowel Dis 2014;20:1416-7.
17. Tibble JA, Sigthorsson G, Bridger S, Fagerhol MK, Bjarnason I. Surrogate markers of intestinal inflammation are predictive of relapse in patients with inflammatory bowel disease. Gastroenterology 2000;119:15-22.
18. D'Haens G, Ferrante M, Vermeire S, et al. Fecal calprotectin is a surrogate marker for endoscopic lesions in inflammatory bowel disease. Inflamm Bowel Dis 2012;18:2218-24.
19. Lin JF, Chen JM, Zuo JH, et al. Meta-analysis: Fecal calprotectin for assessment of inflammatory bowel disease activity. Inflamm Bowel Dis 2014;20:1407-15.
20. Roseth AG, Aadland E, Grzyb K. Normalization of faecal calprotectin: A predictor of mucosal healing in patients with inflammatory bowel disease. Scand J Gastroenterol 2004;39:1017-20.
21. Roseth AG, Aadland E, Jahnsen J, Raknerud N. Assessment of disease activity in ulcerative colitis by faecal calprotectin, a novel granulocyte marker protein. Digestion 1997;58:176-80.
22. Mao R, Xiao YL, Gao X, et al. Fecal calprotectin in predicting relapse of inflammatory bowel diseases: A meta-analysis of prospective studies. Inflamm Bowel Dis 2012;18:1894-9.
23. Costa F, Mumolo MG, Ceccarelli L, et al. Calprotectin is a stronger predictive marker of relapse in ulcerative colitis than in Crohn's disease. Gut 2005;54:364-8.
24. De Vos M, Dewit O, D'Haens G, et al. Fast and sharp decrease in calprotectin predicts remission by infliximab in anti-TNF naive patients with ulcerative colitis. J Crohns Colitis 2012;6:557-62.
25. Wagner M, Peterson CG, Ridefelt P, Sangfelt P, Carlson M. Fecal markers of inflammation used as surrogate markers for treatment outcome in relapsing inflammatory bowel disease. World J Gastroenterol 2008;14:5584-9.
26. Feagan BG, Rutgeerts P, Sands BE, et al. Vedolizumab as induction and maintenance therapy for ulcerative colitis. N Engl J Med 2013;369:699-710.
27. Ho GT, Lee HM, Brydon G, et al. Fecal calprotectin predicts the clinical course of acute severe ulcerative colitis. Am J Gastroenterol 2009;104:673-8.

28. Osterman MT, Aberra FN, Cross R, et al. Mesalamine dose escalation reduces fecal calprotectin in patients with quiescent ulcerative colitis. Clin Gastroenterol Hepatol 2014;12:1887-93,e3.
29. Orlando A, Modesto I, Castiglione F, et al. The role of calprotectin in predicting endoscopic post-surgical recurrence in asymptomatic Crohn's disease: A comparison with ultrasound. Eur Rev Med Pharmacol Sci 2006;10:17-22.
30. Lamb CA, Mohiuddin MK, Gicquel J, et al. Faecal calprotectin or lactoferrin can identify postoperative recurrence in Crohn's disease. Br J Surg 2009;96:663-74.
31. Lobaton T, Lopez-Garcia A, Rodriguez-Moranta F, et al. A new rapid test for fecal calprotectin predicts endoscopic remission and postoperative recurrence in Crohn's disease. J Crohns Colitis 2013;7:e641-51.
32. Scarpa M, D'Inca R, Basso D, et al. Fecal lactoferrin and calprotectin after ileocolonic resection for Crohn's disease. Dis Colon Rectum 2007;50:861-9.
33. Gisbert JP, McNicholl AG. Questions and answers on the role of faecal calprotectin as a biological marker in inflammatory bowel disease. Dig Liver Dis 2009;41:56-66.

43

Characterization of inflammatory bowel disease in hospitalized elderly patients in a large central Canadian health region

Peter Stepaniuk BSc(Hons)[1], Charles N Bernstein MD[1,2], Zoann Nugent PhD[1,2,3], Harminder Singh MD MPH[1,2,4]

P Stepaniuk, CN Bernstein, Z Nugent, H Singh. Characterization of inflammatory bowel disease in hospitalized elderly patients in a large central Canadian health region. Can J Gastroenterol Hepatol 2015;29(5):274-278.

OBJECTIVE: To determine differences in phenotype and treatment among hospitalized elderly and young patients with inflammatory bowel disease (IBD), and the utility of *International Classification of Diseases, 10th Revision* (ICD)-10 codes in hospital discharge abstracts in diagnosing IBD in elderly patients.

METHODS: A large Canadian health region hospitalization discharge database was used to identify elderly (>65 years of age) and young (19 to 50 years of age) patients with IBD admitted between April 1, 2007 and March 31, 2012, and a random sample of elderly patients with other colonic conditions. Medical records were reviewed to confirm IBD diagnosis and extract clinical information. The characteristics of elderly versus young hospitalized IBD patients and accuracy of ICD-10 IBD discharge codes in the elderly were assessed.

RESULTS: One hundred forty-three elderly and 82 young patients with an IBD discharge diagnosis, and 135 elderly patients with other gastrointestinal discharge diagnoses were included. Elderly IBD patients were less likely to have ileocolonic Crohn disease (21.4% versus 50.9%; P=0.001), more likely to be prescribed 5-aminosalicylates (61% versus 43%; P=0.04), and less likely to be prescribed biologics (6% versus 21%; P=0.016) or immunomodulators (21% versus 42%; P=0.01). The sensitivity, specificity and positive predictive value of a single ICD code for CD were 98%, 96% and 94%, respectively, and for ulcerative colitis (UC) were 98%, 92% and 70%, respectively.

CONCLUSIONS: Treatment approaches in elderly patients were different than in younger IBD patients despite having disease sufficiently severe to require hospitalization. While less accurate in UC, a single ICD-10 IBD code was sufficient to identify elderly CD and UC hospitalized patients.

Key Words: *Crohn disease; Elderly; Hospitalized patient outcomes; Ulcerative colitis*

La caractérisation des maladies inflammatoires de l'intestin chez les patients âgés hospitalisés dans une grande région sociosanitaire du centre du Canada

OBJECTIF : Déterminer les différences de phénotype et de traitement entre les personnes âgées et les patients plus jeunes hospitalisés à cause d'une maladie inflammatoire de l'intestin (MII) ainsi que l'utilité des dix codes de la *Classification internationale des maladies, 10e révision* (CIM-10) dans les résumés de congés hospitaliers pour diagnostiquer les MII chez les patients âgés.

MÉTHODOLOGIE : Les chercheurs ont puisé dans une vaste base de données des congés hospitaliers d'une grande région sociosanitaire canadienne pour répertorier les patients âgés (plus de 65 ans) et plus jeunes (de 19 à 50 ans) atteints d'une MII qui avaient été hospitalisés entre le 1er avril 2007 et le 31 mars 2012, de même qu'un échantillon aléatoire de patients âgés atteints d'autres affections du colon. Ils ont examiné les dossiers médicaux pour confirmer le diagnostic de MII et extraire l'information clinique. Ils ont évalué les caractéristiques des personnes âgées hospitalisées à cause d'une MII par rapport à celles des plus jeunes et la précision des codes de congé de la CIM-10 utilisés dans leur dossier.

RÉSULTATS : Ont été inclus dans l'étude 145 patients âgés et 82 patients plus jeunes ayant un diagnostic de MII à leur congé, ainsi que 135 patients âgés ayant un autre diagnostic de maladie intestinale à leur congé. Les patients âgés atteints d'une MII étaient moins susceptibles d'avoir une maladie de Crohn (MC) iléocolique (21,4 % par rapport à 50,9 %; P=0,001), plus susceptibles de se faire prescrire des 5-aminosalicylates (61 % par rapport à 43 %; P=0,04) et moins susceptibles de se faire prescrire des produits biologiques (6 % par rapport à 21 %; P=0,016) ou des immunomodulateurs (21 % par rapport à 42 %; P=0,01). La sensibilité, la spécificité et la valeur prédictive positive d'un seul code de CIM pour la MC s'élevaient à 98 %, 96 % et 94 %, respectivement, et, pour la colite ulcéreuse (CU), à 98 %, 92 % et 70 %, respectivement.

CONCLUSIONS : La conduite thérapeutique chez les patients âgés était différente de celle adoptée chez les patients plus jeunes atteints d'une MII, même si la maladie était assez grave pour exiger une hospitalisation. Bien qu'il soit moins précis en cas de CU, un seul code de CIM-10 suffisait pour distinguer les patients âgés hospitalisés à cause d'une MC et d'une CU.

Inflammatory bowel disease (IBD), including Crohn disease (CD) and ulcerative colitis (UC), are increasing in prevalence in both developed and developing countries around the world. Because IBD has minimal impact on mortality and, as populations are aging, the numbers of elderly individuals living with IBD are increasing (1). In addition, several studies have reported increasing incidence of IBD in the elderly (1-3). Hence, it is important to improve our knowledge base regarding the epidemiology and management of elderly individuals with IBD.

There may be differences in how the elderly and younger patients with IBD manifest with their disease, and how they are managed (4). Furthermore, elderly patients may present with symptoms that can resemble other common gastrointestinal conditions common to the elderly such as diverticulitis, ischemic colitis and infective colitis. More information is required to discern just how elderly and younger patients differ in terms of disease phenotype and treatment, particularly among hospitalized patients. There are limited comparative data regarding the clinical characteristics and current management in routine clinical practices between hospitalized elderly and young patients with confirmed IBD.

International Classification of Diseases (ICD) codes, the primary diagnostic coding system used in hospital settings, are used to identify individuals with chronic diseases and to generate epidemiological data. Very few studies have examined the accuracy of using these diagnostic codes to identify the elderly IBD population, which is important to assess specifically because of the other common gastrointestinal

[1]*Internal Medicine, University of Manitoba;* [2]*University of Manitoba IBD Clinical and Research Centre;* [3]*CancerCare Manitoba, Department of Epidemiology and Cancer Registry;* [4]*Community Health Sciences, University of Manitoba, Winnipeg, Manitoba*

Correspondence: Dr Harminder Singh, Section of Gastroenterology, University of Manitoba, 805-715 McDermot Avenue, Winnipeg, Manitoba R3E 3P4. e-mail harminder.singh@umanitoba.ca

conditions in the elderly. Some studies exploring the epidemiology of IBD in elderly populations are undertaken using health care administrative databases; hence, it is important to determine how robust administrative data definitions are in these populations.

The objectives of the current study were to compare the clinical course, presentation and management of elderly and younger individuals hospitalized with IBD, and to determine the accuracy of hospital discharge diagnosis codes for IBD in the elderly.

METHODS

The present study was approved by the University of Manitoba Health Research Ethics Board (Winnipeg, Manitoba). The present study was performed in Winnipeg (Manitoba), an ethnically diverse city located in the centre of Canada with a metropolitan area containing approximately 780,000 individuals. Manitoba Health is the government agency funding universal health care in the province. The Winnipeg Regional Health Authority (WRHA) administers all six hospitals in the city of Winnipeg. Two of the hospitals (Health Sciences Centre and St Boniface Hospital) are tertiary academic hospitals situated close to the city core, while the four other hospitals (Grace Hospital, Victoria General Hospital, Seven Oaks General Hospital and Concordia Hospital) are community hospitals. All six hospitals submit patient discharge information to the WRHA. The WRHA maintains a Decision Support System (DSS), which includes hospital discharge abstracts containing patient identification, dates of admission and discharge, admitting attending physician, up to 25 diagnoses (coded using ICD, 10th Revision codes, with Canadian Enhancements [ICD-10-CA]) – one for their primary [ie, most responsible] diagnosis and up to 24 additional ICD-10-CA codes) and 20 procedures (coded with Canadian Classification of Interventions codes) performed during the hospitalization. This DSS is used to aid in the planning of the services to the city.

Sample selection

Patients were identified for the present study from those admitted to any of the six hospitals in Winnipeg for at least one overnight stay between April 1, 2007 and March 31, 2012. All elderly (age >65 years of age) patients discharged with a primary diagnosis of IBD (identified from the DSS using ICD-10-CA codes K51 [UC] and K50 [CD]) were included. An approximately 20% random sample of each of the following groups were also included: elderly patients with secondary diagnosis of IBD, young (between 19 and 50 years of age) patients with a primary discharge diagnosis of IBD, and elderly patients with a primary discharge diagnosis of other colonic conditions (K52.3 [indeterminate colitis], K52.9 [noninfective gastroenteritis and colitis, unspecified], K55 [vascular disorders of intestine], K57.3 [diverticular disease of colon NOS], A04.7 [enterocolitis due to *Clostridium difficile*] or A09.9 [gastroenteritis and colitis of unspecified origin]).

Medical records were reviewed and the diagnosis of IBD was determined from review of the index hospitalization, subsequent hospitalization(s), ambulatory care visits with gastroenterologists and/or primary care physicians (including by reviewing the physician clinic records maintained separately from hospital records for patients with unclear diagnosis from hospital records), endoscopy and pathology reports. Follow-up was reviewed to August 2013. For individuals with a definite diagnosis of IBD, information regarding duration of disease, symptoms at presentation, extent of disease, extraintestinal manifestations (arthritis, iritis/uveitis, primary sclerosing cholangitis), IBD comorbidities (abdominal infections, fractures, infections associated with immunomodulator medications, venous thrombosis), hospital length of stay, use of IBD medications (5-aminosalicylates, azathioprine, 6-mercaptopurine, methotrexate, infliximab, adalimumab, cyclosporine, tacrolimus) before and during hospitalization, surgery (including type of surgery before and during hospitalization) and complications during hospitalization was extracted. The extent of disease was ascertained from the admission notes, endoscopy, pathology and/or medical imaging report(s). Medication usage before hospitalization was defined as any record of the patient ever being on the respective medication, including failed medical therapy and past hospitalizations.

Analysis

The accuracy of hospital discharge diagnosis codes for IBD in the elderly was assessed using the hospitalizations for the elderly with and without IBD diagnosis; each hospital admission was considered separately in the present analysis. For comparison of clinical features and management of IBD between the young and elderly with IBD, the first hospital admission during the study time period for those identified with the primary discharge diagnosis of IBD and documented IBD exacerbation on chart review were included. Individuals with mechanical bowel obstructions as determined from the discharge abstracts, medical imaging, endoscopy results and treatment course (conservative management versus treatment with IBD medications) were excluded.

Standard descriptive analysis was undertaken. Categorical data were compared (characteristics of younger and older individuals admitted with IBD) using Fisher's exact test. Logistic regression analyses were performed to assess potential predictors of use of an immunomodulator and/or biological agent in hospital (and before hospitalization in separate models) including age, sex, extent of disease, duration of disease, CD versus UC, comorbidities categorized using the Charlson comorbidity index (CCI), socioeconomic status (SES), admission at tertiary care versus community hospitals, and presence or absence of obstructive symptoms. The SES was determined using a previously validated index (Socioeconomic Factor Index), which is based on the characteristics of the neighbourhood of residence determined from Statistics Canada census data (5). A higher Socioeconomic Factor Index score indicates a more adverse SES. Parametric and nonparametric tests, as appropriate, were used for continuous data. The sensitivity, specificity, positive predictive value (PPV) and negative predictive value (NPV) of the IBD codes (K51 and K50) in the elderly patients were determined. Pearson correlation coefficient was used to assess correlation between primary hospital discharge abstract diagnosis (DAD) of IBD with IBD exacerbation and/or related obstruction on record review among the elderly.

RESULTS

Of the 395 admissions included in the present study, 82 were for individuals 19 to 50 years of age (all with primary discharge diagnosis of IBD) and 313 were for patients >65 years of age. Of the 313 admissions for patients >65 years of age, 178 were for a diagnosis of IBD whereas 135 were in the non-IBD group. In total, the study included 225 patients with IBD (158 CD; 67 UC) who experienced 260 admissions. While all admissions (primary or secondary diagnosis in the DAD) among the elderly were used to assess for accuracy of IBD codes in the DAD (because those are used in administrative database studies), for comparison of clinical characteristics among the elderly and young IBD patients, 138 patients with IBD as the primary discharge diagnosis who were admitted for an IBD exacerbation as determined from the chart reviews were included.

Patient demographics and clinical presentation

The elderly IBD patients were more likely to present with decreased oral intake, while the younger IBD patients were more likely to have inflammatory arthritis as an extraintestinal manifestation of their IBD (Table 1). There was a trend toward family history of IBD being more common in the younger population. There was no significant difference in the duration of IBD before admission. None of the patients admitted with an IBD exacerbation (or intestinal obstruction) died during their hospitalization.

Clinical phenotype

The extent of disease could be determined for 123 CD patients (Table 2) and 53 UC patients (Table 3). Of those with CD, younger patients were significantly more likely to have ileocolonic CD, whereas the elderly patients had disease restricted to either their colon or ileum. Of the younger CD patients, 16% had perianal disease compared with 3% of

TABLE 1
Patient demographics and characteristics of individuals admitted with an inflammatory bowel disease (IBD) exacerbation and IBD primary discharge diagnosis

	Age, years		
Characteristic	>65 (n=62)	19–50 (n=76)	P
Male sex, %	37	45	0.39
Age, years, median (IQR)	73 (69–78)	31 (25–41)	
Crohn disease subjects, %	63	64	0.86
Duration of IBD before admission, years, median	6	7	0.17
IQR	1–29	0–14	
Range	0–59	0–30	
Hospital stay, nights, median (IQR)	6.5 (3–11)	6 (4–9)	0.58
Symptoms at presentation, %			
Abdominal pain	69	74	0.70
Decreased oral intake	45	26	0.03
Diarrhea	68	66	0.86
Failure to thrive	27	28	1
Rectal bleeding	39	47	0.39
Vomiting	31	29	0.85
Family history of IBD, %	15	25	0.14
Extraintestinal manifestations, %			
Arthritis	0	8	0.033
Iritis or uveitis	0	3	0.50
Sclerosing cholangitis	0	3	0.50
IBD complications, %			
Abdominal infections	2	0	0.45
Infections due to immunosuppressant medications	2	0	0.45
Fractures	0	0	
Venous thrombosis	5	1	0.33
Charlson comorbidity index score, %			
0	73	96	<0.001
1	11	3	
2	5	0.0	
≥3	11	1	
Discharge disposition, %			
Home	98.4	98.7	1
Nursing home	1.6	0	
Other hospital	0	1.3	
Death	0	0	

IQR Interquartile range

the >65 years age group (P<0.05). Of the elderly patients with IBD listed as primary discharge diagnosis, 16% presented with an obstruction comapred with 7% of younger patients (P=0.10). Extent of disease was not specified in the charts for 35 CD patients (28 elderly, seven nonelderly) and in 14 UC patients (10 elderly, four nonelderly).

Medication use in IBD patients

The elderly IBD patients were more likely to be prescribed 5-aminosalicylates in hospital or before admission, while the younger IBD patients were more likely to be prescribed biological therapy during hospitalization or before hospitalization (Table 4). Immunomodulators (azathioprine, 6-mercaptopurine, methotrexate) were commonly used in the younger population both before and during hospitalization. Corticosteroid use while in hospital was almost identical, with 77% of the older and 74% of the younger patients receiving them. Immunosuppressants (tacrolimus, cyclosporine) also showed similar usage between the two age groups.

TABLE 2
Extent of disease for Crohn disease patients admitted to hospital

	Age, years			
Disease extent	>65	19–50	Total, n	P
Colonic	31 (41.3)	10 (20.8)	41	0.004
Ileocolonic	22 (29.3)	28 (58.3)	50	
Ileum only	22 (29.3)	10 (20.8)	32	
Total	75	48	123	

Data presented as n (%) unless otherwise indicated

TABLE 3
Extent of disease for ulcerative colitis patients admitted to hospital

	Age, years			
Disease extent	>65	19–50	Total, n	P
Proctosigmoiditis*	6 (20)	2 (8.7)	8	0.24
Left-sided colitis†	4 (13.3)	7 (30.4)	11	
Extensive colitis‡	20 (66.7)	14 (60.9)	34	
Total	30	23	53	

*Data presented as n (%) unless otherwise indicated. *Rectum and/or sigmoid colon only (<30 cm); †Extending to involve descending colon and/or splenic flexure; ‡Extending to involve transverse colon, hepatic flexure, ascending colon and/or cecum*

Comorbidity and IBD

The CCI score and comorbidity burden was calculated from ICD-10-CA codes in the DAD at the index admission. The older cohort was more likely to suffer from chronic pulmonary problems (6.5% in >65 years versus 0% in <50 years; P<0.05), diabetes with complications (8.1% in >65 years versus 0% in <50 years; P<0.05) and have a higher CCI score (Table 1).

Predictors of use of biological and/or immunomodulatory medications

In univariate analyses, the significant predictors of use of biological and/or immunomodulatory medications during the hospitalization were use of these medications before admission and younger IBD patient group. After adjusting for CCI score, these predictors continued to be significant predictors (Table 5). In analysis limited to the elderly with IBD, only previous use of these medications was a significant predictor of use of these medications during hospitalization (Table 5). Use of biologicals was limited to those with a CCI score of 0 and with young versus elderly IBD as the only significant predictor (OR 5.35 [95% CI 1.1.3 to 25.24]). The only predictor of use of biologicals and/or immunomodulatory medications before hospitalization was age (young versus elderly IBD: OR 2.47 [95% CI 1.21 to 5.05], adjusted for CCI score)

Rates of surgery

During the hospital stay, the younger IBD patients were more likely to undergo surgery (37%) compared with the elderly population (18%) (P<0.01). There was no difference in the rates of IBD surgery before admission between the two age groups (34% of elderly versus 33% of younger IBD patients [P=1]).

Accuracy of IBD ICD-10-CA codes

The ICD-10-CA codes for IBD in the elderly had an overall sensitivity of 100%, a specificity of 81%, a PPV of 87% and an NPV of 100%. The sensitivity and specificity of ICD-10 codes were 98% and 96%, respectively, in CD patients. The PPV was 94% and the NPV was 99% in elderly CD patients. The sensitivity and specificity were 98% and 92% in UC patients, respectively. The PPV in UC patients was 70% while the NPV was 99.6%. There was no single predominant diagnosis among the miscoded cases.

TABLE 4
Medication use in hospital and before admission in patients admitted with an inflammatory bowel disease (IBD) exacerbation and IBD primary discharge diagnosis

	Medication use in hospital			Medication use before admission		
	Age, years			Age, years		
Medication	>65 (n=62)	19–50 (n=76)	P	>65 (n=62)	19–50 (n=76)	P
Budesonide	2	1	1	3	4	1
Corticosteroids	77	74	0.69	53	62	0.39
Biologicals*	3	13	0.06	6	21	0.02
Mesalamine	53	29	0.005	61	43	0.04
Immunomodulators†	13	24	0.13	27	43	0.07
Immunosupressants‡	3	7	0.46	2	4	0.63
Biologicals and/or immunomodulators	16	33	0.03	29	47	0.03

*Data presented as % unless otherwise indicated. *Includes adalimumab and infliximab; †Includes azathioprine, 6-mercaptopurine and methotrexate; ‡Includes cyclosporine and tacrolimus*

TABLE 5
Predictors of use of biological and/or immunomodulatory medications during hospitalization

	OR (95% CI)	
	Univariate	Adjusted for CCI
All inflammatory bowel disease patients (n=161)		
Young vs elderly	3.21 (1.46–7.02)	2.82 (1.23–6.44)
Use before hospitalization vs not	9.30 (4.05–21.35)	9.07 (3.93–20.93)
Elderly inflammatory bowel disease patients (n=79)		
Use before hospitalization vs not	23.32 (4.42–122.94)	24.85 (4.59–134.60)

CCI Charlson comorbidity index; vs Versus

TABLE 6
Correlation of discharge abstract inflammatory bowel disease (IBD) primary diagnosis with IBD exacerbation/ bowel obstruction on chart review in the >65 years of age group

	Admission exacerbation/obstruction		
Diagnosis	No	Yes	Total, n
IBD secondary diagnosis	51 (79.7)	13 (20.3)	64
IBD primary diagnosis	8 (10.1)	71 (89.9)	79
Total, n	59	84	143

Data presented as n (%) unless otherwise indicated. Pearson correlation = 0.70

Use of primary discharge diagnosis of IBD in hospital records to detect admissions with IBD exacerbation and/or related intestinal obstruction

Ninety percent of elderly patients coded with IBD as the primary discharge diagnosis (n=79) were admitted for an IBD exacerbation and/or related obstruction, with a correlation of 0.70 between the two variables. Sixty-five percent (52 of 79) of elderly patients were admitted for an exacerbation without an obstruction, 13% for an exacerbation with an obstruction and 11% for an obstruction alone. Five UC and three CD admissions were coded with IBD as the primary discharge diagnosis, but the patient was not admitted due to an IBD exacerbation or bowel obstruction (Table 6). Of the 13 elderly patients with an IBD exacerbation and/or obstruction who were not coded with IBD as the primary discharge diagnosis, 12 were coded with a bowel obstruction (four mechanical) and one with an anal abscess as the primary discharge diagnoses. All 13 had CD and were not included in the comparison of clinical characteristics of young and elderly hospitalized with IBD because records of young patients with IBD as a secondary diagnosis in DAD were reviewed. Of note, the primary discharge diagnosis of IBD plus obstruction coding (among those with IBD as secondary code) identified all elderly patients with IBD admitted with IBD exacerbation and/or intestinal obstruction. All but one case with an IBD exacerbation (anal abscess) was coded with IBD as the primary discharge diagnosis.

DISCUSSION

The present study suggests that hospitalized elderly patients with IBD are less likely to receive immunomodulators and/or biologicals during the hospitalization or before admission compared with younger IBD patients. While there may be a lower threshold to admit the elderly than younger persons with IBD, we were not able to identify an obvious difference in disease severity between elderly and young patients hospitalized with IBD, as indicated by comparable rates of surgery before admission, and corticosteroid use before and during hospitalization. Hence, the difference in utilization of immunomodulators and/or biologicals during hospitalizations may not be due to the differences in severity of IBD in elderly versus younger individuals. Furthermore, the use of a single ICD-10-CA IBD code in the hospitalization discharge abstracts is valid and accurate in identifying elderly (>65 years of age) patients with IBD, although more so for CD than UC. The primary discharge diagnosis IBD coding identifies most elderly IBD patients admitted with IBD exacerbation. Cases with IBD exacerbation and/or obstruction can be identified by including those with IBD listed as primary discharge diagnosis; and those with IBD listed as a secondary diagnosis, concomitantly with obstruction as the primary discharge diagnosis.

Although there are limited data comparing hospitalized elderly and young IBD patients, the results of our comparisons of clinical characteristics of hospitalized IBD patients between the young and the elderly are consistent with the differences found in all (hospitalized and nonhospitalized) IBD patients in other studies. For example, previous studies, including those by Gower-Rousseau et al (1), Lakatos et al (2) and Quezada et al (6), have reported that elderly patients with IBD are less likely to have a family history of IBD. This trend may be due to older patients with IBD being less likely to have a strong genetic component to their disease compared with those diagnosed at a younger age. Environmental influence may play a larger role in those diagnosed with IBD at an older age. In our study, younger hospitalized CD patients were more likely to have ileocolonic disease compared with the older population, who was more likely to have their disease confined to the colon or terminal ileum, similar to that reported in all CD patients by Lakatos et al (1) and Quezada et al (6). Ileocolonic disease has been reported to be even more common in pediatric patients with IBD (7). However, we did not find a difference in the disease extent of UC in hospitalized elderly versus younger UC patients, in contrast to reports of more left-sided colitis in all (hospitalized and nonhospitalized) elderly UC patients (1). This may reflect similar rates of admission among elderly and younger UC patients with UC localized to different parts of the bowel.

We report a higher use of 5-aminosalicylates and lower use (both before admission and during hospitalization) of immunomodulators and/or biologicals among hospitalized elderly IBD patients. In contrast to previous studies, our study was limited to hospitalized patients and, hence, we believe this difference in utilization is not solely due to lower disease severity in the elderly because all patients included in the present study had disease sufficiently severe to warrant admission to hospital. The complications with immunomodulator and/or biological agent use have been reported to be more in the elderly, but complications of alternative forms of treatment, such as surgery and corticosteroids, are also higher in the elderly (8,9). The response rate to the immunomodulator and/or biological agent use in the elderly has been reported to be similar to the younger individuals in observational studies (8,10). Thus, cautious, but increased and earlier use of these more effective medications in the elderly should be considered.

Similarly, the lower rates of surgery among the hospitalized IBD elderly patients in our study may reflect clinician and/or patient preference to nonsurgical treatment. We recently reported that elderly IBD patients undergoing surgery had an increased risk for death compared with younger individuals (9). However, the prognosis of elderly UC patients undergoing emergency surgery is quite poor compared with those undergoing elective surgery (11). Thus, we recommend that surgical management in otherwise healthy elderly IBD patients should be promptly performed when necessary because increased morbidity and mortality can result from untreated or undertreated IBD exacerbations.

In our study, miscoded admissions were due to a wide variety of reasons including diverticulitis, colorectal cancer and surgical complications; however, there was not any particular gastroenterological condition for which ICD-10-CA miscoding was common. The specificity and PPV of the ICD-10-CA codes was higher in elderly CD patients than elderly UC patients, while the sensitivity and NPV were almost identical between the two groups. Thus, the single diagnostic ICD-10-CA code in discharge abstracts is sufficient to identify elderly CD patients, and a single code remains adequate to identify elderly UC patients. This finding is in contrast to studies investigating the use of ICD codes in records of ambulatory care visits. Hou et al (12) determined the accuracy of ICD-9 diagnostic codes for identifying IBD in the American Veterans Affairs health care system, using both inpatient and ambulatory care visits at two Veterans Affairs hospitals. They compared the ICD-9 diagnosis with the clinical diagnosis based on manual chart review and concluded that single ICD-9 codes are inadequate to define IBD patients. Similarly, Benchimol et al (13) concluded that the addition of a pharmacy claim was necessary to improve accuracy and supplement ICD codes, when including ambulatory care visits. Overall, there has been very little research that has examined the accuracy of ICD codes in elderly IBD patients and/or only in the hospital discharge abstracts. Our study suggests that ICD-10 codes can be very accurate in identifying the elderly hospitalized IBD population when used by well-trained coders. Our sample population was selected by ICD-10 codes for IBD (K50 or K51) and other gastrointestinal disorders theoretically more likely to be miscoded as IBD. Hence, the PPV and specificity will likely be higher when assessing value of IBD ICD codes in the general population and not just limited to those with disorders, which are more likely to be miscoded as IBD.

The results of the current study should be interpreted in the context of its strengths and limitations. We evaluated a citywide practice and, hence, the results should be more applicable to usual clinical practice than the studies from specialized centres. We evaluated ICD codes in hospitalized patients and the results may not be applicable to outpatient physician claims datasets. However, other gastrointestinal conditions (eg, ischemic colitis), which are more common in the elderly and have a risk for miscoding as IBD in the discharge abstracts, are more likely to lead to hospitalization and, therefore, may be less of an issue in outpatient settings. Hence, administrative data coding accuracy among young and elderly IBD patients in outpatient settings is likely to be similar. We did not evaluate clinical presentation or management according to age of onset/diagnosis of IBD. We did not assess initiation of immunomodulators and/or biologicals immediately postdischarge from the hospitals, or clinical course after discharge from the hospitals.

CONCLUSION

Our study suggests that immunomodulators and biological agents and surgery are used less often in elderly hospitalized IBD patients. Additional studies should assess approaches to increase the use of immunomodulators and/or biological agents in elderly patients with IBD. Single ICD-10 CA codes can be used to identify elderly hospitalized individuals with IBD, in particular those admitted with IBD exacerbations.

DISCLOSURES: The authors have no financial disclosures or conflicts of interest to declare.

REFERENCES

1. Lakatos P, David G, Pandur T. IBD in the elderly population: Results from a population-based study in Western Hungary, 1977-2008. J Crohns Colitis 2011;5:5-13.
2. Gower-Rousseau C, Vasseur F, Fumery M, et al. Epidemiology of inflammatory bowel diseases: New insights from a French population-based registry (EPIMAD). Dig Liver Dis 2013;45:89-94.
3. Thia KT, Loftus E V, Sandborn WJ, et al. An update on the epidemiology of inflammatory bowel disease in Asia. Am J Gastroenterol 2008;103:3167-82.
4. Gisbert JP, Chaparro M. Systematic review with meta-analysis: Inflammatory bowel disease in the elderly. Aliment Pharmacol Ther 2014;39:459-77.
5. Chateau D, Metge C, Prior H, et al. Learning from the census: The socio-economic factor index (SEFI) and health outcomes in Manitoba. Can J Public Health 2012;103.
6. Quezada SM, Steinberger EK, Cross RK. Association of age at diagnosis and Crohn's disease phenotype. Age Ageing 2013;42:102-6.
7. Israeli E, Ryan JD, Shafer L-A, et al. Younger age at diagnosis is associated with panenteric, but not more aggressive, Crohn's disease. Clin Gastroenterol Hepatol 2014;12:72-9.
8. Ljung T. Infliximab in inflammatory bowel disease: Clinical outcome in a population based cohort from Stockholm County. Gut 2004;53:849-53.
9. Bernstein CN, Nugent Z, Targownik LE, Singh H, Lix LM. Predictors and risks for death in a population-based study of persons with IBD in Manitoba. Gut 2014 (Epub ahead of print).
10. Moleski SM, Lindenmeyer CC, Kozuch PL. The use of infliximab in older inflammatory bowel disease patients. Gastroenterology 2011;142:S-361.
11. Ikeuchi H, Uchino M, Matsuoka H, et al. Prognosis following emergency surgery for ulcerative colitis in elderly patients. Surg Today 2014;44:39-43.
12. Hou JK, Tan M, Stidham RW, et al. Accuracy of diagnostic codes for identifying patients with ulcerative colitis and Crohn's disease in the Veterans Affairs health care system. Dig Dis Sci 2014:7-11.
13. Benchimol EI, Guttmann A, Mack DR, et al. Validation of international algorithms to identify adults with inflammatory bowel disease in health administrative data from Ontario, Canada. J Clin Epidemiol 2014;67:887-96.

Permissions

All chapters in this book were first published in CJGH, by Hindawi Publishing Corporation; hereby published with permission under the Creative Commons Attribution License or equivalent. Every chapter published in this book has been scrutinized by our experts. Their significance has been extensively debated. The topics covered herein carry significant findings which will fuel the growth of the discipline. They may even be implemented as practical applications or may be referred to as a beginning point for another development.

The contributors of this book come from diverse backgrounds, making this book a truly international effort. This book will bring forth new frontiers with its revolutionizing research information and detailed analysis of the nascent developments around the world.

We would like to thank all the contributing authors for lending their expertise to make the book truly unique. They have played a crucial role in the development of this book. Without their invaluable contributions this book wouldn't have been possible. They have made vital efforts to compile up to date information on the varied aspects of this subject to make this book a valuable addition to the collection of many professionals and students.

This book was conceptualized with the vision of imparting up-to-date information and advanced data in this field. To ensure the same, a matchless editorial board was set up. Every individual on the board went through rigorous rounds of assessment to prove their worth. After which they invested a large part of their time researching and compiling the most relevant data for our readers.

The editorial board has been involved in producing this book since its inception. They have spent rigorous hours researching and exploring the diverse topics which have resulted in the successful publishing of this book. They have passed on their knowledge of decades through this book. To expedite this challenging task, the publisher supported the team at every step. A small team of assistant editors was also appointed to further simplify the editing procedure and attain best results for the readers.

Apart from the editorial board, the designing team has also invested a significant amount of their time in understanding the subject and creating the most relevant covers. They scrutinized every image to scout for the most suitable representation of the subject and create an appropriate cover for the book.

The publishing team has been an ardent support to the editorial, designing and production team. Their endless efforts to recruit the best for this project, has resulted in the accomplishment of this book. They are a veteran in the field of academics and their pool of knowledge is as vast as their experience in printing. Their expertise and guidance has proved useful at every step. Their uncompromising quality standards have made this book an exceptional effort. Their encouragement from time to time has been an inspiration for everyone.

The publisher and the editorial board hope that this book will prove to be a valuable piece of knowledge for researchers, students, practitioners and scholars across the globe.

List of Contributors

Emily K Wright MD
Department of Medicine, Division of Gastroenterology, University of Calgary, Foothills Medical Centre, Calgary, Alberta
Department of Gastroenterology, St Vincent's Hospital and The University of Melbourne, Melbourne, Victoria, Australia

Kerri L Novak MD, Cathy Lu MD, Remo Panaccione MD and Subrata Ghosh MD FCAHS
Department of Medicine, Division of Gastroenterology, University of Calgary, Foothills Medical Centre, Calgary, Alberta

Stephanie R Wilson MD
Department of Medicine, Division of Gastroenterology, University of Calgary, Foothills Medical Centre, Calgary, Alberta
Department of Diagnostic Imaging, University of Calgary, Foothills Medical Centre, Calgary, Alberta

Matthew D Sadler MD FRCPC and Pam Crotty MSc
Liver Unit, Division of Gastroenterology and Hepatology

Linda Fatovich MN and Harvey R Rabin MD FRCPC
Adult Cystic Fibrosis Clinic of Calgary and Southern Alberta, Department of Medicine

Stephanie Wilson MD FRCPC
Department of Radiology and Diagnostic Imaging

Robert P Myers MD MSc FRCPC
Liver Unit, Division of Gastroenterology and Hepatology
Department of Community Health Sciences, University of Calgary, Calgary, Alberta

M Eugenia Socías MD
British Columbia Centre for Excellence in HIV/AIDS, St Paul's Hospital
Interdisciplinary Studies Graduate Program

Kate Shannon PhD MPH
British Columbia Centre for Excellence in HIV/AIDS, St Paul's Hospital
Department of Medicine
School of Population and Public Health, University of British Columbia, Vancouver

Julio S Montaner MD, Silvia Guillemi MD and Kathleen Deering PhD
British Columbia Centre for Excellence in HIV/AIDS, St Paul's Hospital
Department of Medicine

Sabina Dobrer MA and Paul Nguyen PhD
British Columbia Centre for Excellence in HIV/AIDS, St Paul's Hospital

Shira Goldenberg PhD
British Columbia Centre for Excellence in HIV/AIDS, St Paul's Hospital
Faculty of Health Sciences, Simon Fraser University, Burnaby, British Columbia

Peter Stepaniuk BScHons
Internal Medicine, University of Manitoba

Charles N Bernstein MD and Laura E Targownik MD MSHS
Internal Medicine, University of Manitoba
University of Manitoba IBD Clinical and Research Centre

Harminder Singh MD MPH
Internal Medicine, University of Manitoba
University of Manitoba IBD Clinical and Research Centre
Community Health Sciences, University of Manitoba, Winnipeg, Manitoba

Khalid Mumtaz MD MSc
Wexner Medical Center, Gastroenterology, Hepatology and Nutrition Division, The Ohio State University, Columbus, Ohio, USA

Nabiha Faisal MD, Max Marquez MD, Alicia Healey CNP, Leslie B Lilly MD MSc and Eberhard L Renner MD FRCP
Liver Transplant Program/Multi-Organ Transplant Program, University Health Network, Toronto General Hospital, University of Toronto, Toronto, Ontario

Maria Ines Pinto-Sanchez MD MSc, Elena F Verdu MD PhD, Paul Moayyedi MD BSc MB ChB PhD MPH and Premysl Bercik MD
Department of Medicine, Farncombe Family Digestive Research Institute, McMaster University

Maria C Gordillo CPA CGA
Kaizen Accounting, Hamilton, Ontario

Julio C Bai MD
Hospital de Gastroenterología Dr C B Udaondo, Buenos Aires, Argentina

Stephen Birch BA MSc DPhil
Centre for Health Economics and Policy Analysis, Department of Clinical Epidemiology & Biostatistics, McMaster University, Hamilton, Ontario

Dorota Koziel PhD and Piotr Lewitowicz MD PhD
Faculty of Health Sciences, Jan Kochanowski University, Kielce

Stanislaw Gluszek Prof and Jaroslaw Matykiewicz Md PhD
Clinical General, Oncological and Endocrinological Surgery, Regional Hospital
Faculty of Health Sciences, Jan Kochanowski University, Kielce

Zuzanna Drozdzak MA
Centre for Evaluation and Analysis of Public Policies, Jagiellonian University, Cracow, Poland

Travis B Murdoch MD MSc and Remo Panaccione MD
Inflammatory Bowel Disease Clinic, Division of Gastroenterology, University of Calgary, Calgary, Alberta

Sarah O'Donnell MD and Mark S Silverberg MD PhD
Zane Cohen Centre for Digestive Diseases, Mount Sinai Hospital, Toronto, Ontario

Neel Malhotra MD and Nilesh Chande MD FRCPC
Department of Medicine and Division of Gastroenterology, Western University/London Health Sciences Centre, London, Ontario

Johan Van Limbergen MD FRCPCH PhD
Division of Gastroenterology & Nutrition, Department of Pediatrics, IWK Health Centre
Department of Microbiology and Immunology, Dalhousie University, Halifax, Nova Scotia
Division of Gastroenterology, Faculty of Medicine, Dalhousie University, Halifax, Nova Scotia

Jennifer Haskett BSNH, PDt
Division of Gastroenterology & Nutrition, Department of Pediatrics, IWK Health Centre

Anne M Griffiths MD and Thomas D Walters MD
Division of Pediatric Gastoenterology, Hepatology and Nutrition, Hospital for Sick Children, Toronto, Ontario

Jeff Critch MD MSc and Elizabeth Sheppard RN BN
Division of Pediatric Gastroenterology, Janeway Child Health Centre, St John's, Newfoundland & Labrador

Hien Huynh MBBS FRCPC and Cheryl Kluthe RN BScN
Division of Pediatric Gastroenterology, Stollery Children's Hospital, University of Alberta, Edmonton, Alberta

Najma Ahmed MD
Division of Pediatric Gastroenterology and Nutrition, Department of Pediatrics, Montreal Children's Hospital, McGill University, Montreal, Quebec

Jennifer C deBruyn MD
Division of Gastroenterology and Nutrition, Department of Paediatrics, University of Calgary, Calgary, Alberta

Robert Issenman MD FRCP
Division of Pediatric Gastroenterology and Nutrition, McMaster Children's Hospital, Hamilton, Ontario

Wael El-Matary MD
Section of Pediatric Gastroenterology, Department of Pediatrics, College of Medicine, Faculty of Health Sciences, University of Manitoba, Winnipeg, Manitoba

Marie-Eve Roy RD
Department of Pediatric Gastroenterology, Hepatology and Nutrition, Centre Hospitalier Universitaire Sainte-Justine, Montréal, Québec

Wallace V Crandall MD
Division of Pediatric Gastroenterology, Hepatology and Nutrition, Nationwide Children's Hospital and The Ohio State University, Columbus, Ohio

Stan Cohen MD
Children's Center for Digestive Health Care, Children's Healthcare of Atlanta, Atlanta, Georgia, USA

Frank M Ruemmele MD PhD
Université Sorbonne Paris Cité, Université Paris Descartes, INSERM U989, Service de Gastroentérologie Pédiatrique, Hôpital Necker Enfants Malades, Paris, France

Arie Levine MD
Pediatric Gastroenterology and Nutrition Unit, Wolfson Medical Center, Sackler School of Medicine, Tel-Aviv University, Tel-Aviv, Israel
Anthony R Otley MD MSc FRCPC
Division of Gastroenterology & Nutrition, Department of Pediatrics, IWK Health Centre
Division of Gastroenterology, Faculty of Medicine, Dalhousie University, Halifax, Nova Scotia

Christopher Ma MD, Chad J Evaschesen MD FRCPC, Grenvil Gracias MD, Vivian W Huang MD FRCPC, Darryl K Fedorak BSc, Karen I Kroeker MD FRCPC, Levinus A Dieleman MD PhD, Brendan P Halloran MD FRCPC and Richard N Fedorak MD FRCPC
Division of Gastroenterology, University of Alberta, Edmonton, Alberta

Bashaar Alibrahim MD
Division of Internal Medicine, Department of Medicine, University of British Columbia, Vancouver, British Columbia
Department of Medicine, King Faisal Special Hospital and Research Centre

Mohammed I Aljasser MD FRCPC DABD,
Department of Medicine, King Saud Bin Abdulaziz University for Health Sciences, King Abdulaziz Medical City, Riyadh, Saudi Arabia

Baljinder Salh MB FRCPC
Division of Gastroenterology, Department of Medicine, University of British Columbia, Vancouver, British Columbia

Meaghan O'Brien MD FRCPC
Department of Internal Medicine, Upper River Valley Hospital, Waterville, New Brunswick

Adara Casselman BSc, Gerry Smart BSc MLT and Ainsley Gretchen BSc
Serology Section, Cadham Provincial Public Health Laboratory

Kelly Kaita MD FRCPC
Section of Hepatology, Department of Internal Medicine, College of Medicine, Faculty of Health Sciences, University of Manitoba

Kamran Kadkhoda PhD FCCM D(ABMM)
Serology Section, Cadham Provincial Public Health Laboratory
Department of Medical Microbiology & Infectious Diseases; and Department of Immunology, College of Medicine, Faculty of Health Sciences, University of Manitoba, Winnipeg, Manitoba

Horacio Vázquez MD, María de la Paz Temprano RD, Emilia Sugai MS, Stella M Scacchi MS, Cecilia Souza MD, Daniel Cisterna MS, Edgardo Smecuol MD, María Laura Moreno MD, Gabriela Longarini MD, Roberto Mazure MD, Andrea González RD, Eduardo Mauriño MD, Julio C Bai MD and María A Bartellini MS
Small Bowel Section, Department of Medicine, Hospital de Gastroenterología C Bonorino Udaondo. Buenos Aires, Argentina

Elena F Verdú MD
Farncombe Family Digestive Health Research Institute, McMaster University, Hamilton, Ontario

Matthew B Lanktree MD PhD and Mark A Crowther MD
Department of Medicine

Bruce B Lanktree MD

Guillaume Paré MD MSc, John S Waye PhD and Bekim Sadikovic PhD
Department of Pathology and Molecular Medicine, McMaster University, Hamilton, Ontario

Alan Hoi Lun Yau MD, Terry Lee PhD, Alnoor Ramji MD FRCPC and Hin Hin Ko MD FRCPC
Department of Medicine, Division of Gastroenterology, University of British Columbia, Vancouver, British Columbia

Victoria Y Kan BSc MHA, Vladimir Marquez Azalgara MD MSc, Jo-Ann E Ford MScN, WC Peter Kwan MD, Siegfried R Erb MD and Eric M Yoshida MD MHSc
Division of Gastroenterology, Vancouver General Hospital, Vancouver, British Columbia

Mamatha Bhat MD MSc, Peter Ghali MD MSc, Philip Wong MD MSc and Marc Deschênes MD
Division of Gastroenterology, McGill University Health Centre

Jonathan M Wyse MD MSc and Nir Hilzenrat MD
Division of Gastroenterology, Sir Mortimer B Davis-Jewish General Hospital

Erica Moodie PhD
Department of Epidemiology, Biostatistics, & Occupational Health, McGill University, Montreal, Quebec

Vivian Wai-Mei Huang MSc MD FRCPC, Neil Dhami BSc, Darryl Fedorak BSc, Karen Ivy Kroeker MSc MD FRCPC, Brendan Phillip Halloran MD FRCPC, Levinus Albert Dieleman MD PhD and Richard Neil Fedorak MD FRCPC
Division of Gastroenterology, Department of Medicine

Connie Prosser PhD and Carol Shalapay MLT
Division of Medical Biochemistry, Department of Laboratory Medicine and Pathology, University of Alberta, Edmonton, Alberta

Omar Kherad MD MPH
Internal Medicine Department, La Tour Hospital and University of Geneva

Sophie Restellini MD
Department of specilaties of Internal Medicine, Divison of Gastroenterology, Geneva University Hospital, Geneva, Switzerland

Myriam Martel Bsc
Division of Gastroenterology

Alan N Barkun MD MSc
Department of Epidemiology and Biostatistics and Occupational Health, McGill University Health Centre, McGill University, Montreal, Quebec

Stephanie Coward MSc, Steven J Heitman MD, James Hubbard MSc, Marie-Claude Proulx MSc, Robert P Myers MD and Gilaad G Kaplan MD MPH
Department of Medicine
Community Health Sciences

Fiona Clement PhD
Community Health Sciences

Scott Zimmer BA
Alberta Health Services, Calgary, Alberta

Remo Panaccione MD, Cynthia Seow MD, Yvette Leung MD, Subrata Ghosh MD Mark Swain MD
Department of Medicine

Neel Datta MD
Surgery, University of Calgary, Calgary, Alberta

Korosh Khalili MD, Ravi Menezes PhD, Tae Kyoung Kim MD, Leyla Kochak Yazdi MD and Hyun-Jung Jang MD
Department of Medical Imaging

Suraj Sharma MD, Jordan Feld MD and Morris Sherman MD
Department of Gastroenterology, University of Toronto, Toronto, Ontario

Jesús K Yamamoto-Furusho MD PhD MSc
Inflammatory Bowel Disease Clinic, Department of Gastroenterology. Instituto Nacional de Ciencias Médicas y Nutrición Salvador Zubirán, Mexico City, Mexico
Pan American Crohns and Colitis Organisation (PANCCO)

Andrea Sarmiento-Aguilar MD
Inflammatory Bowel Disease Clinic, Department of Gastroenterology. Instituto Nacional de Ciencias Médicas y Nutrición Salvador Zubirán, Mexico City, Mexico

Mayur Brahmania MD FRCPC, Madison Young MD, Chetty Muthiah MD FRCPC, Alexandra Ilnyckyj MD FRCPC, Donald Duerksen MD FRCPC, Dana C Moffatt MD FRCPC
Department of Medicine, Division of Gastroenterology, University of Manitoba, St Boniface Hospital, Winnipeg, Manitoba

Noah Switzer MD and Chris de Gara MB MS FRCS FACS FFSTEd
Department of Surgery, University of Alberta, Edmonton

Elijah Dixon MD BSc MScEpi FRCSC FACS
Department of Surgery, University of Calgary, Calgary, Alberta

Jill Tinmouth MD PhD
Division of Gastroenterology, University of Toronto, Toronto, Ontario

Nori Bradley MD
Department of Surgery, University of British Columbia, Vancouver, British Columbia

Melina Vassiliou MD MEd FRCSC
Department of Surgery, McGill University, Montreal, Quebec

Steve Schwaitzberg MD FACS
Department of Surgery, Harvard Medical School, Boston, Massachusetts, USA

Anthony Gomes MD LMCC FRCSC
Division of Surgery, University of Alberta, Lethbridge, Alberta

James Ellsmere MD MSc FRCSC
Department of Surgery, Dalhousie University, Halifax, Nova Scotia

Sergio M Borgia MD MSc
William Osler Health System, Brampton Civic Hospital, Brampton
Division of Infectious Diseases, Department of Medicine, McMaster University, Hamilton, Ontario

Adenike Rowaiye MBA MSc
William Osler Health System, Brampton Civic Hospital, Brampton

Himanish Panda BHSc, Michael Curley MD, Michelle Buresi MD, Lynn Wilsack MSc, Christopher N Andrews MD MSc
Division of Gastroenterology

Philip Mitchell MD
Division of General Surgery, Cumming School of Medicine, University of Calgary, Calgary, Alberta

Gavin Low MBChB MRCS FRCR
Department of Radiology
University of Alberta, Edmonton, Alberta

Graeme JM Alexander MD FRCP
Division of Gastroenterology & Hepatology, Addenbrooke's Hospital, Cambridge University Hospitals NHS Foundation Trust, United Kingdom

David J Lomas MA MB BChir FRCP FRCR
Department of Radiology

Filipe S Cardoso MD, Pedro Fidalgo MD and Sean M Bagshaw MD MSc
Division of Critical Care Medicine

Constantine J Karvellas MD SM
Division of Critical Care Medicine
Division of Gastroenterology, Department of Medicine, Faculty of Medicine and Dentistry

Norman M Kneteman MD MSc
Division of Transplantation, Department of Surgery, University of Alberta

Glenda Meeberg RN MsN
Liver Transplant Program, Alberta Health Services, Edmonton, Alberta

Takafumi Sugimoto MD
Department of Gastroenterology, Kyoundo Hospital
Department of Gastroenterology, Graduate School of Medicine, the University of Tokyo

Yutaka Yamaji MD, Kosuke Sakitani MD, Yoshihiro Isomura MD, Shuntaro Yoshida MD, Atsuo Yamada MD, Yoshihiro Hirata MD and Kazuhiko Koike MD
Department of Gastroenterology, Graduate School of Medicine, the University of Tokyo

Keiji Ogura MD
Department of Gastroenterology, Tokyo Metropolitan Police Hospital

Makoto Okamoto MD
Department of Gastroenterology, JR Tokyo General Hospital, Tokyo, Japan

Michael J Diamant MD MSc
Departments of Medicine

Stephanie Coward MSc and Gilaad G Kaplan MD MPH
Departments of Medicine
Community Health Sciences

W Donald Buie MD, Anthony MacLean MD, Elijah Dixon MD MSc and Chad G Ball MD
Surgery, University of Calgary, Calgary, Alberta

Samuel Schaffer MD
Global Medical Affairs, Shire Canada, Toronto, Ontario

Manuela G Neuman PhD
In Vitro Drug Safety and Biotechnology
Department of Pharmacology and Toxicology, Faculty of Medicine, University of Toronto

Radu M Nanau BSc
In Vitro Drug Safety and Biotechnology

Lawrence B Cohen MD MSc
Division of Gastroenterology, Sunnybrook Health Sciences Centre, Department of Medicine, Faculty of Medicine, University of Toronto, Toronto, Ontario

Mohammed Aljawad MD
Department of Medicine, Western University, London, Ontario
Multiorgan Transplant Unit, King Fahad Specialist Hospital, Dammam, Saudi Arabia

Sanjeev Sirpal MD JD
Division of Gastroenterology, William Osler Health Centre, Brampton
Faculty of Medicine, University of Toronto, Toronto, Ontario

Eric M Yoshida MD
Division of Gastroenterology, University of British Columbia, Vancouver, British Columbia

Natasha Chandok MD MPH
Department of Medicine, Western University, London, Ontario
Division of Gastroenterology, William Osler Health Centre, Brampton

Vivian W Huang MSc MD FRCPC, Hsiu-Ju Chang MSc, Karen I Kroeker MSc MD FRCPC, Karen J Goodman PhD, Levinus A Dieleman MD PhD and Richard N Fedorak MD FRCPC
Division of Gastroenterology, Faculty of Medicine

Kathleen M Hegadoren RN PhD
Faculty of Nursing, University of Alberta, Edmonton, Alberta

Alan N Barkun MD MSc
Division of Gastroenterology
Division of Epidemiology, Biostatistics and Occupational Health, McGill University Health Centre, McGill University, Montreal, Quebec

Viviane Adam MSc and Myriam Martel BSc
Division of Gastroenterology

Khalid AlNaamani MD BSc
Division of Epidemiology, Biostatistics and Occupational Health, McGill University Health Centre, McGill University, Montreal, Quebec
Division of Gastroenterology, Hepatology and Liver Transplantation, The Armed Forces Hospital, Muscat, Oman

Peter L Moses MD
Division of Gastroenterology and Hepatology, University of Vermont, Burlington, Vermont, USA

Robert P Myers MD MSc and Kelly W Burak MD MSc
Liver Unit, Division of Gastroenterology and Hepatology, University of Calgary, Calgary, Alberta

Hemant Shah MD MScCH HPTE and Jordan J Feld MD MPH
Toronto Western Hospital Liver Centre, University of Toronto, Toronto

Curtis Cooper MD
Division of Infectious Diseases, University of Ottawa, Ottawa, Ontario

Dev S Segarajasingam MBBS FRACP
Department of Gastroenterology and Hepatology, Sir Charles Gairdner Hospital and School of Population Health, The University of Western Australia, Crawley, Australia

Stephen C Hanley MD PhD
Division of General Surgery

Alan N Barkun MDCM FRCPC MSc
Division of Gastroenterology
Department of Epidemiology, Biostatistics and Occupational Health, McGill University Health Centre, McGill University, Montreal, Quebec

Kevin A Waschke MDCM FRCPC, Josée Parent MDCM FRCPC and Serge Mayrand MDCM FRCPC, Carlo A Fallone MDCM FRCPC Ernest G Seidman MDCM FRCPC and Myriam Martel BSc
Division of Gastroenterology

Pascal Burtin MD
Endoscopy Unit, Gustave Roussy Institute of Oncology, Villejuif, Greater Paris, France

Gilles Jobin MD
Division of Gastroenterology, Montreal University, Montreal, Quebec

Turki AlAmeel MBBS FRCPC FACP, Bahaa Bseiso MD and Sami AlMomen MRCP(UK)
Division of Gastroenterology

Meteb M AlBugami MBBS FRCPC FACP
Multi-Organ Transplant Center, King Fahad Specialist Hospital-Dammam, Dammam, Saudi Arabia

Lee Stuart Roth MD FRCPC DABIM FACP MSc
William Osler Health System, Toronto, Ontario

Majid A Almadi MBBS, FRCPC, MSc
Division of Gastroenterology, The McGill University Health Centre, Montreal General Hospital, McGill University, Montreal, Quebec
Division of Gastroenterology, King Khalid University Hospital, King Saud University, Riyadh, Saudi Arabia

Maida Sewitch PhD
Division of Clinical Epidemiology, The McGill University Health Centre, Montreal General Hospital, McGill University, Montreal, Quebec

Alan N Barkun MD CM FRCPC MSc
Division of Gastroenterology, The McGill University Health Centre, Montreal General Hospital, McGill University, Montreal, Quebec
Division of Clinical Epidemiology, The McGill University Health Centre, Montreal General Hospital, McGill University, Montreal, Quebec

Myriam Martel BSc
Division of Gastroenterology, The McGill University Health Centre, Montreal General Hospital, McGill University, Montreal, Quebec

Lawrence Joseph PhD
Division of Clinical Epidemiology, The McGill University Health Centre, Montreal General Hospital, McGill University, Montreal, Quebec

Valérie Huynh-Trudeau MD, Stéphanie Maynard MD, Tatjana Terzic MD, Geneviève Soucy MD and Mickael Bouin MD PhD
Hôpital Saint-Luc, Montreal, Quebec

Andrew Szilagyi MD FRCPC
Jewish General Hospital, Department of Medicine, Division of Gastroenterology, McGill University School of Medicine, Montreal, Quebec

Brian Bressler MD
Department of Medicine, Division of Gastroenterology, St Paul's Hospital, Vancouver, British Columbia

Remo Panaccione MD
Department of Medicine, University of Calgary, Calgary

Richard N Fedorak MD
Division of Gastroenterology, University of Alberta, Edmonton, Alberta

Ernest G Seidman MD
Division of Gastroenterology, McGill University Health Centre, Montreal, Quebec

Peter Stepaniuk BSc(Hons)
Internal Medicine, University of Manitoba

Charles N Bernstein MD
Internal Medicine, University of Manitoba
University of Manitoba IBD Clinical and Research Centre

Zoann Nugent PhD
Internal Medicine, University of Manitoba
University of Manitoba IBD Clinical and Research Centre
CancerCare Manitoba, Department of Epidemiology and Cancer Registry

Harminder Singh MD MPH
Internal Medicine, University of Manitoba
University of Manitoba IBD Clinical and Research Centre
Community Health Sciences, University of Manitoba, Winnipeg, Manitoba